BURGER'S
MEDICINAL CHEMISTRY
AND
DRUG DISCOVERY

BURGER'S MEDICINAL CHEMISTRY AND DRUG DISCOVERY

BURGER'S MEDICINAL CHEMISTRY AND DRUG DISCOVERY

Sixth Edition
Volume 2: Drug Discovery and
Drug Development

Edited by

Donald J. Abraham
Department of Medicinal Chemistry
School of Pharmacy
Virginia Commonwealth University
Richmond, Virginia

Burger's Medicinal Chemistry and Drug Discovery
is available Online in full color at
www.mrw.interscience.wiley.com/bmcdd.

**WILEY-
INTERSCIENCE**

A Wiley-Interscience Publication
John Wiley and Sons, Inc.

Cover Description The molecule on the cover is hemoglobin with the allosteric effector RSR 13 attached. Three groups initiated structure-based drug design in the middle to late 1970s. Two of the groups, Peter Goodford's at the Burroughs Wellcome Laboratories in London and the editors' at the School of Pharmacy at the University of Pittsburgh, worked on hemoglobin while David Matthews at Agouron Pharmaceuticals worked on dihydrofolate reductase. Max Perutz and his coworkers' solution of the phase problem produced the first three-dimensional structure of a protein whose coordinates were of interest for drug design. The Editor worked with Max Perutz from 1980 to 1988, attempting to design antisickling agents. One of the active antisickling molecules, clofibric acid, would not lead to a sickle cell drug but to an allosteric effector RSR 13 designed and synthesized at Virginia Commonwealth University, which has been studied clinically for treatment of metastatic brain cancer. Max Perutz, whose work would provide the underpinnings for structure-based drug design, passed away in February 2002. His spirit and love for science that transferred to his students, postdoctoral fellows, visiting scientists, and colleagues worldwide, like Professor Burger's, is the leaven and inspiration for new discoveries.

For general information on our other products and services please contact our Customer Care Department within the U.S. at (877) 762-2974, outside the U.S. at (317) 572-3993 or fax (317) 572-4002.

Wiley also publishes its books in a variety of electronic formats. Some content that appears in print, however, may not be available in electronic format.

For ordering and customer service, call 1-800-CALL-WILEY.

Library of Congress Cataloging-in-Publication Data:

Burger's medicinal chemistry and drug discovery.—6th ed., Volume 2: drug discovery and drug development/ Donald J. Abraham, editor

ISBN 0-471-37028-2 (v. 2: acid-free paper)

Printed in the United States of America.

10 9 8 7 6 5 4 3 2 1

*To Alfred Burger and Max Perutz for their mentorship
and life-long passion for science.*

BURGER MEMORIAL EDITION

The Sixth Edition of Burger's Medicinal Chemistry and Drug Discovery is being designated as a Memorial Edition. Professor Alfred Burger was born in Vienna, Austria on September 6, 1905 and died on December 30, 2000. Dr. Burger received his Ph.D. from the University of Vienna in 1928 and joined the Drug Addiction Laboratory in the Department of Chemistry at the University of Virginia in 1929. During his early years at UVA, he synthesized fragments of the morphine molecule in an attempt to find the analgesic pharmacophore. He joined the UVA chemistry faculty in 1938 and served the department until his retirement in 1970. The chemistry department at UVA became the major academic training ground for medicinal chemists because of Professor Burger.

Dr. Burger's research focused on analgesics, antidepressants, and chemotherapeutic agents. He is one of the few academicians to have a drug, designed and synthesized in his laboratories, brought to market [Parnate, which is the brand name for tranylcypromine, a monoamine oxidase (MAO) inhibitor]. Dr. Burger was a visiting Professor at the University of Hawaii and lectured throughout the world. He founded the *Journal of Medicinal Chemistry, Medicinal Chemistry Research,* and published the first major reference work *"Medicinal Chemistry"* in two volumes in 1951. His last published work, a book, was written at age 90 (*Understanding Medications: What the Label Doesn't Tell You,* June 1995). Dr. Burger received the Louis Pasteur Medal of the Pasteur Institute and the American Chemical Society Smissman Award. Dr. Burger played the violin and loved classical music. He was married for 65 years to Frances Page Burger, a genteel Virginia lady who always had a smile and an open house for the Professor's graduate students and postdoctoral fellows.

PREFACE

The Editors, Editorial Board Members, and John Wiley and Sons have worked for three and a half years to update the fifth edition of Burger's Medicinal Chemistry and Drug Discovery. The sixth edition has several new and unique features. For the first time, there will be an online version of this major reference work. The online version will permit updating and easy access. For the first time, all volumes are structured entirely according to content and published simultaneously. Our intention was to provide a spectrum of fields that would provide new or experienced medicinal chemists, biologists, pharmacologists and molecular biologists entry to their subjects of interest as well as provide a current and global perspective of drug design, and drug development.

Our hope was to make this edition of Burger the most comprehensive and useful published to date. To accomplish this goal, we expanded the content from 69 chapters (5 volumes) by approximately 50% (to over 100 chapters in 6 volumes). We are greatly in debt to the authors and editorial board members participating in this revision of the major reference work in our field. Several new subject areas have emerged since the fifth edition appeared. Proteomics, genomics, bioinformatics, combinatorial chemistry, high-throughput screening, blood substitutes, allosteric effectors as potential drugs, COX inhibitors, the statins, and high-throughput pharmacology are only a few. In addition to the new areas, we have filled in gaps in the fifth edition by including topics that were not covered. In the sixth edition, we devote an entire subsection of Volume 4 to cancer research; we have also reviewed the major published Medicinal Chemistry and Pharmacology texts to ensure that we did not omit any major therapeutic classes of drugs. An editorial board was constituted for the first time to also review and suggest topics for inclusion. Their help was greatly appreciated. The newest innovation in this series will be the publication of an academic, "textbook-like" version titled, "Burger's Fundamentals of Medicinal Chemistry." The academic text is to be published about a year after this reference work appears. It will also appear with soft cover. Appropriate and key information will be extracted from the major reference.

There are numerous colleagues, friends, and associates to thank for their assistance. First and foremost is Assistant Editor Dr. John Andrako, Professor emeritus, Virginia Commonwealth University, School of Pharmacy. John and I met almost every Tuesday for over three years to map out and execute the game plan for the sixth edition. His contribution to the sixth edition cannot be understated. Ms. Susanne Steitz, Editorial Program Coordinator at Wiley, tirelessly and meticulously kept us on schedule. Her contribution was also key in helping encourage authors to return manuscripts and revisions so we could publish the entire set at once. I would also like to especially thank colleagues who attended the QSAR Gordon Conference in 1999 for very helpful suggestions, especially Roy Vaz, John Mason, Yvonne Martin, John Block, and Hugo

Kubinyi. The editors are greatly indebted to Professor Peter Ruenitz for preparing a template chapter as a guide for all authors. My secretary, Michelle Craighead, deserves special thanks for helping contact authors and reading the several thousand e-mails generated during the project. I also thank the computer center at Virginia Commonwealth University for suspending rules on storage and e-mail so that we might safely store all the versions of the author's manuscripts where they could be backed up daily. Last and not least, I want to thank each and every author, some of whom tackled two chapters. Their contributions have provided our field with a sound foundation of information to build for the future. We thank the many reviewers of manuscripts whose critiques have greatly enhanced the presentation and content for the sixth edition. Special thanks to Professors Richard Glennon, William Soine, Richard Westkaemper, Umesh Desai, Glen Kellogg, Brad Windle, Lemont Kier, Malgorzata Dukat, Martin Safo, Jason Rife, Kevin Reynolds, and John Andrako in our Department of Medicinal Chemistry, School of Pharmacy, Virginia Commonwealth University for suggestions and special assistance in reviewing manuscripts and text. Graduate student Derek Cashman took able charge of our web site, *http://www.burgersmedchem.com*, another first for this reference work. I would especially like to thank my dean, Victor Yanchick, and Virginia Commonwealth University for their support and encouragement. Finally, I thank my wife Nancy who understood the magnitude of this project and provided insight on how to set up our home office as well as provide John Andrako and me lunchtime menus where we often dreamed of getting chapters completed in all areas we selected. To everyone involved, many, many thanks.

DONALD J. ABRAHAM
Midlothian, Virginia

Dr. Alfred Burger

Photograph of Professor Burger followed by his comments to the American Chemical Society 26th Medicinal Chemistry Symposium on June 14, 1998. This was his last public appearance at a meeting of medicinal chemists. As general chair of the 1998 ACS Medicinal Chemistry Symposium, the editor invited Professor Burger to open the meeting. He was concerned that the young chemists would not know who he was and he might have an attack due to his battle with Parkinson's disease. These fears never were realized and his comments to the more than five hundred attendees drew a sustained standing ovation. The Professor was 93, and it was Mrs. Burger's 91st birthday.

Opening Remarks

ACS 26th Medicinal Chemistry Symposium

June 14, 1998
Alfred Burger
University of Virginia

It has been 46 years since the third Medicinal Chemistry Symposium met at the University of Virginia in Charlottesville in 1952. Today, the Virginia Commonwealth University welcomes you and joins all of you in looking forward to an exciting program.

So many aspects of medicinal chemistry have changed in that half century that most of the new data to be presented this week would have been unexpected and unbelievable had they been mentioned in 1952. The upsurge in biochemical understandings of drug transport and drug action has made rational drug design a reality in many therapeutic areas and has made medicinal chemistry an independent science. We have our own journal, the best in the world, whose articles comprise all the innovations of medicinal researches. And if you look at the announcements of job opportunities in the pharmaceutical industry as they appear in *Chemical & Engineering News,* you will find in every issue more openings in medicinal chemistry than in other fields of chemistry. Thus, we can feel the excitement of being part of this medicinal tidal wave, which has also been fed by the expansion of the needed research training provided by increasing numbers of universities.

The ultimate beneficiary of scientific advances in discovering new and better therapeutic agents and understanding their modes of action is the patient. Physicians now can safely look forward to new methods of treatment of hitherto untreatable conditions. To the medicinal scientist all this has increased the pride of belonging to a profession which can offer predictable intellectual rewards. Our symposium will be an integral part of these developments.

CONTENTS

BURGER'S
MEDICINAL CHEMISTRY
AND
DRUG DISCOVERY

CHAPTER ONE

Combinatorial Chemistry and Multiple Parallel Synthesis

LESTER A. MITSCHER
APURBA DUTTA
Kansas University
Department of Medicinal Chemistry
Lawrence, Kansas

Contents

Burger's Medicinal Chemistry and Drug Discovery
Sixth Edition, Volume 2: Drug Development
Edited by Donald J. Abraham
ISBN 0-471-37028-2   © 2003 John Wiley & Sons, Inc.

1   INTRODUCTION

The introduction of a new pharmaceutical is a lengthy and expensive undertaking. Methods which promise to shorten the time or the cost are eagerly taken up, and this is clearly the case with combinatorial chemistry and multiple parallel synthesis.

Combinatorial chemistry is somewhat hard to define precisely, but generally speaking, it is a collection of methods that allow the simultaneous chemical synthesis of large numbers of compounds using a variety of starting materials. The resulting compound library can contain all of the possible chemical structures that can be produced in this manner. Multiple parallel synthesis is a related group of methodologies used to prepare a selected smaller subset of the molecules that could in theory have been prepared. The content of libraries prepared by multiple parallel synthesis is more focused and less diverse than those constructed with combinatorial technology.

The primary benefit that combinatorial and multiple parallel chemistry bring to drug synthesis is speed. As with most other human endeavors, uncontrolled speed may be exhilarating but is not particularly useful. Rapid construction of compounds that have no chance of becoming drugs is of little value to the medicinal chemist. After an initial euphoric period when many investigators thought that any novel compound had a realistic chance of becoming a drug, realism has now returned, and libraries are being constructed that reflect the accumulated wisdom of the field of medicinal chemistry. Combinatorial methods have permeated and irreversibly altered most phases of drug seeking that benefits from the attention of chemists. The successful contemporary medicinal chemist must be aware of the strengths and weaknesses of these exciting new methods and be able to apply them cunningly. In the proper hands combining medicinal chemical insight with enhanced speed of synthesis is very powerful.

Libraries are constructed both in solution and on solid supports and the choice between

	A	B	C	D	E
A	AA	BA	CA	DA	EA
B	AB	BB	CB	DB	EB
C	AC	BC*	CC	DC	EC
D	AD	BD	CD	DD	ED
E	AE	BE	CE	DE	EE

Figure 1.1. A combinatorial library constructed from five reacting components.

these techniques is often a matter of personal preference, and they are performed side by side in most laboratories. For very large libraries, however, construction on resins is more practical, whereas for smaller, focused libraries, solution phase chemistry is more practical. Solid phase methods are also specially advantageous for multistep iterative processes and are notable for the comparative ease of purification by simple filtration and the ability to drive reactions to completion by the use of excess reagents. Throughout the previous decade, solid phase organic synthesis (SPOS) has dominated combinatorial chemistry, and many novel methods have been developed as a result.

The main concepts in this field can be summarized in Figs. 1.1 and 1.2. In Fig. 1.1, there is a hypothetical combinatorial compound library of condensation products produced by reacting every possible combination of five starting materials. This results in a library containing 25 (5 × 5) products. The library could be constructed by 25 individual reactions, with each product separate from all of the others. It could also be constructed by run-

	A	B	C	D	E
C	AC	BC*	CC	DC	EC

Figure 1.2. A multiple parallel synthesis library constructed from six participants.

ning reactions simultaneously so that a single mixture of all 25 substances would be obtained.

In this specific example, it is presumed that the reactions were run in a single step so that all compounds were produced simultaneously in a big mixture. Furthermore, it is assumed that the ideal was achieved. That is, all reactions proceeded quantitatively, and that each compound is present in the final compound collection in equal molar concentration. (This ideal is rarely achieved.) It is also assumed that BC is the only active constituent and so it is marked off with an asterisk. If the whole library were tested as a mixture, then it would be seen that it contained an active component, but one would not know which one it was. For this to happen, it is necessary that the components do not interfere with one another so false positives and false negatives are not seen. Many clever means of finding the active component expeditiously have been developed and a number of these will be illustrated later. It is clear that if the components were prepared and tested individually, 25 separate reactions would be required and the identification problem would disappear, but great strain would be placed on the bioassay and the speed of synthesis would be compromised. A perfect combinatorial library for drug discovery would only contain BC, and only a single reaction would be required, but this level of efficiency is rarely achieved by any contemporary medicinal chemical method. The real problem is to construct a compound library that is sufficiently diverse and sufficiently large that there is a high possibility that at least one component will be active in the chosen test system. This example assumes that this has been done. The reader will also readily see that with the library in Fig. 1.1, that instead of testing all of the compounds simultaneously, if one prepared and tested 10 mixtures resulting from combining the five products in each column and each row, then BC would reliably emerge as the active component because only the row C mixture and the column B mixture would be active, and the active component must be the one where the rows and columns intersect. When more than one active component is present then the problem becomes more complex.

Figure 1.2 illustrates a related multiple parallel library. This is much smaller and starts with the assumption that one knows or suspects that the best compound will terminate in component C. Five reactants are chosen to condense with C and the resultant library consists of five components. Each of these products is tested singly and BC* is quickly identified.

The greater efficiency of this library compared with that of Fig. 1.1 for the purpose is obvious. Clearly, the smaller the library that succeeds in solving the problem, the more effective the process is. A perfectly efficient library would only contain BC*. The size of the library produced by either method is the product of the number of variables introduced and the number of steps involved raised exponentially. For example, starting with a given starting material (often called a centroid) and attaching four different groups of 10 side-chains to each product would produce $1 \times 10 \times 10 \times 10 \times 10$ or 10,000 members. The primary advantage of either method is speed, because the products are prepared simultaneously at each step. The efficiency is also enhanced when the condensation steps involve the same conditions.

The use of the library depends on the specific structures included and the purposes for which the library is to be tested. The relevance of speed to drug discovery is easy to explicate. If one knew in advance the particular structure that would satisfy the perceived need, a successful compound library would only need to have one substance in it. If one has a general idea of the type of structure that would be useful, the library will have many promising compounds but still be finite in number. Quantitative bioassay of pure substances would allow one to select the most nearly perfect embodiment. If one has no idea of the type of structure that would give satisfaction, then a successful library must have a larger and more diverse number of compounds in it.

Contemporary drug seeking is a complex, time consuming, and expensive process because a successful drug must not only have outstanding potency and selectivity, but it must also satisfy an increasingly long list of other structure-dependent criteria as well. The elapsed time from initial synthesis to

marketing is estimated to lie on average between 10 and 15 years and to require the preparation and evaluation of a few thousand analogs. The costs are estimated to lie between
$300 and $800 million per agent. Most large
firms now target the introduction of 1–3 novel
drugs per year and target sales at a billion
dollars or more from each. This indicates that
each day of delay in the drug seeking process
not only deprives patients of the putative benefits of the new drug but also represents the
loss of a million dollars or more of sales for the
firm! Added to this is the intense competition
among big pharmaceutical companies for a
winning place in the race and among small
pharmaceutical companies for survival. First
to market in an unserved therapeutic area can
return a great profit if a sufficient number of
sufferers exist who have access to the funds to
pay for their treatment. The next two entries
competing with this agent can also be expected
to do well. After this, success is rather more
problematic because the market grows more
and more fragmented. Being first to finish the
race, therefore, conveys very real survival
value. From an economic standpoint, it is estimated that less than 10% of products introduced repay their development costs. Those
few that do must return a sufficient surplus to
amortize the costs of the rest and sufficient
additional funds to cover the costs of future
projects and to gratify the shareholders. These
imperatives have placed a premium on speed
of discovery and development. The portion of
this time devoted to synthesis and screening in
the drug-seeking campaign is usually about
3–5 years. The enhanced speed of construction
can be expected to decrease the time to market
by perhaps as much as 1 year in favorable
cases. While this is less than was originally
hoped for when these methods were introduced, it is not trivial.

Combinatorial chemistry is now such a per-
vasive phenomenon that comprehensive review of its medicinal chemical features is no
longer possible in less than book length. Full
coverage would require treatment of its impact on all of the phases of drug discovery and
would exceed the space available. Thus, the
remainder of this chapter will illustrate its
main features and applications.

2 HISTORY

Combinatorial chemistry grew out of peptide
chemistry and initially served the needs of biochemists and the subset of medicinal chemists
who specialized in peptide science. Its first decade or so concentrated on oligopeptides and
related molecules. It continued to evolve, however, and now permeates virtually every cor-
ner of medicinal chemistry and a major effort
is underway to discover new, orally active,
pharmaceuticals using these methods.

Many will agree that the path leading to the
present state of combinatorial chemistry essentially started with the solid phase synthetic
experiments on peptides by Bruce Merrifield
in 1962 (1, 2). This work had immediate impact, facilitated in large part because of the
essentially iterative reactions, to completion
by use of reagents in excess, its susceptibility
to automation, and the ease of removing detritus from the products by simple washing and
filtration away from the resins. At first this
extremely useful technology was employed in
a linear fashion. It was probably Furka in
Hungary a decade or so later who realized that
the methodology could lead to simultaneous
synthesis of large collections of peptides and
conceived of the mix and split methods (3).
Geyson made the whole process technically
simpler in 1984 and produced large scale compound collections of peptides (4) and Houghton introduced "tea bag" methodologies in
1985 in which porous bags of resins were suspended in reagents (5). Comparatively few organic chemists undertook the preparation of
ordinary organic substances on solid phases
because the work is rather more complex
when applied to non-oligomeric substances
caused by greater variety of reactants and conditions required, and this work at first failed to
develop a significant following. Solid phase organic chemistry was also comparatively un-
derdeveloped and this held back the field. This
changed in dramatic fashion after the publication of Bunin and Ellman's seminal work on
solid phase organic synthesis (SPOS) of arrays
of 1,4-benzodiazepine-2-ones in 1992 (6). Soon
other laboratories published related work on
this ring system, and work on other drug-like
molecules followed in rapid order and the race
was on. In the initial phases, solid phase or-

ganic synthesis predominated, and this persisted until about 1995, when solution phase combinatorial chemistry began to make serious inroads. Until about 1997, roughly one-half of the libraries reported were either of peptides or peptidomimetics. Subsequently libraries of drug-like small molecules have become increasingly popular.

The work on combinatorial libraries has inspired the rapid development of a wide variety of auxiliary techniques including the use of reagents on solid support, capture resins, chemical and biological analysis of compound tethered to resins, informatics to deal with the huge volume of structural and biological data generated, the synthesis of a wide variety of peptide-like and heterocyclic systems hitherto prepared solely in solution, photolithographic techniques allowing the production of geographically addressed arrays on a "credit card," preparation of gene array chips, attachment of coding sequences, use of robotics, and the preparation of oligonucleotides by Letsinger in 1975 (7) and of oligosaccharides by Hindsgaul in the 1990s (8). At this moment several thousand papers are appearing each year describing the preparation and properties of compound libraries either in mixtures or as individual substances. Several books (9–35) and reviews (36–49) are available for the interested reader. Those of Dolle are particularly recommended because he has undertaken the heroic task of organizing and summarizing each year the world's literature on the topic. That of Thompson and Ellman is especially thorough in reviewing the literature up until 1996 from a chemical viewpoint. A great many other reviews are available, including many in slick-cover free journals that arrive on our desks weekly. In addition to these, at least three specialist journals have been established in the area. These are the *Journal of Combinatorial Chemistry, Molecular Diversity*, and *Combinatorial Chemistry and High Throughput Screening*.

Another important factor leading to the explosion of interest in combinatorial chemical techniques was the development of small firms devoted to the exploitation of genetic discoveries through development of high-throughput screening methods. These firms by and large did not have libraries of com-

pounds to put through these screens and were seeking collections of molecules. Combinatorial chemistry addressed these needs. When these methods were taken up by big pharmaceutical companies, existing libraries quickly proved inadequate for the need and combinatorial methodologies clearly addressed this need as well. Just about 10 years after these seminal events, the face of medicinal chemistry has been irretrievably altered. While combinatorial chemistry has in some respects not lived up to the initial hopes, its value is firmly established and no serious firm today fails to use these methods. By the year 2002, well over 1000 libraries have been reported. Many of these include reports of the biological activity of their contents. This is remarkable considering that the field is scarcely more than a decade old!

3 SOLID PHASE ORGANIC SYNTHESIS OF INFORMATIONAL MACROMOLECULES OF INTEREST TO MEDICINAL CHEMISTS

That the solid phase synthesis of collections of peptides launched this field is not intrinsically surprising. The basic methodology existed because of Merrifield and many others. The peptide linkage has notable advantages for this work because it is relatively chemically stable, non-chiral, constructible by iterative processes amenable to automation, the products are rarely branched, possess a variety of interesting biological properties, and can be constructed in great variety. The counterbalancing defects of these compounds are that they are not easily delivered orally unless they are end capped and rather small in molecular weight, are readily destroyed by enzymatic action, and fail to penetrate into cells. The physiological reason for this is readily understood. Peptides, and other informational macromolecules, function in the body to provide specific structure or to generate signals for cells to respond to according to their sequence and architecture. It would be dangerous if they were absorbed intact from ingestion of other life forms. To prevent cellular disruption they are first digested in the gastrointestinal tract, absorbed as monomers, and then reassembled after our own genetic pattern so that they join or supplement those already present without causing disruption in cellular architecture or function.

Nucleic acids have many of the same advantages and disadvantages, and their compound libraries came into being soon after the peptides. Oligosaccharides, on the other hand, have lagged considerably behind. They possess chiral linkages whose construction must be carefully controlled, often are branched, are fragile in the presence of acid conditions, are often highly polar, and are readily digested. Controlling these processes is much more difficult and it has taken longer to conquer these problems.

3.1 Peptide Arrays

Peptides are prominent among the compounds of interest to biochemists but less so to most medicinal chemists for reasons explicated above. Construction of peptides of specified sequence is an iterative process whose complications largely consist of protection-deprotection steps to ensure that side-chain functionality does not interfere with orderly amide bond formation. When made on resin beads, there are limits to the quantities that can be made in part because of the geometric restrictions caused by bead size and the need to avoid adjacent molecules from interacting with each other, and the comparative lability of the beads to aggressive reagents places limitations on the chemical conditions that can be employed. Porous beads obviously can be loaded more heavily than those whose surface only is accessible after wetting by the reagents. Many different kinds of resins and other solid supports are now available, including some that are solvent soluble depending on the solvent and the temperature.

With their balance of advantages and disadvantages and the present gold standard being oral activity, peptide libraries are currently of primary value in lead seeking, in basic studies on cellular processes, or for the preparation of parenteral medications. Despite intensive study spanning several decades by some of the best minds of this generation, means of delivering thereputically significant blood levels of peptides through the oral route remain elusive. Translating the therapeutic message in peptide leads into oral non-peptide drugs through generally applicable systematic techniques has also been elusive. The several successes that have been achieved have been primarily the result of screening campaigns or serendipitous observations, and the results have so far not revealed an underlying tactical commonality that can be exploited in new cases. Perhaps the best known of these studies has been the translation of snake venom peptides, whose injection by serpents leads to a precipitous fall in blood pressure, into truncated pseudodipeptides like captopril, and then on to enalaprilat and lysinopril, which are pseudotripeptides. The basic lesson learned from all of these studies has been that the resulting agent should be as little peptide-like as possible and not exceed the equivalent of at most four amino acid-like residues. Examination of peptide structures in light of the well-known Lipinski rules provides a rationale for what experience has shown. Beyond about four residues, the molecular weight is becoming too high, the polarity is weighted too much toward water solubility, and the hydrogen-bonding inventory is excessive. Further, the compounds are excessively water soluble so that they do not pass through cellular membranes efficiently by passive diffusion.

An added feature to bear in mind is that the preparation of certain medically important polypeptide drugs, such as human insulin and growth hormone, through genetic engineering methodologies, is well developed and convenient so these substances can be used in parenteral replacement therapy. Their preparation through synthetic peptide chemistry represents important achievements in peptide intellectual technology but does not satisfy a commercial need.

Nonetheless, peptide compound libraries are very convenient for uncovering leads quickly for receptors where natural ligands or serendipitous drugs have not previously been found and large libraries of peptides continue to be made (50–65).

The number of peptides that could in principle be made is stupefying. For example, given that there are approximately 20 common amino acids, and allowing five post-translational modifications (and ignoring the fact that there are more of these and that there are many wholly synthetic amino acids), the avail-

Table 1.1.   Number of Possible Peptide Products as a Function of the Amino Acids Used

Dipeptides	(20×20)	$= 400$
Tripeptides	$(20 \times 20 \times 20)$	$= 8000$
Tetrapeptides	$(20 \times 20 \times 20 \times 20)$	$= 160{,}000$
Pentapeptides	$(20 \times 20 \times 20 \times 20 \times 20)$	$= 3{,}200{,}000$
Hexapeptides	$(20 \times 20 \times 20 \times 20 \times 20 \times 20)$	$= 64{,}000{,}000$
Heptapeptides	$(20 \times 20 \times 20 \times 20 \times 20 \times 20 \times 20)$	$= 1{,}380{,}000{,}000$

able synthons are at least approximately equal to the number of letters in the Western alphabet. By analogy with the number of languages that have been generated using this system, the potential number of peptides that could be made is clearly astronomical. It would require an incredible effort to make a library containing even only one molecule of each, and Furka has estimated that the mass of such a library would exceed that of the universe by more than 200 orders of magnitude (3)!

Were one to use just the common amino acids, the progression of peptides possible is enormous, as is shown in Table 1.1.

The simplest and least ambiguous method for constructing peptide libraries is the spatially separate or spatially addressed method. Here a single peptide is constructed on a single type of resin, and the resin/products are kept separate. No decoding sequence needs to be attached to the beads in this kind of library. This method was introduced by Geysen in 1984. To make 96 peptides at a time and to keep track of the products and facilitate their screening, the reactions were run on resins attached to individual pins so constructed that they fit into individual wells of 96-well plates. (Fig. 1.3) (66). A convenient variation was developed for parallel synthesis in which beads were contained in porous bags and dipped into reagent solutions. These are called "tea bags." The identity of the peptide or peptides contained is recorded on the attached label. Subsequently Fodor et al. developed a very diverse library on silicon wafers using photolitho-

graphic chemistry for forming the peptides and controlled the specific place along an x–y axis, where each peptide would be located through use of variously configured masks (Fig. 1.4.) Photolytic protecting groups were employed followed by coupling the newly revealed "hot spots" with a suitable reactant. After this, the masks are moved as often as desired and the process repeated. In principle this method could produce thousands of individual peptides on a credit card–like surface. Although somewhat laborious, the synthesis can readily be automated. The method requires photosensitive protecting groups and testing methodologies compatible with support-bound assay methods and the libraries are geographically coded by the position of products on the x–y axis (67). These techniques are now widely employed for gene array amplification and identification experiments.

Synthesis of mixtures of peptides further enhanced the speed and convenience of library construction but required development of devolution methods so that active components in the mixtures could be identified. Direct methods of sequence analysis are available. Mass spectrometry is popular as are NMR methods (involving magic angle methods on single beads). Edman degradation of peptides can also be performed. These methods are popular when iterative methods result in linear polymers.

A further complication of simultaneous preparation of peptide mixtures is that individual amino acids differ greatly in their reactivity, so if one simply placed all of the potential reactants in a flask under bond forming conditions, they would not react at the same rate. With each iteration, the disparity between readily formed and poorly formed bonds would widen. One way to deal with this problem is to use less than fully equivalent

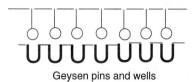

Geysen pins and wells

Figure 1.3. Geysen pins and wells.

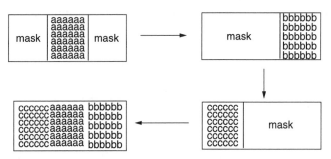

Fodor photolithographic method

Figure 1.4. Fodor photolithographic method.

amounts of each component and to allow the reactions to go to completion. For example, if two components are employed, each should be added in about one-half molar quantities. At the end of the reaction, one should have equimolar amounts of both products. This would allow the subsequent testing results to be quantitatively comparable (68). Alternatively, split synthesis methods were developed contemporaneously by Houghton (69), Furka (70), and Lam (71). In split and pool synthesis (Fig. 1.5), an initial reaction is run on a bead support to attach an amino acid as before, and the resulting beads are then split into equal portions and each of these groups of beads is deprotected and reacted with one of a group of different second amino acids to form dipeptides. These are pooled again and thoroughly mixed. This pool is again separated into equal portions and each is reacted with one of a different group of other amino acids to produce a group of tripeptide mixtures. This is continued until satisfied, and the last group of resin piles is usually not mixed. Although each individual bead will contain a single peptide, the final products from this methodology consist of groups of related materials. By illustration, if run in the manner described, one could start with the same amino acid attached to a bead. If this resulted in 20 piles of beads, each with a different amino acid attached, then each pile

could be reacted in parallel subpiles with a different one of the 20 amino acids and re-pooled. Continuation would lead to 20 collections of all of the possible tripeptides. Detaching and testing would reveal which was the best amino acid to start with (XXB in the example shown in Fig. 1.6). Keeping this one constant in the next series, repeating the process with the remaining 19 would reveal the best second amino acid (XEB in the example). Iteration would lead, after significant labor, to the optimal sequence at all positions and produce a structure-activity relationship based on those other substances approaching its potency (GEB > IEB > HEB in the example). This method of deconvolution is known as positional scanning.

A great diversity of peptides can be constructed in short order using these methods, and very large libraries have been produced. Such collections would be truly combinatorial. Many subvariations of this process can be envisioned. In general it is found that more than one active peptide is obtained. Synthesis of all of these actives as individual pure chemicals (often called "discretes") will allow the development of a structure-activity relationship. No assumptions are made in pursuing the study in this manner. If one has, however, a lead peptide already or wishes to define a re-

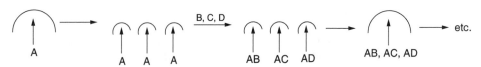

Figure 1.5. Pool and split method.

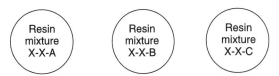

Step 1. Cleave and test each pool separately.

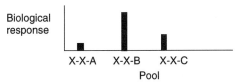

Identify pool X-X-B as the more potent

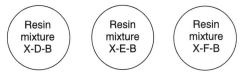

Step 2. Repeat sequence with pool X-X-B however varying the second position.

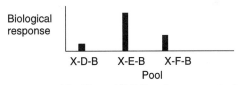

Identify pool X-E-B as the more potent

Step 3. Repeat the sequence with pool X-E-B however varying the first position.

Identify G-E-B as the most potent analog.

Figure 1.6. Split and mix method with positional scanning.

gion in a lead peptide, then one can perform a more limited study by systematically varying all of the individual positions in the region in question in this manner. The general experience is that potency and selectivity increases as each position is optimized and the known sequence grows (72). Although this method is somewhat wasteful, it has no preconceptions, and the residual sequences not of interest in this test series can be archived and examined in future test systems. Another advantage of the method is that the final products are not tethered so are able to assume the solution conformation dictated by their sequence or by the receptor interaction and also to interact with insoluble receptors.

In the substitution/omission method (Fig. 1.7), one starts with a lead sequence whose activity is known and replaces sequentially each amino acid with all possible 20 analogs, keeping the remainder the same. This is similar to the divide-couple-recombine method just described, but it evaluates only a single specific residue at a time. Testing the 20 libraries reveals which amino acid is optimal at that position. This is repeated until all of the amino acids in the sequence being examined have been evaluated and an optimized sequence is at hand. In the illustration, one starts with known sequence A-B-C-D and discovers enhanced potency in the modified sequence of E-F-C-D.

Lead substance = A-B-C-D

Step 1. Vary position A with all 20 amino acids. Detach and test.

Xn-B-C-D   ⟶   E-B-C-D

Detaching from the resin and testing reveals E-B-C-D
to be more active than A-B-C-D.

Step 2. Vary position B with all 20 amino acids. Detach and test.

E-Xn-C-D   ⟶   E-F-C-D

Detaching from the resin and testing reveals E-F-C-D
to be more active than A-B-C-D.

Step 3. Repeat this process with positions C and D

Detaching from the resin and testing reveals E-F-C-D
to be the most active sequence in this illustration.

Figure 1.7. Substitution/omission method.

In the omission method, one deletes one of the amino acids from a given position (most often at the end) or replaces an amino acid residue at any chosen position by an alanine or glycine and finds which omission decreases activity. This is done at each position until the optimal sequence is detected and the relative contribution of each amino acid side-chain is clear.

Devolution by positional scanning is facilitated through testing groups of resins arranged in checkerboard rows and columns as illustrated earlier in Fig. 1.1. Subsequent libraries are much smaller so the process becomes progressively less laborious.

Another popular method of identifying residues is to attach a non-peptide signaling molecule orthogonally to a different attachment arm whose reactivity differs from the first each time an amino acid is attached to its arm (Fig. 1.8). The signaling or coding molecule needs to be attached to its arm using chemistry that does not interfere with the growing peptide on the other arm and does not detach either molecule prematurely. There are a variety of strategies employed in detaching the sequences from the arms (Fig. 1.9). Strategy a represents an internal displacement reaction and leaves no trace behind in the product of the original point of attachment. Strategy b uses an external nucleophile for attachment.

This completes the product structure and usually does not leave a linker trace. Strategy c is a reductive or hydrolytic strategy and sometimes leaves a linker trace in the product but need not do so. The signal sequence carries the history of the bead and therefore codes for the history of the steps involved in the synthesis and thus for the identity of the peptide that one believes one has attached to the bead. The ease of identification of oligonucleotide sequences (by PCR methods) has made these popular for such coding. Various halogenated aromatic residues have also been used for this purpose. Another popular method is to embed an rf generator tuned to individual frequencies in the resin itself so that the substance on the bead can be identified by tuning to the proper frequency. One can also place bar codes on the beads for convenient reading. Much ingenuity has been expended on ingenious methods of deconvolution.

The development of these methods presents the medicinal chemist with the ability to perform a chemical evolution. This would seem to parallel nature's use of biological evolution to produce chemical libraries suitable for particular biological purposes. Chemical evolution of this type is more congenial to the impatient chemist.

Clearly with very large libraries, it is technically not possible to analyze each product, so

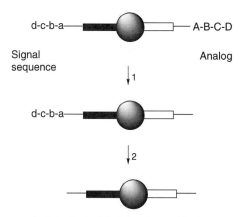

1. Selectively detach analog and test.
2. Selectively detach signal sequence and analyze.

Figure 1.8. Resin with arms containing an analog and a signal sequence.

one must resort to statistical sampling instead or take it as an article of faith that each compound has been successfully prepared. Much work has been devoted to dealing with this problem but a comprehensive treatment of this complex topic is unfortunately too vast to cover here. Clearly careful rehearsal and fanatic attention to detail in the construction of the library helps, but this is conjecture not science. There are relatively few instances in the literature where a careful census has been performed from which to form an opinion on this topic. In one important recent example, a statistical sampling of 7.5% of the contents of a library of 25,200 statin-containing pseudopeptides showed that 85% had the anticipated structure. This is reasonably good for

such complex work, but leaves one with a sense of unease in that about 3800 wrong substances were available. It is important also to note that the wrong structures were not statistically distributed among all of the targeted compounds but rather showed a bias toward certain structures. This is not particularly surprising but underscores the importance of the topic when interpreting the results of screening (73).

Another method of devolution uses a method descended from early work of Pasteur for this work. Here, the resins are poured onto a solid surface previously seeded with a microorganism lawn or a substrate that generates a color. Those resins that contain an active material give a zone of inhibition or a color response. Both of these endpoints can be detected visually and the active resins taken off of the surface with tweezers. The active component can then be detached for analysis and identification.

Initially, combinatorial libraries of peptides consisted primarily of products made from linear combinations of naturally occurring amino acids. Subsequently just about every devise employed in ordinary peptide work has been applied to combinatorial studies, leading to a persistent evolutionary drift away from collections of natural peptides. For example, to enhance the metabolic stability of such libraries, end capping of the amino end and the carboxy end and also cyclization have been employed to stabilize these substances. Before long, the incorporation of unusual amino acids also began (Fig. 1.10). These substances can be termed pseudopeptides

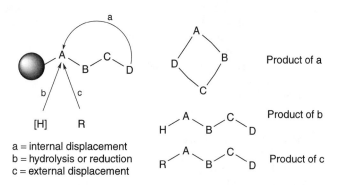

a = internal displacement
b = hydrolysis or reduction
c = external displacement

Figure 1.9. Some detachment strategies for detachment of compounds from resins.

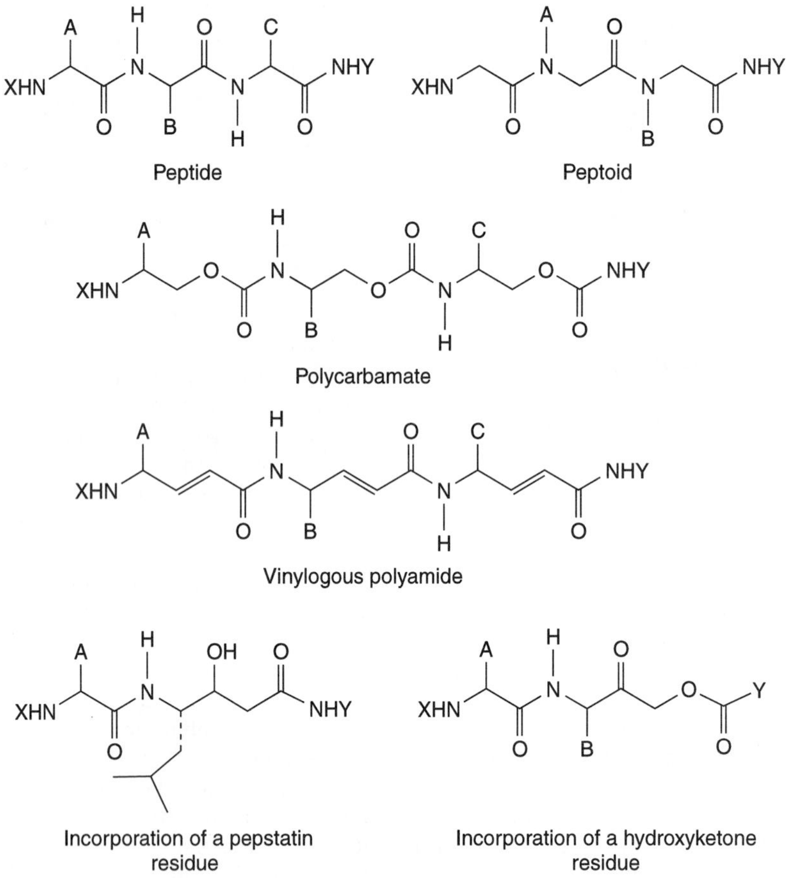

Figure 1.10. Some peptide surrogates employed in libraries.

(74–101). More specifically, such residues as benzamidines (102, 103), phosphinates (104–106), methyleneketones, hydroxyketones, fluoroamides (107), ketoamides, hydroxamates (108, 109), glycols, coumarins (110), boronates, oxazoles (111), nitriles, aldehydes, halogenated ketone hydrates, sulfonamides, and the like progressively appeared. These unusual residues were predominantly incorporated either at the end or at a point where the natural peptides would be cleaved by enzymatic action. These moieties are of special value when the substitution takes place at a site where the processing enzyme must act and employs mechanism-based inhibitory mechanisms. More recent libraries have appeared in which the overall conformation of the peptide has been mimicked so that the resulting heterocycle resembles topographically a β-turn, for example, but may not incorporate any common amino acid components (112, 113). The figure illustrates some of the groups so employed.

Later, the peptide linkage itself began to be modified (Fig. 1.11). For example, the classical mode of stabilization against peptidase cleavage by conversion of the peptide bond NH into N-methyl subsequently evolved into the preparation of peptoids (polyglycine chains with each NH replaced by a variety of N-alkyl groups of the type that resemble the side-chains found in normal amino acids). Libraries of these compounds were very popular for a while but interest has decreased as time passes (114, 115).

Flirtation with other substitutes for normal peptide bonds includes preparation of libraries of polycarbamates (116), vinylogous amides, incorporation of a pepstatin residue

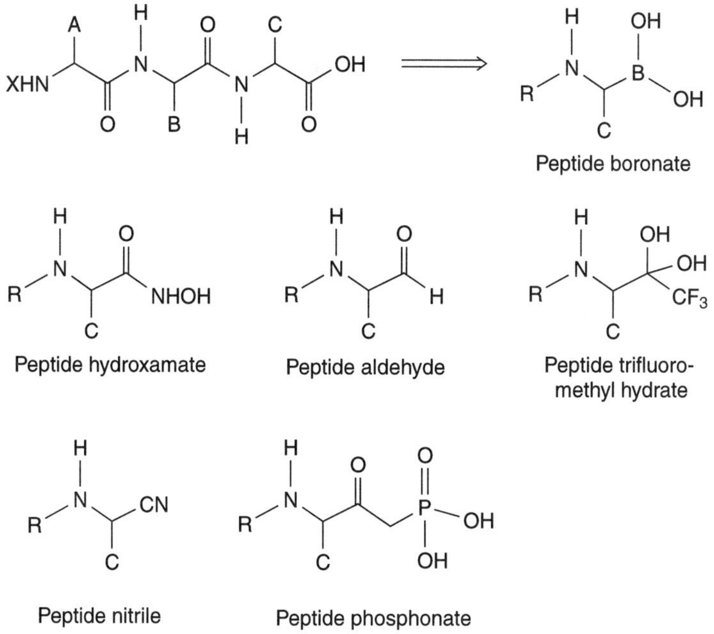

Figure 1.11. Peptides substituted with unusual carboxyl surrogate residues.

(117), and ureas (118) as well. In these libraries, the side-chains project from each fourth rather than each third atom in the chains so these are not close models of amino acids. Such libraries, not surprisingly, are more often antagonist rather than agonist.

In Fig. 1.11, one sees the insertion of unusual peptide bond surrogates for lead seeking. Such residues include peptide boronates, peptide hydroxamates, peptide aldehydes, peptide trifluoromethylketone hydrates, peptide nitriles, peptide phosphonates, and so on. One could also add to this list inclusion of β-turn mimics, β-sheet analogs, and so on. This is at present a very active subfield of medicinal chemistry.

These peptide and pseudopeptide libraries have been replaced progressively by collections containing smaller and more drug-like molecules. These will be covered in their own section below.

3.2 Nucleoside Arrays

The minimum fully realized library of natural peptides would consist of 20^{20} components. An analogous library of nucleic acids would consist of 5^5 components (double this if one used

both ribose and deoxyribose units) and be substantially smaller (Fig. 1.12a). Once again, the use of post-translationally modified bases or wholly unnatural analogs increases the attainable diversity. Conformational effects and self-associations further enhance the diversity. Once again, construction is iterative and bead-based automated procedures are available. Libraries of significant size have been constructed and evaluated (119).

3.3 Oligosaccharide Arrays

Construction of diverse oligosaccharide libraries is much more difficult. The linkage is chiral and relatively hard to control, the bonds are acid fragile, and there are many potentially competing functional groups that can be points of attachment (Fig. 1.12b). Despite these complicating factors, such libraries are beginning to appear. Clearly progress is being made.

3.4 Lipid Arrays

Lipid libraries have largely been neglected. For saturated fatty acids, the construction of carbon–carbon bonds is more difficult. Ste-

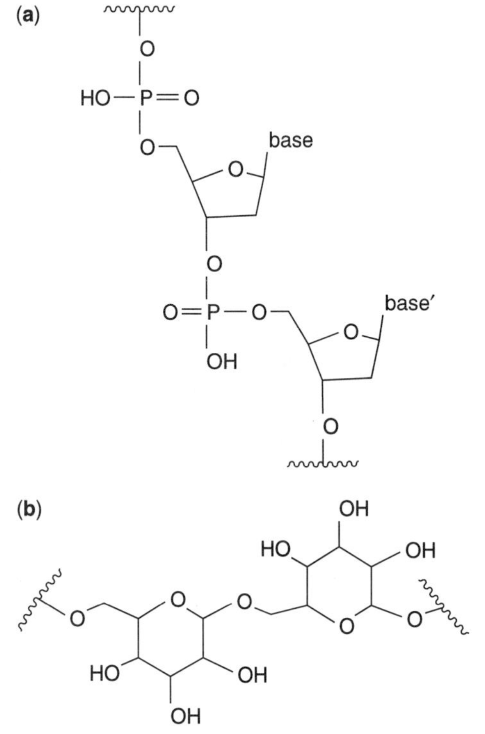

(a)

(b)

Figure 1.12. (a) Oligonucleotide libraries. (b) Oligosaccharide libraries.

roids and other polyisoprenoids are also complex to construct. Mixed triglycerides would seem accessible, and one anticipates developments in this area.

4 SOLID AND SOLUTION PHASE LIBRARIES OF SMALL, DRUGABLE MOLECULES

As noted above, the field of combinatorial chemistry and multiple parallel synthesis started with libraries of peptides. In time, unusual residues crept into the products. While this evolution is still ongoing, it is now accompanied by a major effort to produce libraries of small, drug-like molecules in library form. Many of the methods used for large molecules carry over but the largely non-iterative nature of small molecule synthesis is a significant complication.

The current "gold standard" in small molecule drug seeking is oral activity accompa-

nied by one-a-day dosing. This is a high hurdle. The majority of molecules that have passed have molecular weights of about 500 or so. It has been calculated that the number of small molecules that would fall into this category is approximately 10 raised to the 62nd power! Clearly preparing all of these in reasonable time is beyond the capacity of the entire population of the earth even if they worked tirelessly. The number of compounds that can become satisfactory drugs encompassed in this impossible collection is probably in the range of a few thousand, so most of the effort would be wasted. Hence medicinal chemical skills are still at a premium.

Obviously construction one at a time in the usual iterative or non-iterative fashion results in single molecules of a defined nature. Combination of reactants A and B produces a single product A-B. Reaction of this with another substance produces product A-B-C. In each case, a single reaction produces a single product. The change brought about by combinatorial or multiple parallel synthesis methods is that the reactants are usually linear, but the products can be logarithmic. For example, reacting A with 10 different Bs produces 10 products (A-B$_{1-10}$), and the reaction with 10 different Cs on each of these results in 100 products (A-B$_{1-10}$C$_{1-10}$), either in mixtures or as discrete compounds. Rather large compound collections can be assembled quickly using this scheme.

With non-linear products a different variant is seen than experienced with large molecules (Fig. 1.13). Here a starting material (often called a centroid) with a number of functional groups (preferably with different degrees of reactivity—known as orthogonality) can be reacted with a variety of substituents (often called adornments) to produce a large number of analogs. In the figure, one sees illustrated centroids with two, three, or four such functional groups and given the number of possible variants, this can lead to a very large library of analogs in a brief time (two functional groups with 10 variant adornments quickly results in 100 analogs, three in 1000, and four in 10,000). If the reaction conditions allow, these variations can be run in mixture or in parallel effecting a very significant time savings. It can, however, put a sig-

Two substitutable groups

Three substitutable groups

Four substitutable groups

Etc.

Figure 1.13. Centroid adornment.

nificant strain on purification, analysis, record keeping, budgets, etc. Much work has been expended in addressing these potential limitations.

With 500 as the normal practical upper molecular weight limit, it can be seen that centroid A should be chosen to have the smallest practical molecular weight. It is also helpful if, when fully adorned, it has functional groups remaining that can interact productively with a receptor so the weight devoted to this part of the molecule is not net loss. The molecular weight of the centroid places practical limits on the net weight that the adornments can collectively have. If one adornment is rather large, then this requires one or more of the other adornments to be made compensatingly smaller. The more functional groups present in the centroid, the smaller each adornment can be.

It is particularly helpful if the adornments project into space into quadrants that fit precisely the needs of the receptor if one is optimizing a lead. Alternately, if one is hit seeking, they should project into various quadrants about the centroid so as to allow a fruitful exploration of potential receptor needs. In hit seeking, one often wants molecular flexibility, whereas in lead optimization progressive rigidification is often more effective.

In addition, the adornments must have the usual medicinal chemical characteristics. They should not be chemically reactive, convey toxicity, or be inordinately polar. The product should fit the modified Lipinski rules to allow for the usual molecular weight and lipophilicity creep that often accompanies analoging. The net hydrogen bonding inventories, log P, water solubility, and cell penetrability features set constraints on the individual and the collective nature of the adornments. If one is rather polar, for example, the polarity of another usually must be decreased. Whether the centroid should be tethered to a solid support or free in solution must be considered carefully. If tethered, it is important to consider whether the point of attachment will remain in the final analog, and if so, what affect this may have on its biological properties. One also should consider whether this attachment will prevent the use of one of the potential adornment points.

Just as in one at a time synthesis, linear syntheses are the most risky and produce the lowest yields. Converging methodologies address these limitations successfully, and in combinatorial work, Ugi (four-component) and Passerini (three-component) reactions are very flexible and popular. Generally one has less control over the specific products being produced by such reactions but this is largely compensated for by the molecular diversity available in this way.

Clearly a great deal of thought should go into library design before the work begins.

The first libraries containing heterocycles recognizable as orally active drugs were the 1,4-benzodiazepine-2-ones prepared on resins by Bunin and Ellman in 1992 (Fig. 1.14) (6). A notable chemical feature is the use of amino acid fluorides to drive the amide formation to completion. The choice of benzodiazepines was inspired because of the medicinal importance of these materials and their resemblance to peptides. Here the library was constructed by a combination of three reactants. If each were represented by 10 variations, the library could easily reach 1000 members (10 × 10 × 10) in short order. This would fit the commonly accepted meaning of combinatorial in that all of the possible variants would be constructed. Being selective in the variations actually incorporated, a smaller ("focused") library could be made that answered specific pharmacological questions but at the risk of missing an unexpected discovery. Such a li-

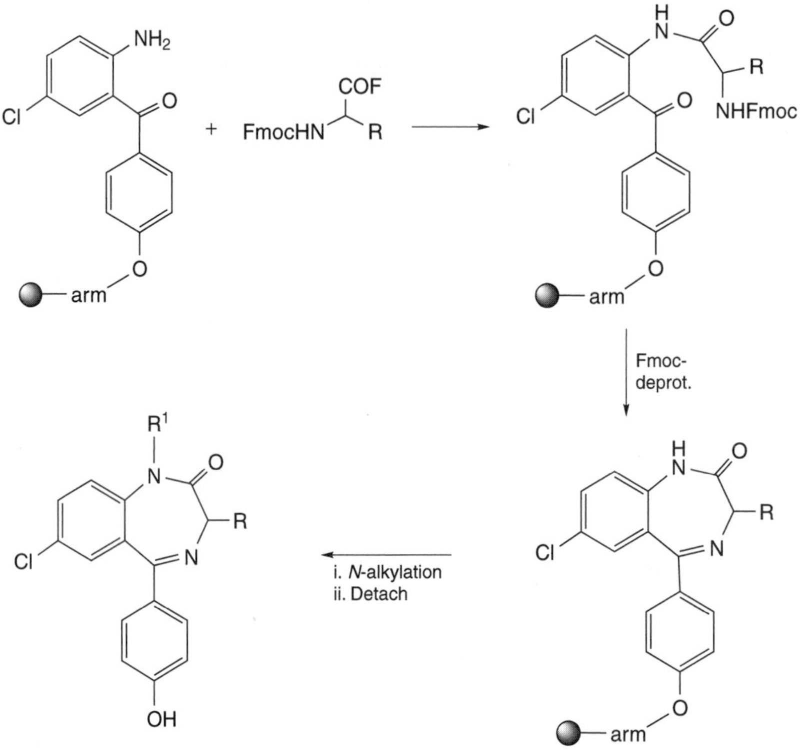

Figure 1.14. Synthesis of benzodiazepine libraries—1.

brary would fit the commonly accepted definition of multiple parallel synthesis. Such focused libraries are often more intellectually satisfying to the practitioner. In this library the attached groups project into space at widely separated compass points around the molecule allowing a systematic exploration of receptor requirements. The centroid has a molecular weight of 160 when all of the available substitution points are occupied by hydrogen atoms. If one accepts an upper molecular weight limit of approximately 500, then four variations can occupy 340 atomic mass units (500–160) so each adornment can have an average of about 85 amu, if the weight is evenly distributed. This gives significant latitude for substitution. When chosen with care to convey drug-like properties and not to exceed collectively the guidelines that Lipinski has developed, the library can contain primarily substances that have a chance to be drug. They also can be so chosen that they allow for sub-

stitution independently of each other and also to be installed without premature separation from the beads. It will also be noted that the centroid chosen has amide and amine linkages that are not involved in adornment attachment and are capable in principle of interacting successfully with a receptor so the molecular weight sacrificed to the centroid may perform pharmacodynamic work. Centroids derived from molecular series that are known to be associated with good pharmacokinetics are often referred to as privileged molecules. Thus, the choice of benzodiazepines to demonstrate the potential power of combinatorial chemistry and multiple parallel synthesis was inspired.

In this pioneering library, the final products were attached to the bead support through a phenolic hydroxyl group that remained as such in the products before testing. Varying the point of attachment of the hydroxyl group would lead to additional multiple

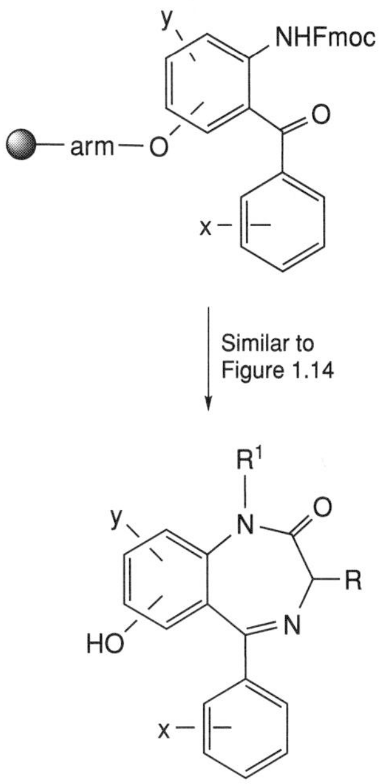

Figure 1.15. Synthesis of benzodiazepine libraries—2.

analogs. Indeed, a variant of this process resulted in a traceable linker in the other aromatic ring (Fig. 1.15).

A somewhat more versatile synthesis of this type using stannanes and palladium acylations (Stille coupling) appeared subsequently (Fig. 1.16) (120). While precedent establishing, this was pharmacologically less

than completely satisfying because agents intended to penetrate well into the CNS should not usually contain such a polar substituent. Despite this, several components in these libraries were bioactive, and the work drew widespread attention to the promise of the methodology and was soon followed by a flood of applications to the preparation of drug-like molecules. In this sequence, attachment to the resin was by an amino acid ester bond. Subsequently this bound intermediate was converted to an imine that cyclized to the benzodiazepine moiety on acid cleavage from the resin (Fig. 1.18) (121). The "traceless linker" technology so introduced has now become standard. One of the additional advantages of this application is that incomplete reaction occurring during the synthesis would lead to products that would not cleave from the resin and could be removed by simple filtration. Furthermore, the products were now indistinguishable from benzodiazepines prepared by usual speed analoging (USA) methods.

Ellman's group also developed a traceless linker sequence of a different type based on HF release of an aryl silicon link to the resin (Fig. 1.17) (122).

As it happens, somewhat gratifyingly, testing of these agents revealed no structural surprises. The intense study of the benzodiazepines in the empirical earlier years had apparently not missed much of significance. Nonetheless, these studies resulted in con-

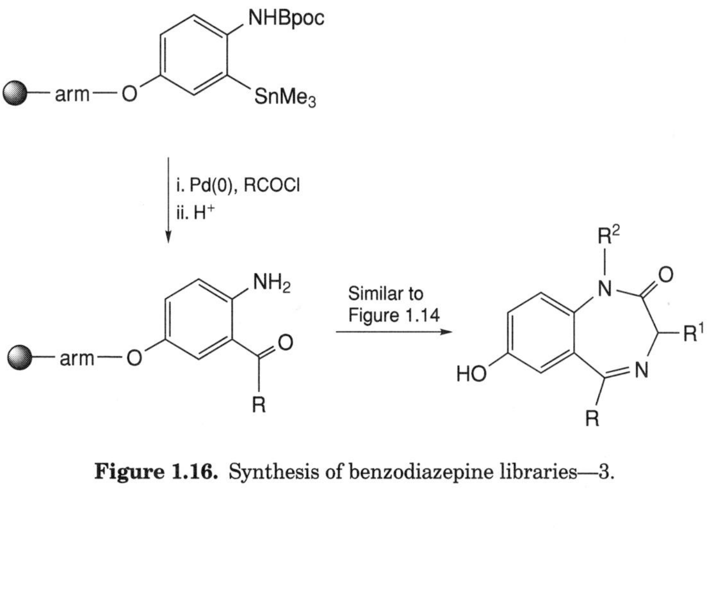

Figure 1.16. Synthesis of benzodiazepine libraries—3.

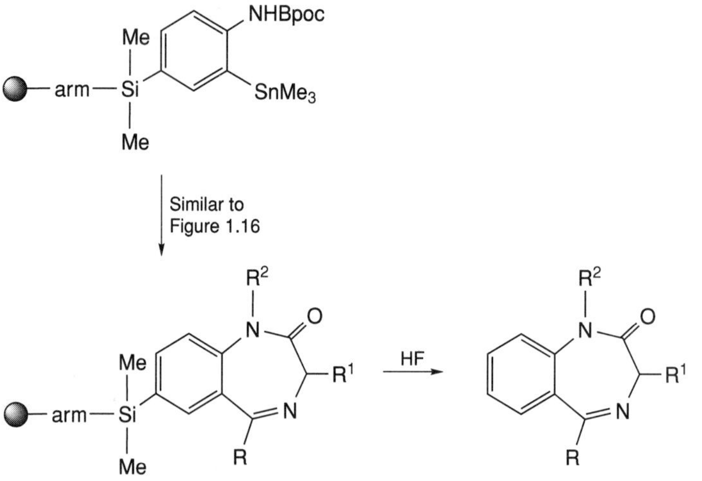

Figure 1.17. Synthesis of benzodiazepine libraries—4.

vincing proof that combinatorial chemical methods would be of dramatic use in preparing agents likely to become orally active drugs.

Benzodiazepine libraries (123) and close analogs such as 1,4-benzodiazepin-2,5-diones (124, 125) continue to be popular. For example, one such library is prepared by a sequence involving a Borch reduction and an internal ester-amide exchange to form the seven-membered ring and then cleavage from the resin. The reaction conditions are mild enough to preserve the optical activity in this library (Fig. 1.19) (126). Use of peptoid starting materials and reductive traceless linker technology are features of the work of Zuckermann et al. (Fig. 1.20) (125).

Representative of another important class of drugs is the 1,4-dihydropyridines. Hantzsch

methodology (without the oxidative step) works efficiently on resins for this purpose. The conditions are mild in this synthesis and the yields are good (Fig. 1.21) (127).

Angiotensin-converting enzyme inhibitors are million dollar molecules. An interesting library of captopril analogs were prepared on resin using a split-mix iterative resynthesis deconvolution procedure. From a collection of about 500 analogs, an analog emerged that was threefold more potent than captopril itself and possessed a K_i of 160 pM (Fig. 1.22) (128)!

Another important class of contemporary drugs that have been made in library form, in this case both in solution and in solid state, are the fluoroquinolone antimicrobial agents (Fig. 1.23) (129, 130). The solution-based yields were superior to those obtained on the resins.

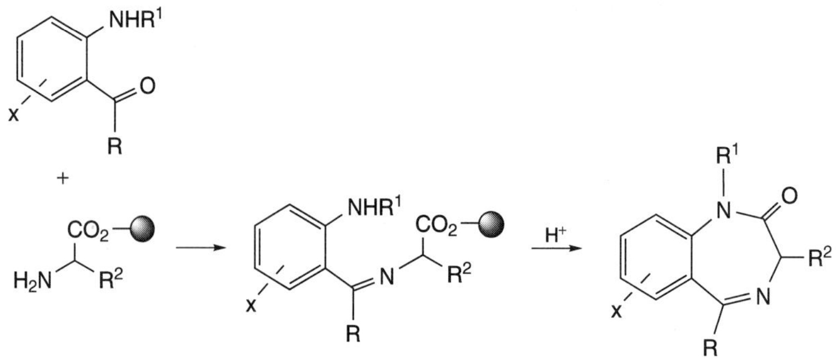

Figure 1.18. Synthesis of benzodiazepine libraries—5.

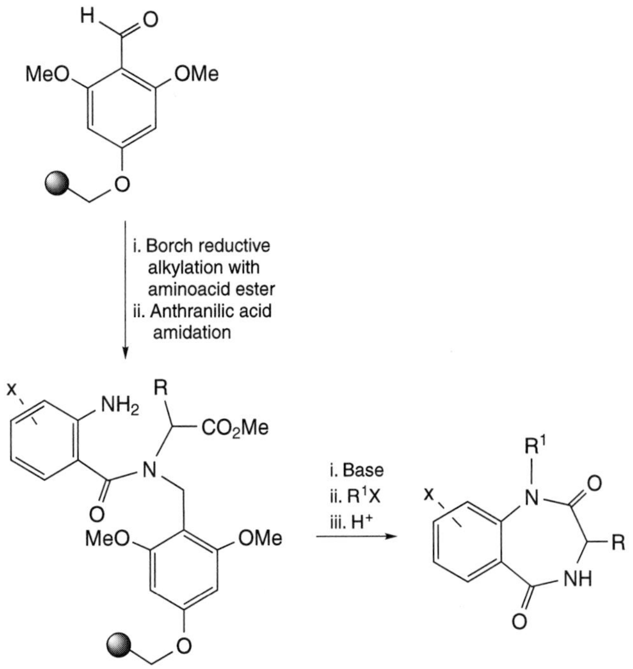

Figure 1.19. Synthesis of benzodiazepine libraries—6.

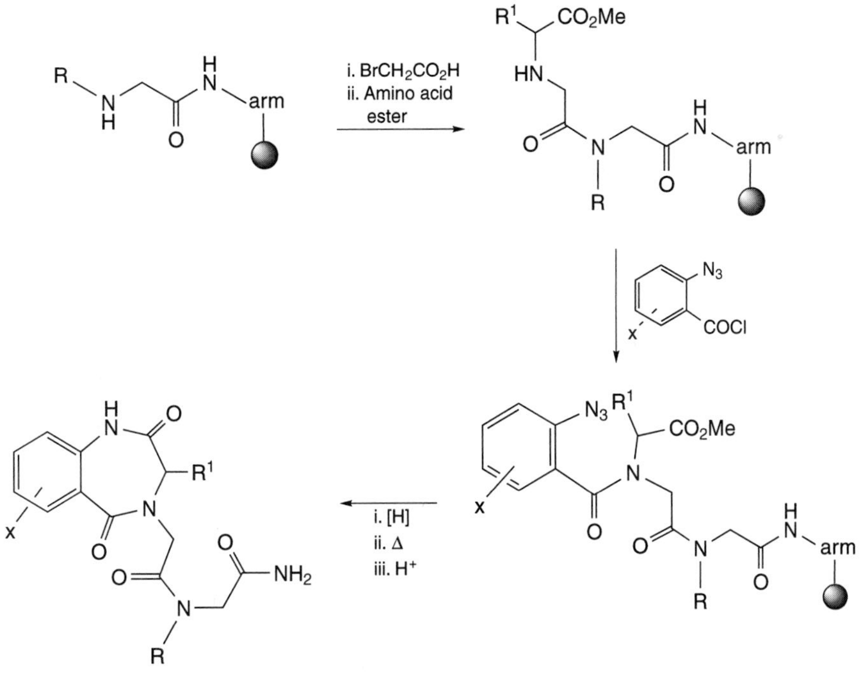

Figure 1.20. Synthesis of benzodiazepine libraries—7.

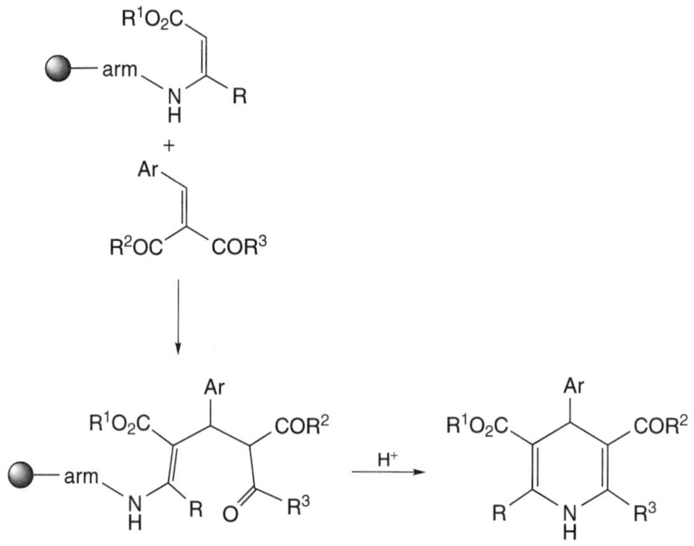

Figure 1.21. Synthesis of dihydropyridine libraries.

Many more examples could be covered, but this gives a representative flavor of the field. By now most of the heterocyclic ring systems have been produced in library form. Particular emphasis has been placed on libraries of molecules with interesting pharmacological properties. Rather extensive reviews of this work exist for the interested reader to consult. [The

reviews of Dolle (36–40) and of Thompson and Ellman (41) are particularly useful in this context.]

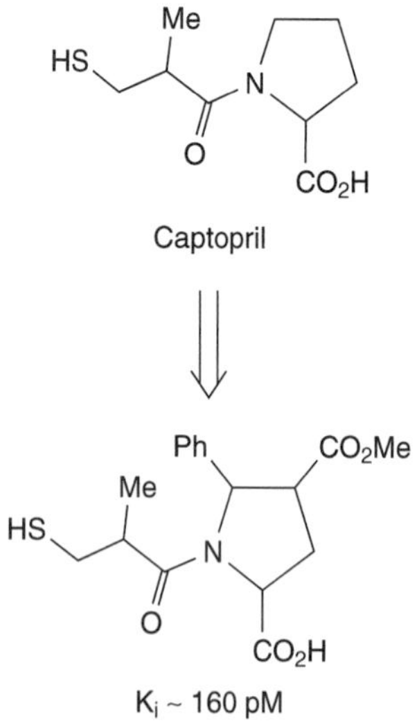

Recently, focused libraries explore the SAR properties of series of contemporary interest where no drugs have yet emerged or where the first of a promising series has been marketed. An example of this is the oxazolidinones. One example of this class, linezolid, has recently been marketed as an orally effective anti-infective, and most large firms have extensive analog programs in progress in attempts to improve on its properties (Fig. 1.24). Combinatorial chemistry plays a significant role in this work. A Pharmacia group has shown that alteration of the morpholine function to a methylated pyrazole moiety produces a broad spectrum analog with oral activity. Palladium coupling of the iodoaromatic moiety of the starting material allows construction of the trimethylsilylacetylene side-chain. Hydrolysis of this group with formic acid produces a methyl ketone, which after a mixed aldol reaction to the dimethylenamine, reacts with methylhydrazine to produce a mixture of the two possible methylated pyrazole isomers. These are separated by chromatography to produce the best of a large series of analogs produced by these and other reaction se-

Figure 1.22. Captopril library results.

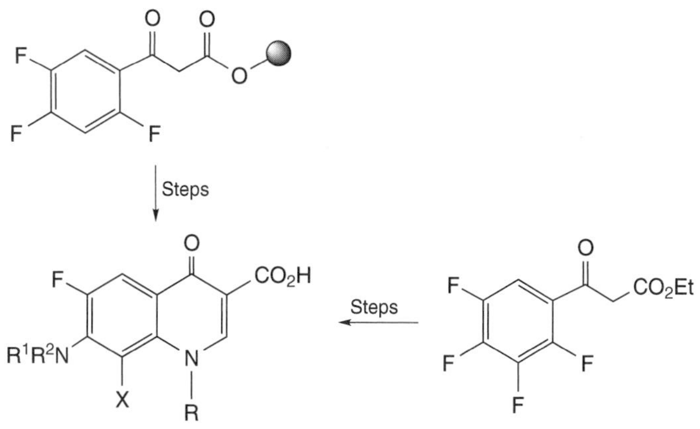

Figure 1.23. Fluoroquinolone libraries.

quences. The product illustrated in Fig. 1.24 has the best combination of *in vitro* and *in vivo* properties in this grouping (131). Reports of related studies have also appeared (132, 133).

A library of cephalosporin antibiotic analogs was made on a solid support (basic alumina) without requiring protection-deprotection. The compounds were prepared in high yield (82–93%) and purity in about 2 min with the aid of microwave irradiation (Fig. 1.25) (134). Microwave acceleration in combinatorial chemistry is a powerful technology for enhancing reactions on solid phases.

The small molecule libraries just exemplified belong to the class called focused. That is, in each case a discrete molecular target was available at the outset and chemical routes were generally available. After some adaptation to the needs of the method and rehearsal of the chemistry, libraries could be generated relatively quickly. Many analogs were then available by comparatively simple variations in the reactants employed. Clearly, in drug seeking, one can operate in much the same manner after identification of a suitable hit molecule.

The strategy required in hit seeking, however, is rather different. Here the initial libraries are usually bigger and more diverse. After the library is screened, and useful molecules are uncovered, subsequent refining libraries are employed that are progressively smaller and more focused. Each succeeding library benefits from the information gained in the previous work so this can be considered

the chemist's equivalent of biological evolution. As the work progresses, the needs for quantities of material for evaluation become more and more so the work usually proceeds back into the larger scale one at a time mode resembling the BC (before combichem) era.

A couple of examples represent the very large amount of work carried out in this manner. First, consider the discovery and progression of OC 144–093, an orally active modulator of P-glycoprotein–mediated multiple drug resistance that has entered clinical studies. First, a 500-membered library of variously substituted imidazoles was prepared on a mixture of aldehyde and amine beads (Fig. 1.26). The choice of materials was based on prior knowledge of the structures of other P-glycoprotein modulators. Screening this library in whole cells led to the identification of two main leads, A, possessing an IC_{50} of 600 nM, and B, possessing an IC_{50} of 80 nM. In addition, B possessed an oral bioavailability in dogs of about 35%. These results were very encouraging.

The third stage involved making a solution-based library based on the structures of A and B. Screening produced leads C, possessing an IC_{50} of 300 nM, and D, with an IC_{50} of 150 nM. Interestingly, D was an unexpected by-product. The chemistry in libraries does not always go as intended. In addition to reasonable potency, D showed enhanced metabolic stability, so it was chosen as the lead for the next phase. Analoging around structure D lead ultimately to OC 144–093, with an IC_{50} of 50 nM and an

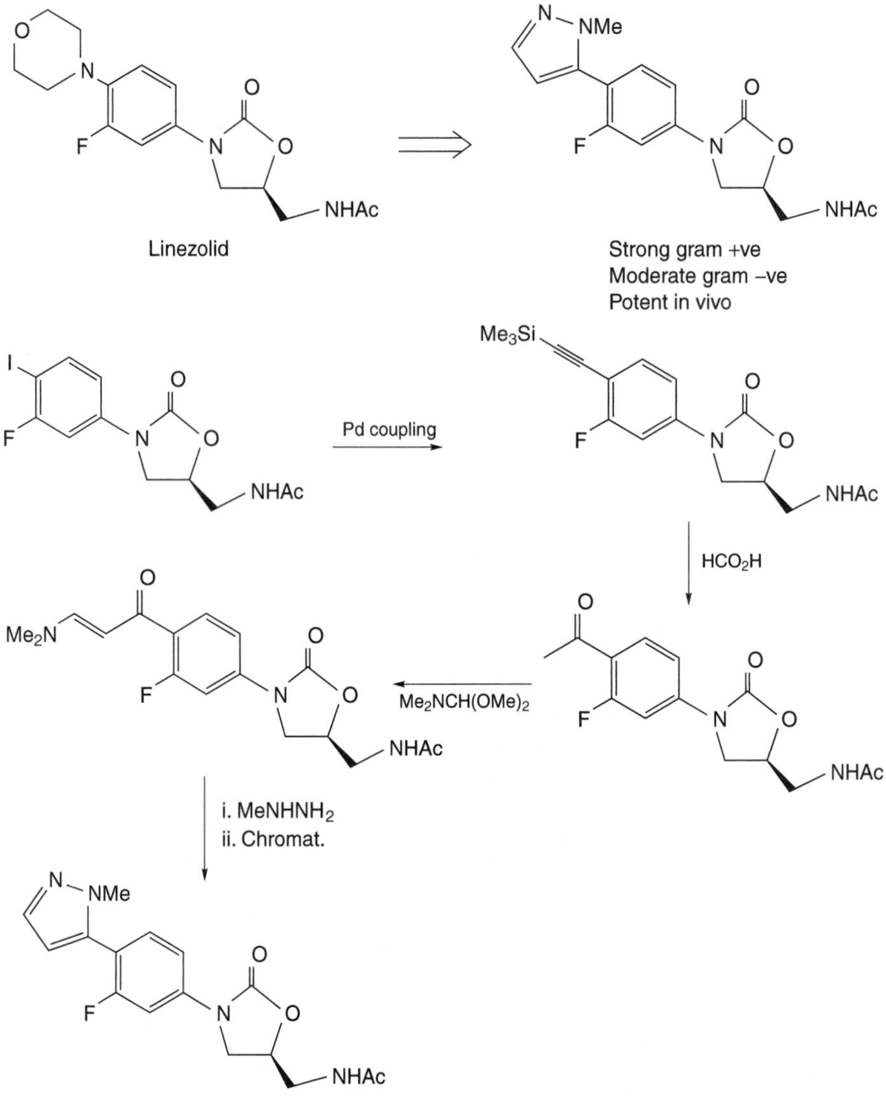

Figure 1.24. Oxazolidinone libraries.

estimated 60% bioavailability after oral administration in humans (135, 136).

Later biological studies *in vitro* and *in vivo* have shown that the agent enhances the activity of paclitaxel by interfering with its export by P-glycoprotein. It is not a substrate for CYP3A and interferes with paclitaxel metabolism only at comparatively high doses. After IV administration, OC 144–093 does not interfere with paclitaxel's pharmacokinetic profile but elevates its area under the curve when given orally. The results are interpreted as

meaning that OC 144–093 interferes with gut P-glycoprotein, enhancing oral bioavailability. Further studies are in progress and it is hoped that a marketed anticancer adjunct will emerge in due course as a result of combinatorial chemistry (137).

In a different study, a search through a company compound collection was made in an attempt to find an inhibitor of the Erm family of methyltransferases. These bacterial enzymes produce resistance to the widely used macrolide-lincosaminide-streptogramin B an-

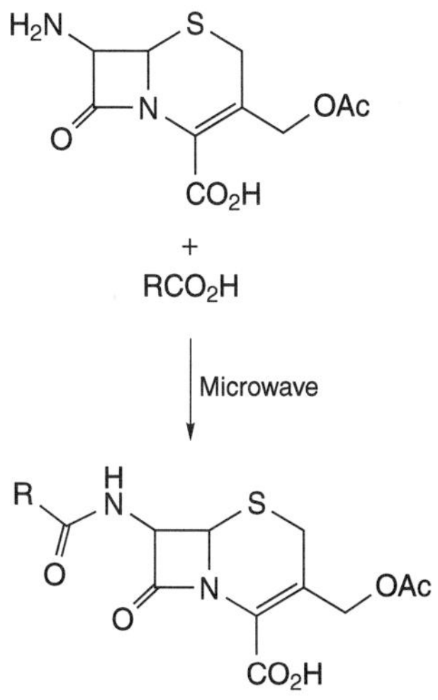

Figure 1.25. Cephalosporin libraries.

tibiotics by catalyzing S-adenosylmethionine–based methylation of a specific adenine residue in 23S bacterial ribosomes. This interferes with the binding of the antibiotics and conveys resistance to them. Using NMR (SAR by NMR) screening, a series of compounds including 1,3-diamino-5-thiomethyltriazine were found to bind to the active site of the enzyme, albeit weakly (1.0 mM for the triazine named) (Fig. 1.27). Analogs were retrieved from the collection, and analogs A, B, and C identified as promising for further work. A solution phase parallel synthesis study was performed from which compound D emerged as being significantly potent. Next a 232-compound library was prepared to discover the best R group on the left side of compound D. From this, compounds E and F emerged. These were now potent in the low micromolar range. The left side of analog E was fixed and the right side was investigated through a 411-membered library. From this, E emerged as the best substance with a K_i of 4 μM against Erm-AM and 10 μM against ErmC. Thus, starting with a very weak lead with a malleable structure, successive libraries produced analogs with quite significant potency for further exploration (138).

It is just a decade after this field became generally active, yet already most of the common drug series and hundreds of different heterocyclic classes have been prepared in library form. Originally the emphasis was on bead-

based chemistry, and this actually slowed general acceptance of the method because few organic and medicinal chemists were familiar with the techniques needed to make small molecules on beads through non-iterative methods, and indeed, much of the needed technology had yet to be developed and disseminated. These problems have largely been overcome, and today the choice of beads or no beads is partly a matter of taste, the size of the libraries being made, and the length of the reaction sequences required.

The remainder of this chapter deals with selected examples that illustrates particular concepts and methodologies.

4.1 Purification

In communicating their results, chemists explicate the route with formulae and often discuss the relative strengths and weaknesses of key reagents but almost never devote time to workup. Even so, the details of the workup require attention to detail in the performance and are sometimes quite challenging. This factor becomes even more demanding in combinatorial work where the need for rapid, effective workup is intensified. Little is gained if one saves much time in construction only to have to give this back by tedious and repetitious purification schemes. Performing chemistry on beads addresses this in that simple filtration and washing often suffices. This is not as useful if the reactions do not go to completion, so considerable excess of reagents and more lengthy times are often employed to drive the reactions further to completion. Separation from solution in solid form or simple evaporation is very convenient, and manifolds for filtration and for solvent removal are commercially available. From a drugability standpoint, there is a danger in this. Compounds that separate readily from polar solvents are often of very low water solubility and present difficulties in testing. A number of commercially available combinatorial screening libraries are peppered with such substances.

Column chromatography is powerful but often labor intensive and solvent consuming. Separation of hundreds of analogs by column chromatography would be a nightmare. With smaller, focused libraries, this is often more manageable.

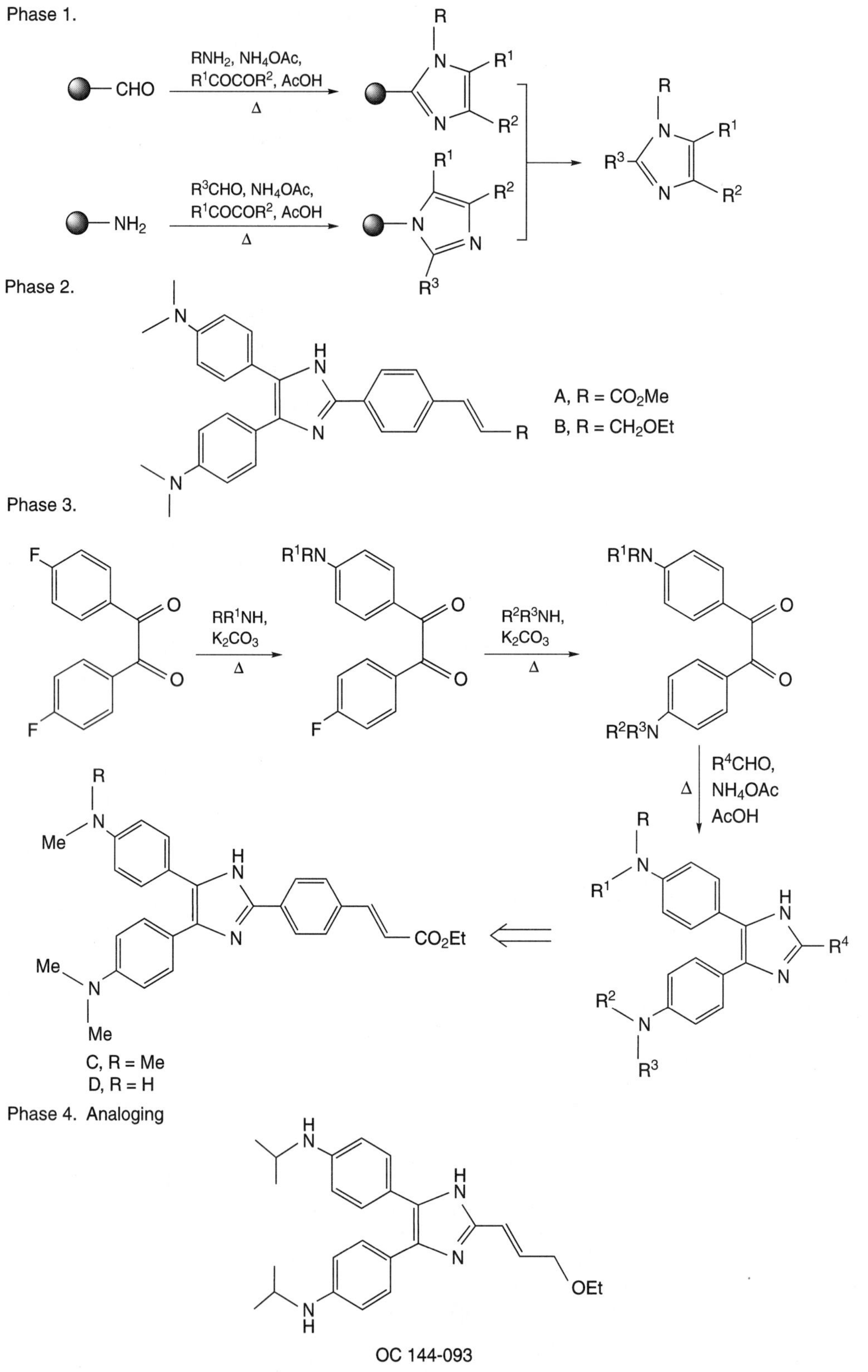

Phase 1.

Phase 2.

A, R = CO$_2$Me
B, R = CH$_2$OEt

Phase 3.

C, R = Me
D, R = H

Phase 4. Analoging

OC 144-093

Figure 1.26. Library-based discovery of clinical candidate C 144–093.

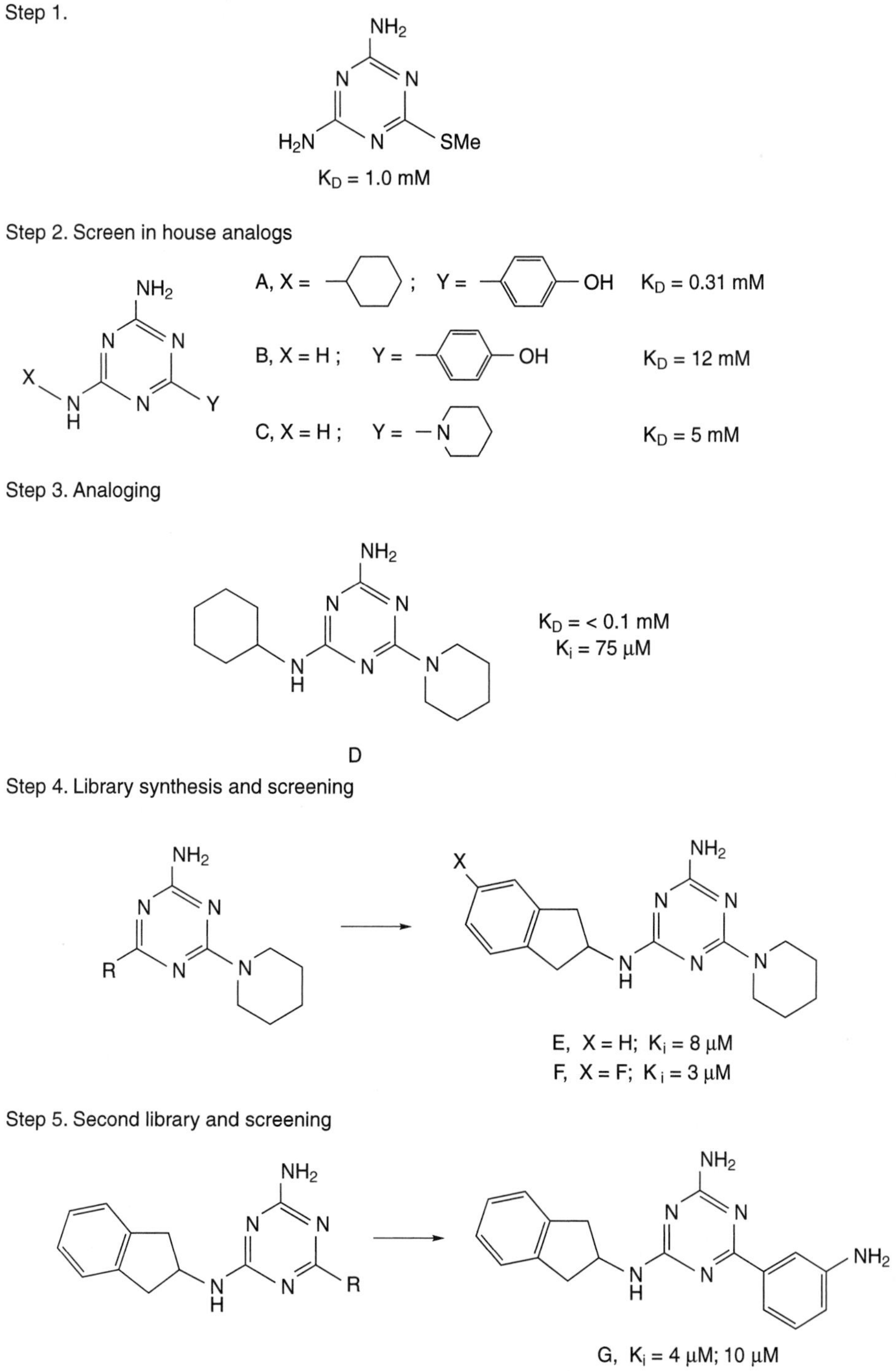

Step 1.

K_D = 1.0 mM

Step 2. Screen in house analogs

A, X = ⬡ ; Y = –◯–OH K_D = 0.31 mM

B, X = H ; Y = –◯–OH K_D = 12 mM

C, X = H ; Y = –N⬡ K_D = 5 mM

Step 3. Analoging

K_D = < 0.1 mM
K_i = 75 μM

D

Step 4. Library synthesis and screening

E, X = H; K_i = 8 μM
F, X = F; K_i = 3 μM

Step 5. Second library and screening

G, K_i = 4 μM; 10 μM

Figure 1.27. A library of Erm inhibitors using SAR by NMR.

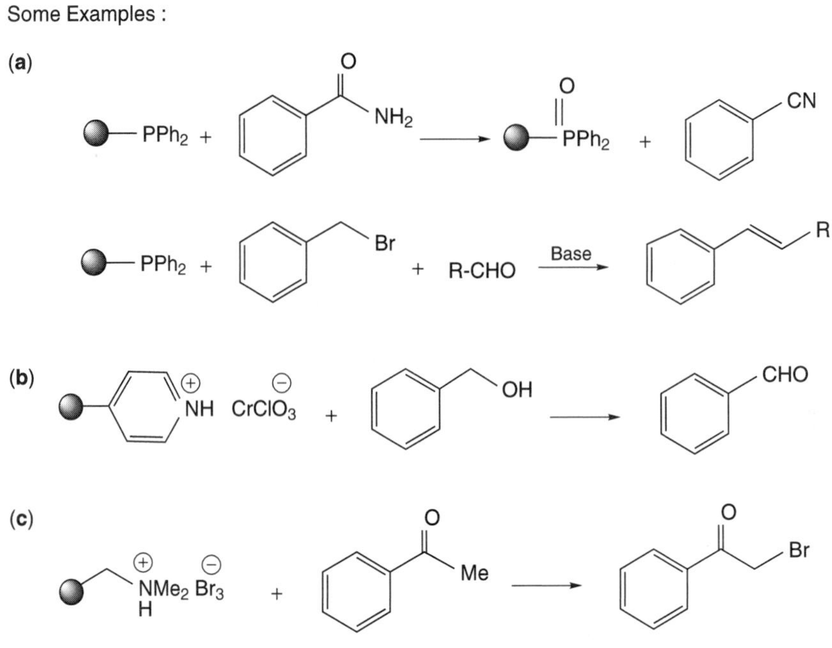

Some Examples :

(a)

(b)

(c)

Figure 1.28. Resin tethered reagents used in combichem.

More frequently automated chromatographic reverse phase methods are employed in which round the clock separations not requiring constant human supervision are available.

Chromofiltration methods are powerful and rapid but often require study for optimization. We have found, for example, that choice of the appropriate solvent for solution-based multiple parallel synthesis can occasionally result in reaction mixtures from which the desired product can be isolated in pure form by suitable choice of absorbent and concentration or evaporation from the eluent (139). However this is generally exceptional. One can sometimes doctor silica gel, for example, to enhance its use for these purposes. After amide formation, when the acid component is used in excess, adding 1% of sodium bicarbonate to silica gel and mixing thoroughly provides a convenient way to remove reaction debris so that filtration produces pure amide on evaporation (139).

More generally applicable has been the development of many reagents tethered to solid supports. These reagents perform their intended role and then the excess reagent and the exhausted reactants can be filtered and washed away for easy product isolation. In this case, the compounds are in solution and the reagents are on the solids. A great many reagents have been prepared for use in this manner and the area has been reviewed extensively (140–142). Whereas ion exchange applications have been around a long time in the medicinal chemist's laboratory, the requirements of combinatorial chemistry have engendered a flowering of additional resins and uses. These are particularly useful in solution-based MPS but clearly find wide applications in other types of chemistry as well. An exhaustive treatment is beyond the scope of this summary, but a few examples help clarify the many uses to which this exciting technology can be put. In Fig. 1.28a, resin bound diphenylphosphine is used to reduce benzamide to benzonitrile. The diphenylphosphine oxide product is often troublesome to remove from nontethered reactions, but here is readily separated leaving the clean product behind. In Fig. 1.28b, tethered chromate cleanly oxidizes benzyl alcohol to benzaldehyde. In Fig. 1.28c, tethered bromine dimethylamino hydrobromide cleanly α-brominates acetophenone.

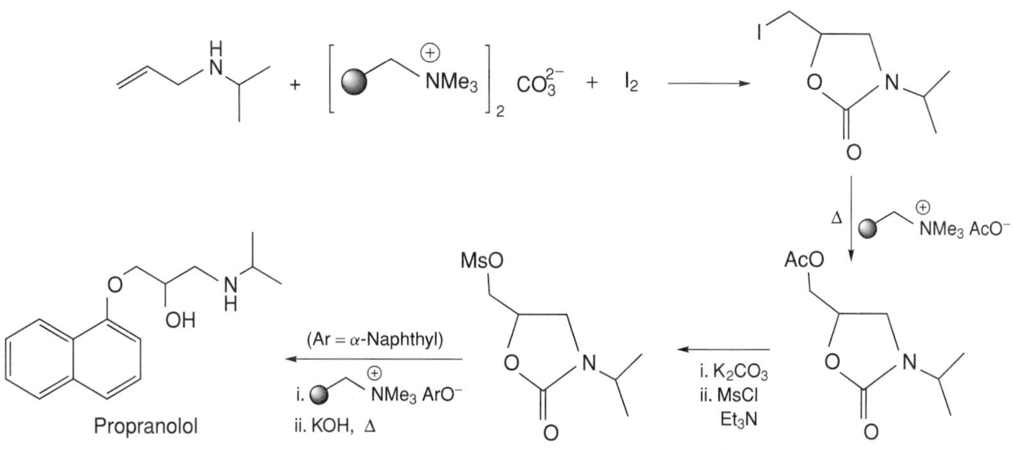

Figure 1.29. Illustration of tethered reagents in preparing propranolol.

An illustrative example of the power of this method is the resin-assisted synthesis of the adrenergic β-blocking agent, propranolol, outlined in Fig. 1.29 (143). In the six chemical steps required for this preparation, three involved the use of resin-based reagents. It is obvious that many possible variants leading to a library of related molecules could be prepared by simple modifications of the reagents and substrates.

Another important and powerful methodology employs capture or scavenger resins. Here resin-tethered bases, for example, are employed to remove acidic reaction products from reaction mixtures, and these can be regenerated for further use. Likewise, these materials can be used to remove acidic reagents or byproducts leaving the desired reaction product in the solution. Isocyanate resins are used to remove primary and secondary amines and alcohols, benzaldehyde tethered to a resin is used to remove primary amines and hydrazines, carbonate resins remove carboxylic acids and phenols, diphenylphosphine resins remove alkyl halides, and tethered trisamine removes acid chlorides, sulfonyl chlorides, and isocyanates. This methodology is similar in concept to the well-established ion exchange methodologies. The use of acid and base resins to remove ionizable products and byproducts from reaction mixtures is familiar. The use of tethered isocyanates to remove excess amine is less familiar but readily comprehended. These and analogous reagents have been

widely employed and reviewed (140–142, 144). An example of the preparation of a library of drug-like molecules in solution employing resin capture methodology is illustrated in Fig. 1.30 wherein tamoxifen analogs are efficiently and cleanly prepared (145). The sequence starts with Suzuki-type coupling with a series of aryl halides. The desired intermediate is present in the product mixture with a variety of reaction detritus including diarylated material. The desired product is captured out of this stew by use of a resin bound aryl iodide which reacts exclusively with it. This device is sometimes called phase switching. The resin bound product is purified by rinsing and the reaction sequence is completed by acid release from the capture resin.

Highly fluorinated organic molecules are often insoluble at room temperature in both water and in organic solvents. At higher temperatures, however, they are soluble. Thus, heating a reaction mixture involving such a molecule to speed the reaction and then cooling on completion often allows the product to separate in pure or at least purified form by phase switching into the fluorinated solvent. This is very convenient for rapid work-up of combinatorial libraries. Recently, silica gel columns with a fluorous phase have been introduced to facilitate separations. Compounds elute from these columns in the order of their decreasing fluorine content. This can be illustrated (Fig. 1.31) by application to a library of 100 mappicine alkaloid analogs. In this case,

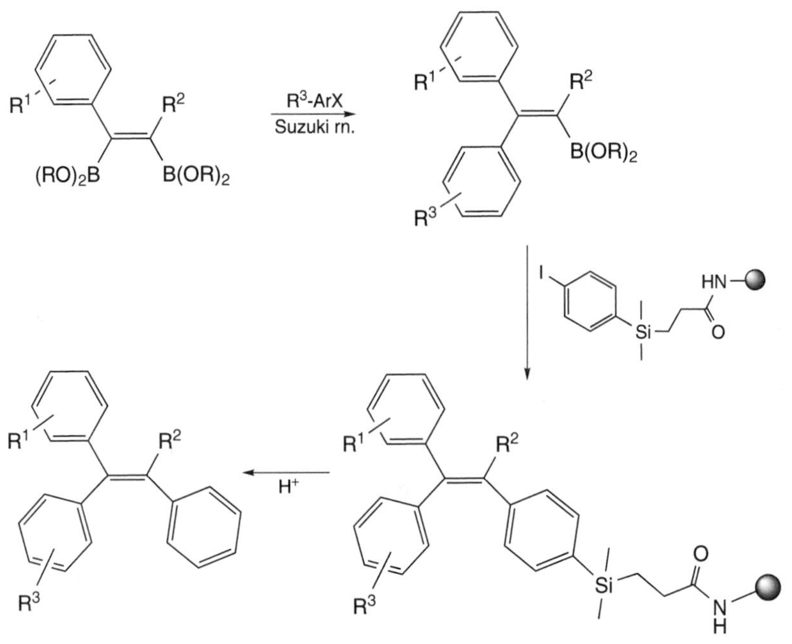

Figure 1.30. Illustration of resin capture use in a library of tamoxiphen analogs.

each analog was tagged with an arm containing one of a series of a perfluorinated hydrocarbons (CnFn+1). Syntheses could then be performed in mixtures and at the end the products were separated based on the number of fluorines in their tags. Detagging produced the individual pure products. The method is not general but is very convenient when used appropriately (146, 147).

It is also possible to remove desired reaction products, if aggressive by-products and reagents are not present, by chromatography over immobilized receptor preparations. Gel filtration is also helpful in sorting out a binding component from a mixture library containing analogs with little affinity for the pharmacological target.

Because of the impact of these newer methods, today one rarely sees a separatory funnel in a combichem laboratory. This chemistry is also comparatively "green" in that solvent needs can be greatly reduced and disposal of unwanted materials is simplified.

4.2 Synthetic Success and Product Purity

Only a relatively few census reports assessing the success rate of combinatorial library con-

struction are available (148), but the consensus is that 85% success is fine. The level of desired purity of the components is also a matter of debate. Actionable quantitative biological data can be obtained from pure samples and uncertainties into selecting the most active constituents to pursue are increasingly introduced by assaying less pure material. Three grades of products can be distinguished. Pure usually means greater than 95% in combinatorial work. A lower but generally acceptable grade of purity is arbitrarily chosen at about 80%, and such compounds can be labeled as "practical grade." Less than this level of purity is generally unacceptable and is sometimes, disparagingly, called "practically." In very large libraries where purity analysis of each component is rarely available, inevitably one has some components in this poor state. Indeed, chemists occasionally report anecdotally finding that an active component in a library has none of the intended compound in it at all. This complicates analoging but is more satisfactory than basing SAR-based design on negative activity data wherein one can be significantly mislead in such cases.

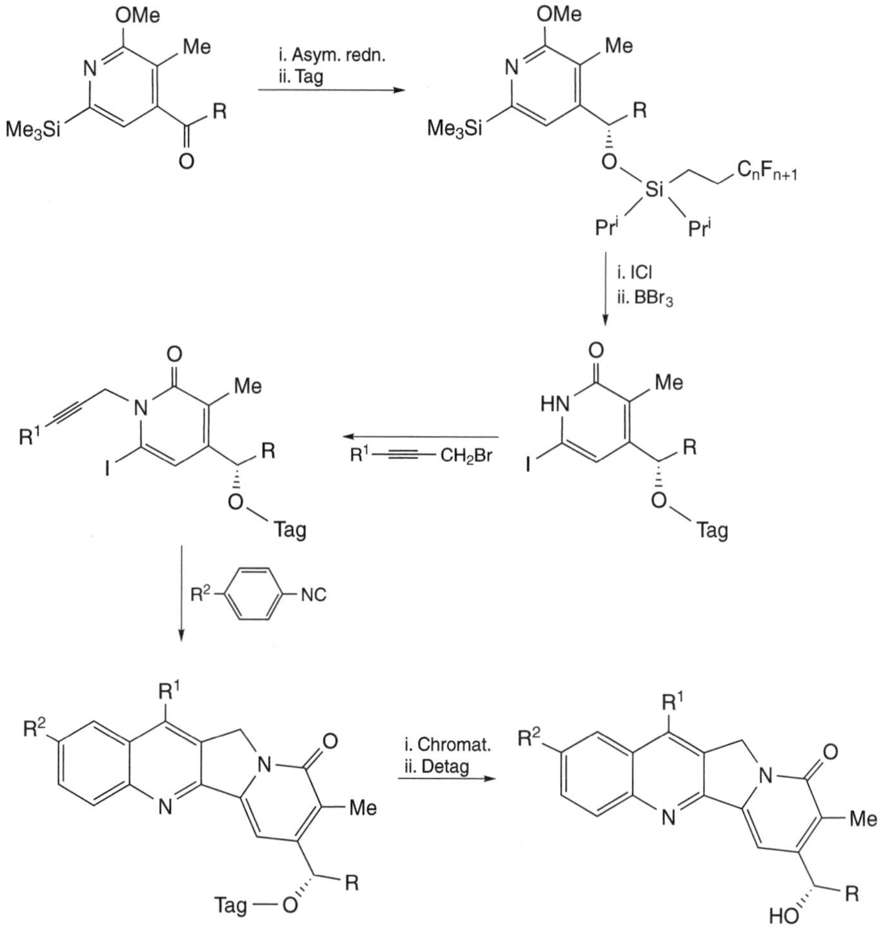

Figure 1.31. Illustration of fluorous phase methods in the synthesis of a mappacine library.

One is disturbed to note also that in a few cases where a census has been taken of very large libraries, the wrong or missing structures are not statistically distributed (149). Thus, such libraries have a structural bias. For example, the chemistry may selectively favor production of more lipophilic substances so that hydrophilic examples are underrepresented. It is hard to see how to get around this conveniently.

4.3   Resins and Solid Supports

A great variety of resins and solid supports are available for combinatorial work (151). Gel-type supports are popular and consist of a flexible polymeric matrix to which is attached functional groups capable of binding small molecules. The particular advantage of this inert support is that the whole volume of the gel is available for use rather than just the surface. Generally these consist of cross-linked polystyrene resins, cross-linked polyacrylamide resins, polyethylene glycol (PEG) grafted resins, and PEG-based resins. Surface functionalized supports have a lower loading capacity and many types are available. These include cellulose fibers, sintered polyethylene, glass, and silica gels. Composite gels are also used. These include treated Teflon membranes, kieselguhr, and the like. Brush polymers consist of polystyrene or the like grafted onto a polyethylene film or tube.

The linking functionality varies. Commonly employed resins are the Merrifield,

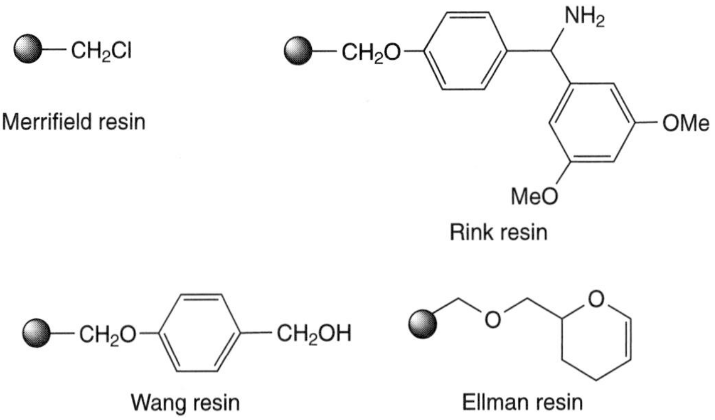

Figure 1.32. Some common resin types.

Wang, Rink, and Ellman types. These are illustrated in Fig. 1.32, and the reader can readily appreciate the kinds of chemistry that they allow.

4.4 Microwave Accelerations

Molecules possessing a permanent dipole align themselves in a microwave apparatus and oscillate as the field oscillates. This rapid motion generates intense homogeneous internal heat greatly facilitating organic reactions, especially in the solid state. For example, heat-demanding Diels-Alder reactions can take days on solid support, hours in solution, and only minutes under microwave. This has been adapted to combinatorial methods and is even compatible with a 96-well plate format (150).

4.5 Analytical Considerations

Analysis of the degree of completeness and the identity of the product is simpler than that seen in solid phase work. This closely parallels general experience in the pre-combichem days with the exception that the work load is greatly magnified. Automation is called for and hplc/tof mass spectrometry is of particular value. Even so, with very large libraries, one is usually restricted by necessity to statistical sampling and compound identification rarely goes beyond ascertaining whether the product has the correct molecular weight. If activity is found then more detailed examination takes place.

With solid-state libraries, the problem is much more complex. An enormous effort has been put into working out analysis of single beads with mass spectrometry, Raman spectroscopy, magic angle NMR, and chemiluminescence techniques being particularly popular.

4.6 Informetrics

Combinatorial synthesis and high-throughput screening generate an enormous amount of data. Keeping track of this is a job for high-speed computers. Many firms have developed their own programs for the data handling, and there are commercial packages that may be useful as well. The best of these have structure drawing capacity also.

4.7 Patents

Patent considerations are complex in combinatorial chemistry. The mass of potential data is hard to compress into a suitable format for this purpose. Commonly, patenting takes place comparatively late in a drug-seeking campaign and so differs little from traditional patenting. One notes however that the comparative speed and ease of molecule construction makes it possible to reduce to practice rather more examples that would have been possible in the one-at-a-time days.

Rather more disturbing is the increasing tendency to patent various means of making and evaluating libraries rather than focusing on their content. The fundamental purpose of patenting is to promote the useful arts and to provide protection for innovative discoveries for a period and then to share them with society in general. Patenting of means of producing libraries, if carried to an extreme, would have a dampening effect on the development of the field and so would inhibit the development of the useful arts. This should be guarded against.

5   SUMMARY AND CONCLUSIONS

Combinatorial chemistry and multiple parallel syntheses have transformed the field of medicinal chemistry for the better. The last decade has seen a revitalization and much dramatically useful technology has been discovered. No laboratory seriously involved in the search for new therapeutic agents can afford not to employ this technology.

From the vantage point of 2002, one can now look back at what has been done in the amazingly short time that this technique has been widely explored and one can see some things more clearly now and use the methodology more cunningly.

In the heady early days of combinatorial chemistry one frequently heard the opinion that existing drugs were only those to which nature or good fortune had laid a clear path. Some believed that there were large numbers of underexplored structural types that could be drugs if only they were prepared and screened. Combichem promised to make this a reality. It would be nice, indeed, if this had turned out to be true! It cannot be denied that there is some justice in this belief; speculative synthesis continues to reveal important drugs. Nonetheless, the cruel restraints that ADME and toxicity considerations place on our chemical imagination have ruined this dream of easy and unlimited progress. The present wedding of combichem with medicinal chemical knowledge is extremely powerful, and we no longer in the main waste time on collections of molecules that have no chance of becoming drugs. Clearly space for chemical diversity is larger than space for medicinal diversity.

BC (before combichem) there was little motivation for enhanced speed of synthesis. Generally, synthesis could be accomplished much more quickly than screening and evaluation of the products. Enhanced speed of construction simply produced a greater backlog of work to be done. The advent of high-through-put screening in the 1980s changed all this. The backlogs emptied rapidly, and there was a demand for more compounds. In addition, new firms were founded to take advantage of newer screening methodologies. These firms had no retained chemical libraries to screen and larger firms were reluctant to allow their libraries to be screened by outsiders. A significant part of these needs were met by the methods in this chapter. With synthesis and screening back in phase, the next choke point in the pipeline has become animal testing, pharmacokinetics, toxicity, solubility, and penetrability. These factors are presently under intensive examination in attempts to elucidate these properties in a similarly rapid fashion or to predict them so that favorable characteristics can be designed into chemical library members from the outset and thus largely avoid having to deal with them. It can readily be seen that further choke points lie distally in the pipeline and these will have to be dealt with in turn. Some time can be saved by speeding things along the way and also by dealing with the remaining constrictions in parallel rather than simultaneously, but it is difficult to see how they can all be resolved in a rapid manner. Fortunately the flow through the pipeline diminishes through increasing failure of leads to qualify for further advancement and this is helpful in reducing the magnitude of the job but the problems remaining will still be vexing. Part of the difficulty is that certain biological phenomena cannot be hurried. For example, no matter how much money and effort one is willing to throw at the problem, producing a baby requires essentially 9 months from conception. Hiring nine women will not result in producing a baby in 1 month. The problem in shortening drug seeking is further compounded in that the problem is not akin to brick laying. To produce a brick wall of a given dimensions more quickly is largely a matter of buying the bricks and hiring and motivating enough skilled labor. In drug seeking, one has to design the bricks first and develop the technology. Combichem does speed the process along but does not remove the elements of uncertainty that must be overcome. Given the strictures placed on clinical studies and their solidification in law and custom, it is hard to see how this phase of the drug seeking sequence can be shortened through chemical effort.

High-throughput screening can be likened to hastening the process of finding a needle in a haystack. Combinatorial chemistry can be likened to the preparation of needles. Ideally one should strive to make a few more useful needles embedded in progressively smaller haystacks. This involves mating as well as is possible productive chemical characteristics with productive biological properties. Combinatorial chemistry and multiple parallel synthesis in the hands of the skillful and lucky chemist rapidly zeros in on the best combination of atoms for a given purpose. This chemist receives approbation for his/her efforts. Those who consistently come up with useless compounds will eventually be encouraged to find other work.

There has been a dramatic increase in investment in drug discovery during the last decade (estimated at 10% annually). Unfortunately this has yet to result in a burst of new introductions. Certainly chemical novelty has largely given way to potential use. Diversity no longer rules. This is perhaps the combinatorial chemist's equivalent of the businessman's mantra that whereas efficiency is doing things properly, effectiveness is doing proper things. As with much of the points being discussed, achieving a proper balance is essential.

It is interesting to note also that a 100-fold increase in screening activity has not yet resulted in a corresponding increase in the introduction of new pharmaceuticals. Part of the explanation for this is that ease of synthesis does not necessarily equate to equivalent value of the products. If each compound in chemical libraries was carefully designed and the data therefrom carefully analyzed, then the disparity would be smaller than the present experience produces. Another exculpatory factor is that much of the low hanging fruit has already been harvested and the remaining diseases are more chronic than acute and are much more complex in their etiology.

Despite all of these considerations, drug seeking is an exciting enterprise calling for the best of our talents and the appropriate use of high speed synthetic methods gives us a powerful new tool to use.

REFERENCES

1. R. B. Merrifield, *J. Am. Chem. Soc.*, **85**, 2149 (1963).

2. R. B. Merrifield, *Science*, **232**, 341 (1986).

3. A. Furka, *Drug Discov. Today*, **7**, 1 (2002).

4. M. H. Geysen, R. H. Meloen, and S. J. Barteling, *Proc. Natl. Acad. Sci. USA*, **81**, 3998 (1984).

5. R. A. Houghton, C. Pinella, S. E. Blondelle, J. R. Appel, C. T. Dooley, and J. H. Cuervo, *Nature*, **354**, 84 (1990).

6. B. A. Bunin and J. A. Ellman, *J. Am. Chem. Soc.*, **114**, 10997 (1992).

7. R. C. Pless and R. L. Letsinger, *Nucl. Acids Res.*, **2**, 773 (1975).

8. Z. G. Wang and O. Hindsgaul, *Adv. Exp. Med. Biol.*, **435**, 219 (1998).

9. J. N. Abelson, *Combinatorial Chemistry*, Academic Press, San Diego, CA, 1996.

10. W. Bannwarth and E. Felder, Eds., *Combinatorial Chemistry: A Practical Approach*, Wiley-VCH, New York, 2000.

11. J. Beck-Sickiner and P. Heber, *Combinatorial Strategies in Biology and Chemistry*, Wiley, New York, 2002.

12. K. Burgess, Ed., *Solid-Phase Organic Synthesis*, Wiley, New York, 1999.

13. B. A. Bunin, *The Combinatorial Index*, Academic Press, San Diego, 1998.

14. I. M. Chaiken and K. D. Janda, Eds., *Molecular Diversity and Combinatorial Chemistry: Libraries and Drug Discovery*, American Chemical Society, Washington, DC, 1996.

15. A. W. Czarnik, Ed., *Solid-Phase Organic Synthesis*, vol. **1**, Wiley, New York, 2001.

16. A. W. Czarnik and S. H. DeWitt, Eds., *A Practical Guide to Combinatorial Chemistry*, American Chemical Society, Washington, DC, 1997.

17. S. El-Basil, *Combinatorial Organic Chemistry: An Educational Approach*, Nova, Huntington, NY, 2000.

18. M. Famulok, E. L. Winnacker, C.-H. Wong, Eds., *Combinatorial Chemistry in Biology*, Springer Verlag, New York, 1999.

19. H. Fenniri, Ed., *Combinatorial Chemistry: A Practical Approach*, Oxford University Press, New York, 2000.

20. A. K. Ghose and V. N. Viswanadhan, Eds., *Combinatorial Library Design and Evaluation*, M. Dekker, New York, 2001.

21. E. M. Gordon and J. F. Kerwin, Eds., *Combinatorial Chemistry and Molecular Diversity in Drug Discovery*, Wiley, New York, 1998.

22. G. Jung, Ed., *Combinatorial Chemistry: Synthesis, Analysis, Screening*, Wiley-VCH, New York, 2000.

23. G. Jung, Ed., *Combinatorial Peptide and Non-Peptide Libraries—A Handbook*, Wiley-VCH, New York, 1997.

24. S. Miertus and G. Fassina, *Combinatorial Chemistry and Technology: Principles, Methods and Applications*, M. Dekker, New York, 1999.

25. W. H. Moos, M. R. Pavia, B. K. Kay, and A. D. Ellington, Eds., *Annual Reports in Combinatorial Chemistry and Molecular Diversity*, vol. **1**, ESCOM Science Publishers, Leiden, The Netherlands, 1997.

26. K. C. Nicolaou, *Handbook of Combinatorial Chemistry*, vols. **1 and 2**, Wiley, New York, 2002.

27. L. Nielsen, *Combinatorial Chemistry in Drug Design*, Blackie Academic and Professional, London, 1998.

28. N. K. Terrett, *Combinatorial Chemistry*, Oxford University Press, New York, 1998.

29. P. Veerapandian, Ed., *Structure-Based Drug Design*, M. Dekker, New York, 1997.

30. S. R. Wilson and A. W. Czarnik, *Combinatorial Chemistry: Synthesis and Application*, Wiley, New York, 1997.

31. P. Seneci, *Solid Phase Synthesis and Combinatorial Technologies*, Wiley-VCH, New York, 2000.

32. M. Swartz, Ed., *Analytical Techniques in Combinatorial Chemistry*, M. Dekker, New York, 2000.

33. N. K. Terrelt, *Combinatorial Chemistry*, Oxford University Press, New York, 1998.

34. B. Yang, *Analytical Methods in Combinatorial Chemistry*, CRC Press, Boca Raton, FL, 2000.

35. F. Z. Doerwald, *Organic Synthesis on Solid Phase—Supports, Linkers and Reactants*, Wiley, New York, 2000.

36. R. E. Dolle, *Mol. Diversity*, **4**, 233 (1998).

37. R. E. Dolle, *Mol. Diversity*, **3**, 199 (1998).

38. R. E. Dolle and K. H. Nelson Jr., *J. Comb. Chem.*, **1**, 235 (1999).

39. R. E. Dolle, *J. Comb. Chem.*, **2**, 383 (2000).

40. R. E. Dolle, *J. Comb. Chem.*, **3**, 477 (2001).

41. L. A. Thompson and J. A. Ellman, *Chem. Rev.* **96**, 555 (1996).

42. M. A. Gallop, R. W. Barrett, W. J. Dower, S. P. Fodor, and E. M. Gordon, *J. Med. Chem.*, **37**, 1233 (1994).

43. E. M. Gordon, M. A. Gallop, and D. V. Patel, *Acc. Chem. Res.*, **29**, 144 (1996).

44. N. Terrett, M. Gardner, E. M. Gordon, R. J. Kobylecki, and J. Steele, *Tetrahedron*, **51**, 8135 (1995).

45. F. Balkenhohl, C. von dem Bussche-Hunnefeld, A. Lansky, and C. Zechel, *Angew. Chem. Int. Educ. English*, **35**, 2288 (1996).

46. J. A. Elolman, *Accts. Chem. Res.*, **29**, 132 (1996).

47. P. H. Hermkens, H. C. Ottenheijm, and D. C. Rees, *Tetrahedron*, **53**, 5643 (1997).

48. P. H. Hermkens, H. C. Ottenheijm, and D. C. Rees, *Tetrahedron*, **52**, 4527 (1996).

49. S. V. Ley, I. R. Baxendale, R. N. Bream, et al., *J. Chem. Soc., Perkin Trans. I*, 3815 (2000).

50. J. D. Reed, D. L. Edwards, and C. F. Gonzalez, *Mol. Plant Microbe Interact.*, **10**, 537 (1997).

51. G. Koppel, C. Dodds, B. Houchins, D. Hunden, D. Johnson, et al., *Chem. Biol.*, **2**, 483 (1995).

52. D. Muller, I. Zeltser, G. Bitan, and C. Gilon, *J. Org. Chem.*, **62**, 411 (1997).

53. A. B. Smith, S. D. Knight, P. A. Sprengeler, and R. Hirschmann, *Bioorg. Med. Chem.*, **4**, 1021 (1996).

54. A. Wallace, K. S. Koblan, K. Hamilton, D. J. Marquis-Omer, P. J. Miller, et al., *J. Biol. Chem.*, **271**, 31306 (1996).

55. J. Eichler and R. A. Houghton, *Mol. Med. Today*, **1**, 174 (1995).

56. J. Blake, J. V. Johnston, K. E. Hellstrom, H. Marquardt, and L. Chen, *J. Exp. Med.*, **184**, 121 (1996).

57. J. Eichler and R. A. Houghton, *Biochemistry*, **32**, 11035 (1993).

58. R. A. Houghton, C. Pinilla, J. R. Appel, S. E. Blondelle, C. T. Dooley, et al., *J. Med. Chem.*, **42**, 3743 (1999).

59. J. A. Camarero, B. Ayers, and T. W. Muir, *Biochemistry*, **37**, 7487, 1998.

60. J. R. Zysk and W. R. Baumbach, *Comb. Chem. High Throughput Screen.*, **1**, 171 (1998).

61. A. Brinker, E. Weber, D. Stoll, J. Voigt, A. Muller, et al., *Eur. J. Biochem.*, **267**, 5085 (2000).

62. S. E. Blondelle and K. Lohner, *Biopolymers*, **55**, 74 (2000).

63. R. A. Houghton, S. E. Blondelle, C. T. Dooley, H. Dorner, J. Eichler, and J. M. Ostresh, *Mol. Divers.*, **2**, 41 (1996).

64. J. Eichler, A. W. Lucks, C. Pinilla, and R. A. Houghten, *Mol. Divers.*, **1**, 233 (1995).

65. J. Eichler, A. W. Lucka, and R. A. Houghten, *Pept. Res.*, **7**, 300 (1994).

66. H. M. Geysen, R. H. Meloen, and S. J. Barteling, *Proc. Natl. Acad. Sci., USA*, **81**, 3998 (1984).

67. S. P. A. Fodor, J. L. Read, M. C. Pirrung, L. Stryer, A. T. Lu, and D. Solar, *Science*, **251**, 767 (1991).

68. F. Sebestyen, G. Dibo, A. Kovacs, and A. Furka, *Biorg. Med. Chem. Lett.*, **3**, 413 (1993).

69. R. A. Houghten, C. Pinilla, S. E. Blondelle, J. R. Appel, C. T. Dooley, and J. H. Cuervo, *Nature*, **354**, 84 (1991).

70. A. Furka, F. Sebestyen, M. Asgedom, and G. Dibo, *Int. J. Pept. Protein Res.* **37**, 487 (1991).

71. K. S. Lam, S. E. Salmon, E. M. Hersh, V. J. Hruby, W. M. Kazmierski, and R. J. Knapp, *Nature*, **354**, 82 (1991).

72. R. A. Houghton, J. R. Appel, S. E. Blondelle, J. H. Cuervo, C. T. Dooley, and C. Pinilla, *Biotechniques*, **13**, 412 (1992).

73. R. E. Dolle, J. Guo, L. O'Brien, Y. Jin, M. Riznik, K. J. Bowman, W. Li, W. J. Egan, C. L. Cavallaro, A. L. Roughton, Q. Zhao, J. C. Reader, M. Orlowski, B. Jacob-Samuel, and C. D. Carroll, *J. Comb. Chem.*, **2**, 716–731 (2000).

74. D. L. Boger, J. K. Lee, J. Goldberg, and Q. Jin, *J. Org. Chem.*, **65**, 1467 (2000).

75. M. B. Andrus, T. M. Turner, Z. E. Sauna, and S. V. Ambudkar, *J. Org. Chem.*, **65**, 4973 (2000).

76. A. Gopalsamy and H. Yang, *J. Comb. Chem.*, **2**, 378 (2000).

77. J. S. Lazo and P. Wipf, *J. Pharmacol. Exp. Ther.*, **293**, 705 (2000).

78. B. Walker, J. F. Lynas, M. A. Meighan, and D. Bromme, *Biochem. Biophys. Res. Commun.*, **275**, 401 (2000).

79. K. C. Ho and C. M. Sun, *Bioorg. Med. Chem. Lett.*, **9**, 1517 (1999).

80. Y. Gong, M. Becker, Y. M. Choi-Sledeski, R. S. Davis, J. M. Salvino, et al., *Bioorg. Med. Chem. Lett.*, **10**, 1033 (2000).

81. K. Illgen, T. Enderle, C. Broger, and I. Weber, *Chem. Biol.*, **7**, 433 (2000).

82. A. P. Ducruet, R. L. Rice, K. Tamura, F. Yokokawa, S. Yokokawa, P. Wipf, and J. S. Lazo, *Biorg. Med. Chem.*, **8**, 1451 (2000).

83. M. Rabinowitz, P. Seneci, T. Rossi, M. Dal Cin, M. Deal, and G. Terstappen, *Bioorg. Med. Chem. Lett.*, **10**, 1007 (2000).

84. Y. Wei, T. Yi, K. M. Huntington, C. Chaudhury, and D. Pei, *J. Comb. Chem.*, **2**, 650 (2000).

85. U. Grabowska, A. Rizzo, K. Farnell, and M. Quibell, *J. Comb. Chem.*, **2**, 475 (2000).

86. D. L. Mohler, G. Shan, and A. K. Dotse, *Bioorg. Med. Chem. Lett.*, **10**, 2239 (2000).

87. G. Bergnes, C. L. Gilliam, M. D. Boisclair, H. Blanchard, K. V. Blake, D. M. Epstein, and K. Pal, *Bioorg. Med. Chem. Lett.*, **9**, 2849 (1999).

88. K. L. Henry Jr., J. Wasicak, A. S. Tasker, J. Cohen, P. Ewing, M. Mitten, et al., *J. Med. Chem.*, **42**, 4844 (1999).

89. W. C. Lumma Jr., K. M. Witherup, T. J. Tucker, S. F. Brady, J. T. Sisko, et al., *J. Med. Chem.*, **41**, 1011 (1998).

90. S. P. Rohrer, E. T. Birzin, R. T. Mosley, S. C. Berk, S. M. Hutchins, D. M. Shea, et al., *Science*, **282**, 737 (1998).

91. J. S. Brady, T. R. Ryder, J. C. Hodges, R. M. Kennedy, and K. D. Brady, *Bioorg. Med. Chem. Lett.*, **8**, 2309 (1998).

92. C. O. Ogbu, M. N. Qabar, P. D. Boatman, J. Urban, J. P. Meara, et al., *Bioorg. Med. Chem. Lett.*, **8**, 2321 (1998).

93. Kundu, B., Bauser, M. Betschinger, W. Kraas, and G. Jung, *Bioorg. Med. Chem. Lett.*, **8**, 1669 (1998).

94. E. Apletalina, J. Appel, N. S. Lamango, R. A. Houghten, and I. Lindberg, *J. Biol. Chem.*, **273**, 26589 (1998).

95. J. S. Warmus, T. R. Ryder, J. C. Hodges, R. M. Kennedy, and K. D. Brady, *Bioorg. Med. Chem. Lett.*, **8**, 2309 (1998).

96. A. Bhandari, D. G. Jones, J. R. Schullek, K. Vo, C. A. Schunk, et al., *Bioorg. Med. Chem. Lett.*, **8**, 2303 (1998).

97. S. Roychoudhury, S. E. Blondelle, S. M. Collins, M. C. Davis, H. D. McKeever, et al., *Mol. Divers.*, **4**, 173 (1998).

98. A. Bianco, C. Brock, C. Zabel, T. Walk, P. Walden, and G. Jung, *J. Biol. Chem.*, **273**, 28759 (1998).

99. J. Ellman, B. Stoddard, and J. Wells, *Proc. Natl. Acad. Sci. USA*, **94**, 2779 (1997).

100. J. R. Schullek, J. H. Butler, Z. J. Ni, D. Chen, and Z. Yuan, *Anal. Biochem.*, **246**, 20 (1997).

101. K. Harada, S. S. Martin, R. Tan, and A. D. Frankel, *Proc. Natl. Acad. Sci. USA*, **94**, 11887 (1997).

102. J. A. Ostrem, F. al-Obeidi, P. Safar, A. Safarova, S. K. Stringer, M. Patek, et al., *Biochemistry*, **37**, 1053 (1998).

103. S. W. Kim, Y. S. Shin, and S. Ro, *Bioorg. Med. Chem. Lett.*, **8**, 1665 (1998).

104. J. Jiracek, A. Yiotakis, B. Vincent, A. Lecoq, A. Nicoalau, et al., *J. Biol. Chem.*, **270**, 21701 (1995).

105. V. Dive, J. Cotton, A. Yiotakis, A. Michaud, S. Vassiliou, et al., *Proc. Natl. Acad. Sci. USA*, **96**, 4330 (1999).

106. J. Jiracek, A. Yiotakis, B. Vincent, B. Checler, and V. Dive, *J. Biol. Chem.*, **271**, 19606 (1996).

107. M. Bastos, N. J. Maeji, and R. H. Abeles, *Proc. Natl. Acad. Sci. USA*, **92**, 6738 (1995).

108. J. M. Salvino, R. Mathew, T. Kiesow, R. Narensingh, H. J. Mason, et al., *Biorg. Med. Chem. Lett.*, **10**, 1637 (2000).

109. S. M. Dankwaardt, R. J. Billedeau, L. K. Lawley, S. C. Abbot, R. L. Martin, et al., *Biorg. Med. Chem. Lett.*, **10**, 2513 (2000).

110. T. A. Rano, T. Timkey, E. P. Peterson, J. Rotonda, D. W. Nicholson, et al., *Chem. Biol.*, **4**, 149 (1997).

111. P. Wipf, A. Cunningham, R. L. Rice, and J. S. Lazo, *Biorg. Med. Chem.*, **5**, 165 (1997).

112. C. Haskell-Luevano, A. Rosenquist, A. Souers, K. C. Khong, J. A. Ellman, and R. D. Cone, *J. Med. Chem.*, **42**, 4380–4387 (1999).

113. C. O. Ogbu, M. N. Qabar, P. D. Batman, J. Urban, J. P. Meara, et al., *Bioorg. Med. Chem. Lett.*, **8**, 2321 (1998).

114. D. S. Wagner, C. J. Markworth, C. D. Wagner, F. J. Schoenen, C. E. Rewerts, et al., *Comb. Chem. High Throughput Screen.*, **1**, 143 (1998).

115. L. Revesz, F. Bonne, U. Manning, and J. F. Zuber, *Bioorg. Med. Chem. Lett.*, **8**, 405 (1998).

116. C. Y. Cho, R. S. Youngquist, S. J. Paikoff, M. H. Beresini, A. R. Hebert, et al., *J. Am. Chem. Soc.*, **120**, 7706 (1998).

117. T. A. Rano, Y. Cheng, T. T. Huening, F. Zhang, W. A. Schleif, et al., *Bioorg. Med. Chem. Lett.*, **10**, 1527 (2000).

118. S. Kaldor, J. E. Fritz, J. Tang, and E. R. McKinney, *Bioorg. Med. Chem. Lett.*, **6**, 3041 (1996).

119. P. W. Davis, T. A. Vickers, L. Wilson-Lingardo, J. R. Wyatt, C. J. Guinosso, et al., *J. Med. Chem.*, **38**, 4363 (1995).

120. B. A. Bunin, M. J. Plunkett, and J. A. Ellman, *Proc. Natl. Acad. Sci. USA*, **91**, 4708 (1994).

121. S. J. DeWitt, J. K. Kiely, C. J. Stankovic, M. C. Schroeder, D. M. R. Cody, and M. R. Pavia, *Proc. Natl. Acad. Sci. USA*, **90**, 6909 (1993).

122. M. J. Plunkett and J. A. Ellman, *J. Org. Chem.*, **60**, 6006 (1995).

123. T. F. Herpin, K. G. Van Kirk, J. M. Salvino, S. T. Yu, and R. F. Labaudiniere, *J. Comb. Chem.*, **2**, 513 (2000).

124. C. G. Boojamara, K. M. Burow, and J. A. Ellman, *J. Org. Chem.*, **60**, 5742 (1995).

125. D. A. Goff and R. N. Zuckermann, *J. Org. Chem.*, **60**, 5744 (1995).

126. J. J. Landi and K. Ramig, *Syn. Commun.*, **21**, 167 (1991).

127. M. F. Gordeev, D. V. Patel, H. P. England, S. Jonnalagadda, J. D. Combs, and E. M. Gordon, *Biorg. Med. Chem.*, **6**, 883 (1998).

128. G. C. Look, M. M. Murphy, D. A. Campbell, and M. A. Gallop, *Tetrahedron Lett.*, **36**, 2937 (1995).

129. A. A. MacDonald, S. H. DeWitt, and R. Ramage, *Chemia*, **50**, 266 (1996).

130. K. E. Frank, P. V. Devasthale, E. J. Gentry, V. T. Ravikumar, A. Keschavarz-Shokri, et al., *Comb. Chem. High Throughput Screen.*, **1**, 89 (1998).

131. C. S. Lee, D. A. Allwine, M. R. Barbachyn, K. C. Grega, L. A. Dolak, et al., *Biorg. Med. Chem.*, **9**, 3243 (2001).

132. M. F. Gordeev, *Curr. Opin. Drug Discov. Dev.*, **4**, 450 (2001).

133. S. C. Bergmeier and S. J. Katz, *J. Comb. Chem.*, **4**, 162 (2002).

134. M. Kidwai, P. Misra, K. R. Bhushan, R. K. Saxena, and M. Singh, *Monatsh. Chem.*, **131**, 937 (2000).

135. C. Zhang, S. Sarshar, E. J. Moran, S. Krane, J. C. Rodarte, K. D. Benbatoul, et al., *Biorg. Med. Chem. Lett.*, **10**, 2603 (2000).

136. S. Sarshar, C. Zhang, E. J. Moran, S. Krane, J. C. Rodarte, et al., *Biorg. Med. Chem. Lett.*, **10**, 2599 (2000).

137. E. S. Guns, T. Denyssevych, R. Dixon, M. B. Bally, and L. Mayer, *Eur. J. Drug Metab. Pharmacokinet.*, **27**, 119 (2002).

138. P. J. Hajduk, J. Dinges, J. M. Schkeryantz, D. Janowick, M. Ksaminski, et al., *J. Med. Chem.*, **42**, 3852 (1999).

139. R. A. Fecik, K. E. Frank, E. J. Gentry, L. A. Mitscher, and M. Shibata, *Pure Appl. Chem.*, **71**, 559 (1999).

140. S. V. Ley, I. R. Baxendale, R. N. Bream, P. S. Jackson, A. G. Leach, et al., *J. Chem. Soc., Perkin Trans. I*, 3815 (2000).

141. S. A. Kates and F. Albericio, Eds., *Solid-Phase Synthesis: A Practical Guide*, M. Dekker, Inc., New York, 2000.

142. D. H. Drewry, D. M. Coe, and S. Poon, *Med. Res. Rev.*, **19**, 97 (1999).

143. G. Cardillo, M. Orena, and S. Sandri, *J. Org. Chem.*, **51**, 713 (1986).

144. D. L. Flynn, *Med. Res. Rev.*, **19**, 408 (1999).

145. S. D. Brown and R. W. Armstrong, *J. Org. Chem.*, **62**, 7076 (1997).

146. D. P. Curran and T. Furukawa, *Org. Lett.*, **4**, 2233 (2002).

147. Z. Luo, Q. Zhang, Y. Oderaotoshi, and D. P. Curran, *Science*, **291**, 1766 (2001).

148. R. Dolle, J. Guo, L. O'Brien, Y. Jin, M. Piznik, et al., *J. Comb. Chem.*, **2**, 716 (2000).

149. D. Cork and N. Hird, *Drug Discov. Today*, **7**, 56 (2002).

150. A. Lew, P. O. Kvatzik, M. E. Hart, and A. R. Chamberlin, *J. Comb. Chem.*, **4**, 95 (2002).

151. R. S. Bohacek and C. McMartin, *J. Am. Chem. Soc.*, **116**, 5560 (1994).

CHAPTER TWO

High-Throughput Screening for Lead Discovery

John G. Houston
Martyn N. Banks
Bristol-Myers Squibb
Pharmaceutical Research Institute
Wallingford, Connecticut

Contents

Burger's Medicinal Chemistry and Drug Discovery
Sixth Edition, Volume 2: Drug Development
Edited by Donald J. Abraham
ISBN 0-471-37028-2 © 2003 John Wiley & Sons, Inc.

1 INTRODUCTION

Drug discovery is one of the most crucial components of the pharmaceutical industry's Research and Development (R&D) process and is the essential first step in the generation of any robust, innovative drug pipeline. Drug candidate pipelines are usually aimed at disease areas with high unmet medical need where the commercial reward for delivering first in class and/or best in class therapies can be enormous. The R&D process is an extremely lengthy, complex, and strongly regulated activity that has increasing attrition rates and escalating costs (1, 2).

The last 5 years has witnessed a dramatic change in the landscape for the pharmaceutical industry. Consolidation in the market place resulting from a series of megamergers has created bigger than ever R&D budgets along with matching expectations and goals for increased productivity and growth (1, 2). Just keeping pace with double digit annual growth rates requires each of the top 10 pharmaceutical companies to launch at least five significant new chemical entities (NCEs) per year with blockbuster potential (greater than $500 million/year). However, this goal is set against a historical track record of less than one NCE per year, per company, of which less than 10% reached sales of greater than $350 million (1, 2). Additionally, in an ambitious attempt to exploit the maximum life of drug patents, companies have also set challenging goals in reducing overall drug development time to market.

An ever increasing regulatory burden and strong emphasis on drug safety ensures a series of extremely tough hurdles before any increased flow of drugs works its way through the industry's pipelines onto the market. Those companies that find the quickest way to put the first and best in class products on the market in a cost-efficient way will become the clear commercial winners in such a competitive arena. The drive to have more innovative and safer drugs on the market, quicker than ever before, has fueled enormous interest in different ways of revolutionizing the process by which drug discovery and development is carried out. This can be most clearly seen in the upstream phase of the R&D process starting from target discovery through early preclinical testing.

From penicillin in 1929 (3) to the β-blockers and 5-hydroxytryptamine (5HT) antagonists of the 1970s (4), pharmaceutical companies have relied on a mix of "accidental" serendipitous discovery, inspired scientific insight, and incremental progression in chemistry to develop drugs.

However, the promise of major improvements in productivity, cost efficiency, and speed of discovery has created radical changes in the way drug discovery is carried out. Leading this wave of change has been the development and implementation of a series of innovative, state-of-the-art technologies aimed at the early lead discovery process.

It is in this area that the overall impact of technological advances in genomics, combinatorial chemistry, compound management, computer-aided drug design (CADD), high-throughput screening (HTS), and bioinformatics has created a new drug discovery paradigm.

In this chapter, we will describe the history and process of lead discovery from initial target identification and disease validation through the lead identification process itself, culminating in lead optimization and pre-clinical candidate selection. The full range of technologies and approaches that have been brought to bear on this complex activity will also be highlighted.

2 HISTORY OF LEAD DISCOVERY AND SCREENING

Serendipity has been the key to the pharmaceutical industry's success over many decades.

The emergence of this "accidental" approach to drug discovery has its origins in early history with traditional natural herbal remedies that were passed from generation to generation in local communities or tribes. In fact, the early history of lead discovery is all about natural products and herbal remedies, the use of which dates back thousands of years.

Ancient Chinese and Indian medicinal recipes, along with African tribal shamen "cures," have included ingredients such as opium, curare, the belladonna alkaloids, and the digitalis alkaloids. The use of these treatments was passed down by each generation through traditional practice and word of mouth.

It was not until the late 18th and early 19th centuries that an analytical investigation of the active components of medicinal plants and herbal remedies was pursued. This resulted in the discovery of alkaloids such as atropine, morphine, codeine, and papaverine, which eventually became the major constituents of many modern analgesic and cardiac medicines. The discovery of aspirin (from Willow bark), the cannabinoids (from *cannabis sativa*), and digitalis (from foxglove leaf) also came many years, in some cases thousands of years, after these remedies were being used for pain relief, sedation, and dropsy (5–9).

The analysis of the active ingredients of these herbal "cures" has supplied an enormous number of the medicines that have been used in the last 100 years. It has been estimated that 119 compounds identified from 90 plants have been used as single entity medicinal agents (10). Significantly, the vast majority of these drugs were obtained as a result of examining the plant based on its ethnomedical use (6, 7, 9, 11).

Of course, plants have not been the only source for these medicines; natural product drugs such as cyclosporin and lovastatin used for immunosuppression and hypocholesterolemia have also been isolated from the secondary metabolites of fungi and bacteria. Additionally, antibiotics such as the β-lactams, echinocandins, and the erythromycin macrolides all originated from microbial and fungal metabolites (5). Even today, natural products continue to be a viable source of new drugs for the pharmaceutical industry (12).

The discovery of penicillin in 1929 by Alexander Fleming was one of the first real documented "chance" observations that eventually led to the development of a drug. He found that penicillin mold could cause the lysis of staphylococcal colonies on agar plates (3). The culture filtrate also possessed activity against pathogens such as Gram-positive bacteria and Gram-negative cocci.

In the 1940s, the wonderful therapeutic efficacy of penicillins was demonstrated by Chain et al. (13), and penicillin and the β-lactam drugs emerged as the first of an incredible series of antibiotic medicines that have saved countless lives.

Through the 1950s and 1960s, the pharmaceutical industry relied heavily on chance discovery to find and develop drugs but made strong attempts to enhance this by developing a drug discovery process with the introduction of compound screening in a concerted fashion.

Scientists such as Paul Ehrlich, searching for an antisyphilitic, have exemplified the idea of serial screening of compounds in an attempt to find a "magic bullet." This approach, although tedious and time consuming, did generate a remarkable number of drugs, such as the benzodiazepines, chlorpromazine, and others (14).

The biggest limitation to this discovery approach was the relatively small number of compounds available for screening (a few hundred) and their lack of chemical diversity. In addition, the lack of *in vitro* assay technology meant that compounds had to be tested in animal models. The resulting high failure rate in these test models reduced considerably the chances of finding a molecule that could then be optimized by chemists.

Compounds found active in these animal models usually had an unknown mechanism of action, making future improvements on the drug very difficult.

In a move from serendipity to rationality, drug companies attempted to develop more useful tools to help facilitate the discovery process. By using specific animal tissues and developing a better understanding of the biological and physiological systems under investigation, scientists were able to piece together a far more detailed picture of the molecular parameters that impact efficacy and affinity.

The development of *in vitro* assays using animal tissues became an essential support tool for tracking structure-activity relationships (SAR) and allowing chemists to optimize structures before whole animal testing. This allowed a wide range of compounds to be made around a SAR hypothesis. If the lead compound subsequently failed in animal testing, new compounds could be made based on the *in vitro* SAR and retested in the animal model. In this iterative process, potency and other drug-like properties could be "built in" to the lead structure in a coordinated, planned fashion and tested.

Rational design of drugs based on knowledge of the biological system being investigated allowed highly specific selective antagonists and agonists to be developed. These molecules could then be developed as drug candidates such as was demonstrated by Black et al. (4) in the development of the H2 receptor antagonist cimetidine. The range of highly specific molecules that became available also helped define a variety of new receptor subtypes that also became candidates for drug intervention. Rational drug design and the pharmacological elucidation of receptor pathways were the preeminent methodology for drug discovery during the 1960s and 1970s. This approach heralded a golden period of drug discovery and development that produced histamine H2 receptor antagonists, β-adrenergic receptor antagonists, and partial agonists (14).

The next major change to the discovery process was seen with the first advances in molecular and cellular biology in the early 1980s. The ability to clone and express human receptors and enzymes radically changed the way drug discovery was carried out. This technology allowed optimization of lead molecules against the human version of a receptor or enzyme and allowed a deeper analysis of their physiological nature. The emergence of functional genomics tools and bioinformatics also allowed us to identify and understand more fully the connections between specific human genes and disease.

By the early 1990s, advances in technologies such as lab automation, detection systems, and data capture systems allowed the first real automated versions of screening labs

to be put in place. Concomitant advances in molecular biology, *in vitro* bioassay design techniques, and microplate technology allowed biological targets to be screened that had proven to be intractable before. The final pieces of the jigsaw were completed with the strides made in automation of compound synthesis and the accumulation of large collections of natural and synthetic compounds in stores available for screening. With all these essential technology ingredients in place, the pharmaceutical laboratory in the early 1990s was set for the first real version of high-throughput screening.

A typical lead discovery organization at that time would have been expected to screen approximately 10–20,000 samples through each of its 15–20 targets per year (15), where each target would be screened for 2–3 years. Throughout the 1990s, the screening capacities rose significantly, and the overall timelines shortened. Now in the early 2000s, some pharmaceutical companies quote that they have capacity to screen over 100 targets per year, each testing over 1 million samples per target, with each screening campaign lasting only a matter of weeks. This is now called ultra-high—throughput screening (UHTS) and represents a significant increase in testing capacity compared with laboratories of the late 1980s (15).

The majority of pharmaceutical companies (16–18) have now reported applications of automation and robotics in activities such as dry compound storage and retrieval, liquid stores, HTS, and combinatorial chemistry libraries. The integration of robotic systems and highly trained scientists with a diverse range of skill sets has allowed the development of the lead discovery process. The manual, labor-intensive process, used in the 1970s and 1980s to screen compounds in low throughput gave way to new processes in the 1990s. Increasing numbers of targets and compounds drove the demand for higher capacity, fully automated screening systems into pharmaceutical organizations.

The modern screening laboratory today is a multifunctional, multiskilled environment that connects a range of discovery functions in a high capacity, integrated process producing a product that consists of a cohort of tractable

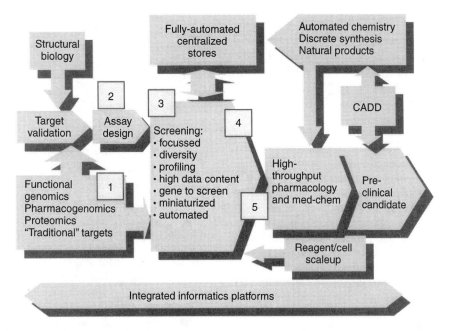

Figure 2.1. A prototypical early drug discovery process and the key activities in lead discovery.

chemical leads against targets of interest. Each component of the lead discovery process is critical to the overall success and impact of a screening campaign. The disease relevance and "druggability" of the biological target and the availability of technologies to supply and test vast arrays of compounds are key factors for success.

The following sections of this chapter describe the generic process of lead discovery and are followed by a more detailed analysis of some of the important functions involved in its execution.

3 LEAD DISCOVERY PROCESS

In contrast to the highly defined and strictly regulated process of drug development and manufacture, the initial discovery of drug candidates can be described as somewhat *ad hoc* and ill defined. Each pharmaceutical company, although agreeing on the ultimate goal, has developed a surprisingly wide array of technologies and approaches to achieve the same endpoint (15–21).

However, in this prototypical early discovery process, we can describe several distinct but overlapping activities that are essential for effective, scalable lead discovery in the modern pharmaceutical industry (Fig. 2.1).

1. Discovering and validating a target of interest

2. Designing a bioassay to measure biological activity

3. Constructing a high-throughput screen

4. Selecting screening decks and screening to find hits

5. Profiling hits and selecting candidates for optimization

Of course, the drug discovery process is a multi-faceted, complex process with numerous iterative sequences and feedback loops, which cannot be fully captured in a chapter such as this. Therefore, we will keep high-throughput screening, the array of ancillary technologies, and processes as the central theme throughout the following section.

Throughout the chapter, we will endeavor to highlight the process by which these distinct activities are brought together in an integrated process.

3.1 Target Discovery and Validation

The drug discovery process starts with the identification, or growing evidence of, biological targets that are believed to be connected to a particular disease state or pathology. Information supporting the role of these targets in disease modulation can come from a variety of sources. Traditionally, the targets have been researched and largely discovered in academic laboratories, and to a lesser extent in the laboratories of pharmaceutical and biotechnology companies. Basic research into understanding the fundamental, essential processes for signaling within and between cells and their perturbation in disease states has been the basic approach for establishing potential targets suitable for drug intervention. By pharmacologically modifying these intra- and intercellular events, it is hoped that particular disease mechanisms and their ensuing pathologies can be modified. This type of approach has generated a number of significant and unique biological targets. More recently, bioinformatics (22), genomics and proteomics techniques (23–25), combined with the huge output from the human genome sequencing project, are helping to identify thousands of potential targets.

This is a significant change to past approaches where the majority of the medicines developed by the pharmaceutical industry had been targeted at only 500 known human targets (26). The emphasis has now changed from finding new drugs that exploit the same well-validated targets to finding drugs against potentially new and innovative biological mechanisms connected to disease. The tough questions that need to be answered at this stage of the process are related to which targets to focus on among the many thousands of potential choices.

Clearly, one of the more important steps in the process of developing a novel pipeline of drugs is the identification of novel genes and assessing their expression and role in various physiological and pathological states. The human genome project has generated a wealth of data around human DNA sequences and gene mapping. There are now billions of base pair DNA sequence data and an estimated 30,000 human genes to be searched and investigated

as potential drug discovery start points. A host of genomics technologies are now available for discovery scientists to find and validate biological targets. Rapid gene sequencing, single nucleotide polymorphism (SNP) identification, differential mRNA display analysis, and bioinformatics data mining tools are just some of the now standard techniques that are available to help identify and analyze novel genes. The ability to search the billions of data points available through the various human genome-related databases has rapidly evolved with the development of bioinformatics tools. Powerful computers and search algorithms allow potential genes to be identified within stretches of DNA sequence. Their DNA coding sequences can be aligned and compared with other known DNA sequences stored in databases to allow comparison with known genes. Programs such as basic local alignment search tool (BLAST) are used for this type of similarity searching. Other methodologies can also be used to identify genes such as using phylogenetic analysis to place the novel gene in context with other genes by molecular evolution.

Cloning, expression, and distribution of the gene of interest are all-important next steps in qualifying the target for further discovery work. A gene that is uniquely expressed in a particular type of cell or tissue will trigger further evaluation. For instance, brain-specific expression of a gene engages neuroscientists in the same way as specific expression of a novel gene in an immune cell would spike the attention of inflammation experts.

Of course, cell and tissue distribution does not solely validate a target, but provides useful circumstantial evidence that this target could be worth investigating in disease models. In fact, differential expression of a candidate gene in disease and non-disease systems helps to focus the interest considerably.

An essential next step in the process is attributing functionality to an unknown gene. The elucidation of gene function and the understanding of its role in the activity of other genes can be critically important.

There are a number of critical methods for determining gene function some of that are similar to those used to initially identify genes.

3.1.1 Gene Function by Homology to Other Defined Genes.

By aligning the sequence of an unknown gene and comparing it with sequences of known genes, homologies are identified that allow a tentative potential function to be assigned. These homologies can be as general as identifying the unknown gene as a target involved in metabolism or the homology could be as specific as identifying a specific function such as fatty acid biosynthesis (27–28).

3.1.2 Gene Function by Gene Subtraction.

When a gene function cannot be identified by direct homology methods or when you want further proof of the importance of a particular candidate gene, it is possible to determine function by deleting the gene (gene knockout) in an *in vivo* model and describing the resulting phenotype (29). An example would be the use of gene knockout studies to establish whether protein tyrosine phosphatase IB is an important target for anti-diabetic drugs (30). Certain animal models such as worms, flies, and mice also have well-characterized phenotypes that will infer function. Yeast models can also be used for this type of analysis. A number of biotechnology companies offer this as a valuable service both for *in vivo* and *in vitro* systems (e.g., Lexicon, Sequitur).

3.1.3 Gene Function by Expression Analysis.

Genes specific for a particular tissue can be expressed differently under a variety of different metabolic conditions or stresses within the cell or organism. Tracking mRNA and protein expression as different indicators of gene activity in normal and disease tissue is now one of the more important methods for gene validation. This type of analysis involves extensive use of DNA microarrays and related chip technologies (31–33).

By using this extensive array of genomics techniques, the discovery scientist can quickly generate a list of novel, tissue-specific genes showing differential expression in disease and non-disease systems. Direct homology or knockout experimentation may have also elucidated their function. These targets may also be members of a known "druggable" target class such as GPCRs or ion channels, which suggests they can be readily integrated into a drug discovery process (34). If the intellectual property (IP) position on this target is also clear, then this target could rise to the top of the prioritized list for entry into the pipeline.

Building a prioritized portfolio of disease targets and subsequently deploying resources is critical in the discovery process. Which disease areas are the foci for the organization? In those disease areas of interest, what target choices are available for drug intervention? What is known about the relationship of this target to disease? All the answers to these questions and more determine the ranking order by which targets may be worked on or not. The fact that a target is known and well validated reduces the discovery and development risk but raises the chances that your competitors are also working on this target.

Selecting a novel target with just the earliest indications of validation, but no real clinical proof, may keep you clear of the competition, but raises the risk of discovery and development failure as the target and drug develop through the pipeline. Entering a biological target into the discovery process is an extremely critical step that can affect the successful discovery of leads and the ultimate success of the drug development process. The greater the knowledge of the biological systems and disease pathologies involved, the greater the chances are that the right type of targets will be selected. Overall, target and disease portfolio decision-making processes are not within the scope of this chapter, but clearly play a direct role in the potential success or failure of any lead discovery program.

By the end of this initial phase of the R&D process, a list of targets will have been selected and passed into the lead discovery phase. At this next stage, the ability to measure the activity and function of the biological target and find compounds that modulate the activity is critical.

3.2 Bioassay Design and Screen Construction

After the identification of a biological target of interest, the next challenge begins with the conversion of the target into a bioassay that can give a readout of biological activity. The range of potential targets is large, from enzymes and receptors to cellular systems that represent an entire biochemical pathway or a disease process. Consequently, the range of assay design techniques and types of assay available have to be correspondingly comprehensive.

Once an assay has been developed that measures the biological activity of the target, by some direct or indirect means, then compounds can be tested in the bioassay to see if they inhibit, enhance, or do nothing to this activity.

This approach is the basis of all compound testing for HTS and structure-activity relationship (SAR) studies in drug discovery programs.

A variety of techniques can be applied to bioassay design depending on the nature of the biological activity being measured. Measurement of product accumulation, measurement of enzyme substrate use, measurement of receptor mediated signaling, receptor antagonism and agonism, cell death, and cell proliferation are just a few of the activities that could be assessed in a bioassay. Each of these bioactivities can be directly or indirectly measured using reporter signals and detection systems such as radioisotopes and scintillation counters, fluorescence and fluorometers, luminescence and luminometers, or voltage changes and patch clamp.

Regardless of the biological activity being measured and the detection system being used, the assay will be optimized against an array of multifactorial parameters to provide an *in vitro* milieu for the bioassay, which most closely resembles the physiological "normal" for the functional activity. In this assay optimization process, factors such as pH, K_d, K_m, and a whole host of other significant parameters are modified to provide the most appropriate bioassay.

If the bioassay is the basis of a high-throughput screen, then other crucial factors are assessed before screen construction. The stability and scalability of the assay's core reagents are just two of the more important factors. An assay format designed to use scarce and time-dependent labile reagents will not make the best design for a HTS environment. Here the reagents have to be made at bulk levels to allow testing of thousands of compounds in conditions where they may have to be stable for hours on an automated screening system.

The process from bioassay design to HTS construction can significantly change the design and final format of an assay. It is important that assay to screen reproducibility is monitored and maintained to ensure consistent results from the different laboratories involved in the discovery process.

3.2.1 Assay Design. Ideally, in the planning phase, a team of medicinal chemists, therapeutic area biologists, and HTS designers define the important objectives for a high-throughput screening campaign. These could include criteria that would initiate a medicinal chemistry program, such as the desired data from the screen, a definition on how to assess the value of the screening hits, etc. The HTS assay designer's responsibility is to select the correct assay methodology to meet the screening objectives. Usually there are numerous potential assay options, and it is often necessary to create a decision matrix to compare the potential methods. A typical matrix of questions is outlined below.

1. What compound characteristics are required, e.g., antagonists or agonist?
2. Do you have the relevant expertise available to build the type of assay? Modification of an existing screening protocol is the natural starting point, especially if one is considering screening a gene family like kinases or nuclear hormone receptors. If at all possible you would like to avoid *de novo* assay design unless there is potential for more targets within this class and the investment can have future impact.
3. Are there any restrictions on the use of the assay technology? For example, are there patent restrictions on the use of any reagents or the actual biochemical technique itself.
4. Does the proposed HTS assay provide the desired information to progress compounds along the drug discovery pipeline? Will the hits need to be assessed in additional assays before progression into medicinal chemistry? For example, if the screen needs to find receptor agonists, there are potential cellular functional assays that directly measure *in vitro* agonist

efficacy. Alternatively, a binding assay could be designed that would find both agonists and antagonists without discriminating between them. A secondary assay would then be needed to distinguish between these two binding activities. If the probability of finding an agonist is perceived to be low, medicinal chemistry may want to modify antagonists to generate agonist activity and the binding assay route would then be preferable.

5. Are there simple ways to detect and eliminate false positive data from the HTS results? Certain compounds will interfere with the readout in the HTS assay; for example, highly colored compounds can quench the signal from a fluorescence-based assay, giving the illusion of inhibition in an enzyme assay.

6. Are all reagents readily available to perform a HTS? Will the reagents be purchased from an external vendor or produced in-house? These reagents include not only substrates and ligands, but also the enzymes/receptors/cells.

7. Is there experience in screening this type of target with the proposed assay technology? Each screening technology has its own challenges, both strengths and weaknesses, and a clear understanding of these is critical to the interpretation of screening results.

8. Is the assay method suitable for miniaturization and/or running on an automated screening system? The decision on whether to use automated screening platforms and miniaturized screening formats will be very dependent on the existing screening infrastructure. This is discussed later in the chapter.

The questions and criteria above are ranked in a generalized order of priority and can direct the assay designer to a particular design choice. For a particular target, a different priority ranking may be more relevant depending on the screen circumstances. For example, if reagents are difficult to produce then the amounts required to run a full deck HTS may indicate that a design applicable to miniaturization should have a higher priority.

For a given target, there are many options available to measure the biological effect of a compound or mixture of compounds in a HTS assay (35–37). The planning phase should conclude with an agreed high-throughput screening assay design, a clear understanding of how the reagents will be procured, and a process for assessing the screening hits.

Having decided on the best methodology option, the construction phase involves the building and optimization of the assay parameters.

3.2.2 Assay Construction. Designing and building a HTS assay can be very time consuming and is often a rate-limiting step in the screening process. It is also the most critical part of the high-throughput screening process. If the screening campaign is perfectly executed and generates hits, but the assay design is flawed (e.g., suboptimal substrate concentration in an enzyme assay), then the value of the results will be questionable. The majority of HTS assay designs are adapted from existing protocols, literature methods, or modifications of assays built in therapeutic area laboratories. Often the existing method is not ideal for high-throughput screening, and the assay will require significant optimization. Generally, homogeneous assay formats are preferred in HTS because they can be readily adapted for automation. In a homogeneous HTS assay, the reagents are added to a microtiter plate already containing compound and the results measured after a suitable incubation period. Examples of these homogeneous assays are described later in the chapter.

The optimization process evaluates all the important assay parameters, such as the following:

1. Buffer type, pH, and ionic strength
2. Reagent purity, stability, protein aggregation, and stabilization additives
3. Additives to improve pipetting accuracy and precision
4. Pipetting method (peristaltic pumps, syringe driven, non-contact piezo, etc.)
5. Incubation conditions (defined, controlled environment, room temperature, etc.)

6. Kinetic parameters, K_m, V_{max}, k_{cat}, or binding constants K_d

7. In cellular assays, cell number, plating conditions, type of media, and passage number are just some of the potential optimization criteria

The objective of the optimization phase is to build an assay that generates a reproducible signal that will allow a statistically significant difference from the background reading. This signal has to be precise; the assay must show acceptable kinetics and respond in a predictable way to known inhibitors or activators. Historically, this optimization process has tended to be a linear process where one or two reagents are optimized, followed by a series of experiments varying other assay parameters until acceptable conditions are achieved. However, optimized assay conditions are often a complex interaction of all the variables in a method and difficult to derive by monitoring each individual component. The assay signal is a composite of both contributions from individual factors as well as combinations of factors. These combination factor contributions to an assay signal can not be explained by the sum of their individual contributions, which means that it is not easy to design assays by the traditional linear approach. It is however possible to use statistical techniques, like experimental design, to optimize multivariate assay conditions. The seminal work in this area was developed by the British biologist/statistician Sir Ronald Fisher over 40 years ago. The interested reader is directed to a more recent publication by Haaland (38) that illustrates numerous examples of the application of experimental design. Specifically with HTS, two publications detail how these statistical techniques have been used to optimize assay conditions (39, 40). In Taylor et al. (39), they describe how one combines the experimental design statistics with automated liquid handling devices to optimize assay conditions.

The major different HTS assays can be conveniently classified into biochemical techniques or cellular techniques. Additionally, there are numerous detection techniques, including radiochemical, absorbance, fluorescence, and luminescence. All of these combinations have advantages and disadvantages, and the critical step in the assay design process is to decide what information is required and match the goal to the available options.

3.2.3 Homogeneous and Non-Homogeneous Biochemical Assays. The majority of high-throughput screening detects the activity of a series of compounds using biochemical or cellular approaches. The advantages and disadvantages of these two approaches have been extensively discussed elsewhere (41, 42). Both biochemical and cellular screening have been automated and miniaturized, and most HTS laboratories maintain capabilities for both approaches.

One of the common critical features of a high-throughput screening method, be it biochemical or cellular, is assay efficiency. Homogeneous assays, where reagents are added, incubated, and the effect of compounds measured without the need to include a separation stage, are the most commonly used and effective formats for achieving high precision, whilst maintaining throughput. However, non-homogenous assays that involve a separation or washing step(s) are still used in several high-throughput screening approaches. A good illustration to compare a homogeneous and non-homogeneous format is an assay that is designed to measure the interaction of a receptor with its ligand using a radioisotopic detection method. In a non-homogeneous format, membranes containing the receptor are incubated with radioactive ligand in the presence of compound in a 96-well microtiter plate. The reaction mixture is then transferred to a second microtiter plate that contains a filter. Under vacuum, the filter traps the receptor—ligand complex. After washing the filter to remove unbound ligand, bound ligand is measured in a conventional scintillation counter. If the compound has blocked binding to the receptor, lower levels of radioactivity are measured by the scintillation counter. An alternative homogeneous assay approach would be to use scintillation proximity assay (SPA) (43).

3.2.3.1 Homogeneous Radioisotopic Assays. In scintillation proximity assays (SPA), a plastic bead containing a fluor and coated with a lectin binds the carbohydrate extracellular

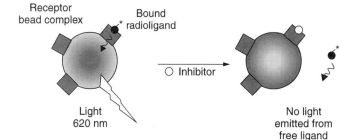

Figure 2.2. The principle behind scintillation proximity assays, e.g., Leadseeker. Only bound radioligand is physically close enough to the fluor-impregnated plastic bead to cause light to be emitted.

surface of a membrane fragment. Consequently, membranes containing a receptor of interest can be bound to the surface of the bead. Radioactive ligand binds to membrane receptor—bead complex at the surface of the bead. During isotopic decay, β particles (3H) or Auger electrons (125I) excite the embedded fluor in the bead and the emitted light is measured. Ligand in solution is too distant from the bead to excite the fluor, and the energy dissipates through the aqueous solution. Therefore, the emitted light is proportional to bound ligand (see Fig. 2.2). In summary, four components are added to a microtiter plate well, SPA bead, membrane, ligand, and the compound under investigation. After a suitable incubation period to allow equilibrium to occur, the emitted light is measured. In a screen designed to find receptor antagonists, a reduction in light is proportional to the activity of the compound tested. One of the significant disadvantages of SPA is the absorption of light by colored assay components or "Quench." Recently the SPA bead has been modified such that the emitted light is redshifted reducing this type of interference.

3.2.3.2 Homogeneous Non-Radioisotopic Assays. Non-radioisotopic homogeneous methods are also widely used in high-throughput screening, e.g., fluorescence polarization (FP) (44), fluorescence correlation spectroscopy (FCS) (45), homogeneous time-resolved fluorometry (HTRF) (46–49), and fluorescence resonance energy transfer (FRET) (50). Direct colorimetric- or absorbance-based assays have been in use in HTS laboratories for many years and are still very popular and effective. A large number of techniques are now available that support high-throughput screening, and the interested reader is directed to a useful high-throughput screening web site,

www.htscreening.net, where a variety of technology companies describe their methodology. Two of the more popular approaches are described below in more detail.

FP is a technique that is popular in high-throughput screening laboratories because of its homogenous format that allows the formation of a stable equilibrium and diffusion controlled kinetics. In FP, the free fluorescent ligand is differentiated from the bound fluorescent ligand based on rotational differences. Fluorescent ligand is excited with polarized light and the emission is measured, first through a polarized emission filter and then through a second filter, perpendicular to the first filter. For small molecular weight ligands, the fluorescent intensity in both planes is essentially identical because ligand rotation will occur during the fluorescence lifetime. Bind this fluorescent ligand to a macromolecule, such as a protein, and the rotation relaxation time significantly increases, and the bound fluorophore will rotate very little during fluorescence lifetime. The intensity of bound fluorescent ligand, now at the two polarized emissions, will be different. This phenomenon then becomes the basis of the high-throughput screening assay (Fig. 2.3).

From a HTS perspective, various considerations should be taken into account before choosing a FP assay. The presence of colored compounds in the screening deck could lead to bioassay interference. This type of inference will be dependent on the type of fluorophore. Typically, there are significant numbers of compounds in a screening deck that absorb at 400–450 nm, therefore, a red shifted dye with absorbance and emission >500 nm would be preferable. Second, the ligand needs to be conjugated with a fluorescent dye. If the ligand is of a relatively low molecular weight, signifi-

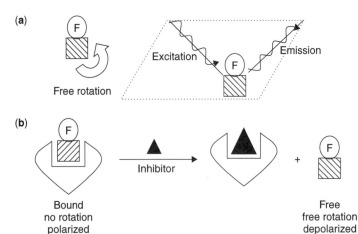

Figure 2.3. This figure shows the theory behind fluorescence polarization-based assays. (a) Ligand in free solution does not display fluorescence polarization. (b) Ligand bound to a macromolecule will show polarization.

cant experimentation will be required to give accurate binding characteristics. Often a fluorescently labeled ligand can be prepared that will compete with a radioligand in a predictable way. However, if the fluorophore on the ligand can still rotate freely even on binding a macromolecule (the propeller effect), a poor polarization signal will result. Nevertheless, FP has been successfully used in a range of assays, e.g., SH3-SH2 binding interactions (51), benzodiazepine receptors (52), and HIV-1 protease inhibitors (53).

Fluorescence is a very versatile detection mode in high-throughput screening assays because of its wide dynamic range and its potential sensitivity. However, assay components can cause interference with the detection of the fluorescent signal because of intrinsic fluorescence from the reagents, light absorption, and fluorescence quench. Time resolved fluorometry (TRF) is one technique that can minimize these types of interference. This methodology is based on the fluorescence of certain lanthanide metal ions, for example, Europium (Eu^{3+}). Under typical biochemical assay conditions, Eu^{3+} is a poor fluorophore and must be complexed in an organic framework that facilitates fluorescence, for example, chelate (54) or cryptate (55).

Eu-cryptates have a relatively long fluorescence lifetime (>0.1 ms) compared with typical assay components (<50 ns). Time gating between excitation and emission reduces intrinsic fluorescence from the signal. The Eu-cryptates also have a relatively large Stokes'

shift, the difference between the emission and excited wavelengths, of 300 nm. Fluorescent assay components have significantly shorter Stokes' shifts, and wavelength discrimination reduces signal interference. These properties have been commercially exploited to produce high-throughput screening assays. Figure 2.4 gives an example of a hTRF assay for a tyrosine kinase. In the presence of ATP and cofactors, tyrosine kinase will phosphorylate an artificial biotinylated peptide substrate. The phosphotyrosyl product is directly quantified using hTRF. This technique is based on fluorescence resonance energy transfer (FRET) from an Eu-cryptate to a red algal protein, allophycocyanin. Streptavidin, a bacterial protein that has a strong affinity for biotin ($K_d = 10^{-15}$ M), is covalently conjugated to allophycocyanin. This conjugate will bind the biotinylated phosphotyrosyl peptide, forming the acceptor partner in the FRET pair. An antibody that specifically recognizes phosphotyrosine, conjugated to Eu-cryptate, forms the donor partner in the FRET pair. Excitation at 337 nm and emission at 665 nm measures the FRET. This assay is homogeneous because the FRET is measured in the presence of all the other reagents. The free Eu-cryptate will emit light at 620 nm and therefore can be discriminated from the FRET emission at 665 nm by wavelength filters. Free streptavidin-allophycocyanin will emit light at 665 nm, but the fluorescence emission is short lived (a few nanoseconds). This can be discriminated from the FRET signal by time gating. As described

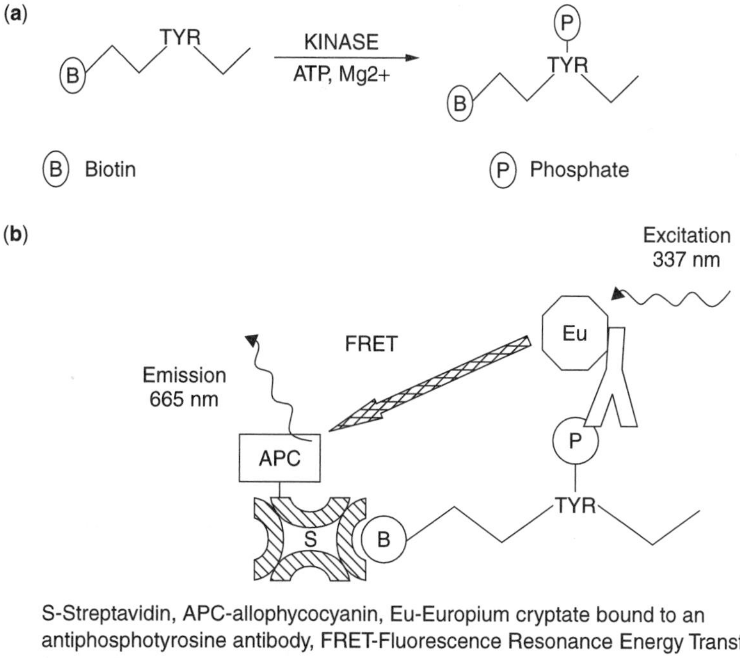
(a)
TYR
KINASE
ATP, Mg2+
P
TYR
B
B Biotin
P Phosphate
(b)
Excitation
337 nm
Eu
FRET
Emission
665 nm
APC
P
TYR
S
B
S-Streptavidin, APC-allophycocyanin, Eu-Europium cryptate bound to an
antiphosphotyrosine antibody, FRET-Fluorescence Resonance Energy Transfer.

Figure 2.4. The schematic shows the basis of a tyrosine kinase hTRF assay. The reaction (a) uses a biotinylated peptide substrate that becomes phosphorylated on tyrosine in the presence of kinase, ATP, and magnesium ions. The reaction product is then detected (b).

earlier in this section, light absorption by assay components is one of the limitations of prompt fluorescence assays. In hTRF, the readout is based on the signal ratio of Eu-cryptate and FRET emissions. In the presence of colored compounds, the two signals will decrease, but the ratio remains constant.

3.2.4 Cellular Assays. Cellular assays are also popular screening formats for identifying lead compounds. These types of assays stand in contrast to the defined, single target biochemical assays described above, because there are multiple points along a network of signaling pathways where compounds can elicit a response. This is a key strength of cellular assays, but can also be their weakness. When the precise biochemical target is unknown, one can use a cell-based "catch all" approach to mitigate against this uncertainty. Cellular screens may also uncover hitherto unknown sites for chemical modulation, thereby potentially identifying new targets. They may also pre-select for compounds with drug-like properties that cross membranes. However, a significant disadvantage of cell-based high-throughput screening appears at the hit assessment stage where the molecular mechanism of a compound may remain ill defined. This requires additional selectivity and profiling assays to help elucidate the molecular target. Most cellular HTS campaigns also include a cytotoxicity assay to help "flag" potentially undesirable compounds.

Cellular assays are generally more complex and time consuming than biochemical assays, requiring investment in generating and maintaining cell lines. To a certain extent, cellular production has been industrialized and new automation options originally designed for biopharmaceuticals are now being adopted by high-throughput screening laboratories, e.g., CELLMATE & SelecT, The Automation Partnership, Cambridge, UK, and ACCELLERATOR RTS Thurnall, Manchester, UK. These approaches minimize contamination and enable a highly precise process to be constructed.

Generating reproducible data over several months can be more challenging for cellular assays compared with biochemical assays, because of the potential variation in cell physiology that may occur with different passage number. Numerous techniques are commonly used for high-throughput cellular assays.

3.2.4.1 Cell Proliferation Assays. Cell proliferation assays are quick and easily automated. The simplest format involves incubating cells in the presence of a suitable stimuli that causes the cells to proliferate and measuring the growth using a vital stain, like Alamar Blue (56), oxygen sensors (57), or radioactive isotope uptake and incorporation. For example, human lymphocytes can be stimulated to proliferate through a variety of stimuli, such as cytokines, bacterial lipopolysaccharides, etc. By measuring the incorporation of radioactive thymidine into DNA, cell proliferation is measured, and compounds that modulate proliferation are readily detected. Variants on this theme have been used for many decades, however, the growth inhibition may just reflect cytotoxicity phenomena, and a range of new cellular assay formats are increasingly being used for high-throughput screening.

3.2.4.2 Reporter Gene Assays. Reporter gene assays are routinely used in high-throughput screening laboratories, and vectors containing reporters are readily available commercially. As the name suggests, a reporter indicates the presence or absence of a particular gene product that in turn reflects changes in a signal transduction pathway. The reporter is usually quantified biochemically; either by measuring its enzymatic activity, as with β-lactamase or luciferase, or by direct measurement, e.g., green fluorescent protein (50). Typically, a reporter gene plasmid is transfected into a desired cell. This plasmid will contain a response element that is capable of initiating transcription, a promoter region, and a coding region for the reporter. The plasmid is stably transfected into cells and the cell clones selected, followed by scale up for the high-throughput screen. Alternatively, transient transfection of cells can be performed just before assay. In high-throughput screening, stable transfection is preferred because of the higher level of reproducibility achieved in readout, which is a critical factor in the overall

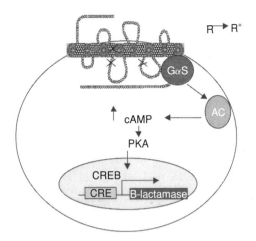

Figure 2.5. A schematic of a β-lactamase reporter cell line that is coupled through a seven transmembrane G-protein—coupled receptor whose second messenger signaling pathway operates through cyclic adenosine monophosphate (cAMP). The G-protein, Gαs, activates adenyl cyclase, increasing the concentration of cAMP, which stimulates the production of β-lactamase through cAMP response element.

quality of the data. However, stable cell line production can be a time-consuming process.

Other types of cell lines are used to assess cytoxicity, selectivity, and other false positive situations. The false positives can arise from a variety of effects like membrane disruption, and indirect inference with the readout, for example, with β-lactamase direct enzyme inhibition. Control cell lines that contain constitutive expression of the reporter are used to assess the hits from a HTS. Figure 2.5 shows a schematic representation of a seven-transmembrane G-protein—coupled receptor (GPCR) that signals through cAMP. In this assay, an agonist stimulates an increase in the intracellular level of cAMP, which cascades down the protein kinase second messenger pathway to trigger the production of β-lactamase. Because β-lactamase is not normally present in mammalian cells, the amount of enzyme produced is proportional to the efficacy of the agonist. A unique fluorescent substrate CCF2 (Fig. 2.6) is ideally suited to high-throughput screening assays. Esterified carboxyl groups allow CCF2 diffusion across the cellular membrane. Inside the cell, esterases rapidly convert the esters into carboxylic ac-

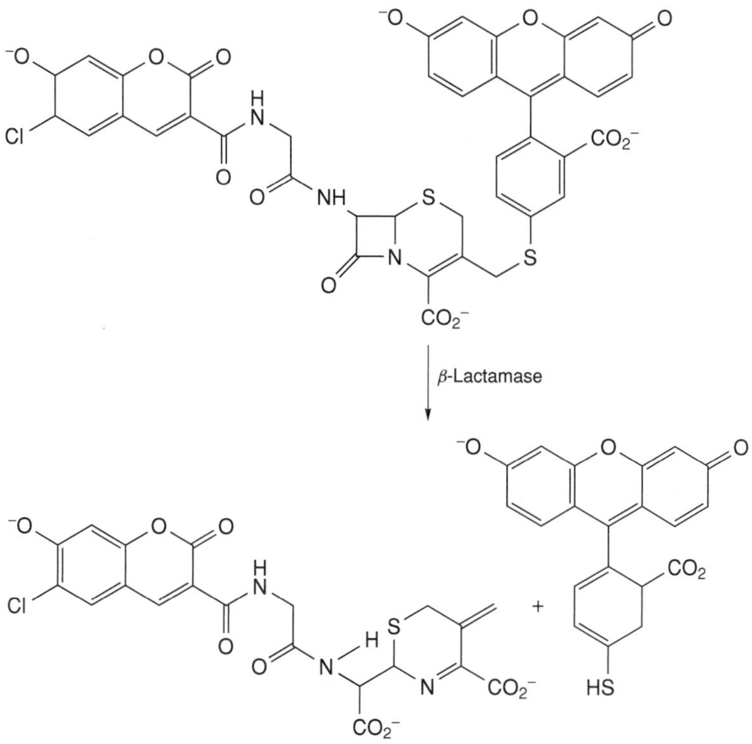

Figure 2.6. Conversion of CCF2 (Aurora Biosciences) from a green fluorescent FRET substrate into a blue fluorescent product by β-lactamase. In the substrate, the coumarin acts as the donor partner and fluorescein acts as the acceptor in the FRET pair. Lactamase cleaves the lactam ring allowing separation of the two fluorescent pairs that in turn prevents FRET occurring. CCF2 has an emission of 518 nm and the product had an emission of 447 nm, allowing simultaneously measurement of the two dyes.

ids, effectively trapping the substrate within the cell. The uncleaved substrate contains two fluorophores that act as a FRET pair and yield an emission at 518 nm. After cleavage by β-lactamase, the FRET pair is broken and the product will produce an emission at 447 nm. Figure 2.7 shows the image of cells before and after treatment with an agonist. By stimulating a GPCR with a known agonist and then test compounds, antagonist activity can be detected. The ratio of the two fluorescence signals allows quantitation of the cellular response and can improve the precision of the assay. Radiometric readouts are popular in cellular HTS assay formats because absolute number of cells is not critical to the assay readout.

Some of the most important cell-based assays are used to find modulators of ion channels or GPCRs. High-throughput screening of GPCRs are either direct binding, using membranes or whole cells, or functional methods. Gi- and Gs-coupled receptors inhibit or stimulate the production of intracellular cyclic AMP that can be quantified directly in an immunoassay or indirectly through a reporter gene. Activation of Gq-coupled receptors leads to an increase in inositol triphosphate (IP3) that can also be quantified directly. Elevated levels of cAMP and IP3 also leads to changes in intracellular calcium for which there are many methods for measuring (58). One useful method for HTS uses calcium-sensitive dyes. These can be measured in the fluorometric imaging plate reader (FLIPR; Molecular Devices, Eugene, OR) (59), and literature examples include the vanilloid and histamine-1 receptor (60, 61). This instrument integrates com-

(a)

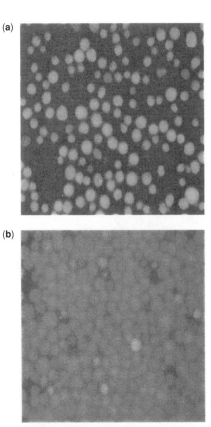

(b)

(c)

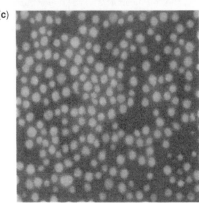

Figure 2.7. Demonstration of a β-lactamase reporter cell line. In a final volume of 0.02 mL, 5000–
10,000 cells were seeded overnight at 37°C before
being stimulated with excess agonist or antagonist
for 4 h. The resultant β-lactamase activity was measured after a 1-h incubation with CCF2 (Aurora Biosciences) at the indicated wavelengths in an LJL
Analyst fluorometer. (a) The unstimulated cels, (b)
the cells incubated with agonist, and (c) cells treated
with agonist and excess antagonist. See color insert.

pound addition to assay plates with direct imaging of the fluorescent event. Ion channel
modulation by compounds can also be measured using FLIPR and a range of voltage-sensitive oxindol dyes for potassium channels
(62), as well as calcium-sensitive dyes for calcium channels.

**3.2.5 Alternate High-Throughput Screening
Techniques.** Sometimes assays for biological
targets cannot be conveniently designed to fit
with standard cellular or biochemical assay
formats. For example, in the search for new
antibacterial agents, genomic experiments
have indicated a large number of proteins that
are essential for the survival of the bacterium,
but their function in the cell is unknown. In
this situation there is no known biological
function that will allow the design a biochemical or cellular screen. To screen these types of
target, an alternative to conventional biochemical and cellular screen needs to be used.

One alternative screening approach that
does not require knowledge or analysis of the
biological function of the target of choice is
direct measurement of compound interaction
with protein. A range of techniques are available to measure the direct binding events such
as NMR and calorimetry (63, 64). These biophysical techniques can yield important binding information, but the current form of the
technology has low throughput and capacity
limits so they cannot be used to screen large
numbers of compounds. In addition, for this
approach, large amounts of relatively pure
protein need to be available. There are a number of biotechnology companies who have developed screening platforms that will detect
direct compound–protein interactions in high-
throughput formats, e.g., Anadys Pharmaceuticals, Neogenesis, Novalon, and 3Dimen-
sional Pharmaceuticals.

The NEOGENESIS (www.neogenesis.
com) (63) approach is based on technologies
that rapidly separate protein-bound ligand
from free ligand using chromatography and
then MS to detect bound ligand. By using mass
encoded libraries, mixtures of compounds can
be screened, yielding a published throughput
of 300,000 compounds in a day. The screening
efficiency is achieved through screening mixtures of compounds.

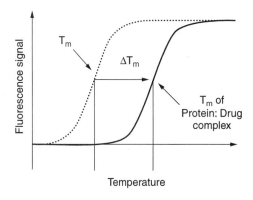

Figure 2.8. The schematic shows the theoretical shift in the melting curve for a protein in the absence of and the presence of a drug binding.

3Dimensional Pharmaceuticals (www.3dp.com) thermofluor (65) approach measures the temperature shift in a melting curve of a protein on ligand binding. The protein's melting curve is measured using a fluorescent indicator dye. A fluorescence dye is chosen such that it binds to the internal hydrophobic domains of a protein. As the temperature increases, the protein melts, exposing more of it internal hydrophobic core and the fluorescence signal increases. By comparing the change in melting transition point with and without ligand, pro-

tein binding can be detected (see Fig. 2.8). An example of a Thermofluor experiment is shown in Fig. 2.9.

These technologies are relatively new, and acceptance in HTS laboratories will depend on their record of accomplishment in discovering viable leads molecules, together with technological improvement to increase their screening efficiency.

3.2.6 Screen Validation and Reagent Scale Up. Having designed and optimized the conditions for a screening assay, the next stage in the process is often termed the screen validation phase. This tests the robustness of an assay in high-throughput screening conditions. The first challenge is to produce sufficient reagents for the screen. For simple enzyme assays, a relatively pure preparation that is devoid of competing enzymatic activity can usually be readily prepared. For example, a tyrosine kinase domain has been requested for high-throughput screening to find enzyme inhibitors. The substrate is a biotinylated peptide that is phosphorylated with ATP and the resulting phosphorylated peptide quantified using a hTRF assay format as described earlier in the chapter. The enzyme preparation should be free of other kinases, competing

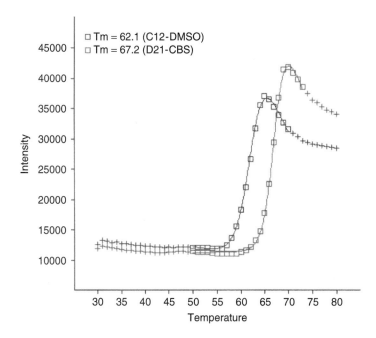

Figure 2.9. An example of a Thermofluor ligand-binding experiment where carboxybenzene sulfonamide (CBS) binds to carbonic anhydrase (CA). The fluorescence intensity was measured over a 30–80°C temperature range using 50 μM CBS (red) or dimethyl sulphoxide (blue) and 0.15 mg/mL CA. See color insert.

phosphatases, and peptidase activity. Usually this is achieved through a combination of purification and the use of inhibitor cocktails. This process should be designed to be scalable and produce sufficient material to support screening as well as additional experiments like compound concentration response curves and any selectivity assays.

For cellular high-throughput screens, a cell production schedule needs to provide a continuous supply of cells for the duration of the screen. It is critical to ensure that cells that have gone through multiple and differing numbers of passages to maintain their pharmacological responsiveness profile.

Screen validation phase tests the screening assay in a more production-like environment. For example, the lab bench results are replicated on the HTS robotic system. Critical quality control experiments are performed at this stage in the process. This involves screen rehearsal with a small number of compounds (i.e., a few thousand), thus validating the screening process. Process precision is measured by repeating the mini-screen on a different day. This procedure allows definition of all the quality control parameters. Typical values include the following:

1. Precision for the maximum signal, usually expressed as the coefficient of variation.

2. Precision for the minimum or background signal expressed as coefficient of variation.

3. Precision of a known inhibitor, agonist, activator, etc., depending on the assay.

4. Analysis of the signal to background, usually expressed as Z' (66): $Z' = 1 - (3s_2 + 3s_1)/(x_2 - x_1)$.

The Z' statistic measures the assay signal window as a fraction of the distance between the means of the distributions, x_1 mean of the low signal, x_2 mean of high signal, s_1 standard deviation of the low signal, and s_2 standard deviation of the high signal. The assay signal window measures the distance between the distributions of the totals and the blanks. For example, in a binding assay, the totals would be the maximum signal in the presence of ligand and the blank measured in the presence of excess competing ligand, the non-specific

binding value. The Z' index is more critical than the signal to background in an assay (see Fig. 2.10). Typically, a screen that yields a Z' greater than 0.5 will yield good screening data.

If a series of biological assays are required to progress a chemical series toward *in vivo* studies, it is important to understand how pharmacological activity translates from one assay to another. This trans-assay precision is critical for the success of an early drug discovery program.

Measuring the effect of a compound library in biological assay can be more complex to interpret because subtle differences exist in different laboratories, on different sites, and even in large pharmaceutical companies and different countries.

If known pharmacological agents are available, then the screening assay is validated across multiple assays. This is more difficult if the target of interest is novel and no pharmacological tools are available. It requires rigorous attention to detail to achieve inter-laboratory precision even down to the way the compounds are solubilized and diluted, the types of equipment, and the detection technology.

3.3 Constructing Compound Decks and Screening for Hits

At this point in the lead discovery process, there are a number of significant compound selection choices that depend on target knowledge and/or the ligands that bind to the target. Some of these choices can be relatively straightforward.

3.3.1 Following the Competition. When analyzing competitor's lead molecules published in the scientific or patent literature, the question to be asked is whether there are clear opportunities available for carving out new IP positions around novel SAR. This is a standard fast follow or "me too" rational design approach to drug discovery and optimization; a good example is described in the development of ranitidine (67). This is also covered in detail in other chapters.

3.3.2 Systems-Based or Focused Discovery. It may be that no specific chemotypes are available in the literature or from competitor

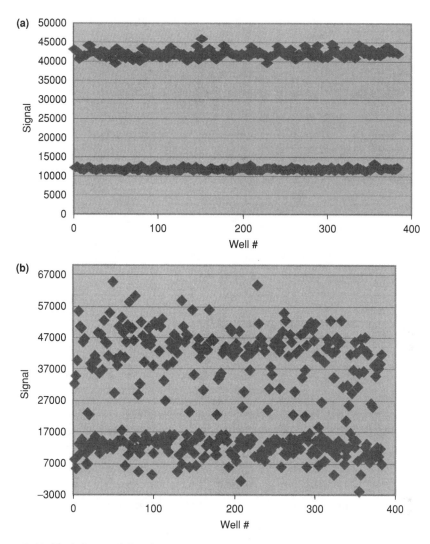

Figure 2.10. Typical control data from two screens. (a) An acceptable assay was designed showing a good separation between signal and background, 3.5-fold and a Z' of 0.86. In contrast, the assay in b performed poorly, with a signal to background of 3.5 and a Z' of -0.2, and it had to be re-designed.

company publications to provide useful starting points for chemistry. However, if there is information on the nature or family of targets that are being worked on, then a number of focused approaches can be taken to discover lead molecules.

If the target is a known "druggable" target such as a GPCR, ion channel, or simple enzyme, then one could take a "targeted" or "systems-based" approach to lead discovery.

By collecting and searching all the available compounds in the company compound collection or external commercial libraries with known activity against these family's of targets, the discovery of active compounds can become more effectively targeted. In some companies, these target class compounds sets can be as small as few hundred compounds or as large as several thousands. At this part of the process, CADD techniques can be used most effectively. Sophisticated computer algorithms matching structural information of a biological target or ligand to search internal and external databases to find structural matches predict active site or receptor binding. Alternately, the technique of "virtual screening" can be used to generate virtual libraries of tens of millions of compounds that can be "docked" into the binding site of a protein target of interest. These virtual compound hits can, if possible, be synthesized and tested in a bioassay to confirm the authenticity of the hit (68–70).

3.3.3 High-Throughput Screening. Of course, when the type of information needed to make smart, focused decisions about lead discovery is not available, most pharmaceutical companies now have at their disposal the powerful technology of HTS.

HTS is the process by which very large numbers of compounds (hundreds of thousands) from a variety of sources such as synthetic compound collections, natural product extracts, and combinatorial chemistry libraries are tested against biological targets. The aim of this exercise is to find compounds active against the target. This will include not only compounds in the screening deck that are expected to be active against the target, but also compounds that would not have been predicted to be active. The underlying premise to high-throughput, random screening is that by sheer scale of numbers, you can bias the serendipitous discovery to occur. If the chances of finding a new chemotype against target X is one in a million, then in the HTS world screening one million compounds would maximize the chance of finding an active hit.

This may seem like a simplistic, anti-intellectual approach to drug discovery, but it has proven to be a very successful lead discovery paradigm for decades. The scale and industrialization of the operation often masks the incredible innovation and smart thinking that has gone into the HTS process over the last 10 years.

High-throughput screening is positioned to have its maximum impact on the earliest part of the drug discovery process, namely lead discovery. The goal of the lead discovery process is to provide a cohort of chemically tractable molecules with sufficiently interesting properties against nominated biological targets. This would trigger further investment of resources in lead optimization programs such as medicinal chemists, disease biology specialists, etc.

A core underpinning technology for the screening process described above is the capability to acquire, store, and rapidly retrieve large numbers of compounds for testing. Most large pharmaceutical companies now use advanced compound management technologies to achieve this goal (71).

The compound management activity covers the acquisition of compounds and libraries, their registration, formatting, QC checking, storage, and retrieval on demand. A typical compound management process would be intimately involved in all aspects of the drug discovery process, supplying compounds in a variety of formats and amounts to lead optimization programs, HTS teams, and biology research teams.

Most pharmaceutical companies have large collections of compounds derived from past and ongoing medicinal chemistry programs and samples acquired from external vendors. In most large pharmaceutical companies, these compounds are stored in large automated storage systems. These compound management systems allowing the storage, reformatting, tracking, and retrieval of compounds, with onward distribution to HTS laboratories or other research laboratories in the company. The size and diversity of the compounds held within a company compound store largely reflects the size and diversity of the medicinal chemistry approaches a company has taken over its' history. Because of the recent spate of mergers, some of these compound collections can be over one million compounds, although the average is more likely to be in the range of 300—500,000. Ideally, the internally acquired and legacy compounds are analyzed before going into the compound store, to select compounds with drug-like properties and eliminate polymers, reactive intermediates, dyes, etc. This refined collection would then form the basis of a "solid" store where compounds exist as dry powders or films in standardized vials. Usually the traditional medicinal chemistry compounds are supplemented with libraries from commercial sources and academic collaborators. Unless an exclusive deal has been negotiated, the compounds available from commercial and academic sources will be available to anyone who wishes to buy them. A competitive edge can be gained by developing tools to select the "best" external compounds to complement the internal compound collection and by acquiring diverse chemotypes that are absent from the current collection. Chemoinformatic resources and a compound acquisition budget are therefore vital to the maintenance of a modern compound store.

Natural products have historically been a significant component of a compound deck (72). Typically, a crude solvent extract of a microbial fermentation, a plant, or a marine organism was provided to the compound store ready for screening. If a particular extract was shown to have the desired biological activity, it was refermented or re-extracted, fractionated, and the pure compound isolated (73). This isolation typically took weeks to months. The protracted timelines invariably became the "Achilles Heel" of natural products screening because the medicinal chemistry program had either advanced beyond a point where the natural product could not add significant value or the program had been terminated. In an attempt to improve this temporal dilemma, some companies have resorted to pre-fractionated crude natural products extracts (74). This tactic involves fractionating natural product extracts before screening to yield a series of fractions that each contains just a few components. By automating this fractionation process, a more efficient overall process can be developed. This approach dramatically decreases the time to reveal an isolated natural product and de-couples the fractionation from the screening process. From our experience, screening less complex natural product extract mixtures tends to reduce the false positive rate in a high-throughput screen. Alternatively, one could purify or purchase purified natural product compounds directly, and then put them through the biological screens as with traditional medicinal chemistry compounds.

A very common source of new compounds for high-throughput screening comes from the prodigious output of combinatorial chemistry laboratories. Large numbers of compounds can be prepared relatively quickly using automated synthesis in either solution or on solid phase. The real art is to build combinatorial libraries that contain "drug-like" molecules that have biological activity. It is beyond the scope of this chapter to detail the various combinatorial chemistries that generate screening libraries.

Having collected and organized a compound deck, the next challenge is to manage the requests from individual investigators both to deposit new compounds from medicinal chemistry laboratories and to ship compounds for biological experiments in a timely fashion. A few tens or hundreds of compounds from the solid compound store can be routinely weighed out using manual or automated systems and delivered to an investigator. However, the supply of the hundreds of thousands required for a high-throughput screen would be totally impractical using the solid store. To overcome this logistical nightmare many compound stores also contain a liquid store. Here, solubilized compounds are stored as individual tubes and in microtiter plates. The solvent of choice is usually dry dimethyl sulfoxide (DMSO). This is a compromise, because not all compounds will readily dissolve in DMSO, but many biological assays can tolerate relatively high concentrations of this solvent. In addition, not all compounds will be stable in DMSO over protracted periods; therefore, there needs to be an appropriate compound refreshment process. It is important to have a high-throughput analytical chemistry capability to monitor the quality of the compounds. The concentration at which compounds are stored very much depends on both the compound handling process, the screening process, and the solubility of the compounds. For example, most cellular assays will not tolerate more than 1% DMSO without adversely affecting the cell physiology, and therefore, a 1 mM stock solution from the compound store will only give 10 μM final concentration in the assay. By storing compounds at a higher concentration, the risk of compounds coming out of solution increases. As a compromise, some groups will routinely store compounds as a 3 mM stock solution in 100% DMSO.

Having compounds available in both tubes and plates is a distinct advantage for screening. The microtiter plates can be readily organized and provided to a high-throughput screening group who will test for biological ac-

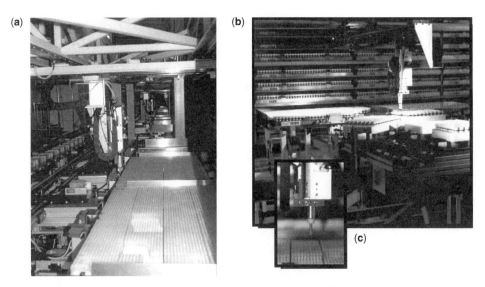

Figure 2.11. This figure shows some of the components within the Haystack compound storage and retrieval system: (a) the microtiter plate handling system, (b) the solid storage system and the robot that handles the compound vials, and (c) the tube picking robot placing a solubilized compound back into a tube rack.

tivity. Certain compounds that yield the desired activity are repeat tested to measure potency, selectivity, and *in vitro* toxicity or safety. Because these compounds are likely to be scattered across hundreds of different microtiter plates, a tube store allows you to prepare a customized set of compounds and supply then for screening. Additionally, focused sets of compounds based around known chemotypes or chemical series are readily assembled from an automatic tube store for lower-throughput screening. Overall, a modern compound inventory management process is highly integrated, requiring a combination of chemistry, information technology, and production engineering skills. Numerous companies now provide very sophisticated compound storage systems that can manage all of the operations described above. An example used at Bristol-Myers Squibb is the Haystack system built by The Technology Partnership, UK (71) (Fig. 2.11).

This particular fully automated storage and retrieval system can store over 750,000 compounds as dry solids and potentially 15 million compounds in a variety of liquid formats.

There have been attempts to specifically engineer combinatorial synthesis approaches to facilitate screening that is more effective by eliminating the need for indirect compound storage and retrieval systems, such as the Haystack (75). In these examples, the compounds were prepared using solid phase synthesis. A compound was synthesized on Tentagel beads using both acid and UV sensitive linkers and encoding tags. The tag encodes the order in which monomers were built onto a scaffold such that the exact monomer sequence could be identified. A mixture of approximately 10–20 beads per microtiter plate well was treated with acid, and the released compound assayed for biological activity. If activity was detected, each of the 10–20 beads were separated into individual microtiter plate wells and exposed to UV. If activity was detected, the tag was decoded, identifying the monomer sequence and hence the structure. Using this format, tens of thousands of compounds could be stored in a relatively small store, i.e., a standard refrigerator.

Compound stores also contain screening decks that are available mainly for the HTS laboratories. These screening decks can represent the full collection of compounds in the store or tactically organized subsets (76, 77). The choice of whether to screen a full compound deck is dependent on the medicinal chemistry insight or structural knowledge of a particular target as discussed earlier.

From a screening perspective, there are significant advantages to a systems based or target class approach. First, similar assays allow parallelization of assay design, the so-called "Plug and Play" approach. Second, the screening data for a family of closely related targets is simultaneously generated for set of compounds.

Full compound deck screening is extremely useful if you wish to find new chemotypes, and there is little information on compounds that effect the target of interest.

3.4 Hit Identification, Profiling, and Candidate Selection

After ordering a copy of the compound deck in the plate format of choice, the screening process can be carried out very rapidly (a matter of weeks). The screening scientist monitors the automated system continuously for both hardware and assay performance. Of all the different stages in the lead discovery process, the actual screening is now the quickest.

3.4.1 Analyzing Screening Hits. The next step is to analyze the primary screening quality control data (66). During a primary screen, a variety of QC plates are inserted into the run, including blank plates, plates containing just DMSO, monitor hardware performance, pipette error, detector misalignments, etc. QC plates containing biological reagents check for drift in the assay over the course of the screen. Typically, panels of known inhibitors, activators, antagonists, or agonists, depending on the assay, are tested at multiple concentrations. Additionally, controls, both negative and positive, are included on each plate. Real-time data analysis allows the screening scientist to continuously monitor the performance of the screen and the robot.

Having analyzed the QC data and eliminated, or repeat tested, any compound plates that failed, the entire screening run is analyzed. It is usual to perform a high-throughput screen with a single replicate of each compound. The results are statistical in nature and interpreted as population data. In the assay validation section (Section 3.2.6), a statistical parameter Z' was introduced. If an assay

had a Z' of 0.7 and a known inhibitor was known to cause 50% inhibition at 10 μm, then in a screen, the percentage of inhibition could vary between 35% and 65% because of population statistics. The initial high-throughput screen is normally viewed as a population frequency histogram and as a scatter plot (see Fig. 2.12, a and b, respectively). Having selected a particular cut-off, any compounds with greater than or equal to this value are retested with replicate determinations. Figure 2.13 (a and b) shows the frequency histogram of the retest values and the associated scatter plot. In this particular example, 73% of the active compounds retested in the second assay. Interestingly, the majority of the weak inhibitors, 20–40% inhibition, were false positive compounds that interfered with the assay readout. Figure 2.14 shows the Z' values for this screen's quality control plates.

The vast majority of compounds from a full deck screen has no effect, and statistical analysis allows one to decide when a compound has had a significant effect, designated as a "hit." A hit that seems to be statistically significant could be explained by a variety of reasons; assay false positives, cytotoxicity effects, etc., as well as true pharmacological response. False positives are compounds under test directly interfere with the detection readout, for example, a fluorescent compound or quencher in a prompt fluorescence assay. False positive results arise from pipetting errors that delivered the incorrect amount of a reagent. A second round of assay(s), to analyze the hits eliminates false positive results. For example, assays that indicate which compounds show the desired selectivity against other biological target or lack of cellular cytotoxicity. A third round of screening generates data from concentration response curves so potency, e.g., K_i or IC50, and or efficacy for agonists, directs further medicinal chemistry.

3.4.2 Profiling Hits. Increasingly, major companies are adding to the value of high-throughput screening by immediately profiling screening hits against a battery of selectivity, toxicity, and safety assays (78–80). The idea behind this type of extensive profiling is

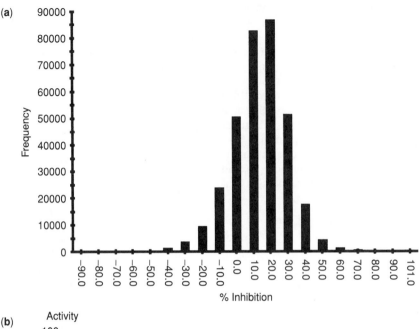

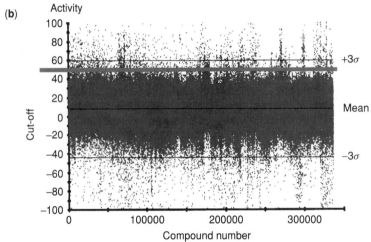

Figure 2.12. (a) A frequency histogram of the data derived from a high-throughput screen run of 340,000 compounds used to find inhibitors of 15-lipoxygenase. (b) A scatter plot of the same data indicating where the cut-off was chosen to retest the compounds that showed some activity.

that the more information scientists have about biologically active compounds, the better the decisions will be on which compounds to progress or to terminate. A further hope is that as compounds fail or succeed in the R&D process, this early profiling may start to indicate predictive *in vitro* profiles for each of these outcomes. This would provide a significant advantage to those companies trying to improve the quality of the drugs in their pipeline and their chances for survival.

After the concentration response phase, decisions on which hits to progress down the drug discovery pipeline need to be made. This is when the team described at the initial stages of the assay design process comes back together to decide which of the active chemotypes display the desired characteristics. The

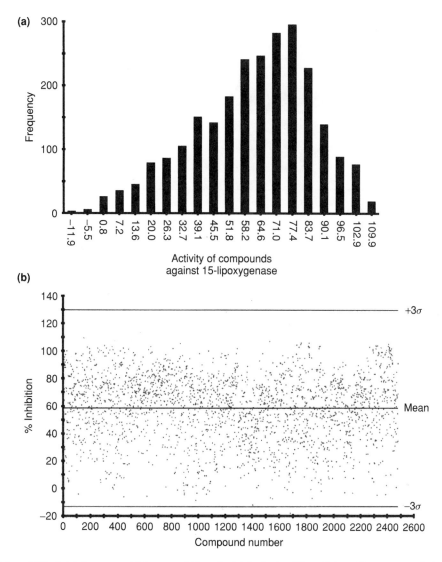

Figure 2.13. (a) A frequency histogram of the retest data derived from the positive compounds from a high-throughput screen run of 340,000 compounds against 15-lipoxygenase. (b) The same data as a scatter plot.

structural integrity of the positive compounds should be determined: LC/MS, as well as a high molecular weight filtration, to remove any compounds that may have polymerized on storage in DMSO. From our experience, approximately 10% of the compounds will not be exactly as described in the database. There are a variety of reasons that include the following:

1. Incorrect structure when the compound was submitted to the compound store

2. The activity was caused by a minor contaminant

3. The compound has degraded on storage

4. Human error anywhere along the process

The choice of compounds to progress can vary, and any additional information can help prioritize the future workflow. HTS processes provide *in vitro* data on all the major cytochrome P450 activities, additional

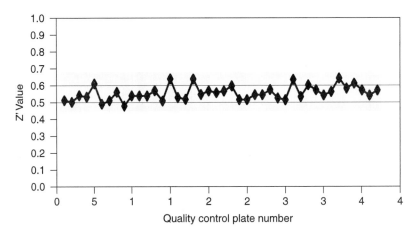

Figure 2.14. These plots show the quality control data that came from the 15-lipoxygenase HTS expressed as Z'. A QC plate was run after every 25th 384-well plate. The data are calculated from the maximum signal and the background where $n = 160$ for each data set.

cytotoxicity indices, cardiac liability reflected through adrenoreceptor or ion channel activity, etc. In Fig. 2.15, a whole matrix of data is supplied to help drive the decision-making process.

- Target potency (μM): activity from the HTS assay
- Target selectivity: search of all the screening databases, both against other targets HTS and lower-throughput screening, to check whether the compound was found to be active against other targets. The number indicates the number of positive results found.
- Human liver enzyme inhibition (μM): the activity in a standard cell-based toxicity model.
- Mutagenesis (μM): the activity in an *in vitro* model of mutagenesis
- Cardiac Liability: activity in essential cardiac ion channels that would cause an adverse side effect in humans.
- Drug-like properties: assessing the compound in a range of drug-like *in silico* models.
- Chemical tractability: a more subjective flag that is based on a medicinal chemist's view on whether this is a good start for a medicinal chemistry program.

- Liability or disadvantage: this is color coded, red = high, yellow = medium, and green = low.

In this example, if the only datum available to the medicinal chemist was the potency of a compound, BMS1 would be the highest priority. BMS5 has the cleanest profile, and this chemical series was the preferred candidate for progression, although it was 10-fold less potent than BMS1. Activity in these liability assays will not stop the progression of a compound, but it helps in understanding how to drive the medicinal chemistry forward. This is the point at which the initial HTS process has ended. The decisions from this point are around compound optimization and continued target validation. HTS approaches and technology are back into play if the project requires further chemotypes.

4 TECHNOLOGY INFRASTRUCTURE IN HIGH-THROUGHPUT SCREENING

Advances in genomics have caused a dramatic increase in the number of potential targets. Together with the growth of combinatorial chemistry, many targets and compounds require HTS. This interdependence has lead to a technological revolution in high-throughput

Candidate	Target potency (IC50)	Target selectivity (activity at other targets)	Human liver enzyme inhibition (IC50)	Toxicity in human liver cells (IC50)	Permeability in human intestine cells (transport rate)	Muta-genesis potential (IC50)	Cardiac liability (IC50)	Integrity of compound (%)	Computer generated drug-like properties (# problem flags)	Chemical tractability fail=red, pass=green
BMS-1	0.6	4	37.8	1.8	122.0	19.1	0.5	98	4	
BMS-2	2.4	2	40	6.3	107.0	22.3	11	53	3	
BMS-3	7.0	0	17.8	4.4	11	78.1	12	31	0	
BMS-4	7.5	0	8.0	>100	125.0	>100	>100	100	2	
BMS-5	8.0	0	>100	51	145.0	>100	>100	92	0	
BMS-6	16.0	3	>100	57	191.1	1.8	33	93	1	
BMS-7	25.0	0	0.5	3.1	34	>100	72.5	95	2	
BMS-8	34.0	3	38.6	33.1	29	>100	0.5	88	0	
BMS-9	53.0	1	5.2	0.2	155.0	61.5	61	93	1	
BMS-10	90.0	0	0.9	120.5	130	>100	0.6	63	3	

Figure 2.15. A compound profile matrix that outlines all the information on a set of compounds that have been identified from a HTS.

screening (15, 81) with significant investment in the automation and miniaturization.

4.1 Automation in High-Throughput Screening

The need to screen large compound libraries that typically range from 10^5–10^6 through a range of biological targets in an efficient and timely manner has been one of the main drivers for automation. From the late 1970s to the mid-1990s, the 96-well microtiter plate reigned supreme in many screening laboratories (82). Manufacturers supplied many variations on the 96-well plate theme, varying the shape of the well, varying the color of the plate, and supplying a range of surface chemistries for specialized assays. Invariably, the 96-well plates were all subtly different, and it was not until 1996 that a standard was recommended by the Society for Biomolecular Screening. Automated screening systems needed accurate plate dimensions. Robotic systems needed higher tolerances and defined dimensions to pick and place plates precisely. Detection instrumentation was also adapted to enable robots to load and unload microplates.

High-throughput screening automation exists at a variety of levels, from manual to semi-automated to fully automated turnkey systems (83). However, the types of equipment tend to be similar, and the way in which the screening process is integrated dictates the level of automation. For a brief discussion on the advantages and disadvantages of automated platforms versus workstations, see refs. 84 and 85.

All the high-throughput screening automation platforms tend to have the same limited number of basic operations; a method moving around microplates, dispensing liquids, a series of detectors, and incubators. The methods

Figure 2.16. Images of the Aurora Bioscience UHTS screening platform that is based around a track that moves the microtiter plates around the different workstations. The 96-well piezo-electric dispensing head is shown in detail.

for moving microplates tend to fall into two general approaches. Movements can be with an articulated robotic arm, picking and placing plates, or through a track that resembles a mini-production line designed to shuttle plates around the system (Fig. 2.16). The liquid handling options can vary between a syringe-based system that gives higher volumetric precision to a peristaltic pump that tends to be more rapid but less precise. The range of potential detectors is dependent on the assay technology discussed earlier in the chapter. The incubators can range from the very simple open racks of shelves to highly environmentally controlled systems. The glue that puts all this together is the scheduling software that controls what goes where and when. In the more sophisticated systems, an operator loads reagents, plates containing test compounds, and disposes of any waste (Fig. 2.17). The scheduling software instructs the articulated arm or track system to move microtiter plates

Figure 2.17. Images of typical turnkey automated screening robots that use an articulated arm to move plates around the screening system. The robot is placing plates into two different liquid handling devices. Top left is a PlateMate Plus and top right is a Multidrop. This system was built internally by Bristol Myers Squibb engineers.

at the calculated times around the various liquid handlers, incubators, and detectors, and eventually to waste. A limited amount of rules-based artificial intelligence can be used to aid the quality of the operation. For example, as data are generated by the detector, online analysis can be used to monitor drift in the assay or whether certain wells fail to meet prescribed quality control parameters, as seen with blocked tips on a dispenser. These types of error alert an operator.

Building these integrated robotic systems requires strong management commitment with time, money, and a willingness to develop the necessary skill sets. A stable, fully automated screening platform does offer continuous operation, a consistency of process that can be verified, automated audit trial of the samples that have been tested, and safety (for further detail see Ref. 86).

Not all laboratories have the resources to build fully integrated screening platforms and support them. Additionally, not all assays can be modified to work on an automated platform. For example, a particular detector may not be available in a format that can be integrated. Most screening laboratories will use workstation approaches in addition to fully automated platforms to enable assay flexibility. Unlike the automated platforms, where plates are processed in a serial manner, plates are batched together, "a stack," in workstation approaches. Workstation approaches replaces the robot with a human, and as long as the number of microtiter plates processed is acceptable, this often works well. The same quality control is incorporated into the workstation process, and from our experience, the workstation data are comparable with that generated by a robot. One real advantage of a fully automated platform is where there is a need to have accurate incubation times, for example, in a kinetic assay.

4.2 Miniaturization of Screening Assays

The increasing operating costs of HTS laboratories have driven a strong interest in implementing more cost-effective ways of carrying out high-throughput screening campaigns. Miniaturizing the plate format is one of the major technology solutions. There has been a stepwise evolution from the glass test tube and plastic Eppendorf tube into microtiter plates containing ever-increasing well densities.

The first tangible step along the miniaturization route was the introduction of the 96-well microtiter plate that replaced the individual tube (82). The 96-well plate became the standard workhorse in academic and industrial laboratories over the last 20 years. However, ever increasing demands to expand testing capacity and improve process efficiency while simultaneously reducing costs have pushed HTS laboratories into using 384-well plates and beyond (Fig. 2.18).

A screening organization that runs 50 screens a year, each testing a 500,000 compound deck with average reagent and plasticware that costs $0.20/well, totals $5 million, excluding waste management costs. This scenario in a 96-well plate format generates 260,000 plates of plastic waste per year. A typical assay volume in the 96-well plate is 100–200 μL, and by reducing this to around 5–10 μL, reagent costs are reduced. Additionally, smaller amounts of compounds are needed for the assay.

In the late 1990s, HTS laboratories adopted the 384-well plate as standard, allowing a fourfold increase in well density and increased screening capacity (87). Instrumentation companies re-invested in designing or adapting liquid handlers, detection systems, and automation to fit the new 384-well plate.

Figure 2.18. There are many different types of microtiter plates that are used in miniaturized assays for HTS. (a) 96-well plate (100 μL assays), (b) 384-well plate (25 μL well assays), (c) 1536-well plate (5 μL well assays), and (d) 3456-well plate (2 μL well assays)

The vast majority of assays have readily miniaturized down to 20–50 μL volumes. The minimum practical volume of 20 μL for the new 384-well plates was defined by the well shape and the need to produce an even layer of liquid at the bottom of the plate.

Even with the 384-well plate, there has been pressure to reduce volumes even further. The 1536-well plate is emerging as the potential next step, with square wells that allow a working volume of 5–10 μL. One advantage of the 1536-well plate in absorbance-based assays is volume reduction while maintaining the path length. Additionally, low-volume 384-well microtiter plates are now commercially available. The 1 μL assay is also now available in the 1536- (Corning Costar Corp., Cambridge, MA and Evotec OAI, Hamburg, Germany) and the 3456-well microtiter plate (Aurora BioSciences, San Diego, CA) (50). In a little over 5 years, we have witnessed a 100-fold reduction in assay volume and a 36-fold increase in the well density. The discussion over well density still causes many debates in screening discussion groups (87) and even higher well densities have been proposed, e.g., 9600-well plate (88).

The move to higher well densities and lower assay volumes has presented significant challenges to instrumentation companies. First, there is the need to detect the results of a particular assay. For example, in the 96-well scintillation proximity assays, the scintillation counter photomultipliers are positioned above each well to measure the emitted light. Using a mask, these machines were adapted to read 384-well microtiter plates. The disadvantage was that it then took four times as long to read a plate. A new solution needed to be found. Imaging technology, using charged-coupled devices, (CCDs) provided the answer (e.g., LEADSeeker; Amersham Pharmacia, Amersham, UK, and CLIPR Molecular Devices, Palo Alto, CA) (89). Imagers take the same time to read a 96-well, 384-well, or 1536-well microtiter plate. A 500,000-compound high-throughput screen using a filter binding assay format consumes approximately 10,400 96-well microtiter and filter-binding plates. For the LEADSeeker format using miniaturized plates, 1536 wells per plate, 325 plates are used. Additionally, it takes approximately 10 min to measure the light from a 96-well plate, and therefore, total time taken to generate the data would be 36 days in a single plate-based scintillation counter. For the 1536-well assay using imaging technology, the reading time is reduced to 27 h. The overall gain in assay efficiency is dramatic. Imagers are now available for fluorescence, time-resolved fluorescence and for measuring light emission.

Another engineering challenge was to dispense volumes in the 20 nL–1μL volume range. At the top end of this scale, a variety of tip-based syringe-driven devices are available, e.g., Matrix Platemate (Matrix Technologies Corp., Lowell, MA). To achieve nanoliter dispensing, two platforms are available: the piezo-electric inkjet dispenser and the solenoid inkjet dispenser (90–92).

As mentioned earlier, the drive to screen more compounds has fueled the need for miniaturization. Additionally, there is a need to rapidly profile and evaluate the selectivity of compounds that are positive in a high-throughput screen. Miniaturization facilitates the parallel processing of numerous targets simultaneously. For example, a GPCR cell reporter assay designed to detect agonists using a β-lactamase reporter can be readily miniaturized to 2 μL in a 3456-well microtiter plate. The hits can be evaluated in this format at multiple concentrations, with null cell lines to remove false positives, and in a range of other cell lines yielding a selectivity index. By combining this with cell toxicity assays and bio-chemical cytochrome P450 assays, a wealth of information is generated in a short period on exactly the same compound solution.

5 SUMMARY

Throughout this chapter we have described how HTS, as a lead discovery tool, fits into the drug discovery process. HTS is a multi-factorial, interactive process that brings together multi-disciplinary teams of chemists, biologists, statisticians, information technology experts, and mechanical and electrical engineers.

From target discovery to lead optimization, an intricate network of processes and decisions are required to produce a successful drug discovery campaign. We have emphasized the high-throughput screening process as part of an integrated approach to hit identification and assessment. The application of industrial automation technology to the process has increased the capacity, speed, and quality of this part of the drug discovery pipeline. In conjunction with corresponding advances in target identification, automated chemistry, and data analysis, the ability of pharmaceutical laboratories to rapidly move from target concept to lead optimization candidates has changed dramatically over the last decade.

The next 10 years will no doubt bring further technological advances and creative insights to improve the drug discovery process even further. This will keep HTS approaches as a mainstream drug discovery tool for years to come.

REFERENCES

1. S. Arlington, *Pharmaceut. Execut.*, **20**, 74–84 (2000).

2. P. K. Banerjee and M. Rosofsky, *Scrip Magazine*, **6**, 35–38 (1997).

3. A. Fleming, *Br. J. Exp. Med.*, **10**, 226–236 (1929).

4. J. W. Black, W. A. M. Duncan, L. J. Durant, J. C. Emmett, C. R. Ganellin, and M. E. Parsons, *Nature*, **236**, 385–390 (1972).

5. A. D. Buss and R. D. Waigh, *Burger's Medicinal Chemistry and Drug Discovery*, 5th ed., vol. **1**, John Wiley & Sons, New York, 1995, pp. 83–1033.

6. W. H. Lewis in H. N. Nigg and D. Seigler, Eds., *Plants Used Medically by Indigenous Peoples*,

Phytochemical Resources for Medicine and Agriculture, Plenum, New York, 1992, pp. 33–74.

7. P. Cox, *The Ethnobotanical Approach to Drug Discovery: Strengths and Limitations in Ethnobotany and the Search for New Drugs*, John Wiley & Sons, Chichester, UK, 1994, pp. 25–41.

8. N. Sneader, *Drug Discovery: The Evolution of Modern Medicines*, John Wiley & Sons, Chichester, UK, 1985.

9. G. A. Cordell, *Phytochemistry*, **55**, 463–480 (2000).

10. N. R. Farnsworth, O. Aberde, A. S. Bingel, D. D. Soejarto, and Z. Guo, *Bull. World Health Organ.*, **63**, 965–981 (1985).

11. G. A. Cordell, N. R. Farnsworth, C. W. W. Beecher, D. D. Soejarto, A. D. Kinghorn, J. M. Pezzuto, M. E. Wall, M. C. Wani, D. M. Brown, M. J. O'Neill, J. A. Lewis, R. M. Tait, and T. J. R. Harris, *Human Medicinal Agents from Plants*, Symposium Series 534, Washington DC, American Chemical Society, 1993, pp. 191–204.

12. Y-Z. Shu, *J. Nat. Prod.*, **61**, 1053–1071 (1998).

13. E. Chain, H. W. Florey, A. D. Gardner, N. G. Heathey, M. A. Jennings, J. Orr-Ewing, and A. G. Sanders, *Lancet*, **2**, 226–228 (1940).

14. E. Ratti and D. Trist, *Farmaco*, **56**, 13–19 (2001).

15. J. Houston and M. Banks, *Curr. Opin. Chem. Biol.*, **8**, 734–740 (1997).

16. W. P. Janzen, *Lab. Robotics Autom.*, **8**, 261–265 (1996).

17. B. Cox, J. C. Denyer, A. Binnie, M. C. Donnelly, B. Evans, D. V. S. Green, J. A. Lewis, T. H. Mander, A. T. Merritt, M. J. Valler, and S. P. Watson, *Prog. Med. Chem.*, **37**, 83–133 (2000).

18. J. J. Burbaum and N. H. Sigal, *Curr. Opin. Chem. Biol.*, **1**, 72–78 (1997).

19. S. Fox, S. Farr-Jones, and M. A. Yund, *J. Biomol. Scrng.*, **4**, 183–186 (1999).

20. S. Sundberg, *Curr. Opin. Biotechnol.*, **11**, 47–53 (2000).

21. S. Fogarty, J. M. Treheme, B. A. Kenny, M. Bushfield, and D. J. Parry-Smith, *Prog. Drug Res.*, **51**, 245–269 (1998).

22. D. B. Searls, *Drug Discov. Today*, **5**, 135–143 (2000).

23. A. R. Dongre and S. A. Hefta, *Drug Discov. World*, **1**, 35–44 (2000).

24. I. Humphery-Smith, S. J. Cordwell, and W. P. Blackstock, *Electrophoresis*, **18**, 1217–1242 (1997).

25. D. A. Jones and F. A. Fitzpatrick, *Curr. Opin. Chem. Biol.*, **3**, 71–76 (1999).

26. J. Drews, *Science*, **287**, 1960–1964 (2000).

27. P. D. Karp, M. Krummenacker, S. Paley, and J. Wagg, *Trends Biotechnol.*, **17**, 275–281 (1999).

28. R. W. King, *Chem. Biol.*, **6**, R327–R333 (1999).

29. E. A. Winzeler, et al., *Science*, **285**, 901–906 (1999).

30. Z. Y. Zhang, *Curr. Opin. Chem. Biol.*, **5**, 416–423 (2001).

31. P. O. Brown and D. Botstein, *Nat. Genet.*, **21**, 33–37 (1999).

32. M. Schena, D. Shalon, R. W. Davis, and P. O. Brown, *Science*, **270**, 467–470 (1995).

33. R. T. Lee, *Drug News Perspect.*, **13**, 403–406 (2000).

34. S. Wilson, D. J. Bergsma, J. K. Chambers, A. I. Muir, K. G. Fantom, C. Ellis, P. R. Murdock, N. C. Herrity, and J. M. Stadel, *Br. J. Pharmacol.*, **125**, 1387–1392 (1998).

35. P. B. Fernandes, *Curr. Opin. Chem. Biol.*, **2**, 597–603 (1998).

36. K. R. Oldenburg, *Annu. Rep. Med. Chem.*, **33**, 301–311 (1998).

37. L. Silverman, R. Campbell, and J. R. Broach, *Curr. Opin. Chem. Biol.*, **2**, 397–403 (1998).

38. P. D. Haaland, Experimental Design in Biotechnology, Statistics: Textbooks and Monographs, vol. **105**, Marcel Dekker, Inc., New York, 1989.

39. P. B. Taylor, F. P. Stewart, D. J. Dunnington, S. Quinn, C. K. Schulz, K. S. Vaidya, E. Kurali, T. R. Lane, W. C. Xiong, T. P. Sherrill, J. S. Snider, N. D. Terpstra, and R. P. Hertzberg, *J. Biomol. Scrng.*, **5**, 213–225 (2000).

40. M. W. Lutz, A. Menius, T. D. Choi, R. G. Lsakody, P. L. Domanico, A. S. Goetz, and D. L. Saussy, *Drug Discov. Today*, **1**, 277–286 (1996).

41. K. Moore and S. Rees, *J. Biomol. Scrng.*, **6**, 69–74 (2001).

42. Z. Parandossh, *J. Biomol. Scrng.*, **2**, 201–204 (1997).

43. N. D. Cook, *Drug Discov. Today*, **1**, 287–294 (1996).

44. P. Banks, M. Gosselin, and L. Prystay, *J. Biomol. Scrng.*, **5**, 159–167 (2000).

45. K. J. Moore, S. Turconi, S. Ashman, M. Ruediger, U. Haupts, V. Emerick, A. J. Pope, *J. Biomol. Scrng.*, **4**, 335–353 (1999).

46. A. J. Pope, *J. Biomol. Scrng.*, **4**, 301–302 (1999).

47. I. Hemmila, *J. Biomol. Scrng.*, **4**, 303–307 (1999).

48. G. Mathis, *J. Biomol. Scrng.*, **4**, 309–313 (1999).

49. L. Upham, *Lab. Robotics Autom.*, **11**, 324–329 (1999).

50. L. Mere, T. Bennett, P. Coassin, P. England, B. Hamman, T. Rink, S. Zimmerman, and P. Negulescu, *Drug Discov. Today*, **4**, 363–369 (1999).

51. T. Gilmer, M. Rodriguez, S. Jordan, K. Crosby, M. Alligood, M. Green, M. Kimery, C. Wagner, D. Kinder, and P. Charifson, *J. Biol. Chem.*, **269**, 31711–31719 (1994).

52. R. T. McCabe, B. R. de Costa, R. L. Miller, R. H. Havunjian, K. C. Rice, and P. Skolnick, *FASEB J.*, **4**, 2934–2940 (1990).

53. J. V. Manetta, M-H. T. Lai, E. E. Osbourne, A. Dee, N. Margolin, J. R. Sportsman, C. J. Vlahos, S. B. Yan, and W. F. Heath, *Anal. Biochem.*, **202**, 10–15 (1992).

54. I. Hemmila, V-M. Mukkala, M. Latva, and P. Kiilhoma, *J. Biochem. Biophys. Methods*, **26**, 283–290 (1993).

55. B. Alpha, J. Lehn, and G. Mathis, *Angew Chem. Int. Ed. Engl.*, **26**, 266–267 (1987).

56. J. O'Brien, I. Wilson, T. Orton, and F. Pognan, *Eur. J. Biochem.*, **267**, 5421–5426 (2000).

57. M. Wodnicka, R. D. Guarino, J. J. Hemperly, M. R. Timmins, D. Stitt, and J. B. Pitner, *J. Biomol. Scrng.*, **5**, 141–152 (2000).

58. A. Takahashi, P. Camacho, J. D. Lechleiter, and B. Herman, *Physiol. Rev.*, **79**, 1089–1125 (1999).

59. E. Sullivan, E. M. Tucker, and I. L. Pale, *Methods Mol. Biol.*, **114**, 125–133 (1999).

60. D. Smart, J. C. Jerman, M. J. Gundhorpe, S. J. Brough, J. Ranson, W. Cairns, P. D. Hayes, A. D. Ranell, and J. B. Davis, *Eur. J. Pharmacol.*, **417**, 51–58 (2001).

61. T. R. Miller, D. G. Witte, L. M. Ireland, C. H. Kang, J. M. Roch, J. N. Masters, T. A. Esbenshade, and A. A. Hancock, *J. Biomol. Scrning.*, **4**, 249–257 (1999).

62. K. I. Whiteaker, S. M. Gopalakrishnan, D. Groebe, C.-C. Shieh, U. Warrior, D. Burns, M. J. Coghlan, V. E. Scott, and M. Gopalakrishnan, *J. Biomol. Scrning.*, **6**, 305–312 (2001).

63. G. R. Lenz, H. M. Nash, and S. Jindal, *Drug Discov. Today*, **5**, 135–143 (2000).

64. G. C. K. Roberts, *Drug Discov. Today*, **5**, 230–240 (2000).

65. M. W. Pantoliano, E. C. Petrella, J. D. Kwasnoski, V. S. Lobanov, J. Myslik, E. Graf, T. Carver, E. Asel, B. A. Springer, P. Lane, and F. R. Salemme, *J. Biomol. Scrning.*, **6**, 429–440 (2001).

66. J. H. Zhang, T. D. Y. Chung, K. R. Oldenburg, *J. Biomol. Scrng.*, **4**, 67–73 (1999).

67. J. Bradshaw, *Chron. Drug Discov.*, **3**, 45–81 (1993).

68. I. D. Kuntz, J. M. Blaney, S. J. Oatley, R. Langridge, and T. E. Ferrin, *J. Mol. Biol.*, **161**, 269–288 (1982).

69. W. P. Walters, M. T. Stahl, and M. A. Murcko, *Drug Discov. Today*, **3**, 160–178 (1998).

70. A. C. Good, S. R. Krystek, and J. S. Mason, *Drug Discov. Today*, **5**, S61–S69 (2000).

71. W. Harrison, *Drug Discov. Today*, **3**, 343–349 (1998).

72. J. P. Devlin, *High Throughput Screening: The Discovery of Bioactive Substances*, Marcel Dekker, Inc., New York, 1997, pp. 3–48.

73. A. Harvey, *Drug Discov. Today*, **5**, 294–300 (2000).

74. I. Schmid, I. Sattler, S. Grabley, and R. Thiericke, *J. Biomol. Scrng.*, **4**, 15–25 (1999).

75. B. Evans, A. Pipe, L. Clark, and M. Banks, *Bioorg. Med. Chem. Lett.*, **11**, 1297–1300 (2001).

76. R. W. Spencer, *Biotech. Engin. (Comb. Chem.)*, **61**, 61–67 (1998).

77. M. J. Valler and D. Green, *Drug Discov. Today*, **5**, 286–293 (2000).

78. M. H. Tarbit and J. Berman, *Curr. Opin. Chem. Biol.*, **2**, 411–416 (1998).

79. D. J. M. Spalding, A. I. Harker, and M. K. Bayliss, *Drug Discov. Today*, **5**, 570–576 (2000).

80. A. P. Watt, D. Morrison, and D. C. Evans, *Drug Discov. Today*, **5**, 17–24 (2000).

81. R. P. Hertzberg and A. J. Pope, *Curr. Opin. Chem. Biol.* **4**, 445–451 (2000).

82. G. R. Nakayama, *Curr. Opin. Drug Discov. Dev.*, **1**, 85–91 (1998).

83. D. Harding, M. Banks, S. Forgarty, and A. Binnie, *Drug Discov. Today*, **2**, 385–390 (1997).

84. K. R. Oldenburg, *J. Biomol. Scrng.*, **4**, 53–54 (1999).

85. M. J. Wildey and C. A. Homon, *J. Biomol. Scrng.*, **4**, 115 (1999).

86. M. Banks in C. K. Alterwill, P. Goldfarb, and W. Purcell, Eds., *Approaches to High Throughput Toxicity Screening*, Pub Yaler and Frances Ltd, New York, 1999, pp. 9–29.

87. J. J. Burbaum, *J. Biomol. Scrng.*, **5**, 5–8 (2000).

88. K. R. Oldenburg, J-H. Zhang, T. Chen, A. Maffia, K. F. Blom, A. P. Combs, and T. C. Chung, *J. Biomol. Scrng.*, **3**, 55–56 (1998).

89. P. Ramm, *Drug Discov. Today*, **4**, 401–410 (1999).

90. D. Rose, *Drug Discov. Today*, **4**, 411–419 (1999).

91. D. A. Dunn and I. Feygin, *Drug Discov. Today*, **5**, S84–S91 (2000).

92. A. V. Lemmo, D. J. Rose, and T. C. Tisone, *Curr. Opin. Chem. Biol.*, **9**, 615–617 (1998).

CHAPTER THREE

Rapid, High Content Pharmacology

STEVEN J. BROWN
IMRAN B. CLARK
BALA PANDI
Axiom Biotechnologies, Inc.
San Diego, California

Contents

Burger's Medicinal Chemistry and Drug Discovery
Sixth Edition, Volume 2: Drug Development
Edited by Donald J. Abraham
ISBN 0-471-37028-2   © 2003 John Wiley & Sons, Inc.

1   INTRODUCTION

Drug discovery program progression from "hit" compounds derived from high-throughput screening to identification of a lead drug candidate suitable for an Investigational New Drug, enabling toxicology studies (hit to lead), requires medicinal chemistry compound optimization driven by relevant pharmacological data. Pharmacological data regarding the mechanism of compound interaction with the molecular target, such as receptor agonists and antagonists, enzyme inhibition kinetics, and protein-ligand binding are clearly important. Quantitative methods for evaluating the effect of pharmacological agents on receptor activation and inhibition, enzyme kinetics, and ligand-receptor binding are reviewed in other chapters. Pharmacological targets adapted to high-throughput methods are often poor predictors of efficacy *in vivo* but are used for ease of testing (Fig. 3.1). Cell-free systems consisting of purified target protein, test compounds, and the necessary substrates and ligands for target protein function and signal generation are often the choice for primary, high-throughput compound screening. Cell-free screening eliminates cell metabolism, cell permeability, and protein binding issues. In the next least complicated system, cells for study are often transfected with the target gene or a reporter system of gene function in a manner that creates a wide assay window but also alters the ratio of the target receptor to cellular components. Such complex interactions, such as those in multiple-protein signaling pathways, are difficult to reconstruct in either cell-free systems or those employing engineered cells. On the other hand, animal and human tissues and whole organism studies cannot be carried out with sufficient throughput to successfully drive hit to lead optimization. Finally, untransfected mammalian cells provide intact signaling pathways and other complex mechanisms not reliably reproduced in cell-free systems or in cells with overexpressed targets. Species-specific compound activities have been observed in drug discovery caused in part by differences in the amino acid

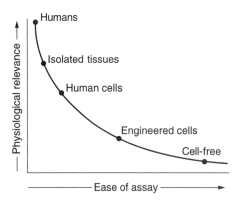

Figure 3.1. The ease compound testing is inversely proportional to the complexity of the system and the physiological relevance of the assay.

sequence of the target protein. An infamous example is provided by the discovery of a mouse G-CSF signal transduction pathway activator (1). This compound was identified in a high-throughput assay in murine cell line. Despite active research, medicinal chemistry efforts have not resulted in a compound active towards human G-CSFR. Thus, untransfected human cells provide the most relevant practical drug-screening systems.

The role of high-throughput pharmacology in the hit to lead compound candidate identification process includes determination of efficacy and evaluation of potential compound liabilities (solubility, adsorption, toxicity) and specificity of action. Other examples include data obtained from pharmacological profiling of lead compounds against a broad panel of natural human cells. It is most useful if these cells have been previously characterized for their response to known pharmacological agents. *In vitro* assays for evaluation of the adsorption and metabolism properties of compounds have been developed. Systems and methods that provide high definition, content-rich information about cellular responses and are robust enough to have proven useful in secondary testing pharmacological properties of compounds are reviewed in this chapter.

2 TARGET-DIRECTED PHARMACOLOGY

2.1 Receptor Pharmacology

High-throughput methods for the determination of receptor-ligand binding and inhibition have been developed. Methods such as radiolabeled binding studies on flash plates (New England Nuclear) and the estimation of the binding of Fluoro-tagged ligands by fluorescence polarization, surface plasmon resonance, and ELISA-type assays have all been used for high-throughput screening and rapid characterization of compounds that interfere with ligand receptor binding. These methods are reviewed elsewhere. Receptors are often screened for ligand binding; however, functional receptor signaling and physiological cell responses are the more relevant endpoints. Examples of receptor pharmacology across the variety of receptor families are covered below.

2.2 Assays and Pharmacology of Rapid Cellular Responses

Probes for the fluorescent detection of rapid cellular responses have been widely used in the characterization of pharmacological agents. Intracellular calcium, detected by Ca-sensitive fluorescent dyes, is perhaps the most commonly measured signal. Other parameters that change within a short time scale include intracellular pH and membrane potential. These endpoints have been widely used in conjunction with appropriate control compounds and/or control cell-lines to both screen compound libraries and evaluate lead dose–response profiles. The high-throughput pharmacology system (HTPS) developed by Axiom Biotechnologies allows for the rapid and high definition characterization of immediate cellular responses. The HTPS uses flow-cell technology to serially introduce compounds to continuously flowing cells in a mixing chamber of the fluidics system (Fig. 3.2). In the configuration shown in Fig. 3.2, the system preincubates antagonist with the cells before introduction of the agonist. In this mode, the relative off-rate of antagonist will determine potency. The agonist and antagonist can be introduced simultaneously to the cells so that both compete for receptors. In both formats, incubation time of agonist with cells has an

effect on the maximal cellular response. The calcium response to GPCR agonists is both time and dose dependent, and typically, agonist potency increases with incubation time. A suitable incubation time to optimize maximal response and agonist receptor potency is selected from initial time-course of response experiments. This flowing system provides several advantages over the static cells in wells approaches. First, dose–response curves are typically generated by introducing different concentrations of test compounds and controls to different wells of cells that have been previously introduced into the microtiter plate. With the HTPS system, a continuous concentration gradient of test compound can be introduced to the cells. Second, for plate-based assays, the cells are often grown in the microtiter plate for 1-3 days in advance of the experiment. This introduces a source of variability because the cells can react to the different growth conditions provided in the individual wells. For example, edges of the microtiter plate experience differences in humidity compared with center wells, and therefore they are more prone to evaporation. In the HTPS system, the cells are kept together until the moment of the test. This greatly reduces variability between testing events. Also, the order of addition of antagonist and agonist is known to change potency and apparent inhibition characteristics. For example, in the HTPS laboratory, Schild analysis with inhibitors pre-incubated with cells demonstrates apparent non-competitive inhibition, whereas the Schild analysis of simultaneous addition of agonist and antagonist demonstrates competitive inhibition. In another example Cascieri et al. found that preincubation of L-742,694, a selective morpholino tachykinin NK1 receptor antagonist, inhibits binding of substance P to the human tachykinin NK1 receptor, increases the apparent EC50 of substance P, and decreases the maximal level of stimulation observed. Simultaneous addition of substance P and L-742,694 to the cells results in an increase in the EC_{50} for substance P with no decrease in the maximal level of stimulation (2). Apparent "non-surmountable antagonism" is observed when the system is preincubated with antagonist (3). These varying results stem from the time-dependent nature of

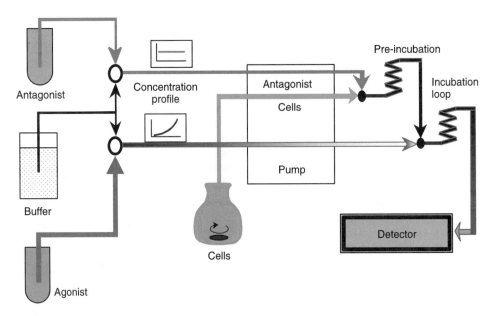

Figure 3.2. A schematic diagram of the high-throughput pharmacology system developed by Axiom Biotechnologies. The open circles (○) represent valves capable of switching between buffer and sample (antagonist or agonist). The closed circles (●) are mixing chambers for the cells and compound. Cells are preloaded with fluorescent dyes sensitive to physiological endpoints such as intracellular calcium or membrane potential. The length of the pre-incubation and incubation loops dictates the incubation times for the antagonists and agonists prior to entering the fluorescent detector.

the physiological responses. Such results are dependent on the kinetic parameters of these systems such as the association and dissociation rates of agonist and antagonist. When an antagonist is preincubated with its receptor, a lowered maximal response is detected if the antagonist dissociates slowly from the receptor compared with the agonist on rate. Indeed, the behavior has been observed before for the kinetics of acetylcholine antagonists in smooth muscle (4). The HTPS system depicted in Fig. 3.2 is very flexible and permits the researcher to readily change incubation times for both agonist and antagonist.

Researchers have turned to molecular biology, gene cloning, and overexpression as a means of facilitating HTS development. A survey of the literature will reveal a vast number of transfected GPCRs, many of these for drug discovery programs. Several advantages are readily apparent. For example, the concentration of a given receptor can be increased, thus increasing the cellular response (Ca flux, cAMP). The limitations are not as readily ap-

parent. Some GPCRs receptors demonstrate constitutive activity when expressed at high levels. These phenomena result from the activation of G-proteins by a small portion of the GPCRs. At high expression levels, this active fraction is large enough to cause a perturbation in the basal signal downstream from the G-protein. It has been suggested that this effect could be used for orphan GPCR screening (5). Expression of receptors in cells containing different G-protein, G-protein receptor kinases, and other proteins that modify GPCR signaling can result in abnormal drug-receptor behavior (6). Such changes result in cellular responses in screening efforts that identify hits that do not translate to nontransfected cells, and most importantly, humans.

3 PHARMACOLOGY OF cAMP ASSAYS

3.1 Introduction

cAMP levels are modulated by GPCRs coupled to Gs and Gi G-proteins, which act to activate or inhibit adenylate cyclase. Many assays for determining cAMP levels produced in cells, tissues, and membrane preparations have been reported, and several are commercially available. These methods are useful for both HTS and follow-up secondary testing. Secondary testing against a panel of cell lines, previously characterized for gene expression, is useful in determining the specificity of action for these compounds.

3.2 Nondestructive Fluorescent Methods

Some nondestructive methods for determining cAMP levels in living cells are known. The first reported involved injection of cAMP-dependent kinase labeled with the fluorescent probes fluorescein and rhodamine on the catalytic and regulatory subunits into cells. Fluorescence resonance energy transfer (FRET) from fluorescein to rhodamine occurs in the holoenzyme confirmation but is ablated on cAMP binding and subunit dissociation. A novel methodology in which the biosensor is generated from an engineered protein has recently been reported. The readout is the cAMP-sensitive FRET signal from differently colored GFP mutant proteins coupled through the cAMP-binding domain. This provides a purely molecular method of detecting intracellular cAMP levels in real time (7, 8). Cyclic nucleotide-gated calcium channels demonstrated use in coupling changes in cAMP levels with intracellular calcium. The changes in intracellular calcium are monitored by Fura-2 fluorescence at 380 nm. While interesting, these molecular methods have not yet proven their worth in the drug discovery process and require stable transfection of the cAMP sensor gene into each cell line before study.

3.3 B-Arrestin Systems

Many GPCRs are down-regulated after ligand binding through the concerted action of several proteins. A family of proteins called G-protein–related kinases phosphorylate activated G-protein coupled receptors. A second molecule, called arrestin, binds to the phosphorylated GPCR, blocking G-protein binding and activation. Arrestin binding also acts as an internalization signal. Arrestin binding to phosphorylated GPCR has been adapted

through several different readouts to provide information about GPCR signaling. B-arrestin-GFP fusion protein binding to activated adrenergic GPCRs monitored by confocal microscopy was reported. A modification of this approach has been used in which Renilla luciferase is attached to the C-terminus of the GPCR. Energy transfer from a hydrolyzed luciferase substrate to GFP is possible on arrestin-GFP binding to luciferase-GPCR fusion protein. This approach, requiring the introduction of two chimeric proteins into cells, provided evidence for GPCR homodimerization and the absence of a B-arrestin interaction with GnRHR (9, 10).

4 ION CHANNEL PHARMACOLOGY

Ion channels are attractive drug targets for several reasons, including the fact that channels represent an important class of drug targets for current therapeutic agents. With the information available from sequencing the human genome, interest in developing new drugs toward specific ion channels is high. Ion channel drug discovery is hampered by the low throughput of conventional patch-clamp methodology. Methods for evaluating ion channel activity based on changes in the fluorescent of dyes in response to changes in membrane potential have been developed. The fast styryl dyes respond to changes in membrane potential on the millisecond time scale. Because the change in their electrochromic properties is limited to 1–10% changes in fluorescence per 100 mV, the use of these dyes is restricted to specialized low-noise equipment (11). Another class of dyes is the negatively charged oxenol dyes. These are the "slow" fluorescent probes, and redistribute within the cytosol, membrane, and intracellular components, forming fluorescent and non-fluorescent dimers and higher aggregates with concomitant changes in spectral properties in response to changes in cellular potential (12). After a change in membrane potential, these dyes can take several minutes to re-equilibrate and, therefore, they are not useful in following channels that desensitize or inactivate after opening. Voltage-sensitive probes based on FRET have also been developed (13).

In this method a voltage-sensing oxonol dye is used as the FRET acceptor. The FRET donor is a courmarin dye linked to a phospholipid, which anchors in the outer leaflet of plasma membrane of cells with the courmarin dye. Under hyperpolarizing conditions, the negative membrane potential will distribute the oxonol toward the courmarin dye in the outer portion of the membrane, resulting in increased FRET. Under depolarizing conditions, the oxonol dye equilibrates toward the cytosolic side of the plasma membrane, and the FRET signal is reduced. FRET assays with response times of only a few seconds provide advantages over oxonol redistribution dyes. These assays are useful not only in primary screening but also for secondary follow-up assay for hit compounds.

5 CELLULAR TOXICOLOGY ASSAYS

Compound toxicity is an issue that is often addressed later in the drug development process than potency. However, because project costs increases rapidly with time, late stage failures caused by unexpected toxicity are undesirably expensive. It would be greatly advantageous to include some measure of compound toxicity earlier in the drug discovery process. Assays that could help rank order "hits" from HTS and early medicinal chemistry efforts would save time and cost derived from pursuing leads with serious cellular toxicity liabilities. In addition, ready access to structure-activity relationships regarding compound-related toxicity would be useful in directing medicinal chemistry efforts to reduce the toxic aspects of a compound as well as improving on other desired properties.

The impairment of critical cellular functions can result in systemic toxicities such as those associated with the neuromuscular system (tremor, cardiac arrhythmia, and paralysis), renal (microtubule function), and vasculature (leaky blood vessels). A noteworthy example is the acquired long QT syndrome (LQTS) associated with blockade of cardiac ion channels (14–16). LQTS results in cardiac arrhythmias, torsade de pointes, ventricular fibrillation, and can lead to sudden death. One such ion channel is the HERG protein, respon-

sible for the rapid component of delayed rectifier K^+ current in the myocardium (17). Several different mutations in HERG are known to cause inherited LQTS (18). A large number of drugs have been found to elicit adverse side effects through inhibition of the HERG channel, thereby inducing LQTS (19). Recent evidence suggests that two key residues, Y652 and F656, may be responsible for the promiscuity of drug binding by the HERG channel (20). This serious and sometimes fatal cardiotoxicity has led to withdrawal of otherwise promising drugs from the marketplace (21), for example, the antipsychotic agent pimozide and the gastric reflux medication cisapride. Some estimates indicate that up to one-half of all compounds under review for market approval may elicit LQTS side effects through the HERG channel. The FDA now recommends that all drugs be screened against the HERG channel before release to market.

Developing useful assays for screening drug effects against the HERG channel has become extremely important to the pharmaceutical industry. Drug screening in humans or animals is most physiologically relevant. However, this method is not always ethical, it is very low throughput, and it is difficult to identify the molecular target of any drug interactions. The patch-clamp method has proven very powerful in its sensitivity and ability to evaluate channels at the single molecule level. However, patch-clamp technology is expensive, difficult, low throughput, and far removed from a physiologically relevant context. Cell culture–based assays have been developed that use fluorescent dyes sensitive to plasma membrane potential in conjunction with plate readers and flow cytometry. These assays have proven to be high throughput but often suffer from decreased sensitivity or difficulty in maintaining cell lines. The need exists for a sensitive, high-throughput, cost effective, and physiologically relevant HERG assay.

Interaction of compounds with a toxicity target can result in a wide variety of in cell phenotypes (22). Deregulation of gene expression can result in inappropriate cell division (neoplasia, teratogenesis), apoptosis, and protein synthesis (e.g., peroxisome proliferation). Cell death can result from stimuli resulting in apoptosis, which has been defined as programmed cell death, and involves a cascade of events, including caspase activation leading to cytochrome C release from the mitochondria and subsequent degradation of chromosomal DNA into distinct fragments, formation of cytoplasmic vacuoles, and plasma membrane and nuclear blebbing (23, 24). Apoptotic cells are removed by macrophages in response to the signals such as the exposure of phosphatidylserine to the outside of the cell before complete loss of plasma membrane integrity. Necrosis is another classical cell death pathway in which cells lose membrane integrity, swell and burst, and spill their contents into the extracellular space. These cellular components often elicit an inflammatory response leading to further damage to surrounding tissues. Cellular factors determining which pathway a cell follows in response to toxic exposure include caspase activity, degree of ATP depletion, extent of intracellular Ca^{2+} increase, levels of reactive oxygen species, and rate and extent of thiol oxidation (24, 25). Additionally, factors such as cell cycle, cell type, duration of exposure, active efflux mechanisms, and compound metabolism may influence the response of cells to toxic agents. Together, the characteristics of these apoptotic, necrotic, and metabolic pathways and environmental factors form a continuum of morphological and biochemical indicators of cell death. Thus, the term "cytotoxicity" is somewhat imprecise, and it would be more useful to be able to readily characterize the toxicity mechanism for hit and medicinal chemistry compounds than simply to classify cells as "alive" or "dead."

Apoptosis can be initiated through TNF receptor family signaling coupled to caspase activation. Other apoptosis triggers included pharmacological agents (staurosporine, Ca^{2+} ionophores, thapsigargin), DNA damaging agents, and a variety of chemical toxicants (24). Assays that measure mitochondrial function and cell viability and growth, cell membrane integrity, membrane potential, intracellular Ca^{2+}, ATP, reduced glutathione concentration, and intracellular pH are useful indicators of cell-based toxicity. The tetrazolium salts such as XTT, MTT, WTS, and others are reductase substrates that are reduced in the mitochondria of living cells to colored formazan dyes readily detected by light absorbance. The colored species generated is proportional to the number of viable cells. Cellular DNA synthesis can be determined by tritium-labeled thymidine (radiometric detection) or bromo-deoxyuridine (antibody detection) incorporation and thus is a direct measure of cell proliferation and can be related to toxic and cytostatic effects of test compounds on proliferating cells. Dyes that do not cross the plasma membrane, such as trypan blue and propidium iodide (PI), are excluded by cells with intact membranes and are an indicator of cell viability. Dye exclusion is readily measured by direct cell counting on a hemocytometer, coulter counter, or flow cytometer.

Flow cytometer-based assays for several other apoptotic indicators have been developed. Changes in intracellular ion levels, in particular Ca^{2+} and H^+, are considered good early indicators of compound-induced cellular toxicity. Elevated Ca^{2+} levels may be important in apoptosis by activating nuclease activity. Intracellular pH and Ca^{2+} levels are readily determined with a variety of H^+- and Ca^{2+}-sensitive fluorescent probes (26–28). Assays for reactive oxygen species and cellular free glutathione content, both indicators of apoptosis, are available. In apoptosis, as part of the signaling to macrophages, phosphytidylserine (PS) "flips" from the inner to outer side of the cell membrane. Annexin V binding to PS on the outer membrane is a characteristic of early stage apoptosis (29). DNA fragmentation assays are also used to discriminate necrotic from apoptotic cells (30, 31). End labeling of the fragmented DNA followed by staining yields signal indicative of the characteristic DNA fragmentation pattern (32). An advantage of measuring several cytotoxicity endpoints simultaneously is that dose and mechanistic properties of moderately and highly toxic compounds are discriminated. These properties are best determined in single-cell analysis as described in the next section.

6 HT-FLOW

Many of the cell-death assays described above are readily performed by flow cytometry

(FCM), and can be combined to yield multiparametric assays that measure several complementary indices of toxicity. The advantage of single cell multiparameteric analyses includes the ability to test mixed cell populations, such as those obtained upon cell differentiation. Simultaneous determination of mitochondrial membrane potential, reactive oxygen species generation, and intracellular glutathione by FCM were used to show that loss of mitochondrial membrane potential and glutathione depletion are nearly simultaneous and could not be uncoupled (33). In particular, while not limited to FCM, determination of light scattering properties of cells yields information that is not available in standard 96-well plate reader assays. Forward scatter is commonly used to measure cell size and decreases early in apoptosis while side-scattered light is proportional to cellular granularity and often increases during apoptosis (34). Thus, FCM has advantages over conventional plate-based assays with regard to multiplexing, readouts such as light scatter, analysis at the single cell level, and high sensitivity. FCM also provides unique advantages for mixtures of cell types. Differential analysis of cellular responses to test compounds is possible because the different cell types or differentiation states can be tagged with specific cell-surface markers. Neuronal cells can be derived through differentiation of precursor cells, and these can be stained for the presence of neural cell adhesion molecule. FCM is therefore well suited to measure the multiplicity of events that occur when cells respond to toxic agents and gives a wealth of information that can be used to identify the mechanism of compound action and direct chemical optimization efforts accordingly.

Drug discovery and development follow-up to a typical HTS program requires the assaying of hundreds to thousands of compounds identified in screening, or cherry picked by structural similarity to original hits and generated by focused combinatorial chemistry around active compounds. One major disadvantage of FCM is the need to handle each sample manually and the resulting low throughput. We have previously reported a high-throughput sample delivery system for FCM (34). The data were collected on a Cyto-

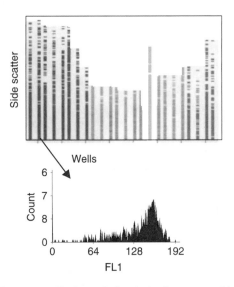

Figure 3.3. Each vertical strip in the top panel is from an individual well of a 96-well plate loaded with fluorescent beads. Two rows, 24 wells, of the plate are shown. The bottom panel represents a one-dimensional histogram for a single well. Data was collected on a Cytomation MoFlo FACS equipped with a Moskito autosampler.

mation MoFlo FCM equipped with a Mosikito autosampler, which is capable of sampling a 96-well plate in ~8 min. Each vertical stripe in the upper panel of Fig. 3.3 shows the side scatter for a single well and the lower panel is the FL1 histogram for a single sample. High-throughput compound screening by flow cytometry provides both a novel set of issues regarding assay cost, compound usage, data handling, and data interpretation and novel cell pharmacology assays. All of these issues are under consideration in the development of the next generation of HTflow instrumentation.

Introduction of compounds to cells in flow provides another dimension to standard flow cytometry. The HTPS system described earlier has been adapted for flow cytometry and increases the information content and expands the repertoire of assays, including high content toxicology data, available to the pharmacology/medicinal chemistry drug development team. Early physiological responses such as changes in pH, intracellular calcium, and membrane potential can be monitored, and this can be coupled to cell-type specific markers, allowing assays to be performed on mixed cell populations. The HT-flow system has been used to measure cellular responses to test compounds at rates of 3–4 compounds per minute as well as determine the relationship between receptor occupancy and cell response (35).

7 HIGH CONTENT METABOLIC ASSAYS

Efforts for the rapid screening of drug adsorption metabolism have intensified as the importance and relevance of the parameters for drug discovery have become more apparent with every failed and recalled drug. The cytochrome P450 (CYP) enzymes metabolize xenobiotics and evolved as a defense against accidentally ingested toxic compounds. Drug pharmacokinetics are modulated in part by CYP metabolism and patients on multi-drug regimens can experience unexpected changes, both positive and negative, in drug mean residence time and maximum serum levels. Drug inhibition of CYPs can lead to increased drug levels and increased risk of toxic side effects. Several high-throughput assays for test compound inhibition of CYPs have been reported (36–38). Most are based on changes in the fluorescence of a known CYP substrate and specific CYP isoform inhibition in the presence of test compounds. Increased CYP activity can result in toxicity as well. The nuclear hormone receptor CAR controls the expression-based response to a class of molecules called the "phenobarbitol-like inducers." In the mouse, these molecules are known to regulate the expression of a particular cytochrome P450 Cyp2b10. Cocaine causes acute liver toxicity in mice previously exposed to CAR agonists, and this toxicity is absent in CAR knockout animals (39). CYP gene induction, modulated at least in part through xenobiotic binding to the nuclear hormone receptors such as CAR, can also result in more rapid drug clearance rates, lower serum concentration, and decreased efficacy. Thus, drug–drug interactions are important in both toxicity and efficacy. In addition to metabolic degradation and inactivation, drug efficacy is reduced by excretion. Another protein specifically targeted for secondary pharmacological screening is the ATP-dependent pump, P-glycoprotein, PGP, (encoded by ABCB1, also known as MDR1), is a broad specificity pump that is responsible for efflux of compounds from the intestine, thus blocking their uptake. P-glycoprotein is overexpressed in some tumor cells and is responsible for cancer drug resistance and is part of the blood-brain barrier where it helps protect the brain from exogenous agents. P-glycoprotein expression is regulated by the orphan nuclear receptor SXR (other names are PXR, PAR, PRR or NR112). Taxol activates the SXR receptor, increasing the expression of both P-glycoprotein and the cytochrome P450 isoforms CYP2C8 and CYP3A4 (40). Taxol's efficacy is decreased by both increased excretion and metabolism through PGP and CYP2C8/CYP3A4 induction, respectively. Reporter assays for lead compound effect on CAR and SXR activity could add useful information to drug development programs by flagging compounds for possible drug–drug interaction and metabolic liabilities.

8 SUMMARY

Pressure to make better decisions earlier in the drug discovery process is high. Rapid, high content methods are available that can facilitate decision-making. Secondary testing in nontranfected cell-lines provides the optimal balance between throughput and disease relevance and target environment. Assays for native expressed GPCRs, coupled through calcium and cAMP, are widely practiced. Alternately, genetically engineered cells provide signal amplification and "universal" methods for detecting receptor activation, but change the ratio of receptor to regulatory molecules and may alter pharmacology. Fluorescence methods for ion channels may yield new lead compounds, but better methods may be needed to over come some of the problems still associated with secondary testing in this class. Flow cytometry methods are becoming more widely used in the pharmacological testing of drug potency and toxicity. High-throughput sampling is made possible with plug-flow coupling of ambient samples and pressurized flow cytometry systems. Assay for compound associated metabolic liabilities, such as cytochrome P450-mediated metabolism or gene induction, can be developed. Finally, when all of these methods are in place, the challenge is the integration of these methods and their data into successful drug development from hit to lead.

REFERENCES

1. S. S. Tian, et al., *Science*, **281**, 257–259 (1998).
2. M. A. Cascieri, et al., *Eur. J. Pharmacol.*, **325**, 253–261 (1997).
3. M. J. Lew, J. Ziogas, and A. Christopoulos. *Trends Pharmacol. Sci.*, **21**, 376–381 (2000).
4. H.P. Rang, *Proc. R. Soc. Lond. B Biol. Sci.*, **164**, 488–510 (1966).
5. G. Chen, et al., *Mol. Pharmacol.*, **57**, 125–134 (2000).
6. J. Bockaert, C. Brand, and L. Journot, *Ann. NY Acad. Sci.*, **812**, 55–70 (1997).
7. M. Zaccolo, et al., *Nat. Cell Biol.*, **2**, 25–29 (2000).
8. Y. Nagai, et al., *Nat. Biotechnol.*, **18**, 313–316 (2000).
9. K. M. Kroeger, et al., *J. Biol. Chem.*, **276**, 12736–12743 (2001).
10. S. Angers, et al., *Proc. Natl. Acad. Sci. USA*, **97**, 3684–3689 (2000).
11. V. Montana, D. L. Farkas, and L. M. Loew, *Biochemistry*, **28**, 4536–4539 (1989).
12. J. Plasek and K. Sigler, *J. Photochem. Photobiol. B*, **33**, 101–124 (1996).
13. J. E. Gonzalez, et al., *Drug Discov. Today*, **4**, 431–439 (1999).
14. A. J. Camm, et al., *Eur. Heart J.*, **21**, 1232–1237 (2000).
15. D. M. Roden, et al., *Circulation*, **94**, 1996–2012 (1996).
16. M. E. Curran, et al., *Cell*, **80**, 795–803 (1995).
17. M. C. Sanguinetti, et al., *Cell*, **81**, 299–307 (1995).
18. M. C. Sanguinetti, et al., *Proc. Natl. Acad. Sci. USA*, **93**, 2208–2212 (1996).
19. I. Zipkin, *BioCentury*, **8**, A10 (2000).
20. J. S. Mitcheson, et al., *Proc. Natl. Acad. Sci. USA*, **97**, 12329–12333 (2000).
21. M. Tonini, et al., *Aliment. Pharmacol. Ther.*, **13**, 1585–1591 (1999).
22. A. Gregus and C. D. Klaassen in C. D. Klaassen, Ed., *Casarett and Doull's Toxicology*, McGraw-Hill, New York, 1996.
23. D. R. Green, *Cell*, **102**, 1–4 (2000).
24. D. J. McConkey, *Toxicol. Lett.*, **99**, 157–168 (1998).
25. P. Nicotera, M. Leist, and E. Ferrando-May, *Biochem. Soc. Symp.*, **66**, 69–73 (1999).
26. R. Y. Tsien, *Methods Cell Biol.*, **30**, 127–156 (1989).
27. R. Y. Tsien, *Trends Neurosci.*, **11**, 419–424 (1988).
28. R. Y. Tsien, *Soc. Gen. Physiol. Ser.*, **40**, 327–345 (1986).
29. J. P. Aubry, et al., *Cytometry*, **37**, 197–204 (1999).
30. W. Gorczyca, M. R. Melamed, and Z. Darzynkiewicz, *Methods Mol. Biol.*, **91**, 217–238 (1998).
31. W. Gorczyca, *Endocr. Relat. Cancer*, **6**, 17–19 (1999).
32. X. Li and Z. Darzynkiewicz, *Cell Prolif.*, **28**, 571–579 (1995).
33. A. Macho, et al., *J. Immunol.*, **158**, 4612–4619 (1997).
34. J. T. Ransom, et al., Flow Cytometry Systems for Drug Discovery and Development. Proceeding of SPIE, **3921**, 90–100 (2000).
35. B. S. Edwards, et al., *J. Biomol. Screen.*, **6**, 83–90 (2000).
36. N. Chauret, et al., *Anal. Biochem.*, **276**, 215–226 (1999).
37. J. H. Kelly and N. L. Sussman, *J. Biomol. Screen.*, **5**, 249–254 (2000).
38. J. Venhorst, et al., *Eur. J. Pharm. Sci.*, **12**, 151–158 (2000).
39. P. Wei, et al., *Nature*, **407**, 920–923 (2000).
40. T. W. Synold, I. Dussault, and B. M. Forman, *Nat. Med.*, **7**, 584–590 (2001).

CHAPTER FOUR

Application of Recombinant DNA Technology in Medicinal Chemistry and Drug Discovery

SOUMITRA BASU
University of Pittsburgh
Center for Pharmacogenetics
Department of Pharmaceutical Sciences
Pittsburgh, Pennsylvania

ADEGBOYEGA K. OYELERE
Rib-X Pharmaceuticals, Inc.
New Haven, Connecticut

Contents

Burger's Medicinal Chemistry and Drug Discovery
Sixth Edition, Volume 2: Drug Development
Edited by Donald J. Abraham
ISBN 0-471-37028-2 © 2003 John Wiley & Sons, Inc.

1 INTRODUCTION

Discovering new drugs has never been a simple matter. From ancient times to the beginning of the last century, treatment for illness or disease was based mainly on folklore and traditional curative methods derived from plants and other natural sources. The isolation and chemical characterization of the principal components of some of these traditional medicines, mainly alkaloids and the like, spawned the development of the modern pharmaceutical industry and the production of drugs in mass quantities. Within the last century, however, the changes the industry has undergone have been profound. As the companion chapters of this volume describe, the emphasis has changed from isolation of active constituents to creation of new, potent chemical entities. This evolution from folklore to science is responsible for the thousands of pharmaceuticals available worldwide at present (1).

1.1 Chemistry-Driven Drug Discovery

The exacting process of discovering new chemical entities that are safe and effective drugs has itself undergone many changes, each of which was prompted by the introduction of some new technology (2, 3). In the 1920s, the first efforts at understanding why and how morphine works in terms of its chemical structure were initiated. During the 1940s, challenges for mass production of medicinally valuable natural products, such as the penicillins, were conquered. By the late 1950s, advances in synthetic organic chemistry enabled the generation of multitudes of novel structures for broad testing into the major focus of the modern pharmaceutical industry. Although serendipitous at best, this approach yielded many valuable compounds, most notably the benzodiazepine tranquilizers chlordiazepoxide and diazepam (4). Even with these successful compounds, however, the process of drug discovery amounted to little more than evaluating available chemical entities in animal models suggestive of human disease.

By the mid-1960s, medicinal chemistry had clearly become the cornerstone technology of modern drug discovery. Systematic development of structure-activity relationships, even to the point at which predictions about activity might be made, became the hallmark of new drug discovery. Even then, however, an understanding of the actions of drugs at the molecular level was often lacking. Receptors and enzymes were still considered as functional "black boxes" whose structures and functions were poorly understood. The first successful attempts at actually designing a drug to work at a particular molecular target happened nearly simultaneously in the 1970s, with the discovery of cimetidine, a selective H_2-antagonist for the treatment of ulcers (5), and captopril, an angiotensin-converting enzyme inhibitor for hypertension (6). The success of these two drugs sparked a realignment of chemistry-driven pharmaceutical research. Since then, the art of rational drug design has undergone an explosive evolution, making use of sophisticated computational and structural methodology to help in the effort (7). During the 1980s, mechanism-targeted design and screening combined to produce a number of novel chemical entities. These include the natural product HMG-CoA reductase inhibitor lovastatin for the treatment of hypercholesteremia (8) and the antihypertensive angiotensin II receptor antagonist losartan, synthetically optimized from a chemical library screening lead (9).

There is little doubt that the task of discovering new therapeutic agents that work potently, specifically, and without side effects has become increasingly important and coincidentally more difficult. Advances in medical research that have provided new clues to the previously obscured etiologies of diseases have revealed new opportunities for therapeutic intervention. This has forced the science of medicinal chemistry, once founded almost solely in near-blind synthesis and screening for *in vivo* effects, to become keenly aware of biochemical mechanisms as an intimate part of the development process. Even with these major advances in the medicinal and pharmaceutical sciences, more fundamental questions remain. What determines a useful biological property? And how is it measured in the discovery process? The answers can determine the success or failure of any drug discovery program, because both the observation of a useful biological property in a novel molecule and the optimization of structure-activity relationships associated with ultimate clinical candidate selection have rightfully relied heavily on practices, and sometimes prejudices, founded in decades of empirical success (10). Although the task of drug development has now been refined into a process without major unidentified obstacles, the challenge to bring the discovery of novel compounds to a comparable state of maturity remains. As in the past, another research avenue synergistic with existing discovery technologies is necessary.

1.2 Advent of Recombinant DNA Technology

The evolution of recombinant DNA technology, from scientific innovation to pharmaceutical discovery process, has occurred in parallel with the development of contemporary medicinal chemistry (11–14). The traditional products of biotechnology research share few of the traits characteristic of traditional pharmaceuticals. These biotechnologically derived therapeutics are large extracellular proteins destined to be, with few exceptions, injectables for use in either chronic replacement therapies or in acute or near-term chronic situations for the treatment of life-threatening indications (15, 16). Many of these products also satisfy urgent and previously unfulfilled therapeutic needs. However, their dissimilarity to traditional medicinal agents does not end there. Unlike most low molecular weight pharmaceuticals, these proteins were developed not because of the novelty of their structures, but because of the novelty of their actions. Their discovery hinged on recognition of a useful biological activity, its subsequent association with an effector protein, and the genetic identification, expression, and production of the effector by the application of recombinant DNA technology (17, 18).

If modulation of biochemical processes by a low molecular weight compound has been the traditional goal of medicinal chemistry, then association of a biological effect with a distinct protein and its identification and production have been considered the domain of molecular genetics. The application of recombinant DNA technology to the identification of proteins and other macromolecules as drugs or drug targets and their production in meaningful quantity as products or discovery tools, respectively, provide an answer to at least one of the persistent problems of new lead discovery. Because a comprehensive review of the genetic engineering of important proteins is well beyond the scope of this volume, this chapter will instead highlight some novel examples of advances in recombinant DNA technology, with respect to both exciting new pharmaceuticals and potential applications of recombinantly produced proteins, be they enzymes, receptors, or hormones, to the more traditional processes of drug discovery.

2 NEW THERAPEUTICS FROM RECOMBINANT DNA TECHNOLOGY

The traditional role of the pharmaceutical industry, i.e. synthesis of new chemical entities as therapeutic agents, was suddenly expanded by the introduction of the first biotechnologically derived products in the 1980s. The approval of recombinant human insulin in 1982 broke important ground for products produced by genetic engineering (19). In 1985 another milestone was achieved when Genentech became the first biotechnology company to be granted approval to market a recombinant product, human growth hormone. These events set an entire industry into motion, to produce not only natural proteins for the treatment of deficiency-associated diseases, but also true therapeutics for both acute and chronic care.

Industry estimates show the upward trend in biotechnologically derived drugs continuing into the new millennium. Over 400 products generated by biotechnology are estimated to be somewhere in the development pipeline or approval process. A comprehensive list of FDA approved biotechnology drugs is available on the web at www.bio.org/er/approveddrugs.asp. The variety of products—from hormones and enzymes to receptors, vaccines, and monoclonal antibodies—seeks to treat a broad range of clinical indications thought untreatable just two decades ago. However, despite this period of phenomenal growth for recombinant DNA-derived therapeutics, the promise of biotechnology, once touted to be limitless, has instead become more realistically defined to include not only the actual recombinant products and the difficulties inherent in their production but also many spin-off technologies, including diagnostics and genetically defined drug discovery tools (20–22).

One particular area of traditional pharmaceutical research in which recombinant DNA technology has made a profound impact has been the engineering of antibiotic-producing organisms (23–25). Always an important source of new bioactive compounds, especially antibiotics (26, 27), fermentation procedures can be directly improved by strain optimization techniques, including genetic recombination and cloning. More exciting is the possibility of producing hybrid antibiotics that combine desirable features of one or more individual compounds for improved potency, bioavailability, or specificity. The art of finding new natural product-based lead compounds by screening fermentation broths, plant sources, and marine organisms by using genetically engineered reagents is becoming of special importance as more of the relevant targets identified by molecular biology operate in obscure or even unknown modes. The structural diversity provided by natural products combined with the ability to test molecular biology driven biochemical hypotheses has already become an important route for the discovery of new therapeutics (28–30).

3 PROTEIN ENGINEERING AND SITE-DIRECTED MUTAGENESIS

Rapid developments in the technique of site-directed mutagenesis have created the ability to change essentially any amino acid, or even substitute or delete whole domains, in any protein, with the goal of designing and constructing new proteins with novel binding, clearance, or catalytic activities (31, 32). The concomitant changes in protein folding and tertiary structure, protein physiology, binding affinities (for a receptor or hormone), binding

specificities (either for substrate or receptor), or catalytic activity (for enzyme active site mutants) are all effects that are measurable against the "wild-type" parent, assuming that expression of the gene and subsequent proper folding have successfully occurred. Several surprising observations have been made during the short period that this technology has been available: amino acid substitutions lead, in general, to highly localized changes in protein structure with few global changes in overall folding; substitutions of residues not involved in internal hydrophobic contacts are extremely well accommodated, leading to few unsynthesizable mutants; and proteins seem extremely tolerant of domain substitution, even among unrelated proteins, allowing often even crude first attempts at producing chimeric proteins to be successful. The implications of this technology for the discovery of new pharmaceuticals lie in two areas: second-generation protein therapeutics and site- or domain-specific mutant proteins for structure-function investigations.

Throughout this chapter, amino acids are denoted by their standard one-letter codes; and site-specific mutations are represented by the code for the wild-type amino acid, the residue number, and the code for the replacement amino acid (31).

3.1 Second-Generation Protein Therapeutics

The cloning, expression, and manufacture of proteins as therapeutics involve the same problems encountered in the development and successful clinical approval of any drug. Potency, efficacy, bioavailability, metabolism, and pharmaceutical formulation challenges presented by the natural protein suggest that second-generation products might be engineered to alleviate the particular problem at hand, producing desired therapeutic improvements. The parent proteins to which this technology has been applied extend across the range of recombinant products already approved and those in advanced stage of clinical evaluation (33, 34a). As an example, for tissue-type plasminogen activator (t-PA), one of the most studied recombinant products (34b–36), four properties functioning in concert [i.e., substrate specificity, fibrin affinity, stimulation of t-PA activity by fibrin and fibrinogen, and sensitivity of the enzyme to inhibition by plasminogen activator inhibitors (PAIs)] are responsible for the localization and potentiation of the lytic reaction at a clot surface and are readily analyzed using molecular variants (37). A consensus structure combining the major domains of t-PA has been predicted based on the significant sequence homology with other serum proteins and serine proteases. The complexity of this structure is reflected in its functional multiplicity: efficient production of plasmin by cleavage of the R560-V561 bond of plasminogen, very low binding to plasminogen in the absence of fibrin, moderately high affinity for fibrin, increase in the efficiency of plasminogen activation by 500-fold in the presence of fibrin, rapid inactivation by PAI-1, and rapid hepatic elimination by receptor-mediated endocytosis (38). BM 06.022, a recombinantly engineered t-PA deletion mutant [t-PA del (V4-E175)], made up of the Kringle 2 and protease domains, has been reported to have the same plasminogenolytic activity but a lower fibrin affinity compared with wild type t-PA (39). Another variant of t-PA (T103N, KHRR 296–299 AAAA) was demonstrated to have the combined desirable properties of decreased plasma clearance, increased fibrin specificity, resistance to PAI-1, and *in vivo* increased potency and decreased systemic activation of plasminogen when administered by bolus dose (40).

Although the systematic changes exemplified by t-PA site-directed mutagenesis studies are the rDNA equivalents of medicinal chemistry [multiple analog synthesis for structure-activity relationship (SAR) development], more recent applications of this technology bear a less straight-forward resemblance to medicinal chemistry–driven drug discovery paradigms. However, these same recombinant techniques can be used to combine domains from different proteins to produce chimeric constructs that incorporate multiple desired properties into a single final product or reagent. For instance, in an effort to overcome the short plasma half-life associated with soluble CD4, chimeric molecules termed *immunoadhesins* (Fig. 4.1) have been recombinantly constructed from the gp120-specific domains of CD4 and the effector domains of various immunoglobulin classes (41, 42). In

addition to dramatically improved pharmacokinetics, these chimeric constructs incorporate functions such as Fc receptor binding, protein A binding, complement fixation, and placental transfer, all of which are imparted by the Fc portion of immunoglobulins. Dimeric constructs from human (CD4–2γl and CD4–4γl) and mouse (CD4-Mγ2a) IgG and a pentameric chimera (CD4-Mµ) from mouse IgM exhibit evidence of retained gpl20 binding and anti-HIV infectivity activity. Both CD4–2γl and CD4–4-γl show significantly increased plasma half-lives of 6.7 and 48 h, respectively, compared with 0.25 h for rCD4. Furthermore, the immunoadhesin CD4–2γl (CD4-IgG) mediates antibody-dependent, cell-mediated cytotoxicity (ADCC) toward HIV-infected cells and is efficiently transferred across the placenta of primates (43).

It is becoming clear that genetic variations play critical roles in patients' response to certain medications. Differential expression of drug targets and or metabolic enzymes has been shown to lead to differences in efficacy and toxicity profiles of drugs in section of population that harbors this genetic variation (44). Molecular biology and its associated techniques feature prominently in bringing to birth an interdisciplinary field, pharmacogenetics, which promises to unravel how genetic make up and variation thereof affect human response to medication (45). It is widely held that advances in pharmacogenetics will revolutionize drug dispensation and drug discovery processes. When fully realized, the gains of pharmacogenetics will positively impact drug discovery processes in numerous ways including the following: (1) identification of new and novel therapeutic targets; (2) an increased understanding of the molecular "uniqueness" of diseases; (3) genetic tagging of diseases with the consequence of developing designer medications that best combat an ailment; and (4) efficient design of clinical trials, with a better chance of success, aided by a genetic pre-screening of candidates. Medications that were judged ineffective by traditional validation methods in a random patient population may be found beneficial to a population of patients having an overexpressed "susceptibility gene(s)."

The role of recombinant DNA technology in medicinal chemistry and drug discovery

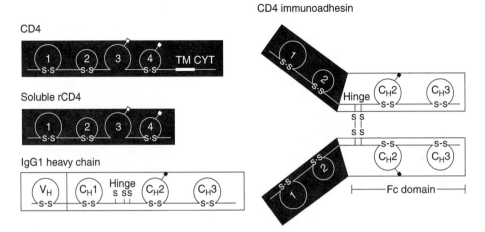

Figure 4.1. Structure of CD4 immunoadhesin, soluble rCD4, and the parent human CD4 and IgG1 heavy-chain molecules. CD4- and IgG1-derived sequences are indicated by shaded and unshaded regions, respectively. The immunoglobulin-like domains are numbered 1 to 4, *TM* is the transmembrane domain, and *CYT* is the cytoplasmic domain. Soluble CD4 is truncated after P368 of the mature CD4 polypeptide. The variable (V_H) and constant (C_H1, Hinge, C_H2, and C_H3) regions of IgG1 heavy chains are shown. Disulfide bonds are indicated by S-S. CD4 immunoadhesin consists of residues 1–180 of the mature CD4 protein fused to IgG1 sequences, beginning at D216, which is the first residue in the IgG1 hinge after the cysteine residue involved in heavy-light chain bonding. The CD4 immunoadhesins shown, which lacks a C_H1 domain, was derived from a C_H1-containing CD immunoadhesin by oligonucleotide-directed deletional mutagenesis, expressed in Chinese hamster ovary cells and purified to >99% purity using protein A-sepharose chromatography (42).

A recent success story in new target identification and validation is seen in the introduction of two new nonsteroidal anti-inflammatory drugs (NSAIDs), Vioxx and Celebrex, by Merck Frosst and Monsanto-Pfizer, respectively. Until recently, the onset of inflammation and pain has been linked to one cyclooxygenase enzyme, COX. Clinically useful NSAIDs such as aspirin, diclofenac, and ibuprofen exhibit their anti-inflammatory and antipyretic activity by inhibiting COX. A prolonged use of most NSAIDs results in gastrointestinal (GI) toxicity (46), which may be debilitating enough to require hospitalization in many patients. Recent progresses in the understanding of the biology of COX, championed by elegant biochemical and recombinant DNA studies, revealed that it exists in two isoforms: constitutive COX-1 and inducible COX-2 (Fig. 4.2). COX-1 is always expressed and mediates the synthesis of prostaglandins that regulate normal cell functions, whereas COX-2 is active only at the onset of inflamma-

tion (47, 48). However, early NSAIDs inhibit both COX isoforms, thereby interfering with the production of the protective prostaglandin products of COX-1 in addition to their pain relieve activity. It was hypothesized that selective inhibitors of inflammation associated COX-2 may produce a better drug profile and possibly avoid many side effects caused by non-selective NSAIDs, especially GI tract toxicity (49). The discovery of the *COX-2* gene stimulated intensive studies aimed at verifying this hypothesis (50, 51). The clinical and commercial success of Vioxx and Celebrex (**1** and **2**), two highly potent COX-2 specific inhibitors, validated COX-2 as a new target for anti-inflammation and antipyretic therapy. Furthermore, the safety profiles of these drug showed that they do not have many toxic side effects of traditional non-selective NSAIDs (52–54).

Alternatively, recombinant DNA technology is making it possible to decipher the roles of disease subtypes and genetic variability in

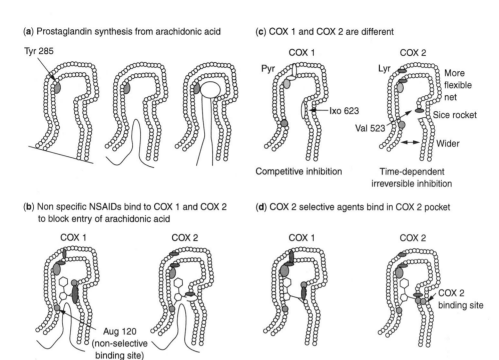

Figure 4.2. Prostaglandin synthesis and inhibition in COX-1 and COX-2. (a) Initial stages of prostaglandin synthesis. (b) Binding stages of standard NSAIDs to arginine 120 to inhibit prostaglandin synthesis by direct blockade of cyclo-oxygenase channel. (c) Differences between COX-1 and COX-2. (d) Specific blockade of COX-2.

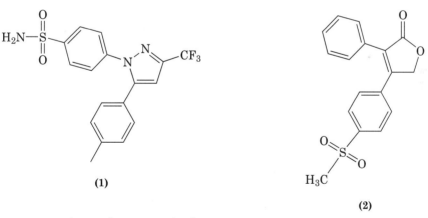

disease progression and response to drugs. New markers that aid disease classification are being identified and explored as targets to develop new and novel therapeutic strategies to better treat or manage diseases. The insights gained from such studies are beginning to yield genetically engineered pharmaceuticals for treating various human disease conditions including diabetes, multiple sclerosis,

rheumatoid arthritis, cancer, and viral diseases. For example, an overexpression of epidermal growth factor receptor-2 (HER-2) is known to occur in about 25–30% of women with breast cancer (55). Herceptin, a humanized anti-HER2 monoclonal antibody, was recently introduced by Genentech to treat metastatic breast cancer, and it is has proven beneficial in patients with metastatic breast cancer in which HER-2 is overexpressed (56).

The fulfillment of the many promises of pharmacogenetics is strongly hinged on the identification of unique markers that correlate genetic make-up to drug response. Molecular biology is and will continue to play frontline roles in the identification of these genetic markers. (Public and private efforts are currently ongoing in identifying such informative markers, details of which are beyond the scope of this chapter).

3.2 Antibody-Based Therapeutics

Antibody therapeutics can potentially treat diseases that can be as diverse as autoimmune disorders to cancer and infectious diseases. Antibodies are currently rated as an important and growing class of biotherapeutics. Other than vaccines, monoclonal antibodies currently in development outnumber all other classes of therapeutics. Recombination technology plays a key role in the development and commercialization of therapeutic antibodies. In fact, eight of nine antibody products on the U.S. market are recombinant products (57, 58).

During the early days of monoclonal antibody use, their therapeutic uses were limited caused by immunogenicity in humans, because they were murine antibodies that induced human antimouse antibodies, leading to allergic reactions and reduced efficacy (59). After the discovery of murine antibodies in 1975, the next generations of antibodies were chimeric in nature with 66% human and 34% mouse produced through genetic engineering. During the 1980s and early 1990s, complementarity-determining region (CDR) grafting and veneering techniques were established, which reduced the mouse portion of the sequence to less than 10%. Lately, genetically engineered transgenic mice that can be used for production of humanized antibodies have been developed. The technology uses the standard hybridoma techniques to produce fully humanized antibodies (59).

3.3 Epitope Mapping

Site-directed mutagenesis technology has also been applied to one of the most challenging problems in structural biochemistry—the nature of the protein–protein interaction. Whereas numerous examples of models of enzyme-ligand complexes have been developed based on active-site modifications, the site-directed mutagenesis method is being extended to the formidable problem of defining the essential elements of a protein–protein (e.g., a protein substrate to a protease or a hormone to its receptor) binding epitope.

An impressive example of a systematic search for a binding epitope is seen in the work used to define the human growth hormone-somatogenic receptor interaction (60, 61). First, using a technique termed homolog-scanning mutagenesis, segments of sequences (7–30 amino acids in length) from homologous proteins known not to bind to the hGH receptor or to hGH-sensitive monoclonal antibodies (Mab) were systematically substituted throughout the hGH structure by using a working model based on the three-dimensional folding pattern found by X-ray crystallographic analysis of the highly homologous porcine growth hormone (62). Using an enzyme-linked immunosorbent assay (ELISA)-based binding assay, which measures the affinity of the mutant hGH for its recombinantly derived receptor, researchers discovered that (63) swap mutations that disrupted binding were found to map within close proximity on the three-dimensional model, even though the residues changed within each subset were usually distant in the primary sequence. By this analysis, three discontinuous polypeptide determinants (the loop between residues 54 and 74, the central portion of helix 4 to the C-terminus, and to a lesser extent, the amino-terminal region of helix 1) were identified as being important for binding to the receptor.

A second technique, termed alanine-scanning mutagenesis was then applied. Single alanine mutations (62 in total) were introduced at every residue within the regions implicated in receptor recognition. The alanine scan revealed a cluster of a dozen large side-chains that when mutated to alanine exhibited more than a fourfold decrease in binding affinity. Many of the residues that constitute the hGH binding epitope for its receptor are altered in close homologues, such as placental lactogen and the prolactins. The overall correct folding of the mutant proteins was deter-

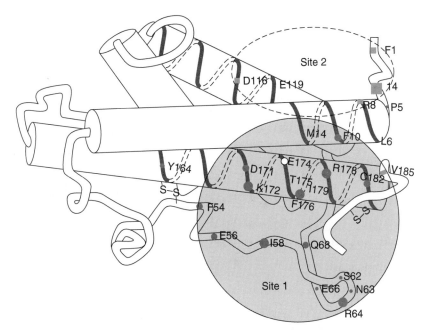

Figure 4.3. Map of alanine substitution in hGH disrupt binding of hGHbp at either site 1 or site 2. The two sites are generally delieated by the large shaded circles. Residues for which alanine mutations reduce site 2 binding are shown: ■, 2- to 4-fold; ■, 4- to 10-fold; ■, 10- to 50-fold; and ■, >50-fold. Sites where alanine mutations in site 1 cause changes in binding affinity for the hGHbp using an immunoprecipitation assay, are shown: ●, 2- to 4-fold reduction; ●, 4- to 10-fold reduction; ●, >10-fold reduction; and ●, 4-fold increase.

mined by cross-reactivity with a single set of conformationally sensitive Mab reagents. Using the receptor-binding determinants identified in these studies, a variant of human prolactin (hPRL) was engineered to contain eight mutations. This variant had an association constant for the hGH receptor that was increased by more than 10,000-fold (64).

Finally, biophysical studies, including calorimetry, size-exclusion chromatography, fluorescence-quenching binding assay (65), and X-ray crystallography (66) revealed the presence of two overlapping binding epitopes (Fig. 4.3) on growth hormone, through which it causes dimerization of two membrane-bound receptors to induce its effect. The crystal structure confirmed both the 1:2 hormone-to-receptor complex structure and the interface residues identified by the scanning mutagenesis mapping technique. These results indicate that the homolog and alanine-scanning mutagenesis techniques should be generally useful starting points in helping to identity amino acid residues important to any protein–protein interaction (67) and that these techniques have potential to provide essential information for rational drug design.

3.4 Future Directions

In an intriguing example of what might be termed *reverse* small molecule design, randomized mutagenesis techniques also have been directly applied to the ever-growing problem of antibiotic resistance. Bacterial resistance to increasingly complex antibiotics has become widespread, severely limiting the useful therapeutic lifetime of most marketed antimicrobial agents (68). Using a mutagenesis technique that randomizes the DNA sequence of a short stretch (3–6 codons) of a gene, followed by determination of the percentage of functional mutants expressed from the randomized gene, localization of the regions of the protein critical to either structure or function can be accomplished. Application of this technique to TEM-1 β-lactamase (the

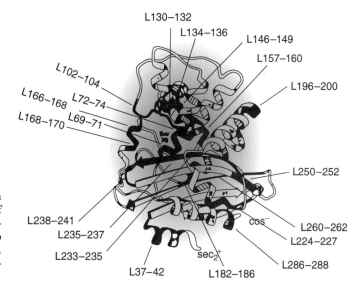

Figure 4.4. Position of random libraries on a ribbon diagram of the homologous *S. aureus* β-lactamase. Dark regions correspond to the position of random libraries. Lines point to the position of individual libraries.

enzyme responsible for bacterial resistance to β-lactam antibiotics such as penicillins and cephalosporins) over a 66-codon stretch revealed that the enzyme is extremely tolerant of amino acid substitutions: 44% of all mutants function at some level and 20% function at the level of the wild- type enzyme (69). The region identified as most sensitive to substitution are located either in the active site or in buried positions that likely contribute to the core structure of the protein (Fig. 4.4). Such a library of functional mutant β-lactamases could, in theory, be used to simulate multiple next generations of natural mutations, but at an accelerated pace. Screening of new synthetic β-lactams against such a mutant library might then be used to discover compounds with the potential for increased useful therapeutic lifetime. The art of rDNA site-directed mutagenesis, although advancing rapidly, is still limited to the repertoire of the 20 natural amino acids encoded by DNA. To effect more subtle changes in proteins, such as increased or decreased acidity, nucleophilicity, or hydrogen-bonding characteristics without dramatically altering the size of the residue and without affecting the overall tertiary structure, it has been proposed that site-directed mutagenesis using unnatural amino acids might offer the needed advantages. In the past, such changes were accomplished semisynthetically on chemically reactive residues such as Cys.

However, methodology for carrying out such mutations recombinantly has been successfully used. The key requirements for successful site specific incorporation of (un)natural amino acids include the following: (*1*) generation of an amber (TAG) "blank" codon in the gene of interest at the position of the desired mutation, (*2*) identification of a suppressor tRNA that can efficiently translate the amber message but that is not a substrate for any endogenous aminoacyl-tRNA synthetases, (*3*) development of a method for the efficient acylation of the tRNA$_{CCA}$ with novel amino acids, and (*4*) availability of a suitable *in vitro* protein synthesis system to which a plasmid bearing the mutant gene or corresponding mRNA and the acylated tRNA$_{CCA}$ can be added. These requirements are individual hurdles that have been crossed in the course of development of this technology, although there is a room for further development. Advances in nucleic acids synthesis have ensured a rapid access to aminoacylated CCA for semisynthesis of tRNAs bearing amino acids of interest (70a–70c). Furthermore, a recent development of wholly *in vitro* translation method promises to dramatically increase the array of unnatural amino acids that can be substituted into proteins (70d). An early successful demonstration of this methodology involved replacement of F66 with three phenylalanine analogues in RTEM β-lactamase and subsequent determination of the kinetic constants k_{cat} and K_m of the mutants (70e). Subsequent applications have centered on the critical issue of the introduction of unnatural amino acid replacements into proteins. The artificial residues can probe effects on stability and folding governed by subtle changes in hydrophobicity and residue side-chain packing to a degree not possible using the 20 natural amino acids (71, 72).

4 GENETICALLY ENGINEERED DRUG DISCOVERY TOOLS

4.1 Reagents for Screening

An increasingly important application of recombinant technology lies not in new protein drug product discovery per se but in the ability to provide cloned and expressed proteins as reagents for medicinal chemistry investigations. The common practice of *in vitro* screening for enzyme activity or receptor binding using animal tissue homogenates (nonhuman, and therefore nontarget) has begun to give way to the use of solid-phase or whole-cell binding assays based on recombinantly produced and isolated or cell-surface expressed reagent quantities of the relevant target protein (73, 74).

The ability to carry out large-scale, high flux screening of chemical, natural product, and recombinantly or synthetically derived diversity libraries (26–29, 75–89) also critically depends on reagent availability and consistency. The inherent differences in these potential sources of drug design information, especially from large combinatorially generated libraries requires that assay variations be reduced to the absolute minimum to ensure the ability to analyze data consistently from possibly millions of assay points.

The discovery of the HIV Tat inhibitor Ro 5–3335 and its eventual development as the analog Ro 24–7429 are successes from screening chemical libraries using recombinant reagents. Tat is a strong positive regulator of HIV expression directed by the HIV-1 long terminal repeat (LTR) and as such constitutes an important and unique target for HIV regulation, because the *tat trans*-activator protein (one of the HIV-1 gene products) has been clearly demonstrated to regulate expression of

the complete genome (90). Assays to detect inhibitors of *tat* function by screening, (91, 92) presented immediate opportunities to control a key step in the HIV-1 viral replication process. In this instance, to screen for Tat inhibitors, two plasmids were cotransfected into COS cells: a gene for either Tat or the reporter gene for secreted alkaline phosphatase (SeAP) was put under the control of the HIV-1 LTR promoter (93). Because Tat is necessary for HIV expression and SeAP expression is under the control of the HIV LTR, an inhibitor of Tat would necessarily lower the apparent alkaline phosphatase activity. This assay was standardized and used in high flux screening to identify structure (3), which was then subjected to medicinal chemistry optimization to produce structure (4) as the ultimate clinical

(3)

(4)

candidate (94). Such rapid lead discovery highlights the continued importance of highly directed screening to drug discovery, now better enabled by use of recombinant reagents and techniques.

Even more to the point of human pharmaceutical discovery and design, however, is the issue of species and/or tissue specificities. Sometimes the differences between tissue isolates and recombinant reagent are small; more frequently, however, the sequence homologies and even functional characteristics can vary greatly, providing a distinct advantage in favor of the recombinant protein. When the possibility of achieving subtype specificity, because of either tissue distribution or differential gene expression, determines a particular isoenzyme as a target for selective drug action, it is of obvious importance to be able to test for the desired specificity. Polymerase chain reaction (PCR), an enzymatic method for the *in vitro* amplification of specific DNA fragments, has revolutionized the search for receptor and enzyme subspecies, making whole families of target proteins available for comparative studies (95). Classic cloning requires knowledge of at least a partial sequence for low stringency screening. This method is unlikely to detect cDNAs corresponding to genes expressed at low levels in the tissue from which the library was constructed. In contrast, the PCR technique can uncover and amplify sequences present in low copy number in the mRNA and offers a greater likelihood of obtaining useful, full-length clones. The selective amplification afforded by PCR can also be used to identify subspecies present in tissue in especially short supply, offering yet another advantage over classic methods. PCR has also been applied to the generation of recombinant diversity libraries of DNA (86), RNA (87, 88), and novel chemical diversity "tagged" for detection and amplification (89).

4.2 Combinatorial Biosynthesis and Microbe Re-Engineering

Many clinically important pharmaceuticals and initial drug candidates are derived from natural sources such as microbes and plants (96). In most cases, the structural complexity of these drugs precludes chemical synthesis as a practical approach to commercially produce them. This consequently contributes to the dearth of derivatives of these compounds for evaluation as potential drug candidates. Also, slow generation time and low quantities of the drugs from their natural producers are usually major obstacles to contend with.

The tools of recombinant DNA technology are now being elegantly applied in pharmaceu-

tical industries to overcome these and other problems. Several examples of biosynthetic pathway engineering, designed to enhance the production of known compounds or generate novel products in microbes, plants and animals have been recently reported (for recent reviews see 97–101). Particularly, major strides have been made in carotenoid and macrocyclic polyketides production (101, 102). Natural product producing organisms have been engineered to produce enhanced levels of the desired compound compared with the wild-type (103). Driven by the realization of the slow growth rate and difficulties in genetic manipulation of natural source-organisms, researchers are making progress in introduction of non-native biosynthetic pathways into genetically amenable organisms (101). For example, the biosynthetic pathways of 6-deoxyerthronolide B(6-dEB), the macrocyclic-aglycon portion of antibiotic erythromycin, has been successfully engineered into *E. coli* (104). The engineered *E. coli* strain was reported to produce 6-dEB in yields comparable with the high-producing mutant of *Saccharopolyspora erythraea*, the source-organism of 6-dEB (104).

Genetic manipulation of gene clusters within the biosynthetic pathways of natural products has been used to rationally design new and novel products (105). Similarly, combinatorial genetic approach has been used to dramatically increase the repertoire of pharmaceutically important metabolites produced by natural producing-organisms (106). Alterations of the erythromycin polyketide synthase genes have been recently demonstrated to generate a mini-library of more than 50 potential pharmaceutically useful macrolides (Fig. 4.5) (107). The structural richness produced by this combinatorial biosynthesis approach is of the sort that will be most tasking to even the best of organic chemist. Potential applications of such combinatorial biosynthesis methods include lead generation and optimization of existing pharmaceuticals (108).

4.3 SELEX or *In Vitro* Evolution

In vitro selection (or SELEX), first reported in 1990, has developed into a powerful technology for drug discovery, diagnostics, structure-function studies, and creating molecules with

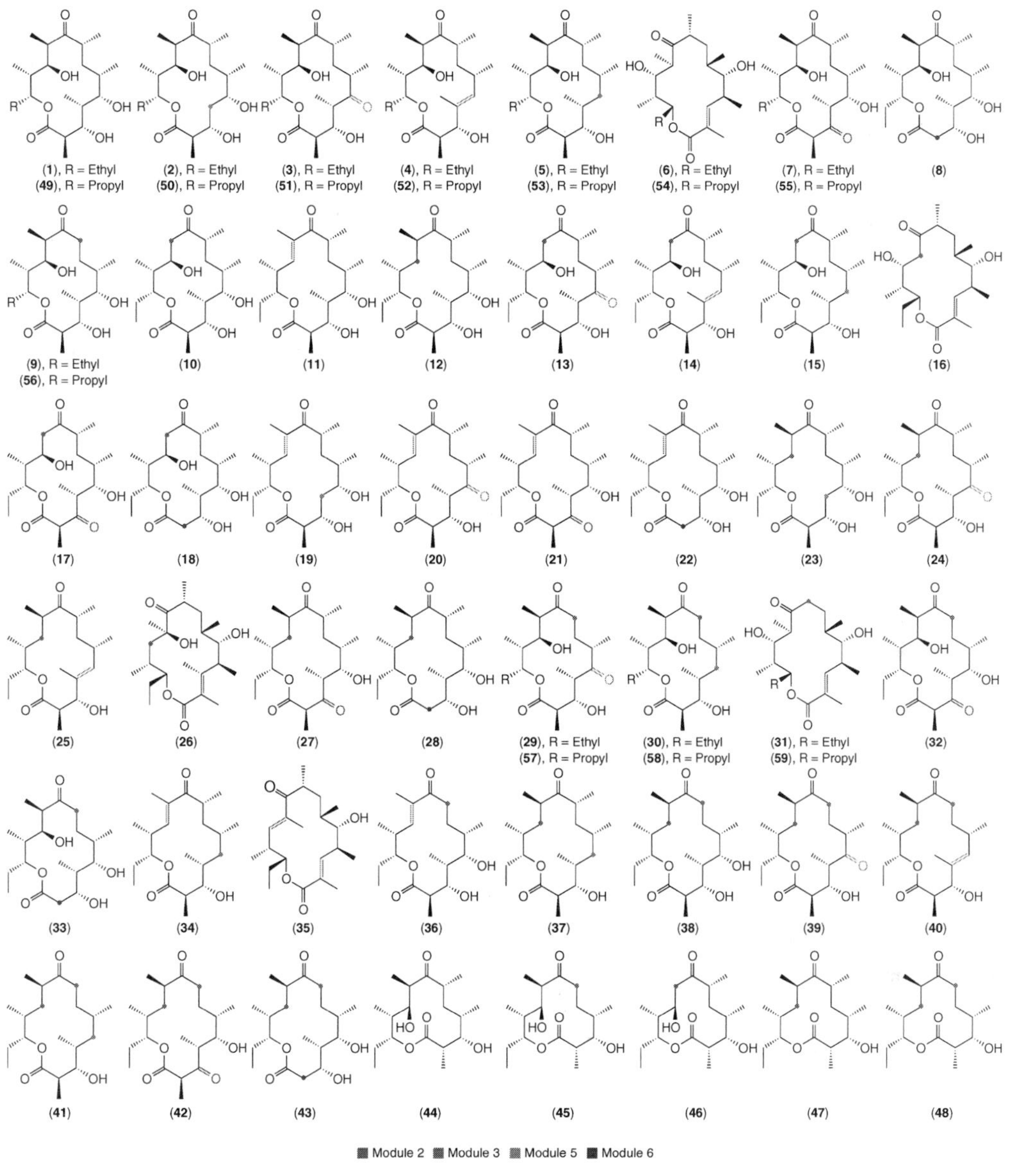

Figure 4.5. DEBS combinatorial library. Colors indicate the location of the engineered carbon(s) resulting from catalytic domain substitutions in module 2 (red), module 5 (green), module 6 (blue), or modules 1, 3, or 4 (yellow). See color insert.

novel catalytic properties (87, 88, 109–113). SELEX involves the selection of RNA or DNA molecules from random sequence pools that can bind to small or large molecule or perform a chosen function. The small or large molecule binders are called aptamers, and they often bind with dissociation constants in the pico-

molar range. A scheme of SELEX is depicted in Fig. 4.6. PCR and subcloning techniques play a major role in SELEX.

4.3.1 Aptamers as Drugs. The success of aptamers as potential drugs relies on finding solutions to many of the similar issues faced

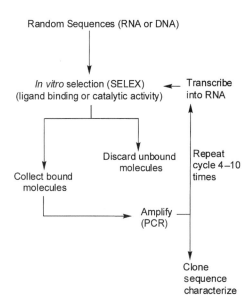

Random Sequences (RNA or DNA)

In vitro selection (SELEX) ← Transcribe
(ligand binding or catalytic activity)   into RNA

Discard unbound   Repeat
molecules   cycle 4–10
            times
Collect bound
molecules

Amplify
(PCR)

Clone
sequence
characterize

Figure 4.6. Schematic of SELEX.

by conventional pharmaceutical agents. The aptamers must have high affinity and specificity to their targets, must be biologically stable, and must reach the target molecule. Aptamers generated by SELEX have been shown to have tight binding and a high degree of specificity. For example, an aptamer designed to bind the vascular endothelial growth factor protein has a dissociation constant of about 100 pM (114). Aptamers have been shown to distinguish between protein kinase C (PKC) isozymes that are 96% identical, reflecting the high level specificity that can be attained by these molecules (115). The issue of biological stability of aptamers has also been addressed. By incorporating chemically modified nucleotide analogs such as 2'-amine or 2'-fluoro, modified pyrimidines can impart stability to nucleases when incorporated into aptamers and increase affinity for the target (114). These analogs are conveniently enough substrates of T7 RNA polymerase and AMV reverse transcriptase, two of the essential enzymes used during the aptamer selection process. The aptamer technology has come a long way as far as realization of therapeutics potential is concerned because several of these molecules are in various stages of clinical trial.

4.3.2 Aptamers as Diagnostics. Aptamers rival antibodies in terms of affinity and speci-

ficity. Aptamers being significantly smaller and simpler molecules than antibodies make them attractive for diagnostic use. In fact some of them surpass the discriminating abilities of antibodies. For example, an RNA aptamer was generated against theophylline with a K_d of 400 nM, and it showed >10,000 times weaker binding to caffeine, which was an order of magnitude higher discrimination than offered by antibodies against theophylline (116).

In vitro evolution or SELEX has proven to be a powerful technology that may have a variety if applications in the development of therapeutics and diagnostics. Chemically modified nucleotides have addressed the concerns about biological stability, but issues related to delivery and mass production still remain as concerns.

4.4 Phage Display

With the identification of new therapeutic targets it has become imperative to identify newer ligands that can bind to them, and one way to identify such molecules is by creating a random library and selecting the members with the desired properties. Coupling recombinant DNA technology with phage biology, the phage display technique has revolutionized the identification of novel peptides that can be used for therapeutic purposes as well as for structural studies such as epitope mapping, identification of critical amino acids responsible for protein–protein interactions, etc. It also provides an opportunity to physically link genotype with the phenotype that has allowed linking DNA encoding novel functions to be selected directly from complex libraries (117, 118).

4.4.1 Preparation of Phage Display Libraries. Phage M13 and other members of the filamentous phage family have been used as expression vectors in which foreign gene products are fused to the phage coat proteins and are displayed on the surface of the phage particle. The probability of finding a ligand in a random peptide library (RPL) is proportional to the affinity of the ligand for the selector molecule and its frequency of occurrence in the library. The frequency can be enhanced by constructing libraries with large sequence diversity and flexibility of the insert, which can be long and unconstrained with the goal of making multiple contacts with the selector molecule. But the libraries that produce higher affinity ligands are created by deliberately introducing predefined structural features or structural constraints like disulfide bridges. Furthermore, it has been established that the chances of finding a good ligand increases by screening more libraries. Various protein and peptides including complex peptide libraries have been displayed on phage. The phage particles that display the desired molecules are mainly isolated by binding to immobilized target molecules or by affinity chromatography. A wide variety of selectors have been used that ranges from organs in whole animals to cell surface proteins and studies with enzyme, antibody, or receptor proteins.

4.4.2 Phage Display Selections against Purified Proteins.
Peptides that will selectively bind to purified cell surface receptor proteins or other therapeutically relevant proteins can be isolated from phage display libraries. One of the classes of therapeutically relevant targets is enzymes. Small molecules are traditionally good binders of active site clefts or allosteric regulatory sites that are often buried within the enzyme structure. Phage display libraries can also effectively identify enzyme inhibitors. Recently, Hyde-Der et al. obtained peptide inhibitors against six of the seven different enzyme classes that they tested (119). Additionally they demonstrated that the isolated peptides can also be useful for identifying small molecules inhibitors of the target enzyme in high-throughput screens.

Selective inhibition of individual proteases of the coagulation cascade may have enormous therapeutic value but is extremely difficult. Dennis et al. isolated a peptide that non-competitively inhibits the activity of serine protease factor VIIa (FVIIa), a key regulator of the coagulation cascade with a high degree of specificity and potency (120). Extensive structure-function characterization showed that the peptide binds to a previously unknown

"exosite" that is distinct from the active site. Apparently the inhibition was via an allosteric mechanism.

The advantage of a peptide-based enzyme inhibitor is that it can theoretically sample the entire exposed surface of the enzyme with more contact points than a small molecule inhibitor. They can therefore act as both drug molecules and as reagents in drug discovery.

4.4.3 Cell-Specific Targeting.
Peptides that can home in on specific cell types have enormous potential both in basic research and therapeutic applications (121). Therapeutic applications include identification of specific cell surface proteins in diseased cells as diagnostic agents and gene delivery agents.

The selection of cell-targeting peptides may involve targeting a known cell surface molecule or screening whole cells without *a priori* knowledge of the chemical nature of the cell surface. The latter approach can lead to identification of hitherto unknown cell surface receptors or target molecules.

Selection from a phage display library of peptides that will bind to a known target has its advantage in that specific ligands can be identified in the presence of a complex milieu of biomolecules. Sometimes this may also pose problem during selection because of the chemical complexity of available targets leading to non-specific binding.

One of the first peptides to be isolated from a selection based on whole cells is an antagonist for the unknown plasminogen activator receptor. This selection was carried out on transfected COS-7 and SF-9 insect cells that overexpressed the receptor. Another example is isolation of ligands for the thrombin receptor. Interestingly, the selection was executed on human platelet cells that naturally express the thrombin receptor. To identify the specific ligand for the receptor, a known agonist for the thrombin receptor was used to elute the bound phages. The selection led to the identification of two peptides, one of which acted as an agonist with the ability to activate the thrombin receptor. The other peptide could immunoprecipitate the thrombin receptor form membrane extract and promoted aggregation of platelets, thereby acting as a true antagonist.

4.4.4 *In Vivo* Phage Display. In a pioneering study, Pasqualini and Ruoslahti demonstrated for the first time that an *in vivo* phage display can specifically target organs (122). Two pools of peptide-phage libraries were injected into mice through the tail vein. The mice were killed, and the two targeted organs, brain and kidney, were isolated. The isolated tissues were homogenized and the bound phages were reamplified in *E. coli*. Several dominant peptide motifs were identified. The isolated phages were injected back into the mice and the selectivity of the phages was measured. The brain-targeting phage accumulated in the brain tissue 4–9 times better compared with the kidney-targeted phage. A synthetic peptide inhibited the uptake of the corresponding phage in the brain showing the specificity of the isolated peptide.

In another exciting study of *in vivo* selection of phage display libraries was used to isolate peptides that home exclusively to tumor blood vessels (123). Phage libraries were injected into the circulation of nude mice bearing breast carcinoma xenografts. Three main peptide motifs were identified that targeted the phages into the tumors. One of the motifs, CDCRGDCFC (embedded RGD motif), homed in to several tumor types (including carcinoma, sarcoma, and melanoma) in a highly selective manner, and the targeting was specifically inhibited by the cognate peptide. By conjugating doxorubicin, a common chemotherapeutic agent, to the identified peptides, the efficacy of doxorubicin was increased and the toxicity was markedly decreased.

Phage display has also been effectively used to identify peptides that will achieve gene delivery by targeting cell surface receptors to augment receptor-mediated gene delivery. This has tremendous potential for gene therapy, where a major hurdle is delivering the gene (124, 125).

The above examples demonstrate vividly the power of recombinant technology for developing techniques and agents that directly advances medicinal chemistry.

4.5 Reagents for Structural Biology Studies

In combination with molecular genetics, structural biology also has used physical techniques—nuclear magnetic resonance (NMR) spectroscopy and X-ray crystallography—to its advantage in the study of proteins as drug targets, models for new drugs, and discovery tools (126). These two techniques can be used independently, or in concert, to determine the complete three-dimensional structure of proteins. Recent advances in NMR techniques, especially multidimensional heteronuclear studies, offer dramatic improvements in spectral resolution and interpretation (127, 128). Identification of differences in the results from comparative studies on the same protein can reveal important structural or dynamic information (129), possibly relevant to the design of synthetic ligands or inhibitors. Inclusion of such structural biology results into the more traditional synthesis-driven discovery paradigm has become a recognized and important component of drug design (130).

The variety of studies undertaken using these structural biology techniques spans the range of proteins of interest, from enzymes and hormones to receptors and antibodies. Recombinantly produced reagents (accessible as either purified, soluble proteins or cell-surface expressed, functional enzymes and receptors) with potential application to drug discovery fall into a number of general categories: enzymes (with catalytic function), receptors (with signal transduction function), and binding proteins (with cellular adhesion properties). Rather than exhaustively catalog further examples, the next sections will highlight instances in which combinations of directed specific assays and structural biology studies have aided in non-protein drug discovery.

4.6 Enzymes as Drug Targets

A large number of enzymes have been cloned and expressed in useful quantities for biochemical characterization. The advent of rational drug design paradigms, in particular the methodology surrounding mechanism-based enzyme inhibition (131a and 131b) and the market success of various enzyme inhibitors such as captopril, have made enzymes of all types more reasonable laboratory tools and therefore accessible targets for medicinal chemistry efforts. Many enzymes linked to pathologies or known to regulate important biochemical pathways have been cloned for subspecies differentiation and/or access to human isotypes. Also, important advances have been made in the molecular biology and target validation of various classes of enzymes including protein kinase C (132, 133), cyclooxygenase (COX) isozymes, phosphodiesterase (134, 135a, 135b), and the phospholipase A_2 (135c, 136, 137a–137c) families.

The rational basis of enzyme-inhibitor interactions, especially to predict or explain specificity, is among the most intensely active areas of structural biology. One of the most studied therapeutic targets is dihydrofolate reductase (DHFR), an enzyme essential for growth and replication at the cellular level. Inhibitors of DHFR, most notably the antifolates methotrexate (MTX) and trimethoprim (TMP), are used extensively in the treatment of neoplastic and infectious disorders. Some of the observed species selectivities for these inhibitors have been explained in terms of distinctive structural differences at the binding sites of the chicken and *E. coli* enzymes (137d, 138), but some of the conclusions made based on the enzyme-inhibitor binding interaction have been challenged by a crystal structure of human recombinant DHFR complexed with folate, the natural substrate (139). Comparisons of the conformations of the conserved human and mouse DHFR side-chains revealed differences in packing, most noticeably the orientation of F31. Site-directed mutagenesis studies confirmed the importance of this observation. The mutant F31L (human F to *E. coli* L mutation) gave equivalent K_i values for inhibition by TMP, but gave a 10-fold increase in K_m for dihydrofolate (140). Similar results were found for the F31S mutant, for which there was also a 10-fold increase in K_m for dihydrofolate and a 100-fold increase in K_d for MTX. The F34S mutant, however, showed greater differences: a 3-fold reduction in K_m for nicotinamide adenine dinucleotide phosphate, reduced form (NADPH), a 24-fold increase in K_m for dihydrofolate, a 3-fold reduction in k_{cat}, and an 80,000-fold increase in K_d for MTX, suggesting that phenylalanines 31

and 34 make different contributions to ligand binding and catalysis in human DHFR (141). These results helped to pinpoint major differences among DHFRs of various species and thus suggest ways to design new and more species-specific inhibitors that would preferentially target pathogen versus host DHFR. Such compounds would be expected to be more potent chemotherapeutics exhibiting less toxicity in humans. The design and refinement of inhibitors of *E. coli* thymidilate synthetase such as (**5**) attests to the viability and potential cost-effectiveness of this rational design approach (142, 143a–143c).

In contrast to the DHFR investigations for which the goal is refinement, problems in *de novo* design of inhibitors require more fundamental help, specifically the availability of the target enzyme in quantity for screening. The ability of rDNA technology to expedite access to quantities of a specific enzyme in a situation in which some indication of specificity would eventually be required of the final inhibitor is no where more evident than in the case of the retroviral aspartic HIV-1 protease (HIV-1 PR) (143d). From among the multitude of potential points of intervention into viral replication of the HIV-1 genome, this enzyme was identified as a viable target for anti-AIDS drugs because mutation of the active site aspartic acid (D25) effectively prevents processing of retroviral polyprotein, producing immature, non-infective virions. In addition to the residues DTG at positions 25 to 27, mutations within the sequence GRD/N (positions 86 to 88 in HIV-1 PR)—a highly conserved domain in the retroviral proteases but not present in cellular aspartic proteases—were found to be completely devoid of proteolytic activity, potentially pinpointing a site critical for design of specific inhibitors capable of recognizing the viral, but not the host proteases.

The search for important tertiary structural differences between HIV-1 PR and known eukaryotic proteases began by determination of the X-ray crystal structure (Fig. 4.7) of recombinantly expressed material at 3 Å resolution (144). Subsequent crystallographic studies on both synthetic (at 2.8 Å) and recombinantly expressed (at 2.7 Å) material helped locate side-chains and resolved

(5)

some ambiguities in the dimer interface region (145, 146). From this information, a model of the substrate binding site was proposed (147). Far more useful for inhibitor design purposes, complexes of four structurally distinct inhibitors bound to HIV-1 PR were solved (148–151), from which a generalized closest contact map (Fig. 4.8) was developed (143). With the functional role and tertiary structure of the protease determined, additional studies with both recombinant and syn-

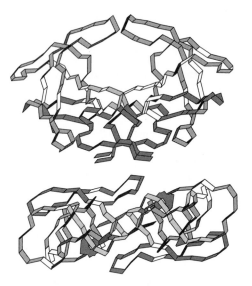

Figure 4.7. The structure of native HIV-1 protease drawn as a ribbon connecting the positions of α-carbons. The upper structure, in which the pseudo-twofold axis relating one monomer to the other is vertical and the plane of the page, represents a view along the substrate binding cleft. The lower structure is a top view, with the pseudo-twofold axis perpendicular to the page.

thetic material have yielded automated robotics assays for screening of chemical libraries, fermentation broths, and designed inhibitors using HIV-1 PR cleavage of synthetic pseudo-substrates. Peptide sequences derived from specific retroviral polyprotein substrates and inhibition by pepstatin and other renin inhibitors identified $(S/T)P_3P_2(Y/F)P$ as a consensus cleavage site for HIV-1 PR. One such inhibitor, $SGN(F\psi[CH2N]P)IVQ$, has been used as an affinity reagent for large-scale purification of recombinant HIV-1 PR (152), whereas $Ac\text{-}TI(nL\psi[CH_2NHl\text{-}nL]Q)R\text{-}NH_2$ was used in co-crystallization studies. From among the large numbers of peptides identified as HIV-1 PR inhibitors, only a limited number have been shown to inhibit effectively viral proteolytic processing and syncytia formation in chronically infected T-cell cultures (153, 154). The most advanced peptidomimetic compound is structure (**6**), which both inhibits HIV-1 PR and exhibits effective and noncytotoxic antiviral activity in chronically infected cells at nanomolar concentrations (155). However, as with other peptidomimetic structures such as inhibitors of another aspartyl protease, renin (156), their transformation into potential drugs will require additional synthetic work. The short interval from identification of the enzyme as a target from among the possibilities presented by the HIV-1 genome to accessing material for assay and structural purposes has obviously hastened the determination of the viability of HIV-1 PR inhibitors as AIDS therapeutics and also has provided an excellent example of structurally driven rational drug design.

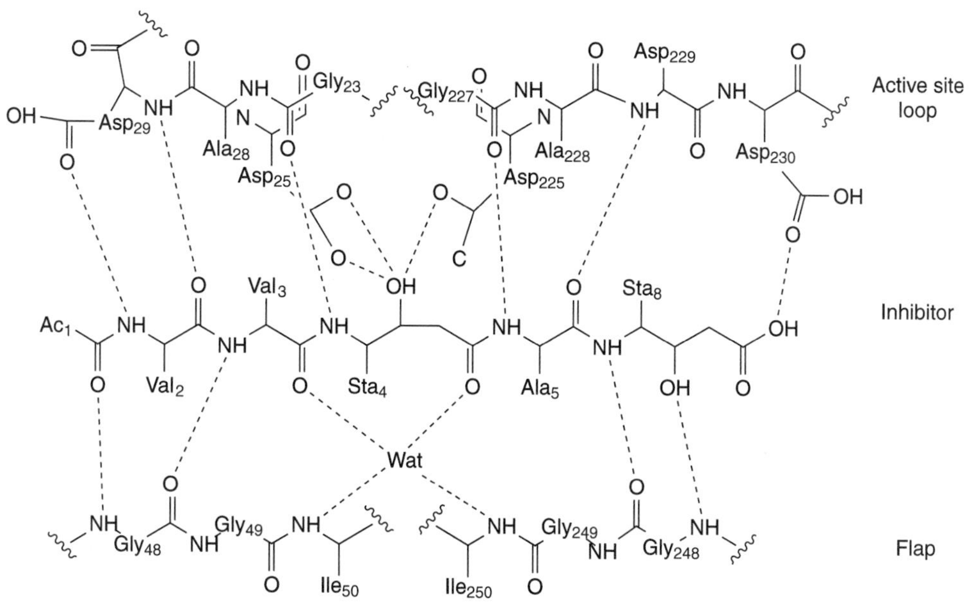

Figure 4.8. Hydrogen bonds between a prototypical aspartic protease inhibitor (acetylpepstatin) and HIV-1 protease. The residues are labeled at the C-β position (C-α for glycine). The residues labeled 25–50 are from monomer A, those labeled 225–250 are from monomer B, and those labeled 1–6 are with acetyl pepstatin.

The search for a common mechanism of action of the immunosuppressive drugs cyclosporin (CsA) (**7**) and FK-506 (**8**) highlights another possibility in which the drug was discovered by screening in cellular or *in vivo* models, but the exact mechanism or site of action is unknown (157). Using the active molecules, cyclophilin (CyP) and the FK binding protein (FKBP) were identified as the specific receptors for cyclosporin and FK-506, respectively. These binding proteins were discovered to be distinct and inhibitor-specific *cis-trans* peptidyl-prolyl isomerases that catalyze the slow *cis-trans* isomerization of proline peptide bonds in oligopeptides and accelerate the rate-limiting steps in the folding of some proteins (158, 159). The biochemical mechanism of inhibition was first proposed to involve a specific covalent adduct between inhibitor and its rotamase (159), but the hypothesis was soon challenged by evidence that showed that the binding interactions are peptide-sequence specific (160) for both proteins. Using recombinant CyP as the standard, four cysteine-to-alanine mutants (C52A, C62A, C115A, and C161A) were shown to retain full affinity for

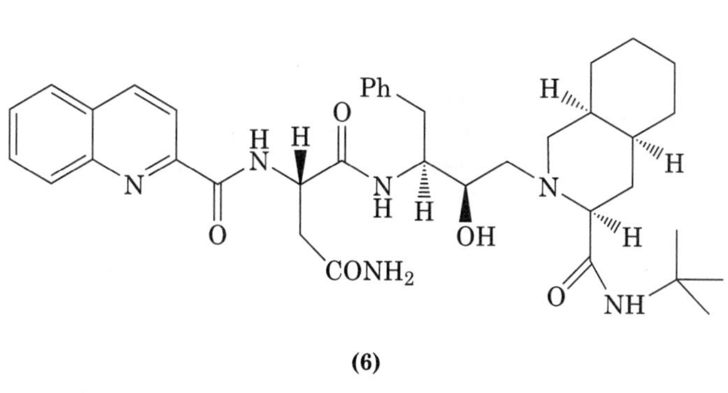

(**6**)

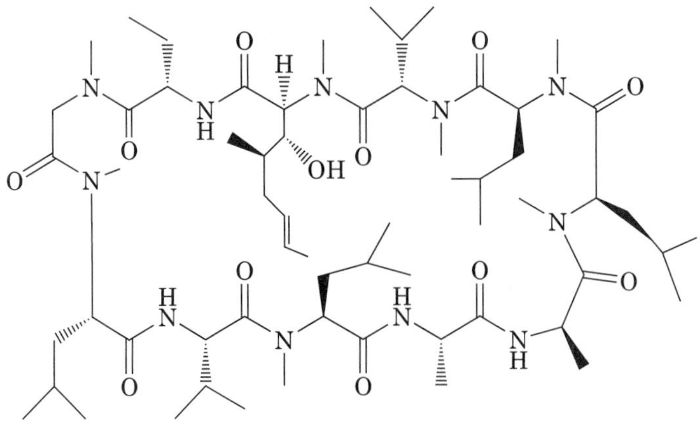

(7)

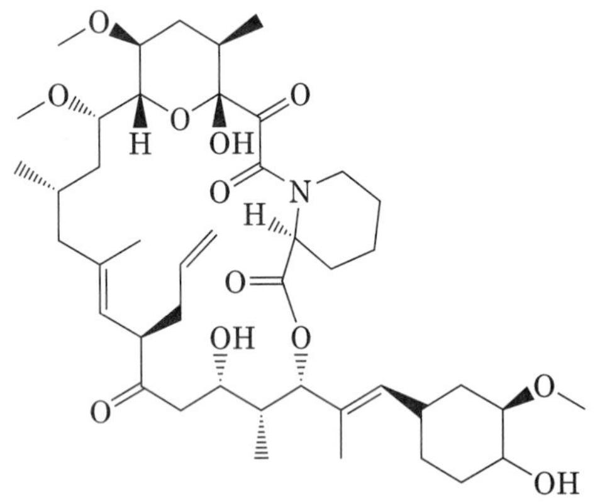

(8)

CsA and equivalent rotamase catalytic activity, indicating that the cysteines play no essential role in binding or catalysis (161). In the case of FKBP, NMR studies of [8,9-[^{13}C]FK-506 bound to recombinant FKBP, wherein the likely mechanism of inhibition also is noncovalent, suggesting that the α-ketoamide of FK-506 serves as an effective surrogate for the twisted amide of a bound peptide substrate (162). Numerous further NMR, X-ray crystallographic and computational modeling studies have been carried out on both CsA/CyP and FK-506/FKBP complexes to attempt to fully determine a structural basis for activity (163, 164).

The exact signal transduction mechanism that triggers the immunosuppressive response in T-cells, however, remained un-

known until an elegant set of experiments identified calcineurin, a calcium- and calmodulin-dependent serine/threonine phosphatase, and a complex of calcineurin with calmodulin as the binding targets of the immunophilin-drug complexes (165). The immunosuppressant, displayed by the immunophilin protein, then effectively functions as the critical element that binds the pentapartite complex together, causing inhibition of the phosphatase (Fig. 4.9). The complex seems also to exert two subsequent effects: halting DNA translation in T-lymphocyte nuclei and inhibition of IgE-induced mast cell degranulation. These *in vitro* observations must necessarily be confirmed by further *in vivo* work, but the elucidation of these molecular-level mechanisms will allow the development of more highly tailored and potent immunosuppressive agents (166, 167a). Toward this goal, experiments that sought to probe the relationship between the immunosuppressive effect and the common side effects, such as hypertension and nephrotoxicity, of CsA were designed by LoRusso et al. (181). It was identified that the immunosuppressive effect of CsA is unconnected with its hypertensive effect; this is vital information that may be of use in the design of CsA derivatives devoid of these side effects.

4.7 Receptors as Drug Targets

Even more so than with enzymes, molecular genetics has been primarily responsible for the identification of functional receptor sub-

Physiological function   Immunosuppression

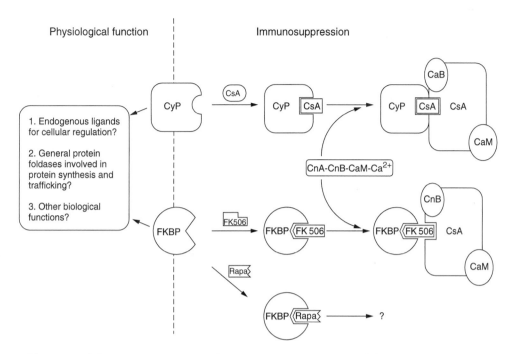

Figure 4.9. Schematic representation of immunosuppressant-immunophilin complex interactions. CyPs bind CsA to form a complex in which both components undergo change in structure. This complex binds to, and inhibits, calcineurin (*CnA*, A subunit; *CnB*, B subunit; *CaM*, camodulin) in a calcium-dependent manner. FK506-binding protein complxes with FK506 or rapamycin (*Rapa*). FKBP-FK506 also binds calcineurin. The target of FKBP-rapamycin is unknown but is presumed to be different from cacineurin.

types. Success in the case of cimetidine, a selective H_2-receptor antagonist, made it clear that the design of specific ligands for at least some receptors is a task amenable to medicinal chemistry. The classic tissue-binding pharmacological methods that made distinctions on the basis of ligand selectivity have been supplemented, and in most cases supplanted, by further subtyping made possible by cross-hybridization cloning using the known receptor genes. Any studies that profile the *in vitro* receptor subtype-specificity of compounds can theoretically help identify potential *in vivo* side effects of the compounds if the association of subtype to effect is known or suspected. At this point, just the indication of a more specific profile with fewer side effects is enough to help choose one compound over another for preclinical development.

One of the earliest examples of the role of molecular biology in discerning receptor subtype roles was the case of the muscarinic cholinergic receptors (MAChRs). The two subtypes M_1 and M_2 had been defined pharmacologically by their affinity, or lack thereof, for pirenzepine and were later confirmed by molecular cloning to be distinct gene products *m1* and *m2*, respectively (167b–169). Three additional muscarinic receptor genes (*m3* to *m5*) were subsequently isolated (170–172). From this work, a subtype-specific heterologous stable expression system in Chinese hamster ovary (CHO) cells suitable for screening potential subtype-specific ligands was developed. From this assay, pirenzepine, previously thought to bind to *M1* only, was found to have only a 50-fold reduced affinity for *M2,* and an almost equivalent (to *M1)* binding affinity for *M3* and *M4,* suggesting that studies using pirenzepine on tissue homogenate have failed to distinguish adequately among the subtypes (173). Similar breakthroughs have been realized across the rest of the family of signal transduction G-protein–coupled receptors,

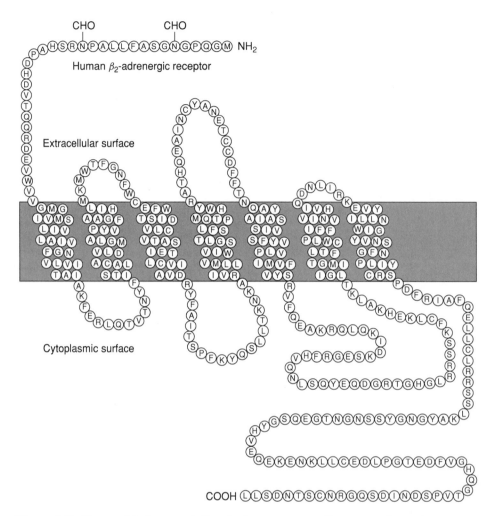

Figure 4.10. Topographical representation of primary sequence of human β_2-adrenergic receptor, a typical G-protein–coupled receptor. The receptor protein is illustrated as possessing seven hydrophobic regions each capable of spanning the plasma membrane, thus creating intracellular and extracellular loops as well as an extracellular amino terminus and a cytoplasmic carboxyl terminal region.

because in addition to the muscarinics, the primary structures of the adrenergic (α_1, α_2, β_1, and β_2), serotonergic, tachykinin, and rhodopsin receptors have been determined (174–176). All of these display the now-familiar homology pattern of seven membrane-spanning domains packed into antiparallel helical bundles (Fig. 4.10). The exceedingly high homology among the large family of G protein-coupled receptors also has allowed the development of three-dimensional models of the proteins to aid in drug refinement (177). For example,

mutagenesis studies on the β_2-adrenergic receptor have localized the intracellular domains involved in (*1*) the coupling of the receptor to G proteins (178); (*2*) homologous desensitization by β-adrenergic receptor kinase (β-ARK) (179), itself cloned and a possible target for down-regulation inhibitors (180); (*3*) heterologous desensitization by cAMP-dependent protein kinase (181); and (*4*) an extracellular domain with conserved cysteine residues implicated in agonist ligand binding (182). A chimeric muscarinic cholin-

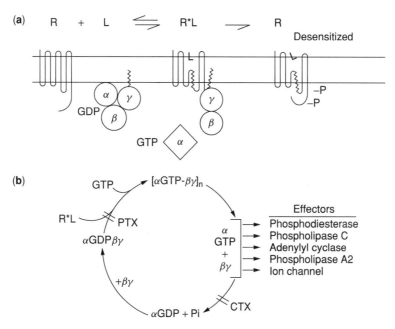

Figure 4.11. Receptor G-protein–mediated signal transduction. (a) Receptor (R) associates with a specific ligand (L), stabilizing an activated form of the receptor (R^*), which can catalyze the exchange of GTP for GDP bound to the α-subunit of a G-protein. The $\beta\gamma$-heterodimer may remain associated with the membrane through a 20-carbon isoprenyl modification of the γ-subunit. The receptor is desensitized by specific phosphorylation (-P). (b) The G protein cycle. Pertussis toxin (PTX) blocks the catalysis of GTP exchange by receptor. Activated α-subunits (aGTP) and $\beta\gamma$-heterodimers can interact with different effectors (E). Cholera toxin (CTX) blocks the GTPase activity of some α-subunits, fixing them in an activated form.

ergic:β-adrenergic receptor engineered to activate adenylyl cyclase (a second messenger system not coupled to MAChR agonism) also has helped identify which intracellular loops may be involved in direct G-protein interactions (183). The diverse signal transduction functional roles of the many G-proteins to which these receptors are coupled (Fig. 4.11) also makes them viable drug targets (184).

The complicated biochemical pharmacology of natriuretic peptides, the regulatory system that acts to balance the renin-angiotensin-aldosterone system (185), has been significantly clarified by the cloning of three receptor subtypes, which revealed the functional characteristics of a new paradigm for second messenger signal transduction through guanylate cyclase. The α-atrial natriuretic peptide (α-ANP) receptor (NPA-R) and the brain natriuretic peptide (BNP) receptor (NPB-R) contain both protein kinase and guanylate cyclase (GC) do-

mains, as determined by both sequence homologies and catalytic activities, whereas the clearance receptor (ANP-C) completely lacks the necessary intracellular domains for signal transduction through the guanylate cyclase pathway (186). This system defines the first example of a cell surface receptor that enzymatically synthesizes a diffusible second messenger system in response to hormonal stimulation (187) (Fig. 4.12). Data from experiments performed with C-ANP$_{4-23}$ indicates that the clearance receptor (NPC-R) may be coupled to the adenylate cyclase/cAMP signal transduction system through an inhibitory guanine nucleotide regulatory protein (188). Because the NPs have differential, but not absolute, affinities for their corresponding receptors (189) and because both agonism (190) and antagonism (191) of the GC activity have been demonstrated *in vitro* using ANP analogs, it may be possible to discriminate among the receptor

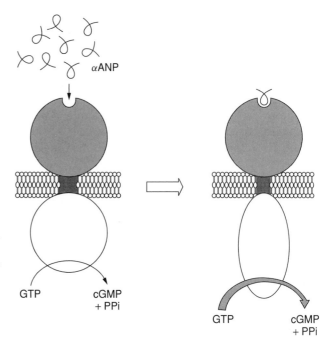

Figure 4.12. Model for ANP-A and ANP-B receptor function. The unoccupied ANP-A receptor is shown on the left with a basal rate of cGMP synthesis (indicated by a thin arrow). The effect of ligand binding to the amino-terminal extracellular domain is shown on the right. Proposed allosteric modulation of guanylate cyclase by a-ANP is schematically illustrated by a change in shape of the intracellular domain and a thicker arrow to denote an increase in guanylate cyclase-specific activity with greater production of the second messenger cGMP.

GCs to obtain more subtle structure-activity information for the design of selective NP analogs. Homology between the NP receptors and another guanylate cyclase firmly identified the latter as the elusive heat-stable enterotoxin receptor St(a)-R (192), which aided in the identification of both the biochemically isolated guanylin (193) and the cloned proguanylin (194) versions of the endogenous natural ligand. This system is presumed to play a role in water retention through cGMP modulation of the CFTR chloride ion channel (195).

The number of receptors of biological significance cloned and expressed for further study continues to grow at an exponential rate (196). These include epidermal growth factor (EGFR), insulin (INSR), insulin-like growth factor-1 (IGF-1R), platelet-derived growth factor (PDGFR) receptors and related tyrosine kinases (197), tumor necrosis factor receptors 1 and 2 (198), subtypes of the $GABA_A$-benzodiazepine receptor complex (199–202), human γ-interferon receptor (203), inositol 1,4,5-triphosphate (IP3)-binding protein P400 (204), kianate-subtype glutamate receptor (205), follicle-stimulating hormone receptor (206), multiple members of the steroid (ER,

PR, AR, GR, MR), thyroid hormone (TRα and β), and retinoid (RARα, β, and γ, and RXRα) receptor superfamily of nuclear transcriptional factors (207–210), multiple members of the interleukin cytokine receptor family (211), and subtypes of the glutamate (212) and adenosine (213) receptor families. The importance of access to human cloned receptors continues to be underscored as receptor binding plays an increasingly critical role in modern drug discovery (214).

To make the case for using cloned human receptors for drug discovery even stronger, dramatic evidence that minor amino acid sequence variations inherent in species variability can produce profoundly different pharmacological effects was recently provided in two instances. In the comparison of rodent versus human analogs of the 5-hydroxytryptamine receptor subtype 5-HT$_{1B}$, the natural receptors were found to bind 5-HT identically, but they differed profoundly in their affinities for many serotonergic drugs. These striking differences could be reversed by change of a single transmembrane domain residue (T355N), which effectively rendered the two receptors pharmacologically identical (215). A similar study comparing human to chicken and hamster progesterone receptors showed that the steroidal abortifacient RU486 (**9**) shows an-

(**9**)

tagonist activity in humans but not in the two other species, because of the presence of a glycine at position 575. Both the chicken and hamster receptors have a cysteine at this position, and replacement by glycine (C575G) generated a mutant receptor that could bind RU486. Likewise, mutation of the human receptor (G575C) abrogated RU486 binding (216). These and many similar findings emphasize the critical importance of the availability of human clones proteins as potential targets for drug action and as sources of structural information in the discovery of potent and selective therapeutic agents.

4.8 Cellular Adhesion Proteins

The understanding of the molecular processes that govern cell localization in various patho-

logical conditions has been significantly expanded because of the cloning and expression of some of the major cellular adhesion proteins, especially in the integrin family, a highly related and widely expressed group of $\alpha\beta$-heterodimeric membrane proteins (217). The interaction in the antigen-receptor crosslinking adhesion of T-cells mediated by the intercellular adhesion molecule (ICAM-1) and the lymphocyte function-associated molecule (LFA-1) was clarified by the cloning and expression of ICAM-1, the major cell surface receptor for rhinovirus (218). A soluble form of human ICAM-1 effectively inhibits rhinovirus infection at nanomolar concentrations (219). As the pivotal interaction in the adhesion of leukocytes to activated endothelium and tissue components exposed during injury (Fig. 4.13), inhibition of ICAM-l/LFA-1 binding represents a potential prime intervention point for new anti-inflammatory drugs.

Other members of the integrin family have been also successfully cloned (220). The availability of the individual members of this heterodimer superfamily—characterized by gross similarities in structure (Fig. 4.13), function, and in some cases, avidity for RGD-containing peptides—will allow their individual roles in specific disease pathophysiologies to be ascertained and provide the means to develop integrin-specific antagonists for a multitude of uses. Of importance to anti-thrombotic drug discovery efforts, the integrin $a_{IIb}b_3$, also known as $GPII_bIII_a$, the plate-

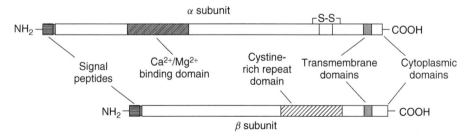

Figure 4.13. General polypeptide structure of integrins. The α-subunit of the integrins is translated from a single mRNA, and in some cases, it is processed into two polypeptides that remain disulfide bonded to one another. The α-subunit and the β-subunit contain typical transmembrane domain that is thought to traverse the cell memebrane and bring COOH-termini of the subunits into the cytoplasmic side of the membrane. The α-subunit contains a series of short-sequence elements homologous to known calcium binding sites in other proteins; the β-subunit is tightly folded by numerous intrachain disulfide bonds.

let fibrinogen receptor (221), was successfully expressed as the functional heterodimer, showing that prior association of the endogenous subunits is necessary to produce the cell surface complexes (222). Surface expression of $GPII_bIII_a$ is the common endpoint in platelet activation, initiating the platelet-platelet cross-linking through fibrinogen that is responsible for thrombus formation. A number of compounds, including disintegrin snake venoms, RGD peptides, and organic mimetics, have been shown to inhibit thrombus formation and platelet aggregation in animal and human clinical trials (223, 224).

Lymphocyte and neutrophil trafficking, the first step in the development of an inflammatory response, is known to occur through specific cell surface receptors and ligands, which match the inflammatory cell to the right target. Different receptors and ligands are expressed at different time points during inflammatory processes, from seconds to hours. These protein recognition signals—previously termed homing receptors (HR) and now uniformly called selectins—are membrane-bound proteins, which target circulating lymphocytes to specialized targets provide one such opportunity for intervention (225, 226). Molecular cloning of the murine HR now called L-selectin revealed that the receptor contains a lectin (carbohydrate binding) domain that is responsible for the binding event (227). The selectin family consists of three cell surface receptors that share this affinity for carbohydrate ligands, specifically the tetrasaccharide sialyl Lewis X (10). The carbohydrates

are likely displayed at multiple O-glycosylation sites on mucin-like glycoproteins such as Spg50, a novel endothelial ligand for L-selectin discovered by a combination of biochemical isolation, sequencing, and cloning techniques (225). The carbohydrate itself, however, offers a viable starting point for drug design as a structural lead for both carbohydrate analogs and noncarbohydrates (228, 229). Molecular modeling of E-selectin based on antibody mapping and homology to mannose binding protein (230) also suggests drug design possibilities based on proposed analogous structural interactions of the selectin with structure (10) (231). Inhibition of selectin-mediated cellular trafficking at specific times might help break a spiraling acute inflammatory cycle, allowing control of acute inflammatory processes, such as shock and adult respiratory distress syndrome (ARDS), and helping in the management of integrin-mediated inflammatory processes that follow (232).

5 FUTURE PROSPECTS

Molecular genetics is only now beginning to identify new targets for drug action. For example, regulation of inducible or tissue-specific gene expression has been an obvious but elusive target for pharmacological intervention (233–235). The tools to monitor such events are now available, as in the case of the low density lipoprotein receptor, for which tissue-specific up-regulation of receptor population may successfully compete with other cholesterol-lowering agents (236, 237). In par-

(10)

allel, another genetic marker for atherosclerotic disease, lipoprotein(a), is a target for selective expression down-regulation (238). An even more direct method to interfere with gene expression is selectively to bind the gene using sequence-specific recognition elements that prohibit transcription. Antisense oligonucleotides are the first sequence-directed molecules designed to inhibit protein expression at the level of translation of mRNA into the undesired protein (239). The first FDA approved drug using antisense technology is Vitravene, introduced by Isis pharmaceuticals in 1998 for local treatment of CMV retinitis in AIDS patients (240a). Genasense, another antisense-based drug, designed to block the production of Bcl-2 protein, a protein overexpressed in most cancer cells, is in late-stage clinical trials as an adjunct cancer-treating drug (240b). The ability to test for inhibition of gene expression or to measure the effects of species-specific agents against relevant pharmacological targets in animal models has also been advanced by the development and use of transgenic (240c) or engineered gene knockout animals, such as the CFTR-defective mouse for cystic fibrosis (241), in screening and evaluation procedures.

The power of molecular genetics to provide unique and valuable tools for drug discovery is beginning to be exploited. The prospects for uncovering the molecular etiology of a disease state or for gaining access to a disease-relevant target enzyme or receptor are already being realized. The recent successful mapping of human genome (242a, 242b) further promises better chances of rationally intervening in disease states at previously inaccessible or unknown points.

The development of recombinant DNA technology into a fully integrated component of the drug discovery process is inevitable (242c, 243). In 1987, Kornberg remarked that "the two cultures, chemistry and biology, [are] growing further apart even as they discover more common ground" (244). However, the broad area of drug development might qualify as one such meeting place for medicinal chemistry and molecular biology where the trend is reversing. The application of genetic engineering techniques to biochemical and pharmacological problems will facilitate the discovery of novel therapeutics with potent and selective actions. There is little doubt among medicinal chemists that effective collaboration between chemistry and biology is not only needed but is actually growing in importance in drug design (245, 246). The eventual extent of the impact that molecular biology will have on the drug discovery process is, and will be for some time, unknown. However, the reality of recombinant protein therapeutics offers the assurance that this same technology, in conjunction with structural biology, computer-assisted molecular modeling, computational analysis, and medicinal chemistry (247), will help make possible better therapies for those diseases already controllable and new therapies for diseases never before treatable.

REFERENCES

1. J. Liebenau, in C. Hansch, P. G. Sammes, and J. B. Taylor, Eds., *Comprehensive Medicinal Chemistry*, vol. 1, Pergamon Press, Oxford, 1990, pp. 81–98.

2. A. Burger, in C. Hansch, P. G. Sammes, and J. B. Taylor, Eds., *Comprehensive Medicinal Chemistry*, vol. 1, Pergamon Press, Oxford, 1990, pp. 1–5.

3. W. Sneader, in C. Hansch, P. G. Sammes, and J. B. Taylor, Eds., *Comprehensive Medicinal Chemistry*, vol. 1, Pergamon Press, Oxford, 1990.

4. L. H. Sterbach, *Prog. Drug Res.*, **22**, 229–266 (1978).

5. C. R. Ganellin in J. S. Bindra and D. Lednicer, Eds., *Chronicles of Drug Discovery*, vol. 1, John Wiley & Sons, Inc., New York, 1982, pp. 1–38.

6. M. A. Ondetti and D. W. Cushman, *J. Med. Chem.*, **24**, 355–361 (1981).

7. J. K. Seydel, in E. Mutschler and E. Winterfeldt, Eds., *Trends in Medicinal Chemistry*, VCH Verlagsgesellschaft, New York, 1987, pp. 83–103.

8. J. M. Henwood and R. C. Heel, *Drugs*, **36**, 429–454 (1986).

9. J. V. Duncia, D. J. Carini, A. T. Chiu, A. L. Johnson, W. A. Price, P. C. Wong, R. R. Wexler, and P. B. M. Timmermans, *Med. Res. Rev.*, **12**, 149–191 (1992).

10. K. R. Freter, *Pharm. Res.*, **5**, 397–400 (1988).

11. J. A. Lowe and P. M. Hobart, *Annu. Rep. Med. Chem.*, **18**, 307–316 (1983).

12. M. C. Venuti, *Annu. Rep. Med. Chem.*, **25**, 201–211 (1990).

13. A. Harvey, *Trends Pharmacol. Sci.*, **12**, 317–319 (1991).

14. M. C. Venuti in T. Friedmann, Ed., *Molecular Genetic Medicine*, vol. **1**, Academic Press, New York, 1991, pp. 133–168.

15. W. Szkrybalo, *Pharm. Res.*, **4**, 361–363 (1987).

16. S. Cometta, *Arzneim. Forsch. Drug Res.*, **39**, 929–934 (1989).

17. P. Swetly in Y. C. Martin, E. Kutter, and V. Austel, Eds., Modern Drug Research: Path to Better and Safer Drugs, Medicinal Research Series, vol. **12**, Marcel Dekker, New York, 1989, pp. 217–241.

18. S. P. Adams in C. Hansch and J. B. Taylor, Eds., *Comprehensive Medicinal Chemistry*, vol. **1**, Pergamon Press, Oxford, 1990, pp. 409–454.

19. I. S. Johnson, *Science*, **219**, 632–637 (1983).

20. C. Hentschel in D. N. Copsey and S. Y. J. Delnatte, Eds., *Genetically Engineered Human Therapeutic Drugs*, Stockton Press, New York, 1988, pp. 3–6.

21. S. L. Gordon in D. N. Delnatte, Ed., *Genetically Engineered Human Therapeutic Drugs*, Stockton Press, New York, 1988, pp. 137–146.

22. P. Bost and A. Jouanneau in C. Hansch and J. B. Taylor, Eds., *Comprehensive Medicinal Chemistry*, vol. **1**, Pergamon Press, Oxford, 1990, pp. 455–479.

23. C. R. Hutchinson, *Med. Res. Rev.*, **8**, 557–567 (1988).

24. C. R. Hutchinson, C. W. Borell, S. L. Otten, K. J. Stutzmann-Engwall, and Y. Wang, *J. Med. Chem.*, **32**, 929–937 (1989).

25. L. Katz and C. R. Hutchinson, *Annu. Rep. Med. Chem.*, **27**, 129–138 (1992).

26. L. J. Nisbet and J. W. Westley, *Annu. Rep. Med. Chem.*, **21**, 149–157 (1986).

27. P. J. Hylands and L. Nisbet, *Annu. Rep. Med. Chem.*, **26**, 259–269 (1991).

28. D. H. Williams, M. J. Stone, P. R. Hauck, and S. K. Rahman, *J. Nat. Prod.*, **52**, 1189–1208 (1989).

29. P. G. Waterman, *J. Nat. Prod.*, **53**, 13–22 (1990).

30. J. D. Coombes, *New Drugs from Natural Sources*, IBC Technical Services, London, 1999.

31. J. R. Knowles, *Science*, **236**, 1252–1258 (1987).

32. E. T. Kaiser, *Angew. Chem. Inc. Edu. Engl.*, **27**, 913–922 (1988).

33. D. J. Livingston, *Annu. Rep. Med. Chem.*, **24**, 213–221 (1989).

34a. H. Kubinyi, *Pharmazie*, **50**, 647–662 (1995).

34b. M. J. Ross, E. B. Grossbard, A. Hotchkiss, D. Higgins, and S. Andersen, *Annu. Rep. Med. Chem.*, **23**, 111–120 (1988).

35. E. Haber, T. Quertermous, G. R. Matsueda, and M. S. Runge, *Science*, **243**, 51–56 (1989).

36. D. L. Higgins and W. F. Bennett, *Annu. Rev. Pharmacol. Toxicol.*, **30**, 91–121 (1990).

37. N. L. Haigwood, et al., *Prot. Eng.*, **2**, 631–620 (1989).

38. J. Krause and P. Tanswell, *Arzneim.- Forsch. Drug Res.*, **39**, 632–637 (1989).

39. U. Kohnert, et al., *Prot. Eng.*, **5**, 93–100 (1992).

40. C. J. Refino, N. F. Paoni, B. A. Keyt, C. S. Pater, J. M. Badillo, F. W. Wurm, J. Ogez, and W. F. Bennett, *Thromb. Haemostas.*, **70**, 313–319 (1993).

41. D. J. Capon, et al., *Nature*, **337**, 525–531 (1989).

42. A. Traunecker, J. Schneider, H. Kiefer, and K. Karjalainen, *Nature*, **339**, 68–70 (1989).

43. R. A. Byrn, et al., *Nature*, **334**, 667–670 (1990).

44. J. A. Rininger, V. A. DiPippo, and B. E. Gould-Rothberg, *Drug Discov. Today.*, **5**, 560–568 (2000).

45. C. M. Henry, *Chem. Eng. News*, **79**, 37–42, (2001).

46. E. Fosslien, *Ann. Clin. Lab. Sci.*, **28**, 67–81 (1998).

47. J. R. Vane, Y. S. Bakhle, and R. M. Botting, *Pharmacol. Toxicol.*, **38**, 97–120 (1998).

48. J. Portanova, Y. Zhang, and G. D. Anderson, *J. Exp. Med.*, **184**, 883–891 (1996).

49. J. Vane, *Nature*, **367**, 215–216 (1994).

50. M. J. Holtzman, J. Turk, and L. P. Sharnick, *J. Biol. Chem.*, **267**, 21438–21445 (1992).

51. T. Hla and K. Nelson, *Proc. Natl. Acad. Sci. USA*, **89**, 7384–7388 (1992).

52. L. S. Simon, A. L. Weaver, and D. Y. Graham, *JAMA*, **282**, 1921–1928 (1999).

53. M. J. Langman, D. M. Jensen, and D. J. Watson, *JAMA*, **282**, 1929–1933 (1999).

54. L. S. Simon, *Arthritis Rheum.*, **41**, 1591–1602 (1998).

55. N. E. Hynes and D. F. Stern, *Biochim. Biophys. Acta.*, **1198**, 165–185 (1994).

56. D. J. Slamon, *New Engl. J. Med.* **344**, 783–792 (2001).

57. R. O. Dillman, *Cancer Invest.*, **19**, 833–841 (2001).

58. E. Drewe and R. J. Powell, *J. Clin. Pathol.*, **55**, 81–85 (2002).

59. H. E. Chadd and S. M. Chamow, *Curr Opinion Biotech.*, **12**, 188–194, (2001).

60. B. C. Cunningham and J. A. Wells, *Science*, **244**, 1081–1085 (1989).

61. B. C. Cunningham, P. Jhurani, P. Ng, and J. A. Wells, *Science*, **243**, 1330–1336 (1989).

62. S. S. Abdel-Meguid, H. S. Shieh, W. W. Smith, H. E. Dayringer, B. N. Violand, and L. A. Bentle, *Proc. Natl. Acad. Sci. USA*, **84**, 6434–6437 (1987).

63. G. Fuh, M. G. Mulkerrin, S. Bass, N. McFarland, M. Brochier, J. H. Bourell, D. R. Light, and J. A. Wells, *J. Biol. Chem.*, **265**, 3111–3115 (1990).

64. B. C. Cunningham, D. J. Henner, and J. A. Wells, *Science*, **247**, 1461–1465 (1990).

65. B. C. Cunningham, M. Ultsch, A. M. D. Vos, M. G. Mulkerrin, K. R. Klausner, and J. A. Wells, *Science*, **254**, 821–825 (1991).

66. A. M. D. Vos, M. Ultsch, and A. A. Kossiakoff, *Science*, **255**, 306–312 (1992).

67. J. A. Wells, B. C. Cunningham, G. Fuh, H. B. Lowman, S. H. Bass, M. G. Mulkerrin, M. Ultsch, and A. M. D. Vos, *Recent Prog. Hormone Res.*, **48**, 253–275 (1992).

68. H. C. Nau, *Science*, **257**, 1064–1073 (1992).

69. T. Palzkill and D. Botstein, *Prot. Struct. Function Genet.*, **14**, 29–44 (1992).

70a. J. Ellman, D. Mendel, S. Anthony-Cahill, C. J. Noren, and P. G. Schultz, *Methods Enzymol.*, **202**, 301–336 (1991).

70b. J. S. Oliver and A. Oyelere, *J. Org. Chem.*, **61**, 4168–4171 (1996).

70c. J. S. Oliver and A. Oyelere, *Tett Lett.*, **38**, 4005–4008 (1997).

70d. Y. Shimizu, A. Inoue, Y. Tomari, T. Suzuki, T. Yokogawa, K. Nishikawa, and T. Ueda, *Nature Biotech*, **19**, 751–755 (2001).

70e. C. J. Noren, S. J. Anthony-Cahill, M. C. Griffith, and P. G. Schultz, *Science*, **244**, 182–188 (1989).

71. J. A. Ellman, D. Mendel, and P. G. Schultz, *Science*, **255**, 197–200 (1992).

72. D. Mendel, J. A. Ellman, Z. Chang, P. A. Kollman, and P. G. Schultz, *Science*, **256**, 1798–1802 (1992).

73. R. M. Burch and D. J. Kyle, *Pharm. Res.*, **8**, 141–147 (1991).

74. M. D. Walkinshaw, *Med. Res. Rev.*, **12**, 317–372 (1992).

75. W. H. Moos, G. D. Green, and M. R. Pavia, *Annu. Rep. Med. Chem.*, **28**, 315–324 (1993).

76. L. Fellows in J. D. Coombes, Ed., *New Drugs from Natural Sources*, IBC Technical Services Ltd., London, 1992, pp. 93–100.

77. H. C. Krebs, *Prog. Chem. Org. Nat. Prod.*, **49**, 151–363 (1986).

78. H. Geysen, R. Meloen, and S. Barteling, *Proc. Natl. Acad. Sci. USA*, **81**, 3998–4002 (1984).

79. S. P. Fodor, J. L. Read, M. C. Pirrung, L. Stryer, A. T. Lu, and D. Solas, *Science*, **251**, 767–773 (1991).

80. K. S. Lam, S. E. Salmon, E. M. Hersch, V. J. Hruby, W. M. Kazmierski, and R. J. Knapp, *Nature*, **334**, 82–84 (1991).

81. J. K. Scott, *Trends Biochem. Sci.*, **17**, 241–245 (1992).

82. R. A. Houghten, C. Pinilla, S. E. Blondelle, J. R. Appel, C. T. Dooley, and J. H. Cuervo, *Nature*, **354**, 84–88 (1991).

83. R. J. Simon, et al., *Proc. Natl. Acad. Sci. USA*, **89**, 9367–9371 (1992).

84. M. E. Wolff and A. McPherson, *Nature*, **345**, 365–366 (1990).

85. J. D. Marks, H. R. Hoogenboom, A. D. Gnffiths, and G. Winter, *J. Biol. Chem.*, **267**, 16007–16010 (1992).

86. L. C. Bock, L. C. Griffin, J. A. Latham, E. H. Vermaas, and J. J. Toole, *Nature*, **355**, 564–566 (1992).

87. A. D. Ellington and J. W. Szostak, *Nature*, **346**, 818–822 (1990).

88. C. Tuerk and L. Gold, *Science*, **249**, 505–510 (1990).

89. S. Brenner and R. A. Lerner *Proc. Natl. Acad. Sci. USA*, **89**, 5381–5383 (1992).

90. W. C. Greene, *Annu. Rev. Immunol.*, **8**, 453–475 (1990).

91. L. T. Batcheler, L. L. Strehl, R. H. Neubauer, S. R. Petteway, and B. Q. Ferguson, *AIDS Res. Hum. Retroviruses*, **5**, 275–278 (1989).

92. J. M. Hasler, T. F. Weighous, T. W. Pitts, D. B. Evans, S. K. Sharma, and W. G. Tarpley, *AIDS Res. Hum. Retroviruses*, **5**, 507–516 (1989).

93. M.-C. Hsu, A. D. Schutt, L. W. Slice, M. I. Sherman, D. D. Richrnan, M. J. Potash, and D. J. Volsky, *Biochem. Soc. Trans.*, **20**, 525–531 (1992).

94. M. Steinmetz in *Rapid Functional Screening for Drug Development*, IBC Press, Southborough, MA, 1992.

95. H. A. Erlich, D. Gelfand, and J. J. Sninsky, *Science*, **252**, 1643–1651 (1991).

96. A. H. Marderosian, *Natural Product Medicine—A Scientific Guide to Foods, Drugs and Cosmetics,* Stickley, G. F. Co., Philadelphia, PA, 1988.

97. M. Chartrain, P. M. Salmon, D. K. Robinson, and B. C. Buckland, *Curr. Opin. Biotech.,* **11**, 209–214 (2000).

98. J. Neilsen, *Appl. Microbiol. Biotech.,* **55**, 263–283 (2001).

99. L. Rohlin, M. K. Oh, and J. C. Liao, *Curr. Opin. Microbiol.,* **4**, 330–335 (2001).

100. R. H. Baltz, *Trends Microbiol.,* **6**, 76–83 (1998).

101. C. Schmidt-Dannert, *Curr. Opin. Biotech.,* **11**, 255–261 (2000).

102. J. T. Kealey, L. Liu, D. V. Santi, M. C. Betlach, and P. J. Barr, *Proc. Natl. Acad. Sci. USA,* **95**, 505–509 (1998).

103. M. Albrecht, N. Misawa, and G. Sandmann, *Biotech. Lett.,* **21**, 791–795 (1999).

104. B. A. Pfeifer, S. J. Admiraal, H. Gramajo, D. E. Cane, and C. Khosla, *Science,* **291**, 1790–1792 (2001).

105. R. McDaniel, S. Ebert-Khosla, D. A. Hopwood, and C. Khosla, *Nature,* **375**, 549–554 (1995).

106. Q. Xue, G. Ashley, C. R. Hutchinson, and D. V. Santi, *Proc. Natl. Acad. Sci. USA,* **96**, 11740–11745 (1999).

107. R. McDaniel, A. Thamchaipenet, C. Gustafsson, H. Fu, M. Betlach, and G. Ashley, *Proc. Natl. Acad. Sci. USA,* **96**, 1846–1851 (1999).

108. D. E. Cane, C. T. Walsh, and C. Khosla, *Science,* **282**, 63–68 (1998).

109. S. E. Osborne, I. Matsumura, and A. D. Ellington, *Curr. Opin. Chem. Biol.* **1**, 5–9, (1997).

110. M. Famulok and A. Jenne, *Curr. Opin. Chem. Biol.,* **2**, 320–327 (1998).

111. S. D. Jayasena, *Clin. Chem.,* **45**, 1628–1650 (1999).

112. D. J. Patel and A. K. Suri, *J. Biotechnol.,* **74**, 39–60 (2000).

113. S. Sun, *Curr. Opin. Mol. Ther.,* **2**, 100–105 (2000).

114. L. S. Green, D. Jellinek, C. Bell, L. A. Beebe, B. D. Feistner, S. C. Gill, F. M. Jucker, and N. Janjic, *Chem. Biol.,* **2**, 683–695 (1995).

115. R. Conrad, L. M. Keranen, A. D. Ellington, and A. C. Newton, *J. Biol. Chem.,* **269**, 32051–32054 (1994).

116. R. D. Jenison, S. C. Gill, A. Pardi, and B. Polisky, *Science,* **263**, 1425–1429 (1994).

117. S. S. Sidhu, *Curr. Opin. Biotech.,* **11**, 610–616 (2000).

118. R. H. Hoess, *Chem. Rev.,* **101**, 3205–3218 (2001).

119. R. Hyde-DeRuyscher, et al., *Chem. Biol.,* **7**, 17–25 (2000).

120. M. S. Dennis, M. Roberge, C. Quan, and R. A. Lazarus, *Biochemistry,* **40**, 9513–9521 (2001).

121. K. C. Brown, *Curr. Opin. Chem. Biol.,* **4**, 16–21 (2000).

122. R. Pasqualini and E. Rouslahti, *Nature,* **380**, 364–367 (1996).

123. W. Arap, R. Pasqualini, and E. Rouslahti, *Science,* **279**, 377–380 (1998).

124. D. Larocca, A. Witte, W. Johnson, G. F. Pierce, and A. Baird, *Hum. Gene Ther.,* **9**, 2393–2399 (1998).

125. D. Larocca, M. A. Burg, K. Jensen-Pergakes, E. P. Ravey, A. M. Gonzalez, and A. Baird, *Curr. Pharm. Biotechnol.* **3**, 45–57 (2002).

126. J. W. Erickson and S. W. Fesik, *Annu. Rep. Med. Chem.,* **27**, 271–289 (1992).

127. G. M. Clore, *Science,* **252**, 1390–1399 (1991).

128. S. W. Fesik, *J. Med. Chem.,* **34**, 2937–2945 (1991).

129. B. Shaanan, A. M. Gronenborn, G. H. Cohen, G. L. Gilliland, B. Veerapandian, D. R. Davies, and G. M. Clore, *Science,* **257**, 961–964 (1992).

130. I. D. Kuntz, *Science,* **257**, 1078–1082 (1992).

131a. J. S. Mason, A. C. Good, and E. J. Martin, *Curr Pharm Des.,* **7**, 567–597 (2001).

131b. R. R. Rando, *Pharmacol. Rev.,* **36**, 111–142 (1984).

132. P. J. Parker, et al., *Science,* **233**, 853–859 (1986).

133. Y. Nishizuka, *Science,* **258**, 607–614 (1992).

134. C. D. Nicholson, R. A. J. Challiss, and M. Shahid, *Trends Pharmacol. Sci.,* **12**, 19–27 (1991).

135a. I. C. Crocker and R. G. Townley, *Drugs Today,* **35**, 519–535 (1999).

135b. C. G. Stief, *Drugs Today,* **36**, 93–99 (2000).

135c. R. M. Kramer, C. Hession, B. Johansen, G. Hayes, P. McGray, E. P. Chow, R. Tizard, and R. B. Pepinsky, *J. Biol. Chem.,* **264**, 5678–5775 (1989).

136. J. L. Seilhamer, W. Pruzanski, P. Vadas, S. Plant, J. A. Miller, J. Kloss, and L. K. Johnson, *J. Biol. Chem.,* **264**, 5335–5338 (1989).

137a. M. Lehr, *Exp. Opin. Thera. Patents,* **11**, 1123–1136 (2001).

137b. E. A. Capper and L. A. Marshall, *Prog. Lipid Res.,* **40**, 167–197 (2001).

137c. M. Murakami and I. Kudo, *J. Biochem. (Tokyo),* **131**, 285–292 (2002).

137d. D. A. Matthews, et al., *J. Biol. Chem.*, **260**, 381–391 (1985).

138. D. A. Matthews, J. T. Bolin, J. M. Burridge, D. J. Filman, W. K. Volz, and J. Kraut, *J. Biol. Chem.*, **260**, 381–391 (1985).

139. C. Oefner, A. D'Arcy, and F. Winkler, *Eur. J. Biochem.*, **174**, 377–385 (1988).

140. N. J. Prendergast, J. R. Appleman, T. J. Delchamp, R. L. Blakley, and J. H. Freisham, *Biochemistry*, **28**, 4645–4650 (1989).

141. B. I. Schweitzer, S. Srirnatkandata, H. Gritsman, R. Sheridan, R. Venkataraghavan, and J. Bertino, *J. Biol. Chem.*, **264**, 20786–20795 (1989).

142. K. Appelt, et al., *J. Med. Chem.*, **34**, 1925–1934 (1991).

143a. T. R. Jones, M. D. Varney, S. E. Webber, K. K. Lewis, G. P. Marzoni, C. L. Palmer, V. Kathardekar, K. M. Welsh, S. Webber, D. A. Matthews, K. Appelt, W. W. Smith, C. A. Janson, J. E. Villafranca, R. J. Bacquet, E. F. Howland, C. L. Booth, S. M. Herrmann, R. W. Ward, J. White, E. W. Moomaw, C. A. Bartlett, and C. A. Morse, *J. Med. Chem.*, **39**, 904–917 (1996).

143b. T. J. Stout, D. Tondi, M. Rinaldi, D. Barlocco, P. Pecorari, D. V. Santi, I. D. Kuntz, R. M. Stroud, B. K. Shoichet, and P. M. Costi, *Biochemistry*, **38**, 1607–1617 (1999).

143c. G. Klebe, *J. Mol. Med.*, **78**, 269–281 (2000).

143d. J. R. Huff, *J. Med. Chem.*, **34**, 2305–2314 (1991).

144. M. A. Navia, et al., *Nature*, **337**, 615–620 (1989).

145. A. Wlodawer, et al., *Science*, **245**, 616–621 (1989).

146. P. Lapatto, et al., *Nature*, **342**, 299–302 (1989).

147. I. T. Weber, M. Miller, M. Jaskolski, J. Leis, A. M. Skalka, and A. Wlodawer, *Science*, **143**, 928–931 (1989).

148. M. Miller, et al., *Science*, **246**, 1149–1152 (1989).

149. P. M. D. Fitzgerald, et al., *J. Biol. Chem.*, **265**, 14209–14219 (1990).

150. J. Erickson, et al., *Science*, **249**, 527–533 (1990).

151. A. L. Swain, M. M. Miller, J. Green, D. H. Rich, J. Schneider, S. B. H. Kent, and A. Wlodawer, *Proc. Natl. Acad. Sci. USA*, **87**, 8805–8809 (1990).

152. J. C. Heimbach, V. M. Garsky, S. R. Michaelson, R. A. F. Dixon, I. S. Sigal, and F. L. Darke, *Biochem. Biophys. Res. Commun.*, **164**, 955–960 (1989).

153. T. D. Meek, et al., *Nature*, **343**, 90–92 (1990).

154. T. J. McQuade, A. G. Tomasseili, V. K. Liu, B. Moss, T. K. Sawyer, R. L. Heinrikson, and W. G. Tarpley, *Science*, **247**, 454–456 (1990).

155. N. A. Roberts, et al., *Science*, **248**, 358–361 (1990).

156. W. Greenlee, *Med. Res. Rev.*, **10**, 173–276 (1990).

157. S. L. Schreiber, *Science*, **251**, 283–287 (1991).

158. N. Takahashi, T. Hayano, and M. Suzuki, *Nature*, **337**, 473–475 (1989).

159. G. Fischer, B. Wittmann-Liebold, K. Lang, T. Kiefhaber, and F. X. Schmid, *Nature*, **337**, 476–478 (1989).

160. R. K. Harrison and R. L. Stein, *Biochemistry*, **29**, 3813–3816 (1990).

161. J. Liu, M. W. Albers, C.-M. Chen, S. L. Schreiber, and C. T. Walsh, *Proc. Natl. Acad. Sci. USA*, **87**, 2304–2308 (1990).

162. M. K. Rosen, S. Standaert, A. Galat, M. Nakatsuka, and S. L. Schreiber, *Science*, **248**, 863–866 (1990).

163. K. Wuthrich, B. von Freberg, C. Weber, G. Wider, R. Traber, H. Widmer, and W. Braun, *Science*, **254**, 953–954 (1991).

164. S. Gallion and D. Ringe, *Prot. Eng.*, **5**, 391–397 (1992).

165. J. Liu, J. D. Farmer, W. S. Lane, J. Friedmann, I. Weissman, and S. L. Schreiber, *Cell*, **66**, 807–815 (1991).

166. F. McKeon, *Cell*, **66**, 823–826 (1991).

167a. A. L. Russo, A. C. Passaquin, P. Andre, M. Skutella, and T. U. Ruegg, *Br. J. Pharmacol.*, **118**, 885–892 (1996).

167b. M. Sokolovsky, *Adv. Drug Res.*, **18**, 431–509 (1989).

168. L. Mei, W. R. Roeske, and H. I. Yamamura, *Life Sci.*, **45**, 1831–1851 (1989).

169. E. C. Hulme, N. J. M. Birdsall, and N. J. Buckley, *Annu. Rev. Pharmacol. Toxicol.*, **30**, 633–673 (1990).

170. E. G. Peralta, A. Ashkenazi, J. W. Winslow, D. H. Smith, J. Ramachandran, and D. J. Capon, *EMBO J.*, **6**, 3923–3929 (1987).

171. T. I. Bonner, N. J. Buckley, A. C. Young, and M. R. Brann, *Science*, **237**, 527–532 (1987).

172. T. I. Bonner, A. C. Young, M. R. Brann, and N. J. Buckley, *Neuron*, **1**, 403–410 (1988).

173. E. G. Peralta, J. W. Winslow, A. Ashkenazi, D. H. Smith, J. Ramachandran, and D. J. Capon, *Trends Pharmacol. Sci.*, **9** (Suppl), 6–11 (1988).

174. A. G. Gilman, *Annu. Rev. Biochem.*, **56**, 615–649 (1987).

175. L. Birnbaumer, *Annu. Rev. Pharmacol. Toxicol.*, **30**, 675–705 (1990).

176. H. G. Dohlman, M. G. Caron, and R. J. Lefkowitz, *Biochemistry*, **26**, 2657–2664 (1987).

177. C. Humblet and T. Mirzadegan, *Annu. Rep. Med. Chem.*, **27**, 291–300 (1992).

178. B. F. O'Dowd, M. Hnatowich, J. W. Regan, W. M. Leader, M. G. Caron, and R. J. Lefkowitz, *J. Biol. Chem.*, **263**, 15985–15992 (1988).

179. J. L. Benovic, A. DeBlasi, W. C. Stone, M. G. Caron, and R. L. Lefkowitz, *Science*, **246**, 235–240 (1989).

180. M. J. Lohse, R. J. Lefkowitz, M. G. Caron, J. L. Benovic, *Proc. Natl. Acad. Sci. USA*, **86**, 3011–3015 (1989).

181. R. B. Clark, J. Friedman, R. A. F. Dixon, and C. D. Strader, *Molec. Pharmacol.*, **36**, 343–348 (1989).

182. C. M. Fraser, *J. Biol. Chem.*, **264**, 9266–9270 (1989).

183. S. K.-F. Wong, E. M. Parker, and E. M. Ross, *J. Biol. Chem.*, **265**, 6219–6224 (1990).

184. M. I. Simon, M. P. Strathmann, and N. Gautam, *Science*, **252**, 802–808 (1991).

185. P. Bovy, *Med. Res. Rev.*, **10**, 115–142 (1990).

186. S. Schultz, P. S. T. Yuen, and D. L. Garbers, *Trends Pharmacol. Sci.*, **12**, 166–120 (1991).

187. D. G. Lowe, M.-S. Chang, R. Hellmiss, E. Cheo, S. Singh, D. G. Garbers, and D. V. Goeddel, *EMBO J.*, **8**, 1377–1384 (1989).

188. M. B. Anand-Srivastava, M. R. Sairam, and M. Cantin, *J. Biol. Chem.*, **265**, 8566–8572 (1990).

189. M.-S. Chang, D. G. Lowe, M. Lewis, R. Hellmiss, E. Chen, and D. V. Goeddel, *Nature*, **341**, 68–72 (1989).

190. P. R. Bovy, et al., *J. Biol. Chem.*, **264**, 20309–20313 (1989).

191. Y. Kambayashi, S. Nakajima, M. Ueda, and K. Inouye, *FEBS Lett.*, **248**, 28–34 (1989).

192. S. Schultz, C. K. Green, P. S. T. Yuen, and D. L. Garbers, *Cell*, **63**, 941–948 (1990).

193. M. G. Currie, K. F. Fok, J. Kato, R. J. Moore, F. K. Hamra, K. L. Duffin, and C. E. Smith, *Proc. Natl. Acad. Sci. USA*, **89**, 947–951 (1992).

194. F. J. deSauvage, R. Horuk, G. Bennett, C. Quan, J. P. Burnier, and D. V. Goeddel, *J. Biol. Chem.*, **267**, 6429–6482 (1992).

195. A. C. Chao, F. J. deSauvage, Y. J. Dong, J. A. Wagner, D. V. Goeddel, and P. Gardner, *EMBO J.*, **13**, 1065–1072 (1994).

196. A. Abbott, *Trends Pharmacol. Sci.*, **13** (Suppl), 169 (1992).

197. A. Ullrich and J. Schlessinger, *Cell*, **61**, 203–212 (1990).

198. L. Tartaglia and D. V. Goeddel, *Immunol. Today*, **13**, 151–153 (1992).

199. R. Sprengel, P. Werner, P. H. Seeburg, A. G. Mukhin, M. R. Santi, D. R. Grayson, A. Guidotti, and K. E. Krueger, *J. Biol. Chem.*, **264**, 20415–20421 (1989).

200. D. B. Pritchett, H. Luddens, and P. H. Seeburg, *Science*, **245**, 1389–1392 (1989).

201. W. Sieghart, *Trends Pharmacol. Sci.*, **10**, 407–411 (1989).

202. R. W. Olsen and A. J. Tobin, *FASEB J.*, **4**, 1469–1480 (1990).

203. V. Jung, C. Jones, C. S. Kumar, S. Stefanos, S. O'Connell, and S. Peska, *J. Biol. Chem.*, **265**, 1827–1830 (1990).

204. T. Furuichi, S. Yoshikawa, A. Miyawaki, K. Wada, N. Maeda, and K. Mikoshiba, *Nature*, **342**, 32–38 (1989).

205. M. Hollmann, A. O'Shea-Greenfield, S. W. Rogers, and S. Heinemann, *Nature*, **342**, 43–48 (1989).

206. R. Sprengel, T. Braun, K. Nikolics, D. L. Segaloff, and P. Seeberg, *Mol. Endocrinol.*, **4**, 525–530 (1990).

207. R. M. Evans, *Science*, **240**, 889–895 (1988).

208. P. J. Godowski and D. Picard, *Biochem. Pharmacol.*, **38**, 3135–3143 (1989).

209. R. F. Power, O. M. Conneely, and B. W. O'Malley, *Trends Pharmacol. Sci.*, **13**, 318–323 (1992).

210. D. P. McDonnell, B. Clevenger, S. Dana, D. Santiso-Mere, M. T. Tzukerman, and M. A. Gleeson, *J. Clin. Pharmacol.*, **33**, 1165–1172 (1993).

211. T. Kishimoto, S. Akira, and T. Taga, *Science*, **258**, 593–597 (1992).

212. S. Nakanishi, *Science*, **258**, 597–603 (1992).

213. P. J. M. Galen, G. L. Stiles, G. Michaels, and K. A. Johnson, *Med. Res. Rev.*, **12**, 423–471 (1992).

214. M. Williams, *Med. Res. Rev.*, **11**, 147–184 (1991).

215. D. Oksenberg, S. A. Marsters, B. F. O'Dowd, H. Jin, S. Havlik, S. J. Peroutka, and A. Ashkenazi, *Nature*, **360**, 161–163 (1992).

216. B. Benhamou, T. Garcia, T. Lerouge, A. Vergezac, D. Gofflo, C. Bigogne, P. Chambon, and H. Gronemeyer, *Science*, **255**, 206–209 (1992).

217. R. O. Hynes, *Cell*, **69**, 11–25 (1992).

218. T. A. Springer, *Nature*, **346**, 425–434 (1990).

219. S. D. Marlin, D. E. Staunton, T. E. Springer, C. Stratowa, W. Sommergruber, and V. J. Merluzzi, *Nature*, **344**, 70–72 (1990).

220. E. Ruoslahti and M. D. Pierchbaucher, *Science*, **238**, 491–497 (1987).

221. D. R. Phillips, I. F. Charo, and R. M. Scarborough, *Cell*, **65**, 359–362 (1991).

222. S. C. Bodary, M. A. Napier, and J. W. McLean, *J. Biol. Chem.*, **264**, 18859–18862 (1989).

223. J. A. Jakubowski, G. F. Smith, and D. J. Sall, *Annu. Rep. Med. Chem.*, **27**, 99–108 (1993).

224. B. K. Blackburn and T. R. Gadek, *Annu. Rep. Med. Chem.*, **28**, 79–88 (1993).

225. L. Lasky, *Science*, **258**, 964–969 (1992).

226. T. A. Yednock and S. D. Rosen, *Adv. Immunol.*, **44**, 313–378 (1989).

227. L. A. Lasky, et al., *Cell*, **56**, 1045–1055 (1989).

228. K. A. Karlsson, *Trends Phamacol. Sci.*, **12**, 265–272 (1991).

229. J. H. Musser, *Annu. Rep. Med. Chem.*, **27**, 301–310 (1992).

230. W. I. Weis, K. Drickamer, and W. A. Hendrickson, *Nature*, **360**, 127–134 (1992).

231. D. V. Erbe, et al., *J. Cell Biol.*, **119**, 215–227 (1992).

232. L. Osborn, *Cell*, **62**, 3–6 (1990).

233. T. Maniatis, S. Goodbourn, and J. A. Fischer, *Science*, **236**, 1237–1245 (1987).

234. A. D. Frankel and P. S. Kim, *Cell*, **65**, 717–719 (1991).

235. R. G. Shea and J. F. Milligan, *Annu. Rep. Med. Chem.*, **27**, 311–320 (1992).

236. A. L. Catapano, *Pharmacol. Ther.*, **43**, 187–219 (1989).

237. W. J. Schneider, *Biochim. Biophys Acta.*, **988**, 303–317 (1989).

238. G. Utermann, *Science*, **246**, 904–910 (1989).

239. M. D. Matteucci and N. Bischofberger, *Annu. Rep. Med. Chem.*, **26**, 287–296 (1991).

240a. D. A. Jabs and P. D. Griffiths, *Am. J. Ophthalmol.*, **133**, 552–556 (2002).

240b. D. Banerjee, *Curr. Opin. Mol. Ther.*, **1**, 404–408 (1999).

240c. J. D. Coombes and M. Evans, *Annu. Rep. Med. Chem.*, **22**, 207–211 (1987).

241. J. N. Snouwaert, K. K. Brigman, A. M. Latour, N. N. Malouf, R. C. Bouchere, O. Smithies, and B. H. Koller, *Science*, **257**, 1083–1088 (1992).

242a. I. H. G. S. Consortium, *Nature*, **409**, 860–921 (2001).

242b. J. C. Venter, et al., *Science*, **291**, 1304–1351 (2001).

242c. L. H. Hurley, *J. Med. Chem.* **30**, 7A–8A (1987).

243. F. J. Zeelen, *Trends Pharmcol. Sci.*, **10**, 472 (1989).

244. A. Kornberg, *Biochemistry*, **26**, 6888–6891 (1987).

245. R. H. Hirschmann, *Angew Chem. Int. Edu. Engl.* **30**, 1278–1301 (1992).

246. S. L. Schreiber, *Chem. Eng. News*, **70**, 22–32 (1992).

247. P. Knight, *Biotechnology*, **8**, 105–107 (1990).

CHAPTER FIVE

Oligonucleotide Therapeutics

STANLEY T. CROOKE
Isis Pharmaceuticals, Inc.
Carlsbad, California

Contents

Burger's Medicinal Chemistry and Drug Discovery
Sixth Edition, Volume 2: Drug Development
Edited by Donald J. Abraham
ISBN 0-471-37028-2 © 2003 John Wiley & Sons, Inc.

1   INTRODUCTION

During the past decade, antisense technology has matured. Today, antisense technology is generally accepted as a broadly useful method for gene functionalization and target validation. Further, with Vitravene's approval by regulatory agencies around the world (making it the first antisense drug to be commercialized), the emerging data showing the activity of a number of antisense drugs in clinical trials, and the overwhelming evidence from studies in animals, the potential of antisense as a therapeutic technology is now better appreciated.

The purposes of this review are to provide a summary of the progress in the technology, to address its role in gene functionalization and target validation as well as therapeutics, and to consider the limitations of the technology and a few of the many questions that remain to be answered.

1.1   Definition

Antisense technology exploits oligonucleotide analogs (typically 15–20 nucleotides) to bind to cognate RNA sequences through Watson-Crick hybridization, resulting in the destruction or disablement of the target RNA. Thus antisense technology represents a "new pharmacology." The receptor, messenger RNA (mRNA), has never before been considered in the context of drug-receptor interactions. Before the advent of antisense technology, no medicinal chemistry had been practiced on the putative "drugs," oligonucleotides. The basis of the drug-receptor interaction, Watson-Crick hybridization, had never been considered as a potential binding event for drugs and put into a pharmacological context. Finally, postbinding events such as recruitment of nucleases to degrade the receptor RNA had never been considered from a pharmacological perspective.

A key to understanding antisense technology is to consider it in a pharmacological context. It is essential to understand the structure, function, and metabolism of the receptors for these drugs. As with any of the class of drugs, it is essential to consider the effects of antisense oligonucleotides in the context of dose-response curves. It is essential to consider the future in the context of advances in antisense biology and medicinal chemistry that result in improved pharmacological behaviors.

2 HISTORY

2.1 Polynucleotides

Before the evolution of effective transfection methods and an understanding of molecular biological techniques, DNA and RNA were administered as potential therapeutic agents. For example, DNA from several sources displayed antitumor activity and the activity was reported to vary as a function of size, base composition, and secondary structure (1–3). However, the molecular mechanisms by which DNA might induce antitumor effects were never defined and numerous other studies failed to demonstrate antitumor activities with DNA (4).

In contrast to studies on DNA as a therapeutic agent, substantially more work has been reported on RNA and polyribonucleotides. Much of the effort focused on the ability of polynucleotides to induce interferon (5) and the most thoroughly studied polynucleotide in this regard is double-stranded polyribonosine:polyribonocytidine (poly rI:poly rC) (6). Poly rI:poly rC was shown to have potent antiviral and antitumor activities *in vitro* and *in vivo* and these activities were generally correlated with interferon induction (6).

Although poly rI:poly rC was shown to have antiviral and antitumor activities in animals, the compound produced substantial toxicities in animals and humans that limited its utility. For example, in mice, anti-RNA antibodies associated with glomerular nephritis and central nervous system (CNS) toxicities were prominent. In rabbits, fever was dose limiting (for review, see Ref. 7). In rats, poly rI:poly rC was lethal at low doses and in mammals, including humans, poly rI:poly rC was immunotoxic, resulting in fever, hypotension, antibody production, and T-cell activation. Pain, platelet depletion, and convulsions were also reported (7–15).

Methods to stabilize, enhance cellular uptake, and alter the *in vivo* pharmacokinetic properties of poly rI:poly rC were extensively studied. Polycationic substances such as DEAE-dextran (diethylaminoethyl dextran), poly L-lysine, and histones were shown to complex with poly rI:poly rC and other polynucleotides and alter all of these properties (7). Poly rI:poly rC, complexed with polylysine in the presence of carboxymethyl cellulose, was studied in patients with cancer and found to be very toxic (12, 13).

In short, poly rI:poly rC and other polynucleotides designed to induce interferon failed to demonstrate substantial antiviral or anticancer activities at doses that did not produce unacceptable toxicities. The mechanisms resulting in the toxicities are still not clearly understood. Complex formation with polycations altered pharmacokinetic properties but did not enhance efficacy and probably exacerbated toxicities.

A second polynucleotide that has been studied extensively is ampligen, a mismatched poly rI:poly rC 12U resulting from mispairing of the duplex (16). This compound has been shown to induce interferon and to activate 2′-5′-adenosine synthetase (17). Ampligen has properties similar to those of poly rI:poly rC, although it has been reported to have broader activities and lower toxicities and is still in development.

Although the initial efforts that focused on polynucleotides that stimulate a variety of immunological events are not directly applicable to the more recent focus on specific effects of oligonucleotides, they have provided a base of toxicological information on the effects of polyanionic compounds and guidance in design of toxicological studies.

2.2 The Antisense Concept

Clearly, the antisense concept derives from an understanding of nucleic acid structure and function and depends on Watson-Crick hybridization (18). Thus, arguably, the demonstration that nucleic acid hybridization is feasible (19) and the advances in *in situ* hybridization and diagnostic probe technology (20) lay the most basic elements of the foundation supporting the antisense concept.

However, the first clear enunciation of the concept of exploiting antisense oligonucleotides as therapeutic agents was in the work of Zamecnik and Stephenson in 1978. In this publication, these authors reported the synthesis of an oligodeoxyribonucleotide, 13 nucleotides long, that was complementary to a sequence in the respiratory syncytial virus genome. They suggested that this oligonucleotide could be stabilized by 3'- and 5'-terminal modifications and showed evidence of antiviral activity. More important, they discussed possible sites for binding in RNA and mechanisms of action of oligonucleotides.

Though less precisely focused on the therapeutic potential of antisense oligonucleotides, the work of Miller and Ts'o and their collaborators during the same period helped establish the foundation for antisense research and reestablish an interest in phosphate backbone modifications as approaches to improve the properties of oligonucleotides (21, 22). Their focus on methyl phosphotriester-modified oligonucleotides as a potential medicinal chemical solution to pharmacokinetic limitations of oligonucleotides presaged a good bit of the medicinal chemistry yet to be performed on oligonucleotides.

Despite the observations of Miller and Ts'o and Zamecnik and colleagues, interest in antisense research was quite limited until the late 1980s, when advances in several areas provided technical solutions to a number of impediments. Because antisense drug design requires an understanding of the sequence of the RNA target, the explosive growth in availability of viral and human genomic sequences provided the information from which "receptor sequences" could be selected. The development of methods for synthesis of research quantities of oligonucleotide drugs then supported antisense experiments on both phosphodiester and modified oligonucleotides (23, 24). The inception of the third key component (medicinal chemistry), forming the foundation of oligonucleotide therapeutics, in fact, is the synthesis in 1969 of phosphorothioate poly rI:poly rC as a means of stabilizing the polynu-

cleotide (25). Subsequently, Miller and Ts'o initiated studies on the neutral phosphate analogs, methylphosphonates (21); and groups at the National Institutes of Health, the Food and Drug Administration, and the Worcester Foundation investigated phosphorothioate oligonucleotides (26–29). With these advances forming the foundation for oligonucleotide therapeutics and the initial studies suggesting *in vitro* activities against a number of viral and mammalian targets (28, 30–33), interest in oligonucleotide therapeutics intensified.

2.3 Strategies to Induce Transcriptional Arrest

An alternative to the inhibition of RNA metabolism by way of an antisense mechanism is to inhibit transcription by interacting with double-stranded DNA in chromatin. Of the two most obvious binding strategies for oligonucleotides binding to double-stranded nucleic acids, strand invasion and triple-strand formation, triple-stranding strategies, until recently, attracted essentially all of the attention.

Polynucleotides were reported to form triple helices as early as 1957. Triple strands can form by non-Watson-Crick hydrogen bonds between the third strand and purines involved in Watson-Crick hydrogen bonding with the complementary strand of the duplex (for review, see Ref. 34). Thus, triple-stranded structures can be formed between a third strand composed of pyrimidines or purines that interact with a homopurine strand in a homopurine-homopyrimidine strand in a duplex DNA. With the demonstration that homopyrimidine oligonucleotides could indeed form triplex structures (35–37), interest in triple-strand approaches to inhibit transcription heightened.

Although there was initially considerable debate about the value of triple-stranding strategies vs. antisense approaches (38), there was little debate that much work remained to be done to design oligonucleotides that could form triple-stranded structures with duplexes of mixed sequences. Pursuit of several strategies has resulted in significant progress (for review, see Ref. 34). Considerable progress in creating chemical motifs capable of binding to duplex DNA with high affinity and specificities supportive of binding to sequences other than polypyrimidine polypurine traits has been reported. For example, peptide nucleic acid (PNA) has been shown to form on triple-stranded structures in isolated DNA and in mouse cell chromosomal DNA (39). Modifications such as 7-deazaxanthine and 2′-aminoethoxy were reported to enhance triplex formation (40, 41). Additionally, 5′-propionyl-modified nucleosides have been shown to enhance triplex formation (42) (for review, see Ref. 43).

In addition to advances in the chemistry of triplex formation, a number of studies have shown results in cells consistent with triplex formation in chromosomal DNA (39, 44–46). Peptide nucleic acids have been shown to be of particular value for triplex interactions because of the relatively high affinity of this modification. They have been shown to inhibit transcription initiation and elongation, to block DNA polymerases (47–49), and to inhibit binary of a number of proteins to DNA (50). Dimeric PNAs have been created that are reported to form PNA/DNA/PNA triplexes and these have been shown to induce gene-targeted mutations in streptolysin-o permeabilized cells (39).

Strand invasion, an alternative approach to obstructing transcription by formation of triple strands, has been shown to be feasible if analogs with sufficient affinity can be synthesized. PNAs have been shown to have very high affinity and be capable of strand invasion of double-stranded DNA under some conditions (51). Additionally, progress in developing sequence-specific minor groove binders has been reported (52) (for review, see Ref. 53). Thus, progress in developing the basic tools with which to evaluate the potential of sequence-specific interactions with DNA has been reported, although much remains to be done.

2.4 Ribozymes

Ribozymes are RNA molecules that catalyze biochemical reactions (54). Ribozymes cleave single-stranded regions in RNA through transesterification or hydrolysis reactions that result in cleavage of phosphodiester bonds (55). To date, several RNA catalytic motifs, group I introns, RNase P, and both hammerhead and hairpin ribozymes, for example, have been identified (56). To achieve potential therapeutic utility, two approaches have been taken. Ribozyme-forming sequences have been incorporated into plasmids and administered, in effect, as ribozyme gene therapy (e.g., see Ref. 57).

In either case, the object is to take advantage of the specificity of hybridization-based interactions and couple that to improved potency that might derive from the ability of the ribozyme to cleave the target RNA. Thus, the value of a ribozyme relative to that of an antisense inhibitor that recruits a cellular enzyme such as RNase H or a double-strand RNase to cleave the target an RNA would be defined as the difference in either specificity or potency achieved by the ribozyme compared to that of the antisense inhibitor vs. the costs and limitations imposed by the structural requirements to effect ribozymic activity. To date, data that address this issue are limited. A comparison of a ribozyme and a phosphorothioate oligodeoxynucleotide to the 5′-transactivator region (TAR) of HIV showed a slight improvement in activity, but thorough comparative dose-response curves were not reported (58). To date no comparative data from *in vivo* experiments have been reported. Perhaps, most important, as discussed later, RNase H has proved to be a remarkably robust mechanism and double-stranded RNase activation, splicing inhibition, and other non-ribozyme antisense-based mechanisms are emerging as alternatives.

Substantial progress has also been reported with regard to the synthesis and testing of nuclease-resistant ribozyme drugs. Modifications including phosphorothioates and nucleoside analogs have been demonstrated to be incorporable in many sites in hammerhead ribozymes, to increase nuclease resistance and support retained ribozyme activity (59–61). In fact, modified relatively nuclease-resistant ribozymes were reported to decrease the target, stromelysin, mRNA levels in knee joints of rabbits after intra-articular injection (62). Further, the pharmacokinetics of a relatively nuclease-stable hammerhead ribozyme were determined after intravenous (i.v.), subcutaneous (s.c.), or intraperitoneal (i.p.) administration to mice. The ribozyme was well absorbed after i.p. or s.c. dosing; distributed to liver, kidney, bone marrow, and other tissues; and displayed an elimination half-life of 33 min (63). This ribozyme designed to inhibit vascular endothelial growth factor (VEGF) receptor synthesis has been reported to be in clinical trials. Additionally, a ribozyme designed to inhibit hepatitis C virus replication is in early clinical trials.

Ribozyme and antisense gene therapy has been used fairly widely *in vitro* to determine the biological roles of various genes (for review, see Ref. 56). In addition to challenges associated with gene therapy, identifying optimal sites in target RNAs for ribozyme binding and the colocalization of the antisense transcript and the target RNA are issues of concern about which progress has been reported (for review, see Ref. 56).

Thus, substantial progress has been reported with regard to both ribozyme gene therapy and synthetic ribozyme drug therapy. Nevertheless, substantial hurdles remain before sufficient data derived from animal and human studies with multiple ribozymes define the potential of the approach. Clearly, for ribozyme gene therapy to be broadly applicable, the challenges gene therapy faces must be met and ribozymes must be shown to be of more value than simply expressing antisense genes. To validate synthetic ribozymes, data in animals and humans for numerous ribozymes, with careful evaluation of mechanism of action, must be generated. Clearly, the 30-min elimination half-life must be lengthened with new modifications and again the value of a ribozyme versus a much simpler antisense approach must be defined.

2.5 Combinatorial Approaches to Oligonucleotide Therapeutics

At least two methods by which oligonucleotides can be created combinatorially have been published (64–66). The potential advantage of a combinatorial approach is that oligonucleotide-based molecules can be prepared to adopt various structures that support binding to nonnucleic acid targets as well as nucleic acid targets. These can then be screened for potential activities without knowledge about the cause of the disease or the structure of the target.

2.6 The Medicinal Chemistry of Oligonucleotides

Because it was apparent almost immediately that native phosphodiester oligodeoxy- or ribonucleotides are unsatisfactory as drugs because of rapid degradation (67), a variety of modifications were rapidly tested. As previously mentioned, perhaps the most interesting of the initial modifications were the phosphate analogs, the phosphorothioates (68) and the methylphosphonates (21). Both fully modified oligonucleotides and oligonucleotides "capped" at the 3'- and/or 5'-termini with phosphorothioate or methylphosphonate moieties were tested (69). However, studies from many laboratories demonstrated that capped oligonucleotides were relatively rapidly degraded in cells (70–72). Nor were point modifications with intercalators that enhanced binding to RNA (73, 74), cholesterol (75), or poly L-lysine (76, 77) sufficiently active or selective to warrant broad-based exploration.

Since the initial approaches to modifications of oligonucleotides, an enormous range of modifications, including novel bases, sugars, backbones, conjugates, and chimeric oligonucleotides have been tested (for review, see Ref. 78). Many of these modifications have proved to be quite useful and are progressing in testing leading to clinical trials. 2'-O-(2-Methoxyethyl) chimeric antisense inhibitors are now in clinical trials.

3 PROOF OF MECHANISM

3.1 Factors That May Influence Experimental Interpretations

Clearly, the ultimate biological effect of an oligonucleotide will be influenced by the local concentration of the oligonucleotide at the target RNA, the concentration of the RNA, the rates of synthesis and degradation of the RNA, type of terminating mechanism, and the rates of the events that result in termination of the RNA's activity. At present, we understand essentially nothing about the interplay of these factors.

3.1.1 Oligonucleotide Purity. Currently, phosphorothioate oligonucleotides can be prepared consistently and with excellent purity (79). However, this has been the case for only the past several years. Before that time, synthetic methods were evolving and analytical methods were inadequate. In fact, our laboratory reported that different synthetic and purification procedures resulted in oligonucleotides that varied in cellular toxicity (72) and that potency varied from batch to batch. Although there are no longer synthetic problems with phosphorothioates, undoubtedly they complicated earlier studies. More important, with each new analog class, new synthetic, purification, and analytical challenges are encountered.

3.1.2 Oligonucleotide Structure. Antisense oligonucleotides are designed to be single stranded. We now understand that certain sequences (e.g., stretches of guanosine residues) are prone to adopt more complex structures (80). The potential to form secondary and tertiary structures also varies as a function of the chemical class. For example, higher affinity 2'-modified oligonucleotides have a greater tendency to self-hybridize, resulting in more stable oligonucleotide duplexes than would be expected based on rules derived from work with oligodeoxynucleotides (Freier, unpublished results, 1990).

3.1.3 RNA Structure. RNA is structured. The structure of the RNA has a profound influence on the affinity of the oligonucleotide and on the rate of binding of the oligonucleotide to its RNA target (81, 82). Moreover, RNA structure produces asymmetrical binding sites that then result in very divergent affinity constants, depending on the position of oligonucleotide in that structure (82–84). This in turn influences the optimal length of an oligonucleotide needed to achieve maximal affinity because in structured RNA the optimal affinity is determined by the difference between the binding energies required for a nucleotide to invade a duplex and those gained per nucleotide by binding of the oligonucleotide. Furthermore, this is only a fraction of the story, given the numerous proteins that interact with RNA that undoubtedly influence binding, and very little is understood about these ternary interactions.

3.1.4 Variations in *In Vitro* Cellular Uptake and Distribution. Studies in several laboratories have clearly demonstrated that cells in tissue culture may take up phosphorothioate oligonucleotides through an active process, and that the uptake of these oligonucleotides is highly variable, depending on many conditions (72, 85). Cell type has a dramatic effect on total uptake, kinetics of uptake, and pattern of subcellular distribution. At present, there is no unifying hypothesis to explain these differences. Tissue culture conditions, such as the type of medium, degree of confluence, and the presence of serum, can all have enormous effects on uptake (85). The oligonucleotide chemical class obviously influences the characteristics of uptake as well as the mechanism of uptake. Within the phosphorothioate class of oligonucleotides, uptake varies as a function of length, but not linearly. Uptake varies as a function of sequence, and stability in cells is also influenced by sequence (85, 86).

Given the foregoing, it is obvious that conclusions about *in vitro* uptake must be very carefully made and generalizations are virtually impossible. Thus, before an oligonucleotide could be said to be inactive *in vitro*, it should be studied in several cell lines. Furthermore, although it may be absolutely correct that receptor-mediated endocytosis is a mechanism of uptake of phosphorothioate oligonucleotides (87), it is obvious that a generalization that all phosphorothioates are taken up by all cells *in vitro* primarily by receptor-mediated endocytosis is simply unwarranted.

Finally, extrapolations from *in vitro* uptake studies to predictions about *in vivo* pharmacokinetic behavior are entirely inappropriate and, in fact, there are now several lines of evidence in animals and humans that, even after careful consideration of all *in vitro* uptake data, one cannot predict *in vivo* pharmacokinetics of the compounds (85, 88–90).

3.1.5 Binding and Effects of Binding to Nonnucleic Acid Targets. Phosphorothioate oligonucleotides tend to bind to many proteins and those interactions are influenced by many factors. The effects of binding can influence cell uptake, distribution, metabolism, and excretion. They may induce non-antisense effects that can be mistakenly interpreted as antisense or complicate the evaluation of whether the pharmaceutical effect is the consequence of an antisense mechanism. By inhibiting RNase H, protein binding may inhibit the antisense activity of some oligonucleotides. Finally, binding to proteins can certainly have toxicological consequences.

Oligonucleotides may interact not only with proteins but also with other biological molecules, such as lipids or carbohydrates, and such interactions like those with proteins will be influenced by the chemical class of oligonucleotide studied. Unfortunately, essentially no data bearing on such interactions are currently available.

An especially complicated experimental situation is encountered in many *in vitro* antiviral assays. In these assays, high concentrations of drugs, viruses, and cells are often coincubated. The sensitivity of each virus to non-antisense effects of oligonucleotides varies, depending on the nature of the virion proteins and the characteristics of the oligonucleotides (91, 92). This has resulted in considerable confusion. In particular, for HIV, herpes simplex viruses, cytomegaloviruses, and influenza virus, the non-antisense effects have been so dominant that identifying oligonucleotides that work through an antisense mechanism has been difficult. Given the artificial character of such assays, it is difficult to know whether non-antisense mechanisms would be as dominant *in vivo* or result in antiviral activity.

3.1.6 Terminating Mechanisms. It has been amply demonstrated that oligonucleotides may employ several terminating mechanisms. The predominant terminating mechanism is influenced by RNA receptor site, oligonucleotide chemical class, cell type, and probably many other factors (93). Obviously, because variations in terminating mechanism may result in significant changes in antisense potency and studies have shown significant variations from cell type to cell type *in vitro*, it is essential that the terminating mechanism be well understood. Unfortunately, at present, our understanding of terminating mechanisms remains rudimentary.

3.1.7 Effects of "Control Oligonucleotides". A number of types of control oligonucleotides have been used including randomized oligonucleotides. Unfortunately, we know little to nothing about the potential biological effects of such "controls" and the more complicated a biological system and test, the more likely that "control" oligonucleotides may have activities that complicate interpretations. Thus, when a control oligonucleotide displays a surprising activity, the mechanism of that activity should be explored carefully before concluding that the effects of the "control oligonucleotide" prove that the activity of the putative antisense oligonucleotide are not the result of an antisense mechanism.

3.1.8 Kinetics of Effects. Many rate constants may affect the activities of antisense oligonucleotides, such as the rate of synthesis and degradation of the target RNA and its protein; the rates of uptake into cells; the rates of distribution, extrusion, and metabolism of an oligonucleotide in cells; and similar pharmacokinetic considerations in animals. Fortunately, in the past several years, many more careful dose-response and kinetic studies have been reported and in general they demonstrated a relatively slow onset of action and a duration of response consistent with the elimination rates of the drugs tested (see below).

Nevertheless, more careful kinetic studies are required and more rational *in vitro* and *in vivo* dose schedules must be developed.

3.2 Recommendations

3.2.1 Positive Demonstration of Antisense Mechanism and Specificity. Until more is understood about how antisense drugs work, it is essential to positively demonstrate effects consistent with an antisense mechanism. For RNase H-activating oligonucleotides, Northern blot, RT-PCR, RNase protection assays, or transcriptional assay analyses showing selective loss of the target RNA are the ideal choices and many laboratories are publishing reports *in vitro* and *in vivo* of such activities (94–97). Ideally, a demonstration that closely related isotypes are unaffected should be included. In brief, then, for proof of mechanism, the following steps are recommended.

1. Perform careful dose-response curves *in vitro* using several cell lines and methods of *in vitro* delivery.

2. Correlate the rank-order potency *in vivo* with that observed *in vitro* after thorough dose–response curves are generated *in vivo*.

3. Perform careful "gene walks" for all RNA species and oligonucleotide chemical classes.

4. Perform careful time courses before drawing conclusions about potency.

5. Directly demonstrate the proposed mechanism of action by measuring the target RNA and/or protein.

6. Evaluate specificity and therapeutic indices through studies on closely related isotypes and with appropriate toxicological studies.

7. Use RNase H protection assays and transcriptional arrays to provide broader analyses of specificity where the assays have been validated.

8. Perform sufficient pharmacokinetics to define rational dosing schedules for pharmacological studies.

9. When control oligonucleotides display surprising activities, determine the mechanisms involved.

4 MOLECULAR MECHANISMS OF ANTISENSE DRUGS

4.1 Occupancy-Only-Mediated Mechanisms

Classic competitive antagonists are thought to alter biological activities because they bind to receptors, thereby preventing agonists from binding the inducing normal biological processes. Binding of oligonucleotides to specific sequences may inhibit the interaction of the RNA with proteins, other nucleic acids, or other factors required for essential steps in the intermediary metabolism of the RNA or its use by the cell.

To create antisense inhibitors that clearly work through non-RNase H mechanisms, the antisense agents must be modified sufficiently not to support RNase H cleavage. Fortunately, numerous analogs have been identified that do not support RNase H cleavage. These can be classified into modifications of sugar moiety, or the phosphate, or replacement of the sugar-phosphate backbone (98). Unfortunately, in a number of earlier publications, conclusions about mechanisms of action were drawn without using appropriately modified oligonucleotides.

4.1.1 Inhibition of 5'-Capping. Conceptually, inhibition of 5'-capping of mRNA could be an effective antisense mechanism. 5'-Capping is critical in stabilizing mRNA and in enabling the translation of mRNA (99). To date, however, no reports of antisense inhibitors of capping have appeared, which may be attributable to the inaccessibility of the 5'-end of mRNA before capping.

4.1.2 Inhibition of Splicing. A key step in the intermediary metabolism of most mRNA molecules is the excision of introns. These "splicing" reactions are sequence specific and require the concerted action of spliceosomes. Consequently, oligonucleotides that bind to sequences required for splicing may prevent binding of necessary factors or physically prevent the required cleavage reactions. This would then result in inhibition of the production of the mature mRNA. Activities have been reported for anti-c-*myc* and antiviral oligonucleotides with phosphodiester, methylphosphonate, and phosphorothioate backbones (31, 99–101). Kole and colleagues (102–104) were the first to use modified oligonucleotides to inhibit splicing. They showed that 2'-MOE phosphorothioate oligonucleotides could correct aberrant beta-globin splicing in a cell-free system. Similar observations were recorded in a cellular system (104).

In our laboratory, we have attempted to characterize the factors that determine whether splicing inhibition is effected by an antisense drug (105). To this end, a number of luciferase-reporter plasmids containing various introns were constructed and transfected into HeLa cells. Then the effects of antisense drugs designed to bind to various sites were characterized. The effects of RNase H-competent oligonucleotides were compared to those of oligonucleotides that do not serve as RNase H substrates. The major conclusions from this study were, first, that most of the earlier studies in which splicing inhibition was reported were probably the result of nonspecific effects. Second, less effectively spliced introns are better targets than those with strong consensus splicing signals. Third, the 3'-splice site and branchpoint are usually the best sites to which to target the oligonucleotide to inhibit splicing.

Several studies have now demonstrated antisense-mediated redirection of slicing of an endogenous cellular mRNA. In one study, 2'-O-methoxyethyl oligonucleotides redirected splicing of IL-5 receptor pre-mRNA (106). Similar results were reported for antisense agents designed to bind to Bclx pre-mRNA (107, 108). Thus, antisense-mediated alternative splicing is a potentially powerful tool with which to investigate this important source of biological diversity, and to create focused therapies for diseases caused by disorders is splicing.

4.1.3 Translational Arrest.

A mechanism for which many oligonucleotides have been designed is translational arrest by binding to the translation initiation codon. The positioning of the initiation codon within the area of complementarity of the oligonucleotide and the length of oligonucleotide used have varied considerably. Again, unfortunately, in only a relatively few studies have the oligonucleotides been shown to bind to the sites for which they were designed, and other data that support translation arrest as the mechanism have not been reported.

Target RNA species that have been reported to be inhibited include HIV (28), vesicular stomatitis virus (VSV) (76), n-*myc*(109), and a number of normal cellular genes (110–113). However, to demonstrate that RNase H is not involved in effects observed, it is necessary to use antisense drugs that do not form duplexes that are RNase H substrates (e.g., fully modified 2'-oligonucleotides).

Studies with PNA analogs have shown that these analogs can inhibit translation in cell-free systems, but to date no data have been reported from cellular studies (114, 115). For morpholino oligomers, antisense activity has been reported in both cell-free and cellular assays (116, 117). Numerous oligonucleotides with 2'-modifications have also been studied and have been shown to inhibit translation when targeted to 5'-UTR or the translation initiation codon (118). However, optimal inhibition is effected by binding at the 5'-cap in RNAs that have significant 5'-untranslated regions (119). In conclusion, translation arrest represents an important mechanism of action for antisense drugs. A number of examples purporting to employ this mechanism have been reported, and recent studies on several compounds have provided data that unambiguously demonstrate that this mechanism can result in potent antisense drugs.

4.1.4 Disruption of Necessary RNA Structure.

RNA adopts a variety of three-dimensional structures induced by intramolecular hybridization, the most common of which is the stem loop. These structures play crucial roles in a variety of functions. They are used to provide additional stability for RNA and as recognition motifs for a number of proteins, nucleic acids, and ribonucleoproteins that participate in the intermediary metabolism and activities of RNA species. Thus, given the potential general activity of the mechanism, it is surprising that occupancy-based disruption RNA has not been more extensively exploited.

As an example, we designed a series of oligonucleotides that bind to the important stem loop present in all RNA species in HIV, the TAR element. We synthesized a number of oligonucleotides designed to disrupt TAR, showed that several indeed did bind to TAR, disrupt the structure, and inhibit TAR-mediated production of a reporter gene (66). Furthermore, general rules useful in disrupting stem-loop structures were developed as well (84).

Although designed to induce relatively nonspecific cytotoxic effects, two other examples are noteworthy. Oligonucleotides designed to bind to a 17-nucleotide loop in Xenopus 28 S RNA required for ribosome stability and protein synthesis inhibited protein synthesis when injected into Xenopus oocytes (120). Similarly, oligonucleotides designed to bind to highly conserved sequences in 5.8 S RNA inhibited protein synthesis in rabbit reticulocyte and wheat germ systems (121).

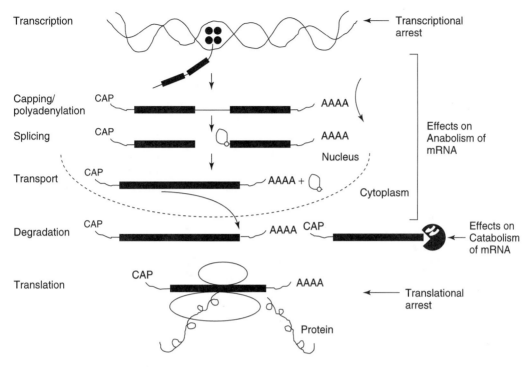

Figure 5.1. Pre-mRNA is transcribed from a gene.

4.2 Occupancy-Activated Destabilization

RNA molecules regulate their own metabolism. A number of structural features of RNA are known to influence stability, various processing events, subcellular distribution, and transport. It is likely that, as RNA intermediary metabolism is better understood, many other regulatory features and mechanisms will be identified.

4.2.1 5′-Capping. A key early step in RNA processing is 5′-capping (Fig. 5.1). This stabilizes pre-mRNA and is important for the stability of mature mRNA. It also is important in binding to the nuclear matrix and transport of mRNA out of the nucleus. Because the structure of the cap is unique and understood, it presents an interesting target. Several oligonucleotides that bind near the cap site have been shown to be active, presumably by inhibiting the binding of proteins required to cap the RNA. For example, the synthesis of SV40 T-antigen was reported to be most sensitive to an oligonucleotide linked to polylysine and targeted to the 5′-cap site of RNA (122). How-

ever, again, in no published study has this putative mechanism been rigorously demonstrated. In fact, in no published study have the oligonucleotides been shown to bind to the sequences for which they were designed.

In our laboratory, we have designed oligonucleotides to bind to 5′-cap structures and reagents to specifically cleave the unique 5′-cap structure (123). These studies demonstrated that 5′-cap-targeted oligonucleotides were capable of inhibiting the binding of the translation initiation factor eIF-4a (124).

4.2.2 Inhibition of 3′-Polyadenylation. In the 3′-untranslated regions of pre-mRNA molecules are sequences that result in the posttranscriptional addition of long (hundreds of nucleotides) tracts of polyadenylate. Polyadenylation stabilizes mRNA and may play other roles in the intermediary metabolism of RNA. Theoretically, interactions in the 3′-terminal region of pre-mRNA could inhibit polyadenylation and destabilize the RNA species. Although there are a number of oligonucleotides that interact in the 3′-untranslated region and display antisense activities, no study to date has reported evidence for alterations in poly-adenylation (125).

4.2.3 Other Mechanisms. In addition to 5′-capping and 3′-adenylation, there are clearly other sequences in the 5′- and 3′-untranslated regions of mRNA that affect the stability of the molecules. Again, there are a number of antisense drugs that may work by these mechanisms. Zamecnik and Stephenson reported that 13-mer targeted to untranslated 3′- and 5′-terminal sequences in Rous sarcoma viruses was active (126). Oligonucleotides conjugated to an acridine derivative and targeted to a 3′-terminal sequence in type A influenza viruses were reported to be active. Against several RNA targets, studies in our laboratories have shown that sequences in the 3′-untranslated region of RNA molecules are often the most sensitive (127–129). For example, ISIS 1939 is a 20-mer phosphorothioate that binds to and appears to disrupt a predicted stem-loop structure in the 3′-untranslated region of the mRNA because the intercellular adhesion molecule (ICAM) is a potent antisense inhibitor. However, inasmuch as the 2′-methoxy analog of ISIS 1939 was much less active, it is likely that, in addition to destabilization to cellular nucleolytic activity, activation of RNase H (see below) is also involved in the activity of ISIS 1939 (94).

4.3 Activation of RNase H

RNase H is a ubiquitous enzyme that degrades the RNA strand of an RNA-DNA duplex. It has been identified in organisms as diverse as viruses and human cells (130). At least two classes of RNase H have been identified in eukaryotic cells. Multiple enzymes with RNase H activity have been observed in prokaryotes (130). Although RNase H is involved in DNA replication, it may play other roles in the cell and is found in both the cytoplasm and the nucleus (131). However, the concentration of the enzyme in the nucleus is thought to be greater and some of the enzyme found in cytoplasmic preparations may be attributed to nuclear leakage.

The precise recognition elements for RNase H are not known. However, it has been shown that oligonucleotides with DNA-like properties as short as tetramers can activate RNase H (132). Changes in the sugar influence RNase H activation as sugar modifications that result in RNA-like oligonucleotides; for example, 2′-fluoro or 2′-methoxy do not appear to serve as substrates for RNase H (133, 134). Alterations in the orientation of the sugar to the base can also affect RNase H activation as α-oligonucleotides are unable to induce RNase H or may require parallel annealing (135, 136). Additionally, backbone modifications influence the ability of oligonucleotides to activate RNase H. Methylphosphonates do not activate RNase H (137, 138). In contrast, phosphorothioates are excellent substrates (74, 139, 140). In addition, chimeric molecules have been studied as oligonucleotides that bind to RNA and activate RNase H (141, 142). For example, oligonucleotides composed of wings of 2′-methoxy phosphonates and a five-base gap of deoxyoligonucleotides bind to their target RNA and activate RNase H (141, 142). Furthermore, a single ribonucleotide in a sequence of deoxyribonucleotides was shown to be sufficient to serve as a substrate for RNase H when bound to its complementary deoxyoligonucleotide (143).

That it is possible to take advantage of chimeric oligonucleotides designed to activate RNase H and have greater affinity for their RNA receptors and to enhance specificity has also been demonstrated (144, 145). In one study, RNase H-mediated cleavage of target transcript was much more selective when deoxyoligonucleotides, composed of methylphosphonate deoxyoligonucleotide wings and phosphodiester gaps, were compared to full phosphodiester oligonucleotides (145).

Despite the information about RNase H and the demonstration that many oligonucleotides may activate RNase H in lysate and purified enzyme assays, relatively little is yet known about the role of structural features in RNA targets in activating RNase H (146–148). In fact, direct proof that RNase H activation is the mechanism of action of oligonucleotides in cells has until very recently been lacking.

Recent studies in our laboratories provide additional, albeit indirect, insights into these questions. ISIS 1939 is a 20-mer phosphorothioate complementary to a sequence in the 3′-untranslated region of ICAM-1 RNA (94). It inhibits ICAM production in human umbilical vein endothelial cells and Northern blots demonstrate that ICAM-1 mRNA is rapidly degraded. A 2'-methoxy analog of ISIS 1939 displays higher affinity for the RNA than that of the phosphorothioate, is stable in cells, but inhibits ICAM-1 protein production much less potently than does ISIS 1939. It is likely that ISIS 1939 destabilizes the RNA and activates RNase H. In contrast, ISIS 1570, an 18-mer phosphorothioate that is complementary to the translation initiation codon of the ICAM-1 message, inhibited production of the protein but caused no degradation of the RNA. Thus, two oligonucleotides that are capable of activating RNase H had different effects, depending on the site in the mRNA at which they bound (94). A more direct demonstration that RNase H is likely a key factor in the activity of many antisense oligonucleotides was provided by studies in which reverse-ligation PCR was used to identify cleavage products from bcr-abl mRNA in cells treated with phosphorothioate oligonucleotides (149).

Given the emerging role of chimeric oligonucleotides with modifications in the 3'- and 5'-wings designed to enhance affinity for the target RNA and nuclease stability and a DNA-type gap to serve as a substrate for RNase H, studies focused on understanding the effects of various modifications on the efficiency of the enzyme(s) are also of considerable importance. In one such study on *E. coli* RNase H, we reported that the enzyme displays minimal sequence specificity and is processive. When a chimeric oligonucleotide with 2'-modified sugars in the wings was hybridized to the RNA, the initial site of cleavage was the nucleotide adjacent to the methoxy-deoxy junction closest to the 3'-end of the RNA substrate. The initial rate of cleavage increased as the size of the DNA gap increased and the efficiency of the enzyme was considerably less against an RNA target duplexed with a chimeric antisense oligonucleotide than against a full DNA-type oligonucleotide (149).

In subsequent studies, we evaluated the interactions of antisense oligonucleotides with both structured and unstructured targets, and the impacts of these interactions on RNase H in more detail (150). Using a series of noncleavable substrates and Michaelis-Menten analyses, we were able to evaluate both binding and cleavage. We showed that *E. coli* RNase H1 is a double-stranded RNA binding protein. The K_d for the RNA duplex was 1.6 μM; the K_d for a DNA duplex was 176 μM; and the K_d for single-stranded DNA was 942 μM. In contrast, the enzyme could cleave RNA only in an RNA-DNA duplex. Any 2'-modification in the antisense drug at the cleavage site inhibited cleavage, but significant charge reduction and 2'-modifications were tolerated at the binding site. Finally, placing a positive charge (e.g., 2'-propoxyamine) in the antisense drug reduced affinity and cleavage. We also examined the effects of antisense oligonucleotide-induced RNA structures on the activity of *E. coli* RNase H1 (150). Any structure in the duplex substrate was found to have a significant negative effect on the cleavage rate. Further, cleavage of selected sites was inhibited entirely, and this was explained by steric hindrance imposed by the RNA loop traversing either the minor or major grooves or the heteroduplex.

Recently, we cloned and expressed human RNase H1. The protein is homologous to *E. coli* RNase H1, but has properties similar to those described for human RNase H type 2 (151, 152). The enzyme is stimulated by low concentrations of Mg^{+2}, inhibited by higher concentrations, and inhibited by Mn^{+2} in the presence of Mg^{+2}. It is a double-stranded, 33-kDa molecular weight RNA binding protein and exhibits unique positional and sequence preferences for cleavage (153). Additionally, human RNase H2 has been cloned, but to date the expressed protein has not been shown to be active (154). Very recently, we reported on the effects of several mutations introduced into human RNase H1 on the activity of the enzyme (155). Thus, we now have the necessary tools to begin to explore the roles of human RNase H proteins in biological and pharmacological processes and to begin to develop drugs designed to interact with them more effectively.

Finally, at least with regard to RNase H-induced degradation of targeted RNA, we recently demonstrated that in the range of 1–150 copies of RNA per cell, the level of target RNA has no effect on the potency of antisense inhibitors (156), a result of the number of mol-

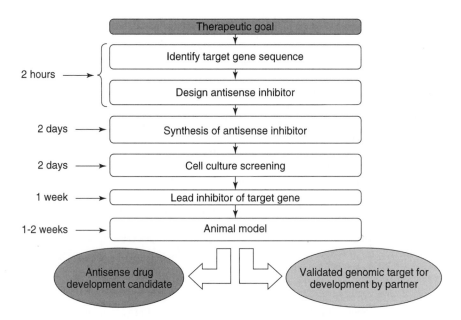

Figure 5.2. Antisense: a drug discovery and genomics tool: identical process. The processes and timelines required to generate optimal antisense inhibitors using automated systems.

ecules of antisense drug per cell exceeding the number of copies of RNA by several orders of magnitude. Thus, other factors must be rate limiting.

4.4 Activation of Double-Strand RNases

By the use of phosphorothioate oligonucleotides with 2'-modified wings and a ribonucleotide center, we have shown that mammalian cells contain enzymes that can cleave double-stranded RNAs (157). This may be an important step forward because it adds to the repertoire of intracellular enzymes that may be used to cleave target RNAs, and because chimeric oligonucleotides 2'-modified wings and oligoribonucleotide gaps have higher affinity for RNA targets than do chimeras with oligodeoxynucleotide gaps.

5   ANTISENSE TECHNOLOGY AS A TOOL FOR GENE FUNCTIONALIZATION AND TARGET VALIDATION

The sequencing of the human and other genomes has resulted in intense interest in the development of tools with which to determine the roles of various gene products and to de-

termine whether they are appropriate targets for drug therapy. With advances in automation, antisense technology has proved to be a versatile, effective tool for these purposes (158). The advances in automation include rapid small-scale synthesizers that can synthesize antisense inhibitors in a 96-well-plate format. These are coupled to in-line automated analytical methods that provide quality assurance. The antisense inhibitors are then screened *in vitro* by the use of an automated reverse transcriptase/polynuclease chain reaction (RT-PCR). With these advances, it is now possible to create antisense inhibitors to hundreds to thousands of genes per year. Figure 5.2 provides a general scheme for rapid creation and evaluation of antisense inhibitors. Figure 5.3 shows a typical 96-well-plate screen designed to determine the optimal site for antisense effects, in this case for TNFα-receptor 1. In this assay, antisense inhibitors to 80 sites in the target RNA are synthesized and screened in TNFα-receptor 1 positive cells. In addition, the effects of a variety of control oligonucleotides are evaluated. In this figure, the RNA is displayed 5' in the translated region to the 3'-untranslated region and each bar represents the level of the target RNA, determined by RT-PCR, clearly interacting in several sites and resulting in significant reductions in the target RNA. Based on these data, several antisense inhibitors can be selected for further study, including *in vivo* evaluation in various animal models.

When one considers the desired attributes for tools for gene functionalization and target validation, it is readily apparent that antisense technology meets these criteria. Antisense inhibitors are gene specific. In the past decade, at Isis Pharmaceuticals alone we have created antisense inhibitors to nearly 1000 genes, so we are confident that the technology can be used for any gene. Having automated antisense, it is rapid, efficient, and cost effective. Antisense inhibitors are versatile in that they can be used for both *in vitro* and *in vivo* studies. Antisense inhibitors are pharmacological agents, so they provide direct insights into the types of responses to be expected from acute interventions with drugs that affect the target. Antisense inhibitors can also be used to dissect the roles of splice variants and to identify novel gene functions.

6 CHARACTERISTICS OF PHOSPHOROTHIOATE OLIGONUCLEOTIDES

Of the first-generation oligonucleotide analogs, the class that has resulted in the broadest range of activities and about which the most is known is the phosphorothioate class. Phosphorothioate oligonucleotides were first synthesized in 1969 when a poly(rIrC) phosphorothioate was synthesized. This modification clearly achieves the objective of increased nuclease stability. In this class of oligonucleotides, one of the oxygen atoms in the phosphate group is replaced with a sulfur. The resulting compound is negatively charged, is chiral at each phosphorothioate phosphodiester, and is much more resistant than the parent phosphorothioate to nucleases (159).

6.1 Hybridization

The hybridization of phosphorothioate oligonucleotides to DNA and RNA has been thoroughly characterized (79, 160–162). The T_m of

a phosphorothioate oligodeoxynucleotide for RNA is approximately 0.5°C less per nucleotide than for a corresponding phosphodiester oligodeoxynucleotide. This reduction in T_m per nucleotide is virtually independent of the number of phosphorothioate units substituted for phosphodiesters. However, sequence context has some influence, given that the ΔT_m can vary from −0.3°C to −1.0°C, depending on sequence. Compared to RNA and RNA duplex formation, a phosphorothioate oligodeoxynucleotide has a T_m value of about −2.2°C lower per unit (163). This means that to be effective *in vitro*, phosphorothioate oligodeoxynucleotides must typically be 17- to 20-mer in length and that invasion of double-stranded regions in RNA is difficult (66, 83, 144, 164).

Association rates of phosphorothioate oligodeoxynucleotide to unstructured RNA targets are typically 10^6–10^7 M^{-1} s^{-1}, independent of oligonucleotide length or sequence (81, 83). Association rates to structured RNA targets can vary from 10^2 to $10^8 M^{-1}$ s^{-1}, depending on the structure of the RNA, site of binding in the structure, and other factors (163). Said another way, association rates for oligonucleotides that display acceptable affinity constants are sufficient to support biological activity at therapeutically achievable concentrations. Interestingly, a recent study using phosphodiester oligonucleotides coupled to fluoroscein showed that hybridization was detectable within 15 min after microinjection into K562 cells (165).

The specificity of hybridization of phosphorothioate oligonucleotides is, in general, slightly greater than that of phosphodiester analogs. For example, a T-C mismatch results in a 7.7 or 12.8°C reduction in T_m, respectively, for a phosphodiester or phosphorothioate oligodeoxynucleotide 18 nucleotides in length with the mismatch centered (163). Thus, from this perspective, the phosphorothioate modification is quite attractive.

6.2 Interactions with Proteins

Phosphorothioate oligonucleotides bind to proteins. The interactions with proteins can be divided into nonspecific, sequence-specific, and structure-specific binding events, each of which may have different characteristics and effects. Nonspecific binding to a wide variety

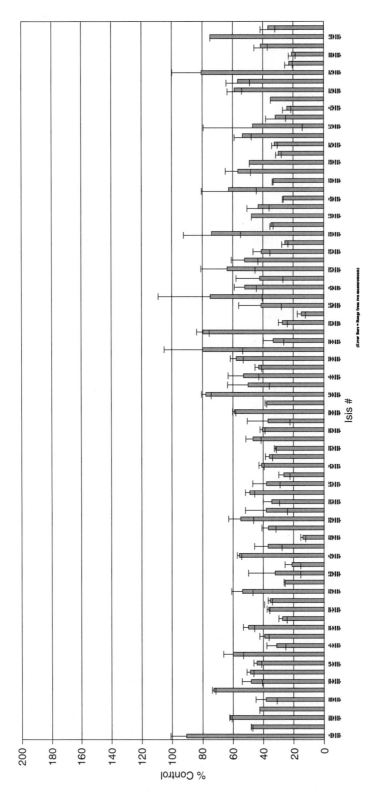

Figure 5.3. TNF R1 screen. Results from a screen to identify the optimal antisense inhibitor to tumor necrosis factor α receptor 1 (TNFα-R1). 2′-Methoxyethyl chimeric antisense inhibitors were synthesized to 80 sites in the target RNA and their effects on TNFα-R1, RNA were evaluated after incubation with TNFα-R1-expressing cells. The results are expressed as percentage control RNA determined by RT-PCR analyses. The effects of the antisense inhibitors are displayed schematically from 5′UTR to 3′UTR right to left on the figure. The antisense inhibitors that result in the maximum reduction in target RNA (smallest bars) are then evaluated with careful dose-response curves and transcriptional arrays along with control oligonucleotides to ensure the effects are specific.

130

of proteins has been demonstrated. Exemplary of this type of binding is the interaction of phosphorothioate oligonucleotides with serum albumin. The affinity of such interactions is low. The K_d value for albumin is approximately 200 μM, in a range similar to that of aspirin or penicillin (166, 167). Furthermore, in this study, no competition between phosphorothioate oligonucleotides and several drugs that bind to bovine serum albumin was observed. In this study, binding and competition were determined in an assay in which electrospray mass spectrometry was used. In contrast, in a study in which an equilibrium dissociation constant was derived from an assay using albumin loaded on a CH-Sephadex column, the K_m value ranged from 1 to 5 $\times$ $10^{-5}\,M$ for bovine serum albumin and 2 to 3 $\times$ $10^{-4}\,M$ for human serum albumin. Moreover, warfarin and indomethacin were reported to compete for binding to serum albumin (168). However, in experiments in our laboratory, we were unable to reproduce the results (for review, see Ref. 169). Clearly, much more work is required before definitive conclusions can be drawn.

Phosphorothioate oligonucleotides can interact with nucleic acid binding proteins such as transcription factors and single-strand nucleic acid binding proteins. However, very little is known about these binding events. Additionally, it has been reported that phosphorothioates bind to an 80-kDa membrane protein that was suggested to be involved in cellular uptake processes (87). However, again, little is known about the affinities, sequences, or structure specificities of these putative interactions. More recently, interactions with 30- and 46-kDa surface proteins in T15 mouse fibroblasts were reported (170).

Phosphorothioates interact with nucleases and DNA polymerases. These compounds are slowly metabolized by both endo- and exonucleases and inhibit these enzymes (160, 171). The inhibition of these enzymes appears to be competitive and this may account for some early data suggesting that phosphorothioates are almost infinitely stable to nucleases. In these studies, the oligonucleotide-to-enzyme ratio was very high and thus the enzyme was inhibited. Phosphorothioates also bind to RNase H when in an RNA-DNA duplex

and the duplex serves as a substrate for RNase H (172). At higher concentrations, presumably by binding as a single strand to RNase H, phosphorothioates inhibit the enzyme (149, 160). Again, the oligonucleotides appear to be competitive antagonists for the DNA-RNA substrate.

Phosphorothioates have been shown to be competitive inhibitors of DNA polymerase α and β with respect to the DNA template, and noncompetitive inhibitors of DNA polymerases γ and δ (172). Despite this inhibition, several studies have suggested that phosphorothioates might serve as primers for polymerases and be extended (140, 173, 174). In our laboratories, we have shown extensions of only 2–3 nucleotides. At present, a full explanation as to why no longer extensions are observed is not available.

Phosphorothioate oligonucleotides have been reported to be competitive inhibitors for HIV-reverse transcriptase and inhibit RT-associated RNase H activity (175, 176). They have been reported to bind to the cell surface protein CD4 and to protein kinase C (PKC) (177). Various viral polymerases have also been shown to be inhibited by phosphorothioates (140). Additionally, we have shown potent, non-sequence-specific inhibition of RNA splicing by phosphorothioates (105).

Like other oligonucleotides, phosphorothioates can adopt a variety of secondary structures. As a general rule, self-complementary oligonucleotides are avoided, if possible, to avoid duplex formation between oligonucleotides. However, other structures that are less well understood can also form. For example, oligonucleotides containing runs of guanosines can form tetrameric structures called G-quartets, and these appear to interact with a number of proteins with relatively greater affinity than that of unstructured oligonucleotides (80).

In conclusion, phosphorothioate oligonucleotides may interact with a wide range of proteins through several types of mechanisms. These interactions may influence the pharmacokinetic, pharmacologic, and toxicologic properties of these molecules. They may also complicate studies on the mechanism of action of these drugs, and may obscure an antisense activity. For example, phosphorothioate oligonucleotides were reported to enhance lipopolysaccharide-stimulated synthesis or tumor necrosis factor (178). This would obviously obscure antisense effects on this target.

6.3 Pharmacokinetic Properties

To study the pharmacokinetics of phosphorothioate oligonucleotides, a variety of labeling techniques have been used. In some cases, 3'- or 5'-^{32}P end-labeled or fluorescently labeled oligonucleotides have been used in either *in vitro* or *in vivo* studies. These are probably less satisfactory than internally labeled compounds because terminal phosphates are rapidly removed by phosphatases and fluorescently labeled oligonucleotides have physicochemical properties that differ from those of the unmodified oligonucleotides. Consequently, either uniformly (179) S-labeled or base-labeled phosphorothioates are preferable for pharmacokinetic studies. In our laboratories, a tritium exchange method that labels a slowly exchanging proton at the C8 position in purines was developed and proved to be quite useful (180). Very recently, a method that added radioactive methyl groups through S-adenosyl methionine was also successfully used (181). Finally, advances in extraction, separation, and detection methods have resulted in methods that provide excellent pharmacokinetic analyses without radiolabeling (182).

6.3.1 Nuclease Stability. The principal metabolic pathway for oligonucleotides is cleavage by endo- and exonucleases. Phosphorothioate oligonucleotides, although quite stable to various nucleases, are competitive inhibitors of nucleases (67, 93, 172, 183, 184). Consequently, the stability of phosphorothioate oligonucleotides to nucleases is probably a bit less than initially thought, given the high concentrations (that inhibited nucleases) of oligonucleotides that were employed in the early studies. Similarly, phosphorothioate oligonucleotides are degraded slowly by cells in tissue culture with a half-life of 12–24 h and are slowly metabolized in animals (183, 185, 186). The pattern of metabolites suggests primarily exonuclease activity, with perhaps modest contributions by endonucleases. However, a number of lines of evidence suggest that, in many cells and tissues, endonucleases play an important role in the metabolism of oligonucleotides. For example, 3'- and 5'-modified oligonucleotides with phosphodiester backbones have been shown to be relatively rapidly degraded in cells and after administration to animals (90, 187). Thus, strategies in which oligonucleotides are modified at only the 3'- and 5'-terminus as a means of enhancing stability have proved to be unsuccessful.

6.4 *In Vitro* Cellular Uptake

Phosphorothioate oligonucleotides are taken up by a wide range of cells *in vitro* (72, 93, 172, 188, 189). In fact, uptake of phosphorothioate oligonucleotides into a prokaryote, *Vibrio parahaemoyticus*, has been reported as has uptake into Schistosoma Mansoni (190, 191). Uptake is time and temperature dependent. It is also influenced by cell type, cell-culture conditions, media and sequence, and length of the oligonucleotide (93). No obvious correlation between the lineage of cells, whether the cells are transformed or are virally infected, and uptake has been identified (93); nor are the factors that result in differences in uptake of different sequences of oligonucleotide understood. Although several studies have suggested that receptor-mediated endocytosis may be a significant mechanism of cellular uptake, the data are not yet compelling enough to conclude that receptor-mediated endocytosis accounts for a significant portion of the uptake in most cells (87).

Numerous studies have shown that phosphorothioate oligonucleotides distribute broadly in most cells once taken up (93, 192). Again, however, significant differences in subcellular distribution between various types of cells have been noted.

Cationic lipids and other approaches have been used to enhance uptake of phosphorothioate oligonucleotides in cells that take up little oligonucleotide *in vitro* (193–195). Again, however, there are substantial variations from cell type to cell type. Other approaches to enhanced intracellular uptake *in vitro* have included streptolysin D treatment of cells and the use of dextran sulfate and other liposome formulations as well as physical means such as microinjections (93, 196, 197).

6.5 *In Vivo* Pharmacokinetics

Phosphorothioate oligonucleotides bind to serum albumin and a-2 macroglobulin. The apparent affinity for albumin is quite low (200–400 μM) and comparable to the low affinity binding observed for a number of drugs (e.g., aspirin, penicillin) (167, 168, 182). Serum protein binding, therefore, provides a repository for these drugs and prevents rapid renal excretion. Because serum protein binding is saturable, at higher doses, intact oligomer may be found in urine (174, 198). Studies in our laboratory suggest that in rats, oligonucleotides administered intravenously at doses of 15–20 mg/kg saturate the serum protein binding capacity (199).

Phosphorothioate oligonucleotides are rapidly and extensively absorbed after parenteral administration. For example, in rats, after an intradermal dose 3.6 mg/kg of ^{14}C-ISIS 2105, a 20-mer phosphorothioate, approximately 70% of the dose was absorbed within 4 h and total systemic bioavailability was in excess of 90% (200). After intradermal injection in humans, absorption of ISIS 2105 was similar to that observed in rats (201). Subcutaneous administration to rats and monkeys results in somewhat lower bioavailability and greater distribution to lymph nodes, as would be expected (202).

Distribution of phosphorothioate oligonucleotides from blood after absorption or i.v. administration is extremely rapid. We have reported distribution half-lives of less than 1 h, and similar data have been reported by others (174, 198, 200, 203). Blood and plasma clearance is multiexponential, with a terminal elimination half-life from 40 to 60 h in all species except humans, where the terminal elimination half-life may be somewhat longer (201).

Phosphorothioates distribute broadly to all peripheral tissues. Liver, kidney, bone marrow, skeletal muscle, and skin accumulate the highest percentage of a dose, but other tissues display small quantities of drug (200, 203). No evidence of significant penetration of the blood-brain barrier has been reported. The rates of incorporation and clearance from tissues vary as a function of the organ studied, with liver accumulating drug most rapidly (20% of a dose within 1–2 h) and other tissues accumulating drug more slowly, and similarly,

elimination of drug relatively rapidly from liver compared to that from many other tissues (e.g., terminal half-life from liver, 62 h; from renal medulla, 156 h). The distribution into the kidney has been studied more extensively and drug was shown to be present in Bowman's capsule, the proximal convoluted tubule, the brush border membrane, and within renal tubular epithelial cells (204). The data suggested that the oligonucleotides are filtered by the glomerulus, then reabsorbed by the proximal convoluted tubule epithelial cells. Moreover, the authors suggested that reabsorption might be mediated by interactions with specific proteins in the brush border membranes. In addition, the oligonucleotide is accumulated in a nonfiltering kidney, suggesting that there is uptake from the basal side also.

Clearance of phosphorothioate oligonucleotides is attributed primarily to metabolism (200, 203, 205). Metabolism is mediated by exo- and endonucleases that result in shorter oligonucleotides and, ultimately, nucleosides that are degraded by normal metabolic pathways. Although no direct evidence of base excision or modification has been reported, these are theoretical possibilities that may occur. In one study, a larger molecular weight radioactive material was observed in urine, but not fully characterized (174). Clearly, the potential for conjugation reactions and extension of oligonucleotides by these drugs serving as primers for polymerases must be explored in more detail. In a very thorough study, 20-nucleotide phosphodiester and phosphorothioate oligonucleotides were administered i.v. at a dose of 6 mg/kg to mice. The oligonucleotides were internally labeled with ^{3}H-CH$_3$ by methylation of an internal deoxycytidine residue using Hha1 methylase and S-[^{3}H]adenosyl methionine (181). The pharmacokinetic properties observed were consistent with our results as described above. Additionally, in this report, autoradiographic analyses showed drug in renal cortical cells (181).

One study of prolonged infusions of a phosphorothioate oligonucleotide to humans has been reported (206). In this study, five patients with leukemia were given 10-day i.v. infusions at a dose of 0.05 mg kg^{-1} h^{-1}. Elimination half-lives reportedly varied from 5.9 to

14.7 days. Urinary recovery of radioactivity was reported to be 30–60% of the total dose, with 30% of the radioactivity being intact drug. Metabolites in urine included both higher and lower molecular weight compounds. In contrast, when GEM-91 (a 25-mer phosphorothioate oligodeoxynucleotide) was administered to humans as a 2-h i.v. infusion at a dose of 0.1 mg/kg, a peak plasma concentration of 295.8 ng/mL was observed at the cessation of the infusion. Plasma clearance of total radioactivity was biexponential with initial and terminal elimination half-lives of 0.18 and 26.71 h, respectively. However, degradation was extensive and intact drug pharmacokinetic models were not presented. Nearly 50% of the administered radioactivity was recovered in urine, but most of the radioactivity represented degrades. In fact, no intact drug was found in the urine at any time (207).

In a more recent study in which the level of intact drug was carefully evaluated using capillary gel electrophoresis, the pharmacokinetics of ISIS 2302, a 20-mer phosphorothioate oligodeoxynucleotide, after a 2-h infusion, were determined. Doses from 0.06 to 2.0 mg/kg were studied and the peak plasma concentrations were shown to increase linearly with dose, with the 2 mg/kg dose resulting in peak plasma concentrations of intact drug of about 9.5 μg/mL. Clearance from plasma, however, was dose dependent, with the 2 mg/kg dose having a clearance of 1.28 mL min^{-1} kg^{-1}, whereas that of 0.5 mg/kg was 2.07 mL min^{-1} kg^{-1}. Essentially, no intact drug was found in urine.

Clearly, the two most recent studies differ from the initial report in several facets. Although a number of factors may explain the discrepancies, the most likely explanation is related to the evolution of assay methodology, not differences between compounds. Overall, the behavior of phosphorothioates in the plasma of humans appears to be similar to that in other species.

In addition to the pharmacological effects that have been observed after phosphorothioate oligonucleotides have been administered to animals (and humans), a number of other lines of evidence show that these drugs enter cells in organs. Autoradiographic, fluorescent, and immunohistochemical approaches have shown that these drugs are localized in endopromal convoluted tubular cells, various bone marrow cells, and cells in the skin and liver (204, 208, 209).

Perhaps more compelling and of more long-term value are studies recently reported showing the distribution of phosphorothioate oligonucleotides in the liver of rats treated intravenously with the drugs at various doses (210). This study showed that the kinetics and extent of the accumulation into Kupffer, endothelial, and hepatocyte cell population varied and that as doses were increased, the distribution changed. Moreover, the study showed that subcellular distribution also varied.

6.5.1 Aerosol Administration. Phosphorothioate oligodeoxynucleotides have been shown to be attractive for inhalation delivery to the lung and upper airway (211–213). Target reduction in the lung has been demonstrated (214) and these drugs have been shown to distribute broadly to all cell types in the lung after aerosol administration. Further, these drugs were shown to be well tolerated at doses up to 12 mg/kg (212).

6.5.2 Topical Administration. Phosphorothioate oligonucleotides formulated in a very simple cream formulation have been shown to penetrate normal mouse, pig, and human skin and to penetrate and accumulate in human psoriatic skin grown on nude mice. Further, in a phase IIa study in patients with plaque psoriasis, ISIS 2302, in ICAM-1 inhibitor, was shown to accumulate throughout the dermis and epidermis after topical administration and to result in a positive trend in the primary endpoint, that is, induration (Kruger, unpublished observations, 2000). Thus, studies have demonstrated a substantial accumulation of drug throughout the dermis and epidermis and reduction of targets such as ICAM-1, B71, and B72 (for review, see Ref. 215).

6.5.3 Summary. In summary, pharmacokinetic studies of several phosphorothioates demonstrate that they are well absorbed from parenteral sites, distribute broadly to many peripheral tissues, do not cross the blood-brain barrier, and are eliminated primarily by nuclease metabolism. In short, once daily or every other day systemic dosing is feasible. Although the similarities between oligonucleotides of different sequences are far greater than the differences, additional studies are required before determining whether there are subtle effects of sequence on the pharmacokinetic profile of this class of drugs. Moreover, they can be delivered to the lung by aerosol and topically administered, and progress in achieving acceptable oral bioavailability continues.

6.6 Pharmacological Properties

6.6.1 Molecular Pharmacology. Antisense oligonucleotides are designed to bind to RNA targets through Watson-Crick hybridization. Because RNA can adopt a variety of secondary structures through Watson-Crick hybridization, one useful way to think of antisense oligonucleotides is as competitive antagonists for self-complementary regions of the target RNA. Obviously, creating oligonucleotides with the highest affinity per nucleotide unit is pharmacologically important, and a comparison of the affinity of the oligonucleotide to a complementary RNA oligonucleotide is the most sensible comparison. In this context, phosphorothioate oligodeoxynucleotides are relatively competitively disadvantaged, in that the affinity per nucleotide unit of oligomer is less than that of RNA ($> -2.0°C\ T_m$ per unit) (216). This results in a requirement of at least 15–17 nucleotides to have sufficient affinity to produce biological activity (164).

Although multiple mechanisms by which an oligonucleotide may terminate the activity of an RNA species to which it binds are possible, examples of biological activity have been reported for only three of these mechanisms. Antisense oligonucleotides have been reported to inhibit RNA splicing, effect translation of mRNA, and induce degradation of RNA by RNase H (28, 100, 125). Without question, the mechanism that has resulted in the most potent compounds and is best understood is RNase H activation. To serve as a substrate for RNase H, a duplex between RNA and a "DNA-like" oligonucleotide is required. Specifically, a sugar moiety in the oligonucleotide that induces a duplex conformation equivalent to that of a DNA-RNA duplex and a

charged phosphate are required (38). Thus, phosphorothioate oligodeoxynucleotides are expected to induce RNase H-mediated cleavage of the RNA when bound. As discussed later, many chemical approaches that enhance the affinity of an oligonucleotide for RNA result in duplexes that are no longer substrates for RNase H.

Selection of sites at which optimal antisense activity may be induced in an RNA molecule is complex, dependent on terminating mechanism and influenced by the chemical class of the oligonucleotide. Each RNA appears to display unique patterns of sites of sensitivity. Within the phosphorothioate oligodeoxynucleotide chemical class, studies in our laboratory have shown antisense activity can vary from undetectable to 100% by shifting an oligonucleotide by just a few bases in the RNA target (94, 160, 217). Although significant progress has been made in developing general rules that help define potentially optimal sites in RNA species, to a large extent, this remains an empirical process that must be performed for each RNA target and every new chemical class of oligonucleotides.

Phosphorothioates have also been shown to have effects inconsistent with the antisense mechanism for which they were designed. Some of these effects are attributed to sequence or are structure specific; others are attributed to nonspecific interactions with proteins. These effects are particularly prominent in *in vitro* tests for antiviral activity because high concentrations of cells, viruses, and oligonucleotides are often coincubated (92, 218). Human immune deficiency virus (HIV) is particularly problematic, in that many oligonucleotides bind to the gp120 protein (80). However, the potential for confusion, arising from the misinterpretation of an activity as being attributed to an antisense mechanism when it is attributed to non-antisense effects, is certainly not limited to antiviral or just *in vitro* tests (219–221). Again, these data simply urge caution and argue for careful dose-response curves, direct analyses of target protein or RNA, and inclusion of appropriate controls before drawing conclusions concerning the mechanisms of action of oligonucleotide-based drugs. In addition to protein interactions, other factors, such as overrepresented sequences of RNA and unusual structures that may be adopted by oligonucleotides, can contribute to unexpected results (80).

Given the variability in cellular uptake of oligonucleotides, the variability in potency as a function of binding site in an RNA target and potential non-antisense activities of oligonucleotides, careful evaluation of dose-response curves and clear demonstration of the antisense mechanism are required before drawing conclusions from *in vitro* experiments. Nevertheless, numerous well-controlled studies have been reported in which antisense activity was conclusively demonstrated. Because many of these studies have been reviewed previously, suffice it to say that antisense effects of phosphorothioate oligodeoxynucleotides against a variety of targets are well documented (79, 140, 160, 173, 222).

6.6.2 *In Vivo* Pharmacological Activities.

A relatively large number of reports of *in vivo* activities of phosphorothioate oligonucleotides have now appeared, documenting activities after both local and systemic administration (for review, see Ref. 78). However, for only a few of these reports have sufficient studies been performed to draw relatively firm conclusions concerning the mechanism of action. Consequently, I will review in some detail only a few reports that provide sufficient data to support a relatively firm conclusion with regard to mechanism of action. Local effects have been reported for phosphorothioate and methylphosphonate oligonucleotides. A phosphorothioate oligonucleotide designed to inhibit c-*myb* production and applied locally was shown to inhibit intimal accumulation in the rat carotid artery (223). In this study, a Northern blot analysis showed a significant reduction in c-*myb* RNA in animals treated with the antisense compound, but no effect by a control oligonucleotide. In a recent study, the effects of the oligonucleotide were suggested to be attributable to a non-antisense mechanism (220). However, only one dose level was studied, so much remains to be done before definitive conclusions are possible. Similar effects were reported for phosphorothioate oligodeoxynucleotides designed to inhibit cyclin-dependent kinases (CDC-2 and CDK-2). Again, the antisense oligonucleotide

inhibited intimal thickening and cyclin-dependent kinase activity, whereas a control oligonucleotide had no effect (224). Additionally, local administration of a phosphorothioate oligonucleotide designed to inhibit N-*myc* resulted in reduction in N-*myc* expression and slower growth of a subcutaneously transplanted human tumor in nude mice (225).

Antisense oligonucleotides administered intraventricularly have been reported to induce a variety of effects in the CNS. Intraventricular injection of antisense oligonucleotides to neuropeptide-y-y1 receptors reduced the density of the receptors and resulted in behavioral signs of anxiety (226). Similarly, an antisense oligonucleotide designed to bind to NMDA-R1 receptor channel RNA inhibited the synthesis of these channels and reduced the volume of focal ischemia produced by occlusion of the middle cerebral artery in rats (226).

In a series of well-controlled studies, antisense oligonucleotides administered intraventricularly selectively inhibited dopamine type 2 receptor expression, dopamine type 2 receptor RNA levels, and behavioral effects in animals with chemical lesions. Controls included randomized oligonucleotides and the observation that no effects were observed on dopamine type 1 receptor or RNA levels (142–144). This laboratory also reported the selective reduction of dopamine type 1 receptor and RNA levels with the appropriate oligonucleotide (227).

Similar observations were reported in studies on AT-1 angiotensin receptors and tryptophan hydroxylase. In studies in rats, direct observations of AT-1 and AT-2 receptor densities in various sites in the brain after administration of different doses of phosphorothioate antisense, sense, and scrambled oligonucleotides were reported (228). Again, in rats, intraventricular administration of phosphorothioate antisense oligonucleotide resulted in a decrease in tryptophan hydroxylase levels in the brain, whereas a scrambled control did not (229).

Injection of antisense oligonucleotides to synaptosomal-associated protein-25 into the vitreous body of rat embryos reduced the expression of the protein and inhibited neurite elongation by rat cortical neurons (230).

Aerosol administration to rabbits of an antisense phosphorothioate oligodeoxynucleotide designed to inhibit the production of adenosine A1 receptor has been reported to reduce receptor numbers in the airway smooth muscle and to inhibit adenosine, house dust mite allergen, and histamine-induced bronchoconstriction (214). Neither a control nor an oligonucleotide complementary to bradykinin B2 receptors reduced adenosine A1 receptors' density, although the oligonucleotides complementary to bradykin in B2 receptor mRNA reduced the density of these receptors.

In addition to local and regional effects of antisense oligonucleotides, a growing number of well-controlled studies have demonstrated systemic effects of phosphorothioate oligodeoxynucleotides. Expression of interleukin 1 in mice was inhibited by systemic administration of antisense oligonucleotides (231). Oligonucleotides to the NF-κB p65 subunit administered intraperitoneally at 40 mg/kg every 3 days slowed tumor growth in mice transgenic for the human T-cell leukemia viruses (232). Similar results with other antisense oligonucleotides were shown in another *in vivo* tumor model, after either prolonged subcutaneous infusion or intermittent subcutaneous injection (233).

Several recent reports further extend the studies of phosphorothioate oligonucleotides as antitumor agents in mice. In one study, a phosphorothioate oligonucleotide directed to inhibition of the *bcr-abl* oncogene was administered i.v. at a dose of 1 mg/day for 9 days to immunodeficient mice injected with human leukemic cells. The drug was shown to inhibit the development of leukemic colonies in the mice and to selectively reduce *bcr-abl* RNA levels in peripheral blood lymphocytes, spleen, bone marrow, liver, lungs, and brain (96). However, it is possible that the effects on the RNA levels were secondary to effects on the growth of various cell types. In the second study, a phosphorothioate oligonucleotide antisense to the protooncogene *myb*, inhibited the growth of human melanoma in mice. Again, *myb* mRNA levels appeared to be selectively reduced (234).

A number of studies from our laboratories that directly examined target RNA levels, target protein levels, and pharmacological effects using a wide range of control oligonucleotides and examination of the effects on closely re-

lated isotypes have been completed. Single and chronic daily administration of a phosphorothioate oligonucleotide designed to inhibit mouse protein kinase C-a (PKC-a), selectively inhibited expression of PKC-a RNA in mouse liver without effects on any other isotype. The effects lasted at least 24 h after a dose and a clear dose-response curve was observed with an i.p. dose of 10–15 mg/kg, reducing PKC-a RNA levels in the liver by 50% 24 h after a dose (235).

A phosphorothioate oligonucleotide designed to inhibit human PKC-a expression selectively inhibited expression of PKC-a RNA and PKC-a protein in human tumor cell lines implanted subcutaneously in nude mice after intravenous administration (236). In these studies, effects on RNA and protein levels were highly specific and observed at doses lower than 6 mg/kg. A large number of control oligonucleotides failed to show activity.

In a similar series of studies, Monia et al. demonstrated highly specific loss of human C-*raf* kinase RNA in human tumor xenografts and antitumor activity that correlated with the loss of RNA (237, 238).

Finally, a single injection of a phosphorothioate oligonucleotide designed to inhibit c-AMP-dependent protein kinase type 1 was reported to selectively reduce RNA and protein levels in human tumor xenografts and to reduce tumor growth (239).

Thus, there is a growing body of evidence that phosphorothioate oligonucleotides can induce potent systemic and local effects *in vivo*. More important, there are now a number of studies with sufficient controls and direct observation of target RNA and protein levels to suggest highly specific effects that are difficult to explain by any mechanism other than antisense. As would be expected, the potency of these effects vary depending on the target, the organ, and the endpoint measured, as well as the route of administration and the time after a dose when the effect is measured.

In conclusion, although it is of obvious importance to interpret *in vivo* activity data cautiously, and it is clearly necessary to include a range of controls and to evaluate effects on target RNA and protein levels and control RNA and protein levels directly, it is difficult to argue with the conclusion today that some effects have been observed in animals that are most likely primarily ascribed to an antisense mechanism.

6.7 Clinical Activities

Vitravene, a phosphorothioate oligonucleotide designed to inhibit cytomegalovirus-induced retinitis, has been shown to be safe and effective in the treatment of this disease after intravitreal injection and has been approved for sale by regulatory agencies in the United States, Europe, and South America (for review, see Ref. 240).

6.7.1 ISIS 2302. ISIS 2302 is a phosphorothioate oligonucleotide inhibitor of human ICAM-1. It has been demonstrated to reduce ICAM-1 levels in a number of organs in a variety of animal models and to result in potent anti-inflammatory effects (for review, see Ref. 241). ISIS 2302 has been evaluated in a number of clinical trials in patients with inflammatory diseases (for review, see Ref. 242). The drug was evaluated in a randomized placebo-controlled phase IIa trial in 43 patients with moderate to severe rheumatoid arthritis (243). In this trial the effects of 0.5, 1.0, and 2.0 mg/kg i.v. three times weekly for 1 month were compared to placebo. The 0.5 and 2.0 mg/kg doses resulted in prolonged improvement in rheumatoid arthritis activity, compared to placebo. All doses were extremely well tolerated. ISIS 2302 was evaluated both by i.v. and topical routes as a potential therapeutic for patients with psoriasis; the dosing for the i.v. study was equivalent to that used for the rheumatoid arthritis study, but only 17 patients were evaluated. Again, all doses were well tolerated. At the end of the treatment period an approximately 30% reduction on psoriasis activity index score (a measure of psoriasis activity) was reported (244). Additionally, several concentrations of ISIS 2302 were evaluated in patients with moderate plaque psoriasis. In this study, the drug was applied topically in a simple cream formulation once a day for 3 months. The patients were followed for a total of 6 months. In approximately one-half the patients, skin biopsies were taken. The study showed that ISIS 2302 achieved very high dermal and epidermal concentrations that increased as the concentration in the cream was increased, with 2% and 4% creams given dermal concentrations similar to those at which pharmacological effects were observed in animals. Further, there was a trend ($P = 0.053$) in favor of increasing ISIS 2302 concentrations with regard to the primary endpoint, induration. Again, the drug was very well tolerated. Additional studies with topical ISIS 2302 are in progress.

ISIS 2302 has been evaluated most extensively in Crohn's disease. In the initial placebo-controlled study in steroid-dependent patients, ISIS 2302 resulted in a substantial reduction in symptoms and a dramatic reduction in steroid use and reduction of ICAM-1 levels in the small intestines of patients treated and evaluated with serial colonoscopies. The duration of effect after a month of dosing was in excess of 6 months. Further, serial colonoscopic biopsies demonstrated reduction of ICAM-1 protein in the small bowels of treated patients (245). In a subsequent 300 patient trial, in which patients were treated with placebo or 2 mg/kg every other day for 2 or 4 weeks, again the drug was shown to be extremely well tolerated. A population pharmacokinetic study demonstrated that heavier patients and women achieved greater exposures to the drug and that in the patients in the upper two quartiles of drug exposure, the drug produced a statistically significant increase in complete remissions compared to placebo (246). Additional phase III trials at higher doses are in progress.

6.7.2 ISIS 3521. ISIS 3521 is a phosphorothioate oligodeoxynucleotide inhibitor of PKCα (for review, see Ref. 247). This drug has been evaluated as a single agent in a variety of solid tumors, using a variety of dose schedules at doses as high as 30 mg kg^{-1} day^{-1} and the drug was well tolerated. Table 5.1 shows the results reported for the drug in these patients with advanced chemoresistant malignancies. ISIS 3521 has also been evaluated in combination in a number of chemotherapeutic agents in a number of solid tumors. The most advanced experience is a phase I/II study in 53 patients with non-small cell carcinoma of the lung in combination with carboplatin and taxol. In this study, with no increase in toxicity, ISIS 3521 was associated with a substantial improvement in survival (16 months) compared to that of historical controls (248). Based on these data, a large phase III program is in progress.

6.7.3 G3139. G3139 is a phosphorothioate oligodeoxynucleotide designed to inhibit BCL_2. It has demonstrated single agent activity in non-Hodgkin's lymphoma. In the non-Hodgkin's lymphoma trial, BCL_2 levels in peripheral blood cells were reduced by G3139 (249). G3139 is also being studied in combination with chemotherapeutic agents. In a Phase I study with decarbazine in patients with malignant melanoma, a 21% response rate in 14 patients was reported (250). This drug is being evaluated in a number of other malignant diseases and in a number of other combinations.

6.7.4 ISIS 2503. ISIS 2503 is a phosphorothioate oligodeoxynucleotide designed to inhibit Ha-*ras*. It is currently completing phase II single-agent and combination studies in patients with a variety of solid malignancies (for review, see Ref. 247). It has been reported to be well tolerated and preliminary suggestions of activity have been reported (see Table 5.1).

6.7.5 ISIS 5132. ISIS 5231 is a phosphorothioate oligodeoxynucleotide designed to inhibit C-*raf* kinase. In phase I studies, this drug was shown to reduce C-*raf* kinase levels in peripheral blood cells and displayed activity in patients with ovarian cancer (251). It too is completing phase II evaluation (for review, see Ref. 247).

6.8 Toxicology

6.8.1 *In Vitro.* In our laboratory, we have evaluated the toxicities of scores of phosphorothioate oligodeoxynucleotides in a significant number of cell lines in tissue culture. As a general rule, no significant cytotoxicity is induced at concentrations below 100 μM oligonucleotide. Additionally, with a few exceptions, no significant effect on macromolecular synthesis is observed at concentrations below 100 μM (188, 192).

Polynucleotides and other polyanions have been shown to cause release of cytokines (8). Also, bacterial DNA species have been reported to be mitogenic for lymphocytes *in vitro* (259). Furthermore, oligodeoxynucleotides (30–45 nucleotides in length) were reported to induce interferons and enhance natural killer cell activity (260). In the latter study, the oligonucleotides that displayed natural killer cell (NK)-stimulating activity contained specific palindromic sequences and tended to be guanosine rich. Collectively, these observations indicate that nucleic acids may have broad immunostimulatory activity.

It has been shown that phosphorothioate oligonucleotides stimulate B-lymphocyte proliferation in a mouse splenocyte preparation (analogous to bacterial DNA), and the response may underlie the observations of lymphoid hyperplasia in the spleen and lymph nodes of rodents caused by repeated administration of these compounds (see below) (261). We also have evidence of enhanced cytokine release by immunocompetent cells when exposed to phosphorothioates *in vitro* (262). In this study, both human keratinocytes and an *in vitro* model of human skin released interleukin 1-a when treated with 250 μM to 1 mM of phosphorothioate oligonucleotides. The effects seemed to be dependent on the phosphorothioate backbone and independent of sequence or 2'-modification. In a study in which murine B-lymphocytes were treated with phosphodiester oligonucleotides, B-cell activation was induced by oligonucleotides with unmethylated CpG dinucleotides (263). This has been extrapolated to suggest that the CpG motif may be required for immune stimulation of oligonucleotide analogs such as phosphorothioates. This clearly is not the case with regard to release of IL-1a from keratinocytes (262), nor is it the case with regard to *in vivo* immune stimulation (see below).

6.8.2 Genotoxicity. As with any new chemical class of therapeutic agents, concerns about genotoxicity cannot be dismissed, given that little *in vitro* testing has been performed and no data from long-term studies of oligonucleotides are available. Clearly, given the limitations in our understanding about the basic mechanisms that might be involved, empirical data must be generated. We performed mutagenicity studies on five phosphorothioate oligonucleotides, ISIS 2302, ISIS 2105, ISIS

Table 5.1 Single-Agent Antitumor Activity in Clinical Trials of Antisense Oligonucleotides

Oligo/Target	Schedule	Tumor	Efficacy Results	Ref.
ISIS 3521/PKC-α	2 h i.v. infusion tiw, 3 weeks out of 4	Lymphoma	Complete response in 2 of 2 patients with low grade disease: One continuing 30+ months after start of treatment. One with skin recurrence 15 months after start of treatment	11
		Non-small cell lung	Stable disease for 6 months	12
	2 day c.i.v., 3 weeks out of 4	Ovarian	Partial response: time to progression (TTP): 11 months Marker only disease with 75% ↓ in CA-125. TTP: 7 months Marker only disease with 40% ↓ in CA-125. TTP: 7 months 80% ↓ in CA-125 with stable measurable disease	
ISIS 5132/C-raf	2 h i.v. infusion tiw, 3 weeks out of 4	Colon	Stable disease for 7 months; 31% decrease of CEA Concurrent ↓ C-raf expression in peripheral blood mononuclear cells	41, 42
		Renal	Stable disease for 9 months. Concurrent ↓ C-raf expression in peripheral blood mononuclear cells	
	21 day c.i.v., 3 weeks out of 4	Ovarian	97% decrease in CA-125 with stable evaluable disease. TTP: 10 months	40
		Pancreatic	Stable disease for 10 months	
		Renal	Stable disease for 9 months	
ISIS 2503/Ha-ras	14 day c.i.v., 2 weeks out of 3	Sarcoma	Stable disease for 10 cycles	31
		Pancreatic	Stable disease for 9 cycles	
		Colon	Stable disease for 8 cycles	
		Mesothelioma	Stable disease for 6 cycles	
	24 h c.i.v., repeated weekly	Melanoma	Melanoma: stable disease for 7 cycles	32
G3139/bcl-2	14-day s.c. infusion	Lymphoma	Complete response: 1/17 patients 11 stable disease (including 2 minor responses) 6/17 patients with improved lymphoma symptoms 7/16 patients with ↓ Bcl-2 protein in PBMC, bone marrow, or lymph node	48

5132, ISIS 14803, and Vitravene, and found them to be nonmutagenic at all concentrations studied (264, 265).

Two mechanisms of genotoxicity that may be unique to oligonucleotides have been considered. One possibility is that an oligonucleotide analog could be integrated into the genome and produce mutagenic events. Although integration of an oligonucleotide into the genome is conceivable, it is likely to be extremely rare. For most viruses, viral DNA integration is itself a rare event and, of course, viruses have evolved specialized enzyme-mediated mechanisms to achieve integration. Moreover, preliminary studies in our laboratory have shown that phosphorothioate oligodeoxynucleotides are generally poor substrates for DNA polymerases, and it is unlikely that enzymes such as integrases, gyrases, and topoisomerases (that have obligate DNA cleavage as intermediate steps in their enzymatic processes) will accept these compounds as substrates. Consequently, it would seem that the risk of genotoxicity resulting from genomic integration is no greater, and probably less, than that of other potential mechanisms (e.g., alteration of the activity of growth factors, cytokine release, nonspecific effects on membranes that might trigger arachidonic acid release, or inappropriate intracellular signaling). Presumably, new analogs that deviate significantly more from natural DNA would be even less likely to be integrated.

A second concern that has been raised about possible genotoxicity is the risk that oligonucleotides might be degraded to toxic or carcinogenic metabolites. However, metabolism of phosphorothioate oligodeoxynucleotides by base excision would release normal bases, which presumably would be nongenotoxic. Similarly, oxidation of the phosphorothioate backbone to the natural phosphodiester structure would also yield nonmutagenic (and probably nontoxic) metabolites. Finally, it is possible that phosphorothioate bonds could be hydrolyzed slowly, releasing nucleoside phosphorothioates that presumably would be rapidly oxidized to natural (nontoxic) nucleoside phosphates. However, oligonucleotides with modified bases and/or backbones may pose different risks.

6.8.3 In Vivo
6.8.3.1 Acute and Transient Toxicities
6.8.3.1.1 Complement Activation. Rapid infusion of phosphorothioate oligodeoxynucleotides to nonhuman primates can result in cardiovasulcar collapse (266–268). These hemodynamic effects are associated with complement activation, but complement activation is necessary but not sufficient to produce the observed cardiovascular effects (for review, see Ref. 169). The other factors contributing to cardivascular collapse are not yet fully elucidated. However, it has been suggested that dosing monkeys that are restrained may result in exacerbating cardiovascular events. Activation of the complement cascade, attributed to activation of the alternate pathway, is relatively insensitive to sequence, but is absolutely related to peak plasma concentration (269). The threshold concentration for activation of complement in the monkey is 40–50 μg/mL phosphorothioate oligodeoxynucleotide and, once the threshold is reached, variable but potentially dangerous levels of complement activation are observed.

Studies in humans have avoided peak plasma concentrations that would induce complement. However, it appears that monkeys are substantially more sensitive than humans to complement activation. A comparison of the effects on complement activation in human vs. monkey serum demonstrates a dramatic difference. In monkey serum, phosphorothioate oligodeoxynucleotides activate complement in a concentration-dependent fashion. It human serum, complement activation is actually inhibited at higher concentrations of oligonucleotide (169). This is thought to be attributable to the sensitivity of human serum to inhibition of complement activators. The mechanism of complement activation is currently believed to be attributable to an interaction with factor H (169). These effects can be reduced by chemical modifications and formulations that reduce plasma protein binding (169).

6.8.3.1.2 Inhibition of Clotting. In all species studies, phosphorothioate oligodeoxynucleotides induce a transient, apparently self-limited, peak plasma concentration-related inhibition of clotting, manifested as an increase in activated partial thromboplastin time (aPTT) (270–273). Increases in aPTT are minimally affected by sequence, although the effects are directly proportional to the length of the phosphorothioate oligodeoxynucleotide (169). The mechanism of aPTT increase appears to be an interaction with the intrinsic tenase complex (274, 275). This interaction is complex and involves effects on multiple clotting factors including factors VIIIa, IXa, and X. The effects on clotting are transient, appear to be self-limited, and have not resulted in bleeding diatheses in either animals or humans. Effects on clotting can be ameliorated by chemical modifications such as 2'-O-methoxyethyl substitution (169).

6.8.3.2 Subchronic Toxicities

6.8.3.2.1 Immune Stimulation. In the rodent, the most prominent toxicity is immune stimulation. This effect is also frequently a confounding variable that must be evaluated with regard to the mechanism of action of pharmacological effects (for review, see Ref. 265). Subchronic administration of doses as low as 10 mg kg^{-1} day^{-1} in rodents results in splenomegaly, lymphoid hyperplasia, and diffuse multiorgan mixed mononuclear cell infiltrates (276–279). These effects are reminiscent of stimulator effects observed on isolated splenocytes (261, 280, 281). In fact, these *in vitro* studies are reasonably predictive of the relative potencies of phosphorothioates in moving *in vivo* immune stimulation. Immune stimulation in the rodent is a property common to all phosphorothioate oligodeoxynucleotides, but potency varies substantially as a function of sequence. Rodent immune stimulatory motifs include palindromic sequences and CG motifs (282, 283).

In primates, immune stimulation is far less prominent than that in the rodent. In the monkey, doses necessary to produce immune stimulation have not been identified, despite evaluation of numerous oligonucleotides at a variety of doses and schedules (169). At least one factor contributing to these species differences is that the optimal rodent immune stimulator sequence is simple and immune stimulation is induced by all sequences. In contrast, in the primate, the optimal sequences are different and more complex (284) and the response is much less promiscuous with regard to sequence. Modifications of the base, the

sugar, and the backbone can reduce immune stimulation in the rodent. For example, 5'-methycytosine and 2'-O-methoxyethyl containing oligonucleotides display substantially reduced potency for immune stimulation in the rodent (169).

6.8.3.2.2 Other Toxicities. Occasionally, in the rodent but not the monkey, single-cell hepatolyte neurons is observed and this has been related to immunostimulation (265). Also, on occasion, in the monkey transient thrombocytopenia is observed, perhaps associated with complement activation (for review, see Ref. 169). Other toxicities noted in animals are mild and occur infrequently at therapeutic doses. For example, occasional increases in liver function enyzmes are noted, but these occur at high doses and are not associated with histopathological changes (169).

6.8.4 Human Safety. At Isis, we have had the opportunity to study approximately 10 phosphorothioate oligonucleotides in humans. We have studied antisense drugs administered intravitreally, intradermally, subcutaneously, intravenously, and topically. Vitravene, an intravitreally administered drug, is commercially available around the world.

After systemic administration, in clinical trials, we have studied more than 2000 patients treated with more than 70,000 doses. Intravenous doses have ranged from 0.5 mg/kg every other day to as much as 30 mg kg^{-1} day^{-1} as a continuous infusion. Most patients have been treated with multiple doses and many have been treated for prolonged periods (i.e., multiple months).

6.8.4.1 Complement Activation. With 2-h infusions at a dose of 2 mg/kg, no increases in complement-split products were observed in more than 300 patients with inflammatory diseases treated with ISIS 2302, a 20-nucleotide phosphorothioate antisense inhibitor of ICAM-1. Similarly, ISIS 5132, an inhibitor of C-*raf* kinase was dosed from 0.5 to 6.0 mg/kg and no meaningful increases in complement-split products were observed. ISIS 3521, a PKCα inhibitor gave equivalent data and both of these drugs were studied in patients with various malignant diseases. Similar data were observed with all the phosphorothioate oligodeoxynucleotides administered on this schedule in a variety of patients (for review, see Refs. 285 and 286). Three phosphorothioate oligodeoxynucleotides, ISIS 3521, a PKCα inhibitor; ISIS 5132, a C-*raf* kinase inhibitor; and ISIS 2503, a Ha-*ras* inhibitor, have been thoroughly characterized in patients with malignant diseases and dosed with long-term infusions. When given as 21-day antitumor infusions at doses as high as 10 mg kg^{-1} day^{-1}, no increases in complement-split products were observed (257, 287, 288). In contrast, when these drugs were given as 24-h infusions, significant increases in complement-split products were observed at doses of 18 mg kg^{-1} day^{-1} and greater (285). Despite these increases in complement-split products, only a small number of patients experienced mild fevers and myalgias at these very high doses.

6.8.4.2 Cytokines. At very high doses (i.e., 24 mg kg^{-1} day^{-1}) as a continuous i.v. infusion, significant increases in IL-6, IL-IR$_\alpha$, and TNFα were observed, once these often correlated with flulike syndromes (285). At present, the precise roles of each of the cytokines and complement activation in the clinical signs and symptoms (myalgia and fever) at high doses are not defined.

6.8.4.3 Coagulation. All phosphorothioate oligodeoxynucleotides studied in normal volunteers and in patients have resulted in transient, self-limited increases in activated partial thromboplastin time (aPTT). The effects are dose and peak plasma concentration dependent (285). These effects appear to be more prominent after the first dose. In none of the patients treated did we observe any evidence of bleeding, so the effects on aPTT have not proved to be a problem in the clinic and humans appear to behave similarly to other animals with regard to this side effect.

6.8.4.4 Platelet Effects. When phosphorothioate oligodeoxynucleotides have been administered by continuous i.v. infusion in patients with malignant diseases, transient thrombocytoplastin has been observed in a few patients (285). These effects were more frequent during the first course of therapy, were not obviously dose-related, and were not associated with bone marrow effects (253, 289). Most of the time, platelet counts returned to normal while dosing was continued and no bleeding was observed. The mechanisms for these effects are not clear, but probably involve margination.

6.8.4.5 Other Toxicities. At no dose studied have significant effects on liver, kidney, or other organ performance been observed (285).

6.9 Conclusions

Phosphorothioate oligonucleotides have perhaps outperformed many expectations. They display attractive parenteral pharmacokinetic properties. They have produced potent systemic effects in a number of animal models and, in many experiments, the antisense mechanism has been directly demonstrated as the hoped-for selectivity. Further, these compounds appear to display satisfactory therapeutic indices for many indications.

Nevertheless, phosphorothioates clearly have significant limits. Pharmacodynamically, they have relatively low affinity per nucleotide unit. This means that longer oligonucleotides are required for biological activity and that invasion of many RNA structures may not be possible. At higher concentrations, these compounds inhibit RNase H as well. Thus, the higher end of the pharmacological dose-response curve is lost. Pharmacokinetically, phosphorothioates must be dosed every other day, do not cross the blood-brain barrier, are not significantly orally bioavailable, and may display dose-dependent pharmacokinetics. Toxicologically, clearly the release of cytokines, activation of complement, and interference with clotting limit the dosing and may limit the use of these agents in the treatment of some diseases. Further, with chronic subcutaneous administration, local lymphoadenopathy has been dose limiting (285).

7 THE MEDICINAL CHEMISTRY OF OLIGONUCLEOTIDES

7.1 Introduction

The core of any rational drug discovery program is medicinal chemistry. Although the synthesis of modified nucleic acids has been a subject of interest for some time, the intense focus on the medicinal chemistry of oligonucleotides dates perhaps to no more than 5 years before the preparation of this chapter.

Consequently, the scope of medicinal chemistry has recently expanded enormously, although the biological data to support conclusions about synthetic strategies are only beginning to emerge.

Modifications in the base, sugar, and phosphate moieties of oligonucleotides and oligonucleotide conjugates have been reported. The subjects of medicinal chemical programs include approaches to create enhanced affinity and more selective affinity for RNA or duplex structures; the ability to cleave nucleic acid targets; enhanced nuclease stability; cellular uptake and distribution; and *in vivo* tissue distribution, metabolism, and clearance.

7.2 Heterocycle Modifications

7.2.1 Pyrimidine Modifications. A relatively large number of modified pyrimidines have been synthesized and are now incorporated into oligonucleotides and evaluated. The principal sites of modification are C2, C4, C5, and C6. These and other nucleoside analogs have recently been thoroughly reviewed (290, 291). Consequently, a very brief summary of the analogs that displayed interesting properties is incorporated here. Inasmuch as the C2 position is involved in Watson-Crick hybridization, oligonucleotides containing C2 alkyl-modified pyrimidines have shown unattractive hybridization properties. However, an oligonucleotide containing 2-thiothymidine was found to hybridize well to DNA and, in fact, even better to RNA with a ΔT_m value of 1.5°C/modification (Swayze et al., unpublished results, 1995). In a different study, oligoribonucleotides with 2'-O-methyl-2-thiouridine exhibited a ΔT_m value of +5.5°C/modification when hybridized against RNA (292), resulting from a highly preorganized RNA-like C3'-endo conformation (attributed to the combination of 2-thio modification and 2'-O-Me substituent). Oligonucleotides with this modification also exhibit better hybridization discrimination for the wobble U-G base pair formation compared to the normal U-A base pair. This selectivity is a result of weaker hydrogen bonding and increased steric bulk of the 2-thiocarbonyl group (293, 294) (Fig. 5.4).

In contrast, several modifications in the 4-position that have interesting properties have been reported. 4-Thiopyrimidines have been incorporated into oligonucleotides with no significant negative effect on hybridization (295). However, recent studies have shown destabilization in the normal U-A base pair formation and stabilization of the wobble U-G base pair for 4-thiouridine (294). A bicyclic and an N4-methoxy analog of cytosine were shown to hybridize with both purine bases in DNA with T_m values approximately equal to that of natural base pairs (296). Additionally, a fluorescent base has been incorporated into oligonucleotides and shown to enhance DNA-DNA duplex stability (297).

A large number of modifications at the C5 position have also been reported, including halogenated nucleosides. Although the stability of duplexes may be enhanced by incorporating 5-halogenated uracil-containing nucleosides, the occasional mispairing with G and the potential that the oligonucleotide might degrade and release toxic nucleosides analogs cause concern (290). Oligonucleotides containing 5-propynylpyrimidine modifications have been shown to enhance the duplex stability (ΔT_m = 1.6°C/modification), and support RNase H activity. The 5-heteroarylpyrimidines were also shown to increase the stability of duplexes (218, 298). A more dramatic influence was reported for the tricyclic 2'-deoxycytidine analogs, termed phenoxazine, exhibiting an enhancement of 2–5°C/modification, depending on the positioning of the modified bases (299). Further, phenoxazine, phenothiazine, and tetrafluorophenoxazine have been synthesized and evaluated. When incorporated into oligonucleotides, these base modifications were shown to hybridize with guanine and to enhance thermal stability by extended stacking interactions and increased hydrophobic interactions. Tetrafluorophenoxazine is less selective in hybridization because it also hybridizes with adenine (300). Phenoxazine has been reported to increase the cellular permeation of oligonucleotides and alter RNase H cleavage sites (301). These helix-stabilizing properties have been shown to be further improved with the G-clamp, another tricyclic cytosine analog in which an aminoethoxy moiety is attached to the rigid phenoxazine scaffold. Binding studies demonstrated that a single incorporation of the "G-clamp" enhanced the binding affin-

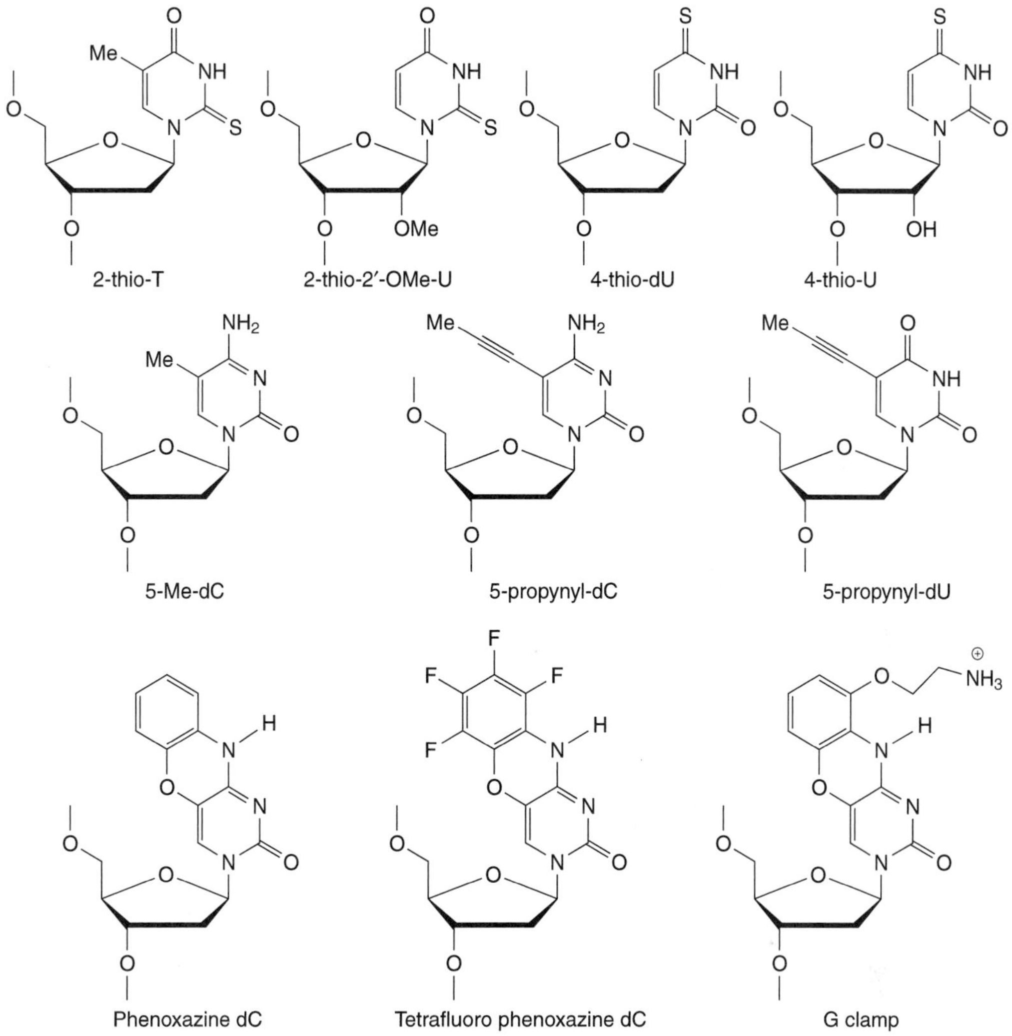

Figure 5.4. Selected pyrimidine modifications.

ity of a model oligonucleotide to its complementary target DNA or RNA with a ΔT_m value of up to 18°C relative to that of 5-methyl cytosine (302). This is the highest affinity enhancement attained so far with a single modification. Importantly, the gain in helical stability does not compromise the specificity of the oligonucleotide for complementary RNA. The tethered amino group may act as a hydrogen bond donor and interact with the O6 of guanine on the Hoogsteen face of the base pair. Thus the increased affinity of a G-clamp is presumably attributed to the combination of extended base stacking and an additional

specific hydrogen bond. There is no structural proof yet available for this hypothesis. The RNase H activation properties of eleven 5-propynyl modifications compared to a single G-clamp substitution have been assessed in another oligonucleotide sequence in terms of percentage cleavage of complementary RNA with nuclear extracts of HeLa cells. The propynyl oligonucleotide was four times less active than the control oligonucleotide, whereas the G-clamp oligonucleotide maintained similar activity (303).

As expected, modifications in the C6 position of pyrimidines are highly duplex destabi-

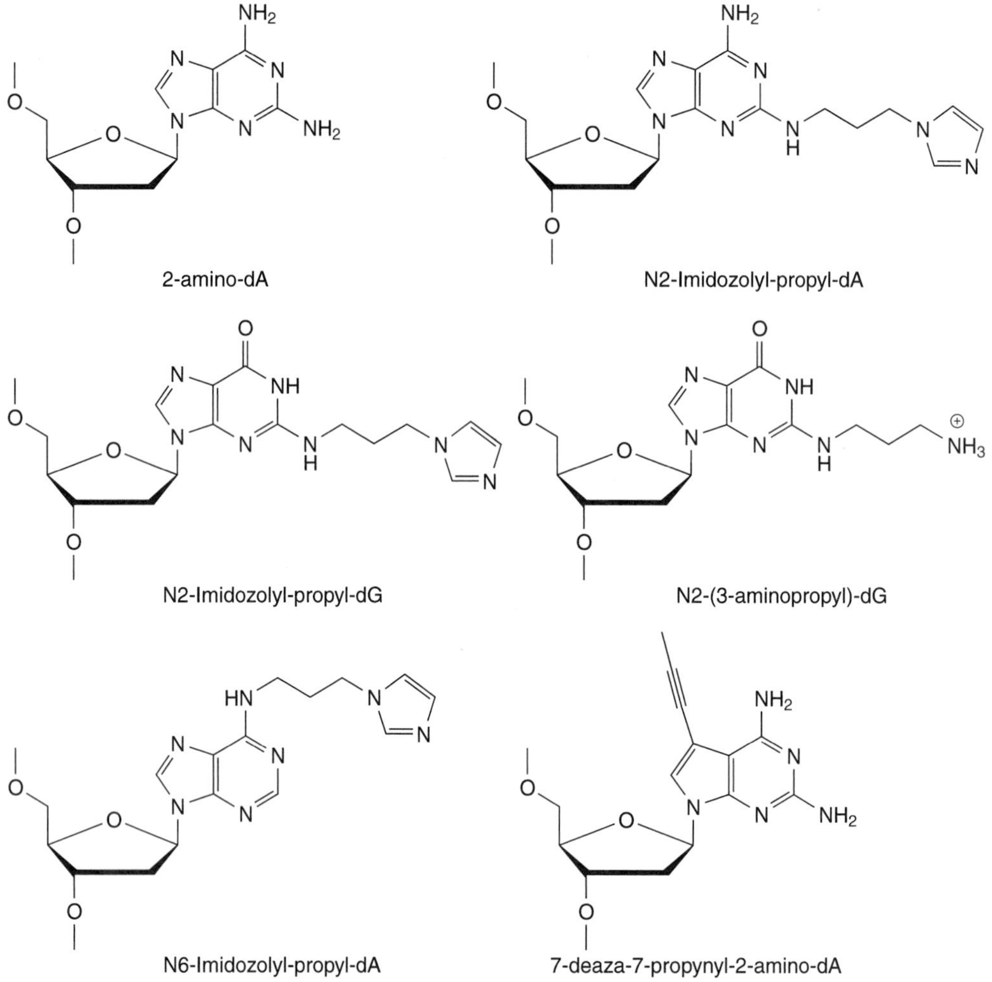

Figure 5.5. Selected purine modifications.

lizing (304). Oligonucleotides containing 6-aza pyrimidines have been shown not only to reduce the T_m value by 1–2°C per modification, but also to enhance the nuclease stability of oligonucleotides and to support *E. coli* RNase H-induced degradation of RNA targets (304).

7.2.2 Purine Modifications. Although numerous purine analogs have been synthesized, when incorporated into oligonucleotides, they usually have resulted in destabilization of duplexes. However, there are a few exceptions where a purine modification had a stabilizing effect. A brief summary of some of these analogs is discussed below (Fig. 5.5).

Generally, N1 modifications of the purine moiety have resulted in destabilization of the duplex (305). Similarly, C2 modifications have usually resulted in destabilization. However, oligonucleotides containing N2-imidazolylpropylguanine and N2-imidazolylpropyl-2-aminoadenine moieties showed a remarkable enhancement of binding affinity when hybridized to complementary DNA (306). This modification reduced *E. coli* RNase H activity by 2.5 times, possibly because of increased steric crowding in the minor groove (149). Oligonucleotides with an aminopropyl group placed at the N2-position of 2′-deoxyguanosine have also been studied. This modification showed enhanced binding properties against both DNA and RNA targets and was used for conjugating other functionalities (307). In the preceding instances, neither the structural reasons for increased binding affinity nor the sequence dependency of the binding affinity enhancement was explored. 2–6-Diaminopurine has been reported to enhance hybridization by approximately 1°C per modification when paired with T (308). This increase in binding affinity is independent of either the presence or the absence of any 2'-modification in the oligonucleotides. A convenient method of synthesizing this modification has been reported from our laboratories (309). Of the 3-position-substituted bases reported to date, only the 3-deaza adenosine analog has been shown to have no negative effect on hybridization.

Modifications at the C6 and C7 positions have likewise resulted in only a few interesting bases from the viewpoint of hybridization and other antisense-related properties. N6-imidazolylpropyl-adenine-modified oligodeoxynucleotides enhanced E. coli RNase H activity (149). Inosine has been shown to have little effect on duplex stability, but because it can pair and stack with all four normal DNA bases, it behaves as a universal base and creates an ambiguous position in an oligonucleotide (310). Incorporation of 7-deazainosine into oligonucleotides was destabilizing, and this was considered to be attributed to its relatively hydrophobic nature (311). 7-Deazaguanine was similarly destabilizing, but when 8-aza-7-deazaguanine was incorporated into oligonucleotides, it enhanced hybridizations (312). Thus, on occasion, introduction of more than one modification in a nucleobase may compensate for destabilizing effects of some modifications. Interestingly, the 7-iodo 7-deazaguanine residue was incorporated into oligonucleotides and shown to enhance the binding affinity dramatically ($\Delta T_m = 10.0°C$/ modification compared to that of 7-deazaguanine) (312). The increase in T_m value was attributed to (1) the hydrophobic nature of the modification, (2) increased stacking interaction, and (3) favorable pK_a of the base. Oligodeoxynucleotides with 7-propynyl-, 7-iodo-, and 7-cyano-7-deaza-2-amino-2'-deoxyadenosines showed increases of 3–4°C per modification in T_m against complementary RNA for single substitutions and smaller increases per incorporation for multiple substitutions relative to those of unmodified control sequences. When the 7-propyne and 7-iodo nucleosides were incorporated into antisense sequences targeting the 3'-UTR of murine C-raf mRNA, the sequences with three and four substitutions of the 7-propyne-7-deaza-2-amino-2'-deoxyadenosine exhibited a two- to threefold increase in potency over that of unmodified controls in vitro (313). In contrast, some C8-substituted bases have yielded improved nuclease resistance when incorporated in oligonucleotides, but seem to be somewhat destabilizing (290).

7.3 Oligonucleotide Conjugates

Although conjugation of various functionalities to oligonucleotides has been reported (314) to achieve a number of important objectives, the data supporting some of the claims are limited and generalizations are not possible based on the data presently available (Fig. 5.6).

7.3.1 Nuclease Stability. Numerous 3'-modifications have been reported to enhance the stability of oligonucleotides in serum (305). Both neutral and charged substituents have been reported to stabilize oligonucleotides in serum and, as a general rule, the stability of a conjugated oligonucleotide tends to be greater as bulkier substituents are added. Inasmuch as the principal nuclease in serum is a 3'-exonuclease, it is not surprising that 5'-modifications have resulted in significantly less stabilization. Internal modifications of base, sugar, and backbone have also been reported to enhance nuclease stability at or near the modified nucleoside (305). Such nuclease-resistant modifications include 3'-conjugates (305, 315), cationic modifications (316–318), zwitterionic modifications in the 2'-position (319, 320), geometrically altered linkages such as 2'-5' linkages (321), L-nucleotides and analogs (322, 323), α-anomeric oligonucleotides (324), 3'-3' terminal linkages (325, 326), 3'-3' linked oligonucleotides (327), and 3'-loop oligonucleotides (Khan, 1993, No. 4579; Agrawal, 1991, No. 2687). In many cases, these nuclease-resistant modifications and conjugates require

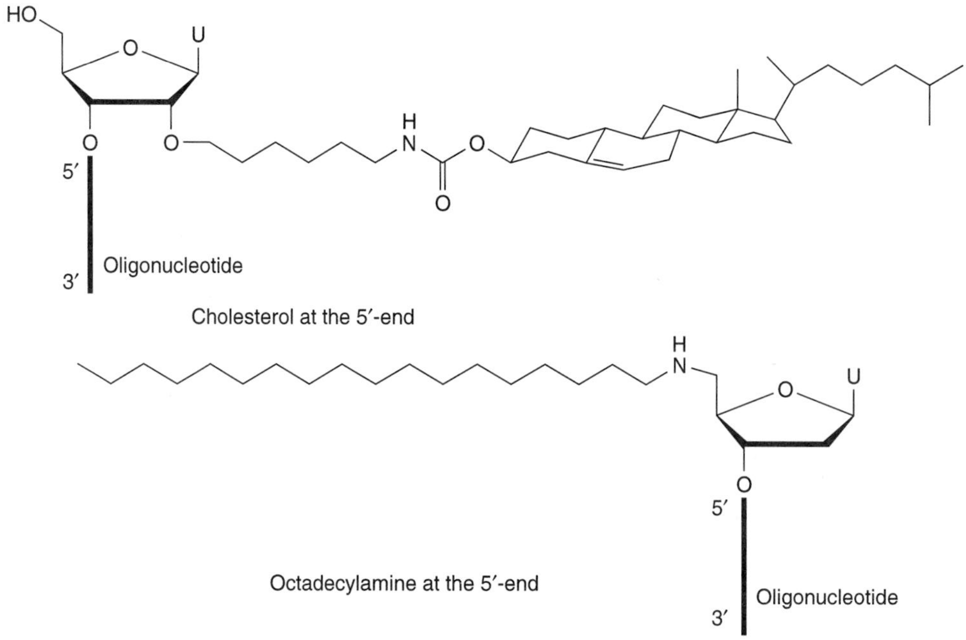

Figure 5.6. Selected 5'-oligonucleotide conjugates.

additional synthetic steps and modifications to the oligonucleotide synthesis protocol or monomer synthesis. In addition, there may be a loss in binding affinity to the target RNA because of these changes with geometrically altered linkages. Thiono triester (adamantyl, cholesteryl, and others) modified oligonucleotides have been reported to improve nuclease stability, cellular association, and binding affinity (328).

The demonstration that modifications may induce nuclease stability sufficient to enhance activity in cells in tissue culture and in animals has proved to be much more complicated because of the presence of 5'-exonucleases and -endonucleases. In our laboratory, 3'-modifications and internal point modifications have not provided sufficient nuclease stability to demonstrate pharmacological activity in cells (183). In fact, even a 5-nucleotide-long phosphodiester gap in the middle of a phosphorothioate oligonucleotide resulted in sufficient loss of nuclease resistance to cause complete loss of pharmacological activity (164). In mice, neither a 5'-cholesterol nor 5'-C18 amine conjugate altered the metabolic rate of a phosphorothioate oligodeoxynucleotide in liver, kid-

ney, or plasma (166). Furthermore, blocking the 3'- and 5'-termini of a phosphodiester oligonucleotide did not markedly enhance the nuclease stability of the parent compound in mice (90). However, 3'-modification of a phosphorothioate oligonucleotide was reported to enhance its stability in mice relative to that of the parent phosphorothioate (329). Moreover, a phosphorothioate oligonucleotide with a 3'-hairpin loop was reported to be more stable than its parent in rats (328). Thus, 3'-modifications may enhance the stability of the relatively stable phosphorothioates sufficiently to be of value.

7.3.2 Enhanced Cellular Uptake. Although oligonucleotides have been shown to be taken up by a number of cell lines in tissue culture, with perhaps the most compelling data relating to phosphorothioate oligonucleotides, a clear objective has been to improve cellular uptake of oligonucleotides (72, 264). Inasmuch as the mechanisms of cellular uptake of oligonucleotides are still very poorly understood, the medicinal chemistry approaches have been largely empirical and based on many unproven assumptions.

Because phosphodiester and phosphorothioate oligonucleotides are hydrophilic, the conjugation of lipophilic substituents to enhance membrane permeability has been a subject of considerable interest. Unfortunately, studies in this area have not been systematic and, at present, there is precious little information about the changes in physicochemical properties of oligonucleotides actually effected by specific lipid conjugates. Phospholipids, cholesterol and cholesterol derivatives, cholic acid, and simple alkyl chains have been conjugated to oligonucleotides at various sites in the oligonucleotide. The effects of these modifications on cellular uptake have been assessed using fluorescent, or radiolabeled, oligonucleotides or by measuring pharmacological activities. From the perspective of medicinal chemistry, very few systematic studies have been performed. The activities of short alkyl chains, adamantine, daunomycin, fluorescein, cholesterol, and porphyrin-conjugated oligonucleotides were compared in one study (330). A cholesterol modification was reported to be more effective than the other substituents at enhancing uptake. It also seems likely that the effects of various conjugates on cellular uptake may be affected by the cell type and target studied. For example, we have studied cholic acid conjugates of phosphorothioate deoxyoligonucleotides or phosphorothioate 2'-O-methyl oligonucleotides and observed enhanced activity against HIV and no effect on the activity of ICAM-1-directed oligonucleotides (331).

Additionally, polycationic substitutions and various groups designed to bind to cellular carrier systems have been synthesized. Although many compounds have been synthesized, the data reported to date are insufficient to draw firm conclusions about the value of such approaches or structure-activity relationships (SARs) (332).

7.3.3 RNA Cleaving Groups. Oligonucleotide conjugates were recently reported to act as artificial ribonucleases, albeit in low efficiencies (333). Conjugation of chemically reactive groups such as alkylating agents, photoinduced azides, porphyrins, and psoralens have been used extensively to effect a crosslinking of oligonucleotide and the target

RNA. In principle, this treatment may lead to translation arrest. In addition, lanthanides and complexes thereof have been reported to cleave RNA by way of a hydrolytic pathway. Recently, a novel europium complex was covalently linked to an oligonucleotide and shown to cleave 88% of the complementary RNA at physiological pH (334).

7.3.4 RNase H Activity. Oligonucleotide conjugates may play a distinct role in improving RNase H activity because conjugation of small molecules, especially at the oligonucleotide termini, may not affect the factors required for enzyme activity. Acridine-conjugated oligonucleotides targeted to β-globin mRNA were shown to be better inhibitors of β-globin synthesis than the unmodified oligonucleotides (335) in a microinjected Xenopus oocytes assay as a result of RNase H activity. Cholesterol-conjugated oligonucleotides targeted to an mRNA fragment of Ha-*ras* oncogene were able to promote a greater extent of target RNA hydrolysis, nearly three- to five-fold, by RNase H compared to the nonconjugated parent (336). The 3'-conjugate performed better than the 5'-conjugate. More recently, polycyclic aromatic chromophores such as phenazine (PZN) and dipyridophenazine (DPPZ) have been evaluated in *E. coli* RNase H cleavage of the target strand when conjugated to oligonucleotides (337, 338). The 3'-conjugated antisense oligomers promoted faster hydrolysis of the target RNA than the unmodified compound and a stabilizing interaction between the ligands and the enzyme has been proposed.

7.3.5 In Vivo Effects. To date, relatively few studies have been reported *in vivo*. The properties of 5'-cholesterol and 5'-C18 amine conjugates of a 20-mer phosphorothioate oligodeoxynucleotide have been determined in mice. Both compounds increased the fraction of an i.v. bolus dose found in the liver. The cholesterol conjugate resulted in more than 80% dose accumulation in the liver. Neither conjugate enhanced stability in plasma, liver, or kidney (166). Interestingly, the only significant change in the toxicity profile was a slight increase in effects on serum transamineses and histopathological changes, indicative of

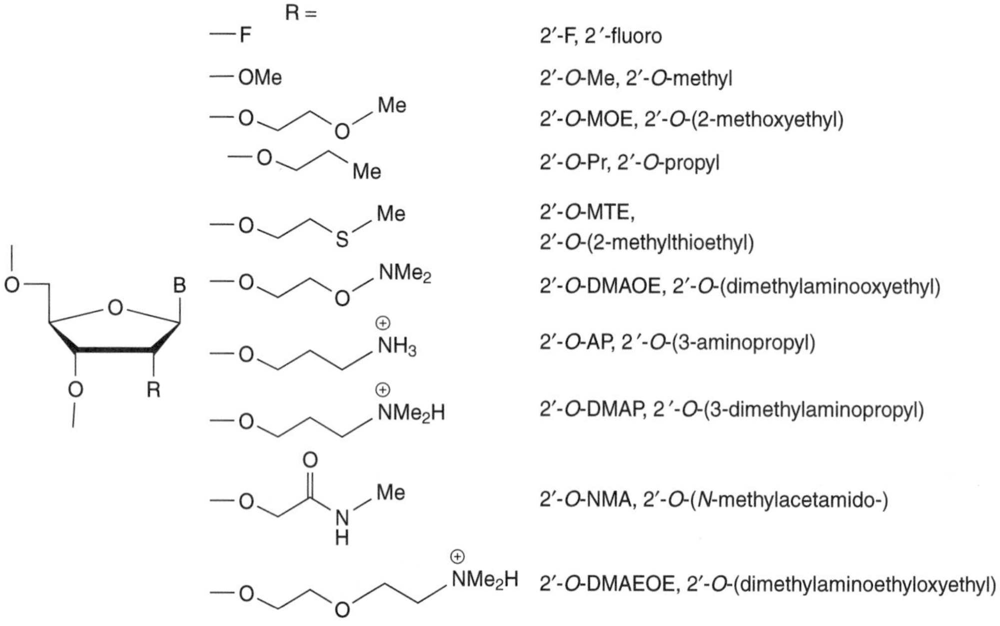

R =

—F   2'-F, 2'-fluoro

—OMe   2'-O-Me, 2'-O-methyl

—O⌒⌒O⌒Me   2'-O-MOE, 2'-O-(2-methoxyethyl)

—O⌒⌒Me   2'-O-Pr, 2'-O-propyl

—O⌒⌒S⌒Me   2'-O-MTE,
                                            2'-O-(2-methylthioethyl)

—O⌒⌒O⌒NMe₂   2'-O-DMAOE, 2'-O-(dimethylaminooxyethyl)

—O⌒⌒⌒N⊕H₃   2'-O-AP, 2'-O-(3-aminopropyl)

—O⌒⌒⌒N⊕Me₂H   2'-O-DMAP, 2'-O-(3-dimethylaminopropyl)

—O⌒C(=O)N(H)⌒Me   2'-O-NMA, 2'-O-(N-methylacetamido-)

—O⌒⌒O⌒⌒N⊕Me₂H   2'-O-DMAEOE, 2'-O-(dimethylaminoethyloxyethyl)

Figure 5.7. Selected 2' modifications.

slight liver toxicity associated with the cholesterol conjugate (339). A 5'-cholesterol phosphorothioate conjugate was also recently reported to have a longer elimination half-life, to be more potent, and to induce greater liver toxicity in rats at a high dose of 50 mg/kg (340).

7.4 Sugar Modifications

The focus of second-generation oligonucleotide modifications has centered on the sugar moiety. In oligonucleotides, the pentofuranose sugar ring occupies a central connecting manifold that also positions the nucleobases for effective stacking. Recently, a symposium series has been published on the carbohydrate modifications in antisense research, which covers this topic in great detail (341), and another review has been published, which summarizes the antisense properties (342). Therefore, the content of the following discussion is restricted to a summary of the main events in this area (Fig. 5.7).

A growing number of oligonucleotides in which the pentofuranose ring is modified or replaced have been reported (343). Uniform modifications at the 2'-position have been shown to enhance hybridization to RNA, and

in some cases, to enhance nuclease resistance (343). Chimeric oligonucleotides containing 2'-deoxyoligonucleotide gaps (for RNase H activation) with 2'-modified wings have been shown to be more potent than parent molecules (144). Other sugar modifications include α-oligonucleotides, carbocyclic oligonucleotides, and hexapyranosyl oligonucleotides (343). Of these, α-oligonucleotides have been most extensively studied. They hybridize in parallel fashion to single-stranded DNA and RNA and are nuclease resistant. However, they have been reported to be oligonucleotides designed to inhibit Ha-*ras* expression. All these oligonucleotides support RNase H and, as can be seen, there is a direct correlation between affinity and potency.

A growing number of oligonucleotides in which the C2'-position of the sugar ring is modified have been reported (332, 333). These modifications include lipophilic alkyl groups, intercalators, amphipathic amino-alkyl tethers, alkoxy alkyl groups, acetamide groups, positively charged polyamines, highly electronegative fluoro or fluoro alkyl moieties, and sterically bulky methylthio derivatives. 2'-O-Aminopropyl (2'-O-AP) modification exhib-

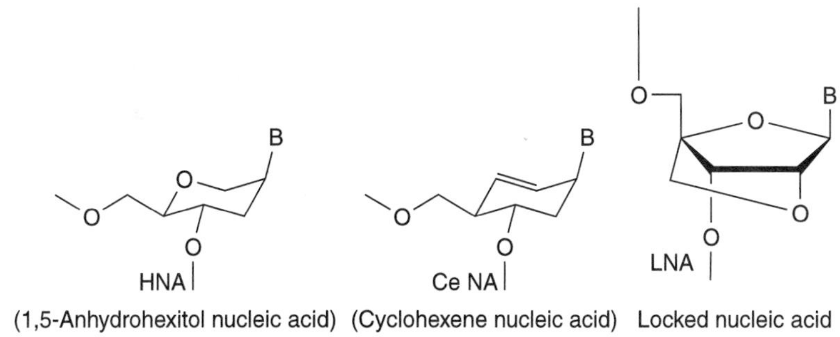

HNA   Ce NA   LNA

(1,5-Anhydrohexitol nucleic acid) (Cyclohexene nucleic acid) Locked nucleic acid

Figure 5.8. Some new sugar modifications.

ited very high nuclease resistance in *in vitro* studies, and crystallographic studies involving this modification suggest that the interference of the positively charged 2'-O-substituent with the metal ion binding site of the exonuclease may be the reason for effective slow down of the nuclease degradation. The beneficial effects of a C2'-substitution on the antisense oligonucleotide cellular uptake, nuclease resistance, and binding affinity have been well documented in the literature. In addition, excellent review articles have appeared in the last few years on the synthesis and properties of C2'-modified oligonucleotides (333, 344–347).

Other modifications of the sugar moiety have also been studied, including other sites such as the 1'-position and the 4'-position as well as more substantial modifications. However, much less is known about the antisense effects of these modifications (93). Oligodeoxynucleotides containing 4'-α-C-aminoalkyl-thymidines have been reported to have acceptable RNA binding properties and the hybrids formed between this class of oligomers and their complementary RNAs were reported to be substrates for *E. coli* RNase H and human RNase H present in HeLa cell nuclear extracts. However, the rates of RNase H cleavage were found to be slower than those of natural substrates (348).

2'-O-Methyl-substituted phosphorothioate oligonucleotides have recently been reported to be more stable than their parent compounds in mice and to display enhanced oral bioavailability (328, 349). The analogs displayed tissue distribution similar to that of the parent phosphorothioate. Similarly, we have compared the pharmacokinetics of 2'-O-pro-

pyl-modified phosphodiester and phosphorothioate deoxynucleotides (182). As expected, the 2'-O-propyl modification increased lipophilicity and nuclease resistance. In mice the 2'-O-propyl phosphorothioate was too stable in liver or kidney to measure an elimination half-life. Interestingly, the 2'-O-propyl phosphodiester was much less stable than the parent phosphorothioate in all organs, except the kidney, in which the 2'-O-propyl phosphodiester was remarkably stable. The 2'-O-propyl phosphodiester did not significantly bind to albumin, whereas the affinity of the phosphorothioate for albumin was enhanced. The only difference in toxicity between the analogs was a slight increase in renal toxicity associated with the 2'-O-propyl phosphodiester analog (339).

Incorporation of 2'-O-methoxyethyl into oligonucleotides increased the T_m value by 1.1°C/modification when hybridized to the complement RNA. In a similar manner, several other 2'-O-alkoxy modifications have been reported to enhance the affinity (350). The increase in affinity with these modifications was attributed to (1) the favorable gauche effect of side chain, (2) additional solvation of the alkoxy substituent in water, and (3) preorganization of the sugar moiety into a C3'-endo conformation (351).

More substantial carbohydrate modifications have also been studied (Fig. 5.8). The 4'-oxygen has been replaced with sulfur. Although a single substitution of a 4'-thio-modified nucleoside resulted in destabilization of a duplex, incorporation of two 4'-thio-modified nucleosides increased the affinity of the duplex (352). Also, hexose-containing oligonucleotides were created and found to have very low affinity for RNA (353). However, 1,5-anhydro-hexitol-based nucleic acids (HNA) exhibit greater than 2°C/modification for RNA hybridization but fail to activate Rnase H in RNA-HNA hybrids. A closely related analog with the cyclohexene-containing oligonucleotides, called cyclohexene nucleic acids (CeNA) (354), exhibit T_m stabilization of 1.2°C/modification when hybridized to both the complement RNA and DNA. This modification can interconvert easily between C3'-endo-like and C2'-endo-like conformations because of a low energy barrier associated with the cyclohexene ring system. Because of this process, this analog is able to activate *E. coli* RNase H, resulting in cleavage of the RNA strand. However, the rates of RNase H compared to those of the natural analog have not been evaluated. Bicyclic sugars have been synthesized with the hope that preorganization into more rigid structures would enhance hybridization. Several of these modifications have been reported to enhance hybridization (355). A bicyclic sugar analog connecting 2'-oxygen to 4'-carbon by way of a methylene bridge has been termed locked nucleic acids (LNA) and shown to have improved hybridization (greater than 4–5°C/modification) to both DNA and RNA (356). This modification has also been claimed to activate RNase H (357) as a mixed oligomer with DNA and not as a chimera involving deoxy gaps. This claim is questionable on the basis of conformational and steric factors associated with this modification.

7.4.1 2'-Modified Oligonucleotides with Ability to Activate RNase H.

Uniformly 2'-modified oligonucleotides are not capable of activating RNase H in an oligonucleotide/RNA hybrid. However, recent reports show that hybrids of RNA and arabinonucleic acids [the 2'-stereoisomer of RNA based on D-arabinose instead of the natural D-ribose (2'-arabino-OH, ANA) and 2'-deoxy-2'-fluoro-D-arabinonucleic acid (2'F-ANA)] are substrates of RNase H (358, 359) (Fig. 5.9). However, ANA/RNA hybrids had a lower T_m value, compared to that of the control DNA/RNA hybrids. This destabilization ($\Delta T_m = -1.0$°C/modification) is presumed to derive from steric interference by the -C2'-OH group, which is oriented into the

major groove of the helix, causing local deformation (unstacking). Replacing the *ara*-2'-OH group by a 2'-F atom resulted in an increase in duplex melting temperature. The 2'F-ANA/RNA duplex had higher values of T_m compared to those of the corresponding hybrids formed by ANA, phosphorothioate DNA, and DNA ($\Delta T_m = +0.5$°C/modification compared to that of DNA). Both ANA and 2'-F-ANA are expected to have C2'-endo conformations. When hybridized to RNA they form an A-type helix similar to that of RNA-DNA hybrids, as shown by circular dichroism studies. 2'-F-ANA has been shown to activate RNase H when hybridized to target RNA and cleave it. However, ANA/RNA hybrids are poor substrates for RNase H, possibily because of the poor binding affinity of ANA for RNA. The nuclease stability of ANA and 2'-F-ANA oligonucleotide phosphodiesters is not sufficient to use them as antisense drugs. Furthermore, the toxicity properties of arabino nucleosides may be a matter of concern.

7.5 Backbone Modifications

Substantial progress in creating new backbones for oligonucleotides that replace the phosphate or the sugar-phosphate unit has been made. The objectives of these programs are to improve hybridization by removing the negative charge, enhance stability, and potentially improve the pharmacokinetics (Fig. 5.10). For a review of the backbone modifications reported to date, see Refs. 93 and 355. Suffice it to say that numerous modifications have been made that replace phosphate, retain hybridization, alter charge, and enhance stability. One of the important advantages of the high affinity backbone modifications is their use in RNase H-independent mechanisms of action, such as inhibition or modulation of splicing, translation arrest, and disruption of necessary RNA structure. Because these modifications are now being evaluated *in vitro* and *in vivo*, a preliminary assessment should be possible shortly.

Replacement of the entire sugar-phosphate unit has also been accomplished and the oligomeric compounds produced have displayed very interesting characteristics. PNA oligomers have been shown to bind to single-stranded DNA and RNA with extraordinary

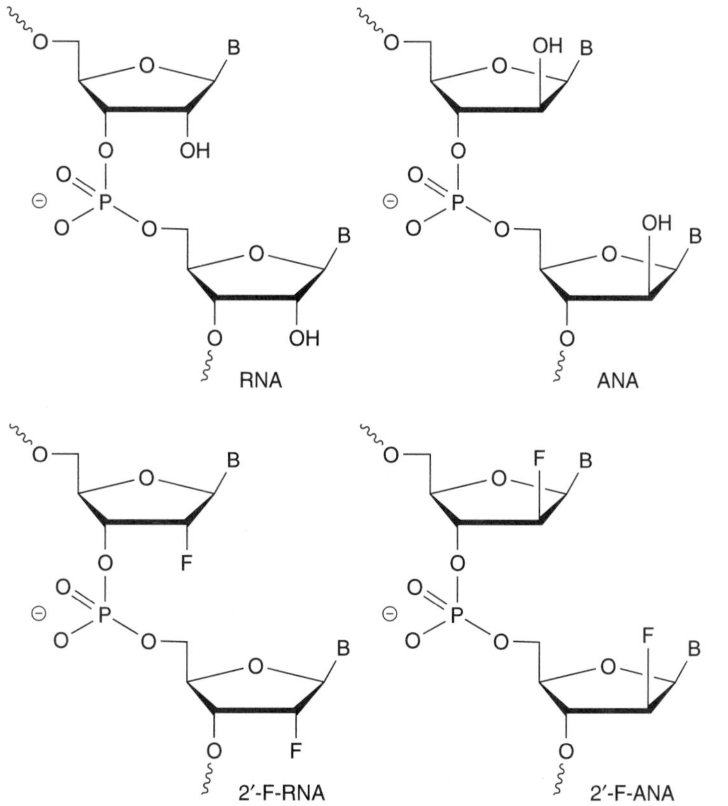

Figure 5.9. Structures of RNA, ANA, 2'-F-RNA, and 2'-F-ANA.

affinity and high sequence specificity. They have been shown to be able to invade some double-stranded nucleic acid structures. PNA oligomers can form triple-stranded structures with DNA or RNA. PNA oligomers with lysine ends have better water solubility, less aggregation, and higher target binding affinity. Peptide nucleic acid oligomers were shown to be able to act as antisense and transcriptional inhibitors when microinjected in cells (47). PNA oligomers appear to be quite stable to nucleases and peptidases as well. A recent study demonstrated that antisense PNAs alter splicing of the IL-5R-α pre-mRNA efficiently in a fashion similar to their 2'-*O*-MOE-modified counterparts of the same sequence when electroporated in cells. Moreover, using PNA as the splicing modulator, the length of the antisense oligomer could be shortened from 20 to 15 nucleobase units to obtain a comparable effect (360).

7.6 Summary

In summary, then, in the past 5 years, enormous advances in the medicinal chemistry of oligonucleotides have been reported. Modifications at nearly every position in oligonucleotides have been attempted and numerous potentially interesting analogs have been identified. Although it is far too early to determine which of the modifications may be most useful for particular purposes, it is clear that a wealth of new chemicals is available for systematic evaluation and that these studies should provide important insights into the SAR of oligonucleotide analogs.

8 2'-*O*-(2-METHOXYETHYL) CHIMERAS: SECOND-GENERATION ANTISENSE DRUGS

The 2'-*O*-(2-methoxyethyl) (Fig. 5.11) substitution represents a significant advance in an-

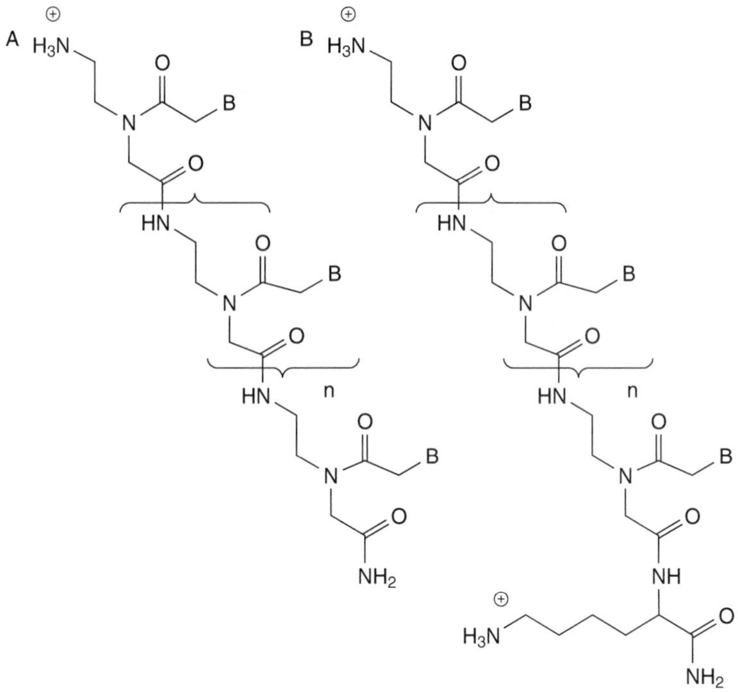

Figure 5.10. PNA peptide nucleic acid without (A) and with Lys end (B).

tisense therapeutics and the culmination of more than a decade of progress in antisense technology. Although oligonucleotides in which every nucleotide contains a 2'-*O*-(2-methoxyethyl) modification have proved of great value when used for occupancy-only-mediated mechanisms (e.g., induction of alternative splicing; for review, see Ref. 361). They

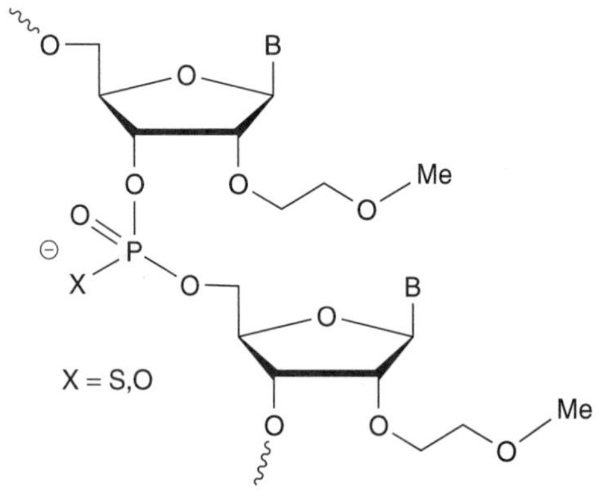

Figure 5.11. Structure of 2'-*O*-MOE: 2'-*O*-(2-methoxyethyl) RNA.

have been used most broadly in chimeric structures designed to serve as substrates for RNase H. Numerous chimeric 2'-*O*-(2-methoxyethyl) oligonucleotides have not been studied extensively. Although such oligonucleotides typically display affinity for target RNA that is several orders of magnitude greater than that for phosphorothioate oligodeoxynucleotides, they are less attractive substrates for RNase H (for review, see ref. 362). Consequently, they typically display only 5- to 15-fold increases in potency both *in vitro* and *in vivo*. For example, a direct comparison of two antisense inhibitors of C-*raf* kinase with the same sequence, one of which is a phosphorothioate oligodeoxynucleotide, the other a 2'-*O*-(2-methoxyethyl) chimera, showed that *in vitro* the 2'-*O*-(2-methoxyethyl) chimera was approximately fivefold more potent. Similarly, after i.v. administration in a rat heart allograph model, the modified oligonucleotide was at least fivefold more potent (363). Similar data were reported from a direct comparison of antisense inhibitors of survivin both *in vitro* and *in vivo* in a human tumor xenograft model (364). Pharmacological studies of antisense oligomers modified with 2'-O-(2-methoxyethyl) chemistry have recently been reviewed (365).

Perhaps more important, 2'-O-(2-methoxyethyl) chimeras are substantially more stable than phosphorothioate oligonucleotides. In mice, rats, and monkeys, the elimination half-life is nearly 30 days in plasma and several tissues (314, 366–368). Furthermore, in the liver, elegant correlations between pharmacokinetic and pharmacodynamic effects have been reported to show a duration of pharmacological action of nearly 20 days in the mouse (369).

Studies in animals have recently been extended to humans. ISIS 104838, a TNFα antisense inhibitor, a 2'-O-(2-methoxyethyl) chimera, was shown to reduce TNFα secretion at doses at least 10 times lower than would be expected in humans by first-generation antisense drugs (Dorr, unpublished observations, 2001) and to have an elimination half-life in plasma of approximately 30 days (Geary, unpublished observations, 2001). These studies demonstrate the feasibility of once-a-month dosing.

Although significant progress in achieving acceptable oral bioavailability of antisense inhibitors has been reported, much remains to be accomplished and key clinical studies in progress will help determine how feasible oral delivery is (for review, see Ref. 215). Based on the progress, two key barriers to oral bioavailability have been identified and partially overcome: presystemic metabolism and penetration across the gastrointestinal (GI) mucosa.

Because the gut has a very high level of nucleases contributed by both the host and bacteria resident in the GI tract, metabolism of phosphorothioate oligodeoxynucleotides in the gut occurs much too rapidly to support adequate oral bioavailability (370). Chimeric modifications of the oligonucleotides have been shown to enhance stability and oral bioavailability (328, 371). With the development of 2'-O-MOE phosphorothioates, oral bioavailability is potentially feasible (215).

In rodents, dogs, and monkeys, the permeability of the GI tract to antisense inhibitors has been significantly enhanced by using formulations containing penetration enhancers such as bile acid salts and fatty acids. After intrajejunal administration of several 2'-O-MOE-modified oligonucleotides in the presence of penetration enhancer-containing formulations, systemic bioavailability in excess of 20% was observed in all three species (215). Initial studies with solid dose forms containing penetration enhancers resulted in significantly less systemic bioavailability (5.5%); clearly therefore more progress is required before solid dose forms with adequate (10–20%) oral bioavailability are available (215, 366, 367). Initial clinical trials in which a variety of solid dose forms containing ISIS 104803, a 2'-O-MOE chimeric antisense inhibitor of TNFα, and various penetration enhancers are in progress, so within the next few years we should have some sense of the near-term potential for oral delivery of antisense inhibitors in humans.

Finally, 2'-O-(2-methoxyethyl) chimeras have been shown to be less proinflammatory (169, 372). After repeated subcutaneous dosing in humans, ISIS 104838 produced dramatically lower local inflammation than that of first-generation antisense drugs. Thus, more convenient and better locally tolerated subcutaneous administration may be feasible.

9 CONCLUSIONS

Although many more questions about antisense remain to be answered than are currently answered, progress has continued to be gratifying. Clearly, as more is learned, we will be in the position to perform progressively more sophisticated studies and to understand more of the factors that determine whether an oligonucleotide actually works through an antisense mechanism. We should also have the opportunity to learn a great deal more about this class of drugs as additional studies are completed in humans.

10 ABBREVIATIONS

aPTT	activated thromboplatin
AT-1	angiothromin receptor 1
AT-2	angiothrombin 2
c-AMP	cyclic AMP
CeNA	cyclohexene nucleic acid
CMV	cytomegalovirus

Ha-*ras* Harvey *ras*
HIV human immune deficiency virus
HPV human papillomavirus
HSV herpes simplex virus
ICAM intercellular adhesion molecule
IL interleukin
IL-1a interleukin 1a
LNA locked nucleic acid
mRNA messenger RNA
NK natural killer
PKC protein kinase C
ply rC poly ribocytosine
PNA peptide nucleic acid
poly rI poly ribinosine
PTHrP parathyroid hormone-related peptide
T_m thermal melting temperature
ΔT_m change in T_m relative to DNA
TAR transactivator response element
VSV vesicular stomatitis virus

11   ACKNOWLEDGMENTS

I thank Drs. Art Levin and Mano Manoharan for thoughtful review and additions to several sections of the manuscript. Thanks also to Bella Aguilar and Marcia Southwell for converting the manuscript into readable material.

REFERENCES

1. J. L. Glick and A. R. Goldberg, *Science*, **149**, 997–998 (1965).

2. J. L. Glick, *Cancer Res.*, **27**, 2338–2341 (1967).

3. J. L. Glick and A. P. Salim, *Nature*, **213**, 676–678 (1967).

4. S. T. Crooke, *Preliminary Studies on Genetic Engineering: The Uptake of Oligonucleotides and RNA by Novikoff Hepatoma Ascites Cells*, School of Graduate Studies, Baylor College of Medicine, Houston, TX, 1972.

5. A. Isaacs, R. A. Coo, and Z. Rotem, *Lancet*, **2**, 113 (1963).

6. C. Colby, M. J. Chamberlin, P. H. Duesberg, and M. I. Simon, Proceedings of the Symposium on Molecular Biology, University of California, La Jolla, CA, Springer-Verlag, New York, 1971.

7. R. F. Beers, Proceedings of the Symposium on Molecular Biology, University of California, La Jolla, CA, Springer-Verlag, New York, 1971.

8. C. J. Colby, *Prog. Nucleic Acid Res. Mol. Biol.*, **11**, 1–32 (1971).

9. W. A. Carter and P. M. Pitha, Proceedings of the Symposium on Molecular Biology, University of California, La Jolla, CA, Springer-Verlag, New York, 1971.

10. M. C. Yu, P. A. Young, and W. H. Yu, *Am. J. Pathol.*, **64**, 305–320 (1971).

11. E. R. Homan, R. P. Zendzian, H. B. Levy, and R. H. Adamson, *Toxicol. Appl. Pharmacol.*, **23**, 579–588 (1972).

12. H. C. Stevenson, P. G. Abrams, C. S. Schoenberger, R. B. Smalley, R. B. Herberman, and K. A. Foon, *J. Biol. Response Mod.*, **4**, 650–655 (1985).

13. B. C. Lampkin, A. S. Levine, H. Levy, W. Krivit, and D. Hammond, *J. Biol. Response Mod.*, **4**, 531–537 (1985).

14. E. Schlick, F. Bettens, R. Ruffmann, M. A. Chirigos, and P. Hewetson, *J. Biol. Response Mod.*, **4**, 628–633 (1985).

15. S. E. Krown, D. Kerr, W. E. D. Stewart, A. K. Field, and H. F. Oettgen, *J. Biol. Response Mod.*, **4**, 640–649 (1985).

16. W. A. Carter, D. R. Brodsky, and M. G. Pellegrino, *Lancet*, **1**, 1286–1292 (1987).

17. R. J. Suhadolnik, N. L. Reichenbach, C. Lee, D. R. Strayer, I. Brodsky, and W. A. Carter, *Prog. Clin. Biol. Res.*, **202**, 449–456 (1985).

18. J. Watson and F. Crick, *Nature*, **171**, 737 (1953).

19. D. Gillespie and S. Spiegelman, *J. Mol. Biol.*, **12**, 829–842 (1965).

20. J. D. Thompson and D. Gillespie, *Clin. Biochem.*, **23**, 261–266 (1990).

21. P. O. Ts'o, P. S. Miller, and J. J. Greene, *Development of Target-Oriented Anticancer Drugs*, Raven Press, New York, 1983, pp. 189.

22. J. C. Barrett, P. S. Miller, and P. O. P. Ts'o, *Biochemistry*, **13**, 4897–4906 (1974).

23. M. H. Caruthers, *Science*, **230**, 281–285 (1985).

24. G. Alvarado-Urbina, G. M. Sathe, W. C. Liu, M. F. Gillen, P. D. Duck, R. Bender, and K. K. Ogilvie, *Science*, **214**, 270–274 (1981).

25. E. De Clercq, F. Eckstein, and T. C. Merigan, *Science*, **165**, 1137–1139 (1969).

26. C. J. Marcus-Sekura, A. M. Woerner, G. Zon, and G. V. Quinnan, *Nucleic Acids Res.*, **15**, 5749–5763 (1987).

27. M. Matsukura, K. Shinozuka, G. Zon, H. Mitsuya, M. Reitz, J. S. Cohen, and S. Broder, *Proc. Natl. Acad. Sci. USA*, **84**, 7706–7710 (1987).

28. S. Agrawal, J. Goodchild, M. P. Civeira, A. H. Thornton, P. S. Sarin, and P. C. Zamecnik, *Proc. Natl. Acad. Sci. USA*, **85**, 7079–7083 (1988).

29. J. Goodchild, S. Agrawal, M. P. Civeira, P. S. Sarin, D. Sun, and P. C. Zamecnik, *Proc. Natl. Acad. Sci. USA*, **85**, 5507–5511 (1988).

30. W. Gao, C. A. Stein, J. S. Cohen, G. E. Dutschman, and Y.-C. Cheng, *J. Biol. Chem.*, **264**, 11521–11526 (1989).

31. C. C. Smith, L. Aurelian, M. P. Reddy, P. S. Miller, and P. O. P. Ts'o, *Proc. Natl. Acad. Sci. USA*, **83**, 2787–2791 (1986).

32. R. Heikkila, G. Schwab, E. Wickstrom, S. L. Loke, D. H. Pluznik, R. Watt, and L. M. Neckers, *Nature*, **328**, 445–449 (1987).

33. E. L. Wickstrom, T. A. Bacon, A. Gonzalez, G. H. Lyman, and E. Wickstrom, *In Vitro Cell Dev Biol.*, **25**, 297–302 (1989).

34. C. Helene, *Antisense Research and Applications*, CRC Press, Boca Raton, FL, 1993, pp. 375–385.

35. H. E. Moser and P. B. Dervan, *Science*, **238**, 645–650 (1987).

36. T. Le Doan, L. Perrouault, D. Praseuth, N. Habhoub, J. L. Decout, and N. T. Thuong, *Nucleic Acids Res.*, **15**, 7749–7760 (1987).

37. M. Cooney, G. Czernuszewicz, E. H. Postel, S. J. Flint, and M. E. Hogan, *Science*, **241**, 456–460 (1988).

38. C. K. Mirabelli and S. T. Crooke, *Antisense Research and Applications*, CRC Press, Boca Raton, FL, 1993, pp. 7–35.

39. A. F. Faruqi, M. Egholm, and P. M. Glazer, *Proc. Natl. Acad. Sci. USA*, **95**, 1398–1403 (1998).

40. M. J. Blommers, F. Natt, W. Jahnke, and B. Cuenoud, *Biochemistry*, **37**, 17714–17725 (1998).

41. A. F. Faruqi, S. H. Krawczyk, M. D. Matteucci, and P. M. Glazer, *Nucleic Acids Res.*, **25**, 633–640 (1997).

42. L. Lacroix, J. Lacoste, J. F. Reddoch, J. L. Mergny, D. D. Levy, M. M. Seidman, M. D. Matteucci, and P. M. Glazer, *Biochemistry*, **38**, 1893–1901 (1999).

43. M. A. Macris and P. M. Glazer, *Antisense Drug Technology: Principles, Strategies, and Applications*, Marcel Dekker, New York, 2001, pp. 859–886.

44. G. Wang, M. M. Seidman, and P. M. Glazer, *Science*, **271**, 802–805 (1996).

45. G. Wang, D. D. Levy, M. M. Seidman, and P. M. Glazer, *Mol. Cell. Biol.*, **15**, 1759–1768 (1995).

46. G. Wang and P. M. Glazer, *J. Biol. Chem.*, **270**, 22595–22601 (1995).

47. J. C. Hanvey, N. C. Peffer, J. E. Bisi, S. A. Thomson, R. Cadilla, J. A. Josey, D. J. Ricca, C. F. Hassman, M. A. Bonham, K. G. Au, S. G. Carter, D. A. Bruckenstein, A. L. Boyd, S. A. Noble, and L. E. Babiss, *Science*, **258**, 1481–1485 (1992).

48. U. Koppelhus, V. Zachar, P. E. Nielsen, X. Liu, J. E. Olsen, and P. Ebbesen, *Nucleic Acids Res.*, **25**, 2167–2173 (1997).

49. D. Praseuth, M. Grigoriev, A.-L. Guieysse, L. L. Pritchard, A. Marel-Bellan, P. E. Nielsen, and C. Helene, *Biochim. Biophys. Acta*, **1309**, 226–238 (1996).

50. P. E. Nielsen, M. Egholm, R. H. Berg, and O. Buchardt, *Nucleic Acids Res.*, **21**, 197–200 (1993).

51. P. E. Nielsen, M. Egholm, R. H. Berg, and O. Buchardt, *Antisense Research and Applications*, CRC Press, Boca Raton, FL, 1993, pp. 363–373.

52. S. White, J. W. Szewczyk, J. M. Turner, E. E. Baird, and P. B. Dervan, *Nature*, **391**, 468–471 (1998).

53. R. W. Burli and H. E. Moser, *Antisense Technology: Principles, Strategies and Applications*, Marcel Dekker, New York, 2001, pp. 833–857.

54. T. R. Cech, *Annu. Rev. Biochem.*, **59**, 543–568 (1990).

55. O. Uhlenbeck, *Antisense Research and Applications*, CRC Press, Boca Raton, FL, 1993, p. 83.

56. J. J. Rossi, *Antisense Technology: Principles, Strategies and Applications*, Marcel Dekker, New York, 2001, pp. 887–905.

57. R. Parthasarathy, G. J. Cote, and R. F. Gagel, *Cancer Res.*, **59**, 3911–3914 (1999).

58. M. Ventura, P. Wang, N. Franck, and S. Saragosti, *Biochem. Biophys. Res. Commun.*, **203**, 889–898 (1994).

59. O. Heidenreich, F. Benseler, A. Fahrenholz, and F. Eckstein, *J. Biol. Chem.*, **269**, 2131–2138 (1994).

60. M. Koizumi and E. Ohtsuka, *Biochemistry*, **30**, 5145–5150 (1991).

61. L. Prasmickaite, A. Hogset, G. Maelandsmo, K. Berg, J. Goodchild, T. Perkins, O. Fodstad, and E. Hovig, *Nucleic Acids Res.*, **26**, 4241–4248 (1998).

62. C. M. Flory, P. A. Pavco, T. C. Jarvis, M. E. Lesch, F. E. Wincott, L. Beigelman, S. W.

HuntIII, and D. J. Schrier, *Proc. Natl. Acad. Sci. USA*, **93**, 754–758 (1996).

63. J. A. Sandberg, K. S. Bouhana, A. M. Gallegos, A. B. Agrawal, S. L. Grimm, F. E. Wincott, M. A. Reynolds, P. A. Pavco, and T. J. Parry, *Antisense Nucleic Acid Drug Dev.*, **9**, 271–277 (1999).

64. C. Tuerk and L. Gold, *Science*, **249**, 505–510 (1990).

65. A. D. Ellington and J. W. Szostak, *Nature*, **346**, 818–822 (1990).

66. T. Vickers, B. F. Baker, P. D. Cook, M. Zounes, R. W. Buckheit, Jr., J. Germany, and D. J. Ecker, *Nucleic Acids Res.*, **19**, 3359–3368 (1991).

67. E. Wickstrom, *J. Biochem. Biophys. Methods*, **13**, 97–102 (1986).

68. E. De Clercq, F. Eckstein, and T. C. Merigan, *Science*, **165**, 1137–1140 (1969).

69. D. M. Tidd and H. M. Warenius, *Br. J. Cancer*, **60**, 343–350 (1989).

70. J. M. Dagle, J. A. Walder, and D. L. Weeks, *Nucleic Acids Res.*, **18**, 4751–4757 (1990).

71. J. M. Dagle, D. L. Weeks, and J. A. Walder, *Antisense Res. Dev.*, **1**, 11 (1991).

72. R. M. Crooke, *Anticancer Drug Des.*, **6**, 609–646 (1991).

73. U. Asseline, M. Delarue, G. Lancelot, F. Toulme, N. T. Thuong, T. Montenay-Garestier, and C. Helene, *Proc. Natl. Acad. Sci. USA*, **81**, 3297–3301 (1984).

74. C. Cazenave, C. A. Stein, N. Loreau, N. T. Thuong, L. M. Neckers, C. Subasinghe, C. Helene, J. S. Cohen, and J.-J. Toulme, *Nucleic Acids Res.*, **17**, 4255–4273 (1989).

75. R. L. Letsinger, G. R. Zhang, D. K. Sun, T. Ikeuchi, and P. S. Sarin, *Proc. Natl. Acad. Sci. USA*, **86**, 6553–6556 (1989).

76. M. Lemaitre, B. Bayard, and B. Lebleu, *Proc. Natl. Acad. Sci. USA*, **84**, 648–652 (1987).

77. J. P. Leonetti, B. Rayner, M. Lemaitre, C. Gagnor, P. G. Milhaud, J. L. Imbach, and B. Lebleu, *Gene*, **72**, 323–332 (1988).

78. S. T. Crooke, *Handbook of Experimental Pharmacology: Antisense Research and Application*, Springer-Verlag, Berlin, 1998, pp. 1–50.

79. S. T. Crooke and C. K. Mirabelli, *Antisense Research and Applications*, CRC Press, Boca Raton, FL, 1993, pp. 7–35.

80. J. R. Wyatt, T. A. Vickers, J. L. Roberson, R. W. Buckheit, Jr., T. Klimkait, E. DeBaets, P. W. Davis, B. Rayner, J. L. Imbach, and D. J. Ecker, *Proc. Natl. Acad. Sci. USA*, **91**, 1356–1360 (1994).

81. S. M. Freier, *Antisense Research and Applications*, CRC Press, Boca Raton, FL, 1993, pp. 67–82.

82. D. J. Ecker, *Antisense Research and Applications*, CRC Press, Boca Raton, FL, 1993, pp. 386–400.

83. W. F. Lima, B. P. Monia, D. J. Ecker, and S. M. Freier, *Biochemistry*, **31**, 12055–12061 (1992).

84. D. J. Ecker, T. A. Vickers, T. W. Bruice, S. M. Freier, R. D. Jenison, M. Manoharan, and M. Zounes, *Science*, **257**, 958–961 (1992).

85. S. T. Crooke, *Antisense Res Dev.*, **4**, 145–146 (1994).

86. S. T. Crooke, *Drug Target. Deliv.*, **4**, 297–320 (1995).

87. S. L. Loke, C. A. Stein, X. H. Zhang, K. Mori, M. Nakanishi, C. Subasinghe, J. S. Cohen, and L. M. Neckers, *Proc. Natl. Acad. Sci. USA*, **86**, 3474–3478 (1989).

88. P. A. Cossum, H. Sasmor, D. Dellinger, L. Truong, L. Cummins, S. R. Owens, P. M. Markham, J. P. Shea, and S. T. Crooke, *J. Pharmacol. Exp. Ther.*, **267**, 1181–1190 (1993).

89. P. A. Cossum, L. Troung, S. R. Owens, P. M. Markham, J. P. Shea, and S. T. Crooke, *J. Pharmacol. Exp. Ther.*, **269**, 89–94 (1994).

90. H. Sands, L. J. Gorey-Feret, S. P. Ho, Y. Bao, A. J. Cocuzza, D. Chidester, and F. W. Hobbs, *Mol. Pharmacol.*, **47**, 636–646 (1995).

91. L. M. Cowsert, *Antisense Research and Applications*, CRC Press, Boca Raton, FL, 1993, pp. 521–533.

92. R. F. Azad, V. B. Driver, K. Tanaka, R. M. Crooke, and K. P. Anderson, *Antimicrob. Agents Chemother.*, **37**, 1945–1954 (1993).

93. S. T. Crooke, *Therapeutic Applications of Oligonucleotides*, R. G. Landes, Austin, TX, 1995, p. 138.

94. M. Y. Chiang, H. Chan, M. A. Zounes, S. M. Freier, W. F. Lima, and C. F. Bennett, *J. Biol. Chem.*, **266**, 18162–18171 (1991).

95. N. M. Dean, R. McKay, T. P. Condon, and C. F. Bennett, *J. Biol. Chem.*, **269**, 16416–16424 (1994).

96. T. Skorski, M. Nieborowska-Skorska, N. C. Nicolaides, C. Szczylik, P. Iversen, R. V. Iozzo, G. Zon, and B. Calabretta, *Proc. Natl. Acad. Sci. USA*, **91**, 4504–4508 (1994).

97. N. J. Hijiya, M. Z. Zhang, J. A. Ratajczak, K. Kant, M. DeRiel, G. Z. Herlyn, and A. M. Gewirtz, *Proc. Natl. Acad. Sci. USA*, **91**, 4499–4503 (1994).

98. B. F. Baker and B. P. Monia, *Biochim. Biophys. Acta*, **1489**, 3–18 (1999).

99. M. E. McManaway, L. M. Neckers, S. L. Loke, A. A. Al-Nasser, R. L. Redner, B. T. Shiramizu, W. L. Goldschmidts, B. E. Huber, K. Bhatia, and I. T. Magrath, *Lancet*, **335**, 808–811 (1990).

100. M. Kulka, C. C. Smith, L. Aurelian, R. Fishelevich, K. Meade, P. Miller, and P. O. P. Ts'o, *Proc. Natl. Acad. Sci. USA*, **86**, 6868–6872 (1989).

101. P. C. Zamecnik, J. Goodchild, Y. Taguchi, and P. S. Sarin, *Proc. Natl. Acad. Sci. USA*, **83**, 4143–4146 (1986).

102. Z. Dominski and R. Kole, *Proc. Natl. Acad. Sci. USA*, **90**, 8673–8677 (1993).

103. Z. Dominski and R. Kole, *Mol. Cell. Biol.*, **14**, 7445–7454 (1994).

104. H. Sierakowska, M. J. Sambade, S. Agrawal, and R. Kole, *Proc. Natl. Acad. Sci. USA*, **93**, 12840–12844 (1996).

105. D. Hodges and S. T. Crooke, *Mol. Pharmacol.*, **48**, 905–918 (1995).

106. J. G. Karras, K. McGraw, R. A. McKay, S. R. Cooper, D. Lerner, T. Lu, C. Walker, N. M. Dean, and B. P. Monia, *J. Immunol.*, **164**, 5409–5415 (2000).

107. D. R. Mercatante, C. D. Bortner, J. A. Cidlowski, and R. Kole, *J. Biol. Chem.*, **276**, 16411–16417 (2001).

108. J. K. Taylor, Q. Q. Zhang, J. R. Wyatt, and N. M. Dean, *Nat. Biotechnol.*, **17**, 1097–1100 (1999).

109. A. Rosolen, L. Whitesell, N. Ikegaki, R. H. Kennett, and L. M. Neckers, *Cancer Res.*, **50**, 6316–6322 (1990).

110. G. Vasanthakumar and N. K. Ahmed, *Cancer Commun.*, **1**, 225–232 (1989).

111. A. R. Sburlati, R. E. Manrow, and S. L. Berger, *Proc. Natl. Acad. Sci. USA*, **88**, 253–257 (1991).

112. H. Zheng, B. M. Sahai, P. Kilgannon, A. Fotedar, and D. R. Green, *Proc. Natl. Acad. Sci. USA*, **86**, 3758–3762 (1989).

113. J. A. Maier, P. Voulalas, D. Roeder, and T. Maciag, *Science*, **249**, 1570–1574 (1990).

114. H. Knudsen and P. E. Nielsen, *Nucleic Acids Res.*, **24**, 494–500 (1996).

115. C. Gambacorti-Passerini, L. Mologni, C. Bertazzoli, P. Le Coutre, E. Marchesi, F. Grignini, and P. Nielsen, *Blood*, **88**, 1411–1417 (1996).

116. M. Taylor, J. Paulauskis, D. Weller, and L. Kobzik, *J. Biol. Chem.*, **271**, 17445–17452 (1996).

117. J. Summerton, D. Stein, S. B. Huang, P. Matthews, S. Weller, and M. Partridge, *Antisense Nucleic Acid Drug Dev.*, **7**, 63–70 (1997).

118. B. F. Baker, S. S. Lot, T. P. Condon, S. Cheng-Flournoy, E. A. Lesnik, H. M. Sasmor, and C. F. Bennett, *J. Biol. Chem.*, **272**, 11994–12000 (1997).

119. S. T. Crooke, *Antisense Nucleic Acid Drug Dev.*, **8**, 133–134 (1998).

120. S. K. Saxena and E. J. Ackerman, *J. Biol. Chem.*, **265**, 3263–3269 (1990).

121. K. Walker, S. A. Elela, and R. N. Nazar, *J. Biol. Chem.*, **265**, 2428–2430 (1990).

122. P. Westermann, B. Gross, and G. Hoinkis, *Biomed. Biochim. Acta*, **48**, 85–93 (1989).

123. B. F. Baker, *J. Am. Chem. Soc.*, **115**, 3378–3379 (1993).

124. B. F. Baker, L. Miraglia, and C. H. Hagedorn, *J. Biol. Chem.*, **267**, 11495–11499 (1992).

125. C. H. Hagedorn, M. Y. Chiang, M. A. Zounes, S. M. Freier, W. F. Lima, and C. F. Bennett, *J. Biol. Chem.*, **266**, 18162–18171 (1991).

126. P. C. Zamecnik and M. L. Stevenson, *Proc. Natl. Acad. Sci. USA*, **75**, 280–284 (1978).

127. A. Zerial, N. T. Thuong, and C. Helene, *Nucleic Acids Res.*, **15**, 9909–9919 (1987).

128. N. T. Thuong, U. Asseline, and T. Monteney-Garestier, *Oligodeoxynucleotides: Antisense Inhibitors of Gene Expression*, CRC Press, Boca Raton, FL, 1989, p. 25.

129. C. Helene and J. Toulme, *Oligodeoxynucleotides: Antisense Inhibitors of Gene Expression*, CRC Press, Boca Raton, FL, 1989, p. 137.

130. R. J. Crouch and M.-L. Dirksen, *Nucleases*, Cold Spring Harbor Laboratory Press, Cold Spring Harbor, NY, 1985, pp. 211–241.

131. C. Crum, J. D. Johnson, A. Nelson, and D. Roth, *Nucleic Acids Res.*, **16**, 4569–4581 (1988).

132. H. Donis-Keller, *Nucleic Acids Res.*, **7**, 179–192 (1979).

133. A. M. Kawasaki, M. D. Casper, S. M. Freier, E. A. Lesnik, M. C. Zounes, L. L. Cummins, C. Gonzalez, and P. D. Cook, *J. Med. Chem.*, **36**, 831–841 (1993).

134. B. S. Sproat, A. I. Lamond, B. Beijer, P. Neuner, and U. Ryder, *Nucleic Acids Res.*, **17**, 3373–3386 (1989).

135. F. Morvan, B. Rayner, and J. L. Imbach, *Anticancer Drug Des.*, **6**, 521–529 (1991).

136. C. Gagnor, B. Rayner, J. P. Leonetti, J. L. Imbach, and B. Lebleu, *Nucleic Acids Res.*, **17**, 5107–5114 (1989).

137. L. J. MaherIII, B. Wold, and P. B. Dervan, *Science*, **245**, 725–730 (1989).

138. P. S. Miller, *Oligodeoxynucleotides: Antisense Inhibitors of Gene Expression*, CRC Press, Boca Raton, FL, 1989, p. 79.

139. C. K. Mirabelli, C. F. Bennett, K. Anderson, and S. T. Crooke, *Anticancer Drug Des.*, **6**, 647–661 (1991).

140. C. A. Stein and Y.-C. Cheng, *Science*, **261**, 1004–1012 (1993).

141. R. S. Quartin, C. L. Brakel, and J. G. Wetmur, *Nucleic Acids Res.*, **17**, 7253–7262 (1989).

142. P. J. Furdon, Z. Dominski, and R. Kole, *Nucleic Acids Res.*, **17**, 9193–9204 (1989).

143. P. S. Eder and J. A. Walder, *J. Biol. Chem.*, **266**, 6472–6479 (1991).

144. B. P. Monia, E. A. Lesnik, C. Gonzalez, W. F. Lima, D. McGee, C. J. Guinosso, A. M. Kawasaki, P. D. Cook, and S. M. Freier, *J. Biol. Chem.*, **268**, 14514–14522 (1993).

145. R. V. Giles and D. M. Tidd, *Nucleic Acids Res.*, **20**, 763–770 (1992).

146. R. Y. Walder and J. A. Walder, *Proc. Natl. Acad. Sci. USA*, **85**, 5011–5015 (1988).

147. J. Minshull and T. Hunt, *Nucleic Acids Res.*, **14**, 6433–6451 (1986).

148. C. Gagnor, J. R. Bertrand, S. Thenet, M. Lemaitre, F. Morvan, B. Rayner, C. Malvy, B. Lebleu, J. L. Imbach, and C. Paoletti, *Nucleic Acids Res.*, **15**, 10419–10436 (1987).

149. S. T. Crooke, K. M. Lemonidis, L. Neilson, R. Griffey, E. A. Lesnik, and B. P. Monia, *Biochem. J.*, **312**, 599–608 (1995).

150. W. F. Lima and S. T. Crooke, *Biochemistry*, **36**, 390–398 (1997).

151. H. Wu, W. F. Lima, and S. T. Crooke, *Antisense Nucleic Acid Drug Dev.*, **8**, 53–61 (1998).

152. P. Frank, S. Albert, C. Cazenave, and J. J. Toulme, *Nucleic Acids Res.*, **22**, 5247–5254 (1994).

153. H. Wu, W. F. Lima, and S. T. Crooke, *J. Biol. Chem.*, **274**, 28270–28278 (1999).

154. P. Frank, C. Braunshofer-Reiter, U. Wintersberger, R. Grimm, and W. Busen, *Proc. Natl. Acad. Sci. USA*, **95**, 12872–12877 (1998).

155. H. Wu, H. Xu, L. J. Miraglia, and S. T. Crooke, *J. Biol. Chem.*, **275**, 36957–36965 (2000).

156. L. Miraglia, A. T. Watt, M. J. Graham, and S. T. Crooke, *Antisense Nucleic Acid Drug Discov.*, **10**, 453–461 (2000).

157. W. F. Lima and S. T. Crooke, *J. Biol. Chem.*, **272**, 27513–27516 (1997).

158. C. F. Bennett and L. M. Cowsert, *Biochim. Biophys. Acta*, **1489**, 19–30 (1999).

159. J. S. Cohen, *Antisense Research and Applications*, CRC Press, Boca Raton, FL, 1993, p. 205.

160. S. T. Crooke, *BioTechnology*, **10**, 882–886 (1992).

161. R. M. Crooke, *Antisense Research and Applications*, CRC Press, Boca Raton, FL, 1993, pp. 427–499.

162. S. T. Crooke, *Ann. Rev. Pharmacol. Toxicol.*, **32**, 329–376 (1992).

163. S. M. Freier, *Antisense Research and Applications*, CRC Press, Boca Raton, FL, 1993, pp. 67–82.

164. B. P. Monia, J. F. Johnston, D. J. Ecker, M. A. Zounes, W. F. Lima, and S. M. Freier, *J. Biol. Chem.*, **267**, 19954–19962 (1992).

165. D. L. Sokol, X. Zhang, P. Lu, and A. M. Gewirtz, *Proc. Natl. Acad. Sci. USA*, **95**, 11538–11543 (1998).

166. S. T. Crooke, M. J. Graham, J. E. Zuckerman, D. Brooks, B. S. Conklin, L. L. Cummins, M. J. Greig, C. J. Guinosso, D. Kornburst, M. Manoharan, H. M. Sasmor, T. Schleich, K. L. Tivel, and R. H. Griffey, *J. Pharmacol. Exp. Ther.*, **277**, 923–937 (1996).

167. R. W. Joos and W. H. Hall, *J. Pharmacol. Exp. Ther.*, **166**, 113 (1969).

168. S. K. Srinivasan, H. K. Tewary, and P. L. Iversen, *Antisense Res. Dev.*, **5**, 131–139 (1995).

169. A. A. Levin, S. Henry, D. Monteith, and M. Templin, *Antisense Technology: Principles, Strategies and Applications*, Marcel Dekker, New York, 2001, pp. 201–267.

170. P. Hawley and I. Gibson, *Antisense Nucleic Acid Drug Dev.*, **6**, 185–195 (1996).

171. R. M. Crooke, M. J. Graham, M. E. Cooke, and S. T. Crooke, *J. Pharmacol. Exp. Ther.*, **275**, 462–473 (1995).

172. W.-Y. Gao, F.-S. Han, C. Storm, W. Egan, and Y.-C. Cheng, *Mol. Pharmacol.*, **41**, 223–229 (1992).

173. S. T. Crooke, *Burger's Medicinal Chemistry and Drug Discovery*, Vol. **1**, John Wiley & Sons, New York, 1995, pp. 863–900.

174. S. Agrawal, J. Temsamani, and J. Y. Tang, *Proc. Natl. Acad. Sci. USA*, **88**, 7595–7599 (1991).

175. C. Majumdar, C. A. Stein, J. S. Cohen, S. Broder, and S. H. Wilson, *Biochemistry*, **28**, 1340–1346 (1989).

176. Y. Cheng, W. Gao, and F. Han, *Nucleosides Nucleotides*, **10**, 155–166 (1991).

177. C. A. Stein, M. Neckers, B. C. Nair, S. Mumbauer, G. Hoke, and R. Pal, *J. Acquir. Immune Defic. Syndr. Hum. Retrovirol.*, **4**, 686–693 (1991).

178. G. Hartmann, A. Krug, K. Waller-Fontaine, and S. Endres, *Mol. Med.*, **2**, 429–438 (1996).

179. L. M. Cowsert, M. C. Fox, G. Zon, and C. K. Mirabelli, *Antimicrob. Agents Chemother.*, **37**, 171–177 (1993).

180. M. J. Graham, S. M. Freier, R. M. Crooke, D. J. Ecker, R. N. Maslova, and E. A. Lesnik, *Nucleic Acids Res.*, **21**, 3737–3743 (1993).

181. H. Sands, L. J. Gorey-Feret, A. J. Cocuzza, F. W. Hobbs, D. Chidester, and G. L. Trainor, *Mol. Pharmacol.*, **45**, 932–943 (1994).

182. S. T. Crooke, M. J. Graham, J. E. Zuckerman, D. Brooks, B. S. Conklin, L. L. Cummins, M. J. Greig, C. J. Guinosso, D. Kornbrust, M. Manoharan, H. M. Sasmor, T. Schleich, K. L. Tivel, and R. H. Griffey, *J. Pharmacol. Exp. Ther.*, **277**, 923–937 (1996).

183. G. D. Hoke, K. Draper, S. M. Freier, C. Gonzalez, V. B. Driver, M. C. Zounes, and D. J. Ecker, *Nucleic Acids Res.*, **19**, 5743–5748 (1991).

184. J. M. Campbell, T. A. Bacon, and E. Wickstrom, *J. Biochem. Biophys. Methods*, **20**, 259–267 (1990).

185. P. Cossum, H. Sasmor, D. Dellinger, L. Truong, L. Cummins, S. Owens, P. Markham, J. Shea, and S. Crooke, *J. Pharmacol. Exp. Ther.*, **267**, 1181–1190 (1993).

186. S. T. Crooke, *Curr. Opin. Invest. Drugs*, **2**, 1045–1048 (1993).

187. T. Miyao, Y. Takakura, T. Akiyama, F. Yoneda, H. Sezaki, and M. Hashida, *Antisense Res. Dev.*, **5**, 115–121 (1995).

188. R. M. Crooke, *Antisense Research and Applications*, CRC Press, Boca Raton, FL, 1993, pp. 471–492.

189. L. M. Neckers, *Antisense Research and Applications*, CRC Press, Boca Raton, FL, 1993, pp. 451–460.

190. L. A. Chrisey, S. E. Walz, M. Pazirandeh, and J. R. Campbell, *Antisense Res Dev.*, **3**, 367–381 (1993).

191. L. F. Tao, K. A. Marx, W. Wongwit, Z. Jiang, S. Agrawal, and R. M. Coleman, *Antisense Res Dev.*, **5**, 123–129 (1995).

192. R. M. Crooke, *Antisense Research and Applications*, CRC Press, Boca Raton, FL, 1993, pp. 427–449.

193. C. F. Bennett, M. Y. Chiang, H. Chan, and S. Grimm, *J. Liposome Res.*, **3**, 85–102 (1993).

194. C. F. Bennett, M. Y. Chiang, H. Chan, J. E. E. Shoemaker, and C. K. Mirabelli, *Mol. Pharmacol.*, **41**, 1023–1033 (1992).

195. A. Quattrone, L. Papucci, N. Schiavone, E. Mini, and S. Capaccioli, *Anticancer Drug Des.*, **9**, 549–553 (1994).

196. R. V. Giles, D. G. Spiller, and D. M. Tidd, *Antisense Res. Dev.*, **5**, 23–31 (1995).

197. S. Wang, R. J. Lee, G. Cauchon, D. G. Gorenstein, and P. S. Low, *Proc. Natl. Acad. Sci. USA*, **92**, 3318–3322 (1995).

198. P. Iversen, *Anticancer Drug Des.*, **6**, 531–538 (1991).

199. R. Z. Yu, R. S. Geary, J. M. Leeds, T. Watanabe, M. Moore, J. Fitchett, J. Matson, T. Burckin, M. V. Templin, and A. A. Levin, *J. Pharm. Sci.*, **90**, 182–193 (2000).

200. P. A. Cossum, L. Truong, S. R. Owens, P. M. Markham, J. P. Shea, and S. T. Crooke, *J. Pharmacol. Exp. Ther.*, **269**, 89–94 (1994).

201. S. Crooke, I. Grillone, A. Tendolkar, A. Garrett, M. Fratkin, J. Leeds, and W. Barr, *Clin. Pharmacol. Ther.*, **56**, 641–646 (1994).

202. J. M. Leeds, S. P. Henry, R. Geary, T. Burckin, and A. A. Levin, *Antisense Nucleic Acid Drug Dev.*, **10**, 435–441 (2000).

203. P. A. Cossum, H. Sasmor, D. Dellinger, L. Truong, L. Cummins, S. R. Owens, P. M. Markham, J. P. Shea, and S. Crooke, *J. Pharmacol. Exp. Ther.*, **267**, 1181–1190 (1993).

204. J. Rappaport, B. Hanss, J. B. Kopp, T. D. Copeland, L. A. Bruggeman, T. M. Coffman, and P. E. Klotman, *Kidney Int.*, **47**, 1462–1469 (1995).

205. P. Iverson, *Anticancer Drug Des.*, **6**, 531–538 (1992).

206. E. Bayever, P. L. Iversen, M. R. Bishop, J. G. Sharp, H. K. Tewary, M. A. Arneson, S. J. Pirruccello, R. W. Ruddon, A. Kessinger, and G. Zon, *Antisense Res. Dev.*, **3**, 383–390 (1993).

207. R. Zhang, J. Yan, H. Shahinian, G. Amin, Z. Lu, T. Liu, M. S. Saag, Z. Jiang, J. Temsamani, R. R. Martin, P. J. Schechter, S. Agrawal, and R. B. Diasio, *Clin. Pharmacol. Ther.*, **58**, 44–53 (1995).

208. Y. Takakura, R. I. Mahato, M. Yoshida, T. Kanamaru, and M. Hashida, *Antisense Nucleic Acid Drug Dev.*, **6**, 177–183 (1996).

209. M. Butler, K. Stecker, and C. F. Bennett, *Lab. Invest.*, **77**, 379–388 (1997).

210. M. J. Graham, S. T. Crooke, D. K. Monteith, S. R. Cooper, K. M. Lemonidis, K. K. Stecker, M. J. Martin, and R. M. Crooke, *J. Pharmacol. Exp. Ther.*, **286**, 447–458 (1998).

211. J. W. Nyce, *Expert Opin. Invest. Drugs*, **6**, 1149–1156 (1997).

212. M. V. Templin, A. A. Levin, M. J. Graham, P. M. Aberg, B. I. Axelsson, M. Butler, R. S. Geary, and C. F. Bennett, *Antisense Nucleic Acid Drug Dev.*, **10**, 359–368 (2000).

213. J. W. Nyce, *Antisense Technology: Principles, Strategies and Applications*, Marcel Dekker, New York, 2001, pp. 679–685.

214. J. W. Nyce and W. J. Metzger, *Nature*, **385**, 721–725 (1997).

215. G. Hardee, S. Weinbach, and L. Tillman, *Antisense Technology: Principles, Strategies and Applications*, Marcel Dekker, New York, 2001, pp. 795–832.

216. P. D. Cook, *Antisense Research and Applications*, CRC Press, Boca Raton, FL, 1993, pp. 149–187.

217. C. F. Bennett and S. T. Crooke, *Therapeutic Modulation Cytokines*, CRC Press, Boca Raton, FL, 1996, pp. 171–193.

218. R. W. Wagner, M. D. Matteucci, J. G. Lewis, A. J. Gutierrez, C. Moulds, and B. C. Froehler, *Science*, **260**, 1510–1513 (1993).

219. C. M. Barton and N. R. Lemoine, *Br. J. Cancer*, **71**, 429–437 (1995).

220. T. L. Burgess, E. F. Fisher, S. L. Ross, J. V. Bready, Y. Qian, L. A. Bayewitch, A. M. Cohen, C. J. Herrera, S. F. Hu, T. B. Kramer, F. D. Lott, F. H. Martin, G. F. Pierce, L. Simonet, and C. L. Farrell, *Proc. Natl. Acad. Sci. USA*, **92**, 4051–4055 (1995).

221. M. Hertl, L. M. Neckers, and S. I. Katz, *J. Invest. Dermatol.*, **104**, 813–818 (1995).

222. K. M. Nagel, S. G. Holstad, and K. E. Isenberg, *Pharmacotherapy*, **13**, 177–188 (1993).

223. M. Simons, E. R. Edelman, J.-L. DeKeyser, R. Langer, and R. D. Rosenberg, *Nature*, **359**, 67–70 (1992).

224. J. Abe, W. Zhou, J. Taguchi, N. Takuwa, K. Miki, H. Okazaki, K. Kurokawa, M. Kumada, and Y. Takuwa, *Biochem. Biophys. Res. Commun.*, **198**, 16–24 (1994).

225. L. Whitesell, A. Rosolen, and L. M. Neckers, *Antisense Res. Dev.*, **1**, 343–350 (1991).

226. C. Wahlestedt, E. M. Pich, G. F. Koob, F. Yee, and M. Heilig, *Science*, **259**, 528–531 (1993).

227. S.-P. Zhang, L.-W. Zhou, and B. Weiss, *J. Pharmacol. Exp. Ther.*, **271**, 1462–1470 (1994).

228. P. Ambuhl, R. Gyurko, and M. I. Phillips, *Regul. Pept.*, **59**, 171–182 (1995).

229. M. M. McCarthy, D. A. Nielsen, and D. Goldman, *Regul. Pept.*, **59**, 163–170 (1995).

230. A. Osen-Sand, M. Catsicas, J. K. Staple, K. A. Jones, G. Ayala, J. Knowles, G. Grenningloh, and S. Catsicas, *Nature*, **364**, 445–448 (1993).

231. R. M. Burch and L. C. Mahan, *J. Clin. Invest.*, **88**, 1190–1196 (1991).

232. W. F. Lima and S. T. Crooke, *Biochemistry*, **36**, 390–398 (1997).

233. K. A. Higgins, J. R. Perez, T. A. Coleman, K. Dorshkind, W. A. McComas, U. M. Sarmiento, C. A. Rosen, and R. Narayan, *Proc. Natl. Acad. Sci. USA*, **90**, 9901–9905 (1993).

234. N. Hijiya, J. Zhang, M. Z. Ratajczak, J. A. Kant, K. DeRiel, M. Herlyn, G. Zon, and A. M. Gewirtz, *Proc. Natl. Acad. Sci. USA*, **91**, 4499–4503 (1994).

235. N. M. Dean and R. McKay, *Proc. Natl. Acad. Sci. USA*, **91**, 11762–11766 (1994).

236. N. Dean, R. McKay, L. Miraglia, R. Howard, S. Cooper, J. Giddings, P. Nicklin, L. Meister, R. Ziel, T. Geiger, M. Muller, and D. Fabbro, *Cancer Res.*, **56**, 3499–3507 (1996).

237. B. P. Monia, J. F. Johnston, T. Geiger, M. Muller, and D. Fabbro, *Nat. Med.*, **2**, 668–675 (1995).

238. B. P. Monia, J. F. Johnston, H. Sasmor, and L. L. Cummins, *J. Biol. Chem.*, **271**, 14533–14540 (1996).

239. M. Nesterova and Y. S. Cho-Chung, *Nat. Med.*, **1**, 528–533 (1995).

240. L. Grillone, *Antisense Technology: Principles, Strategies and Applications*, Marcel Dekker, New York, 2001, pp. 725–738.

241. F. Bennett, *Antisense Therapeutics: Principles, Strategies and Applications*, Marcel Dekker, New York, 2001, pp. 291–318.

242. J. Shanahan, *Antisense Drug Technology: Principles, Strategies and Applications*, Marcel Dekker, New York, 2001, pp. 773–794.

243. W. Maksymowych, W. Blackburn, E. Hutchings, L. Williams, J. Tami, K. Wagner, and W. Shanahan, *Arthritis Rheum.*, **42**, S170 (1999).

244. T. Fredriksson and U. Pettersson, *Dermatologica*, **157**, 238–244 (1978).

245. B. R. Yacyshyn, M. B. Bowen-Yacyshyn, L. Jewell, J. A. Tami, C. F. Bennett, D. L. Kisner, and W. R. Shanahan, Jr., *Gastroenterology*, **114**, 1133–1142 (1998).

246. B. Yacyshyn, B. Woloschuk, M. Yacyshyn, D. Martini, K. Doan, J. Tami, F. Bennett, D. Kisner, and W. Shanahan, Proceedings of the Annual Meetings of American Gastroenterological Association and American Association for the Study of Liver Diseases, Washington, DC, 1997.

247. J. Holmlund, *Antisense Technology: Principles, Strategies and Applications*, Marcel Dekker, New York, 2001, pp. 739–759.

248. A. Yuen, R. Advani, G. Fisher, J. Halsey, B. Lum, R. Geary, T. J. Kwoh, J. Holmlund, A. Dorr, and B. I. Sikic, Proceedings of the American Society of Clinical Oncology, San Francisco, CA, 2001.

249. J. Waters, A. Webb, D. Cunningham, P. Clarke, F. Raynaud, F. di Stefano, and F. Cotter, *J. Clin. Oncol.*, **18**, 1812–1823 (2000).

250. B. Jansen, V. wacheck, and E. Heere-Ress, Proceedings of AACR, 2000.

251. P. O'Dwyer, J. Stevenson, M. Gallagher, A. Cassella, I. Vasilevskaya, B. Monia, J. Holmlund, F. Dorr, and K.-S. Yao, *Clin. Cancer Res.*, **5**, 3977–3982 (1999).

252. J. Nemunaitis, J. Holmlund, M. Kraynak, D. Richards, J. Bruce, N. Ognoskie, T. Kwoh, R. Geary, A. Dorr, D. Von Hoff, and S. Eckhardt, *J. Clin. Oncol.*, **17**, 3586–3595 (1999).

253. A. Yuen, J. Halsey, G. Fisher, J. Holmlund, R. Geary, T. Kwoh, A. Dorr, and B. Sikic, *Clin. Cancer Res.*, **5**, 3357–3363 (1999).

254. J. P. Stevenson, K.-S. Yao, M. Gallagher, D. Friedland, E. P. Mitchell, A. Cassella, B. Monia, T. J. Kwoh, R. Yu, J. Holmlund, F. A. Dorr, and P. J. O'Dwyer, *J. Clin. Oncol.*, **17**, 2227–2236 (1999).

255. P. O' Dwyer, J. Stevenson, M. Gallagher, A. Cassella, I. Vasilevskaya, B. Monia, J. Holmlund, A. Dorr, and K. Yao, *Clin. Cancer Res.*, **5**, 3977–3982 (1999).

256. C. Cunningham, J. Holmlund, J. Schiller, R. Geary, T. Kwoh, A. Dorr, and J. Nemunaitis, *Clin Cancer Res.*, **6**, 1626–1631 (2000).

257. A. Dorr, J. Nemunaitis, J. Bruce, B. Monia, J. Johnston, R. Geary, T. Kwoh, and J. Holmlund, Proceedings of the 35th Annual Meeting of the American Society of Clinical Oncology, Atlanta, GA, 1999.

258. M. S. Gordon, A. B. Sandler, J. T. Holmlund, A. Dorr, L. Battiato, K. Fife, R. Geary, T. J. Kwoh, and G. W. J. Sledge, Proceedings of the 35th Annual Meeting of the American Society of Clinical Oncology, Atlanta, GA, 1999.

259. J. P. Messina, G. S. Gilkeson, and D. S. Pisetsky, *J. Immunol.*, **147**, 1759–1764 (1991).

260. E. Kuramoto, O. Yano, Y. Kimura, M. Baba, T. Makino, S. Yamamoto, T. Yamamoto, T. Kataoka, and T. Tokunaga, *Jpn. J. Cancer Res.*, **83**, 1128–1131 (1992).

261. D. S. Pisetsky and C. F. Reich, *Life Sci.*, **54**, 101–107 (1994).

262. R. M. Crooke, S. T. Crooke, M. J. Graham, and M. E. Cooke, *Toxicol. Appl. Pharmacol.*, **140**, 85–93 (1996).

263. A. M. Krieg, A.-K. Yi, S. Matson, T. J. Waldschmidt, G. A. Bishop, R. Teasdale, G. A. Koretzky, and D. M. Klinman, *Nature*, **374**, 546–549 (1995).

264. S. T. Crooke, L. R. Grillone, A. Tendolkar, A. Garrett, M. J. Fratkin, J. Leeds, and W. H. Barr, *Clin. Pharmacol. Ther.*, **56**, 641–646 (1994).

265. A. A. Levin, D. K. Monteith, J. M. Leeds, P. L. Nicklin, R. S. Geary, M. Butler, M. V. Templin, and S. P. Henry, *Antisense Research and Application*, Springer-Verlag, Berlin/Heidelberg, 1998, pp. 169–214.

266. W. M. Galbraith, W. C. Hobson, P. C. Giclas, P. J. Schechter, and S. Agrawal, *Antisense Res. Dev.*, **4**, 201–206 (1994).

267. S. P. Henry, P. C. Giclas, J. Leeds, M. Pangburn, C. Auletta, A. A. Levin, and D. J. Kornbrust, *J. Pharmacol. Exp. Ther.*, **281**, 810–816 (1997).

268. K. G. Cornish, P. Iversen, L. Smith, M. Arneson, and E. Bayever, *Pharmacol. Commun.*, **3**, 239–247 (1993).

269. S. P. Henry, D. Monteith, and A. A. Levin, *Anticancer Drug Des.*, **12**, 395–408 (1997).

270. S. P. Henry, W. Novotny, J. Leeds, C. Auletta, and D. J. Kornbrust, *Antisense Nucleic Acid Drug Dev.*, **7**, 503–510 (1997).

271. T. L. Wallace, S. A. Bazemore, D. J. Kornbrust, and P. A. Cossum, *J. Pharmacol. Exp. Ther.*, **278**, 1306–1312 (1996).

272. L. Griffin, G. Tidmarsh, L. Bock, J. Toole, and L. Leung, *Blood*, **81**, 3271–3276 (1993).

273. P. L. Nicklin, J. Ambler, A. Mitchelson, D. Bayley, J. A. Phillips, S. J. Craig, and B. P. Monia, *Nucleosides Nucleotides*, **16**, 1145–1153 (1997).

274. J. P. Sheehan and H.-C. Lan, *Blood*, **92**, 1617–1625 (1998).

275. S. P. Henry, R. Larkin, W. F. Novotny, and D. J. Kornbrust, *Pharm. Res.*, **11**, S351–S490 (1994).

276. D. K. Monteith, R. S. Geary, J. M. Leeds, J. Johnston, B. P. Monia, and A. A. Levin, *Toxicol. Sci.*, **46**, 365–375 (1998).

277. S. P. Henry, H. Bolte, C. Auletta, and D. J. Kornbrust, *Toxicology*, **120**, 145–155 (1997).

278. U. M. Sarmiento, J. R. Perez, J. M. Becker, and R. Narayanan, *Antisense Res. Dev.*, **4**, 99–107 (1994).

279. S. P. Henry, D. Monteith, and A. A. Levin, *Anticancer Drug Des.*, **12**, 395–408 (1997).

280. H. Liang, Y. Nishioka, C. F. Reich, D. S. Pisetsky, and P. E. Lipsky, *J. Clin. Invest.*, **98**, 1119–1129 (1996).

281. K. Lemonidis-Farrar, M. Templin, M. D. Graham, and R. M. Crocke, *Toxicologist*, **42**, 120 (2000).

282. A. M. Krieg, *Antisense Research and Application*, Springer-Verlag, Berlin/Heidelberg, 1998, pp. 243–262.

283. H. L. Davis, R. Weeranta, T. J. Waldschmidt, L. Tygrett, J. Schorr, and A. M. Kreig, *J. Immunol.*, **160**, 870–876 (1998).

284. G. Harmann and A. Krieg, *J. Immunol.*, **164**, 944–953 (2000).

285. F. A. Dorr, J. M. Glover, and J. Kwoh, *Antisense Technology: Principles, Strategies and Applications*, Marcel Dekker, New York, 2001, pp. 269–290.

286. J. Nemunaitis, J. T. Holmlund, M. Kraynak, D. Richards, J. Bruce, N. Ognoskie, T. J. Kwoh, R. Geary, A. Dorr, D. Von Hoff, and S. G. Eckhardt, *J. Clin. Oncol.*, **17**, 3586–3595 (1999).

287. J. Holmlund, J. Nemunaitis, J. Schiller, A. Dorr, and D. Kisner, Proceedings of the American Society of Clinical Oncology, Los Angeles, CA, 1998.

288. G. I. Sikic, A. R. Yuen, J. Halsey, G. A. Fisher, J. P. Pribble, R. M. Smith, R. Geary, and A. Dorr, Proceedings of the 33rd Annual Meeting of the American Society of Clinical Oncology, Denver, CO, 1998.

289. J. Nemunaitis, J. Holmlund, M. Kraynak, D. Richards, J. Bruce, N. Ognoskie, T. Kwoh, R. Geary, A. Dorr, D. Von Hoff, and S. Eckhardt, *J. Clin. Oncol.*, **17**, 3586–3595 (1999).

290. Y. S. Sanghvi, *Antisense Research and Applications*, CRC Press, Boca Raton, FL, 1993, pp. 273–288.

291. P. Herdewijn, *Antisense Nucleic Acid Drug Dev.*, **10**, 297–310 (2000).

292. K. Shohda, I. Okamoto, T. Wada, K. Seio, and M. Sekine, *Bioorg. Med. Chem. Lett.*, **10**, 1795–1798 (2000).

293. R. K. Kumar and D. R. Davis, *Nucleic Acids Res.*, **25**, 1272–1280 (1997).

294. S. M. Testa, M. D. Disney, D. H. Turner, and R. Kierzek, *Biochemistry*, **38**, 16655–16662 (1999).

295. T. T. Nikiforov and B. A. Connolly, *Tetrahedron Lett.*, **32**, 3851–3854 (1991).

296. P. K. T. Lin and D. M. Brown, *Nucleic Acids Res.*, **17**, 10373–10383 (1989).

297. H. Inoue, A. Imura, and E. Ohtsuka, *Nucleic Acids Res.*, **13**, 7119–7128 (1985).

298. A. J. Gutierrez, T. J. Terhorst, M. D. Matteucci, and B. C. Froehler, *J. Am. Chem. Soc.*, **116**, 5540–5544 (1994).

299. K.-Y. Lin, R. J. Jones, and M. Matteucci, *J. Am. Chem. Soc.*, **117**, 3873–3874 (1995).

300. T. Wang, H. An, T. A. Vickers, R. Bharadwaj, and P. D. Cook, *Tetrahedron*, **54**, 7955–7976 (1998).

301. W. M. Flanagan, R. W. Wagner, D. Grant, K.-Y. Lin, and M. D. Matteucci, *Nat. Biotechnol.*, **17**, 48–52 (1999).

302. K.-Y. Lin and M. D. Matteucci, *J. Am. Chem. Soc.*, **120**, 8531–8532 (1998).

303. W. M. Flanagan, J. J. Wolf, P. Olson, D. Grant, K. Y. Lin, R. W. Wagner, and M. D. Matteucci, *Proc. Natl. Acad. Sci. USA*, **96**, 3513–3518 (1999).

304. Y. S. Sanghvi, G. D. Hoke, S. M. Freier, M. C. Zounes, C. Gonzalez, L. Cummins, H. Sasmor, and P. D. Cook, *Nucleic Acids Res.*, **21**, 3197–3203 (1993).

305. M. Manoharan, *Antisense Research and Applications*, CRC Press, Boca Raton, FL, 1993, pp. 303–349.

306. K. S. Ramasamy, M. Zounes, C. Gonzalez, S. M. Freier, E. A. Lesnik, L. Cummins Lendell, R. H. Griffey, B. P. Monia, and P. D. Cook, *Tetrahedron Lett.*, **35**, 215–218 (1994).

307. M. Manoharan, K. S. Ramasamy, V. Mohan, and P. D. Cook, *Tetrahedron Lett.*, **37**, 7675–7678 (1996).

308. B. S. Sproat, A. M. Iribarren, R. G. Garcia, and B. Beijer, *Nucleic Acids Res.*, **19**, 733–738 (1991).

309. B. S. Ross, R. H. Springer, R. Bharadwaj, A. M. Symons, and M. Manoharan, *Nucleosides Nucleotides*, **18**, 1203–1204 (1999).

310. F. H. Martin, M. M. Castro, F. Aboul-ela, and I. J. Tinoco, *Nucleic Acids Res.*, **13**, 8927–8938 (1985).

311. J. SantaLucia, Jr., R. Kierzek, and D. H. Turner, *J. Am. Chem. Soc.*, **113**, 4313–4322 (1991).

312. F. Seela, N. Ramzaeva, and Y. Chen, *Bioorg. Med. Chem. Lett.*, **5**, 3049–3052 (1995).

313. G. Balow, V. Mohan, E. A. Lesnik, J. F. Johnston, B. P. Monia, and O. L. Acevedo, *Nucleic Acids Res.*, **26**, 3350–3357 (1998).

314. M. Manoharan, *Antisense Technology: Principles, Strategies and Applications*, Marcel Dekker, New York, 2001, pp. 391–470.

315. H. B. Gamper, M. W. Reed, T. Cox, J. S. Virosco, A. D. Adams, A. A. Gall, J. K. Scholler, and R. B. Meyer, Jr., *Nucleic Acids Res.*, **21**, 145–150 (1993).

316. M. A. Maier, A. P. Guzaev, and M. Manoharan, *Org. Lett.*, **2**, 1819–1822 (2000).

317. J. G. Zendegui, K. M. Vasquez, J. H. Tinsley, D. J. Kessler, and M. E. Hogen, *Nucleic Acids Res.*, **20**, 307–314 (1992).

318. M. Manoharan, K. L. Tivel, L. K. Andrade, and P. D. Cook, *Tetrahedron Lett.*, **36**, 3647–3650 (1995).

319. R. H. Griffey, B. P. Monia, L. L. Cummins, S. Freier, M. J. Greig, C. J. Guinosso, E. Lesnik, S. M. Manalili, V. Mohan, S. Owens, B. R. Ross, H. Sasmor, E. Wancewicz, K. Weiler, P. D. Wheeler, and P. D. Cook, *J. Med. Chem.*, **39**, 5100–5109 (1996).

320. T. P. Prakash, M. Manoharan, A. M. Kawasaki, E. A. Lesnik, S. R. Owens, and G. Vasquez, *Org. Lett.*, **2**, 3995–3998 (2000).

321. M. Wasner, D. Arion, G. Borkow, A. Noronha, A. H. Uddin, M. A. Parniak, and M. J. Damha, *Biochem. J.*, **37**, 7478–7486 (1998).

322. M. J. Damha, P. A. Giannaris, and P. Marfey, *Biochemistry*, **33**, 7877–7885 (1994).

323. H. Urata, H. Miyagoshi, T. Yumoto, K. Mori, R. Teramichi, and M. Akagi, *Nucleic Acids Symp. Ser.*, **42**, 45–46 (1999).

324. S. Vichier-Guerre, A. Pompon, I. Lefebvre, and J. L. Imbach, *Antisense Res. Dev.*, **4**, 9–18 (1994).

325. J.-P. Shaw, K. Kent, J. Bird, J. Fishback, and B. Froehler, *Nucleic Acids Res.*, **19**, 747–750 (1991).

326. H. Seliger, A. Froehlich, M. Montenarh, J. F. R. Ortigao, and H. Roesch, *Nucleosides Nucleotides*, **10**, 469–477 (1991).

327. C. Chaix, R. P. Iyer, and S. Agrawal, *Bioorg. Med. Chem. Lett.*, **6**, 827–832 (1996).

328. R. Zhang, Z. Lu, H. Zhao, X. Zhang, R. B. Diasio, I. Habus, Z. Jiang, R. P. Iyer, D. Yu, and S. Agrawal, *Biochem. Pharmacol.*, **50**, 545–556 (1995).

329. J. Temsamani, J. Tang, A. Padmapriya, M. Kubert, and S. Agrawal, *Antisense Res. Dev.*, **3**, 277–284 (1993).

330. A. Boutorine, C. Huet, and T. Saison, *Nucleic Acid Ther.*, **24**, 273 (1991).

331. M. Manoharan, L. K. Johnson, C. F. Bennett, T. A. Vickers, D. J. Ecker, L. M. Cowsert, S. M. Freier, and P. D. Cook, *Bioorg. Med. Chem. Lett.*, **4**, 1053–1060 (1994).

332. M. Manoharan, *Antisense Research and Applications*, CRC Press, Boca Raton, FL, 1993, pp. 303–349.

333. A. De Mesmaeker, R. Haener, P. Martin, and H. E. Moser, *Acc. Chem. Res.*, **28**, 366–374 (1995).

334. J. Hall, D. H. usken, U. Pieles, H. E. Moser, and R. Haner, *Chem. Biol.*, **1**, 185–190 (1994).

335. C. Cazenave, N. Loreau, N. T. Thuong, J.-J. Toulme, and C. Helene, *Nucleic Acids Res.*, **15**, 4717–4736 (1987).

336. G. Godard, A. S. Boutorine, E. Saison-Behmoaras, and C. Helene, *Eur. J. Biochem.*, **2**, 404–410 (1995).

337. N. V. Amirkhanov, E. Samaratski, and J. Chattopadhyaya, *Tetrahedron Lett.*, in press.

338. E. Zamaratski, D. Ossipov, P. I. Pradeepkuman, N. Amirkhanov, and J. Chattopadhyaya, *Tetrahedron*, **57**, 593–606 (2001).

339. S. P. Henry, J. E. Zuckerman, J. Rojko, W. C. Hall, R. J. Harman, D. Kitchen, and S. T. Crooke, *Anticancer Drug Des.*, **12**, 1–14 (1997).

340. J. Desjardins, J. Mata, T. Brown, D. Graham, G. Zon, and P. Iversen, *J. Drug Target.*, **2**, 477–485 (1995).

341. Y. S. Sanghvi and P. D. Cook, *Carbohydrate Modifications in Antisense Research*, ACS Symposium Series **580**, American Chemical Society, Washington, DC, 1994.

342. M. Manoharan, *Biochim. Biophys. Acta*, **1489**, 117–130 (1999).

343. K. J. Breslauer, R. Frank, H. Blocker, and L. A. Marky, *Proc. Natl. Acad. Sci. USA*, **83**, 3746–3750 (1986).

344. A. I. Lamond and B. S. Sproat, *FEBS Lett.*, **325**, 123–127 (1993).

345. B. S. Sproat and A. I. Lamond, *Antisense Research and Applications*, CRC Press, Boca Raton, FL, 1993, pp. 351–362.

346. G. Parmentier, G. Schmitt, F. Dolle, and B. Luu, *Tetrahedron*, **50**, 5361–5368 (1994).

347. M. Manoharan, *Biochim. Biophys. Acta*, **1489**, 117–130 (1999).

348. M. Kanazaki, Y. Ueno, S. Shuto, and A. Matsuda, *J. Am. Chem. Soc.*, **122**, 2422–2432 (2000).

349. S. Agrawal, X. Zhang, Z. Lu, H. Zhao, J. M. Tamburin, J. Yan, H. Cai, R. B. Diasio, I. Habus, Z. Jiang, R. P. Iyer, D. Yu, and R. Zhang, *Biochem. Pharmacol.*, **50**, 571–576 (1995).

350. P. Martin, *Helv. Chim. Acta*, **78**, 486–504 (1995).

351. M. Teplova, G. Minasov, V. Tereshko, G. B. Inamati, P. D. Cook, M. Manoharan, and M. Egli, *Nat. Struct. Biol.*, **6**, 535–539 (1999).

352. L. Bellon, C. Leydier, and J. L. Barascut, *Carbohydrate Modifications in Antisense Research*, American Chemical Society, Washington, DC, 1994, pp. 68–79.

353. S. Pitsch, R. Krishnamurthy, M. Bolli, S. Wendeborn, A. Holzner, M. Minton, C. Lesueur, I. Schloenvogt, B. Jaun, et al., *Helv. Chim. Acta*, **78**, 1621–1635 (1995).

354. J. Wang, B. Verbeure, I. Luyten, E. Lescrinier, M. Froeyen, C. Hendrix, H. Rosemeyer, F. Seela, A. Van Aerschot, and P. Herdewijn, *J. Am. Chem. Soc.*, **122**, 8595–8602 (2000).

355. Y. S. Sanghvi and P. D. Cook.*Carbohydrate Modifications in Antisense Research*, ACS Symposium Series**580**, American Chemical Society, Washington, DC, 1994.

356. J. Wengel, *Acc. Chem. Res.*, **32**, 301–310 (1999).

357. C. Wahlestedt, P. Salmi, L. Good, J. Kela, T. Johnsson, T. Hokfelt, C. Broberger, F. Porreca, J. Lai, K. Ren, M. Ossipov, A. Koshkin, N. Jakobsen, J. Skouv, H. Oerum, M. H. Jacobsen, and J. Sengel, *Proc. Natl. Acad. Sci. USA*, **97**, 5633–5638 (2000).

358. M. J. Damha, C. J. Wilds, A. Noronha, I. Brukner, G. Borkow, D. Arion, and M. A. Parniak, *J. Am. Chem. Soc.*, **120**, 12976–12977 (1998).

359. A. M. Noronha, C. J. Wilds, C.-N. Lok, K. Viazovkina, D. Arion, M. A. Parniak, and M. J. Damha, *Biochemistry*, **39**, 7050–7062 (2000).

360. J. G. Karras, M. A. Maier, T. Lu, A. Watt, and M. Manoharan, *Biochemistry*, **40**, 7853–7859 (2001).

361. R. Kole and D. Mercatante, *Antisense Technology: Principles, Strategies and Applications*, Marcel Dekker, New York, 2001, pp. 517–539.

362. S. T. Crooke, *Antisense Technology: Principles, Strategies and Applications*, Marcel Dekker, New York, 2001, pp. 1–28.

363. S. M. Stepkowski, X. Qu, M.-E. Wang, L. Tian, W. Chen, E. V. Wancewicz, J. F. Johnston, C. F. Bennett, and B. P. Monia, *Transplantation*, **70**, 656–661 (2000).

364. T. Zellweger, H. Miyake, S. Cooper, K. Chi, B. S. Conklin, B. P. Monia, and M. E. Gleave, *J. Pharmacol. Exp. Ther.*, **298**, 1–7 (2001).

365. N. M. Dean, M. Butler, B. P. Monia, and M. Manoharan, *Antisense Technology: Principles, Strategies and Applications*, Marcel Dekker, New York, 2001, pp. 319–338.

366. R. S. Geary, O. Khatsenko, K. Bunker, R. Crooke, M. Moore, T. Burchkin, L. Truong, H. Sasmor, and A. A. Levin, *J. Pharmacol. Exp. Ther.*, **296**, 898–904 (2001).

367. R. S. Geary, T. A. Watanabe, L. Truong, S. Freier, E. A. Lesnik, N. B. Sioufi, H. Sasmor, M. Manoharaon, and A. A. Levin, *J. Pharmacol. Exp. Ther.*, **296**, 890–897 (2001).

368. R. Crooke and M. Graham, *Antisense Technology: Principles, Strategies and Applications*, Marcel Dekker, New York, 2001, pp. 155–182.

369. Z. Zhang, J. Cook, J. Nickel, R. Yu, K. Stecker, K. Myers, and N. M. Dean, *Nat. Biotechnol.*, **18**, 862–867 (2000).

370. P. L. Nicklin, S. J. Craig, and J. A. Phillips, *Antisense Research and Application*, Springer-Verlag, Berlin/Heidelberg, 1998, pp. 141–168.

371. R. Zhang, R. B. Diasio, Z. Lu, T. Liu, Z. Jiang, W. M. Galbraith, and S. Agrawal, *Biochem. Pharmacol.*, **49**, 929–939 (1995).

372. S. P. Henry, K. Stecker, D. Brooks, D. K. Monteith, B. Conklin, and C. F. Bennett, *J. Pharmacol. Exp. Ther.*, **292**, 468–479 (2000).

CHAPTER SIX

Therapeutic Agents Acting on RNA Targets

Jason P. Rife
Virginia Commonwealth University
Richmond, Virginia

Contents

Burger's Medicinal Chemistry and Drug Discovery
Sixth Edition, Volume 2: Drug Development
Edited by Donald J. Abraham
ISBN 0-471-37028-2 © 2003 John Wiley & Sons, Inc.

1 INTRODUCTION

Structure-aided drug design began in earnest in the 1970s. An early example at that time was against dihydrofolate reductase (1). However, X-ray crystallography was being used to investigate enzyme/inhibitor complexes a decade before that (2). In addition to those studies, numerous enzymes were available in sufficient quantity for most of the 20th century to allow biochemists to design and test inhibitors of protein function. Therefore, the design of inhibitors of protein function has been aided by the structural probing of inhibitor/protein complexes since the late 1960s, and lessons learned from those structures continue to contribute to the design of drugs in the present day. The advent of recombinant DNA technologies and fast computers provided the last pieces of the infrastructure necessary to permit structural biology to play a regular role both in explaining drug action and in development of new drugs. Modern day drug discovery relies on structural biology and molecular modeling, natural product screening, high-throughput screening, genomics, and proteomics to identify drug candidates against existing and novel targets (3).

Consider the maturity of drug design against protein targets compared with similar efforts against RNA targets. The first RNA-containing enzyme to be discovered was the ribosome in the early 1950s. Its complexity has served as a tantalizing biochemical problem for hundreds of scientists over several decades and it is arguably the most studied enzyme (4). Identification of discrete ribosome particles occurred in the early 1950s, and definitive proof that ribosomes carried out protein synthesis came in 1955. Study of ribosome structure, function, and mechanism has continued unabated to the present day, progress that has recently been punctuated with reports of numerous crystal structures of both subunits at atomic resolution (5–8), medium resolution structures of whole ribosomes (9), and a variety of drug-ribosomes complexes at good resolution (10–14).

The antibiotic streptomycin provides an excellent example of why the design of novel RNA binding therapeutics has lagged behind the drug discovery process in general. Isolated and shown to have strong antibiotic properties (15), streptomycin entered clinical use in 1949 as a treatment for gram-negative infections, and initially served as a complement to the penicillins, which are useful against gram-positive infections (16). As early as 1948, it was clear that streptomycin inhibited protein synthesis (17), but not until 1961 were suggestions made that streptomycin inhibited ribosome function (18). In 1987, clear evidence that streptomycin interacted extensively with RNA was provided (19), and in 2000, more than 50 years after the discovery of streptomycin, the details of streptomycin binding were finally revealed with the report of the crystal structure of the streptomycin/30S complex (10). Since then, there have been several other structural reports of antibiotics that bind to ribosomal RNA, many of which are human therapeutics, in complex with ribosomal subunits (10–14, 20). These structures now provide the basic knowledge needed to explain the molecular function of those drugs and to design novel drugs similar to them.

Drugs that bind to RNA are half a century old, but the intentioned design of RNA-targeted drugs is new. Some 25 years after structural biology was initially brought to bear on the problem of drug design, medicinal chemists have in hand similar toolkits with which to aid in their search for novel drugs that target RNA structure and function. Several questions remain to be answered. (1) Do small ligands bind to RNA in fundamentally different ways than they do to proteins? (2) Are there enough drug/RNA complexes available to serve as a foundation for future drugs design? (3) Are there qualities unique to rRNA that qualify it as a drug target that would preclude other RNAs from being suitable drug targets? (4) Can any RNA motif confer sufficient specificity to be the sole *in vivo* target.

2 RNA AS A DRUG TARGET

Can drugs be found that specifically target RNA? For rRNA in the ribosome the answer is easy; it is a validated target that has been exploited therapeutically for more than 50 years. Arguments in favor of targeting other RNA molecules have been made elsewhere and only the most salient points will be recapitulated here (21–23). RNA function in biology is diverse (24). It serves to preserve genetic information (retroviruses) or to transport it temporarily (mRNA), it is the core of the ribosome (rRNA), it serves as a co-factor in protein synthesis (tRNA), and it is integral to other RNA/protein enzymes (e.g., the spliceosome, telomerase), to give a few examples.

The search for RNA-binding drugs that inhibit vital protein-RNA interactions in HIV replication has been underway for about a decade (25). Although no compound has yet been elevated to the status of clinical drug candidate, many structurally diverse compounds have been identified that bind tightly to their RNA targets. Drug discovery efforts are underway against other viral targets as well (26).

The quest for drugs that specifically downregulate expression of individual genes, such as oncogenes, is ongoing at the proof-of-principle stage. For example, within untranslated regions of a given mRNA there exist numerous structured elements integral to the expression of that mRNA. In theory, the expression of a particular mRNA could be regulated through the binding of drug-like compounds to those elements (27–29). At least two pieces of evidence suggest that this is possible *in vivo*. First, when a neomycin-specific binding element was engineered into the 5′ untranslated region of a reported gene, the expression of that gene could be regulated by neomycin, whereas a similar aminoglycoside, tobramycin, failed to affect gene expression (30). Second, it has recently been suggested that in bacteria mRNA can act as a direct sensor of small molecules such as thiamin to control their expression (31). In this example, expression of mRNA that encodes for thiamin-synthesizing proteins is down-regulated by the presence of thiamin itself, which binds directly to those mRNAs. Considerable effort is underway to characterize the function of the numerous mRNA structural elements known to exist and to identify others.

Drug development is a multi-factorial endeavor, of which binding affinity is only a single component. Toxicology and pharmacokinetics of drug candidates are equally important. Nevertheless, binding remains the defining event in drug action. This chapter will address the binding of four classes of drugs: streptomycin, neomycin B/gentamicin-like aminoglycosides, tetracyclines, and macrolides (e.g., erythromycin). Other important considerations pertaining to these drugs are addressed elsewhere in this series and will not be addressed here (32). Each of these drugs had similar beginnings; they are all natural products that were discovered as fermentation products from soil microbes. Exhaustive searches have been conducted for semisynthetic analogs with improved therapeutic qualities; in all cases except for streptomycin these efforts have led to clear improvements. In large part the success or failure of many of those analogs can be explained by the recent crystal structures of antibiotic/ribosome complexes. Once key drug-target interactions have been identified, it is then possible to search for other analogs that retain those key interactions yet also satisfy other necessary drug requirements.

3 RIBOSOME STRUCTURE AND FUNCTION

Ribosomes are found in all cells and are responsible for translating mRNA into protein. Ribosomes are composed of two subunits. In bacteria, the large subunit, whose sedimentation velocity is 50S, is about twice the mass of the small ribosomal subunit, whose velocity is 30S. In *E. coli*, the 50S subunit is composed of two RNA molecules, 5S rRNA, which is 120 nucleotides in length, and 23S rRNA, which is 2904 nucleotides in length, and 31 proteins (L1–L31). The 30S subunit (hereafter referred to as 30S) is composed of a single 1542-nucleotide RNA molecule, 16S rRNA, and 21 proteins (S1–S21). Both subunits together have a molecular weight of about 2.5×10^6 Da and are composed of roughly one-third protein and two-thirds RNA. Eukaryotic ribosomes are essentially identical to prokaryotic ribosomes, except that they are slightly larger, because they have more components and the average size of all components is larger. Some parts of ribosome structure are highly conserved across all domains of life, while others are not. Certain eukaryotic organelles (mitochondria, chloroplasts, and plastids) also have ribosomes, which tend to be stripped-down versions of prokaryotic ribosomes. No matter the source, a common convention is to number the rRNA nucleotide sequence and helices according to those found in *E. coli*, and that is the convention that will be followed here.

Ribosome function is complex; numerous cofactors are required for initiation, elongation, and termination (for a detailed description please see ref. 33). Independent functions can be ascribed to the two subunits. Peptide bond formation takes place on the 50S subunit within the peptidyl transferase center, whereas decoding of the mRNA takes place on the 30S subunit within the decoding A- and P-sites. tRNAs in the P-site and the A-site span both subunits and couple the two events. After each round of peptide bond formation, a translocation step takes place that involves the movement of the mRNA through the ribosome, transfer of the P-site tRNA to the E (or exit)-site, and transfer of the A-site tRNA to the P-site. Most antibiotics target one of the listed steps: decoding at either the A-site or the P-site, peptide bond formation within the peptidyl transferase center, or translocation.

4 STREPTOMYCIN

Streptomycin and other antibiotics that are close analogs (Fig. 6.1) all contain a streptamine aminocyclitol, which is linked to a streptose ring, which in turn is linked to an *N*-methylglucosamine. Of the 10 natural streptomycin-like compounds listed in Fig. 6.1, all have guanidino groups at the C1 and C3 position except for bluensomycin, which contains a carbamoyl group at the C1 position instead. Some variation is tolerated within the streptose ring: the aldehyde can be reduced to an alcohol, and methyl alcohol can substitute for the methyl at the C4′ position (34). Considerable modifications of the glucosamine moiety are also tolerated.

Shortly after its discovery, streptomycin was brought into service to treat Gram-negative and some Gram-positive infections, with its greatest contribution coming as the primary agent to treat tuberculosis (35). Its use has waned over the years because bacterial resistance against streptomycin has emerged and less toxic agents have become available. Nonetheless, it is still frequently used as a component of a multi-drug mixture to treat multi-drug resistant cases of tuberculosis.

4.1 Mechanism of Action

Streptomycin binds to the 30S ribosomal subunit and disrupts normal protein synthesis both by weakening proofreading and by interfering with initial tRNA selection. The streptomycin/30S complex has recently been solved by X-ray crystallography (10). As predicted by chemical protection, crosslinking, and mutagenesis data, streptomycin binds tightly to a pocket involving nucleotides from four regions of 16S rRNA. The binding of streptomycin is stabilized by a variety of salt bridges and hydrogen bonds between streptomycin and backbone phosphates and through hydrogen bonds to Lys45 of protein S12.

30S exists in one of two conformations termed the *ram* state and the restrictive state (10, 36). The *ram* state (or *r*ibosomal *am*biguity) was identified many years ago by correlating the presence of certain 16S rRNA and ribosomal protein mutations with error-prone translation. Conversely, mutations that support the restrictive state exhibit hyper-accurate translation. A balance between sloppy, fast translation and hyper-accurate, slow translation must be struck for organisms to be successful. The mechanics of this two-state switch have been described (10). In the *ram* state, nucleotides 912–910 in helix 27 of 16S rRNA base pair with nucleotides 885–887, whereas in the restrictive state 912–910 base pair with nucleotides 888–890. Mutations or ligands that stabilize the 912–910/885–887 pairing promote indiscriminant tRNA selection. Mutations or ligands that stabilize the 912–910/888–890 pairing restrict tRNA binding, leading to hyper-accurate translation. Streptomycin binds to 30S in the *ram* state and consequently stabilizes that state, which results in error-prone translation and consequently the accumulation of miscoded proteins and the reduction of proteins with the correct sequence.

4.2   Structure-Activity Relationship

Initial understanding of streptomycin structure-activity relationships can be gleaned from comparing the activities of the component rings in groupings. Rings I and II alone are sufficient for antibiotic activity,[1] but rings II and III (streptobiosamine) together are inactive; this in part explains the high tolerance of ring III for modifications (37).

Figure 6.2 highlights permissible and nonpermissible alterations to streptomycin and analogs. Within the streptamine ring, any change to the guanidino group at C3 abolishes all activity (38–40). However, methylation of the C1 guanidino group or substitution of that group with chemically similar moieties is usually tolerated. Phosphorylation or adenylation of the hydroxyl group at C6 abolishes activity (41); removal of that group greatly reduces activity (42). The consequences of modifying the hydroxyl group at either C2 or C5 are unknown.

Inspection of the streptomycin/30S crystal structure and sequence comparisons of 16S sequences from numerous organisms generally explain these observations (10). The C3 guanidino group, which cannot be altered, forms a bifurcated hydrogen bond with O2′ and O3′ of A914, a universally conserved nucleotide. Methylation of the C3 guanidine would result in the disruption of one of these hydrogen bonds. Other modifications would similarly disrupt the multiple hydrogen bonds involving the C3 guanidino group.

The C1 guanidino group forms a salt bridge with the phosphate of C1490. When this guanidino group is replaced with a carbamoyl group, as in bluensomycin, an analogous interaction can take place and is therefore moderately tolerated; bluensomycin is about 10-fold less active than streptomycin for most streptomycin-sensitive organisms. Methylation of

the C1 guanidino group is also tolerated and the streptomycin/30S crystal structure supports this, because a salt bridge to the phosphate of C1490 can be retained. Synthetic analogs in which the C1 guanidino group has been replaced with 1-N-[(S)-4-amino-2-hydroxybutyryl] or 1-N-[(S)-4-guanidino-2-hydroxybutyryl] have been evaluated (38). These compounds were tested on a variety of organisms, some of which were resistant to streptomycin. In all cases, the reported activity was lower than for dihydrostreptomycin (which has activity similar to streptomycin) and similar to or somewhat lower than for bluensomycin. Replacement of the C1 guanidino group with a 2-imidazolin-2-ylamino substituent was inactive (40). Overall, however, most of the C1 guanidino modifications investigated are tolerated.

Modifications of the C6 hydroxyl group are poorly tolerated. The 6-deoxydihydrostreptomycin analog was synthesized and tested in hopes of circumventing the action of C6- OH modifying enzymes found in some streptomycin-resistant bacteria, but this compound was only $^1/_{10}$–$^1/_{80}$ as active as dihydrostreptomycin (42). Phosphorylation, adenylation, or removal of the C6 hydroxyl group causes a strong reduction in activity; the reason is apparent from the streptomycin/30S structure, in which this hydroxyl group forms hydrogen bonds with Lys45 of the ribosomal protein S12 and the phosphate of G1491. These bonds clearly must be maintained for full activity.

Although no analogs of streptomycin exist in which the C5 position is altered, the structure of the complex suggests that very few, if any, modifications would be tolerated, because that hydroxyl group is also involved in a hydrogen bond with Lys45 of S12. Similar restrictions are not expected for C2 modification, as there are no direct contacts between its hydroxyl group and the ribosome.

The site most often exploited for streptomycin modification is the 3′ aldehyde moiety of ring II, the streptose ring. The first semisynthetic analog of streptomycin, dihydrostreptomycin, contained an alcohol in place of the streptose aldehyde (43). This analog was later isolated from fermentation sources (*Streptomyces humidus*) (44). Figure 6.2b illustrates many of the aldehyde conversions

---

[1] Unless otherwise stated, activity refers to the ability of an agent to stop bacterial growth *in vitro*. Important measurements such as binding affinity for 30S were rarely carried out. Therefore, in general, the observed activity (or lack of activity) is assumed to result from a mechanism of action identical to that of streptomycin. Examples described in the text illustrate that such assumptions are not always valid.

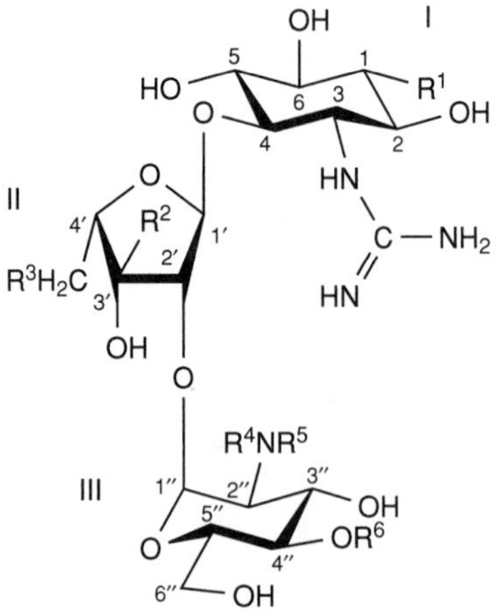

	R^1	R^2	R^3	R^4	R^5	R^6
Streptomycin	NH–CNH–NH$_2$	CHO	CH$_3$	CH$_3$	H	H
Dihydro-Streptomycin	NH–CNH–NH$_2$	CH$_2$–OH	CH$_3$	CH$_3$	H	H
5′-hydroxy-Streptomycin	NH–CNH–NH$_2$	CHO	CH$_2$–OH	CH$_3$	H	H
N-Demethyl-Streptomycin	NH–CNH–NH$_2$	CHO	CH$_3$	H	H	H
Mannosido-hydroxystrepto-mycin	NH–CNH–NH$_2$	CHO	CH$_2$–OH	CH$_3$	H	α-D-mannoside
Dimannosido-streptomycin	NH–CNH–NH$_2$	CHO	CH$_3$	CH$_3$	H	α-D-mannose-1,6-α-D-Mannose
Bluensomycin	O–CO–NH$_2$	CHO	CH$_3$	CH$_3$	H	H
Ashimycin A	NH–CNH–NH$_2$	CHO	CH$_3$	CH$_3$	H	Ashimose
Ashimycin B	NH–CNH–NH$_2$	CHO	CH$_3$	CH$_3$	CO–CH$_2$OH	α-D-mannose-1,6-α-D-Mannose
AC4437	= 5′-hydroxystreptomycin lacking ring III					

Figure 6.1. The structures of streptomycin and several natural analogs of streptomycin.

which have been synthesized and tested. The aldehyde oxygen forms a hydrogen bond with a phosphate oxygen of G527.

Conversion of the aldehyde to an acid or reduction to a methyl abolishes all activity (45, 46). Numerous groups have investigated conversion of the aldehyde to its amino derivative and a variety of alkylamine derivatives, with surprising results (47–50). The amino deriva-tive and short-chain alkylamine derivatives remained active up to the hexylamine analog. However, activity diminished with increasing chain length; the hexylamine derivative is only about $^1/_{100}$ as active as dihydrostrepto-mycin. The heptaamine derivative was not in-vestigated. Longer alkylamines (octyl and above) were nearly as active as dihydrostrep-tomycin. This sharp inflection prompted one

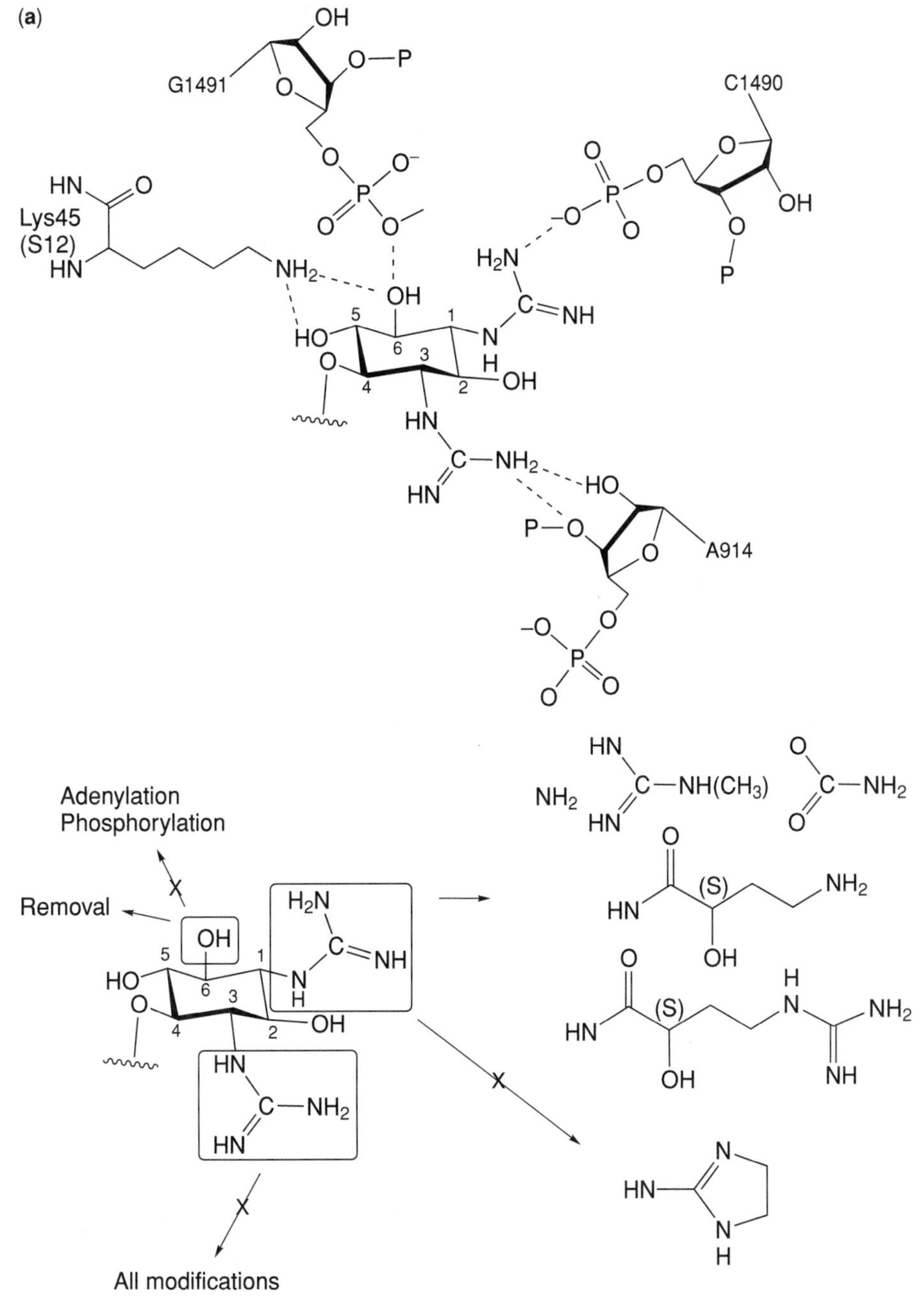

Figure 6.2. Molecular interactions between streptomycin and 30S (*E. coli* numbering, top) with various modifications tested for activity (bottom). Dashed lines indicate possible hydrogen bonds (some of which are salt bridges when suitably reinforced with favorable electrostatic potentials). Arrows point to permissible modifications; arrows with an X point to non-permissible modifications. (a) Ring I. (b) Ring II. Dashed arrow points to modifications that results in compounds that are active by an unknown mechanism. (c) Ring III.

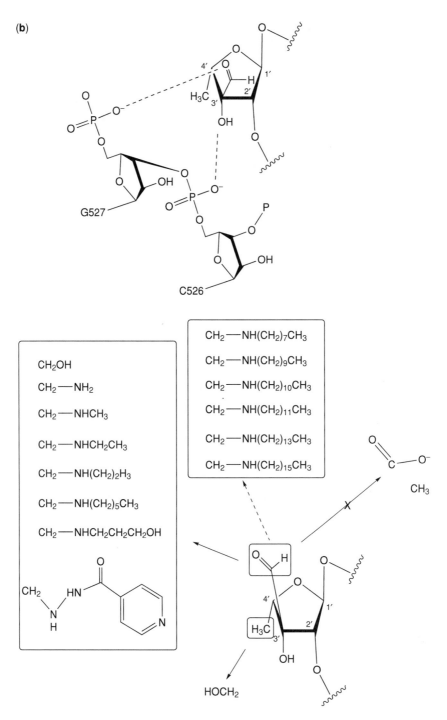

Figure 6.2. (Continued.)

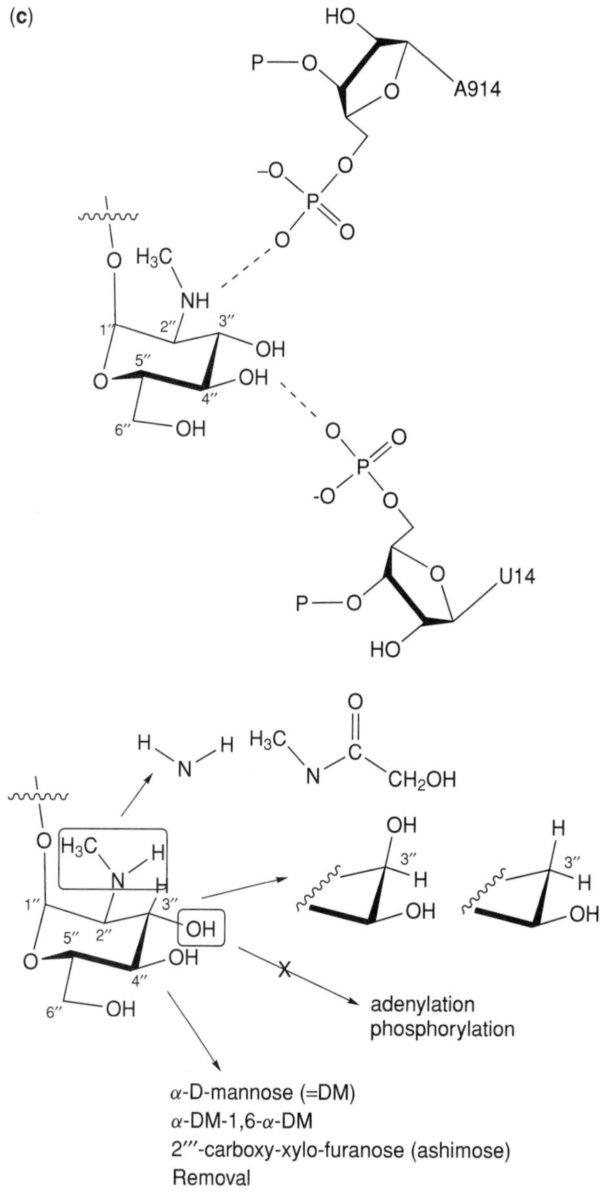

(c)

Figure 6.2. (Continued.)

α-D-mannose (=DM)
α-DM-1,6-α-DM
2‴-carboxy-xylo-furanose (ashimose)
Removal

group to consider the mechanism of action of the alkylamine derivatives. The activity data suggested that the short-chain alkylamine analogs have a mechanism of action identical to that of streptomycin, whereas the long-chain alkylamine analogs operate by an unknown but ribosomally unrelated mechanism; the long-chain alkylamine analogs no longer bind to 30S.

Another alkylamino derivative tested was a conjugate of streptomycin and isoniazid, another prominent anti-tuberculosis drug. This compound termed streptohydrazid, was synthesized and found to be at least as active as combined therapy using both streptomycin and isoniazid (51). Streptohydrazid was tested long before the mechanism of action of streptomycin was known (the mechanism of isoniazid is not fully understood); it was reasoned that a conjugate of the two might act synergistically. The mechanism by which streptohydrazid works is not known, but presumably it has streptomycin-like function, isoniazid-like function, or some combination of the two.

Correlation of the streptomycin/30S crystal structure with the various streptomycin analogs that involve aldehyde modification suggests that only a hydrogen bond between the aldehyde oxygen and a protonated phosphate oxygen (or a salt bridge for the amino derivatives) must be maintained. Reduction in binding only occurs when the modification becomes too large to be accommodated within the binding pocket. The only exception to this is when streptomycin is oxidized to streptomycinic acid. Although the possibility of forming the required hydrogen bond exists, the analog is inactive, presumably because of the electrostatically unfavorable close approach of the carboxylic acid to a phosphate that would occur on binding.

Several active natural streptomycin analogs, such as 5'-hydroxystreptomycin and AC4437, are hydroxylated at C5' (34, 52). Semisynthetic derivatives of this position are absent. A cursory inspection of the streptomycin/30S structure suggests that the methyl group at C4' contributes little to ribosome binding, suggesting that modifications at this position might be tolerated.

Ring III (glucosamine) makes two direct contacts with 30S, neither of which are essential to activity (37, 52). Indeed, all of ring III can be dispensed with and cause only modest reductions in activity. However, deleterious modifications to this ring are possible. Some streptomycin-resistant bacteria harbor genes that encode proteins that either phosphorylate or adenylate the C3'' hydroxyl group of streptomycin (41). Two semisynthetic analogs, 3''-epidihydrostreptomycin and 3''-deoxydihydrostreptomycin, were synthesized to circumvent common streptomycin resistance (53, 54). The logic is sound: both analogs should retain streptomycin-like binding affinity, yet not be substrates for inactivating enzymes. As expected, these analogs worked well against common bacterial strains and better than streptomycin against many streptomycin resistant strains. Yet, for reasons that are not clear, they never reached clinical status.

Some glycosylations have been observed at C4'' in natural streptomycin analogs, yielding somewhat less active antibiotics, yet remain active against bacteria that express enzymes that phosphylate and adenylate the C3''- OH. A limited number of natural modifications occur at C2'', none of which abolish activity.

5 NEOMYCIN AND GENTAMICIN TYPE AMINOGLYCOSIDES

Shortly after the discovery of streptomycin, members of the 4,5-linked and 4,6-linked 2-deoxystreptamine aminoglycoside antibiotics were found in fermentation products of actinomycetes (Fig. 6.3) (55, 56). (Although many antibiotics contain aminoglycoside rings and are often referred to as aminoglycosides, the general term "aminoglycoside" will be used here to specifically refer to the 4,5- and the 4,6-subclasses and will be abbreviated AG.) Neomycin B is the prototypical 4,5-linked AG and is composed of four rings, because it is the most effective agent in this class. The gentamicins are the most commonly used 4,6-linked AGs and are composed of only three rings. The 4,6-linked subclass constitutes the majority of the clinically useful agents. The two subclasses have their first two rings in common and work by the identical mechanism of binding to the decoding A-site of 30S, thus causing misreading of mRNA (57). Biochemical probing has firmly established the binding of these agents to the major groove of an asymmetric loop composed in part of several absolutely conserved nucleotides (19). Other poorly conserved nucleotides within this loop also form part of the AG binding pocket and provide the basis for organismal specificity (58).

The structure of paromomycin, a close analog of neomycin B, in complex with rRNA, has been solved multiple times, once by NMR and twice by X-ray crystallography (10, 13, 59). These structures clearly describe the important modes of binding for the 4,5-linked subclass. NMR was also used to solve the structure of gentamicin C1a in complex with its target RNA sequence (60). The orientation of binding is the same—rings I and II from

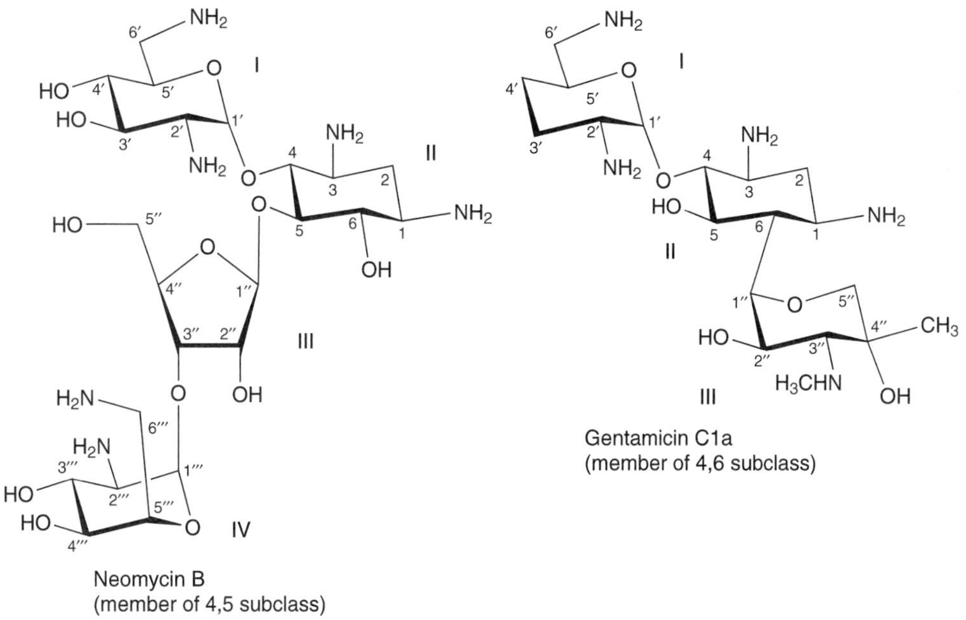

Figure 6.3. The structures of neomycin B and gentamicin C1a.

both subclasses bind identically; however, because of alternate linkages that exist between rings II and III, the remaining portions of the two subclasses take different trajectories. The ultimate consequence of AG binding remains the same; 30S is no longer able to discriminate between cognate and noncognate tRNAs when the tRNA binds to the A-site (57).

5.1 Mechanism of Action

The molecular details by which AGs cause miscoding were recently elucidated (61). Nucleotides A1492 and A1493, along with G517, play pivotal roles in discriminating between cognate codon/anticodon interactions and non-cognate interactions. When the correct tRNA occupies the A-site, the tRNA and mRNA nucleotides involved in the codon/anticodon interaction form a regular helix conformation, and consequently normal minor groove geometry. In the correct presentation a helical minor groove stabilizes the "looped-out" conformation of A1492, A1493, and G517, the conformation which serves as a signal for peptide bond formation to occur. Incorrect tRNAs do not elicit the same response and eventually diffuse away from the A-site with-

out the formation of a new peptide bond. However, when either a 4,5 or 4,6 AG binds to 16S rRNA, it displaces A1492 and A1493 into the "looped-out" conformation. Therefore, peptide bond formation will occur regardless of whether or not the correct tRNA occupies the A-site.

5.2 Structure-Activity Relationship of Rings I and II of Both Subclasses

Although the two subclasses of the A-site binding aminoglycosides are chemically distinct, they do share rings I and II in common and up to that point bind 30S in identical ways (60). Therefore, discussion of ring I and ring II modifications of both subclasses will be treated at the same time, without regard for whether or not a particular modification was observed in the 4,5 subclass or the 4,6 subclass (Fig. 6.4a).

For both classes, ring I is the only essential ring. Removal of ring I from all aminoglycosides results in inactive compounds; ring I of neomycin alone is minimally active as an antibiotic (41). Although this class is grouped by the inclusion of 2-deoxystreptamine, this ring alone is insufficient for antibiotic activity. Sev-

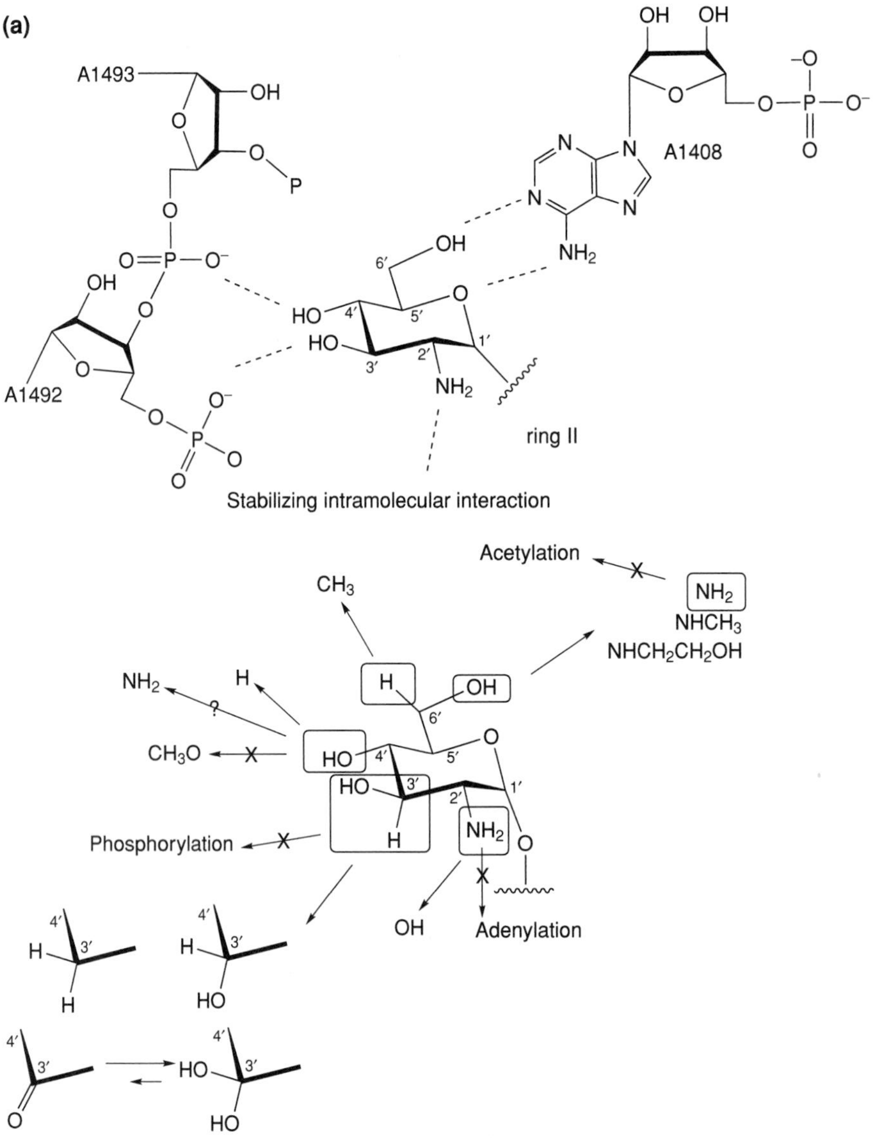

Figure 6.4. Molecular interactions between 4,5 and 4,6 linked aminoglycosides and 30S (*E. coli* numbering, top) with various modifications tested for activity (bottom). Dashed lines indicate possible hydrogen bonds (some of which are salt bridges when suitably reinforced with favorable electrostatic potentials). Arrows point to permissible modifications; arrows with an X point to non-permissible modifications. (a) Paromomycin ring I. The arrow with a ? points to a modification that was shown to be inactive, but is not readily explained by the crystal structure. (b) Paromomycin ring II. (c) Paromomycin ring III. (d) Paromomycin ring IV. (e) Gentamicin C1a ring III.

eral naturally occurring antibiotics contain only ring I and ring II (paromamine and neamine are examples), but all clinically active agents contain four rings for the 4,5 subclass and three rings for the 4,6 subclass.

Some AGs have an NH_2 at C6', which when acetylated, abolishes antibiotic activity (41). Replacement of C6'-OH with an amine or a short-chain alkylamine is tolerated or improves activity. The ring oxygen and C6'-OH

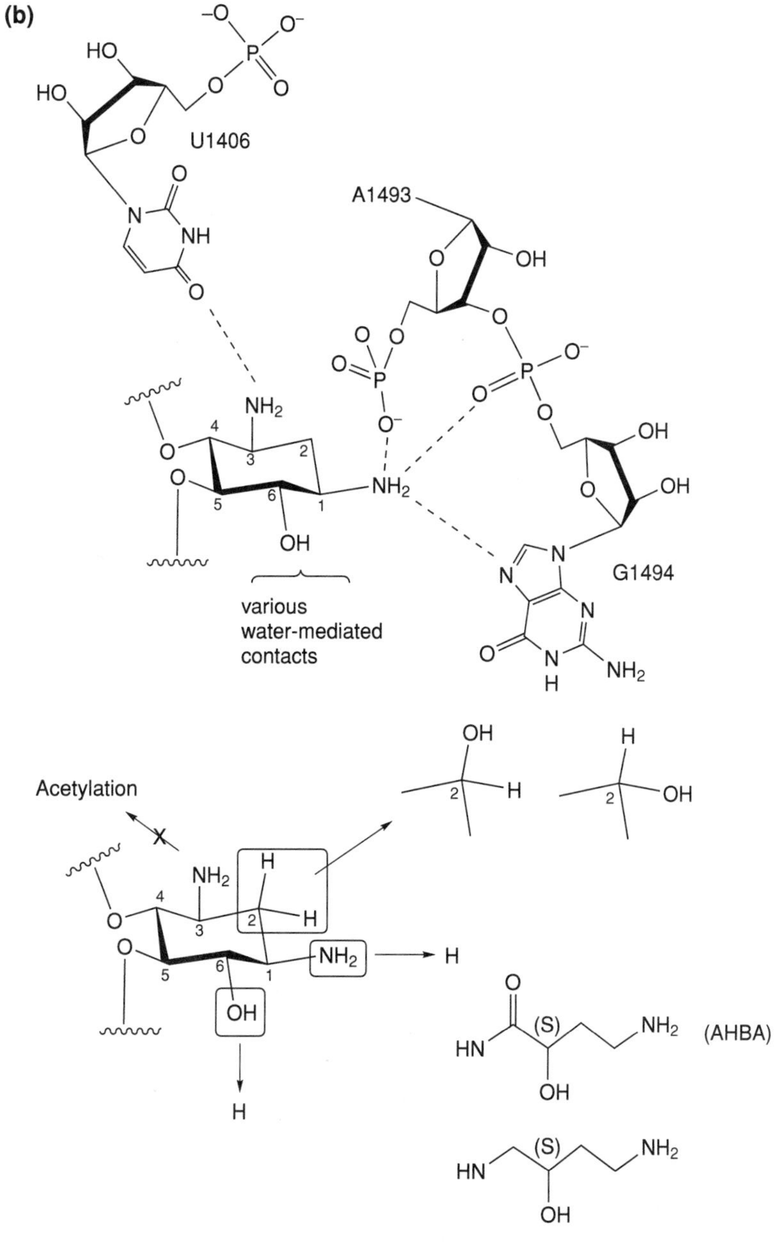

Figure 6.4. (Continued.)

(or C6'-NH$_2$) form a psuedo-*trans* Watson-Crick pair with A1408 (13). In eukaryotes, G substitutes for A at this position, and this largely explains the prokaryotic specificity of aminoglycoside antibiotics (58). With a G in position 1408, an analogous pseudo-pairing isn't possible. An important observation is that paromomycin remains active against protozoa despite the fact that in protozoa nucleotide 1408 is a guanosine, because at least one hydrogen bond is preserved. But an amino group at C6' (as in neomycin B) isn't active, because neither of the original hydrogen bonds can be satisfied. Finally, branching at

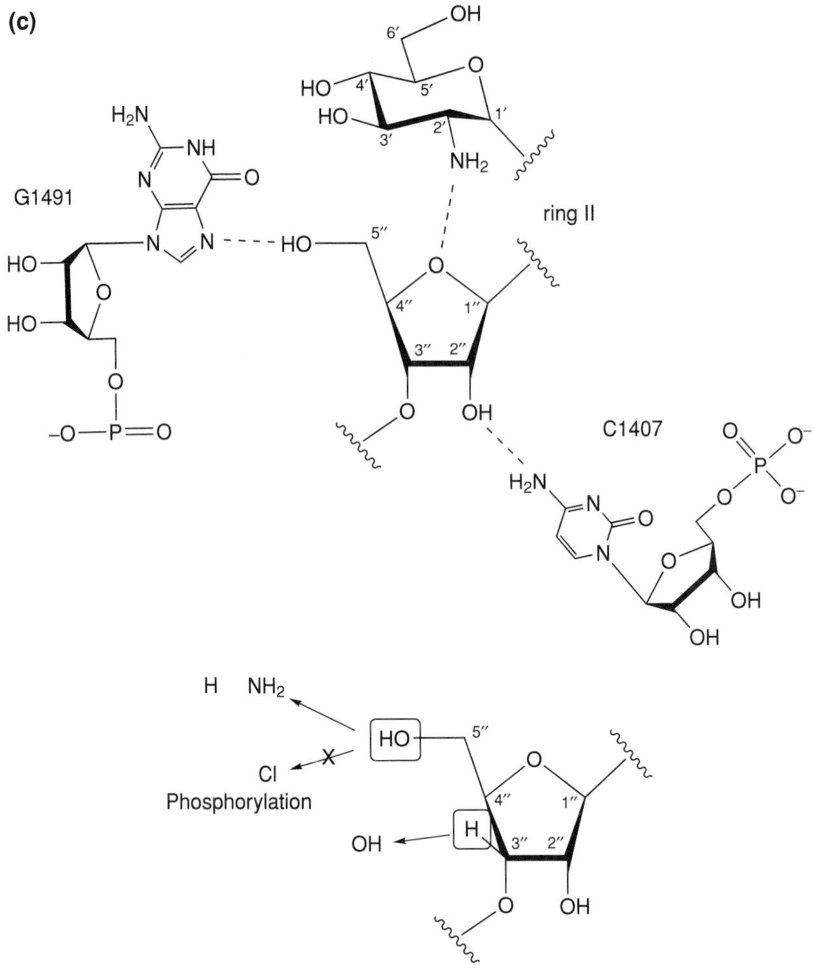

Figure 6.4. (Continued.)

C6′ is permissible, at least up to one carbon, because branching at this position doesn't disrupt the pseudo-*trans* pairing to G1408 (41).

All active aminoglycoside compounds have a hydroxyl at C4′, are C4′-deoxy, or are unsaturated at C4′ (41); hydroxyl groups at C4′ form a hydrogen bond to the phosphate of A1493. The methoxy derivative is not tolerated at C4′ (41), presumably because of steric clash with the phosphate of A1493. An amino at the C4′ position surprisingly abolishes activity. Because an amino at this position should make a productive salt bridge, the particular C4′- NH₂ derivative tested may be inactive for other reasons.

The 3′ hydroxyl group in paromomycin forms a hydrogen bond to the phosphate of A1492. Phosphorylation at this position abolishes all activity and forms a mode of resistance (41, 62); however, 3′-deoxy and 3′-*epi* analogs are not substrates for such resistance enzymes, and they remain active against some bacteria resistant to aminoglycosides (41). Recognition that all 3′-deoxyderivatives lose a productive interaction that presumably lowers their affinities for 16S rRNA prompted the design of 3′ ketokanamycin A, in an effort to preserve the hydroxyl phosphate hydrogen bond while eliminating the possibility of inactivation by phosphorylating enzymes (63). It is

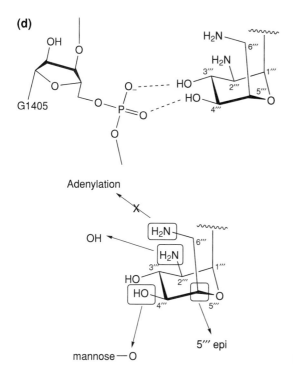

(d)

Adenylation

mannose—O

5''' epi

Figure 6.4. (Continued.)

expected that the 3' keto derivative will exist in equilibrium with a hydrated variant (3'-gemdiol analog) and serve as a substrate for phosphorylation. However, the phosphorylated product would undergo spontaneous elimination of the phosphate moiety, thereby regenerating the 3' ketokanamycin A analog, which retains the same hydrogen bonding to A1492. Although the keto analog was substantially less active than kanamycin A against *E. coli*, it was more active than kanamycin A against *E. coli* harboring the gene for APH(3')-Ia, a resistance gene which, when expressed, phosphorylates C3'-OH.

Only hydroxyl and amino groups are tolerated at C2' because an intramolecular hydrogen bond must be satisfied between ring I and ring III of the 4,5 subclass and ring I and ring II of the 4,6 subclass to stabilize the conformation of the antibiotics in the active state. In both subclasses other substitutions at the C2' position would disrupt this hydrogen bond. Adenylation of this site by resistance enzymes or conversion to the 2'-*epi* configuration abolishes activity (41, 62).

Ring II structurally defines this class of antibiotics as the 2-deoxystreptamine aminogly-

cosides and has given way to modification more frequently than the other rings (Fig. 6.4b). Conversion of kanamycin A to include 4-amino-2(S)-hydroxylbutyrylamide (AHBA) at C1 yielded an aminoglycoside (amikacin) that was effective against many aminoglycoside resistant bacteria with little reduction in activity against aminoglycoside sensitive bacteria (41). Success with amikacin prompted an exhaustive search for other C1 derivatives with improved efficacy over parental aminoglycosides. Several important commercial aminoglycosides emerged from such efforts. It is impossible to catalogue all of the reported C1 modifications, therefore only the most illustrative examples will be described here.

N-acylation of C1-NH$_2$ is the most explored type of modification at this position. Although only a limited number of highly active derivatives were found, a large variety of modifications are tolerated at C1-NH$_2$ (41, 64, 65). Using AHBA as a basis of comparison for all *N*-1 acyl modifications, clear structure-activity relationships of other acylated products are observed. Shortening or lengthening the carbon chain by more than one carbon unit drastically reduces activity, as does moving the hydroxyl

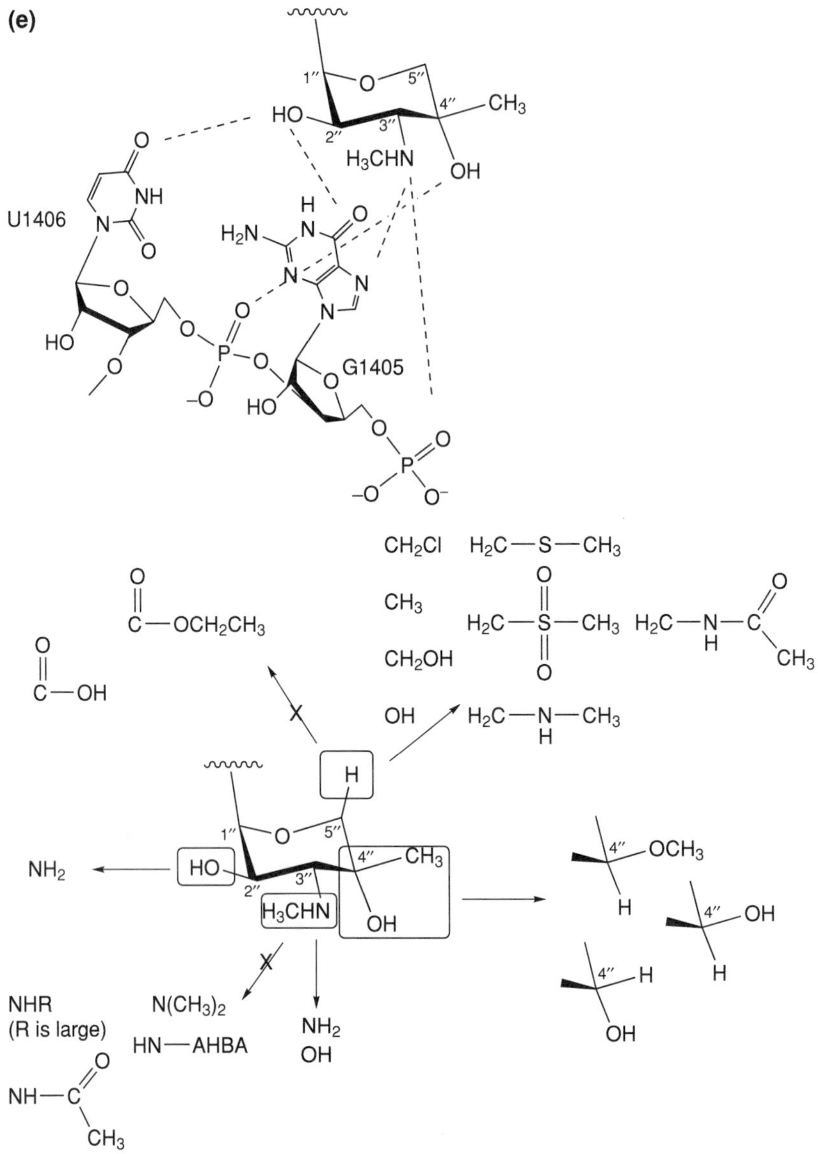

(e)

Figure 6.4. (Continued.)

group away from the α position or inverting the stereochemistry of the α carbon. Substituting the α-hydroxyl group with an amino group abolished all activity, whereas a fluorine replacement retains full activity. The α-deoxy derivative retains partial activity. With the exception of unsubstituted amidino and guanidino groups, any change to the terminal amine generally reduces activity or abolishes activity altogether. Taken together, it is clear

that AHBA makes highly selective, yet unknown contacts with 30S.

The positive qualities of C1 AHBA modification may be universal to all 2-DOS containing aminoglycosides that bind in the decoding A-site pocket, because the same moiety is found in the naturally occurring aminoglycoside butirosin and was successfully introduced into gentamicin B, yielding the active compound isepamicin (41). (The observation that the same modification is optimal in both the 4,5-subclass (butirosin) and in the 4,6-subclass (amikacin and isepamicin) gives further weight to the supposition that rings I and II of both subclasses bind in the same manner.)

Considerably fewer C1-NH_2 alkyl derivatives have been synthesized and tested for activity, but it seems that considerable structural diversity can be tolerated. For example, both AHBA-like alkane and C1-NH_2 dihydroxypropyl kanamycin analogs showed near equal efficacy despite being structurally dissimilar (66, 67).

The direct interactions between C1-NH_2 and 30S include a hydrogen bond with N7 of G1494 and stabilizing electrostatic interactions with the phosphates of G1494 and A1493. These interactions are presumably maintained in the above derivatives. The nucleotides A1492 and A1493 are displaced on drug binding, which provides a large pocket adjacent to C1-NH_2 (10); this explains why so many modifications are easily tolerated at this position. What isn't known is why some modifications are better than others. A co-structure of amikacin (or one of the other N-1 AHBA aminoglycosides) bound to 30S or extensive molecular modeling is required before more insight into the positive attributes of AHBA can be realized.

Acetylation of C3-NH_2 by resistance enzymes abolishes activity, as do virtually all other modifications at this position (62). The hydrogen bond that this amino group forms with U1406 is evidently a crucial interaction. The only information that exists for modifications at C2 regards substitution with a hydroxyl group (41); substitution of this type in either configuration is tolerated.

In the 4,5 subclass there is generally a free hydroxyl at C6, whereas for the 4,6 subclass there is generally a free hydroxyl at C5. In both subclasses, conversion to the deoxy analog (at either position C5 or C6) is tolerated (41). The NMR structure of the gentamicin 1a/RNA complex reports a hydrogen bond between the C5 hydroxyl group and ring I, which cannot be satisfied by deoxy analogs of the 4,6 subclass, suggesting that this intramolecular interaction isn't necessary for activity (41).

5.3 Structure-Activity Relationship of Rings III and IV of the 4,5 Subclass

The chemical structure and the trajectories of ring III in each of the two subclasses diverge, so from this point the structure-activity relationships of the two subclasses will be treated separately. In the 4,5 subclass, ring III is usually a ribose (41). Only two direct interactions between 30S and ring III of paromomycin are observed in the crystal structure: hydrogen bonds between the 5″-hydroxyl group and N7 of G1491 and between the 2″-hydroxyl group and N4 of C1407. Chlorination or phosphorylation of C5″ abolishes activity, whereas C5″ deoxy and C5″ amino derivatives are active, suggesting that these latter groups are too large (Fig. 6.4c) (41).

Connection to ring IV is made from C3″ of ring III to C1‴ of ring IV. Extensive modifications are possible within ring IV of the 4,5 subclass of aminoglycosides in agreement with the limited number of ribose contacts. Ring IV has only a few natural modifications: connection to mannose through C4‴ (e.g., lividomycin) (41), the observation of both stereoisomers at C5‴ (41), and substitution at C2‴ (41), which can be either an amino or a hydroxyl group (41).

Group 4,5-aminoglycosides are inactivated on adenylation of the C6‴ amino group (Fig. 6.4d) (41). Conversion of the C6″ amine to a hydroxyl is tolerated (68). Replacement of ring IV with a short alkyldiamino tail yielded a compound with activity identical to that of neomycin B (68). Related efforts to find novel aminoglycoside analogs sought to replace both rings III and IV; however, the best compounds only had intermediate activities between neamine (ring I and II only) and neomycin B (69).

5.4 Structure-Activity Relationship of Ring III of the 4,6 Subclass

Ring III of the 4,6 subclass is an alternative to rings III and IV of the 4,5 subclass, and it makes numerous direct contacts with 30S (Fig. 6.4c). The 2″ hydroxyl is within hydrogen bonding distance of O6 of G1405 and O4 of U1406; this doesn't preclude the 2″ hydroxyl to amine substitution, which is permissible (41). The 3″ amino group forms hydrogen

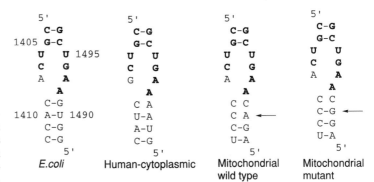

Figure 6.5. 30S A-site sequences of ribosomes from various sources. The arrows indicate the point of variation between the 1555 polymorphic mitochondrial sequences.

bonds to the N7 and a phosphate oxygen of G1405 that must be maintained for activity. At this position restrictions on the size and geometry of a substituent limits other allowable modifications. A hydrogen bond exists between the hydroxyl at C4″ and a phosphate oxygen of U1406. Some natural variations exist at C4″; a hydroxyl group (in either configuration) is usually present but sometimes is replaced by a methoxy (41).

A wide variety of natural and synthetic variations is permissible at C5″, and at C6″ for C5″ branched compounds (41). An extensive study of C6″ modifications of kanamycin B revealed that halogeno, azido, amino, alkylthio, and alkoxy substitutions are all well tolerated until they become very bulky (70). For example, only the N-hexanoyl-N-butylamino derivative was essentially inactive against all bacteria tested, but the smaller substituents retained good activity. The NMR structure of gentamicin C1a (which is 5″ deoxy) indicates that any modification at C5″ would be directed away from 30S (60). The predicted lack of interaction between C5″ substituents and 30S explains why a variety of small substitutions is permissible.

6 AMINOGLYCOSIDE TOXICITY

All aminoglycosides, including streptomycin, have the potential to produce irreversible vestibular and cochlear intoxication (16). Non-streptomycin aminoglycosides are also renally toxic (16). The underlying causes of aminoglycoside ototoxicity are unresolved (reviewed in Ref. 71). Multiple hypotheses exist to explain aminoglycoside ototoxicity, but none are com-

pletely satisfactory; indeed, multiple mechanisms may be at work. One of the two most complete theories states that redox-active aminoglycoside/iron complexes catalyze the formation of free radicals, which are destructive to the hair cells of the inner ear (71). The other theory holds that for at least some individuals aminoglycosides bind to mitochondrial ribosomes and shut down protein synthesis (72). Because of its ribosome involvement only the second theory will be discussed here.

Mechanisms of the severe ototoxicity associated with aminoglycoside use are only partially known, but circumstantial data suggest that aminoglycoside ototoxicity results from unintended interactions with human mitochondrial ribosomes. (1) Reports of several pedigrees exist, which correlate hypersensitivity to aminoglycosides with a single polymorphism in the mitochondrial genome (A1555G) (72–76). This polymorphism conspicuously occurs in the A-site of 12S rRNA, which is analogous to the gentamicin/neomycin binding site in bacterial ribosomes. It is speculated that this polymorphism makes the A-site of 12S rRNA (mitochondrial) look more like the A-site of 16S rRNA (bacterial) (Fig. 6.5). (2) The phenotype of sensoneural hearing loss or deafness due to aminoglycoside toxicity is the same as that found in most mitochondrial encephalomyopathies, suggesting mitochondrial involvement (77). (3) Mitochondrial DNA from individuals with the A1555G polymorphism can be co-transfected into ρ° HeLa cells, which lack mitochondrial DNA altogether, causing these cells to become sensitive to streptomycin intoxication (78). (4) Significant uptake of tritium labeled aminoglycosides has been observed in proximal tubular kidney cells and within sensory hair cells, with significant localization to mitochondria (79). (5) *In vitro* measurements demonstrate that gentamicin/ neomycin aminoglycosides bind poorly to the human mitochondrial decoding A-site, but bind to the RNA containing the A1555G polymorphic sequence with an affinity similar to that of RNA with the bacterial sequence (80). (6) Paromomycin has been observed to shut down mitochondrial activity in *Leishmania* (81).

A direct model of aminoglycoside ototoxicity exists for neomycin-like and gentamicin-like compounds, in which aminoglycosides bind to mitochondrial ribosomes containing the A1555G mutation more tightly than they do to the wild-type ribosomes. Ring I of both subclasses stacks atop the Watson-Crick base pair formed by C1409 and G1491. This base pair and the adjacent base pair form mismatched pairings (C1409 · C1491 and C1410 · A1490) in human mitochondrial ribosomes, and the second pair is thought to disrupt the stabilizing stacking interaction between ring I and nucleotides 1409 and 1491. In individuals with the A1555G polymorphism (analogous to A1490 in the *E. coli* small subunit rRNA), one of the two mismatches is converted to a Watson-Crick base pair (C1410-G1490), which may help restore the stacking interaction between ring I and 1409 and 1491. *In vitro* binding studies of common aminoglycosides and both mitochondrial RNA sequences confirm that aminoglycosides bind the A1555G polymorphic sequence more tightly than the wild-type (79). If the direct model is correct, then there is optimism that agents with enhanced specificity for the bacterial sequence will necessarily be less ototoxic.

One obvious weakness in the theory of mitochondrial involvement in aminoglycoside intoxication is that streptomycin doesn't share a binding site with gentamicin/neomycin-like aminoglycosides. However, streptomycin does interact with the backbone of A1491 (10). It is possible that the A1555G polymorphism potentiates streptomycin biding against mitochondrial ribosomes, either directly or indirectly. A second weakness is that tissue specificity for aminoglycoside toxicity is poorly understood, but that is true for all competing theories of aminoglycoside ototoxicity.

7 TETRACYCLINE AND ANALOGS

Tetracyclines were the first broad-spectrum antibiotics and have been used successfully for decades to treat both gram-positive and gram negative bacterial infections (82). Chlortetracycline was the first tetracycline to be isolated, in 1948, from *Streptomyces aureofaciens* (83). Other common tetracyclines, such as oxytetracycline and tetracycline, were isolated from *Streptomyces* sources in subsequent years. The abundance of natural, active tetracyclines, coupled with extensive synthetic alterations, provides a rich collection of compounds from which to build meaningful structure-activity relationships. As was found for the aminoglycosides, previous observations of tetracycline structure and activity can be rationalized from recent tetracycline/30S crystal structures (11, 12).

Although tetracycline was not the first agent of its class discovered, it nonetheless provides the basis for tetracycline nomenclature (Fig. 6.6). Numerous reviews exist that report on the biosynthesis and use of tetracyclines (82, 84–86).

Extensive efforts to identify tetracycline analogs from both natural and semisynthetic sources that possess greater efficacy, lower toxicity, and greater chemical stability have produced superior drugs (82), such as minocycline and doxycycline, but newer semisynthetic analogs may be on the way. Members of a new tetracycline class, known as glycylcyclines, are superior to all existing clinically useful agents against several types of tetracycline-resistant bacteria (87–89).

7.1 Mechanism of Action

Biochemical probing identified multiple tetracycline binding sites within 30S, one of which resides in the decoding A-site and seems to be responsible for the antibiotic effect of tetracycline (19). Two groups have independently solved tetracycline/30S structures by X-ray crystallography (11, 12). The Berlin group refined the positions of six tetracycline binding sites, ranging in occupancies from 0.41 to 1.0,

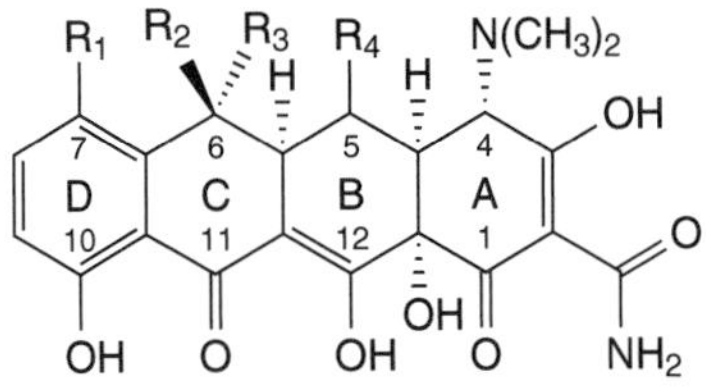

	R^1	R^2	R^3	R^4
Tetracycline	H	OH	CH$_3$	H
Chlortetracycline	Cl	OH	CH$_3$	H
Demeclocycline	Cl	OH	H	H
6-Demethyltetracycline (Demecycline)	H	OH	H	H
Oxytetracycline	H	OH	CH$_3$	OH
6-Deoxytetracycline	H	H	CH$_3$	H
Minocycline	N(CH$_3$)$_2$	H	H	H
Doxycycline	H	H	CH$_3$	OH

Figure 6.6. The structures of tetracycline and several analogs of tetracycline.

to a resolution of 4.5 Å (12), while the Cambridge group identified two tetracycline sites, which they refined to 3.4Å (11). The two sites identified by the Cambridge group are a subset of the six sites reported by the Berlin group, with both groups identifying tetracycline bound in a pocket formed by 16S rRNA helices 31 and 34 as the dominant and most relevant site. Only the structure of the H31/H34 tetracycline/30S complex will be discussed here, although other tetracycline binding sites could contribute to ribosome inhibition.

It has been accepted for some time that tetracycline interferes with proper tRNA binding to the A-site (57). A refinement of this model has the elongation factor Tu/tRNA complex docking correctly into the A-site, allowing the correct codon-anticodon interaction to take place, and the subsequent hydrolysis of GTP by EF-Tu (11). At this point a necessary rotation of the A-site tRNA is blocked by tetracycline, leading to the ejection of tRNA from the A-site without peptide bond formation. In this scenario, tetracycline extracts two payments from the cell, inhibition of protein synthesis and the unproductive hydrolysis of GTP.

7.2 Structure-Activity Relationship

Tetracyclines contain four fused rings (A-D), one of which, D, is aromatic; rings A and B contain sites of unsaturation (Fig. 6.7). Essentially all alterations to the general tetracycline ring skeleton are deleterious; breaking any ring, disrupting aromaticity in ring D, or aromatization of ring A or C destroys all tetracycline activity (85). Modifications to the functional groups on ring A and the bottoms of rings B, C, and D are generally not tolerated, but extensive modification is possible elsewhere. Maintenance of the ketones at C1 and C11 is essential, as are the hydroxyls at C3, C10, C12, and C12a. All of these groups are involved in an extensive network of RNA binding interactions composed of hydrogen bonds and shared coordination to Mg^{2+}. Neither the hydroxyl nor the methyl at C6 is essential for activity; no direct ribosome interaction is observed for the C6 substituents. Removal of the dimethylamino group at C4 diminishes *in vitro* activity and abolishes *in vivo* activity (85). Reduced binding is expected with the removal of this group; it interacts electrostatically with the phosphate of G966. The amido group at C2 is rarely substituted, but a 2-acetyl moiety has been observed (85). Either chemical entity is able to hydrogen bond to the ribose of C1195.

Tetracycline occupies a pocket formed by helices 31 and 34 of 16S rRNA (11). As with most ribosome targeting antibiotics, the con-

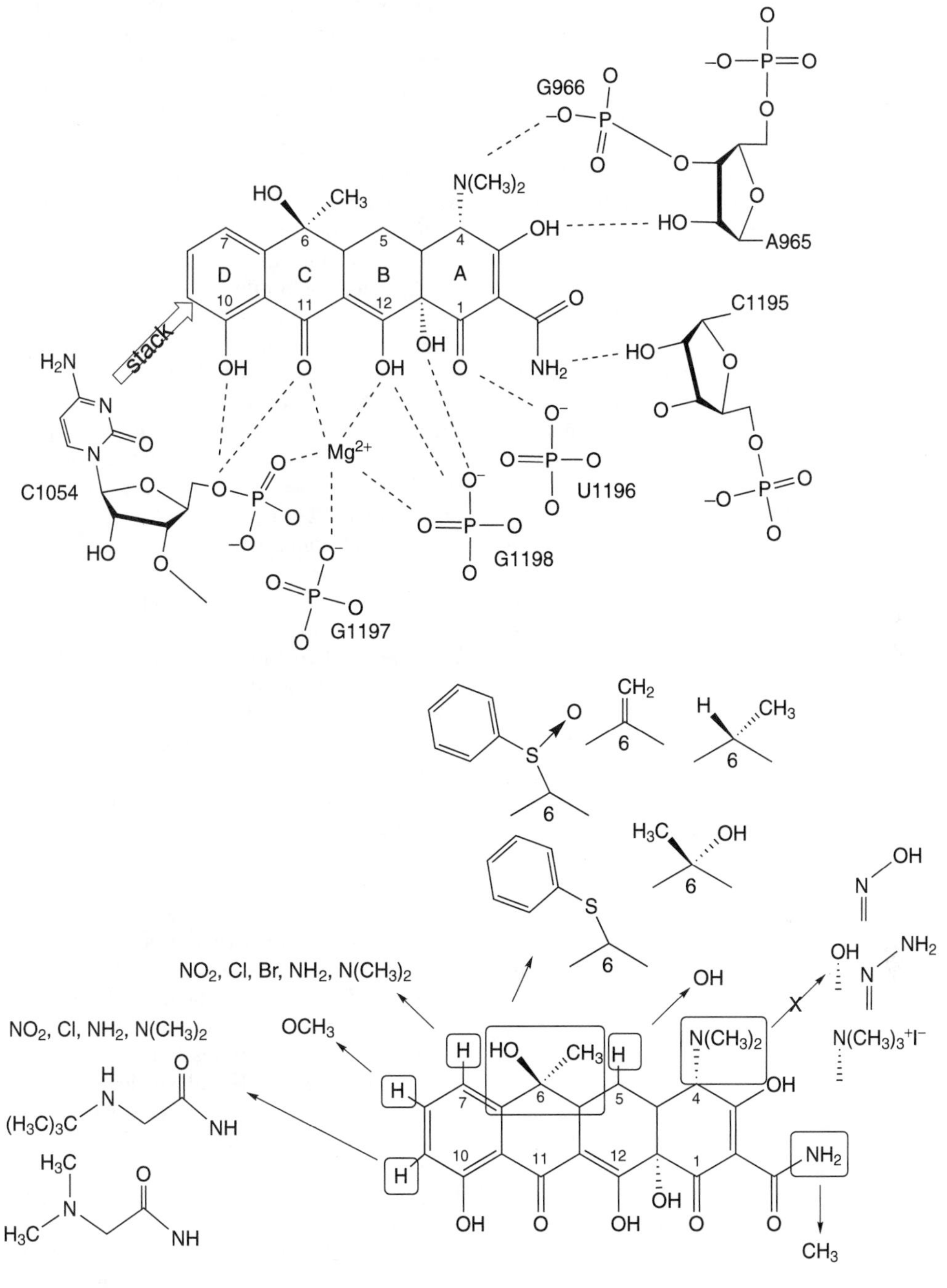

Figure 6.7. Molecular interactions between tetracycline and 30S (*E. coli* numbering, top) with various modifications tested for activity (bottom). Dashed lines indicate possible hydrogen bonds or metal coordination. Arrows point to permissible modifications; arrows with an X point to non-permissible modifications.

tacts are to RNA only. The structures explain virtually all of the absolutely required tetracycline groups. Extensive hydrogen bonding networks are observed between the drug and backbone phosphate and ribose oxygen atoms. Similar to streptomycin, no hydrogen bonds are formed between tetracycline and any nucleobases—sequence selective binding comes from the proper presentation of sequence-independent groups. Tetracycline binding has two interactions that are unusual drug/RNA interactions: binding through a shared Mg^{2+}, which coordinates several oxygens from tetracycline and phosphate groups from the ribosome, and aromatic stacking between ring D and C1054. All attempts to modify tetracycline groups involved in the intricate interactions detailed above are deleterious. The remaining perimeter of tetracycline makes no close interactions with 30S. Indeed, the C5–C9 edge of tetracycline is directed away from any nearby 30S, which explains why extensive modifications at C6, C7, and C9, and the C8 methoxy derivative are possible.

Removal of one or both of the common C6 substituents, hydroxyl and methyl, is common, and virtually all substitutions at C6 are permitted, even for groups as large as benzylmercaptan and glycosides (85). Both electron donating and electron withdrawing groups are tolerated at C7 (85). C8 methoxy derivatives are occasionally observed and are active (85). Numerous substitutions at C9 are also permissible. Recently, C9 amido substitutions yielded compounds with enhanced ribosome binding that escape recognition by certain efflux pumps (87–89). These derivatives, known as glycylcylines, remain active against a wide variety of tetracycline-resistant bacteria. Of particular interest here is the fact that many of these derivatives have greatly enhanced ribosome binding, indicating that the amido substitutions are making additional unique, productive interactions with 30S. More work will be needed before the identities of these new interactions are revealed.

8 ERYTHROMYCIN AND ANALOGS

Isolation of macrolides began in the 1950s with the discovery of pikromycin (90); the iso-

lation of numerous other macrolides quickly followed and the search for others continues today. Natural macrolides contain a highly substituted 12- to 16-member monocycle lactone ring, referred to as an aglycone, to which one or more deoxysugars are attached. Erythromycin, the first clinically used macrolide, was isolated in 1953 from *Saccharoplyspora erythraea* (formerly *Streptomyces erythreus*) (91). Erythromycin is a member of the 14-atom macrolide family and contains two sugar moieties: desosamine, an amino sugar, and cladinose (Fig. 6.8). It is a relatively broad-spectrum antibiotic with low toxicity, but it is also characterized by poor chemical stability, uncertain pharmacokinetics, and widespread bacterial resistance (32, 82). These negative qualities provided the impetus to search for erythromycin alternatives, both from natural sources and from semisynthetic modifications, that were improvements over erythromycin in one or more of the above categories. Several analogs (i.e., azithromycin and clarithromycin) were noticeable improvements in all regards except resistance; bacteria were generally cross-resistant to erythromycin and its analogs. A new class of macrolide analogs, referred to as the ketolides, show great promise because of their increased activity both overall and against many erythromycin resistant bacteria (92–94).

8.1 Mechanism of Action

Macrolides bind to a region of 50S near the peptidyl transferase center and block the progress of the nascent peptide through the exit channel (14). Drug binding encompasses interactions to nucleotides from domain V (directly) and domain II (allosterically) of 23S rRNA (95). The dominant mechanism of resistance is through methylation of A2058 in domain V by the *erm* family of methyltransferases, which confers resistance to macrolides, lincosamide, and group B streptogramines, a collection of structurally diverse antibiotics that are collectively referred to as MLS_B antibiotics (57, 93). Cross-resistance arises from the sharing of a common binding site that includes interaction with A2058. The second most common form of erythromycin resistance comes from active efflux encoded for by *mef* genes (93). Also, clinical isolates

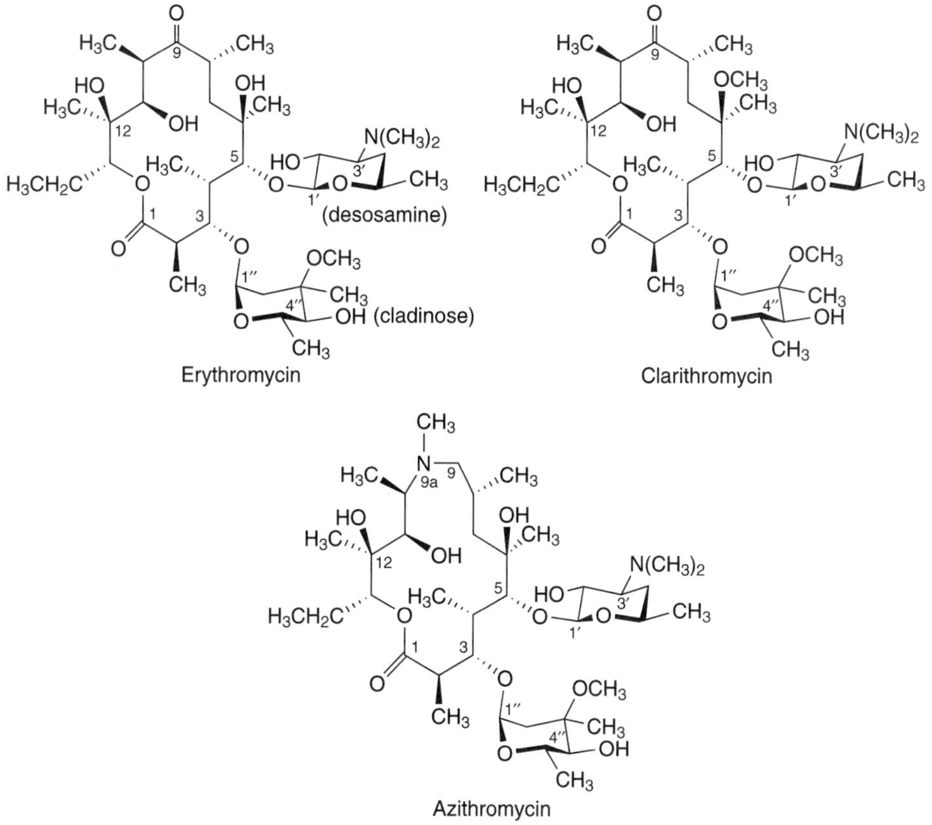

Figure 6.8. The structures of the macrolides: erythromycin, clarithromycin, and azithromycin.

resistant to erythromycin have recently been reported to contain 23S rRNA and ribosomal protein mutations (93).

8.2 Structure-Activity Relationship

Macrolides have a rich assortment of chemical features through which tight, selective molecular recognition of 50S can exist; however, the crystal structure and structure-activity relationships of erythromycin demonstrate that only a handful of these potential contact points are actually involved in direct contact (Fig. 6.9a-c). This limited engagement by some regions of erythromycin has allowed the construction of diverse subgroups of macrolide antibiotics. Notably, extensive modification of the ring skeleton is possible.

Isolated macrolides typically have 12, 14, or 16 atom aglycones, with the 14-atom class serving as the basis for most clinically relevant mac-

rolide antibiotics. The 12-atom aglycone macrolides exhibit poor activity, whereas 16-member aglycones, such as josamycin and tylosin, have both medical and veterinary use. Tylosin and erythromycin compete for the same binding site on 50S, and both drugs block the exit channel; by virtue of a disaccharide in place of desosamine, a ring common to 14-membered macrolides, tylosin and other 16-membered macrolides also inhibit peptide bond formation (96). Structural information exists only for the 14-membered macrolides; although there is extensive evidence to support a similar mode of binding for the 16-membered ring macrolides, enough differences between the mechanisms of action of the two classes exist to caution against extrapolating the 16-membered macrolides from existing 14-membered macrolide/50S structures. Therefore, only the 14-membered macrolides will be addressed here in detail.

(a)

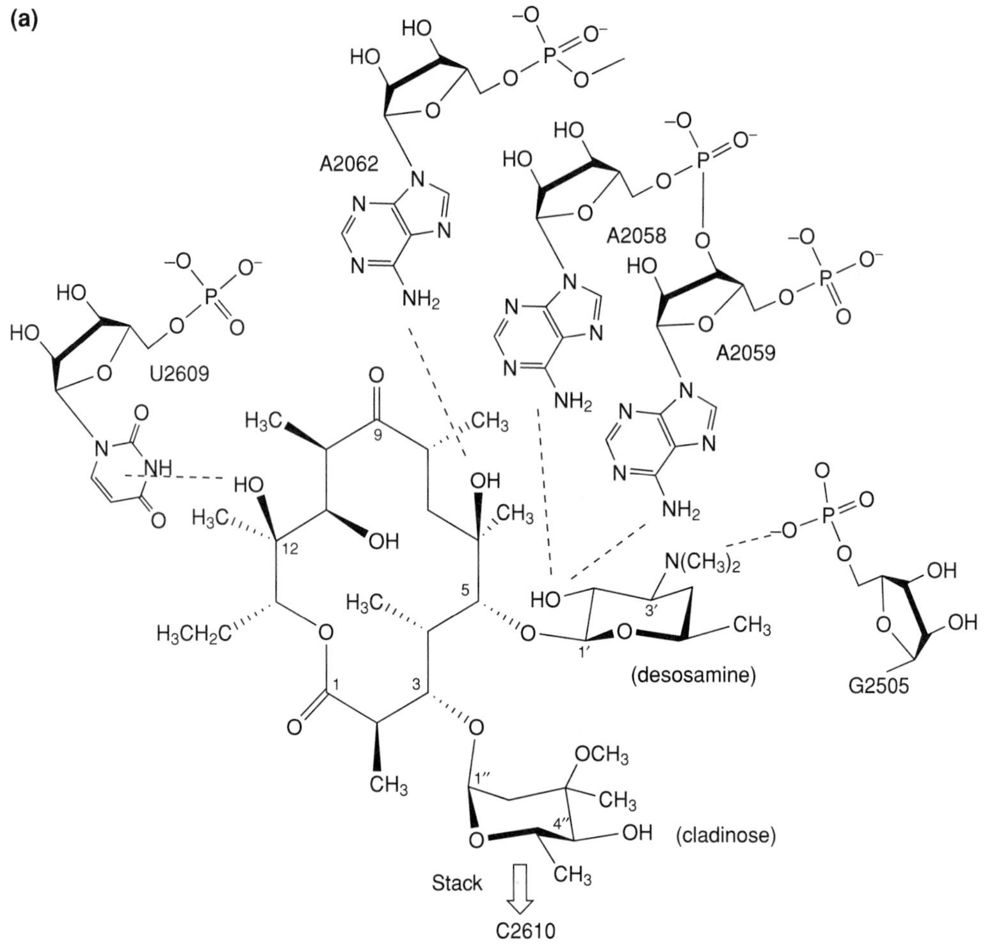

Figure 6.9. Molecular interactions between erythromycin and 50S (*E. coli* numbering) with various modifications tested for activity. (a) Erythromycin/50S interactions. Dashed lines indicate possible hydrogen bonds (some of which are salt bridges when suitably reinforced with favorable electrostatic potentials). (b and c) Several modifications tested. Arrows point to permissible modifications; arrows with an X point to non-permissible modifications. Dashed arrow points to a modification that is not permissible unless corresponding changes are made elsewhere.

Erythromycin seems to be anchored to domain V of 23S rRNA by only a few hydrogen bonds and a salt interaction: the C6 hydroxyl forms a hydrogen bond to N6 of A2062; the C2′ desosamine hydroxyl group forms a hydrogen bond to N1 of A2058 and either N1 or N6 of A2059; the C3′ dimethylamino group is close to the phosphate of G2505; and the C12 hydryoxyl group forms a hydrogen bond to the π system of U2609 (14). Clearly, other important non-bonding interactions must contribute in a large way to erythromycin binding.

For example, the cladinose ring seems to make hydrophobic contacts with 30S. Nevertheless, there are also regions of erythromycin that make only minor contributions to binding, such as the region encompassed by C7-C11.

Under acidic conditions erythromycin can undergo spontaneous intramolecular ring cyclizations, the products of which are inactive (96–99). All contain either the 6,9-hemiketal or the 8,9-anhydro-6,9-hemiketal; they may also contain the 6,9;9,12-spiroketal. The synthesis and isolation of the biologically active

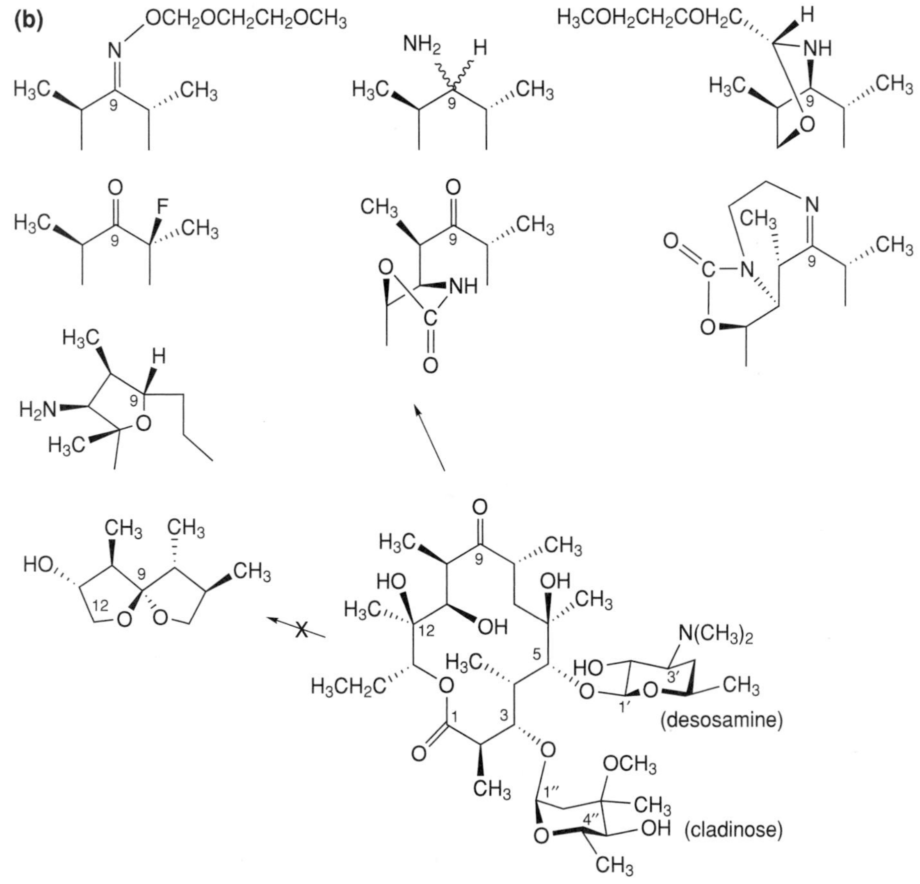

Figure 6.9. (Continued.)

9,12-bicyclic-epoxy macrolides established that it is the presence of the 6,9-bicycle that leads to inactive macrolides (100). Conversion of the C9 ketone to O-substituted oximes and amino groups yields active compounds and removes the possibility of forming the above inactive cyclization products (101, 102). Erythromycin 9-oxime undergoes a Beckman rearrangement to give the ring-expanded 15-member aglycone, which is further converted to the 9-"azalide" (azithromycin) on reduction and N-methylation (103, 104). The synthesis of 11,12-carbonate and 11,12-N-substituted carbamates has generated many new useful compounds currently under clinical investigation (105). Tricyclic analogs are also active (106). The above modifications are varied and sometimes extreme, yet all of them are tolerated except for the formation of 6,9-bicyclic

derivatives; all can be rationalized through inspection of the erythromycin/50S structure.

The two sugars that form part of erythromycin are important for activity in most macrolide antibiotics; cladinose can be replaced or dispensed with when other compensating modifications are made that restore ribosome binding (107, 108). Hydrogen bonding and strong electrostatic interactions are absent between cladinose and 50S, but the ring itself stacks atop the base equivalent to C2610 in $E.$ $coli$. This interaction must account for a significant amount of binding energy, because removal of this ring without other compensating changes elsewhere abolishes activity.

The lack of directional interactions, such as hydrogen bonds involving 2″-O and 4″-O, agrees with the observation that methylation of ring hydroxyls is of little consequence to

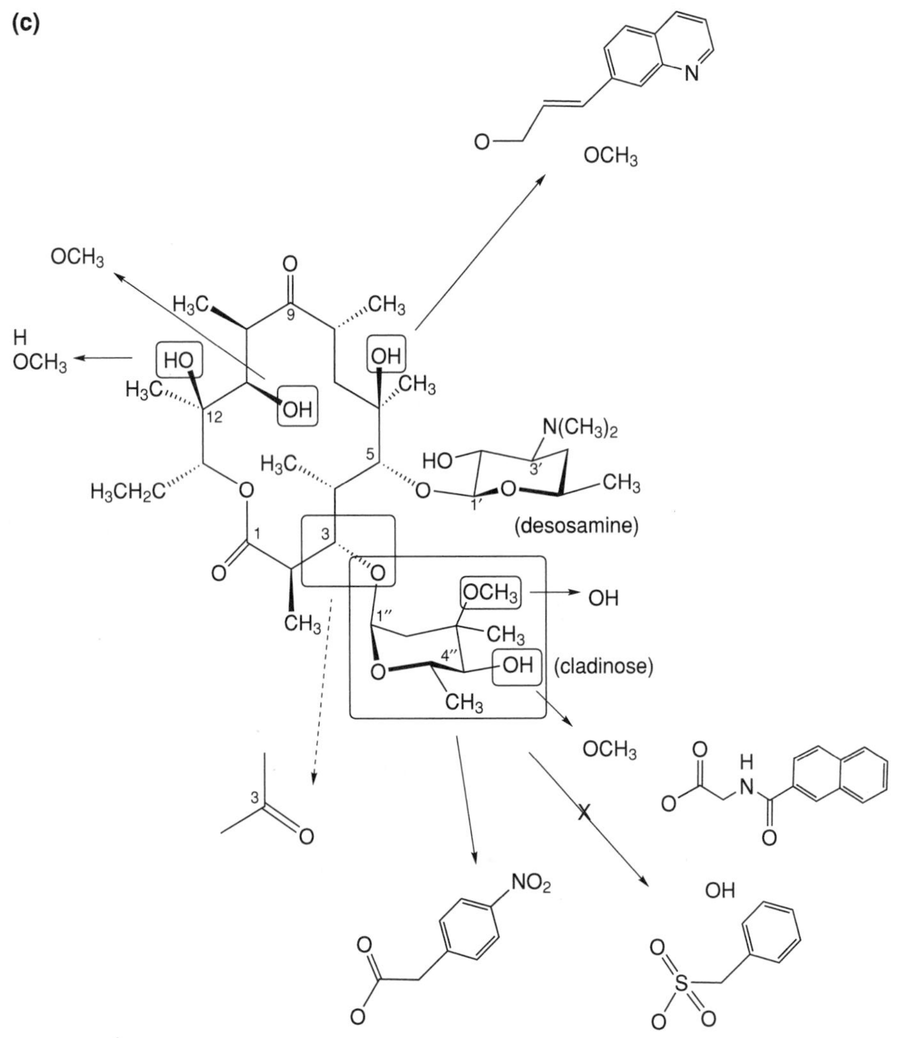

(c)

Figure 6.9. (Continued.)

activity. Other modifications to cladinose are also possible; the C2″ methoxy derivative is active, as are several C4″ derivatives (109, 110).

The cladinose ring seems to be an essential determinant for the induction of the inducible form of MLS$_B$ resistance, an observation that prompted the investigation of analogs in which the cladinose was removed or replaced by other moieties. Numerous 3-O-acyl derivatives of erythromycin were synthesized and observed to be active antibiotics (111). Some of those antibiotics remained active against MLS$_B$-resistant *Staphylococcus aureus* and ef-

flux-resistant *Streptococcus pneumoniae*, illustrating that some of the derivatives compensated for the cladinose ring in terms of ribosome binding but not its ability to induce antibiotic resistance. Substitutions that would be expected to stack most strongly (i.e., aromatic rings) showed the best activity, further supporting the notion that the primary binding energy of cladinose is from stabilizing nonbonding interactions (111).

Like the cladinose ring, the desosamine ring is important for activity; unlike cladinose, no synthetic alternatives have been identified. In complex with 50S, desosamine makes numerous specific interactions: the C2' hydroxyl is within hydrogen bonding distance of N1 and N6 of A2058 and N6 of A2059, and the dimethylamine forms a salt bridge to the phosphate of G2505. MLS$_B$-resistant strains of bacteria produce a methyltranseferase that converts A2058 to an N6,N6-dimethyladenosine (112), which when present would disrupt hydrogen bonding between the C2' hydroxyl and the adenosine (14). Because disruption of the hydroxyl/A2058 interaction is enough to reduce the efficacy of erythromycin by several orders of magnitude, that interaction coupled with any negative steric clash that may be present must account for a significant portion of the overall binding energy. Clinical isolates of *Helicobacter pylori* and other pathogenic bacteria with a limited number of rRNA operons resistant to clarithromycin sometimes contain either a common A2058 → G mutation or a rarer A2058 → C mutation in 23S rRNA (113). Ribosomes containing these mutations bind macrolide antibiotics less well, again illustrating the importance of the C2' hydroxyl interactions. A pikromycin biosynthetic analog was produced that was β-glycosylated at the 2' position of desosamine, yielding a compound devoid of antibiotic activity; this also affirms the importance of the C2' hydroxyl group (114).

Numerous 16-membered macrolides (e.g., tylosin) have mycaminose, a disaccharide that contains desosamine, instead of the monosaccharide desosamine alone (115). If the structure-activity relationships of the 14- and 16-membered macrolides prove to be identical, it is clear that the pocket into which the desosamine binds needs to be explored further before adequate structure-activity relationships can be realized.

Modification of C6-OH of the aglycone is not only possible, it is preferable. Methylation of this hydroxyl group yields an antibiotic (clarithromycin) with superior pharmacokinetics (82). Numerous 6-O-aryl derivatives yielded agents with superior antibiotic activity for erythromycin, ketolide, and 11,12-carbamate derivatives (106, 116). The authors of this study speculated that the newly introduced 6-O-aryl groups enhance the overall interaction of those drugs with domain II of 23S rRNA. If true, this means that these modifications may occupy the same site as N-linked

moieties from recent 11,12-carbamate analogs that have been shown to have improved binding to domain II nucleotides, which means that both types of modifications could not exist simultaneously.

In both the erythromycin/50S and the clarithromycin/50S structures the C6-O is within hydrogen bonding distance of the N6 of A2062. Inspection of both of those structures suggests that the same hydrogen bond can be preserved for the 6-O-aryl derivatives. Nevertheless, more work will needs to be done before definitive conclusions can be drawn about where the aryl arm extends with respect to 50S.

The ketolides, such as the recently approved drug telithromycin, are a new class of erythromycin analogs (Fig. 6.10) (108). The cladinose ring is absent and the C3-OH is oxidized to a ketone. Removal of the cladinose ring, as discussed above, reduces overall binding affinity, which can be more than compensated by the inclusion of N-linked arms that extend from either the 11,12-carbamate functional groups (telithromycin) or the C6-O-aryl groups (ABT 773) (107, 108). Biochemical probing implicates nucleotide 752 of domain II as being involved with ketolide binding, specifically with either of the extensions; laboratory isolates resistant to kelotides but not to erythromycin suggest additional domain V involvement as well (108, 117, 118). The erythromycin/50S crystal structure reveals a close interaction (3.2 Å) between A752 of domain II and U2609 of domain V, which suggests either a direct mechanism by which the 11,12-carbamate arm could interact simultaneously with both nucleotides or an indirect mechanism by which ketolide interaction with one nucleotide subsequently affects the internucleotide interaction.

The hydroxyl groups from both C11 and C12 are within hydrogen bonding distance of U2609, with the C12- OH/U2609 putative interaction seeming to be more favorable. Methylation of both hydroxyl groups is tolerated, but is not ideal (111).

8.3 Refined View of Macrolide Binding

At the resolution limit of the erythromycin/50S structure, ordered water molecules, which help mediate many RNA-ligand interac-

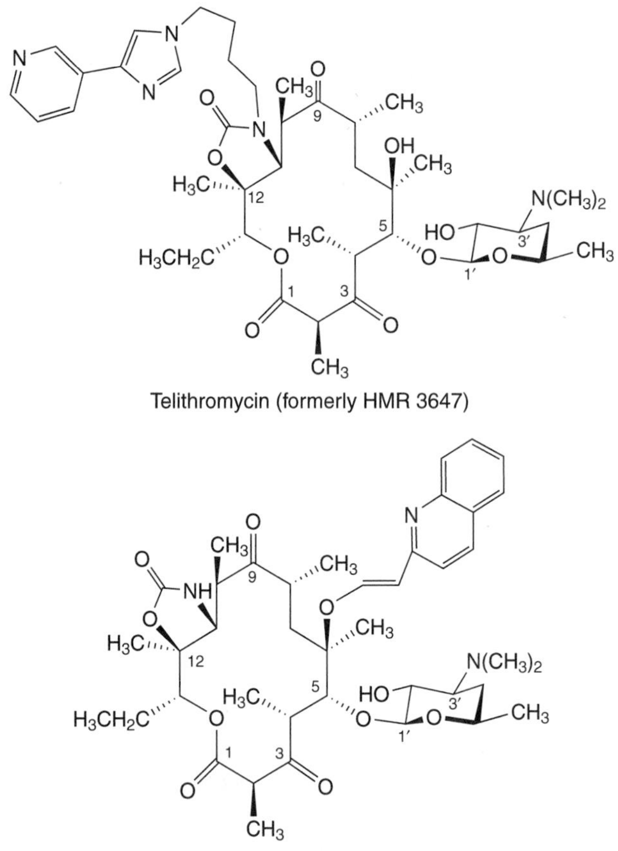

Telithromycin (formerly HMR 3647)

ABT 773

Figure 6.10. The structures of the ketolides: telithromycin and ABT 773.

tions, are not observed, but placement of large features like the core aglycone ring and the external cladinose and desosamine rings can be positioned with confidence (14). Therefore, gross features of erythromycin binding are sound, but the finer details, such as intricate hydrogen bonding networks, remain to be identified. And while a quantitative accounting of the interactions that govern macrolide binding is absent at this time, a global model of binding is emerging.

Macrolides such as erythromycin bind 50S by occupying one large pocket formed by nucleotides within the central loop of domain V and then branch out into other neighboring pockets by way of external sugar groups or, in the case of the ketolides, by extensions from either the 11,12-carbamate or the C6-OH. With the possible exception of the 11,12-carbamate and C6-OH extensions, interactions with these satellite pockets are essentially in-

dependent from each other; pruning of one interaction can be compensated for by growth into another pocket. At least three satellite pockets exist: the desosamine pocket, the cladinose pocket, and the 11,12-extension/6-O-aryl pocket. It seems that extensions from the C6-OH occupy the same pocket as do extensions from the 11,12-carbamate moiety (118). If they do not, then there are at least four pockets, all of which can be maximized for antibiotic activity (Fig. 6.11).

9   LESSONS LEARNED FROM RIBOSOME/DRUG COMPLEXES

Reports of the individual drug/ribosome complexes are important both in terms of describing the mechanism of action and binding of those particular drugs and in terms of cataloguing and comparing the various modes of

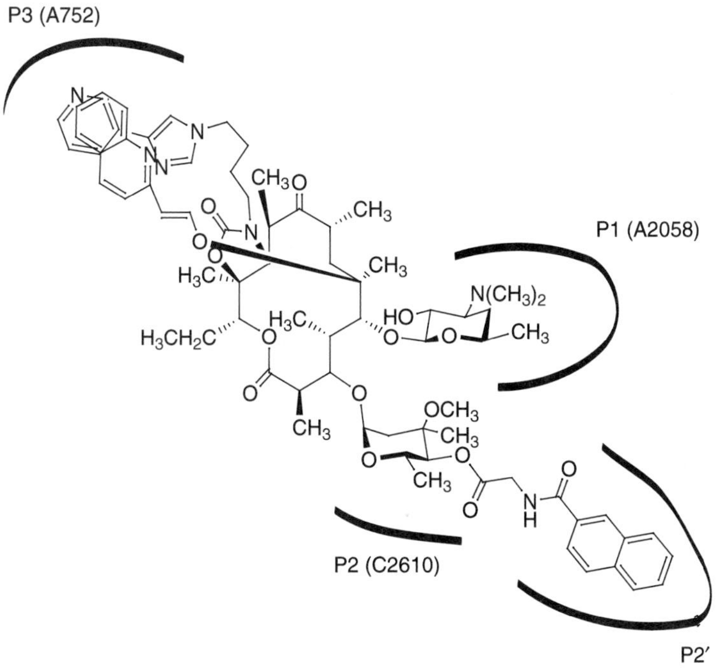

Figure 6.11. An overall view of macrolide/ketolide binding. Several arms extend from the aglycone ring. Desosamine occupies the P1 binding pocket and makes interactions to multiple nucleotides including A2058. Cladinose occupies the P2 binding pocket and stacks with C2610. Extensions from cladinose extend into another pocket, P2′. Ketolide antibiotics have either 11,12 carbamate extensions or aryl extensions from C6, both extensions seem to interact with a pocket involving A752. Therefore, a macrolide analog presumably cannot have both types of extensions.

binding involved (10–14). Earlier NMR structures of aminoglycoside/RNA complexes confirmed the predicted importance of electrostatic interactions between drug amino groups and RNA phosphate groups, but also illustrated other important features of drug binding, such as hydrogen bonds to nucleobases and stacking of drug sugars atop base pairs (59, 60). Previous sections described the binding of four classes of drugs in detail; given their level of chemical diversity, it shouldn't surprise anyone that the observed modes of binding were very different. Other antibiotic/RNA complexes have been solved (8–12), adding further to the collective knowledge of drug/RNA interactions; like the drugs discussed above, each brings knowledge of previously unknown modes of interaction. Their structures plus the structure of linezolid, a recently approved synthetic antibiotic, are drawn in Fig. 6.12.

Streptomycin binds to a site that ties together multiple helices almost exclusively through hydrogen bonds and salt bridges to backbone phosphates, whereas compounds such as hygromycin B bind to a pocket within an expanded major groove through several hydrogen bonds to nucleotide bases but with no

phosphate contacts (10). Paromomycin and gentamicin share the same pocket within a major groove and make numerous contacts to both backbone phosphates and nucleotide bases (10, 59, 60).

Some antibiotics, such as erythromycin and chloramphenicol, seem to rely heavily on hydrophobic interactions to provide necessary binding energy, but there is substantial variation in how this is accomplished. For example, aromatic rings from tetracycline and pactamycin stack with nucleotide bases, whereas paromomycin, gentamicin, and erythromycin contain sugars that stack atop the aromatic nucleotide bases (10, 14, 60). Pactamycin illustrates the latter phenomenon in reverse: a ribose sugar from 16S rRNA stacks atop one of pactamycin's two aromatic rings (11).

Two drugs, chloramphenicol and tetracycline, interact with rRNA through Mg^{2+} atoms that are coordinated by both the drugs and the RNA phosphate oxygens (11, 14). Shared metal coordination is not a common interaction between drug and proteins; it remains to be seen if this is a common phenomenon for RNA/antibiotic complexes or not.

Table 6.1 lists several ribosome binding antibiotics and one novel compound, linezolid

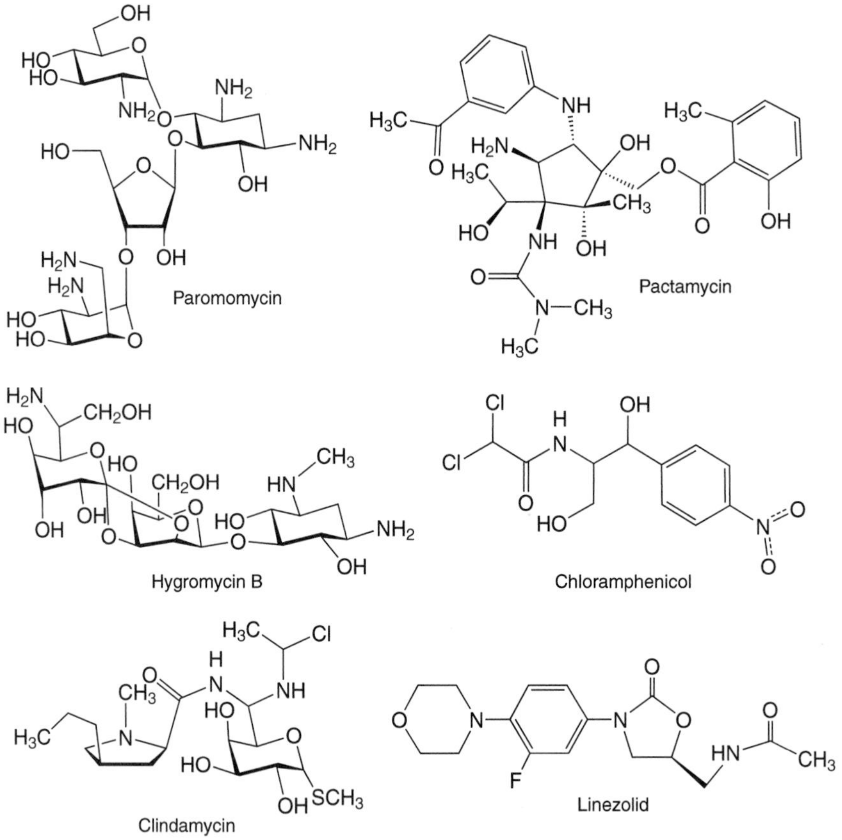

Figure 6.12. Structures of several other antibiotics whose 3D structures have been solved in complex with either 30S or 50S, and the novel synthetic antibiotic linezolid.

(119), which also inhibits ribosome function (120–123), and the log of their estimated *n*-octanol/water partition coefficients (LogP). Estimated LogP values range from −7.2 for paromomycin to 2.8 for erythromycin for fully protonated drugs. The values for the same compounds in their neutral forms range from −4.1 to 4.6. It is interesting and not surprising to note that the two most hydrophobic compounds, erythromycin and clindamycin, seem from their crystal structures to engage in mostly hydrophobic interactions with the ribosome, whereas the most hydrophilic compounds, paromomycin, gentamicin, hygromycin B, and streptomycin, all interact extensively with backbone phosphates through amino groups. The structure of linezolid in complex with 50S has not been solved, but if this trend holds true, then linezolid should interact with the ribosome pri-

marily through hydrophobic interactions. As a group these LogP values tend to be lower (more negative) than the LogP values for most drugs, which may reflect the type of interactions that are observed between antibiotics and RNA, but may also reflect the more hydrophilic nature of the surrounding RNA. RNA, even in tightly packed structures, is extensively hydrated, whereas the core of proteins is not. In a more electrophilic environment the magnitude of some non-bonded interactions, such as Coulombic forces, will be dampened, but others, such as the hydrophobic effect, will be enhanced, hinting that the relative values that contribute to ligand/receptor associations may be different for RNA and protein targets. Numerous high-resolution protein/RNA complexes exist in the crystallographic database. Comparison between forces that mediate protein/protein association

Table 6.1 "Rule of 5" Values for Selected RNA Binding Antibiotics

Compound	LogP[a]	NH + OH[b]	MW	N + O[c]	Alert[d]
Chloramphenicol	−1.689	3	323	6	0
Clindamycin	0.826	4	410	7	0
Erythromycin	2.874	5	734	14	1
Gentamicin 1Ca	−5.622	8	449	12	1
Hygromycin B	−5.619	11	527	16	1
Linezolid	2.011	1	337	7	0
Pactamycin	0.495	7	531	12	1
Paromomycin	−7.212	13	600	18	1
Streptomycin	−5.26	14	581	19	1
Tetracycline	−2.688	6	444	10	1
Average:	−2.1884	7.2	493	12.1	

[a]LogP calculated using HINT (135). All alkylamines were treated as ammonium ions, which at least for paromomycin and gentamicin 1Ca underestimates LogP.

[b]Sum of OH and NH H-bonds donors (132).

[c]Sum of O and N H-bond acceptors (132).

[d]Computational alert according to rule of 5:0, no problem:1, poor absorption or permeation likely (132).

and those that mediate protein/RNA association may help in explaining global features of RNA molecule recognition (124–131).

Ample evidence exists to confirm that RNA-binding ligands can be found that have drug-like affinity and specificity against a variety of RNA targets. A question that remains to be answered is whether similar agents can be applied to a wide range of targets, including cellular targets where a balance between drug solubility and cellular permeability is more important. Membrane permeability and drug solubility are intricately linked; the more permeable a membrane is to a drug, the less soluble the drug needs to be (132). The rule of 5 states that poor absorption or permeability is more likely when there are more than 5 H-bond donors, more than 10 H-bond acceptors, the molecular weight is greater than 500, and the calculated LogP is greater than 5 (133). This simple mnemonic is a set of criteria which most, but certainly not all, drugs obey. In the drug discovery process they are often used as a filter to reduce the number of potential candidates before costs of synthesis and testing are incurred (134). Here the rule of 5 will be used to predict whether or not any of the drugs in Table 6.1 would be acceptable drugs for a cellular target. Or put another way, are the chemical properties required for RNA binding mutually exclusive of the chemical properties required for good membrane permeability, absorption, and solubility?

Three of the antibiotics listed in Table 6.1, chloramphenicol, clindamycin, and the synthetic drug linezolid, are expected to have absorption and permeation qualities similar to the vast majority of drugs. All of the compounds that failed had either too many hydrogen bond acceptors, too many hydrogen bond donors, or both. Only two compounds in the failed set, gentamicin C1a and tetracycline, had molecular weights under 500. Many compounds were borderline; in particular, tetracycline has six hydrogen bond donors instead of the limit of five, but otherwise satisfies the rule of 5. Note that many active tetracycline derivatives satisfy the rule of 5.

Can drugs be developed that act on nonribosomal RNA? Several facts highlighted within this chapter should provide optimism that such drugs can be found. The structural data that exist for antibiotic/ribosome complexes illustrate a wide range of intermolecular interactions that are sufficient to selectively inhibit RNA function; other tightly binding ligands against RNA targets, such as ligands against HIV genomic RNA, demonstrate similar specificity. And on balance, antibiotic/RNA binding is more hydrophilic in nature than drug/protein complexes. However, some agents, such as erythromycin, make use of extensive hydrophobic interactions, and for three compounds, chloramphenicol, clindamycin, and linezolid, chemical features required for RNA binding seem to intersect with chemical properties required for drug-like absorption and permeability. The semisynthetic library, assembled largely in the absence of structural information, is rich for several classes of antibiotics; in most cases improvements in drug performance were possible, indicating that, even in the absence of structural data, promising lead compounds can be optimized enough to afford bona fide therapeutic agents.

Note added in proof: In the final stages of preparing this chapter, an article has appeared from the groups of Moore and Steitz describing the structures of several macrolide drugs in complex with the 50S ribosomal subunit (136). Readers interested in macrolide antibiotics are encouraged to visit this article. It contains new information not presented within this chapter.

10 ACKNOWLEDGMENTS

The author thanks the Deafness Research Foundation and the American Cancer Society (IN-105), which supported some of this work. I am grateful to Ms. Heather O'Farrell for proofreading the manuscript and to Prof. Glen Kellogg for helpful discussions.

REFERENCES

1. D. A. Matthews, R. A. Alden, J. T. Bolin, S. T. Freer, R. Hamlin, N. Xuong, J. Kraut, M. Poe, M. Williams, and K. Hoogsteen, *Science*, **197**, 452–455 (1977).

2. D. J. Abraham in T. J. Perun and C. L. Propst, Eds., *Computer-Aided Drug Design: Methods and Applications*, Marcel Dekker, New York, 1989, pp. 93–132.

3. D. J. Abraham, Ed., *Burger's Medicinal Chemistry*, 6th ed., vol. **1A**, Wiley-Interscience, New York, 2002.

4. P. B. Moore in R. F. Gesteland, T. R. Cech, and J. F. Atkins, Eds., *The RNA World*, Cold Spring Harbor Laboratory Press, New York, 1993, pp. 119–135.

5. N. Ban, P. Nissen, J. Hansen, P. B. Moore, and T. A. Steitz, *Science*, **289**, 905–920 (2000).

6. B. T. Wimberly, D. E. Brodersen, W. M. J. Clemons, R. Morgan-Warren, C. von Rhein, T. Hartsch, and V. Ramakrishnan, *Nature*, **407**, 327–339 (2000).

7. F. Schuenzen, A. Tocilij, R. Zarivach, J. Harms, M. Gluehmann, D. Janell, A. Bashan, H. Bartels, I. Agmon, F. Franceschi, and A. Yonath, *Cell*, **102**, 615–623 (2000).

8. J. Harms, F. Schuenzen, R. Zarivach, H. Bashan, S. Gat, I. Agmon, A. Bartels, F. Franceschi, and A. Yonath, *Cell*, **107**, 679–688 (2001).

9. J. H. Cate, M. M. Yusupov, G. Z. Yusupov, T. N. Earnest, and H. F. Noller, *Science*, **285**, 2095–2104 (1999).

10. A. P. Carter, W. M. Clemons, D. E. Brodersen, R. J. Morgan-Warren, B. T. Wimberly, and V. Ramakrishnan, *Nature*, **407**, 340–348 (2000).

11. D. E. Brodersen, W. M. Clemons, A. P. Carter, R. J. Morgan-Warren, B. T. Wimberly, and V. Ramakrishnan, *Cell*, **103**, 1143–1154 (2000).

12. M. Pioletti, F. Schlunzen, J. Harms, R. Zarivach, M. Gluhmann, H. Avila, A. Bashan, H. Bartels, T. Auerbach, C. Jacobi, T. Hartsch, A. Yonath, and F. Franceschi, *EMBO J.*, **20**, 1829–1839 (2001).

13. Q. Vicens and E. Westhof, *Structure*, **9**, 647–658 (2001).

14. F. Schlunzen, R. Zarivach, J. Harms, A. Bashan, A. Tocilj, R. Albrecht, A. Yonath, and F. Franceschi, *Nature*, **413**, 814–821 (2001).

15. A. Schatz, E. Bugie, and S. A. Waksman, *Proc. Soc. Exp. Biol. Med.*, **55**, 66–69 (1944).

16. H. F. Chambers and M. A. Sande in J. G. Hardman, L. E. Limbird, P. B. Molinoff, R. W. Rudden, and A. G. Gilman, Eds., *Goodman & Gilman's The Pharmacological Basis of Therapeutics*, 9th ed., McGraw-Hill, New York, 1996, pp. 1103–1153.

17. R. J. Fitzgerald, F. Bernheim, and D. B. Fitzgerald, *J. Biol. Chem.*, **175**, 195 (1948).

18. C. R. Spots and R. Y. Stanier, *Nature*, **192**, 633 (1961).

19. D. Moazed and H. F. Noller, *Nature*, **327**, 389–394 (1987).

20. P. Nissen, J. Hansen, N. Ban, P. B. Moore, and T. A. Steitz, *Science*, **289**, 920–930 (2000).

21. N. D. Pearson and C. D. Prescott, *Chem. Biol.*, **4**, 409–414 (1997).

22. T. Hermann, *Angew Chem. Int. Edu.*, **39**, 1890–1905 (2000).

23. J. Gallego and G. Varani, *Acc. Chem. Res.*, **34**, 836–843 (2001).

24. J. A. Doudna, *Nature Struct. Biol.*, **7**(Suppl), 954–956 (2000).

25. M. L. Zapp, S. Stern, and M. R. Green, *Cell*, **74**, 969–978 (1993).

26. A. J. Collier, J. Gallego, R. Klink, P. T. Cole, S. J. Harris, G. P. Harrison, F. Aboul-Ela, G. Varani, and S. Walker, *Nat. Struct. Biol.*, **9**, 375–380 (2002).

27. S. J. Suscheck, W. A. Greenberg, T. J. Tolbert, and C.-H. Wong, *Angew Chem. Int. Edu.*, **39**, 1080–1084 (2000).

28. J. B.-H. Tok, J. Cho, and R. R. Rando, *Biochemistry*, **38**, 199–206 (1999).

29. L. Varani, M. G. Spillantini, M. Goedert, and G. Varani, *Nucleic Acids Res.*, **28**, 710–719 (2000).

30. G. Werstuck and M. R. Green, *Science*, **282**, 296–298 (1998).

31. G. D. Stormo and Y. Ji, *Proc. Natl. Acad. Sci. USA*, **98**, 9465–9467 (2001).

32. G. D. Wright in D. J. Abraham, Ed., *Burger's Medicinal Chemistry*, 6th ed., vol. **4**, Wiley-Interscience, New York, 2002.

33. V. Ramakrishnan, *Cell*, **108**, 557–572 (2002).

34. W. Piepersberg in W. R. Strohl, Ed., *Biotechnology of Antibiotics*, 2nd ed., Marcel-Dekker, New York, 1997.

35. G. L. Mandell and W. A. Petri in J. G. Hardman, L. E. Limbird, P. B. Molinoff, R. W. Rudden and A. G. Gilman, Eds., *Goodman & Gilman's The Pharmological Basis of Therapeutics*, 9th ed., McGraw-Hill, New York, 1996, pp. 1155–1174.

36. O. Jerinic and S. Joseph, *J. Mol. Biol.*, **304**, 707–713 (2000).

37. R. L. Peck, R. P. Graber, A. Walti, E. W. Peel, C. E. Hoffhine, and K. Folkers, *J. Am. Chem. Soc.*, **68**, 29–31 (1946).

38. S. Umezawa and T. Tsuchiya in H. Umezawa and I. R. Hooper, Eds., *Aminoglycoside Antibiotics*, Springer-Verlag, Berlin, 1982, pp. 37–110.

39. T. Usui, T. Tsuchiya, and S. Umezawa, *J. Antibiotics*, **31**, 991–996 (1978).

40. D. L. Delaware, M. S. Sharma, B. S. Iyengar, W. A. Remers, and T. A. Pursiano, *J. Antibiotics*, **34**, 251–258 (1986).

41. K. E. Price, J. C. Godfrey, and H. Kawaguchi in D. Perlman, Ed., *Structure Activity Relationships Among the Semisynthetic Antibiotics*, Academic Press, New York, 1977, pp. 239–395.

42. T. Tsuchiya, T. Kishi, S. Kobayashi, Y. Kobayashi, S. Umezawa, and H. Umezawa, *Carb. Res.*, **104**, 69–77 (1982).

43. R. L. Peck, C. E. Hoffhine, and K. Folkers, *J. Am. Chem. Soc.*, **68**, 1390–1391 (1946).

44. S. Tatsuoka, T. Kusaka, A. Miyake, M. Inoue, H. Hitomi, Y. Shiraishi, H. Iwasaki, and M. Imanishi, *Chem. Pharm. Bull.*, **5**, 343–349 (1957).

45. J. Fried and O. Wintersteiner, *J. Am. Chem. Soc.*, **69**, 79–86 (1947).

46. H. Heding, *Tet. Lett.*, **1969**, 2831–2832 (1969).

47. W. A. Winsten, C. I. Jarowski, F. X. Murphy, and W. A. Lazier, *J. Am. Chem. Soc.*, **72**, 3969–3972 (1950).

48. H. P. Treffers, N. O. Besler, and D. C. Alexander, *Antibiotics Ann.*, **1953**, 595–603 (1953).

49. H. Heding, *Acta Chem. Scand.*, **26**, 3251–3256 (1972).

50. H. Heding and A. Diedrichsen, *J. Antibiotics*, **28**, 312–316 (1975).

51. F. C. Pennington, P. A. Guerico, and I. A. Solomons, *J. Am. Chem. Soc.*, **75**, 2261 (1953).

52. M. Awata, N. Muto, M. Hayashi, and S. Yaginuma, *J. Antibiotics*, **39**, 724–726 (1986).

53. T. Tsuchiya, S. Sakamoto, T. Yamasaki, and S. Umezawa, *J. Antibiotics*, **35**, 639–641 (1982).

54. H. Sano, T. Tsuchiya, S. Kobayashi, M. Hamada, S. Umezawa, and H. Umezawa, *J. Antibiotics*, **25**, 978–980 (1976).

55. S. A. Waksman and H. A. Lechevalier, *Science*, **109**, 305–307 (1949).

56. M. Murase, T. Wakazawa, M. Abe, and S. Kawaji, *J. Antibiotics*, **A14**, 156–157 (1961).

57. E. F. Gale, E. Cundliff, P. E. Reynolds, M. H. Richmond, and M. J. Waring, *The Molecular Basis of Antibiotic Action*, 2nd ed., Wiley-Interscience, London, 1981, pp. 402–547.

58. M. I. Recht, S. Douthwaite, and J. D. Puglisi, *EMBO J.*, **18**, 3133–3138 (1999).

59. D. Fourmy, M. I. Recht, S. C. Blanchard, and J. D. Puglisi, *Science*, **274**, 1367–1371.

60. S. Yoshizawa, D. Fourmy, and J. D. Puglisi, *EMBO J.*, **17**, 6437–6448 (1998).

61. J. M. Ogle, D. E. Brodersen, W. M. Clemons, M. J. Tarry, A. P. Carter, and V. Ramakrishnan, *Science*, **292**, 897–902 (2001).

62. J. Davies and G. D. Wright, *Trends Microbiol.*, **5**, 234–240 (1997).

63. J. Haddad, S. Vakulenko, and S. Mobashery, *J. Am. Chem. Soc.*, **121**, 11922–11923 (1999).

64. T. Yamasaki, Y. Narita, H. Hoshi, S. Aburaki, H. Kamei, T. Naito, and H. Kawaguchi, *J. Antibiotics*, **44**, 646–653 (1991).

65. H. Hoshi, S. Aburaki, S. Iimura, T. Yamasaki, T. Naito, and H. Kawaguchi, *J. Antibiotics*, **43**, 858–872 (1990).

66. K. Richardson, K. W. Brammer, S. Jevons, R. M. Plews, and J. R. Wright, *J. Antibiotics*, **32**, 973–977 (1979).

67. K. Richardson, S. Jevons, J. W. Moore, B. C. Ross, and J. R. Wright, *J. Antibiotics*, **30**, 843–846 (1977).

68. P. B. Alper, M. Hendrix, P. Sears, and C.-H. Wong, *J. Am. Chem. Soc.*, **120**, 1965–1978 (1998).

69. W. A. Greenberg, E. S. Priestley, P. S. Sears, P. B. Alper, C. Rosenbohm, M. Hendrix, S.-C. Hung, and C.-H. Wong, *J. Am. Chem. Soc.*, **121**, 6527–6541 (1999).

70. A. Van Schepdael, J. Delcourt, M. Mulier, R. Busson, L. Verbist, H. J. Vanderhaeghe, M. P. Mingeot-Leclercq, P. M. Tulkens, and P. J. Chaes, *J. Med. Chem.*, **34**, 1468–1475 (1991).

71. J. Schacht, *Ann. NY Acad. Sci.*, **884**, 125–130 (1999).

72. R. A. Casano, Y. Bykhovskaya, D. F. Johnson, M. Hamon, F. Torricelli, M. Bigozzi, and N. Fischel-Ghodsian, *Am. J. Med. Genet.*, **79**, 388–391 (1998).

73. X. Estivill, N. Govea, E. Barcelo, C. Badenas, E. Romero, L. Moral, R. Scozzri, L. D'Urbano, M. Zeviani, and A. Torroni, *Am. J. Hum. Genet.*, **62**, 27–35 (1998).

74. A. Pandya, X. Xia, J. Radnaabazar, J. Batsuuri, B. Dangaansuren, N. Fischel-Ghodsian, and W. E. Nance, *J. Med. Genet.*, **34**, 169–172 (1997).

75. M. Shohat, N. Fischel-Ghodsian, C. Legum, and G. J. Halpern, *Am. J. Otolaryngol.*, **20**, 64–67 (1999).

76. S. Kupka, T. Toth, M. Wrobel, U. Zeissler, W. Syzfter, K. Szyfter, G. Niedzielska, J. Bal, H. P. Zenner, I. Sziklai, N. Blin, and M. Pfister, *Hum. Mutat.*, **19**, 308–309 (2002).

77. D. C. Wallace, *Annu. Rev. Biochem.*, **61**, 1175–1212 (1992).

78. K. Inoue, D. Takai, A. Soejima, K. Isobe, T. Yamasoba, Y. Oka, Y. Goto, and J. Hayashi, *Biochem. Biophys. Res. Commun.*, **223**, 496–501 (1996).

79. D. Beauchamp, P. Gourde, and M. G. Bergeron, *Antimicrob. Agents Chemother.*, **35**, 2173–2179 (1991).

80. K. Hamasaki and R. R. Rando, *Biochemistry*, **36**, 12323–12328 (1997).

81. M. Maarouf, Y. de Kouchkovsky, S. Brown, P. X. Petit, and M. Robert-Gero, *Exp. Cell Res.*, **232**, 339–348 (1997).

82. J. E. Kapusnik, M. A. Sande, and H. F. Chambers in J. G. Hardman, L. E. Limbird, P. B. Molinoff, R. W. Rudden and A. G. Gilman, A. G., Eds., *Goodman & Gilman's The Pharmological Basis of Therapeutics*, 9th ed., McGraw-Hill, New York, 1996, pp. 1123–1153.

83. J. H. Boothe and J. J. Hlavka in Kirk-Othmer Encyclopedia of Chemical Technology, vol. **3**, 4th ed., New York, Wiley, 1992, p. 331.

84. I. S. Hunter and R. A. Hill in W. R. Strohl, Ed., *Biotechnology of Antibiotics*, 2nd ed., Marcel-Dekker, New York, 1997, pp. 659–682.

85. R. K. Blackwood and A. R. English in D. Perlman, Ed., *Structure-Activity Relationships Among the Semisynthetic Antibiotics*, Academic Press, New York, 1977, pp. 397–426.

86. D. Schnappinger and W. Hillen, *Arch. Microbiol.*, **165**, 359–369.

87. J. Bergeron, M. Ammirati, D. Danley, L. James, M. Norcia, J. Retsema, C. A. Strick, W.-G. Su, J. Sutcliffe, and L. Wondrack, *Antimicrob. Agents Chemother.*, **40**, 2226–2228.

88. P. J. Petersen, N. V. Jacobus, W. J. Weiss, P. E. Sum, and R. A. Tests, *Antimicrob. Agents Chemother.*, **43**, 738–744.

89. P.-E. Sum and P. Petersen, *Bioorg. Med. Chem. Lett.*, **9**, 1459–1462.

90. H. Brockman and W. Henkel, *Naturwissenschaften*, **37**, 138–139 (1950).

91. J. M. McGuire, R. L. Bunch, R. C. Anderson, H. E. Boaz, E. H. Flynn, H. M. Powell, and J. W. Smith, *Antibiot. Chemother.*, **2**, 281–283 (1952).

92. A. Bonnefoy, A. M. Girard, C. Agouridas, and J. F. Chantot, *J. Antimicrob. Chemother.*, **40**, 85–90 (1997).

93. R. Leclerq, *J. Antimicrob. Chemother.*, **48**, 9–23 (2001).

94. D. J. Hoban, A. K. Wierzbowski, K. Nichol, and G. G. Zhanel, *Antimicrob. Chemother.*, **45**, 2147–2150 (2001).

95. M. Dam, S. Douthwaite, T. Tenson, and A. S. Mankin, *J. Mol. Biol.*, **259**, 1–6 (1996).

96. S. M. Poulsen, C. Kofoed, and B. Vester, *J. Mol. Biol.*, **304**, 471–481 (2000).

97. P. Kurath, P. H. Jones, R. S. Egan, and T. J. Perun, *Experientia*, **27**, 362 (1971).

98. C. Vinckier, R. Hauchecorne, T. Cachet, G. Van der Mooter, and J. Hoogmartens, *Int. J. Pharmaceutics*, **55**, 67–76 (1989).

99. R. J. Pariza and L. A. Freiberg, *Pure Appl. Chem.*, **66**, 2365–2368 (1994).

100. D. J. Hardy, R. N. Swanson, N. L. Shipkowitz, L. A. Freiberg, P. A. Lartey, and J. J. Clement, *Antimicrob. Agents Chemother.*, **35**, 922–928 (1991).

101. J.-C. Gasc, S. G. d'Ambières, A. Lutz, and J.-F. Chantot, *J. Antibiotics*, **44**, 313–330 (1991).

102. E. H. Massey, B. S. Kitchell, L. D. Martin, and K. Gerzon, *J. Med. Chem.*, **17**, 105–107 (1974).

103. G. M. Bright, A. A. Nagel, J. Bordner, K. A. Desai, J. N. Dibrino, J. Nowakowska, L. Vincent, R. M. Watrous, F. C. Sciavolino, A. R. English, J. A. Retsema, M. R. Anderson, L. A. Brennan, R. J. Borovoy, C. R. Cimochowski, J. A. Faiella, A. E. Girard, D. Girard, C. Herbert, M. Manousos, and R. Mason, *J. Antibiotics*, **41**, 1029–1047 (1988).

104. S. Djokic, G. Kobrehel, N. Lopotar, B. Kamenar, A. Nagl, and D. Mrvos, *J. Chem. Res.*152–153 (1988).

105. Y. J. Wu and W. G. Su, *Curr. Med. Chem.*, **14**, 1727–1758 (2001).

106. Z. Ma, R. F. Clark, A. Brazzale, S. Wang, M. J. Rupp, L. Li, G. Griesgraber, S. Zhang, H. Yong, L. T. Phan, P. A. Nemoto, D. T. W. Chu, J. J. Plattner, X. Zhang, P. Zhong, Z. Cao, A. M. Nilius, V. D. Shortridge, R. Flamm, M. Mitten, J. Meulbroek, P. Ewing, J. Alder, Y. S. Or, *J. Med. Chem.*, **44**, 4137–4156 (2001).

107. C. Agouridas, A. Denis, J.-M. Auger, Y. Benedetti, A. Bonnefoy, F. Bretin, J.-F. Chantot, A. Dussarat, C. Fromentin, S. G. D'Ambrières, S. Lachaud, P. Laurin, O. Le Martret, V. Loyau, and N. Tessot, *J. Med. Chem.*, **41**, 4080–4100.

108. S. Douthwaite and W. S. Champney, *J. Antimicrob. Chemother.*, **48**, 1–8 (2001).

109. P. B. Fernandes, W. R. Baker, L. A. Freiberg, D. J. Hardy, and E. J. McDonald, *Antimicrob. Agents Chemother.*, **33**, 78–81 (1989).

110. S. T. Waddell, G. M. Santorelli, T. A. Blizzard, A. Graham, and J. Occi, *Bioorg. Med. Chem. Lett.*, **8**, 549–554 (1998).

111. T. Tanikawa, T. Asaka, M. Kashimura, Y. Misawa, K. Suzuki, S. Morimoto, and A. Nishida, *J. Med. Chem.*, **44**, 4027–4030.

112. C. Abad-Zapatero, P. Zhong, D. E. Bussiere, K. Stewart, and S. W. Muchmore in X. Cheng and R. Blumenthal, Ed., *S-Adenosylmethionine-Dependent Methyltransferases: Structures and Functions*, World Scientific Publishing, River Edge, NJ, 1999, pp. 199–225.

113. B. Vester and S. Douthwaite, *Antimicrob. Agents Chemother.*, **45**, 1–12 (2001).

114. L. Zhao, D. H. Sherman, and H. Lui, *J. Am. Chem. Soc.*, **120**, 9374–9375 (1998).

115. S. M. Poulsen, C. Kofoed, and B. Vester, *J. Mol. Biol.*, **304**, 471–481 (2000).

116. L. T. Phan, R. F. Clark, M. Rupp, Y. Or, D. T. W. Chu, and S. Ma, *Organic Lett.*, **2**, 2951–2954 (2000).

117. S. Douthwaite, L. H. Hansen, and P. Mauvais, *Mol. Microbiol.*, **36**, 183–193 (2000).

118. G. Garza-Ramos, L. Xiong, P. Zhong, and A. Mankin, *J. Bacteriology*, **183**, 6898–6907 (2001).

119. S. J. Brickner, D. K. Hutchinson, M. R. Barbachyn, P. R. Manninen, D. A. Ulanowicz, S. A. Garmon, K. C. Grega, S. K. Hendges, D. S. Troops, C. W. Ford, and G. E. Zurenko, *J. Med. Chem.*, **39**, 673–679 (1996).

120. P. Kloss, L. Xiong, D. L. Shinabarger, and A. S. Mankin, *J. Mol. Biol.*, **294**, 93–101 (1999).

121. L. Xiong, P. Kloss, S. Douthwaite, N. M. Andersen, S. Swaney, D. L. Shinabarger, and A. S. Mankin, *J. Bacteriology*, **182**, 5325–5331 (2000).

122. U. Patel, Y. P. Yan, F. W. Hobbs, J. Kaczmarczyk, A. M. Slee, D. L. Pompliano, M. G. Kurilla, and E. Y. Bobkova, *J. Biol. Chem.*, **276**, 37199–37205 (2001).

123. W. S. Champney and M. Miller, *Curr. Trends Microbiol.*, **44**, 350–356 (2002).

124. K. Nagai, *Curr. Opin. Struct. Biol.*, **6**, 53–61 (1996).

125. R. N. De Guzman, R. B. Turner, and M. F. Summers, *Biopolymers*, **48**, 181–195 (1998).

126. A. R. Ferre-D'Amare and J. A. Doudna, *Ann. Rev. Biophys. Biomol. Struct.*, **28**, 57–73 (1999).

127. Y. Muto, C. Oubridge, and K. Nagai, *Curr. Opin. Biol.*, **10**, R19–R21 (2000).

128. J. R. Williamson, *Nat. Struct. Biol.*, **7**, 834–837 (2000).

129. J. M. Perez-Canadillas and G. Varani, *Curr. Opin. Struct. Biol.*, **11**, 53–58 (2001).

130. P. B. Moore, *Biochemistry*, **40**, 3243–3250 (2001).

131. V. Ramakrishnan and P. B. Moore, *Curr. Opin. Struct. Biol.*, **11**, 144–145.

132. C. A. Lipinski, *J. Pharmacol. Toxicol. Methods*, **44**, 235–249 (2000).

133. C. A. Lipinski, F. Lombardo, B. W. Dominy, and P. J. Feeney, *Adv. Drug Deliv. Rev.*, **46**, 3–26 (2001).

134. W. P. Walters and M. A. Murcko, *Adv. Drug Deliv. Rev.*, **54**, 255–271 (2002).

135. G. E. Kellogg and D. J. Abraham, *Eur. J. Med. Chem.*, **35**, 651–661 (2000).

136. J. L. Hansen, J. A. Ippolito, N. Ban, P. Nissen, P. B. Moore, and T. A. Steitz, *Mol. Cell*, **10**, 117–128 (2002).

CHAPTER SEVEN

Carbohydrate-Based Therapeutics

JOHN H. MUSSER
Pharmagenesis, Inc.
Palo Alto, California

Contents

Burger's Medicinal Chemistry and Drug Discovery
Sixth Edition, Volume 2: Drug Development
Edited by Donald J. Abraham
ISBN 0-471-37028-2   © 2003 John Wiley & Sons, Inc.

1 INTRODUCTION

A decade has elapsed since we first wrote—"by the 21st century, research on carbohydrates will emerge as significant new approach to drug discovery (1)." Indeed, significant progress has been made toward the realization of this statement. The pharmaceutical impact of this class of biomolecules can be seen in the carbohydrate-based therapeutics recently approved for marketing. For example, the trend toward producing lower molecular weight heparin-based products has culminated in the approval of a pentasaccharide for preventing venous thromboembolism after major orthopedic surgery. Also, in the antiviral arena that has traditionally resisted the introduction of new medicines, not one but two new monosaccharide glycomimetic drugs that block the ability of influenza viruses to exit infected cells by binding to neuraminidase were registered for the treatment of flu.

The pharmaceutical impact of carbohydrate-based therapeutics can also be seen in the agents now under evaluation in pre-clinical and clinical development. For example, synthetic carbohydrate chains, modeled on tumor cell antigens but modified to stimulate an immune response, are being used to develop novel cancer vaccines.

In research, incremental increases in the capability of research tools have made the localization and functional characterization of glyconjugates possible. Hints at how to solve the many carbohydrate synthesis problems can be gleaned from the developments made in solid and solution phase as well enzymatic chemistry. Most importantly, carbohydrate chemists and glycobiologist have forged links that have allowed carbohydrate-based therapeutics to enter the mainstream drug discovery process. Thus, all the above factors have combined to clearly demonstrate the importance of carbohydrates as a source for new drug discovery.

2 CARBOHYDRATE-BASED DRUGS

The carbohydrate-based drug section is divided into three parts. The first part contains drugs that have made it to the market in the last decade. An emphasis is placed on new product introductions both because there has been significant progress in the commercialization of carbohydrate-based therapeutics and the product approval process is a very stringent measure of success. The second part of this section deals with therapeutics in development. Although not as stringent as marketed products, the introduction of new therapeutics into clinical trials is not trivial. Also, showing safety first, then efficacy, is a demanding task, and there are failures. A review of carbohydrate-based therapeutics that failed in development is as informative as those that have succeeded. The final part is focused on carbohydrate agents in research. Indeed, it is interesting times for carbohydrate research because there are both promising ideas and naïve approaches, and only time will tell which research will lead to new carbohydrate-based drugs.

2.1 To Market—Last Decade

What marketed carbohydrate-based drugs means in the context of this chapter is that the indicated drugs have been approved by government regulatory agencies during the last decade in one or more countries as safe and efficacious medicines for human use. It also means the drug's chemical structure or the drug's mechanism of action is related to carbohydrate chemistry or glycobiology.

In the last decade, a number of new carbohydrate-based therapeutics have been approved for human use including both new chemical entities (usually low molecular weight) and new biological entities (usually high molecular weight). This section is divided by molecular weight and organized into therapeutic classes. The therapeutic classes are grouped as follows: diabetes, gastrointestinal, central nervous system, infection, thrombosis, inflammation, and enzyme replacement.

2.1.1 Diabetes. Diabetes is a condition where there is too much glucose in the blood. Therefore, it is not surprising that carbohydrate-based drugs would be active in a disease like diabetes.

The tetrasaccharide acarbose (**1**) is used as an adjunctive in non–insulin-dependent type 2 diabetes and is also of benefit in insulin-dependent type 1 diabetes when used in conjunction with insulin (**2**). Acarbose (**1**) is derived from the fermentation process of the fungus *Actinoplanes utahensis* and is orally active. It works by inhibiting intestinal α-glucosidases, resulting in lower blood glucose levels after meals. Thus, acarbose (**1**) is an effective treatment for post-prandial hyperglycemia.

The mechanism of action for acarbose (**1**) is informative as to how carbohydrate-based drugs may work effectively. In a typical reaction in the intestine, the polymeric sugar chains are broken down into individual monosaccharide units by the action of α-glucosidase. This enzyme recognizes specific residues in the polymeric sugar chain and hydrolyzes the inter-glycoside linkage through the formation of an oxonium ion intermediate. A subsequent hydrolysis reaction of the intermediate releases the non-reducing end monosaccharide that becomes available for quick absorption in the intestine and release in the blood.

Acarbose (**1**) is a stable transition state inhibitor of this reaction. Its structure resembles the transition state structure of the polysaccharide, although it lacks the ability to form a stable oxonium ion intermediate. A carbose (**1**) has a 105-fold higher affinity than typical substrates. Thus, it binds the α-glucosidase enzyme, incapacitating it from its action on its normal substrates. This prevents the release of monosaccharides from polysaccharides in the stomach and intestine. The normal polysaccharides that are usually broken down in the gastrointestinal tract are eliminated in feces.

There are gastrointestinal side effects with acarbose (**1**) treatment including pain, diarrhea, and flatulence. The drug is not recommended for people with limited kidney function, and high doses cause abnormal liver enzyme values. Therefore, the maximum dosage should not exceed 100 mg, three times per day.

Voglibose (**2**) is anti-diabetic carbohydrate-based drug, and like acarbose (**1**), it is an α-glucosidase inhibitor for prevention of postprandial hyperglycemia. Clinical studies have shown that voglibose (**2**) limits daily fluctua-

(1)

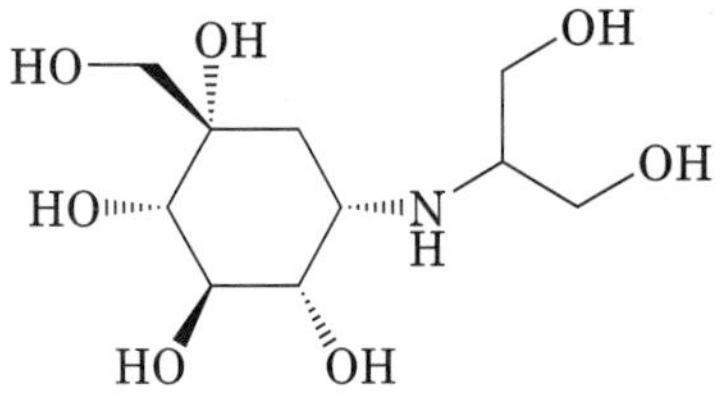

(2)

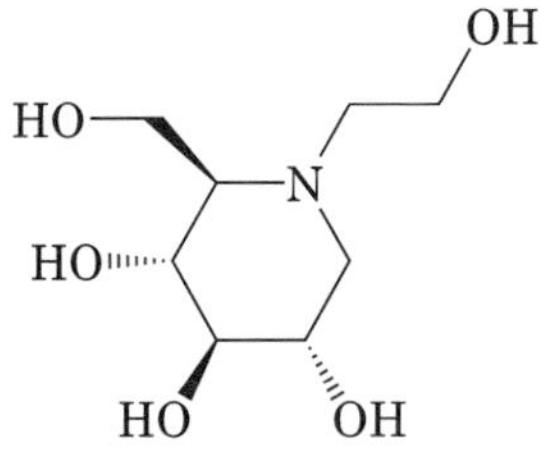

(3)

tions in blood glucose level and decreases the amount of insulin produced by the pancreas. Voglibose (**2**) is well tolerated and has been proven to be safe for further clinical use in combination with other drugs for diabetes, such as glibenclamide (3). In rats with type 2 diabetes, the combination of voglibose and a pioglitazone seems to control blood glucose levels and improves the ability to regulate future glucose levels by restoring the function of damaged islet cells in the pancreas (4). Structurally, voglibose (**2**) is a synthetic monosaccharide that may offer advantages over the tetrasaccharide, acarbose (**1**). However, when compared side by side clinically, 1-day administration of acarbose (**1**) and voglibose (**2**) at currently recommended clinical doses demonstrated that acarbose (**1**) was more effective in sparing endogenous insulin secretion than was voglibose (**2**) (5).

Another azasugar that has been used for diabetes treatment is miglitol (**3**) (6). Like acarbose (**1**) and voglibose (**2**), it is an oral α-glucosidase inhibitor for use in the management of non–insulin-dependent diabetes mellitus. Structurally, miglitol (**3**) is a desoxynojirimycin derivative, and this family of

compounds has antiviral and anticancer properties (vide infra). Nevertheless, this agent has anti-diabetic properties that are similar to both volglibose (**2**) and acarbose (**1**).

2.1.2 Gastrointestinal System. Because historically carbohydrates were considered to function as merely energy storage, such as starch, it is ironic that some newly approved therapeutics are used to heal the gastrointestinal system.

Dosmalfate (**4**) is an oral cytoprotective agent that is used for healing gastric, duodenal, and esophageal ulcers (7). An additional indication is to prevent lesions caused by the chronic intake of non-steroid anti-inflammatory drugs. Its mechanism of gastric protection is dependent on multiple factors including the modulation of prostaglandins. Dosmalfate (**4**) is reported to have a coating effect, including antiacid diffusion activity, but there are not any anti-secretory properties. It was well tolerated in the clinic.

2.1.3 Central Nervous System. Carbohydrate-based drugs acting on the central ner-

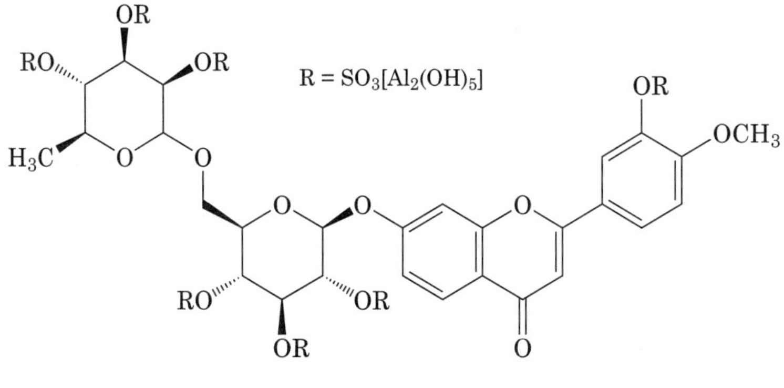

R = $SO_3[Al_2(OH)_5]$

(**4**)

vous system (CNS) is surprising because, in general, saccharides are water-soluble, and drugs that are CNS active usually have appreciable lipid solubility. Apparently, a way to solve the problem is to mask the free hydroxyl groups of the saccharide with alkyl groups.

Topiramate (**5**) is a monosaccharide diacetonide with a sulfamate group that has activ-

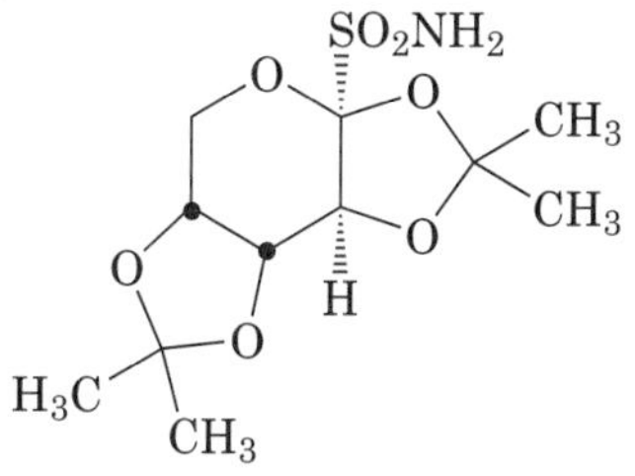

(**5**)

ity as a novel anti-convulsant (8), and more recently, it was approved to treat epilepsy. As one of the new generation of anti-epileptics, it is used to treat partial-onset seizures as adjunctive therapy in children. In 2001, topiramate (**5**) was also approved as adjunctive treatment for primary, generalized, tonic-clonic seizures in which a temporary loss of consciousness and muscle control occurs in young adults and children. Off-label prescribing of topiramate (**5**) as a migraine preventive is known, and this is of concern especially in light of a side effect consisting of acute myopia-associated glaucoma that has been reported in patients receiving topiramate (**5**) (9).

Symptoms include acute onset of decreased visual acuity and/or ocular pain.

2.1.4 Infection. Carbohydrate-based drugs that treat infection include antibiotics and antivirals. In this section are agents to treat sepsis and those that act as vaccines. Also, no attempt is made to cover all carbohydrate-based anti-infectives because of the number involved and because antibiotics are covered more thoroughly elsewhere (10).

Aminoglycosides are a diverse group of carbohydrate-based antibiotics that are used to treat a wide range of human infections. Aminoglycosides include streptomycin, kanamycin, and dibekacin. These compounds inhibit protein synthesis in susceptible bacteria, although resistance and the risk of serious side effects, such as ear and kidney damage, have lessened their use. In view of the current concerns over the global rise in antibiotic-resistant microorganisms, there is a renewed interest in the new aminoglycosides that are approved for the effective treatment of resistant microorganism infections.

The aminoglycoside, arbekacin (**6**), is a derivative of dibekacin and is effective against most resistant strains. It showed remarkable antimicrobial effects on methicillin-resistant *Staphylococcus aureus* (MRSA), a leading cause of nosocomial infections. Since its introduction, arbekacin (**6**) has been clinically used as one of the most effective antibiotics in the treatment of MRSA infections (11).

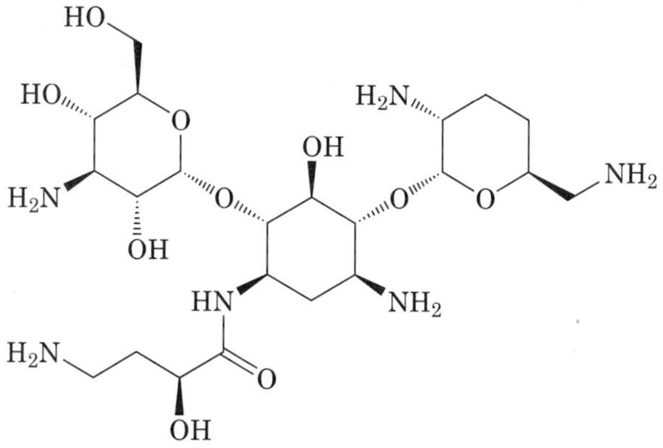

(**6**)

In contrast to antibiotics, viral diseases present an especially difficult challenge to the pharmaceutical industry. This is evident simply because there are not enough good antiviral drugs available. The discovery and development of novel antiviral therapeutics with new mechanisms of action is a notable event. Thus, it is especially significant that the carbohydrate chemistry of sialic acid was used in combination with neuriminidase glycobiology to produce not one but two new antiviral therapeutic drugs for the treatment of influenza.

Neuraminidase cleaves terminal sialic acid (**7**) residues from carbohydrate moieties on

(**7**)

the surfaces of host cells and influenza virus envelopes. The enzymatic process promotes the release of progeny viruses from infected cells (12).

Neuraminidase inhibitors are analogs of sialic acid (**7**). They work by binding to the active, catalytic site of neuraminidase that protrudes from the surface of influenza viruses. Viral hemagglutinin binds to the intact sialic acid residues, which results in viral aggregation at the host cell surface and a reduction in the amount of virus that is released to infect other cells (13).

The sialic acid analogs or glycomimetics, zanamivir (**8**) and oseltamivir (**9**), are members of a new class of antiviral agents that selectively inhibit the neuraminidase of both influenza A and B viruses.

Zanamivir (**8**) is an inhaled anti-viral drug for the treatment of uncomplicated types A and B influenza, the two types most responsible for flu epidemics (14). Patients need to start treatment within 2 days of the onset of symptoms, and the drug is less effective in patients whose symptoms do not include a fever. Zanamivir (**8**) is a powder that is inhaled twice

(**8**)

(**9**)

a day for 5 days, and it is not recommended in patients with severe asthma or chronic obstructive pulmonary disease.

In contrast to zanamivir (**8**), oseltamivir (**9**) is an oral anti-viral drug for the treatment of uncomplicated influenza in patients whose flu symptoms have not lasted more than 2 days (15, 16). This product treats types A and B influenza; however, the majority of patients in the United States are infected with type A. Efficacy of oseltamivir (**9**) in the treatment of influenza in subjects with chronic cardiac disease and/or respiratory disease has not been established. Oseltamivir (**9**) is also approved for the prevention of influenza in adults and adolescents older than 13 years. Efficacy of oseltamivir (**9**) for the prevention of influenza has not been established in immune-compromised patients.

In the treatment of influenza, further comments on the mechanism of action of neuraminidase inhibitors are warranted. Neuraminidase inhibitors occupy the active site of neuraminidase by binding to it more readily than sialic acid (**7**), the sugar residue it normally cleaves (17). Based on structure-activity relationship of analogs, it is apparent that sialic acid (**7**) is held in the active site through its glycerol and carboxylate groups, which form hydrogen bonds with amino acids in the active site. Zanamivir (**8**) adds other hydrogen bonds by replacing the hydroxyl group of a sialic acid derivative with a large, positively charged guanidine moiety, which forms strong attachments to two negatively charged amino acids at the bottom of the target protein cleft.

Oseltamivir (**9**) is a prodrug that is converted to free acid in the body, and the resulting molecule retains the carboxylate bonds found in sialic acid (**7**). It also makes use of a hydrophobic group that apparently binds better than a glycerol of sialic acid (**7**) that it replaces. The antibiotics zanamivir (**8**) and oseltamivir (**9**) are outstanding examples of new carbohydrate-based therapeutics.

Historically, the only therapeutics for sepsis were antibiotics. People suffering from sepsis are unable to regulate inflammation and coagulation, leading to tissue damage, organ failure, and sometimes death. Trauma, surgery, burns, or illnesses like cancer and pneumonia can trigger the condition. Sepsis is second to heart attack as the leading cause of death with intensive care unit patients.

Drotrecogin alfa is a genetically engineered version of the human activated protein C molecule, which helps balance the forces behind sepsis, including inflammation, coagulation, and suppression of fibrinolysis (18). Drotrecogin alfa differs from the natural form of activated protein C by specialized chemical attachments composed of complex carbohydrates that are not found in other biotech treatments such as insulin. Protein C is a serine protease and naturally occurring anticoagulant that plays a role in the regulation of homeostasis through its ability to block the generation of thrombin production by inactivating factors Va and VIIIa in the coagulation cascade. Human protein C is made *in vivo* primarily in the liver as a single polypeptide of 461 amino acids. The protein C precursor molecule undergoes multiple post-translational modifications.

The reasons why drotrecogin alfa is included as a carbohydrate-based therapeutic is not only because it is a glycoprotein but also because to be fully active the core protein C must be glycosylated by the addition of four Asn-linked oligosaccharides. Other post-translational modifications required for activity include the addition of nine γ-carboxy-gluta-

mates and one erythro-β-hydroxy-Asp, and the removal of the leader sequence and the dipeptide Lys 156-Arg 157. Without such post-translational modifications, protein C is non-functional. Drotrecogin alfa is a good example of the role saccharides may play in the activity of glycoproteins.

Several new glycoconjugate vaccines have recently been approved, including those made from *Pneumococcal*, *Haemophilus* B, and *Typhoid* Vi capsular polysaccharides. This type of carbohydrate-based drug involves carbohydrate natural products isolation and serves as an example for synthetic oligosaccharide glycoconjugates in research and development.

Hib vaccines as a class have virtually eliminated bacterial meningitis from the United States, most European countries, and a growing number of developing nations (19). The bacterium primarily responsible is *Haemophilus influenza* type b (Hib).

Four different vaccines for Hib are licensed in the United States. All these vaccines are composed of Hib capsular polysaccharide conjugated to a protein carrier. All four vaccines use a different carrier. Polyribosyl ribitol phosphate (PRP) capsule is an important virulence factor that renders Hib resistant to phagocytosis by neutrophils in the absence of specific anti-capsular antibody. Antibody directed against PRP capsule of Hib is primarily responsible for host resistance to infection. The Hib vaccine is composed of the purified, capsular polysaccharide of Hib. Antibodies to this antigen correlate with protection against invasive disease.

Prevnar is a glycoconjugate heptavalent pneumococcal vaccine for young children made from the polysaccharide capsule of the pneumococcus (20). Although older vaccines made from pneumococcus capsular polysaccharide are available for adults, young children cannot make antibodies to the capsule. By coupling the pneumococcal capsular polysaccharide with a protein carrier, a vaccine was created that will trigger an infant's immune system to produce antibodies.

2.1.5 Thrombosis. Heparin was the first polysaccharide-based drug to find widespread application in humans. Isolated from mammalian tissues as a complex mixture of glycosaminoglycan (GAG) polysaccharides, heparin has been used clinically since 1937 to treat thrombosis because of its powerful anti-coagulant activity. Heparin acts by enhancing the ability of antithrombin (AT) to inactivate thrombin and factor Xa, enzymes that promote coagulation (21).

Low molecular weight (LMW) heparins are produced by chemical or enzymatic depolymerization of heparin, and have a chain length of 13–22 saccharides with a mean molecular weight of 4–6.5 kDa. LMW heparins act through AT inhibition of factor Xa, where only the pentasaccharide high-affinity binding sequence is required (22).

Although lacking full anti-coagulant activity, LMW heparins are at least as effective as heparin in the prevention of deep venous thrombosis (DVT) and subsequent venous thromboembolism (VTE) (23). LMW heparins have more predictable activity than heparin, longer duration of action, do not require the measurement of anticoagulation, and can be self administered subcutaneously. However, LMW heparins cannot be effectively reversed by protamine, and duration of action is prolonged in renal failure. In higher doses, they have been shown to be equally effective with intravenous heparin in unstable angina, pulmonary embolism, and ischemic stroke. LMW heparins include dalteparin (24), enoxaparin (25), nadroparine (26), ardeparin (27), and danaparoid (28, 29).

The progression from the isolated mixture heparin to the depolymerized LMW heparins has culminated in a synthesized single chemical entity that can act like heparin and the LMW heparins with fewer side effects.

Fondaparinux (10) is the first synthetic, selective factor Xa inhibitor. This pentasaccharide agent is significantly better than the

LMW heparin, enoxaparin, for preventing VTE after major orthopedic surgery (30). The results from a meta-analysis show that fondaparinux (10) is associated with an overall risk reduction versus enoxaparin for the prevention of venous thromboembolism. Of all the LMW heparins, enoxaparin is widely regarded as the treatment standard for VTE prophylaxis. Fondaparinux (10) is an injectable solution for the prevention of DVT and is the only antithrombotic agent approved in the United States for hip fracture surgery.

DVT is caused by a blood clot forming in a deep vein in the leg. It may cause pain and swelling and enlargement and discoloration of the veins. The clot can grow in size and block other veins. In the most serious thrombotic event, portions of the clot may break away and travel through the veins to the lungs, leading to a life-threatening pulmonary embolism.

Fondaparinux (10) exhibits antithrombotic activity, which is the result of AT-mediated selective inhibition of factor Xa. By selectively binding to AT, fondaparinux (10) activates the innate neutralization of factor Xa by AT. This neutralization interrupts the blood coagulation cascade and thus inhibits thrombin formation and thrombus development.

Note that heparin, LMW heparins, and fondaparinux (10) all consist of long linear sequences of alternating hexosamine (D-glucosamine or D-galactosamine) and hexuronic acid (D-glucuronic acid or L-iduronic acid) units carrying sulfate substituents at various positions.

2.1.6 Inflammation. Chemically related to heparin is the GAG, hyaluronic acid. It is a natural component of the extra-cellular matrix found in connective tissues. Clinical investigations have demonstrated the safety and ef-

(10)

ficacy of hyaluronic acid in the treatment of osteoarthritis of the knee and other large joints. In addition to restoring the elasticity and viscosity of the synovial fluid, hyaluronic acid modulates acute and chronic inflammation processes both in animals and human beings (31). Specifically, hyaluronic acid interacts with endogenous receptors such as CD 44, ICAM-1, and RHAMM and may play an important role in controlling a variety of cellular behaviors, such as the migration, adhesion, and activation of pro-inflammatory cells, chondrocyte maturation or differentiation, and matrix synthesis in the cartilage microenvironment (32).

A highly purified, high molecular weight, high viscosity injectable form of hyaluronic acid (Orthovisc) is available to improve joint mobility and range of motion in patients suffering from osteoarthritis of the knee (33). Orthovisc is injected into the knee to restore the elasticity and viscosity of the synovial fluid.

2.1.7 Enzyme Replacement. In contrast to all the drugs discussed thus far, carbohydrate enzyme replacement therapeutics is different in that the enzyme drugs are not primarily structurally related to carbohydrates, but play an important role in saccharide or glycoconjugate processing. Coincidentally, the replacement strategy requires high affinity receptor-mediated uptake and delivery to lysosomes that is regulated by specific oligosaccharide side-chains of these enzymes.

Carbohydrate enzyme replacement therapy is the administration of exogenous enzymes to patients that have defective saccharide-processing enzymes that result in an accumulation of harmful metabolic products. Metabolic diseases often have genetic causes and some of these diseases involve saccharide or glycoconjugate metabolites. Two examples are Gaucher's and Fabry's disease.

Gaucher's disease is a glycolipid storage disease and results from genetic mutations that can either slow or prevent the breakdown of certain glycolipids. Patients with Gaucher's disease are born with a deficiency in the enzyme glucocerebrosidase that results harmful quantities of a fatty substance called glucocerebroside to accumulate in the spleen, liver, lungs, and bone marrow.

Fabry's disease is an inherited genetic disorder caused by deficient activity of the lysosomal enzyme α-galactosidase A. In patients with Fabry's disease, harmful quantities of globotriaosylceramide accumulate in the kidney, heart, nervous system, and blood vessels.

Imiglucerase is a recombinant form of the enzyme glucocerebrosidase and it is approved to treat type 1 Gaucher's disease. The therapy is highly effective, but it requires a 2-h intravenous infusion as often as three times a week and is expensive.

Agalsidase alfa is a human α-galactosidase A produced by genetic engineering technology in a human cell line. Patients can receive agalsidase alfa every other week in a 40-min intravenous infusion at home rather than in a hospital setting.

Although Table 7.1 does not contain all the new carbohydrate-based drugs introduced into the market in the last decade, it should be representative. The choice of what to include in Table 7.1 is also arguable. Nevertheless, it is impressive that in the last decade, over 20 carbohydrate-based drugs have been approved.

2.2 Therapeutics in Development

In the past decade there has been significant progress in development of carbohydrate-based therapeutics. Also, new methods for the large-scale production of carbohydrates and their analogs are allowing the thorough evaluation of their safety and efficacy in human trials.

Challenges that face the development of carbohydrate-based therapeutics continue to include the need to procure sufficient amounts of complex carbohydrates, which is often difficult and expensive, whether they involve synthesis or isolation from natural sources. Additional challenges include the low bioavailability of orally delivered carbohydrates and the inability of many animal models to provide data relevant to human disease. In contrast to the challenges, advantages in favor of carbohydrate-based therapeutics include low toxicity and high structural diversity. This section, like marketed drugs above, is divided by molecular weight and organized into therapeutic classes.

The therapeutic classes are grouped as follows: gastrointestinal system, central nervous

Table 7.1 To Market—Last Decade

Generic Name (Brand Name)	Indication	Company	Approval Country and Year
Acarbose (**1**) (Precose, Glucobay, Prandase)	Diabetes	Bayer AG	U.S. in 1995
Voglibose (**2**) (Basen, Glustat) (AO-128)	Diabetes	Takeda and Abbott	Japan in 1994
Miglitol (**3**) (Glyset, BAY m 1099)	Diabetes	Bayer	U.S. in 1996
Dolsamate (**4**) (Flavalfate, F3616M)	Gastric ulceration	Faes	Spain in 2000
Topiramate (**5**) (Topamax)	Anti-convulsant, anti-epileptic	Ortho-McNeil, Johnson & Johnson	U.S. in 1996, 1999
Arbekacin (**6**) (Habekacin)	Anti-microbial	Meiji Seika	Japan in 1997
Zanamivir (**8**) (Relenza) GG167	Anti-viral	GlaxoSmithKline	U.S. in 1999
Oseltamivir (**9**) (Tamiflu)	Anti-viral	Hoffmann-La Roche Gilead	U.S. in 1999
Pneumococcal heptavalent vaccine (Prevnar)	Conjugate vaccine	Wyeth	U.S. in 2000
Haemophilus b (ActHIB, OmniHIB)	Conjugate vaccine	SmithKline Beecham, Pasteur Merieux	France in 1996
Typhoid Vi (Typhim Vi)	Conjugate vaccine	Pasteur Merieux	U.S. in 1995
Drotrecogin alfa (Xigris)	Sepsis	Lilly	U.S. in 2001
Dalteparin (Fragmin)	Anti-coagulant, antithrombotic	Pharmacia Upjohn	U.S. in 2000
Enoxaparine (Lovenox)	Anti-coagulant Antithrombotic	Aventis	U.S. in 1998
Nadroparine (Fraxiparine)	Anti-coagulant, antithrombotic	Sanofi-synthelabo	France in 1986
Ardeparin (Normiflo)	Anti-coagulant, antithrombotic	Wyeth	U.S. in 1997
Danaparoid (Orgaran)	Anti-coagulant; antithrombotic	Organon	U.S. in 2000
Fondaparinux (**10**) (Arixtra)	Anti-coagulant; antithrombotic	Organon and Sanofi-Synthelabo	U.S. in 2001
Hyaluronic acid (Orthovisc)	Viscoelastic supplement	Anika, Zimmer Europe	Europe and Canada in 1998
Imiglucerase (Cerezyme)	Gaucher's disease	Genzyme	U.S. in 1994
Agalsidase alfa (Replagal)	Fabry's disease	Transkaryotic Therapies	Europe in 2001

system, infection, cancer, inflammation, and lysosomal storage. The major difference from the marketed drugs is that the diabetes class is not represented. With diabetes, it is not clear why no new mechanism drug is being developed to follow up the success of α-glucosidase inhibition. In place of diabetes, cancer therapeutics is added with a focus on cancer vaccines that are in clinical development.

2.2.1 Gastrointestinal System. Oral cytoprotective agents like dosmalfate (**4**) have not been followed in development presumably because the mechanism of action is not well understood. In place of cytoprotection in gastrointestinal drug development is motilin antagonism.

Mitemcinal (**11**) is an agonist of motilin, a peptide hormone that plays a role in the con-

(11)

tractile movement of the gastrointestinal tract (34). Mitemcinal (**11**) is an erythromycin A derivative that is being studied in phase II clinical trials to assess its effect on the recovery of gastrointestinal motility. Gastrointestinal motor-stimulating activity is a side effect of erythromycin, and other compounds based on erythromycin, such as EM574, are in development (35).

2.2.2 CNS. Because CNS active drugs tend to have appreciable lipid solubility, it is not

surprising that a glycolipids-based drug would be in development. What seems to be the problem, however, is glycolipid drug supply.

The development the glycolipid ganglioside GM1, called Sygen (**12**), for treating Parkinson's disease was reported (36). This complex molecule is currently isolated from natural sources and studies into its large-scale synthesis are underway. A phase II clinical trial for acute spinal cord injured patients was started in 1997. Results from single center studies, as well as efforts in experimental spinal cord injury and stroke, indicated that Sygen (**12**) promoted improvement in neurological status. However, there were problems. Early forms of the drug were often impure and there were several severe complications from the use of the European version of the drug. Also, higher doses of Sygen (**12**) were associated with increased risks in that there was a small incidence of allergies and an increase in cholesterol that was related to being on the drug for a period of time (37).

2.2.3 Infection. Anti-infective carbohydrate-based therapeutics in development include antibiotics and antivirals. The emergence of multi-drug resistant bacteria has increased the need for new antibiotics or modifications of older antibiotic agents. The modification of known antibiotics may be a productive way to combat multi-drug resistant bacteria.

(12)

(13)

Everninomycin (**13**), an intravenous antibiotic, was in development for hospitalized patients with resistant gram-positive infections. Phase III clinical studies, however, showed that the balance between efficacy and safety did not justify further development of the product.

Everninomycin (**13**), although promising, may have been a victim of antibiotic overuse. Another oligosaccharide, avilamycin, which is structurally similar, has been used as a growth promoter for food animals in the European Union. The use of avilamycin as a growth promoter has created a reservoir of isolates with decreased susceptibility to everninomycin (**13**) before this antibiotic was finally approved for human use (38).

The total synthesis everninomycin may allow semi-synthetic variants of this antibiotic to be made that are active versus multi-resistant bacteria (39).

Development of the pediatric ear infection drug, NE-1530, was terminated based on a phase II clinical trial data in which NE-1530 showed no significant benefit compared with placebo (40). Another candidate, NE-0080, was being evaluated as a treatment for stomach ulcers and other gastric complaints caused by the bacterium *Helicobacter pylori*, but after phase I trial indicated the safety (41), no further information became available.

Shiga toxin, which is produced by *Shigella dysenteriae*, and the homologous Shiga-like toxins (SLTs) of *E. coli*, can cause serious clinical complications in humans infected by these organisms. The functional toxin receptor on mammalian cells is the glycolipid Gb3 [α-D-Gal(1 → 4) β-D-Gal(1 → 4) β-D-Glc(1 → O-ceramide)], and the high incidence of complications such as acute kidney failure in children correlates with the expression of Gb3 in the pediatric renal glomerulus. Synthetic Gb3 analogs covalently attached to insoluble silica particles can competitively adsorb toxin from the gut and can serve as a starting point for the adsorbent carbohydrate drugs.

Indeed, two orally delivered, carbohydrate-based, silica-linked, gastrointestinal anti-infective therapies were advanced to clinical trials. Synsorb Pk was developed for the treatment of *E. coli* infections to prevent hemolytic uremic syndrome. The target *E. coli* variant produces verotoxin, the toxin in "hamburger disease." The other silica-linked oligosaccharide drug, Synsorb Cd, was for the treatment of *C. difficile*-associated diarrhea. Both compounds were discontinued in development.

Peramivir (**14**) is a neuraminidase inhibitor that is reported to have both an excellent safety profile and effective against influenza A and B. It is orally active, given in a once-a-day dosage, and can be made into a liquid form, allowing for use by the elderly and young children.

In a clinical challenge study, a clear dose response is seen for the primary endpoint (area under the curve versus viral titer). Doses of peramivir (**14**) ranging from 100 to 400 mg/day produced a dose-dependent, virological efficacy response compared with placebo. Peramivir (**14**) at 400 mg once daily for 5 days results in a highly significant reduction in mean viral titer area under the curve and duration of viral shedding compared with placebo.

Despite promising early data on activity, further development of SC-48334 (**15**) (*N*-butyl-deoxynojirimycin) for HIV infection was discontinued. Trials in AIDS/ARC patients indicated that the drug had no significant effect on p24 antigen levels or CD4 cell counts (42). SC-48334 (**15**) was not found to be of any ma-

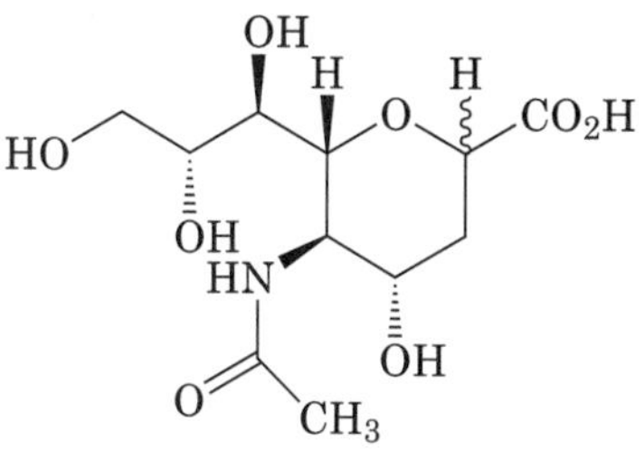

(7)

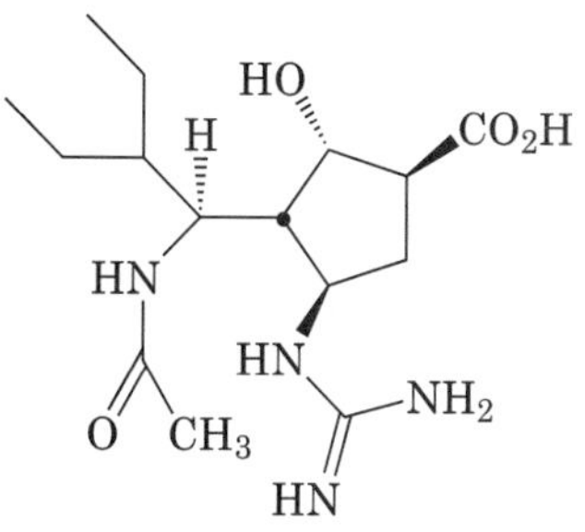

(14)

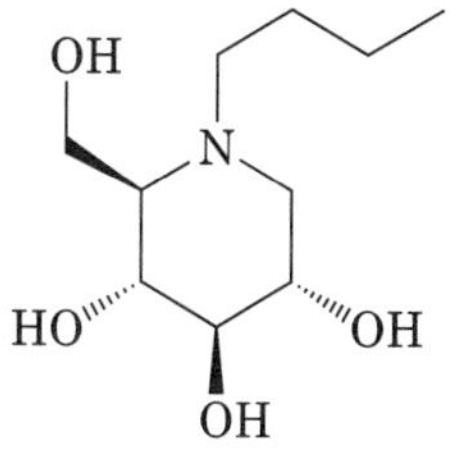

(15)

jor benefit in reducing viral load in a phase II trial (43). Although (15) did not find use in HIV, it is being developed under a different name, vevesca, for the treatment of lysosomal storage disease (see Section 2.2.7). This is an example of how understanding the glycobiology and using this knowledge can be used to discover efficacy in another indication.

Celgosivir (16) demonstrates inhibitory effect on glucosidases and has antiviral activity

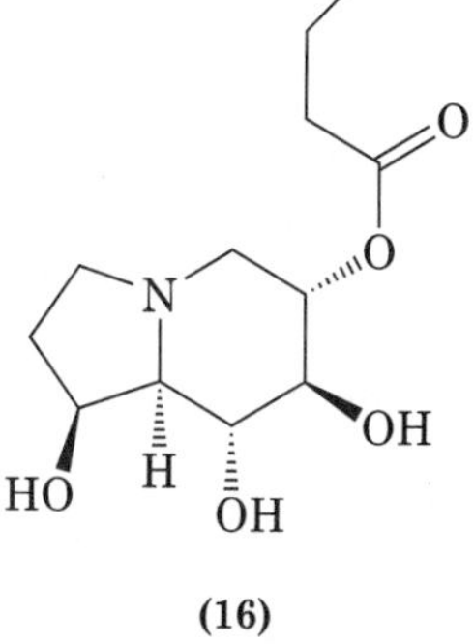

(16)

against HIV-1 and against murine leukemia virus. Celgosivir (16) was tested alone or in combination with AZT orally in AIDS patients with CD4$^+$ counts between 100 and 500 cells/

μL. Preliminary efficacy measurements showed good results for one-third of the patients, stable results for one-third, and rather poor results for the other one-third of the patients (44). No further clinical development has been forthcoming with Celgosivir (16).

A prototype vaccine against E. coli O157 triggered a strong protective response in healthy volunteers (45). Within a week of vaccination, 80% of the 87 volunteers had developed large numbers of antibodies that killed laboratory cultures of E. coli O157. The response may be rapid enough to protect people once an outbreak has begun, particularly the elderly and young children who are most vulnerable to infection.

The vaccine has two components, a carbohydrate called the O-specific polysaccharide, which appears on the surface coat of E. coli O157, and a protein from Pseudomonas aeruginosa, the bacterium that causes pneumonia.

2.2.4 Thrombosis. The enzymatic cleavage at specific sites in heparin could be important for anticoagulation therapeutics. By cleaving heparin, heparinase I inactivates heparin's anticoagulant effects. Neutralase (heparinase I) is being developed as a replacement to protamine that is currently used to reverse the effects of heparin. In contrast to protamine, which binds to heparin to form large macromolecular complexes, neutralase enzymatically breaks heparin into small, mostly inactive, fragments.

Pre-clinical studies indicate that neutralase can also inactivate the anticoagulant effects of the low LMW heparins and the synthetic pentasaccharide fondaparinux (10) (see

Section 2.1.5). Neutralase is able to neutralize these agents because it cleaves the recognition site for anti-thrombin.

OP2000 is an oligosaccharide product derived from heparin with antithrombotic and anti-inflammatory properties and is in development for inflammatory bowel disease (IBD).

IBD is a group of chronic inflammatory disorders of the intestine of unknown cause, often causing recurrent abdominal pain, cramps, diarrhea with or without bleeding, fever, and fatigue. Two forms of IBD are Crohn's disease, which affects the lowest portion of the small intestine, and ulcerative colitis, which results in inflammation of the large intestine.

Clinical observations suggest that IBD may result from increased clotting activity. Investigators have observed evidence of increased clotting in the bowel and other organs during flares of IBD. Clinical pharmacology of heparin in ulcerative colitis support the idea that heparin can safely induce remission in IBD patients. OP2000 is a product of the chemical cleavage of heparin and has the comparatively low molecular weight of 2.5 kDa. OP2000 is a potent anti-clotting agent like other LMW heparins.

2.2.5 Inflammation. Cylexin was studied for preventing reperfusion injury in infants (46). Clinical trials with cylexin were halted based on disappointing results from its ongoing phase II/III clinical trial. Cylexin showed no benefit over placebo in a 138-patient trial for the treatment of reperfusion injury in infants undergoing cardiopulmonary bypass to facilitate the surgical repair of life-threatening heart defects. Until then, cylexin was the most advanced selectin-binding inhibitor.

P-selectin is a major component in the early interaction between platelets and neutrophils, and endothelial cells, in the initial inflammatory response. A major ligand for P-selectin is P-selectin glycoprotein ligand-1 (PSGL-1), and this ligand is expressed on the surface of monocyte, lymphocyte, and neutrophil membranes.

PSGL-1 is a 240-kDa homodimer consisting of two 120-kDa polypeptide chains. The structure of functional PSGL-1 includes the sLe^x oligosaccharide. The *PSGL-1* gene en-

codes a transmembrane polypeptide rich in proline, serine, and threonine residues typical of mucin glycoproteins. The *O*-linked glycans displayed by PSGL-1 must have two specific post-translational modifications, α-1,3-fucosylation and α-2,3-sialylation, to be active as a counter-receptor for P-selectin. Bonds between P-selectin and PSGL-1 primarily mediate the rolling phase of the adhesion cascade.

A partial form of recombinant human PSGL-1 has been covalently linked to IgG to give rPSGL-Ig (47). This fusion peptide is active as a competitive inhibitor of PSGL-1, and in an animal model, pretreatment with rPSGL-Ig reduces both the thrombo-inflammatory and neointimal proliferative responses (48). As an inhibitor of neutrophil-endothelial cell adherence, rPSGL-Ig was in early clinical development for the treatment of ischemia-reperfusion injury.

A double-blind randomized, multi-center placebo-controlled, dose-ranging phase II study on the efficacy and safety of recombinant rPSGL-Ig in patients with acute myocardial infarction (MI) was assessed by positron emission tomography. The trial was negative, and product development has been discontinued for MI.

Inhibiting P-selectin may prevent not only ischemia reperfusion injury but also the subsequent clotting events that can re-occlude a vessel that's just been opened. PSGL-1 binds to a selectin-P subtype on platelet cells, causing red blood cells to stick to leukocytes and create blood clots.

In addition, inhibiting P-selectin may also prove to be useful for stopping the spread of tumors. Metastasizing tumor cells recognize P-selectin on platelets and use them as a protective shield against immune system cells. The development indications for rPSGL-Ig are not only in inflammation but also cancer.

2.2.6 Cancer. The development of carbohydrate-based vaccines designed to activate T-cells targeting of cancer cells has been pursued for some time. The goal of a cancer vaccine is to make the host immune system responsive to antigens characteristic of cancer cells. Several carbohydrate-based therapeutics that have now reached clinical trials seek to establish the molecular vaccine design in which a cell surface antigen activates T-cells (49).

Theratope is a cancer vaccine that consists of modified tumor glycopeptide antigens for patients with advanced breast cancer. The antigen is attached to Keyhole Limpet Hemocyanin (KLH) protein to increase the immune response to the vaccine. The vaccine uses a carbohydrate antigen known as STn, part of a larger antigen, mucin-1, found on breast cancer cells. Theratope is in phase III human trial, which has as an endpoint of increased survival.

Earlier studies in humans provide strong evidence of the development of an anti-tumor T-cell response after immunization with a cancer-associated carbohydrate antigen (50).

The hexasaccharide, globo H, is found on cancer cells and is an antigen for vaccine design. Globo H occurs on cell surface glycoproteins of breast, prostate, and other epithelial cancers, and the expressed levels of these antigens are magnified in the context of such tumors. Chemists have synthesized globo H, and, as in the case of Theratope, attached it to KLH to augment the antigen's immunogenicity (51).

The globo H vaccine is being tested in clinical trials to determine whether it will block the reappearance or progression of cancer. Thus far, in a group of 27 women with metastatic breast cancer, globo H raised blood levels of IgM antibodies. In 16 of these patients, IgM antibodies bound to the cancer cells, and some patients showed evidence of cancer cell death (52).

Similarly, prostate cancer patients who relapsed after surgery or radiation therapy are being treated with the globo H tumor vaccine (53). The vaccine was established as safe and capable of inducing specific antibody response to globo H. More importantly, the immune response to the vaccine based on synthetic globo H-protein conjugate also recognized naturally occurring cancer-derived glycoprotein.

An effective treatment for the deadly skin cancer melanoma is clearly needed. Current best therapy is surgical removal of high-risk melanoma and treatment with Interferon alfa. Melanoma patients have a 50–80% chance of relapse without adjuvant treatment. The GMK vaccine is in development for melanoma

and contains the ganglioside GM2 linked to the carrier protein KLH and combined with the adjuvant QS-21.

GMK vaccine is capable of producing a specific immune response against a defined and chemically purified melanoma antigen. Anti-GM2 antibodies in melanoma patients induced by the GMK demonstrate binding to melanoma cells, complement-mediated cytotoxicity, and antibody-dependent cellular cytotoxicity (54).

GMK vaccine was compared with high-dose Interferon alfa 2b in high-risk melanoma patients who had undergone surgery. Interferon-treated patients were 50% less likely to experience a return of their disease and 52% less likely to die than were GMK-treated patients (55).

Another vaccine, MGV, is being developed for several other cancer indications, including colorectal, lymphoma, small-cell lung, sarcoma, gastric, and neuroblastoma. The cell surface gangliosides GM2 and GD2 are expressed on several cancers, including melanoma. GM2 is immunogenic, and induction of anti-GM2 antibodies correlates with improved survival in stage III melanoma patients with no evidence of disease after surgery. GD2 is immunogenic when coupled to KLH and mixed with the adjuvant QS21. The development of GMK and MGV cancer vaccines seems to be on hold.

Other carbohydrate-based drugs in development to treat cancer include a sulfated oligosaccharide, a pectin derivative, and an azasugar. PI-88 is a sulfated oligosaccharide, GAG mimetic that inhibits growth factors and heparanase produced by tumors and has anti-angiogenic and anti-metastatic properties. In pre-clinical studies, PI-88 retarded the growth of primary tumors by inhibiting angiogenesis, and in a rat model, it showed a dose-dependent inhibition of tumor growth in the BC1 syngeneic rat mammary adenocarcinoma (56). This inhibition is thought to be the result of its ability to inhibit endothelial cell proliferation by preventing basic fibroblast growth factor (bFGF) and vascular endothelial growth factor (VEGF) from binding to the receptors on endothelial cells. PI-88 also inhibits heparinase secreted by a variety of cells, thus preventing the degradation of extra-cellular matrix and the release of growth factors.

A multi-center phase II clinical trial to further investigate the efficacy and safety of PI-88 in patients with multiple myeloma was started. The trial will also gather information about PI-88's ability to inhibit the blood vessels formation in tumors.

Carbohydrates have demonstrated an increasing role in tumor biology, with a known correlation between the expression of the galectin 3 receptor and the propensity for metastatic spread in pre-clinical models (57). The carbohydrate GCS-100 is a pectin derivative that binds to the galectin 3 receptor. Pectin is the methylated ester of polygalacturonic acid.

In pre-clinical models, GCS-100 is able to inhibit the metastatic spread of tumor lines to the lung and is able to shrink large tumors. In pre-clinical toxicology studies, the dose limiting effect of GCS-100 is pulmonary insufficiency.

GCS-100 is in phase II clinical trials for pancreatic and colorectal cancer. Phase IIa data in failed metastatic pancreatic and colorectal cancer patients warranted proceeding into larger studies.

Oligosaccharide moieties of cell-surface glycoproteins are involved in recognition events associated with cancer metastasis. GD0039 (**17**), an inhibitor of Golgi α-manno-

(17)

(15)

sidase II, is used to examine the role of glycan structures in metastasis (58). GD0039 (**17**) blocks pulmonary colonization by tumor cells and stimulates components of the immune system, such as macrophages and splenocytes. Furthermore, GD0039 (**17**) abrogates much of the toxicity of commonly used chemotherapeutic agents in both healthy and tumor-bearing mice. GD0039 (**17**) is in phase II clinical

trials in metastatic renal cancer and 5-FU resistant colorectal cancer.

2.2.7 Lysosomal Storage. Lysosomal storage diseases (LSDs) are genetic errors of metabolism resulting primarily from the absence of an enzyme whose target is a substance to be lowered in cellular tissues. The build-up of these substances causes a loss of function in one or several crucial areas of the body and may result in mental and physical disability, or in most cases, shortened lifespan. A partial list of LSDs includes glycolipidosis disorders, such as Fabry's and Gaucher's diseases (discussed in Section 2.1.7), and mucopolysaccharidosis (MPS), such as MPS I and VI diseases. Both small molecule enzyme inhibitors and large molecule enzyme replacement approaches are being taken to treat LSDs.

OGT-918 (15) (vevesca, N-butyldeoxynojirimycin) is in clinical development for the treatment of Gaucher's and Fabry's diseases. OGT 918 is a small molecule drug and as such can be given orally. Analysis of a 6-month study to investigate the potential of OGT 918 treatment in Fabry's patients who had been receiving enzyme therapy (see agalsidase alfa, Section 2.1.7) indicates that the patients were successfully maintained on oral therapy alone during the study period.

It will be of interest in the future to contrast the success of enzyme replacement that requires infusion with orally delivered drugs for the treatment of Gaucher's and Fabry's diseases. Oral drugs are clearly more convenient; however, the side effects and long-term toxicity must be determined.

Long-term data from type 1 Gaucher patients at 24 months show a progressive improvement on the results previously obtained after 12 and 18 months of treatment with OGT-918. Those Gaucher patients that reached the 24-month time point have continued into their third year of therapy. Chitotriosidase activity, as a surrogate of activity in patients, is a biochemical marker related to the disease.

Aldurazyme is an enzyme replacement therapy for the treatment of MPS I disease. Phase III results with aldurazyme warranted filing for marketing approval in 2002. MPS I is a genetic disease caused by the deficiency of

α-L-iduronidase, an enzyme normally required for the breakdown of GAGs. The normal breakdown of GAGs is incomplete or blocked if the enzyme is not present in sufficient quantity. The accumulation of GAGs in the lysosomes of the cell cause MPS I.

Patients with MPS I are usually diagnosed in childhood and get progressively worse, leading to severe disability and early death. During the course of the disease, the build-up of GAGs results in inhibited growth and mental development, impaired vision and hearing, reduced cardiovascular and pulmonary function, and joint deformities. About 3000–4000 patients in developed countries have MPS I.

Arylsulfatase B (N-acetylgalactosamine 4-sulfatase) is a deficient enzyme in patients with the genetic defect disease MPS VI. Like MPS I, GAGs are only partially broken down, and carbohydrate residues build relentlessly in the lysosomes of cells. Also like α-L-iduronidase replacement, the exogenous supply of arylsulfatase could breakdown the stored GAGs and cellular function could be restored.

Aryplase is a specific form of the recombinant human enzyme, arylsulfatase B. In a phase I trial, the enzyme was well tolerated, and there were no drug-related serious adverse events and no significant allergic reactions to the infusions. Urinary GAG excretion was reduced by a mean of 70% in the high dose group and 55% in the low dose group. Urinary GAG excretion is a biomarker for MPS.

Long-term observation of the development process in the pharmaceutical industry indicates that roughly one of every six therapeutics (17%) submitted as investigational new drug applications go on to be filed successfully as New Drug Applications with the U.S. Food and Drug Administration. In Table 7.2 , 15 of 26 carbohydrate-based therapeutics (58%) are still in development. Clearly, more carbohydrate-based therapeutics in development will be discontinued; however, if the industrial average holds, at least four of the therapeutics in development will go on to be successful drugs.

2.3 Agents in Research

Research agents are the source of new drugs. Research on carbohydrate-based agents that may have use as drugs is increasing; however, getting specific structural information is more difficult. Large pharmaceutical companies that are a traditional source of new drugs, including the early publication of structures, have not wholly endorsed carbohydrate chemistry or glycobiology. In contrast, small biotechnical companies that have focused on these technologies aggressively advertise "products in the R&D pipeline" without disclosing specific molecular information that describe the potential products.

Areas of special interest in carbohydrate drug research include the treatment of *Helicobacter pylori* gastritis and prion-induced CNS disease. The following is organized in the approximate order as Sections 2.1 and 2.2, starting with diabetes research that continues to attract new carbohydrate-based agents.

2.3.1 Diabetes. DYN 12 is a small molecule that reduces plasma concentrations of 3-deoxyglucosone (3DG) in diabetic rats by 50%. 3DG is a highly reactive molecule that can cross-link proteins, causing them to change or lose function, including the inactivation of certain enzymes. 3DG is also a proven factor in the development of advanced glycation end products (AGEs), which are implicated in the development of diabetic kidney disease, arteriosclerosis, and other diseases. Amadorase or the closely related fructoseamine-3-kinase is the putative enzyme responsible for the formation of most of the 3DG in the body (59).

2.3.2 Gastrointestinal System. The human specific gastric pathogen, *H. pylori*, has emerged as the causative agent in chronic active gastritis and peptic ulcer disease (60). The bacterium is unique in its carbohydrate-binding complexity. The complete genome sequence of *H. pylori* reveals a surprisingly large part coding for outer surface proteins, reflecting a complex interrelation with the environment. Specific binding molecules are many and include polyglycosylceramide (61), and sialic acid conjugates (62). Recently, a sialylated receptor analog was evaluated in clinical trials with *H. pylori*-infected patients with promising results (63).

2.3.3 CNS. Prions are composed exclusively of a single sialoglycoprotein called PrP. They contain no nucleic acid, have a mass of 30

Table 7.2   Therapeutics in Development

Name Generic (Code)	Indication	Company	Development
Mitemcinal (**11**) (GM-611)	Gastroparesis; gastroesophageal reflux	Chugai	Phase II
EM-574	Chronic gastritis gastroesophogeal reflux	Takeda, Kitazato	Phase II
Sygen (**12**) (GM-1)	Parkinson's disease, spinal cord injury	Fidia	Phase II
Everninomicin (**13**) (Ziracin, SCH 27899)	Antibiotic	Schering-Plough	Discontinued (phase III)
NE-1530	Pediatric ear infection	Neose	Discontinued (phase II)
NE-0080	*Helicobacter pylori* infection	Neose	Discontinued (phase I)
Synsorb Pk	*E. coli* O157:H7 Hemolytic uremic syndrome	Synsorb	Discontinued (phase III)
Synsorb Cd	*C. difficile* associated diarrhea	Synsorb	Discontinued (phase III)
Peramivir (**14**) (RWJ-270201)	Anti-viral	Johnson & Johnson, BioCryst	Phase III
SC 48334 (**15**) (N-Butyl-deoxynojirimycin)	HIV/ARC	Searle	Discontinued (phase II)
Celgosivir (**16**) (MDL 28574, DRG-0202)	HIV	Aventis	Discontinued (phase II)
E. coli O157 vaccine,	*E. coli* O157 infection	NICHHD	Phase I
Heparinase I (Neutralase)	Anti-coagulation reversal	BioMarin	Phase II
Deligoparin (OP2000)	Antithrombosis, Inflammatory bowel disease	Opocrin, Incara, Elan	Phase III
Cylexin	Reperfusion injury	Cytel	Discontinued (phase II)
rPSGL-Ig, PSGL-1	Heart attack	Wyeth	Discontiued (phase II)
Theratope (Sialyl Tn Ag conjugate vaccine)	Metastatic colorectal, breast cancer	Biomira	Phase III
Globo H conjugate vaccine	Breast, prostate cancer	Memorial Sloan-Kettering Cancer Center	Phase I
GMK (GM2 KLH/QS-21 conjugate vaccine)	Malignant melanoma	Progenics, Bristol-Myers Squibb	Discontinued (phase III)
MGV (GM2/GD2 KLH QS21 conjugate vaccine)	Colorectal, gastric, small-cell lung, cancer	Progenics, Bristol-Myers Squibb	Discontinued (phase II)
PI-88	Multiple myeloma, Tumor angiogenesis	Progen, Medigen	Phase II
GCS 100, (GBC-590)	Pancreatic cancer, adenocarcinoma	SafeScience	Phase II
GD0039 (**17**) (Swainsonine)	Renal, colorectal, breast cancer	GlycoDesign	Phase II
Vevesca 15 (OGT 918, N-butyldeoxy-nojirimycin)	Fabry's, Gaucher's disease	Oxford GlycoSciences	Phase III
Alpha -L- iduronidase (Aldurazyme)	MPS I	BioMarin, Genzyme	Phase III
Arylsulfatase B (Aryplase)	MPS VI	BioMarin	Phase II

kDa, and are composed of 145 amino acids. This protein polymerizes into rods possessing the structural characteristics of amyloid. Amyloid protein is deposited in many human CNS diseases including Alzheimer's, Creutzfeldt-Jakob, and Down's syndrome.

PrP undergoes several post-translational events to become the prion protein including glycosylation at amino acids 181 and 197, and an addition of a phosphatidylinositol glycolipid at the C-terminal. A conformational change in the PrP from an α-helix to a β-sheet induces the same α-helix to β-sheet conformation change in the other PrP protein. This is a permanent conformational change and induces replication. The β-sheet–forming peptides aggregate to form amyloid fibrils, and the amyloid fibrils kill thalamus neurons through apoptosis.

In the absence of evidence for any significant covalent modifications that can distinguish PrPC (α-helix) from PrPSc (β-sheet), focus has shifted to the difference in secondary and tertiary structure between the two forms of the prion protein.

Glycosylation may modify the conformation of PrPC (64). Glycosylation could also affect the affinity of PrPC for a particular conformer of PrPSc, thereby determining the rate of PrPSc formation and the specific patterns of PrPSc deposition. Variations in the complex carbohydrate structure in the CNS could account for selective targeting in particular areas of the brain.

Sulphated GAGs, such as heparin, keratin, and chondroitin, inhibit the neurotoxicity of amyloid fibrils (65). This seems to be mediated by the inhibition of the polymerization of the PrP peptide into fibrils. These findings provide a basis for using sulphated polysaccharides for CNS therapy.

2.3.4 Infection. Vancomycin (**18**) is a relatively small glycoprotein derived from *Nocardia orientalis*, which is well known for its activity against Gram-positive bacteria including *Streptococci*, *Corynebacteria*, *Clostridia*, *Listeria*, and *Bacillus*.

Modifying vancosamine residue on vancomycin (**18**) could increase the antibiotic's activity against vancomycin-resistant strains. One chemical approach is to sequentially protect the functional groups in the molecule according to their reactivities, and then selectively cleave the glycosidic linkage to vancosamine. The result is a protected pseudoaglycon. The goal is to use sulfoxide glycosylation chemistry to reattach vancosamine derivatives and produce glycosylated vancomycin (**18**) analogs with increased antibiotic activity and activity against resistant microorganisms (66).

Olivomycin A (**19**) is a member of the aureolic acid family and is derived from *Streptomyces*. A recent synthesis of olivomycin A (**19**), involving a convergent synthesis in which preassembled sugar units are attached in specific order to the aglycone, has renewed interest in this class of agents (67). The methodology can be used for synthesis of aureolic acid analogs that may prove useful as cancer chemotherapeutic agents. It may provide tools in studies of the molecular basis of the binding to the minor groove of DNA. Also important is the development of methods for synthesizing and manipulating the unusual monosaccharides found in the aureolic acid family, which lack one of the hydroxyl groups found in most sugars.

Pathogens from the *Staphylococcus* genus are found in both the hospital and the community, and the most prevalent species, *aureus*, causes illnesses that range from minor skin abscesses to severe pneumonia, meningitis, and infections of the heart, bloodstream, bone, and joint. Multiple strains of *S. aureus* are now antibiotic resistant, including a few strains that are partially resistant to vancomycin (**18**), the last effective antibiotic against *S. aureus*. New treatments for *S. aureus* infections are clearly needed.

One approach is to discover an *S. aureus* vaccine. Usually, antigens isolated from bacteria grown in the laboratory are new vaccine candidates. However, the consideration that bacteria grown under laboratory conditions may have different antigens than the same bacteria in actual infection, led to the isolation of a promising vaccine antigen, poly-*N*-succinyl-β-1–6-glucosamine (PSG) (68).

Purified PSG injected into rabbits produced large amounts of PSG antibodies that persisted for at least 8 months. When PSG antibodies were injected into mice and ex-

(18)

posed to eight different strains of *S. aureus*, including strains partially resistant to vancomycin (**18**), none of the animals developed an infection.

An interesting observation is that small cell lung cancer (SCLC) and *N. meningitidis* both have polysialic acid (polySA) cell surface antigens, and research is being directed toward the generation of vaccines that may have use in these diseases. SCLC is responsive to therapy; however, relapses are common and the use of immunization against polySA antigenic targets on SCLC cells could improve outcomes. Although serogroup A and C vaccines exists, the polySA expressed by serogroup B meningococci makes this pathogen invisible to immune surveillance.

Replacing the *N*-acetyl chemical moiety on polySA with an *N*-propionyl moiety gives nppolySA, and it is able to stimulate an immune response.

PolySA or the modified version, nppolySA, was conjugated to the immunogenic carrier protein KLH and mixed with the immune adjuvant QS-21 (69). After achieving a major response on standard therapy, SCLC patients were vaccinated. Serologic analysis indicated

no reactivity in the unmodified polySA vaccine; however with nppolySA, all patients developed antibodies against polySA. Some patients developed strong reactivity against the polySA positive SCLC cell line H69. Interestingly, sera from five of six patients in the nppolySA vaccine group also exhibited reactivity against serogroup B meningococci.

Hepatitis B infection occurs worldwide, and the resulting morbidity and mortality is significant. Current therapy includes interferon and nucleoside inhibitors (lamivudine) treatment.

In vitro, *N*-nonyl deoxynojirimycin (NN-DNJ; **20**) inhibits hepatitis B virus ability to construct its M envelope protein in human liver cells. Apparently, minor cellular carbohydrate processing disruption (6%) results in a greater than 99% reduction in the secretion of hepatitis B virus. *In vivo*, using the standard woodchuck model for HBV with the woodchuck hepatitis virus, treatment with NN-DNJ (**20**) induced a loss of viremia (70). Not unexpectedly, after treatment ceased, viral titers returned to pretreatment levels.

NN-DNJ (**20**), in a dose-dependent manner, does alter the ability of the virus to be

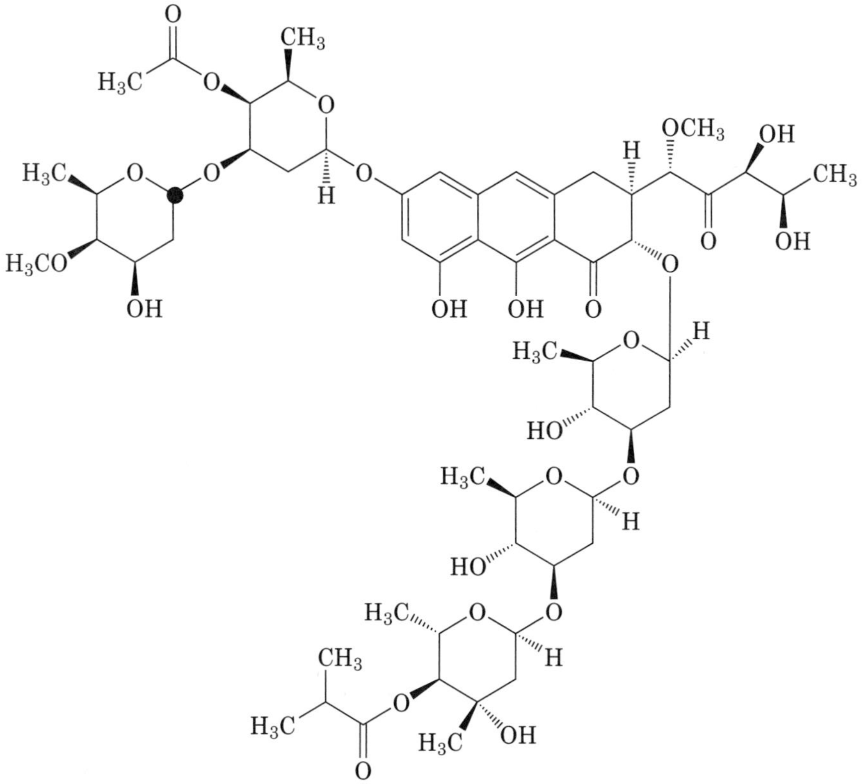

(19)

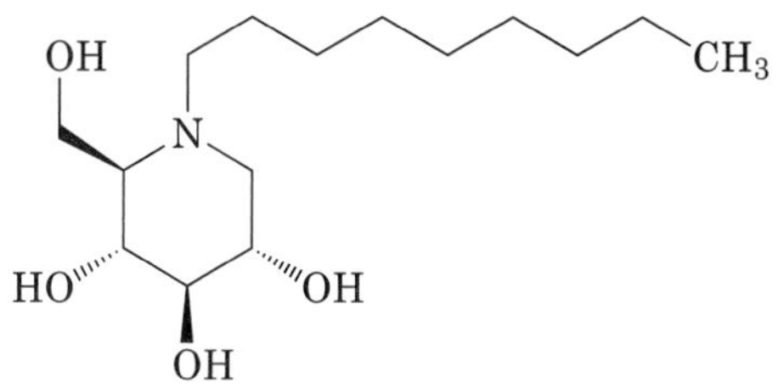

(20)

secreted in an infectious form. Presumably the mechanism involves an interaction with a cellular trafficking protein and an improper folding of the altered hyper-glycosylated envelope protein.

2.3.5 Inflammation. Adhesion of cells in the immune system is clinically important when considering such indications as cancer, microbial infection, and inflammatory, allergic, and autoimmune disease, including arthritis, allergies, and reperfusion injury. Selectins are a family of glycoproteins that are involved in the adhesion of leukocytes to platelets or vascular endothelium (71). Adhesion is an early step in leukocyte extravasation in which sequelae includes thrombosis, recirculation, and inflammation. Three protein receptors, E-, L-, and P-selectins, are assigned to the selectin family based on their cDNA sequences. Each contains a domain similar to calcium-dependent lectins or C-lectins, an epidermal growth factor-like domain, and several complement binding protein-like domains (72).

Research to define the native carbohydrate ligands for each selectin receptor continues (73). The carbohydrate ligand for E-selectin on leukocytes contains the sLe[x] epitope (74). Because the C-type lectin domains of E- and P-selectins are highly homologous, it was no surprise that sLe[x] serves as a ligand also for P-selectin (75).

Another P-selectin glycoprotein ligand is PSGL-1, a 220-kDa glycoprotein found on leukocytes (see Section 2.2.5) (76). PSGL-1 has a sLe^x group on its carbohydrate side-chains and contains at least one sulfated tyrosine residue near the N-terminus. The sLe^x group and sulfated tyrosine residue are necessary for P-selectin to bind the cells, whereas E-selectin requires presence of only sLe^x group.

L-selectin is the smallest of the vascular selectins, a 74- to 100-kDa molecule, and is constitutively expressed on granulocytes, monocytes, and many circulating lymphocytes. L-selectin is important for lymphocyte homing and adhesion to high endothelial cells of post-capillary venules of peripheral lymph nodes. Moreover, this adhesion molecule contributes greatly to the capture of leukocytes during the early phases of the adhesion cascade (77). After capture, L-selectin is shed from the leukocyte surface after chemoattractant stimulation. L-selectin interacts with three known counter receptors or ligands, MAdCAM-1, GlyCAM-1, and CD-34.

Another major ligand for L-selectin is a sulfated form of sLe^x, sialyl 6-sulfo Lewis X (78). This determinant is expressed on HEV in human lymph nodes where it recruits naive T-lymphocytes expressing L-selectin. The cell adhesion mediated by L-selectin and sialyl 6-sulfo Lewis X triggers the action of chemokines and activates lymphocyte integrins, leading the cells to home into lymph node parenchymas.

Considering the forgoing discussion, the blocking of selectin-ligand binding with glycomimetics is a seductive approach to the discovery of novel therapeutic agents.

Structure-activity data on sLe^x analogs have suggested important chemical features for selectin binding. By combining the structural data obtained from the identification of the natural selectin ligands with synthetic structure-activity information, important chemical features for ligand binding to E-, P-, and L-selectin can be recognized.

In general, sulfated oligosaccharides have a higher affinity for proteins than the corresponding non-sulfated precursors. In fact, sulfation of sLe^x does increase its affinity for L-selectin (79, 80).

Similarly, a chemoenzymatic route was used to generate sLe^x -substituted glycopeptides to elucidate the critical binding epitope of PSGL-1, the physiological ligand for P-selectin (81). In a P-selectin inhibition assay, macrocyclic sLex analog exhibits a dramatic enhancement (100 times) relative to sLe^x (82).

With knowledge of the importance of sLe^x functional groups for complexation, mimics of the tetrasaccharide have been synthesized

(21)

that are potent selectin antagonists (83). For example, a tetrasaccharide mimic is greater than 50 times more active at blocking E-selectin than sLe^x (84). These data underscore the tremendous progress made in generating efficacious glycomimetics.

A rational design and synthesis of (3-O-carboxymethyl)-β-D-galactopyranosyl-α-D-mannopyranoside (**22**) provided a disaccha-

(**22**)

ride that is five times as active as sLe^x in binding to E-selectin and is also effective against P- and L-selectin (85). A new method for the 1,1-glycosidic bond formation through coupling of protected trimethylsilyl-β-D-galactoside and α-mannosyl fluoride in the presence of boron trifluoride-etherate is also described.

Certain tetravalent sLe^x-glycans are extremely potent inhibitors of L-selectin-dependent lymphocyte traffic to sites of inflammation with IC$_{50}$ values in the nanomolar range. Large-scale synthesis of the lead compounds is needed for *in vivo* experiments in models if acute inflammation such as reperfusion injury and rejection of organ transplants.

2.3.6 Cancer. Basic research on the structure of antigens has stimulated new concepts that are being converted into experimental vaccines. In general, malignant transformation of cells is accompanied by an increase in cell surface carbohydrates, resulting not only in unusual multiple antigens, but also in a serial clustering of antigens.

The observation of unusual carbohydrate antigens led to the synthesis of glycoconjugates that mimic unusual carbohydrate structural antigens on tumor cell surface. Synthetic

oligosaccharides and glycoconjugates have been shown to trigger humoral responses in murine and human immune systems. The potential of inducing active immunity with a fully synthetic carbohydrate-based vaccine is now possible, particularly if vaccine compounds can be synthesized that resemble the surface environment of transformed cancerous cells.

For example, this strategy is the basis for the convergent total synthesis of a tumor-associated mucin motif (86).

In the pursuit of optimal carbohydrate-based anticancer vaccines, a multi-antigenic, unimolecular glycopeptide containing the Tn, MBR1, and Lewis y antigens was prepared. The three carbohydrate antigens were linked together to form a single molecule that in turn was conjugated to a KLH carrier protein (87). However, in the initial development of such a vaccine, serious regulatory issues have been raised.

The effects of epitope clustering, carrier structure, and adjuvant on antibody responses to Lewis y conjugates were examined in mice (88). It is thought that tumor-associated glycoproteins are often presented in clustered form. That is, an antigenic carbohydrate appendage is attached to a serine or threonine, and the resulting carbohydrate-amino acid motif is repeated in a series of residues. To more closely approximate a tumor cell surface, a carbohydrate construct was synthesized that presents Lewis y on three consecutive amino acids.

The results in mice showed that the clustered presentation is strongly antigenic, and that the presentation mode of the vaccine construct could be simplified because a KLH carrier was not needed.

Like other carbohydrate-based antigens, production of the carbohydrate β-1,6-GlcNAc-branched N-glycans is elevated in human malignancies of breast, colon, and skin cancers (89). The gene Mgat5 encodes N-acetylglucosaminyltransferase V (GlcNAc-TV), the Golgi enzyme required in the biosynthesis of β-1,6-GlcNAc-branched N-glycans.

With Mgat5 knockout mice, there is observed a suppression of cancerous tumor growth and the spread of tumor cells to the lung (90). The mutant mice deficient in Mgat5 seem normal; however, they react differently

Table 7.3   Agents in Research

Name Generic (Code)	Indication	Mechanism	Company/Institution
DYN 12	Diabetes	Amadorase/ fructoseamine-3- kinase inhibition	Dynamis Therapeutics
Vancomycin (**18**) analogs	Anti-bacterial	Cell wall biosynthesis Inhibition	Princeton University
Olivomycin A (**19**)	Anti-bacterial, anti-cancer	DNA minor groove binding	University of Michigan
S. aureus vaccine, (NSG, poly-N-succinyl-β- 1–6 glucosamine)	Anti-bacterial	Vaccine	Harvard Medical School NIAID
Polysialic acid, nppolySA, KLH QS-21	Meningitis, cancer	Vaccine	Memorial Sloan-Kettering Cancer Center
N-nonyl deoxynojirimycin (**20**) (NN-DNJ)	Anti-viral	Inhibition of glycoprotein folding	IgX Oxford Hepatitis
Tn MBR1 Lewis y KLH	Anti-cancer	Multi-antigenic, unimolecular vaccine	Memorial Sloan-Kettering Cancer Center
Galactopyranosyl-α D- mannopyranoside (**21**)	Anti-inflammatory	Leukocyte traffic	UC San Diego
Lewis y epitope	Anti-cancer	Lewis y clustering vaccine	Memorial Sloan-Kettering Cancer Center
β-1,6-GlcNAc- branched N-glycans	Anti-cancer	GlcNAc-TV modulation, Mgat5 gene	University of Toronto, GlycoDesign
ManLev (**22**)	Anti-cancer	Toxins or immune recognition	UC Berkeley

from other mice when exposed to a powerful gene that causes cancer. The Mgat5-deficient mice are resistant to the induction in breast cancer growth and metastasis to the lungs compared with other mice, indicating that this carbohydrate structure plays a role in promoting the growth and spread of cancer. The *Mgat5* gene in cancers promotes cell movement and directs involvement of carbohydrate chains in cancer growth.

New drugs designed to mimic β-1,6-GlcNAc-branched N-glycans or modulate GlcNAc-TV could be used to either to dampen T-cells in autoimmune disease or to sensitize the immune system in the treatment of cancer and infections.

Experimental work on a tumor-targeted, keto-carbohydrate–based technology is ad-vancing from *in vitro* to *in vivo* experiments (91). Ketone-containing monosaccharides can serve as substrates in the oligosaccharide bio-synthesis (92). Briefly, cells incubated with a ketone-containing monosaccharide produce ketone-tagged cell-surface oligosaccharides. Because cell surface ketones are rare in cells, conjugation with keto-selective nucleophiles is possible. For example, sialic acids are abundant terminal components of oligosaccharides on mammalian cell-surface glycoproteins and are synthesized from the six-carbon precursor N-acetylmannosamine. When HeLa cells in culture are incubated with N-levulinoyl-D-mannosamine (ManLev **23**), the ketone ends up on a cell surface. On conjugation with the ricin toxin, the cell dies. The killing of ManLev (**23**)-labeled HeLA cells with ricin is depen-

(23)

dent on the number of ketones expressed on the cell surface. Using this method, it may be possible to make cells susceptible to toxins or immune recognition.

Saccharides of glycoconjugates found on cell surfaces can modulate cellular interactions and are under investigation as research agents. In a similar fashion, saccharide agents or their glycomimetics can modulate carbohydrate biosynthetic and processing enzymes that can block the assembly of specific saccharide structures. Modulators of carbohydrate recognition and biosynthesis can reveal the biological functions of the carbohydrate epitope and its target receptors. Carbohydrate biosynthetic pathways are often amenable to interception with synthetic saccharides, corresponding glycomimetics or xenobiotic natural products. Table 7.3 summarizes carbohydrate based agents in research.

3 CARBOHYDRATE CHEMISTRY

Carbohydrate-based therapeutics is now established, and the research and development pipeline for these agents is well outlined, all of which promises a robust future for carbohydrate drugs.

3.1 Chemistry

To fully realize the potential of carbohydrate-based therapeutics, new chemical methods to discover, characterize, and supply carbohydrates is needed.

3.1.1 Challenges and Opportunities. The one characteristic of carbohydrates that most vividly demonstrates both the opportunities and challenges of carbohydrate chemistry is the potential of this class of biomolecules to form diverse structures.

With oligosaccharides, carbohydrates of low-to-intermediate complexity, the shear number of possible isomers becomes difficult to calculate. For example, considering a relatively small tetrasaccharide and using only nine common monosaccharides found in humans, there is greater than 15 million possible isomers that can be assembled (93).

Traditionally, carbohydrates have been prepared by synthesis or natural product isolation.

3.1.2 Solution Phase. In general, saccharides are a challenge to synthesize because monosaccharides have both regiochemistry and stereochemistry in both open and closed ring forms. If monosaccharides are combined, either α- or β-covalent bonds can link the resulting molecules. And if many monosaccharides are covalently bonded, the linkages may be linear or branched, including multiple branch points with a given monosaccharide. Also, oligosaccharide synthesis requires multiple selective protection and deprotection steps.

The inevitable comparison of polysaccharides to other biopolymer types helps put the challenge into perspective. Nucleic acids are made in a linear manner through chemical and biological synthetic techniques. Likewise, protein sequences are also linear and can be easily determined, produced, and manipulated through a combination of recombinant DNA technology and solid phase synthesis.

Nature has evolved very sophisticated mechanisms for making exact polypeptides and polynucleotides. In contrast, saccharides are made in an apparent random fashion with a diverse set of enzymes competing to produce very diverse products. Although random in appearance, carbohydrates do possess precise structures that are on the surface of complex large molecules and that affect function. These precise carbohydrate structures are targets of isolation and synthesis.

The creation of the glycosidic linkage between carbohydrate monomers is a strategically important step in oligosaccharide synthesis and requires stereocontrol of the newly formed linkage. Glycosides that have a *trans* relationship of the C-1 and C-2 substituents are synthesized by taking advantage of neigh-

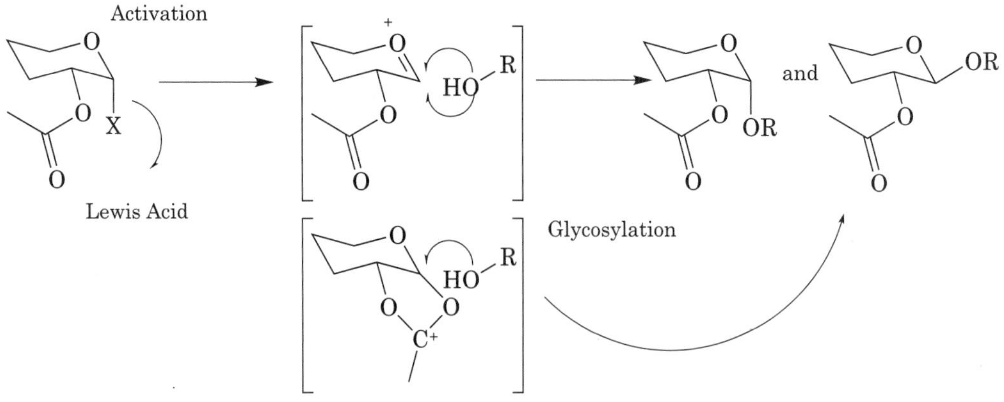

Scheme 7.1.

boring group participation from the 2 position. With acyl type protecting groups at C-2, the reaction proceeds through an acyloxonium cation intermediate that reacts with a nucleophile to give 1,2-*trans* stereochemistry in the product, whereas the oxocarbonium ion is expected to give mixtures of the 1,2-*cis* and 1,2-*trans* products (Scheme 7.1). For the synthesis of 1,2-*cis* glycosides, a prerequisite is the use of non-participating blocking groups at the 2 position. Using a non-participating substituent, however, does not necessarily result in the formation of 1,2-*cis* glycosides; the stereoselectivity depends on the chemical reactivity of the glycosyl donor and acceptor, the strength of the promoter used for the reaction, the solvent, and the reaction temperature. Note that the protecting groups used influence the chemical reactivity of the glycosyl donors and acceptors, and in some cases, changing a single protecting group at a remote position may result in changes in stereoselectivity and yields.

The Koenigs-Knorr method of coupling glycosyl halides (X = halogen in Scheme 7.1) is one of the most widely used techniques (94). The relative instability of the sugar halide necessitates the construction of the saccharide from the reducing end. Alternative leaving groups have been developed that are more stable.

Trichloroacetimidates and glycosyl sulfoxides are now commonly used for coupling (95, 96). Other coupling functional groups include phosphites, phosphates, thioglycosides, and pentenyl glycosides (97–99).

The multi-functional nature of the carbohydrate molecules requires the use of protecting groups on hydroxyls or other functional groups not involved in the reaction to provide an unambiguous synthesis. Blocking groups are required for functional groups that are not involved in the whole synthetic scheme, whereas temporary protecting groups are required on functional groups that must be manipulated at a later stage of the synthesis. Benzyl ethers are most commonly used as blocking groups in combination with temporary acyl protecting groups.

Carbohydrates with an *N*-acetyl group at the 2 position are found in many important biological structures; however, methods for synthesizing such sugars require many steps and often yield complex mixtures of stereoisomers. A single vessel method that allows easier access to this key class of oligosaccharides is now available (100). The method uses a new sulfonium reagent that activates glycals for direct addition of the acetamido group and coupling with a variety of glycosyl acceptors.

Note that *N*-silylated amide must be used to carry out the nitrogen-transfer process and use of unprotected primary amides as nitrogen nucleophiles does not seem to be effective.

Two major advances used to streamline the chemical synthesis of oligosaccharides are one-pot reactions (101) and polymer-supported synthesis (see Section 3.1.4) (102).

One promising approach at developing automated synthesis uses single-vessel reaction schemes that take advantage of the reactivity

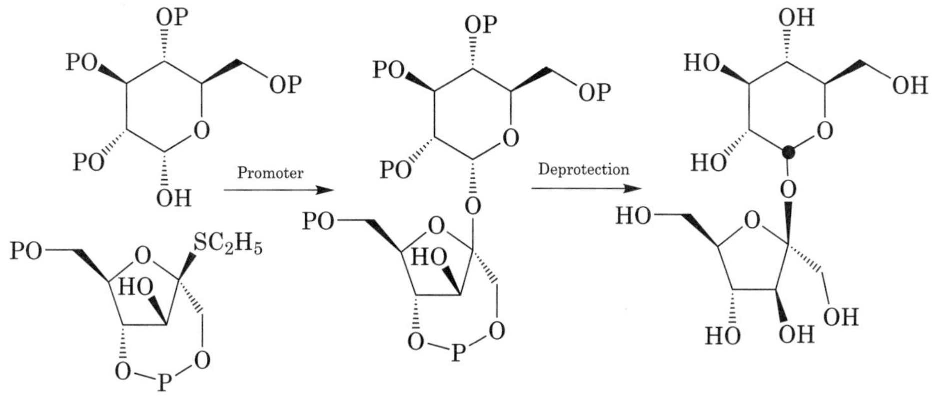

Scheme 7.2.

profile of different protected monosaccharides to determine the outcome (103). The reactivity of a monosaccharide is highly dependent on the protecting groups and the anomeric-activating group used. By adding substrates in sequence from most reactive to least reactive, the predominance of a desired target compound is assured.

The key to this approach is to have extensive quantitative data regarding the relative reactivities of different protected monosaccharides. Thus, a reactivity database on protected *p*-methylphenyl thioglycosides was generated and used as the basis of a computer program that selects the best reactants for one-vessel synthesis of oligosaccharides (104).

This approach has been used in the synthesis of a large number of oligosaccharides, including the cancer antigen globo H hexasaccharide (see Section 2.2.6) (105). Work is needed to design a complete set of building blocks for use in the synthesis of most bioactive saccharides. By using disaccharide thioglycosides as reactants in the linear scheme, branch points are incorporated. These reactions are typically performed in solution.

A novel exception to generalities above is demonstrated in the stereospecific synthesis of sucrose (Scheme 7.2) (106). The problems faced by the sucrose synthesis are that the two monosaccharides are connected through tertiary anomeric centers, and one anomeric center has the capability of isomerization. The synthesis was accomplished by remote tethering the two ends of the donor part so that only one possible coupling stereochemistry is pos-

sible. However, the generality of remote tethering for ordinary oligosaccharide syntheses is not clear.

Chemical synthesis and enzyme-based routes are complementary. Chemical synthesis offers exceptional flexibility. Natural and non-natural saccharide building blocks can be assembled with natural or non-natural linkages. In comparison, enzymes can be used to effect glycosylation with absolute regio- and stereo-control. If the necessary enzyme is available, the desired bond can be formed, often with high efficiency. Although some enzymes will act on alternative substrates, chemical synthesis provides the means to generate any oligosaccharide, glycoconjugate, or glycomimetic.

3.1.3 Enzymatic. Enzymes are useful reagents in oligosaccharide synthesis (107, 108). Many reagent enzymes accept unnatural substrates, operate at room temperature under neutral aqueous conditions, and can form stereochemical and regiochemical reaction products with rate acceleration. Of particular value, enzymatic transformations are performed on unprotected carbohydrates. Both glycosyltransferases and glycosidases are employed in the enzymatic synthesis of oligosaccharides.

As an exception, mutant glycosidases can be employed in connection with novel donor substrates, such as glycosyl fluorides of the opposite anomeric configuration, and are used to produce both oligosaccharides (109) and polysaccharides (110).

The major problem with enzymatic synthesis is generating the required catalysts and substrates. Few glycosyltransferases are readily available, and the substrates for these enzymes are the nucleotide-activated monosaccharides that are also not readily available and must be prepared from monosaccharides through enzymatic or biological methods (111). Although glycosidases use available substrates, such as monosaccharide halides and p-nitrophenyl glycosides, yields are lower. Other novel donor substrates used in enzymatic reactions include oxazolines for the synthesis of artificial chitin (112) and 6-oxo-glycosides for the synthesis of N-acetyl-D-lactosamine derivatives (113).

Other novel enzymatic examples include the synthesis of cyclic imine saccharides through the use of the fructose diphosphate aldolase (114). This enzyme has also been used for the synthesis of bicyclic carbohydrates (115).

Reaction on solid phase requires soluble enzymes that are either recovered for recycling or discarded. Many resins can be used, including Sepharose, polyethylene-based resins, polyacrylamide, silica, and polystyrene (116, 117). Solution phase supports have the problem of product recovery. An alternative approach is to couple the substrate to a water-soluble polymer that can be removed from solution by precipitation of the polymer or size exclusion chromatography. Water-soluble supports have been used in the enzymatic synthesis of pseudo-GM3 (118).

The endoglycosidase, ceramide glycanase, is used to transfer an oligosaccharide group from a water-soluble polymer to ceramide. This enzymatic tool provides an efficient new method for polymer-supported synthesis of glycosphingolipids (119).

3.1.4 Solid Phase. Solid phase synthesis of oligosaccharides and glyconjugates similarly offers powerful advantages for oligosaccharide synthesis in comparison with conventional methods because it circumvents multiple purification steps needed in traditional solution syntheses (120). Newly developed glycosylation methods and advances in solid phase organic chemistry increased interest in solid phase oligosaccharide synthesis in the 1990s

(121). The productivity from these efforts is evident from the diversity and complexity of oligosaccharides that have been synthesized, including sequences containing up to 12 residues (122).

Although of limited scope, one approach for solid phase carbohydrate synthesis has actually incorporated coupling chemistry into an automated solid phase oligosaccharide synthesizer (123). In addition, there is the use of glycosyltransferase-regenerating beads that are potentially applicable to automated oligosaccharide synthesis (124).

Many oligosaccharides are linked covalently to peptides, and the integration of oligosaccharides into glycopeptides has started. Several laboratories have used protected glycosylated amino acids as building blocks for automated solid phase peptide synthesis to generate glycosylated peptide fragments reminiscent of natural glycoproteins (125). Modern 9-fluorenyl methoxy carbonyl (FMOC)-based peptide synthesis methods are sufficiently mild that the oligosaccharide remains intact throughout the synthesis. For example, a glycopeptide derived from the mucin-like leukocyte antigen leukosialin was made using trisaccharide-amino acid as a building block (126). Chemical linkage of saccharides to peptides is affected in several ways. The acceptor amino acid can be part of the evolving chain in solid phase peptide synthesis and reacted with a saccharide to produce the target glycopeptide. For example, glycosylated amino acids containing one to three sugars have been incorporated in solid phase synthesis of glycopeptides (127), and mucin glycopeptides containing common core structures have been synthesized (128).

Another example, glycopeptide segments prepared on solid phase, can be condensed using serine proteases. Incorporation of additional monosaccharides using glycosyltransferases is a novel method of glycoprotein synthesis (129).

A problem with attempting to couple large oligosaccharide side-chains to a polypeptide is low yield, which may be attributed to steric factors. A method that overcomes the steric problem was developed in the case of coupling a high mannose-type pentasaccharide to a variety of tripeptides (130). Alternatively, amino acids containing high mannose-type oligosaccharides with as many as 11 primary sugars in a triantennary structure were prepared (131).

3.1.5 Glycoprotein.
The degree and heterogeneity of glycosyl groups on glycoproteins produced in nature is not only a challenge in synthetic methodology but also is a problem in characterization and in target product specification.

Currently, cell fermentation is used to glycosylate proteins. Unfortunately, fermentation produces many different protein glycoforms. The resulting mixture, however, can be used in subsequent digestion reactions where the sugar chains are cleaved to produce a near homogeneous glycoprotein that, in turn, can be further reacted enzymatically. For example, glycans on N-glycosylated proteins are digested to the core N-acetylglucosamine by endoglycosidases, and then elaborated enzymatically to increase size and complexity of the glycan by using enzyme-catalyzed transglycosylation (132).

3.1.6 Carbohydrate Library.
Although supply continues to be a major challenge to carbohydrate chemists, a new tool for drug discovery, library synthesis, is now becoming available. Efficient methods for the combinatorial synthesis of carbohydrates on the solid support and in solution are known (133). Progress on this front has resulted in the generation of oligosaccharide libraries (134, 135). Thus, complex oligosaccharides and oligosaccharide libraries are becoming accessible to non-chemists.

Early on, random glycosylation of unprotected disaccharides, involving the nonselective coupling of a fucosyl donor, was used to produce three sub-libraries of α-fucosylated disaccharides in one step (136). Nearly statistical mixtures of all possible trisaccharides were obtained after chromatography. Linear and branched trisaccharide libraries were prepared by the latent-active glycosylation (137, 138). This concept made use of the convenient access to both glycosyl donors and acceptors from common allyl glycoside precursors.

Solution phase split-and-pool methods are capable of generating diverse libraries containing large numbers of compounds. Problems include hit identification and isolation, especially with oligosaccharide libraries that contain identical molecular weight compounds that only differ in the stereochemistry of the anomeric center. A split-and-pool approach was used to generate a small library of chondroitin sulfate disaccharides containing all eight possible sulfated forms (139). A 1000-compound library of acylated amino di- and trisaccharides were produced by a split-and-pool protocol employing glycosyl sulfoxides as donors for solid phase glycosylations and reacting with C2 azidosugars first, and then mono- and disaccharides second. Reductive conversion of the azides into amines allowed for further differentiation using acyl groups. Screening against a bacterial lectin, two carbohydrates that exhibited a higher affinity than the known natural ligand were identified (140).

Monosaccharides are ideal scaffolds for the preparation of libraries because they are enantiomerically pure, often possess a rigid conformation, and exhibit a high degree of functionalization. Monosaccharides used as scaffolds include D-glucose (141), D- and L-glucose and L-mannose (142), 2-deoxy-2-aminoglucuronic acid and 3-deoxy-3-aminoglucuronic acid (143), and tunicamycin (144).

In preparing a large number of resin-bound carbohydrates in a combinatorial fashion, it was observed that the bead-bound carbohydrates exhibited multivalent behavior. When screening a resin-bound library against a carbohydrate binding protein, it was found that the protein bound with high selectivity to only one carbohydrate among more than 1300 closely related compounds in the library (145).

Combinatorial methods for the identification of biologically active carbohydrates are evolving. Generation of random glycosylation-based libraries as heterogeneous mixtures have given way to focused libraries that employ discrete saccharide cores.

Progress in solid phase saccharide chemistry (120), in combination with techniques employed in glycomimetic chemistry, will advance carbohydrate library methodology and be a source for promising new therapeutic agents.

3.2 Analysis

To gain an insight into the biological and physicochemical functions of complex carbohydrates at the molecular level, knowledge of their primary and secondary structures is a required. The complete structural characterization of complex carbohydrates involves determining the type, number, and primary sequence of the constituent saccharide residues. This includes the occurrence of branch points and the location of appended non-carbohydrate groups and the three-dimensional conformation(s) and dynamics in solution.

3.2.1 Mass Spectrometry.
Soft ionization techniques of fast atom bombardment (FAB), electrospray ionization (ES), or matrix-assisted laser desorption ionization (MALDI) have advanced carbohydrate analysis (146, 147).

Of the three ionization technologies, MALDI-Q-TOF is arguably the most sensitive and useful type instrument for complex carbohydrate analysis (148).

Chemical derivatization methods as an aid to MS analysis continue to be important. Derivatization can be divided into tagging of reducing ends and protection of most or all of the functional groups. Permethylation is still an important type of full derivatization.

Devising strategies to optimize fragment ion information include the following: inducing fragmentation by collisional activation, monitoring natural ionization-induced fragmentation, and selecting derivatives that enhance and direct fragmentation.

Highly sensitive mass spectrometric methods for defining the primary structures of glycoprotein glycoforms is advancing structural carbohydrate chemistry (149).

Linkage and monosaccharide analysis by a combination of derivitization and LC/MS continues to provide useful carbohydrate characterization, especially in development, but where more structural detail is needed, NMR still remains a powerful tool.

3.2.2 NMR.
Nuclear Magnetic Resonance (NMR) spectroscopy is critical in determining the complete structure of a carbohydrate. The natural allocation of hydrogen and carbon atoms in carbohydrate molecules lends itself to the revelation of their full primary structure through H-1 and C-13 NMR analysis.

The main emphasis of current carbohydrate structural analysis is the applicability of modern multi-dimensional NMR for solving the two crucial problems in complex carbohydrate structural analysis, namely, the elucidation of the sequence of glycosyl residues and the solution conformation and dynamics of a carbohydrate (150). Techniques include 2D Total Correlation Spectroscopy (TOCSY), Nuclear Overhauser effect spectroscopy (NOESY), rotational nuclear Overhauser effect spectroscopy (ROESY), heteronuclear single quantum coherence (HSQC), heteronuclear multiple quantum correlation (HMQC), heteronuclear multiple bond correlation (HMBC), and (pseudo) 3D and 4D extensions.

NMR techniques used in combination with databases is helpful for carbohydrate analysis. For example, SUGABASE is a carbohydrate-NMR database that combines CarbBank complex carbohydrate structure data (CCSD) with proton and carbon chemical shift values (151).

3.3 Natural Products Isolation

Natural products isolated from higher plants and microorganisms have provided novel, clinically active drugs. The key to the success of discovering naturally occurring therapeutic agents rests on bioassay-guided fractionation and purification procedures. A traditional area of carbohydrate isolation is bacterial capsular polysaccharide chemistry for vaccine production (see Section 2.1.4). Although not a major focus for research, unique discoveries are made in carbohydrate natural products isolation.

In the study of the structure of the cell-wall polysaccharides of a poorly characterized *Proteus* microorganism, a new type of disaccharide linkage, structure (**24**), in bacterial polysaccharides was discovered (152). Presumably, this is the first example of an open-chain glycosidic linkage; however, it is not clear whether this new linkage has important biological meaning.

With the potential structural diversity of carbohydrates, it is curious that novel linkages, and for that matter novel monosacchar-

(24)

ides, are not found more often in nature. Perhaps a study of the carbohydrate chemistry of microbes that exist in extreme conditions may yield additional novel carbohydrate-based structures and perhaps new therapeutics.

3.4 Multivalency

Carbohydrate function is often contingent on multiple repeat saccharide presentation and subsequent multidentate binding. The binding of monosaccharides to target sites on proteins is generally weak, yet the affinity and specificity required for effecting a physiological response is high. Carbohydrate multivalency is the simultaneous attachment of two or more binding sites on one oligosaccharide, polysaccharide, or glycoconjugate to multiple receptor sites on another molecule, which is usually a protein.

Naturally occurring carbohydrate multivalent displays are widespread in nature and include mucins, pathogen, and mammalian cell surfaces. Lectins, carbohydrate-binding proteins, often present multiple repeat carbohydrate binding domains. The interaction of multivalent presentations can result in the formation of numerous simultaneous non-covalent bonds that proceed to afford the observed high avidity of carbohydrates (153).

Because successful drugs usually bind tightly to target receptors with binding affinities often in the nanomolar range and monomeric carbohydrates generally have low receptor binding with binding affinities in the millimolar range at best, employing multiva-

lent presentations found in nature is a seductive strategy for drug discovery. Indeed, the use of multivalency is also a new approach to drug design (154).

3.4.1 Viral Inhibitors. A multivalency approach was used to inhibit influenza virus infectivity. Several sialic acid–based polymers have been synthesized that inhibit flu virus receptor-binding activity, which in turn impedes flu viruses from sticking to cell surfaces and subsequent viral infection of cells (see also Section 2.1.4) (155, 156).

Flu viruses attach to cell surfaces through multivalent interactions. Sialic acid–based polymeric agents can inhibit the interaction of trimers of hemagglutinin and sialic acid. Significant affinity enhancements are observed in these studies. For example, in hemagglutination or the cross-linking of erythrocytes by a virus, the monomer shows inhibition in the millimolar range; however, when presented on a polymer, effective inhibition is observed in the picomolar range (157).

Multivalent sialic acid–containing liposomes have been shown to inhibit flu virus (158). Indeed, a combination of a sialic acid liposome and a lysoganglioside/poly-L-glutamic acid conjugate act as inhibitors of flu virus hemagglutinin at picomolar concentrations (159). The multivalent liposome inhibits not only hemagglutinination but also the viral surface protein neuraminidase activity. The liposome's potency is about 1000-fold higher than that of the corresponding monomeric sialic acid ligand, and the potency of the lysoganglioside is a million-fold higher than that of the monomeric ligand.

3.4.2 RNA Target. Multivalent inhibitors can be used to target not only proteins but also RNA. A bifunctional dimeric aminoglycoside antibiotic that binds to bacterial RNA can be made to bind three orders of magnitude more tightly than the monomeric antibiotic. The linked dimer also inhibits the bacterial enzyme-catalyzed modification of aminoglycoside drugs that is an antibiotic-resistance mechanism (160). This new antibiotic is active against several bacteria, including tobramycin-resistant strains.

3.4.3 Toxin Inhibitors. Multivalent inhibitors are being designed to inhibit disease-causing bacterial agents including Shiga-like toxin and enterotoxin. Shiga-like toxins from *E. coli* can be classified into two subgroups, SLT-I and SLT-II (161). One tailored multivalent carbohydrate ligand, starfish, is reported to neutralize Shiga-like toxins (162). The structure of starfish (**25**) consists of two trisaccharide units at the tips of five tethers that radiate from a central glucose core yielding a decavalent presentation. The resulting inhibitor binds to multiple sites on a target bacterial toxin that is closely related to cholera toxin. The binding prevents the toxin from attaching itself to carbohydrates on cell surfaces, which is the biological mode of action. This multivalent carbohydrate ligand exhibits more than six orders of magnitude increase in inhibition over trisaccharide unit. This sub-nanomolar activity is within the range desired for an anti-toxin therapeutic.

An X-ray crystallographic study of the complex formed between the B-subunit pentamer of SLT-I and starfish (**25**) revealed the mode of binding. Instead of binding to the theoretical two sites per monomer, it binds only to one.

That is, only five of starfish's trisaccharide arms are engaged, leaving the other five free to bind another B-subunit. Thus, starfish (**25**) is sandwiched between two subunits.

A bacterial enterotoxin related to cholera toxin is also a target for a multivalent inhibitor. When ingested, the enterotoxin's five identical binding sites recognize and bind to ganglioside carbohydrate groups in the gut lining (163).

Another approach is to develop synthetic strategies for attaching carbohydrate residues to different parts of cyclodextrin molecules. Photoaddition of thiols to allyl ethers in an anti-Markovnikov fashion is used as a key step in the multiple attachment of carbohydrates to cyclodextrin cores (164).

3.4.4 Mechanism of Multivalency. Mechanisms that contribute to the high avidity observed for multivalent ligands include neighboring site participation, target protein clustering, a chelate effect, and a precipitation process. In neighboring site participation, many target proteins possess multiple or secondary sites adjacent to the initial binding site (165). The avidity of many multivalent carbo-

(25)

hydrate molecules is caused by their ability to cluster target receptors (166). When a chelate effect operates, multiple interactions occur after formation of the first contact, which are facilitated because of the high effective concentration of the intramolecular binding groups (167).

Another model is that most of the enhancement in avidity observed in multivalent binding arises from an aggregation and precipitation process that follows the initial intermolecular binding step (168) If the avidity increases because of aggregation, apparent avidities will be strongly dependent on concentration, and if multivalency effects have aggregation events as their origin, there will be no avidity at biologically relevant concentrations. Thus, the use of multivalent compounds as systemic drugs will be limited.

An additional problem with multivalent presenting molecules is that by their very nature they are high molecular weight molecules. Divalent and trivalent compounds could have low enough molecular weights (less than a 1000 Da) to be orally active.

Polymeric multivalent structures will have high molecular weights and as such will not likely to be orally active and would require other routes of administration such as parenteral delivery. Also, high molecular weight materials tend to be mixtures that are difficult to characterize and to control in manufacturing.

3.5 Heparin

3.5.1 Synthesis. The coagulation of blood is the result of a biochemical cascade catalyzed by enzymes called activated blood-coagulation factors, which include thrombin. The protein antithrombin (AT) is the main physiological inhibitor of these factors, and its action is amplified after binding to heparin. This accounts for the anticoagulant and antithrombotic activities of heparin. Heparin is a complex mixture of heterogeneously sulphated polymers consisting of alternating L-iduronic and D-glucosamino sugars and is produced and stored in mast cells present throughout animal tissues. Heparin has numerous biological activities that can be attributed to binding interactions between these anionic groups and the positively charged groups on proteins.

Ideally, heparin mimetics should discriminate between thrombin and other proteins, particularly platelet factor-4 (PF4), which is related to heparin-induced thrombocytopenia (see also Section 2.1.5).

The total synthesis of a series single glycomimetics of heparin was recently announced (169). One structurally optimized oligosaccharide heparin mimetic is 10 times more potent *in vivo* than both standard heparin and low molecular weight heparins and is also devoid of the undesired side effect, thrombocytopenia, that is associated with heparin treatment. Thus, chemical synthesis was employed to optimize the length and the charge of the synthetic oligosaccharides.

Synthetic heparin mimetics consisting of 20 monosaccharides (20-mers) exhibited antithrombotic activity similar to that obtained for heparin. However, the cross-reactivity of these compounds with PF4 prompted the exploration of related molecules structurally closer to heparin.

In a synthetic series that explored chain length, it was clear that reducing the size to the minimum that still allowed thrombin inhibition was not sufficient to abolish the undesired interaction with PF4. The next step was to reduce the charge of the molecule for antithrombin, and that provided the structurally optimized oligosaccharide heparin mimetic.

Indeed, the optimized oligosaccharide heparin mimetic has 10 times the antithrombotic potency relative to that obtained for heparin but is unable to activate platelets in the presence of plasma from patients with heparin-induced thrombocytopenia and is devoid of other side effects attributed to the anionic charges of heparin, such as hemorrhage.

3.5.2 FGF. An X-ray structure of a heparin-linked biologically active dimer of fibroblast growth factor was described (170). The fibroblast growth factors (FGFs) form a large family of structurally related, multifunctional proteins that mediate cellular functions by binding to transmembrane FGF receptors (171, 172).

Complexes of the decasaccharides with both native and selenomethionyl aFGF were purified and crystallized, providing two crystal forms each containing a total of four independent heparin-linked aFGF dimers. In the structures of all four dimers, the decasaccharide chains are ordered and bind through sulphate groups to residues in the heparin-binding loop of aFGF residues.

The crystal structure of the biologically active, heparin-decasaccharide–linked dimer of aFGF exhibits a unique mode of ligand-induced protein dimerization. This aFGF dimer is bridged by heparin and does not have a protein–protein interface. Instead, the protomers are positioned to bind sulphate groups of five to six monosaccharide units on opposite sides of the heparin helix axis.

3.5.3 Heparin Analytical Tool.
Recently, a heparin sequencing technology called property-encoded nomenclature/MALDI sequencing was developed (173).

To sequence a heparin-related compound, using property encoded nomenclature/MALDI method, the polysaccharide is first fragmented by enzymatic and/or chemical means. Mass signatures of the fragments are then obtained by MALDI-MS, and the signatures are used as constraints to search a database of possible sequences for the correct match.

The database was constructed as follows. First, a unique code for each of the 32 GAG building blocks was generated. Each code is read by computers and also indicates structural information. The next step is a compositional analysis in which the sample is fragmented into its individual building blocks and counted as to how many of 32 GAG building blocks it contains. Then, each building block is identified by its mass, which is in turn determined by MS.

With the list of building blocks and the corresponding codes for a given sample in hand, a computer analysis generates a list of all possible sequences that satisfy the overall composition. The correct sequence is among the ones listed and is defined by a process of elimination.

The process involves the use enzymatic and chemical techniques to cut a second, intact sample into smaller and smaller pieces. Because these techniques cut heparin fragments at specific sites, each piece provides structural information at reducing and non-reducing ends. In addition, the mass of each of these smaller pieces is determined, which gives the number and kind of building blocks in each. Repeating this process for smaller and smaller pieces eventually gives the correct sequence.

Carbohydrate chemistry is contributing both to the understanding of biological functions of saccharides and glycoconjugates and to the discovery and development of carbohydrate-based therapeutics. Advances in carbohydrate chemistry have solved some problems associated with carbohydrate research and have provided new strategies for solving challenges in glycobiology. Technical problems that hinder carbohydrate research still exist. A crucial milestone in the maturation of carbohydrate chemistry will be the development of robust automated systems for the synthesis of oligosaccharides and glycoconjugates. The maturation of carbohydrate chemistry systems that are easily accessible to both biologists and chemists will have an important impact on our understanding of glycobiology and on the discovery of carbohydrate-based therapeutics. Collectively, chemical tools have proven indispensable for studies in glycobiology (174).

4 GLYCOBIOLOGY

Glycobiology is now an established discipline that is expanding rapidly, and an understanding of the major advances in this discipline is essential for the discovery of new carbohydrate-based therapeutics. What follows is a discussion of a few of the interesting developments and trends that have occurred in this field over the past 10 years.

4.1 Driving Forces

Several forces drive the surge in interest in the field of glycobiology. They include the sequencing of the human genome, glycoconjugate structure-function studies, and immunogen and lectin research.

4.1.1 Human Genome.
One force is the sequencing of the human genome. As it becomes clear that the number of genes encoded by the genome cannot alone explain the complexity of life, post-translation modifications have become obvious explanation for the source of

this complexity. The complexity of saccharide side-chains could more than sufficiently provide micro-specificity between proteins. For instance, it has been calculated that a hexa-saccharide has 1012 isomeric permutations. Furthermore, the glycans are readily modified after synthesis of the core structure (175). Additionally, gradients of carbohydrate side-chains could play key roles in processes such as development.

4.1.2 Therapeutic Glycoproteins.

Second, as the use of glycoproteins as therapeutic agents increases, so does the need for understanding the role that sugars play in their function. For example, correct glycosylation of the top-selling drug, erythropoietin (EPO), is critical to maintaining a reasonable circulating half-life, and Amgen regularly rejects batches of EPO because of incorrectly attached sugars (176). Glycosylation of monoclonal antibodies is important for their recognition by Fc γ receptors (177), and removal of glycosylation sites in therapeutic monoclonal antibodies could result in reduced immunogenicity and "first-dose" cytokine release syndrome (178). Modification of glycosylation sites has also led to second generation drugs, such as reteplase, a tissue plasminogen activator (TPA) molecule in which the glycosylation sites have been removed, resulting in an enzyme with a slightly higher half-life and lower fibrin-binding affinity than recombinant TPA (179).

4.1.3 Structure-Function.

Third, there have been successes within the last 10 years in making effective drugs that are based on carbohydrate structure, including glycomimetics and an understanding of glycobiology. The use of neuriminidase inhibitors to prevent the escape of the influenza virus from its host cell and the conversion of heparin that is a complex mixture with poly-pharmacy into a single chemical entity with a well-defined pharmacology are real drug discovery success stories. Future success stories will depend in part on an understanding of glycobiology including structure-function of the principal biomolecular conjugate structures.

4.2 Glycoconjugates

Glycoconjugates are expressed primarily as three different forms: glycolipids, glycoproteins, and the very complex proteoglycans found in the extracellular matrix. In this section, the glycobiology of glycolipids and glycoproteins and their ligands and proteoglycans are reviewed, focusing on how the basic research may effect the discovery on new carbohydrate-based therapeutics.

4.2.1 Glycolipids.

Glycolipids are made up of oligosaccharides linked covalently to a fatty acid portion by means of inositol or sphingosine moieties. The association of the non-polar function with the cell membranes effectively anchors these molecules to the extracellular surface. Glycolipids are subdivided into (1) glycosphingolipids, (2) glycerol-type glycolipids or glycosylated glycerols having fatty acid substituents on the glycerol, (3) ester-type glycolipids or esters of carbohydrates with fatty acids, (4) steryl glycosides or glycosylated derivatives of sterol, (5) polyisoprenyl-type glycolipids in which the sugar is normally linked through a phosphate bridge to the polyisoprenols, and (6) glycosylphosphatidyl-inositols.

The role of glycolipids extends well beyond the essential structural role they play in cell membrane lipid bilayers, including cell–cell and cell–matrix interactions. The examples are numerous, including glycophosphatidyl inositol anchors, which act as sites of attachment for proteins to the cell membrane, and gangliosides, a form of glycosphingolipids, which are thought to be crucial in the development of nervous tissue. In the last 10 years, there has been a great deal of research and high expectations that glycolipid-based therapeutics would be successful. For example, the glycosphingolipid GM1 was in clinical trials to determine whether it could play a role in repairing damaged nervous tissue caused by stroke or spinal injury (see Section 2.2.2). Whereas its mechanism of action was unclear at that time, animal studies have suggested this use. It was also expected that glycosphingolipids would have use in the diagnosis and treatment of cancer, because it had been established that almost all human cancers exhibit aberrant glycosylation patterns.

4.2.2 Proteoglycans.

Proteoglycans are very large extracellular glyconjugates (up to millions of Daltons) composed of a protein core with polysaccharide chains extending out perpendicularly in brush-like structures. In fact, the polysaccharides account for 90–95% of the proteoglycan's size by weight. With GAGs for example, chains are built up of long linear sequences of alternating hexamine (D-glucosamine or D-galactosamine) and hexuronic acid (D-glucuronic acid or L-iduronic acid) units, and despite the regularity in this alternating sequence, GAGs show substantial structural heterogeneity. Their attachment to the polypeptide backbone, in general, occurs through a trisaccharide linker to an O-glycosidic bond at serine residues. GAGs provide viscosity, and as a result, impart low compressibility, which makes proteoglycans ideal molecules for the joint extracellular matrix, including cartilage. The predominant proteoglycan in the cartilaginous matrix is aggrecan (180) and is degraded in osteoarthritis (OA), a progressive disease in which the joint cartilage is gradually eroded. The GAGs on aggrecan consist of chondroitin sulfate and keratin sulfate and play an important role in the hydration of the cartilage. Clinical trials are ongoing to determine whether chondroitin sulfate can be used as a treatment for OA (181).

Heparin is now known to play a key role in packaging proteases in mast cells granules (see Section 2.1.5) (182). In mice, a gene necessary for the production of fully functional heparin was disrupted or knocked out. The gene encodes glucosaminyl N-deacetylase/N-sulphotransferase-2, an enzyme that adds sulfur trioxide to heparin. Loss of this enzyme results in improper packaging of key components of the storage granules in mast cells, which serve as the first line of defense for the immune system. The components affected are several enzymes called mouse mast-cell proteases. These proteases are used by the immune system to destroy foreign proteins. The knockout mice thus indicate a physiologic function for heparin unrelated to its ability to prevent blood clots.

4.2.3 Glycoproteins.

Glycoproteins are by far the most complex glycoconjugates, and there are two major classes of glycoproteins, O-linked and N-linked, depending on whether the oligosaccharide chain is linked to the protein through threonine or serine side-chains (O-linked) or aspargine (N-linked). Attachment and maturation of the carbohydrate side-chains occurs within the endoplasmic reticulum (ER) and Golgi in a highly ordered manner. In the case of N-linked oligosaccharides, the process begins with the attachment, through a dolichol lipid carrier, of Glc3Man9-GlcNAc2 structures containing trimannosyl-chitobiose cores to polypeptides as they enter the ER. Specifically, the carbohydrate precursor is attached to asparagine residues within the recognition sequence Asn-X-Ser/Thr. It has been estimated that 90% of all of these sequences are glycosylated (183). After the transfer from the dolichol lipid, the sugar side-chains are "trimmed" by exoglycosidases.

Trimming by α-mannosidases, without subsequent glycosyl addition, results in ManxGlcNAc2 side-chains called "high mannose" structures. In the case of "complex-type" carbohydrate side-chains, "trimming" is followed by the stepwise addition of new sugar residues by glycosyltransferases. The biosynthetic processes of elongation and capping of oligosaccharides varies, resulting in carbohydrate moieties of differing size and terminal sugars. In addition to high mannose and complex-type N-linked oligosaccharides, there is a third, less common family referred to as "hybrid." Members of this family share structural features of the other two families (184).

O-linked oligosaccharides are attached to the hydroxyl component of serines and threonines, and unlike N-linked glycans, no consensus sequence is known. Attachment is initiated in the Golgi and the initiating sugar is usually a GlcNAc. O-linked side-chains can vary in size from a single GlcNAc residue to structures as large as complex-type N-linked oligosaccharides. Unlike N-linked oligosaccharides, there are at least seven O-linked oligosaccharide core structures; however, many of the processing steps are similar, resulting in similar terminal sugars between the two oligosaccharide types.

The carbohydrates groups on glycoproteins confer important physical properties such as conformational stability, protease resistance, and charge and water-binding capacity. Additionally, the carbohydrates can act as epitopes for ligand binding and as sinks for carbohydrate-binding growth factors (185). Most of the protein therapeutics on the market today are glycoproteins. Understanding the functions of the carbohydrate in these molecules could lead to improved second-generation protein products with improved efficacy and potency and reduced side effects. One of the areas where the role of carbohydrates and glycoproteins is best understood and is most likely to stimulate new carbohydrate-based therapeutic discovery is in the immune system (186). This may not be surprising given that many of the biologics on the market are immune modulators, and almost all the key molecules involved in the innate and adaptive immune system are glycoproteins.

4.3 Immunogens

With the ever-increasing number of monoclonal antibody (mAb) therapies entering the clinic, the interest for understanding the role of carbohydrates in antigen recognition and in mAb function has also increased. Over the last several decades, there has been sustained interest in developing therapeutic mAbs that bind specific carbohydrate epitopes on diseased cells such as cancer. For example, two antibodies, A33 and R24, which recognize a novel palmitoylated surface glycoprotein (187) and the GD3 ganglioside (188), respectively, were taken into clinical trials (189). Unfortunately to date, none of these efforts has resulted in a therapeutic agent.

A newer strategy has emerged in which people are immunized with carbohydrates immunogens for certain diseases. A real success story in this area was the development of the Hib vaccine, in which the immunogen is the Hib capsular polyribosylribitolphosphate polysaccharide (190). Additionally, clinical trials are underway to determine whether vaccinating patients with conjugates of carbohydrate antigens expressed on tumor cells and keyhole limpet hemocyanin (KLH) can be effective therapeutics at breaking immune tolerance to cancer cells (see Section 2.1.4) (191).

4.3.1 Immunoglobulins. There is also much interest in understanding role of glycosylation in monoclonal antibody and Fc γ receptor (Fc γRI, Fc γRIIA, Fc γRIIB, Fc γRIIIA, and Fc γRIIIB) interactions. Binding of the Fc portion of the IgG to the Fc γ receptors leads to release of inflammatory mediators, endocytosis of immune complexes, phagocytosis of microorganisms, antibody-dependent cellular cytotoxicity (ADCC), and regulation of immune cell activation (192–195). Manipulation of this interaction could lead to designer antibodies with varying levels of antibody effector functions. IgG and the Fc γRIIIA have been cocrystallized, and the solved structure is at the binding interface (196). Furthermore, the binding sites on the Fc portion of IgG for all classes of Fc γ receptors has been mapped to high resolution (197). The CH_2 region of the Fc domain is important for both the dimerization of the immunoglobulins and for binding of the IgGs to Fc γ receptors. Crystallographic studies showed that within the CH_2 domains, the oligosaccharides attached to Asn297, and not protein–protein interactions, help to drive the dimerization of the IgG monomers. While the crystal structure suggests that the carbohydrates are on the periphery of the binding interface between IgGs and the Fc γ receptors, deglycosylation studies demonstrated that carbohydrates are required for this interaction. Gradual truncation of the carbohydrates in the CH_2 domain gradually reduced the CH_2 domain thermal stability and binding to a soluble form Fc γRIIb (198), suggesting these oligosaccharides play a role in maintaining the proper geometry of that the Fc domain.

Reducing the ability of therapeutic mAbs to bind to Fc γ receptor could be an effective way to reduce the side effect of "first dose" cytokine release syndrome often experienced by patients. In fact, an Asn297-mutated variant of the anti-CD3 mAb, OKT3 (muronomonab-CD3), an approved therapeutic for the prevention of transplant rejection, has been tested in phase I studies with some success. Conversely, engineering in improved Fc γ receptor-binding activity could be beneficial for anticancer mAbs in which recruitment of immune system to kill target cells could be desired.

4.3.2 Hemagglutinins. Increasingly, carbohydrate-based therapeutics are being explored for anti-viral therapies. There are already two marketed drugs, zanamivir and oseltamivir, which are neuraminidase inhibitors, available for the treatment of influenza A and B. The neuraminidase enzyme, which cleaves sialic acid, is one of the coat protein spikes on these viruses and is required for release of the new virions from the host cells. In the presence of these neuraminidase inhibitors, which are sialic acid analogs, the new virions remain aggregated at the host cell surface through their coat hemagglutinin, thereby reducing infectivity (199). Similarly, mimetics of N-acetyl-neuraminic acid thioglycosides, which seem to block rotavirus adhesion to host cells, are being investigated for their ability to inhibit rotavirus infection (see Section 2.1.4) (200).

Other anti-viral strategies involve the use of glycosylation inhibitors to interfere with the folding of viral envelope proteins, and hence production of viable viruses. There is increasing evidence that N-butyl deoxynojirimycin (**15**), which inhibits the ER resident α-glycosidases I and II and is important for trimming terminal glucoses in the maturation pathway of precursor N-glycans (201), could be an effective treatment for both hepatitis B virus (HBV) and HIV (202). In HBV, N-butyl deoxynojirimycin (**15**) treatment resulted in hyperglucosylated M coat proteins that were misfolded, and as a result, not incorporated into the viral coat. This prevented viral secretion.

In the case of HIV, treatment with N-butyl deoxynojirimycin (**15**) did not prevent viral coat assembly, but it did greatly impair infectivity. This is believed to be because of conformation changes in gp120 coat protein, which prevented its shedding after CD4 binding, and as a result, prevented gp41 exposure, precluding fusion and entry into the cells (see Sections 2.2.3 and 2.2.6) (203).

4.4 Lectins

A discussion on proteins that specifically recognize the carbohydrate-containing molecules, the lectins, is of special interest to the design and discovery of carbohydrate-based therapeutics. In recent years there has been a surge in identification and clarification of the biological role of the lectin family of proteins. Lectins are defined as those proteins that specifically bind to carbohydrate ligands. They are often complex, multi-domain proteins, but the sugar-binding activity can usually be ascribed to a single protein module within the lectin polypeptide. Such a module is designated a carbohydrate-recognition domain (CRD). For the most part, lectins do not contain enzymatic activity, and they are classified by their carbohydrate-specificities and defined protein domains. Lectins are found throughout multiple organisms; however, in this review, the discussion will be limited to the animal lectins. The animal lectin family can be loosely divided into seven subfamilies based on their binding specificities and requirements: (*1*) The Ca^{2+}-dependent (C-type) lectins, which are a diverse group of proteins containing a highly conserved CRD and the requirement of Ca^{2+} for binding, (*2*) the galectins, which are metal independent in their binding activity and bind galactose through globular galectin-type CRDs, (*3*) the annexins, which were originally defined as Ca^{2+} and phosholipid binding proteins; however, recent studies have determined they also bind to glycoconjugates, (*4*) siglecs (I-type lectins), which are sialic acid–binding proteins that contain I-type CRDs derived from the immunoglobulin fold, (*5*) calnexin and calreticulin, which bind to N-linked oligosaccharides and retain misfolded glycoproteins in the endoplasmic reticulum, (*6*) the P-type lectins, which bind mannose-6-phosphate moieties, and (*7*) the R-type lectins, which contain CRDs similar to those found in ricin.

4.4.1 C-Type Lectins. Currently, the largest subfamily of lectins is the Ca^{2+}-dependent (C-type) lectins. The Ca^{2+} is not directly involved with the CRD binding, but is important for the maintenance of a structure that permits receptor/ligand interactions. The C-type lectins are further subdivided as those family members that are membrane proteins (selectins, type II receptors, and transmembrane lectins with tandem CRDs) and that are soluble proteins (lecticans and collectins). Perhaps the best understood among C-type lectins are the selectins, consisting of L-, P-, and E-selectin. The selectins are type I membrane proteins, with a single CRD, and are involved in the leukocyte–endothelial cell interactions in the initial homing of leukocytes to sites of inflammation. L-selectin is expressed on leukocytes and binds to ligands on activated endothelial cells. Conversely, E- and P-selectins are expressed on activated endothelial cells and bind ligands on leukocytes. Selectins bind to oligosaccharides with sLe^x and sialylated Lewis a oligosaccharide determinants. Early on, molecule glycomimetics that could inhibit selectin interactions were tested for efficacy in treating of a number of inflammatory diseases. Unfortunately, they were unsuccessful. Recently efomycine M, a macrolide antibiotic produced by *Streptomyces* with a molecular structure similar to sLe^x, showed selective selectin inhibition (204). Another approach for targeting selectin-mediated inhibition is through the glycosylases that modify their ligands, such as the carbohydrate sulfotransferases (205) and the fucosyltransferases (206). For example, HEC-GlcNAc6ST, whose expression is restricted to high endothelial venules, is a sulfotransferase responsible for addition of the critical binding determinant of GlcNAc-6-sulfate to L-selection ligands. Mice null for HEC-GlcNAc6ST activity showed markedly reduced L-selectin activity (207).

PSGL-1 (P-selectin glycoprotein ligand-1), originally identified as a ligand for P-selectin, has been shown to bind to all the selectins. Interestingly, the important T-cell homing receptor to the skin, cutaneous lymphocyte antigen (CLA), is slight variant of PSGL-1 containing a carbohydrate element that has likely been modified by fucosyltransferase VII (208). Evidence strongly suggests that CLA is important for T-cell homing to the skin in psoriatic patients (209), suggesting that blocking this selectin interaction could be an effective psoriasis treatment. In a similar strategy, rPSGL-Ig was advanced to the clinic for ischemia/reperfusion injury. It is hoped that this strategy will be more effective than the previous attempts using drugs that mimicked the sLe^x and sialylated Lewis epitopes. Perhaps this molecule will be more potent; however, no data has been published on the affinity of

PSGL-1 for the selectins, making it hard to predict whether this molecule will be any more clinically effective than its carbohydrate-based predecessors (see Section 2.2.4).

4.4.2 Mannan-Binding Lectin. Mannan-binding lectin (MBL) is a soluble C-type lectin in the collectin family. The collectins (collagen-like lectins) consist of a carboxyl terminal C-type lectin domain and an amino-terminal collagen domain (Gly-Xaa-Yaa triplet), which acts to trimerize the molecule. The collectins bind to the terminal non-reducing sugar residues mannose, GlcNAc, fucose, and glucose and play important roles in innate immunity (210). MBL is part of the complement system and plays a key role in immune defense. In the presence of Ca^{2+}, it recognizes a wide spectrum of oligosaccharides (211) and binds through multiple lectin domains to the repeated arrays of carbohydrate residues present on the cell surface of many microorganism. This results either in the recruitment of the complement system through an associated serum protease, MASP-2 (212), or binding to phagocyte cell surface receptors. Interestingly, a number of immune diseases have been correlated with low MBL activity.

Mutations in codon 52, 54, or 57 have been reported to occur naturally in humans and disrupt MBL secondary structure. There seems to be a higher frequency of these mutations and low MBL serum levels in patients experiencing high numbers of unexplained infections or those with systemic lupus erythematosus. There is also evidence suggesting that HIV-positive men harboring these MBL mutations experience more rapid AIDS progression. Furthermore, MBL has been implicated in contributing to the pathology of rheumatoid arthritis (RA).

In RA, galactosyl transferase levels are decreased, resulting in increased concentrations of the agalactosyl glycoforms of IgG (IgG0) in the serum, synovial fluid, and synovial tissues of RA patients. MBL recognizes these truncated carbohydrate side-chains, which may allow for inappropriate activation of the innate immune system. Conceivably, inhibiting MBL binding activity or increasing galactosyl transferase activity in RA patients could reduce the severity of disease.

4.4.3 Siglecs. Perhaps the lectin family that has seen the biggest growth in the past couple of years is the siglec family. Siglecs are sialic acid–binding Ig-like lectins, which are expressed abundantly on cells within the immune system. There are at least 11 family members, and 6 of these have been identified only within the last 4 years. While the exact mechanisms of action of these molecules are unknown, it is hoped that these molecules will have therapeutic value because a number of them seem to play a role in both the adaptive and innate immune system. For example, siglec-2 (CD22) is an inhibitory receptor for B-cell receptor (BCR) signaling. High levels of siglec-1 (sialoadhesin/CD169) are expressed on tumor infiltrating macrophages in breast cancer, and these macrophages bind to carcinomas expressing MUC-1, a membrane mucin that can be bound specifically by siglec-1 (213). Additionally, like other inhibitory receptors in the immune system, many siglecs contain immune receptor tyrosine-based inhibitory motifs (ITIMs) in the membrane proximal region of their cytoplasmic domains.

Studies examining binding specificities have determined that siglecs bind with varying preferences to at least four different terminal sugars: α-2–3-linked sialylactosamine, α-2–6-linked sialylactosamine, α-2–8-linked sialic acid, and the sialyl Tn epitope (Neu5Acα2–6GlcNAc), a tumor marker associated with poor prognosis. With the exception of the sialyl Tn antigen, these sialic acid structures are commonly found on the cell surface and in extracellular spaces. Different sialic acid side-chains, underlying sugars, and the presence of α-1–3-linked frucose all influence siglec binding to their ligands (214). Conversely, glycosylation on the siglecs themselves also affect binding (215). Sialic acid binding also serves to regulate siglec-mediated cell–cell interactions.

All siglecs, with the exception of siglec-1, are "masked" by *cis* interactions with sialic acids on the same cell surface, thereby preventing cell–cell interactions. On cell activation, by an unknown mechanism, the siglecs seem to be "unmasked." Modulation of siglec binding can also be controlled by post-translational modifications in their cytoplasmic domains. For example, protein kinase C phosphorylation of serine/threonine residues in the siglec-3 cytoplasmic domain modulated it sialic acid binding, and presumably this mechanism could apply to other siglecs, which also contain putative serine/threonine phosphorylation sites (216). While the number of therapeutics based on siglecs is limited, there is clinical trial data suggesting that humanized antibodies against siglec-3 (CD33) could be used for selective ablation of acute myeloid leukemia (217), and anti-CD22 (siglec-2) monoclonal antibodies have shown promise as a lymphoma therapy (218).

4.5 Health and Disease

The role of carbohydrates in health and disease is increasingly being appreciated, and while initially it looked like carbohydrate therapeutics would not be feasible drugs, there are currently at least several biotechnical companies taking carbohydrate-based therapies into the clinic. Most of the biological therapeutics already on the market are glycoproteins, and there is a number of examples of successful glycomimetics. For many of these glycoproteins, as well as monoclonal antibody therapies, the complete role of their carbohydrate structures is not fully understood. Furthermore, production of these glycoproteins is still not perfected, and improving cell lines so that they consistently glycosylate these therapeutics correctly will be an important step for future developments. As scientists dissect the heterogeneity of carbohydrates and understand their role in protein/lipid structure, function, and ligand interactions, there should be a greater ability to produce "designer" drugs with potentially longer half-lifes and fewer side effects.

5 SUMMARY—CONTINUE THE PACE

Now that we are in the 21st century, it is clear that carbohydrates have emerged as a significant new approach to drug discovery. At least two new carbohydrate-based drugs were registered on an annual basis as therapeutics during the last decade (Table 7.1). Considering the therapeutics now in development (Table 7.2), this pace should continue into the near future. A sustained launch of two new products annually is an enviable accomplishment even to a large, global pharmaceutical company.

For new carbohydrate-based therapeutics to continue at this pace of product registration in the intermediate and long term, several objectives must be reached. Technologies supplying carbohydrates in quality as well as in quantity must be refined or discovered. This includes any combination of automated, solid and solution phase, and enzymatic synthesis. Work to determine whether carbohydrate multivalency is a successful approach to new drug discovery, or at best, a way to make interesting research tools needs to be done. Likewise, examples that illustrate how carbohydrate-based libraries can actually accelerate drug discovery need to be shown.

Despite the failures of carbohydrate companies, the venture community should continue to fund the carbohydrate-based technologies and drug discovery efforts because there will be a significant return on investment. Also needed to continue the pace of product introduction is the application of a multi-disciplinary approach to drug discovery. An important factor for the success of carbohydrate-based therapeutics in the last decade is the combination of glycobiology and carbohydrate chemistry.

REFERENCES

1. J. H. Musser, P. Fugedi, and M. B. Anderson in M. A. Wolff, Ed., *Burger's Medicinal Chemistry*, 5th ed., Vol. **1**, John Wiley and Sons, New York, 1994, p. 901.

2. J. A. Balfour and D. McTavish, *Drugs*, **46**, 1025–1054 (1993).

3. P. Kleist, A. Ehrlich, Y. Suzuki, W. Timmer, N. Wetzelsberger, P. W. Lücker, and H. Fuder, *Eur. J. Clin. Pharmacol.*, **53**, 149 (1997).

4. H. Ishida, S. Kato, M. Nishimura, N. Mizuno, S. Fujimoto, E. Mukai, M. Kajikawa, Y. Yamada, H. Odaka, H. Ikeda, and Y. Seino, *Horm. Metab. Res.*, **30**, 673–678 (1998).

5. S. Kageyama, N. Nakamichi, H. Sekino, and S. Nakano, *Clin. Ther.*, **19**, 720–729 (1997).

6. P. J. Kingma, P. P. C. A. Menheere, J. P Sels, and A. C. Nieuwenhuijzen-Kruseman, *Diabetes Care*, **15**, 478–483 (1992).

7. R. Mosquera, L. Labeaga, and A. Orjales, *II Congresso Congiunto Italiano-Spagnolo di Chimica Farmaceutica*, August 30–September 3, (1995).

8. E. Faught, B. J. Wilder, R. E. Ramsay, R. A. Reife, L. D. Kramer, G. W. Pledger, and R. M. Karim, *Neurology*, **46**, 1684–1690 (1996).

9. K. Edwards, J. Hulihan, L. Kraut, et al., *Updates Clin. Neurology*, June 22–23, 2001, Milano, Italy.

10. A. Kucers, S. Crowe, M. L. Grayson, and J. Hoy, *The Use of Antibiotics: A Clinical Review of Antibacterial, Antifungal and Antiviral Drugs*, 5th ed., Butterworth-Heinemann, Burlington, MA, 1997.

11. T. Watanabe, K. Ohashi, K. Matsui, and T. Kubota, *J. Antimicrob. Chemother.*, **39**, 471–476 (1997).

12. D. P. Calfee and F. G. Hayden, *Drugs*, **56**, 537–553 (1998).

13. P. Palese and R. W. Compans, *J. Gen. Virol.*, **33**, 159–163 (1976).

14. MIST (Management of Influenza in the Southern Hemisphere Trialists), *Lancet*, **352**, 1877–1881 (1998).

15. J. J. Treanor, P. S. Vrooman, F. G. Hayden, N. Kinnersley, P. Ward, and R. G. Mills, *Final Program and Exhibits Addendum of the 38th Interscience Conference on Antimicrobial Agents and Chemotherapy*. American Society for Microbiology, Washington, DC, 1998.

16. F. Aoki, A. Osterhaus, G. Rimmelzwaan, N. Kinnersley, and P. Ward, *Final Program and Exhibits Addendum of the 38th Interscience Conference on Antimicrobial Agents and Chemotherapy*. American Society for Microbiology, Washington, DC, 1998.

17. W. G. Laver, N. Bischofberger, and R. G. Webster, *Sci. Am.*, **Jan 06**, 1–10 (1999).

18. G. R. Bernard, J.-L. Vincent, P.-F. Laterre, S. P. LaRosa, J.-F. Dhainaut, A. Lopez-Rodriguez, J. S. Steingrub, G. E. Garber, J. D. Helterbrand, E. W. Ely, and C. J. Fisher, *N. Engl. J. Med.*, **344**, 699–709 (2001).

19. W. G. Adams, et al., *J. Am. Med. Assoc.*, **269**, 221–226 (1993).

20. M. B. Rennels, K. M. Edwards, H. L. Keyserling, K. S. Reisinger, D. A. Hogerman, D. V. Madore, I. Chang, P. R. Paradiso, F. J. Malinoski, and A. Kimura, *Pediatrics*, **101**, 604–611 (1998).

21. J. Hirsh, R. Raschke, T. E. Warkentin, et al., *Chest*, **108**, 258S–275S (1995).

22. J. I. Weitz, *N. Engl. J. Med.*, **337**, 688–698 (1997).

23. M. T. Nurmohamed, H. ten Cate, and J. W. ten Cate, *Drugs*, **53**, 736–751 (1997).

24. J. J. Goy, *Lancet*, **354**, 694–695 (1999).

25. M. Cohen, C. Demers, E. P. Gurfinkel, A. G. G. Turpie, G. J. Fromell, S. Goodman, A. Langer, R. M. Califf, K. A. A. Fox., J. Premmereur, F. Bigonzi, J. Stephens, and B. Weatherley, *N. Engl. J. Med.*, **337**, 447 (1997).

26. P. Mismett, A. Kher, and S. Laporte-Simitsidis, *Sang Thrombose Vaisseaux*, **12**, 76–82 (2000).

27. Z. S. Goldhaber, R. B. Morrison, L. L. Diran, M. A. Creager, and T. H. Lee Jr., *Arch. Intern. Med.*, **158**, 21 (1998).

28. H. N. Magnani, *Thromb. Haemost.*, **70**, 554–561 (1993).

29. J. W. ten Cate, *Haemostasis*, **22**, 109–111 (1992).

30. K. A. Bauer, B. I. Eriksson, M. R. Lassen, and A. G. G. Turpie, *N. Engl. J. Med.*, **18**, 345 (2001).

31. K. L. Goa and P. Benfield, *Drugs*, **47**, 536–566 (1994).

32. R. D. Altman, R. Moskowitz, and R. Hyalgan, *J. Rheumatol.*, **25**, 2203–2012 (1998).

33. J. Peyron, *J. Rheumatol.*, **20**(suppl 39), 10–15 (1993).

34. H. Takanashi, K. Yogo, K. Ozaki, M. Ikuta, M. Akima, H. Koga, and H. Nagata, *Gastroenterology*, **106**, A575 (1994).

35. M. Satoh, T. Sakai, I. Sano, K. Fujikura, H. Koyama, K. Ohshima, Z. Itoh, and S. Omura, *J. Pharm. Exp. Therap.*, **271**, 574–579 (1994).

36. J. S. Schneider, D. P. Roeltgen, D. S. Rothblat, J. Chapas-Crilly, L. Seraydarian, and J Rao, *Neurology*, **45**, 1149–1154 (1995).

37. F. H. Geisler Gangliosides in P. L. Peterson and J. W. Phillis, Eds., *Traumatic and Ischemic Injuries to the Brain and Spinal Cord: Prevention and Repair*, CRC Press, Boca Raton, FL, 1995, pp. 291–310.

38. F. M. Aarestrup, *Microb. Drug Resist.*, **4**, 137–141 (1998).

39. K. C. Nicolaou, H. Suzuki, R. M. Rodríguez, K. C. Fylaktakidou, and H. J. Mitchell, *Angew Chem. Int. Ed.*, **38**, 3334–3345 (1999).

40. K. Pihl-Carey, *BioWorld Today*, **Dec. 21** (1999).

41. Neose Technologies, Inc., Press Release, November 21, 1996.

42. M. A. Fischl, L. Resnick, R. Coombs, et al., *J. Acquir. Immune Defic. Syndr. Hum. Retrovirus*, **7**, 139–147 (1994).

43. http://www.aegis.com.

44. P. S. Sunkara, M. S. Kang, T. L. Bowlin, P. S. Liu, A. S. Tyms, and A. Sjoerdsma, *Ann. NY Acad. Sci.*, **616**, 90–96 (1990).

45. E. Y. Konodu, J. C. Parke, H. T. Tran, D. A. Bryla, J. B. Robbins, and S. C. Szu, *J. Infect. Dis.*, **177**, 383 (1998).

46. J. Shrine, *BioWorld Today*, (1999).

47. A. Kumar, XIVth World Congress of Pharmacology, San Francisco, CA, July 7–12, 2002.

48. J.-F. Tanguay, P. Geoffroy, M. G. Sirois, J.-F. Theoret, R. G. Schaub, A. Kumar, and Y. Merhi, *Canadian Cardiovascular Congress*, Halifax, Nova Scotia, 2001.

49. T. F. Greten and E. M. Jaffee, *J. Clin. Oncol.*, **17**, 1047 (1999).

50. B. M. Sandmaier, D. V. Oparin, L. A. Holmberg, M. A. Reddish, G. D. MacLean, and B. M. Longenecker, *J Immunother.*, **22**, 54–66 (1999).

51. G. Ragupathi, S. F. Slovin, S. Adluri, D. Sames, I. J. Kim, H. M. Kim, M. Spassova, W. G. Bornmann, K. O. Lloyd, H. I. Scher, P. O. Livingston, and S. J. Danishefsky, *Angewandte Chemie Intern. Ed.*, **38**, 563–566 (1999).

52. T. Gilewski, G. Ragupathi, S. Bhuta, L. J. Williams, C. Musselli, X.-F. Zhang, K. P. Bencsath, K. S. Panageas, J. Chin, C. A. Hudis, L. Norton, A. N. Houghton, P. O. Livingston, and S. J. Danishefsky, *Proc. Natl. Acad. Sci. USA*, **98**, 3270–3275 (2001).

53. S. F. Slovin, G. Ragupathi, S. Adluri, G. Ungers, K. Terry, S. Kim, M. Spassova, W. G. Bornmann, M. Fazzari, L. Dantis, K. Olkiewicz, K. O. Lloyd, P. O. Livingston, S. J. Danishefsky, and H. I. Scher, *Proc. Natl. Acad. Sci. USA*, **96**, 5710–5715 (1999).

54. P. B. Chapman, D. Morrissey, J. Ibrahim, T. Richards, D. Lawson, R. J. Israel, M. B. Atkins, and J. M. Kirkwood, *Am. Soc. Clin. Oncol.*, Orlando, FL, 1999.

55. J. M. Kirkwood, *J. Clin. Oncol.*, **19**, 2370–2380 (2001).

56. N. Pavlakis, C. Parish, R. Davey, D. Podger, and H. Wheeler, *Clin. Cancer Res.*, **6**, abstract 290 (2000).

57. H. Inohara and A. Raz, *Glycoconj. J.*, **11**, 527–532 (1994).

58. J. Roberts, J. Klein, R. Palmantier, S. Dhume, M. George, and K. Olden, *Cancer Detect. Prev.*, **20**, 514 (1996).

59. G. Delpierre, M. H. Rider, F. Collard, V. Strool-bant, F. Vanstapel, H. Santos, and E. Van Schaftingen, *Diabetes*, **49**, 1627–1634 (2000).

60. A. Dubois, *Emerg. Infect. Dis.*, **1**, 79 (1995).

61. H. Miller-Podraza, M. Abul Milh, J. Berg-ström, and K.-A. Karlsson, *Glycoconj. J.*, **13**, 453 (1996).

62. H. Miller-Podraza, J. Bergström, M. Abul Milh, and K.-A. Karlsson, *Glycoconj. J.*, **14**, 467 (1997).

63. P. M. Simon, P. L. Goode, A. Mobasseri, et al., *Infect. Immun.*, **65**, 750 (1997).

64. S. J. DeArmond, H. Sanchez, Y. Qiu, A. Nin-chak-Casey, V. Daggett, A. Paminiano-Cam-erino, J. Cayetano, F. Yehiely, M. Rogers, D. Groth, M. Torchia, P. Tremblay, M. R. Scott, F. E. Cohen, and S. B. Prusiner, *Neuron*, **19**, 1337–1348 (1997).

65. M. Pérez, F. Wandosell, C. Colaço, and J. Avila, *Biochem. J.*, **335**, 369–374 (1998).

66. D. Kahne and M. Ge, *J. Am. Chem. Soc.*, **120**, 11014 (1998).

67. W. Roush, R. A. Hartz, and D. J. Gustin, *J. Am. Chem. Soc.*, **121**, 1990 (1999).

68. D. McKenney, K. L. Pouliot, Y. Wang, V. Mur-thy, M. Ulrich, G. Döring, J. C. Lee, D. A. Gold-mann, and G. B. Pier, *Science*, **284**, 1523–1527 (1999).

69. K. K. Ng, M. G. Kris, V. A. Miller, L. M. Krug, N. Michaelson, P. O. Livingston, G. Ragupathi, and H. Jennings, *Am. Soc. Clin. Oncol.*, **20**, 2644 (2001).

70. T. M. Block, X. Lu, A. S. Mehta, et al., *Nat. Med.*, 610–614 (1998).

71. L. Osborn, *Cell*, **62**, 3 (1990).

72. L. A. Lasky, M. S. Singer, T. A. Yednock, D. Dowbenko, C. Fennie, H. Rodriguez, T. Nguyen, S. Stachel, and S. D. Rosen, *Cell*, **56**, 1045 (1989).

73. R. D. Cummings in A. Varki, R. D. Cummings, J. Esko, H. Freeze, G. Hart, and J. Marth, Eds., *Essentials of Glycobiology*, Cold Spring Harbor Laboratory Press, New York, 1999, pp. 391–415.

74. B. K. Brandley, S. J. Sweidler, and P. W. Rob-bins, *Cell*, **63**, 861 (1990).

75. M. J. Pauley, M. L. Phillips, E. Warner, E. Nudleman, A. K. Singhal, S. I. Hakomori, and J. C. Paulsen, *Proc. Natl. Acad. Sci. USA*, **88**, 6224 (1991).

76. D. Sako, X. J. Chang, K. M. Barone, G. Vachino, H. M. White, G. Shaw, G. M. Veld-man, K. M. Bean, T. J. Ahern, and B. Furie, *Cell*, **75**, 1179–1186 (1993).

77. K. Ley, D. C. Bullard, M. L. Arbones, R. Bosse, D. Vestweber, T. F. Tedder, and A. L. Beaudet, *J. Exp. Med.*, **181**, 669–675 (1995).

78. R. Kannagi and A. Kanamori, *Trends Glycosci. Glycotechnol.*, **11**, 329–344 (1999).

79. S. Hemmerich, H. Leffler, and S. D. Rosen, *J. Biol. Chem.*, **270**, 12035 (1995).

80. C. R. Bertozzi, S. Fukuda, and S. D. Rosen, *Biochemistry*, **34**, 14271 (1995).

81. A. Leppanen, S. P. White, J. Helin, R. P. McEver, and R. D. Cummings, *J. Biol. Chem.*, **275**, 39569 (2000).

82. C.-Y. Tsai, X. Huang, and C.-H. Wong, *Tetra-hedron Lett.*, **41**, 9499 (2000).

83. E. E. Simanek, G. J. McGarvey, J. A. Jab-lonowski, and C.-H. Wong, *Chem. Rev.*, **98**, 833 (1998).

84. R. Banteli, P. Herold, C. Bruns, J. T. Patton, J. L. Magnani, and G. Thoma, *Helv. Chim. Acta*, **83**, 2893 (2000)

85. K. Hiruma, T. Kajimoto, G. Weitz-Schmidt, I. Ollmann, and C.-H. Wong, *J. Am. Chem. Soc.*, **118**, 9265–9270 (1996).

86. D. Sames, X. T. Chen, and S. J. Danishefsky, *Nature*, **389**, 587–591 (1997).

87. J. R. Allen, C. R. Harris, and S. J. Danishefsky, *J. Am. Chem. Soc.*, **123**, 1890–1897 (2001).

88. V. Kudryshov, P. W. Glunz, S. Hintermann, S. J. Danishefsky, and K. O. Lloyd, *Proc. Natl. Acad. Sci. USA*, **98**, 3264 (2001).

89. J. W. Dennis, M. Granovsky, and C. E. Warren, *Biochim. Biophys. Acta*, **1473**, 21–34 (1999).

90. J. Dennis, M. Granovsky, J. Pawling, J. Fata, R. Khokha, and W. J. Muller, *Nat. Med.*, **6**, 3 (2000).

91. S. L. Shames, E. S. Simon, C. W. Christopher, W. Schmid, G. M. Whitesides, and L. L. Yang, *Glycobiology*, **1**, 187–191 (1991).

92. L. K. Mahal, K. J. Yarema, and C. R Bertozzi, *Science*, **276**, 1125–1128 (1997).

93. P. Sears and C.-H. Wong, *Science*, **291**, 2344–2350 (2001).

94. H. Paulsen, *Angew Chem. Int. Ed. Engl.*, **21**, 155 (1982).

95. R. Liang, L. Yan, J. Loebach, M. Ge, Y. Uo-zumi, K. Sekanina, N. Horan, J. Gildersleeve, C. Thompson, A. Smith, K. Biswas, W. C. Still, and D. Kahne, *Science*, **274**, 1520–1522 (1996).

96. J. Rademann, A. Geyer, and R. R. Schmidt, *Angew Chem. Int. Ed. Engl.*, **37**, 1241 (1998).

97. B. Fraser-Reid, J. R. Merritt, A. L. Handlon, and C. W. Andrews, *Pure Appl. Chem.*, **65**, 779 (1993).

98. T. J. Martin and R. R. Schmidt, *Tetrahedron Lett.*, **33**, 6123 (1992).

99. S. Hashimoto, T. Honda, and S. Ikegami, *J. Chem. Soc. Chem. Commun.*, 685 (1989).

100. D. Y. Gin, V. Di Bussolo, J. Liu, and L. G. Huffman, *Angew Chem. Int. Ed. Engl.*, **39**, 204 (2000).

101. K. M. Koeller and C. H. Wong, *Chem. Rev.*, **100**, 4465 (2000).

102. P. H. Seeberger and W.-C. Haase, *Chem. Rev.*, **100**, 4549 (2000).

103. N. L. Douglas, S. V. Ley, U. Lücking, and S. L. Warriner, *J. Chem. Soc. Perkin Trans.*, **1**, 51 (1998).

104. Z. Zhang, I. R. Ollmann, X.-S. Ye, R. Wischnat, T. Baasov, and C.-H. Wong, *J. Am. Chem. Soc.*, **121**, 734 (1999).

105. P. Sears and C.-H Wong, *Science*, **291**, 2344–2350 (2001).

106. S. Oscarson and F. W. Sehgelmeble, *J. Am. Chem. Soc.*, **122**, 8869 (2000).

107. K. M. Koeller and C.-H. Wong, *Nature*, **409**, 232–240 (2001).

108. K. M. Koeller and C.-H. Wong, *Chem. Rev.*, **100**, 4465 (2000).

109. L. F Mackenzie, Q. Wang, R. A. J. Warren, and S. G. Withers, *J. Am. Chem. Soc.*, **120**, 5583–5584 (1998).

110. S. Fort, V. Boyer, L. Greffe, G. J. Davies, O. Moroz, L. Christiansen, M. Schulein, S. Cottaz, and H Driguez, *J. Am. Chem. Soc.*, **122**, 5429–5437 (2000).

111. J. E. Heidlas, K. W. Williams, and G. M. Whitesides, *Acc. Chem. Res.*, **25**, 307 (1992).

112. S . Kobayashi, T. Kiyosada, and S. Shoda, *J. Am. Chem. Soc.*, **118**, 13113–13114 (1996).

113. T. Kimura, S. Takayama, H. Huang, and C.-H. Wong, *Angew Chem. Int. Ed. Engl.*, **35**, 2348–2350 (1996).

114. T. D. Machajewski and C.-H. Wong, *Angew Chem. Int. Ed. Engl.*, **39**, 1352–1374 (2000).

115. M. T. Zannetti, C. Walter, M. Knorst, and W.-D. Fessner, *Chem. Eur. J.*, **5**, 1882–1890 (1999).

116. K. Witte, O. Seitz, and C.-H. Wong, *J. Am. Chem. Soc.*, **120**, 1979 (1998).

117. P. M. St. Hilaire and M. Meldal, *Angew Chem. Int. Ed. Engl.*, **39**, 1162 (2000).

118. K. Yamada and S.-I. Nishimura, *Tetrahedron Lett.*, **36**, 9493 (1995).

119. S.-I. Nishimura and K. Yamada, *J. Am. Chem. Soc.*, **119**, 10555–10556 (1997).

120. P. H. Seeberger, Ed., *Solid Support Oligosaccharide Synthesis and Combinatorial Carbohydrate Libraries*, J. C. Wiley, West Sussex, UK, 2001.

121. D. Kahne, *Curr. Opin. Chem. Biol.*, **1**, 130 (1997).

122. K. C. Nicolaou, N. Winssinger, J. Pastor, and F. DeRoose, *J. Am. Chem. Soc.*, **119**, 449 (1997).

123. O. J. Plante, E. R. Palmacci, and P. H. Seeberger, *Science*, **291**, 1523–1527 (2001).

124. X. Chen, J. Fang, J. Zhang, Z. Liu, J. Shao, P. Kowal, P. Andreana, and P. G. Wang, *J. Am. Chem. Soc.*, **123**, 2081–2082 (2001).

125. H. Herzner, T. Reipen, M. Schultz, and H. Kunz, *Chem. Rev.*, **100**, 4495 (2000).

126. D. Sames, X.-T. Chen, and S. J. Danishefsky, *Nature*, **389**, 587 (1997).

127. H. Herzner, T. Reipen, M. Schultz, and H. Kunz, *Chem. Rev.*, **100**, 4495 (2000).

128. N. Mathieux, H. Paulsen, M. Meldal, and K. Bock, *J. Chem. Soc. Perkin Trans.*, **1**, 2359 (1997).

129. K. Witte, O. Seitz, and C.-H. Wong, *J. Am. Chem. Soc.*, **120**, 1979–1989 (1998).

130. S. J. Danishefsky, S. Hu, P. F. Cirillo, M. Eckhardt, and P. H. Seeberger, *Chem. Eur. J.*, **3**, 1617 (1997).

131. E. Meinjohanns, M. Meldal, H. Paulsen, R. A. Dwek, and K. Bock, *J. Chem. Soc. Perkin Trans.*, **1**, 549 (1998).

132. L.-X. Wang, M. Tang, T. Suzuki, K. Kitajima, Y. Inoue, S. Inoue, J.-Q. Fan, and Y. C. Lee, *J. Am. Chem. Soc.*, **119**, 11137–11146 (1997).

133. P. H. Seeberger and W.-C. Haase, *Chem. Rev.*, **100**, 4349–4394 (2000).

134. R. Liang, L. Yan, J. Loebach, M. Ge, Y. Uozumi, K. Sekanina, N. Horan, J. Gildersleeve, C. Thompson, A. Smith, K. Biswas, W. C. Still, and D. Kahne, *Science*, **274**, 1520 (1996).

135. M. J. Sofia, et al., *J. Med. Chem.*, **42**, 3193 (1999).

136. Y. Ding, J. Labbe, O. Kanie, and O. Hindsgaul, *Bioorg. Med. Chem.*, **4**, 683–692 (1996).

137. G.-J. Boons, B. Heskamp, and F. Hout, *Angew Chem. Int. Ed. Engl.*, **35**, 2845–2847 (1996).

138. G.-J. Boons and S. J. Isles, *J. Org. Chem.*, **61**, 4262–4271 (1996).

139. A. Lubineau and D. Bonnaffé, *Eur. J. Org. Chem.*, 2523–2532 (1999).

140. M. H. J. Ohlmeyer, R. N. Swanson, L. W. Dillard, J. C. Reader, G. Asouline, R. Kobayashi,

M. Wigler, and W. C. Still, *Proc. Natl. Acad. Sci. USA*, **90**, 10922–10926 (1993).

141. R. Hirschmann, K. C. Nicolaou, S. Pietranico, E. M. Leahy, J. Salvino, B. Arison, M. A. Cichy, P. G. Spoors, W. C. Shakespeare, et al.; *J. Am. Chem. Soc.*, **115**, 12550–12568 (1993).

142. R. Hirschmann, et al., *J. Med. Chem.*, **41**, 1382–1391 (1998).

143. M. J. Sofia, R. Hunter, T. Y. Chan, A. Vaughan, R. Dulina, H. Wang, and D. Gange, *J. Org. Chem.*, **63**, 2802–2803 (1998).

144. D. J. Silva and M. J. Sofia, *Tetrahedron Lett.*, **41**, 855–858 (2000).

145. R. Liang, L. Yan, J. Loebach, M. Ge, Y. Uozumi, K. Sekanina, N. Horan, J. Gildersleeve, C. Thompson, A. Smith, K. Biswas, W. C. Still, and D. E. Kahne, *Science*, **274**, 1520 (1996).

146. J. B. Fenn, M. Mann, C. K. Meng, S. F. Wong, and C. M. Whitehouse, *Mass Spectrom. Rev.*, **9**, 37 (1990).

147. M. Karas, A. Ingendoh, U. Bahr, and F. Hillenkamp, *Biomed. Environ. Mass Spectrom.*, **18**, 841 (1989).

148. D. J. Harvey, R. H. Bateman, R. S. Bordoli, and R. Tyldesley, *Rapid Commun. Mass Spectrom.*, **14**, 2135 (2000).

149. A. Dell and H. R. Morris, *Science*, **291**, 2351–2356 (2001).

150. A. S. Serianni in S. Hecht, Ed., *Bioorganic Chemistry: Carbohydrates*, Oxford University Press, Oxford, UK, 1999.

151. A. van Kuik, Experimental Sugabase, available online at http://www.boc.chem.uu.nl/static/sugabase/sugabase.html, accessed on September 25, 2002.

152. E. Vinogradov and K. Bock, *Angew Chem. Int. Ed. Engl.*, **38**, 671–674 (1999).

153. M. Mammen, S.-K. Choi, and G. M. Whitesides, *Angew Chem. Int. Ed. Engl.*, **37**, 2754 (1998).

154. G. M. Whitesides and S.-K. Choi, *Angew Chem. Int. Ed. Engl.*, **37**, 2754 (1998).

155. G. M Bovin, et al., *FEBS Lett.*, **272**, 209 (1990).

156. A. Spaltenstein and G. M. Whitesides, *J. Am. Chem. Soc.*, **113**, 686–687 (1991).

157. G. M Whitesides, *Chem. Biol.*, **3**, 757 (1996).

158. J. E. Kingery-Wood, K. W. Williams, G. B. Sigal, and G. M. Whitesides, *J. Am. Chem. Soc.*, **114**, 7303–7305 (1992).

159. C.-H. Wong, *Angew Chem. Int. Ed. Engl.*, **37**, 1524 (1998).

160. S. J. Sucheck, A. L. Wong, K. M. Koeller, D. D. Boehr, K.-A. Draker, P. Sears, G. D. Wright, and C.-H. Wong, *J. Am. Chem. Soc.*, **122**, 5230–5231 (2000).

161. A. D. O'Brien, et al., *Curr. Top. Microbiol. Immunol.*, **180**, 65–94 (1992).

162. P. I. Kitov, J. M. Sadowska, G. Mulvey, G. D. Armstrong, H. Ling, N. S. Pannu, R. J. Read, and D. R. Bundle, *Nature*, **403**, 669–672 (2000).

163. E. Fan, Z. Zhang, W. E. Minke, Z. Hou, C. L. M. J. Verlinde, and W. G. J. Hol, *J. Am. Chem. Soc.*, **122**, 2663–2664 (2000).

164. D. A. Fulton and J. F. Stoddart, *Org. Lett.*, **2**, 1113–1116 (2000).

165. W. I. Weis and K. Drickamer, *Annu. Rev. Biochem.*, **65**, 441 (1996).

166. S. D. Burke, Q. Zhao, M. C. Schuster, and L. L. Kiessling, *J. Am. Chem. Soc.*, **122**, 4518 (2000).

167. M. I. Page and W. P. Jencks, *Proc. Natl. Acad. Sci. USA*, **68**, 1678 (1971).

168. E. J. Toone, *Tetrahedron*, **11**, 95 (2000).

169. M. Petitou, J.-P. Hérault, A. Bernat, P.-A. Driguez, P. Duchaussoy, J.-C. Lormeau, and J.-M. Herbert, *Nature*, **398**, 417–422 (1999).

170. A. D. Digabriele, et al. *Nature*, **393**, 812–817 (1998).

171. C. Basilico, D. Moscatelli, *Adv. Cancer Res.*, **59**, 115–165 (1992).

172. J. M. Schlessinger and J. C. Dionne, *Biochem. Biophys. Acta*, **1135**, 185–199 (1992).

173. R. Sasisekharan, *Science*, **286**, 537 (1999).

174. C. R. Bertozzi and L. L. Kiessling, *Science*, **291**, 2357–2364 (2001).

175. S. Roseman, *J. Biol. Chem.*, **276**, 41527–41542 (2001).

176. J. Alper, *Science*, **291**, 2338–2343 (2001).

177. S. Radaev and P. Sun, *Mol. Immunol.*, **38**, 1073–1083 (2002).

178. P. J. Friend, G. Hale, L. Chatenoud, et al., *Transplantation*, **68**, 1632–1637 (1999).

179. M. B. Wooster and A. B. Luzier, *Ann. Pharmacother.*, **33**, 318–324 (1999).

180. C. B. Knudson and W. Knudson, *Semin. Cell Dev. Biol.*, **12**, 69–78 (2001).

181. B. Mazieres, B. Combe, A. Van Phan, et al., *J. Rheumatol.*, **28**, 173–181 (2001).

182. D. E. Humphries, et al., *Nature*, **400**, 769 (1999).

183. Y. Gavel and H. G. von Heijne, *Protein Eng.*, **3**, 433–442 (1990).

184. A. Dell and H. R. Morris, *Science*, **291**, 2351–2356 (2001).

185. K. L. Bennett, D. G. Jackson, J. C. Simon, et al., *J. Cell Biol.*, **128**, 687–698 (1995).

186. P. M. Rudd, T. Elliott, P. Cresswell, I. A. Wilson, and R. A. Dwek, *Science*, **291**, 2370–2376 (2001).

187. G. Ritter, L. S. Cohen, E. C. Nice, et al., *Biochem. Biophys. Res. Commun.*, **236**, 682–686 (1997).

188. H. C. Maguire Jr., D. Berd, E. C. Lattime, et al., *Cancer Biother. Radiopharm.*, **13**, 13–23 (1998).

189. J. Tschmelitsch, E. Barendswaard, C. Williams Jr., et al., *Cancer Res.*, **57**, 2181–2186 (1997).

190. J. O. Hendley, J. G. Wenzel, K. M. Ashe, et al., *Pediatrics*, **80**, 351–354 (1987).

191. C. Musselli, P. O. Livingston, and G. Ragupathi, *J. Cancer Res. Clin. Oncol.*, **127**(suppl 2), R20–R26 (2001).

192. J. E. Gessner, H. Keiken, A. Tamm, et al., *Ann. Hematol.*, **76**, 231–248 (1998).

193. A. Gavin, M. Hulett, and P. M. Hogarth, *The Immunoglobulin Receptors and Their Physiological and Pathological Roles in Immunity*, Kluwer Academic Publishers Group, Dordrecht, The Netherlands, 1998, pp. 11–35.

194. C. Sautes, *Cell-mediated Effects of Immunoglobulins*, R. G. Landes Co., Austin, TX, 1997.

195. M. Da'ron, *Annu. Rev. Immunol.*, **15**, 203–234 (1997).

196. P. Sondermann, R. Huber, V. Oosthulzen, et al., *Nature*, **406**, 267–273 (2000).

197. R. L. Shields, A. K. Namenuk, K. Hong, et al., *J. Biol. Chem.*, **276**, 6591–6604 (2001).

198. Y. Mimura, P. Sondermann, R. Ghirlando, et al., *J. Biol. Chem.*, **276**, 45539–45547 (2001).

199. P. Palese and R. W. Compans, *J. Gen. Virol.*, **33**, 159–163 (1976).

200. A. Fazli, S. J. Bradley, M. J. Kiefel, et al., *J. Med. Chem.*, **44**, 3292–3301 (2001).

201. R. Kornfeld and S. Kornfeld, *Ann. Rev. Biochem.*, **54**, 631–664 (1985).

202. A. Mehta, N. Zitzmann, P. M. Rudd, et al., *FEBS Lett.*, **430**, 17–22 (1998).

203. P. B. Fischer, G. B. Karlsson, R. A. Dwek, et al., *J. Virol.*, **70**, 7153–7160 (1996).

204. M. P. Schon, T. Krahn, M. Schon, et al., *Nat. Med.*, **8**, 366–372 (2002).

205. S. Hemmerich, *Drug Discov. Today*, **6**, 27–35 (2001).

206. W. Weninger, L. H. Ulfman, G. Cheng, et al., *Immunity*, **12**, 665–676 (2000).

207. S. Hemmerich, A. Bistrup, M. S. Singer, et al., *Immunity*, **15**, 237–247 (2001).

208. R. C. Fuhlbrigge, J. D. Kieffer, D. Armerding, et al., *Nature*, **389**, 978–981 (1997).

209. H. Sigmundsdottir, J. E. Gudjonsson, I. Jonsdottir, et al., *Clin. Exp. Immunol.*, **126**, 365–369 (2001).

210. J. Lu, *Bioessays*, **19**, 509–518 (1997).

211. M. W. Turner, *Immunobiology*, **199**, 327–339 (1998).

212. M. W. Turner and R. M. Hamvas, *Rev. Immunogenet.*, **2**, 305–322 (2000).

213. P. R. Crocker and A. Varki, *Immunology*, **103**, 137–145 (2001).

214. E. C. Brinkman-Van der Linden and A. Varki, *J. Biol. Chem.*, **275**, 8625–8632 (2000).

215. S. Freeman, H. C. Birrell, K. D'Alessio, et al., *Eur. J. Biochem.*, **268**, 1228–1237 (2001).

216. K. Grob and L. D. Powell, *Blood*, **99**, 3188–3196 (2002).

217. I. D. Bernstein, *Clin. Lymphoma*, **2**(suppl 1), S9–S11 (2002).

218. R. O. Dillman, *Cancer Pract.*, **9**, 71–80 (2001).

CHAPTER EIGHT

Membrane Transport Proteins and Drug Transport

PETER W. SWAAN
Division of Pharmaceutics
Ohio State Biophysics Program
OSU Heart & Lung Institute Core Laboratory
    for Bioinformatics and Computational Biology
The Ohio State University
Columbus, Ohio

Burger's Medicinal Chemistry and Drug Discovery
Sixth Edition, Volume 2: Drug Development
Edited by Donald J. Abraham
ISBN 0-471-37028-2   © 2003 John Wiley & Sons, Inc.

Contents

1 INTRODUCTION

To exert their desired pharmacological activity, drugs must reach their sites of action with certain minimal effective concentration. With the exception of a few drug classes, such as general anesthetics and osmotic diuretics, most therapeutic agents produce their effects by acting on specific membrane proteins or intracellular enzymes. To gain access to these cellular targets, drugs must first reach the systemic circulation by penetrating the epithelial barriers covering the absorptive surfaces of the body, such as skin, intestine, and lung.

In most biological epithelia, drug molecules confront two obstacles in reaching the systemic circulation: (*1*) a biochemical barrier resulting from enzymatic degradation; and (*2*) a physical barrier originating from the lipid bilayer. The first obstacle can be overcome by changing the route of administration or drug formulation, for example, by encapsulating drugs in vehicles impenetrable to metabolic enzymes. However, for hydrophilic drugs and drugs with high molecular weight, especially macromolecules, epithelial membranes still impose a formidable barrier to drug entry.

Epithelia may vary in thickness or functions among different tissues, but the general transepithelial transport mechanisms for

drug molecules are similar (Fig. 8.1). Based on the route that drug molecules penetrate, epithelial transport can be classified into two pathways: paracellular and transcellular. In paracellular transport, molecules move across the epithelium through intercellular junctions *between* adjacent cells, whereas molecules cross the epithelium *through* the cells when they use the transcellular pathway. Depending on the nature of the driving force, transcellular transport can be further categorized into passive diffusion and either receptor-mediated or active carrier–mediated translocation. In passive diffusion, movement of drug molecules is derived by its concentration gradient. In active carrier–mediated transport, a membrane-embedded transport protein transports molecules against a concentration gradient using an energy supply provided by either adenosine 5′-triphosphate (ATP) hydrolysis or cotransport of ions moving down their concentration gradient, often Na^+ or H^+. Originally thought to play only a minor role in the overall drug absorption process, transport proteins have been appreciated recently to be involved in all aspects of drug absorption, distribution, and elimination. The significance of their role in drug transport is discussed later in this chapter.

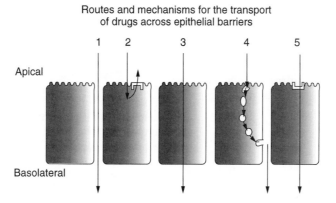

Routes and mechanisms for the transport
of drugs across epithelial barriers

Figure 8.1. Routes and mechanisms of solute transport across epithelial membranes. In general, routes 2–5 are transcellular pathways (i.e., compounds move *through* the cells), whereas route 1 is considered a paracellular pathway (i.e., a compound moves *between* the cells). (1) Tight junctional pathway; (2) drug efflux pathway (e.g., P-glycoprotein mediated); (3) passive diffusion; (4) receptor-mediated endocytosis and/or transcytosis pathways; (5) carrier-mediated route. Note that receptor and carrier proteins in epithelial cells are expressed on both the apical and basolateral surfaces.

2 STRATEGIES TO ENHANCE DRUG PERMEABILITY

The permeability of drugs that are poorly absorbed because of their hydrophilic character can be influenced by manipulating either the drug or the membrane. Therefore, most strategies for absorption enhancement either change the permeability properties of the epithelial cells or alter the physicochemical properties of the compound itself. Currently, three general strategies can be employed to increase the transport of a solute across an epithelial membrane:

1. Opening of the tight junctions or changing the epithelial lipid bilayer membrane (penetration enhancement)
2. Oral lipid-based formulations
3. Design of prodrugs with increased membrane permeability by:
   — lipophilization
   — targeting to a carrier-mediated transport system

Changing the permeability of the membrane by penetration enhancers is an aspecific approach that has met limited success in the drug development process and is discussed only briefly here.

2.1 Penetration Enhancers

Permeation-enhancing agents, including compounds such as surfactants, bile salts, salicylates, chelating agents (e.g., EDTA), or short-chained fatty acids, have been used to improve transport of poorly absorbed drugs. Most agents enhance uptake of drugs by compromising the integrity of the cell membrane. Generally, the increased absorption results from either disrupting the tight junctions or altering the membrane fluidity or both. The mechanism behind the increased permeability observed for various compounds after the co-administration of salicylates is thought to be an opening of the tight junctions. The aspecific nature of this opening of the tight junction could possibly lead to severe side effects when applied *in vivo*. The intestinal barrier becomes permeable not only for the drugs studied, but it can also become more permeable for toxic xenobiotics or even antigens, the latter leading to severe immunological side reactions. Furthermore, morphological studies indicate that salicylates cause epithelial cell damage and widen both the intercellular junctional spaces and the pores of the epithelial cell (1). Although the basement membrane and subepithelial structures are not damaged, it is not yet completely understood whether this tissue damage is reversible.

Notwithstanding its obvious disadvantages, the opening of tight junctions between enterocytes by chemical modifiers is still a promising technique to increase epithelial permeability of macromolecules, including insulin. Despite a surge in research in the late 1980s and early 1990s, this approach has found rather limited clinical application because of the nonspecific behavior of many tight junctional perturbants. A greater understanding of the factors that govern junction control has led to the discovery and development of more specific and potent permeation enhancers.

Vibrio cholera, which infects the intestinal tract and causes severe diarrhea, produces a protein known as the zonula occludens toxin (ZOT), which is able to increase the permeability of tight junctions. ZOT specifically targets the actin filaments associated with the tight junction without compromising the overall intestinal integrity or function. Interestingly, ZOT protein seems to be effective only at receptors in the jejunum and ileum but not the colon. As a result, this regulation of the paracellular pathway has been shown to be a safe, reversible, and time- and dose-dependentstrategy, and limited to intestinal tissue. This controlled permeation enhancement would be preferred to the nonspecific disruptions caused by fatty acids, bile salts, and chelators. *In vivo*, ZOT has been shown to increase the absorption of insulin by 10-fold in rabbit ileum and jejunum without affecting colonic absorption (2). In diabetic animals, those treated with ZOT and insulin orally showed comparable survival and decreases in blood glucose to those diabetic animals treated with insulin parenterally. These early findings are very promising, in that ZOT may be used to enhance the intestinal absorption of proteins by safely modulating the paracellular pathway. Furthermore, ZOT in combination with other targeted delivery techniques such as nanosystems could become a highly versatile approach.

Another promising class of permeation enhancers is the biocompatible polymer chitosan and its analogs (3), which were shown to substantially increase the bioavailability of several macromolecules across the intestinal epithelium.

2.2 Lipid-Based Oral Drug Delivery

In recent years there has been an increased interest in the utility of lipid-based delivery systems to enhance oral bioavailability (4). It is generally known that membrane permeability is directly correlated to a drug's water–lipid partition coefficient; however, the systemic availability of highly lipophilic drugs is impeded by their low aqueous solubility. In an effort to improve this solubility-limited bioavailabiliy, formulators have turned to the use of lipid excipients to solubilize the compounds before oral administration. Several formulations are currently on the market, for example, Sandimmun/Neoral (cyclosporin microemulsion), Norvir (ritonavir), and Fortovase (saquinavir).

From a mechanistic point of view, the interaction of lipid-based formulations with the gastrointestinal system and associated digestive processes is not completely understood and appears to be more complex than mere solubility enhancement. For example, an increasing body of evidence has shown that certain lipid excipients can inhibit both presystemic drug metabolism and intestinal drug efflux mediated by P-glycoprotein (PGP). Furthermore, it is well known that lipids are capable of enhancing lymphatic transport of hydrophobic drugs, thereby reducing drug clearance resulting from hepatic first-pass metabolism. As more mechanistic studies emerge we can expect more extensive application of this flexible oral drug delivery approach.

2.3 Prodrugs

The dominant factor governing passive drug transport is the lipophilicity of the compound, generally described by the oil/water partition coefficient (P) or related parameters (e.g., distribution coefficient D, which takes into account partitioning with respect to aqueous pH). Transport across a lipid bilayer membrane involves a number of steps: diffusion across a stagnant aqueous boundary layer, interfacial transfer into the membrane, passage across the membrane, interfacial transfer out of the membrane, and diffusion across the second stagnant aqueous layer. It is now generally accepted that the overall transfer rate across the membrane increases initially with

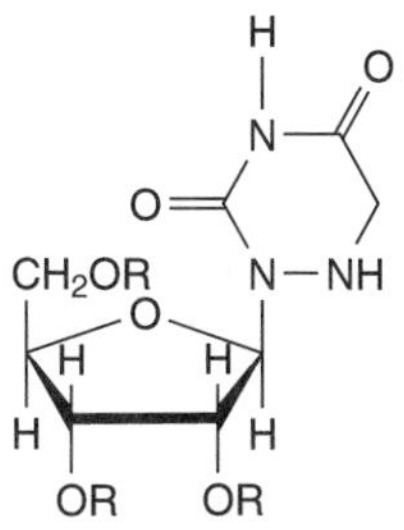

R = —H   6-azauridine (6-AZA)

R = —COCH₃   2′, 3′, 5′-triacetyl-6-AZA

R = — COC₆H₅   2′, 3′, 5′-tribenzoyl-6-AZA

Figure 8.2. Structural formulas of 6-azauridine and its prodrug analogs.

lipophilicity, but eventually reaches a maximum as diffusion across the aqueous boundary layer becomes a rate-limiting step (see below). Whereas many nutrients and drugs are readily transported across the intestinal membrane, there are a large number of highly water soluble compounds whose transfer across the intestinal membrane is limited as a result of extremely low P values.

An approach to increase epithelial permeation of these compounds is to alter the lipophilicity of drugs through chemical modifications. In general, a prodrug is synthesized from the parent compound by converting its hydrophilic residues into less polar moieties. During or after absorption, the parent drugs are then released from their lipophilic derivatives by hydrolysis or specific enzymatic actions. This strategy has successfully improved absorption of numerous drugs, especially for compounds with relatively small molecular weight. However, not all poorly absorbed drugs can be subjected to structural modifications. For example, macromolecules such as polypeptides and proteins, whose structures are closely related to their biological activities, can potentially aggregate on chemical modification. Upon hydrolysis, the aggregated parent molecule may no longer be biologically active.

2.3.1 Definition of Prodrugs. Historically, the term *prodrug* was first introduced by Albert (5), who used the word *prodrug* or *proagent* to describe compounds that undergo biotransformation before exhibiting their pharmacological effects. Consequently, he suggested that this concept could be used for many different purposes.

Nowadays, the term "prodrug" is used to describe a compound that is converted to the pharmacologically active substance *after* ad-

ministration (6). Although many drugs are known to be inactive until biotransformed, their utility as prodrugs was based on serendipity. Today, however, rational prodrug design is widely used to overcome problems of absorption, distribution, and biotransformation associated with certain drug molecules. The prodrug concept has been most successfully applied to:

- facilitate absorption and distribution of drugs with poor lipid solubility
- stabilize against metabolism during oral absorption
- increase the duration of action of drugs that are rapidly eliminated
- overcome problems of poor product acceptance by patients
- eliminate stability and other formulation problems
- promote site-specific delivery of a drug
- increase the aqueous solubility
- lower the toxicity of a drug

In fact, the vast majority of prodrugs are designed to increase the intestinal absorption of polar drugs—that is, to increase lipophilicity.

2.3.2 Prodrug Design by Increasing Lipophilicity. An example of highly polar, nonlipophilic molecules with resulting poor permeability characteristics and therefore low bioavailability is the structural analogs of the natural purine and pyrimidine nucleosides. The regular use of the nucleoside analog 6-azauridine (Fig. 8.2) in the treatment of psoriasis and neoplastic diseases was impractical because of its poor oral bioavailability. This low bioavailability can be mainly attributed to poor permeability characteristics.

The synthesis of various ester prodrugs of 6-azauridine (6-AZA), such as 2′,3′,5′-triacetyl-6-AZA, 2′,3′,5′-tribenzoyl-6-AZA, as well as other mono- and polyacyl derivatives, was carried out in an effort to obtain orally bioavailable 6-AZA. It was shown that the triacetyl ester can be administered every 8 h and is absorbed completely. On oral dosing, the triacetyl ester is converted for 80% into 6-AZA and 17% as its 5′-monoacetyl derivative. Orally administered 2′,3′,5′-triacetyl-6-AZA caused the same clinical effects as an equivalent intravenously administered dose of 6-AZA (7).

Numerous other examples of successful ester prodrug design are presented by Yalkowski and Morozowich (8). The pivaloyl ester prodrug of the antibiotic drug ampicillin, pivampicillin, shows about 90% bioavailability upon oral administration in humans, whereas ampicillin shows about 33%; the ester has a $\log P$ value about 2.7 log units higher than that of the parent compound. Psicofuranine is not absorbed upon oral administration in humans, having a $\log P$ of -1.95; the triacetate ester $\log P = 0.72$ is well absorbed.

Evidently, lipophilization seems straightforward enough to create orally absorbable prodrugs of many polar substances.

2.3.3 Rationale and Considerations for the Use of Prodrugs.
In the rational design and synthesis of the ideal prodrug derivative, several factors should also be considered.

1. *Prodrugs should be easily synthesized and purified*. Elaborate synthetic schemes should be avoided because of increased costs. Multistep syntheses increase operator time and decrease yield.
2. *Prodrugs must be stable in bulk form and dosage form*.
3. *Neither the prodrug nor its metabolic derivatives should be toxic*. Not only the newly created prodrug but also the derivatives formed after bioconversion should be nontoxic. Relatively safe moieties include amino acids, short to medium length alkyl esters and many inorganic and organic acid and base salt combinations. Nowadays, pivaloyl (e.g., in pivampicillin, pivcephalexin),

palmitoyl, propyl, and ethyl ester groups are frequently used in prodrug design.

As discussed below, the lipophilicity of a drug cannot be increased indefinitely to improve epithelial absorption. Because of the low permeation at both low and high $\log P$ values, an optimal lipophilicity (e.g., $-1 < \log P < 2$) has to be considered in prodrug design.

2.4 Absorption Enhancement by Targeting to Membrane Transporters

Several strategies have been developed to enhance the bioavailability of drugs with limited membrane permeability. These approaches can be categorized into methods that manipulate the membrane barrier properties or increase drug solubility.

Whereas charged, hydrophilic compounds and pharmaceutical macromolecules encounter difficulties in permeating the cell membrane, the systemic absorption of many water-soluble nutrients (e.g., sugars, vitamins), endogenous proteins (e.g., insulin, growth factors), and toxins (e.g., ricin, cholera toxin) appears to be highly efficient. The effective transcellular movement of these molecules is facilitated by specialized transport processes in the epithelia, that is, carrier-mediated transport and receptor-mediated endocytosis/transcytosis (discussed in more detail in subsequent sections). Both processes are operated by specific membrane-associated proteins and share common features of active—that is, concentration-, energy-, and temperature-dependent—transport mechanisms, and subject to structural analog inhibition. On the other hand, they also differ significantly in ways that transporter proteins are anchored in the membrane and in their ligand internalization mechanisms.

Apart from naturally occurring substrates, it is now well recognized that many drugs can be selectively taken up by active transport processes. For pharmaceutical scientists, these membrane transporters provide alternative routes for the delivery of drugs that would normally be impermeable to the biological barriers. Using a method similar to the conventional prodrug approach, absorption is enhanced by formation of conjugates between drugs and the endogenous ligands of the membrane transporters. Consequently, through specific interaction of ligands between the moiety and its transporter, drug candidates can be shuttled across or into the cells and eventually be released from the ligands.

Taking advantages of recent advances in molecular biology and computer modeling, scientists are now starting to design prodrugs based on the structural requirements of the transporter systems. In general, prodrug strategies involving carrier-mediated pathways have the advantage of high uptake capacity. However, the size of drug conjugates is relatively limited ($\sim$1000 Da), probably because larger conjugates fail to be shuttled through the restricted space within the carrier protein. For peptide and protein delivery, carrier-mediated pathways could facilitate peptides only up to four amino acids.

Compared to active carriers, receptor-mediated endocytosis (RME) systems have a rather limited uptake capacity, which in some cases is insufficient to elicit pharmacological activities. Yet, because of the endocytic pit formation (up to several hundred nanometers) and the vesicular internalization mechanism, RME pathways are perfectly suited to accommodate large molecular weight peptide and/or protein conjugates. More important, recent success in transport of RME ligand–drug vehicle conjugates (e.g., nanoparticles, liposomes) by way of RME pathways opens new possibilities for macromolecular delivery across biological barriers. First of all, formulating pharmaceuticals in drug vehicle systems compensates the limited capacity of RME systems, resulting in 10^3- to 10^6-fold increase in uptake. Second, drug vehicle systems also protect drug molecules from possible enzymatic degradation in the biological membrane. Furthermore, this type of conjugation avoids direct chemical reaction between drug molecules and ligands, allowing incorporation of drugs with more diverse structural properties.

3 PASSIVE DIFFUSION

In general, drug transport across any epithelium is dictated by the characteristics of the cell membrane and the physicochemical properties of drugs. In absorptive epithelia such as enterocytes, the intercellular space is sealed by tight junctions or *zonula occludens* (ZO). The junctional proteins ZO-1 and ZO-2 play essential roles in epithelial barrier function; they not only maintain cell polarity by confining surface proteins to their appropriate membrane domains but also prevent diffusion of water-soluble molecules and backflow of the absorbed nutrients.

Studies in rat small intestine have shown that only water and small hydrophilic solutes with molecular radius smaller than 25 Å can move across the paracellular pathway (9). Despite intense interest in the structure and regulation of the human intestinal tight junction, its functional dimensions have remained poorly defined. Madara and Dharmsathaphorn (10) suggested a pore size for human colonic T84 cells in the range 3.6–15 Å based on the ability of two probes of widely different size, mannitol and inulin, to cross T84 monolayers. Similarly, Ma and colleagues (11) proposed that the paracellular pore in rat colon was accessible to molecules with a radius >11 Å on the basis of significant permeation of inulin. Knipp and coworkers (12) used a group of structurally unrelated compounds of known hydrodynamic radii and estimated the pore radius of human Caco-2 cells to be 5.2 Å, although the fact that all of the probes used were smaller than the estimated pore size was an acknowledged limitation. Using a series of polyethyleneglycol (PEG) beads of known radii that span that of the restrictive paracellular pore, Watson and colleagues (13) calculated a pore radius of 4.5 and 4.3 Å for Caco-2 and T84 cell monolayers, respectively. *In vivo* PEG absorption profiles in rat, dog, and human (14) show a significant molecular mass cutoff at about 600 Da, corresponding to a hydrodynamic radius of about 5.3 Å, which compares well with data obtained in cell culture.

The overall surface area available for transcellular transport is significantly greater than that of the paracellular pathway. Therefore, the transcellular pathway is n aturally the preferred route of transport for most molecules. Lipophilic and small amphiphilic compounds can traverse the epithelium efficiently by partitioning into and out of the lipid bilayers. In constrast, large hydrophilic molecules cannot diffuse freely through the cells, even when thermodynamic conditions (e.g., concentration gradient) favors such action. Factors influencing the transcellular passive diffusion of drugs have been thoroughly characterized. Analyzing the physicochemical properties and permeability characteristics of several thousand drug molecules, Lipinski (15) deduced that only compounds with a molecular weight lower than 500, a $\log P$ less than 5, and less than 5 hydrogen bond donors and 10 hydrogen bond acceptors are likely to permeate efficiently across the cell membrane by passive diffusion. This set of characteristics for well-permeating molecules is now popularly known as Lipinski's "rule of five" and has served in the drug industry as an extremely helpful screening mechanism for recognizing drug permeability issues early on in the drug discovery process (15).

3.1   Kinetics of Passive Diffusion

Passive diffusion refers to movement of a solute along its concentration gradient. As long as the diffusing molecule does not interact with elements of the membrane, the driving force behind the diffusion of a molecule through the lipid bilayer is the electrochemical potential difference of the compound on both sides of the membrane. The change in mass (M) of a solute as a function of time (t) during its diffusion through a membrane barrier with area S is known as the flux J:

$$J = \frac{dM}{S \cdot dt} \qquad (8.1)$$

The flux, in turn, is proportional to the concentration gradient dC/dx:

$$J = -D\frac{dC}{dx} \qquad (8.2)$$

where D is the diffusion coefficient of the solute in cm^2/s, C is its concentration in mol/cm^3, and x is the distance in cm of movement perpendicular to the surface of the barrier. The diffusion constant D or diffusivity does not ordinarily remain constant and may change at

higher concentrations, and is affected by temperature, pressure, solvent properties, and the chemical nature of the solute.

Equation 8.2 is known commonly as Fick's first law. Important boundary conditions to the first law are (1) steady state (i.e., a constant rate of diffusion) and (2) sink conditions (i.e., homogeneously mixed compartments on both sides of the barrier). Without these boundary conditions, Fick's second law applies:

$$\frac{\partial C}{\partial t} = -\frac{\partial J}{\partial x} \qquad (8.3)$$

which is usually differentiated to express changes in concentration in three dimensions (x, y, z) in the general form

$$\frac{\partial C}{\partial t} = D\left(\frac{\partial^2 C}{\partial x^2} + \frac{\partial^2 C}{\partial y^2} + \frac{\partial^2 C}{\partial z^2}\right) \qquad (8.4)$$

The reader should appreciate that this form of Fick's law does not have to be taken into consideration if the aforementioned boundary conditions are met (steady state, i.e., $dC/dt = 0$).

In experimental situations, a barrier (e.g., epithelial tissue) usually separates two compartments of a diffusion cell of cross-sectional area S and thickness h. If the concentrations in the membrane on the donor and receptor sides are C_1 and C_2, respectively, Fick's first law may be written as

$$J = \frac{dM}{S \cdot dt} = D\left(\frac{C_1 - C_2}{h}\right) \qquad (8.5)$$

where $(C_1 - C_2)/h$ is an approximation of dC/dx. The concentrations C_1 and C_2 within the membrane generally are not known but can be replaced by the partition coefficient K multiplied by the concentration C_d on the donor side or C_r on the receiver side. The distribution or partition coefficient K is expressed by

$$K = \frac{C_1}{C_d} = \frac{C_2}{C_r} \qquad (8.6)$$

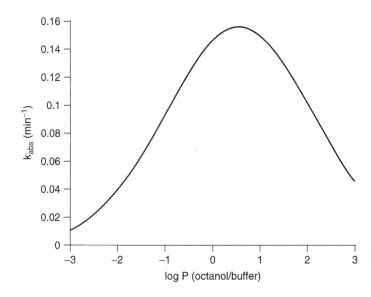

Figure 8.3. Empirical relationship between *in situ* intestinal absorption rate constants (k_{abs}) and apparent partition coefficients. The curve is described by $\log k_{abs} = 0.103 \log P - 0.09(\log P)^2 - 0.833$ (16).

Thus,

$$J = DK\left(\frac{C_d - C_r}{h}\right) \qquad (8.7)$$

In general, under physiological conditions, sink conditions will apply in the receptor compartment, $C_r \approx 0$,

$$J = \frac{DKC_d}{h} = P \cdot C_d \qquad (8.8)$$

The variables D, K, and h cannot always be determined independently and commonly are lumped together to provide the permeability coefficient P with units of linear velocity (cm/s). P can be assessed in experimental systems and provides a relative parameter to classify solute penetration through a lipid bilayer.

Numerous studies support the idea that drug molecules are absorbed through lipid bilayer membranes in the un-ionized state by the process of passive diffusion. The rate of absorption, the pK_a of the diffusing solute, and the pH at the absorption site are interrelated. Equation 8.8 demonstrates that the absorption rate of solutes through biological membranes is directly proportional to the value of the oil/water partition coefficient. Houston et al. (16) studied the absorption of a series of carbamate esters through rat everted intestine and observed a bell-shaped relationship between the absorption rate and the partition coefficient (Fig. 8.3). The reader should be aware that P in this case is used to denote the partition coefficient and not permeability. It is common to express the partition coefficient in terms of $\log P$. At low P values ($\log P < -2$), the compound cannot penetrate the lipid membrane because of an excessive thermodynamic barrier. Conversely, at high P values ($\log P > 3$), the compound becomes so lipid soluble that the diffusion through the unstirred water layers flanking both sides of the membrane becomes the rate-limiting step in the overall absorption process. Moreover, a decrease in water solubility takes place that would make the compound highly insoluble; this, in turn, would prevent the compound from reaching the membrane surface. Because of the low intestinal permeation at both low and high P values, an optimal P value (e.g., $-1 < \log P < 2$) has to be considered in drug design.

4 FACILITATED AND ACTIVE TRANSPORT PATHWAYS

4.1 Receptor-Mediated Transport

A distinct difference between carrier- and receptor-mediated systems should be pointed out. Carrier-mediated systems involve transport proteins that are anchored to the membrane by multiple membrane-spanning fragments or protein loops, whereas receptor-mediated systems use receptor proteins that span the membrane only once. Carriers operate by shuttling their substrates across the membrane by means of an energy-dependent (ATP or cotransport) flip-flop mechanism and receptors are internalized in vesicles after binding to their substrate, as explained in more detail below.

4.1.1 Receptor-Mediated Endocytosis.

Mammalian cells have developed an assortment of mechanisms to facilitate the internalization of specific substrates and target these to defined locations inside the cytoplasm. Collectively, these processes of membrane deformations are termed "endocytosis," consisting of phagocytosis, pinocytosis, receptor-mediated endocytosis (clathrin-mediated), and potocytosis [nonclathrin (caveolin)-mediated RME]. The emphasis of this section is receptor-mediated endocytosis in the intestinal tract, but the interested student may consult alternative reviews covering the complete spectrum of endocytotic processes in other cell types (17, 18).

RME is a highly specific cellular biologic process by which, as its name implies, various ligands bind to cell surface receptors and are subsequently internalized and trafficked within the cell. In many cells the process of endocytosis is so active that the entire membrane surface is internalized and replaced in less than 0.5 h (19).

RME can be dissected into several distinct events. Initially, exogenous ligands bind to specific externally oriented membrane receptors. Binding occurs within 2 min and is followed by membrane invagination until an internal vesicle forms within the cell [the early endosome, "receptosome," or CURL (compartment of uncoupling receptor and ligand) (20)]. Localized membrane proteins, lipids, and extracellular solutes are also internalized during this process. When the ligand binds to its specific receptor, the ligand–receptor complex accumulates in coated pits. Coated pits are areas of the membrane with high concentrations of endocellular clathrin subunits. The assembly of clathrin molecules on the coated pit is believed to aid the invagination process.

Specialized coat proteins, which are actually a multisubunit complex, called adaptins, trap specific membrane receptors (which move laterally through the membrane) in the coated pit by binding to a signal sequence (Tyr-X-Arg-Phe, where X = any amino acid) at the endocellular carboxy terminus of the receptor. This process ensures that the correct receptors are concentrated in the coated pit areas and minimizes the amount of extracellular fluid that is taken up in the cell. RME appears to require the GTP-binding protein dynamin, but the process by which dynamin is recruited to clathrin-coated pits remains unclear (21).

After the internalization process, the clathrin coat is lost through the help of chaperone proteins and proton pumps lower the endosomal pH to approximately 5.5, which causes dissociation of the receptor–ligand complex (22). CURL serves as a compartment to segregate the recycling receptor (e.g., transferrin) from the receptor involved in transcytosis (e.g., transcobalamin) (23). Endosomes may then move randomly or by saltatory motion along the microtubules (24) until they reach the trans-Golgi reticulum, where they are believed to fuse with Golgi components or other membranous compartments and convert into tubulovesicular complexes and late endosomes or multivesicular bodies. The fate of the receptor and ligand are determined in these sorting vesicles. Some ligands and receptors are returned to the cell surface, where the ligand is released into the extracellular milieu and the receptor is recycled. Alternatively, the ligand is directed to lysosomes for destruction, whereas the receptor is recycled to the cell membrane. Figure 8.4 presents an overview of the existing possibilities in the fate of ligands and receptors.

The endocytotic recycling pathways of polarized epithelial cells (Fig. 8.5), such as enterocytes, are generally more complex than those in nonpolarized cells. In these enterocytes there is a common recycling compartment (CRC) that receives molecules from both apical and basolateral membranes and is able to correctly return them to the appropriate membrane or membrane recycling compartment (ARC or BLRC) (25). The signals required for this sorting step have not been defined as of yet, but are presumably similar to

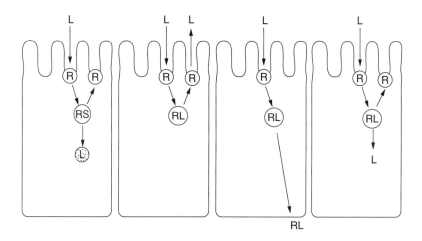

Receptor recycles, ligand degraded	Receptor recycles, ligand recycles	Receptor transported, ligand transported	Receptor recycles, ligand released
L = low-density lipoprotein	L = transferrin	L = IgA	L = viruses, toxins

Figure 8.4. Intracellular sorting pathways of RME. The initial binding and uptake steps [including receptor clustering in coated or noncoated pits, internalization of the receptor–ligand complex into coated vesicles (noncoated in the case of potocytosis), and fusion of vesicles to form endosomes] are common to all pathways. After entry into acidic endosomes, ligand and receptors are sorted and trafficked independently, which may result in degradation, recycling or transcytosis of either molecule (see text). L, ligand; R, receptor; lysosomes are depicted as shaded circles. (Adapted from Ref. 10.)

the peptide sequences required for proper sorting in the *trans*-Golgi network.

4.1.2 Structure of Cell Surface Receptors. Our general understanding of RME receptor structure and related structure-function relationships has been significantly enhanced by ongoing efforts to clone mRNA sequences coding for endocytotic receptors. It appears that most RME receptors share several structural features, such as an extracellular ligand binding site, a single hydrophobic transmembrane domain (unless the receptor is expressed as a dimer), and a cytoplasmic tail encoding endocytosis and other functional signals (26). Two classes of receptors are proposed based on their orientation in the cell membrane: the amino terminus of Type I receptors is located on the extracellular side of the membrane, whereas Type II receptors have this same protein tail in the intracellular milieu. Although protein orientation may appear trivial, it strongly influences the eventual endocytotic mechanism (26).

4.1.3 Transcytosis. One of the least understood aspects in vesicular trafficking and sorting, and possibly one of the most important aspects for successful oral drug delivery by RME, is the transport of endocytotic vesicles to the opposite membrane surface, more commonly referred to as transcytosis. Recent studies in the area of cellular biology have reported specific proteins, named TAPs (transcytosis-associated protein), that are particularly found on transcytotic vesicles and are believed to be required for fusion with target membrane (27, 28). Other methods to stimulate transcytosis have been recently explored. Transcytosis of transferrin (Tf) was found to be stimulated in the presence of Brefeldin A (BFA), a fungal metabolite that has profound effects on the structure and function of the Golgi apparatus. Shah and coworkers (29) showed in Caco-2 cell monolayers that BFA causes a marked decrease in the number of basolateral Tf receptors (TfR) along with a slight increase in the number of apical TfR. BFA enhanced the TfR-mediated transcytosis

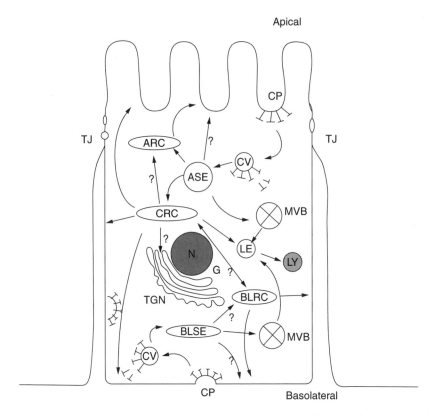

Figure 8.5. Schematic representation of endocytotic pathways in polarized cells. Question marks along an arrow indicate that only circumstantial evidence exists for those pathways at the present time. Double-headed arrows indicate that similar magnitudes of transfer occur in either direction. Ligand and receptor complex is initially taken up from coated pits (CP) into coated vesicles (CV). The CV loses its clathrin coat as described in the text and transforms into an apical sorting endosome (ASE). The fate of receptor and ligand can be: a degradative pathway where the material is passaged to multivesicular bodies (MVB) and eventually late endosomes (LE), which fuse with lysosomes (LY); a recycling pathway to the central recycling compartment (CRC) and/or the apical recycling compartment (ARC), which eventually exocytose the contents of the vesicle; or direct exocytosis of the ASE to apical membrane. In polarized cells, similar pathways occur at the basolateral membrane. The CRC, in combination with the *trans*-Golgi network (TGN), interconnects the apical and basolateral sorting pathways and enables the potential for transcytosis. BLRC, basolateral recycling compartment; BLSE, basolateral sorting endosome; G, Golgi apparatus; N, cell nucleus; TJ, tight junction.

of both 125I–Tf and the horseradish peroxidase–Tf conjugate across Caco-2 cells in both apical-to-basolateral and basolateral-to-apical directions. Prydz and colleagues found that BFA treatment rapidly increased apical endocytosis of both ricin and HRP in MDCK cells, whereas basolateral endocytosis was unaffected (30).

4.1.4 Potocytosis. It was not until recently that potocytosis has been accepted as a dis-

tinct RME pathway (31–33). Potocytosis, or non-clathrin-coated endocytosis takes place through caveolae, which are uniform omega- or flask-shaped membrane invaginations (50–80 nm diameter) (17) and was first described as the internalization mechanism of the vitamin folic acid in a mouse keratinocyte cell line (34). Years before the term "potocytosis" was coined, various ligands had been reported to localize in non-clathrin-coated membrane regions, including cholera and tetanus toxins (35).

Morphological studies have implicated caveolae in (*1*) the transcytosis of macromolecules across endothelial cells; (*2*) the uptake of small molecules by potocytosis involving GPI-linked receptor molecules and an unknown anion transport protein; (*3*) interactions with the actin-based cytoskeleton; and (*4*) the compartmentalization of certain signaling molecules involved in signal transduction, including G-protein–coupled receptors. Caveolae are characterized by the presence of an integral 22-kDa membrane protein termed VIP21-caveolin, which coats the cytoplasmic surface of the membrane (36, 37).

From a drug delivery standpoint, the advantage of potocytosis pathways over clathrin-coated RME pathways lies in the absence of the pH-lowering step, thereby circumventing the classical endosomal/lysosomal pathway (32). This may be of invaluable importance to the effective delivery of pH-sensitive macromolecules.

4.2 Receptor-Mediated Oral Absorption Systems

RME in Enterocytes versus M-Cells. According to Walker and Sanderson (38) the preferred route of intestinal uptake of low concentrations of antigens is through the M-cells (microfold or membranous), located in Peyer's patches, although at higher concentrations the regular enterocytes are also involved. M-cells are specialized epithelial cells of the gut-associated lymphoid tissues (GALT) that transport antigens from the lumen to cells of the immune system, thereby initiating an immune response or tolerance. Soluble macromolecules, small particles (39), and also entire microorganisms are transported by M-cells. The importance of M-cells in the uptake of particles is still a point of discussion. Recently, Hussain and colleagues (40) deduced from their own work and the work of others that "the importance of Peyer's patches as the principal site of particulate absorption may have been overemphasized, and that normal epithelial cells can also be induced, with appropriate ligands such as plant lectins and bacterial adhesins, to absorb particulate matter." In this light, it should be stressed that the surface area of M-cells is only 10% compared to the surface area of normal epithelial cells.

4.2.1 Immunoglobulin Transport

4.2.1.1 Maternal and Neonatal IgG Transport. Receptor-mediated transcytosis of immunoglobulin G (IgG) across the neonatal small intestine serves to convey passive immunity to many newborn mammals (41). In rats, IgG in milk selectively binds to neonatal Fc receptors (FcRn) expressed on the surface of the proximal small intestinal enterocytes during the first 3 weeks after birth. FcRn binds IgG in a pH-dependent manner, with binding occurring at the luminal pH (6–6.5) of the jejunum and release at the pH of plasma (7.4). The Fc receptor resembles the major histocompatibility complex (MHC) class I antigens, in that it consists of two subunits: a transmembrane glycoprotein (gp50) in association with β2-microglobulin (41). In mature absorptive cells both subunits are colocalized in each of the membrane compartments that mediate transcytosis of IgG. IgG administered *in situ* apparently causes both subunits to concentrate within endocytic pits of the apical plasma membrane, suggesting that the ligand causes redistribution of receptors at this site. These results support a model for transport in which IgG is transferred across the cell as a complex with both subunits. Interestingly, Benlounes and coworkers (42) recently showed that IgG is effectively transcytosed at lower concentrations (<300 $\mu g/mL$), whereas a degradative pathway dominates at higher mucosal IgG concentrations.

Site-directed mutagenesis of a recombinant Fc hinge fragment has been used to localize the site of the mouse IgG1 (mIgG1) molecule that is involved in the intestinal transfer of recombinant Fc hinge fragments in neonatal mice. These studies definitively indicate that the neonatal Fc receptor, FcRn, is involved in transcytosis across both yolk sac and neonatal intestine, in addition to the regulation of IgG catabolism (43, 44). Continued binding to vesicle membranes appears to be required for successful transfer, given that unbound proteins are removed from the transport pathway before exocytosis. These results favor the proposal that IgG is transferred across cells as an IgG–receptor complex (45).

Drug carriers such as liposomes are not readily transported intact across epithelial barriers. Patel and Wild (46) showed that coating liposomes with appropriate IgG enhances their transport across rabbit yolk sac endoderm and enterocytes of suckling rat gut proximal small intestine. They measured the effect of liposomal transcytosis both by radiolabel assay of entrapped [^{125}I]PVP and [^{3}H]inulin, and by the hypoglycemic effect of entrapped insulin. These results suggested that transported liposomes followed a pathway of transcytosis in clathrin-coated vesicles, thus escaping lysosomal degradation.

4.2.1.2 Polymeric IgA and IgM Transport.
Polymeric IgA is produced by plasma cells and found in all external excretions, including bile and saliva (47, 48). In the small intestine, polymeric IgA and IgM bind to the polymeric immunoglobulin receptor (pIgR), which is located on the basolateral surface of the cell. pIgR expression can be upregulated by cytokines (49). The pIgR–IgA complex is internalized into endosomes, where it is sorted into vesicles that transcytose it to the apical surface. At the apical surface the pIgR is proteolytically cleaved, and the large extracellular fragment (known as secretory component) is released together with the ligand. The pIgR contains a cytoplasmic domain of 103 amino acids that contains several sorting signals. Targeting from the *trans*-Golgi network to the basolateral surface is determined by the membrane-proximal 17 residues of this domain. For endocytosis there are two signals, both of which contain tyrosines. Transcytosis of the pIgR is signaled by serine phosphorylation and may be regulated by the heterotrimeric Gs protein, protein kinase C, and calmodulin. IgG is transcytosed from the apical to basolateral surface in several epithelial tissues such as the placenta and the small intestine of newborn rats. The receptor for intestinal transport of IgG is structurally similar to class I MHC molecules (Mostov, 1994, No. 2735; Mostov, 1993, No. 2734).

4.2.2 Bacterial Adhesins and Invasins.
For many bacterial species, adherence to host cells is the initial key step toward colonization and establishing an infectious disease. Two components are necessary for the adherence process: a bacterial "adhesin" (adherence or colonization factor) and a "receptor" on the host (eukaryotic) cell surface. Bacteria usually ex-

press various cell adherence mechanisms, depending on the environmental conditions and nature of the adhesins as well as receptors. In a study on the colonization mechanism of *Klebsiella*, *Enterobacter*, and *Serratia* strains, Livrelli and coworkers (50) found no relationship between the adhesive pattern and the production of specific fimbriae, suggesting that several unrecognized adhesive factors are involved that remain to be identified.

Bacteria causing gastrointestinal infection need to penetrate the mucus layer before attaching themselves to the epithelial surface. This attachment is usually mediated by bacterial fimbriae or pilus structures, although other cell surface components may also take part in the process. Adherent bacteria colonize intestinal epithelium by multiplication and initiation of a series of biochemical reactions inside the target cell through signal transduction mechanisms (with or without the help of toxins) (51).

Several adhesin and invasin molecules have been identified, such as a mannose-specific adhesin in *Lactobacillus plantarum* (52) and *V. cholerae* (53). Metcalfe and coworkers (54) found that adherence of *Escherichia coli* K-12(K88ab) to immobilized porcine small intestine mucus was caused by a 40- to 42-kDa glycoprotein K88-specific receptor. Using monoclonal antibodies against fimbrial adhesins of porcine enterotoxigenic *E. coli*, K99 and K88 adhesin were detected, but not F41 and 987P adhesins (55).

The colonization mechanism of the enteropathogenic bacterium *Yersinia pseudotuberculosis* has been studied in great detail. In contrast to other infective agents, such as Salmonella strains or enteroinvasive *E. coli* (EIEC), invasion and transcytosis of *Y. pseudotuberculosis* is mediated by a single 986 amino acid protein, invasin, on the bacterial surface that binds to α5β1 integrin (56). This single factor is sufficient to promote entry of inert particles by binding multiple integrin receptors during cellular uptake (57). This phenomenon has also recently been found by Mengaud and coworkers, who identified E-cadherin as the ligand for internalin, a *Listeria monocytogenes* protein essential for entry into epithelial cells. The internalization process of many microorganisms is impressively fast: it was recently shown that within 45 min after introduction of *Y. pseudotuberculosis* into the lumen of BALB/C mice, wild-type bacteria can be found in the Peyer's patch (58). Mutants expressing defective invasin derivatives were unable to promote efficient translocation into the Peyer's patch and instead colonized on the luminal surface of the intestinal epithelium.

The study of bacterial adhesins and invasins for the application in drug delivery strategies has recently become the focus of much attention. Paul and colleagues used an invasin fusion protein system for gene delivery strategies (59) and Easson and coworkers (60, 61) used a similar approach for intestinal delivery of nanoparticles. The latter group found that latex microspheres up to 1 μm coupled to maltose-binding protein, which was fused with invasin, can be internalized by MDCK cell monolayers (60, 61).

4.2.3 Bacterial and Plant Toxins.

After reaching early endosomes by RME, diphtheria toxin (DT) molecules have two possible fates. A large pool enters the degradative pathway, whereas a few molecules become cytotoxic by translocating their catalytic fragment A (DTA) into the cytosol (62).

The B subunit of the *E. coli* heat-labile toxin binds to the brush border of intestinal epithelial cells in a highly specific, lectinlike manner. Uptake of this toxin and transcytosis to the basolateral side of the enterocytes was observed both *in vivo* (63) and *in vitro* (64).

Fisher and coworkers expressed the transmembrane domain of diphtheria toxin in *E. coli* as a maltose-binding fusion protein and coupled it chemically to high molecular weight poly-L-lysine. The resulting complex was successfully used to mediate the internalization of a reporter gene *in vitro* (65).

Staphylococcus aureus produces a set of proteins [e.g., staphylococcal enterotoxin A (SEA), SEB, toxic shock syndrome toxin 1 (TSST-1)], which act both as superantigens and toxins. Hamad and coworkers (66) found dose-dependent, facilitated transcytosis of SEB and TSST-1, but not SEA, in Caco-2 cells. They extended their studies in mice *in vivo* by showing that ingested SEB appears in the blood more efficiently than does SEA.

Various plant toxins, mostly ribosome-inactivating proteins (RIPs), have been identified that bind to any mammalian cell surface expressing galactose units and are subsequently internalized by RME (67). Toxins such as nigrin b (68), α-sarcin (69), ricin and saporin (70), viscumin (71), and modeccin (72) are highly toxic upon oral administration (i.e., are rapidly internalized). The possibility exists, therefore, that modified and, most important, less toxic subunits of these compound can be used to facilitate the uptake of macromolecular compounds or microparticulates.

4.2.4 Viral Hemagglutinins.

The initial step in many viral infections is the binding of surface proteins (hemagglutinins) to mucosal cells. These binding proteins have been identified for most viruses, including rotaviruses (73), varicella zoster virus (74), Semliki Forest virus (75), adenoviruses (76), potato leafroll virus (77), and reovirus (78).

Recently, Etchart and colleagues (79) compared the immune response to a vaccinia virus recombinant, expressing the measles virus hemagglutinin (VV-HA), after parenteral or mucosal immunizations in mice. Oral immunizations with 10^8 pfu (plaque-forming units) of VV-HA generated low numbers of HA-specific IgA-producing cells in the lamina propria of the gut, whereas oral coimmunization with VV-HA and cholera toxin greatly enhanced the level of HA-specific spot-forming cells (IgA > IgG). Interestingly, intrajejunal immunizations with 10^8 pfu VV-HA alone induced high levels of anti-HA IgG-producing cells in the spleen and anti-HA IgA-secreting cells in the lamina propria of the gut. This study shows that VV-A can induce measles-specific immunity in the intestine, provided that it is protected from degradation in the gastrointestinal tract, or that cholera toxin is used as an adjuvant.

4.2.5 Lectins (Phytohemagglutinins).

Lectins are plant proteins that bind to specific sugars, which are found on the surface of glycoproteins and glycolipids of eukaryotic cells. Such binding may result in specific hemagglutinating activity. Because lectins are relatively heat stable, they are abundant in the human diet (e.g., cereals, beans, and other seeds).

Concentrated solutions of lectins have a "mucotractive" effect as a result of irritation of the gut wall, which explains why so-called high fiber foods (rich in lectins) are thought to be responsible for stimulating bowel motility (80, 81).

In another study demonstrating the rapid RME uptake of lectins, Weaver and colleagues (82) directly infused concanavalin A, conjugated with 10 nm colloidal gold particles, into the lumen of the jejunum in neonatal guinea pigs. Within 60 min, both villous and crypt epithelial cells contained gold particles, demonstrating the rapid accessibility of crypt cells to the lectin.

Hussain and coworkers showed that the uptake mechanism for lectins can be used for intestinal drug targeting *in vivo* (40). They covalently coupled polystyrene nanoparticles (500 nm) to tomato lectin and observed 23% systemic uptake after oral administration to rats. Control animals exerted a systemic uptake of <0.5%, indicating a 50-fold increase in oral absorption. Interestingly, they showed the intestinal uptake of tomato lectin-conjugated nanoparticles through the villous tissue to be 15 times higher than uptake by gut-associated lymphoid tissue (GALT) (40).

Although lectins are generally believed to be transported by means of an RME mechanism, there is substantial evidence that these compounds have significant affinity to intestinal M-cells (83, 84). Binding studies have revealed that M-cells exhibit pronounced regional and species variation in glycoconjugate expression. Sharma and colleagues (85) studied the nature of cell-associated carbohydrates in the human intestine that may mediate transepithelial transport of bacterial and dietary lectins and their processing by the lymphoid cells of Peyer's patches. Upon comparison of human and mouse glycoconjugates of follicle-associated epithelium and GALT, they found a distinct difference in glycosylation between mouse and human Peyer's patches and their associated lymphoid cells. Thus, choosing the appropriate lectin is apparently important when considering cell surface glycoconjugates as target molecules for intestinal drug delivery strategies. In the future, knowledge of the site and species-related variations in M-cell surface glycoconjugate expression may allow lectins to be used to selectively target an-

tigenic material and oral vaccines to the mucosal immune system at specific locations (84). Again, it should be pointed out that the overall contribution of M-cells to the absorptive surface area of the gastrointestinal tract is minimal, which could jeopardize the widespread application of drug targeting to these specialized cells.

4.3 RME of Vitamins and Metal Ions

4.3.1 Folate. The cellular uptake of free folic acid is mediated by the folate receptor and/or the reduced folate carrier. The folate receptor is a glycosylphosphatidylinositol (GPI)-anchored 38-kDa glycoprotein clustered in caveolae mediating cell transport by photocytosis (32). Whereas the expression of the reduced folate carrier is ubiquitously distributed in eukaryotic cells, the folate receptor is principally overexpressed in human tumors. Two homologous isoforms (α and β) of the receptor have been identified in humans. The α-isoform is found to be frequently overexpressed in epithelial tumors, whereas the β-isoform is often found in nonepithelial lineage tumors (86). Consequently, this receptor system has been used in drug-targeting approaches to cancer cells (87), but also in protein delivery (88), gene delivery (89), and targeting of antisense oligonucleotides (90) to a variety of cell types. Although considerable success has been met in other areas of drug targeting, to our knowledge there are currently no reports in the literature describing the use, or attempt, of this system for intestinal drug delivery purposes. This may, in part, be attributable to the low expression level of the receptor in (healthy) enterocytes. However, given that the α-isoform of the folate receptor is overexpressed in epithelial cell lines, local targeting to intestinal cancer cells (e.g., colon carcinoma) appears to be a fertile approach.

4.3.2 Riboflavin. Although not as extensively investigated as the transferrin and folate pathways, it was recently shown by Low and coworkers (91) that serum albumin coupled to riboflavin showed RME-mediated uptake in distal lung epithelium. In this paper-study the authors speculate that a similar uptake process exists in the small intestine.

Indeed, Huang and Swaan (92, 93) recently showed that riboflavin is taken up in human small intestinal and placental epithelial cells by a riboflavin-specific RME process.

4.3.3 Vitamin B_{12}. Vitamin B_{12}, the colloquial name for cobalamin (Cbl), is a large polar molecule that must be bound to specialized transport proteins to gain entry into cells. After oral administration it is bound to intrinsic factor (IF), a protein released from the parietal cells in the stomach and proximal cells in the duodenum. The Cbl–IF complex binds to an IF-receptor located on the surface of the ileum, which triggers a yet undefined endocytotic process. After internalization, the fate of the IF–Cbl complex has yet to be clarified. It was reported that IF–Cbl complex dissociates at acidic pH, and Cbl is transferred to transcobalamin II by Ramasamy and coworkers (94), whereas Dan and Cutler (95) found evidence of free Cbl in endosomes and the basolateral side of the membrane after administration to the apical surface of Caco-2 cell monolayers. It is clear, however, that Cbl is transported into all other cells only when bound to transcobalamin II.

The vitamin B_{12} RME system is probably the most extensively studied system for the oral delivery of peptides and proteins. In humans, the uptake of cobalamin is approximately 1 nmol per intestinal passage, with a potential for multiple dosing (2–3 times per hour) (96). Russell-Jones and coworkers have shown that this particular system can be employed for the intestinal uptake of luteinizing hormone releasing factor (LHRH) analogs (97), granulocyte colony-stimulating factor (G-CSF, 18.8 kDa), erythropoietin (29.5 kDa), α-interferon (98, 99), and the LHRH antagonist ANTIDE (96). More recently, they showed (both *in vitro* and *in vivo*) the intestinal uptake of biodegradable polymeric nanoparticles coupled to cobalamin to be two- to threefold higher compared to that of control (nonspecific uptake of nanoparticles) (98, 100). Thus far, the universal application of this transport system for the oral delivery of peptides and proteins seems to be hampered only by its limited uptake capacity: 1 nmol per dose. Even though this amount of uptake may be adequate for molecules such as LHRH or

erythropoietin, it is clearly not sufficient for the delivery of insulin or G-CSF. However, the recent successes with cobalamin-conjugated nanoparticles *in vivo* (100, 101) are promising and eliminate the requirement of covalently coupling cobalamin and the substrate to be delivered. This, in turn, would permit the delivery of any macromolecule through the vitamin B_{12} uptake mechanism.

4.3.4 Transferrin. Transferrin, an 80-kDa iron-transporting glycoprotein, is efficiently taken up into cells by the process of carrier-mediated endocytosis. Transferrin receptors are found on the surface of most proliferating cells, in elevated numbers on erythroblasts, and on many kinds of tumors. According to current knowledge of intestinal iron absorption, transferrin is excreted into the intestinal lumen in the form of apotransferrin and is highly stable to attacks from intestinal peptidases. In most cells, diferric transferrin binds to transferrin receptor (TfR), a dimeric transmembrane glycoprotein of 180 kDa (102), and the ligand–receptor complex is endocytosed within clathrin-coated vesicles. After acidification of these vesicles, iron dissociates from the transferrin/TfR complex and enters the cytoplasm, where it is bound by ferritin (Fn). The role that transferrin, TfR, and Fn have in regulating dietary iron uptake and maintaining total body iron stores is unknown. Both the TfR and Fn genes can be detected on small intestinal mRNA (102), and both proteins have been isolated from intestinal enterocytes (103, 104). The uptake of iron in the intestinal tract amounts up to 20 mg/day and is primarily mediated by transferrin. Recently, Shah and Shen (105) showed that insulin covalently coupled to transferrin, was transported across Caco-2 cell monolayers by RME. More recently, they showed that oral administration of this complex to streptozotocin-induced diabetic mice significantly reduced plasma glucose levels (−28%), which was further potentiated by BFA pretreatment (−41%) (106).

5 TRANSPORT PROTEINS

Membrane transporters in general are a large group of membrane proteins that have one (bitopic) or more (polytopic) hydrophobic transmembrane segments. These transporters are involved in almost all facets of biological processes in the cell. Their involvement in cellular function can be classified as follows [after Saier (107)]:

1. Mediate entry of all essential nutrients into the cytoplasmic compartment and subsequently into organelles, thus facilitating the metabolism of exogenous sources of carbon, nitrogen, sulfur, and phosphorus.

2. Provide a means for regulation of metabolite concentrations by catalyzing the efflux of end products of metabolic pathways from organelles and cells.

3. Mediate the active extrusion of drugs and other toxic substances from either the cytoplasm or the plasma membrane.

4. Mediate uptake and efflux of ion species that must be maintained at concentrations dramatically different from those in the external milieu.

5. Participate in the secretion of proteins, complex carbohydrates, and lipids into and beyond the cytoplasmic membrane.

6. Transfer of nucleic acids across cell membranes, allowing genetic exchange between organisms and thereby promoting species diversification.

7. Facilitate the uptake and release of pheromones, hormones, neurotransmitters, and a variety of other signaling molecules that allow a cell to participate in the biological experience of multicellularity.

8. Transporters allow living organisms to conduct biological warfare, secreting, for example, antibiotics, antiviral agents, antifungal agents, and toxins of humans and other animals that may confer upon the organisms producing such an agent a selective advantage for survival purposes. Many of these toxins are themselves channel-forming proteins or peptides that serve a cell-disruptive transport function.

Polytopic membrane proteins are indispensable to the cellular uptake and homeostasis of many essential nutrients. During the past decade it has become clear that a vast number of drugs share transport pathways with nutrients. Moreover, a critical role has been recognized for transport proteins in the absorption, excretion, and toxicity of drug molecules, as well as in their pharmacokinetic and pharmacodynamic (PK/PD) profiles. Because cellular transporter expression is often regulated by nuclear orphan receptors that simultaneously regulate the translation and expression of metabolic enzymes in the cell (e.g., P-glycoprotein and cytochrome P450 regulation by the pregnane X receptor), they indirectly control drug metabolism. Thus, transport proteins are involved in all facets of drug ADME and ADMET (absorption, distribution, metabolism, excretion, and toxicology), conferring an important field of study for pharmaceutical scientists involved in these areas. As a result, in-depth knowledge of membrane transport systems may be extremely useful in the design of new chemical entities (NCE). After all, it is now well appreciated that the most critical parameter for a new drug to survive the drug development pipeline on its way to the market is its ADMET profile.

Despite the involvement of solute transporters in fundamental cellular processes, most are poorly characterized at the molecular level. As a result, we are unable to predict the interaction of drugs with this important class of membrane proteins *a priori*, and detection of drug–transporter interactions remains unacceptably serendipitous.

This section aims to give an overview of current strategies for modeling transporter systems illustrated by three well-characterized transport systems: (*1*) the P-glycoprotein efflux pump, a prototypical ABC-transporter and a product of the multidrug resistance (MDR-1, ABC-B1) gene, which exports metabolites as well as drugs from various cell types; (*2*) the small peptide transporter (PepT1, SLC15A1), which transports di- and tri-peptide as well as numerous therapeutic compounds; and (*3*) the apical sodium-dependent bile acid transporter (ASBT, SLC10A2), which plays a key role in intestinal reabsorption and enterohepatic recycling of bile salts, cholesterol homeostasis, and as a therapeutic target for hypocholesterolemic agents (108–110).

5.1 The ATP-Binding Cassette (ABC) and Solute Carrier (SLC) Genetic Superfamilies

Organic solutes such as nutrients (amino acids, sugars, vitamins, and bile acids), neurotransmitters, and drugs are transferred across cellular membranes by specialized transport systems. These systems encompass integral membrane proteins that shuttle substrates across the membrane by either a passive process (channels, facilitated transporters) or an active process (carriers), the latter energized directly by the hydrolysis of ATP or indirectly by coupling to the cotransport of a counterion down its electrochemical gradient (e.g., Na^+, H^+, Cl^-).

Our understanding of the biochemistry and molecular biology of mammalian transport proteins has significantly advanced since the development of expression cloning techniques. Initial studies in *Xenopus leavis* oocytes by Hediger and colleagues resulted in the isolation of the intestinal sodium-dependent glucose transporter SGLT1 (111, 112). To date, sequence information and functional data derived from numerous transporters have revealed unifying designs, similar energy-coupling mechanisms, and common evolutionary origins (107). The plethora of isolated transporters motivated the Human Gene Nomenclature Committee to classify these proteins into a distinct genetic superfamily named SLC (for SoLute Carrier). Currently, the SLC class contains 37 families with 205 members and is rapidly expanding (http://www.gene.ucl.ac.uk/nomenclature/). The ABC superfamily contains 7 families with 48 members (113); class B contains the well-known multidrug-resistance gene (MDR1) derived P-glycoprotein (ABCB1), whereas class C is composed of members of the multidrug-resistance protein (MRP) subfamily. With the completion of the Human Genome Project, it can be anticipated that a vast number of membrane transport proteins will be identified without known physiological function.

Paulsen and colleagues (114, 115) determined the distribution of membrane transport proteins for all organisms with completely sequenced genomes and identified 81 distinct families. Two superfamilies, the ATP-binding cassette (ABC) and major facilitator (MFS) superfamilies account for nearly 50% of all transporters in each organism. The other half of these genes will be members of the SLC superfamily. Furthermore, Paulsen predicts that 15% of all genes in the human genome will code for transport proteins. With a current number of estimated sequence-tagged sites (STS) of 30,000 (116), we can expect an additional 4500 membrane transporters to emerge. Thus, the SLC superfamily is anticipated to consist of at least 2300 members; at this moment, only a fraction (10%) has been characterized in certain detail (i.e., membrane topology, substrate specificity, organ expression pattern). Table 8.1 present a concise overview of ABC and SLC members that have been classified and characterized.

The current status of transporter nomenclature and classification is in a similar state of disarray to that which the cytochrome P450 enzyme field found itself in during the early 1980s. However, the Human Gene Nomenclature Committee (HUGO) has taken on the task of directing gene classification and defining distinct subclasses in the ABC and SLC families. Eventually, this may eradicate the rampant use of trivial names for these classes of proteins.

5.2 Therapeutic Implications of Membrane Transporters

It has been generally acknowledged that transporters play an important role in clinical pharmacology (Table 8.2). Several classes of pharmacologically active compounds share transport pathways with nutrients (117). A substantial role has been recognized for transport proteins in oral absorption and drug bioavailability (118); drug resistance [e.g., efflux of antineoplastic compounds from tumor cells mediated by multidrug resistance (MDR) gene products (119, 120)]; excretion of drugs and their metabolites, mediated by transporters in the kidney and liver; drug toxicity (121); and drug pharmacokinetics and pharmacodynamics (122–124). Furthermore, the pathophysiology of several hereditary diseases (i.e., clearly defined phenotypes shown to be inherited as monogenic Mendelian traits) has been attributed to mutations in transport proteins. Most of these mutations in human genes and genetic disorders have been recorded and can

Table 8.1   Overview of the ABC and SLC Genetic Superfamilies: Nomenclature and Expression

Gene Symbol	Name/Substrate	Trivial Names	Organ Expression
	ATP-Binding Cassette Family		
ABC-A (12)	Cholesterol efflux regulatory proteins, photoreceptor proteins	CERP, RMP	
ABC-B (11)	Efflux transporters, multiple drug resistance	MDR, TAP, P-gp	Epithelial cells, tumor cells
ABC-C (12)	Cytic fibrosis transconductance regulators, multiple drug resistance associated proteins	CFTR, MRP	Lung (CFTR), liver (MRP)
ABC-D (4)	Cholesterol, fatty acid transporters, adrenoleukodystrophy-associated	ALD, ALDL	Brain
ABC-E (1)	Rnase L inhibitor	OABP, RNS4I	Ubiquitous
ABC-F (3)	TNF-α-stimulated ABC-member, non-membrane-bound	ABC50, GCN20	Synoviocytes
ABC-G (6)	Eye pigment transmembrane permeases	WHITE	Eye
	Solute Carrier Family		
SLC1 (7)	High affinity glutamate transporter	EAAT	Neurons, kidney, intestine, brain, retina
SLC2 (11)	Facilitated glucose transporters 1	GLUT	Most cells
SLC3 (2)	Cystine, dibasic, and neutral amino acid transporters	ATR, RBAT	Kidney and intestine
SLC4 (10)	Anion exchangers (1–3), sodium bicarbonate cotransporter (4–10)	EPB, AE1, NBC	Erythrocytes, brain, GI tract, kidney, reproductive organs
SLC5 (7)	Sodium/glucose cotransporter, iodide transporter (A5), vitamin transporters (A6), choline (A7)	SGLT, SMVT, NIS	Intestine, kidney, brain, thyroid
SLC6 (16)	Neurotransmitter transporters, GABA	GAT, GLYT	Brain
SLC7 (11)	Cationic amino acid transporter, y+	LAT, CAT	Ubiquitous (intestine, kidney)
SLC8 (3)	Sodium/calcium exchangers	NCX	Heart, brain, retina skeletal and smooth muscles
SLC9 (6)	Sodium/hydrogen exchangers	NHE	Ubiquitous (intestine, kidney)
SLC10 (2)	Sodium/bile acid cotransporters	Ntcp, ASBT	Liver, Ileum
SLC11 (3)	Proton-coupled divalent metal ion transporters	NRAMP, DCT	Ubiquitous
SLC12 (1–7)	Sodium/potassium/chloride transporters	NKCC	Ubiquitous (kidney)
SLC13 (2)	Sodium/sulfate and dicarboxylate symporters	NaSi, NADC	
SLC14 (2)	Urea transporters	UT	Kidney and red cells
SLC15 (2)	Oligopeptide transporters	PepT	Intestine, kidney
SLC16 (7)	Monocarboxylic acid transporters	MCT	Erythrocytes, muscle, intestine, and kidney
SLC17 (5)	Sodium phosphate transporters	NPT	Kidney, intestine
SLC18 (3)	Vesicular monoamine transporters	VAT	Brain
SLC19 (2)	Folate/thiamine transporters	FOLT, THTR	Placenta, small intestine and other tissues
SLC20 (2)	Phosphate transporters	GLVR	
SLC21 (14)	Organic anion transporters	OAT, OATP	Liver, kidney, intestine

Table 8.1 (*Continued*)

Gene Symbol	Name/Substrate	Trivial Names	Organ Expression
SLC22 (8)	Organic cation transporters	OCT	Kidney, liver, intestine
SLC23 (2)	Nucleobase transporters	SVCT	Brain, eye, intestine, kidney, liver
SLC24 (4)	Sodium/potassium/calcium exchangers	NCKX	Ubiquitous
SLC25 (21)	Mitochondrial carriers (citrate, adenine, carnitine)	CTP, ANT	Liver, gut, heart
SLC26 (11)	Solute carrier family 26 (sulfate, mostly undefined)	SAT, DTD	Cartilage and intestine
SLC27 (6)	Fatty acid transporters	FATP	Adipocytes, skeletal muscle, heart, and fat
SLC28 (2)	Sodium-coupled nucleoside transporters	CNT	
SLC29 (2)	Nucleoside transporters	ENT	Ubiquitous
SLC30 (4)	Zinc transporters	ZNT	Ubiquitous
SLC31 (2)	Copper transporters	COPT	Most organs
SLC32 (1)	GABA vesicular transporter	VGAT	
SLC34 (2)	Sodium phosphate	NaPi3, NPT	Kidney, lung, pancreas
SLC35 (3)	CMP-sialic acid and UDP-galactose transporter	CST, UGALT	Ubiquitous
SLC37 (1)	Glycerol-3-phosphate transporter		

be found online at the National Institute for Biotechnology Information (http://www.ncbi.nlm.nih.gov) OMIM database (Online Mendelian Inheritance in Man). For example, congenital glucose-galactose malabsorption syndrome is caused by a defect (D28N, D28G) in SGLT1 (125) and bile acid malabsorption syndrome can be attributed to a single C to T transition in the ASBT gene, resulting in a P290S mutation that abolishes bile acid reuptake (126).

In summary, solute transporters play an invaluable role in fundamental cellular processes in health and disease, and function as important mediators governing all aspects of drug therapy. Despite the apparent clinical importance of SLC proteins, the knowledge of their structure and mechanism of action has lagged far behind the knowledge of these properties of proteins in general; however, this

Table 8.2 Top-Selling 100 FDA-Approved Drugs

Molecular Target	Market (%)
Transporters/channels	30
Membrane receptors	25
Enzymes	20
Nuclear receptors	15
Foreign molecules (pathogens)	5

might change in the future if more research and technologies are applied to this area.

5.3 Structural Models of Transport Proteins and Methods to Design Substrates

The primary and, to a lesser extent, secondary structures of many transport systems are known. Because of the absence of suitable crystallization methods, however, only a few polytopic membrane proteins have yielded to X-ray crystallographic analyses (127–130), resulting in high resolution three-dimensional structural information. Recently, combinatorial approaches have been applied to successfully crystallize membrane proteins in a high throughput approach, resulting in the structure of a homolog of the multidrug-resistant ATP-binding cassette transporters to a resolution of 4.5 Å (131). In this study, approximately 96,000 crystallization conditions using about 20 detergents were tested to yield crystals with good quality for X-ray structure determination. Unfortunately, the resulting protein was not functionally active, but the structural information corroborated several previously obtained experimental results. The combinatorial crystallization approach used by Chang and Roth (131) will likely stimulate other researchers to attempt crystallization of other important membrane proteins.

Meanwhile, we base our views of solute transport on molecular models that provide working models of transport systems (132–134). It is well recognized that any two proteins that show sequence homology (i.e., share sufficient primary structural similarity to have evolved from a common ancestor) will prove to exhibit strikingly similar three-dimensional structures (135). Furthermore, the degree of tertiary structural similarity correlates well with the degree of primary structural similarity. Phylogenetic analyses allow application of modeling techniques to a large number of related proteins and additionally allow reliable extrapolation from one protein family member of known structure to others of unknown structure. Thus, once three-dimesnional structural data are available for any one family member, these data can be applied to all other members within limits dictated by their degrees of sequence similarity.

5.4 Techniques for Studying Integral Membrane Protein Structure

There are various approaches to study the topography of integral membrane proteins in the absence of a resolved crystal structure. *In vitro* and *in vivo* translation of constructs containing one or more transmembrane sequences, epitope localization, reaction with sided reagents, and proteolysis of the purified membrane-inserted protein have all been used individually or in combination. (Table 8.3). Each has advantages and disadvantages, and only rarely does one method enable conclusive topographic analysis of the membrane-embedded segments of a polytopic integral membrane protein.

Considerable progress has been made over the past 10 years by the group of Kaback at UCLA, who have taken numerous biophysical approaches toward studying the structure of the bacterial transporter lactose permease (136). Table 8.3 lists the techniques that this and other groups have used to solve the structure and topology of membrane transporters in the absence of a crystal structure.

5.4.1 Membrane Insertion Scanning. Membrane topology of transporter proteins can be

Table 8.3   Techniques for Studying Polytopic Membrane Protein Structure

1. Site-directed Ala scanning
2. Membrane insertion scanning
3. Site-directed thiol crosslinkers
4. Excimer fluorescence
5. Engineered divalent metal-binding sites
6. EPR
7. Metal–spin-labeling interactions
8. Site-directed chemical cleavage
9. Identification of discontinuous monoclonal antibody probes
10. *N*-glycosylation site engineering

predicted by various computer algorithms; however, none of the existing methods can definitively define these regions, often producing a "sliding window" of groups of hydrophobic amino acid residues that are suitable candidates for the formation of a membrane-spanning segment. Examples of these software programs are TopPred II (137) (http://www.biokemi.su.se), the PHD Topography neural network system developed by Rost and coworkers (138) (http://www.embl-heidelberg.de/predictprotein), or the hidden Markov model recently described by Tusnády and Simon (139) (http://www.enzim.hu/hmmtop). Sachs and colleagues at UCLA pioneered a technique aptly named "membrane insertion scanning" to determine whether a sequence of amino acids is capable of spanning a membrane. They tested the ability of individual protein segments to function as signal anchors for membrane insertion by placement between a cytoplasmic anchor encompassing the first 101 amino acids of the rabbit H^+,K^+-ATPase (HK M0) subunit and a glycosylation flag sequence consisting of five N-linked glycosylation sites located in the C-terminal 177 amino acids of the rabbit HK M0 subunit (140). Stop transfer properties of hydrophobic sequences can be examined in a similar manner using the first 139 N-terminal amino acids of the rabbit H^+,K^+-ATPase subunit (HK M1) containing the first membrane sequence of the H^+,K^+-ATPase as a signal anchor upstream of individual predicted transport transmembrane sequences linked to the N-glycosylation flag. The membrane insertion scanning technique has been used to determine the membrane topology of various transporters, channels, and receptors. A common criticism of membrane insertion studies is the placement of a hydrophobic amino acid sequence out of its physiological or environmental context, thus "forcing" an abbreviated sequence through the membrane. Transmembrane regions in polytopic membrane proteins may require flanking topogenic information to become integrated into the lipid bilayer and, thus, topology determination by this technique alone may not be definitive.

5.4.2 Cys-Scanning Mutagenesis. Studying the lactose permease of *Escherichia coli*, a polytopic membrane transport protein that catalyzes β-galactoside/H^+ symport, Frillingos and colleagues (141) used Cys-scanning mutagenesis to determine which residues play an obligatory role in the mechanism and to create a library of mutants with a single-Cys residue at each position of the molecule for structure/function studies. In general, this type of study will define amino acid side chains that play an irreplaceable role in the transport mechanism and positions where the reactivity of the Cys replacement is altered upon ligand binding. Furthermore, helix packing, helix tilt, and ligand-induced conformational changes can be determined by using the library of mutants in conjunction with a battery of site-directed techniques.

5.4.3 N-Glycosylation and Epitope Scanning Mutagenesis. A technique that can be used in addition to membrane insertion scanning is based on the fact that *N*-glycosylation occurs only on the luminal side of the endoplasmic reticulum. This method has been successfully used to determine which domains of the receptors/channels are located extracellularly. In general, a glycosylation-free mutant is first designed and subsequently, *N*-glycosylation consensus sequences (*NXS/T*) are engineered into hydrophilic regions of an aglycomutant. Based on the positioning of the glycosylation groups within the extracellular membrane, a molecular weight shift may indicate successful glycosylation and reveals definitive information on protein topology. A drawback of this technique is the dependency of glycosylation on efficiency and accessibility of the concensus sequence to the glycosylation machinery;

thus, nonglycosylated mutants do not provide conclusive information about the topology and these data should be interpreted carefully. Regardless, the topology of various proteins has been successfully solved using this technique, including the sodium-dependent glucose transporter (SGLT1) (142) and the γ-aminobutyric acid transporter GAT-1 (143).

Alternatively, small peptide epitopes can be inserted into the extramembranous parts of an SLC prote in that are recognized by a well-characterized monoclonal antibody. Covitz and colleagues (144) used this technique to determine the topology of the peptide transporter, PepT1. An epitope tag, EYMPME, was inserted into different extramembranous locations of hPEPT1 by site-directed mutagenesis. The membrane topology was solved by labeling reconstituted, functionally active, EYMPME-tagged hPEPT1 mutants with an anti-EYMPME monoclonal antibody in nonpermeabilized and permeabilized cells.

5.4.4 Excimer Fluorescence. Site-directed excimer fluorescence (SDEF) and site-directed spin labeling (SDSL) are two particularly useful techniques to study proximity relationships in membrane helices. The experiments are based on site-directed pyrene labeling of combinations of paired Cys replacements in a mutant devoid of Cys residues. Because pyrene exhibits excimer fluorescence if two molecules are within about 3.5 Å, the proximity between paired labeled residues can be determined. Moreover, interspin distances in the range of 8–25 Å between two spin-labeled Cys residues can be measured in the frozen state. Using this technique, Kaback and coworkers showed that ligands of the lac permease cause a dramatic increase in reactivity that is consistent with the notion that the mutated amino acid positions are transferred into a more hydrophobic environment (145, 146).

5.4.5 Site-Directed Chemical Cleavage. The insertion of short reporter sequences (e.g., factor Xa protease cleavage sites) into hydrophilic loops has proved to be a useful alternative to *N*-glycosylation scanning mutagenesis. However, this approach requires isolation of homogeneous preparations of intact membranes and many tedious control experiments, and experimental difficulties associated with protease accessibility are well documented (147). In general, in-frame factor Xa protease sites are inserted into a target sequence at positions within the NH_2- and COOH-terminal domains, and into hydrophilic loops. The factor Xa protease recognizes the tetrapeptide motif IEGR and specifically cleaves the protein sequence COOH-terminal of the arginine residue (148). Generally, the recognition motif is tandomly (IEGRIEGR) inserted to increase the probability of cleavage (149). After digestion of purified protein vesicles with the factor Xa enzyme, fragments are isolated on SDS–PAGE and can be analyzed to further determine membrane topology.

5.5 Case Studies

5.5.1 P-Glycoprotein: Understanding the Defining Features of Regulators, Substrates, and Inhibitors.
The ABC efflux transporter P-glycoprotein (P-gp) is a large 12 transmembrane-domain bound protein initially noted to be present in certain malignant cells associated with the multidrug resistance (MDR) phenomenon that results from the P-gp–mediated active transport of anticancer drugs from the intracellular to the extracellular compartment (150). However, P-gp is normally expressed at many physiological barriers including the intestinal epithelium, canalicular domain of hepatocytes, brush border of proximal tubule cells, and capillary endothelial cells in the central nervous system (CNS) (150). Expression of P-gp in such locations results in reduced oral drug absorption and enhanced renal and biliary excretion of substrate drugs (151). Moreover, P-gp expression at the blood–brain barrier is a key factor in the limited CNS entry of many drugs. The expressed level of P-gp as well as altered functional activity of the protein attributed to genetic variability in the MDR1 gene also appears to impact the ability of this transporter to influence the disposition of drug substrates (152).

Interestingly, cytochrome P450 3A4 (CYP3A4), a drug-metabolizing enzyme with broad substrate specificity, appears to coexist with P-gp in organs such as the intestine and liver. These observations led to the hypothesis that there may be a relationship between these two proteins in the drug disposition process. Wacher et al. have described the overlapping substrate specificity and tissue distribution of CYP3A4 (153) and P-gp. Schuetz et al. found that modulators and substrates coordinately upregulate both proteins in human cell lines (154). Similarly, P-gp–mediated transport was found to be important in influencing the extent of CYP3A induction in the same cell lines and also in mice (155). More recent data have suggested that there may be a dissociation of inhibitory potencies for molecules against these proteins. Although some molecules can interact with CYP3A4 and P-gp to a similar extent, for the most part the potency of inhibition for CYP3A4 did not predict the potency of inhibition for P-gp, and vice versa (58). Moreover, not all CYP3A substrates such as midazolam and nifedipine are P-gp substrates (156). The key to this linkage between P-gp and CYP3A4 appears to be their coregulation by the pregnane-X-receptor at the level of transcription (157, 158). Recently, with the description of the X-ray crystal structure of this protein with SR12813 bound, we may be closer to understanding the structural features necessary for PXR ligands. In addition, a pharmacophore for PXR ligands may also enable us to select new drugs that are less likely to induce P-gp and CYP3A4 (159).

To account for the observed broad substrate specificities for both CYP3A4 and P-gp, the presence of multiple drug binding sites has been proposed (160–163). The first elegant experimentally determined signs of a complex behavior for P-gp appeared in 1996 when cooperative, competitive, and noncompetitive interactions between modulators were found to interact with at least two binding sites in P-gp (164). The multiple site hypothesis was confirmed by other groups (165–167). Subsequent results have indicated there may be three or more binding sites (168). Steady-state kinetic analyses of P-gp–mediated ATPase activity using different substrates indicate that these sites can show mixed-type or noncompetitive inhibition indicative of overlapping substrate specificities (169). Other researchers have determined that immobilized P-gp demonstrates competitive behavior between vinblastine and doxorubicin, cooperative allosteric interactions between cyclosporin and vinblastine or ATP, and anticooperative allosteric interactions between ATP, vinblastine, and verapamil (170). Clearly allosteric behavior by multiple substrates, inhibitors, or modulators of CYP3A4 or P-gp complicates predicting the behavior and drug–drug interactions of new molecules *in vivo* and has important implications for drug discovery.

In terms of understanding P-gp structure–activity relationships, photoaffinity experiments have been valuable in defining the cyclosporin binding site in hamster P-gp (171) and indicating that trimethoxybenzoylyohimbine (TMBY) and verapamil bind to a single or overlapping sites in a human leukemic cell line (172). Additional studies have shown that TMBY is a competitive inhibitor of vinblastine binding to P-gp (173). The P-gp modulator LY 335979 has been shown to competitively block vinblastine binding (173), whereas vinblastine itself can competitively inhibit verapamil stimulation of P-gp-ATPase (172). With the growth in knowledge derived from these and other studies, it would be valuable to use structural information to define whether unrelated molecules are likely to interact with P-gp.

Interestingly, there has been over a decade's worth of studies that have identified pharmacophores for this efflux pump that could have been used to help predict P-gp–related bioavailability issues. Early computational studies using P-gp modulators such as verapamil, reserpine, 18-epireserpine, and TMBY showed that they could be aligned, suggesting the importance of aromatic rings and a basic nitrogen atom in P-gp modulation (174, 175). A subsequent, more extensive study with 232 phenothiazines and structurally related compounds indicated that molecules with a carbonyl group that is part of an amide bond plus a tertiary amine were active P-gp inhibitors (176). A model built with 21 molecules of various structural classes that modulate P-gp-ATPase activity suggested these molecules competed for a single binding site (177). Similarly, 19 propafenone-type P-gp inhibitors were then used to confirm the requirement for a carbonyl oxygen, suggested to form a hydrogen bond with P-gp (178).

Others have used MULTICASE to determine important substructural features like

$CH_2–CH_2–N–CH_2–CH_2$ (179), and linear discriminant analysis with topological descriptors (180). In 1997 the first 3D-QSAR (quantitative structure-activity relationships) analysis of phenothiazines and related drugs known to be P-gp inhibitors was described (181). This was followed by Hansch-type QSAR studies with propafenone analogs (182), CoMFA studies of phenothiazines and related drugs (183), and simple regression models of propafenone analogs (184, 185). These latter models confirmed the relevance of hydrogen bond acceptors and the basic nitrogen for inhibitors (184, 185) and multiple hydrogen bond donors in substrates 2.5–4.6 Å apart (186). One study using a diverse array of inhibitors with P-gp-ATPase activity noted that size of the molecular surface, polarizability, and hydrogen bonding had the largest impact on the ATPase activity (187). A number of computational approaches and models of P-gp have yielded useful information that is usually derived from a series of structurally related molecules. This is not surprising, given that *in vitro* studies with P-gp inhibitors frequently take this approach. A recent example suggested P-gp inhibitors with high lipophilicity and polarizability were found to be more likely to be high affinity ligands for the verapamil-binding site (188). However, some complexity arises if one considers more structurally diverse molecules because they may bind to different sites within P-gp. This hypothesis derives from experimental results describing a complex behavior for P-gp, such that cooperative, competitive, and noncompetitive interactions between modulators may occur (164), indicative of multiple binding sites within P-gp (165–167).

Recently, specific models addressing the individual P-gp binding sites using a diverse array of inhibitors have been described. The use of a computational approach to model *in vitro* data derived from structurally diverse inhibitors of digoxin transport in Caco-2 cells, vinblastine and calcein accumulation in P-gp expressing LLC-PK1 (L-MDR1) cells, vinblastine binding in vesicles derived from CEM/VLB100 cells, and verapamil binding in Caco-2 cells (159) suggested there may be some overlap in the binding sites for these four substrates. The inhibitor pharmacophores were suggested to be mainly large with hydrophobic and hydrogen bonding features. It is hypothesized that vinblastine, digoxin, and verapamil are likely to bind a single site, given that strong correlations between these models were observed. A simple P-gp substrate pharmacophore was generated using the alignment of verapamil and digoxin, onto which vinblastine was aligned (159). This model also contained many hydrophobic and hydrogen bond acceptor features.

As models are generated for the other binding sites and substrates, they may help in elucidating important structural differences between ligands for each site. Such models will be of value for modulating the bioavailability of drugs and understanding the locations of the P-gp binding site(s). The latter may be of particular importance because we are now in an era in which membrane proteins are being crystallized. The 22-Å resolution X-ray structure of the monomer for the related ABC family member (MRP1/ABCC1) is suggested to be structurally similar to P-gp, in that it appears to possess a pore region ringed by protein (189). The recent determination of the 4.5-Å resolution X-ray structure of the lipid flippase MsbA from *E. coli*, another ABC family member, provides a further structure on which to build homology models for P-gp because its amino acid sequence is similar (73%) (131). It is suggested that P-gp pharmacophores could be docked into these structures and the amino acids likely to correspond on both monomers defined. The question of whether P-gp behaves like this flippase or like more traditional pores still remains, but computational techniques will be at the forefront of continuing to provide some enlightenment and aid in drug discovery.

5.5.2 The Intestinal Peptide Transporter (PepT1)

5.5.2.1 Introduction. Over the past two decades, targeting delivery to nutrient transport systems in the gastrointestinal tract has emerged as a strategy to increase oral availability of poorly absorbed drugs and it has met with several successes. In this approach, drug penetration is enhanced by coupling a drug molecule to natural ligands for nutrient transporters. The toxicity and immunogenicity of

drug–ligand conjugates is expected to be inherently low because of the use of endogenous substrates. Alternatively, an approach named "substrate mimicry" can be used; this approach aims to create novel drug entities that mimic the three-dimensional features of natural ligands. Both approaches should result in compounds that are recognized by a specific transport protein embedded in the enterocyte brush-border membrane, thereby facilitating transport across the intestinal wall.

5.5.2.2 Physiology of PepT1. In general, the intestinal small peptide carrier (PepT1) is a proton-coupled, low affinity, active, oligopeptide transport system with a broad substrate specificity. In addition to transporting its natural substrates, di- and tripeptides occurring in food products (190–194), it shows affinity toward a broad range of peptidelike pharmaceutically relevant compounds, such as β-lactam antibiotics (195, 196) and angiotensin-converting enzyme (ACE) inhibitors (197–207). In fact, these molecules can oftentimes be viewed as "peptidelike" in their molecular composition (Fig. 8.6). For this reason, the transporter has been recognized as an important intermediate in the oral bioavailability of peptidomimetic compounds (199). However, the lack of knowledge regarding structural specificity toward its substrates has prevented the use of this transporter on a more rational basis. In addition, its cellular localization to the apical membrane does not provide a mechanism for cellular exit into the basal compartment (208, 209). Currently there is a keen interest in understanding the structural determinants for substrates and inhibitors of this transport protein.

PepT1 belongs to a larger family of oligopeptide transporters, the proton/oligopeptide transporter (POT) family, where currently only two members have been identified in humans, hPepT1 and hPepT2. POT family members have two characteristic protein signatures assigned by Steiner and Becker, known as the PTR2 family signatures (210, 211). The first of these signature sequences includes the end of transmembrane span 2 (TMD2), the intracellular loop, and TMD3. The second PTR2 signature corresponds to the core of TMD5. A third consensus sequence (GTGGIKPXV), proposed by Fei and cowork-

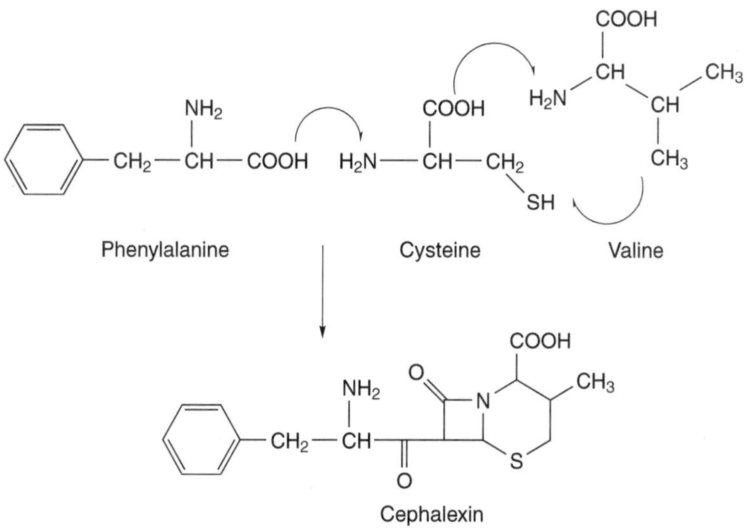

Figure 8.6. Molecular building blocks of the β-lactam antibiotic cephalexin. The three amino acids phenylalanine, valine, and cysteine are shown to merge through (pseudo)peptide bonds, to form a peptidomimetic compound with the molecular features of a tripeptide.

ers (212), is well conserved between mammalian and *C. elegans* peptide transporters CPTA and CPTB, but is not specific to or well conserved in other members of the POT superfamily.

5.5.2.3 Structure–Transport Relationship of PepT1. Structural information on PepT1 has been limited to its primary sequence and predicted structural membrane topology. Hydropathy analysis of the human, rabbit, and rat PepT1 isoforms have predicted the presence of 12 transmembrane domains (TMD) in each isoform (212). This model has been partially proved by other investigators (144, 213).

The human PepT1 sequence has been predicted to contain a 2127 base-pair (bp) open reading frame that encodes a 708 amino acid, 79-kDa protein, with an estimated pI of 8.58 (212). Hydropathic analysis also predicts that the N- and C-termini are located on the cytoplasmic side of the membrane. Site-directed mutagenesis, using EYMPME (EE)-epitope tag insertion at different locations of the human PepT1 transmembrane regions have identified the COOH-terminal to be intracellular, whereas epitope tags at positions 106 and 412 showed that the predicted loops between 3TM and 4TM, and 9TM and 10TM, respectively, were extracellularly localized

(144). These results support the predicted loop and TMD numbers and orientations from TMD4 through the C-terminus (144). The results using EE epitope insertions in the amino terminal region were inconclusive, possibly because of EE effects on function, leaving ambiguity to the predicted structure from the N-terminus to TMD3.

Because no three-dimensional structure of the transporter is available, design of new substrates for PepT1 has relied on indirect structure–affinity relationships. Early studies were directed to define a pharmacophoric pattern for this transporter (214–216). A pharmacophore or "recognition site" is defined based on a common arrangement of essential atoms or groups of atoms appearing in each active molecule. In many published studies the active analog approach (AAA), originally developed by Marshall and coworkers (217), has proved to be useful in rationalizing and predicting pharmacological data of active and inactive substrates. In the AAA, the structural requirements common to a set of compounds showing affinity for the same transporter may be used to define a pharmacophoric pattern of atoms or groups of atoms mutually oriented in space, which are necessary for binding to this transporter. Two factors have to be considered for the successful application of this technique: the included compounds should be structurally homogeneous and the individual compounds should have a low level of conformational flexibility. Because the natural substrates, di- and tripeptides, contain many freely rotatable bonds, most studies have concentrated on substrates with more rigid backbones, such as a β-lactam nucleus (214, 215, 218, 219). Therefore, the AAA is limited, in that it requires structurally similar rigid molecules. One study used β-lactams and showed that a carboxylic carbon (likely to position in a positively charged pocket), two carbonyl oxygen atoms (hydrogen bond acceptors), a hydrophobic site, and finally an amine nitrogen atom (hydrogen bonding region) were important features of substrates (218).

Another study examined the three-dimensional structural features of three structurally closely related PepT1 substrates: enalapril, enalaprilat, and lisinopril (Fig. 8.7) (220). Enalapril is an ester prodrug of the pharmacologically active enalaprilat. After oral administration of enalapril, the active compound (enalaprilat) is formed by bioconversion of enalapril. Enalaprilat, a diacid, binds slowly and tightly to ACE, producing well-defined clinical effects, but is poorly absorbed from the gastrointestinal tract (3–12% bioavailability) (221). The prodrug approach of esterifying enalaprilat to enalapril is required to enhance the oral bioavailability to 60–70%. By use of a combined *in vitro* and molecular modeling approach, we showed that intramolecular hydrogen bond formation between the lysyl side chain in enalaprilat and its carboxylic acid groups may explain decreased affinity for PepT1 (220).

Structural studies of PepT1 substrates were extended using comparative molecular field analysis (CoMFA) of 10 known substrates for the peptide transporter with data derived from an *in situ* rat model (222). The CoMFA approach required manual alignment of relatively rigid molecules; however, it allowed explanation of the variation in the permeability with respect to steric and electrostatic interaction energies of the molecules (222). This 3D-QSAR produced by our laboratory provided a valuable starting point for future prediction of

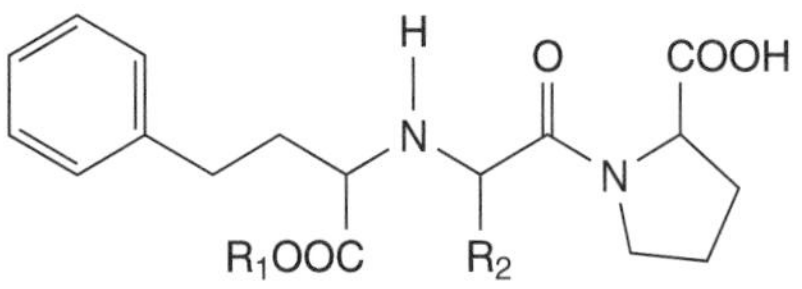

	R_1	R_2
Enalapril	—C_2H_5	—CH_3
Enalaprilat	—H	—CH_3
Lisinopril	—H	—$(CH_2)_4NH_2$

Figure 8.7. Molecular structure of the ACE-inhibitors enalapril, enalaprilat, and lisinopril. Enalapril and lisinopril are substrates for PepT1.

the affinity of substrates for this transporter and at the same time confirmed the earlier findings from AAA studies.

Studies by other research groups have sought to address the requirements for binding and transport by PepT1. The allowable backbone distance between the amine and the carboxylic acid terminals for binding to and transport by PepT1 have been demonstrated in part with a series of ω-amino fatty acids (ω-AFA) (223). These studies clearly demonstrated that 4–10 methylene carbon groups (CH_2 units) could be accommodated between the termini and allow direct binding with PepT1. Binding of ω-AFAs was demonstrated to inhibit D-Phe-Ala uptake competitively in transiently transfected hPepT1 cells, with the competitive inhibition (EC_{50} values) appearing not to change substantially with chain lengths from 5–11 CH_2 units between the terminal functional groups (223). The conclusions of this study indicate that a peptide bond is not essential for substrates of PepT1, and that two ionized amino or carboxyl groups with at least four CH_2 units, providing a distance of 500 to 635 pm between them, are required for transport by PepT1 (223). This study also raised considerable concerns about conformation during presentation because the CH_2—CH_2 midsection has considerable flexibility in contrast to that of a peptide bond.

The distance between the charged functionalities for PepT1 recognition has been demonstrated to be critical (223). Insights into the affinity of β-amino groups were also elucidated in these studies, where 8-amino-octanoic acid effectively inhibited D-Phe-Ala up-take, whereas 2-amino-octanoic acid had no inhibitory effect (223). Interestingly, further studies have demonstrated that the α- or β-amino carbonyl functionality may affect PepT1 substrate affinity, and serve as the key determinant in PepT2 affinity preference over PepT1, where coexpressed (224). For example, the presence of a β-amino group resulted in the substrate having a higher affinity for transport by PepT2 than by PepT1 (224). However, anionic β-lactam antibiotics, lacking an α-amino group, demonstrated a preference for PepT1.

A recent study has taken this understanding of PepT1 further by using a meta-analysis of K_i data for 42 substrates across eight classes of molecules (225). The findings by these authors were in agreement with previous studies and provided a template consisting of an N-terminal NH_3 site, a hydrogen bond to the carbonyl group of the first peptide bond, a hydrophobic pocket, and a carboxylate binding site (225). This model was also used to classify the compounds as possessing high, medium, or low affinity for PepT1. Another group has generated K_i values for 23 β-lactams and suggested their data supported the need for all of the structural features previously described for recognition by the transporter (226). Recent studies have used Caco-2 cells and assessed the inhibition (K_i) of Gly-[^{3}H]-L-Pro transport by ACE inhibitors as a means to understand dipeptide transport. A two-dimensional model was constructed that indicated the types of functional groups favored for inhibition (197).

Other studies have shown the importance of certain amino acid residues in determining PepT1 peptide transport activity (227, 228). Site-directed mutations of single amino acids located within the PepT1 transmembrane domains (TMDs) showed that Trp at position 294 and Glu at position 595 reduced significantly the glycylsarcosine uptake by human embryonic kidney cells (HEK293), whereas Tyr at position 167 (TMD5) inactivated the transporter completely (227). PepT1 has shown to present another important structural characteristic: the presence of conserved histidyl residues (H) in several transmembrane regions. Sequence alignment demon-strated the presence of H57, H121, and H260 residues (corresponding positions in the PepT1 sequence) in PepT1 and PepT2. This group is speculated to be involved in the H^+-binding, and thus being fundamental for transport function. Histidyl residues at positions 57 and 121 have been demonstrated to be essential to maintain PepT1 function, whereas the residue at position 260 remains to be completely defined (228). The molecular mechanism(s) by which these residues may interact with H^+ remains to be elucidated.

5.5.3 The Apical Sodium-Dependent Bile Acid Transporter (ASBT)

5.5.3.1 Introduction. The "apical sodium-dependent bile acid transporter" (ASBT) has been targeted for oral drug delivery as well. The high efficacy of ASBT combined with its high capacity makes this system an interesting target for drug delivery purposes, including local drug targeting to the intestine and improving the intestinal absorption of poorly absorbable drugs. ASBT has recently received much attention because of its pivotal role in cholesterol metabolism. Inhibition of intestinal bile acid reabsorption elicits increased hepatic bile acid synthesis from its precursor cholesterol, thereby lowering plasma cholesterol levels (229). Application of this approach has met considerable clinical success using unspecific bile acid sequestrants, such as cholestyramine and colestipol (230). Currently, several novel ASBT inhibitors are in clinical trials for the treatment of hypercholesterolemia (231).

5.5.3.2 Physiology of ASBT. Bile acids mediate the digestion and absorption of fat and fat-soluble vitamins (232). The total bile acid pool in humans, 3–5 g, circulates 6–10 times a day, giving rise to a daily bile acid turnover of 20–30 g in humans. Only 0.2–0.5 g of bile acids is lost in feces per day and this amount is repleted by the *de novo* synthesis of bile acids (233). A detailed description of the enterohepatic circulation of bile acids may be found in several reviews (234–239). The primary bile acids cholate and chenodeoxycholate are synthesized from cholesterol in the liver mediated by the enzyme cholesterol 7α-hydroxylase (CYP7A) (Fig. 8.8). The activity of this enzyme is under inhibitory feedback control by bile

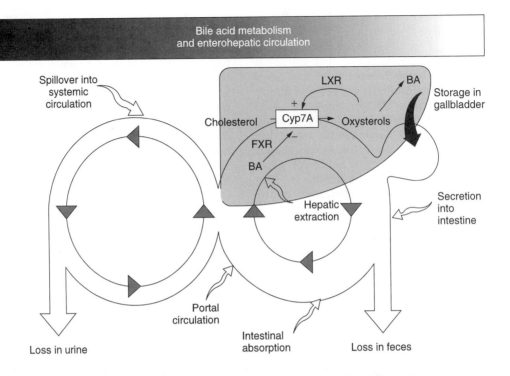

Figure 8.8. Biosynthesis of bile acids and the enterohepatic circulation. Bile acids are synthesized from cholesterol in the liver under feedback regulation of the nuclear orphan receptors farnesoid X receptor (FXR) and lignane X receptor (LXR). They are stored in the gallbladder and released through the bile duct into the duodenum, where they aid in the digestion of dietary fats. Intestinal uptake of bile acids takes place along the entire length of the small intestine, but active reabsorption is confined to the distal ileum to minimize loss of bile salts in the feces. The portal circulation carries bile acids from the intestine to the liver, where they are actively absorbed by hepatocytes and secreted into bile.

acids. In light of the importance of this pathway for the removal of cholesterol, the mechanisms for transcription regulation of the CYP7A gene have been extensively studied. In the ileum, conjugated bile acids are reabsorbed by ASBT (also named IBAT) gene SLC10A2.

Within the cytoplasm of the enterocyte, a second 14- to 15-kDa soluble protein binds the absorbed bile acid and mediates its transfer across the cell. This protein is generally known as ileal bile acid binding protein (IBABP) (240), although it was originally named gastrotropin (241, 242). The exact interaction between ASBT and IBABP is currently unknown, but there has been speculation that IBABP may bind to ASBT at the cytosolic surface and may play a role in mediating bile acid affinity for ASBT (243). In addition to IBABP, 20-, 35-, and 43-kDa proteins that have remained unidentified appear to bind bile acids in the cytosol (244), although the specific affinity of these proteins for bile acids is probably low. For example, the putative 43-kDa protein was shown to share high homology with actin, which plays a role in many transport processes in the cell (245).

Expression of IBABP and ASBT is regulated by the farnesoid X receptor (FXR), a nuclear orphan receptor (246, 247). A specific binding site for FXR was found on the promotor region of the IBABP gene, named bile acid response element (BARE) (248). The response was greatest in the presence of chenodeoxycholic acid (CDCA), whereas cholic acid was much less effective and the secondary bile acids deoxycholic and lithocholic acid had variable responses.

Molecular regulation of CYP7A appears to be regulated by the liver X receptor (LXR), a member of the steroid receptor superfamily that, in turn, appears to be regulated by oxysterols (249). Two isoforms of LXR, α and β, have been identified; the α isoform is responsible for regulating CYP7A and some SAR and pharmacophore studies suggest the 24-oxo ligands acting as hydrogen bond acceptors bind tightly to LXRα and may be candidates for the natural ligand (250, 251). ASBT is primarily expressed on the apical membranes of ileal enterocytes in mammals, although weak translation has been observed in the jejunum; however, these observations remain controversial and may have resulted from (patho)physiologic induction. Furthermore, ASBT has been detected in rat bile duct epithelial cells (cholangiocytes) (252), where it probably functions as part of a bile acid secretion feedback mechanism. Microinjection of guinea pig and rabbit ileal mucosal Poly (A+) mRNA into *Xenopus laevis* oocytes resulted in translation of a functional Na$^+$/bile acid carrier that is expressed in the surface membrane (253). As expected, incubation with rabbit jejunal–mucosa Poly (A+) mRNA did not result in expression of the bile acid transporter. Timed photoaffinity labeling techniques revealed the brush-

border (254), cytosolic (244), and basolateral proteins (255) involved in the transfer of bile acids across the rat ileal enterocyte. A putative transport protein involved in coupled Na$^+$-bile acid transfer across the ileal brush-border membrane had a molecular weight of 99 kDa (254, 256). Dawson and coworkers (257, 258) were the first to clone and characterize a 43-kDa protein as the rat ileal bile acid transporter protein. The 99-kDa protein that was detected previously can be explained as a dimeric form of the glycosylated transport protein (Fig. 8.9).

Subsequent electrophsyiological characterization revealed that ASBT is an electrogenic, active cotransporter driven by a Na$^+$-gradient across the apical membrane of ileocytes with a 2:1 Na$^+$:bile acid coupling stoichiometry (259). To date, structural information on ASBT has been limited to its primary (sequence) and secondary (membrane topology) structures. Topography models of ASBT predict seven transmembrane (7TM) regions (258), but eight or nine membrane-spanning regions have been suggested by topology/hydropathy analysis (260). Experimental evidence indicates a *trans* position of the N- and C-terminal protein domains (N$_{exo}$/C$_{cyt}$) (126, 258, 261), which effectively eliminates the putative 8TM model. Based on membrane insertion scanning techniques, a 9TM model was recently suggested by Hallén and coworkers (260); however, this model suggests the presence of two very short TM domains, presumably in the β-helix conformation. Overall, the 7TM model appears to be the most feasible topology for ASBT to date (Fig. 8.10). Studies with antibodies raised against the terminal amino acid sequences verify that the amino terminal is located extracellularly, whereas the carboxy terminal is located at the cytoplasmic side of the membrane (262). This unique topology makes ASBT different from other members of the SLC family in that it does not adhere to the "positive-inside" rule (263). Human ASBT contains 13 cysteine residues, of which 12 are conserved in the hamster, rat, rabbit, and mouse. Only four of these 12 cysteines are conserved in the human hepatic bile acid transporter (Ntcp), which has 37% homology and 48% similarity with ASBT, and only two cysteines are conserved in the P3-protein, a related orphan transporter.

Because bile acids are biosynthesized from

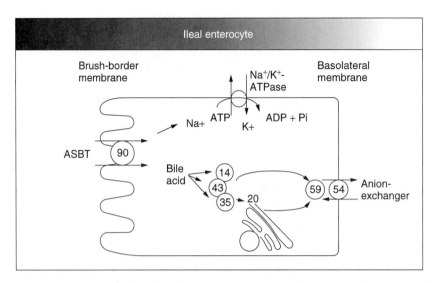

Figure 8.9. Physiology and molecular biology of intestinal bile acid transport. Bile acids are actively absorbed in enterocytes through a sodium-dependent cotransporter, ASBT. The sodium gradient is maintained by the sodium–potassium ATPase, located at the basolateral membrane. In the cytosol, bile acids are shuttled through the cell by the aid of various proteins, most importantly the ileal bile acid binding protein, iBABP. An anion exchanger transports bile acids across the basolateral membrane into the portal circulation.

cholesterol and are the only forms for cholesterol to be excreted *in vivo*, a specific nonabsorbable inhibitor of the ileal bile acid transporter system would lower plasma cholesterol level by blocking the intestinal reabsorption of bile acids and consequently raising the conversion of cholesterol to bile acids. Recently, several compounds have been shown to be able to lower serum cholesterol in animal studies by specifically inhibiting ASBT (264–267). These findings substantiate the feasibility of using ASBT as the new pharmacological target for cholesterol-lowering therapy.

5.5.3.3 Structure-Activity Relationships for ASBT. The studies with endogenous bile acids have led to physiological understanding of the function of bile acids and the enterohepatic circulation. The search for a deeper understanding of the molecular mechanism behind the affinity and recognition of molecules by the bile acids carriers in both ileum and liver has led researchers to modify bile acids and study the carrier affinity of these compounds. These modifications typically entail either the substitution of the hydroxyl groups at the 3, 7, or 12 positions by other functionalities or the

addition to or alterations at the C-17 side chain (Fig. 8.11).

In a series of seminal experiments, Lack and Weiner were the first to establish a basic structure–activity relationship for intestinal bile acid transport using the rat everted sac model (268). Important additional findings were recorded by the group of Kramer. The following general observations were reported:

1. The K_m is related to whether the bile acid is conjugated or unconjugated (269, 270). In general, conjugated bile acids have a two-fold higher K_m value than that of their unconjugated parent compounds (271).

2. The V_{max} is independent of conjugation but appears to be related to the number of hydroxyl groups attached to the sterol nucleus: trihydroxy $\gg$ dihydroxy $>$ monohydroxy (271, 272). However, no single hydroxy group is essential for transport. Triketo bile acids, such as taurodehydrocholic acid, are devoid of all hydroxyl groups and show considerably less transport capacity (273, 274). Krag and Phillip (275) found similar structure–V_{max} rela-

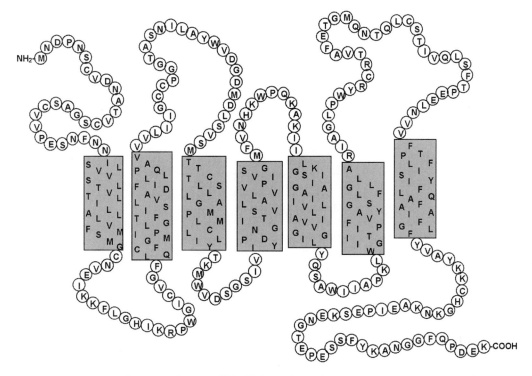

Figure 8.10. Membrane topology of ASBT. Hydropathy analysis proposes a seven transmembrane topology for ASBT. Studies with antibodies to terminal epitopes have confirmed the inside-out orientation of this membrane protein.

tionships in humans; however, no correlation was observed between K_T values and conjugation.

3. For efficient transport, a bile acid molecule must possess a single negative charge (276) that should be located on the C-17 side-chain of the sterol nucleus (277).

4. Although bile acids with two negative charges around the C-17 position give minimal active transport, the addition of an extra negative charge in the form of sulfonation at the C-3 position does not preclude active transport (278). Replacing the anionic moiety from the C-17 side-chain to position 3 results in a complete loss of affinity (277).

5. Stereospecificity of the hydroxyl groups on the sterol nucleus was shown by substitution of this group at the 3-position with a hydroxyethoxy moiety (279). The 3α-isomer was able to inhibit transport of [³H]taurocholate in rabbit ileal brush-border membrane vesicles, whereas the 3β-iso-

mer showed very weak affinity and was unable to inhibit taurocholic acid transport.

Lack and coworkers have proposed that the recognition site for carrier-mediated bile acid transport is a hydrophobic pocket on the membrane surface that consists of three components: a recognition site for interaction with the steroid nucleus; a cationic site for coulombic interaction with the negatively charged side chain; and an anionic site for interaction with Na^+. Supposedly, this anionic site could be responsible for the reduced affinity of bile acid derivatives with a dianionic side chain (280). Generalizations on the structural requirements for ASBT affinity include:

1. The presence of at least one hydroxyl group on the steroid nucleus at position 3, 7, or 12.

2. A single negative charge in the general vicinity of the C-17 side-chain; however, an

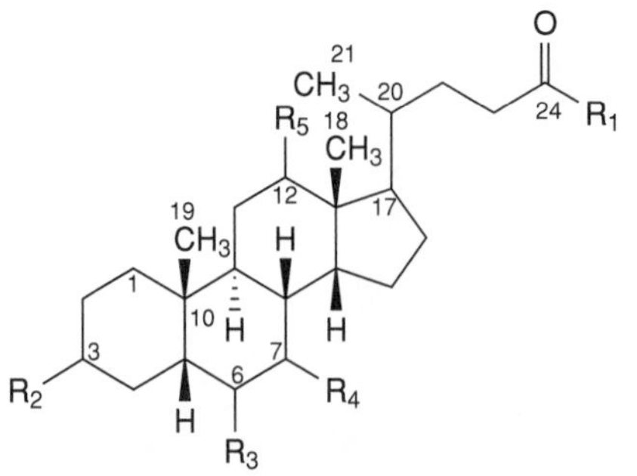

R_1	R_2	R_3	R_4	R_5	Prefix
-OH	-OH	-H	-OH	-OH	–
-OH	-OH	-H	-OH	-H	Chenodeoxy-
-OH	-OH	-H	-H	-OH	Deoxy-
-OH	-OH	-H	-H	-H	Litho-
-OH	-OH	-H	-OH (β)	-OH	Urso-
-OH	-OH	-H	-OH (β)	-H	Ursodeoxy-
-OH	-OH (β)	-H	-OH (β)	-H	Isoursodeoxy-
-OH	-OH	-H	-H	-OH (β)	Lagodeoxy-
-OH	-OH	-OH	-OH	-H	Hyo-
-NHCH$_2$COOH	–		–	–	Glyco-[1]
-NH(CH$_2$)$_2$SO$_3$H	–		–	–	Tauro-[1]

[1]The prefixes glyco- and tauro- prevail all others, i.e. glyco-hyocholic acid.

Figure 8.11. Structure and nomenclature of bile acids.

additional negative charge at the C-3 position does not prevent active transport.

3. Substitutions at the C-3 position do not interfere with active transport.

4. Substitutions at the C-17 position do not interfere with transport as long as a negative charge around the C-24 position remains present.

5. A *cis* configuration of rings A and B within the sterol nucleus.

More recently, our group and the joint laboratories of Baringhaus and Kramer have reported on the three-dimensional structural requirements of ASBT that will be of specific use in the development of novel substrates for this transport protein. By use of a training set of 17 chemically diverse inhibitors of ASBT, Baringhaus and colleagues (281) developed an enantiospecific catalyst pharmacophore that mapped the molecular features essential for ASBT affinity: one hydrogen bond donor, one hydrogen bond acceptor, and three hydrophobic features. For natural bile acids they found that (*1*) ring D in combination with methyl-18 mapped one hydrophobic site and methyl-21 mapped a second; (*2*) an α-OH group at position 7 or 12 constituted a hydrogen donor; (*3*) the negatively charged side chain constituted the hydrogen bond acceptor; and (*4*) the 3α-OH group does not necessarily map a hydrogen bond functionality. The ASBT pharmacophore model is in good agreement with the 3D-QSAR model we previously developed using a series of 30 ASBT inhibitors and substrates (282). In this study, the electrostatic and steric fields around bile acids were mapped using comparative molecular field analysis (CoMFA) to identify regions of putative interaction with ASBT. This model enabled the *in silico* design of substrates for ASBT, especially for conjugation at the C-17 position. It should be pointed out that our model was silent on biological diversity around the C-3 position because of limited molecular variability in this region; however, alterations at that position have been described in detail by Kramer and coworkers (283). The two indirect models outlined above should facilitate the rational design of (pro)drugs for targeting to ASBT. There may also be the possibility of combining models from different groups, although this may present its own problems.

One serious limitation of indirect structure–activity models is the inherent lack of information on substrate–transport protein interactions at the molecular level. Currently, there is no crystal structure for ASBT and it is anticipated that suitable methods for crystallizing intrinsic membrane proteins will not be available for several years. As an alternative to high resolution molecular information, we have recently constructed a model for the ligand binding of the ASBT transport protein using knowledge-based homology modeling (284). Using the transmembrane domains of bacteriorhodopsin as a scaffold, the extracellular loop regions were superposed and optimized with molecular dynamics simulations. By probing the protein surface with cholic acid we identified five binding domains, three of which are located on the outer surface of the protein (Fig. 8.12). Another binding domain was located between two extracellular loops in close enough proximity to the first binding region to accommodate dicholic acid conjugates; this observation explains the extreme inhibitory capacity of this class of compounds (110). In addition to the previously described pharmacophore and 3D-QSAR models, the molecular representation of ASBT may provide a powerful tool in the design of novel substrates or inhibitors for this transporter.

5.5.3.4 Use of ASBT for Enhancing Intestinal Absorption. Bile acids display a physiologic organotropism for the liver and the small intestine, which suggests that bile acid transport pathways can be exploited in two ways: to shuttle compounds across the intestinal epithelium or to target them to the liver and the hepatobiliary system. This approach was first envisioned in 1948 by Berczeller in a U.S. patent application (No. 2,441,129), describing the synthesis of sulphonylcholylamide, which was claimed to be particularly suitable for the treatment of "germ and virus diseases which attack the liver." It was not until the mid-

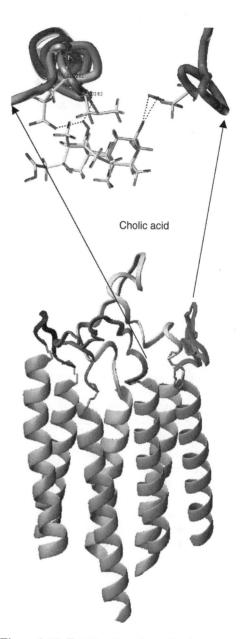

Cholic acid

Figure 8.12. Putative three-dimensional structure of ASBT and its binding sites.

1980s when Ho (285) suggested several potential therapeutic applications of using the bile acid transport systems for: (*1*) the improvement of the oral absorption of an intrinsically, biologically active, but poorly absorbed hydrophilic drug; (*2*) liver site-directed delivery of a drug to bring about high therapeutic concentrations in the diseased liver with the minimization of general toxic reactions elsewhere in the body; and (3) gallbladder-site delivery systems of cholecystographic agents and cholesterol gallstone dissolution accelerators. His experimental evidence was based on bile acid analogs with minor structural modifications (i.e., 3-iodo, 3-tosyl, and 3-benzoylcholate) that were handled like natural bile acids by the liver and intestine during *in situ* perfusion experiments. The feasibility and viability of using the bile acid transport pathway for targeting of drug–bile acid derivatives was more definitively demonstrated by the laboratory of Kramer, who primarily used the 3-position for drug conjugation. Our recent studies with small peptides attached to the C-24 position further demonstrate the general applicability of this transport system for drug targeting and delivery purposes (286).

From a structural perspective, largely based on the above-outlined structure–activity relationships, the following drug-targeting strategies are feasible with bile acids:

1. Attachment of drugs to the bile acid side-chain (C-17) with positional retention of the negatively charged group.
2. Attachment of drugs to the steroid nucleus at hydroxyl positions 3, 7, or 12 with conservation of the C-17 side-chain.

5.5.3.5 Therapeutic Applications of ASBT.
The uses of carrier-mediated bile acid for drug delivery purposes can be divided in three groups: liver- and gallbladder-directed delivery, oral absorption enhancement, and lowering serum cholesterol. Oral absorption enhancement can be directed to either the liver or gallbladder or systemic delivery.

5.5.3.5.1 Liver and Gallbladder Delivery. So far, most studies have focused on using the bile acid transport system for liver targeting. The research groups of Kramer (287–289) and Stephan (290) have successfully shown specific hepatic delivery of chlorambucil, HMG-CoA reductase inhibitors, and L-T$_3$, respectively. These studies prove that the coupling of drug entities to bile acids does not cause a loss of affinity for the hepatic bile acid carrier. Apart from the necessity of a negatively

charged group around the C-24 position, Kim and colleagues (291) have shown that some size restrictions apply when compounds are coupled to the C-3 position. Both the 3-position and the 24-position appear to be usable coupling points for a prodrug strategem. The 24-position appears to be an attractive site in the bile acid molecule for coupling purposes. The carboxylic acid moiety is easily linked to an amine using conventional peptide-synthesizing techniques (292–294), making the synthesis of these compounds relatively easy. It should be stressed, however, that the need for a negatively charged group around the C-24 position is required.

5.5.3.5.2 Systemic Delivery. When one compares the maximal transport flux (J_{max}) values of taurine conjugated bile acids measured in rat liver and distal ileum, a trend for relatively higher hepatic maximal transport rates can be observed. This observation can have important consequences when using the bile acid transporter for oral drug delivery. By use of a prodrug approach, with a bile acid molecule as a shuttle, liver targeting is easily accomplished, whereas systemic drug delivery needs to address the problem of rapid biliary excretion. Thus far, no single study has unequivocally shown the release of the parent compound from the conjugate after passage across the intestinal wall. It has to be mentioned, however, that no studies so far have attempted to develop a prodrug approach in which the drug will be released before arrival in the liver. In that case, the drug moiety must be released from the bile acid it is coupled to, within either the enterocyte or the portal vein. Only if these conditions are met, will systemic delivery using a prodrug approach be successful. Although promising, the suitability of this transport system for systemic drug delivery remains to be demonstrated.

5.5.3.5.3 Cholesterol-Reducing Agents. Hypercholesterolemia is well known as a major risk factor for coronary heart disease. In clinical practice, two main hypocholestrolemic agents are commonly used. One is the 3-hydroxy-3-methylgrutaryl coenzyme A (HMG-CoA) reductase inhibitors (such as Lipitor); another is the bile acid sequestrants, such as cholestyramine and colestipol (97), which bind bile acid in the intestinal lumen and thus increase their excretion. The main drawback of

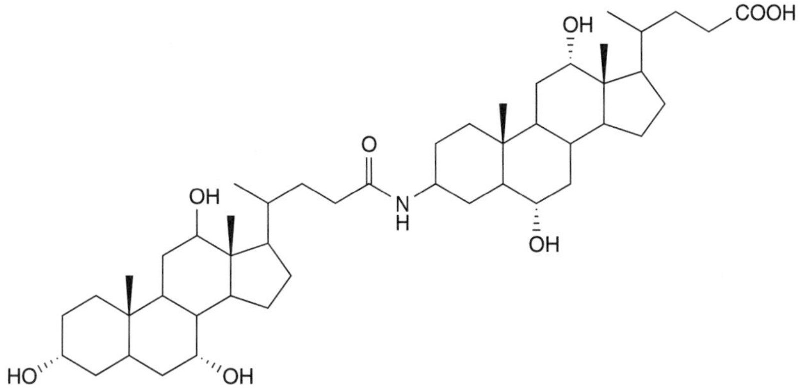

Figure 8.13. Structure of specific bile acid transporter inhibitors. See text for additional details.

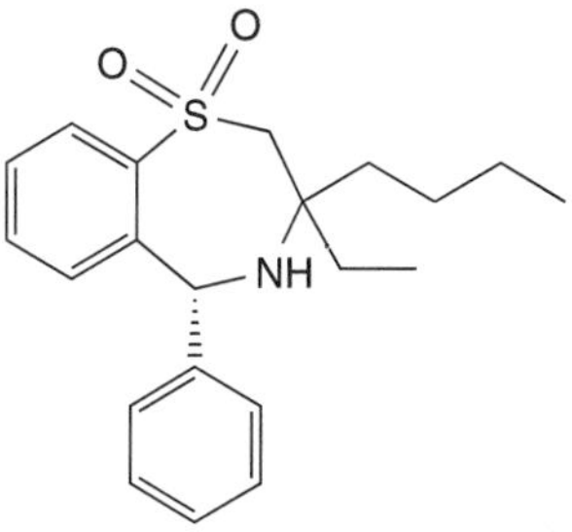

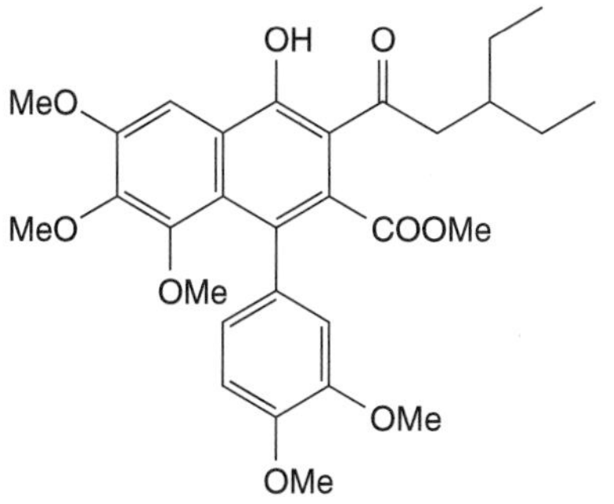

these agents is poor compliance of patients stemming from adverse side effects, such as high dosages of 10–30 g per day, constipation, maldigestion, and malabsorption syndromes. As an alternative method to bile acid sequestrants, any reagent that can inhibit the bile acid active transport system could block the reabsorption of bile acids and consequently reduce the serum cholesterol level. So far, several molecules have been found to possess this effect in animal studies (108–110, 295).

The first such inhibitors consisted of the coupling of two bile acid molecules by means of a spacer to allow simultaneous interaction with more than one transporter site, resulting in an efficient inhibition of bile acid reabsorption without or with only low absorption of the inhibitor itself (110) (Fig. 8.13). Recently, it was shown that a benzothiazepine derivative, 2164U90, was able to selectively inhibit active ileal bile acid absorption in rats, mice, mon-

keys, and humans (296, 297). Similarly, another compound, S-8921 (Fig. 8.13), a lignan derivative, was able to reduce serum cholesterol in hamsters, mice, and rabbits. The inhibition of the intestinal bile acid transport system is thought to be the underlying mechanism for an increased fecal bile acid excretion and lower plasma LDL cholesterol levels after oral administration of these drugs.

6 CONCLUSIONS AND FUTURE DIRECTIONS

It is clear that over the past decade we have gained a great deal of knowledge around the functioning of SLC proteins and how they can be manipulated as drug targets for maximizing absorption and bioactivity through prodrug approaches. Although we are still without a crystal structure for the SLC proteins, there has been an explosion in interest in structure-activity relationships. This has been seen largely as part of a wider growth in using computational approaches for understanding *in vitro* data, particularly in the fields of drug design, optimization, and ADME properties (298). The resolution of the three-dimensional structures of solute transporter protein models is presumably low; however, they can be justified by their ability to confirm biologically relevant phenomena. As with all models, the continued input of novel experimental data and revalidation of the model may eventually lead to highly predictive systems capable of *in silico* detection and design of novel solute transporter substrates. Furthermore, we can expect that a combination of indirect (3D-QSAR) and direct (homology models) techniques may lead to newer, higher resolution screening systems. It is likely that the integration of bioinformatics and both computational and *in vitro* models will drive our understanding of SLC proteins to new levels and afford an opportunity for further possible therapeutic targets and drug design opportunities.

7 ACKNOWLEDGMENTS

I thank Se-Ne Huang and Yongheng (Eric) Zhang for their contributions. Dr. Sean Ekins (Concurrent Pharmaceuticals, Cambridge, MA) contributed to the P-glycoprotein section in this chapter.

REFERENCES

1. J. G. Kingham, P. J. Whorwell, and C. A. Loehry, *Gut*, **17**, 354 (1976).

2. A. Fasano and S. Uzzau, *J. Clin. Invest.*, **99**, 1158 (1997).

3. I. M. van der Lubben, J. C. Verhoef, G. Borchard, and H. E. Junginger, *Adv. Drug Deliv. Rev.*, **52**, 139 (2001).

4. C. J. Porter and W. N. Charman, *Adv. Drug Deliv. Rev.*, **50**, S127 (2001).

5. A. Albert, *Nature*, **182**, 421 (1958).

6. V. Stella in V. Stella, Ed., *Prodrugs as Novel Drug Delivery Systems*, ACS Symposium Series **14**, American Chemical Society, Washington, DC, 1975, p. 1.

7. A. D. Welch, *Cancer Res.*, **21**, 1475 (1961).

8. S. H. Yalkowski and W. Morozowich in E. J. Ariens, Ed., *Drug Design*, Vol. **9**, Academic Press, New York, 1980, p. 121.

9. J. L. Madara and J. R. Pappenheimer, *J. Membr. Biol.*, **100**, 149 (1987).

10. J. L. Madara and K. Dharmsathaphorn, *J. Cell. Biol.*, **101**, 2124 (1985).

11. T. Y. Ma, D. Hollander, R. A. Erickson, H. Truong, H. Nguyen, and P. Krugliak, *Gastroenterology*, **108**, 12 (1995).

12. G. T. Knipp, N. F. Ho, C. L. Barsuhn, and R. T. Borchardt, *J. Pharm. Sci.*, **86**, 1105 (1997).

13. C. J. Watson, M. Rowland, and G. Warhurst, *Am. J. Physiol. Cell Physiol.*, **281**, C388 (2001).

14. Y. L. He, S. Murby, G. Warhurst, L. Gifford, D. Walker, J. Ayrton, R. Eastmond, and M. Rowland, *J. Pharm. Sci.*, **87**, 626 (1998).

15. C. A. Lipinski, *J. Pharmacol. Toxicol. Methods*, **44**, 235 (2000).

16. J. B. Houston, D. G. Upshall, and J. W. Bridges, *J. Pharmacol. Exp. Ther.*, **195**, 67 (1975).

17. S. Mukherjee, R. N. Ghosh, and F. R. Maxfield, *Physiol. Rev.*, **77**, 759 (1997).

18. C. M. Lehr in A. G. d. Boer, Ed., *Drug Absorption Enhancement: Concepts, Possibilities, Limitations and Trends*, Harwood, Switzerland, 1994, p. 325.

19. M. Marsh and A. Helenius, *J. Mol. Biol.*, **142**, 439 (1980).

20. A. Pol, D. Ortega, and C. Enrich, *Biochem. J.*, **323**, 435 (1997).

21. P. Wigge, Y. Vallis, and H. T. McMahon, *Curr. Biol.*, **7**, 554 (1997).

22. D. J. Yamashiro and F. R. Maxfield, *J. Cell. Biochem.*, **26**, 231 (1984).

23. H. J. Geuze, J. W. Slot, G. J. Strous, J. Peppard, K. von Figura, A. Hasilik, and A. L. Schwartz, *Cell*, **37**, 195 (1984).

24. B. Herman and D. F. Albertini, *Nature*, **304**, 738 (1983).

25. E. J. Hughson and C. R. Hopkins, *J. Cell Biol.*, **110**, 337 (1990).

26. A. L. Schwartz, *Pediatr. Res.*, **38**, 835 (1995).

27. M. Barroso, D. S. Nelson, and E. Sztul, *Proc. Natl. Acad. Sci. USA*, **92**, 527 (1995).

28. P. P. Lemons, D. Chen, A. M. Bernstein, M. K. Bennett, and S. W. Whiteheart, *Blood*, **90**, 1490 (1997).

29. D. Shah and W. C. Shen, *J. Drug Target.*, **2**, 93 (1994).

30. K. Prydz, S. H. Hansen, K. Sandvig, and B. van Deurs, *J. Cell Biol.*, **119**, 259 (1992).

31. N. J. Severs, *J. Cell Sci.*, **90**, 341 (1988).

32. R. G. Anderson, B. A. Kamen, K. G. Rothberg, and S. W. Lacey, *Science*, **255**, 410 (1992).

33. R. G. Anderson, *Proc. Natl. Acad. Sci. USA*, **90**, 10909 (1993).

34. H. Matsue, K. G. Rothberg, A. Takashima, B. A. Kamen, R. G. Anderson, and S. W. Lacey, *Proc. Natl. Acad. Sci. USA*, **89**, 6006 (1992).

35. R. Montesano, J. Roth, A. Robert, and L. Orci, *Nature*, **296**, 651 (1982).

36. T. V. Kurzchalia, P. Dupree, and S. Monier, *FEBS Lett.*, **346**, 88 (1994).

37. K. G. Rothberg, J. E. Heuser, W. C. Donzell, Y. S. Ying, J. R. Glenney, and R. G. Anderson, *Cell*, **68**, 673 (1992).

38. W. A. Walker and I. R. Sanderson, *Ann. N. Y. Acad. Sci.*, **664**, 10 (1992).

39. M. P. Desai, V. Labhasetwar, G. L. Amidon, and R. J. Levy, *Pharm. Res.*, **13**, 1838 (1996).

40. N. Hussain, P. U. Jani, and A. T. Florence, *Pharm. Res.*, **14**, 613 (1997).

41. M. Berryman and R. Rodewald, *J. Cell Sci.*, **108**, 2347 (1995).

42. N. Benlounes, R. Chedid, F. Thuillier, J. F. Desjeux, F. Rousselet, and M. Heyman, *Biol. Neonate*, **67**, 254 (1995).

43. C. Medesan, C. Radu, J. K. Kim, V. Ghetie, and E. S. Ward, *Eur. J. Immunol.*, **26**, 2533 (1996).

44. C. Medesan, D. Matesoi, C. Radu, V. Ghetie, and E. S. Ward, *J. Immunol.*, **158**, 2211 (1997).

45. D. R. Abrahamson and R. Rodewald, *J. Cell Biol.*, **91**, 270 (1981).

46. H. M. Patel and A. E. Wild, *FEBS Lett.*, **234**, 321 (1988).

47. J. Mestecky, *J. Clin. Immunol.*, **7**, 265 (1987).

48. R. P. Hirt, G. J. Hughes, S. Frutiger, P. Michetti, C. Perregaux, O. Poulain-Godefroy, N. Jeanguenat, M. R. Neutra, and J. P. Kraehenbuhl, *Cell*, **74**, 245 (1993).

49. K. R. Youngman, C. Fiocchi, and C. S. Kaetzel, *J. Immunol.*, **153**, 675 (1994).

50. V. Livrelli, C. De Champs, P. Di Martino, A. Darfeuille-Michaud, C. Forestier, and B. Joly, *J. Clin. Microbiol.*, **34**, 1963 (1996).

51. A. C. Ghose, *Indian J. Med. Res.*, **104**, 38 (1996).

52. I. Adlerberth, S. Ahrne, M. L. Johansson, G. Molin, L. A. Hanson, and A. E. Wold, *Appl. Environ. Microbiol.*, **62**, 2244 (1996).

53. S. Crennell, E. Garman, G. Laver, E. Vimr, and G. Taylor, *Structure*, **2**, 535 (1994).

54. J. W. Metcalfe, K. A. Krogfelt, H. C. Krivan, P. S. Cohen, and D. C. Laux, *Infect. Immun.*, **59**, 91 (1991).

55. C. J. Thorns, G. A. Wells, J. A. Morris, A. Bridges, and R. Higgins, *Vet. Microbiol.*, **20**, 377 (1989).

56. R. R. Isberg, *Mol. Biol. Med.*, **7**, 73 (1990).

57. R. R. Isberg and J. M. Leong, *Cell*, **60**, 861 (1990).

58. A. Marra and R. R. Isberg, *Infect. Immun.*, **65**, 3412 (1997).

59. R. W. Paul, K. E. Weisser, A. Loomis, D. L. Sloane, D. LaFoe, E. M. Atkinson, and R. W. Overell, *Hum. Gene Ther.*, **8**, 1253 (1997).

60. J. H. J. Easson and C.-M. D. Lehr, *Proc. Int. Symp. Controlled Rel. Bioact. Mater.*, 1997.

61. J. H. J. Easson and C.-M. D. Lehr, *Pharm Res.*, **14**, S21 (1997).

62. E. Lemichez, M. Bomsel, G. Devilliers, J. vanderSpek, J. R. Murphy, E. V. Lukianov, S. Olsnes, and P. Boquet, *Mol. Microbiol.*, **23**, 445 (1997).

63. J. Lindner, A. F. Geczy, and G. J. Russell-Jones, *Scand. J. Immunol.*, **40**, 564 (1994).

64. L. Lazorova, A. Sjolander, G. J. Russell-Jones, J. Linder, and P. Artursson, *J. Drug Target.*, **1**, 331 (1993).

65. K. J. Fisher and J. M. Wilson, *Biochem. J.*, **321**, 49 (1997).

66. A. R. Hamad, P. Marrack, and J. W. Kappler, *J. Exp. Med.*, **185**, 1447 (1997).

67. F. Stirpe and L. Barbieri, *FEBS Lett.*, **195**, 1 (1986).

68. M. G. Battelli, L. Citores, L. Buonamici, J. M. Ferreras, F. M. de Benito, F. Stirpe, and T. Girbes, *Arch. Toxicol.*, **71**, 360 (1997).

69. I. D. Sylvester, L. M. Roberts, and J. M. Lord, *Biochim. Biophys. Acta*, **1358**, 53 (1997).

70. M. G. Battelli, L. Buonamici, L. Polito, A. Bolognesi, and F. Stirpe, *Virchows Arch.*, **427**, 529 (1996).

71. F. Stirpe, K. Sandvig, S. Olsnes, and A. Pihl, *J. Biol. Chem.*, **257**, 13271 (1982).

72. S. Sperti, L. Montanaro, M. Derenzini, A. Gasperi-Campani, and F. Stirpe, *Biochim. Biophys. Acta*, **562**, 495 (1979).

73. J. E. Ludert, F. Michelangeli, F. Gil, F. Liprandi, and J. Esparza, *Intervirology*, **27**, 95 (1987).

74. J. K. Olson and C. Grose, *J. Virol.*, **71**, 4042 (1997).

75. A. Irurzun, J. L. Nieva, and L. Carrasco, *Virology*, **227**, 488 (1997).

76. B. S. Worthington and J. Syrotuck, *J. Nutr.*, **106**, 20 (1976).

77. A. Garret, C. Kerlan, and D. Thomas, *Arch. Virol.*, **131**, 377 (1993).

78. J. L. Wolf, R. S. Kauffman, R. Finberg, R. Dambrauskas, B. N. Fields, and J. S. Trier, *Gastroenterology*, **85**, 291 (1983).

79. N. Etchart, F. Wild, and D. Kaiserlian, *J. Gen. Virol.*, **77**, 2471 (1996).

80. S. E. Bardocz, S. W. Grant, and G. Pasztai, in S. P. Bardocz, Ed., *Lectins: Biomedical Perspectives*, Taylor & Francis, London, 1995, p. 103.

81. C.-M. P. Lehr in S. P. Bardocz, Ed., *Lectins: Biomedical Perspectives*, Taylor & Francis, London, 1995, p. 117.

82. L. T. Weaver and D. S. Bailey, *J. Pediatr. Gastroenterol. Nutr.*, **6**, 445 (1987).

83. A. Gebert, H. J. Rothkotter, and R. Pabst, *Int. Rev. Cytol.*, **167**, 91 (1996).

84. M. A. Jepson, M. A. Clark, N. Foster, C. M. Mason, M. K. Bennett, N. L. Simmons, and B. H. Hirst, *J. Anat.*, **189**, 507 (1996).

85. R. Sharma, E. J. van Damme, W. J. Peumans, P. Sarsfield, and U. Schumacher, *Histochem. Cell. Biol.*, **105**, 459 (1996).

86. J. F. Ross, P. K. Chaudhuri, and M. Ratnam, *Cancer*, **73**, 2432 (1994).

87. J. Fan, N. Kureshy, K. S. Vitols, and F. M. Huennekens, *Oncol. Res.*, **7**, 511 (1995).

88. C. P. Leamon and P. S. Low, *Biochem. J.*, **291**, 855 (1993).

89. K. A. Mislick, J. D. Baldeschwieler, J. F. Kayyem, and T. J. Meade, *Bioconjug. Chem.*, **6**, 512 (1995).

90. S. Wang, R. J. Lee, G. Cauchon, D. G. Gorenstein, and P. S. Low, *Proc. Natl. Acad. Sci. USA*, **92**, 3318 (1995).

91. O. D. Wangensteen, M. M. Bartlett, J. K. James, Z. F. Yang, and P. S. Low, *Pharm. Res.*, **13**, 1861 (1996).

92. S. N. Huang and P. W. Swaan, *J. Pharmacol. Exp. Ther.*, **298**, 264 (2001).

93. S. N. Huang and P. W. Swaan, *J. Pharmacol. Exp. Ther.*, **294**, 117 (2000).

94. M. Ramasamy, D. H. Alpers, C. Tiruppathi, and B. Seetharam, *Am. J. Physiol. Gastrointest. Liver Physiol.*, **257**, G791 (1989).

95. N. Dan and D. F. Cutler, *J. Biol. Chem.*, **269**, 18849 (1994).

96. G. J. Russell-Jones in M. D. A. Taylor, Ed., *Peptide-Based Drug Design: Controlling Transport and Metabolism*, American Chemical Society, Washington, DC, 1995, p. 181.

97. G. J. Russell-Jones, S. W. Westwood, P. G. Farnworth, J. K. Findlay, and H. G. Burger, *Bioconjug. Chem.*, **6**, 34 (1995).

98. A. D. Habberfield, K. Ralph, S. W. Westwood, and G. J. Russell-Jones, *Int. J. Pharm.*, **145**, 1 (1996).

99. G. J. Russell-Jones, S. W. Westwood, and A. D. Habberfield, *Bioconjug Chem.*, **6**, 459 (1995).

100. G. J. Russell-Jones, L. Killinger, and S. W. Westwood, *Proc. Intl. Symp. Controlled Rel. Bioact. Mater.*, 1997.

101. G. J. Russell-Jones, S. W. Westwood, and A. D. Habberfield, *Proc. Int. Symp. Controlled Rel. Bioact. Mater.*, 1996.

102. A. Pietrangelo, E. Rocchi, G. Casalgrandi, G. Rigo, A. Ferrari, M. Perini, E. Ventura, and G. Cairo, *Gastroenterology*, **102**, 802 (1992).

103. G. J. Anderson, L. W. Powell, and J. W. Halliday, *Gastroenterology*, **98**, 576 (1990).

104. G. P. Jeffrey, K. A. Basclain, and T. L. Allen, *Gastroenterology*, **110**, 790 (1996).

105. D. Shah and W.-C. Shen, *J. Pharm. Sci.*, **85**, 1306 (1996).

106. J. S. Wang, D. Shah, and W.-C. Shen, *Pharm. Res.*, **14**, S469 (1997).

107. M. H. Saier Jr., *Int. Rev. Cytol.*, **190**, 61 (1999).

108. W. Kramer and G. Wess, *Eur. J. Clin. Invest.*, **26**, 715 (1996).

109. M. B. Tollefson, W. F. Vernier, H. C. Huang, F. P. Chen, E. J. Reinhard, J. Beaudry, B. T. Keller, and D. B. Reitz, *Bioorg. Med. Chem. Lett.*, **10**, 277 (2000).

110. G. Wess, W. Kramer, A. Enhsen, H. Glombik, K. H. Baringhaus, G. Boger, M. Urmann, K. Bock, H. Kleine, G. Neckermann, A. Hoffmann, C. Pittius, E. Falk, H. W. Fehlhaber, H. Kogler, and M. Friedrich, *J. Med. Chem.*, **37**, 873 (1994).

111. M. A. Hediger, M. J. Coady, T. S. Ikeda, and E. M. Wright, *Nature*, **330**, 379 (1987).

112. M. A. Hediger, T. Ikeda, M. Coady, C. B. Gundersen, and E. M. Wright, *Proc. Natl. Acad. Sci. USA*, **84**, 2634 (1987).

113. M. Dean, A. Rzhetsky, and R. Allikmets, *Genome Res.*, **11**, 1156 (2001).

114. I. T. Paulsen, M. K. Sliwinski, B. Nelissen, A. Goffeau, and M. H. Saier Jr., *FEBS Lett.*, **430**, 116 (1998).

115. I. T. Paulsen, M. K. Sliwinski, and M. H. Saier Jr., *J. Mol. Biol.*, **277**, 573 (1998).

116. G. D. Schuler, M. S. Boguski, E. A. Stewart, L. D. Stein, G. Gyapay, K. Rice, R. E. White, P. Rodriguez-Tome, A. Aggarwal, E. Bajorek, S. Bentolila, B. B. Birren, A. Butler, A. B. Castle, N. Chiannilkulchai, A. Chu, C. Clee, S. Cowles, P. J. Day, T. Dibling, N. Drouot, I. Dunham, S. Duprat, C. East, and T. J. Hudson, *Science*, **274**, 540 (1996).

117. P. W. Swaan, S. Oie, and F. C. Szoka Jr., *Adv. Drug Del. Rev.*, **20**, 1 (1996).

118. S. A. Adibi, *Gastroenterology*, **113**, 332 (1997).

119. D. M. Bradshaw and R. J. Arceci, *J. Clin. Oncol.*, **16**, 3674 (1998).

120. A. Krishan, C. M. Fitz, and I. Andritsch, *Cytometry*, **29**, 279 (1997).

121. W. O. Berndt, *Toxicol. Pathol.*, **26**, 52 (1998).

122. M. J. Owens and C. B. Nemeroff, *Depress. Anxiety*, **8**, 5 (1998).

123. M. E. Reith, C. Xu, and N. H. Chen, *Eur. J. Pharmacol.*, **324**, 1 (1997).

124. D. L. Baly and R. Horuk, *Biochim. Biophys. Acta*, **947**, 571 (1988).

125. E. M. Wright, E. Turk, B. Zabel, S. Mundlos, and J. Dyer, *J. Clin. Invest.*, **88**, 1435 (1991).

126. M. H. Wong, P. Oelkers, and P. A. Dawson, *J. Biol. Chem.*, **270**, 27228 (1995).

127. J. Deisenhofer and H. Michel, *Annu. Rev. Biophys. Biophys. Chem.*, **20**, 247 (1991).

128. J. Deisenhofer and H. Michel, *Annu. Rev. Cell Biol.*, **7**, 1 (1991).

129. N. Unwin, *Nature*, **373**, 37 (1995).

130. T. Tsukihara, H. Aoyama, E. Yamashita, T. Tomizaki, H. Yamaguchi, K. Shinzawa-Itoh, R. Nakashima, R. Yaono, and S. Yoshikawa, *Science*, **272**, 1136 (1996).

131. G. Chang and C. B. Roth, *Science*, **293**, 1793 (2001).

132. H. R. Kaback, J. Voss, and J. Wu, *Curr. Opin. Struct. Biol.*, **7**, 537 (1997).

133. V. C. Goswitz and R. J. Brooker, *Protein Sci.*, **4**, 534 (1995).

134. M. F. Varela and T. H. Wilson, *Biochim. Biophys. Acta*, **1276**, 21 (1996).

135. R. F. Doolittle, M. S. Johnson, I. Husain, B. Van Houten, D. C. Thomas, and A. Sancar, *Nature*, **323**, 451 (1986).

136. J. Voss, J. Wu, W. L. Hubbell, V. Jacques, C. F. Meares, and H. R. Kaback, *Biochemistry*, **40**, 3184 (2001).

137. M. G. Claros and G. von Heijne, *Comput. Appl. Biosci.*, **10**, 685 (1994).

138. B. Rost, P. Fariselli, and R. Casadio, *Protein Sci.*, **5**, 1704 (1996).

139. G. E. Tusnady and I. Simon, *J. Mol. Biol.*, **283**, 489 (1998).

140. K. Bamberg and G. Sachs, *J. Biol. Chem.*, **269**, 16909 (1994).

141. S. Frillingos, M. Sahin-Toth, J. Wu, and H. R. Kaback, *FASEB J.*, **12**, 1281 (1998).

142. E. Turk, C. J. Kerner, M. P. Lostao, and E. M. Wright, *J. Biol. Chem.*, **271**, 1925 (1996).

143. E. R. Bennett and B. I. Kanner, *J. Biol. Chem.*, **272**, 1203 (1997).

144. K. M. Covitz, G. L. Amidon, and W. Sadee, *Biochemistry*, **37**, 15214 (1998).

145. Q. Wang, J. Voss, W. L. Hubbell, and H. R. Kaback, *Biochemistry*, **37**, 4910 (1998).

146. M. Zhao, K. C. Zen, W. L. Hubbell, and H. R. Kaback, *Biochemistry*, **38**, 7407 (1999).

147. R. C. Hresko, H. Murata, B. A. Marshall, and M. Mueckler, *J. Biol. Chem.*, **269**, 32110 (1994).

148. K. Nagai and H. C. Thogersen, *Nature*, **309**, 810 (1984).

149. M. Sahin-Toth, R. L. Dunten, and H. R. Kaback, *Biochemistry*, **34**, 1107 (1995).

150. C. Wandel, R. B. Kim, S. Kajiji, P. Guengerich, G. R. Wilkinson, and A. J. Wood, *Cancer Res.*, **59**, 3944 (1999).

151. A. George, *Curr. Opin. Drug Discov. Dev.*, **2**, 286 (1999).

152. S. Hoffmeyer, O. Burk, O. von Richter, H. P. Arnold, J. Brockmoller, A. Johne, I. Cascorbi, T. Gerloff, I. Roots, M. Eichelbaum, and U. Brinkmann, *Proc. Natl. Acad. Sci. USA*, **97**, 3473 (2000).

153. V. J. Wacher, C. Y. Wu, and L. Z. Benet, *Mol. Carcinog.*, **13**, 129 (1995).

154. E. G. Schuetz, W. T. Beck, and J. D. Schuetz, *Mol. Pharmacol.*, **49**, 311 (1996).

155. E. G. Schuetz, A. H. Schinkel, M. V. Relling, and J. D. Schuetz, *Proc. Natl. Acad. Sci. USA*, **93**, 4001 (1996).

156. R. B. Kim, C. Wandel, B. Leake, M. Cvetkovic, M. F. Fromm, P. J. Dempsey, M. M. Roden, F. Belas, A. K. Chaudhary, D. M. Roden, A. J. Wood, and G. R. Wilkinson, *Pharm. Res.*, **16**, 408 (1999).

157. E. G. Schuetz, S. Strom, K. Yasuda, V. Lecureur, M. Assem, C. Brimer, J. Lamba, R. B. Kim, V. Ramachandran, B. J. Komoroski, R.

Venktaramanan, H. Cai, C. J. Sinal, F. J. Gonzalez, and J. D. Schuetz, *J. Biol. Chem.*, **276**, 39411 (2001).

158. T. W. Synold, I. Dussault, and B. M. Forman, *Nat. Med.*, **7**, 584 (2001).

159. S. Ekins and J. A. Erickson, *Drug Metab. Dispos.*, **30**, 96 (2002).

160. S. Ekins, B. J. Ring, S. N. Binkley, S. D. Hall, and S. A. Wrighton, *Int. J. Clin. Pharmacol. Ther.*, **36**, 642 (1998).

161. K. R. Korzekwa, N. Krishnamachary, M. Shou, A. Ogai, R. A. Parise, A. E. Rettie, F. J. Gonzalez, and T. S. Tracy, *Biochemistry*, **37**, 4137 (1998).

162. J. B. Houston and K. E. Kenworthy, *Drug Metab. Dispos.*, **28**, 246 (2000).

163. L. M. Greenberger, C. P. Yang, E. Gindin, and S. B. Horwitz, *J. Biol. Chem.*, **265**, 4394 (1990).

164. S. Ayesh, Y. M. Shao, and W. D. Stein, *Biochim. Biophys. Acta*, **1316**, 8 (1996).

165. A. B. Shapiro and V. Ling, *Eur. J. Biochem.*, **250**, 130 (1997).

166. S. Scala, N. Akhmed, U. S. Rao, K. Paull, L. B. Lan, B. Dickstein, J. S. Lee, G. H. Elgemeie, W. D. Stein, and S. E. Bates, *Mol. Pharmacol.*, **51**, 1024 (1997).

167. S. Dey, M. Ramachandra, I. Pastan, M. M. Gottesman, and S. V. Ambudkar, *Proc. Natl. Acad. Sci. USA*, **94**, 10594 (1997).

168. A. B. Shapiro, K. Fox, P. Lam, and V. Ling, *Eur. J. Biochem.*, **259**, 841 (1999).

169. E. J. Wang, C. N. Casciano, R. P. Clement, and W. W. Johnson, *Biochim. Biophys. Acta*, **1481**, 63 (2000).

170. L. Lu, F. Leonessa, R. Clarke, and I. W. Wainer, *Mol. Pharmacol.*, **59**, 62 (2001).

171. M. Demeule, A. Laplante, G. F. Murphy, R. M. Wenger, and R. Beliveau, *Biochemistry*, **37**, 18110 (1998).

172. R. L. Shepard, M. A. Winter, S. C. Hsaio, H. L. Pearce, W. T. Beck, and A. H. Dantzig, *Biochem. Pharmacol.*, **56**, 719 (1998).

173. A. H. Dantzig, R. L. Shepard, J. Cao, K. L. Law, W. J. Ehlhardt, T. M. Baughman, T. F. Bumol, and J. J. Starling, *Cancer Res.*, **56**, 4171 (1996).

174. H. L. Pearce, M. A. Winter, and W. T. Beck, *Adv. Enzyme Regul.*, **30**, 357 (1990).

175. H. L. Pearce, A. R. Safa, N. J. Bach, M. A. Winter, M. C. Cirtain, and W. T. Beck, *Proc. Natl. Acad. Sci. USA*, **86**, 5128 (1989).

176. A. Ramu and N. Ramu, *Cancer Chemother. Pharmacol.*, **30**, 165 (1992).

177. M. J. Borgnia, G. D. Eytan, and Y. G. Assaraf, *J. Biol. Chem.*, **271**, 3163 (1996).

178. P. Chiba, G. Ecker, D. Schmid, J. Drach, B. Tell, S. Goldenberg, and V. Gekeler, *Mol. Pharmacol.*, **49**, 1122 (1996).

179. G. Klopman, L. M. Shi, and A. Ramu, *Mol. Pharmacol.*, **52**, 323 (1997).

180. G. A. Bakken and P. C. Jurs, *J. Med. Chem.*, **43**, 4534 (2000).

181. M. Wiese and I. K. Pajeva, *Pharmazie*, **52**, 679 (1997).

182. C. Tmej, P. Chiba, M. Huber, E. Richter, M. Hitzler, K. J. Schaper, and G. Ecker, *Arch. Pharm. (Weinheim)*, **331**, 233 (1998).

183. I. Pajeva and M. Wiese, *J. Med. Chem.*, **41**, 1815 (1998).

184. D. Schmid, G. Ecker, S. Kopp, M. Hitzler, and P. Chiba, *Biochem. Pharmacol.*, **58**, 1447 (1999).

185. G. Ecker, M. Huber, D. Schmid, and P. Chiba, *Mol. Pharmacol.*, **56**, 791 (1999).

186. A. Seelig, *Int. J. Clin. Pharmacol. Ther.*, **36**, 50 (1998).

187. T. Osterberg and U. Norinder, *Eur. J. Pharm. Sci.*, **10**, 295 (2000).

188. S. Neuhoff, P. Langguth, C. Dressler, T. B. Andersson, C. G. Regardh, and H. Spahn-Langguth, *Int. J. Clin. Pharmacol. Ther.*, **38**, 168 (2000).

189. M. F. Rosenberg, Q. Mao, A. Holzenburg, R. C. Ford, R. G. Deeley, and S. P. Cole, *J. Biol. Chem.*, **276**, 16076 (2001).

190. G. K. Grimble and D. B. A. Silk, *Nutr. Res. Rev.*, **2**, 87 (1989).

191. D. M. Mathews and S. A. Adibi, *Gastroenterology*, **71**, 151 (1976).

192. S. A. Adibi, *Metabolism*, **36**, 1001 (1987).

193. P. Furst, S. Albers, and P. Stehle, *Proc. Nutr. Soc.*, **49**, 343 (1990).

194. G. K. Grimble, *Annu. Rev. Nutr.*, **14**, 419 (1994).

195. N. J. Snyder, L. B. Tabas, D. M. Berry, D. C. Duckworth, D. O. Spry, and A. H. Dantzig, *Antimicrob. Agents Chemother.*, **41**, 1649 (1997).

196. H. Jiang, L. B. Tabas, and A. H. Dantzig, submitted.

197. V. A. Moore, W. J. Irwin, P. Timmins, P. A. Lambert, S. Chong, S. A. Dando, and R. A. Morrison, *Int. J. Pharm.*, **210**, 29 (2000).

198. F. H. Leibach and V. Ganapathy, *Annu. Rev. Nutr.*, **16**, 99 (1996).

199. H. K. Han, J. K. Rhie, D. M. Oh, G. Saito, C. P. Hsu, B. H. Stewart, and G. L. Amidon, *J. Pharm. Sci.*, **88**, 347 (1999).

200. K. M. Covitz, G. L. Amidon, and W. Sadee, *Pharm. Res.*, **13**, 1631 (1996).

201. J. S. Kim, R. L. Oberle, D. A. Krummel, J. B. Dressman, and D. Fleisher, *J. Pharm. Sci.*, **83**, 1350 (1994).

202. J. P. Bai and G. L. Amidon, *Pharm. Res.*, **9**, 969 (1992).

203. M. Hu, P. Subramanian, H. I. Mosberg, and G. L. Amidon, *Pharm. Res.*, **6**, 66 (1989).

204. D. M. Oh, H. K. Han, and G. L. Amidon, *Pharm. Biotechnol.*, **12**, 59 (1999).

205. M. Boll, D. Markovich, W. M. Weber, H. Korte, H. Daniel, and H. Murer, *Pflügers Arch.*, **429**, 146 (1994).

206. M. E. Ganapathy, M. Brandsch, P. D. Prasad, V. Ganapathy, and F. H. Leibach, *J. Biol. Chem.*, **270**, 25672 (1995).

207. M. E. Ganapathy, W. Huang, H. Wang, V. Ganapathy, and F. H. Leibach, *Biochem. Biophys. Res. Commun.*, **246**, 470 (1998).

208. H. Shen, D. E. Smith, T. Yang, Y. G. Huang, J. B. Schnermann, and F. C. Brosius 3rd, *Am. J. Physiol. Renal Fluid Electrolyte Physiol.*, **276**, F658 (1999).

209. H. Ogihara, T. Suzuki, Y. Nagamachi, K. Inui, and K. Takata, *Histochem. J.*, **31**, 169 (1999).

210. H. Y. Steiner, F. Naider, and J. M. Becker, *Mol. Microbiol.*, **16**, 825 (1995).

211. D. Meredith and C. A. Boyd, *Cell. Mol. Life Sci.*, **57**, 754 (2000).

212. Y. J. Fei, V. Ganapathy, and F. H. Leibach, *Prog. Nucleic Acid Res. Mol. Biol.*, **58**, 239 (1998).

213. S. Nussberger, A. Steel, D. Trotti, M. F. Romero, W. F. Boron, and M. A. Hediger, *J. Biol. Chem.*, **272**, 7777 (1997).

214. P. W. Swaan and J. J. Tukker, *Pharm. Weekbl. Sci. Ed.*, **14F**, 62 (1992).

215. P. W. Swaan and J. J. Tukker, *Pharm. Weekbl. Sci. Ed.*, **14M**, 4 (1992).

216. J. J. Tukker and P. W. Swaan, *Pharm. Res.*, **9**, S180 (1992).

217. C. Humblet and G. R. Marshall, *Ann. Rep. Med. Chem.*, **15**, 267 (1980).

218. J. Li and I. J. Hidalgo, *J. Drug Target.*, **4**, 9 (1996).

219. P. W. Swaan and J. J. Tukker, *J. Pharm. Sci.*, **86**, 596 (1997).

220. P. W. Swaan, M. C. Stehouwer, and J. J. Tukker, *Biochim. Biophys. Acta*, **1236**, 31 (1995).

221. S. H. Kubo and R. J. Cody, *Clin. Pharmacokinet.*, **10**, 377 (1985).

222. P. W. Swaan, B. C. Koops, E. E. Moret, and J. J. Tukker, *Receptors Channels*, **6**, 189 (1998).

223. F. Doring, J. Will, S. Amasheh, W. Clauss, H. Ahlbrecht, and H. Daniel, *J. Biol. Chem.*, **273**, 23211 (1998).

224. T. Terada, K. Sawada, M. Irie, H. Saito, Y. Hashimoto, and K. Inui, *Pflügers Arch.*, **440**, 679 (2000).

225. P. D. Bailey, C. A. R. Boyd, J. R. Bronk, I. D. Collier, D. Meredith, K. M. Morgan, and C. S. Temple, *Angew. Chem. Int. Ed. Engl.*, **39**, 506 (2000).

226. B. Bretschneider, M. Brandsch, and R. Neubert, *Pharm. Res.*, **16**, 55 (1999).

227. M. B. Bolger, I. S. Haworth, A. K. Yeung, D. Ann, H. von Grafenstein, S. Hamm-Alvarez, C. T. Okamoto, K. J. Kim, S. K. Basu, S. Wu, and V. H. Lee, *J. Pharm. Sci.*, **87**, 1286 (1998).

228. X. Z. Chen, A. Steel, and M. A. Hediger, *Biochem. Biophys. Res. Commun.*, **272**, 726 (2000).

229. H. Buchwald, D. K. Stoller, C. T. Campos, J. P. Matts, and R. L. Varco, *Ann. Surg.*, **212**, 318 (1990).

230. M. Ast and W. H. Frishman, *J. Clin. Pharmacol.*, **30**, 99 (1990).

231. K. Takashima, T. Kohno, T. Mori, A. Ohtani, K. Hirakoso, and S. Takeyama, *Atherosclerosis*, **107**, 247 (1994).

232. M. D. Carey, D. M. Small, and C. M. Bliss, *Annu. Rev. Physiol.*, **45**, 651 (1983).

233. F. A. Wilson in R. A. Frizzell, Ed., *The Gastrointestinal System*, Vol. **IV**, American Physiological Society, Bethesda, MD, 1991, p. 389.

234. R. H. Dowling, *Gastroenterology*, **62**, 122 (1972).

235. A. F. Hofmann, *Adv. Intern. Med.*, **21**, 501 (1976).

236. S. Ewerth, *Acta Chir. Scand.*Suppl., **513**, 1 (1982).

237. R. H. Erlinger in L. R. Johnson, Ed., *Physiology of the Gastrointestinal Tract*, 2nd ed., Raven Press, New York, 1987, p. 1557.

238. C. D. Klaassen, *Toxicol. Pathol.*, **16**, 130 (1988).

239. A. F. Hofmann in J. Forte, Ed., *The Gastrointestinal System*, Vol. **III**, American Physiological Society, Bethesda, MD, 1989, p. 567.

240. S. Stengelin, S. Apel, W. Becker, M. Maier, J. Rosenberger, U. Bewersdorf, F. Girbig, C.

Weyland, G. Wess, and W. Kramer, *Eur. J. Biochem.*, **239**, 887 (1996).

241. A. D. Vodenlich, Y.-Z. Gong, K. F. Geoghegan, M. C. Lin, A. J. Lanzetti, and F. A. Wilson, *Biochem. Biophys. Res. Commun.*, **177**, 1147 (1991).

242. S. Iseki, O. Amano, T. Kanda, H. Fujii, and T. Ono, *Mol. Cell. Biochem.*, **123**, 113 (1993).

243. W. Kramer, D. Corsiero, M. Friedrich, F. Girbig, S. Stengelin, and C. Weyland, *Biochem. J.*, **333**, 335 (1998).

244. M. C. Lin, W. Kramer, and F. A. Wilson, *J. Biol. Chem.*, **265**, 14986 (1990).

245. M. C. Lin, E. Mullady, and F. A. Wilson, *Am. J. Physiol. Gastrointest. Liver Physiol.*, **265**, G56 (1993).

246. J. Grober, I. Zaghini, H. Fujii, S. A. Jones, S. A. Kliewer, T. M. Willson, T. Ono, and P. Besnard, *J. Biol. Chem.*, **274**, 29749 (1999).

247. J. Y. Chiang, R. Kimmel, C. Weinberger, and D. Stroup, *J. Biol. Chem.*, **275**, 10918 (2000).

248. M. Crestani, A. Sadeghpour, D. Stroup, G. Galli, and J. Y. Chiang, *J. Lipid Res.*, **39**, 2192 (1998).

249. J. M. Lehmann, S. A. Kliewer, L. B. Moore, T. A. Smith-Oliver, B. B. Oliver, J. L. Su, S. S. Sundseth, D. A. Winegar, D. E. Blanchard, T. A. Spencer, and T. M. Willson, *J. Biol. Chem.*, **272**, 3137 (1997).

250. B. A. Janowski, M. J. Grogan, S. A. Jones, G. B. Wisely, S. A. Kliewer, E. J. Corey, and D. J. Mangelsdorf, *Proc. Natl. Acad. Sci. USA*, **96**, 266 (1999).

251. T. A. Spencer, D. Li, J. S. Russel, J. L. Collins, R. K. Bledsoe, T. G. Consler, L. B. Moore, C. M. Galardi, D. D. McKee, J. T. Moore, M. A. Watson, D. J. Parks, M. H. Lambert, and T. M. Willson, *J. Med. Chem.*, **44**, 886 (2001).

252. K. N. Lazaridis, L. Pham, P. Tietz, R. A. Marinelli, P. C. deGroen, S. Levine, P. A. Dawson, and N. F. LaRusso, *J. Clin. Invest.*, **100**, 2714 (1997).

253. S. Sorscher, J. Lillienau, J. L. Meinkoth, J. H. Steinbach, C. D. Schteingart, J. Feramisco, and A. F. Hofmann, *Biochem. Biophys. Res. Commun.*, **186**, 1455 (1992).

254. W. Kramer, G. Burckhardt, F. A. Wilson, and G. Kurz, *J. Biol. Chem.*, **258**, 3623 (1983).

255. M. C. Lin, S. L. Weinberg, W. Kramer, G. Burckhardt, and F. A. Wilson, *J. Membr. Biol.*, **106**, 1 (1988).

256. G. Burckhardt, W. Kramer, G. Kurz, and F. A. Wilson, *J. Biol. Chem.*, **258**, 3618 (1983).

257. B. L. Shneider and M. S. Moyer, *J. Biol. Chem.*, **268**, 6985 (1993).

258. B. L. Shneider, P. A. Dawson, D.-M. Christie, W. Hardikar, M. H. Wong, and F. J. Suchy, *J. Clin. Invest.*, **95**, 745 (1995).

259. S. A. Weinman, M. W. Carruth, and P. A. Dawson, *J. Biol. Chem.*, **273**, 34691 (1998).

260. S. Hallen, M. Branden, P. A. Dawson, and G. Sachs, *Biochemistry*, **38**, 11379 (1999).

261. C. J. Sippel, P. A. Dawson, T. Shen, and D. H. Perlmutter, *J. Biol. Chem.*, **272**, 18290 (1997).

262. M. H. Wong, P. Oelkers, A. L. Craddock, and P. A. Dawson, *J. Biol. Chem.*, **269**, 1340 (1994).

263. G. von Heijne and Y. Gavel, *Eur. J. Biochem.*, **174**, 671 (1988).

264. J. Higaki, S. Hara, N. Takasu, K. Tonda, K. Miyata, T. Shike, K. Nagata, and T. Mizui, *Arterioscler. Thromb. Vasc. Biol.*, **18**, 1304 (1998).

265. T. Ichihashi, M. Izawa, K. Miyata, T. Mizui, K. Hirano, and Y. Takagishi, *J. Pharmacol. Exp. Ther.*, **284**, 43 (1998).

266. M. W. Love and P. A. Dawson, *Curr. Opin. Lipidol.*, **9**, 225 (1998).

267. A. F. Hofmann, *Arch. Intern. Med.*, **159**, 2647 (1999).

268. L. Lack, *Environ. Health Perspect.*, **33**, 79 (1979).

269. M. P. Tyor, J. T. Garbutt, and L. Lack, *Am. J. Med.*, **51**, 614 (1971).

270. S. Walker, A. Stiehl, R. Raedsch, P. Kloters, and B. Kommerell, *Digestion*, **32**, 47 (1985).

271. R. Aldini, A. Roda, P. L. Lenzi, G. Ussia, M. C. Vaccari, G. Mazzella, D. Festi, F. Bazzoli, G. Galletti, S. Casanova, et al., *Eur. J. Clin. Invest.*, **22**, 744 (1992).

272. E. R. Schiff, N. C. Small, and J. M. Dietschy, *J. Clin. Invest.*, **51**, 1351 (1972).

273. K. W. Heaton and L. Lack, *Am. J. Physiol.*, **214**, 585 (1968).

274. L. Lack and I. M. Weiner, *Am. J. Physiol.*, **210**, 1142 (1966).

275. E. Krag and S. F. Phillips, *J. Clin. Invest.*, **53**, 1686 (1974).

276. K. Gallagher, J. Mauskopf, J. T. Walker, and L. Lack, *J. Lipid Res.*, **17**, 572 (1976).

277. L. Lack, A. Tantawi, C. Halevy, and D. Rockett, *Am J Physiol. Gastrointest. Liver Physiol.*, **246**, G745 (1984).

278. T. S. Low-Beer, M. P. Tyor, and L. Lack, *Gastroenterology*, **56**, 721 (1969).

279. G. Wess, W. Kramer, W. Bartmann, A. Enhsen, H. Glombik, S. Müllner, K. Bock, A. Dries, H. Kleine, and W. Schmitt, *Tetrahedron Lett.*, **33**, 195 (1992).

280. L. Lack, J. T. Walker, and G. D. Singletary, *Am. J. Physiol.*, **219**, 487 (1970).

281. K. H. Baringhaus, H. Matter, S. Stengelin, and W. Kramer, *J. Lipid Res.*, **40**, 2158 (1999).

282. P. W. Swaan, F. C. Szoka Jr., and S. Oie, *J. Comput.-Aided Mol. Des.*, **11**, 581 (1997).

283. W. Kramer, S. Stengelin, K. H. Baringhaus, A. Enhsen, H. Heuer, W. Becker, D. Corsiero, F. Girbig, R. Noll, and C. Weyland, *J. Lipid Res.*, **40**, 1604 (1999).

284. F. Helsper and P. W. Swaan, submitted.

285. N. F. H. Ho, *Ann. N. Y. Acad. Sci.*, **507**, 315 (1987).

286. P. W. Swaan, K. M. Hillgren, F. C. Szoka Jr., and S. Oie, *Bioconjug. Chem.*, **8**, 520 (1997).

287. W. Kramer, G. Wess, G. Schubert, M. Bickel, F. Girbig, U. Gutjahr, S. Kowalewski, K.-H. Baringhaus, A. Enhsen, H. Glombik, S. Müllner, G. Neckermann, S. Schulz, and E. Petzinger, *J. Biol. Chem.*, **267**, 18598 (1992).

288. W. Kramer, G. Wess, A. Enhsen, K. Bock, E. Falk, A. Hoffmann, G. Neckermann, D. Gantz, S. Schulz, L. Nickau, et al., *Biochim. Biophys. Acta*, **1227**, 137 (1994).

289. G. Wess, W. Kramer, X. B. Han, K. Bock, A. Enhsen, H. Glombik, K. H. Baringhaus, G. Boger, M. Urmann, A. Hoffmann, and E. Falk, *J. Med. Chem.*, **37**, 3240 (1994).

290. Z. F. Stephan, E. C. Yurachek, R. Sharif, J. M. Wasvary, R. Steele, and C. Howes, *Biochem. Pharmacol.*, **43**, 1969 (1992).

291. D.-C. Kim, A. W. Harrison, M. J. Ruwart, K. F. Wilkinson, J. F. Fisher, I. J. Hidalgo, and R. T. Borchardt, *J. Drug Target*, **1**, 347 (1993).

292. C. O. Mills, G. H. Martin, and E. Elias, *Biochim. Biophys. Acta*, **876**, 677 (1986).

293. M. S. Anwer, E. R. O'Maille, A. F. Hofmann, R. A. DiPietro, and E. Michelotti, *Am. J. Physiol. Gastrointest. Liver Physiol.*, **249**, G479 (1985).

294. C. O. Mills, S. Iqbal, and E. Elias, *J. Hepatol.*, **1**, 199 (1985).

295. W. F. Caspary and W. Creutzfeldt, *Diabetologia*, **11**, 113 (1975).

296. C. Root, C. D. Smitth, D. A. Winegar, L. E. Brieaddy, and M. C. Lewis, *J. Lipid Res.*, **36**, 1106 (1995).

297. M. C. Lewis, L. E. Brieaddy, and C. Root, *J. Lipid Res.*, **36**, 1098 (1995).

298. S. Ekins, C. L. Waller, P. W. Swaan, G. Cruciani, S. A. Wrighton, and J. H. Wikel, *J. Pharmacol. Toxicol. Methods*, **44**, 251 (2000).

CHAPTER NINE

Allosteric Proteins and Drug Discovery

J. Ellis Bell
Department of Chemistry
Gottwald Science Center
University of Richmond
Richmond, Virginia

James C. Burnett
School of Pharmacy and Department of Medicinal Chemistry
Institute for Structural Biology and Drug Discovery
Virginia Commonwealth University
Richmond, Virginia

Jessica K. Bell
Department of Pharmaceutical Chemistry
University of California at San Francisco
San Francisco, California

Peter S. Galatin
Donald J. Abraham
School of Pharmacy and Department of Medicinal Chemistry
Institute for Structural Biology and Drug Discovery
Virginia Commonwealth University
Richmond, Virginia

Contents

Burger's Medicinal Chemistry and Drug Discovery
Sixth Edition, Volume 2: Drug Development
Edited by Donald J. Abraham
ISBN 0-471-37028-2   © 2003 John Wiley & Sons, Inc.

1 INTRODUCTION

The concept of allostery for the regulation of
protein function has been known for approximately half a century. However, medicinal
chemists have only recently been successful in
developing allosteric effectors as therapeutic
agents. Allosteric modifiers do not act as agonists or antagonists; instead they regulate
substrate binding or release, or catalysis. Allosteric proteins regulate some of the most important biochemical pathways in living systems. Allosteric effectors represent a new
mechanistic approach in drug therapy. Pharmacologists have explored similar and closely
related areas using different terminology such
as inverse agonists. The purpose of this chapter is to integrate the underpinnings grounded
in structural biology and mechanistic biochemistry with drug discovery. Because an increasing number of human illnesses are now
know to result from mutations affecting the
allosteric regulation of various enzymes (1), it
is safe to assume that a number of new drug
entities will emerge in the near future that
modulate allosteric interactions. The major
barrier to more rapid discovery of allosteric
effectors has been the lack of detailed structural information about the different allosteric states that an allosteric protein assumes.
Until recently, hemoglobin represented the
only well-defined allosteric protein with a
plethora of structural and mechanistic studies. However, those involved in unraveling the
mechanistic details of the allosteric transition
still argue over what triggers it, and what is
the order of tertiary and quaternary conformational changes.

Despite the difficulty in obtaining structural information for each allosteric state in
any given protein or receptor, drug discovery
with allosteric systems is well under way and
bearing fruit. The purpose of this chapter is to
review the underlying principles of allosteric
proteins and how these may be used to design
"allosteric" drugs.

2 ALLOSTERIC THEORY AND MODELS

Enzymes function as monomers, multido-
mained protomers, or complex oligomeric species. They bind substrates, cofactors, and/or
effectors, and stabilize transition states to increase the rate of catalysis and release product. Aside from regulation arising at the transcriptional, translational, or posttranslational
levels, the control of an enzymatic reaction relies on the enzyme's ability to regulate substrate binding or product release (so called K-
type allosteric effects), or the rate of catalysis
(so-called V-type allosteric effects) in response
to cellular and whole organism changes.

Most proteins exhibit hyperbolic saturation curves; however, a number of important
biological processes involve multi-subunit
proteins whose overall activities are regulated
by allosteric effects. These proteins exhibit
saturation curves that deviate from the normal hyperbolic curve (Fig. 9.1), and may involve effects where the activity increases dis-
proportionately with saturation [a so-called
positive cooperativity, giving a sigmoidal saturation curve (Fig. 9.1)] or decreases relatively as saturation is increased (a so-called
negative cooperativity), giving rise to a saturation curve (Fig. 9.1) that is proportionally
too steep below half saturation, and less steep
relative to a normal saturation curve. Above
half saturation a further deviation from a normal (hyperbolic) saturation curve (Fig. 9.2 a)
that is often seen is the phenomenon of substrate inhibition, which also (among other alternatives) can have its root in allosteric interactions. In a Lineweaver-Burk plot, substrate
inhibition results in a distinct upturn at high
substrate concentrations (Fig. 9.2b).

Mathematically, cooperativity observed in
ligand binding has been described by both the
Hill and Adair equations. The derivations of
these equations are modeled on the classic
textbook example of first ligand binding, and
then specifically the Hill and Adair equations
(2, 3). First, consider the simple binding of L

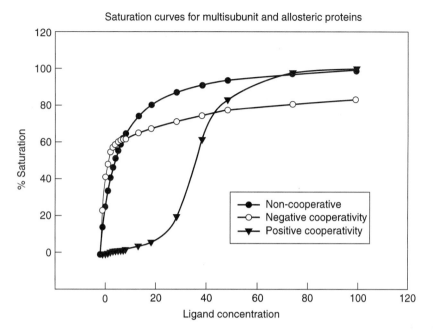

Figure 9.1. Saturation curves for a protein exhibiting no cooperativity (●), negative cooperativity (○), and positive cooperativity (▼).

[ligand] to P [protein], $P + L \rightleftharpoons PL$ with a dissociation constant of K_d given by

$$K_d = [P][L]/[PL] \qquad (9.1)$$

Rearrangement of the preceding equation gives

$$[PL] = [P][L]/K_d \qquad (9.2)$$

The concentration of bound sites divided by the total number of sites available,

$$Y = [PL]/[P] + [PL] \qquad (9.3)$$

represents the fractional saturation Y. Substituting Equation 9.2 into Equation 9.3 gives

$$Y = [L]/K_d + [L] \qquad (9.4)$$

A normal saturation curve is observed when Y is plotted as a function of [L].

In 1910 Archibald Hill (4) derived an equation similar to Equation 9.4 to describe the hemoglobin: O_2 sigmoidal curve of Y versus [L] plots. Hill assumed that a protein has n sites to bind ligand. Upon binding one site, the remaining $n - 1$ sites are immediately occupied. This all-or-none binding does not allow for accumulation of intermediates and may be represented as $P + nL = PL_n$, with the dissociation constant of L given by

$$K_d = [P][L]^n/[PL_n] \qquad (9.5)$$

The fractional saturation Y is now

$$Y = [PL_n]/[P] + [PL_n] \qquad (9.6)$$

Substituting Equation 9.5 for [PL] gives

$$Y = [L]^n/K_d + [L]^n \qquad (9.7)$$

Equation 9.7 can be rearranged to give

$$Y/(1 - Y) = [L]^n/K_d \qquad (9.8)$$

where n represents the Hill coefficient. The log of Equation 9.8 gives

$$\log[Y/(1 - Y)] = n \log[L] - \log K_d \qquad (9.9)$$

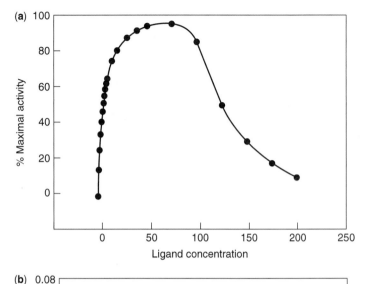

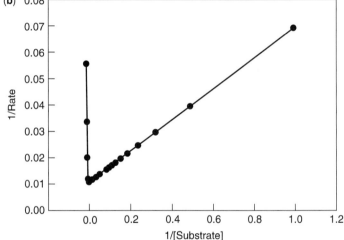

Figure 9.2. Substrate inhibition manifested in (a) a saturation plot and (b) a Lineweaver-Burk plot of data.

The plot of $\log[Y/(1 - Y)]$ versus $\log[L]$ is linear (Fig. 9.3) with a slope of n. The all-or-none binding assumes extreme cooperativity in ligand binding. Under such conditions n should equal the number of sites within the protein. Few proteins exhibit extreme cooperativity. Therefore, the value for n is usually less than the number of sites but, when positive cooperativity occurs, n is greater than 1. In the case of negative cooperativity n is less than 1. An n value of 1 reduces the fractional saturation for the Hill equation (Equation 9.7) to that of simple ligand binding (Equation 9.4).

The Hill derivation accounts for only two states, liganded and unliganded protein. The observation of partially oxygenated hemoglobin led Gilbert Adair (5) in 1924 to propose that ligand binding could occur in a sequential fashion:

$$P + L \leftrightarrows PL$$

$$PL + L \leftrightarrows PL_2$$

$$PL_2 + L \leftrightarrows PL_3$$

$$PL_{n-1} + L \leftrightarrows PL_n$$

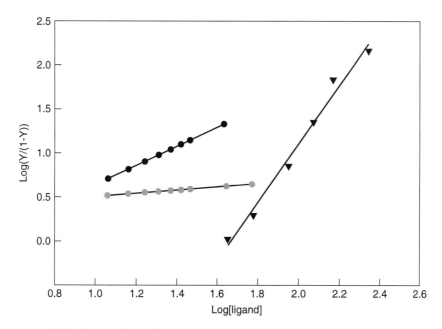

Figure 9.3. Hill plots of data indicating no cooperativity, negative cooperativity, and positive cooperativity.

The individual binding steps could be described by a dissociation constant:

$$K_{dn} = [PL_{n-1}][L]/[PL_n] \qquad (9.10)$$

The fractional saturation, defined as the concentration L bound divided by total concentration of sites for n sites, would give the equation

$$Y = [PL] + 2[PL_2] + \cdots$$
$$+ [PL_n]/n[P] + [PL] + [PL_2] \qquad (9.11)$$
$$+ \cdots + [PL_n]$$

Substituting Equation 9.10 into Equation 9.11 and factoring out [P] results in the Adair equation:

$$Y = ([L]/K_{d1} + 2[L]^2/K_{d1}K_{d2} + \cdots$$
$$+ n[L]^n/K_{d1}K_{d2} \cdots K_{dn})/$$
$$n(1 + [L]/K_{d1} + [L]^2/K_{d1}K_{d2} \qquad (9.12)$$
$$+ \cdots + [L]^n/K_{d1}K_{d2} \cdots K_{dn})$$

An advantage to assessing ligand binding data with the Adair equation is that this derivation does make assumptions about the type or presence of cooperativity. Instead, cooperativity is evaluated using K_{dn}, the macroscopic or apparent dissociation constants, and the statistically related microscopic or intrinsic dissociation constants $[K_n{}']$, which describe the individual dissociation constants for the ligand binding sites. For a protein containing four binding sites, the relationship between the macroscopic and microscopic dissociation constants is as follows:

$$K_{d1} = K_1{}'/4$$

$$K_{d2} = 2K_2{}'/3$$

$$K_{d3} = 3K_3{}'/2$$

$$K_{d4} = 4K_4{}'$$

No cooperativity is observed when $K_1{}' = K_2{}' = K_3{}' = K_4{}'$; the Adair equation reduces to the fractional saturation equation of simple ligand

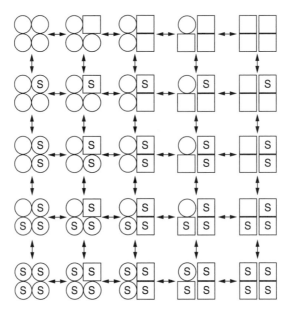

Figure 9.4. Schematic of the MWC, KNF/DE, and unified mechanistic models of cooperativity. The first and fifth columns represent the nonexclusive MWC model. In this model an equilibrium between the T and R states exists before ligand binding. Ligand may bind to either the T or R state but will bind preferentially to the R state. Upon ligand binding the equilibrium is reestablished, causing a shift from the T to the R state. The newly formed, high affinity sites are quickly bound by ligand, resulting in positive cooperativity. The diagonal represents the KNF/DE sequential model. As ligand binds it induces a conformation change within that subunit, which is relayed as either increased or decreased affinity for ligand to surrounding binding sites/subunits. The MWC and KNF/DE models are two restrained examples of a much more general scheme that is represented by the entire figure. Because of the complexity of the general scheme and its mathematical derivations, systems are generally evaluated as following a concerted or sequential pathway.

binding. When $K_1' > K_2' > K_3' > K_4'$, positive cooperativity occurs as each ligand binds. If the inverse, $K_1' < K_2' < K_3' < K_4'$, is true, ligand binding proceeds with negative cooperativity.

2.1 The Monod-Wyman-Changeux and Koshland-Nemethy-Fimer/Dalziel-Engel Models

With the mathematical evaluation of cooperativity based on the Hill and Adair equations, the question remains, How does the enzyme convey ligand binding between sites? Historically, the mechanisms by which proteins convey cooperativity in ligand binding have been described by two models: Monod-Wyman-Changeaux (6) (MWC) and Koshland-Nemethy-Filmer (7)/Dalziel-Engel (8) (KNF/DE). These two models serve as starting points to

describe what conformational changes within an enzyme may lead to the observed cooperativity.

2.1.1 Monod-Wyman-Changeaux: The Concerted Model.
In 1965 Monod, Wyman, and Changeaux (MWC) proposed the first model to explain the mechanism by which cooperativity may be transmitted within an oligomer. The MWC model begins with two unliganded states, (1) the low affinity, taut state T, and (2) the high affinity, relaxed state R, at equilibrium predominated by the T state (refer to Fig. 9.4). Each complete oligomer may contain either the T or R but not a mixture such that the molecule is symmetric. A ligand molecule binds to the R state, following Le Chatelier's principle; the equilibrium between the T and R states is reestablished, creating another R

state molecule. Ligand will rapidly bind to the newly formed, unoccupied high affinity sites, resulting in positive cooperativity. The MWC model explains positive cooperativity by the transition from low to high affinity states. Heterotropic effectors, activators, or inhibitors fit into the model by stabilizing/destabilizing the T or R state. The major disadvantage to the MWC model is its inability to describe negative homotropic cooperativity. Because the MWC model begins with a preexisting equilibrium of the T and R state and the binding of ligand stabilizes the R state, the constraints of the model (9) do not allow an increase in the proportion of the T state, negative cooperativity.

The model is described mathematically by the equation:

$$Y = \frac{\alpha(1 + \alpha)^{n-1} + L\alpha C(1 + \alpha C)^n}{(1 + \alpha)^n + L(1 + \alpha C)^n} \quad (9.13)$$

In this equation, L is the apparent conformational equilibrium constant in the absence of substrate:

$$L = T/R \quad (9.14)$$

where n is the number of equivalent, independent binding sites for the ligand and corresponds to the number of identical subunits, and α and C are constants related to the intrinsic dissociation constants of the two states (K_R and K_T) and the free ligand concentration (F) by

$$C = K_R/K_T \quad (9.15)$$

$$\alpha = F/K \quad (9.16)$$

2.1.2 Koshland-Nemethy-Filmer/Dalziel-Engel: The Sequential Model.
A second model to describe cooperativity within systems was proposed independently by both the Koshland and Dalziel groups. Unlike the concerted model, the sequential model does not constrain the oligomer to either the T or R state and therefore no preequilibrium of states is present. The sequential model, also known as the induced-fit model, hypothesizes that when a ligand binds, a conformational change occurs within its binding site. This change is then transmitted to adjacent protomers, inducing either increased or decreased affinity for the binding of the next ligand. The KNF/DE model, because it is not locked into a preexisting equilibrium, can adequately describe both positive and negative homotropic cooperativity. Heterotropic effectors, again, stabilize one form of the enzyme. The sequential model is a mechanistic explanation of the Adair equation and the observed intrinsic dissociation constants.

The MWC and KFN/DE models need not be mutually exclusive. M. Eigen (10) took these two models as limiting cases of a general scheme, represented by Fig. 9.4, as a whole for a tetrameric protein. The mathematical analysis for so many components is complex. Under most circumstances, therefore, cooperativity is generally evaluated under the simplified cases of either the MWC or KNF/DE models.

2.2 Hemoglobin: Classic Example of Allostery

The classic example of allostery and cooperativity is the binding of oxygen to hemoglobin. Hemoglobin (Hb) is a heterotetramer composed of two α- and two β-chains. An α- and β-chain associate to form a dimer, $\alpha_1\beta_1$ and $\alpha_2\beta_2$; the dimers then associate to form the tetramer. Figure 9.5 illustrates the topology of both the tetramer and the subunit. Within a subunit the heme is located in a hydrophobic pocket between helices E and F. The coordination of the Fe(II) atom depends on the oxygenation state. In the deoxy state the Fe(II) is 5-coordinated by the four pyrrole nitrogen atoms and the proximal histidine vs. the oxygenated state, where the Fe(II) is 6-coordinated with the sixth site occupied by the O_2. The main subunit-subunit interactions in the tetramer occur between the $\alpha_1\beta_1$ and $\alpha_1\beta_2$ interfaces and their symmetry-related counterparts. A central cavity, resulting from the arrangement of the subunits, has been shown to accommodate allosteric effectors such as 2,3-bisphosphoglycerate.

From crystallographic studies Hb has been shown to exist in at least two structurally distinct states, referred to as the T (deoxy) and R (oxy) states. In the T or constrained state the

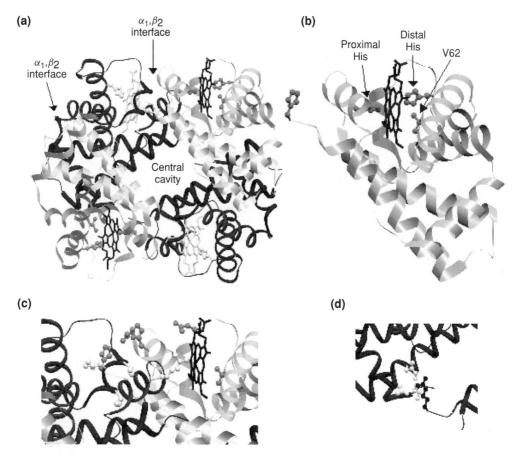

Figure 9.5. Structural overview of hemoglobin. (a) The T-state of the heterotetramer of Hb is composed of two α (black) and two β (white) subunits. The packing of the subunits creates a central cavity, which is the binding site of the allosteric effector, 2,3-bisphosphoglycerate. (b) Both subunits are α-helical. The β-subunit shows the position of the porphyrin ring, located in a hydrophobic pocket between two helices with a proximal and distal histidine capable of coordinating the Fe(II). Valine-67 of the β subunit in the deoxy state partially blocks O_2 from binding. (c) The two-way switch between the R and T state involves His-97 of the β_2 chain and Thr-41 and Thr-38 of the α_1 chain. In the deoxy state a salt bridge exists between H97 and T41. When the Fe(II) moves into the plane of the heme upon binding O_2, H92 and helix F must also move closer to the heme. To allow this movement, the H97:T41 bridge is broken and moved down one turn of the helix to T38. A transition between T41 and T38 is unfavored because of steric clashes. A salt bridge that stabilizes the T state is also depicted: His-146 at the C-terminus of β_2 coordinates its carboxyl to the amino group of Lys-40 in the α_1 chain and the Nε of His-146 is coordinated to the COδ of Asp-94. With the rearrangement of the H97 when O_2 is bound, the salt bridge between H146 and K40 is broken. (d) Salt bridges between the α-subunits also stabilize the T state. Arg-141 of the α1 chain forms a hydrogen bond with the amino group of Lys-127 of the α_2 chain through its carboxy terminus. The guanidino group of Arg-141 forms bonds with COδ of Asp-126, CO of Val-34, and the C-terminus of Val-1, all in the α_2 subunit. Again, these bonds are disrupted when O_2 binds.

Fe(II) is about 0.6 Å out of the heme plane toward the proximal histidine. In the β-subunits Val-62 partially occludes O_2 binding on the distal side of the heme ring. The T state is stabilized by salt bridges between the α_1 and α_2 chains (R141 of the α_1 to D126α_1, CO of V34β_2, and C-terminus of V1α_2) and the β_1- and β_2-chains (Y145β_2 to COV98β_2, H146β_2 to

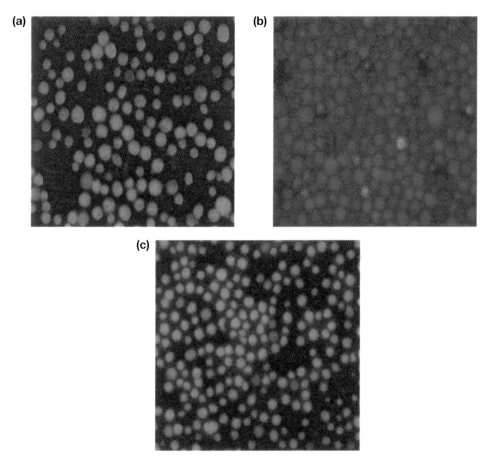

Figure 2.7. Demonstration of a β-lactamase reporter cell line. In a final volume of 0.02 mL, 5000–10,000 cells were seeded overnight at 37°C before being stimulated with excess agonist or antagonist for 4 h. The resultant β-lactamase activity was measured after a 1-h incubation with CCF2 (Aurora Biosciences) at the indicated wavelengths in an LJL Analyst fluorometer. (a) The unstimulated cells, (b) the cells incubated with agonist, and (c) cells treated with agonist and excess antagonist.

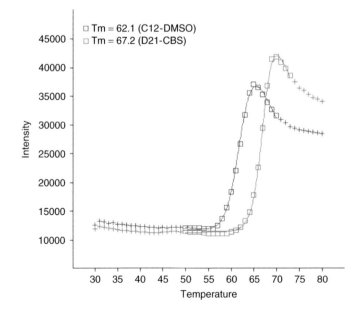

□ Tm = 62.1 (C12-DMSO)
□ Tm = 67.2 (D21-CBS)

Figure 2.9. An example of a Thermofluor ligand-binding experiment where carboxybenzene sulfonamide (CBS) binds to carbonic anhydrase (CA). The fluorescence intensity was measured over a 30–80°C temperature range using 50 μM CBS (red) or dimethyl sulphoxide (blue) and 0.15 mg/mL CA.

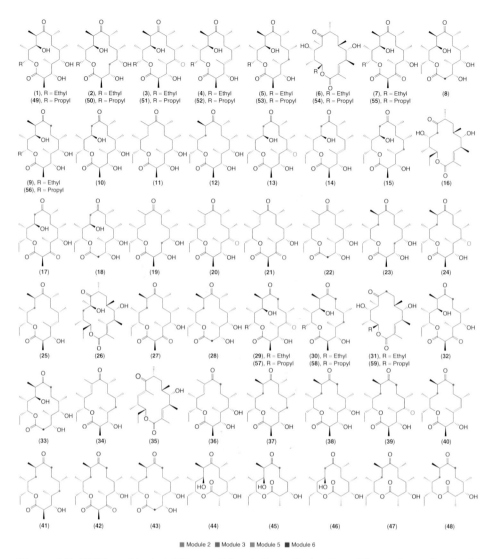

Figure 4.5. DEBS combinatorial library. Colors indicate the location of the engineered carbon(s) resulting from catalytic domain substitutions in module 2 (red), module 5 (green), module 6 (blue), or modules 1,3, or 4 (yellow).

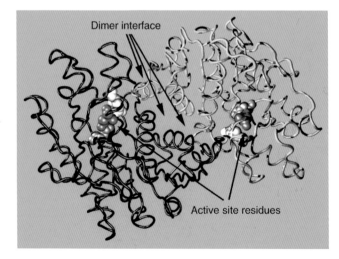

Figure 9.8. The malate dehydrogenase dimer, indicating the location of the active sites in each protein plus the dimer interface. Malate dehydrogenase demonstrates substrate inhibition that has been attributed to subunit interactions and allosteric regulation by citrate, although the crystal structure of the protein reveals the absence of a separate allosteric site for citrate.

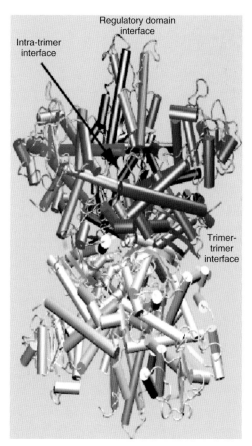

Figure 9.10. The quaternary structure of gluta-mate dehydrogenase reveals a complex array of subunit interfaces.

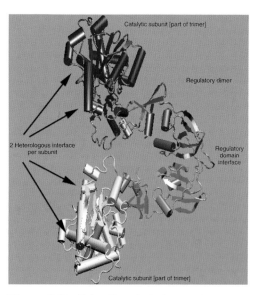

Figure 9.12. Subunit contacts in aspartate tran-scarbamoylase show heterologous contacts in the catalytic trimers, isologous contacts between regu-latory dimers, and regulatory subunit-catalytic subunit interactions.

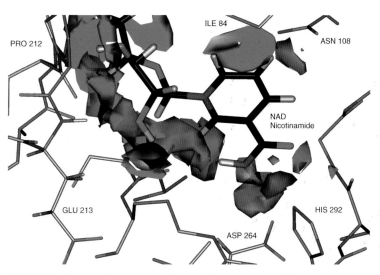

Figure 9.13. HINT analysis of interactions between the confactor NAD nicotinamide moiety and surrounding 3-phosphoglycerate dehydrogenase residues. Protein carbons are light gray and NAD carbons are black. Nitrogen atoms are blue and oxygen atoms are red. HINT contours are color coded to represent the different types of noncovalent interactions as follows: blue = favorable hydrogen bonding and acid-base interactions; green = favorable hydrophobic-hydrophobic inter-actions; and red = unfavorable acid-acid and base-base interactions. The volume of the HINT con-tour represents the magnitude of the interactions.

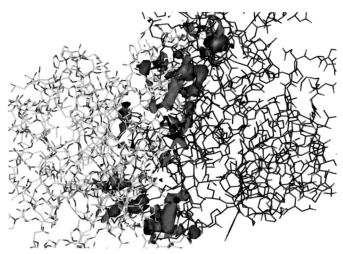

Figure 9.14. HINT analysis of interactions between residues at the interface between two subunits of 3-phosphoglycerate dehydrogenase. The carbons of the two distinct subunits are colored light and dark gray, respectively. Nitrogen atoms are blue, oxygen atoms are red, and sulfur atoms are yellow. HINT contour maps visually displaying interactions between residues at the subunit interface are color coded according to the type of interaction: blue represents hydrogen bonding and favorable acid-base interactions, green displays favorable hydrophobic-hydrophobic contacts, red indicates unfavorable base-base and acid-acid interactions, and purple indicates unfavorable hydrophobic-polar interactions. The volume of the HINT contour represents the magnitude of the interaction. For the analyzed phosphoglycerate dehydrogenase subunits, HINT indicates that a balance between favorable and unfavorable interactions characterizes the interface.

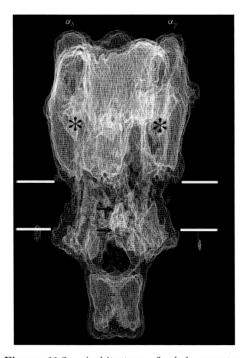

Figure 11.2. Architecture of whole receptor, emphasizing the external surface and openings to the ion-conducting pathway on the outer (extracellular) and inner (cytoplasmic) sides of the membrane. The positions of the two α subunits, the binding pockets (asterisks), gate of the closed channel (upper arrow), and the constricting part of the open channel (lower arrow) are indicated.

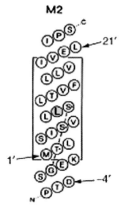

Figure 11.3. Helical net plot of the amino acid sequence around the membrane-spanning segment M2 (*Torpedo* α subunit). The leucine residue near the middle of the membrane (yellow) is the conserved leucine L251 (at the 9′ position), which may be involved in forming the gate of the channel. The dots denote other residues that have been shown to affect the binding affinity of an open channel blocker (17, 18) and ion flow through the open pore (19). The numbers shown refer to the numbering scheme for M2 residues used in the text.

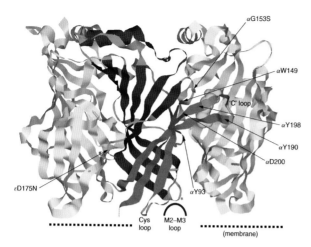

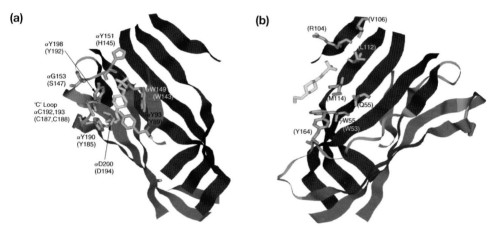

Figure 11.4. Three subunits of the *Lymnaea* AChBP viewed perpendicular to the fivefold axis of symmetry and from the outside of the pentamer. The inner and outer sheets of the β-sandwich are blue and red, respectively, whereas the putative ligand HEPES is purple. The approximate positions of the α carbons of residues discussed in the text are marked with arrows on the foremost subunit. The approximate positions of the cell membrane and of the M2-M3 loop are shown diagrammatically. The inner β-sheet (blue) is thought to rotate after agonist binding and to interact with the M2-M3 loop (as indicated by the asterisk).

Figure 11.6. The binding site of AChBP. The ligand (HEPES) is in yellow. Blue and red regions denote the inner- and outer-sheet parts of the β-sandwich (30). The views in (a) and (b) are with the fivefold axis vertical, and the "membrane" at the bottom. (a) The structure shown is analogous to the binding site of the *Torpedo* or human α1 subunit, to which the numbering of identified residues refers. Numbers in parentheses refer to AChBP. (b) Surface of the neighboring protomer that faces the binding site of the AChBP. In the receptor this would be part of the γ, ε, or δ subunit. Numbers in parentheses refer to the AChBP. W53 aligns with W55 in mouse γ, ε, and δ nicotinic receptor subunits, but most of the residues have no obvious analog in the γ, ε, or δ subunits of the nicotinic receptor. For example, Q55 aligns with mouse γE57, εG57, and δD57; L112 aligns with mouse γY117 or δT119, and Y164 is A or G in the nicotinic subunits.

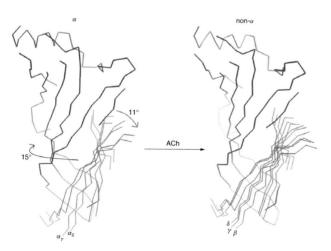

Figure 11.9. Arrangements of the inner (blue) and outer (red) β-sheet parts of the (a) α and (b) non-α subunits (see Figs. 11.4 and 11.6), after fitting to the densities in the 4.6-Å map of the receptor. The arrangement of the sheets in the α subunits switches to that of the non-α sheets when ACh binds. The views are in the same direction, toward the central axis from outside the pentamer. Arrows and angles in A denote the sense and magnitudes of the rotations relating the α to the non-α sheets. The traces are aligned so that the inner sheets are superimposed. [Adapted from Unwin et al. (30).]

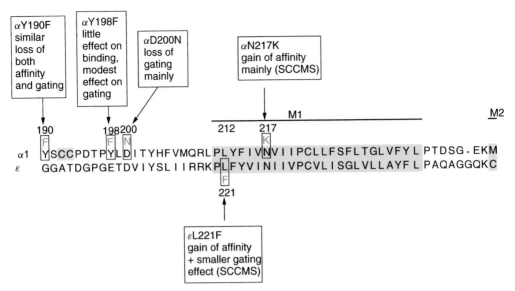

Figure 11.12. Aligned sequences of the human α1 and ε subunits, to show the position of some of the mutations that are discussed in the text. The (approximate) positions of the M1 region and the beginning of M2 are shown.

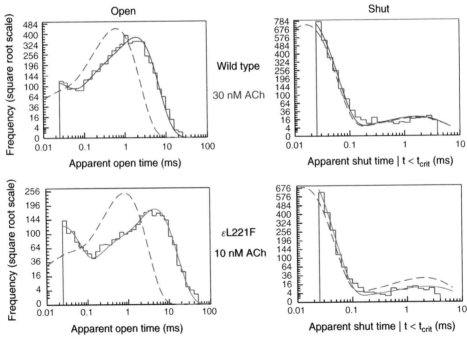

Figure 11.14. Distributions of (logarithms of) the apparent open time (left) and apparent shut time (right) for wild-type human receptors (top) and for mutant εL221F receptors (bottom). The histogram shows the experimental observations. The continuous lines were not fitted directly to the data in the histograms, but were calculated from the rate constants for the mechanism that was fitted (Fig. 11.10, scheme B, with the two sites constrained to be independent). The distributions were calculated with appropriate allowances for missed events (HJC distributions) (65, 66). The fact that they superimpose well on the histograms shows that the mechanism was a good description of the observations. The dashed lines show the distributions calculated from the fitted rate constants in the conventional way (45), without allowance for missed events, so they are our estimate of the true distributions of open and shut times (296).

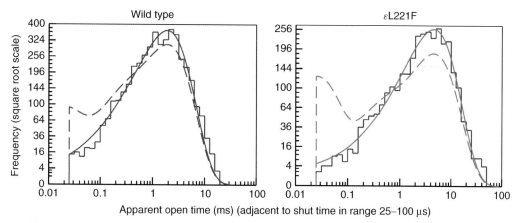

Figure 11.15. Conditional open time distributions. The histograms show the observed open times only for openings that were adjacent to short shut times (25–100 μs in duration). The continuous lines were not fitted to these histograms, but are the HJC conditional distributions of apparent open times calculated from the fitted rate constants (67). The fact that they superimpose well on the histograms shows that the fit predicts correctly the negative correlation between adjacent apparent open and shut times. The dashed lines show the calculated HJC distributions of *all* open times (as shown as solid lines in Fig. 11.14, left).

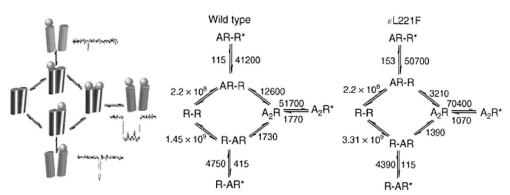

Figure 11.16. (Left) Schematic representation of reaction scheme B in Fig. 11.10, with examples of the sort of channel activations that are produced by the wild-type receptor with either one or both of the binding sites occupied by ACh. (Middle and right) Reaction scheme B (Fig. 11.10), with the results of a particular HJCFIT fitting marked on the arrows (the binding sites were assumed to be independent).

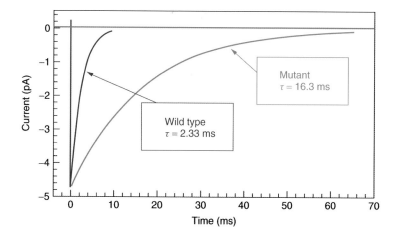

Figure 11.17. The predicted macroscopic current. The rate constants that have been fitted to results from equilibrium recordings (see Figs. 11.14–11.16) were used to calculate the macroscopic response to a 0.2-ms pulse of ACh (1 m*M*), as in Colquhoun and Hawkes (44). This calculation predicts that the mutation will cause a sevenfold slowing of the decay of the synaptic current, much as observed (102).

(a)

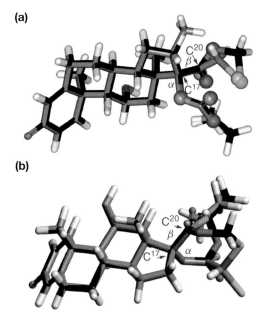

(b)

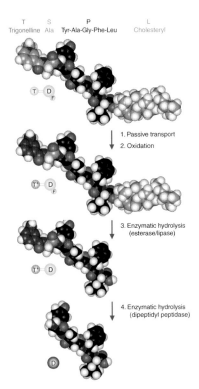

T   S   P   L
Trigonelline Ala Tyr-Ala-Gly-Phe-Leu Cholesteryl

1. Passive transport
2. Oxidation

3. Enzymatic hydrolysis
(esterase/lipase)

4. Enzymatic hydrolysis
(dipeptidyl peptidase)

Figure 15.11. Overlapping pharmacophore structures of corticosteroids. (a) Clobetasol propionate (in gray) and the soft corticosteroid loteprednol etabonate (**26**) (in black). The view is from the α side, from slightly below the steroid ring system. (b) Loteprednol etabonate (in black) and a 17α-dichloroester soft steroid (in gray). This view is from the β side, from above the steroid ring system.

Figure 15.40. CPK structures illustrating the sequential metabolism of the molecular package used for brain delivery of a leucine enkephalin analog.

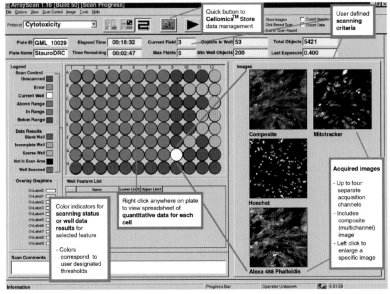

Figure 16.7. Graphical user interface of ArrayScan HCS System. Staurosporine treatment of BHK cells in 96-well microplate. Drug concentrations increase from left to right (controls in first two columns). Well color coding reflects (*1*) well status (current well being scanned is shown in white whereas unscanned wells are shown in grey) and (*2*) well results (pink wells have insufficient cell number, blue wells are below user-defined threshold range, red wells are above range, and green wells are in range for the selected cell parameter). Images correspond to current well, with three monochrome images reflecting three respective fluorescence channels, and composite (three-channel) color overlay in upper left panel. Nuclei are identified by the Hoechst dye (blue), microtubules by fluorescently labeled phalloidin (green), and mitochondria by Mitotracker stain (red).

D94, C-terminus β_2 to K40α_1). Upon oxygenation the Fe(II) atom moves into the plane of the heme. The proximal His also reorients and moves with the Fe(II) as does the F helix (of which the His is a part). For the F helix to move, a rearrangement at the $\alpha_1\beta_2$ interface is necessary. The contact between the H97 of the β-chain to the T41 of the α-chain must dissociate. H97 then is able to form a contact with T38, one turn down the helix. This movement is much like the knuckles of two hands sliding over one another. The result of these tertiary movements is a gross 15° rotation relative to the $\alpha_1\beta_1$ dimer to the $\alpha_2\beta_2$ dimer and a narrowing of the central solvent-filled cavity. Physiologically, the cooperativity of O_2 binding allows Hb to take up and release O_2 over a small range of O_2 pressures, comparable to pressure changes between the lungs and target tissues. (For a complete review of hemoglobin, refer to Ref. 16 and references therein.)

The binding of O_2 to hemoglobin can be mathematically explained by both the MWC and KNF/DE models. Given the crystallographic snapshots of the R and T states and abundant biochemical data, which model best explains the cooperativity in Hb? Both.

The quaternary transition of Hb between the T and R state is consistent with the concerted or MWC model. The two potential interactions between His-97 of the β-chain and either T41 or T38 of the α-chain sterically does not allow intermediary states within the tetramer. Coupled with the constraints of the more rigid $\alpha_1\beta_1$ and $\alpha_2\beta_2$ interfaces, this binary switch (T41 or T38) forces the transition of the entire molecule concomitantly, T → R.

The X-ray structure of human hemoglobin, in which only the α-subunits are oxygenated, offers some evidence for an induced-fit model (11). The partially oxygenated structure resembles the T state, with the exception of the α-chain Fe(II) atoms, which have moved about 0.15 Å closer to the heme plane. Perutz had postulated that the movement of the Fe(II) atom into/out of the heme plane in the oxy/deoxy states would exert a tension on the Fe(II) atom proximal His bond. This tension was measured spectroscopically by setting up Hb with NO, which binds with higher affinity than O_2, to pull Fe(II) into the plane of the heme and IHP, which acts as a mimic of BPG

and induces the T state, to pull Fe(II) out of the plane. The opposing forces on the Fe(II), if it triggers the transition from T → R, would exceed the strength of the bonds and they would break. Therefore, the observation of the human Hb intermediary position of the Fe(II) atom suggests that as more O_2 binds to the T state, this tension builds within the tertiary structure until the molecule is forced to undergo the T → R transition.

In part Hb fits both the MWC and KNF/DE models and underlies the premise that the models provide a starting point for interpreting the mechanistic machinations of a protein, but cannot be exclusive of one another or other potential intermediates. The examination of Hb's transition between its "T" and "R" states exemplifies the means by which proteins may transmit and communicate within and between subunit's ligand binding events. To summarize, Hb utilizes salt bridge formation/disruption at the C terminus, movement of secondary structure (F helix) induced by ligand coordination to a metal, and changes in a subunit-subunit interface ($\alpha_1\beta_2$) that, although not changing the nature of the hydrogen bonding pattern, does trigger gross quaternary movement. The knowledge of Hb's T → R transition and other proteins, such as ATCase, form the basis of structural changes that can be used to identify triggering mechanisms in other allosterically regulated and cooperative processes in proteins.

Although the above discussion has focused primarily on ligand affinity, it should be pointed out that the induced conformational changes or the differences between preexistent states may be reflected in altered catalytic activity, not just ligand affinity. This latter point illustrates another aspect of allosteric regulation, which can best be described in the concepts of K-type and V-type effects. A K-type effect is one that affects ligand affinity, whereas a V-type effect is one that affects the catalytic activity of the protein. Although the majority of allosteric enzymes are K-type systems, in recent years a number of V-type enzymes have been identified, and in many ways may be the more important as potential targets for drug design.

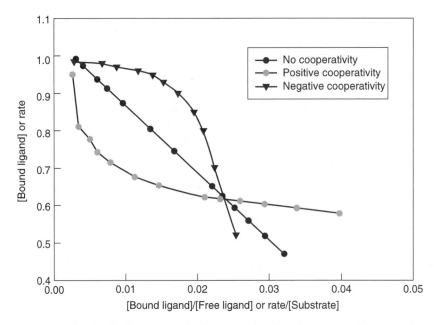

Figure 9.6. Scatchard or Eadie-Hofstee plots of data indicative of no cooperativity, negative cooperativity and positive cooperativity.

2.3 Common Features of Allosteric Proteins

Allosteric proteins are composed of multiple subunits and are regulated by both homotropic (substrate) and heterotropic ligands. In the case of heterotropic effects, this is usually a binding site quite different from that of the substrates or functional ligands of the protein, and may involve a separate subunit or domain of the overall protein. With homotropic effects, the spatially separate site is another chemically identical active site in the oligomer. The substrate affinity of allosteric proteins is cooperative in nature—the binding of substrate to one subunit affects the substrate binding affinity of other subunits. Cooperativity may be either positive or negative and is indicated experimentally by a curved ligand binding plot of protein activity versus substrate concentration (for comparison, a straight line is observed for non-allosteric proteins) (Fig. 9.6). A concave downwards curve (plot) indicates positive interactions/cooperativity, while a concave upwards plot indicates negative interactions/cooperativity. In an alternative analysis of the data, where the enzyme velocity is measured (Lineweaver-Burk plot) or the saturation with ligand is deter-

mined (Klotz Plot), a double-reciprocal representation (Fig. 9.7 of 1/Rate (or 1/[Bound Ligand] vs. 1/[Substrate] or 1/[Free Ligand]), which is linear for an enzyme showing no allosteric effects, is concave downward for negative cooperativity or concave upward for positive cooperativity. A coefficient of 1 indicates no cooperativity, whereas $n > 1$ equals positive cooperativity and $n < 1$ equals negative cooperativity.

While either ligand binding or kinetic observations are often used to invoke allosteric interactions, it is critically important to recognize that such manifestations of cooperative interactions are not unique. Kinetic plots showing either "positive" or "negative" cooperativity can simply be a manifestation of a complex kinetic mechanism, whereas binding studies indicative of "negative cooperativity" can result from multiple independent species capable of binding the same ligand with different affinities. Sigmoidal saturation curves of binding data, however, can be attributed only to "positive" cooperative effects. As has been extensively discussed (12–14), a clear demonstration of an allosteric effect requires demonstration of an induced conformational change

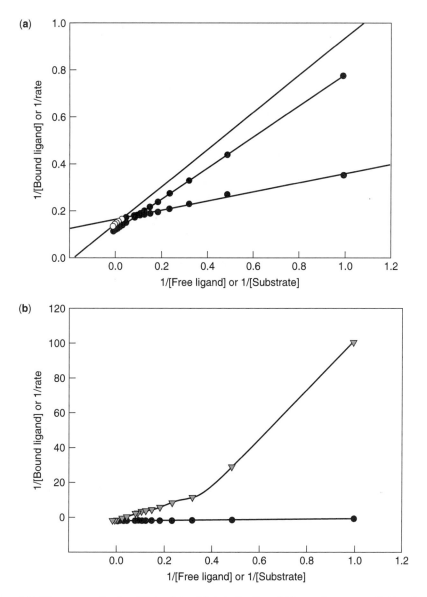

Figure 9.7. Lineweaver-Burk or Klotz plots of data indicative of (a) negative cooperativity, where at least two distinct linear regions are seen (as shown here), or (b) positive cooperativity.

affecting a distant site (for the Koshland/ Dalziel-type models) or the demonstration of a preexistent equilibrium between R and T states of the protein for Monod-type models.

2.4 Understanding Allosteric Transitions at the Molecular Level

Hemoglobin (Hb) was one of the first protein structures to be solved using X-ray crystallog-

raphy, and, as it displays positive cooperativity for oxygen binding, has served as the standard for correlating atomic level interactions and allostery (15, 16).

Structurally, Hb is a tetrameric protein composed of four subunits—two identical α subunits (referred to as $\alpha 1$ and $\alpha 2$) and two identical β subunits (referred to as $\beta 1$ and $\beta 2$). The subunits are arranged such that there are

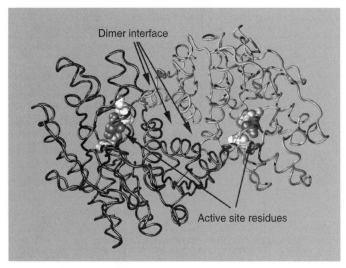

Figure 9.8. The malate dehydrogenase dimer, indicating the location of the active sites in each protein plus the dimer interface. Malate dehydrogenase demonstrates substrate inhibition that has been attributed to subunit interactions and allosteric regulation by citrate, although the crystal structure of the protein reveals the absence of a separate allosteric site for citrate. See color insert.

six intersubunit interfaces: $\alpha1\alpha2$, $\beta1\beta2$, $\alpha1\beta1$, $\alpha2\beta2$, $\alpha1\beta2$, and $\alpha2\beta1$. In dilute solutions, the Hb tetramer dissociates into two $\alpha\beta$ dimers: $\alpha1\beta1$ and $\alpha2\beta2$. Hence this protein is also referred to as a dimer of dimers.

Two structural end states of Hb have been determined through X-ray crystallography—a low affinity deoxy state (tense or T-state) (17, 18) and a high affinity oxy state (relaxed or R-state) (19, 20). Analyses of these two structures indicate that the allosteric transition from the non-liganded deoxy state to the fully liganded oxy state is accompanied by both tertiary and quaternary conformational changes. A 15° rotation of the $\alpha1\beta1$ dimer with relation to the $\alpha2\beta2$ dimer is observed on ligation. This rotation is triggered by structural modification of the heme groups during oxygen binding and is accompanied by numerous changes in noncovalent contacts between residues at the dimer–dimer interface. The T-state dimer–dimer interface possesses more stabilizing noncovalent interactions, and is substantially more stable than the R-state dimer-dimer interface.

Advances in X-ray crystallography and nuclear magnetic resonance spectroscopy have facilitated the solution of a range of other allosteric proteins. These structures have contributed significantly to our understanding of the nature of allostery and the complex atomic level changes that accompany transitions from active/high affinity conformations to in-

active/low affinity conformations. Examples include: fructose-1,6-bisphosphatase (21), glucosamine-6-phosphate deaminase (22), chorismate mutase (23), pyruvate kinase (24), hemocyanin (25), D-3-phosphoglycerate dehydrogenase (26), glutamine-5-phospho-1-pyrophosphatase (27), and lactate dehydrogenase (28).

2.5 The Molecular Mechanisms of Allosteric Changes

It is instructive to consider the various types of subunit structures that have been observed to be associated with allosteric behavior. Although such structures are usually discussed in terms of homo-polymers (consisting of two or more chemically identical subunits) or hetero-polymers (consisting of two or more chemically distinct subunits, one often "catalytic" and the other "regulatory"; see examples later), it is perhaps better to consider the types of interfaces that exist in allosteric proteins, particularly when it comes to discussing how potential drugs may be designed to influence protein function at the level of allosteric regulation. It is across these interfaces that the conformational effects associated with allosteric proteins must occur.

Consider first three "homo-polymeric" enzymes, all shown to involve allosteric effects: malate dehydrogenase, 3-phosphoglycerate dehydrogenase, and glutamate dehydrogenase. Malate dehydrogenase (Fig. 9.8), a dimer, has a single interface (29); 3-phospho-

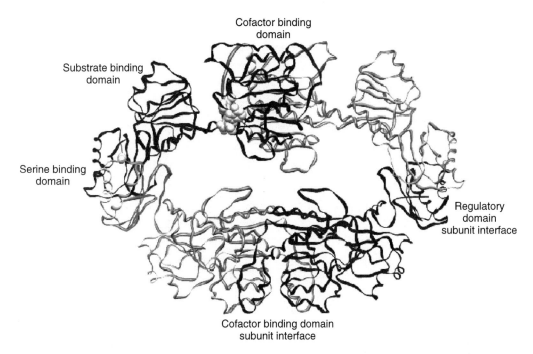

Figure 9.9. Quaternary structure of 3-phosphoglycerate dehydrogenase (PGDH). The ribbon diagram represents the homotetramer of PGDH. Two subunits are shown in black vs. gray to distinguish each subunit. The NADH and serine are shown for one subunit in van der Waal's surfaces. PGDH is unique in that it does not form a compact globular protein but rather a flexible donut-shaped moiety. The subunits form a string of three domains connected by hinge regions. The active site is located in a cleft between two of these domains.

glycerate dehydrogenase (30) (Fig. 9.9), a tetramer, has two quite distinct interfaces; whereas glutamate dehydrogenase (Fig. 9.10), a hexamer, has an even more complex situation, with three quite distinct interface regions (31, 32). In a dimer such as MDH, only isologous contacts are possible; the regions of the protein involved are the same on both sides of the interface. The same is true of the tetramer in 3-phosphoglycerate dehydrogenase, although there are clearly two quite distinct types of contacts; both, however, are isologous. In the hexamer structure of glutamate dehydrogenase a more complex situation is in effect. Although the interaction between the top and bottom halves (trimers) of the molecule are clearly isologous, the interactions between subunits in each of the trimers are heterologous, as shown schematically in Fig. 9.11; otherwise, there would be enforced asymmetry in the trimer. Superimpose on these pictures the functional aspects of the various regions of

each protein other than malate dehydrogenase. Of the two types of contacts in 3-phosphoglycerate dehydrogenase, one is composed of regions of the protein directly involved in substrate binding and catalysis and the other is composed of regions (domains) of the protein involved in regulator binding, in this case the binding of the V-type regulator serine. In glutamate dehydrogenase the isologous interface involves substrate binding domains; of the two heterologous interfaces, the one in the plane of the trimer also involves substrate binding domains, whereas the other involves regulator binding domains.

This discussion of functional domains within an allosteric protein can be extended to the case of heteropolymeric allosteric enzymes. Consider aspartate transcarbamoylase (33) (Fig. 9.12). This enzyme consists of catalytic and regulatory subunits as shown, and ligand binding must trigger conformational

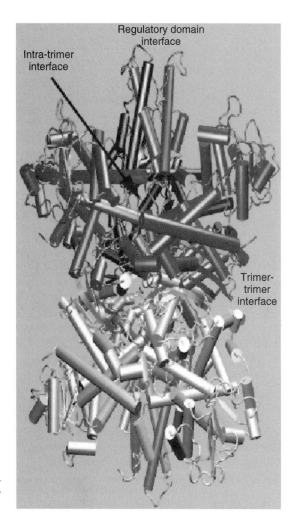

Figure 9.10. The quaternary structure of glutamate dehydrogenase reveals a complex array of subunit interfaces. See color insert.

changes that are transmitted through the regulatory subunits to other catalytic subunits.

How strong are such interfaces and what governs how the protein will change "shape" during an allosteric conformational change? Although the exact details of the overall "strength" of an interaction at an interface depend on the specific nature of the side chains and contacts between the two surfaces, to a first approximation the area of contact is directly proportional to the strength of the interactions across that interface. If one wants to rotate subunits relative to one another, or induce a conformational effect across an interface, these changes are most likely to involve the weakest link of the various interfaces in a given molecule. If one wants to block the in-

duction of an allosteric conformational change (i.e., the R to T transition) there would be a need to strengthen the "weakest link" in the molecule to the extent that the conformational change induced by ligand binding (whether heterotropic or homotropic) was no longer possible.

Identifying the region of the molecule involved in the direct transmission of allosteric effects is clearly important for understanding how allosteric transitions might be manipulated for therapeutic purposes, particularly if one wished to block an allosteric transition. Understanding the structure-function relationships of the ligands binding to their various sites in an allosteric protein is also critically important. For a homotropic interaction

(a)

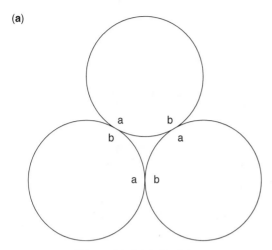

Trimer showing heterologous
subunit interfaces

(b)

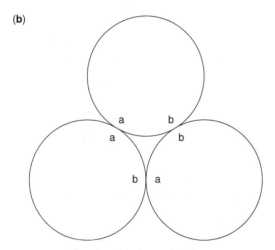

Creation of isologous interfaces
leads to asymmetry

Figure 9.11. Heterologous interactions between subunits in a trimer.

part of the ligand must interact with the active site of the protein and "trigger" the interaction between subunits. In a heterotropic interaction parts of the regulatory ligand must provide direct binding energy for the binding and one part must again act as a trigger for the induced conformational change. In either case dissecting which parts of the ligand contribute to which types of interaction is critical in the design of an allosteric drug.

Over the last decade, major improvements in computer and graphics technology have facilitated the generation of numerous compu-

tational chemistry programs that allow investigators not only to view protein and small molecule structures, but also to use semiempirical parameters to quantitate atomic-level interactions between species of biomolecules (protein–protein, protein–ligand, protein–ligand–water). One such program, HINT (Hydropathic INTeractions; eduSoft, LC, Richmond, VA), allows for the analysis of all possible noncovalent biomolecular atom-atom interactions, including hydrogen bonding, coulombic, acid-base, and hydrophobic forces. In brief, HINT employs thermodynamic,

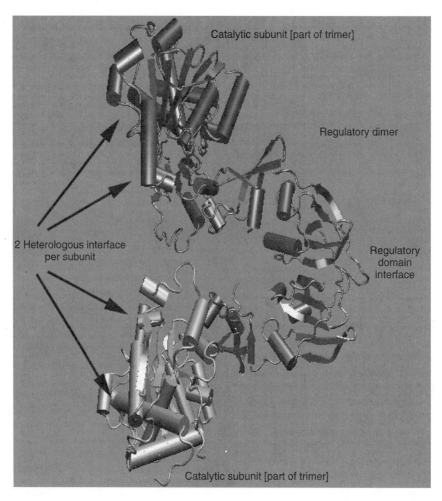

Figure 9.12. Subunit contacts in aspartate transcarbamoylase show heterologous contacts in the catalytic trimers, isologous contacts between regulatory dimers, and regulatory subunit-catalytic subunit interactions. See color insert.

atom-based hydropathy values derived from solvent partitioning measurements of organic molecules, to quantitatively "score" interactions between the atoms of two species. These interactions can then be examined as color-coded contours.

If the crystal structure of the protein-ligand complex is available, HINT analysis can yield valuable insight into specific contributions to the interaction. For example, in the binding of cofactor to the allosteric protein 3-phosphoglycerate dehydrogenase, HINT analysis, as shown in Fig. 9.13 indicates regions of both the protein surface and the co-

factor, which make either strong positive or negative contributions to the overall interaction. Use of this type of analysis as a starting point can lead to the design of analogs of the cofactor, or mutants of the protein, to further dissect contributions to binding and induced conformational changes. An understanding of the nature of ligand binding to an allosteric protein at this level is critical to the successful design of an "allosteric" drug (see next section). Of equal importance is understanding the overall mechanism of transmission of the allosteric effect through the protein conformation to an adjacent subunit. As with the anal-

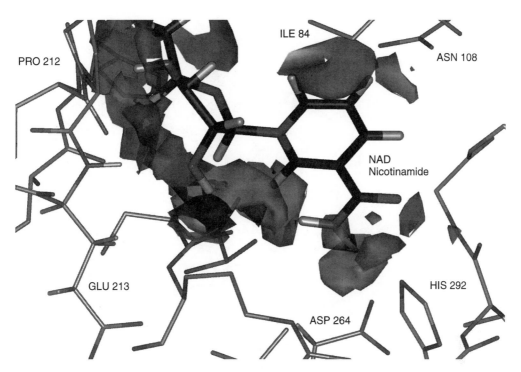

Figure 9.13. HINT analysis of interactions between the cofactor NAD nicotinamide moiety and surrounding 3-phosphoglycerate dehydrogenase residues. Protein carbons are light gray and NAD carbons are black. Nitrogen atoms are blue and oxygen atoms are red. HINT contours are color coded to represent the different types of noncovalent interactions as follows: blue = favorable hydrogen bonding and acid-base interactions; green = favorable hydrophobic-hydrophobic interactions; and red = unfavorable acid-acid and base-base interactions. The volume of the HINT contour represents the magnitude of the interactions. See color insert.

ysis of ligand interactions with a protein, discussed above, a combination of computational and experimental approaches will be necessary to fully understand the mechanism of the transmission of an allosteric effect through a protein.

HINT analysis, in addition to being useful for the analysis of ligand-protein interactions, can be used to examine potential contributions to subunit interfaces in allosteric proteins. An example of such an analysis is shown in Fig. 9.14, where the cofactor binding domain subunit interface has been subjected to HINT analysis. Different regions of the subunits contribute either positively or negatively to the overall strength of the interface. Although in Hb, as discussed earlier, much detail is known about how the allosteric changes are triggered, this is not true for most allosteric proteins. A few have had their three-dimensional structures determined in both the so-called R and T states, but even this has failed to give significant insight into the nature of the trigger or transmission of allosteric changes. As discussed above, computational analysis of crystal structures of oligomeric proteins and of ligand–protein complexes can give insight that can lead to tests by direct experimentation. Similar approaches are being developed to examine the ability of ligands to induce conformational changes. These involve docking substrates with the binding sites of the nonliganded form and using dynamics calculations to assess how the protein may readjust its conformation. The use of snapshots of these dynamic simulations are suggestive evidence of how the protein adjusts its conformation. If such approaches can be

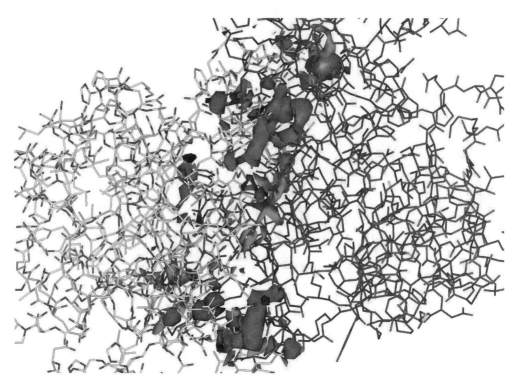

Figure 9.14. HINT analysis of interactions between residues at the interface between two subunits of 3-phosphoglycerate dehydrogenase. The carbons of the two distinct subunits are colored light and dark gray, respectively. Nitrogen atoms are blue, oxygen atoms are red, and sulfur atoms are yellow. HINT contour maps visually displaying interactions between residues at the subunit interface are color coded according to the type of interaction: blue represents hydrogen bonding and favorable acid-base interactions, green displays favorable hydrophobic-hydrophobic contacts, red indicates unfavorable base-base and acid-acid interactions, and purple indicates unfavorable hydrophobic-polar interactions. The volume of the HINT contour represents the magnitude of the interaction. For the analyzed phosphoglycerate dehydrogenase subunits, HINT indicates that a balance between favorable and unfavorable interactions characterizes the interface. See color insert.

validated by direct experimentation (e.g., using site-directed mutagenesis to change seemingly important residues in a predictable way), new insights into the triggers and transmission of allosteric changes will be obtained. With detailed information about both the triggers and transmission of allosteric conformational changes, the rational design of allosteric drugs will become more feasible.

2.6 The Basis for Allosteric Effectors as Potential Therapeutic Agents

Types of considerations suggesting how "allosteric" drugs of various types might be designed are schematically illustrated in Figs.

9.15 and 9.16. For heterotropic effects a drug might bind but not trigger the necessary allosteric change. For either homotropic or heterotropic effects a drug might be designed that stabilizes the region of the molecule across which the allosteric change must be transmitted, thus blocking the required allosteric transition. Finally, if an MWC type model is in effect, yet a third type of "allosteric" drug can be envisaged, one that stabilizes either the R or T state as appropriate without directly impacting either the normal allosteric regulator binding or the regions of the molecule involved in the normal conformational change. Such a drug would bind to some aspect of the confor-

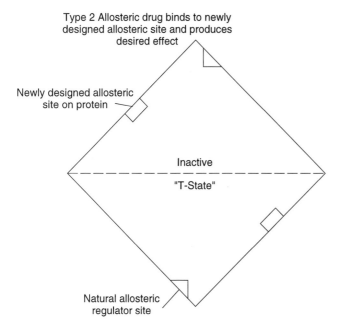

Type 2 Allosteric drug binds to newly
designed allosteric site and produces
desired effect

Newly designed allosteric
site on protein

Inactive

"T-State"

Natural allosteric
regulator site

Type 1 Allosteric drug binds to
normal allosteric site and produces desired effect

Figure 9.15. Schematic outline of the basics of allosteric drug design in ligand agonists and antagonists.

mation of either the R or T state distinct for that state and displace the equilibrium between R and T toward the appropriate state, as shown in Fig. 9.16.

3 ALLOSTERIC TARGETS FOR DRUG DISCOVERY AND FUTURE TRENDS

Small molecule allosteric effectors of allosteric proteins represent a growing class of potential therapeutic agents. With continued progress in the elucidation of the three-dimensional structures of both the high and low affinity states of allosteric proteins, opportunities to design state-specific synthetic effectors will become increasingly feasible. As opposed to designing drugs that bind directly to an active site, which can often be limited by the need to generate molecules with receptor affinity that is severalfold that of the natural substrate, allosteric effectors offer the unique opportunity of binding to a location that is removed from the substrate binding site, and hence will elicit an effect (e.g., destabilizing an allosteric protein's active state), regardless of the concentration of endogenous substrate present. In addition, allosteric binding sites offer a potential for greater specificity for target proteins

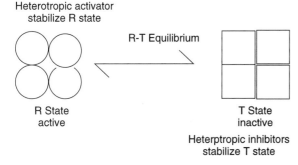

Heterotropic activator
stabilize R state

R-T Equilibrium

R State
active

T State
inactive

Heterptropic inhibitors
stabilize T state

Figure 9.16. Allosteric drugs could stabilize either the R or T state in a Monod-Wyman-Changeux type allosteric enzyme.

Table 9.1 Allosteric Protein Targets and Small Molecule Effectors

Allosteric Targets	Description	Allosteric Effector(s)
Hemoglobin	Hemoglobin (Hb) is a tetrameric protein composed of two identical alpha subunits and two identical beta subunits. X-ray crystal structures of the low (18) and high (20) affinity states of this protein have been solved at high resolution.	Hb effectors of both the low and high affinity forms of the protein have been generated. Effectors such as vanillin (**1**), which bind to the high affinity (R) state (35), are potential therapeutics for sickle-cell anemia; effectors such as RSR-13 (**2**), which bind to the low affinity state (36, 37), are under clinical evaluation for the treatment of ischemic conditions.
HIV reverse transcriptase	HIV reverse transcriptase (RT) is a heterodimeric protein. The structure of HIV RT has been elucidated by X-ray crystallography in several forms, including the unliganded enzyme (38), in complex with nonnucleoside RT inhibitors (NNTRIs) (39, 40), and bound to template primer with (41) and without dNTP substrate (42).	NNRTIs are chemically diverse compounds that are largely hydrophobic (43). Many NNRTIs fall into the HEPT (**3**), TIBO, a-APA, and nevirapine (**4**) families of compounds (44).
HIV fusion proteins	Viral fusion with host cell membranes is mediated by the HIV protein gp41, which upon interaction with a host cell receptor undergoes a conformational change, leading to membrane fusion and infection (45, 46)	The normal "allosteric effector" in this system is the host cell receptor. A variety of "fusion blocking" drugs are under development, which bind to gp41 and "allosterically" prevent (i.e., stabilize the inactive conformation) the induced conformational change from occurring.
Proteases and protease zymogens	Although most protease inhibitors are active site directed, mutants such as the "flap" mutants of HIV protease (47) suggest that "allosteric" drugs can be developed that inhibit by blocking conformational changes necessary to reveal the active site of some proteases (48). Many proteases undergo a zymogen activation process that, like the protease involved in steroid signaling (49–50), may be allosterically triggered, indicating possible "allosteric" drug development targeted at preventing or promoting zymogen activation as necessary.	
P53	P53 can exist as multiple-weight aggregates (34). P53's tetramerization domain has been elucidated by NMR and X-ray crystallography (51–56).	Researchers at Pfizer have discovered two compounds [CP-31398 (**5**) and CP-25704] that are allosteric activators of P53 mutants (57). A potential allosteric inhibitor of P53, pifithrin-α (**6**) has also been reported (58).
Ion channels/ neuroreceptors	Several ion channels and neuroreceptors are allosterically modulated by small molecules. These include calcium channels, sodium channels, GABA receptors, and nicotinic receptors, among others.	Anxiolytic compounds classified as benzodiazepines [e.g., diazepam (**7**)] bind to an allosteric site on GABA receptors (59). The compound galanthamine (**8**), which is a cholinesterase inhibitor, has been found to modulate nicotinic receptors (60). Dihydropyridines act upon calcium channels (61).
Glycogen phosphorylase	Glycogen phosphorylase (GP) is a dimeric enzyme that catalyzes the degradation of glycogen to glucose. Its structure in both the resting (GPb) (62) and activated (GPa) (26) forms has been solved in complex with allosteric effectors.	The allosteric effector W1807 (**9**) has been structurally shown to bind to and stabilize the active form (GPa) of the protein (63). Compound CP320626 (**10**) binds to an allosteric site of the resting (GPb) form of the enzyme and may find therapeutic application as an antidiabetic drug (64).

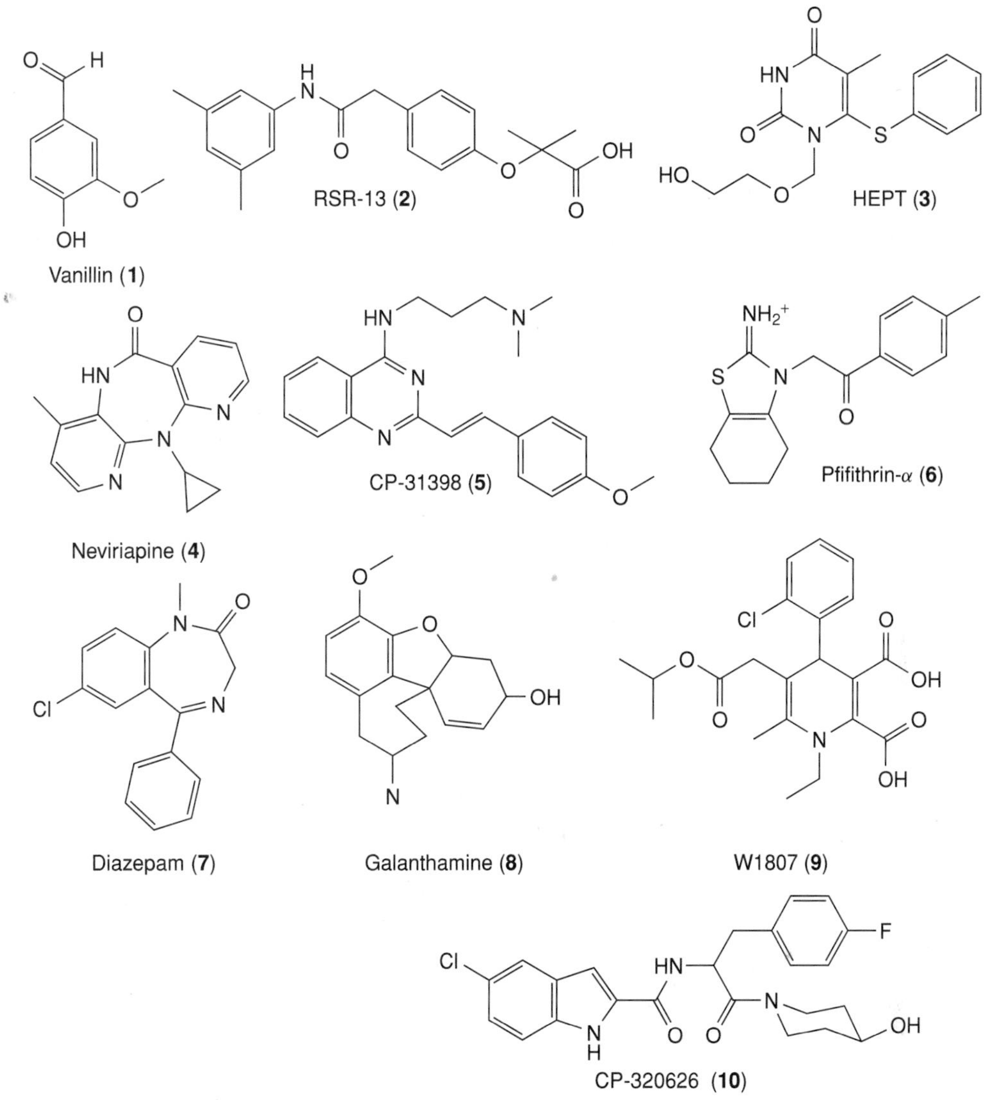

Figure 9.17. Examples of small molecule effectors of allosteric proteins (discussed in Table 9.1).

(34). Thus, modulating the activity of allosteric proteins by degree, vs. all or none, leaves open the possibility of tailoring drug activity to the severity of the disease state.

Several allosteric protein targets are outlined in Table 9.1. Many of these targets are covered in greater detail in other volumes and chapters of this edition. The chemical structures of a select number of small molecule allosteric effectors covered in Table 9.1 are shown in Fig. 9.17.

REFERENCES

1. J. Fang, B. Y. Hsu, C. M. MacMullen, M. Poncz, T. J. Smith, and C. A. Stanley, *Biochem. J.*, **363**, 81–87 (2002).

2. G. G. Hammes and C. W. Wu, *Ann. Rev. Biophys. Bioeng.*, **3**, 1–33 (1974).

3. J. E. Bell and E. Bell, *Proteins and Enzymes*, Prentice-Hall, Englewood Cliffs, NJ, 1988.

4. A. V. Hill, *J. Physiol.*, **40**, 4–11 (1910).

5. G. S. Adair, *J. Biol. Chem.*, **63**, 529 (1925).

6. J. Monod, J. Wyman, and J. P. Changeux, *J. Mol. Biol.*, **12**, 88–118 (1965).

7. D. E. Koshland Jr., G. Nemethy, and D. Filmer, *Biochemistry*, **5**, 365–385 (1966).

8. P. C. Engel and K. Dalziel, *Biochem. J.*, **115**, 621–631 (1969).

9. K. Dalziel and P. C. Engel, *FEBS Lett.*, **1**, 349–352 (1968).

10. M. Eigen, *Nobel Symp.*, **5**, 333 (1967).

11. B. Luisi and N. Shibayama, *J. Mol. Biol.*, **206**, 723–736 (1989).

12. J. E. Bell and K. Dalziel, *Biochim. Biophys. Acta*, **309**, 237–242 (1973).

13. L. O'Connell, E. T. Bell, and J. E. Bell, *Arch. Biochem. Biophys.*, **263**, 315–322 (1987).

14. S. Alex and J. E. Bell, *Biochem. J.*, **191**, 299–304 (1980).

15. M. F. Perutz, *Nature*, **228**, 726–739 (1970).

16. M. F. Perutz, A. J. Wilkinson, M. Paoli, and G. G. Dodson, *Ann. Rev. Biophys. Biomol. Struct.*, **27**, 1–34 (1998).

17. G. Fermi, *J. Mol. Biol.*, **97**, 237–256 (1975).

18. G. Fermi, M. F. Perutz, B. Shaanan, and R. Fourme, *J. Mol. Biol.*, **175**, 159–174 (1984).

19. M. F. Perutz, H. Muirhead, J. M. Cox, and L. C. G. Goaman, *Nature*, **219**, 131–139 (1968).

20. B. Shaanan, *J. Mol. Biol.*, **171**, 31–59 (1983).

21. Y. Zhang, J.-Y. Liang, S. Huang, and W. N. Lipscomb, *J. Mol. Biol.*, **244**, 609–624 (1994).

22. G. Oliva, M. R. M. Fontes, R. C. Garratt, M. M. Altamirano, M. L. Calcagno, and E. Horjales, *Structure*, **3**, 1323–1332 (1995).

23. N. Strater, K. Hakansson, G. Schnappauf, G. Braus, and W. N. Lipscomb, *Proc. Natl. Acad. Sci. USA*, **93**, 3330–3334 (1996).

24. A. Mattevi, G. Valentini, M. Rizzi, M. L. Speranza, M. Bolognesi, and A. Coda, *Structure*, **3**, 729–741 (1995).

25. B. Hazes, K. A. Magnus, C. Bonaventura, J. Bonaventura, Z. Dauter, K. H. Kalk, and W. G. J. Hol, *Protein Sci.*, **2**, 597–619 (1993).

26. G. A. Grant, D. J. Schuller, and L. Banaszak, *Protein Sci.*, **5**, 34–41 (1996).

27. J. L. Smith, *Curr. Opin. Struct. Biol.*, **5**, 752–757 (1995).

28. S. Iwata, K. Kamata, S. Yoshida, T. Minowa, and T. Ohta, *Nat. Struct. Biol.*, **1**, 176–185 (1994).

29. M. D. Hall, D. G. Levitt, and L. J. Banaszak, *J. Mol. Biol.*, **226**, 867–882 (1992).

30. D. J. Schuller, G. Grant, and L. Banaszak, *Nat. Struct. Biol.*, **2**, 69–76 (1995).

31. P. E. Peterson and T. J. Smith, *Structure*, **769**, 7 (1999).

32. T. J. Smith, T. Schmidt, J. Fang, J. Wu, G. Siuzdak, and C. A. Stanley, *J. Mol. Biol.*, **318**, 777 (2002).

33. M. K. Williams, B. Stec, and E. R. Kantrowitz, *J Mol Biol.*, **281**, 121 (1998).

34. B. S. DeDecker, *Chem. Biol.*, **7**, R103–R107 (2000).

35. D. J. Abraham, A. S. Mehanna, F. C. Wireko, J. Whitney, R. P. Thomas, and E. P. Orringer, *Blood*, **77**, 1334–1341 (1991).

36. D. J. Abraham, J. Kister, G. S. Joshi, M. C. Marden, and C. Poyart, *J. Mol. Biol.*, **248**, 845–855 (1995).

37. M. K. Safo, C. M. Moure, J. C. Burnett, G. S. Joshi, and D. J. Abraham, *Protein Sci.*, **10**, 951–957 (2002).

38. Y. Hsiou, J. Ding, K. Das, A. D. Clark, S. H. Hughes, and E. Arnold, *Structure*, **4**, 853–860 (1996).

39. L. A. Kohlstaedt, J. Wang, J. M. Friedman, P. A. Rice, and T. A. Steitz, *Science*, **256**, 1783–1790 (1992).

40. K. Das, J. Ding, Y. Hsiou, A. D. Clark Jr., H. Moereels, L. Koymans, K. Andries, R. Pauwels, P. A. Janssen, and P. L. Boyer, *J. Mol. Biol.*, **264**, 1085–1100 (1996).

41. H. Huang, R. Chopra, G. L. Verdine, and S. C. Harrison, *Science*, **282**, 1669–1675 (1998).

42. A. Jacobo-Molina, J. Ding, R. G. Nanni, A. D. Clark Jr., X. Lu, C. Tantillo, R. L. Williams, G. Kamer, A. L. Ferris, and P. Clark, *Proc. Natl. Acad. Sci. USA*, **90**, 6320–6324 (1993).

43. G. Tachedjian, M. Orlova, S. G. Sarafianos, E. Arnold, and S. P. Goff, *Proc. Natl. Acad. Sci. USA*, **98**, 7188–7193 (2001).

44. H. Tanaka, R. T. Walker, A. L. Hopkins, J. Ren, E. Y. Jones, K. Fujimoto, M. Hayashi, T. Miyasaka, M. Baba, D. K. Stammers, and D. I. Stuart, *Antiviral Chem. Chemother.*, **9**, 325–332 (1998).

45. Y. H. Chow, O. L. Wei, S. Phogat, I. A. Sidorov, T. R. Fouts, C. C. Broder, and D. S. Dimitrov, *Biochemistry*, **41**, 7176–7182 (2002).

46. K. Morris, *Lancet*, **349**, 1227 (1997).

47. W. Shao, L. Everitt, M. Manchester, D. D. Loeb, and C. A. Hutchison, *Proc. Natl. Acad. Sci. USA*, **94**, 2243–2248 (1997).

48. S. Jeonghoon, P. Jaume, and C. S. Craik, *Proc. Natl. Acad. Sci. USA*, submitted.

49. J. L. Goldstein and M. S. Brown, *Science*, **292**, 1310–1312 (2001).

50. M. Matsuda, B. S. Korn, R. E. Hammer, Y. A. Moon, R. Komuro, J. D. Horton, J. L. Goldstein, M. S. Brown, and I. Shimomura, *Genes Dev.*, **15**, 1206–1216 (2001).

51. G. M. Clore, J. G. Omichinski, K. Sakaguchi, N. Zambrano, H. Sakamoto, E. Appella, and A. M. Gronenborn, *Science*, **265**, 386–391 (1994).

52. W. Lee, T. S. Harvey, Y. Yin, P. Yau, D. Lichtfield, and C. H. Arrowsmith, *Struct. Biol.*, **1**, 877–890 (1994).

53. P. D. Jeffrey, S. Gorina, and N. P. Pavletich, *Science*, **267**, 1498–1502 (1995).

54. G. M. Clore, J. G. Omichinski, K. Sakaguchi, N. Zambrano, H. Sakamoto, E. Appella, and A. M. Gronenborn, *Science*, **267**, 1515–1516 (1995).

55. G. M. Clore, J. Ernst, R. Clubb, J. G. Omichinski, W. M. Poindexter, Kennedy, K. Sakaguchi, E. Appella, and A. M. Gronenborn, *Struct. Biol.*, **2**, 321–333 (1995).

56. M. Miller, J. Lubkowski, J. K. Mohana Rao, A. T. Danishefsky, J. G. Omichinski, K. Sakaguchi, H. Sakamoto, E. Appella, A. M. Gronenborn, and G. M. Clore, *FEBS Lett.*, **399**, 166–170 (1996).

57. B. A. Foster, H. A. Coffey, M. J. Morin, and F. Rastinejad, *Science*, **286**, 2507–2510 (1999).

58. P. G. Komarov, E. A. Komarova, R. V. Kondratov, K. Christov-Tselkov, J. S. Coon, M. V. Chernov, and A. V. Gudkov, *Science*, **285**, 1733–1737 (1999).

59. E. Sigel and A. Buhr, *Trends Pharmacol. Sci.*, **18**, 425–429 (1997).

60. J. Coyle and P. Kershaw, *Biol. Psychiatry*, **49**, 289–299 (2001).

61. G. H. Hockerman, B. Z. Peterson, E. Sharp, T. N. Tanada, T. Scheuer, and W. A. Catterall, *Proc. Natl. Acad. Sci. USA*, **94**, 14906–14911 (1997).

62. N. G. Oikonomakos, V. T. Skamnaki, K. E. Tsitsanou, N. G. Gavalas, and L. N. Johnson, *Struct. Fold. Des.*, **8**, 575–584 (2000).

63. K. E. Tsitsanou, V. T. Skamnaki, and N. G. Oikonomakos, *Biochem. Biophys.*, **384**, 245–254 (2000).

64. N. G. Oikonomakos, S. E. Zographos, V. T. Skamnaki, and G. Archontis, *Bioorg. Med. Chem.*, **10**, 1313–1319 (2002).

CHAPTER TEN

Receptor Targets in Drug Discovery and Development

MICHAEL WILLIAMS
Department of Molecular Pharmacology and Biological Chemistry
Northwestern University Medical School
Chicago, Illinois

CHRISTOPHER MEHLIN
Molecumetics Inc.
Bellevue, Washington

DAVID J. TRIGGLE
State University of New York
Buffalo, New York

Burger's Medicinal Chemistry and Drug Discovery
Sixth Edition, Volume 2: Drug Development
Edited by Donald J. Abraham
ISBN 0-471-37028-2   © 2003 John Wiley & Sons, Inc.

Contents

1 INTRODUCTION

Drug discovery, the process of identifying and developing novel compounds to treat human disease states, has entered a new century with an unheralded wealth of sophisticated technologies and information generation platforms that theoretically will allow the more rapid development of medicines with improved selectivity and safety profiles. The successes of previous generations of products from the pharmaceutical industry, both drugs and vaccines, have led to an increase in human life expectancy with a concomitant anticipation of new drugs for an aging population that will both improve the quality of life and add to its span. The draft map of the human genome (1) has led to the possibility of understanding diseases at a level of precision never before possible, with predictions of new drugs that will address the specific molecular lesion(s) of a given disease, theoretically avoiding interactions with targets that are responsible for the side effects that limit the use of currently available drugs. In the last decade, there has been an exponential increase in the understanding of the pathophysiologies associated with disorders like pain, diabetes, and neurodegenerative processes (Alzheimer's and Parkinson's diseases). There is also a realistic expectation of novel therapeutic treatments that will effectively treat cancer by targeting the mutated cells rather than agents that are promiscuously cytotoxic.

The challenge in using the information now available for disease-related gene identification and molecular targets involved in the cause(s) of various diseases is how to handle and productively focus the bewildering flow of data to cost effectively discover and develop new drugs—a problem that has given rise to the disciplines of bio- and chemo-informatics. Irrespective of compound or drug target sources, more productive high-throughput screening (HTS) processes, and more detailed structural information, success in drug discovery will depend on the iteration and integration of lead compound identification and optimization through the hierarchical complexity of *in vitro* and *in vivo* assays that measure efficacy, selectivity, side effect liability, absorption, distribution, metabolism and excretion (ADME), and toxicology. To do this in an effective manner, it is critical to understand how existing knowledge of drug targets has evolved and what the realistic potential is for identifying, validating, and prioritizing new ones.

2 CELLULAR COMMUNICATION

The transfer of information between cells and the subsequent cellular integration of multiple information sources that is necessary to maintain cellular homeostasis and tissue viability—under both normal and adverse, disease-related conditions—involves a variety of different external signaling modalities. These include temperature, membrane potential, mechanical distention and stress, alterations in ion (H^+/K^+) concentrations, and pheromones and oderants, as well as the more traditional classes of neurotransmitters, neuromodulators, and hormones. These physical stimuli and endogenous chemicals elicit their effects through interactions with cell surface targets, usually proteins, that are classified as receptors (Table 10.1). Once receptors are ac-

Table 10.1 Drug Recognition Sites Classifiable as Receptors

Recognition Site	Location	Example
Receptor	Cell surface GPCR	Muscarinic, dopamine
		β-adrenoceptor
		$GABA_B$
	Cell surface LGIC	NMDA, nicotinic
		TRPV, P2X, $GABA_A$
	Steroid	GR, ER
	Intracellular	RXR, RAR
		PPAR, NFκB
	Cytokine	IL-2, GM-CSF
		Toll, EPO
Enzyme	Cell surface/extracellular	ACE, NOS, acetylcholinesterase
		HMG CoA reductase
	Intracellular	Caspase
		DNA polymerase

tivated by an agonist ligand, a compound capable of producing a cellular response, they transduce or couple the energy associated with the binding event to an effect within the cell through a signal transduction process(es) involving protein–protein interactions or second messenger signaling systems, e.g., G-proteins, protein kinase modulation, lipid metabolism, to produce acute or more long-term effects on cell and tissue function. Thus, neurotransmitters, neuromodulators, and hormones can produce transient increases in second messengers like cyclic AMP or inositol triphosphate (IP_3) or more long-term changes that involve changes in gene expression through activation of transcription factors.

Alterations in receptor function can occur in the following ways: (1) as the result of a functional overstimulation of receptors with their consequent desensitization resulting from excess ligand availability or an enhanced coupling of the ligand-activated receptor to second messenger systems, or (2) a reduction in stimulation resulting from decreased ligand availability or dysfunctional receptor coupling processes. These two events are described, respectively, as the "gain of function" and "loss of function" molecular lesions that are thought to underlie many disease pathophysiologies.

Drugs that effectively treat human disease states by restoring disease-associated defects in signal transduction can act either by replacing endogenous transmitters, e.g., l-dopa treatment to replace the dopamine (DA) lost

in Parkinson's disease, or blockade of excess agonist stimulation, e.g., the histamine H_2 receptor antagonist, cimetidine, which blocks histamine-induced gastric acid secretion and thus reduces ulcer formation. Of the approximately 450 targets through which known drugs act, 71% of these are receptors (2). Of these drugs, approximately 90% produce their effects by antagonizing the actions of endogenous agonists (3).

Enzymes, by producing products that regulate second messenger availability, e.g., adenylyl cyclase-catalyzed production of cyclic AMP or phosphodiesterase-mediated degradation of cAMP, or act by modifying protein targets, e.g., by adding (protein kinases) or removing (protein phosphatases) phosphates on serine or threonine residues, serve a similar role in cellular homeostasis, and as such, also represent key drug targets that can be conceptually included in the category of receptors.

The seminal concept that therapeutic agents produce their effects by acting as "magic bullets" at discrete molecular targets on tissues within the body is attributed (4) to Ehrlich and Langley at the beginning of the 20th century who, independently, described experimental data that led to the evolution of the "lock and key" hypothesis. A ligand (L) thus acted as the unique "key" to selectively modulate receptor (R) activity, the latter functioning as the "lock" for the "entry" of external signals into the cell. In this model, an agonist ligand would form an RL complex and have the ability to "turn" the lock (Equation

10.1), whereas a receptor antagonist would simply occupy the lock and prevent agonist access.

$$\text{Receptor} + \text{ligand} \underset{k-1}{\overset{k+1}{\rightleftharpoons}}$$
$$\text{RL} \rightarrow \text{Signaling event} \qquad (10.1)$$

Despite quantal advances in the technology used to study receptor interactions that have emerged over the past century, the receptor–ligand (RL) concept and the similar enzyme–substrate (ES) complex have remained the conceptual foundations for understanding receptor and enzyme function, disease pathophysiology, and medicinal chemistry-driven approaches to drug discovery. Over the past decade, however, with the explosion in the number and diversity of receptors (5) driven by receptor cloning approaches and the mapping of the human genome (1, 6), there has been an increased appreciation of the inherent complexity of their function.

At one time it was thought that a receptor-mediated response was a predictable, linear process that involved ligand-induced activation of a protein monomer and its signal transduction pathway independently, or with very little influence, from other membrane proteins. It has become increasingly evident, however, that receptors can physically interact both with one another and other membrane proteins (7). Numerous examples exist of receptor co-expression and interactions, e.g., $GABA_BR1$ and $GABA_BR2$ (8–10), dopamine with somatostatin (11), and $GABA_A$ (12) receptors and opioid receptors with β_2-adrenoceptors (13). These are necessary to obtain functional cell surface receptors and allow permissive interactions to modulate the effects of one another. This can occur through modulation of function through allosteric sites (the benzodiazepine site in the $GABA_A$ receptor (14), the glycine receptor on the NMDA receptor complex (15), or by integration of the functional effects of signal transduction pathways (receptor cross talk) (11–13). There also exist several discrete classes of receptor-associated proteins including receptor-activity-modifying proteins (RAMPs) (16) and trafficking chaperones (17), which currently number

in the many hundreds and play key roles in cellular events like receptor trafficking from the endoplasmic reticulum to the cell surface (17) and modulation of cell surface responses. Examples of the complexity of receptor signaling at the postsynaptic level include the NMDA receptor, where proteomic analysis demonstrated more than 70 proteins other than the receptor potentially involved in the function of the receptor complex (18), and the ATP-sensitive $P2X_7$ receptor, where a signaling complex comprised of 11 proteins including laminin α-3, integrin β2, β-actin, supervillin, MAGuK, three heat shock proteins, phosphatidylinositol 4-kinase, and the receptor protein tyrosine phosphatase-β (RPTP-β) was identified (19). RPTP-β seems to modulate $P2X_7$ receptor function through control of its phosphorylation state.

3 RECEPTOR AND ENZYME CONCEPTS

While the major focus of this chapter is on cell surface receptors, the theoretical concepts involved can be applied enzymes and the various classes of intracellular receptors. Thus, for the purposes of this review, any endogenous target at which a drug can act is designated a receptor.

Receptor theory is based on the classical Law of Mass Action as developed by Michaelis and Menten (20) for the study of enzyme catalysis. The extrapolation of classical enzyme theory to receptors is, however, an approximation. In an enzyme–substrate (ES) interaction, the substrate S undergoes an enzyme-catalyzed conversion to a product or products. Because of the equilibrium established, product accumulation has the ability to reverse the reaction process. Alternatively, the latter can be used in other cellular pathways and is thus removed from the equilibrium situation or can act as a feedback modulator (21) to alter the ES reaction either positively or negatively (Equation 10.2).

$$\text{Enzyme} + \text{substrate} \rightleftharpoons \text{ES} \rightleftharpoons \text{E} + \text{Product}$$
$$(10.2)$$

For the receptor–ligand interaction, binding of the ligand to the receptor to form the RL

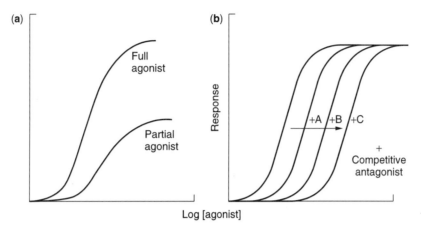

Figure 10.1. Dose-response curve. The addition of increasing concentrations of an agonist ligand causes an increase in a biological response. Plotted on a logarithmic scale, a sigmoidal curve is obtained. (a) A full agonist produced a maximal response, whereas a partial agonist reaches a plateau that is only part of the response seen with a full agonist. (b) In the presence of antagonist concentrations A, B, C, the dose-response curve is progressively moved to the right. Increasing agonist concentrations overcome the effects of the antagonist.

complex results in a response driven by the thermodynamics of the binding reaction that leads to functional changes in the target cell (Equation 10.1). Whereas conformational changes occur in either the ligand, the receptor, or both, there is no chemical change in the ligand resulting from the RL interaction such that there is no chemical product derived from the ligand that results from the RL interaction. Theoretically, despite events such as receptor internalization, receptor phosphorylation, second messenger system activation, etc., the ligand is chemically unchanged by the binding event, and thus, there is no equilibrium established between the RL complex and the consequences of receptor activation.

After the formation of an RL complex, a functional response to receptor activation can be related to the concentration of the ligand present (Fig. 10.1a). *Occupancy theory*, developed by Clark in 1926 (22), is based on a dose/concentration-response relationship. This theory has undergone continuous refinement based on experimental data to aid in further delineating the increasingly complex concept of ligand efficacy (23) and to accommodate allosteric site modulation of receptor function (21, 24, 25), ternary complex models, (26, 27) and their extension to constitutively active re-

ceptor systems (23, 28), the latter being those that are functional in the absence of an identified ligand.

3.1 Occupancy Theory

From observing that acetylcholine (ACh) receptor antagonists such as atropine caused a rightward shift in the dose-response (DR) curve in muscle preparations when plotted logarithmically, Clark introduced the concept of occupancy theory (22). Its basic premise, based on Michaelis-Menten theory (20), was that the effect produced by an agonist was dependent on the number of receptors occupied by that agonist, a reflection of the agonist concentration present. The basic tenets of occupancy theory were as follows: (1) the RL complex was assumed to be reversible; (2) the association of the receptor with the ligand to form the RL complex was defined as a bimolecular process with dissociation being a monomolecular process; (3) all receptors in a given system were assumed to be equivalent to one another and able to bind ligand independently of one another; (4) formation of the RL complex did not alter the free (F) concentration of the ligand or the affinity of the receptor for the ligand; (5) the response elicited by receptor occupancy was directly proportional to the number of receptors occupied; and (6) the

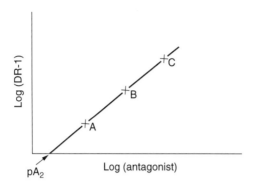

Figure 10.2. Schild plot regression. Data from Fig. 10.1b for antagonist concentrations A, B, and C can be plotted by the method of Schild (29) to yield a pA_2 value, a measure of antagonist activity. A slope of unity indicates a competitive antagonist.

biological response was dependent on attaining an equilibrium between R and L according to Equation 10.1.

Although it is not always possible to determine the concentration of free ligand (F) or that of the RL complex, rearranging the latter, the equilibrium dissociation constant, K_d, which equals $k - 1/k + 1$, can be derived from the equation:

$$K_d = ([R][L]/[RL]) \qquad (10.3)$$

and is equal to the concentration of L that occupies 50% of the available receptors.

Antagonist interactions with the receptor were further defined by Gaddum (29) as resulting in receptor occupancy without the ability to elicit a functional response. Antagonists thus block agonist actions. By increasing their concentrations, agonists can overcome the effects of a competitive (e.g., reversible/surmountable) antagonist (Fig. 10.1b) such that in the presence of increasing fixed concentrations of a competitive antagonist, a series of parallel agonist DR curves can be generated that shift progressively to the right (Fig. 10.1b). A Schild regression relationship (30), a plot of log (DR-1) versus the log concentration of antagonist (Fig. 10.2), can then be derived with the pA_2 value for the antagonist being determined from the intercept of the abscissa. The pA_2 value is the negative logarithm of the affinity of an antagonist for a given receptor in

a defined biological system and is equal to $-\log_{10} K_B$ where K_B is the dissociation constant for a competitive antagonist with a slope of near unity. Not all antagonists act in a competitive manner. Non- or uncompetitive antagonists that act at sites distinct from the agonist binding site or that bind irreversibly to the agonist site have slopes that are significantly less than unity. The Schild plot can thus be used to determine the mechanism by which an antagonist produces its effects.

Ariens (31) modified occupancy theory based on experimental data showing that not all cholinergic agonists were able to elicit a maximal response in a skeletal muscle preparation, even when administered at supramaximal concentrations. This led to the introduction of the concept of the *intrinsic activity*, α, of a ligand. A full agonist was defined as having an α value of 1.0 with the α value for an antagonist being 0. Intrinsic activity was thus originally defined on a zero to unity scale. However, many compounds were subsequently identified that could bind to the receptor but produce only a portion of the response seen with a full agonist. These were defined as partial agonists (Fig. 10.1a). By definition, these compounds were also partial antagonists.

Other agonists have also been identified that produce a response greater than unity. These have been termed "super agonists." One example is the muscarinic cholinergic receptor agonist, L 670,207 (Fig. 10.3), a bioisostere of arecoline that had 70% greater activity than that observed with arecoline and thus had an intrinsic activity of 1.7 (32). From the activity seen in response to L 670,207, the system used to characterize these muscarinic agonists was obviously capable of a greater response than that seen with arecoline, making the latter compound a partial agonist.

The partial agonist concept was additionally refined by Stephenson (33), who introduced the concept of *efficacy*, ε, which differed from intrinsic activity in that the latter was defined as a proportion of the maximal response [effect = α (RL)]. This concept was extended to situations where a maximal response to an agonist could occur when only a small proportion of the total number of receptors on a tissue were occupied, as in the situa-

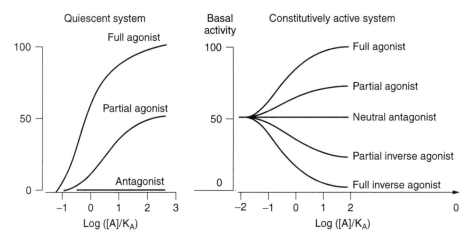

Figure 10.3. Efficacy in quiescent and constitutively active systems. In a quiescent system, three types of ligand can be defined, full agonist, partial agonist, and antagonist, depending on the response elicited. In a constitutively active system (51), an antagonist from a quiescent system is defined as a neutral antagonist with ligands that inhibit the activity of the constitutively active system and are defined as full and partial inverse agonists depending on the degree of inhibition.

tions when receptors were inactivated by alkylating agents using the method of Furchgott (34). This resulted in a non-linear occupancy relationship with the response then being defined as the stimulus, S, a product of the fraction of receptors occupied and the ligand efficacy (33). A non-linear functional response clearly complicates data interpretation, especially when spare receptor or receptor reserve concepts are introduced to rationalize individual data sets. An additional issue in defining efficacy was the degree to which the receptor activation event and its blockade by antagonists was measured through events that were spatially and temporally removed from the receptor activation event and also the degree to which the response could be amplified through cofactor and signal transduction cascades.

Kenakin (35) has described ligand-mediated responses in a given tissue as being determined by four parameters: (1) receptor density; (2) the efficiency of the transductional process; (3) the equilibrium dissociation constant of the RL complex; and (4) the intrinsic efficacy of the ligand at the receptor. Receptor occupancy for exogenous ligands is primarily dependent on pharmacokinetic parameters, whereas that for native endogenous ligands is most probably under intrinsic homeostatic controls, leading Tallarida (36) to introduce a

feedback function defined as Φ that relates to the control of RL complex stability and is dependent on the relative concentrations of free and bound ligand. Using chaos theory, Tallarida defined the stable RL complex as an "attractor," with situations that lead to loss of control for a given RL interaction leading to unstable points or "repellers." In addition to being both provocative and intellectually challenging, chaos theory as related to classical receptor occupancy theory accommodates pulsatile and chronotropic transmitter availability and may also provide viable models for disease states resulting from receptor dysfunction within a tissue.

Classical receptor theory assumed that affinity and efficacy were independent parameters (35). Thus, there was thought to be no consistent relationship between the affinity of a ligand and its ability to elicit a full response. A ligand with relatively low affinity ($<10^{-6} M$) could still be a full agonist when a sufficient concentration interacted with the receptor. More recently, with the discovery of constitutive receptor activity, it seems that this lack of a consistent relationship is more reflective of an inability to measure receptor-mediated activity than a potency disconnect (23). Thus, one may conclude that all ligands have efficacy if the appropriate system is used to measure this parameter.

The relationship between the receptor and ligand in the classical lock and key model with the RL interaction resulting in no change in the receptor conformation essentially described a static situation. In 1958, however, Koshland (37) described an *induced fit* model of the ES complex, where substrate binding to the enzyme caused a change in the three-dimensional structure of the protein, leading to a change in activity. Together with the pioneering work of Hill on hemoglobin allosterism (38), this led to the concept of allosteric modulation of receptor function.

3.2 Rate Theory

Based on experimental data showing the persistence of antagonist-mediated responses and agonist "fade" where maximal responses occur transiently to be followed by lesser responses of longer duration and agonist-mediated blockade of agonist effects, Paton (39) modified the concept of occupancy to include a chemically based rate term. According to rate theory, it was not only the number of receptors occupied by a ligand that determined the tissue response, but also the *rate* of RL formation. The resultant effect, E, was considered equal to a proportionality factor, φ, that included an efficacy component and the velocity of the RL interaction, V. Thus,

$$E = \varphi V eq \qquad (10.4)$$

The rate of RL formation, like that of neurotransmitter release, was measured in quantal terms with discrete "all or none" changes in receptor-mediated events. Pharmacokinetic considerations also play a major role determining the rate of RL formation from exogenously applied ligands, which most drugs are, again invoking some consideration of aspects of Tallarida's chaos theory modeling of the RL complex (36).

The primary factor delineating occupancy and rate theory seemed to be the dissociation rate constant. Thus, if this factor was large, the ligand was an agonist; if the factor was small, reducing the quantal response to receptor occupancy, the ligand functioned as an an-

tagonist. The kinetic aspects of rate theory did not, however, take into account the efficacy of transductional coupling and the potential for amplification after the initial binding event leading to its description by Limbird (40) as "a provocative conceptualization . . . with limited applicability," a phrase that might equally be applied to many of the newer aspects of receptor theory.

3.3 Inactivation Theory

Receptor inactivation theory, initially proposed by Gosselin in 1977 has been widely disseminated by Kenakin (35) and to some degree is based on the two-state model originally proposed by Katz and Thesleff (41) for ion channels, specifically the *Torpedo* nicotinic receptor, where the multimeric receptor exists in active and inactive states, with ligand binding altering the equilibrium between these two states. Receptor inactivation theory reflects a synthesis of both occupancy and rate theories providing an alternative consideration for the study of the RL interaction.

$$[R] + [L] \underset{k-1}{\overset{k+1}{\rightleftharpoons}} [RL]$$
$$k_3 \diagdown \qquad \diagup k_2 \qquad (10.5)$$
$$[R'L]$$

Inactivation theory assumes that the RL complex is an intermediate "active" state that gives rise to an inactive form of the receptor, R', which is part of an RL complex termed R'L. The rate term, k_{+1}, is the rate of association and k_{-1} is the rate of dissociation of the RL complex (Equation 10.5). k_2 is the rate constant for the transition from RL to R'L with the rate constant for the regeneration of the active form of the receptor, R being k_3. The response is proportional to the rate of R' formation which is equal to K_3 (R'L), a variable that is dependent on the number of receptors occupied and the rate of R' formation. Unequivocal experimental data to support receptor inactivation theory has been difficult to obtain as has data to distinguish between occupancy, rate and inactivation theories of the RL interaction. Nonetheless, the inclusion of an additional step in terms of the active recep-

Table 10.2 Allosteric Receptor Models

Monod, Wyman, Changeux Concerted Model (20)

Receptor complex is a multicomponent oligomer comprised of a finite number of identical binding sites.

Subunits are symmetrically arranged each having a single ligand binding site.

Receptor complex exists in two conformational states, one of which has a preference for ligand binding.

Conformational transition state involves a simulataneous shift in the state of all subunits.

No hybrid states exist implying cooperativity.

Koshland Nemethy Filmer Sequential Model (23)

Receptor complex is a multicomponent oligomer with symmetrically arranged protomers each with a single ligand-binding site.

Protomers exist in two conformational states with transition induced by ligand binding.

Receptor symmetry is lost on ligand binding.

Hybrid states of the receptor complex can be stabilized by protomers.

Stabilization is equivalent to negative cooperativity.

tor generation inherent in receptor inactivation theory provides a further nuance related to the potential mechanisms by which antagonists and allosteric modulatory ligands may elicit their effects on the RL interaction, the signal transduction process and receptor function.

4 RECEPTOR COMPLEXES AND ALLOSTERIC MODULATORS

Hill's studies (38) on the binding of oxygen by hemoglobin demonstrated that identical binding sites on protein oligomers could influence one another such that the binding of the first ligand (in the case of hemoglobin in oxygen) facilitated the binding of a second, identical ligand and so on for sequentially bound ligands such that the saturation curve describing the interaction of the ligand with its recognition site is steeper than that which would be predicted from classical Michaelis-Menten kinetics (20). The process of one ligand, homologous or heterologous, interacting with the binding of another is thought to occur by a cooperative, conformational change in the binding protein for the second ligand from a site adjacent to the ligand recognition site. Koshland's induced fit model (37) built on these findings, leading to the development of two key models of cooperativity or allosterism. These were the *sequential or induced fit* model described by Koshland et al. (24) and the *concerted model* of Monod et al. (21), the key elements of which are summarized in Table 10.2.

Both models assume the existence of oligomeric protein units existing in two states that are in equilibrium with one another in the absence of ligand. Ligand binding induces a conformational change in the protein(s), moving the equilibrium of the two states to favor that with the higher affinity for the ligand. This in turn alters the kinetic and functional properties of the oligomeric complex. This model has been further refined in terms of *ligand-stabilizing conformational ensembles* (42). Conformational changes can also occur independently of the ligand, being driven by random thermal fluctuations (43).

The site on a receptor that defines its pharmacology and membership of a particular receptor superfamily, e.g., 5HT, nicotine, etc. is termed the *orthosteric site*. Ligands that bind to this receptor have a spatial overlap for the binding site such that their binding is mutually exclusive (unless an antagonist covalently binds to the orthosteric site). In contrast, the *allosteric site* (of which there may be more than one associated with a single orthosteric site and which can affect that site) is distinct from the latter in that ligands that bind to the allosteric site(s) can produce effects on ligand binding and efficacy to the orthosteric site through an indirect, conformational modulation of this site that probably involves alterations in either the association or dissociation rates of the orthosteric ligand. While much of the early work on allosterism derived from studies on enzymes and ligand-gated ion channels (LGICs), it is now clear that GPCRs that were once considered as monomeric proteins with only an orthosteric site now are known to contain allosteric sites and can form oligomeric complexes (44). The concept of allosterism increases in complexity when considering multiple ligand sites on a receptor that have different pharmacology, e.g., the ligand recognition sites are totally heterogenous.

The identification of allosteric ligands that can have both positive and negative effects on the function of the orthsteric site has occurred in a largely serendipitous manner. The first drug identified as an allosteric modulator was the BZ, diazepam, which has anxiolytic, hypnotic, and muscle relaxant activities and produces its effects by facilitating the actions of the $GABA_A$ receptor (14). Unlike directly acting $GABA_A$ receptor agonists, diazepam has a relatively safe side effect profile. In contrast, $GABA_A$ receptor agonists have not advanced beyond early stage clinical trials because of marked side effect limitations.

Allosteric modulators are viewed (25) as having three advantages over drugs that act through orthosteric sites: their effects are saturable such that there is a ceiling effect to their activity that results in a good margin of safety in human use; their effects are selective, and as mentioned, frequently "use-dependent." Thus, the actions of an allosteric modulator occur only when the endogenous orthosteric ligand is present. In the absence of the latter, an allosteric modulator is theoretically quiescent and may thus represent an ideal prophylactic treatment for disease states associated with sporadic or chronotropic occurrence. Finally, allosteric modulators are considered to be more specific in their effects partly because of the nature of their binding sites that are distinct from the orthosteric site and partly because of the extent to which their effects depend on the degree of cooperativity between the allosteric and orthosteric sites. This is exemplified by the muscarinic allosteric modulator, N-chloromethylbrucine (45). While having identical affinity for muscarinic M_3 and M_4 receptors, this compound has a positive cooperative effect on the M_3 but is neutral, having little effect, on the M_4 receptor.

The "cys-loop" (46) family of LGICs including the $GABA_A$, glycine, $5HT_3$, and nicotinic cholinergic receptors are among the best characterized of the allosterically regulated receptors. GPCRs are also subject to allosteric modulation. The families identified to date

demonstrating this property include adenosine (P1), α- and β-adrenergic, dopamine, chemokine, $GABA_B$, endothelin, metabotropic glutamate, neurokinin-1, P2Y, and muscarinic cholinergic, as well as some members of the 5HT superfamiliy (25). Interestingly, the diuretic amiloride and some of its analogs are active at adenosine A_1, A_{2A}, α_1, α_{2A}, and α_B receptors, suggesting a common allosteric motif of these receptors. Cone snail conotoxins including ω-GVIA and ρ-TIA are allosteric modulators of the $P2X_3$ (47) and the α_{1B} adrenoceptor (48), respectively. Changes in GPCR function resulting from alterations in the ionic milieu also reflect the potential for allosteric modulation of receptor function (25).

5   TERNARY COMPLEX MODELS

The ternary complex model (TCM) (26, 27) describes allosteric interactions between orthosteric and allosteric sites present on a single protein monomer and can also be extended to reflect other two-state interactions involving sites on adjacent proteins and the effect of signaling proteins, e.g., G-proteins on the function of the receptor. As noted by Christopoulos (25), allosteric interactions are reciprocal such that the effects of a ligand A on the binding properties of ligand B also imply an effect on the binding of ligand B on the properties of ligand A. Similarly, because GPCRs alter the conformation of G-proteins to elicit transductional events and an alteration in cell function, changes in G-protein conformation and interactions alter receptor function, and this may be reflected in desensitization.

6   CONSTITUTIVE RECEPTOR ACTIVITY

A basic principle of receptor theory is that when a receptor is activated by a ligand, the effect produced by the ligand is proportional to the concentration of the ligand, e.g., it follows the Law of Mass Action (20). It is now becoming apparent that receptors spontaneously form active complexes as a result of interactions with other proteins. This is especially true when receptor cDNA is expressed in cell systems such that the relative abundance of a receptor is in excess of that normally occurring in the native state or associates with proteins that reflect the host cell milieu in which the receptor is expressed, rather than an intrinsic property of the receptor in its natural environment. This is a major issue in the characterization of ligand efficacy (49). A spontaneous interaction between receptor and effector can occur more frequently in a system where the proteins are in excess and where factors normally present that control such interactions are absent. This is shown graphically in Fig. 10.3, where constitutive activity is shown in the range of 0–50 and where the theoretical effects of inverse agonists, full and partial, are shown. A quiescent system that is more reflective of classical receptor theory shows a full agonist, partial agonist, and what is now defined as a neutral antagonist. Constitutive receptor activation has been described in terms of *protein ensemble* theory (23) and in terms of *allosteric transition* (25), where changes in receptor conformation can occur through random thermal events (42, 43) and may also be described in terms of chaos theory (36).

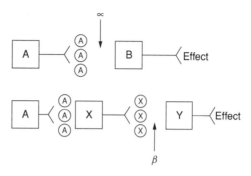

Figure 10.4. Pharmacological versus functional antagonism. In the top panel, neurotransmitter A is released from neuron A, directly interacting with neuron B to produce a functional response. Antagonist α blocks the effects of A, a direct pharmacological antagonism of the effects of A. In the bottom panel, neurotransmitter A is released from neuron A, directly interacting with neuron X, which in turn releases neurotransmitter X, which acts on cell Y to produce a functional response. Antagonist β blocks the effects of X on cell Y, but in the absence of other data on the actions of antagonist β, seems to block the functional effects of A because of the circuitry involved. Antagonist β thus acts as a functional antagonist.

7 EFFICACY CONSIDERATIONS

Historical receptor theory describes a ligand efficacy continuum, with full agonism at one end and full antagonism at the other. Between the two ends of this continuum lie partial agonists that, as already noted, imply that ligands can also be partial antagonists. Antagonism *per se* implied that a ligand could bind to a receptor without producing any effect and limiting access to the native agonist—block receptor activation. Such a compound is now called a *neutral antagonist*.

With the ability to measure constitutive receptor activity, some compounds like the β2-adrenoceptor antagonist, ICI 118551, were found to inhibit constitutive activity, thus functioning as an *inverse agonist* or *negative antagonist* (50). The concept of an inverse agonist was first proposed by Braestrup et al. in their studies of the GABA$_A$/BZ receptor complex (51).

Compounds can also have different efficacy properties depending on the system in which they are examined (52) and in an intact tissue

preparation can often have distinct agonist and antagonist properties at different receptor subtypes.

Whereas the actions of a competitive antagonism can be surmounted by the addition of increasing concentrations of the agonist ligand, resulting in a functional dose-response curve that undergoes a rightward shirt with approximately the same shape and maximal effect (Fig. 10.1), noncompetitive or uncompetitive antagonists interact at sites distinct from the agonist recognition site and can modulate agonist binding either by proximal interactions with this site from a site adjacent to the recognition site or by allosteric modulation. The effects of noncompetitive antagonists are usually not reversible by the addition of excess agonist. This type of antagonism, whether competitive or noncompetitive, occurring at a distinct molecular target is known as *pharmacological antagonism* and involves the interactions between ligands and the receptor site (Fig. 10.4). In contrast, *functional antagonism* refers to a situation in which an antagonist that does not interact with a given receptor can still block the actions of an agonist of that receptor and is typically measured in intact tissue preparations of whole animal models.

In the hypothetical example shown in Fig. 10.4, neurotransmitter A released from neuron A interacts with A-type receptors on neuron B. Antagonist α can block the effects of A on cell B by interacting with A receptors. Antagonist α is thus a pharmacological antagonist of A receptors. In the second example in Fig. 10.4, neuron A releases neurotransmitter A, which interacts with A-type receptors located on neuron X. In turn, neuron X releases neurotransmitter X that interacts with X-type receptors on neuron Y. Antagonist β is a competitive antagonist that interacts with X receptors to block the effect of neurotransmitter X, and in doing so, indirectly blocks the actions of neurotransmitter A. Antagonist β is thus a pharmacological antagonist of receptors for the neurotransmitter X, but a functional antagonist for neurotransmitter A. In interpreting functional data in complex systems, it is always important to consider the possibility that a ligand has more than one effect mediated through a single class of receptor. For this reason, in advancing new ligands from *in vitro* evaluation to more complex tissue systems or animal models, it is extremely helpful to have a ligand-binding profile, e.g., the activity of a compound at a battery of 70 or more receptors and enzymes (a Cerep profile), to fully understand any new findings. For instance, when a ligand for a new receptor is advanced to animal models and found to elicit changes in blood pressure, it would be extremely helpful to know whether in addition to its activity at the new receptor, it has some other properties that relate to the blood pressure effects rather than assuming that some unknown mechanism related to activation of the new receptor has cardiovascular-related liabilities.

8 RECEPTOR DYNAMICS

Receptors are very dynamic in their active presence at the cell surface. Both ligand binding and alterations in gene function can alter the number, half-life, and responsiveness of

receptors and channels. Receptors turn over as a normal consequence of cell growth, with half-lives that vary between hours and days. Ligand binding can result in receptor internalization through phosphorylation-dependent events often initiated as a result of the ligand-binding process, exposing serine and threonine residues in the receptor protein. Binding of substance P to the NK-1 receptors can lead to receptor internalization removing the RL complex from interaction with an antagonist. Histamine H_2 antagonists acting as inverse agonists in a constitutively active system can up-regulate their cognate receptors, potentially increasing cell sensitivity (53). Given what is known about the processes of neurotransmission and neuromodulation, it is reasonable to assume that the target cells for endogenous effector agents, neurotransmitters, neuromodulators, and neurohormones are under tonic control, an implicit assumption of chaos theory (36). This tonicity may be chronotropic, varying with the circadian rhythm of the organism. In contrast, the effects of exogenously administered ligands, e.g., drugs, are rarely under normal homeostatic control, and as a result, their effects frequently become blunted on repeated administration or when they are administered in controlled release forms. It should not be a surprise, therefore, that the majority of effective therapeutic agents are antagonists of receptor function.

The molecular basis of many disease states reflects changes in cell surface receptor function (54). In Parkinson's disease, the presynaptic nerve cells in the *sustantia nigra* that produce dopamine die as the result of an as yet unknown disease etiology. This transmitter defect results in a decrease in dopamine levels and a consequent hypersensitivity of postsynaptic responses as the homeostatic processes in the target cell attempt to compensate for a lack of endogenous ligand. Cells with the NGF receptor, p75, die in the absence of nerve growth factor (55). Familial hypercholesterolemia and cystic fibrosis involve defects in the trafficking of their cognate receptors (54). Alzheimer's disease is characterized by a loss of cholinergic innervation in the basal forebrain (56) that leads to a generalized neuronal loss, cognitive dysfunction, and death.

Agonists that rapidly desensitize a receptor, e.g., ATP at the P2X$_3$ receptor, can seem to be antagonists because their net effect is to attenuate the normal receptor response.

9 RECEPTOR NOMENCLATURE

Before the publication of the draft maps of the human genome, the techniques of molecular biology had already resulted in an explosion in the number of putative receptor families and subtypes within these as well as classes or receptors known as *orphan receptors,* which were structurally members of receptor classes but for which the endogenous ligand and associated function was unknown. The latter are discussed further below.

Because of the speed with which new receptors were identified and because different laboratories frequently identified the identical receptor almost simultaneously and gave it their own unique name, leading to considerable confusion in the literature, the *International Union of Pharmacology* (IUPHAR) has undertaken the development of a systematic nomenclature system (57). The deliberations of the various committees enlisted to devise a systematic nomenclature are published on a periodic basis in *Pharmacological Reviews* and in *Trends in Pharmacological Sciences*. Compendiums are also published at regular intervals in *The Sigma-RBI Handbook* (5) and the annual *Trends in Pharmacological Sciences* Receptor Nomenclature Supplement. Various Internet websites and a palm-based PDA database (58) are also valuable sources for keeping abreast of new developments in receptor identification and classification.

9.1 Receptor Classes

Receptors are divided into five major classes (Table 10.1): the heptahelical, 7-transmembrane (7-TM) G-protein–coupled receptor (GPCR) class, the ion channel class, the steroid receptor superfamily, intracellular receptors, and the non–GPCR-linked cytokine receptors. Of these, the 7-TM receptors have historically represented the most fertile class for drug discovery (2), perhaps primarily because they have been the most studied. Ion channels can be further subdivided into ligand-gated (LGIC), voltage-sensitive calcium and potassium (VSCC, Kir), and ion-modulated (ASIC) subtypes, all of which have similar but very distinct structural motifs. Ion channels can also be modulated by temperature, e.g., vanilloid receptors (59). This latter family, TRPV, has now been reclassified as part of the transient receptor potential (TRP) ion channel family. The steroid receptor superfamily comprises glucocorticoid (GR), progesterone (PR), mineralocorticoid (MR), androgen (AR), thyroid hormone (TR), and vitamin D$_3$ (VDR) receptors (60). Another diverse class of targets that mediates or modulates the effects of drugs are the intracellular receptors, which includes the cytochrome P450 (CYP) family, the SMAD family of tumor suppressors, the retinoic acid receptor (RXR, RAR) superfamilies (61), receptor-activated transcription factors (RAFTs), and signal transducers and activators of transcription (STATs). The latter encompass AP-1, NFκB, NF-AT, STAT-1, PPARs, the hormone responsive elements on DNA and RNA promoters, and ribozymes. The interferon, tumor necrosis factor (TNF), and receptor kinase families are grouped together in the cytokine receptor class because of similarities in their signal transduction mechanisms.

In addition to this multitude of receptor classes, further complexity in conceptualizing receptors as distinct, classifiable entities is exemplified by recent findings related to GPCRs. By implicit definition, these receptors produce their physiological effects by coupling through the G-protein family. However, there are several instances where ion channels can produce their effects by coupling through G-proteins and also examples where GPCRs can function independently of G-proteins (62). Similarly, HCN-1 and HCN-3, members of the hyperpolarizing activated, cyclic nucleotide–regulated receptor family are insensitive to cyclic nucleotides. Receptor classification is thus a very dynamic process with few absolutes.

9.1.1 G-Protein–Coupled Receptors.

The 7TM motif of GPCRs based on the X-ray structure of rhodopsin is a relatively simple, single protein comprised of approximately 300–500 amino acids with discrete amino acid motifs in the transmembrane regions and on the C-terminal extracellular loop, determining the ligand specificity of the receptor and those on the amino terminal designating G-protein interactions (63) Postgenomic alternative splicing can alter the composition of the GPCR to create isoforms that may be species, tissue, and disease-state dependent (54). RL interactions are thought to take place within a pocket in the 7TM motif that can be generically designated to lie between TMs III, IV, and VI. Much of the current knowledge related to the structure of GPCRs is based on the high resolution structure of crystalline bovine rhodopsin, a 7TM protein (64), which has provided proof of concept that GPCR-like structures can be isolated, purified, and crystallized, and has stimulated efforts to solve the crystal structures of GPCRs (65).

The number of GPCRs present in the human genome, including orphan receptors, has been estimated to be 1000–2000 (66, 67), with over 1000 of these coding for odorant and pheromone receptors.

GPCRs can be organized into six main families (66). There are approximately 150 GPCRs, 18 amine receptors, 50 peptide receptors, and 80 orphan GPCRs in family 1. This family also contains receptors for oderants with subfamilies: 1a that includes rhodopsin, β-adrenoceptors, thrombin, and the adenosine A_{2A} receptor, with a binding site localized within the 7TM motif; 1b that includes receptors for peptides with the ligand-binding site in the extracellular loops, the N-terminal, and the superior regions of the TM motifs: and 1c that comprises receptors for glycoproteins. Binding to this receptor class is mostly extracellular. Family 2 is morphologically, but not sequence, related to the 1c family and consists of four GPCRs activated by hormones like glucagon, secretin, and VIP-PACAP. Family 3 contains four metabotropic glutamate receptors and three $GABA_B$ receptors. Family 4 is a pheromone (VN) family and family 5 is a group of GPCRs that includes "frizzled" and "smoothened," both involved in embryonic development. The sixth receptor family is a group of cAMP receptors that to date has only been identified in the slime mold, $D.$ $discoidium.$

Interactions with G-proteins and other associated signaling molecules have the potential for considerable complexity (67) because

there are four major G-protein families that interact with GPCRs: (1) $G_{\alpha s}$ that activates adenylyl cyclase; (2) $G_{\alpha i/o}$ that inhibits adenylyl cyclase and can also regulate ion channels and activation of cGMP PDE; (3) $G_{\alpha q}$ that activates phospholipase C; and (4) $G_{\alpha 12}$ that regulates Na^+/H^+ exchange. G-proteins are heteromers formed from 20 or more α-, 5 β-, and 10 γ-isoforms that offer a considerable variety of potential functional G-proteins. $Ras,$ $Rac,$ and Rho are low molecular weight G-proteins involved in mitogenic signaling. Receptor-associated guanylyl cyclase activity is also regulated by G-proteins.

9.1.2 Other G-Protein–Associated or –Modulating Proteins. The cyclic nucleotide phosphodiesterases (PDEs) responsible for hydrolytic degradation of the cyclic nucleotides, cAMP and cGMP, exist in more than 15 isoforms (68), whereas the protein kinases, PKA and PKC, are responsible for protein phosphorylation, the GPCR kinases, GRK 1–6, are responsible for GPCR phosphorylation (69), and the protein phosphatases are responsible for dephosphorylation, the latter potentially numbering in excess of 300, significantly increasing the complexity of GPCR-associated signaling processes.

Superimposed on these signaling proteins are the calmodulins that mediate calcium modulation of receptor function, the β-arrestins (70), involved in inactivation of phosphorylated receptors (6 members), a group of 15 proteins termed RGS (regulators of G-protein signaling), RAMPs (15 receptor-activated modulating proteins), and a protein known as Sst2p that is involved in receptor desensitization.

The number of GPCRs present in the human genome, including orphan receptors, has been estimated to vary between 600 and 800. With 35 G-protein isoforms, 300 phosphatases, and the various GPCR-associated signaling proteins described above, there is obviously considerable scope for complexity in cell signaling associated with the GPCR family.

As noted, there is a considerable body of data showing that GPCRs can form homo- and heteromeric forms (e.g., $GABA_B$, adenosine A_1 and A_{2A}, angiotensin, bradykinin, chemokine, dopamine, metabotropic glutamate, muscarinic, opioid, serotonin, and somatostatin), increasing the potential complexity of ligand-driven GPCR signaling processes and offering an opportunity to explore new targets in medicinal chemistry.

Applying an evolutionary trace method (ETM) to assess potential protein–protein interactions, functionally important residue clusters have been identified on transmembrane (TM) helices 5 and 6 in over 700 aligned GPCR sequences (44). Similar clusters have been found on TMs 2 and 3. TM 5 and 6 clusters were consistent with 5,6-contact and 5,6-domain swapped dimer formation. Additional application of ETM to 113 aligned G-protein sequences identified two functional sites: one associated with adenylyl cyclase, β/γ, and regulator of G-protein signaling (RGS) binding, and the other extending from the ras-like to helical domain that seems to be associated with GPCR dimer binding. From such findings, it was concluded that GPCR dimerization and heterodimerization occur in all members of the GPCR superfamily and its subfamilies.

From these findings, potential new approaches to ligand design include the following: (1) antagonists that can act by inhibiting dimer formation, e.g., transmembrane peptide mimics; (2) bivalent compounds/binary conjugates; and (3) compounds targeting the GPCR–G-protein interface.

9.2 Ligand-Gated Ion Channels

Ion channels consist of homo-or heteromeric complexes numbering between three (P2X) and eight (Kirs) subunits. Examples of these are the GABA$_A$/benzodiazepine and NMDA/glycine receptor, neuronal nicotinic receptors (nAChR), and P2X receptors.

The GABA$_A$/benzodiazepine (BZ) receptor (14) is an LGIC that is the target site for numerous clinically effective anxiolytic, anticonvulsant, muscle relaxant, and hypnotic drugs that produce their therapeutic effects by enhancing the actions of the inhibitory neurotransmitter, GABA. It is a pentameric LGIC, the constituent subunits of which are formed from a family of six α, four β, one δ, and two ρ subunits, leading to the potential existence of several thousand different pentamers. The functional receptor complex contains a

GABA$_A$ receptor, a BZ recognition site, and by virtue of its pentameric structure, a central chloride channel. Allosteric recognition sites on this complex include those for ethanol, avermectin, barbiturates, picrotoxin, and neurosteroids like allopregnanolone. The pharmacology and function of the allosteric sites depends on the actual subunit composition with receptor knock-in studies showing that the $\gamma2$ subunit is critical for BZ recognition (71) and that the α subunits code for key features of this drug class (72). The $\alpha1$ subunit, which is present in nearly 60% of GABA$_A$ receptors, in mouse brain mediates the sedative, anticonvulsant, and amnestic effects of diazepam; the $\alpha2$, the anti-anxiety effects of diazepam; and the $\alpha5$, associative temporal memory. With this knowledge, rather than screening new ligands for the GABA$_A$ receptor in various animal models to derive a profile for a non-sedating anxiolytic, by understanding the structural determinants of BZ interactions with $\alpha1$ and $\alpha2$ subunits and designing compounds that preferentially interact with the latter, the process of designing novel ligands can be considerably enhanced.

The NMDA LGIC is a member of the glutamate receptor superfamily that mediates the effects of the major excitatory transmitter, glutamate (15). It is composed of an NMDA receptor, a central ion channel that binds magnesium, the dissociative anesthetics, ketamine and phencyclidine (PCP), the non-competitive NMDA antagonist, dizocilpine (MK 801), glycine and polyamine binding sites, activation of which can marked alter NMDA receptor function, and some 70 other ancillary proteins, the physiological function of which remains to be determined (18). The activation state of the receptor can define the effects of the allosteric modulators. Thus, some are termed "use-dependent," reflecting modulatory actions only when the channel is opened by glutamate.

The nicotinic cholinergic receptor, nAChR (73, 74), is another pentameric LGIC comprised of distinct α, β, γ, and δ subunits (75). The subunit composition of the receptor varies, imparting different functionality when the channel is activated by ACh (70). Allosteric modulators of neuronal nAChRs include dizocilpine, avermectin, steroids, barbiturates, and ancillary proteins.

The P2X receptor family is another LGIC family that is responsive to ATP, functioning as a trimer formed from a family of seven subunits (76).

Little is known regarding the structural elements involved in the ligand pharmacology and function of LGICs, such that signaling transduction pathways have not been extensively characterized. For nAChRs, it is known that the recognition site for acetylcholine (ACh) is formed between two subunits. Thus multiple orthosteric sites for ACh are possible, depending on the types of subunit forming the receptor. Many new classes of LGIC are still in the process of being discovered.

9.3   Steroid Receptor Superfamily

The steroid receptor superfamily comprises the glucocorticoid (GR), progesterone (PR), mineralocorticoid (MR), androgen (AR), thyroid hormone (TR), and vitamin D_3 (VDR) receptors (60). These receptors bind to steroid hormones and are translocated to the nucleus where they bind to hormone responsive elements on DNA promotor regions to alter gene expression. While steroids are very effective anti-inflammatory agents, they have a multiplicity of serious side effects that limit their full use.

The anti-estrogen, tamoxifen, is the most commonly used hormonal therapy for breast cancer and has demonstrated positive effects on the cardiovascular and skeletal systems of postmenopausal women but is associated with an increased risk of uterine cancer. Tamoxifen is described as a SERM, a selective estrogen receptor modulator with a tissue selective profile that is caused by the different distribution of the α- and β-subtypes of the estrogen receptor (ERα and ERβ) that activate and inhibit transcription respectively (77). These selective effects have been ascribed to differential interactions with gene promotor elements and coregulatory proteins depending on whether the ERα interacts directly, or in a tethered manner with DNA (78). In uterine tissue, tamoxifen interacts with a specific coactivator, SRC1, that is abundant in uterine tissue.

9.4   Intracellular Receptors

Members of the intracellular receptor family include the cytochrome P450 (CYP) family, the SMAD family of tumor suppressors, intracellular kinases and phosphatases (67, 79–81), nitric oxide synthases (82), caspases (83), the retinoic acid receptor (RXR, RAR) superfamilies (61), receptor activated transcription factors (RAFTs), and signal transducers and activators of transcription (STATs) such as AP-1, NFκB, NF-AT, STAT-1, PPARs (84, 85), various hormone responsive elements on DNA and RNA promoters, and ribozymes. These represent a bewildering number of potential drug targets especially in the metabolic disease and cancer areas. Given their roles in normal cell function, it will be a challenge to ensure specificity in ligands interacting with these targets.

9.5   Non–GPCR-Linked Cytokine Receptors

Cytokines are polypeptide mediators and are involved in the inflammatory/immune response (86). There are three cytokine families: hematopoietin, which includes IL-2–IL7, L-9–13; IL-15–IL-17; IL-19, 1L-21, IL-22, GMCSF, GCSF, EPO, LIF, OSM, and CNTF, with primary signal transduction through the Jak/STAT pathway; tumor necrosis factor, comprising the receptors, TNFRSF1–18 that signal through NFκB, TRAF and caspases; and the interleukin 1/TIR family that includes IL-1RI and IL-1RII, IL-1Rrp2, and IL1RAPL. TIGGR-1, ST2, IL-18, and Toll 1–9 also signal through NFκB and TRAF. Some of these cytokine receptors mediate inflammatory responses, whereas others are anti-inflammatory, making this a highly complex as well as multimembered family (87).

9.6   Orphan Receptors

Orphan receptors have been generally defined as proteins with a receptor motif that lack both a ligand and function (88). Approximately 160–300 orphan GPCRs are thought to be in the draft sequence of the human genome and intense efforts are currently ongoing to identify the ligands for these and their function as novel intellectual property for the drug discovery process. Orphan receptor validation can be done using expression profiling to identify tissues rich in the expression of the receptor of interest and a technique known as reverse pharmacology that can be used to identify a ligand for the orphan receptor. In the latter, the orphan receptor is used as "bait" to bind selective ligands. These can then be used to further characterize receptor function (6, 88). Nearly 30 orphan GPCRs have been validated in this manner. While most of the current interest on orphan receptors is focused on GPCRs because of the considerable body of existing knowledge about this receptor class, it is anticipated that orphan receptors will also be discovered for other receptor classes.

The orphan receptor approach to drug discovery is exemplified by the orphanin/FQ receptor, ORL1, a structural homolog of the opioid receptor family (89). Identified in 1995 using a homology-based screening strategy, ORL1 had low affinity for known opioid ligands. A novel heptadecapeptide ligand for the ORL1 receptor, orphanin/FQ, was subsequently isolated from brain regions rich in ORL1 (90), which provided the key tool to validate the target and identify a functional role for the receptor in stress-related situations in animal models. In turn, after an intensive screening program, an antagonist of this receptor was identified, Ro 64–6198 (91), that represented a novel anxiolytic/antidepressant drug candidate.

9.7 Neurotransmitter Binding Proteins

A binding protein for corticotrophin releasing factor (CRF) was identified in the early 1990s and exploited as a potential drug discovery target (92). This protein, which acted as a reservoir for CRF, seemed to be unique to the CRF neurotransmitter system. However, a soluble acetylcholine-binding protein (AChBP) has recently been described that modulates synaptic transmission in the mollusk, *L. stagnalis* (93), suggesting that neurotransmitter binding proteins may play a more general role in buffering synaptic message transfer. Equally importantly, the structure of this soluble AChBP has also provided important information about the ligand-binding domain of nicotinic receptors (94) and is being used to model the receptor itself (75).

9.8 Drug Receptors

Most drugs interact with receptors (or enzymes) for which the natural ligand (or substrate) is known. There are, however, a number of receptors, distinct from the evolving class of orphan receptors, for which the synthesized drug is the only known ligand. The best example of this is the central benzodiazepine (BZ) receptor present on the $GABA_A$ ion channel complex that was originally identified using radiolabeled diazepam (14). Because this is the site of action of the widely used BZ anxiolytic drug class, this is a *bona fide* receptor with clinical relevance. However, despite considerable efforts and a number of interesting candidate compounds, no endogenous ligand has yet been unambiguously identified that would represent an endogenous modulator of anxiety acting through this site. Other examples of drug receptors are the cannabinoid receptor family, comprised of two members, CB_1 and CB_2, through which the following compounds act: Δ9-tetrahydrocannibol, the active ingredient of the psychoactive recreational drug/analgesic/antiemetic drug marijuana acts (95); the vanilloid receptor, VR-1, the known ligand for which is capsaicin, the ingredient in red pepper that evokes the sensation of heat (59); and the opioid receptor family (96), the site of action of morphine and other derivatives of the poppy. For each of these drug receptors, endogenous mammalian ligands, anandamide (95), the endovanilloid, *N*-arachidonyl-dopamine (NADA) (96) enkephalins (97), and orphanin/FQ (90) have been identified.

10   RECEPTOR MOLECULAR BIOLOGY

Recombinant DNA methods have been extensively used to isolate and analyze the sequence of numerous receptors from various mammalian complimentary DNA (cDNA) libraries using the polymerase chain reaction (PCR) technique (98) to clone receptors that can then be expressed in various pro- and eukaryotic cell lines and *Xenopus oocyte*. Conserved regions in receptors that are involved in ligand binding, coupling to transductional systems and ion channel formation, have thus been identified. Receptor cloning, sequencing, and expression are now an intimate part of the drug discovery process.

From the receptor/ligand modeling standpoint, the ability to specifically alter the structure of cloned receptors through the modification of a small number of nucleotides, the process known as site-directed mutagenesis (99), the removal of specific regions of the receptor gene "deletion mutagenesis," and the construction of hybrid (chimeric) receptor proteins containing the recognition site for α-adrenergic agonists and the G-protein region for the β-adrenoceptor (100) can provide additional clues on the relative importance of individual amino acids in the process of ligand recognition and receptor function.

11 TARGET VALIDATION AND FUNCTIONAL GENOMICS

The identification of novel receptor targets from the human genome and their subsequent use in the drug discovery process requires that the target be validated. Validation requires that the ligand and function of the receptor be known, and this topic has been addressed in relation to orphan receptors above.

Target validation, especially that related to novel, genomically derived targets, is a highly complex and resource intense process and has added significantly to the cost of drug discovery (6, 101). In addition to orphan receptors, identified on the basis of their homology to members of known receptor classes (88), novel genomic targets can be identified by drug-related differential gene expression analysis (102, 103), by population genetic approaches (101, 104), or a combination of both. Once a gene is associated with a disease, its protein product needs to be characterized and a biochemical function, e.g., receptor, enzyme, or transporter, ascribed to it. The probable function of the protein can then be assessed and its cognate ligand or substrate identified (5). Not all proteins identified in this manner are obviously involved in mammalian cell function. The protein product of a novel lithium-related (NLR) gene identified in mice exposed to lithium has sequence homology to a bacterial nitrogen permease (105). The potential role of this protein in the etiology of bipolar affective disorder remains unclear. Patients can also be genotyped for susceptibility related to a drug target or drug side effects to identify "responsive" patients for more efficient and safer drug use (104).

To identify what are termed "druggable" gene products, the still somewhat mystical process of functional genomics can be used to construct protein–protein interaction maps to identify other proteins in a pathway/network that can represent a more facile entry points to the protein cluster associated with the genetically identified target associated with the disease state (106, 107). Using techniques like yeast two-hybrid and gal-pull down, the function of an uncharacterized protein can be assigned on the basis of the known function of its interacting partners, involving the extensive use of bioinformatics tools.

Targeted gene disruption, antisense, and ribozyme inhibition (RNAi) are other techniques for assessing the phenotype of a given gene. Antisense to the rat $P2X_3$ receptor had marked hyperalgesic activity showing an unambiguous role for this receptor in chronic inflammatory and neuropathic pain states (108). Transgenic animals, where the function of a gene is knocked out by genetic techniques, are less useful in understanding the function of a protein. In addition to being restricted to the mouse, this approach is high in cost and time (taking 1–2 years to generate sufficient animals for evaluation), and in many instances, the absence of a given protein has no overt effect on the phenotype of an animal because of compensatory changes during the developmental phase. Alternatively, the gene knockout may lead to an animal that has a limited, if any, life span or reproductive capability.

Another target validation approach involves the use of ligands to define the function of a new target. Already discussed in regard to orphan receptors (88), other ligand-directed target validation models fall under the rubric of *chemical genomics* (109). More driven by acronyms than novelty, such technologies include ATLAS (any target affinity ligand screen), ALIS (automated ligand identification system), RAPTV (rapid pharmacological target validation), and SCAN (screen for compounds with affinity for nucleic acids), as well as well as NeoMorph (6); these technologies are designed to screen new identified proteins with unknown function against compound libraries to identify high affinity binders. The latter can then be used to see what effects they produce on intact cellular systems to deduce the function of the protein to which they bind. A key to identifying useful ligands that selectively interact with proteins of unknown function to define their function is a sufficient level pharmacophore diversity to ensure success (110).

Assuming that a novel protein target is amenable to high-throughput screening approaches to identify its cognate ligand, that the ligand is identified, and that other approaches are able to define a putative function for the target, the next step in target validation is to identify a "druggable" molecule that can be advanced to the clinic. For ORL1, an excellent example of this approach, this molecule is Ro 64–6198 (91).

This raises the key test of target validation. Does a compound that has the appropriate potency, selectivity, and ADME properties, that is active in "predictive" animal models, and is free of other systems toxicology work in the targeted human disease state? This is the ultimate test of a target-based drug discovery program. With the considerable compound attrition rates in moving from animals to diseased humans, this has not proven to be a predictable transition. A case in point is that of the NK1 receptor activated by the peptide, substance P.

Data accumulated over the past decade from both animals and humans has implicated this receptor in a variety of human pain states including migraine and neuropathic pain (111). As the result of several successful HTS campaigns run in parallel at Pfizer, Lilly, Merck, and Sanofi, a number of highly efficacious and selective antagonists of the NK1 receptor were identified that were then chemically optimized for use as drug candidates. These compounds, with varying degrees of potency, were active in animal models of pain and were free of overt side effects in phase I human clinical trials. But they uniformly failed in phase II studies as novel analgesic agents in patients with various pain conditions (112).

The reasons for this remain unknown and have focused the relevance of animal pain models to the human condition, e.g., a lack of understanding of the true human disease condition and various nuances of substance P signaling pathways. However, before the results with NK1 antagonists, drug discovery in the analgesia area was considered one of the most robust. All known analgesics, e.g., aspirin and morphine, were active in one or more of the animal pain models; new receptors could be mapped in pain pathways and their function assessed using knockout and antisense procedures, and the occupancy of receptors in human brain and pain-sensing pathways could be non-invasively imaged. The only limitation was whether the side effect profile of a putative analgesic agent acting on a novel target or suboptimal ADME characteristics would limit human exposure.

Based on the available data, the NK1 receptor has not been validated as a target for pain treatment despite an overwhelming body of robust preclinical data (112). This example thus serves to underline the many significant challenges of validating targets in the drug discovery process at the present time.

A second example, more directly related to the human genome, is the search for genes that are associated, and by inference may be causative, in producing the psychiatric disorder schizophrenia. This disease affects nearly 1% of the population and has a strong genetic association. From studies comparing schizophrenic patients with individuals lacking symptoms of the disorder, l.o.d. scores of 2.4–6.5 on markers between or close to markers D1S1653 and D1S1679 on chromosome 1q have been reported (113, 114). A l.o.d. score of above a numerical value of 3 is considered by experts in genetics to be indicative of a relevant association akin to a significant P value ($P < 0.05$) in statistics. With this information, a search for the gene on chromosome 1q and the delineation of its function in humans would be a logical approach to finding novel genomic targets that could lead to new drugs that would more effectively and safely treat schizophrenia. However, a subsequent study (115) using eight individually collected schizophrenia populations and a sample set that was "100% powered to detect a large genetic effect under the reported recessive model," showed no evidence for a linkage between chromosome 1q and schizophrenia; a finding that could not be ascribed to ethnicity, statistical approach, or population size. Another group (116) using two prefrontal cortex tissue from two separate schizophrenic populations showed an up-regulation of apolipoprotein L1 gene expression that was not seen in tissue from patients with bipolar disorder or depression. The genes related to apolipoprotein L1 gene expression are clustered on the chromosome locus 22q12, providing another target for the functional genomic approach to target discovery and validation. Yet another locus at chromosome 3q has been reported (117).

To the researcher embarking on a well-funded, science-driven approach to new targets for schizophrenia in 1997, these new findings in 2002, as well as others showing gene association with schizophrenia on chromosomes 6 and 13, would give pause to wonder how best to proceed. Do all five locations represent *bona fide* targets for functional genomics? Should one continue work on the original chromosome 1 findings in light of the failure to replicate? Is the only real validation, in light of conflicting findings, to proceed to the identification of multiple compounds that can be tested in schizophrenic populations—to validate the genome-based approach? Given that the cost of initiating a project, finding leads, optimizing these, and running a single clinical candidate to phase IIa clinical proof of principle is in the range of $24–28 million, how many organizations can afford the $120 million+ cost to undertake scientifically logical yet financially prohibitive strategies? These are difficult questions that can be applied to many other psychiatric and neurological diseases, as well as any other disease state with a potential genetic causality, One answer to these questions obviously points to multifactorial genetic causes in the genesis of many diseases and thus negates the overly simplistic "one gene, one disease" mantra that heralded the age of drug discovery based on the human genome map (114, 118). Strategies resulting from the answers to these questions also make the process of genome-based target and drug discovery a much more costly endeavor (6), with good evidence for two to three or more

potential targets for each disease that need to be examined in parallel. Hopefully, the quality of the target finally validated using compounds in the human disease state will be a magnitude of order superior to currently existing drugs and will thus justify the cost of this approach.

12 COMPOUND PROPERTIES

Once a suitable drug discovery target has been identified, the task is then to identify compounds that interact with the target and can be used as the basis of a lead optimization program to identify potential drug candidates.

The ideal target compound must have the appropriately unique recognition characteristics to impart affinity and selectivity for its target, have the necessary efficacy to alter the assumed deficit in cell function associated with the targeted disease state, be bioavailable, metabolically, and chemically stable, be chirally pure, and be easy and cost effective to synthesize.

12.1 Structure-Activity Relationships

The structure-activity relationship (SAR) of a compound series is a means to relate changes in chemical diversity to the biological activity of the compound *in vitro* and *in vivo*, as well as the pharmacokinetic (gut/blood brain barrier transit, liver metabolic stability, plasma protein binding, etc.) and toxicological properties of the molecule (119). These SARs, when known, are frequently distinct such that changes that improve the bioavailability of a compound often decrease its activity and/or selectivity. Compound optimization is thus a highly iterative and dynamic process.

For the purposes of the present chapter, however, SAR will be used to describe classical compound efficacy unless otherwise stated. Quantitative SAR (QSAR) involves a more mathematical approach involving neural networks and computer assisted design (120–123).

With the characterization of the SAR and the documentation of the effects of different pharmacophores and various substituents on biological activity, it is possible to theoretically model the way in which the ligand interacts with its target and thus derive a two- or three-dimensional approximation of the active site of the receptor or enzyme (119–123). Computer-assisted molecular design (CAMD) techniques can then be used to predict—sometimes successfully—the key structural requirements for ligand binding, thus defining those regions of the receptor target that are necessary for ligand recognition and/or functional coupling to second messenger system to permit the design of new pharmacophores.

The SAR can be further delineated in terms of the type of activity measured as the readout. *In vitro*, this can be displacement of a radiolabeled ligand from a receptor, receptor activation as measured in a functional assay, blockade of receptor function by an antagonist ligand, etc. An additional ligand property is that of selectivity, the degree to which a ligand interacts with the target of choice compared with related structural targets. The degree of selectivity typically determines the side effect profile of a new compound, given that the targeted mechanism itself does not produce untoward effects when stimulated beyond the therapeutic range. As already noted, the development of a ligand binding profile for a ligand active at a new target is of considerable use in assessing its effects in more complex tissue systems.

12.2 Defining the Receptor–Ligand Interaction

The complex physiochemical interactions that describe the interaction of a small molecule with a protein despite the many sophisticated technologies used to study this interaction are still highly empirical, being implicitly defined by the SAR for a series of active and less active compounds. In an increasing number of instances however, the ability to clone, express, and readily derive crystals of a receptor or enzyme and analyze the interaction of a ligand or substrate with these using X-ray crystallographic, NMR, and amino acid point mutation approaches has provided information on the actual topography of the selected drug target that can then be used in *de novo* ligand design.

There are a variety of approaches to deriving information on the RL interaction for use in understanding the key features required in ligand necessary to dictate a potent and selective interaction with its target. Analysis of the interactions of a series of structurally distinct pharmacophores, agonists, and antagonists with the target can be combined with point mutation changes in the target to elucidate key amino acids involved in compound recognition. This can be complimented by X-ray crystallographic data (123) and NMR-derived data (6, 124) to design novel pharmacophores. However, many proteins of interest, especially membrane-bound receptors and ion channels, are not available in the soluble form amenable to the use of these techniques such that the design of new compounds is based on the conceptualization of the target, e.g., virtual receptors (125). There are also limitations to structure-based design approaches, e.g., approximations of the hydration state of the isolated protein, the impact of removal from its native environment, and three-dimensional structural issues with recombinantly derived proteins. Nonetheless, these technologies have added to knowledge regarding compound design. The secondary structure motif developed by Kahn et al. (126) has been successfully used to design totally novel, non-protonated opioid pharmacophores (127). Similarly, the technique of SAR by NMR (124) has provided important insights into ligand design. Small molecules that bind to protein subsites are identified, optimized, and linked together to generate high-potency ligands. The process combines random screening of low molecular-weight compounds whose binding is measured by NMR shifts using two-dimensional techniques with ^{15}N-enriched proteins. Using SAR by NMR, two small molecules were detected that bound to FK binding protein with micromolar and millimolar affinities. Combination of these two molecules, determined by modeling techniques, led to molecules with nanomolar affinities.

More recently, techniques have been developed to allow automation and HTS techniques to be applied to X-ray crystallography, thus providing a more rapid means to generate information on multiple ligand interactions with a given crystal (128). This technology has been used as the basis of three companies: Astex, Structural GenomiX, and Syrrx.

12.2.1 Receptor Binding Assays. Until the 1950s, newly synthesized compounds and compounds isolated from natural sources were assessed by a mixture of *in vivo*, whole animal screens, and classical tissue assays. While many useful therapeutic agents were identified by this approach (3), the cost in terms of compound as well as time and animal use were significant. *In vivo* test paradigms also suffered from the possible elimination of interesting compounds on the basis of unknown pharmacokinetic properties. Thus, test paradigms were usually rigid in terms of timing, and as a result, many potentially interesting compounds that underwent rapid elimination or exhibited short plasma half-lives were considered "inactive" because data on their actions was sought after their peak plasma concentration. This type of screening approach also provided little information to the chemist regarding the discrete interaction of the drug with its target, limiting useful information that could be used in the design of analogs. The specificity of the response, ignoring caveats related to pharmacokinetics, was not ideal because the mechanism inducing the overt response and potential points of intervention to block the response were unknown.

This empirical approach would predict, in the absence of any data related to interactions with a molecular target, that because β-adrenoceptor antagonists lower blood pressure, then any compound that lowers blood pressure is by definition a β-adrenoceptor antagonist, an absurd conclusion, but one that happened nonetheless. The 1960s saw the development of a number of *in vitro* biochemical screens that moved the measurement of the RL interaction a little closer to the molecular level. Nonetheless, the major challenge was to develop assays that measured the RL interaction independently of "downstream" events such as enzyme activation and second and third messenger systems. By such means, the ability of a compound to bind to a receptor could be determined on the basis of the SAR and thus provide the chemist with a more direct means to model the RL interaction.

Snyder, building on pioneering work by Roth (129), Cuatrecasas (26), and Rang (130) established radioligand binding as a valuable tool in the drug discovery process in 1973 (131). This led to an explosion in the identification and characterization of new receptors and their subtypes, that in turn, was enhanced by the application of tools of recombinant DNA technology to the process.

While the technique of radioligand binding has been largely supplanted by activity-based assays in many high-throughput settings, it remains the gold standard for compound characterization and SAR development. It is done by measuring the RL interaction (Equation 10.1) *in vitro* using a radioactive ligand, R*, to bind with highly affinity and selectivity to receptor sites. The interaction of unlabeled ("cold") ligands with the receptor competes with the radioligand, decreasing its binding. This simple technique revolutionized compound evaluation in the 1970s, allowing SARs to be determined with milligram amounts of compound in a highly cost- and resource-effective manner. At steady state, the RL* complex can either be separated out from free radioligand using filtration or assayed using a scintillation proximity assay.

The parameters measured in a binding assay are the dissociation constant, K_d, the reciprocal of the affinity constant, K_a. The K_d is a measure of the affinity of a radioligand for the target site: the B_{max}, usually measured in moles per milligram protein, a measure of the concentration of binding sites in a given tissue source and the IC_{50} value. The K_d and B_{max} values can be determined using a saturation curve where the concentration of radioligand is increased until all the ligand recognition sites are occupied or by measuring radioligand association and dissociation kinetics; the K_d is the ratio of the dissociation and association rate constants (132).

The IC_{50} value is the concentration of unlabeled ligand required to inhibit 50% of the specific binding of the radioligand. This value is determined by running a competition curve (Fig. 10.5) with a fixed concentration of radioligand and tissue and varying concentrations of the unlabeled ligand. To accurately determine the IC_{50}, it is essential that sufficient data points be included. As shown in Fig. 10.5, if the data used to derive the IC_{50} value are clustered over a range that reflects 40–60% of the competition curve, much useful information is lost. Ideally, the competition curve should encompass the range of 10–90% of the competition curve and include a minimum of

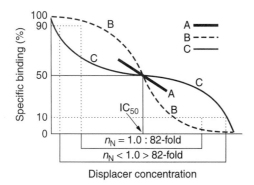

Figure 10.5. Measuring ligand interactions in a receptor-binding assay. Binding of the radioligand is 100%. In curve A, close to the 50% point, an IC_{50} value can be obtained but ignores complexities of the displacement curve. More complete displacement, as in curves B and C, provides more information on the ligand. Displacement curve B has a Hill coefficient of unity requiring an 82-fold difference in displacer concentration to displace 10–90% of binding. Displacement curve C has a Hill coefficient of less than one, requiring a greater than 82-fold difference in displacer concentration to displace 10–90% of binding and indicating the presence of more than one site. By extending the range of concentrations used, curves B and C can assess the possibility of multiple sites being present.

20 data points. Based on Michalis-Menten kinetics (20), when binding is the result of the interaction of the displacer with one recognition site, 10–90% of the radioligand is inhibited over an 82-fold concentration range of the displacer. The slope of a competition curve can then be analyzed to assess the potential cooperation of the RL interactions. When binding is complex resulting in the interaction of the displacer with more than one recognition site, a greater than 82-fold concentration of displacer is required to inhibit the same 10–90% of specific radioligand binding.

The IC_{50} value for a given compound is dependent on the assay conditions: the concentration of the radioligand used, the receptor density, and the affinity of the receptor, the K_d, for the radioligand. To compare the activity of a "cold" ligand across different radioligand binding assays, the Cheng-Prusoff equation (133) is used to compensate for differences in K_d and the radioligand concentra-

tion to obtain a K_i value derived by the relationship:

$$K_i = IC_{50}/1 + [L]/K_d \qquad (10.6)$$

where [L] is the concentration of radioligand used and K_d is the dissociation constant for the radioligand at the receptor. This relationship corrects for inherent differences in assay conditions.

Binding assays can be rapidly used to assess compound recognition characteristics but are generally limited in their ability to delineate agonists from antagonists, especially in a high-throughput setting.

12.2.2 Functional Assays. Biochemical assays involving the measurement of cAMP production or phosphatidylinositol turnover have given way in HTS scenarios to reporter systems where receptor activation or inhibition can be measured using a fluorescence based readout. This depends on the use of sensitive calcium sensing dyes coupled with real time measurement using coupled charge device (CCD) cameras and data capture. A widely used system is the fluorescence imaging plate reader (FLIPR) from Molecular Devices (134). Using a 96- or 384-well microtiter plate format, the throughput on a FLIPR is such that a compound library of 0.5 million distinct compounds can be assayed in less than a month.

Other approaches to functional assay include various reporter gene constructs where formation of the RL complex leads to the expression of a gene that produces a response that can be read immediately. These include various luciferase reporter gene or aequorin-based assays that produce light that can be measured by multiple photodiode arrays and a β-galactosidase reporter gene that leads to a colorimetric readout (135). GPCR-transfected frog melanocytes represent yet another approach to determining whether a ligand is an agonist or antagonist that is independent of a reporter gene construct (136). Addition of melatonin to a transfected oocyte reduces intracellular cAMP concentrations resulting in the aggregation of the pigment in the cells. Agonist stimulation of the transfected cell increases cAMP resulting in pigment dispersal, a reaction that can he immediately determined visually on a plus/minus basis but can also be quantified in terms of light transmission.

The ability to measure co-localization of two cellular components in living cells is an increasingly powerful technique in understanding ligand–receptor and protein–protein interactions. Two such approaches are fluorescence resonance energy transfer (FRET) that measures the proximity of two proteins through the use of a luciferase tag on one partner and green fluorescent protein (GFP) on the other (137), and fluorescence polarization (FP) that relies on a polarized excitation light source to illuminate a binding reaction mixture (138). The smaller partner (e.g., the ligand) must be fluorescently labeled, and if this is unbound then it will, by tumbling in solution, emit depolarized fluorescence. This is thus a means of looking at the amount of unbound ligand in the binding mixture and has the benefit of not requiring any washing steps. Current systems do not work well with turbid solutions or fluorescent compounds, although the latter has largely been dealt with through the use of red fluorescent dyes.

For ion channels, patch-clamping electrophysiological techniques, FRET, and ^{86}Rb+ efflux assays are used by have limited throughput. The FLIPR has proven useful for nicotinic and P2X receptors (134).

12.3 Receptor Sources

The choice of a tissue or cell line as a receptor source has a significant impact on the data generated. The natural receptor concentration in most tissues is in the femtomole to picomole range; brain tissue has a much higher density of receptors because of more extensive nerve innervation. Expression of the drug target in a cell line can, however, lead to differences in the number of receptors expressed per clone, a factor dependent on the relative proportion of transient to stable expressed cells and the passage number of the transfectants. Thus, the number of receptors can vary affecting the apparent activity of unknown ligands that compete for binding with the radioligand (49).

As the drug targets of greatest interest are those in the human, the use of a human receptor or enzyme would seem ideal in defining the

SAR of a potential drug series. The drawback, however, is that nearly all the toxicology and safety studies done in preparing a compound for clinical trials are conducted in rodents, dogs, and nonhuman primates. If there are no species differences between rat and human, this testing becomes a moot point. If on the other hand, the human target is substantially different from that in rat or dog or monkey, and there are many examples of this, the safety and toxicology studies may be conducted on a compound that has limited interactions with the drug target in species other than human. One approach is to incorporate human receptor orthologs into mice (139).

The use of transfected cell lines in compound evaluation can lead to a number of potential artifacts. Activation of transfected receptor may result in an increase in cAMP, a second messenger effect that may already be known to be a consequence of ligand activation of the receptor in its natural state. It is also possible that the transfected receptor may activate a cell signaling pathway in the transfected cell that is not linked to the receptor in its normal tissue environment. In this instance, the second messenger readout actually functions as a "reporter," a G-protein–linked phenomenon that results from the introduction of the cDNA for a GPCR and the generic or promiscuous interaction of the receptor with the G-protein systems. The introduction of the cDNA for any GPCR may then act to elicit a similar response. It is then advisable to use caution in extrapolating events occurring in the transfected cell to the physiological milieu of the intact tissue (49), and there is at least one case where a compound identified as an antagonist in a cell system overexpressing the receptor of interest was subsequently found to be a partial agonist after it exacerbated disease symptomatology in phase II clinical trials.

12.4 ADME

In evaluating new compounds, it is only in the past decade that the ability of a compound to reach its putative site of action has been a priority in the discovery phase of compound identification. For many years, it was naively assumed that there were some generic approaches that could be used on compound with poor bioavailability that would turn them into drug candidates. With attrition rates of lead compounds in the clinical development process of 50–60% range, with one estimate of greater than 90% (140), this was clearly not the case. As Hodgson (141) has noted, "a chemical cannot be a drug, no matter how active nor how specific its action, unless it is also taken appropriately into the body (absorption), distributed to the right parts of the body, metabolized in a way that does not instantly remove its activity and eliminated in a suitable manner—a drug must get in, move about, hang around and then get out." The factors involved in defining *in vivo* activity form the basis of Lipinski's "rule of 5" (142). In this widely read, retrospective study, a number of compounds have been assessed and used to design new molecules. The physical properties that were determined as limiting bioavailability were as follows: molecular weight greater than 500; more than 5 hydrogen bond donors; more than 10 hydrogen bond acceptors; and a $C\log P$ value less than 5.

A retrospective evaluation of the oral bioavailability of 1100 novel drug candidates with an average molecular weight of 480 from SmithKline Beecham's drug discovery efforts in rats by Veber et al. (143) established that reduced molecular flexibility, measured by the number of rotatable bonds and low polar surface area or total hydrogen bond count (sum of acceptors and donors), were important predictors of good (>20–40%) oral bioavailability, independent of molecular weight. A molecular weight cutoff of 500 did not significantly separate compounds with acceptable oral bioavailability from those with poor oral bioavailability, the predictive value of molecular weight was more correlated with molecular flexibility than molecular weight *per se*. From this retrospective analysis, Veber et al. suggested that compounds with 10 or fewer rotatable bonds and a polar surface area equal to or less than 140 Å^2 (representing 12 or fewer H-bond acceptors and donors) have a high probability of having good oral bioavailability in rats. Reduced polar surface area was a better predictor of artificial membrane permeation that lipophilicity ($C\log P$), with increased ro-

tatable bond count having a negative effect on permeation and having no correlation with *in vivo* clearance.

ADME is now a routine part of the efforts of a drug discovery team with the use of a number of *in vitro* approaches, e.g., Caco 2 intestinal cell lines (144, 145), human liver slices, or homogenates to assess potential metabolic pathways (119), in addition to classical rat, dog, and nonhuman primate *in vivo* studies, none of which has yet shown reliable predictability for the human situation. More recently, the use of proteomics to derive potential toxicological profiles has been assessed (146) as has increasing the throughput of compound evaluation in conjunction with increased computational support (147).

12.5 Compound Databases

With the exponential increase in information flow resulting from the ability to make many more compounds and test these in multiple assays, the capture, analysis, and management of data is a critical success factor in drug discovery.

Many databases used in drug discovery are ISIS/Oracle-based using MDL structural software and a variety of data entry and analysis systems, some are PC/Mac-based, and some are on a server. The ability to capture data and then reassess its value through *in silico* approaches (148) has the potential to be a vast improvement over the "individual memory" systems that many drug companies used for the better part of the last century. Thus, with the retirement of a key scientist, the whole history of a project or even a department disappeared, and whatever folklore existed regarding unexplained findings with compounds 10 or 20 years before was lost with the individual.

13 LEAD COMPOUND DISCOVERY

As evidenced by the compound code numbers used by pharmaceutical companies, many thousands of new chemical entities have been made since the industry began in the late 19th century. Until the advent of combinatorial chemistry, the chemical libraries at most of the major pharmaceutical companies numbered from 50,000–800,000 compounds comprised of newly synthesized compounds as well as those from fermentation and natural product sources. Approximately 2–5 million compounds were identified in the search for new drugs to treat human disease states over the past century. This number has obviously leaped to the billions with combinatorial approaches. Given decomposition and/or depletion of compounds, the 2–5 million compounds could be rounded down to 1 million.

The 2001 edition of the *Merck Index*, the compendium of drugs and research tools, lists a total of 10,250 compounds (149). Thus, from a hypothetical million compounds, only 1% have proven to be of sustained interest as either therapeutic agents or research tools.

In the early 1980s, with the not yet materialized promise that compounds could be created and tested on a computer screen, the screening of large numbers of compounds against selected targets came to be viewed as irrational. Indeed, the head of research at one of the top 20 pharmaceutical companies told the medicinal chemists at that company in the early 1980s that the demand for their skills was becoming less and would cease by the 1990s, a viewpoint somewhat akin to the apocryphal story of the head of the U.S. Patent Office in the early 1900s who recommended its closure because "everything that could be discovered had been discovered."

The breakthrough for screening came in 1984 with the identification by Chang et al. at Merck of the CCK_2 antagonist, asperlicin (150), and its subsequent use as a lead structure to discover the clinical candidate, MK 329 (151). Considerable effort has been expended in the pharmaceutical industry worldwide to capitalize on this approach with significant successes that have enhanced the search for novel chemical entities as well as providing research tools to better understand receptor and enzyme function.

13.1 High-Throughput Screening

To identify compounds, it is imperative that a rapid, economical, and information-rich evaluation of biological activities be available. The term high-throughput screening describes a set of techniques designed to permit rapid and automated (robotic) analysis of a library of compounds in a battery of assays that generate specific receptor- or enzyme-based signals. These signals may be membrane-based (radioligand binding, enzyme catalysis) or cell-based (flux, fluorescence). The primary purpose of HTS is not to identify candidate drugs, but rather to identify lead structures, preferably containing novel chemical features, which may serve as a guide for more tailored iterative optimization. HTS should generate as few false-positive leads as possible because exploitation of leads is an expensive component of the drug discovery process. Presently, most HTS are designed to give information principally about potency, and a combination of screens may provide information about selectivity and specificity. The increasing use of cell-based assays provides additional information, including agonist/antagonist characteristics, and a biological "read-out" under physiological or nearly physiological conditions. Additionally, cell-based assays can provide information about the cytotoxicity and bioavailability of molecules. The increased use of "designer cells" with visual and fluorescent signal read-outs will continue to facilitate the screening process, and in the future, will doubtlessly include measures of metabolism, toxicity, bioavailability, and other important pharmacokinetic parameters.

Using FLIPR and related techniques in 96-, 384-, 1536-, and greater numbers of microtiter plate arrays, 100,000 or more compounds can be screened through 10–20 targeted assays in less than a week, and complete libraries numbering in the millions can be assessed at single concentration points in less than 6 weeks. The issue is the success rate, typically in the 0.1% range, and the ability to capture and store the data for posterity. Clearly, the quality and diversity of the compounds and the robustness of the assays play a key role in a successful screening program.

13.2 Compound Sources

New chemical entities (NCEs) are discovered or developed/optimized from the following: (*1*) natural products and biodiversity screening; (*2*) exploitation of known pharmacophores; (*3*) rationally planned approaches, e.g., computer-assisted molecular design; (*4*) combinatorial or focused library chemistry approaches; and (5) evolutionary chemistry.

13.2.1 Natural Product Sources. Approximately 70% of the drugs currently in human use originate from natural sources (3, 152). These include morphine, pilocarpine, physostigmine, theophylline, cocaine, digoxin, salicylic acid, reserpine, and a host of antibiotics (3). Medicinal and herbal extracts form the basis for the health care of approximately 80% of the world's population; some 21,000 plant species are used world-wide. Screening of natural products led to the discovery of the immunosuppressants, cyclosporin, rapamycin, and FK 506 (153), and there is a continued search for new compounds, even in relatively well-explored areas such as China (154).

The continued destruction of habitat with the accompanying loss of animal and plant species may impede further natural product-based drug discovery. Many interesting drugs have vanished (155). For example, a plant called silphion by the Greeks and sylphium by the Romans grew around Cyrene in North Africa. It may have been an extremely effective anti-fertility drug in the ancient world but was harvested to extinction (156).

Whereas 100% of the world's mammals are known (155), as few as 1–5% of other species, notably bacteria, viruses, fungi, and most invertebrates, are well characterized. It has been estimated that only 0.00002% to 0.003% of the world's estimated $3–500 \times 10^6$ species are used as a source of modern drugs (157). Exploration of environments previously assumed to be hostile to life has revealed bacterial species living at extreme depths, at extraordinary temperatures, and in the presence of high concentrations of heavy metals.

The sea covers almost three-quarters of the earth's surface and contains a broader genetic variation among species relative to the terrestrial environment (155, 158). Although a number of important molecules have been derived from marine sources, including arabinosyl nucleotides, didemnin B, and bryostatin 1, there has been an inadequate focus on this potentially chemically productive biosphere. Sea snails, often called "nature's combinatorial chemistry factories," produce a bewildering array of novel conotoxins active at mamma-

lian drug targets (159, 160) that are being exploited for ion channel HTS by Xenome in Australia in collaboration with ICAgen, Ionix, and Antalium. These toxins are typically 10–30 amino acid residues in length, contain several disulfide bridges, and are rigid in structure (160). There are several hundred varieties of cone snails, and each may secrete more than 100 toxins. Therefore, there are likely to be several tens of thousands of these toxins, representing a library of substantial structural and functional diversity. These peptides are first synthesized as larger precursors from which the mature peptide is cleaved (159). In the mature peptide there is a constant N-terminus region and a hypervariable C-terminus region from which the biological diversity is derived. Conus toxins have proven to be invaluable as molecular probes for a variety of ion channels and neuronal receptors and as templates for drug design (160).

Poison frogs of the *Dendrobatidae* family contain a wide variety of skin-localized poisonous alkaloids that are presumably secreted for defensive purposes (161). Among the chemical structures present are the batrachotoxins, pumiliotoxins, histrionicotoxins, gephyrotoxins, and decahydroquinolines (Fig. 10.6) that target both voltage- and ligand-gated ion channels. The alkaloid epibatidine (Fig. 10.6), present as a trace entity in *Epipedobates tricolor,* is of particular interest because it has powerful analgesic activities, being 200 times more potent than morphine (162). The alkaloid is a potent neuronal nicotinic channel agonist selective for the $\alpha 4\beta 2$ subtype. Isolated by Daly and Myers in 1974, the structure of epibatidine was not elucidated until 1992. The discovery of epibatidine as a novel and potent analgesic led to the identification of ABT-594, which had equivalent analgesic efficacy to epibatidine but with reduced side effect liabilities (163). Despite continued successes in isolating new compounds with pharmaceutical potential from natural sources, the pharmaceutical industry has tended to loose interest in this approach limiting an important aspect of chemical diversity (164).

13.2.2 Pharmacophore-Based Ligand Libraries. The majority of pharmaceutical companies have relatively large chemical libraries

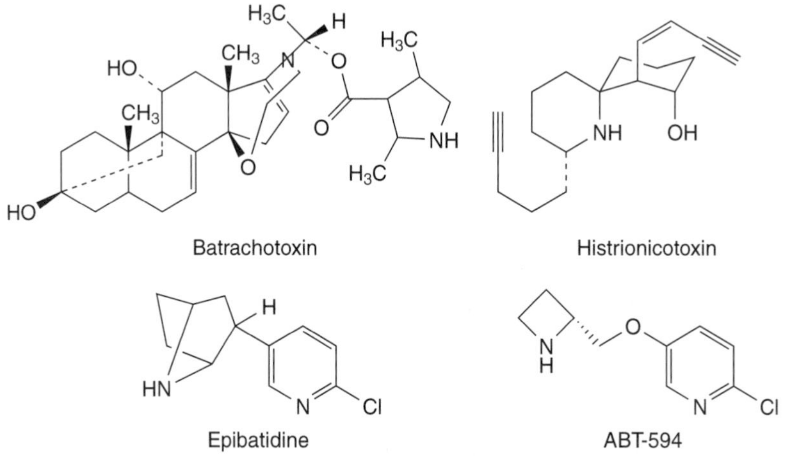

Figure 10.6. Alkaloids and related structures.

representing the cumulative synthetic efforts of the medicinal chemists within the company. Typically, the chemical diversity in these libraries is not extreme because the synthetic approach to drug design revolves around defined pharmacophores in lead series and the rational and systematic development of the SAR. Certain companies will thus have a large number of similar compounds based on the approaches and successes attendant to a therapeutic area.

The systematic modification of existing structures, both natural and synthetic, is an approach to compound optimization with improvements in potency, selectivity, efficacy, and pharmacokinetics being linked to discrete changes in molecular constituents on the basic pharmacophore.

Angiotensin converting enzyme (ACE) inhibitors are important cardiovascular drugs that block the conversion of angiotensin I, formed by the action of renin on substrate angiotensinogen, to angiotensin II (a powerful pressor and growth factor agent). Until 1973, peptide inhibitors from the venom of the Brazilian viper were the only known inhibitors of this enzyme. The nonapeptide, teprotide, was an orally active and competitive inhibitor of ACE (165). Benzylsuccinic acid, a potent inhibitor of carboxypeptidase A (166), an enzyme that had structural and mechanistic similarities to ACE, led Ondetti et al. to select N-succinyl-L-proline as a lead (IC$_{50}$ = 330 pM).

This was subsequently optimized based on the presence of a Zn^{2+} in the active site of both ACE and carboxypeptidase and the likely presence of hydrophobic pockets. An SH group to coordinate Zn^{2+} was incorporated into the stereoselectively active α-methyl analog, which eventually led to captopril (165) and the analogs, enalapril, cilazapril, and lisinopril (Fig. 10.7).

The Ca^{2+}-channel antagonists, the 1,4-dihydropyridines (DHPs), represent a remarkably successful group of cardiovascular drugs that have antihypertensive, antianginal, and antiarrhythmic properties (167). Nifedipine (Fig. 10.8) was the first member of the DHP family; a structure that embraces both Ca^{2+}-channel antagonists and activators (168), which was followed by the synthesis of a number of analogs with a prolonged duration of action and enhanced vascular selectivity (167).

13.2.2.1 Molecular Modeling. Rationally planned approaches to structure design are an increasingly important part of the drug discovery process, whether planned *ab initio*, derived from structural knowledge of a putative ligand-binding site on a biological target in the absence of ligand information, or derived by rational exploitation of an existing chemical lead (121–125).

13.2.2.2 Privileged Pharmacophores. It is increasingly apparent that a small number of common structures—"basic pharmacophores," "templates," or "scaffolds"—are associated

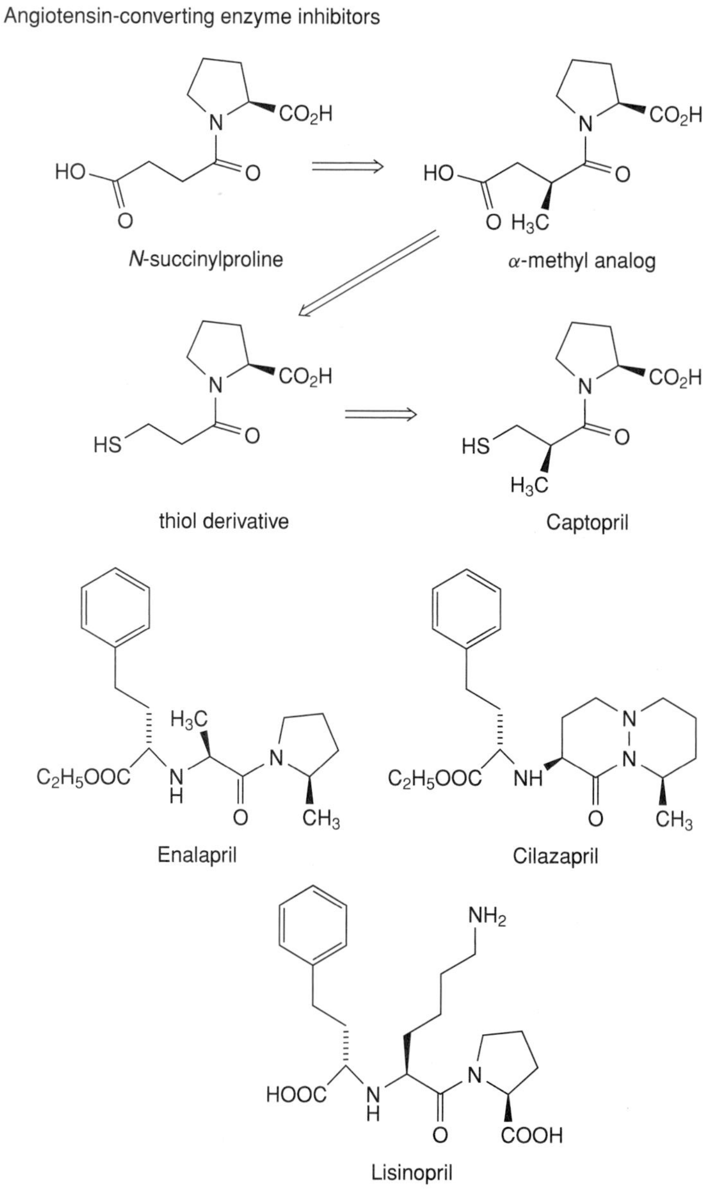

Angiotensin-converting enzyme inhibitors

Figure 10.7. ACE Inhibitors.

with a multiplicity of diverse biological activities. These structures represent facile starting points for combinatorial chemical approaches to ligand diversity.

13.2.2.2.1 Benzodiazepines. The benzodiazepines (BZs) are well established as anxiolytics, hypnotics, and muscle relaxants as represented by diazepam, clonazepam, midazolam, and triazolam (Fig. 10.9). The BZ nucleus also occurs in natural products. Asperlicin is a

naturally occurring ligand that is a weak, albeit selective, antagonist at cholecystokinin receptors and contains a benzodiazepine nucleus (150). From this lead was derived a series of potent and selective benzodiazepine ligands, active at CCK_1 and CCK_2 receptors, one of which is L-364,718 (151). Other BZs (Fig. 10.9) with activity at receptors distinct from the BZ receptor are as follows: tifluadom (opioid receptor) (169); somatostatin (170);

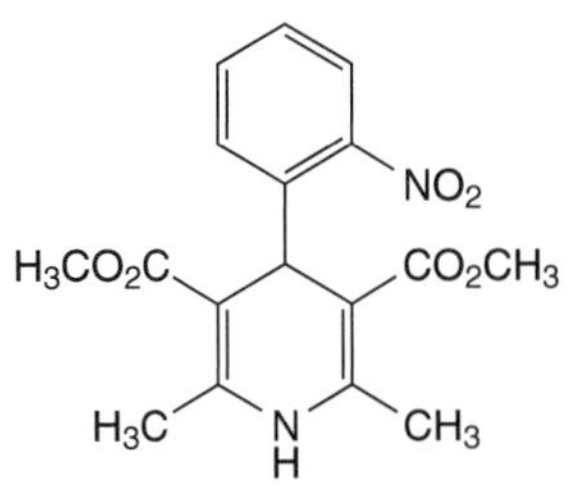

Nifedipine

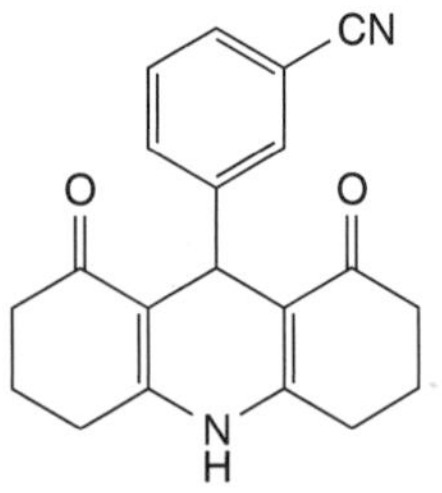

ZM 244085

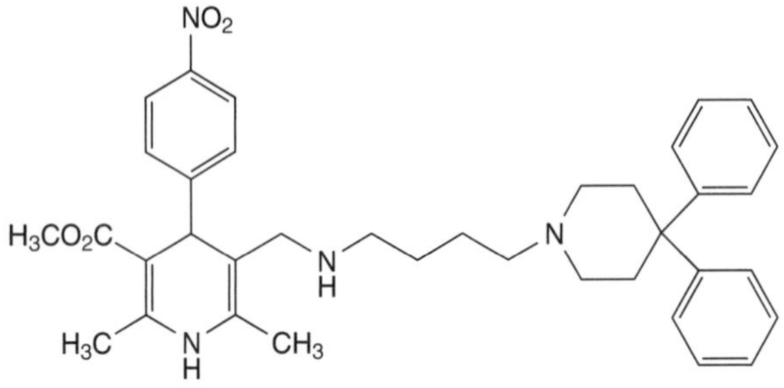

SNAP 5089

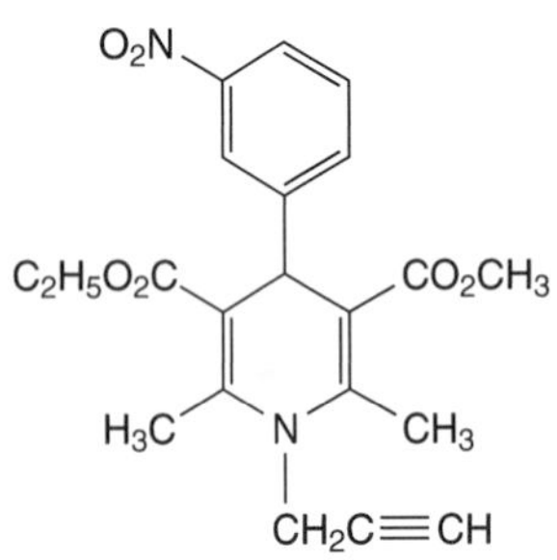

MRS 1845

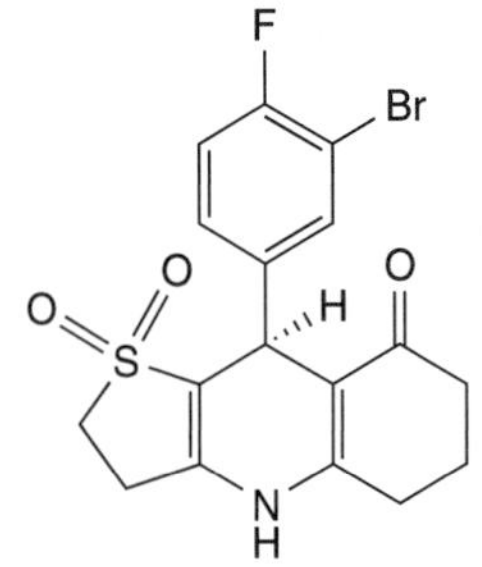

A-278637

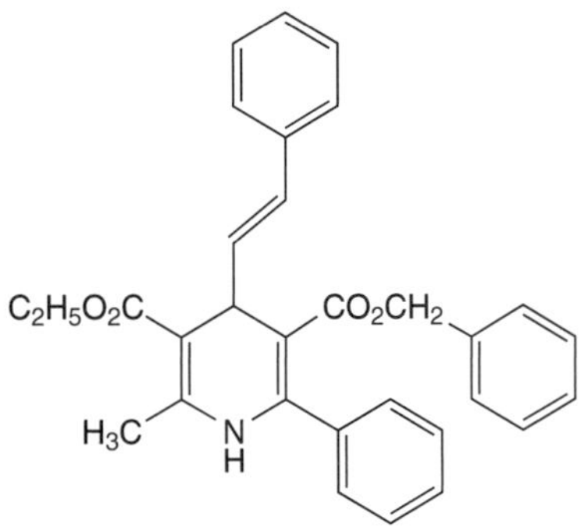

MRS 1191

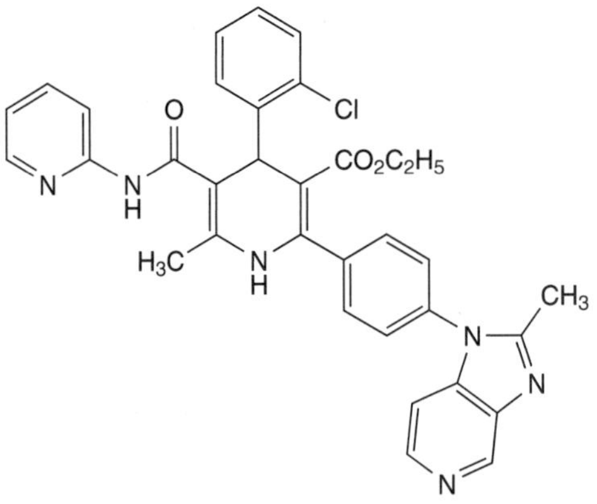

UK 74,505

Figure 10.8. Dihydropyridines; calcium channel, and other receptor antagonists.

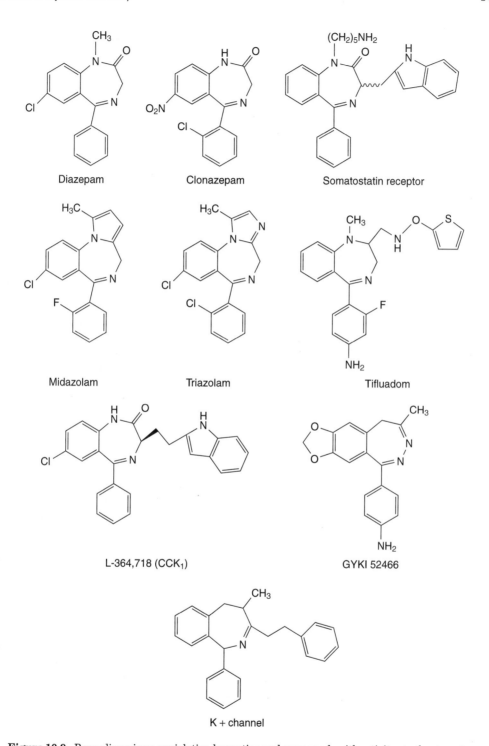

Figure 10.9. Benzodiazepines: anxiolytics, hypnotics, and compounds with activity at other targets.

GYKI 52466 (glutamate receptor) (171); and inward-rectifying potassium channels (172).

13.2.2.2.2 1,4-Dihydropyridines. The DHP pharmacophore is a well-established chemical entity derived from the classic chemistry of Hantzsch (173). Nifedipine (Fig. 10.8) and several related DHPs, including amlodipine, felodipine, nicardipine, nimodipine, and nisoldipine, are well-established antihypertensive and vasodilating agents that act through voltage-gated L-type Ca^{2+} channels in the vasculature (174). However, DHPs also interact with lower activity at other classes of Ca^{2+} channels, including N-type channels (175), T-type channels (176), and "leak" channels (177). DHPs can block delayed rectifier K^+ channels (178) and cardiac Na^+ channels (179). Other DHP analogs (Fig. 10.8) are active at PAF receptors (UK 74,505) (180), adenosine A_3 receptors (MRS 1191) (181), K^+ ATP channels (ZM 24405 and A-278367) (182, 183), capacitative SOC channels (MRS 1845) (184), and α_{1A}-adrenoceptors (SNAP 5089) (185).

13.2.3 Diversity-Based Ligand Libraries.
The issue of diversity reflects the need to enhance the scope of the library beyond those available within a company. This can be done by compound exchange with other companies, by acquiring compounds from university departments and commercial sources (Sigma, Bader, Mayhew, Cookson, etc.). Additional sources of novel synthetic compounds include the libraries of major present or former chemical/agricultural companies and their successors like Eastman Kodak, Stauffer, FMC, and Shell, and more recently, chemicals made in the former Eastern Bloc countries that are now being brokered to pharmaceutical companies. In conjunction with the use of computerized cluster programs, the selection of compounds based on diverse structures can be considerably enhanced to provide maximum coverage of molecular space. This can be done by generating libraries of approximately 2000 compounds that is used to rapidly identify potential leads for a new drug target using the SAR generated in an HTS assay. Such libraries are assembled from compounds that are available in relatively large supply and may not necessarily be proprietary to the company.

Their value is in rapidly eliminating unlikely structures in a systematic manner for each new target.

Combinatorial chemistry has provided the means to synthesize literally billions of molecules, a major step forward in the exploration of "molecular space." This random or "brute-force" approach to the search for new leads has been a major disappointment in its contribution to product pipelines (186) and the sacrifice of quality for quantity (187). However, combinatorial exploration around lead compounds or pharmacophores to generate dedicated libraries, when coupled with high-throughput screening, clearly provides an economical and efficient way to rapidly generate lead compounds (188).

Combinatorial diversity can be exploited in nature in the repertoire of antibodies with their remarkable combination of high affinity and selectivity and is also reflected in the conotoxins (159, 160).

The polyketides are a family of natural products containing many important pharmaceutical agents that are synthesized through the multienzyme complex, polyketide synthase, which can display substantial molecular diversity with respect to chain length, monomer incorporated, reduction of keto groups, and stereochemistry at chiral centers (189). This variability, together with the existence of several discrete forms of polyketide synthase, allows the generation of diverse structures like erythromycin, avermectin, and rapamycin. This biochemical diversity has been considerably expanded by the introduction of new substrate species that were used by the enzymes to produce new or unnatural polyketides (190).

Such methods are likely to be more extensively used in the future to provide a highly focused combinatorial approach to generating molecular diversity. The multienzyme pathway responsible for converting simple linear unsaturated allylic alcohols to sterols, carotenoids, and terpenes is incompletely characterized, but offers excellent potential for genetic manipulation to provide directed bio-combinatorial chemistry (191, 192).

13.2.4 Evolutionary Chemistry.
An alternative pathway to the directed synthesis of

molecules, whether individually or by the millions, is to establish conditions where spontaneous mutation can occur accompanied by selection for a specific property (193). Using this approach, ribozymes, from *Tetrahymena* that catalyzed sequence-specific cleavage of RNA were directed through selective amplification with error-prone PCR to cleave DNA *in vitro*. Cycles of amplification and selection eventually produced the conversion of an RNA-cleaving to a DNA-cleaving enzyme in just 10 generations (194). While this process of *in vitro*, or directed, molecular evolution has thus far been employed in the synthesis and characterization of macromolecules, it can be extended the synthesis of small molecules like the polyketides and isoprenoids discussed above.

13.3 Biologicals and Antisense

The use of the naturally occurring hormones as drugs is not new as evidenced by the use of insulin and epinephrine. The techniques of molecular biology allow the production of an increasing number of native and modified hormones and their soluble forms. Erythropoietin (EPO) is a classic example of a successful drug taken from the human body. Soluble receptors like Enbrel and humanized antibodies, e.g., herceptin and D2E7, are additional examples of how cloning techniques have altered the concept of rational drug design and what can be considered as a drug. Anti-cytokine therapies are particularly amenable to this approach (195).

Antisense oligonucleotides and RNA interference (RNAi) approaches may similarly provide novel, specific to disease treatment (196, 197) by the suppression/modification of a gene product. The fundamental concept is that an oligonucleotide complementary to a disease-causing gene will anneal to that gene and prevent its transcription. In a similar manner, RNAi approaches use the fact that double-stranded RNA is rapidly degraded, and double-stranded RNA oligomers can trigger degradation of the genes encoded by them. While delivery issues remain a concern, one topically active antisense drug, Vitravene (Isis) is approved for the treatment of CMV-induced retinitis, and several promising antisense constructs, e.g., Genasense (Genta/Aventis), an 18-mer against the first six codons of *bcl*2 that

is being investigated for the potential treatment of melanoma, prostate, breast, and colon cancer, are now in clinical trials.

14 FUTURE DIRECTIONS

The gradual evolution of the receptor concept as the basis for drug discovery over the past century has led to major advances in the understanding of biological systems and human disease states.

While the development of ever more sophisticated structure-based technologies and ultra, micro-HTS that encompass activity, ADME, toxicity, and structural data generation has resulted in an ever increasing body of knowledge, it does not seem to have added significantly to success as measured by increased numbers of quality INDs (186), but rather, together with genomics and proteomics, significantly added to the cost of the search for new medicines (6). The medicinal chemist and the pharmacologist play key roles in integrating and interpreting the data flow that constitutes much of the day-to-day workings of the drug discovery environment. Their challenge is to provide the intellectual framework to use this data to find drugs rather than play a technology-driven numbers game. As noted nearly a decade ago (198), the compact disk is a technological marvel that has replaced fragile magnetic tape and vinyl as a recording media. There continues to be little evidence that this digital media, however much oversampled or upsampled, has improved either the quality, innovation or longevity of the recorded music.

15 ACKNOWLEDGMENTS

The authors thank Walter Moos for critical input, Rob Jones, Tony Evans, Jan Urban, and Doug Boatman for stimulating discussions.

REFERENCES

1. E. Pennisi, *Science*, **291**, 1177–1180 (2001).
2. J. Drews, *Science*, **287**, 1960–1964 (2000).
3. W. Sneader, *Drug Discovery: The Evolution of Modern Medicines*, Wiley, Chichester, UK, 1985.
4. R. Flower, *Nature*, **415**, 587 (2002).

5. K. J. Watling, *The RBI Handbook*, 4th ed, Sigma-RBI,Natick, MA, 2001.

6. G. R. Lenz, H. M. Nash, and S. Jinfal, *Drug Discov. Today*, **5**, 145–156 (2000).

7. G. Milligan, *J. Cell. Sci.*, **114**, 1265–1271 (2001).

8. K. A. Jones, B. Borowsky, J. A. Tamm, et al., *Nature*, **396**, 674–679 (1997).

9. J. H. White, A. Wise, M. J. Main, et al., *Nature*, **396**, 679–682 (1997).

10. M. Margeta-Mitrovic, Y. N. Jan, and L. Y. Jan., *Proc. Natl. Acad. Sci. USA*, **98**, 14643–14648 (2001).

11. M. Rocheville, D. C. Lange, U. Kumar, et al., *Science*, **288**, 154–157 (2000).

12. F. Liu, Q. Wan, Z. B. Pristupa, et al., *Nature*, **403**, 274–280 (2001).

13. B. A. Jordan, N. Trapadize, I. Gomes, et al., *Proc. Natl. Acad. Sci. USA*, **98**, 343–348 (2001).

14. H. Mohler, J. M. Fritschy, and U. Rudolph, *J. Pharmacol. Exp. Ther.*, **300**, 2–8 (2002).

15. T. Yamakura and K. Shimoji, *Prog. Neurobiol.*, **59**, 279–298 (1999).

16. P. M. Sexton, A. Albiston, M. Morfis, and N. Tilakartne, *Cell Signal.*, **13**, 73–83 (2001).

17. J. C. Bermak and Q.-Y. Zhou, *Mol. Interven.*, **1**, 282–287 (2001).

18. H. Husi, M. A. Ward, J. S. Choudhary, et al., *Nature Neurosci.*, **3**, 661–669 (2000).

19. M. Kim, L.-H. Jiang, H. L. Wilson, et al., *EMBO J.*, **20**, 6347–6358 (2001).

20. L. Michaelis and M. L. Menten, *Biochem. Z.*, **49**, 333–369 (1913).

21. J. Monod, J. Wyman, and J.-P. Changeux, *J Mol. Biol.*, **12**, 88–118 (1965).

22. A. J. Clark, *The Mode of Action of Drugs on Cells*, Arnold, London, UK, 1933.

23. T. P. Kenakin, *Nature Rev. Drug Discov.*, **1**, 103–110 (2002).

24. D. E. Koshland, G. Nemethy, and D. Filmer, *Biochemistry*, **6**, 365–387 (1966).

25. A. Christopoulos, *Nature Rev. Drug Discov.*, **1**, 198–210 (2002).

26. P. Cuatrecasas, *Ann. Rev. Biochem.*, **43**, 169–214 (1974).

27. A. DeLean, J. M. Stadel, and R. J. Lefkowitz, *J. Biol. Chem.*, **255**, 7108–7117 (1980).

28. T. P. Kenakin, *FASEB J.*, **15**, 598–611 (2001).

29. J. H. Gaddum, *Pharmacol. Rev.*, **9**, 211–218 (1957).

30. O. Arunlakshana and H. O. Schild, *Br. J. Pharmacol.*, **14**, 48–58 (1959).

31. E. J. Ariens, *Arch. Int. Pharmacodyn.*, **99**, 32–49 (1954).

32. J. Saunders and S. B. Freedman, *Trends Pharmacol. Sci. Suppl.* 70–75 (1989).

33. R. P. Stephenson, *Br. J. Pharmacol.*, **11**, 379–392 (1956).

34. R. F. Furchgott, *Ann Rev. Pharmacol.*, **4**, 21–38 (1964).

35. T. P. Kenakin, *Pharmacologic Analysis of Drug-Receptor Interaction*, 3rd ed, Lippincott, Philadelphia, PA, 1997.

36. R. J. Tallarida, *Drug Dev. Res.*, **19**, 257–274 (1990).

37. D. E. Koshland, *Proc. Natl. Acad. Sci. USA*, **44**, 98 (1958).

38. V. Hill, *J. Physiol. (Lond.)*, **39**, 361–373 (1909).

39. W. Paton, *Proc. Roy. Soc. Ser. B*, **154**, 21–69 (1961).

40. L. Limbird, *Cell Surface Receptors*, Nijhoff, Boston, MA, 1986.

41. B. Katz and S. Thesleff, *J. Physiol. (Lond.)*, **138**, 63–80 (1957).

42. H. O. Onaran, A. Scheer, S. Cotecchia, and T. Costa, *Handbook Exp. Phamacol.*, **148**, 217–280 (2000).

43. H. Frauenfelder, S. G. Sigar, and P. G. Wolynes. *Science*, **254**, 1598–1603 (1991).

44. M. K. Dean, C. Higgs, R. E Smith, et al., *J. Med. Chem.*, **44**, 4595–4614 (2001).

45. S. Larareno, P. Gharagozloo, D. Kuonen, et al., *Mol. Pharmacol.*, **53**, 573–589 (1998).

46. N. Le Novere and J. P. Changeux, *J. Mol. Ecol.*, **40**, 155–172 (1995).

47. U. V. Lalo, Y. V. Pankratov, D. Arndts, and O. A. Krishtal, *Brain Res. Bull.*, **54**, 507–512 (2001).

48. A. Sharpe, J. Gehrmann, M. L. Loughnan, et al., *Nature Neurosci.*, **4**, 902–907 (2001).

49. T. P. Kenakin, *Pharmacol. Rev.*, **48**, 413–463 (1996).

50. P. Samama, G. Pei, T. Costa, et al., *Mol. Pharmacol.*, **45**, 390–394 (1994).

51. C. Braestrup and M. Nielsen, in R. W. Olsen and J. C. Venter, Eds., *Benzodiazepine/GABA Receptors and Chloride Channels: Structure and Functional Properties*, Wiley-Liss, New York, 1986, pp. 167–184.

52. M. Lutz and T. P. Kenakin, *Quantitative Molecular Pharmacology and Informatics In Drug Discovery*, Wiley, Chichester, UK, 1999.

53. M. J. Smit, R. Leurs, A. E. Alewijnse, et al., *Proc. Natl. Acad. Sci. USA*, **93**, 6802–6807 (1996).

54. S. W. Edwards, C. M. Tan, and L. L. Limbird, *Trends Pharmacol. Sci.*, **21**, 304–308 (2000).

55. G. Vantini, *Psychoneuroendocrinol.*, **17**, 410–410 (1992).

56. J. T. Coyle, D. L. Price, and M. L. Delong, *Science*, **219**, 1184–1190 (1983).

57. T. P. Kenakin, R. A. Bond, and T. I. Bonner, *Pharmacol. Rev.*, **44**, 351–362 (1992).

58. Available online at http//:biomedpda.com.

59. M. J. Gunthorpe, C. D. Benham, A. Randall, and J. B. Davis, *Trends Pharmacol. Sci.*, **23**, 183–191 (2002).

60. R. M. Evans, *Science*, **240**, 889–895 (1988).

61. D. M. Mangelsdorf and R. M. Evans, *Cell*, **83**, 841–850 (1995).

62. J. A. Brzostowski and A. R. Kimmel, *Trends Biochem. Sci.*, **26**, 291–297 (2001).

63. J. M. Herz and W. J. Thomsen, *Encyclopedia Pharmaceutical Technology*, Dekker, New York, 2002, pp. 2375–2395.

64. K. Palczewski, et al., *Science*, **289**, 739–745 (2000).

65. K. Lundstrom, *Curr. Drug Discov.* 2002, 29–33 (2002).

66. J. Bockaert and J. P. Pin, *EMBO J.*, **18**, 1723–1729 (1999).

67. D. M. Berman and A. G. Gilman, *J. Biol. Chem.*, **273**, 1269–1272 (1998).

68. M. Conti and S. L. C. Jin, *Prog. Nucleic Acid Res. Mol. Biol.*, **63**, 1–38 (2000).

69. M. Bunemann and M. M. Hosey, *J. Physiol.*, **517**, 5–23 (1999).

70. R. J. Lefkowitz, *J. Biol. Chem.*, **273**, 18677–18680 (1998).

71. E. Costa, *Ann. Rev. Pharmacol. Toxicol.*, **38**, 321–350 (1998).

72. U. Rudolph, F. Crestani, D. Benke, et al., *Nature*, **401**, 796–800 (2001).

73. G. K. Lloyd and M. Williams, *J. Pharmacol. Exp. Ther.*, **292**, 461–467 (2000).

74. J.-P. Changeux and S. Edelstein, *Neuron*, **21**, 959–980 (1998).

75. M. Le Noviere, T. Grutter, and J.-P. Changeux, *Proc. Natl. Acad. Sci. USA*, **99**, 3210–3215 (2002).

76. K. A. Jacobson, M. F. Jarvis, and M. Williams, *J. Med. Chem.*, **45**, 4057–4093 (2002).

77. E. Enmark, M. Pelto-Huikko, M. Grandien, et al., *J. Clin. Endocrinol. Metab.*, **82**, 4258–4265 (1997).

78. Y. Shang and M. Brown, *Science*, **295**, 2465–2468 (2002).

79. R. Davies, *Cell*, **103**, 239–252 (2000).

80. S. M. Keyse, *Curr. Opin. Cell Biol.*, **12**, 186–193 (2000).

81. S. Jaken and P. J. Parker, *Bioessays*, **22.3**, 245–254 (2000).

82. D. S. Bredt and S. H. Snyder, *Ann. Rev. Biochem.*, **63**, 175–195 (1994).

83. B. Wolf and D. R. Green, *J. Biol. Chem.*, **274**, 20049–20052 (1999).

84. S. Schoonbroodt and J. Piette, *Biochem. Pharmacol.*, **60**, 1075–1083 (2000).

85. P. Escher and W. Wahli, *Mut. Res.*, **448**, 121–138 (2000).

86. S. K. Dower, *Nature Immunol.*, **1**, 367–368 (2000).

87. J. N. Ihle, *Nature*, **377**, 591–594 (1995).

88. A. Wise, K. Gearing, and S. Rees, *Drug Discov. Today*, **7**, 235–246 (2002).

89. C. Mollereau, M. Parmentier, P. Mallieux, et al., *FEBS Lett.*, **341**, 33–38 (1994).

90. J. C. Meunier, C. Mollereau, L. Toll, et al., *Nature*, **377**, 532–535 (1995).

91. W. Jenck, J. Wichmann, F. M. Dautzenberg, et al., *Proc. Natl. Acad. Sci. USA*, **97**, 4938–4943 (2000).

92. E. De Souza and D. E. Grigoriadis in K. L. Davis, D. Charney, J. T. Coyle, and C. Nemeroff, Eds., *Psychopharmacology: Fifth Generation of Progress*. Lippincott Williams and Wilkins, Philadelphia, PA, 2002, pp. 91–107.

93. B. Smit, N. I. Syed, D. Schaap, et al., *Nature*, **411**, 261–268 (2001).

94. K. Brejc, W. J. van Dijk, and R. V. Klaasen, et al., *Nature*, **411**, 269–276 (2001).

95. C. Felder and M. Glass, *Ann. Rev. Pharmacol. Toxicol.*, **38**, 179–200 (1998).

96. S. M. Huang, T. Bisogno, T. Trevisani, et al., *Proc. Natl. Acad. Sci. USA*, **99**, 8400–8405 (2002).

97. N. Dahwan, F. Sesselin, R. Raghubir, et al., *Pharmacol. Rev.*, **48**, 567–592 (1996).

98. K. Mullis and F. A. Faloona, *Methods Enzymol.*, **155**, 335–350 (1987).

99. R. Oliphant and K. Struhl, *Proc. Natl. Acad. Sci. USA*, **86**, 9094–9098 (1989).

100. S. B. Liggett, N. J. Freedman, D. A. Schwinn, and R. J. Lefkowitz, *Proc. Natl. Acad. Sci. USA*, **90**, 3665–3669 (1993).

101. L. Essioux, B. Destenaves, P. Jais, and F. Thomas in J. Licinio and M.-L. Wong, Eds.,

Pharmacogenomics, Wiley-VCH, Weinheim, Germany, 2002, pp. 57–82.

102. L. Shiue, *Drug Dev. Res.*, **41**, 142–159 (1997).

103. M. J. Marton, J. L. DiRisi, H. A. Bennett, et al., *Nature Med.*, **4**, 1293–1301 (1998).

104. A. D. Roses, *Nature Rev. Drug Discov.*, **1**, 541–549 (2002).

105. J. F. Wang, B. Chen, and L. T. Young, *Mol Brain Res.*, **70**, 66–73 (1999).

106. P. Legrain, J. Wojcik, and J-M. Gauthier, *Trends Genetics*, **17**, 346–352 (2001).

107. A. R. Mendelsohn and R. Brent, *Science*, **284**, 1948–1950 (1999).

108. P. Honore, K. Kage, J. Mikusa, et al., *Pain*, **99**, 19–27 (2002).

109. S. L. Schreiber, *Science*, **287**, 1964–1969 (2000).

110. G. M. Makara, *J. Med. Chem.*, **44**, 3563–3571 (2001).

111. C. J. Woolf and R. J. Mannion, *Lancet*, **353**, 1959–1964 (1999).

112. R. G. Hill, *Trends Pharmacol. Sci.*, **21**, 244–246 (2000).

113. L. M. Brzustowicz, K. A. Hodginkson, W. C. Chow, et al., *Science*, **288**, 678–682 (2000).

114. A. Sawa and S. H. Snyder, *Science*, **296**, 692–695 (2002).

115. D. F. Levinson, P. A. Holmans, C. Laurent, et al., *Science*, **296**, 739–741 (2002).

116. M. L. Mimmack, M. Ryan, H. Baba, et al., *Proc. Natl. Acad. Sci. USA*, **99**, 4680–4685 (2002).

117. U. Bailer, F. Leisch, K. Meszaros, et al., *Biol. Psychiatr.*, **52**, 40–52 (2002).

118. M. Williams, J. T. Coyle, S. Shaikh, and M. W. Decker, *Ann. Rep. Med. Chem.*, **36**, 1–10 (2001).

119. A. P. Li, *Drug Discov. Today*, **6**, 357–366 (2001).

120. Ajay, *J. Med. Chem.*, **36**, 3565–3571 (1993).

121. R. S. Bohachek, C. M. McMartin, and W. C. Guida, *Med. Res. Rev.*, **16**, 3–50 (1996).

122. M. A. Navia and P. R. Chaturverdi, *Drug Discov. Today*, **1**, 179–189 (1996).

123. T. L. Blundell, H. Jhoti, and V. Abell, *Nature Rev. Drug Discov.*, **1**, 45–54 (2002).

124. S. B. Shuker, P. J. Hajduk, R. P. Meadows, and S. W. Fesik, *Science*, **274**, 1531–1534 (1996).

125. W. P. Walters, M. T. Stahl, and M. A. Murko, *Drug Discov. Today*, **3**, 160–178 (1998).

126. M. Kahn, *Tetrahedron*, **49**, 3433–3689 (1993).

127. M. Eguchi, R. Y. W. Shen, J. P. Shea, et al., *J. Med. Chem.*, **45**, 1395–1398 (2002)

128. H. Jhoti, *Trends Biotech.*, **19** (suppl), S67–S71 (2001).

129. J. Roth, *Metabolism*, **22**, 1059–1073 (1973).

130. D. Colquhoun, *Mol. Interven.*, **2**, 128–131 (2002).

131. S. H. Snyder, *J. Med. Chem.*, **26**, 1667–1672 (1983).

132. M. McKinney in S. J. Enna and M. Williams, Eds., *Current Protocol for Pharmacology*, Wiley, New York, 1998, pp. 1.3.1–1.3.33.

133. Y.-C. Cheng and W. C. Prusoff, *Biochem. Pharmacol.* **22**, 3099–3108 (1972).

134. K. L. Whiteaker, J. P. Sullivan, and M. Gopalkrishnan in S. J. Enna, M. Williams, Eds., *Current Protocol for Pharmacology*, Wiley, New York, 2000, pp. 9.2.1–9.2.23.

135. S. J. Hill, J. G. Baker, and S. Rees, *Curr. Opin. Pharmacol.*, **1**, 526–532 (2001).

136. G. F. Graminski, C. K. Jayawickreme, M. N. Potenza, and M. R. Lerner, *J. Biol. Chem.*, **268**, 5957–5964 (1993).

137. K. Truong and M. Ikura, *Curr. Opin. Structural Biol.*, **11**, 573–578 (2001).

138. A. J. Pope, U. M. Haupt, and K. J. Moore, *Drug Discov. Today*, **4**, 350–362 (1999).

139. H. M. Prosser, D. G. Cooper, I. T. Forbes, et al., *Drug Dev. Res.*, **55**, 197–209 (2002).

140. R. A. Coleman, W. P. Bowen, I. A. Baines, et al., *Drug Discov. Today*, **6**, 1116–1126 (2001).

141. A. Hodgson, *Nature Biotech.*, **19**, 722–726 (2001).

142. C. A. Lipinski, F. Lombardo, B. W. Dominy, and P. J. Feeney, *Adv. Drug. Deliv. Rev.*, **23**, 3–25 (1997).

143. D. F. Veber, S. R. Johnson, H. Y. Cheng, et al., *J. Med. Chem.*, **45**, 2615–2623 (2002).

144. S. Yee, *Pharm. Res.*, **14**, 763–766 (1997).

145. A. K. Mandagere, T. N. Thompson, and K.-K. Hwang, *J. Med. Chem.*, **45**, 304–311 (2002).

146. L. R. Bandara and S. Kennedy, *Drug Discov. Today*, **7**, 411–418 (2002).

147. A. P. Beresford, H. E. Selick, and M. H. Tarbit, *Drug Discov. Today*, **7**, 109–116 (2002).

148. P. Finn, *Drug Discov. Today*, **1**, 363–370 (1996).

149. M. J. O'Neil, A. Smith, P. E. Heckelman, et al., *The Merck Index*, 13th Edn, Merck, Rahway, NJ, 2001.

150. R. S. L. Chang, V. J. Lotti, R. L. Monaghan, et al., *Science*, **230**, 177–179 (1985).

151. B. E. Evans, K. E. Rittle, M. G. Bock, et al., *J. Med. Chem.*, **31**, 2235–2246 (1988).

152. A. Harvey, *Drug Discov. Today*, **5**, 294–300 (2000).

153. T. R. Brazelton and R. E. Morris, *Curr. Opin. Immunol.*, **8**, 710–720 (1996).

154. D. De-Zai, *Drug Dev. Res.*, **39**, 123–200 (1996).

155. D. J. de Vries and M. R. Hall, *Drug Dev. Res.*, **33**, 61–173 (1994).

156. J. M. Riddle and J. X. Estes, *Ant. Sci.*, **80**, 226–234 (1992).

157. A. T. Bull, M. Goodfellow, and J. H. Slater, *Ann. Rev. Microbiol.*, **46**, 219–252 (1992).

158. E. F. DeLong, *Curr. Opin. Microbiol.*, **4**, 290–295 (2001).

159. G. S. Shen, R. T. Layer, and R. T. McCabe, *Drug Discov. Today*, **5**, 98–106 (2000).

160. B. M. Olivera, D. R. Hillyard, M. Marsh, and D. Yoshikami, *Trends Biotechnol.*, **13**, 422–426 (1995).

161. J. W. Daly, H. M. Garraffo, and C. W. Myers, *Pharm. News*, **4**, 9–14 (1997).

162. B. Badio, H. M. Garraffo, T. R. Spande, and J. W. Daly, *Med. Chem. Res.*, **4**, 440–448 (1994).

163. A. W. Bannon, M. W. Decker, M. W. Holladay, et al., *Science*, **289**, 77–80 (1998).

164. A. L. Demain, *Nature Biotech.*, **20**, 331 (2002).

165. M. Ondetti, *Ann. Rev. Pharmacol. Toxicol.*, **34**, 1–16 (1994).

166. L. D. Byers and R. Wolfenden, *Biochemistry*, **12**, 2070–2078 (1973).

167. D. J. Triggle in M. Epstein, Ed., *Calcium Antagonists in Clinical Medicine*, 2nd ed., Hanley & Belfus, Philadelphia, PA, 1997, pp. 1–26.

168. S. Goldmann and J. Stoltefuss, *Angew. Chemie Int. Ed.*, **30**, 1559–1578 (1991).

169. D. Romer, H. H. Buscher, and R. C. Hill, et al., *Nature*, **298**, 759–760 (1982).

170. C. Papaageorgiou and X. Borer, *Bioorg. Med Chem. Lett.*, **6**, 267–272 (1996).

171. S. D. Donevan, S.-I. Yamaguchi, and M. A. Rogawski, *J. Pharmacol. Exp. Ther.*, **271**, 25–29 (1994).

172. R. E. Johnson, E. R. Baizman, E. R. C. Becker, et al., *J. Med. Chem.*, **36**, 3361–3370 (1993).

173. D. M. Stout and A. I. Meyers, *Chem. Rev.*, **82**, 223–242 (1982).

174. D. J. Triggle, *Cleveland Clin. J. Med.*, **59**, 617–627 (1992).

175. T. Furukawa, T. Aukida, K. Suzuki, et al., *Br. J. Pharmacol.*, **121**, 1136–1140 (1997).

176. C. J. Cohen, S. Spires, and D. van Skiver, *J. Gen. Physiol.*, **100**, 703–728 (1992).

177. W. Hopf, R. Reddy, J. Hong, and R. A. Steinhardt, *J. Biol. Chem.*, **271**, 22358–22367 (1996).

178. X. Zhang, J. W. Anderson, and D. Fedida, *J. Pharmacol. Exp. Ther.* **281**, 1247–1256 (1997).

179. H. Miyawaki, F. Yamakazi, T. Furata, et al., *Drug Dev. Res.*, **22**, 293–298 (1991).

180. K. Cooper, M. J. Frayk, M. H. Parry, et al., *J. Med. Chem.*, **35**, 3115–3129 (1992).

181. A. M. van Rhee, J.-L. Wang, N. Melman, et al. *J. Med. Chem.*, **39**, 2980–2989 (1996).

182. S. Trivedi, L. Potter-Lee, M. W. McConville, et al., *Res. Commun. Mol. Path. Pharmacol.*, **88**, 137–151 (1995).

183. M. Goplakrishnan, S. A. Buckner, K. L. Whitaker et al., *J. Pharmacol. Exp. Ther.* (2002) in press.

184. J. L. Harper, C. S. Cameriini-Otero, A.-H. Li, et al., *Biochem. Pharmacol.*, (2002) in press.

185. J. M. Wetzel, S. W. Miao, C. Forray, et al., *J. Med. Chem.*, **38**, 1579–1581 (1995).

186. F. Horrobin, *Nature Biotech.*, **19**, 1099–1100 (2001).

187. J. Everett, M. Gardner, F. Pullen, et al., *Drug Discov. Today*, **6**, 779–785 (2001).

188. A. W. Czarnik and J. A. Ellman, *Acc. Chem. Res.*, **29**, 112–170 (1996).

189. J. R. Khosla and R. J. X. Zawada, *Trends Biotechnol.*, **14**, 137–142 (1996).

190. J. R. Jacobsen, C. R. Hutchinson, D. E. Cane, and C. Khosla, *Science*, **277**, 367–369 (1997).

191. C. A. Lesburg, G. Zhai, D. E. Cone, and D. W. Christianson, *Science*, **277**, 1820–1824 (1997).

192. J. C. Sacchettini and C. D. Poulter, *Science*, **277**, 1788–1789 (1997).

193. F. Flam, *Science*, **265**, 1032–1033 (1994)

194. A. A. Beaudry and G. F. Joyce, *Science*, **257**, 635–641 (1992).

195. X.-Y. R. Song, T. J. Torphy, D. E. Griswold, and D. Shealy, *Mol. Interven.*, **2**, 36–46 (2002).

196. B. Opalinska and A. M. Gerwitz, *Nature Rev. Drug Discov.*, **1**, 503–514 (2002).

197. P. Estibeiro and J. Godfray, *Trends Neurosci.*, **24**, S56–S62 (2001).

198. M. Williams, T. Giordano, R. A Elder, H. J. Reiser, and G. L. Neil, *Med. Res. Rev.*, **13**, 399–448 (1993).

CHAPTER ELEVEN

Nicotinic Acetylcholine Receptors

DAVID COLQUHOUN
CHRIS SHELLEY
CHRIS HATTON
Department of Pharmacology
University College London
London, United Kingdom

NIGEL UNWIN
Neurobiology Division
MRC Laboratory of Molecular Biology
Cambridge, United Kingdom

LUCIA SIVILOTTI
Department of Pharmacology
The School of Pharmacy
London, United Kingdom

Contents

Burger's Medicinal Chemistry and Drug Discovery
Sixth Edition, Volume 2: Drug Development
Edited by Donald J. Abraham
ISBN 0-471-37028-2 © 2003 John Wiley & Sons, Inc.

1 INTRODUCTION

Nicotinic acetylcholine (ACh) receptors are responsible for transmission of nerve impulses from motor nerves to muscle fibers (muscle types) and for synaptic transmission in autonomic ganglia (neuronal types). They are also present in the brain, where they are presumed to be responsible for nicotine addiction, although little is known about their normal physiological function there. Nicotinic receptors form cation-selective ion channels. When a pulse of ACh is released at the nerve-muscle synapse, the channels in the postsynaptic membrane of the muscle cell open, and the initial electrochemical driving force is mainly for sodium ions to pass from the extracellular space into the interior of the cell. However, as the membrane depolarizes, the driving force increases for potassium ions to go in the opposite direction. Nicotinic channels (particularly some of the neuronal type) are also permeable to divalent cations, such as calcium.

Nicotinic receptors are the most intensively studied type of neurotransmitter-gated ion channel, and in this review we summarize what is known about their structure and function.

2 STRUCTURE AND TOPOLOGY

2.1 Genes

All of the nicotinic receptors are oligomers, composed of a ring of five subunits encircling a central pathway for the ions. The genes for the known subunit types are shown in Table 11.1.

2.2 Structure of Muscle-Type Nicotinic Receptors

Far more is known about the muscle-type nicotinic receptor than about neuronal receptors. One of the reasons for our greater understanding is that a receptor similar to the muscle-type nicotinic receptor is present in great abundance in the electric organ of the *Torpedo* ray. A large proportion of the surface in the modified muscle tissue in the electric organ is occupied by postsynaptic membrane that contains densely packed, partially crystalline arrays of nicotinic receptors. This has allowed purification (with the help of high affinity ligands such as α-bungarotoxin), partial sequencing of the receptor subunits, and hence cloning of the genes for these subunits. Tubular crystals of *Torpedo* receptors embedded in their native lipids can be grown from isolated postsynaptic membranes, and these have been

Table 11.1 Nicotinic Subunit Genes in Man

	Chromosomal Location (OMIM)	Number of Amino Acids (including signal peptide)	Gene Name	Swiss-Prot Entry Name (primary accession number)	Notes
Muscle subunits					
α1	2q24–q32	457*	*CHRNA1*	ACHA_HUMAN (P02708)	*Isoform 1; a splice variant with an additional 25 amino acids is known (isoform 2): this does not form functional channels)
β1	17p12–p11	501	*CHRNB*	ACHB_HUMAN (P11230)	
γ	2q33–q34	517	*CHRNG*	ACHG_HUMAN (P07510)	Embryonic
δ	2q33–q34	517	*CHRND*	ACHD_HUMAN (Q07001)	
ε	17p13–p12	493	*CHRNE*	ACHE_HUMAN (Q04844)	Adult
Neuronal subunits					
α2	8p21	529	*CHRNA2*	ACH2_HUMAN (Q15822)	
α3	15q24	503	*CHRNA3*	ACH3_HUMAN (P32297)	
α4	20q13.2–q13.3	627	*CHRNA4*	ACH4_HUMAN (P43681)	
α5	15q24	468	*CHRNA5*	ACH5_HUMAN (P30532)	
α6		494	*CHRNA6*	ACH6_HUMAN (Q15825)	
α7	15q14	502	*CHRNA7*	ACH7_HUMAN (P36544)	The human α7 gene is partially duplicated in the same chromosomal region
α9	8p11.1	479	*CHRNA9*	ACH9_HUMAN (P43144)	
α10	11p15.5	450	*CHRNA10*	ACH10_HUMAN (Q9GZZ6)	
β2	1p21	502	*CHRNB2*	ACHN_HUMAN (P17787)	
β3	8p11.2	458	*CHRNB3*	ACHO_HUMAN (Q05901)	
β4	15q24	498	*CHRNB4*	ACHP_HUMAN (P30926)	

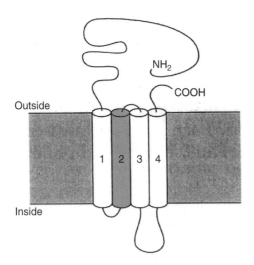

Figure 11.1. Diagram of topology of a single receptor subunit. Each subunit is thought to cross the cell membrane four times.

used extensively in investigations of receptor structure and function by electron microscopy (see below).

The muscle-type ACh receptor is a glycoprotein complex (~290 kDa), which consists of five subunits arranged around a central membrane-spanning pore. Nicotinic subunits are similar in amino acid sequence and have the same topology (Fig. 11.1): each subunit consists of a large extracellular amino-terminal domain, four predicted membrane-spanning segments (M1-M4), and a long cytoplasmic loop between M3 and M4.

These characteristics are shared with subunits that form other ion channels/receptors and thus define a receptor superfamily, usually referred to as the nicotinic family. All members in this superfamily function as either cation- or anion-selective channels, thereby mediating fast excitatory or inhibitory synaptic transmission. In mammalian cells, the cation-selective members include nicotinic and $5HT_3$ receptors, whereas the anion-selective members include $GABA_A$, $GABA_C$, and glycine receptors. Anion-selective channels in this family are also found in invertebrates: these channels are gated by glutamate, 5-HT, histidine, and acetylcholine (1).

The muscle-type receptor has the composition $(\alpha1)_2$, β, γ, and δ in embryonic (or denervated) muscle, but in the adult the γ subunit is replaced by an ϵ subunit. The adult receptor is found, at high density, only in the end-plate region of the muscle fiber, but before innervation embryonic receptors are distributed over the whole muscle fiber. The electric organ contains only the embryonic γ-form of the receptor. All the subunits share a high degree of homology (typically 31–41% pairwise identity to the α subunits, depending on the species).

The properties of and interactions between individual subunits have been explored extensively by a range of biochemical, molecular genetic, and electrophysiological techniques [for recent reviews, see Karlin (2); Corringer et al. (3)]. Their order around the pore is most likely to be α, γ, α, β, and δ going in the clockwise sense and viewed from the direction of the synaptic cleft. Opening of the channel occurs upon binding of ACh to both α subunits (α_γ and α_δ) at sites that are at, or close to, the interfaces made with neighboring γ and δ subunits (4–6). These sites are shaped by three separate regions of the polypeptide chain (3) and include the so-called C-loop (see below).

2.2.1 Molecular Architecture. The tubular crystals from the *Torpedo* ray form the basis of almost all quantitative three-dimensional studies of the whole receptor [e.g., see Kistler and Stroud (7); Miyazawa et al. (8)]. Tubes are built from tightly packed ribbons of receptor dimers and intervening lipid molecules (9). They grow naturally from the isolated postsynaptic membranes, retaining a curvature similar to that at the crests of the junctional folds. Apparently, there is a close structural correspondence between the tubes, which are simply elongated protein-lipid vesicles, and the receptor-rich membrane as it exists *in vivo*.

Ice-embedded tubes, imaged with the electron microscope, can be made to retain their circular cross section and be analyzed as helical particles (10). At low resolution, using this approach, the receptor appears as an approximately 70 × 160-Å (diameter × length) cylinder composed of five similar rod-shape subunits arranged around the central axis and aligned approximately normal to the membrane plane. The ion-conducting pathway, delineated by the symmetry axis, appears as a narrow (unresolved) pore across the mem-

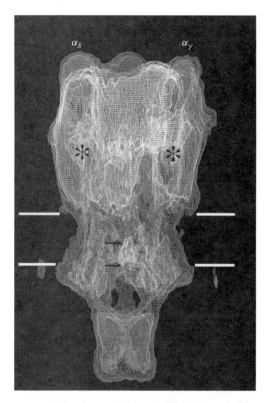

Figure 11.2. Architecture of whole receptor, emphasizing the external surface and openings to the ion-conducting pathway on the outer (extracellular) and inner (cytoplasmic) sides of the membrane. The positions of the two α subunits, the binding pockets (asterisks), gate of the closed channel (upper arrow), and the constricting part of the open channel (lower arrow) are indicated. See color insert.

brane, bounded by two large (~20-Å diameter) vestibules. Further development of this approach has led to resolutions of 9 Å (11) and, more recently, 4.6 Å (8), being achieved.

Figure 11.2 shows the appearance of the whole receptor at 4.6-Å resolution. In the extracellular portion, the subunits form a pentagonal wall around the central axis and make the cylindrical outer vestibule of the channel. The outer vestibule is about 20×65 Å (width × length). About halfway up this portion are the ACh-binding regions in the two α subunits (asterisks). In the cytoplasmic portion, the subunits form an inverted pentagonal cone, which comes together on the central axis at the base of the receptor, so shaping a spherical inner vestibule of the channel. This is about 20

Å in diameter. The only aqueous links between the inner vestibule and the cell interior are the narrow (<8–9 Å wide) "windows" between the subunits lying directly under the membrane surface. The gate of the channel, made by the pore-lining segments, M2, is near the middle of the membrane (upper arrow), and the constriction zone (the narrowest part of the open channel) is at the cytoplasmic membrane surface (lower arrow).

2.2.2 The Vestibules. One likely physiological role of the vestibules is to serve as preselectivity filters for ions, making use of charged groups at their mouths and on their inner walls, to concentrate the ions they select for (cations), while screening out the ions they discriminate against (anions). In this way, the ionic environment would be modified close to the narrow membrane-spanning pore, increasing the efficiency of transport of the permeant ions and enhancing the selectivity arising from their direct interaction with residues and/or backbone groups lining the constriction zone. A more direct means of increasing the cation conductance may be achieved by rings of negative charge located at the mouth of the pore. These rings (at positions $-4'$, $-1'$, and $20'$ of M2, using the numbering system for M2 residues defined in Fig. 11.3) are significant in that they have been shown to influence channel conductance (12).

Consistent with a screening role, the cylindrical shape and about 10-Å radius of the outer (extracellular) vestibule provide a route that is narrow enough for charged groups on the inner wall to influence ions at the center, but not too narrow to restrict their diffusion. The design of this portion of the receptor might therefore have some parallels with the fast-acting enzyme acetylcholinesterase, where the whole protein surface plays a role in producing an electrostatic field that guides the positively charged ACh substrate to the active site (13). The inner (cytoplasmic) vestibule is architecturally distinct from the outer vestibule, yet presumably plays a similar functional role in concentrating the cations, given that electrophysiological experiments on the muscle-type receptor have shown that there is no marked preference for cations to go in one direction across the membrane (i.e., rectifica-

M2

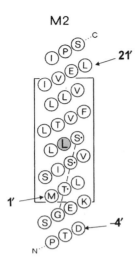

Figure 11.3. Helical net plot of the amino acid sequence around the membrane-spanning segment M2 (*Torpedo* α subunit). The leucine residue near the middle of the membrane (yellow) is the conserved leucine L251 (at the 9′ position), which may be involved in forming the gate of the channel. The dots denote other residues that have been shown to affect the binding affinity of an open channel blocker (17, 18) and ion flow through the open pore (19). The numbers shown refer to the numbering scheme for M2 residues used in the text. See color insert.

tion). Negatively charged groups framing the windows would have a strong local effect, given that the windows are not significantly wider than the diameter of an ion, including its first hydration shell.

The large proportion of mass (~70%) that is not within the membrane, and shapes these vestibules is also needed for other purposes, such as making the (complex) ACh-binding pockets, and providing sites of attachment for regulatory molecules and other proteins (such as rapsyn) that are concentrated at the synapse.

2.2.3 Membrane-Spanning Pore.
The membrane-spanning portion of the receptor has not yet been completely resolved by direct structural methods, although the pore-lining segments are partially visible as a ring of five rod-shape densities, consistent with an α-helical configuration. This helix is the part of the structure closest to the axis of the receptor

and therefore must correspond to M2, the stretch of sequence shown by chemical labeling (14, 15) and by site-directed mutagenesis/electrophysiology experiments (12, 16, 17) to be lining the pore. In the shut-channel form of the receptor, this helix is bent inward, toward the central axis, making the lumen of the pore narrowest near the middle of the membrane. This is the most constricted region of the whole ion pathway and therefore presumably corresponds to the gate of the channel.

A tentative alignment can be made between the three-dimensional densities and the amino acid sequence of M2 (Fig. 11.3) (11). This alignment places the charged groups at the ends of M2 symmetrically on either side of the lipid bilayer and a highly conserved leucine residue (*Torpedo* αLeu251) at the level of the bend. It seems likely that the leucine side chains, by side-to-side interactions with neighboring M2 segments, are involved in making the gate of the channel. Site-directed mutagenesis experiments, combined with electrophysiological study of function, have highlighted the uniqueness of the conserved leucine residue in relation to the gating mechanism. The profound effects of mutating this leucine to a hydrophilic amino acid on the agonist sensitivity of the receptor and its desensitization properties were first reported for the recombinant homomeric α7 neuronal nicotinic receptor by Revah et al. (20). In the muscle-type receptor, progressive replacement of leucines by serines (21) or by threonines (22) increases, by roughly uniform increments, the sensitivity of the channel (i.e., decreases the EC_{50} value for ACh). A similar effect is seen in neuronal nicotinic receptors that contain α3, β4, and β3 subunits (23). However, other experiments [e.g., Wilson and Karlin (24)] have been interpreted to indicate that the gate is located closer to the cytoplasmic membrane surface.

2.2.4 The Snail ACh-Binding Protein.
Before going on to discuss the agonist binding site, we next discuss the snail acetylcholine-binding protein (AChBP). Glial cells in the snail, *Lymnea stagnalis*, produce and secrete this protein, which is a homopentamer having structural homology with the large *N*-terminal portion of the extracellular domain of ion

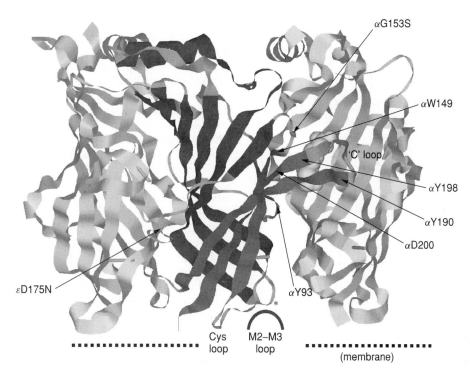

Figure 11.4. Three subunits of the *Lymnaea* AChBP viewed perpendicular to the fivefold axis of symmetry and from the outside of the pentamer. The inner and outer sheets of the β-sandwich are blue and red, respectively, whereas the putative ligand HEPES is purple. The approximate positions of the α carbons of residues discussed in the text are marked with arrows on the foremost subunit. The approximate positions of the cell membrane and of the M2-M3 loop are shown diagrammatically. The inner β-sheet (blue) is thought to rotate after agonist binding and to interact with the M2-M3 loop (as indicated by the asterisk). See color insert.

channels in the nicotinic superfamily. The protomer of AChBP is composed of 210 amino acids and has 20–23% sequence identity with the muscle-type ACh receptor subunits. It contains most of the residues that were previously suspected to be involved in ACh binding to the receptor. Its crystal structure was solved recently to 2.7-Å resolution (25), revealing the protomer to be organized around two sets of β-strands, forming Greek key-like motifs, folded into a curled β-sandwich. The β-sandwich can be divided into inner- and outer-sheet parts, shown respectively in blue and red in Figure 11.4, which are covalently linked together through a disulfide bond. The "cys-loop" disulfide bond (C128-C142 in *Torpedo* and human α1 subunits) plays an important structural role in stabilizing the three-dimensional fold (25) and is absolutely

conserved among all members of the ion channel superfamily.

AChBP has been crystallized only with HEPES, rather than ACh, bound, and "owing to low occupancy and limited resolution, the precise orientation of the HEPES molecule cannot be definitely resolved" (25). In overall appearance, it is very similar to the extracellular domain of the receptor (Fig. 11.5). The C-loop is particularly prominent both in AChBP (Fig. 11.4) and in the receptor, where it is shown as the projection labeled C in Fig. 11.5. The C-loop contains several conserved residues that are thought to be part of the acetylcholine binding site: two adjacent cysteines that are characteristic of α subunits, homologs of αC192, αC193, two tyrosines (αY190, αY198), and an aspartic acid (αD200). It is orientated more tightly against the neighboring

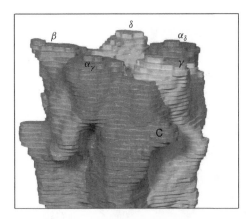

Figure 11.5. Wooden model of the extracellular part of the ACh receptor, based on the 4.6-Å map of the shut channel (8). The membrane surface is at the bottom of the figure. C denotes the C-loop region of the receptor (see also AChBP, Figs. 11.4 and 11.6), which makes part of the ACh-binding site. The width of each wooden slab corresponds to 2 Å. [From Unwin et al. (30).]

subunit in AChBP than it is in the (unliganded) receptor. These residues on the C-loop, and other nearby aromatic residues (see below) that form part of the binding site in AChBP, are homologs of residues that have been postulated to form the binding site of nicotinic receptors, on the basis of mutational studies and/or photoaffinity labeling studies. One exception is H145 (see Fig. 11.6a), which aligns with αY151 and which had not been thought to be important. Also, some residues postulated to be part of the binding site by other methods do not appear to be in the AChBP binding site (e.g., αY86).

2.3 ACh Binding Region of the Receptor

The two ACh binding sites are located in the extracellular domain, about 30 Å from the membrane surface, or 45 Å from the gate. Although the actual ligand-binding site has not yet been identified definitively within the three-dimensional structure of the receptor, ACh is expected to bind through cation-π interactions, where the positive charge of its quaternary ammonium moiety interacts with electron-rich aromatic side chains (26). The recently solved structure of AChBP (25), discussed above, shows that the "signature"

aromatic residues lie in a pocket next to the interface with the anticlockwise-positioned protomer, as seen from the "synaptic cleft." The pocket identified in AChBP would lie behind the protruding densities, labeled C in Fig. 11.5, near the α/γ and α/δ interfaces. The densities at C can also be identified with the C-loop structure in AChBP (see Fig. 11.4), although they do not curve around so tightly toward the neighboring subunits, making a more open cleft in the (unliganded) receptor.

The key aromatic residues at the binding site are most likely αY93, αW149, αY190, and αY198. These residues are located in three separate loops of the polypeptide chain (3), designated A (Y93), B (W149), and C (Y190 and Y198). All were identified as being near the agonist binding site by labeling with a small photoactivatable ligand that covalently reacts with the receptor upon UV-irradiation and acts as a competitive antagonist (27), and the first three are highly conserved in aligned positions of all muscle and neuronal α subunits. Experiments in which a series of unnatural tryptophan derivatives were substituted in place of the natural residues have suggested that the side chain of αW149 is in van der Waals contact with the quaternary ammonium group of ACh in the bound state of the receptor (26). Chemical labeling has also shown that the pair of adjacent cysteines (αC192 and αC193) is likely to be close to the binding site (28).

Figure 11.6a shows the binding site region for the AChBP with the ligand (HEPES), to show the position of the residues mentioned above. The "plus" side of the interface (Fig. 11.6a) is analogous with the α subunit of the nicotinic receptor.

The most important residues in neighboring subunits that influence ACh binding are W57 of the δ subunit and the homologous W55 of the γ subunit (6, 29). The "minus" side of the AChBP interface, shown in Fig. 11.6b, would be the γ/ε or the δ subunit in the receptor, but in the AChBP it is another identical subunit. The residues shown to be in contact with the ligand, labeled in Fig. 11.6b, mostly have no obvious analog in the ACh receptor. The one important exception is W53, which corresponds to the tryptophan residues mentioned above. Consequently, the snail protein

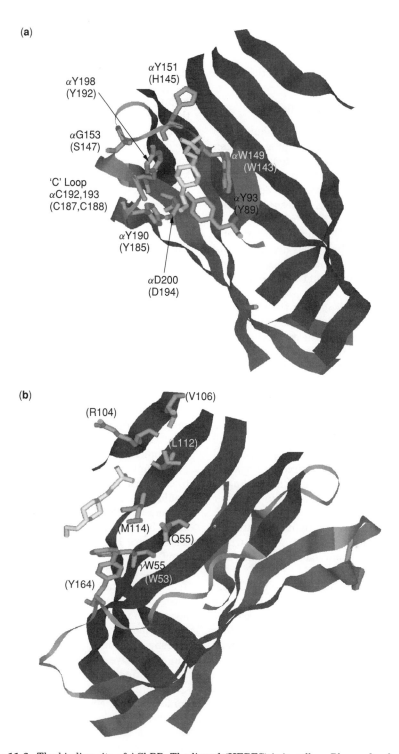

Figure 11.6. The binding site of AChBP. The ligand (HEPES) is in yellow. Blue and red regions denote the inner- and outer-sheet parts of the β-sandwich (30). The views in a and b are with the fivefold axis vertical, and the "membrane" at the bottom. (a) The structure shown is analogous to the binding site of the the *Torpedo* or human α1 subunit, to which the numbering of identified residues refers. Numbers in parentheses refer to AChBP. (b) Surface of the neighboring protomer that faces the binding site of AChBP. In the receptor this would be part of the γ, ε, or δ subunit. Numbers in parentheses refer to the AChBP. W53 aligns with W55 in mouse γ, ε, and δ nicotinic receptor subunits, but most of the residues have no obvious analog in the γ, ε, or δ subunits of the nicotinic receptor. For example, Q55 aligns with mouse γE57, εG57, and δD57; L112 aligns with mouse γY117 or δT119, and Y164 is A or G in the nicotinic subunits. See color insert.

model is rather less helpful about the non-α side of the receptor interface.

2.4 The Mechanism of Activation

The structural transition from the shut- to the open-channel form of the receptor has been analyzed at 9-Å resolution by comparing the three-dimensional map of the shut form, as described above, with that of the open form, obtained by spraying ACh onto the tubes and then freezing them rapidly within 5 ms of spray impact (31). The rapid freezing combined with minimal delay was needed to trap the activation reaction and minimize the number of receptors that would become desensitized.

A detailed comparison of the two structures indicated that the binding of ACh initiates two interconnected events in the extracellular domain. One is a local disturbance, involving all five subunits, in the region of the binding sites, and the other an extended conformational change, involving predominantly the two α subunits, which communicates to the transmembrane portion. These experiments give a picture of the receptor in either of the two states (i.e., the shut- and open-channel forms), and thus provide no direct information relating to the possibility that the binding to one site might affect the binding to the other *before* the channel opens [see Hatton et al. (32)]. However, there is a tight association of α_γ with the neighboring γ subunit (30), which is next to α_δ; thus some coupling is quite possible.

In the membrane, the exposure to ACh did not bring about any obvious alteration to the outer structure facing the lipids, whereas the M2 helices switched quite dramatically to a new configuration in which the bends, instead of pointing toward the axis of the pore, had rotated (clockwise) over to the side, as shown in Fig. 11.7.

This rearrangement had the effect of opening up the pore in the middle of the membrane, and making it narrowest at the cytoplasmic membrane surface, where the α-helices now came close enough to associate by side-to-side interactions around the ring. Thus there appear to be two alternative configurations of M2 helices around the pore: one (the shut configuration) stabilized by side-to-side interactions near the middle of the membrane, and the other (the open configuration) stabilized by side-to-side interactions close to the cytoplasmic membrane surface. These limited sets of interactions, combined with the rigid α-helical folds, might be important in ensuring the precise permeation and fast-gating kinetics that characterize acetylcholine-gated channels.

A tentative alignment of the M2 sequence with the densities in the cytoplasmic leaflet suggests that a line of small polar (serine or threonine) residues would lie almost parallel to the axis of the pore when the channel opens (Fig. 11.7), an orientation that should stabilize the passing ions by providing an environment of high polarizability. The threonine residue at the point of maximum constriction (*Torpedo* αT244), when substituted by other residues of different volume, has a pronounced effect on ion flow, as if it were at the narrowest part of the open pore (33). The diameter of this most constricted portion of the channel, based on permeability measurements made with small uncharged molecules of different size, is about 10 Å (34, 35). This value is similar to that indicated by the structural results.

A simple mechanistic picture of the structural transition, derived from these studies, would be as illustrated in Fig. 11.8. First, ACh triggers a localized disturbance in the region of the binding sites. Second, the effect of this disturbance is communicated by axial rotations, involving mainly the α subunits, to the M2 helices in the membrane (see Fig. 11.4). Third, the M2 helices transmit the rotations to the gate-forming side-chains, drawing them away from the central axis; the mode of association near the middle of the membrane is thereby disfavored, and the helices switch to the alternative side-to-side mode of association, creating an open pore.

A more precise description of the extended conformational change, linking to the transmembrane portion, has recently been derived by comparing the 4.6-Å structure of the extracellular domain with the crystal structure of the AChBP. It is found that, to a good approximation, there are two alternative extended conformations of the receptor subunits (one characteristic of either α subunit before activation, and the other characteristic of the

(a)

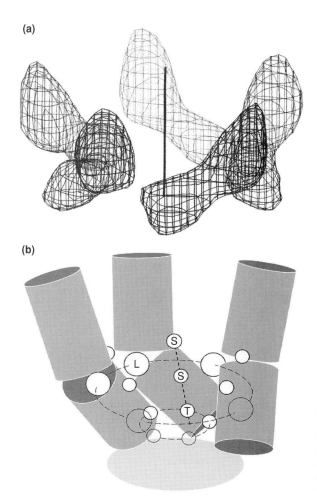

(b)

Figure 11.7. Transient configuration of M2 helices around the open pore. (a) A barrel of α-helical segments, having a pronounced twist, forms in the cytoplasmic leaflet of the bilayer, constricting the pore maximally at the cytoplasmic membrane surface. The bend in the rods is at the same level as for the closed pore, but instead of pointing inward has rotated over to the side. (b) Schematic representation of the most distant three rods. A tentative alignment of the amino acid sequence with the densities suggests that a line of polar residues (serines and threonine; see Fig. 11.3) should be facing the open pore. [From Unwin (31).]

three non-α subunits) and that the binding of ACh converts the structures of the two α subunits to the non-α form (30). Evidently, the α subunits are distorted initially by their interactions with neighboring subunits, and the free energy of binding overcomes these distortions, making the whole assembly more symmetrical, analogous to the ligand-bound AChBP.

This transition to the activated conformation of the receptor involves relative movements of the inner and outer parts of the β-sandwich, which compose the core of the α subunit (see Figs. 11.4-11.6), around the *cys*-loop disulfide bond, as shown in Fig. 11.9.

Most strikingly, there are 15–16° clockwise rotations of the polypeptide chains on the inner surface of the vestibule next to the mem-

brane-spanning pore. The M2 segments and also the M2-M3 loops lie directly under these rotating elements. The importance of the M2-M3 loop for gating was first suggested by the group of Schofield as a plausible interpretation of the mechanism by which startle mutations in this area impair the agonist sensitivity of another member of the nicotinic superfamily, the glycine receptor (37). Thus, Lynch et al. (38), on the basis of a scanning alanine mutagenesis study and macroscopic dose-response curves, suggested that *both* M1-M2 and M2-M3 loops are involved in gating. The single channel work of Lewis et al. (39), based on a preliminary plausible model of the glycine receptor activation mechanism, confirmed that a mutation in M2-M3 (αK276E) predominantly changes gating. In

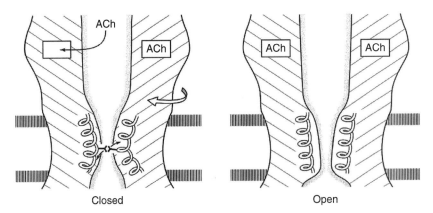

Figure 11.8. Simplified model of the channel-opening mechanism suggested by time-resolved electron microscopic experiments. Binding of ACh to both α subunits initiates a concerted disturbance at the level of the binding pockets, which leads to small (clockwise) rotations of the α subunits at the level of the membrane. The rotations destabilize the association of bent α-helices forming the gate and favor the alternative mode of association (Fig. 11.7), in which the pore is wider at the middle of the membrane and most constricted at the cytoplasmic membrane surface. [Adapted from Unwin (36).]

the nicotinic receptor, too, there is evidence that the M2-M3 region is important in coupling the binding reaction to gating (40). Thus it seems that coupling does not occur directly through M1, which does not appear to move significantly, but through an interaction of the M2-M3 loop with a part of the extracellular chain associated with the inner sheet, probably the loop between the $\beta1$ and $\beta2$ strands [see asterisk in Fig. 11.4 and Brejc et al. (25)] and/or the *cys* loop.

How does binding of ACh bring about the extended conformational change in the α subunits, converting them into a non-α form? A likely possibility, consistent with the three-dimensional maps and also with the results of biochemical experiments using binding-site reagents (4), is that the C-loop is drawn inward by the bound ACh, bringing it closer to its location in the (non-α) protomer of AChBP. The joined outer sheet could in this way be reorientated and stabilized in the configuration it would have in the absence of subunit interactions, hence favoring a switch toward the relaxed, non-α form of the subunit. Whatever the precise details, the movements that result in the open/shut transition must be fast because it is known that the whole transition from shut to open takes less than 3 μs (how much less is not known) to complete once it

has started (41), and the channel often shuts briefly (average 12 μs) and reopens (see below).

3 FUNCTION AND STRUCTURE IN MUSCLE RECEPTORS

The high resolving power of single ion channel measurements means that the function of ion channels is probably understood better than that of most enzymes. On the other hand, the lack of detailed crystal structures has hampered efforts to relate these functional measurements to the protein structure. That situation is improving rapidly.

3.1 The Nature of the Problem: Separation of Binding and Gating

Methods for recording the currents through single ion channels were developed by Neher and Sakmann (42) and Hamill et al. (43). The theoretical basis for their interpretation was developed initially by Colquhoun and Hawkes (44, 45). Far more is known about the muscle-type nicotinic acetylcholine receptor than about any other. The first attempts to investigate the mechanism of action of acetylcholine (ACh) itself were made by Colquhoun and Sakmann (46, 47). By that time it was already

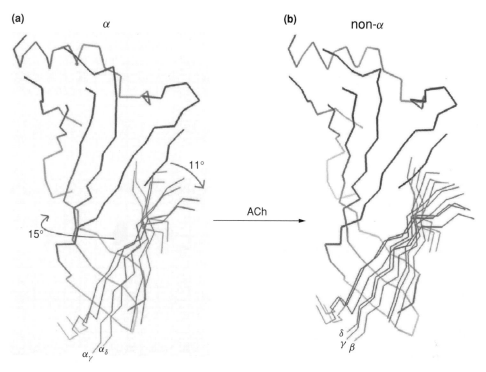

Figure 11.9. Arrangement of the inner (blue) and outer (red) β-sheet parts of the (a) α and (b) non-α subunits (see Figs. 11.4 and 11.6), after fitting to the densities in the 4.6-Å map of the receptor. The arrangement of the sheets in the α subunits switches to that of the non-α sheets when ACh binds. The views are in the same direction, toward the central axis from outside the pentamer. Arrows and angles in a denote the sense and magnitudes of the rotations relating the α to the non-α sheets. The traces are aligned so that the inner sheets are superimposed. See color insert. [Adapted from Unwin et al. (30).]

well known that there were two binding sites for ACh. Single-channel measurements made it clear that the channel *can* open with only one ACh molecule bound (though much less efficiently than with two bound). This is something that is essentially impossible to detect unambiguously from whole-cell measurements. Perhaps more important, these measurements allowed a distinction to be made between the initial binding of the agonist and the subsequent conformational change. This distinction is absolutely crucial for understanding the action of agonists in terms of classical ideas of affinity and efficacy, and it is the crucial logical basis for the use of mutation studies to identify the position of the ligand-binding site. This "binding-gating" problem has been reviewed by Colquhoun (48). The problem has been solved most fully for the muscle-type nicotinic acetylcholine receptor, and there are some reasonably good estimates for GABA and glycine receptors. For other ion channels, and for all G-protein–coupled receptors, the problem is still unsolved.

The binding-gating problem in its simplest form can be described in terms of the del Castillo and Katz (49) mechanism. This describes the binding of a single agonist molecule (A) to a receptor (R), with an equilibrium constant $K = k_{-1}/k_{+1}$. The binding may be followed by a conformation change from the shut to the open state, with an equilibrium constant $E = \beta/\alpha$; thus

$$\text{R} \underset{k_{-1}}{\overset{k_{+1}}{\rightleftharpoons}} \text{AR} \underset{\alpha}{\overset{\beta}{\rightleftharpoons}} \text{AR*}. \qquad (11.1)$$

The fraction of channels that are open at equilibrium is related hyperbolically to the concentration of agonist, with maximum $E/(1 + E)$. The fraction of receptors that have an agonist bound to them at equilibrium is also related hyperbolically to the concentration of agonist (with maximum 1). The concentration that is needed to produce 50% of the maximum effect, for both response *and* binding, is $K_{eff} = K/(1 + E)$. Thus the effects of binding (K) and gating (or efficacy, E) cannot be separated by either functional or binding experiments at equilibrium.

Binding experiments do not measure agonist binding (in any sense that is useful for learning about the binding site, or for elucidating structure-function relationships). In the context of a Monod-Wyman-Changeux-type mechanism (50), it is true that an agonist that is very selective for the open state will give, in a macroscopic binding experiment, an equilibrium constant for binding that approaches the true (microscopic) affinity for the *open* state. However, to get from the shut to the open state requires a change in conformation that is potentially affected by mutations in any part of the molecule that moves. This change in conformation is what determines the microscopic affinity for the open state. Thus, if we want to know about the binding site itself (as opposed to other regions that change shape on opening), then we need to know the microscopic affinity for the *shut* state, and this cannot be obtained from a ligand-binding experiment with an agonist. For a more detailed discussion of this question, see Colquhoun (48).

3.2 Methods for Measurement of Function

Solving the binding-gating problem is equivalent to finding a reaction mechanism that describes the actual reaction mechanism of the receptor (to a sufficiently good approximation), and then estimating values for the rate constants in that mechanism. If anything has been learned in the last 40 years it is that it is futile to imagine that firm conclusions can be drawn about channel function without a physically realistic mechanism. Null methods that circumvent the need for detailed knowledge of mechanism work well for antagonists [the Schild method (51)], but they do *not* work for

agonists (48, 52). In many ways the qualitative step of identifying the mechanism is harder that the quantitative problem of estimating rate constants, though the latter can be hard enough. Mechanisms like that in Equation 11.1 can be ruled out straightaway for the nicotinic receptor because it is known that two agonist molecules must be bound to open the channel efficiently. Schemes A and B in Fig. 11.10 are the two that have been most commonly used.

In scheme A of Fig. 11.10, the two binding sites for ACh are supposed to be the same, although the possibility is allowed that there may be cooperativity in the process of binding to the shut receptor, so binding of the second ACh molecule may not have the same rate constants as binding of the first. In scheme B of Fig. 11.10 (which in most cases is the more realistic) the two binding sites for ACh are assumed to be different from the start, so two distinguishable monoliganded states exist. In its most general form, this mechanism also allows for cooperativity of the binding reaction, in the sense that the rates for binding to site a may depend on whether site b is occupied (and vice versa). In almost all work it has been assumed that such cooperativity is absent, so the following constraints are applied:

$$k_{-2a} = k_{-1a}, \quad k_{-2b} = k_{-1b}, \quad k_{+2b} = k_{+1b}. \quad (11.2)$$

These, together with the microscopic reversibility constraint, also ensure that

$$k_{+1a} = k_{+2a} \quad (11.3)$$

In addition to the states that are shown in these schemes, we need to add, at higher agonist concentrations, states that represent open channels that have become blocked by ACh molecules. Block by ACh is fast, the blockages lasting only for 15–20 μs or so on average (similar to the spontaneous short shuttings of the receptor). The affinity for block of the open channel is highly dependent on membrane potential but is usually around 1 mM (53, 54). All other agonists and antagonists that have been tested can also produce block, some with much higher affinity than that of ACh [e.g., (+)-tubocurarine (55); suxa-

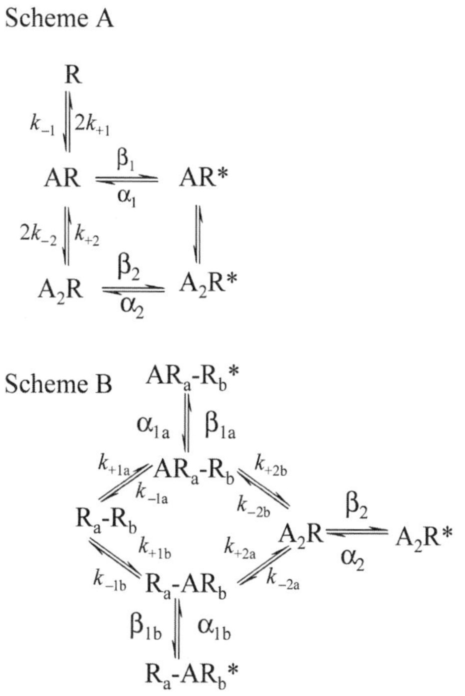

Scheme A

$$R$$
$$k_{-1} \left\|\, 2k_{+1} \right.$$
$$AR \underset{\alpha_1}{\overset{\beta_1}{\rightleftharpoons}} AR^*$$
$$2k_{-2} \left\|\, k_{+2} \right. \qquad \left\| \right.$$
$$A_2R \underset{\alpha_2}{\overset{\beta_2}{\rightleftharpoons}} A_2R^*$$

Scheme B

$$AR_a\text{-}R_b{}^*$$
$$\alpha_{1a} \left\|\, \beta_{1a} \right.$$
$$\overset{k_{+1a}}{\underset{k_{-1a}}{\nearrow}} AR_a\text{-}R_b \overset{k_{+2b}}{\underset{k_{-2b}}{\searrow}}$$
$$R_a\text{-}R_b \qquad\qquad \overset{\beta_2}{\underset{\alpha_2}{\rightleftharpoons}} A_2R \overset{}{\rightleftharpoons} A_2R^*$$
$$\overset{k_{+1b}}{\underset{k_{-1b}}{\searrow}} R_a\text{-}AR_b \overset{k_{+2a}}{\underset{k_{-2a}}{\nearrow}}$$
$$\beta_{1b} \left\|\, \alpha_{1b} \right.$$
$$R_a\text{-}AR_b{}^*$$

Figure 11.10. Two reaction schemes that have been widely used to represent the activation of the nicotinic receptor. R represents the inactive (shut) receptor; R^*, the active (open) receptor; and A, the agonist. The rate constants for the individual reaction steps are denoted k (for association and dissociation) or α (for shutting) and β (for opening). Scheme A: In this case the two binding sites for ACh are supposed to be the same, although the possibility is allowed that there may be cooperativity in the process of binding to the shut receptor, so binding of the second ACh molecule may not have the same rate constants as binding of the first. Scheme B: The two binding sites for ACh are assumed to be different from the start, so two distinguishable monoliganded states exist. In its most general form, this mechanism also allows for cooperativity of the binding reaction, in the sense that the rates for binding to site a may depend on whether site b is occupied (and vice versa).

methonium (56)]. The channel block by (+)-tubocurarine has a particularly high affinity, which seems strange in that it is too big to fit in the pore. Presumably, it must bind further out (although still within the electric field), where the channel/vestibule is wider, and in such a way that ions cannot pass it or it might prevent opening of the pore.

3.3 Desensitization

The schemes in Fig. 11.10 do not contain desensitized states, that is, states in which the agonist is still bound to the channel, but the channel is closed in a conformation distinct from that of the (unliganded) shut-channel form. This omission can be justified by the facts that (1) we can measure the things that we need without having to include desensitized states; (2) desensitization is probably of no physiological importance for nicotinic receptors (57), and, in that sense, is of peripheral interest; and (3) desensitization is a complex and ill-understood phenomenon, so it is hard to describe an adequate reaction mechanism for it. It has been known for some time that desensitization is a complex phenomenon that develops on many different time scales, from milliseconds to minutes [e.g., Katz and Thesleff (58); Cachelin and Colquhoun (59); Butler et al. (60); Franke et al. (61)]. Recently, the extent of this complexity has been shown elegantly by single-channel methods (62). There is clearly not just one desensitized state, but many, although nothing is known yet about the structural differences between these states. It is simply not feasible to fit the many rate constants, and fortunately it is not necessary to do so to learn about channel activation.

3.4 Fitting Rate Constants

In earlier work, inferences about rate constants were made from single-channel recordings by *ad hoc* methods, and only rough corrections were possible for the fact that brief events (too fast for the bandwidth of the recording) are not detectable in single-channel recordings. Inferences were made from distributions of quantities such as the distribution of apparent open and shut times, the distribution of the apparent number of openings per burst, and the distribution of burst length. Use of these univariate distributions (i.e., obtaining the time constants by fitting each distribution separately) does not make the best use of the information in the record because the lengths of openings and shuttings are correlated in just about every sort of ion channel in which the question has been examined, and use of bivariate distributions is necessary to

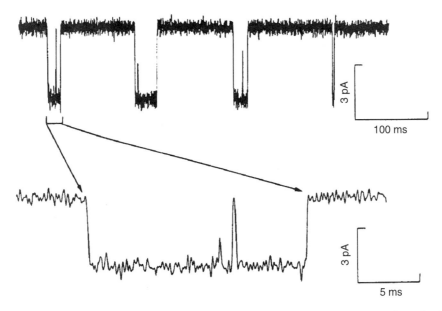

Figure 11.11. Four individual activations of the muscle nicotinic receptor by acetylcholine. The first activation is enlarged to show that it is not a single opening, but (at least) three openings in quick succession. In the enlarged part the durations of the openings are 10.7, 1.0, and 5.7 ms. These are separated by shuttings of 0.061 and 0.289 ms (filtering of the record is such that the 61-μs shut time is too short to reach the baseline). Recording from frog muscle (*cutaneus pectoris*) end plate (ACh 100 n*M*, −100 mV) (Colquhoun and Sakmann, unpublished).

extract all the information (63). A method that extracts all of the information in the record was first proposed by Horn and Lange (64), but it could not be used in practice because of the missed event problem.

Recordings from muscle-type nicotinic receptors contain many brief closures [see Figs. 11.11 and 11.13 (below)], with a mean lifetime of around 15 μs at 20°C, and because the shortest event that can be detected reliably is around 20–30 μs, the majority of these are missed (because the durations are exponentially distributed, it is possible to estimate the mean even when observations as short as the mean are missed). Methods have improved since then. Now an exact method for allowing for missed events is available, so it is possible to analyze an entire observed recording by maximum likelihood methods that extract all of the information and that incorporate missed event correction. There are some other methods under development, particularly methods based on the theory of Hidden Markov processes, but none are in routine use,

and none apart from maximum likelihood methods has had any systematic investigation of the properties of the estimators. A brief description of the maximum likelihood approach will be given next.

3.5 Maximum Likelihood Estimation of Rate Constants from Single-Channel Records

"Likelihood" means the probability (density) of the observations, given a hypothesis concerning the reaction mechanism and the values of the rate constants in it. Two programs are available for doing such calculations: MIL from Buffalo (the lab of Auerbach and Sachs) (http://www.qub.buffalo.edu/index.html) and HJCFIT (http://www.ucl.ac.uk/Pharmacology/dc.html) from University College London (UCL). Both programs work on similar principles, both can fit several data sets simultaneously, and MIL may be faster, although the UCL version has a number of advantages over MIL in other respects. For example, (*1*) it uses exact missed event correction rather than an approximation; (*2*) it uses exact "start and end of burst" vectors that improve accuracy when low concentration records have to be fitted in bursts (because of lack of knowledge of the number of channels in the patch); (3) the final fit can be tested not only by plotting open and shut time distributions but also by conditional distributions and dependency plots, which show how well the fit predicts correlations in the observations; and (4) it is the only method for which the quality of the estimates has been tested by repeated simulations.

The method of maximum likelihood allows the rate constants in a specified reaction mechanism to be estimated *directly* from an idealized single-channel record. There is no need to plot open and shut time distributions and so forth beforehand. The principles of the methods currently in use have been described by Hawkes et al. (65, 66), Colquhoun et al. (67), Qin et al. (68) [see also Colquhoun and Hawkes (69)]. The HJCFIT approach has been tested by Colquhoun et al. (70) and used by Hatton et al. (32) and by Beato et al. (71). The MIL program has been used in many publications from the Auerbach and Sine labs.

Many mutants of nicotinic receptors, both naturally occurring and artificial, have been investigated. The need for methods such as those just described is emphasized by the fact that only a minority of these mutants have been investigated by methods that make a serious attempt to distinguish binding and gating effects (many of the recent studies come from the labs of Auerbach and Sine). There is little point in trying to relate the results to receptor structure if this has not been done, so we next discuss some of the better characterized mutants, as well as the wild-type receptor. In particular, several mutations that were found to reduce the potency of ACh were initially guessed to have affected ACh binding, and therefore the mutated residues were presumed to be in the ACh binding site. However, reexamination has shown that some of these mutations actually affect conformation change rather than binding and, conversely, some residues that were not thought to be part of the binding site seem to have a significant effect on binding. Some reassessment of structure-activity relationships therefore seems desirable.

3.6 The Wild-Type Nicotinic Acetylcholine Receptor (Muscle Type)

Scheme B shown in Fig. 11.10 seems to be adequate to describe the behavior of the wild-type human receptor, although the results of Hatton et al. (32) suggest that the entire concentration range can be fitted well only if either (1) the rate constants for binding to site a depend on whether site b is occupied, or (2) an extra shut state with a lifetime of around 1 ms is added to the right of the open state (72).

3.7 Diliganded Receptors

From the point of view of the structure-function relationships of proteins, the difference between the two ACh binding sites is of great interest. However, from the physiological point of view it is not very important. Singly liganded openings are brief and, except at very low concentrations, rare. They contribute next to nothing to the end-plate current that is responsible for neuromuscular transmission. From the physiological point of view, the rates that matter most are the opening and shutting rate constants for the doubly liganded channel (α_2 and β_2) and the total rate at which agonist dissociates from the doubly liganded receptor ($k_{-2a} + k_{-2b}$; see Fig. 11.10). After exposure to the transient high concentration of ACh released from a nerve ending, most receptor molecules will be in the doubly liganded states, and these three values determine the length of each individual opening, the number of reopenings, and the lengths of the short shut periods that separate each opening. In other words, they are sufficient to determine the characteristics of the predominant doubly liganded bursts of openings (channel "activations") that are responsible for neuromuscular transmission. These three rates are, fortunately, the easiest to determine [see Colquhoun et al. (70)].

For the wild-type receptor of most species, the channel opening rate constant β_2 is 50,000 to 60,000 s^{-1} at 20°C [e.g., Salamone et al. (72); Hatton et al. (32)]. This is not much different from the value originally found in frog muscle (30,000 s^{-1} at 11°C) (47). In the lab of Auerbach and Sachs, this value is usually estimated by extrapolation to infinite concentration of the reciprocal of the durations of shut times within bursts. This method has the drawback that the extrapolation has to be done with the wrong equation (the Hill equation), and that it is often hard to get close to saturation. In any case, it is not necessary because β_2 is easy to estimate directly, simultaneously with the other parameters, by the maximum likelihood method (32, 70). In practice, however, there is not much disagreement about the value of β_2. Most values are in the range 40,000–60,000 s^{-1}. Fast concentration jump methods give similar values to those found by single-channel analysis [e.g., 15,000 s^{-1} (73); 30,000 s^{-1} (61); or 30,000–100,000 s^{-1} (74)]. These measurements provide valuable confirmation that the proposed interpretation of the single-channel observations is essentially right. The channel-shutting rate α_2 is about 1500 to 2000 s^{-1} at 20°C, so individual openings last about 0.5–0.7 ms on average. The total dissociation rate from doubly liganded receptors is 14,000–15,000 s^{-1}. These numbers imply that an average doubly liganded channel activation consists of about 4.8 openings (each of 0.6 ms), separated by 3.8 brief shuttings (each of 14.4 μs), so the mean length of the activation is about 2.9 ms. These numbers seem to be similar in human, rat, and frog, but some species variation is possible. It is hard to compare values for mean open times in the literature because most are not corrected for missed brief shuttings. For this reason, it is safer to compare values for burst lengths, although they are often not given.

3.8 Monoliganded Receptors

It is clear that the two ACh binding sites differ, and this has been found to be the case for most subtypes of the receptor. However, there has been little unanimity about the extent to which they differ [see also Edmonds et al. (75)]. There seems to be a particularly large difference for the *Torpedo* receptor (76). Binding of a single ACh molecule is sufficient to produce brief openings of the channel (though with very low efficacy) (46), and in the adult human receptor the most obvious sign that the sites differ lies in the fact that two classes of singly liganded openings are detectable (see Fig. 11.16 below), one much briefer than the other (32). The shorter one is barely resolvable, but has been demonstrated more clearly

in ultralow noise recordings from mouse embryonic receptors (in myoballs) by Parzefall et al. (77). Colquhoun and Sakmann (47) found little evidence for a difference in frog muscle receptors. Through the use of more recent methods, it has been suggested that there is little difference between the two sites in the adult form of the mouse receptor (72). These authors (72) found a significantly better fit if the sites were not assumed to be equivalent, but had similar equilibrium constants for binding. However, this study used only high agonist concentrations and ignored singly liganded openings, so it is unlikely to be very sensitive to differences between the sites. The extent of the difference between the two sites for ACh may be species dependent, or perhaps the inconsistent reports merely reflect the difficulty of the electrophysiological experiments. The fact is that it is hard to distinguish all of the separate rate constants for the two sites by electrophysiological methods (the shortest open times are on the brink of resolvability), and even with the best forms of analysis now available, it is not possible to resolve the 13 free parameters in scheme B in Fig. 11.10, without imposing the (possibly untrue) constraint of independence of the sites (see Equations 11.1 and 11.2) (70).

Binding experiments have given very convincing evidence for the two binding sites being different in their ability to bind *antagonists*. Although studies with antagonists do not tell us directly about how ACh will behave, they do have the enormous advantage that there is no binding-gating problem with antagonists and their microscopic binding affinities can be measured much more directly than for agonists.

For example, the small peptide α-conotoxin MI binds with much higher affinity to the α/δ site than to the α/γ site of mouse muscle receptors (5); see also the discussion on conotoxins below. This interaction seems to involve particularly Y198 on the α subunit and S36, Y113, and I178 on the δ subunit (78). Site selectivity is opposite for tubocurarine and its derivative metocurine (dimethyltubocurarine), which have higher affinity for the α/γ (or α/ϵ) site than for the α/δ site of mouse muscle or *Torpedo* receptors (79–81). Note that site selectivity is species dependent for α-conotoxin MI, which targets α/γ (rather than α/δ) in *Torpedo* receptors (82).

3.9 Some Well-Characterized Mutations

In the enzyme acetylcholinesterase, there is evidence that the charged quaternary ammonium group of ACh interacts with an aromatic residue, tryptophan (83). The evidence concerning this interaction is much less certain in the nicotinic receptor, but the putative binding site region contains a number of aromatic amino acids that have been investigated. Several of them are labeled by photoaffinity reagents (84).

αW149. An ingenious study by Zhong et al. (26) used unnatural amino acids to conclude that αW149 was "the primary cation-π binding site" in the nicotinic receptor. This sort of study cannot, however, distinguish between effects on binding and gating because it relies on macroscopic EC_{50} data only. A single-channel study by Akk (85) found an 80-fold increase in EC_{50} values for the αW149F mutation, but only a 12-fold weakening in binding affinity. The remainder of the potency reduction resulted from impairment of gating, the opening rate constant β_2 of the doubly occupied channel being reduced 93-fold.

In the AChBP, the analogous residue W143 forms part of the wall of the HEPES binding site (Fig. 11.6a).

αY93. Properties of the αY93F mutation have been described by Auerbach et al. (86), Akk and Steinbach (87), and Akk (85). This mutation increases the EC_{50} value for ACh by 39-fold, with a fourfold increase in the dissociation equilibrium constant estimated on the assumption that the two binding sites are equivalent (a 13-fold reduction in the association rate constant, plus a threefold reduction in the dissociation rate constant). Gating was quite strongly affected: there was a 50-fold decrease in the channel opening rate constant (β_2) and a twofold increase in the closing rate constant (α_2), and therefore about 100-fold reduction in the gating equilibrium constant, $E = \beta_2/\alpha_2$. Even allowing for the fact that the EC_{50} depends roughly on $\sqrt{E}$ [see Colquhoun (48)], the major effect is on gating rather than on binding.

In the AChBP, the analogous residue Y89 forms part of the bottom half of the HEPES binding site (Fig. 11.6a).

αY190. This tyrosine is on the C-loop and is close to the pair of adjacent cysteine residues (C192 and C193) that characterizes α subunits. The mutation Y190F (like Y93F and W149F) also decreases macroscopic ACh binding (88) and increases the EC_{50} value by 184-fold (in embryonic mouse muscle receptor) (89). These effects were originally attributed to changes in the binding. Chen et al. (89) found that the equilibrium dissociation constant for binding to the *shut* receptor was indeed increased about 70-fold (they fitted a mechanism with two sequential bindings, so this factor refers to the product of the two binding equilibrium constants, which is what matters for doubly occupied receptors). However, they also found large effects on gating: a 400-fold decrease in the channel-opening rate (β_2) and a twofold increase in the shutting rate (α_2), so there was an 800-fold reduction in the gating equilibrium constant, $E = \beta_2/\alpha_2$. Qualitatively similar mixed effects were seen when Y190 was replaced by W, S, or T (89).

In the AChBP, the analogous residue Y185 forms part of the bottom half of the HEPES binding site (Fig. 11.6A).

αY198. Although αY198 (which is also in the C-loop) has been proposed to interact directly with ACh in its binding site (90, 91), single-channel analysis of αY198F shows hardly any effect on the dissociation equilibrium constant (and only twofold slowing of the rates), with a larger but still modest effect on gating (92).

In the AChBP, the analogous residue Y192 forms part of the wall of the HEPES binding site (Fig. 11.6a).

αD200. O'Leary and White (90) suggested that gating changes account for the modest increase in EC_{50} values observed in αD200N. This conclusion was based on the complete loss of the efficacy of partial agonists in mutant receptors. This interpretation was confirmed by the single-channel study of Akk et al. (93), who observed a profound decrease in β_2 (100- and 400-fold for the adult and embryonic mouse muscle receptor, respectively). This was accompanied by a small (threefold) increase in the closing rate α_2, which further

decreased E_2 to a value between 0.1 and 0.2 for both embryonic and adult receptors. The binding of ACh was investigated by use of scheme A in Fig. 11.10, with omission of the singly liganded open state (a reasonable approximation, given that most experiments were at high enough concentration to make them rare). This scheme describes cooperativity of binding rather than nonequivalence of sites, but it should give reasonable estimates of the total dissociation rate from the doubly liganded state. The results suggested a slight *increase* in affinity for ACh for adult mouse receptor and little change in embryonic mouse receptor. The effects of the mutation are virtually entirely the result of impaired gating.

In the AChBP there is also aspartate at the equivalent position (D194) (see Fig. 11.6a). It is at the end of the C-loop and almost 10 Å from the nearest part of the HEPES ligand. This residue forms a hydrogen bond with K139, and this may be important for keeping the C-loop in an appropriate position for ACh binding. The C-loop seems to be mobile in the unbound structure but comes inward upon binding, moving the outer sheet with it to initiate the extended conformational change. One might expect residues on the C-loop (Y190, Y198, D200), residues near the inner-/outer-sheet interface (especially G153), and residues at the α/γ or the α/δ subunit interfaces (Y93, W149), to be important for coupling the ACh-binding reaction to gating, given that they are in locations where relative movements occur. Also, the pair of Cys residues, which can switch conformations about their disulfide bond, might play a role in the conformational change by stabilizing alternative configurations of the C-loop.

αG153S. This is a naturally occurring "gain-of-function" mutation in humans, in which it causes a slow channel congenital myasthenic syndrome (SCCMS) (94). It is also interesting because it is one of the mutations whose effects are almost entirely on binding. It leads to prolonged decay of miniature endplate currents (MEPCs) as a result of the channel activations (bursts) being about 15-fold longer than those in wild-type, on average (95, 96). Single-channel analysis indicates that the main reason for the prolonged bursts of openings in receptors that contain αG153S,

is slowed dissociation of ACh from the doubly liganded shut state, so the channel reopens more often (72, 95). The fits in the study by Salamone et al. (72) suggest a 30-fold decrease in equilibrium dissociation constant for ACh binding, attributable mainly to a reduced dissociation rate from the doubly liganded shut state, with relatively little effect on gating (the opening rate β_2 is hardly changed by the mutation, but openings become somewhat shorter, resulting in about threefold increase in E_2).

Despite this rather selective effect on binding, the residue (S147) in AChBP that is analogous with αG153, does not seem to form part of the binding site (see Figs. 11.4 and 11.6), but is separated from it by at least 10 Å. Nevertheless, this residue seems to be in a critical location in terms of linking binding to gating. It is on the loop connecting the inner sheet to the outer sheet part of the β-sandwich, and the relative movements are greatest in this region when ACh binds.

αN217K. This is another naturally occurring SCCMS mutant in humans; and like αG153S, it is a "gain-of-function" mutation. MEPCs recorded with intracellular microelectrodes decayed biexponentially, the slower component being approximately sevenfold longer than that of the control (97). Wang et al. (98) used single-channel analysis (with the MIL program), to fit the rate constants in schemes A and B (Fig. 11.10). The potency of ACh is increased 20-fold (EC$_{50}$ value is reduced 20-fold) in adult human receptors that contain αN217K. This appears to result almost entirely from an increase in the microscopic binding affinity, and in particular from a slowing of dissociation from the doubly liganded receptor. The gating effects were even smaller than those for αG153S. The main channel opening rate β_2 was slowed by approximately 40% in the mutant, whereas α_2 was slowed by approximately 50%, so there was hardly any change in the main efficacy term E_2.

This result is somewhat surprising, given that αN217 is nowhere near the area that is normally considered to be the binding site. Indeed, it is not even in the extracellular region at all, but is buried several amino acids down in the predicted first transmembrane domain

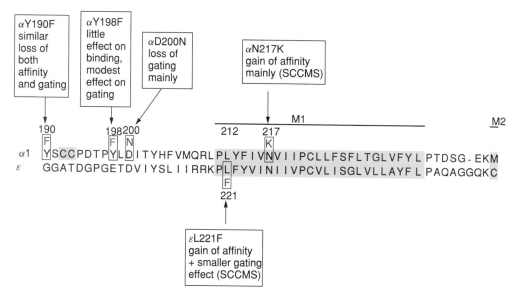

Figure 11.12. Aligned sequences of the human α1 and ε subunits, to show the position of some of the mutations that are discussed in the text. The (approximate) positions of the M1 region and the beginning of M2 are shown. See color insert.

(M1), as indicated in Fig. 11.12. It has no analog in the AChBP (which is only 210 amino acids long).

However, there is still some uncertainty about which residues in M1 are actually within the lipid membrane, and it could be that the boundary of M1 is actually closer to α217 than the conventional position shown in Fig. 11.12. It is possible (but not proved) that this mutation in M1 points to, and interacts with, the M2 region, which moves during gating (whereas M1 is not thought to move).

3.10 Mutations in the Epsilon Subunit

Much evidence has suggested that the ACh binding sites are close to subunit interfaces, and the AChBP structure seems to confirm this view. Most of the residues of the α subunit that were thought to be closely involved in binding do indeed appear in or near the binding site for HEPES (shown in Fig. 11.6a), which is on one side of the subunit-subunit interface. The other side of the interface (see Fig. 11.6b) is thought to be formed from the γ or ε subunit for one site, and from the δ subunit for the other.

In the AChBP, all the subunits are the same, and for the most part their interface

residues do not align in any very convincing way with the γ, ε, or δ subunits. The one exception is W53 (see Fig. 11.2a), which aligns with W55 in the mouse γ, ε, and δ subunits. The γ W55 residue contributes to the binding of the *Naja naja* α-toxin (99), but no binding-gating studies have been done on mutations at this position. An additional problem on this side of the interface is that stated by Brejc et al. (25): "the loop F region has an unusual conformation, but as it is relatively weakly resolved, its precise analysis is difficult."

εD175N. Position 175 in the mouse ε subunit was thought to be of interest because of studies by Czajkowski et al. on the homologous position in the δ subunit, δD180 (100, 101). They mutated all of the aspartate and glutamate residues to asparagine or glutamine, respectively, in a region known to be proximal to the binding site from crosslinking studies. The greatest effect seen was an 80-fold increase in the EC_{50} value for ACh produced by the δD180N mutation.

Akk et al. (92) constructed the homologous mouse mutation εD175N to test the effects on the rate constants of binding and gating. By use of high concentrations of ACh and P_{open} curves, the EC_{50} value of εD175N was found to

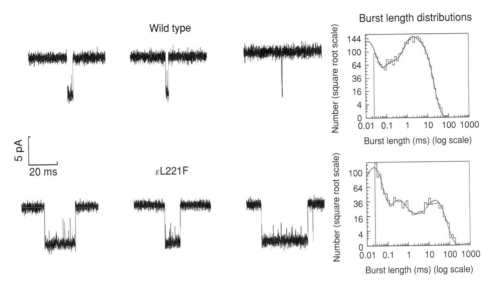

Figure 11.13. Typical activations of wild-type and εL221F human muscle nicotinic receptors by ACh (10–30 nM). It is obvious that the bursts of openings (activations) are longer on average for the mutant receptor. This is shown by the typical distributions of the durations of bursts shown on the right.

be increased 10-fold relative to that of the wild-type receptor. Through the use of a simple kinetic scheme that assumed equivalent binding sites, maximum likelihood fitting was used to demonstrate that the mutation decreased the efficacy E_2 nearly 80-fold. There were binding effects, too, *both* the association and dissociation rates being reduced more than 10-fold, so the equilibrium dissociation constant for binding was essentially unchanged. The authors suggested that the mutation affects the mobility of ACh around the binding site but *not* the affinity of the binding site for ACh.

The AChBP contains no obvious analog of εD175, and so casts no light on its role in binding.

εL221F. This is another gain-of-function SCCMS mutation that has been found in two unrelated families and that causes myasthenic symptoms (102). Ultrastructural studies revealed degenerated junctional folds and diffusely thickened end-plate basal lamina, as in other forms of SCCMS.

The εL221 residue is located near the *N*-terminal end of M1 and is therefore presumably very close to, if not actually within, the cell membrane (although, as discussed above,

there is some uncertainty about where M1 starts). Like other SCCMS mutants, it causes prolongation of the single-channel activations (bursts) produced by ACh and consequent slowing of the decay of MEPCs (103). The time constant for decay of MEPCs increased from 1.7 ms to approximately 15 ms.

Typical activations of wild-type and mutant receptor are shown in Fig. 11.13.

It is obvious that the bursts of openings (activations) are longer on average for the mutant receptor. This is shown by the typical distributions of the durations of bursts shown on the right. What is *not* obvious to the naked eye is whether the activations are longer because the individual openings are longer or because the channel reopens more often (in fact the latter is the predominant effect).

Hatton et al. (32) used maximum likelihood estimation of rate constants from single-channel records, by use of HJCFIT. Various mechanisms were tested, including those shown in Fig. 11.10, and variants of these that included either channel block or the extra shut state that Auerbach and his colleagues found to be necessary to fit some records. This postulates an extra shut state, connected directly to the doubly liganded open state that describes isomerization into a short-lived (about 1 ms) shut state that could (although this is merely semantic) be described as a very short lived desensitized state. A wide range of ACh concentrations was tested (30 nM to 30 μM for wild-type; 1 nM to 30 μM for ϵL221F).

When low concentrations were fitted separately, good fits could be obtained with the mechanism in scheme B (Fig. 11.10), with the assumption that the binding to site a was the same whether site b was occupied or not occupied. This assumption of independent binding to the two different sites was effected by using the constraints in Equations 11.2 and 11.3. The same was true when high concentration records were fitted alone, although somewhat different values were obtained for some rates. As might be expected as a result of this, it did not prove possible to fit both high and low concentrations simultaneously with this scheme. To achieve this it was necessary to either (1) relax the constraints in Equations 11.2 and 11.3, or (2) add the extra short-lived shut state (see above) to scheme B. At present it is not possible to tell which of these schemes is closer to the truth. The first option requires that the rates of association and dissociation for binding to site a depended on whether site b was occupied (and vice versa), and this implies that two binding sites, which are quite a long way apart, should be able to interact *before* the occurrence of the major conformation change that accompanies the opening of the channel. This is, physically, a somewhat unattractive idea, although by no means impossible. The second option also has an unattractive feature, in that it involves postulating a rather arbitrary shut state without any good physical or structural reason.

Fortunately, it is not necessary to decide between these options to obtain estimates of the main doubly liganded parameters, α_2 and β_2, and total dissociation rate from the doubly liganded shut channel.

Figures 11.14 and 11.15 show the results of using HJCFIT to estimate the rate constants in scheme B of Fig. 11.10 from low concentration single-channel recordings (see legends for details). Figure 11.16 shows a schematic representation of the mechanism, with examples of activations caused by occupancy of either

site alone, or of both sites. The numbers shown in Fig. 11.16 are estimates of the rate constants for typical fits.

The effect of the ϵL221F mutation was, surprisingly (given its location), mainly on binding. The total dissociation rate was decreased from 15,000 s^{-1} for wild-type to 4000 s^{-1} for the mutant receptor. A small gating effect was also seen. The doubly liganded opening rate β_2 was increased from 50,000 to 73,000 s^{-1}, whereas α_2 decreased from 1900 to 1200. This combination leads to a twofold increase in E_2, the efficacy of doubly liganded gating.

3.11 Macroscopic Currents

Once an appropriate mechanism has been chosen, and values for its rate constants have been estimated, then programs exist to calculate the time course of macroscopic currents under any specified conditions (e.g., SCALCS from http://www.ucl.ac.uk/Pharmacology/dc.html). Thus it can be seen whether the fitted mechanism is capable of predicting the time course of synaptic currents. The resolution of single-channel experiments is so much greater than that for measurements of macroscopic currents that it is not possible to go the other way. Figure 11.17 shows the calculated response to a 0.2-ms pulse of 1 mM ACh for both wild-type and the ϵL221F mutant. Despite the complexity of the mechanism (the curve has six exponential components), the decay phase is very close to a single exponential curve in both cases. This calculation predicts that the mutation will cause a sevenfold slowing of the decay of the synaptic current, much as observed for miniature end-plate currents measured on a biopsied muscle fiber from a patient with the ϵL221F mutation (102).

The relationship between single-channel currents and macroscopic currents has been considered both experimentally and theoretically by Wyllie et al. (104), who give the general relationship that relates the two sorts of measurement.

3.12 Summary of Effects of Mutations

Before considering the effect of mutations, it is first necessary to ask how accurately it is possible to distinguish binding from gating. The results of single-channel analysis of the sort

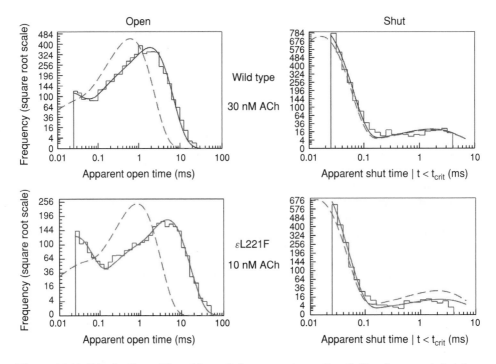

Figure 11.14. Distributions of (logarithms of) the apparent open time (left) and apparent shut time (right) for wild-type human receptors (top) and for mutant εL221F receptors (bottom). The histogram shows the experimental observations. The continuous lines were not fitted directly to the data in the histograms, but were calculated from the rate constants for the mechanism that was fitted (Fig. 11.10, scheme B, with the two sites constrained to be independent). The distributions were calculated with appropriate allowance for missed events (HJC distributions) (65, 66). The fact that they superimpose well on the histograms shows that the mechanism was a good description of the observations. The dashed lines show the distributions calculated from the fitted rate constants in the conventional way (45), without allowance for missed events, so they are our estimate of the true distributions of open and shut times (296). See color insert.

described are consistent with those of macroscopic jumps, and simulations (with HJCFIT) show that this method can clearly give good estimates, at least of the main doubly liganded rate constants. For the method to give misleading results the reaction mechanisms on which it is based (like those in Fig. 11.12) would have to be in some way seriously wrong.

Obviously, it is to be expected that mutations in residues that form part of the physical binding site will affect agonist binding, and most of those discussed above do so. They mostly affect gating, too, but this is not unreasonable. Because the act of binding has to be transmitted to other parts of the molecule to trigger the large conformation change that occurs on opening, it is perhaps not at all surprising that residues such as Y190, which al-

most certainly form part of the binding site itself, also influence the gating process. It is harder to explain why αY198F shows hardly any effect on binding, although according to the analogy with AChBP it is part of the binding site. On the other hand, the lack of effect of αD200N on binding, yet strong effect on gating, is consistent with its position some distance from the binding site, as predicted from the AChBP structure, but at the base of the C-loop where it could play a role in initiating the extended conformational change.

It is salutary, though, to notice that one of the purest binding effects is produced by the αG153S mutation, and it is unlikely that this residue is part of the binding site. Of course, it is usually easy to produce *post hoc* rationalizations of such results. A glycine to serine muta-

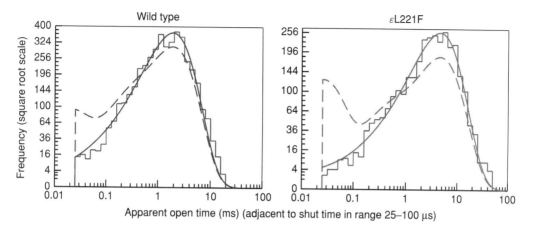

Figure 11.15. Conditional open time distributions. The histograms show the observed open times only for openings that were adjacent to short shut times (25–100 μs in duration). The continuous lines were not fitted to these histograms, but are the HJC conditional distributions of apparent open times calculated from the fitted rate constants (67). The fact that they superimpose well on the histograms shows that the fit predicts correctly the negative correlation between adjacent apparent open and shut times. The dashed lines show the calculated HJC distributions of *all* open times (as shown as solid lines in Fig. 11.14, left). See color insert.

tion would be expected to reduce the rotational freedom around the peptide bond in this position. This increase in rigidity as well as the introduction of a larger more polar side chain may disrupt the structure of the binding site without actually being in the binding site.

Even more extreme examples are provided by the αN217K and εL221F mutations. Both seem to have greater effects on binding than on gating, yet they are about 30 Å from the binding site, and close to, or buried in, the cell membrane. No sense can be made of this at present, beyond making the obvious statement that it appears that the binding site can be influenced at a distance.

The optimistic way to look at the outcome of the work that has gone into mutational studies of the binding site is that a reasonably

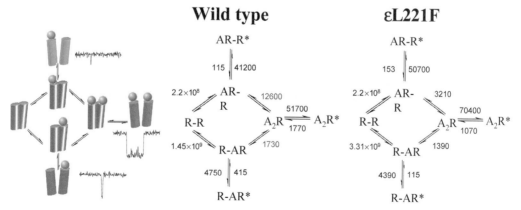

Figure 11.16. (Left) Schematic representation of reaction scheme B in Fig. 11.10, with examples of the sort of channel activations that are produced by the wild-type receptor with either one or both of the binding sites occupied by ACh. (Middle and right) Reaction scheme B (Fig. 11.10), with the results of a particular HJCFIT fitting marked on the arrows (the binding sites were assumed to be independent). See color insert.

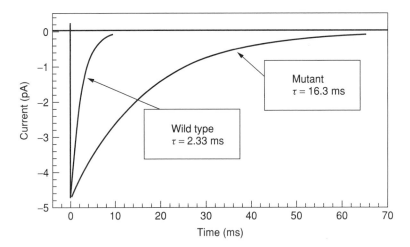

Figure 11.17. The predicted macroscopic current. The rate constants that have been fitted to results from equilibrium recordings (see Figs. 11.14–11.16) were used to calculate the macroscopic response to a 0.2-ms pulse of ACh (1 mM), as in Colquhoun and Hawkes (44). This calculation predicts that the mutation will cause a sevenfold slowing of the decay of the synaptic current, much as observed (102). See color insert.

accurate picture of the site appears to have emerged, even when the arguments have been somewhat irrational (like looking only at the EC_{50} value of agonists). However, as expected, this approach has also led to some wrong conclusions. The less optimistic conclusion that has to be drawn is that, even when results are analyzed by the best available methods, it is not possible to infer that a residue is part of the binding site, as exemplified by the cases of the αG153S, αN217K, and ϵL221F mutations. Perhaps now that we are just beginning to understand the physical movements of chains and residues that accompany the process by which binding is translated into a conformation change, it may become possible to explain, and even predict, some of the effects that are seen. For the moment, though, it is still necessary to engage in quite a lot of *post hoc* rationalization. As always, comparisons with enzymes are interesting. Functional assessment of enzymes has to be done at a much cruder level than can be done with channels. However, enzymes have the advantage that complete structures of many mutants have been determined. Despite this, the ability to rationalize the effect of mutations is still limited (105).

3.13 Antagonists of the Muscle-Type Nicotinic Receptor

There are two main types of antagonist action, competitive block and ion channel block. Every antagonist that has been tested (and indeed every agonist too) can produce block of the ion channel, apparently by plugging the pore. Despite this fact, the competitive action is much more important for their clinical effects than the channel block action (unlike some of the antagonists of autonomic neuronal nicotinic receptors). This is still true even when the antagonist has a higher affinity for the open channel pore than it has for the receptor binding site. For example, (+)-tubocurarine (TC) has an equilibrium constant of 0.34 μM for competition at the frog receptor (not voltage dependent), but an equilibrium constant of 0.12 μM for the open ion channel at -70 mV (or 0.02 μM at -120 mV) (55). Like most channel blockers it can block the channel only when it is open (or at least faster when it is open; the selectivity for the open channel is far from complete for many channel blockers). However, at the concentrations of TC that are used, the *rate* at which block develops is quite slow, and under physiological conditions the channels are open (and so susceptible to block) for a very short time only. This means that during normal neuromuscular transmission, the equilibrium level of channel block cannot be attained and the competitive action is far more important (106).

It is relatively easy to obtain equilibrium constants for the binding of antagonists, either by equilibrium binding methods (there is no binding-gating problem for antagonists), or from measurements of responses by the Schild method. Nevertheless, it is still unfortunately common for nothing but IC_{50} values to be given, and because these must inevitably depend on the nature of the tissue and on concentration of agonist, this is not very helpful. A "Cheng-Prusoff"-type correction cannot be applied to responses for an ion channel that show cooperativity. An additional complication is that, to understand the action of antagonists under physiological conditions, under which the application of agonist is very brief, we need the association and dissociation rate constants, not just their ratio, the equilibrium constant.

3.13.1 Rates of Action of Channel Blockers.
There is a lot of information available about the rate of action of channel blocking antagonist because this is relatively easy to determine by single-channel methods and also can often be obtained by macroscopic methods like voltage jump relaxations [e.g., Colquhoun et al. (55)] or noise analysis (107).

Which method is best will depend on the nature of the channel blocker. The average duration of a blockage may be 10 μs (for carbachol) up to 3 s (for TC). If the mean length of a blockage is in the range of tens of microseconds up to several milliseconds, then blockages produce obvious interruptions in the single-channel record that can be measured as a component of the shut time distribution. If, on the other hand, the blockages are very long, then it becomes hard to distinguish which shut times correspond to blockages, and macroscopic relaxation methods are better.

3.13.2 Rates of Action of Competitive Blockers.
Knowledge of the competitive mechanism of action is far older than knowledge of channel block. Competitive block was already

formulated quantitatively in 1937 (108), in contrast with channel-blocking agents, which were discovered only in the 1970s (109–111). It was soon discovered how to make robust estimates of the equilibrium constants for competitive antagonists from measurements of responses (51, 112), and soon after by direct measurement of ligand binding (113). Despite this, there have been very few measurements of the *rate constants* for the association and dissociation of competitive antagonists.

Attempts to measure the rate constants for competitive block have a long history. Probably the first was by Hill (114), in the famous study in which he gave the first derivation of the Langmuir binding equation. He got it wrong, as most studies since have done. The usual reason for getting it wrong is that the observed rates are limited by diffusion rather that by receptor association and dissociation. The fact that the molecules are bound tightly as they diffuse makes the diffusion far slower than simple calculations might predict (when TC is present in a synaptic cleft at a concentration equal to its equilibrium constant, so half the sites are occupied, there is only one free molecule for every 200 or so that are bound). Worse still, under these conditions, diffusion plus binding may slow association and dissociation rates to roughly similar extents (115), so the test that is often applied, that the ratio of the rate constants should be similar to the equilibrium constant determined independently, may give quite misleading agreement. Single-channel methods are almost totally useless for solving this problem because a competitive blocker produces long shut times in the record, even when they dissociate and associate rapidly (long periods are spent shuttling, possibly rapidly, between various shut states; unpublished observations). When, as is usual, we do not know how many channels are present in the patch, these long shut periods cannot be distinguished from those caused by, for example, desensitization. Channel activations are not changed by the competitive antagonist; they are just rarer. The problem, for nicotinic antagonists, seems to have been solved at last by Wenningmann and Dilger (116) and Demazumder and Dilger (117). They used concentration jump methods (exchange time 100–200 μs on bare pipette) on the embryonic ($\alpha_2\beta\gamma\delta$) mouse receptor in the BC3H1 cell line. They worked at -50 mV to minimize the effects of channel block. They measured occupancy of the receptors by TC and pancuronium by applying a pulse of ACh at various times after addition or removal of the antagonist.

They did not try to describe quantitatively the response to the test pulse, but rather assumed that it is sufficiently fast that at its peak there will have been essentially no change in occupancy by antagonist. Insofar as this is true, the peak response to the test pulse (as fraction of control) should be a reasonable measure of the fraction of channels not occupied by antagonist (and in this case the IC_{50} value should be close to the equilibrium constant for antagonist binding, K_B). They also assume that block of *one* site (the high affinity one mostly, if they differ) is sufficient to block the response.

For (+)-tubocurarine they found the association rate constant $k_{+B} = 1.2 \times 10^8 M^{-1} s^{-1}$, and the dissociation rate constant to be $k_{-B} = 5.9$ s^{-1}, so $\tau_{off} = 170$ ms. The ratio of these gives $K_B = 50$ nM, which was not greatly different from the estimate of K_B obtained from the IC_{50} (41 nM). Despite the size of the TC molecule, the association rate constant is quite fast, similar to that for ACh. At a clinical concentration that blocks, say, 90% of receptors (0.44 μM), there will be 59 associations per second (per free receptor), each occupancy lasting, on average, 170 ms, and the time constant for the association reaction will be about 17 ms.

For pancuronium, Wenningmann and Dilger (116) found an association rate constant of $2.7 \times 10^8 M^{-1} s^{-1}$, and the dissociation rate constant to be $k_{-B} = 2.1$ s^{-1}, so $\tau_{off} = 476$ ms. The ratio of these gives $K_B = 7.8$ nM, similar to the estimate of K_B obtained from the IC_{50} (5.5 nM). Again, the association rate constant is quite fast, similar to that for ACh. At a concentration that blocks 90% of receptors (0.09 μM) there will be 21 associations per second (per free receptor), each occupancy lasting, on average, 480 ms, and the time constant for the association reaction will be about 48 ms.

4   NEURONAL NICOTINIC RECEPTORS

The first striking difference between muscle and neuronal nicotinic receptors lies in the sheer number of genes that code for neuronal subunits, namely nine α ($\alpha2$-$\alpha10$) and three β subunits ($\beta2$–$\beta4$; see Table 11.1). The important questions then become first of all what combination of subunits *can* form functional receptors when expressed together and, even more important, which of these combinations are present in neuronal cells and matter to the physiological and pharmacological role of neuronal nicotinic receptors.

A first approach to determine the rules that govern assembly of the neuronal subunits is to find which combinations form functional receptors in heterologous expression systems.

- Some α subunits can form homomeric receptors, that is, give functional responses when expressed alone. Among mammalian subunits these are $\alpha7$ and $\alpha9$.
- Other α subunits ($\alpha2$, $\alpha3$, $\alpha4$, and $\alpha6$) can form a functional receptor only if expressed with a β subunit ($\beta2$ or $\beta4$); these heteromeric α/β receptors have a stoichiometry of 2 α to 3 β (23, 118, 119) and, by analogy with muscle receptors, are likely to have a topology of $\alpha\beta\alpha\beta\beta$.
- $\alpha10$ can participate in the formation of a receptor only if expressed with $\alpha9$ (120, 121) (the stoichiometry is as yet unknown).
- $\alpha5$ and $\beta3$ can form a receptor only if expressed together with a pair of "classical" α and β subunits (i.e., $\alpha2$-$\alpha4$ plus $\beta2$ or $\beta4$); these receptors are formed by two copies each of the "classical" α and β subunits plus one copy of $\alpha5$ or $\beta3$ [thus $\alpha5$ or $\beta3$ takes the place of a "classical" β subunit (23, 122, 123)].

These rules of assembly seem to hold broadly for both *Xenopus* oocytes and mammalian cell lines [for a review of the differences, see Sivilotti et al. (124)].

The information from the heterologous expression work must be viewed in the context of the actual pattern of expression of different subunits in central and peripheral neurons. *In situ* hybridization data show that $\alpha4$ and $\beta2$

are the most widespread and abundant of the "heteromeric-type" subunits in the CNS, whereas $\alpha3$ and $\beta4$ are the most important subunits in autonomic ganglia and chromaffin cells. This distinction is not absolute [for a recent review, see Sargent (125)]. Other subunits may have a more discrete localization, appearing in specific CNS areas: this is the case for $\alpha6$ [which is concentrated in catecholaminergic nuclei of the brain (126)] and $\beta3$ [present in substantia nigra, striatum, cerebellum, and retina (125, 127)]. Transcripts for the $\alpha7$ subunit are present both in the CNS and in the peripheral nervous system, whereas the expression of $\alpha9$ and $\alpha10$ is confined to the cochlea.

The restrictions in the subunit distribution go some way to offset the most important limitation of the data from heterologous expression experiments, that is, that they tell us only what the *minimal* subunit requirement is for a functional nicotinic receptor. Nevertheless, it is not uncommon for a native cell to express many of the known subunits. Does this mean that the receptors on the surface of such a neuron are a mosaic of "minimal" combination receptors or that more complex combinations are possible and/or favored in those circumstances? For instance, what will be the subunit composition of a nicotinic receptor on a cell that expresses the $\alpha3$, $\beta2$, and $\beta4$ subunits? Will the majority of receptors on the cell surface consist of all the subunits or will such a complex $\alpha3\beta2\beta4$ heteromer be rare relative to $\alpha3\beta2$ and $\alpha3\beta4$ receptors?

There is indeed evidence that subunit combinations even more complex than the ones outlined above do form in native neurons (the list that follows is by no means exhaustive). For instance, immunoprecipitation showed that a fraction of the major type of nicotinic receptor in ciliary ganglion neurons ($\alpha3\beta4\alpha5$) also contains $\beta2$ subunits (128). Evidence for the formation of nicotinic receptors containing four different subunits ($\alpha4$, $\beta2$, $\beta3$, and $\beta4$) also comes from work on CNS-type subunits by Forsayeth and Kobrin (127). Finally, in autonomic neurons, another major type of nicotinic receptor is based on $\alpha7$ subunits and consists of more than one pharmacological class of receptors: one class resembles the recombi-

nant homomeric $\alpha7$ receptor, whereas the other(s) may result from coassembly with other subunits (129–131).

This problem is recognized by the current nomenclature convention for neuronal nicotinic receptors, which allows referring to a receptor type as, for example, $\alpha7^*$, where the asterisk means that $\alpha7$ may not be the only subunit present.

4.1 General Properties of Neuronal Receptors

It is difficult to relate the numerous receptor subtypes that are possible on the basis of cloning and heterologous expression data to the functional properties of actual native receptors. Biophysical properties and the pharmacological tools available so far allow us to make only broad distinctions between classes of native receptor subtypes. They cannot distinguish all the native receptor subtypes that are in principle possible. One reason for this limitation is that the doubt remains that heterologous expression of three or more subunits may lead to the expression of heterogeneous receptor mosaics rather than the complex combination in a form pure enough to be characterized.

The problem is further compounded by the difficulty of carrying out kinetic studies on neuronal nicotinic receptors (mostly because of rundown): this means, for instance, that we do not know anything about agonist binding affinities or efficacy and how these change with subunit composition (or indeed with inherited mutations).

The first broad functional distinction is between homomeric-type and heteromeric-type receptors (the edges are blurred when we take into account the existence of $\alpha7^*$ receptors, although the distinction is clear for recombinant receptors). The most distinctive properties for homomeric receptors in this context are their sensitivity to the antagonist effects of α-bungarotoxin and methyllycaconitine (the latter selective at concentrations up to approximately 1 nM) and their faster desensitization and higher calcium permeability than those of heteromeric α/β receptors.

More detailed distinctions among the different types of recombinant heteromeric α/β combinations can be done on the basis of differences in agonist sensitivity. For instance, cytisine is both potent and efficacious as an agonist on receptors that contain the $\beta4$ subunit but is poorly efficacious on $\beta2$-containing receptors [Luetje and Patrick (132); see discussion of agonists below]. It is hard to predict what the sensitivity to cytisine would be for a complex heteromeric receptor (i.e., a receptor containing both $\beta2$ and $\beta4$ subunits). Help in this respect may come from the increasing availability of competitive antagonists selective for the different ligand-binding interfaces, such as conotoxins (see below). At present the most useful ones are α-conotoxin MII (specific for $\alpha3\beta2$ interfaces) and AuIB (specific for $\alpha3\beta4$ interfaces). It is reasonable to assume that a complex heteromeric receptor would be blocked, even if only one interface were occupied by the antagonist, but there is no proof of that at present.

It is hard to identify a distinctive and consistent contribution of subunits such as $\alpha5$ or $\beta3$ to nicotinic receptor properties. This may be because these subunits may not participate in the formation of the receptor binding site as either classical α- or β-type subunits. In the case of $\alpha5$, the most consistent change appears to be a speeding of desensitization. Increases in calcium permeability and single-channel conductance have also been reported, but the changes in agonist potency observed (in absolute or relative terms) depend on the specific combination expressed and are therefore not very useful as a diagnostic tool for identifying subunit combinations in native receptors [for a review, see Sivilotti et al. (124)]. Similarly, recombinant $\beta3$-containing receptors have higher single-channel conductance, but differ little in other receptor properties (Beato, Boorman, and Groot-Kormelink, personal communication). It is worth noting that single-channel conductance, calcium permeability, and speed of apparent desensitization are not distinctive enough to be useful in receptor classification in most cases (see below).

Finally, coexpressing $\alpha10$ with $\alpha9$ in oocytes is reported to increase receptor expression, speed of desensitization, and sensitivity to external calcium (120) and to change the receptor sensitivity to α-bungarotoxin and (+)-tubocurarine (121).

4.2 Biophysical Properties

4.2.1 Calcium. Calcium has multiple effects on nicotinic receptors. Not only is it to some extent permeant, but it affects the receptor single-channel conductance and modulates the agonist response of neuronal-type receptors.

All nicotinic receptors are somewhat calcium permeable: the most permeable are neuronal homomeric receptors ($\alpha7$, $\alpha9$) and the least permeable, embryonic muscle receptors. It must be noted that the measurement of relative calcium permeability by the simplest technique (reversal potential shift induced by changes in extracellular calcium concentrations) is error-prone for neuronal nicotinic receptors because their extreme inward rectification makes it difficult to measure reversal potentials accurately. A further technical difficulty (for recombinant receptors) arises from the presence in *Xenopus* oocytes of a calcium-dependent chloride conductance that has to be suppressed or minimized by either intracellular calcium chelation or chloride depletion. Some of these problems can be overcome by expressing the receptors in mammalian cell lines and using ratiometric measurements of intracellular calcium and coupled with whole-cell recording, to obtain a measure of what proportion of the nicotinic current is carried by calcium (a measure that also has the advantage of being physiologically more relevant).

Bearing in mind these cautions, the permeability to calcium is around 1/10 to 1/5 of that to sodium or cesium ions for mammalian muscle receptors of the embryonic type (native or recombinant), and is much higher (0.5–0.9) for the adult form of the receptor (133–135). This is in agreement with estimates that calcium ions carry approximately 2% and 4% of the total nicotinic current through embryonic and adult muscle receptors, respectively [at physiological calcium concentrations and holding potentials (136–138)].

The calcium permeability of ganglion-type nicotinic receptors is reported to be similar to that of adult muscle in experiments on superior cervical ganglion, intracardiac ganglia, and chromaffin cells, with values for calcium permeability (relative to sodium or cesium) between 0.5 and 1 and fractional current measurement between 2.5% and 4.7% (133, 137, 139–142). This is in broad agreement with recombinant work on heteromeric receptors or native receptors likely to be heteromeric (134, 138, 143). Some reversal potential method studies suggest much higher calcium permeability (vs. adult muscle receptors) for superior cervical ganglion neurons (144) or heteromeric recombinant receptors (145), particularly if the α5 subunit is expressed (146).

Nicotinic receptors formed by α7 or α9 are by far the most calcium permeant. Thus for recombinant or native α7-like receptors reported values for relative calcium permeability range from 6 to 20 (147–150), with fractional current carried by calcium as high as 20% (151); for work on the M2 determinants of this high calcium permeability, see Bertrand et al. (147). This would mean that α7 receptors are almost as calcium permeable as the NMDA-type of glutamate receptor. Equally high calcium permeability was reported for α9-type receptors (152, 153), expressed alone or with α10 (121).

Finally, what matters most from the physiological point of view is that direct calcium entry through nicotinic receptors can be sufficient to act as a postsynaptic signal, for instance, activating calcium-dependent SK potassium channels in outer hair cells of the cochlea (154) or contributing to apamin-sensitive hyperpolarization in rat otic ganglion (155).

Note that increasing the concentration of extracellular calcium reduces the single-channel conductance of both muscle and neuronal nicotinic receptors (137, 143, 156–159).

Finally, changes in extracellular calcium (in the low millimolar range) can modulate nicotinic responses. Increases in calcium concentration strongly enhance macroscopic responses of either native or recombinant heteromeric nicotinic receptors to low ACh concentrations, decreasing the EC_{50} value of ACh and increasing the Hill slope of the curve (133, 140, 160, 161). This effect is not seen with muscle embryonic channels (133). In native α7* receptors the modulation has been reported to be biphasic, with potentiation at submillimolar calcium concentrations and depression at higher concentrations (162). The sequence determinants for this effect have been investigated for chick recombinant α7/5HT3 chimeric receptors by Galzi et al. (163), who have identified residues α7 161–172 as particularly important: in the AChBP these residues are on the minus face, at the end of loop 9, which is near the extracellular end of the pore. Le Novère et al. (164) proposed, on the basis of their modeling of the α7 subunit on the AChBP, that the binding site for calcium is formed at the subunit interface by residues belonging to different subunits. These residues include some identified by Galzi et al. (163), such as E44 and E172, but also D43 and D41. Of these, E44 and D43 would be on the (+) side and E172 and D41 on the (−) side.

4.2.2 Single-Channel Conductance. The main structure determinants for the single-channel conductance of neuronal nicotinic receptors are likely to be (as for muscle receptors) the residues in positions −4′, −1′, and 20′ of M2 (this numbering system for M2 residues is defined in Fig. 11.3) [for review, see Buisson et al. (161)]. The total charge on each of these rings of charges has an important effect on conductance. It is worth noting that the M2 sequence is well conserved across neuronal nicotinic subunits. In particular, the residue in −4′ is negatively charged in all except the α5 and α9 subunits (which have a neutral residue in this position) and −1′ is always negatively charged. A difference is seen in the 20′ residue, which is negatively charged for all subunits except β2 and β4. This may be the reason for the conductance increase observed in channels containing α5 or β3 (122, 165; and Beato, Boorman, and Groot-Kormelink, personal communication): if these subunits replace a classical β subunit, they will produce a −2 change in the charge on the external ring.

Although single-channel conductance is a useful diagnostic criterion for the classification of other ionotropic receptors (for instance native NMDA receptors), this does not apply to neuronal nicotinic channels. Conductance levels are not very distinctive, given that even recombinant receptors that should in principle be homogeneous have multiple conductance levels, and these levels overlap considerably for different combinations. In addition, the same combination expressed in different heterologous system can give rise to different conductances (166) and conductances are exquisitely sensitive to divalent ions concentrations, making it difficult to compare data that have not been obtained in identical recording solutions. Characterization of channel conductances is also hindered by the phenomenon of "rundown" in the excised patch configuration (i.e., the disappearance of channel activity), which appears to be mostly agonist independent and may be triggered by patch excision. These factors make it very difficult to use single-channel conductance (and worse still kinetics) as a criterion for the identification of specific subunit combinations in native receptors.

4.2.3 Inward Rectification. One striking property of neuronal nicotinic channels is the extreme inward rectification of the macroscopic current-voltage relation. In contrast muscle-type receptors show only modest inward rectification that can be accounted for almost entirely by the fact that the main channel shutting rate becomes slower as the membrane is hyperpolarized (about e-fold per 60–100 mV) (47). In neuronal receptors, rectification is so extreme that there is hardly any whole-cell current at all at potentials between -10 and $+60$ mV: in functional terms, neuronal nicotinic channels could legitimately be described as "discordance detectors" because they pass little current at depolarized potentials. Inward rectification has been reported for a variety of native and recombinant neuronal nicotinic receptors [with the notable exception of $\alpha9^*$ receptors from the cochlea (153)]. A clue for the understanding of its cause came from the absence of rectification in excised patches and the linear current-voltage relationship of the single-channel conductance (when the artificial intracellular medium does not contain magnesium ions). This suggested that rectification is caused either by channel block by intracellular components (that are not present in artificial intracellular solutions) and/or by voltage dependence of the channel kinetics (167). We now know that both are important: the major role is played by channel block by micromolar concentration of the intracellular polyamine spermine (168), with a minor contribution by intracellular magnesium ions (167, 169, 170) and by the

voltage dependence of the channel P_{open}. Although this work was carried out on native neuronal nicotinic receptors of autonomic ganglia (and on recombinant $\alpha4\beta2$ and $\alpha3\beta4$ channels), it is likely that similar mechanisms underlie the rectification of other neuronal nicotinic receptors (171).

The M2 determinants for inward rectification have been investigated in recombinant chick $\alpha7$ receptors by Forster and Bertrand (172).

4.3 Native Neuronal Nicotinic Receptors: Physiological Role

Neuronal nicotinic receptors are found on a variety of classes of neurons, both in the peripheral and in the central nervous system, and on nonneuronal cells [for a review of the latter, see Wessler et al. (173)].

4.3.1 Peripheral Nervous System. In the peripheral nervous system, these receptors mediate fast synaptic transmission at autonomic ganglia and at efferent cholinergic synapses onto cochlear outer hair cells.

4.3.1.1 Autonomic Ganglia

$\alpha3^*$ Receptors. The pattern of subunit expression, immunoprecipitation, and antisense data all agree in recognizing a major role of $\alpha3\beta4^*$-type receptors in autonomic ganglion neurons, including chromaffin cells [128, 174–176; but see Skok et al. (177) and Klimaschewski et al. (178)]. In chick ciliary ganglia these receptors also contain the $\alpha5$ subunit: additionally, a significant fraction of them contains both the $\alpha5$ and the $\beta2$ subunit [reviewed in Berg et al. (179)].

This class of receptor has traditionally been thought to be the major or indeed the only type of receptor involved in synaptic transmission in ganglia because of the resistance of synaptic transmission to α-bungarotoxin (180) and the subsynaptic location of these receptors [reviewed in Temburni et al. (181)]. These data were confirmed by the profound autonomic defect observed in mice, in which the $\alpha3$ had been knocked out (182). Similar problems were observed in mice lacking *both* the $\beta2$ and the $\beta4$ subunits, whereas mice lacking only one of these β subunits had a relatively normal autonomic phenotype (183). It is worth noting that the range of subunit expressed in ganglia [and the variability of the range from one neuron to the other (184)] offers scope for considerable heterogeneity within this class of receptors [see, for instance, Britt and Brenner (185)].

α7* Receptors. It is common for autonomic ganglion neurons to express α7-type receptors, too. In embryonic chick ciliary ganglion, these receptors have a location distinct from α3* receptors, in that they are specifically targeted to spine mats on the soma of the postsynaptic neuron [for a review, see Berg et al. (179)]. It is not clear to what extent postsynaptic densities are localized on these spines, but there is good evidence that blocking α7 receptors damages the reliability with which most neurons (two-thirds of the neurons in E13-E14 chick ciliary ganglion) follow frequencies of presynaptic stimulation equal to or greater than 1 Hz (186). The situation for adult *mammalian* ganglion neurons is less clear, but these cells do express both classical α7 receptors (i.e., homomeric) and α7* receptors, which differ in their slower desensitization and quicker recovery from α-bungarotoxin block [for intracardiac and superior cervical ganglia, see Cuevas and Berg (130) and Cuevas et al. (131)]. Transcripts for the α7 subunit are also present in chromaffin cells (187), but functional studies have so far failed to demonstrate functional α-bungarotoxin-sensitive receptors on these cells (188–191), although binding sites for α-bungarotoxin can be revealed by autoradiography (192). Although α7 null mice are viable and do not display gross phenotypic defects (193), they do have a subtle autonomic deficit, manifest as an impairment of baroreflex responses. This impairment is limited to responses, such as tachycardia, mediated by the sympathetic nervous system (194).

Additionally, functional nicotinic receptors have been found on the preganglionic terminals of embryonic chick ganglia (195) and on the axon terminals of postganglionic superior cervical ganglion neurons in culture, where it was found that receptors on terminals differ in agonist sensitivity from the somatic ones (196).

4.3.1.2 Cochlea. Outer hair cells receive cholinergic input from the olivary complex. Synaptic transmission here is mediated by α9

(or possibly α9α10) nicotinic receptors: their high calcium permeability and their coupling to a calcium-dependent potassium conductance mean that the cholinergic inward current is swamped by the outward potassium current [see Ashmore (197); Fuchs and Murrow (198)]. Both the native receptors on outer hair cells and recombinant α9 and α9α10 nicotinic receptors are unusual, in that they are insensitive to nicotine. These receptors are blocked by the glycine receptor antagonist strychnine (likely to be competitive; a reported IC_{50} value around 20 nM) and by the GABA receptor antagonist bicuculline, albeit less potently [IC_{50} value around 1 μM (120, 199, 200)].

4.3.2 Central Nervous System. Although nicotinic receptors are much less abundant in the CNS than acetylcholine muscarinic receptors, they are nevertheless widespread and found (for instance) in cortical areas, including the hippocampus, in the thalamus, basal ganglia, cerebellum, and retina. Particularly high levels of high affinity receptor binding are found in the habenula and interpeduncular nucleus [reviewed by Sargent (125)].

Despite the widespread presence of neuronal nicotinic receptors, for a long time their only known physiological role in the CNS was in mediating the excitation of Renshaw cells by motoneuron axon collaterals (201). The situation has now changed, particularly with respect to the demonstration of a synaptic role, especially for α7-like receptors.

4.3.2.1 Receptor Types. In many areas of the CNS, the most represented type of heteromeric nicotinic receptor is likely to be α4β2 or α4β2*: these receptors correspond to high affinity binding sites for [³H]nicotine. This is likely to be the synaptic receptor on Renshaw cells, which are immunopositive for the α4 and β2, but not the α7 subunit, in agreement with the insensitivity of their synaptic responses to methyllycaconitine (202).

Judging from the distribution of the different subunits, other receptor types may be important in discrete locations of the CNS. A case in point is that of the α6 subunit, which is represented in basal ganglia and catecholaminergic neurons (126, 203–205). Combinations of the α3β4 type are also thought to be important in discrete CNS areas (i.e., in the habenula and interpeduncular nucleus). Homomeric receptors of the $\alpha7$ type are also present in the CNS and correspond to the binding sites for $[^{125}I]$-α-bungarotoxin.

Identifying nicotinic subunit combinations in the central nervous system is an area of intense research, which makes use of all the tools available, such as biophysical and pharmacological characterization of functional receptor responses, *in situ* hybridization and single-cell RT-PCR, and both antisense and transgenic techniques.

In particular, knockout transgenic mice lacking the $\alpha3$ (182), $\alpha4$ (206), $\alpha7$ (193), $\alpha9$ (207), $\beta2$ (208), and $\beta4$ subunits (183) have been bred. In addition, mice bearing the L9'T mutation in the $\alpha4$ or the $\alpha7$ subunits have been obtained (209, 210): in both strains homozygous mice die soon after birth. A discussion of the implications of these data for our understanding of the diversity and the physiological and pharmacological roles of nicotinic receptors can be found in Cordero-Erausquin et al. (211) and Zoli et al. (212).

We shall focus our review of receptor types to a specific area, the mammalian hippocampus and to electrophysiological evidence.

The first descriptions of nicotinic responses in hippocampus refer to agonist responses recorded in long-term primary cultures (213, 214). Three main types of agonist responses were described, broadly corresponding to $\alpha7$, $\alpha4\beta2$, and $\alpha3\beta4$-like responses [reviewed in Albuquerque et al. (215)].

Of these the pure $\alpha7$-type (type IA) is by far the most common. Type I current was described by other groups in acute slices of rat hippocampus (CA1 or dentate gyrus) in response to pressure-applied ACh (216–218). It is a fast-desensitizing response, distinctive in its marked sensitivity to the antagonists α-bungarotoxin (10 nM), MLA (1 nM), and α-conotoxin ImI and its resistance to mecamylamine and dihydro-β-erythroidine (both 1 μM). Another property that links these receptors to recombinant $\alpha7$ homomers is the sensitivity to the agonist effect of choline (1 mM) [for the choline selectivity for $\alpha7$, see Papke et al. (219)].

In acute hippocampal slices, this current is abolished by 100 nM α-bungarotoxin. It is

rarely if at all present on principal neurons, but common in interneurons [50% of all interneurons have a pure type IA (216–218)] and may be especially important in interneurons that control input onto the pyramidal cell dendrites. Note that fast $\alpha7$-like responses have been described in pyramidal cells in culture (215) or in mouse CA1 pyramidal neurons (acute slices) (220). Other properties of this current include sensitivity to other classical $\alpha7$ antagonists such as MLA [2 nM (218); 10 nM (216, 217)] and α-conotoxin ImI (200–500 nM), and resistance to mecamylamine (0.5–1 μM) and dihydro-β-erythroidine (100–150 nM). The $\alpha7$ involvement is confirmed by the disappearance of type I currents in $\alpha7$ knockout mice (193).

In hippocampal cultures, other less common types of nicotinic responses were the slower α-bungarotoxin resistant types II and III. Of these, type II is the most common (10% of neurons in primary culture) and may correspond to $\alpha4\beta2^*$. It is very sensitive to dihydro-β-erythroidine (10 nM) and is decreased by high concentrations of methyllycaconitine (100 nM). It is also blocked by mecamylamine (1 μM). The most rare and slow responses were termed type III (2% of hippocampal neurons in culture): these may correspond to $\alpha3\beta4^*$ receptors, are sensitive to 1 μM mecamylamine or 20 μM tubocurarine, and resistant to 100 nM methyllycaconitine. This classification is likely to hold outside the hippocampus as well, as shown by results in normal and $\beta2$-knockout mice [see, for instance, Zoli et al. (212), who distinguish a fourth type of nicotinic response, similar to type III, but with faster desensitization at high nicotine doses and different properties in equilibrium binding assays with agonists].

In acute slices, a mixed response, which consists of fast $\alpha7$-like and slow components, has been described in interneurons of the *stratum oriens* (36% of all interneurons). The slow response is sensitive to mecamylamine (1 μM) and to a certain extent to dihydro-β-erythroidine (100 nM), but resistant to the $\alpha3\beta2$ antagonist α-conotoxin MII (200 nM). In its moderate sensitivity to dihydro-β-erythroidine, this response may resemble more $\alpha3\beta4$-like responses (type III) than $\alpha4\beta2$ type II responses (218), although it has been suggested that sensitivity to all antagonists is lower in slices than in cultured dissociated neurons (221): the apparent difference could be simply attributed to access problems, given that no K_d data are available [for a review of the difficulty in comparing IC_{50} values, see Sivilotti et al. (124)]. Note that it is still controversial whether *pyramidal* cells do have nicotinic responses: there are reports of both fast $\alpha7$-type responses (215, 220) and slow responses (221).

Recent work shows that of these receptors, the $\alpha7$ type seems to be the most important for synaptic transmission in the hippocampus. $\alpha7$ immunoreactivity is present at nearly all the synapses in CA1 *stratum radiatum* (including GABAergic and glutamatergic ones) (222) and $\alpha7$-like receptors mediate fast synaptic transmission onto CA1 interneurons (223, 224). The identification was based on the sensitivity of synaptic currents (both evoked and spontaneous) to the antagonists methyllycaconitine (50–150 nM) and α-bungarotoxin (100 nM) and to desensitizing concentrations of the selective $\alpha7$ agonist choline. Such cholinergic currents were relatively rare, being found only in 17/125 stratum radiatum interneurons, but represented 10% of the total evoked synaptic current in these cells. It is possible that the rarity of these currents is a result of the difficulty in recruiting cholinergic afferents, given that in most interneurons $\alpha7$ responses to ACh application could be detected [see also Buhler and Dunwiddie (225)].

Nicotinic synaptic currents of the $\alpha7$ type were also detected in pyramidal cells (226) in acute or organotypic hippocampal slices. Here the contribution of nicotinic currents to the total postsynaptic current was very modest (less than 3% of total).

Although it is natural that one should look for nicotinic fast synaptic transmission in the central nervous system, in analogy to the peripheral role of these receptors, central nicotinic receptors are often present at a presynaptic level [reviewed by McGehee and Role (227); Wonnacott (228); Kaiser and Wonnacott (229)]. Thus in many brain areas pharmacological activation of these channels produces an increase in spontaneous release of a variety of transmitters, including catecholamines, GABA, 5-HT, glutamate, and ACh itself. This may be a result of either to direct calcium en-

try through nicotinic channels located on presynaptic terminals or to the firing of sodium-dependent action potentials (which eventually reach terminals) by depolarization produced by preterminal nicotinic receptors. These two mechanisms can be distinguished on the basis of the tetrodotoxin sensitivity of the latter [see, for instance, Léna et al. (230)].

In most preparations direct electrophysiological recording from presynaptic structures is not possible and hence characterization of presynaptic (or preterminal) nicotinic receptors has to rely on neurochemical measurements of transmitter release, postsynaptic recording of the effects of such release, or presynaptic intracellular calcium measurement. There is evidence that both $\alpha7$ and non-$\alpha7$ receptor types can play a presynaptic role. For instance, a predominant $\alpha7$ involvement was reported by McGehee et al. (231) for glutamate release in chick habenula/interpeduncular nucleus cocultures and by Gray et al. (232) for mossy fiber terminals in rat hippocampus slices. Nevertheless, the type of nicotinic receptor involved depends both on the brain region and on the nature of the terminal (i.e., on the transmitter released). A particularly striking example is that of the rat dorsal raphe nucleus, where the nicotinic receptors involved in noradrenaline release are sensitive to 100 nM methyllycaconitine (i.e., $\alpha7$-like), whereas those involved in 5-HT release are not (233). Indeed, depending on the pattern of nicotine application, *both* $\alpha7$ and non-$\alpha7$ receptors may enhance glutamate release in rat hippocampal microisland cultures (234). Nicotinic enhancement of GABA release in hippocampus is likely to be mediated by both $\alpha7$ and non-$\alpha7$ receptors (221, 235).

Catecholaminergic terminals bear non-$\alpha7$ nicotinic receptors, which are thought to be of either the $\beta2^*$ or the $\beta4^*$ type in the case of dopaminergic and noradrenergic terminals, respectively. A thorough review of the pharmacology of these receptors can be found in Kaiser and Wonnacott (229).

Because nicotinic receptors can both modulate the release of a variety of transmitters and directly depolarize postsynaptic neurons, the functional consequences of their pharmacological activation can be both subtle and widespread. It has been reported that activation of presynaptic nicotinic receptors results in increases in the amplitude of submaximal glutamatergic synaptic currents (234, 236–238) or in the increase in the non-NMDA component and decrease in the NMDA component (239). Important modulatory effects on synaptic plasticity have also been described (220, 238).

The widespread presence of nicotinic receptors in the central nervous system coupled with the relative rarity of a classical synaptic role means that it is still difficult to describe a clear physiological role for these receptors. This is particularly true for receptors in presynaptic locations, for which the level and temporal pattern of exposure to the neurotransmitter are unknown. Approaches that are casting light on this problem include mouse knockout models and transmitter depletion by blockers of vesicular transport processes such as vesamicol: the combination of these techniques has recently shown that normal evoked dopamine release in striatal slices is strongly dependent on endogenous cholinergic mechanisms that involve the activation of $\beta2$-containing nicotinic receptors (240).

Central neuronal nicotinic receptors are the target for the pharmacological actions of nicotine in tobacco. Plasma levels of nicotine in smokers go through short-lasting peaks superimposed onto a sustained lower concentration that rises steadily through the day. This is likely to result in a complex pattern of activation and desensitization, accompanied by long-term regulatory effects on the number of nicotinic receptors [reviewed in Dani et al. (241); Hyman et al. (242)]. Knockout models suggest that it is the $\beta2$-containing receptor type that has a primary role in sustaining nicotine self-administration in mice (243).

4.4   Autosomal-Dominant Nocturnal Frontal Lobe Epilepsy: A Central Nicotinic Defect

Clues on the physiological role of neuronal nicotinic receptors come from the identification of a form of human epilepsy that can be caused by mutations in either the $\alpha4$ or the $\beta2$ nicotinic subunits. This rare syndrome, autosomal dominant nocturnal frontal lobe epilepsy (ADNFLE), was the first idiopathic epilepsy to be identified as a monogenic disorder (244), and consists of seizures that occur during light

sleep. Although these can be quite distinctive, it is not uncommon for the seizures to be misdiagnosed as nightmares or other sleep-related disturbances [for a review, see Sutor and Zolles (245)]. As other Mendelian forms of epilepsy, this syndrome is very heterogeneous; thus it is unclear whether it is attributable exclusively to defects in central neuronal nicotinic receptors. In at least one family this syndrome is linked to 15q24, a chromosomal locus that does not contain either $\alpha4$ or $\beta2$ genes, but rather a cluster of *peripheral* neuronal nicotinic subunits (i.e., $\alpha3$, $\beta4$, and $\alpha5$) (246), which have a restricted expression pattern in the CNS. Furthermore, the autosomal dominant inheritance pattern is obscured by incomplete penetrance (estimated at 75%) and the actual symptoms are extremely variable from one patient to the other *within* the same family (245).

Five mutations have been identified so far, all either in the pore-lining domain, M2, or in the short linker that connects it with M3. The characterization of the functional consequences of the mutations (by heterologous expression in *Xenopus* oocytes) has been carried out through the use of mostly macroscopic techniques because of the difficulty in obtaining excised patch recordings for neuronal receptors. Again, intrinsic technical difficulties mean that (with few exceptions) the electrophysiological data are from "all mutant" receptors (i.e., *not* the sort of receptor that patients who are heterozygous would have). Furthermore, the data come from expression of $\alpha4\beta2$ receptors and may not be entirely predictive of the behavior of native receptors, which may also contain other nicotinic subunits. Finally, it must also be said that there is considerable divergence in the reported effects of the same mutation between one lab and the other and between the effects of different mutations that in humans produce similar phenotypes.

Three mutations have been reported for the $\alpha4$ subunit, all in the M2 region. Two are missense mutations [at 6' S248F (244)] or at 10' [S252L; this mutation has not been characterized electrophysiologically (247)], whereas one is the insertion of a Leu after 17' (248). The two ADNFLE mutations known for the $\beta2$ subunit are at the same residue (22') of the M2-M3 linker, which is a region likely to be important in receptor gating. These are V287L (249) and V287M (250).

Studies that used recombinant expression of mutant $\alpha4\beta2$ receptors described a variety of effects for these mutations, encompassing increases in ACh EC_{50} values for S248F and V287M (145, 250, 251) and decreases for the 776ins3 Leu insertion (251). A reduction in the maximum current elicited by ACh was reported for S248F, but not for the 776ins3 Leu insertion mutant (251, 252). A potential gain-of-function effect was seen for the $\alpha4$ mutations and consisted of a "wind-up" or increase in the response to low agonist concentrations upon repeated application (145, 252).

It is difficult to interpret the varied biophysical effects of the epilepsy mutations on recombinant receptors *in vitro* with the pathogenesis of the actual disease. Is this disease attributable to loss or gain of nicotinic receptor function? Which is the most important of these changes? The uncertainty is inevitable, given that at present we cannot tell whether the major physiological role of central nicotinic receptors is postsynaptic or presynaptic. If the postsynaptic role is the most important, we must try to argue on the effect of mutations on the receptor response to brief saturating transients of ACh. On the other hand, presynaptic receptors may be activated by concentrations of transmitter that are lower and in conditions closer to equilibrium. Furthermore, it is entirely possible that the same central nicotinic receptors play both roles to different extents in different CNS areas.

Finally, it must also be borne in mind the case of muscle receptors, both gain of function and loss of function mutations, can produce congenital myasthenia.

4.5 Agonists for Neuronal Nicotinic Receptors

The different combinations of recombinant neuronal nicotinic receptors differ in their sensitivity to agonists. The list of nicotinic agonists is long and consists of both natural compounds such as choline, nicotine, cytisine, lobeline, epibatidine, anabaseine, and synthetic compounds such as tetramethylammo-nium (TMA), 1,1-dimethyl-4-phenylpiperazinium (DMPP), and carbachol, to name but a few.

We next consider the different combinations in groups, that is homomeric $\alpha7$ receptors on one hand and heteromers on the other (i.e., central type $\alpha4\beta2$ receptors and ganglionic type $\alpha3\beta4$ receptors), highlighting the compounds that are most useful for receptor classification purposes, irrespective of species differences by use of functional assays. It is worth noting that classical pharmacology shows that use of agonists for receptor classification is fraught with problems. This is because the functional EC_{50} value of an agonist on the same receptor will depend on several experimental variables, and especially on the rate of application [discussed in Sivilotti et al. (124)]. The usefulness of the technique is greater if agonist potency ranks are determined, especially at their low concentration limit [to reduce the confounding effects of agonist self-block and desensitization; see Covernton et al. (253)]. Particular caution is also needed because our knowledge of the relative potency of agonists comes from recombinant expression of pure homomeric or "pair" heteromeric receptors. We do not know how agonist potency would be changed in a complex heteromeric receptor (i.e., one that contained two different interface binding sites).

In the case of neuronal nicotinic receptors, the choice to use agonists is dictated by the paucity of suitably selective competitive antagonists, a situation that may change with the increased availability of an increased range of conotoxins (see below).

Broadly speaking, choline is the most useful agonist for $\alpha7$ receptors (219, 254, 255) because it is a full agonist on these receptors (EC_{50} values of 1.6 versus 0.13 mM for ACh) (255) but is ineffective or a very poor partial agonist on $\alpha3\beta4$- and $\alpha4\beta2$-type receptors.

On the other hand, cytisine is both efficacious and potent as an agonist on heteromers containing the $\beta4$ subunits (i.e., ganglion-type receptors), but is only a partial agonist on $\beta2$ heteromers (132, 253, 256). On the latter receptor type, the maximum current to cytisine is no more than 25% of that produced by ACh (the precise value depends on the α subunit, the species from which the clones are derived and the nature of the functional assay; the range of values reported is 1–25%).

Most other agonists do not show such a great level of selectivity as the ones discussed above. It is nevertheless worth mentioning the nonselective agonists 1,1-dimethyl-4-phe-nylpiperazinium (DMPP), for its widespread use, and epibatidine (from the skin of an Ec-uadorean tree frog, *Epidobates tricolor*), for its extreme potency, orders of magnitude greater than that of other agonists, especially for the heteromeric receptors.

4.6 Antagonists

There is a great paucity of data for the affinity of competitive antagonists of neuronal nico-tinic receptors. Functional studies usually re-port antagonist IC_{50} values: an IC_{50} value may be (marginally) quicker to obtain than dissoci-ation constants in a Schild-type design, but an IC_{50} value depends on the agonist concentra-tion used. Given that comparison of IC_{50} val-ues across different preparations is difficult, differences in receptor types can be argued only if relatively large IC_{50} differences are ob-served and if the agonist concentration in-volved is similar. This consideration is impor-tant, especially if synaptic responses are studied and compared to agonist applications, given that synaptic currents are likely to be produced by very short (submillisecond) in-creases in ACh concentration to very high lev-els. Furthermore, IC_{50} experiments do not tell us anything about the actual mechanism of action of the antagonist (i.e., competitive vs. open channel block; see the discussion of an-tagonists of muscle receptors). This means that much potentially valuable information on the binding site is lacking. This is true not only for functional studies, but also for binding studies. In most cases, binding assays for neu-ronal nicotinic antagonists use displacement of a labeled *agonist* by the antagonist; this is because there are no selective antagonists of sufficiently high affinity that can be used in such work (S. Wonnacott, personal communi-cation). The resulting K_i value is the equiva-lent of an IC_{50} (i.e., not a true dissociation constant). The exception is $\alpha 7$-type receptors, for which labeled α-bungarotoxin and methyl-lycaconitine can be used (257).

Because of the limitations in the data avail-able in the literature, we direct our focus on the antagonists that are most useful for recep-tor classification. Traditionally, the main group is that of competitive antagonists that selectively block $\alpha 7$ and other homomeric re-ceptors, that is, α-bungarotoxin, methyllyca-conitine (at low concentrations), and α-cono-toxin ImI. For a recent review of toxin antagonists of neuronal nicotinic receptors, see McIntosh (258).

4.6.1 Antagonists for α_7-Type Receptors. α-Bungarotoxin is one of the components of the venom of the banded krait, *Bungarus mul-ticinctus* (74 amino acids, MW 8000). The af-finity of this toxin for the homomeric $\alpha 7$ recep-tor appears to be high in binding assays (1–2 nM) (257), but considerably lower than that for muscle-type receptors. In practice, α-bun-garotoxin is used at concentrations between 10 and 100 nM to block $\alpha 7$-type receptors. This block is nearly irreversible for "pure" $\alpha 7$ receptors. Indeed, quick reversal of the block by removal of the antagonist has been taken to indicate the presence of a different receptor (i.e., $\alpha 7^*$), which may contain subunits other than $\alpha 7$ (131). Other homomeric-type recep-tors, such as $\alpha 9$ and $\alpha 9/\alpha 10$ are also sensitive to nanomolar concentrations of α-bungaro-toxin (120, 121, 200, 259).

Another toxin from the banded krait is κ-bungarotoxin (66 amino acids): this is a competitive blocker of neuronal receptors, particularly potent (nearly irreversible) on $\alpha 3\beta 2$ receptors. The main difficulty in using κ-bungarotoxin lies in its limited availability, mostly because of the difficulty in eliminating contaminant α-bungarotoxin in the purifica-tion from crude venom (κ-bungarotoxin is not commercially available to our knowledge). Re-combinant expression in yeast may improve availability of the pure toxin.

It is interesting to note that an endogenous molecule related to snake neurotoxins, *lynx1*, is present in the rodent CNS, where it is sur-face anchored and expressed by neurons posi-tive for nicotinic $\alpha 4\beta 2$ and $\alpha 7$ receptors. In recombinant systems, the effects of coexpress-ing this molecule with $\alpha 4\beta 2$ receptors are com-plex: increases in EC_{50} and in the relative frequency of the greatest conductance and speeding of desensitization during sustained agonist application have been reported (260, 261).

Methyllycaconitine, an alkaloid derived from *Delphinium brownii* (262), is a competitive antagonist selective for $\alpha7$ and $\alpha7^*$ receptors, effective at low nanomolar concentrations (2–5 nM) (257, 263). Work on the macroscopic kinetics of the onset and offset of antagonist action on recombinant $\alpha7/5HT3$ chimeras suggests methyllycaconitine affinity may be an order of magnitude higher than that for pure $\alpha7$ receptors (264). Additionally, this study was consistent with the idea that homomeric receptors have indeed got five binding sites for the alkaloid (i.e., up to one per subunit), if it was assumed that binding of one antagonist molecule is enough to block the response. Heteromeric receptors are also blocked by methyllycaconitine, but at much higher concentrations (tens of nanomoles/liter); recombinant $\alpha4\beta2$ receptors recover from methyllycaconitine block with a time course consistent with the presence of two antagonist binding sites (264).

It is worth mentioning that the glycine receptor antagonist strychnine is also a good competitive antagonist at homomeric receptors, both of the $\alpha7$ and $\alpha9/\alpha10$ type; strychnine is effective on these receptors at submicromolar concentrations, such as are commonly used to suppress glycine receptor activity in native preparations (200, 265, 266).

When it comes to heteromeric receptors, many antagonists have little useful selectivity. The exceptions are dihydro-β-erythroidine and the rapidly growing family of the α-conotoxins (see below). An important point is that many of the antagonists available have channel-blocking properties (see, for instance, mecamylamine and hexamethonium).

4.6.2 Dihydro-β-Erythroidine.
Dihydro-β-erythroidine is an alkaloid obtained from the seeds of several species of the genus *Erythrina*. Its mechanism of action on neuronal nicotinic receptors is likely to be competitive (267), although to our knowledge no K_d values have been reported in the literature. Even with the limitations of the IC$_{50}$ approach, it is clear that the compound has a marked selec-

tivity for some types of heteromeric receptors. Thus it is a poor antagonist of both $\alpha3\beta4$ (IC$_{50}$ range 14–23 μM) (268, 269) and $\alpha7$ receptors (IC$_{50}$ range 2–20 μM) (266, 268, 270). On the other hand, dihydro-β-erythroidine is effective at submicromolar concentrations on recombinant $\alpha4\beta2$ and $\alpha4\beta4$ receptors [IC$_{50}$ values below 0.4 μM (268, 271–273)] and is perhaps slightly less potent on $\alpha3\beta2$ receptors (IC$_{50}$ values 0.4–1.6 μM) (268, 269); all these data were obtained at equilibrium, against agonist concentrations between EC$_{20}$ and EC$_{50}$, depending on the study. For a mutagenesis study of the α subunit residues that determine the difference in sensitivity between $\alpha3\beta2$ and $\alpha3\beta4$ combinations, see Harvey and Luetje (269).

4.6.3 Conotoxins.
Recent work on this family of compounds has given rise to some of the most useful nicotinic antagonists because of the exquisite selectivity of some of these peptides for individual binding interfaces and because of their likely competitive mechanism of action.

The *Conus* genus of marine snails provides an enormous variety of small peptide toxins (estimated at 200–500 per species). These are active on disparate voltage- and ligand-gated ion channels [for reviews, see McIntosh et al. (274); McIntosh (258); McIntosh and Jones (275)]. The venom is used by the snail to hunt its prey, which, depending on the snail species, can be worm, mollusc, or fish. The active principles are small peptides that are maintained in a specific configuration by one or more disulfide bonds. The peptides with more than one disulfide bond are called conotoxins, and they are further subdivided into superfamilies on the basis of the pattern of disulfide bonds in the molecule. Conotoxins that are competitive antagonists of nicotinic receptors belong to the A superfamily [see McIntosh et al. (274)]. Interestingly, the exquisite specificity of these compounds means that the data reported for receptors of a particular species cannot be extrapolated, even to a related mammalian species, given that small differences in the binding domain sequences can markedly change the sensitivity to a conotoxin (258).

4.6.3.1 Conotoxins that Act on Muscle Receptors.
A first grouping is that of toxins that act only on muscle-type nicotinic receptors,

which are derived from the fish-hunting species *C. geographus, ermineus, magus*, and *striatus*. Chemically, most of these toxins belong to the α group and the 3/5 subfamily [see McIntosh et al. (274)].

Of this group, the toxins that are available commercially (as of June 2002) are α-conotoxins GI and MI (which are the most extensively characterized), SI, and SIA.

α-Conotoxin GI, α-conotoxin MI, and α-conotoxin SIA have a strong selectivity for the α/δ interface of mouse embryonic muscle receptor. The highest affinity is observed for GI and MI, and the range of equilibrium constants reported is 1–5 nM for the α/δ site versus 8–58 μM for the α/γ site (5, 82, 276). Note that the selectivity is reversed for *Torpedo* (80). No effect is reported for either α-conotoxin GI and MI at 5 μM on a variety of rat homomeric and heteromeric neuronal nicotinic receptors expressed in oocytes (259).

A peptide from another fish-hunting snail, *C. ermineus*, α-conotoxin EI (belonging to the 4/7 subfamily of the αA group), has a similar preference for mouse muscle α/δ interfaces, but is much less selective than α-conotoxin GI or MI. Interestingly, this conotoxin also targets α/δ in *Torpedo* receptors (277), contrary to the behavior of α-conotoxins GI, MI, and SIA.

4.6.3.2 Conotoxins that Act on Neuronal Receptors.

Note that IC_{50} values in this section were obtained in oocytes against very brief applications of near-maximal concentrations of ACh; as discussed for tubocurarine in the section on muscle receptors, this method would give acceptable estimates of the true equilibrium constant of the antagonist if its receptor occupancy cannot equilibrate with the agonist during the agonist application.

A worm-eating species, *C. imperialis*, produces α-conotoxin ImI, which is an effective antagonist of homomeric-type neuronal nicotinic receptors, such as rat α7 or α9 (reported IC_{50} values of 0.22 and 1.8 μM, respectively). This commercially available 12 amino acid amidated peptide is a very weak blocker of mouse muscle receptors (IC_{50} = 51 μM) and is ineffective at 5 μM on heteromeric rat neuronal receptors (259).

Another commercially available conotoxin, selective for neuronal nicotinic receptors, is α-conotoxin MII, a 16 amino acid α4/7 peptide from *C. magus*. This toxin is selective for rat α3β2 (IC_{50} = 0.5 nM) versus other heteromeric and homomeric combinations of rat neuronal subunits. Only at much higher concentration (200 nM) did this toxin have a small effect on rat recombinant α4β2, α3β4, α4β4 or muscle receptors (20–30% reduction in the maximal ACh response). Rat homomeric α7 receptors are also resistant to α-conotoxin MII (IC_{50} = 200 nM) (278).

Finally, α-conotoxin AuIB (15 amino acids, from *C. aulicus*, at present not commercially available) is selective for the rat α3β4 interface, but is not very potent (IC_{50} = 0.5–0.75 μM). Rat α7 receptors are approximately 10-fold less sensitive than α3β4, whereas other heteromeric neuronal combinations and muscle receptors are at least 100-fold less sensitive (279).

4.6.4 Trimetaphan.

This sulfonium ganglion blocker still has a limited use in clinical practice (as i.v. infusion) to induce controlled hypotension in surgery. Trimetaphan produces a voltage-independent block in a variety of autonomic ganglion preparations and is likely to have a competitive mechanism of action (280–282). The approximate K_d value reported is 1.44 μM (280). Our own work with human recombinant α3β4 receptors indicates a K_d in the region of 70 nM (Boorman, Groot-Kormelink, and Sivilotti, in preparation). Little is known of the selectivity of trimetaphan on different receptor combinations: the only data available (283) suggest that α3β4 receptors are more sensitive than α3β2. Trimetaphan is known to be a poor antagonist of nicotinic receptors on outer hair cells in the cochlea (284), now known to be of the α9/α10 type.

4.6.5 Methonium Compounds.

The polymethylene bistrimethyl ammonium series has been investigated since the nineteenth century [see Colquhoun (285)]. This series of compounds consists of two quaternary ammonium groups joined by a polymethylene chain of variable length. They work on both muscle-type and neuronal-type nicotinic receptors, some as agonists and others as antagonists. Their actions were characterized by Paton and

Zaimis (286, 287). In fact these studies were perhaps the first to give a clear demonstration of how different the muscle and neuronal types of receptor really are.

On the muscle receptor, hexamethonium was a weak antagonist, although the most potent member of the series, decamethonium, worked, albeit unusually, as an agonist. The mechanism of block by depolarization was elucidated by Burns and Paton (288). Their observations can now be explained in more detail as the result of inexcitability of the muscle fiber membrane, close to the neuromuscular synapse, brought about by inactivation of perijunctional sodium channels, caused by prolonged depolarization of the end-plate region. On the neuronal receptor (in peripheral ganglia) hexamethonium was the most potent of the series, but it worked in a quite different way, as an antagonist. Subsequent work has shown that all of these compounds, including those that are agonists, can block the open ion channel to a greater or lesser extent. In fact, as first envisaged by Blackman (109), all the compounds that are antagonists work primarily by channel block; none of them is a good competitive antagonist (though it is still not unknown for textbooks to describe them as "competitive," simply because they are not agonists). Decamethonium also blocks the neuronal receptor channel, although it is a weaker antagonist than hexamethonium because it dissociates more rapidly (280). The slower dissociation (and higher potency) of hexamethonium on ganglion receptors was shown by Gurney and Rang (281) to be a result of the fact that hexamethonium (but not decamethonium) was small enough to be trapped in the channel. It was as if, the blocker having entered the channel while it was open, the channel could then shut again, trapping the hexamethonium inside, and slowing its dissociation. In fact, the hexamethonium can barely escape at all unless the channel is opened again by an agonist; recovery from block required both agonist application and membrane depolarization.

4.6.6 Mecamylamine. This secondary amine compound merits a mention because it is the most used antagonist in behavioral studies, thanks to its ability to cross the blood-brain barrier [for a review, see Young et al. (289)].

Mecamylamine was originally developed as a ganglion blocker and antihypertensive, but, like trimetaphan, its clinical use is now very limited, although it has been suggested as a possible therapeutic agent in Tourette's syndrome.

Heterologous expression data show that mecamylamine is not selective for the different receptor types [see, for instance, Chavez-Noriega et al. (268, 273)] and is effective at low micromolar concentrations. At concentrations greater than 1 μM, mecamylamine is an open channel blocker on recombinant $\alpha 4\beta 2$ receptors (290), and on nicotinic receptors of intracardiac ganglia (139) and chromaffin cells (282, 291). Indeed, there is now evidence that, like hexamethonium, mecamylamine gives rise to a persistent block because it is trapped in the channel. Thus, recovery is sped up by combining agonist application with membrane depolarization: modeling of use dependency and time course of recovery suggests that channels that have trapped blocker open more slowly (291). Nevertheless, when mecamylamine is applied at very much lower concentrations to rat parasympathetic ganglion neurons, it is more effective against *low* agonist concentrations (139, 280). This observation is inconsistent with channel block and suggests a competitive mechanism at these low concentrations, with a K_d of the order of 25–50 nM.

4.6.7 Chlorisondamine. This compound acts as open channel blocker on neuronal nicotinic receptors [see, for instance, for autonomic ganglia (292)]. It is often used for *in vivo* studies because of its very long lasting effects (293), which are probably attributable to trapped channel block [as shown at the frog neuromuscular junction (294)].

4.6.8 (+)-Tubocurarine. Neuronal nicotinic receptors are blocked by micromolar concentrations of (+)-tubocurarine, which is an effective, slow dissociating open channel blocker on ganglionic receptors (280). Nevertheless, there is no evidence that the block involves "trapping" of tubocurarine (291). This compound is likely to have additional effects, other than simple open channel block, given that effects compatible with a partial agonist action have been described for both ganglionic and recombinant $\alpha3\beta4$ receptors (282, 295).

REFERENCES

1. V. Raymond and D. B. Sattelle, *Nat. Rev. Drug Discov.*, **1**, 427–436 (2002).

2. A. Karlin, *Nat. Rev. Neurosci.*, **3**, 102–114 (2002).

3. P. J. Corringer, N. Le Novère, and J.-P. Changeux, *Ann. Rev. Pharmacol. Toxicol.*, **40**, 431–458 (2000).

4. A. Karlin, *Curr. Opin. Neurobiol.*, **3**, 299–309 (1993).

5. S. M. Sine, H. J. Kreienkamp, N. Bren, R. Maeda, and P. Taylor, *Neuron*, **15**, 205–211 (1995a).

6. Y. Xie and J. B. Cohen, *J. Biol. Chem.*, **276**, 2417–2426 (2001).

7. J. Kistler and R. M. Stroud, *Proc. Natl. Acad. Sci. USA*, **78**, 3678–3682 (1981).

8. A. Miyazawa, Y. Fujiyoshi, M. Stowell, and N. Unwin, *J. Mol. Biol.*, **288**, 765–786 (1999).

9. A. Brisson and P. N. T. Unwin, *J. Cell Biology.*, **99**, 1202–1211 (1984).

10. C. Toyoshima and N. Unwin, *J. Cell Biol.*, **111**, 2623–2635 (1990).

11. N. Unwin, *J. Mol. Biol.*, **229**, 1101–1127 (1993).

12. K. Imoto, C. Busch, B. Sakmann, M. Mishina, T. Konno, J. Nakai, H. Bujo, Y. Mori, K. Fukuda, and S. Numa, *Nature*, **335**, 645–648 (1988).

13. D. R. Ripoll, C. H. Faerman, P. H. Axelsen, I. Silman, and J. L. Sussman, *Proc. Natl. Acad. Sci. USA*, **90**, 5128–5132 (1993).

14. F. Hucho, W. Oberthur, and F. Lottspeich, *FEBS Lett.*, **205**, 137–142 (1986).

15. J. Giraudat, M. Dennis, T. Heidmann, J. Y. Chang, and J.-P. Changeux, *Proc. Natl. Acad. Sci. USA*, **83**, 2719–2723 (1986).

16. K. Imoto, C. Methfessel, B. Sakmann, M. Mishina, Y. Mori, T. Konno, K. Fukuda, M. Kurasaki, H. Bujo, Y. Fujita, and S. Numa, *Nature*, **324**, 670–674 (1986).

17. R. J. Leonard, C. G. Labarca, P. Charnet, N. Davidson, and H. A. Lester, *Science*, **242**, 1578–1581 (1988).

18. P. Charnet, C. Labarca, R. J. Leonard, N. J. Vogelaar, L. Czyzyk, A. Gouin, N. Davidson, and H. A. Lester, *Neuron*, **2**, 87–95 (1990).

19. A. Villarroel, S. Herlitze, M. Koenen, and B. Sakmann, *Proc. R. Soc. Lond. Ser. B*, **243**, 69–74 (1991).

20. F. Revah, D. Bertrand, J. L. Galzi, A. Devillers-Thiéry, C. Mulle, N. Hussy, S. Bertrand, M. Ballivet, and J.-P. Changeux, *Nature*, **353**, 846–849 (1991).

21. C. Labarca, M. W. Nowak, H. Zhang, L. Tang, P. Deshpande, and H. A. Lester, *Nature*, **376**, 514–516 (1995).

22. G. N. Filatov and M. M. White, *Mol. Pharmacol.*, **48**, 379–384 (1995).

23. J. P. Boorman, P. J. Groot-Kormelink, and L. G. Sivilotti, *J. Physiol. (London)*, **529**, 565–577 (2000).

24. G. G. Wilson and A. Karlin, *Neuron*, **20**, 1269–1281 (1998).

25. K. Brejc, W. J. van Dijk, R. V. Klaassen, M. Schuurmans, O. J. van Der, A. B. Smit, and T. K. Sixma, *Nature*, **411**, 269–276 (2001).

26. W. Zhong, J. P. Gallivan, Y. Zhang, L. Li, H. A. Lester, and D. A. Dougherty, *Proc. Natl. Acad. Sci. USA*, **95**, 12088–12093 (1998).

27. M. Dennis, J. Giraudat, F. Kotzyba-Hibert, M. Goeldner, C. Hirth, J. Y. Chang, C. Lazure, M. Chretien, and J.-P. Changeux, *Biochemistry*, **27**, 2346–2357 (1988).

28. P. N. Kao and A. Karlin, *J. Biol. Chem.*, **261**, 8085–8088 (1986).

29. D. C. Chiara and J. B. Cohen, *J. Biol. Chem.*, **272**, 32940–32950 (1997).

30. N. Unwin, A. Miyazawa, J. Li, and Y. Fujiyoshi, *J. Mol. Biol.*, **319**, 1165–1176 (2002).

31. N. Unwin, *Nature*, **373**, 37–43 (1995).

32. C. J. Hatton, C. Shelley, D. Beeson, and D. Colquhoun, *J. Physiol.* (submitted).

33. A. Villarroel and B. Sakmann, *Biophys. J.*, **62**, 196–208 (1992).

34. T. M. Dwyer, D. J. Adams, and B. Hille, *J. Gen. Physiol.*, **75**, 469–492 (1980).

35. B. N. Cohen, C. Labarca, N. Davidson, and H. A. Lester, *J. Gen. Physiol.*, **100**, 373–400 (1992).

36. N. Unwin, *J. Struct. Biol.*, **121**, 181–190 (1998).

37. S. Rajendra, R. J. Vandenberg, K. D. Pierce, A. M. Cunningham, P. W. French, P. H. Barry, and P. R. Schofield, *EMBO J.*, **14**, 2987–2998 (1995).

38. J. W. Lynch, S. Rajendra, K. D. Pierce, C. A. Handford, P. H. Barry, and P. R. Schofield, *EMBO J.*, **16**, 110–120 (1997).

39. T. M. Lewis, L. G. Sivilotti, D. Colquhoun, R. M. Gardiner, R. Schoepfer, and M. Rees, *J. Physiol. (London)*, **507**, 25–40 (1998).

40. C. Grosman, F. N. Salamone, S. M. Sine, and A. Auerbach, *J. Gen. Physiol.*, **116**, 327–339 (2000).

41. D. J. Maconochie, G. H. Fletcher, and J. H. Steinbach, *Biophys. J.*, **68**, 483–490 (1995).

42. E. Neher and B. Sakmann, *Nature*, **260**, 799–802 (1976).

43. O. P. Hamill, A. Marty, E. Neher, B. Sakmann, and F. J. Sigworth, *Pflügers Arch.*, **391**, 85–100 (1981).

44. D. Colquhoun and A. G. Hawkes, *Proc. R. Soc. Lond.B*, **199**, 231–262 (1977).

45. D. Colquhoun and A. G. Hawkes, *Philos. Trans. R. Soc. Lond. Ser. B*, **300**, 1–59 (1982).

46. D. Colquhoun and B. Sakmann, *Nature*, **294**, 464–466 (1981).

47. D. Colquhoun and B. Sakmann, *J. Physiol. (London)*, **369**, 501–557 (1985).

48. D. Colquhoun, *Br. J. Pharmacol.*, **125**, 923–948 (1998).

49. J. del Castillo and B. Katz, *Proc. R. Soc. Lond. Ser. B*, **146**, 369–381 (1957).

50. J. Monod, J. Wyman, and J.-P. Changeux, *J. Mol. Biol.*, **12**, 88–118 (1965).

51. O. Arunlakshana and H. O. Schild, *Br. J. Pharmacol.*, **14**, 47–58 (1959).

52. D. Colquhoun in J. W. Black, Ed., *Perspectives on Hormone Receptor Classification*, Alan R. Liss, New York, 1987, pp. 103–114.

53. S. M. Sine and J. H. Steinbach, *Biophys. J.*, **46**, 277–284 (1984).

54. D. C. Ogden and D. Colquhoun, *Proc. R. Soc. Lond. Ser. B*, **225**, 329–355 (1985).

55. D. Colquhoun, F. Dreyer, and R. E. Sheridan, *J. Physiol. (London)*, **293**, 247–284 (1979).

56. C. G. Marshall, D. C. Ogden, and D. Colquhoun, *J. Physiol. (London)*, **428**, 154–174 (1990).

57. K. L. Magleby and B. S. Pallotta, *J. Physiol. (London)*, **316**, 225–250 (1981).

58. B. Katz and S. Thesleff, *J. Physiol. (London)*, **138**, 63–80 (1957).

59. A. B. Cachelin and D. Colquhoun, *J. Physiol. (London)*, **415**, 159–188 (1989).

60. J. Butler, C. Franke, V. Witzemann, J. P. Ruppersberg, S. Merlitze, and J. Dudel, *Neurosci. Lett.*, **152**, 77–80 (1993).

61. C. Franke, H. Parnas, G. Hovav, and J. Dudel, *Biophys. J.*, **64**, 339–356 (1993).

62. S. Elenes and A. Auerbach, *J. Physiol. (London)*, **541**, 367–383 (2002).

63. D. R. Fredkin, M. Montal, and J. A. Rice in L. M. Le Cam, Ed., Proceedings of the Berkeley Conference in Honor of Jerzy Neyman and Jack Kiefer, Wadsworth, Monterey, CA, 1985, pp. 269–289.

64. R. Horn and K. Lange, *Biophys. J.*, **43**, 207–223 (1983).

65. A. G. Hawkes, A. Jalali, and D. Colquhoun, *Philos. Trans. R. Soc. Lond. A..*, **332**, 511–538 (1990).

66. A. G. Hawkes, A. Jalali, and D. Colquhoun, *Philos. Trans. R. Soc. Lond. Ser. B*, **337**, 383–404 (1992).

67. D. Colquhoun, A. G. Hawkes, and K. Srodzinski, *Philos. Trans. R. Soc. Lond. A*, **354**, 2555–2590 (1996).

68. F. Qin, A. Auerbach, and F. Sachs, *Biophys. J.*, **70**, 264–280 (1996).

69. D. Colquhoun and A. G. Hawkes in B. Sakmann and E. Neher, Eds., *Single Channel Recording*, Plenum Press, New York, 1995, pp. 397–482.

70. D. Colquhoun, C. J. Hatton, and A. G. Hawkes, *J. Physiol. (London)* (submitted).

71. M. Beato, P. J. Groot-Kormelink, D. Colquhoun, and L. G. Sivilotti, *J. Gen. Physiol.*, **119**, 443–466 (2002).

72. F. N. Salamone, M. Zhou, and A. Auerbach, *J. Physiol. (London)*, **516**, 315–330 (1999).

73. Y. Liu and J. P. Dilger, *Biophys. J.*, **60**, 424–432 (1991).

74. D. J. Maconochie and J. H. Steinbach, *J. Physiol. (London)*, **506**, 53–72 (1998).

75. B. Edmonds, A. J. Gibb, and D. Colquhoun, *Ann. Rev. Physiol.*, **57**, 469–493 (1995).

76. S. M. Sine, T. Claudio, and F. J. Sigworth, *J. Gen. Physiol.*, **96**, 395–437 (1990).

77. F. Parzefall, R. Wilhelm, M. Heckmann, and J. Dudel, *J. Physiol. (London)*, **512**, 181–188 (1998).

78. N. Bren and S. M. Sine, *J. Biol. Chem.*, **275**, 12692–12700 (2000).

79. S. M. Sine, *Proc. Natl. Acad. Sci. USA*, **90**, 9436–9440 (1993).

80. R. M. Hann, O. R. Pagán, and V. A. Eterovic, *Biochemistry*, **33**, 14058–14063 (1994).

81. N. Bren and S. M. Sine, *J. Biol. Chem.*, **272**, 30793–30798 (1997).

82. D. R. Groebe, J. M. Dumm, E. S. Levitan, and S. N. Abramson, *Mol. Pharmacol.*, **48**, 105–111 (1995).

83. I. Silman, M. Harel, P. Axelsen, M. Raves, and J. L. Sussman, *Biochem. Soc. Trans.*, **22**, 745–749 (1994).

84. A. Devillers-Thiéry, J. L. Galzi, J. L. Eiselé, S. Bertrand, D. Bertrand, and J.-P. Changeux, *J. Membr. Biol.*, **136**, 97–112 (1993).

85. G. Akk, *J. Physiol. (London)*, **535**, 729–740 (2001).

86. A. Auerbach, W. Sigurdson, J. Chen, and G. Akk, *J. Physiol. (London)*, **494**, 155–170 (1996).

87. G. Akk and J. H. Steinbach, *J. Physiol. (London)*, **527**, 405–417 (2000).

88. G. F. Tomaselli, J. T. McLaughlin, M. E. Jurman, E. Hawrot, and G. Yellen, *Biophys. J.*, **60**, 721–727 (1991).

89. J. Chen, Y. Zhang, G. Akk, S. Sine, and A. Auerbach, *Biophys. J.*, **69**, 849–859 (1995).

90. M. E. O'Leary and M. M. White, *J. Biol. Chem.*, **267**, 8360–8365 (1992).

91. S. M. Sine, P. Quiram, F. Papanikolaou, H. J. Kreienkamp, and P. Taylor, *J. Biol. Chem.*, **269**, 8808–8816 (1994).

92. G. Akk, M. Zhou, and A. Auerbach, *Biophys. J.*, **76**, 207–218 (1999).

93. G. Akk, S. Sine, and A. Auerbach, *J. Physiol. (London)*, **496**, 185–196 (1996).

94. A. G. Engel, E. H. Lambert, D. M. Mulder, C. F. Torres, K. Sahashi, T. E. Bertorini, and J. N. Whitaker, *Ann. Neurol.*, **11**, 553–569 (1982).

95. S. M. Sine, K. Ohno, C. Bouzat, A. Auerbach, M. Milone, J. N. Pruitt, and A. G. Engel, *Neuron*, **15**, 229–239 (1995b).

96. R. Croxen, C. Newland, D. Beeson, H. Oosterhuis, G. Chauplannaz, A. Vincent, and J. Newsom-Davis, *Hum. Mol. Genet.*, **6**, 767–774 (1997).

97. A. G. Engel, K. Ohno, M. Milon, H. L. Wang, S. Nakano, C. Bouzat, J. N. Pruitt II, D. O. Hutchinson, J. M. Brengman, N. Bren, J. P. Sieb, and S. M. Sine, *Hum. Mol. Genet.*, **5**, 1217–1227 (1996).

98. H. L. Wang, A. Auerbach, N. Bren, K. Ohno, A. G. Engel, and S. M. Sine, *J. Gen. Physiol.*, **109**, 757–766 (1997).

99. H. Osaka, S. Malany, B. E. Molles, S. M. Sine, and P. Taylor, *J. Biol. Chem.*, **275**, 5478–5484 (2000).

100. C. Czajkowski and A. Karlin, *J. Biol. Chem.*, **266**, 22603–22612 (1991).

101. C. Czajkowski, C. Kaufmann, and A. Karlin, *Proc. Natl. Acad. Sci. USA*, **90**, 6285–6289 (1993).

102. H. J. Oosterhuis, J. Newsom-Davis, J. H. Wokke, P. C. Molenaar, T. V. Weerden, B. S. Oen, F. G. Jennekens, H. Veldman, A. Vincent, and D. W. Wray, *Brain*, **110**, 1061–1079 (1987).

103. R. Croxen, C. J. Hatton, C. Shelley, M. Brydson, G. Chauplannaz, H. Oosterhuis, A. Vincent, A. Newsom-Davis, D. Colquhoun, and D. Beeson, *Neurology*, **59**, 162–168 (2002).

104. D. J. A. Wyllie, P. Béhé, and D. Colquhoun, *J. Physiol. (London)*, **510**, 1–18 (1998).

105. D. Shortle, *Q. Rev. Biophys.*, **25**, 205–250 (1992).

106. K. L. Magleby, B. S. Pallotta, and D. A. Terrar, *J. Physiol. (London)*, **312**, 97–113 (1981).

107. D. Colquhoun and R. E. Sheridan, *Proc. R. Soc. Lond. Ser. B*, **211**, 181–203 (1981).

108. J. H. Gaddum, *J. Physiol. (London)*, **89**, 7P–9P (1937).

109. J. G. Blackman, *Proc. Univ. Otago Med. School*, **48**, 4–5 (1970).

110. C. M. Armstrong and B. Hille, *J. Gen. Physiol.*, **59**, 388–400 (1972).

111. P. R. Adams, *J. Physiol. (London)*, **260**, 531–552 (1976).

112. H. O. Schild, *Br. J. Pharmacol.*, **2**, 189–206 (1947).

113. W. D. M. Paton and H. P. Rang, *Proc. R. Soc. Lond. Ser. B*, **163**, 1–44 (1965).

114. A. V. Hill, *J. Physiol. (London)*, **39**, 361–373 (1909).

115. D. Colquhoun and J. M. Ritchie, *Mol. Pharmacol.*, **8**, 285–292 (1973).

116. I. Wenningmann and J. P. Dilger, *Mol. Pharmacol.*, **60**, 790–796 (2001).

117. D. Demazumder and J. P. Dilger, *Mol. Pharmacol.*, **60**, 797–807 (2001).

118. R. Anand, W. G. Conroy, R. Schoepfer, P. Whiting, and J. Lindstrom, *J. Biol. Chem.*, **266**, 11192–11198 (1991).

119. E. Cooper, S. Couturier, and M. Ballivet, *Nature*, **350**, 235–238 (1991).

120. A. B. Elgoyhen, D. E. Vetter, E. Katz, C. V. Rothlin, S. F. Heinemann, and J. Boulter, *Proc. Natl. Acad. Sci. USA*, **98**, 3501–3506 (2001).

121. F. Sgard, E. Charpantier, S. Bertrand, N. Walker, D. Caput, D. Graham, D. Bertrand, and F. Besnard, *Mol. Pharmacol.*, **61**, 150–159 (2002).

122. J. Ramirez-Latorre, C. R. Yu, X. Qu, F. Perin, A. Karlin, and L. Role, *Nature*, **380**, 347–351 (1996).

123. P. J. Groot-Kormelink, J. P. Boorman, and L. G. Sivilotti, *Br. J. Pharmacol.*, **134**, 789–796 (2001).

124. L. G. Sivilotti, D. Colquhoun, and N. Millar in F. Clementi, D. Fornasari, and C. Gotti, Eds., *Neuronal Nicotinic Receptors*, Springer-Verlag, Berlin/Heidelberg, 2000, pp. 379–416.

125. P. B. Sargent in F. Clementi, D. Fornasari, and C. Gotti, Eds., *Neuronal Nicotinic Receptors*, Springer-Verlag, Berlin/Heidelberg, 2000, pp. 163–192.

126. N. Le Novère, M. Zoli, and J.-P. Changeux, *Eur. J. Neurosci.*, **8**, 2428–2439 (1996).

127. J. R. Forsayeth and E. Kobrin, *J. Neurosci.*, **17**, 1531–1538 (1997).

128. W. G. Conroy and D. K. Berg, *J. Biol. Chem.*, **270**, 4424–4431 (1995).

129. C. R. Yu and L. W. Role, *J. Physiol. (London)*, **509**, 651–665 (1998).

130. J. Cuevas and D. K. Berg, *J. Neurosci.*, **18**, 10335–10344 (1998).

131. J. Cuevas, A. L. Roth, and D. K. Berg, *J. Physiol. (London)*, **525**, 735–746 (2000).

132. C. W. Luetje and J. Patrick, *J. Neurosci.*, **11**, 837–845 (1991).

133. S. Vernino, M. Amador, C. W. Luetje, J. Patrick, and J. A. Dani, *Neuron*, **8**, 127–134 (1992).

134. A. C. S. Costa, J. W. Patrick, and J. A. Dani, *Biophys. J.*, **67**, 395–401 (1994).

135. A. Villarroel and B. Sakmann, *J. Physiol. (London)*, **496**, 331–338 (1996).

136. E. R. Decker and J. A. Dani, *J. Neurosci.*, **10**, 3413–3420 (1990).

137. S. Vernino, M. Rogers, K. A. Radcliffe, and J. A. Dani, *J. Neurosci.*, **14**, 5514–5524 (1994).

138. D. Ragozzino, B. Barabino, S. Fucile, and F. Eusebi, *J. Physiol. (London)*, **507**, 749–757 (1998).

139. L. A. Fieber and D. J. Adams, *J. Physiol. (London)*, **434**, 215–237 (1991).

140. Z. Zhou and E. Neher, *Pflügers Arch.-Eur. J. Physiol.*, **425**, 511–517 (1993).

141. T. J. Nutter and D. J. Adams, *J. Gen. Physiol.*, **105**, 701–723 (1995).

142. M. Rogers and J. A. Dani, *Biophys. J.*, **68**, 501–506 (1995).

143. C. Mulle, D. Choquet, H. Korn, and J.-P. Changeux, *Neuron*, **8**, 135–143 (1992a).

144. J. Trouslard, S. J. Marsh, and D. A. Brown, *J. Physiol. (London)*, **468**, 53–71 (1993).

145. A. Kuryatov, V. Gerzanich, M. Nelson, F. Olale, and J. Lindstrom, *J. Neurosci.*, **17**, 9035–9047 (1997).

146. V. Gerzanich, F. Wang, A. Kuryatov, and J. Lindstrom, *J. Pharmacol. Exp. Ther.*, **286**, 311–320 (1998).

147. D. Bertrand, J.-L. Galzi, A. Devillers-Thiéry, S. Bertrand, and J.-P. Changeux, *Proc. Natl. Acad. Sci. USA*, **90**, 6971–6975 (1993).

148. S. B. Sands, A. C. S. Costa, and J. W. Patrick, *Biophys. J.*, **65**, 2614–2621 (1993).

149. N. G. Castro and E. X. Albuquerque, *Biophys. J.*, **68**, 516–524 (1995).

150. S. Fucile, E. Palma, A. M. Mileo, R. Miledi, and F. Eusebi, F. *Proc. Natl. Acad. Sci. USA*, **97**, 3643–3648 (2000).

151. O. Delbono, M. Gopalakrishnan, M. Renganathan, L. M. Monteggia, M. L. Messi, and J. P. Sullivan, *J. Pharmacol. Exp. Ther.*, **280**, 428–438 (1997).

152. E. Katz, M. Verbitsky, C. V. Rothlin, D. E. Vetter, S. F. Heinemann, and A. B. Elgoyhen, *Hearing Res.*, **141**, 117–128 (1999).

153. D. J. Jagger, C. B. Griesinger, M. N. Rivolta, M. C. Holley, and J. F. Ashmore, *J. Physiol. (London)*, **527**, 49–54 (2000).

154. D. Oliver, N. Klocker, J. Schuck, T. Baukrowitz, J. P. Ruppersberg, and B. Fakler, *Neuron*, **26**, 595–601 (2000).

155. R. J. Callister, J. R. Keast, and P. Sah, *J. Physiol. (London)*, **500**, 571–582 (1997).

156. P. D. Bregestovski, R. Miledi, and I. Parker, *Nature*, **279**, 638–639 (1979).

157. C. A. Lewis, *J. Physiol. (London)*, **286**, 417–445 (1979).

158. A. Mathie, S. G. Cull-Candy, and D. Colquhoun, *Proc. R. Soc. Lond. Ser. B, Biol. Sci.*, **232**, 239–248 (1987).

159. R. Neuhaus and A. B. Cachelin, *Proc. R. Soc. Lond. Ser. B, Biol. Sci.*, **241**, 78–84 (1990).

160. C. Mulle, C. Léna, and J.-P. Changeux, *Neuron*, **8**, 937–945 (1992b).

161. B. Buisson, Y. F. Vallejo, W. N. Green, and D. Bertrand, *Neuropharmacology*, **39**, 2561–2569 (2000).

162. R. Bonfante-Cabarcas, K. L. Swanson, M. Alkondon, and E. X. Albuquerque, *J. Pharmacol. Exp. Ther.*, **277**, 432–444 (1996).

163. J.-L. Galzi, S. Bertrand, P.-J. Corringer, J.-P. Changeux, and D. Bertrand, *EMBO J.*, **15**, 5824–5832 (1996).

164. N. Le Novère, T. Grutter, and J.-P. Changeux, *Proc. Natl. Acad. Sci. USA*, **99**, 3210–3215 (2002).

165. L. G. Sivilotti, D. K. McNeil, T. M. Lewis, M. A. Nassar, R. Schoepfer, and D. Colquhoun, *J. Physiol. (London)*, **500**, 123–138 (1997).

166. T. M. Lewis, P. C. Harkness, L. G. Sivilotti, D. Colquhoun, and N. Millar, *J. Physiol. (London)*, **505**, 299–306 (1997).

167. A. Mathie, D. Colquhoun, and S. G. Cull-Candy, *J. Physiol. (London)*, **427**, 625–655 (1990).

168. A. P. Haghighi and E. Cooper, *J. Neurosci.*, **18**, 4050–4062 (1998).

169. C. K. Ifune and J. H. Steinbach, *Proc. Natl. Acad. Sci. USA*, **87**, 4794–4798 (1990).

170. C. K. Ifune and J. H. Steinbach, *J. Physiol. (London)*, **457**, 143–165 (1992).

171. M. Alkondon, S. Reinhardt, C. Lobron, B. Hermsen, A. Maelicke, and E. X. Albuquerque, *J. Pharmacol. Exp. Ther.*, **271**, 494–506 (1994).

172. I. Forster and D. Bertrand, *Proc. R. Soc. Lond. Ser. B, Biol. Sci.*, **260**, 139–148 (1995).

173. I. Wessler, C. J. Kirkpatrick, and K. Racke, *Pharmacol. Ther.*, **77**, 59–79 (1998).

174. M. Listerud, A. B. Brussaard, P. Devay, D. R. Colman, and L. W. Role, *Science*, **254**, 1518–1521 (1991) [Published erratum appears in Science 1992 Jan 3;255(5040):12].

175. G. Rust, J. M. Burgunder, T. E. Lauterburg, and A. B. Cachelin, *Eur. J. Neurosci.*, **6**, 478–485 (1994).

176. A. Campos-Caro, F. I. Smillie, E. Domínguez del Toro, J. C. Rovira, F. Vicente-Agulló, J. Chapuli, J. M. Juíz, S. Sala, F. Sala, and J. J. Ballesta, *J. Neurochem.*, **68**, 488–497 (1997).

177. M. V. Skok, L. P. Voitenko, S. V. Voitenko, E. Y. Lykhmus, E. N. Kalashnik, T. I. Litvin, S. J. Tzartos, and V. I. Skok, *Neuroscience*, **93**, 1427–1436 (1999).

178. L. Klimaschewski, S. Reuss, R. Spessert, C. Lobron, A. Wevers, C. Heym, A. Maelicke, and H. Schröder, *Mol. Brain Res.*, **27**, 167–173 (1994).

179. D. K. Berg, R. D. Shoop, K. T. Chang, and J. Cuevas in F. Clementi, D. Fornasari, and C. Gotti, Eds., *Neuronal Nicotinic Receptors*, Springer-Verlag, Berlin/Heidelberg, 2000, pp. 247–267.

180. D. A. Brown and L. Fumagalli, *Brain Res.*, **129**, 165–168 (1977).

181. M. K. Temburni, R. C. Blitzblau, and M. H. Jacob, *J. Physiol. (London)*, **525**, 21–29 (2000).

182. W. Xu, S. Gelber, A. Orr-Urtreger, D. Armstrong, R. A. Lewis, C. N. Ou, J. Patrick, L. Role, M. DeBiasi, and A. L. Beaudet, *Proc. Natl. Acad. Sci. USA*, **96**, 5746–5751 (1999a).

183. W. Xu, A. Orr-Urtreger, F. Nigro, S. Gelber, C. Ballard Sutcliffe, D. Armstrong, J. W. Patrick, L. W. Role, A. L. Beaudet, and M. De Biasi, *J. Neurosci.*, **19**, 9298–9305 (1999b).

184. K. Poth, T. J. Nutter, J. Cuevas, M. J. Parker, D. J. Adams, and C. W. Luetje, *J. Neurosci.*, **17**, 586–596 (1997).

185. J. C. Britt and H. R. Brenner, *Pflügers Arch.-Eur. J. Physiol.*, **434**, 38–48 (1997).

186. K. T. Chang and D. K. Berg, *J. Neurosci.*, **19**, 3701–3710 (1999).

187. M. García-Guzmán, F. Sala, S. Sala, A. Campos-Caro, W. Stühmer, L. M. Gutierrez, and M. Criado, *Eur. J. Neurosci.*, **7**, 647–655 (1995).

188. J. M. Nooney, J. J. Lambert, and V. A. Chiappinelli, *Brain Res.*, **573**, 77–82 (1992a).

189. R. Afar, J.-M. Trifaró, and M. Quik, *Brain Res.*, **641**, 127–131 (1994).

190. J. M. Nooney and A. Feltz, *Br. J. Pharmacol.*, **114**, 648–655 (1995).

191. E. Tachikawa, K. Mizuma, K. Kudo, T. Kashimoto, S. Yamato, and S. Ohta, *Neuroscience Lett.*, **312**, 161–164 (2001).

192. M. Criado, E. Domínguez del Toro, C. Carrasco-Serrano, F. I. Smillie, J. M. Juíz, S. Viniegra, and J. J. Ballesta, *J. Neurosci.*, **17**, 6554–6564 (1997).

193. A. Orr-Urtreger, F. M. Göldner, M. Saeki, I. Lorenzo, L. Goldberg, M. De Biasi, J. A. Dani, J. W. Patrick, and A. L. Beaudet, *J. Neurosci.*, **17**, 9165–9171 (1997).

194. D. Franceschini, A. Orr-Urtreger, W. Yu, L. Y. Mackey, R. A. Bond, D. Armstrong, J. W. Patrick, A. L. Beaudet, and M. De Biasi, *Behav. Brain Res.*, **113**, 3–10 (2000).

195. J. S. Coggan, J. Paysan, W. G. Conroy, and D. K. Berg, *J. Neurosci.*, **17**, 5798–5806 (1997).

196. D. Kristufek, E. Stocker, S. Boehm, and S. Huck, *J. Physiol. (London)*, **516**, 739–756 (1999).

197. J. F. Ashmore, *Exp. Physiol.*, **79**, 113–134 (1994).

198. P. A. Fuchs and B. W. Murrow, *Proc. R. Soc. Lond. Ser. B, Biol. Sci.*, **248**, 35–40 (1992).

199. G. D. Housley and J. F. Ashmore, *Proc. R. Soc. Lond. Ser. B, Biol. Sci.*, **244**, 161–167 (1991).

200. A. B. Elgoyhen, D. S. Johnson, J. Boulter, D. E. Vetter, and S. Heinemann, *Cell*, **79**, 705–715 (1994).

201. J. C. Eccles, P. Fatt, and K. Koketsu, *J. Physiol. (London)*, **126**, 524–562 (1954).

202. M. Dourado and P. B. Sargent, *J. Neurophysiol.*, **87**, 3117–3125 (2002).

203. F. M. Göldner, K. T. Dineley, and J. W. Patrick, *NeuroReport*, **8**, 2739–2742 (1997).

204. R. Klink, A. de Kerchove d'Exaerde, M. Zoli, and J.-P. Changeux, *J. Neurosci.*, **21**, 1452–1463 (2001).

205. C. Léna, A. D. d'Exaerde, M. Cordero-Erausquin, N. Le Novère, M. d. M. Arroyo-Jiménez, and J.-P. Changeux, *Proc. Natl. Acad. Sci. USA*, **96**, 12126–12131 (1999).

206. L. M. Marubio, M. d. M. Arroyo-Jiménez, M. Cordero-Erausquin, C. Léna, N. Le Novère, A. de Kerchove d'Exaerde, M. Huchet, M. I. Damaj, and J.-P. Changeux, *Nature*, **398**, 805–810 (1999).

207. D. E. Vetter, M. C. Liberman, J. Mann, J. Barhanin, J. Boulter, M. C. Brown, J. Saffiote-Kolman, S. F. Heinemann, and A. B. Elgoyhen, *Neuron*, **23**, 93–103 (1999).

208. M. R. Picciotto, M. Zoli, C. Léna, A. Bessi, Y. Lallemand, N. Le Novère, P. Vincent, E. Merlo Pich, P. Brulet, and J.-P. Changeux, *Nature*, **374**, 65–67 (1995).

209. C. Labarca, J. Schwarz, P. Deshpande, S. Schwarz, M. W. Nowak, C. Fonck, R. Nashmi, P. Kofuji, H. Dang, W. Shi, M. Fidan, B. S. Khakh, Z. Chen, B. Bowers, J. Boulter, J. M. Wehner, and H. A. Lester, *Proc. Natl. Acad. Sci. USA*, **98**, 2786–2791 (2001).

210. A. Orr-Urtreger, R. S. Broide, M. R. Kasten, H. Dang, J. A. Dani, A. L. Beaudet, and J. W. Patrick, *J. Neurochem.*, **74**, 2154–2166 (2000).

211. M. Cordero-Erausquin, L. M. Marubio, R. Klink, and J.-P. Changeux, *Trends Pharmacol. Sci.*, **21**, 211–217 (2000).

212. M. Zoli, C. Léna, M. R. Picciotto, and J.-P. Changeux, *J. Neurosci.*, **18**, 4461–4472 (1998).

213. C. F. Zorumski, L. L. Thio, K. E. Isenberg, and D. B. Clifford, *Mol. Pharmacol.*, **41**, 931–936 (1992).

214. M. Alkondon and E. X. Albuquerque, *J. Pharmacol. Exp. Ther.*, **265**, 1455–1473 (1993).

215. E. X. Albuquerque, E. F. R. Pereira, M. Alkondon, H. M. Eisenberg, and A. Maelicke in F. Clementi, D. Fornasari, and C. Gotti, Eds., *Neuronal Nicotinic Receptors*, Springer-Verlag, Berlin/Heidelberg, 2000, pp. 337–358.

216. S. Jones and J. L. Yakel, *J. Physiol. (London)*, **504**, 603–610 (1997).

217. C. J. Frazier, Y. D. Rollins, C. R. Breese, S. Leonard, R. Freedman, and T. V. Dunwiddie, *J. Neurosci.*, **18**, 1187–1195 (1998a).

218. A. R. McQuiston and D. V. Madison, *J. Neurosci.*, **19**, 2887–2896 (1999).

219. R. L. Papke, M. Bencherif, and P. Lippiello, *Neurosci. Lett.*, **213**, 201–204 (1996).

220. D. Ji, R. Lape, and J. A. Dani, *Neuron*, **31**, 131–141 (2001).

221. M. Alkondon, E. F. Pereira, H. M. Eisenberg, and E. X. Albuquerque, *J. Neurosci.*, **19**, 2693–2705 (1999a).

222. R. Fabian-Fine, P. Skehel, M. L. Errington, H. A. Davies, E. Sher, M. G. Stewart, and A. Fine,. *J. Neurosci.*, **21**, 7993–8003 (2001).

223. M. Alkondon, E. F. Pereira, and E. X. Albuquerque, *Brain Res.*, **810**, 257–263 (1998).

224. C. J. Frazier, A. V. Buhler, J. L. Weiner, and T. V. Dunwiddie, *J. Neurosci.*, **18**, 8228–8235 (1998b).

225. A. V. Buhler and T. V. Dunwiddie, *Neuroscience*, **106**, 55–67 (2001).

226. S. Hefft, S. Hulo, D. Bertrand, and D. Muller, *J. Physiol. (London)*, **515**, 769–776 (1999).

227. D. S. McGehee and L. W. Role, *Nature*, **383**, 670–671 (1996).

228. S. Wonnacott, *Trends Neurosci.*, **20**, 92–98 (1997).

229. S. Kaiser and S. Wonnacott, *Mol. Pharmacol.*, **58**, 312–318 (2000).

230. C. Léna, J.-P. Changeux, and C. Mulle, *J. Neurosci.*, **13**, 2680–2688 (1993).

231. D. S. McGehee, M. J. S. Heath, S. Gelber, P. Devay, and L. W. Role, *Science*, **269**, 1692–1696 (1995).

232. R. Gray, A. S. Rajan, K. A. Radcliffe, M. Yakehiro, and J. A. Dani, *Nature*, **383**, 713–716 (1996).

233. X. Li, D. G. Rainnie, R. W. McCarley, and R. W. Greene, *J. Neurosci.*, **18**, 1904–1912 (1998).

234. K. A. Radcliffe and J. A. Dani, *J. Neurosci.*, **18**, 7075–7083 (1998).

235. L. Maggi, E. Sher, and E. Cherubini, *J. Physiol. (London)*, **536**, 89–100 (2001).

236. D. S. McGehee and L. W. Role, *Ann. Rev. Physiol.*, **57**, 521–546 (1995).

237. A. Bordey, P. Feltz, and J. Trouslard, *J. Physiol. (London)*, **497**, 175–187 (1996).

238. H. D. Mansvelder and D. S. McGehee, *Neuron*, **27**, 349–357 (2000).

239. J. L. Fisher and J. A. Dani, *Neuropharmacology*, **39**, 2756–2769 (2000).

240. F.-M. Zhou, Y. Liang, and J. A. Dani, *Nat. Neurosci.*, **4**, 1224–1229 (2001).

241. J. A. Dani, D. Ji, and F.-M. Zhou, *Neuron*, **31**, 349–352 (2001).

242. S. E. Hyman, S. E. Hyman, and R. C. Malenka, *Nat. Rev. Neurosci.*, **2**, 695–703 (2001).

243. M. R. Picciotto, M. Zoli, R. Rimondini, C. Léna, L. M. Marubio, E. Merlo Pich, K. Fuxe, and J.-P. Changeux, *Nature*, **391**, 173–177 (1998).

244. O. K. Steinlein, J. C. Mulley, P. Propping, R. H. Wallace, H. A. Phillips, G. R. Sutherland, I. E. Scheffer, and S. F. Berkovic, *Nat. Genet.*, **11**, 201–203 (1995).

245. B. Sutor and G. Zolles, *Pflügers Arch.-Eur. J. Physiol.*, **442**, 642–651 (2001).

246. H. A. Phillips, I. E. Scheffer, K. M. Crossland, K. P. Bhatia, D. R. Fish, C. D. Marsden, S. J. Howell, J. B. Stephenson, J. Tolmie, G. Plazzi, O. Eeg-Olofsson, R. Singh, I. Lopes-Cendes, E. Andermann, F. Andermann, S. F. Berkovic, and J. C. Mulley, *Am. J. Hum. Genet.*, **63**, 1108–1116 (1998).

247. S. Hirose, H. Iwata, H. Akiyoshi, K. Kobayashi, M. Ito, K. Wada, S. Kaneko, and A. Mitsudome, *Neurology*, **53**, 1749–1753 (1999).

248. O. K. Steinlein, A. Magnusson, J. Stoodt, S. Bertrand, S. Weiland, S. F. Berkovic, K. O. Nakken, P. Propping, and D. Bertrand, *Hum. Mol. Genet.*, **6**, 943–947 (1997).

249. M. De Fusco, A. Becchetti, A. Patrignani, G. Annesi, A. Gambardella, A. Quattrone, A. Ballabio, E. Wanke, and G. Casari, *Nat. Genet.*, **26**, 275–276 (2000).

250. H. A. Phillips, I. Favre, M. Kirkpatrick, S. M. Zuberi, D. Goudie, S. E. Heron, I. E. Scheffer, G. R. Sutherland, S. F. Berkovic, D. Bertrand, and J. C. Mulley, *Am. J. Hum. Genet.*, **68**, 225–231 (2001).

251. S. Bertrand, S. Weiland, S. F. Berkovic, O. K. Steinlein, and D. Bertrand, *Br. J. Pharmacol.*, **125**, 751–760 (1998).

252. A. Figl, N. Viseshakul, N. Shafaee, J. Forsayeth, and B. N. Cohen, *J. Physiol. (London)*, **513**, 655–670 (1998).

253. P. J. O. Covernton, H. Kojima, L. G. Sivilotti, A. J. Gibb, and D. Colquhoun, *J. Physiol. (London)*, **481**, 27–34 (1994).

254. A. Mandelzys, P. De Koninck, and E. Cooper, *J. Neurophysiol.*, **74**, 1212–1221 (1995).

255. M. Alkondon, E. F. Pereira, W. S. Cortes, A. Maelicke, and E. X. Albuquerque, *Eur. J. Neurosci.*, **9**, 2734–2742 (1997).

256. R. L. Papke and S. F. Heinemann, *Mol. Pharmacol.*, **45**, 142–149 (1994).

257. A. R. L. Davies, D. J. Hardick, I. S. Blagbrough, B. V. L. Potter, A. J. Wolstenholme, and S. Wonnacott, *Neuropharmacology*, **38**, 679–690 (1999).

258. J. M. McIntosh in F. Clementi, D. Fornasari, and C. Gotti, C., Eds., *Neuronal Nicotinic Receptors*, Springer-Verlag, Berlin/Heidelberg, 2000, pp. 455–476.

259. D. S. Johnson, J. Martinez, A. B. Elgoyhen, S. F. Heinemann, and J. M. McIntosh, *Mol. Pharmacol.*, **48**, 194–199 (1995).

260. J. M. Miwa, I. Ibañez-Tallon, G. W. Crabtree, R. Sánchez, A. Sali, L. W. Role, and N. Heintz, *Neuron*, **23**, 105–114 (1999).

261. I. Ibañez-Tallon, J. M. Miwa, H. L. Wang, N. C. Adams, G. W. Crabtree, S. M. Sine, and N. Heintz, *Neuron*, **33**, 893–903 (2002).

262. M. Alkondon, E. F. R. Pereira, S. Wonnacott, and E. X. Albuquerque, *Mol. Pharmacol.*, **41**, 802–808 (1992).

263. L. Yum, K. M. Wolf, and V. A. Chiappinelli, *Neuroscience*, **72**, 545–555 (1996).

264. E. Palma, S. Bertrand, T. Binzoni, and D. Bertrand, *J. Physiol. (London)*, **491**, 151–161 (1996).

265. P. Séguéla, J. Wadiche, K. Dineley-Miller, J. Dani, and J. W. Patrick, *J. Neurosci.*, **13**, 596–604 (1993).

266. X. Peng, M. Katz, V. Gerzanich, R. Anand, and J. Lindstrom, *Mol. Pharmacol.*, **45**, 546–554 (1994).

267. D. Bertrand, S. Bertrand, and M. Ballivet, *Neurosci. Lett.*, **146**, 87–90 (1992).

268. L. E. Chavez-Noriega, J. H. Crona, M. S. Washburn, A. Urrutia, K. J. Elliott, and E. C. Johnson, *J. Pharmacol. Exp. Ther.*, **280**, 346–356 (1997).

269. S. C. Harvey and C. W. Luetje, *J. Neurosci.*, **16**, 3798–3806 (1996).

270. C. Virginio, A. Giacometti, L. Aldegheri, J. M. Rimland, and G. C. Terstappen, *Eur. J. Pharmacol.*, **445**, 153–161 (2002).

271. S. C. Harvey, F. N. Maddox, and C. W. Luetje, *J. Neurochem.*, **67**, 1953–1959 (1996).

272. B. Buisson, M. Gopalakrishnan, S. P. Arneric, J. P. Sullivan, and D. Bertrand, *J. Neurosci.*, **16**, 7880–7891 (1996).

273. L. E. Chavez-Noriega, A. Gillespie, K. A. Stauderman, J. H. Crona, B. O. Claeps, K. J. Elliott, R. T. Reid, T. S. Rao, G. Veliçelebi, M. M. Harpold, E. C. Johnson, and J. Corey-Naeve, *Neuropharmacology*, **39**, 2543–2560 (2000).

274. J. M. McIntosh, A. D. Santos, and B. M. Olivera, *Ann. Rev. Biochem*, **68**, 59–88 (1999).

275. J. M. McIntosh and R. M. Jones, *Toxicon*, **39**, 1447–1451 (2001).

276. R. M. Hann, O. R. Pagán, L. M. Gregory, T. Jácome, and V. A. Eterovic, *Biochemistry*, **36**, 9051–9056 (1997).

277. J. S. Martinez, B. M. Olivera, W. R. Gray, A. G. Craig, D. R. Groebe, S. N. Abramson, and J. M. McIntosh, *Biochemistry*, **34**, 14519–14526 (1995).

278. G. E. Cartier, D. Yoshikami, W. R. Gray, S. Luo, B. M. Olivera, and J. M. McIntosh, *J. Biol. Chem.*, **271**, 7522–7528 (1996).

279. S. Luo, J. M. Kulak, G. E. Cartier, R. B. Jacobsen, D. Yoshikami, B. M. Olivera, and J. M. McIntosh, *J. Neurosci.*, **18**, 8571–8579 (1998).

280. P. Ascher, W. A. Large, and H. P. Rang, *J. Physiol. (London)*, **295**, 139–170 (1979).

281. A. M. Gurney and H. P. Rang, *Br. J. Pharmacol.*, **82**, 623–642 (1984).

282. J. M. Nooney, J. A. Peters, and J. J. Lambert, *J. Physiol. (London)*, **455**, 503–527 (1992b).

283. A. B. Cachelin and G. Rust, *Pflügers Arch.-Eur. J. Physiol.*, **429**, 449–451 (1995).

284. C. Erostegui, C. H. Norris, and R. P. Bobbin, *Hearing Res.*, **74**, 135–147 (1994).

285. D. Colquhoun in A. T. Birmingham and D. A. Brown, Eds., *Landmarks in Pharmacology* [Suppl. to Vol. 120(4) of *Br. J. Pharmacol.*], Macmillan, London, 1997, pp. 57–59.

286. W. D. M. Paton and E. J. Zaimis, *Br. J. Pharmacol.*, **4**, 381–400 (1949).

287. W. D. M. Paton and E. J. Zaimis, *Br. J. Pharmacol.*, **6**, 155–168 (1951).

288. B. D. Burns and W. D. M. Paton, *J. Physiol. (London)*, **115**, 41–73 (1951).

289. J. M. Young, R. D. Shytle, P. R. Sanberg, and T. P. George, *Clin. Ther.*, **23**, 532–565 (2001).

290. D. Bertrand, M. Ballivet, and D. Rungger, *Proc. Natl. Acad. Sci. USA*, **87**, 1993–1997 (1990).

291. R. A. Giniatullin, E. Sokolova, S. Di Angelantonio, A. Skorinkin, M. Talantova, and A. Nistri, *Mol. Pharmacol.*, **58**, 778–787 (2000).

292. M. Amador and J. A. Dani, *J. Neurosci.*, **15**, 4525–4532 (1995).

293. P. B. Clarke, I. Chaudieu, H. el Bizri, P. Boksa, M. Quik, B. A. Esplin, and R. Capek, *Br. J. Pharmacol.*, **111**, 397–405 (1994).

294. A. Neely and C. J. Lingle, *Biophys. J.*, **50**, 981–986 (1986).

295. A. B. Cachelin and G. Rust, *Mol. Pharmacol.*, **46**, 1168–1174 (1994).

296. C. Hatton, J. Chen, C. Shelley, R. Croxen, D. Beeson, and D. Colquhoun, *J. Physiol. (London)*, 110P, 527P (2000).

CHAPTER TWELVE

Large-Scale Synthesis

FRANK GUPTON
Boehringer Ingelheim Chemicals, Inc.
Ridgefield, Connecticut

KARL GROZINGER
Boehringer Ingelheim Pharmaceuticals, Inc.
Ridgefield, Connecticut

Contents

Burger's Medicinal Chemistry and Drug Discovery
Sixth Edition, Volume 2: Drug Development
Edited by Donald J. Abraham
ISBN 0-471-37028-2   © 2003 John Wiley & Sons, Inc.

1 INTRODUCTION

The ability to produce active pharmaceutical ingredients (APIs) to support the various disciplines of the drug development process is an enabling element of pharmaceutical product development. In the initial stages of drug development, bulk active materials are typically supplied from bench-scale laboratory synthesis. However, API requirements can quickly exceed the capacity of normal laboratory operations, thus providing the need to carry out the synthesis of the drug candidate on a larger scale. The first portion of this chapter is devoted to providing a general overview of the issues and requirements associated with the scale-up of chemical processes from the laboratory to pilot and commercial-scale operations.

The second portion of this chapter describes the process development of nevirapine, a novel nonnucleoside reverse transcriptase (NNRT) inhibitor used in the treatment of AIDS. This case study details the evolution of the nevirapine process from conception in medicinal chemistry through process development, pilot plant scale-up, and commercial launch of the bulk active drug substance. Restricting the case study to nevirapine allows the process and rationale to be described in more detail. The authors are aware of the vast amount of excellent process development that has been performed in the commercialization of other drug products. The processes described herein are not necessarily a unique solution to this particular synthesis. To some extent, they reflect the culture, philosophy, raw materials, equipment, and synthetic tools available during this period of time (1990–1996) as well as the initiatives of the process chemists.

2 SCALE-UP

2.1 General

The process development and scale-up of APIs require a multidisciplinary cooperation between organic chemists, analytical chemists, quality control, quality assurance, engineers, and plant operations. Furthermore, the development of a drug candidate requires collaboration with pharmaceutics for formulation studies, drug metabolism and pharmacokinetics, toxicology, clinical, purchasing, and marketing. Outsourcing specialists, working in concert with purchasing, also play a key role in the identification, coordination, and procurement of key raw materials in support of the scale-up effort. This particular function has gained greater importance in recent years as a result of the increasing emphasis in the pharmaceutical industry to improve the overall efficiency of the drug development process.

In the early stages of process development, the chemist must often balance the need to optimize each synthetic step with the API delivery requirements for toxicology, formulation, and clinical trials. To fulfill these requirements, the process chemist may often scale-up a process in the pilot plant with less than optimal process conditions. As a result, the first quantities of API produced in the pilot plant can be the most time consuming to prepare. However, as the drug candidate passes through the various stages of drug development, the probability of commercialization increases and the need to address the commercial viability of the process becomes more important. This section presents an overview of the issues associated with the preparation of multikilogram quantities of APIs throughout the drug development process.

2.2 Synthetic Strategy

The types of development activities that are associated with the large-scale synthesis of a drug candidate can be divided into a series of discrete functions. Although the terminology used to describe these activities may vary, for the purpose of these discussions the specific functions of the drug development process related to chemical synthesis will be divided into the following three categories: (*1*) chemical de-

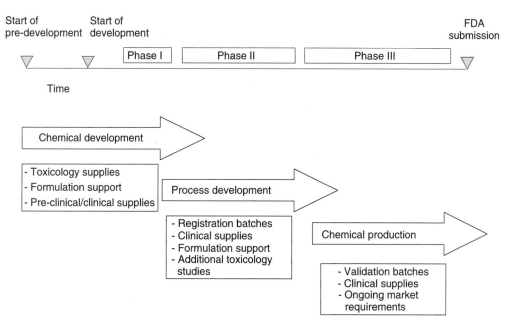

Figure 12.1. Large-scale synthesis requirements for drug.

velopment, (2) process development, and (3) commercial production.

Figure 12.1 indicates the specific areas of the drug development process where each of these activities occurs. Although each function has specific requirements and outputs from its respective activities, the overlap that is indicated between these activities is critical to the successful implementation of the project.

In the initial stages of *chemical development*, the focus of the effort is to supply materials to assess the viability of the drug candidate. The emphasis of this effort is on the expeditious supply of these materials rather than the commercial viability of the process used to produce the compound. Unique raw materials, reagents, solvents, reaction conditions, and purification techniques can and will be employed in this phase of the process to produce the desired compound in a timely fashion. The initial transition from the laboratory to pilot-scale operations typically takes place during this portion of the drug development process to supply larger quantities of the bulk active material for toxicology, formulation, and preclinical evaluations. As the project proceeds through drug development,

chemical development personnel continue to evaluate potential improvements to the synthesis. The insights obtained from these efforts provide the platform for future process development investigation.

The role of *process development* is to balance the timeline and material requirements of the project with the need to develop a commercially viable method for the preparation of the drug candidate. This stage of the drug development process will concentrate on such issues as (1) synthetic strategy, (2) improvement of individual reaction yields, (3) identification and use of commercially available raw materials and reagents, (4) evaluation of alternative solvent systems, (5) compatibility of process conditions with existing manufacturing assets, (6) identification and quantification of potential process safety hazards, (7) simplification of purification methods, (8) evaluation of process waste streams, and (9) the improvement of the overall process economics.

Both chemical and process development activities typically require that the drug candidate be prepared on a pilot plant scale. Although the batch size may vary depending on the drug substance requirements, these oper-

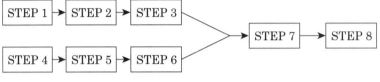

Case I

STEP 1 → STEP 2 → STEP 3 → STEP 4 → STEP 5 → STEP 6 → STEP 7 → STEP 8

Case II

STEP 1 → STEP 2 → STEP 3

STEP 4 → STEP 5 → STEP 6

→ STEP 7 → STEP 8

Case III

STEP 1

STEP 2

→ STEP 5

STEP 3

STEP 4

→ STEP 6

→ STEP 7

ations are usually conducted in 100- to 2000-L reactors. The scale-up factor from the laboratory to the pilot plant is quite large (1 to 200 or more), and particular emphasis is placed on detailed safety analysis of the scale-up. The outcome of these efforts is a documented process that is included in the drug submission package to the U.S. Food and Drug Administration (FDA).

The overall objective of *chemical production* activities is to reproduce the process that has been transferred from process development to meet the current and future market requirements for the drug product. Particular emphasis is placed on issues related to process safety, environmental issues, equipment requirements, and production economics. The scale-up factor from the pilot plant to commercial production is usually rather small (approximately 1–20). As a result, the informa-

tion obtained from the process development efforts can be quite valuable in the successful implementation of the commercial process. The reproducibility of the process is confirmed and documented as part of the process validation package, which in turn is part of the transfer process.

2.2.1 Route Selection. When considering the merits of alternate synthetic pathways to produce a specific molecule, the route that incorporates the most convergent subroutes is generally the most advantageous option, provided yields for the individual steps are essentially equivalent (1). For example, an 8-step linear synthesis, in which each step has an 85% yield, results in a 27% overall yield (Case I). However, if the eight steps can be divided into two 3-step converging pathways leading into two final steps, as in Case II, the overall process yield is increased to 44%, which is a 63% improvement over the Case I scenario. Furthermore, if the process is broken down to even shorter converging pathways, as in Case III, the overall yield improves by 125%, from Case I, to 61%.

In addition to the obvious yield advantages, an important benefit of a convergent approach is the proximity of the starting materials to the product. In Case II, the raw materials are only five steps away from the product, and only three steps away in Case III. This can significantly reduce the time required to respond to any need for additional product. Also, the value of each intermediate in a linear synthesis becomes greater with each additional step as a result of the resources required to produce material from that step. In a convergent synthesis, the cost is spread over two or more intermediates, thus reducing the overall risk in the event that material losses occur. In many cases, putting the major components to the molecule together can also simplify the regulatory filing requirements because these intermediates may be classified as starting materials under current FDA guidelines.

2.2.2 Chiral Requirements. Over the last several decades, drug development efforts have placed increasing emphasis on the development of the biologically active stereoisomers of drug products. Chiral APIs offer the opportunity to provide higher drug potency while reducing the metabolic burden and risk of undesirable side effects to the patient (2). It has been estimated that over half of the best-selling drugs worldwide are single enantiomers (3). As a result, the process chemist is presented with the challenge of developing commercially viable processes for the production and isolation of these chiral compounds. Several approaches can be used to produce enantiomerically enriched bulk active pharmaceutical products. The resolution of racemic mixtures with chiral adjuvants has been a common approach in the past to isolate the desired optical isomer of drug products. Chiral amines and acids are typically used to isolate an enantiomer by crystallization of the diastereomeric salt. The major drawback with this approach is the significant loss of material as the undesired enantiomer. This can be miti-

gated by racemization of the off isomer followed by recycling of the racemate back into the resolution. However, the equipment requirements to execute this procedure can be significant and must be justified economically.

An alternative approach for the preparation of chiral APIs is the use of chiral raw materials. The increased availability of functionalized chiral raw materials from both synthetic as well as natural sources has made this a more viable option in recent years. In the event that the desired chiral precursors are not commercially available, asymmetric synthetic techniques may be employed to introduce one or more stereogenic centers into the molecule. Many elegant techniques have been developed using chiral induction, chiral templates, and chiral catalysts to produce enantiomerically enriched drug substances, and this area of research continues to be at the forefront of organic chemistry.

Regardless of the approach that is used to introduce the stereogenic center into the molecule, a significant cost is incurred in achieving this objective. For this reason, it is important to introduce the chiral component later in the synthesis and employ the principles of convergent synthesis (Section 2.2.1) to effectively minimize the impact of this cost to the overall process economics.

2.3 Bench-Scale Experimentation

To scale up a chemical process to pilot or commercial-scale operations, a significant laboratory effort is required to define the operating ranges of the critical process parameters. A critical process parameter is any process variable that may potentially affect the product quality or yield. This information is required to prepare a Process Risk Analysis, which is an FDA prerequisite for process validation. Process parameters that are often evaluated as part of the risk analysis include reaction temperature, solvent systems, reaction time, raw material and reagent ratios, rate and orders of addition, agitation, and reaction concentration. If catalysts are employed as part of the process, additional laboratory evaluation may also be required to further define the process limits. Experimental design is often used for the evaluation of critical process parameters to minimize the total laboratory effort (4).

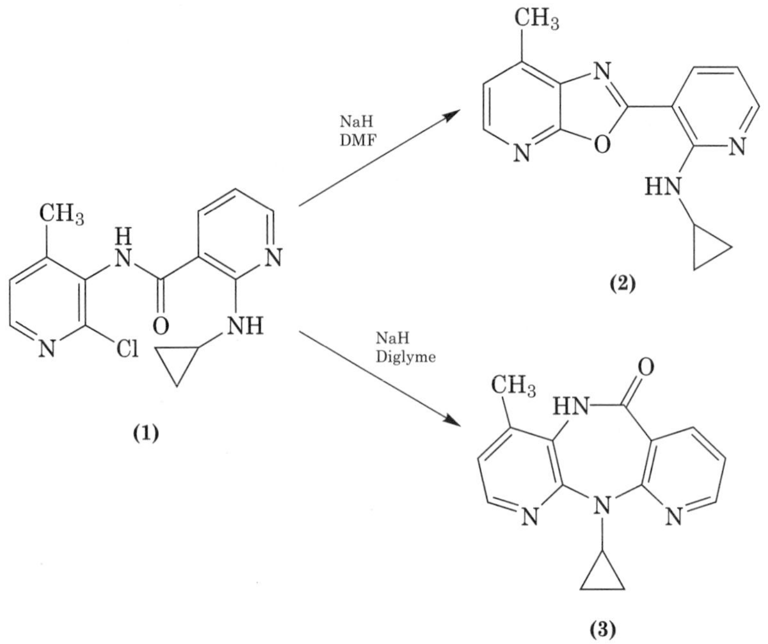

Scheme 12.1.

This technique is equally important in identifying interdependent process parameters that can have a synergistic impact on product yield and quality. In-process controls (IPCs) are also defined during this phase of the development process. All of these bench-scale activities help to provide a better understanding of the capabilities and limitations of the process and are discussed in further detail in this section.

2.3.1 Selection of Reaction Solvents. Solvents are generally used to promote the solubility of reagents and starting materials in a reaction mixture. Reactants in solution typically undergo conversion to product at a higher rate of reaction and are generally easier to scale up because of the elimination of mass transfer issues. For this reason, the solubility properties of the reagents and raw materials are a major consideration in the solvent selection process for scale-up. In addition, the solvent must be chemically compatible with the reagents and raw materials to avoid adverse side reactions. For example, an alcohol solvent would be a poor choice for a reaction when a strong base such as butyl lithium is

being employed as a reagent. Information pertaining to the physical properties of solvents is available to assist in the solvent selection process (5).

Solvents can also be used to promote product isolation and purification. An ideal solvent system is one that exhibits high solubility with the reagents and starting materials but only limited solubility with the reaction product. Precipitation of the reaction product from the mixture can increase the reaction rate, drive reactions in equilibrium to completion, and isolate the product in the solid state to minimize the risk of undesirable side reactions. Solvents can also aid in the regio control of the reaction pathway. It was found in the preparation of nevirapine (**3**) that, when diglyme was used as the reaction solvent with sodium hydride, the ring closure of (**1**) (**Scheme 12.1**) proceeded by the desired reaction pathway (6). However, when dimethyl formamide was used for this reaction, the exclusive product was the oxazolopyridine (**2**). In this particular case, the solvation effects may have helped to stabilize the transition state of the desired product.

One of the most challenging aspects of solvent selection is the avoidance of certain classes of solvent that are routinely used in laboratory operations but are inappropriate for pilot and commercial-scale applications. Solvents such as benzene and 1,4-dioxane can present significant health risks to employees (7) handling large quantities of these materials. Toluene is routinely used as a commercial substitute for benzene and other aromatic solvents. Likewise, solvents that promote peroxide formation such as diethyl ether and tetrahydrofuran present significant safety hazards in scale-up operations (8). Methyl *tert*-butyl ether is a good commercial substitute for these materials. The autoignition temperature of the solvent should also be considered against the process operating conditions and electrical classifications of the equipment being used. With regard to environmental issues, several chlorinated solvents have been identified as priority pollutants (9) and can present permitting issues if adequate environmental containment capabilities are not incorporated in the scale-up facility. Although specific health, safety, and environmental issues for a given solvent can usually be addressed, it is important to evaluate the advantages of using an undesirable solvent against the additional cost and operational constraints that are imposed on the process.

2.3.2 Reaction Temperature. Before conducting a reaction temperature profile experiment, it is important to understand the temperature limitations of the specific scale-up equipment that is to be used. For example, the typical operating temperature range for a pilot or production facility that employs a silicone-based heat transfer system is $-20-180°C$. It is also important to understand the capabilities of the temperature control system used in the scale-up facility. The selected reaction temperature range must also be consistent with the accuracy and precision limits of the equipment. Given these constraints, the objective of this effort is to identify the optimal temperature range that gives the maximum conversion of starting materials to product in the shortest period of time and with the minimum amount of impurity formation. A general rule for the evaluation of reaction temperature is that increasing the reaction temperature by $10°C$ will generally double the reaction rate.

However, this will also increase the potential for by-product formation that could adversely impact both product yield and quality. The optimal temperature range is typically a balance between these three dependent variables.

2.3.3 Reaction Time. In a laboratory environment, reactions are often run overnight with limited concern for the actual time requirements to complete the reaction. When selecting the reaction time for a specific process step to be scaled up, consideration should be given both to the potential reaction yield improvement and to the equipment utilization requirements. In many cases doubling the reaction time will result in only a small percentage increase in yield. The cost of the additional equipment time can more than offset the potential yield benefit in cases in which the raw material costs are low. However, in cases where raw materials of high cost and chemical complexity are employed, the additional reaction time may be easily justified on an economic basis.

Consideration should also be given to the quantification of potential adverse effects from extending the reaction time beyond the optimum condition. Product decomposition and by-product formation are often observed under these circumstances. This information can be beneficial in scale-up operations when reaction times are extended beyond the specified period because of unforeseen circumstances. This information is also important in evaluation of this variable as a potential critical process parameter for the process risk assessment.

2.3.4 Reaction Stoichiometry and Order of Addition. Reaction rates, product yields, and by-product formation can often be effectively managed by the selection of the appropriate ratios of reactants and raw materials as well as by the rate and order of addition of these materials. A fundamental mechanistic understanding of the process is essential for the effective evaluation of these parameters. Reaction kinetic information can be beneficial in defining the limiting reagent for the reaction under evaluation. More often, the financial impact of specific raw materials will be a key driver of the overall process economics and, as a result, optimization efforts will focus on the minimization of these materials. This issue has become of increasing importance because of the chemical complexity of advanced starting materials in bulk pharmaceutical production. Likewise, a statistical design of experiments can assist in the evaluation of multiple process parameters and also identify interactions between multiple process variables.

The minimization of by-product formation can be a particularly difficult task because of the high degree of chemical functionality in bulk pharmaceutical intermediates and products. Oligomerization reactions are a major mode of impurity formation in these types of chemical processes and can often be effectively minimized by the control of addition rates. Characterization of these impurities can also provide valuable insights into the control of these side reactions. The order and rate of addition are also frequently used to control extremely exothermic reactions. Chlorinating reagents such as thionyl chloride and phosphorous oxychloride, as well as strong bases such as butyl lithium, lithium diisopropylamide, and sodium hydride are usually added in a controlled manner to limit both heat and by-product formation in these reactions.

2.3.5 Solid-State Requirements.

The solid-state properties of active pharmaceutical ingredients can have a dramatic impact on critical dosage form parameters such as bioavailability and product stability. For this reason, FDA filing requirements include the definitive characterization of drug substance physical properties as part of the NDA information package. Formulation activities during the drug development process are directly linked to these parameters, and control of these physical properties during laboratory, pilot, and commercial-scale operations can be challenging.

The particle size distribution of the API can affect the dissolution rate of the drug product and thus the bioavailability of the product. Once particle size requirements have been defined from formulation studies, the process must be capable of routinely meeting these requirements. One of the ways that particle size distribution can be controlled is by the conditions under which the product is crystallized. Typically for cooling crystallizations, the particle size distribution is dependent on the rate of cooling. Generally, smaller size particles are formed under rapid cooling conditions, whereas larger crystal growth is experienced with slower cooling rates. Milling and grinding techniques can also control particle size. However, these methods exclusively result in particle size reduction. Both the milling conditions and the solid-state characteristics of the bulk active material being charged to the mill thus determine the particle size distribution of the API. Milling parameters are discussed in further detail in Section 2.4.4.

Bulk drug products often exist in different crystalline or polymorphic forms. Because the polymorphs of a specific API can exhibit distinguishably different bulk stability properties as well as bioavailability characteristics as a result of the differences in surface area between the different crystalline forms, specification of the polymorphic form is recommended for FDA submission. Products such as ranitidine (10), lorazepam (11), and natamycin (12) serve as examples of APIs that exist in several different polymorphic forms. The solvent system and the crystallization conditions generally determine the specific crystallization form that is isolated. Polymorph selection for regulatory submission is usually based on the ability to reliably produce and process the material in the same crystalline form. In many cases this is the thermodynamically most stable polymorphic form. In the event that a less stable polymorphic form is desired, because of stability or bioavailability issues, seeding techniques can be used to control the crystallization selectivity of a specific polymorph.

2.4 Scale-Up from Bench to Pilot Plant

Bulk active pharmaceutical ingredients are most often produced at the pilot scale under batch-mode operations with multipurpose equipment. In contrast, continuous operations are typically reserved for high volume products that can be produced in dedicated facilities. For this reason, these discussions are restricted to issues associated with batch operations. From a procedural perspective, batch operations more closely resemble bench-scale operations. However, the successful transformation of bench-scale experiments in laboratory glassware to pilot and commercial-scale operations requires a more detailed understanding of the physical issues related to scale-up, such as heating and cooling requirements, agitation, liquid-solid separation techniques, and solids handling requirements. Particular emphasis is placed on understanding the thermal requirements because this can often be the area of greatest perceived risk. This can influence the rate of by-product formation, which has an impact on both the impurity profile and the yield. Fortunately, reactions proceed by the same mechanism regardless of the scale, and problems in scale-up are typically restricted to physical parameters.

2.4.1 Heating and Cooling.

A pilot plant is generally outfitted with multipurpose vessels that can obtain an operating temperature range of -20 to $+150°C$. Broader temperature ranges can be obtained with silicone-based heat-transfer fluids such as Syltherm. Temperatures lower than $-20°C$ are sometimes required in API production and can be achieved with liquid nitrogen cooling systems.

The heating and cooling capabilities of a reactor system are determined by several factors. Variables such as reactor surface area, materials of construction, the temperature of the heating and cooling media, and the heat capacity of the reactor contents contribute to the thermal properties of the reactor system. The effects of these parameters on heating and cooling are greatly magnified upon scale-up from the bench to the pilot plant. For example, a 250-mL round-bottom flask in the laboratory has a large surface area to volume ratio. As a result, the flask can be heated and cooled quickly. In comparison, the surface area to volume ratio of a 100-L glass-lined steel reactor is drastically reduced and may influence the ability to control the reactor contents effectively. In general, from a 250-mL flask to a 100-L reactor, the surface area vs. volume is reduced by a factor of 10. Likewise, the surface heat constant (k) of a stainless steel reactor is much greater than that of a laboratory reaction flask, which could result in a thermal transfer that is much more rapid than that of the laboratory experience.

This effect of heating and cooling can be calculated as follows (13):

$$T = t_s - (t_s - t^0)e^{-kF/C}$$

where T is the temperature of the vessel in °C, t^0 is t at the beginning of the heating, t_s is the temperature of the heat-exchange fluid, F is the reactor surface, k is the heat constant on the surface (kcal m^{-2} h^{-1} C^{-1}), kF is the heat surface, and C is the heat capacity of the reaction vessel with contents.

2.4.2 Agitation.

The key function of agitation is to ensure homogeneity of the reactor contents. The major factors that affect reactant homogeneity are both the reactor-agitator configuration and the physical properties of the reactor contents. Miscible liquids of low viscosity, such as ethanol and water, represent mixtures with which one can easily attain homogeneity with minimal agitation. As one might expect, biphasic mixtures require more vigorous agitation than miscible solutions. The extent of the additional agitation requirement is dependent on the viscosities of the individual phases. Liquid-solid mixtures also require greater agitation to maintain homogeneity. In many cases the solid is formed later in the process, resulting in different agitation requirements over the duration of the reaction.

Catalytic hydrogenations can represent some of the most challenging agitation issues. A typical hydrogenation reaction will require the dispersion of a heterogeneous catalyst and hydrogen gas throughout a specific solution containing the material that is to undergo the reduction. Hydrogenation agitators are often specifically designed to maximize the dispersion of the hydrogen gas throughout the liquid phase.

The ability to transfer heat to the reaction mixture is also a function of agitation. A typical agitation heat-transfer correlation is as follows:

$$k \propto \frac{L^{4/3}N^{2/3}}{D}$$

where k is the surface heat constant, L is the agitator impeller length, N is the agitator speed, and D is the vessel diameter.

2.4.3 Liquid-Solid Separations. In the majority of drug syntheses, the reaction product is a solid. The isolation of the solid product from the reaction mixture is often accomplished in bench-scale operations by rotary evaporation of the volatile components of the reaction mixture, leaving a solid residue that is easily recovered. This technique is clearly not amenable to scale-up, and therefore alternate methods of solids isolation are required. Crystallization of the desired product from the reaction mixture is the most desirable approach as the first step to product isolation. Laboratory, pilot, and commercial-scale crystallizations are typically carried out by cooling, evaporative concentration, or by pH adjustment to precipitate the salt form of the product. However, the use of cosolvents to reduce the product solubility can also be effective in promoting dissolution. Typical liquid-solid slurries are manageable in the 20–30% solids range in a pilot plant or commercial operation. At higher solids concentrations the transfers become more difficult.

Separation of the solid product from the liquid phase is usually accomplished at the bench scale by vacuum filtration through a single-stage filter such as a Buchner funnel. Although pilot and commercial-scale facilities are equipped with similar types of equipment, centrifugation is commonly used for liquid-solid separations. This is particularly true for commercial-scale operations. One of the major advantages of centrifuge systems is their ability to effectively remove liquid from a product cake. This can result in a significant reduction in both the product drying time requirements and the impurity content. For example, the residual solvent content of solids isolated by centrifugation is typically in the 5–10% range, whereas solids isolated by vacuum filtration can be in the 20–30% range. Measurement of filtration rates and cake compressibility at the bench scale can provide valuable insights into the commercial feasibility of the isolation conditions and the selection of appropriate equipment.

2.4.4 Drying and Solid Handling. Drying operations under laboratory conditions are typically restricted to the use of vacuum ovens. Similar types of equipment are often used in pilot operations and are commonly referred to as tray dryers. These types of dryers fall into a specific FDA class of dryer systems referred to as indirect conduction heating static solid-bed dryers and are very versatile when processing wet solids that are difficult to dry. One of the drawbacks with these systems is the static nature of the drying operation that limits the ability for heat transfer to occur across the solid mass. In addition, these units are very labor intensive and can present significant industrial hygiene and validation challenges on a commercial scale. For these reasons, pilot plants are often equipped with a variety of types of dryers to make an effective transition between the laboratory and commercial-scale operations.

The most commonly used commercial drying systems are rotary tumble dryers. This type of dryer falls into the FDA classification of indirect conduction, moving solids bed dryers. These units work well for free-flowing solids that have high volume requirements but are less effective with solids that have a tendency to agglomerate and cake while drying. Agitated drying systems such as paddle and spherical dryers are another type of solids drying system that are of the same FDA dryer class as the rotary tumble dryers. These units typically have a fixed heated surface and internal agitation to maximize heat transfer while breaking up any agglomerated solids. Agitated dryers are often outfitted with chopper attachments to the agitation system that can also effect particle size reduction and avoid an additional milling step. As a result, these units can provide high throughput drying of a variety of difficult-to-handle materials, are applicable for both pilot and commercial applications, and are commonly found in more modern installations. Fluidized bed dryers represent a second FDA classification of drying system. These units use a hot inert gas flowing at a high velocity to suspend and dry the solid in a finely divided state. This type of dryer equipment falls into the FDA classification of direct heating, dilute solids bed, and flash dryers and has been used for both batch and continuous drying operations on a more limited basis.

Whenever possible, particle size distribution is controlled by crystallization parameters such as agitation and cooling rate. Once the solid material has been isolated and dried, particle size reduction can be achieved by various milling techniques. The particle size requirements as well as the physical properties of the solid dictate the type of milling equipment used for a specific application. The particle size requirements are usually defined during drug formulation development and impact the bioavailability of the drug candidate. Some of the physical properties of the solid that can affect the selection of milling equipment and conditions include hardness, crystal morphology, and thermal stability. The stability of the solid is a critical issue with regard to milling operations because of the energy applied by the milling equipment. Fluid impact mills, such as jet mills, are one type of milling equipment that is used in both development and commercial applications. These mills promote particle size reduction through high speed particle-to-particle collisions. In contrast, impact mills, such as hammer and pin mills, impart particle size reduction by particle-to-mill surface as well as particle-to-particle collisions. These units are also used routinely in pilot and production environments. Other milling techniques such as compression milling and particle size classification can also be applied, depending on the particle size specifications and the physical properties of the solid.

2.4.5 Safety. When transferring processes from the laboratory to the pilot plant, it is important to identify and address potential safety issues as early as possible in the transfer process. Typically, calorimetry studies and process hazard reviews are carried out to meet this requirement. Calorimetry experiments can assist in the identification and quantification of reaction exotherms associated with the process. This information can then be used to determine the capability of the pilot equipment to control the reaction.

Process hazard reviews are conducted to identify potential hazards that could occur as a result of operational failures such as loss of power or cooling capacity. The process hazard review is conducted subsequent to the calorimetry studies to obtain the benefit of this additional information in the assessment of

risk. The compatibility of the pilot plant electrical classification with the process solvents, reagents, and solids is also evaluated as part of this process. Predictive evaluations are often made during the process hazard review by using information obtained from reactions carried out using similar reaction conditions and raw materials. In addition, the pilot plant materials of construction should also be evaluated against the reaction conditions that are to be employed as part of the scope of the review process.

Because these drug candidates have potential biological activity, precautions should be taken to limit worker exposure during scale-up operations. Personal protective equipment requirements and adequate containment and ventilation provisions should also be defined as part of the safety review process. Often this assessment can be difficult because the material produced from the pilot plant will be used for toxicology evaluation purposes. In these cases, structure-activity relationship evaluations with regard to the relative toxicity of the compound may be appropriate to estimate the extent of risk.

2.5 Commercial-Scale Operations

The commercial implementation of a new process is primarily dependent on three factors: (1) the quality of the information obtained from laboratory and piloting efforts, (2) the effective transfer of the knowledge gained from these efforts, and (3) the ability to match the process requirements with the production capabilities. The production capabilities may be new and/or existing but in all cases should incorporate the effective utilization of existing assets while meeting the process requirements. Fortunately, the scale-up factor from the pilot plant to commercial operations is usually 1 to 20 or less, so that pilot information can be easily transferred to commercial practice. Likewise, the information transfer can be facilitated by the participation of production personnel in process development scale-up operations. The issues of equipment requirements, implementation of in-process controls, and validation requirements, as well as safety and environmental matters related to commercial production, are addressed in this section.

2.5.1 Identification of Processing Equipment Requirements. When transferring processes from pilot to commercial-scale operations, a comparative analysis is usually made between the equipment used in the pilot operation with the proposed commercial facility. Process flow diagrams (PFDs) that include material balances from pilot plant experiments can facilitate this analysis. Specifications and requirements for agitation, filtration, drying, and milling devices are established based on experimental results that support these specifications and are documented.

Vent treatment requirements are also established during the evaluation of the process equipment requirements. The compatibility of the existing vent gas treatment system is evaluated against the process information obtained from the pilot runs and the existing environmental permit constraints. Permissible levels of venting are then established based on this assessment and the design requirements are documented.

Process streams from pilot plant experiments are usually retained to evaluate the compatibility of various types of materials of construction. Mass balance information is often sufficient to determine these requirements based on pH, halide content, solvents used, and process temperature. Corrosion testing of pilot plant process streams through the use of electrochemical techniques is often recommended. These results are compiled and documented along with the specifications for the materials of construction.

The process requirements for both temperature and pressure should also be evaluated against the production equipment capabilities as part of the production equipment assessment. Normal operating conditions are used as a base case, but upset conditions should also be included as part of the evaluation. If venting is chosen to control unintended reactions, vent sizing calculations must be performed and peripheral equipment selected as needed. Experiments and simulations to determine consequences of unintended reactions and the interpretation of these experiments are documented as part of the production process safety review.

The ignition prevention requirements for the process must also be defined based on electrical classification of solvents used in the process as well as the ignition characteristics of dry powders used in the process. Preferably, dry powder characteristics such as minimum ignition energy and temperature can be established based on testing of the solid materials. However, dry powder characteristics may also be estimated based on experience with similar materials. Experiments to establish ignition characteristics and the interpretation of experimental results should also be documented as part of the production process safety review.

2.5.2 In-Process Controls. In-process controls (IPCs) are used in both pilot and commercial operations to confirm that the process is in control and that the reactions and unit operations have been carried out to their expected completion point. Other process control points such as pH measurement, reactor content volume, distillation end points, filter cake washings, and drying end points are often considered to be critical process parameters that are also incorporated into IPCs. In pilot operations, numerous IPCs are taken to establish benchmarks for various process parameters, such as reaction time, drying time, distillation end points, and many other process variables. In a production environment, these benchmarks have been established and fewer IPCs are typically used to control the process. It is important when transferring a process from the pilot plant to review, identify, and separate the IPCs that were used only to establish process benchmarks in the pilot operations from those IPCs that would be appropriate for routine commercial operations. The appropriate commercial IPCs should then be documented and incorporated into the production procedure. This is important because the New Drug Application (NDA) will define the IPCs in the Chemistry, Manufacturing, and Controls (CMC) section of the submission, and the elimination of an unnecessary control at this point can be quite time consuming.

2.5.3 Validation. The purpose of a validation program is to establish documented evidence that provides a high degree of assurance that specified processes consistently produce product that meet predetermined specifications and quality attributes. The validation program ensures that all systems, instruments, and equipment that impact the quality or integrity of the product have been validated.

The validation program is composed of several different elements and is designed to ensure that all validation requirements are addressed. General validation requirements for each of these elements are outlined in a master plan. Descriptions of the various validation elements are as follows:

Equipment/Systems Qualification. Qualification of the equipment in which the product is manufactured, the support services, and computer systems supporting the process.

Process Validation. Validation of the manufacturing process through the execution of production batches to establish that all product performance criteria have been met.

Cleaning Validation. Validation of the cleaning procedures used to clean the product from production equipment.

Method Validation. Validation of the analytical methods used to support the process validation, cleaning validation, in-process testing, and release testing of the product.

All FDA-regulated products should be validated. Validation of each product is performed in a phased approach, encompassing all of the elements mentioned. The typical sequence for these activities is shown in the following timeline:

In addition to validating these elements, several support programs must be in place to ensure that, once validated, manufacturing processes will be maintained in a validated state on an ongoing basis. These programs include calibration, preventive maintenance (PM), personnel training, and change control. The *personnel training* program should be designed so that all operational procedures and requirements are defined and communicated. It is also important that documentation of the training activities is completed and readily retrievable.

The *calibration* program ensures that instruments associated with manufacturing processes are calibrated and maintained. Instruments in the facility are typically classified as either critical, noncritical, or reference. Those instruments that are deemed either critical or noncritical are typically calibrated using NIST or other applicable standards on a routine basis.

A *preventive maintenance* (PM) program should be in place to support the ongoing qualification requirements for all production and support facilities. The objective of the program is to ensure that preventive maintenance requirements for the equipment are carried out throughout the operational life of the equipment. The PM requirements are established using the equipment manufacturer's recommendations and any additional requirements established by the operation site.

Activity		
Support Services Qualification		
Process Equipment Qualification		
Method Validation—Process		
Process Validation		
Method Validation—Cleaning		
Cleaning Validation		

Validation Activity Timeline

A *change control* program is used to regulate the alteration of systems and changes to processes. The program should outline the methods to be followed when a system, process, or equipment change is proposed. The change control program ensures that proposed changes are reviewed and approved by the quality unit and other appropriate departmental representatives before initiating changes. The program should also ensure implemented changes are reviewed before use in manufacturing. It is through this review that any required testing and documentation is defined to verify that the proposed change is acceptable and that the equipment remains in a validated state. The review also ensures that governing regulatory issues for the affected process/operation are addressed.

Validation should be performed in accordance with preapproved written protocols. The execution of the approved validation protocol will generate documentation that supports the intended use of the equipment and demonstrates compliance with current good manufacturing practices (cGMPs). A validation package that contains all documentation relevant to the validation study should be prepared at the completion of each protocol. The validation package should include the summary report that documents the results, observations, and conclusions from the implementation of the protocol, a copy of the protocol, and completed data sheets corresponding to each section of the protocol. The combination of the summary report and the protocol with completed attachments serves as a permanent document of the validation study.

2.5.4 Chemical Safety in Production. Before scaling the process from the pilot plant to commercial scale, a process hazard review is performed based on the additional data obtained during the pilot campaign. This review should include the evaluation of reaction calorimetry data, powder ignitability, and the results of the acute and chronic toxicity testing of all raw materials, intermediates, and the product. The battery of toxicity testing can include mutagenicity, teratogenicity, and carcinogenicity; acute dermal and ocular irritation testing results; absorption routes for raw ma-

terials, intermediates, and products; and potential sensitizers in the process. Antidotes to acutely toxic materials should also be identified as part of this evaluation. Personnel protective equipment such as gloves, goggles, and protective garments should also be reevaluated at this point based on experience gained from pilot plant operations. Any industrial hygiene sampling data obtained from various operations during the pilot studies should also be included in this evaluation.

2.5.5 Environmental Controls in Production. Environmental permit requirements should be evaluated based on the commercial-scale material balance and new equipment specifications. Testing requirements for environmental evaluation should include acute fish and invertebrate toxicity for raw materials, intermediates, and products; biodegradation of raw materials, intermediates, and products; microbial growth inhibition of raw materials, intermediates, and products; water coefficients (KOW) and water solubility for raw materials, intermediates, and products; and waste treatability test results. Particular emphasis should be placed on the evaluation of the compatibility of the new process waste streams with the existing waste-treatment systems. If any process waste streams require off-site disposal into regulated hazardous waste landfills, leaching experiments may also be required.

3   NEVIRAPINE

3.1   Background

In 1986 Boehringer Ingelheim Pharmaceuticals initiated an antiviral HIV research program that focused on the identification of potential nonnucleosidic reverse transcriptase inhibitors with the specific intent to develop an AIDS drug with reduced adverse side effects. A high throughput screening method was established using AZT as a standard. Promising candidates were screened for mammalian DNA polymerase as well as other enzymes and receptors. Nine months after the first lead compound was identified, nevirapine (**3**) was approved as a development candidate.

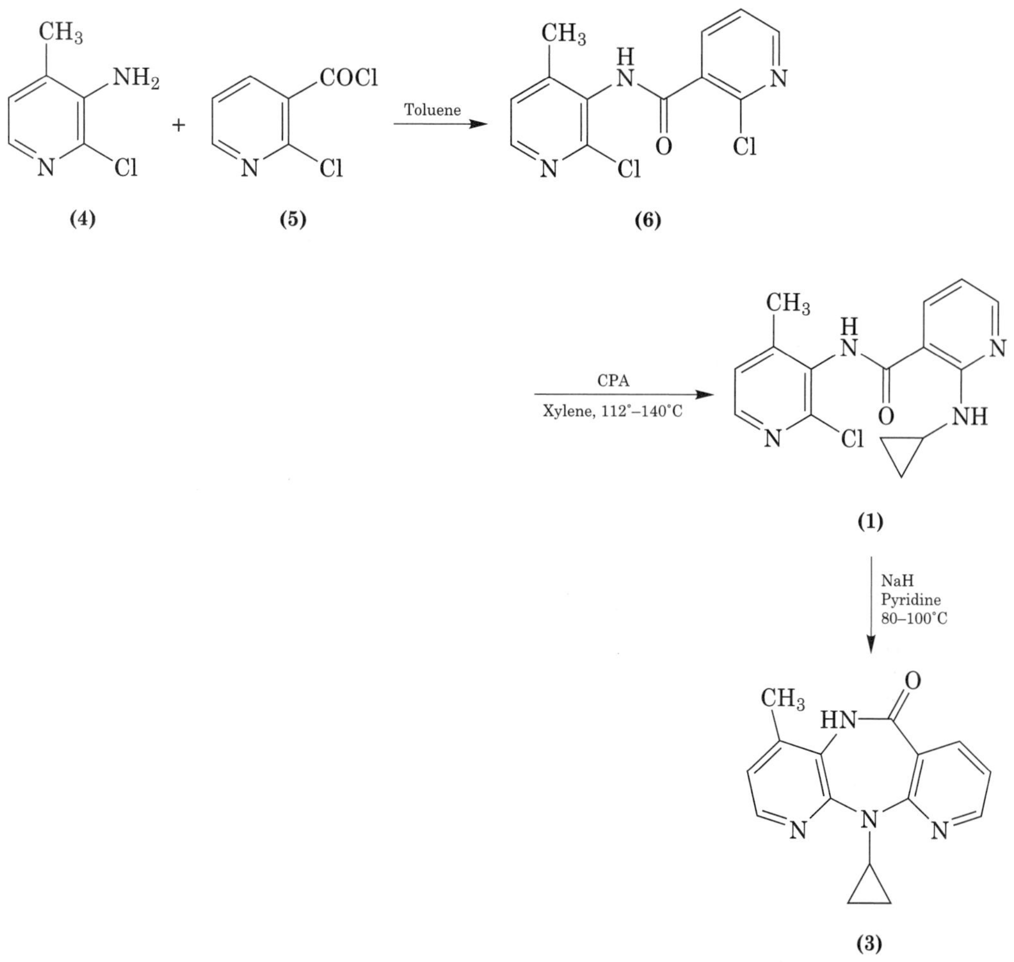

Scheme 12.2.

The nevirapine clinical program focused on both single and combination drug therapies. Clinical results indicated that this material not only was effective in the treatment of HIV-related illness but also was found to be well tolerated and safe. Boehringer Ingelheim submitted an NDA (New Drug Application) in February 1996 and the new AIDS drug was approved in July 1996.

3.2 Evolution of the Nevirapine Synthesis

3.2.1 Medicinal Chemistry Synthetic Route. The initial nevirapine synthesis developed by the Medicinal Chemistry Group entailed the condensation of 2-chloro-3-amino-4-methyl-

pyridine (CAPIC, **4**) with 2-chloronicotinoyl-chloride (**5**), to give the 2,2′-dihaloamide (**6**). Treatment of (**6**) with four equivalents of cyclopropylamine in xylene at 120–140°C under autogenous pressure produced the 2′-alkyl-amino adduct (**1**) followed by ring closure with sodium hydride in pyridine at 80–100°C, to give nevirapine (**3**), as shown in **Scheme 12.2**.

The basic synthetic strategy developed during this phase of drug development was quite sound and provided an excellent starting point for future process development efforts. The synthesis exhibited significant elements of convergence, starting with two functionalized pyridine precursors, which were only three

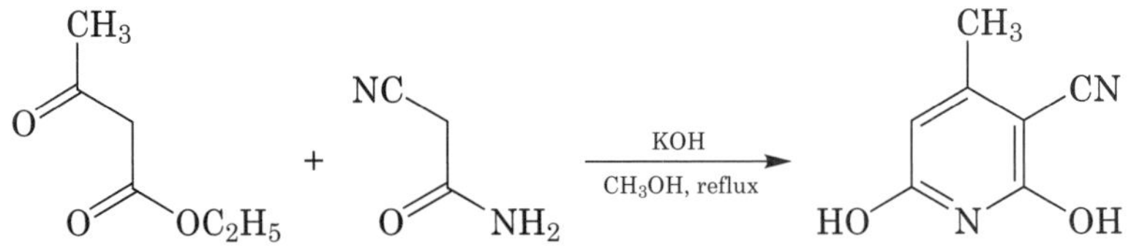

92% Yield

Scheme 12.3.

chemical steps away from the target molecule. However, technical barriers existed that impeded the ability to meet the short-term API requirements for toxicology and clinical supplies, and additional process issues would need to be addressed to obtain a commercially viable synthesis from this point in the nevirapine process development. The most critical short-term issue was with the ability to obtain significant quantities of raw materials to meet the bulk active needs to support the drug evaluation efforts.

The 2-chloronicotinoylchloride (**5**) was easily prepared from 2-chloronicotinic acid, which was commercially available in multi-ton quantities. The most significant initial concern was with the ability to obtain pilot-scale quantities of CAPIC (**4**). Gram quantities of this material were initially obtained by the reduction of 2-chloro-4-methyl-3-nitropyridine. Small quantities of this material were initially obtained from laboratory supply houses, but significant scale-up quantities were not commercially available.

Attempts were made to nitrate both 2-hydroxy-4-methylpyridine and 2-amino-4-methylpyridine, which are commercially available, by conventional synthetic methods (14). However, the major product from both of these reactions was the 5-nitro adduct, with less than 30% of the desired 3-nitro isomer present in the reaction mixture. The yield of the desired isomer was improved to 82% by adding the nitric acid/sulfuric acid premix to the respective substrates at 0–5°C, followed by heating at 60–80°C for 1 h (15). However, the physical separation of the products proved to be industrially impractical because of the similarity in physical properties of the respective isomers.

Based on the information obtained from

these experiences, the initial nitration approach was abandoned. The thrust of chemical and process development activities was redirected toward the development of a CAPIC process that could, as a minimum requirement, be scaled up to produce pilot plant quantities of this raw material to support, toxicology, formulation, and clinical studies.

3.2.2 Chemical Development and Pilot Plant Scale-Up. Having benefited from the experience gained during the development of the nevirapine medicinal chemistry route, chemical development technical efforts initially shifted to the identification and evaluation of synthetic alternatives for the preparation of CAPIC. As previously stated, 2-chloro-3-nitro-4-methylpyridine could be readily converted to CAPIC by catalytic reduction. However, the ability to obtain significant quantities of this material was problematical because of the lack of reaction selectivity. Alternative approaches to the introduction of the 3-amino group were examined and found to be quite promising. Functionalized nicotinonitriles have been produced in high yield from readily available acyclic precursors by the Guareschi-Thorpe condensation (16). The 3-cyano substituent could be readily converted to the corresponding amine by hydrolysis to the amide followed by a Hofmann rearrangement (17). A process for the preparation of 2,6-dihydroxy-4-methyl-3-cyanopyridine (**9**) through use of this method was identified, which employed ethyl acetoacetate (**7**) and cyanoacetamide (**8**) as relatively inexpensive and readily available raw materials (18) (**Scheme 12.3**). Further investigation revealed that (**9**) is commercially available in multi-ton quantities.

Conditions were established to chlorinate

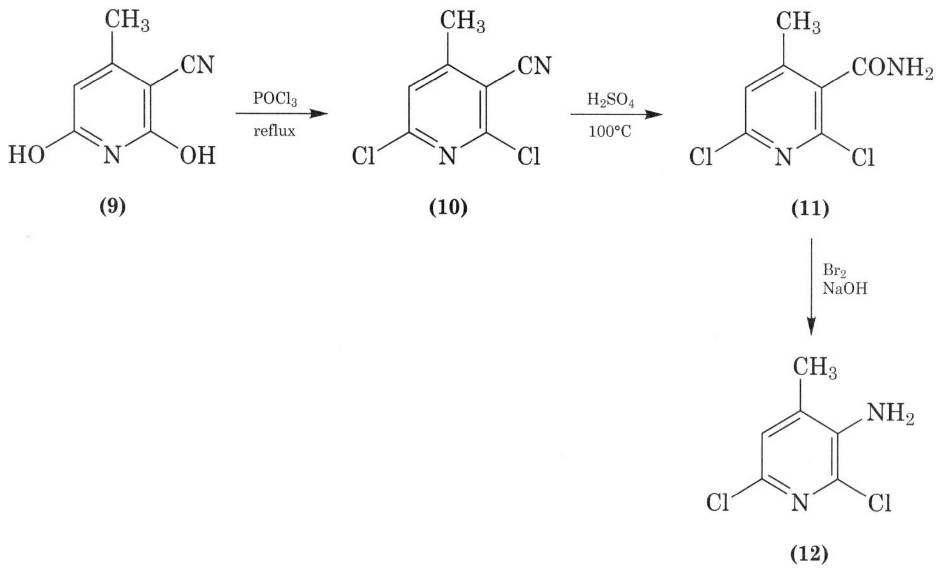

Scheme 12.4.

this intermediate by use of phosphorous oxychloride, to give 2,6-dichloro-4-methyl-3-cyanopyridine (**10**), followed by acid hydrolysis of the 3-cyano substituent and conversion to the amine under Hofmann rearrangement conditions (**Scheme 12.4**).

Efforts to selectively remove the 6-chloro substituent from either (**10**), (**11**), or (**12**) were unsuccessful. However, removal of both chlorine atoms by catalytic dechlorination (19) followed by selective rechlorination in the 2-position gave the desired product, as shown in **Scheme 12.5** (20).

This method was used to produce the nevirapine API requirements through Phase III clinical trials as well as for commercial launch and production. Although this synthetic approach lacked atom economy with respect to the removal and addition of chlorine atoms, it provided the opportunity to meet the short-

term API supply needs and established a synthetic strategy upon which further development activities would benefit. It should also be noted that all process steps from the commercially available raw material 2,6-dihydroxy-4-methyl-3-cyanopyridine (**9**) are carried out in aqueous media, making this option also environmentally attractive.

An alternative synthetic option was also examined that would eliminate the dechlorination/rechlorination process steps for the preparation of CAPIC, which is shown in **Scheme 12.6**. In this approach the chloride was removed in the last chemical step from (**17**), eliminating the dechlorination of (**12**) and selective rechlorination of (**13**). Although this option provided significant synthesis advantages over the existing process, this alternative method produced a different impurity profile from that of the original process

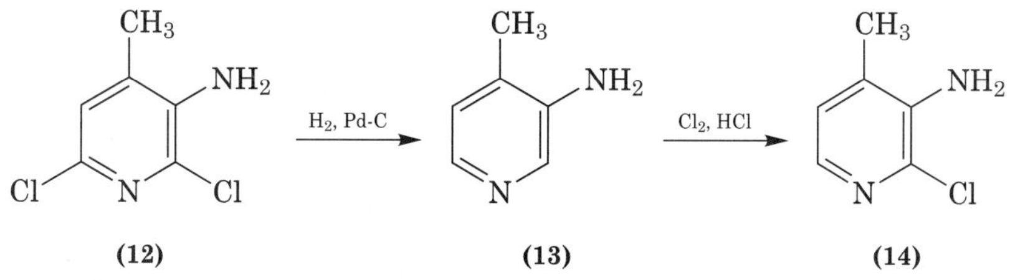

Scheme 12.5.

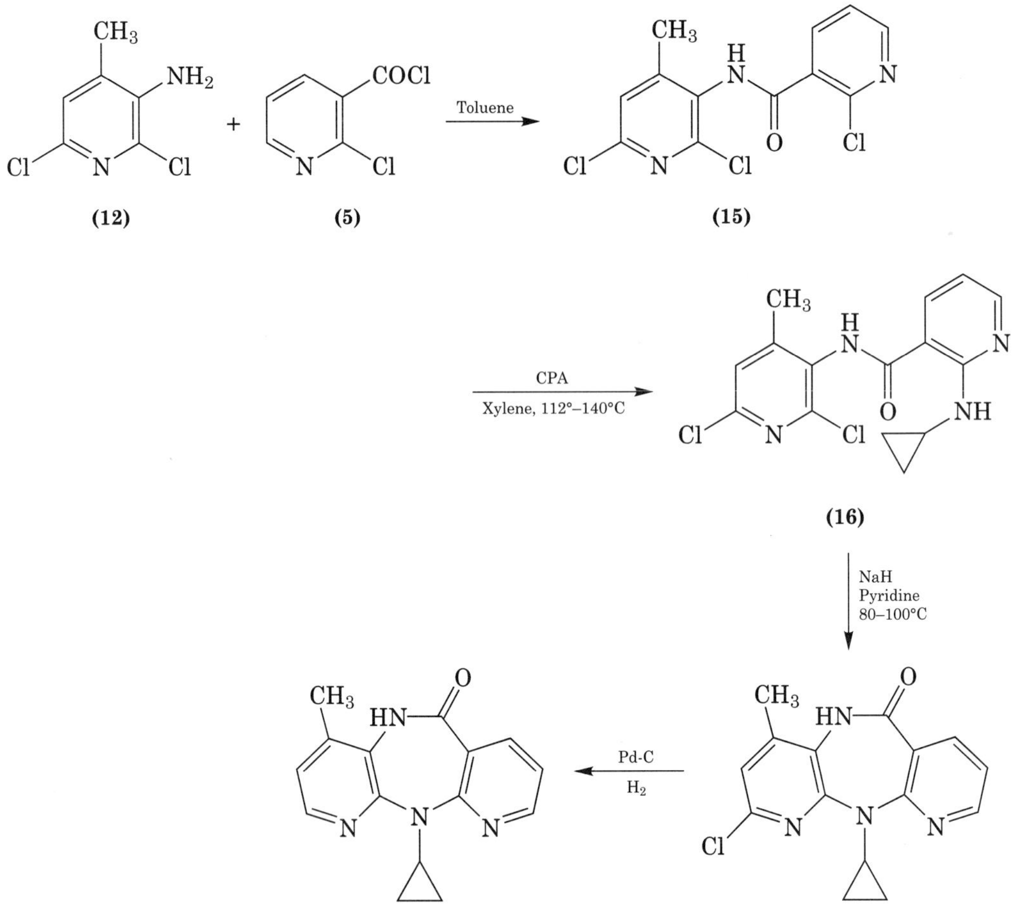

Scheme 12.6.

(**Scheme 12.2**). For this reason, revalidation of the API impurity profile, toxicology, and other pharmacological and regulatory issues would be required. Because this option was identified late in the chemical development process, it was decided that the potential process benefits would be more than offset by the additional efforts required to requalify the alternative process, and this option was eliminated from further commercial consideration.

3.2.3 Process Development and Pilot Plant Scale-Up. The nevirapine process scheme used during chemical development (**Scheme 12.2**) provided the basis on which to begin pro-

cess development studies, with the objective of defining reaction conditions that would allow production to be carried out on a routine commercial basis. In this process, the basic elements of the molecule are introduced with the condensation of CAPIC (**4**) and 2-chloronicoulinoylchloride (**5**). Using FDA guidelines (21) for defining the starting point in the synthesis for regulatory purposes, CAPIC was considered a raw material in the synthesis. This provided the opportunity to implement further CAPIC process improvements in the preparation of CAPIC after the product launch with limited regulatory impact. With this in mind, priority was given to the conden-

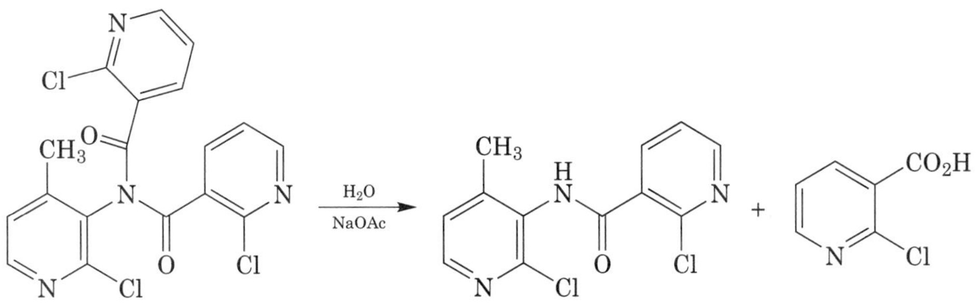

Scheme 12.7.

sation ring closure and purification steps that were filed in the New Drug Application (NDA).

In the manufacture of CAPIC, the reaction conditions were found to be quite acceptable for commercial operations, with minimal process modifications required to maximize the reactor utilization. However, reaction selectivity problems were encountered with Step 4 during pilot runs that had not been observed in previous work. An alternative set of reaction conditions was established that employed hydrochloric acid and hydrogen peroxide to selectively chlorinate the 2-position of (13). Process research established that a very narrow temperature range is required for this step, and the reaction temperature is controlled by the rate of addition of the hydrogen peroxide.

The condensation reaction required significant process development modifications from the procedure used to produce the initial drug development requirements. 2-Chloronicotinoylchloride (5) was prepared *in situ* during medicinal and chemical development runs by adding 2-chloronicotinic acid to a 5 molar excess of thionyl chloride as a neat reaction mixture. Upon completion of the reaction, excess thionyl chloride was removed by distillation. The residue was then redissolved in toluene, followed by the addition of CAPIC in toluene and sodium carbonate to neutralize the excess HCl liberated from the condensation. An alternative procedure was developed with the use of a 10% molar excess of thionyl chloride in toluene to produce (5). The excess thionyl chloride was removed by distillation and to the

resulting mixture was added to the CAPIC/ toluene solution, to give (14).

The work-up conditions for the condensation step (Scheme 12.6) were also modified to accommodate commercial operations. Sodium carbonate was used in the initial chemical development pilot plant batches to absorb the by-product HCl from the reaction. The quantities of carbon dioxide produced from the neutralization made this approach impractical in a commercial plant. To complicate matters, the amide bond formed during the condensation was subject to hydrolysis under strongly acidic conditions. Solid sodium acetate was added to the reaction mixture as a buffer to address this issue. A significant quantity of the diacetylation product (18) was also detected in the reaction mixture before work-up. However, this material rapidly hydrolyzes to the condensation product (6) and 2-chloronicotinic acid upon exposure to water (Scheme 12.7).

The use of cyclopropylamine for the preparation of (1) (Scheme 12.6) also presented a significant process optimization opportunity. In the initial pilot studies, four molar equivalents of cyclopropylamine (CPA) were used in the reaction medium. Although CPA appears to be a simple building block, it is rather expensive on a per kilogram basis and represents a significant cost contribution to the overall drug substance. One mole of CPA was initially used to absorb the by-product HCl from the reaction. Calcium oxide was found to be a much more cost-effective substitution as a neutralizing agent. To efficiently remove the

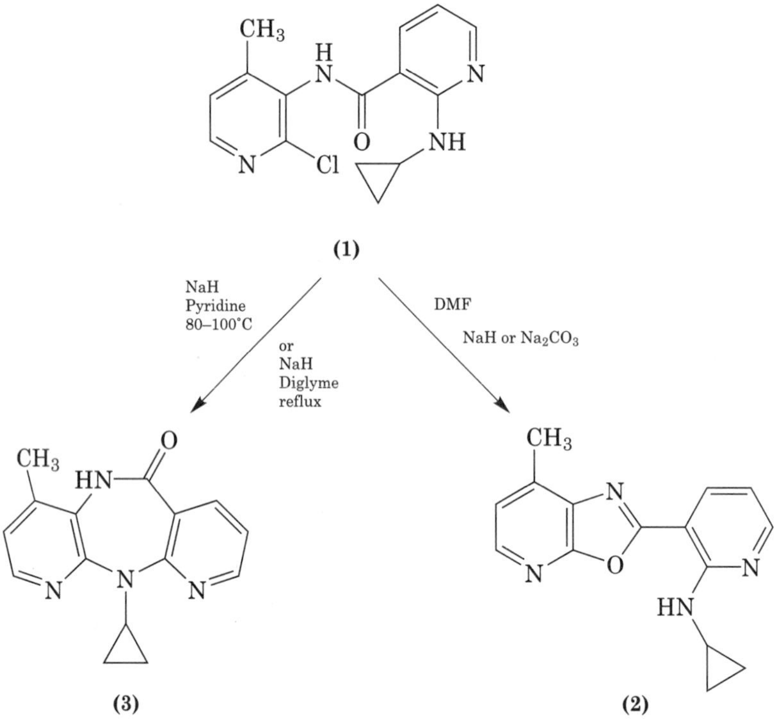

Scheme 12.8.

calcium salts before further processing, a centrifugation step was added to the work-up. However, even with calcium oxide present, a 2.5 molar excess of cyclopropylamine was still required to carry the reaction to completion. Efforts to telescope this operation into the ring-closure step were successful and (**1**) was treated as a nonisolated intermediate in process development pilot runs as well as on a commercial scale after removal of the calcium salts. Reaction temperature profile studies of this step indicated that the cyclopropylamine addition reaction occurred between 125°C and 145°C. However, a significant exothermic side reaction was observed above 145°C. Although this side reaction was not observed in any pilot trials, redundant cooling and ventilation systems were installed to pilot and commercial equipment to ensure safe operation of this process step.

One of the most critical issues to be addressed from these development activities concerned the specific reaction conditions employed in the final cyclization step (**Scheme**

12.8). The medicinal chemistry route used pyridine as the reaction solvent medium and sodium hydride as the base. It was later recognized from solvent screening studies that the reaction pathway for the ring closure was solvent dependent. When dimethylformamide (DMF) was used as the solvent, an alternate cyclization pathway was observed (**Scheme 12.8**). The oxazolo[5,4-]pyridine (**2**) is the exclusive product under these conditions. This product arises from the displacement of the chlorine atom by the amide carbonyl oxygen. A 2.8 molar excess of sodium hydride is required to carry the reaction to completion. The first mole of sodium hydride is consumed with the deprotonation of the more acidic amide protone.

If a base of insufficient strength is used in this reaction, the ring closure to the oxazol (**2**) primarily occurs. Although no industrially practical substitute could be found for sodium hydride as a reagent base, diglyme was found to be an effective alternative solvent to promote the conversion of (**1**) to nevirapine (**3**).

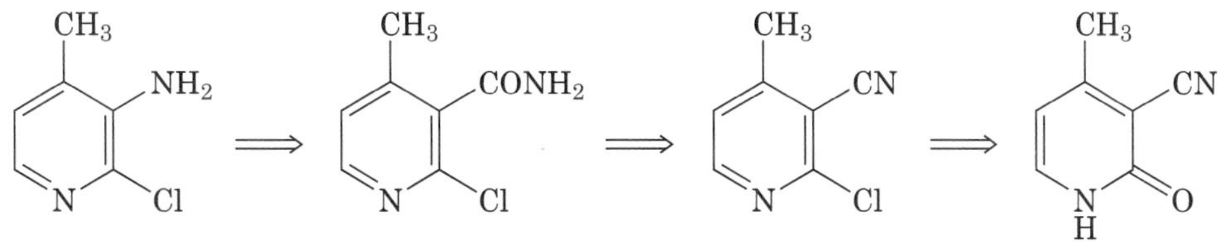

Scheme 12.9.

This was accomplished through an effective collaboration between members of the respective chemical and process development teams. Because of the low autoignition temperature of diglyme, significant equipment modifications were required to upgrade the pilot and production facilities to meet the more stringent electrical code requirements.

One issue with the use of sodium hydride as a reagent in pilot and commercial operations is the storage and handling requirements for this material. Sodium hydride is typically obtained commercially as a 60% amalgam in mineral oil to stabilize the reagent. In the nevirapine process, the mineral oil tends to agglomerate with the product upon precipitation from the reaction mixture. An intermediate purification step was developed through use of DMF as a crystallization medium. The crude product was dissolved in hot DMF followed by charcoal treatment to absorb the residual mineral oil associated with the product. The charcoal was then removed by filtration followed by evaporative crystallization of the product. A final aqueous crystallization was carried out to remove residual quantities of DMF from the product by acidification with hydrochloric acid followed by treatment with caustic to precipitate the product.

3.2.4 Commercial Production and Process Optimization. On February 23, 1996, Boehringer Ingelheim Pharmaceuticals submitted the NDA for nevirapine to the FDA. Production of the nevirapine API launch batches began within weeks after the submission. The company received regulatory approval for the product in July of that same year. A priority review of the NDA was initiated by the Agency based on the nature of the drug indication.

Because of the accelerated drug development timeline, the procedure used in the final process development piloting campaign was transferred to the production unit virtually unchanged. Only minor modifications to the existing production equipment were made to address electrical code requirements.

As previously noted, having developed a relatively converging synthesis for nevirapine provided the opportunity to define CAPIC as a raw material rather than a registered intermediate in the process. This in turn provided Boehringer Ingelheim with the flexibility to manage the CAPIC manufacturing requirements more effectively. This can be a particularly important issue with new product launches in general, given the high level of uncertainty in initial market forecasts.

As it turned out, nevirapine was well received in the marketplace as an effective AIDS treatment and post-launch sales consistently exceeded the market projections. With this rapid growth came an increasing awareness of the need to improve the synthesis of CAPIC (4) to meet the growing drug substance demands. The linear nature of the CAPIC synthesis as well as the lack of atom economy in the method were recognized as the major process deficiencies. A retrosynthetic analysis of (4) (**Scheme 12.9**) was carried out to evaluate alternative options for the preparation of this material. The goal of this effort was to limit the number of chemical transformations in the synthesis by constructing a pyridine ring with the optimal functionalization from acyclic precursors. The conditions used in the existing commercial process to introduce the amino group in the 4-position by the hydrolysis and Hofmann rearrangement appeared to be an effective approach. The

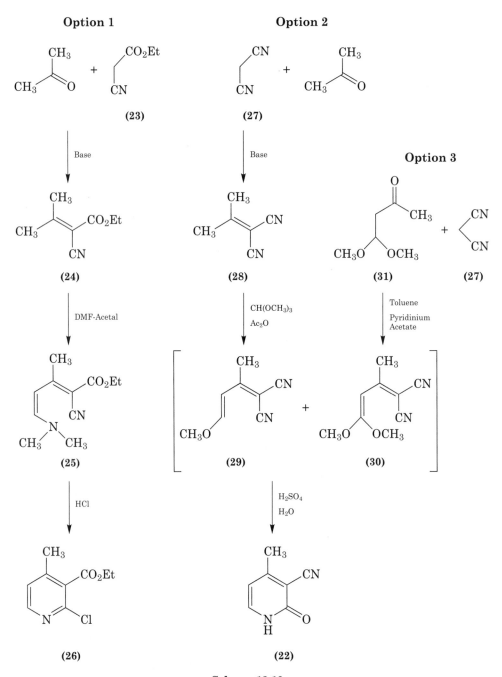

Scheme 12.10.

2-chloro-3-cyano-4-picoline (**21**) could be readily obtained from 3-cyano-4-methyl-2-pyridone (**22**) by chlorination with phosphorous oxychloride.

Research efforts were directed toward the evaluation of options for the preparation of (**22**) from commercially available starting materials. Several approaches were examined, all with the common feature to use a Knovenagel condensation reaction to establish the desired regiochemistry for the target molecule.

Option 1 (22), as shown in **Scheme 12.10**, employs acetone and 2-cyanoethyl acetate (**23**) in the initial Knovenagel condensation. The resulting α-β-unsaturated cyanoacetate (**24**) is reacted with DMF-acetal to produce (**25**). The ring-closure step was conducted under Pinner reaction conditions, to give ethyl 2-chloro-4-methylnicotinate (**26**), which is converted to (**4**) in three steps. However, low yields were observed in the ring-closure steps in this route. An alternative approach (23) was examined using an alternative synthetic approach (Option 2). In this procedure, acetone was reacted with malononitrile (**27**), to produce (**28**), followed by reaction with trimethylorthoformate, to give a mixture of (**29**) and (**30**). Although (**30**) is the predominant product from this reaction, both compounds are readily converted to (**22**) upon treatment with sulfuric acid. Low yields observed in the formulation step led to the development of Option 3. In this procedure, the formulation step is avoided by using a protected β-ketoaldehyde (**31**) in the Knovenagel condensation with malononitrile (**27**). The protected β-ketoaldehyde (**31**) is prepared from acetone, methylformate, and sodium methoxide, and is readily available in commercial quantities. The Knovenagel intermediate (**30**) was converted into (**22**) under acidic reaction conditions. Upon completion of an economic evaluation of these procedures, Option 3 (24) was selected for commercialization.

Other efforts to improve the commercial nevirapine process have been primarily driven by equipment modifications rather than amendments to the chemical process.

3.3 Summary

The FDA approval of nevirapine for the treatment of AIDS was granted less than 7 years after the submission of the IND. During this period of time, many technical and regulatory barriers were overcome to bring this product to the marketplace. From a process development perspective, the challenge, as always, is in ensuring the uninterrupted supply of bulk active drug substance in support of the overall drug development effort without sacrificing the ability to deliver a commercially viable chemical process. Although these issues represent a common theme in most drug development case studies, the accelerated pace of the nevirapine project significantly magnified the complexity of the drug development effort. Fortunately, the major elements of the original synthesis remained intact throughout the various phases of process development and provided the opportunity to conduct these activities in parallel with minimal regulatory impact.

REFERENCES

1. S. Warren, *Organic Synthesis: The Disconnection Approach*, John Wiley & Sons, New York, 1982.

2. A. N. Collins, G. N. Sheldrahe, and J. Crosby, *Chirality in Industry*, John Wiley & Sons, New York, 1982.

3. S. C. Stinson, *Chem. Eng. News*, **76**, 83 (1998).

4. C. Hicks, *Fundamental Concepts in the Design of Experiment*, Oxford, New York, 1999.

5. C. Reichardt, *Solvents and Solvent Effects in Organic Chemistry*, 2nd ed., VCH, Weinheim, Germany, 1990.

6. M. H. Norman, D. J. Minick, and G. E. Martin, *J. Heterocyclic Chem.*, **30**, 771 (1993).

7. N. G. Anderson, *Practical Process Research*, Academic Press, San Diego, CA, 2000.

8. R. J. Kelly, *Chem. Health Safety*, **September/October**, 28 (1996).

9. Federal Water Pollution Control Act (Clean Water Act), 33, U.S.C.A. Section 1251 et seq.

10. K. Murthy, B. Radatus, and S. Kanwarpal, U.S. Pat. 5,523,423 (1996).

11. K. Flory, Analytical Profiles of Drug Substances, Vol. **23**, Academic Press, New York, 1988.

12. S. R. Chemburkar, J. Bauer, K. Deming, H. Spi, S. Spanton, W. Dzjlki, W. Porter, J. Quick, I. M. Soldani, D. Riley, and K. McFarland, *Org. Process Res. Dev.*, **4**, 413–417 (2000).

13. R. H. Perry and D. W. Green, *Perry's Chemical Engineers' Handbook*, 7th ed., McGraw-Hill, New York, 1997.

14. K. Grozinger, V. Fuchs, K. Hargrave, S. Mauldin, J. Vitous, and S. A. Campbell, *J. Heterocyclic Chem.*, **32**, 259 (1995).

15. A. G. Burton, P. J. Halls, and A. R. Katritzky, *Tetrahedron Lett.*, **24**, 20211 (1971).

16. M. Bobbitt and D. A. Scala, *J. Org. Chem.*, **25**, 560 (1960).

17. R. Wallis and G. Lane, *Org. React.*, **3**, 267–306 (1946).

18. K. G. Grozinger and K. D. Hargrove, U.S. Pat. 5,200,522 (1993).

19. K. G. Grozinger, K. D. Hargrave, and J. Adams, U.S. Pat. 5,668,287 (1997).

20. K. G. Grozinger, K. D. Hargrave, and J. Adams, U.S. Pat. 5,571,912 (1996).

21. *Guidance for Industry BACPAC I: Intermediates in Drug Substance Synthesis Bulk Active Post Approval Changes: Chemistry, Manufacturing, and Controls Documentation*, U.S. Department of Health and Human Services Food and Drug Administration Center for Drug Evaluation and Research (CDER), Center for Veterinary Medicine (CVM), February 2001.

22. K. G. Grozinger, U.S. Pat. 6,136,982 (2000).

23. K. G. Grozinger, U.S. Pat. 6,111,112 (2000).

24. B. F. Gupton, U.S. Pat. 6,399,781 (2002).

CHAPTER THIRTEEN

Principles of Drug Metabolism

BERNARD TESTA
University of Lausanne
Institute of Medicinal Chemistry, School of Pharmacy
Lausanne, Switzerland

WILLIAM SOINE
Virginia Commonwealth University
Department of Medicinal Chemistry
Richmond, Virginia

Contents

Burger's Medicinal Chemistry and Drug Discovery
Sixth Edition, Volume 2: Drug Development
Edited by Donald J. Abraham
ISBN 0-471-37028-2 © 2003 John Wiley & Sons, Inc.

1 INTRODUCTION

Xenobiotic metabolism, which includes drug metabolism, has become a major pharmacological science with particular relevance to biology, therapeutics, and toxicology. Drug metabolism is also of great importance in medicinal chemistry because it influences (in qualitative, quantitative, and kinetic terms) the deactivation, activation, detoxification, and toxification of the vast majority of drugs. As a result, medicinal chemists engaged in drug discovery (lead finding and lead optimization) should be able to integrate metabolic considerations into drug design. To do so, however, requires a good knowledge of xenobiotic metabolism.

This chapter, which is written by medicinal chemists for medicinal chemists, offers knowl-

edge and understanding rather than encyclopedic information. Readers wanting to go further in the study of xenobiotic metabolism may consult various classic or recent books (1–10).

1.1 Definitions and Concepts

Drugs are but one category among the many xenobiotics (Table 13.1) that enter the body but have no nutritional or physiological value (11). The study of the disposition—or fate—of xenobiotics in living systems includes the consideration of their absorption into the organism, how and where they are distributed and stored, the chemical and biochemical transformations they may undergo, and how and by what route(s) they are finally excreted and returned to the environment. As for "metabolism," this word has acquired two meanings, being synonymous with disposition (i.e., the

Table 13.1   Major Categories of Xenobiotics (modified from Ref. 11)

Drugs

Food constituents devoid of physiological roles

Food additives (preservatives, coloring and flavoring agents, antioxidants, etc.)

Chemicals of leisure, pleasure, and abuse (ethanol, coffee and tobacco constituents, hallucinogens, doping agents, etc.)

Agrochemicals (fertilizers, insecticides, herbicides, etc.)

Industrial and technical chemicals (solvents, dyes, monomers, polymers, etc.)

Pollutants of natural origin (radon, sulfur dioxide, hydrocarbons, etc.)

Pollutants produced by microbial contamination (e.g., aflatoxins)

Pollutants produced by physical or chemical transformation of natural compounds (polycyclic aromatic hydrocarbons by burning, Maillard reaction products by heating, etc.)

---

sum of the processes affecting the fate of a chemical substance in the body) and with biotransformation as understood in this chapter (12).

In pharmacology, one speaks of pharmacodynamic effects to indicate what a drug does to the body and pharmacokinetic effects to indicate what the body does to a drug; two aspects of the behavior of xenobiotics that are strongly interdependent. Pharmacokinetic effects will obviously have a decisive influence on the intensity and duration of pharmacodynamic effects, whereas metabolism will generate new chemical entities (metabolites) that may have distinct pharmacodynamic properties of their own. Conversely, by its own pharmacodynamic effects, a compound may affect the state of the organism (e.g., hemodynamic changes, enzyme activities) and therefore the organism's capacity to handle xenobiotics. Only a systemic approach can help one appreciate the global nature of this interdependence (13).

1.2 Types of Metabolic Reactions Affecting Xenobiotics

A first discrimination that can be made among metabolic reactions is based on the nature of the catalyst. Reactions of xenobiotic metabolism, like other biochemical reactions, are catalyzed by enzymes. However, while the vast majority of reactions of xenobiotic metabolism are indeed enzymatic ones, some nonenzymatic reactions are also well documented. This is because a variety of xenobiotics have been found to be labile enough to react nonenzymatically under biological conditions of pH and temperature (14). But there is more. In a normal enzymatic reaction, metabolic intermediates exist en route to the product(s) and do not leave the catalytic site. However, many exceptions to this rule are known, with the metabolic intermediate leaving the active site and reacting with water, an endogenous molecule or macromolecule, or a xenobiotic. Such reactions are also of a nonenzymatic nature but are better designated as postenzymatic reactions (14).

The metabolism of drugs and other xenobiotics is often a biphasic process in which the compound may first undergo a functionalization reaction (phase I reaction) of oxidation, reduction, or hydrolysis. This introduces or unveils a functional group such as a hydroxy or amino group suitable for coupling with an endogenous molecule or moiety in a second metabolic step known as a conjugation reaction (phase II reaction). In a number of cases, phase I metabolites may be excreted before conjugation, whereas many xenobiotics can be directly conjugated. Furthermore, reactions of functionalization may follow some reactions of conjugation, e.g., some conjugates are hydrolyzed and/or oxidized before their excretion.

Xenobiotic biotransformation thus produces two types of metabolites, namely functionalization products and conjugates. But with the growth of knowledge, biochemists and pharmacologists have progressively come to recognize the existence of a third class of metabolites, namely xenobiotic-macromolecule adducts, also called macromolecular conjugates (15). Such peculiar metabolites are formed when a xenobiotic binds covalently to a biological macromolecule, usually following metabolic activation (i.e., postenzymatically). Both functionalization products and conjugates have been found to bind covalently to biological macromolecules; the reaction is often toxicologically relevant.

1.3 Specificities and Selectivities in Xenobiotic Metabolism

The words "selectivity" and "specificity" may not have identical meanings in chemistry and biochemistry. In this chapter, the specificity of an enzyme is taken to mean an ensemble of properties, the description of which makes it possible to specify the enzyme's behavior. In contrast, the term selectivity is applied to metabolic processes, indicating that a given metabolic reaction or pathway is able to select some substrates or products from a larger set. In other words, the selectivity of a metabolic reaction is the detectable expression of the specificity of an enzyme. Such definitions may not be universally accepted, but they have the merit of clarity.

What, then, are the various types of selectivities (or specificities) encountered in xenobiotic metabolism? What characterizes an enzyme from a catalytic viewpoint is first its chemospecificity, i.e., its specificity in terms of the type(s) of reaction it catalyzes. When two or more substrates are metabolized at different rates by a single enzyme under identical conditions, substrate selectivity is observed. In such a definition, the nature of the product(s) and their isomeric relationship are not considered. Substrate selectivity is distinct from product selectivity, which is observed when two or more metabolites are formed at different rates by a single enzyme from a single substrate. Thus, substrate-selective reactions discriminate between different compounds, whereas product-selective reactions discriminate between different groups or positions in a given compound.

The substrates being metabolized at different rates may share various types of relationships. They may be chemically dissimilar or similar (e.g., analogs), in which case the term of substrate selectivity is used in a narrow sense. Alternatively, the substrates may be isomers such as positional isomers (regioisomers) or stereoisomers, resulting in substrate regioselectivity or substrate stereoselectivity. Substrate enantioselectivity is a particular case of the latter (see Section 5.1.2).

Products formed at different rates in product-selective reactions may also share various types of relationships. Thus, they may be analogs, regioisomers, or stereoisomers, resulting in product selectivity (narrow sense), product regioselectivity or product stereoselectivity (e.g., product enantioselectivity). Note that the product selectivity displayed by

two distinct substrates in a given metabolic reaction may be different; in other words, the product selectivity may be substrate-selective. The term substrate-product selectivity can be used to describe such complex cases, which are conceivable for any type of selectivity but have been reported mainly for stereoselectivity.

1.4 Pharmacodynamic Consequences of Xenobiotic Metabolism

The major function of xenobiotic metabolism can be seen as the elimination of physiologically useless compounds, some of which may be harmful as witnessed by the tens of thousands of toxins produced by plants. The function of toxin inactivation justifies the designation of detoxification originally given to reactions of xenobiotic metabolism. However, the possible pharmacological consequences of biotransformation are not restricted to detoxification. In the simple case of a xenobiotic having a single metabolite, four possibilities exist:

1. Both the xenobiotic and its metabolite are devoid of biological effects (at least in the concentration or dose range investigated); such a situation has no place in medicinal chemistry.
2. Only the xenobiotic elicits biological effects; a situation which in medicinal chemistry is typical of, but not unique to, soft drugs.
3. Both the xenobiotic and its metabolite are biologically active; the two activities being comparable or different either qualitatively or quantitatively.
4. The observed biological activity is caused exclusively by the metabolite; a situation which in medicinal chemistry is typical of prodrugs.

When a drug or another xenobiotic is transformed into a toxic metabolite, the reaction is one of toxification (16). Such a metabolite may act or react in a number of ways to elicit a variety of toxic responses at different biological levels (17, 18). However, it is essential to stress that the occurrence of a reaction of toxification (i.e., toxicity at the molecular level) does *not* necessarily imply toxicity at the levels

Table 13.2 Metabolism-Related Questions

Answered in lead discovery and optimization
  Susceptibility to metabolism?
  Expected rate of metabolism?
  Nature of major metabolites?
  Enzymes/isozymes involved?
  Potential for enzyme inhibition?
Answered in the preclinical and clinical phases
  Nature and relative formation of major and
    minor metabolites?
  Enzymes/isozymes and tissues involved?
  Influence of genetic factors? Influence of other
    factors?
  Distribution and elimination of metabolites?
  Activities and toxicities of metabolites?
  Activity of drug and metabolites as inducer,
    autoinducer, and/or *inhibitor?*
  Potential for and occurrence of drug-drug
    interactions?

Table 13.3 Aspects of Drug Metabolism of Major Interest to Medicinal Chemists (21)

The chemistry and biochemistry of metabolic
  reactions
Predictions of drug metabolism based on
  quantitative structure-metabolism relationships
  (QSMRs), expert systems, and molecular
  modeling of enzymatic sites
The consequences of such reactions on activation
  and inactivation, toxification, and detoxification
Prodrug and soft drug design
Changes in physicochemical properties (pKa,
  lipophilicity, etc.) resulting from
  biotransformation
The potential for drug–drug interactions
  (inhibition and/or induction)
The potential for genetic polymorphism

of organs and organisms. This will be discussed later in this text.

1.5 Setting the Scene

In drug research and development, metabolism is of pivotal importance because of the interconnectedness between pharmacokinetic and pharmacodynamic processes (Table 13.2). *In vitro* metabolic studies are now initiated very early during lead optimization to assess the overall rate of oxidative metabolism, to identify the metabolites, to obtain primary information on the enzymes involved, and to postulate metabolic intermediates. Based on these findings, the metabolites must be synthetized and tested for their own pharmacological and toxicological effects. In preclinical and early clinical studies, many pharmacokinetic data must be obtained and relevant criteria must be satisfied before a drug candidate can enter large-scale clinical trials (19, 20). As a result of these demands, the interest of medicinal chemists for drug metabolism has grown remarkably in recent years (Table 13.3) (21).

As will become apparent, the approach followed in this chapter is an analytical one, meaning that the focus is on metabolic reactions, the target groups they affect, and the enzymes by which they are catalyzed. This information provides the foundations of drug metabolism, but it must be complemented by a synthetic view to allow a broader understanding and meaningful predictions. Two steps are required to approach these objectives, namely (*1*) the elaboration of metabolic schemes where the competitive and sequential reactions (Sections 2 and 3) undergone by a given drug are ordered, and (*2*) an assessment of the various biological factors (Section 4) that influence such schemes both quantitatively and qualitatively. As an example of a metabolic scheme, Fig. 13.1 presents the biotransformation of propranolol (**1**) in humans (22). There are relatively few studies as comprehensive and clinically relevant as this one, which remains as current today as it was when published in 1985. Indeed, over 90% of a dose was accounted for and consisted mainly of products of oxidation and conjugation. The missing 10% may represent other minor and presumably quite numerous metabolites, e.g., those resulting from ring hydroxylation at other positions or from the progressive breakdown of glutathione conjugates.

A large variety of enzymes and metabolic reactions are presented in Sections 2 and 3. As will become clear, some enzymes catalyze only a single type of reaction (e.g., *N*-acetylation), whereas others use a basic catalytic mechanism to attack a variety of moieties and produce different types of metabolites (e.g., cytochromes P450). As an introduction to these enzymes and reactions, we present an estimate of their relative importance in drug metabolism (Table 13.4). In this table, the correspondence between the number of substrates

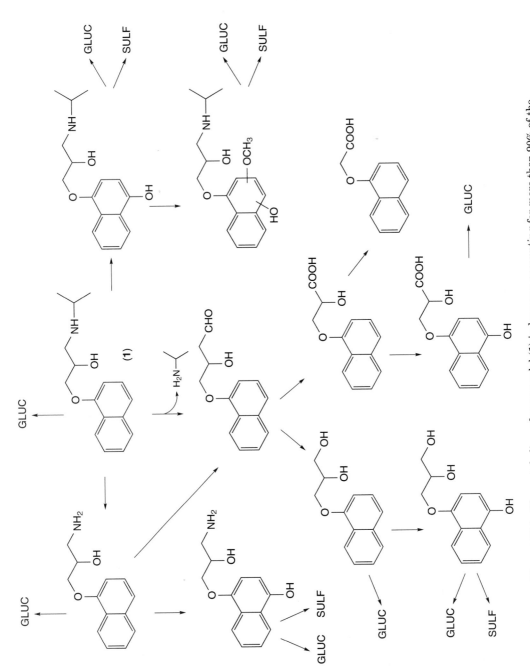

Figure 13.1. The metabolism of propranolol (**1**) in humans, accounting for more than 90% of the dose. GLUC, glucuronide(s); SULF, sulfate(s) (22).

Table 13.4   Estimate of the Relative Contributions of Major Drug-Metabolizing Enzymes[a]

Enzymes	Number of Substrates[b]	Overall Contribution to Drug Metabolism
Cytochromes P450 (Section 2.2.1)	****	****
Dehydrogenases and reductases (Section 2.2.2)	***	***
Flavin-containing monooxygenases (Section 2.2.2)	* to **	*
Hydrolases (Section 2.2.3)	** to ***	***
Methyltransferases (Section 3.2)	*	*
Sulfotransferases (Section 3.3)	**	*
Glucuronyltransferases (Section 3.4)	***	***
N-Acetyltransferases (Section 3.5)	*	*
Acyl-coenzyme A synthetases (Section 3.6)	*	*
Glutathione S-transferases (Section 3.7)	**	**
Phosphotransferases (Section 3.8)	*	*

[a]* low, ** intermediate, *** high, **** very high.
[b]Including drug metabolites.

and the overall contribution to drug metabolism does not need to be perfect, because some enzymes show a limited capacity (e.g., sulfotransferases), whereas others make a significant contribution to the biotransformation of their substrates (e.g., hydrolases).

2   FUNCTIONALIZATION REACTIONS

2.1   Introduction

Reactions of functionalization are comprised of oxidations (electron removal, dehydrogenation, and oxygenation), reductions (electron addition, hydrogenation, and removal of oxygen), and hydrations/dehydrations (hydrolysis and addition or removal of water). The reactions of oxidation and reduction are catalyzed by a very large variety of oxidoreductases, whereas various hydrolases catalyze hydrations. A large majority of enzymes involved in xenobiotic functionalization are briefly reviewed in Section 2.2 (23). Metabolic reactions and pathways of functionalization constitute the main body of Section 2.

2.2   Enzymes Catalyzing Functionalization Reactions

2.2.1   Cytochromes P450.   Monooxygenation reactions are of major significance in drug metabolism and are mediated by various enzymes that differ markedly in their structure and

properties. Among these, the most important as far as xenobiotic metabolism is concerned are the cytochromes P450 (EC 1.14.14.1, 1.14.15.1, and 1.14.15.3–1.14.15.6), a very large group of enzymes belonging to hemecoupled monooxygenases (7, 24–28). The cytochrome P450 enzymes (CYPs) are encoded by the *CYP* gene superfamily and are classified in families and subfamilies as summarized in Table 13.5. Cytochrome P450 is the major drugmetabolizing enzyme system, playing a key role in detoxification and toxification, and is of additional significance in medicinal chemistry because several CYP enzymes are drug targets, e.g., thromboxane synthase (CYP5) and aromatase (CYP19). The three CYP families mostly involved in xenobiotic metabolism are CYP1–CYP3, whose relative importance is shown in Table 13.6.

Examples of the many drugs interacting with cytochromes P450 as substrates, inhibitors, or inducers will be considered later (see Table 13.9 in Section 4), whereas this section focuses on the metabolic reactions. An understanding of the regiospecificity and broad reactivity of cytochrome P450 requires a presentation of its catalytic cycle (Fig. 13.2). This cycle involves a number of steps that can be simplified as follows:

1. The enzyme in its ferric (oxidized) form exists in equilibrium between two spin states,

Table 13.5 The Human *CYP* Gene Superfamily: A Table of the Families and Subfamilies of Gene Products (7, 24–28)

Families	Subfamilies	Representative Gene Products
P450 1 Family (*Aryl hydrocarbon hydroxylases; xenobiotic metabolism; inducible by polycyclic aromatic hydrocarbons*)	P450 1A Subfamily P450 1B Subfamily	CYP1A1, CYP1A2 CYP1B1
P450 2 Family (*Xenobiotic metabolism; constitutive and xenobiotic-inducible*)	P450 2A Subfamily P450 2B Subfamily (*Includes phenobarbital-inducible forms*)	CYP2A6, CYP2A13 CYP2B6
	P450 2C Subfamily (*Constitutive forms; includes sex-specific forms*)	CYP2C8, CYP2C9, CYP2C18, CYP2C19
	P450 2D Subfamily P450 2E Subfamily (*Ethanol-inducible*)	CYP2D6 CYP2E1
	P450 2F Subfamily P450 2J Subfamily	CYP2F1 CYP2J2
P450 3 Family (*Xenobiotic and steroid metabolism; steroid-inducible*)	P450 3A Subfamily	CYP3A4, CYP3A5, CYP3A7 (*fetal CYP enzyme*), CYP3A43
P450 4 Family (*Peroxisome proliferator-inducible*)	P450 4A Subfamily	CYP4A11 (*Fatty acid ω- and (ω-1)-hydroxylases*)
	P450 4B Subfamily P450 4F Subfamily	CYP4B1 CYP4F2, CYP4F3, CYP4F8, CYP4F12
P450 5 Family	P450 5A Subfamily	CYP5A1 (*TXA synthase*)
P450 7 Family (*Steroid 7-hydroxylases*)	P450 7A Subfamily P450 7B Subfamily	CYP7A1 CYP7B1
P450 8 Family	P450 8A Subfamily P450 8B Subfamily	CYP8A1 (*Prostacyclin synthase*) CYP8B1
P450 11 Family (*Mitochondrial steroid hydroxylases*)	P450 11A Subfamily P450 11B Subfamily (*Steroid hydroxylases*)	CYP11A1 (*Cholesterol side-chain cleavage*) CYP11B1, CYP11B2
P450 17 Family (*Steroid 17α-hydroxylase*)		CYP17
P450 19 Family (*Steroid aromatase*)		CYP19
P450 21 Family (*Steroid 21-hydroxylases*)		CYP21
P450 24 Family (*25-Hydroxyvitamin D 24-hydroxylase*)		CYP24
P450 26 Family	P450 26A Subfamily	CYP26A1
P450 27 Family (*Mitochondrial steroid hydroxylases*)	P450 27A Subfamily	CYP27A1
	P450 27B Subfamily	CYP27B1
P450 39 Family		CYP39
P450 46 Family		CYP46
P450 51 Family		CYP51 (*Lanosterol 14α-demethylase*)

This list reports all human CYPs with known substrate(s) and/or inhibitor(s). At the time of writing, human CYPs of unknown function were 2A7, 2R1, 2S1, 2U1, 2W1, 4A20, 4A22, 4F11, 4F22, 4V2, 4X1, 20, 26B1, 26C1, and 27C1 (28).

Table 13.6 Levels and Variability of Human CYP Enzymes Involved in Drug Metabolism (27, 28)

CYP	Level of Enzyme in Liver (% of Total)	Variability Range	Percent of Drugs (or Other Xenobiotics) Interacting with Enzyme		
			As Substrates	As Inhibitors	As Inducers/ Activators
1A1			3 (12)	3 (13)	6 (33)
1A2	~13	~40-fold	10 (15)	12 (18)	3 (25)
1B1	<1		1 (10)	1 (7)	1 (9)
2A6	~4	~30- to 100-fold	3 (10)	2 (6)	2 (1)
2B6	<1	~50-fold	4 (9)	3 (5)	13 (no data)
2C	~18	25- to 100-fold	25 (13)	27 (17)	21 (6)
2D6	up to 2.5	>1000-fold	15 (2)	22 (8)	2 (2 activators)
2E1	up to 7	~20-fold	3 (16)	4 (10)	7 (7)
3A4	up to 28	~20-fold	36 (13)	26 (16)	45 (17)

an hexa-coordinated low spin form that cannot be reduced and a penta-coordinated high spin form. Binding of the substrate to enzyme induces a shift to the reducible high spin form (reaction a).

2. A first electron enters the enzyme-substrate complex (reaction b).

3. The enzyme in its ferrous form has a high affinity for diatomic gases such as CO (a strong inhibitor of cytochrome P450) and dioxygen (reaction c).

4. Electron transfer from Fe^{2+} to O_2 within the enzyme-substrate-oxygen ternary complex reduces the dioxygen to a bound molecule of superoxide. Its possible liberation in the presence of compounds with good affinity but low reactivity (uncoupling) can be cytotoxic (reaction d).

5. The normal cycle continues with a second electron entering through either NADPH-cytochrome P450 reductase (F_{P1}) or NADH-cytochrome b_5 reductase (F_{P2}) and reducing the ternary complex (reaction e).

6. Electron transfer within the ternary complex generates bound peroxide anion (O_2^{2-}).

7. The bound peroxide anion is split, liberating H_2O (reaction f).

8. The remaining oxygen atom is an oxene species. This is the reactive form of oxygen that will attack the substrate.

9. The binary enzyme-product complex dissociates, thereby regenerating the initial state of cytochrome P450 (reaction h).

Oxene is a rather electrophilic species; it is neutral but has only six electrons in its outer shell. Its detailed reaction mechanisms are beyond the scope of this chapter, but some indications will be given when discussing the various reactions of oxidation catalyzed by cytochromes P450.

2.2.2 Other Oxidoreductases. Alcohol dehydrogenases (ADH; alcohol:NAD^+ oxidoreductase; EC 1.1.1.1) are zinc enzymes found in the cytosol of the mammalian liver and in various extrahepatic tissues. Mammalian liver alcohol dehydrogenases (LADHs) are dimeric enzymes. The human enzymes belong to three different classes: class I (ADH1), comprising the various isozymes that are homodimers or heterodimers of the α (ADH1A, ADH2A, ADH3A), β, and γ subunits (e.g., the $\alpha\alpha$, $\beta_1\beta_1$, $\alpha\beta_2$, and $\beta_1\gamma$ isozymes); class II (ADH2), comprising the $\pi\pi$ enzyme; and class III (ADH3), comprising the $\chi\chi$ enzyme (29, 30).

Enzymes categorized as aldehyde reductases include alcohol dehydrogenase ($NADP^+$; aldehyde reductase [NADPH]; alcohol:$NADP^+$ oxidoreductase; EC 1.1.1.2), aldehyde reductase [alditol:$NAD(P^+)$ 1-oxidoreductase; aldose reductase; EC 1.1.1.21], and many others

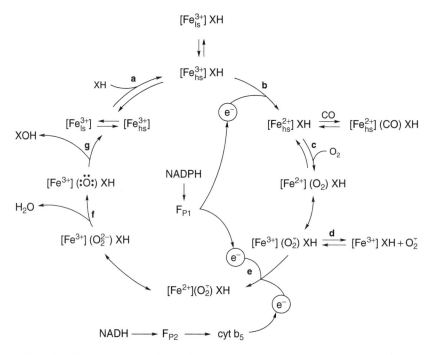

Figure 13.2. Catalytic cycle of cytochrome P450 associated with monooxygenase reactions. (Fe^{3+}), ferricytochrome P450; hs, high spin; ls, low spin; (Fe^{2+}), ferrocytochrome P450; F_{P1}, flavoprotein 1–NADPH–cytochrome P450 reductase; F_{P2}, NADH–cytochrome b_5 reductase; cyt b_5, cytochrome b_5; XH, substrate (modified from Ref. 6).

of lesser relevance (31). Aldehyde reductases are widely distributed in nature and occur in a considerable number of mammalian tissues. Their subcellular location is primarily cytosolic, and in some instances is also mitochondrial. The so-called ketone reductases include α- and β-hydroxysteroid dehydrogenases (e.g., EC 1.1.1.50 and EC 1.1.1.51), various prostaglandin ketoreductases (e.g., prostaglandin-F synthase, EC 1.1.1.188; prostaglandin-E_2 9-reductase, EC 1.1.1.189), and many others that are comparable with aldehyde reductases. One group of particular importance are the carbonyl reductases (NADPH; EC 1.1.1.184). Furthermore, the many similarities (including some marked overlap in substrate specificity) between monomeric, NADPH-dependent aldehyde reductase (AKR1), aldose reductase (AKR2), and carbonyl reductase (AKR3) have led to their designation as aldo-keto reductases (AKRs) (32).

Other reductases that have a role to play in drug metabolism include glutathione reductase (NADPH:oxidized-glutathione oxidoreductase; EC 1.6.4.2) and quinone reductase [NAD(P)H:(quinone acceptor) oxidoreductase; DT-diaphorase; EC 1.6.99.2].

Aldehyde dehydrogenases [ALDHs; aldehyde:NAD(P)$^+$ oxidoreductases; EC 1.2.1.3 and EC 1.2.1.5] exist in multiple forms in the cytosol, mitochondria, and microsomes of various mammalian tissues. It has been proposed that ALDHs form a superfamily of related enzymes consisting of class 1 ALDHs (cytosolic), class 2 ALDHs (mitochondrial), and class 3 ALDHs (tumor-associated and other isozymes). In all three major classes, constitutive and inducible isozymes exist. In a proposed nomenclature system, the human ALDHs are designated as 1A1, 1A6, 1B1, 2, 3A1, 3A2, 3B1, 3B2, 4A1, 5A1, 6A1, 7A1, 8A1, and 9A1 (33–35).

Dihydrodiol dehydrogenases (*trans*-1,2-dihydrobenzene-1,2-diol:NADP$^+$ oxidoreductase; EC 1.3.1.20) are cytosolic enzymes; several of which have been characterized. Although the isozymes are able to use NAD^+, the preferred cofactor is $NADP^+$.

Other oxidoreductases that can play a major or less important role in drug metabolism are hemoglobin, monoamine oxidases (EC 1.4.3.4; MAO-A and MAO-B), which are essentially mitochondrial enzymes, the cytosolic molybdenum hydroxylases (xanthine oxidase, EC 1.1.3.22; xanthine dehydrogenase, EC 1.1.1.204; and aldehyde oxidase, EC 1.2.3.1), and the broad group of copper-containing amine oxidases (EC 1.4.3.6) (36–39).

Besides cytochromes P450, other monooxygenases of importance are the flavin-containing monooxygenases (EC 1.14.13.8; FMO1 to FMO5) and dopamine β-hydroxylase (dopamine β-monooxygenase; EC 1.14.17.1).

Various peroxidases are progressively being recognized as important enzymes in drug metabolism. Several cytochrome P450 enzymes have been shown to have peroxidase activity (40). Protaglandin-endoperoxide synthase (EC 1.14.99.1) is able to use a number of xenobiotics as cofactors in a reaction of cooxidation (41). And finally, a variety of other peroxidases may oxidize drugs, e.g., catalase (EC 1.11.1.6) and myeloperoxidase (donor:hydrogen-peroxide oxidoreductase; EC 1.11.1.7) (42).

2.2.3 Hydrolases.

Hydrolases constitute a very complex ensemble of enzymes; many of which are known or suspected to be involved in xenobiotic metabolism. Relevant enzymes among the serine hydrolases include carboxylesterases (carboxylic-ester hydrolase; EC 3.1.1.1), arylesterases (aryl-ester hydrolase; EC 3.1.1.2), cholinesterase (acylcholine acylhydrolase; EC 3.1.1.8), and a number of serine endopeptidases (EC 3.4.21). The role of arylsulfatases (EC 3.1.6.1), aryldialkylphosphatases (EC 3.1.8.1), β-glucuronidases (EC 3.2.1.31), and epoxide hydrolases (EC 3.3.2.3) is worth noting. Some cysteine endopeptidases (EC 3.4.22), aspartic endopeptidases (EC 3.4.23), and metalloendopeptidases (EC 3.4.24) are also of potential interest (9).

2.3 Reactions of Carbon Oxidation and Reduction

When examining reactions of carbon oxidation (oxygenations and dehydrogenations) and carbon reduction (hydrogenations), it is convenient from a mechanistic viewpoint to distinguish between sp^3-, sp^2-, and sp-carbon atoms.

2.3.1 sp^3-Carbon Atoms.

Reactions of oxidation and reduction of sp^3-carbon atoms are shown in Fig. 13.3 and discussed sequentially below. In the simplest cases, a nonactivated carbon atom in an alkyl group undergoes cytochrome P450–catalyzed hydroxylation. The penultimate position is a preferred site of attack (reaction 1-B), but hydroxylation can also occur in the terminal position (reaction 1-A) or in another position in the case of steric hindrance or with some specific cytochromes P450. Dehydrogenation by dehydrogenases can then yield a carbonyl derivative (reactions 1-C and 1-E) that is either an aldehyde or a ketone. Note that reactions 1-C and 1-E act not only on metabolites, but also on xenobiotic alcohols, and are reversible (i.e., reactions 1-D and 1-F) because dehydrogenases catalyze the reactions in both directions. And whereas a ketone is seldom oxidized further, aldehydes are good substrates for aldehyde dehydrogenases or other enzymes and lead irreversibly to carboxylic acid metabolites (reaction 1-G). A classical example is that of ethanol, which in the body exists in redox equilibrium with acetaldehyde; this metabolite is rapidly and irreversibly oxidized to acetic acid.

For a number of substrates, the oxidation of primary and secondary alcohol and of aldehyde groups can also be catalyzed by cytochrome P450. A typical example is the C10-demethylation of androgens and analogs catalyzed by aromatase (CYP19).

A special case of carbon oxidation, recognized only recently and probably of underestimated significance, is desaturation of a dimethylene unit by cytochrome P450 to produce an olefinic group (reaction 2). An interesting example is provided by testosterone, which among many cytochrome P450-catalyzed reactions, undergoes allylic hydroxylation to 6β-hydroxytestosterone and desaturation to 6,7-dehydrotestosterone (43).

There is a known regioselectivity in cytochrome P450-catalyzed hydroxylations for carbon atoms adjacent (α) to an unsaturated system (reaction 3) or an heteroatom such as N, O, or S (reaction 4-A). In the former cases, hydroxylation can easily be followed by dehy-

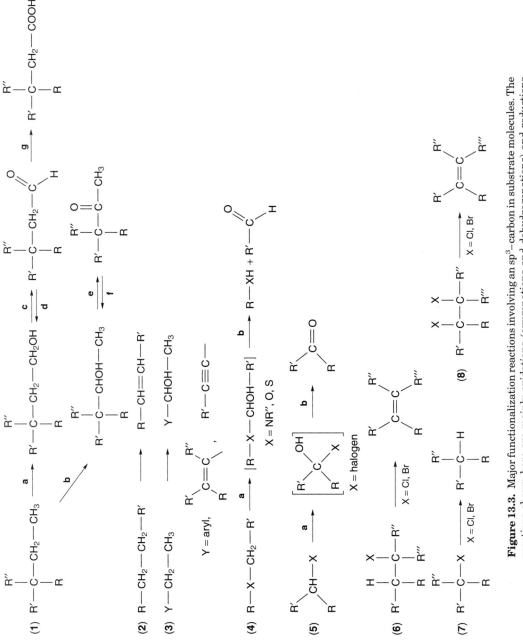

Figure 13.3. Major functionalization reactions involving an sp³–carbon in substrate molecules. The reactions shown here are mainly oxidations (oxygenations and dehydrogenations) and reductions (hydrogenations), plus some postenzymatic reactions of hydrolytic cleavage.

drogenation (not shown). In the latter cases, however, the hydroxylated metabolite is usually unstable and undergoes a rapid, postenzymatic elimination (reaction 4-B). Depending on the substrate, this pathway produces a secondary or primary amine, an alcohol or phenol, or a thiol, while the alkyl group is cleaved as an aldehyde or a ketone. Reaction 4 constitutes a very common and frequent pathway as far as drug metabolism is concerned, because it underlies some well-known metabolic reactions of *N*-C cleavage discussed later. Note that the actual mechanism of such reactions is usually more complex than shown here and may involve intermediate oxidation of the heteroatom.

Aliphatic carbon atoms bearing one or more halogen atoms (mainly chlorine or bromine) can be similarly metabolized by hydroxylation and loss of HX to dehalogenated products (reactions 5-A and 5-B; see below). Dehalogenation reactions can also proceed reductively or without change in the state of oxidation. The latter reactions are dehydrohalogenations (usually dehydrochlorination or dehydrobromination) occurring nonenzymatically (reaction 6). Reductive dehalogenations involve replacement of a halogen by a hydrogen (reaction 7) or *vic*-bisdehalogenation (reaction 8). Some radical species formed as intermediates may have toxicological significance.

Reactions 1-A, 1-B, 3, 4-A, and 5-A are catalyzed by cytochromes P450. Here, the iron-bound oxene (Section 2.2.1) acts by a mechanism known as "oxygen rebound, " whereby a H atom is exchanged for a OH group. In simplified terms, the oxene atom attacks the substrate by cleaving a C—H bond and removing the hydrogen atom (hydrogen radical). This forms an iron-bound HO· species and leaves the substrate as a C-centered radical. In the last step, the iron-bound HO· species is transferred to the substrate.

Halothane (**2**) offers a telling example of the metabolic fate of halogenated compounds of medicinal interest. Indeed, this agent undergoes two major pathways, oxidative dehalogenation leading to trifluoroacetic acid (**3**) and reduction producing a reactive radical (**4**) (Fig. 13.4).

2.3.2 sp^2- and sp-Carbon Atoms. Reactions at sp^2-carbons are characterized by their own

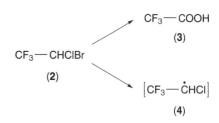

Figure 13.4. Halothane (**2**) and two of its metabolites, namely trifluoroacetic acid (**3**) produced by oxidation and a reactive radical (**4**) produced by reduction.

pathways, catalytic mechanisms, and products (Fig. 13.5). Thus, the oxidation of aromatic rings generates a variety of (usually stable) metabolites. Their common precursor is often a reactive epoxide (reaction 1-A), which can either be hydrolyzed by epoxide hydrolase (reaction 1-B) to a dihydrodiol or rearranged under proton catalysis to a phenol (reaction 1-C). The production of a phenol is a very common metabolic reaction for drugs containing one or more aromatic rings. The *para*-position is the preferred position of hydroxylation for unsubstituted phenyl rings, but the regioselectivity of the reaction becomes more complex with substituted phenyl or with other aromatic rings.

Dihydrodiols are seldom observed, as are catechol metabolites produced by their dehydrogenation catalyzed by dihydrodiol dehydrogenase (reaction 1-D). It is interesting to note that this reaction restores the aromaticity that had been lost on epoxide formation. The further oxidation of phenols and phenolic metabolites is also possible, the rate of reaction and the nature of products depending on the ring and on the nature and position of its substituents. Catechols are thus formed by reaction 1-E, whereas hydroquinones are sometimes also produced (reaction 1-F).

In a few cases, catechols and hydroquinones have been found to undergo further oxidation to quinones (reactions 1-G and 1-I). Such reactions occur by two single-electron steps and can be either enzymatic or nonenzymatic (i.e., resulting from autoxidation and yielding as by-product the superoxide anion-radical $O_2^{·-}$). The intermediate in this reaction is a semiquinone. Both quinones and semiquinones are reactive, particularly to-

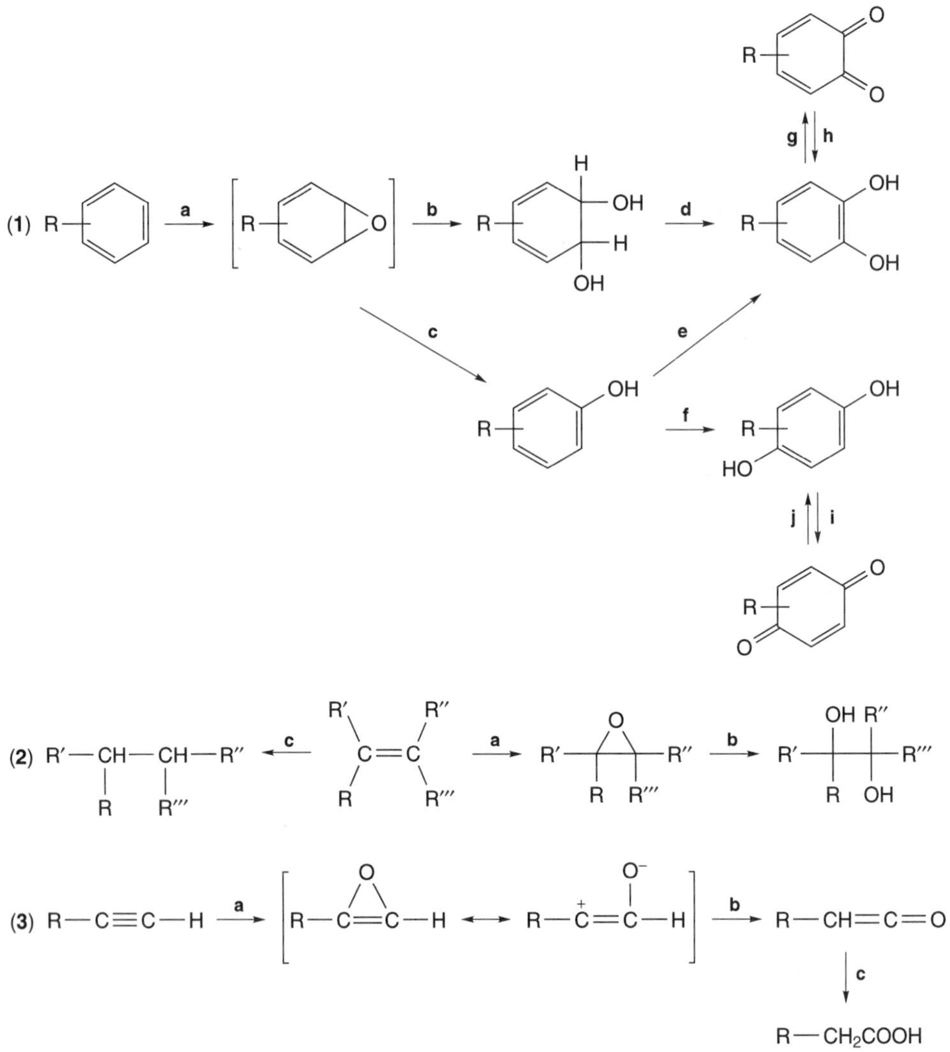

Figure 13.5. Major functionalization reactions involving an sp^2– or sp–carbon in substrate molecules. These reactions are oxidations (oxygenations and dehydrogenations), reductions (hydrogenations), and hydrations, plus some postenzymatic rearrangements.

ward biomolecules, and have been implicated in many toxification reactions. For example, the high toxicity of benzene for bone marrow is believed to be a result of the oxidation of catechol and hydroquinone catalyzed by myeloperoxidase.

The oxidation of diphenols to quinones is reversible (reactions 1-H and 1-J); a variety of cellular reductants are able to mediate the reduction of quinones either by a two-electron mechanism or by two single-electron steps.

The two-electron reduction can be catalyzed by carbonyl reductase and quinone reductase, whereas cytochrome P450 and some flavoproteins act by single-electron transfers. The nonenzymatic reduction of quinones can occur, for example, in the presence of $O_2^{\cdot-}$ or some thiols such as glutathione.

Olefinic bonds in xenobiotic molecules can also be targets of cytochrome P450-catalyzed epoxidation (reaction 2-A). In contrast to arene oxides, the resulting epoxides are fairly

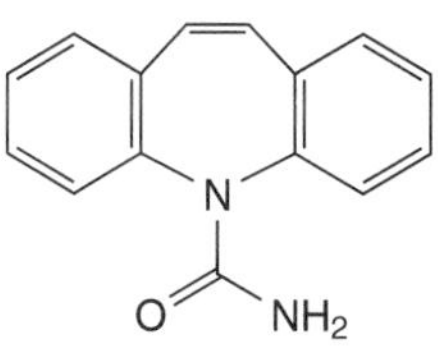

Figure 13.6. Carbamazepine (**5**) and its 10,11–epoxide metabolite (**6**).

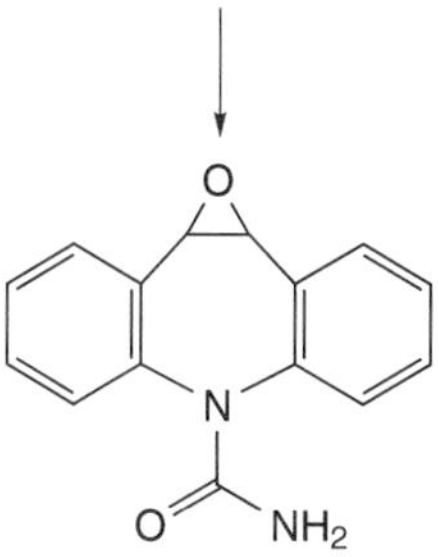

stable and can be isolated and characterized. But like arene oxides, they are substrates of epoxide hydrolase yielding dihydrodiols (reaction 2-B). This is exemplified by carbamazepine (**5**), whose 10,11-epoxide (**6**) is a major and pharmacologically active metabolite in humans and is further metabolized to the inactive dihydrodiol (44) (Fig. 13.6).

The reduction of olefinic groups (reaction 2-C) is documented for a few drugs bearing an α,β-ketoalkene function. The reaction is thought to be catalyzed by various NAD(P)H oxidoreductases.

The few drugs that contain an acetylenic moiety are also targets for cytochrome P450-catalyzed oxidation. Oxygenation of the triple bond (reaction 3-A) yields an intermediate that, depending on the substrate, can react in a number of ways, for example, binding covalently to the enzyme or forming a highly reactive ketene whose hydration produces a substituted acetic acid (reactions 3-B and 3-C).

2.4 Reactions of Nitrogen Oxidation and Reduction

The main metabolic reactions of oxidation and reduction of nitrogen atoms in organic molecules are summarized in Fig. 13.7. The functional groups involved are amines and amides

and their oxygenated metabolites, as well as 1,4-dihydropyridines, hydrazines, and azo compounds. In many cases, the reactions can be catalyzed by cytochrome P450 and/or flavin-containing monooxygenases. The first oxygenation step in reactions 1–4 and 6 have frequently been observed.

Nitrogen oxygenation is a (apparently) straightforward metabolic reaction of tertiary amines (reaction 1-A), whether they are aliphatic or aromatic. Numerous drugs undergo this reaction and the resulting N-oxide metabolite is more polar and hydrophilic than the parent compound. Identical considerations apply to pyridines and analogous aromatic azaheterocycles (reaction 2-A). Note that these reactions are reversible; a number of reductases are able to deoxygenate N-oxides back to the amine (i.e., reactions 1-B and 2-B).

Secondary and primary amines also undergo N-oxygenation and the first isolable metabolites are hydroxylamines (reactions 3-A and 4-A, respectively). Again, reversibility is documented (reactions 3-B and 4-B). These compounds can be aliphatic or aromatic amines, and the same metabolic pathway occurs in secondary and primary amides (i.e., R = acyl), whereas tertiary amides seem to be resistant to N-oxygenation. The oxidation of secondary amines and amides usually stops at the hydroxylamine/hydroxylamide level, but formation of short-lived nitroxides (not shown) has been reported.

As opposed to secondary amines and amides, their primary analogs can be oxidized to nitroso metabolites (reaction 4-C), but further oxidation of the latter compounds to nitro compounds does not seem to occur *in vivo*. In contrast, aromatic nitro compounds can be reduced to primary amines through reactions 4-E, 4-D, and finally 4-B. This is the case for numerous chemotherapeutic drugs such as metronidazole.

Note that primary aliphatic amines having a hydrogen on the alpha-carbon can display additional metabolic reactions, shown as reaction 5 in Fig. 13.5. Indeed, N-oxidation may also yield imines (reaction 5-A), whose degree of oxidation is equivalent to that of hydroxylamines (45). Imines can be further oxidized

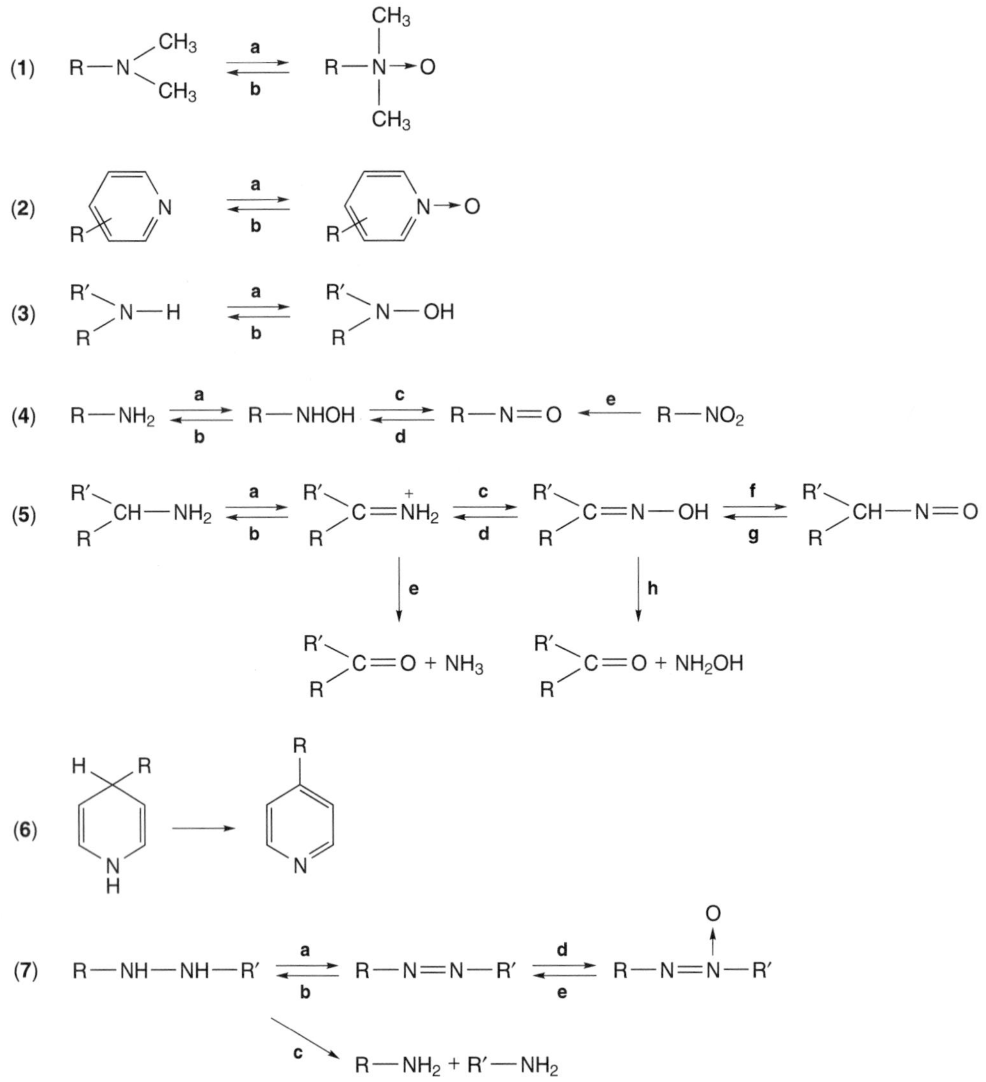

Figure 13.7. Major functionalization reactions involving nitrogen atoms in substrate molecules. The reactions shown here are mainly oxidations (oxygenations and dehydrogenations) and reductions (deoxygenations and hydrogenations).

to oximes (reaction 5-C), which are in equilibrium with their nitroso tautomer (reactions 5-F and 5-G).

1,4-Dihydropyridines, and particularly calcium channel blockers such as nivaldipide (**7**) (Fig. 13.8), are effectively oxidized by cytochrome P450. The reaction is one of aromatization (reaction 6 in Fig. 13.7), yielding the corresponding pyridine.

Dinitrogen moieties are also targets of oxidoreductases. Depending on their substituents, hydrazines are oxidized to azo compounds (reaction 7-A), some of which can be oxygenated to azoxy compounds (reaction 1-D). Another important pathway of hydrazines is their reductive cleavage to primary amines (reaction 7-C). Reactions 7-A and 7-D are reversible and the corresponding reduc-

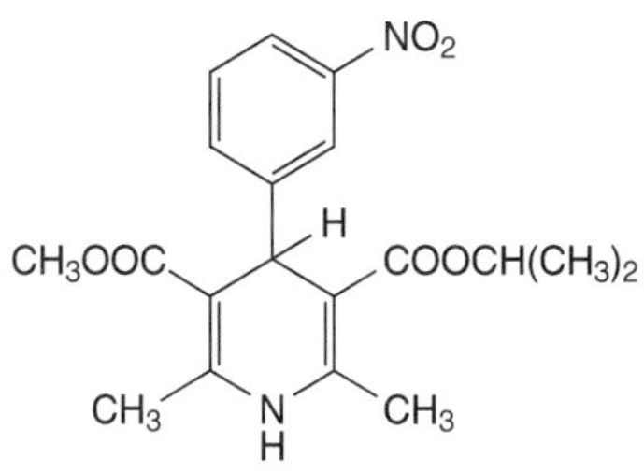

Figure 13.8. Nivaldipine (**7**).

tions (reactions 7-B and 7-E) are mediated by cytochrome P450 and other reductases. A toxicologically significant pathway thus exists for the reduction of some aromatic azo compounds to potentially toxic primary aromatic amines (reactions 7-B and 7-C).

2.5 Reactions of Oxidation and Reduction of Sulfur and Other Atoms

A limited number of drugs contain a sulfur atom, usually as a thioether. The major redox reactions occurring at sulfur atoms in organic compounds are summarized in Fig. 13.9.

Thiol compounds can be oxidized to sulfenic acids (reaction 1-A), then to sulfinic acids (reaction 1-E), and finally to sulfonic acids (reaction 1-F). Depending on the substrate, the pathway is mediated by cytochrome P450 and/or flavin-containing monooxygenases. Another route of oxidation of thiols is to disulfides either directly (reaction 1-C through thiyl radicals) or by dehy-

dration between a thiol and a sulfenic acid (reaction 1-B). However, our understanding of sulfur biochemistry is largely incomplete, and much remains to be learned. This is particularly true for reductive reactions. Whereas it is well known that reaction 1-C is reversible (i.e., reaction 1-D), the reversibility of reaction 1-A is unclear and reduction of sulfinic and sulfonic acids seems unlikely.

The metabolism of sulfides (thioethers) is rather straightforward. Besides the S-dealkylation reactions discussed earlier, these compounds can also be oxygenated by monooxygenases to sulfoxides (reaction 2-A) and then to sulfones (reaction 2-C). Here, it is known with confidence that reaction 2-A is indeed reversible, as documented by many examples of reduction of sulfoxides (reaction 2-B), whereas the reduction of sulfones has never been found to occur.

Thiocarbonyl compounds are also substrates of monooxygenases, forming S-monoxides (sulfines, reaction 3-A) and then S-dioxides (sulfenes, reaction 3-C). As a rule, these metabolites cannot be identified as such because of their reactivity. Thus, S-monoxides rearrange to the corresponding carbonyl by expelling a sulfur atom (reaction 3-D). This reaction is known as oxidative desulfuration and occurs in thioamides and thioureas (e.g., thiopental). As for the S-dioxides, they react very rapidly with nucleophiles and particularly with nucleophilic sites in biological macromolecules. This covalent binding results in the formation of adducts of toxico-

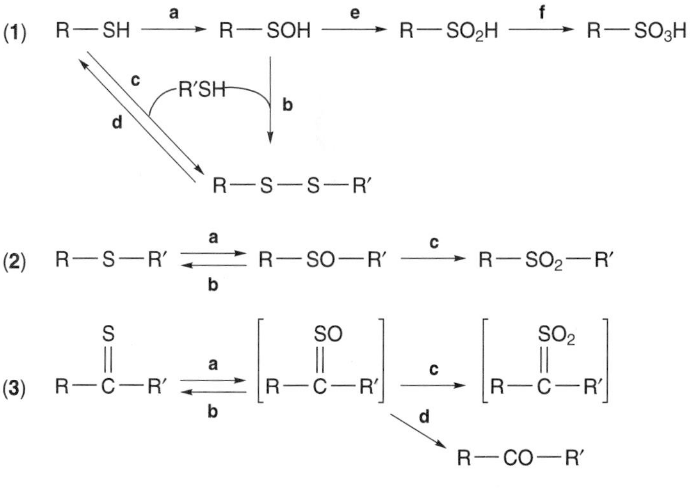

Figure 13.9. Major reactions of oxidation and reduction involving sulfur atoms in organic compounds.

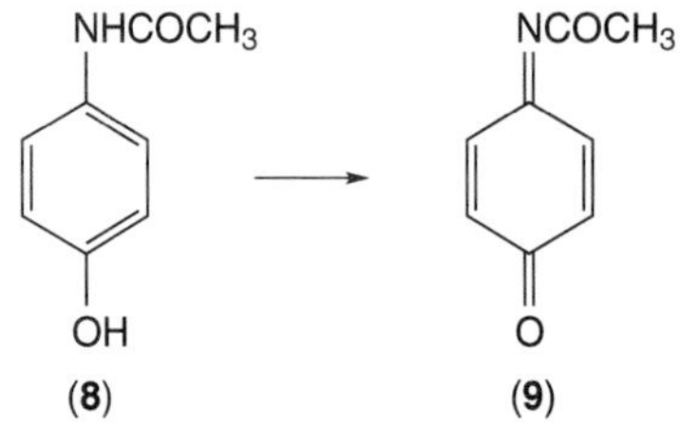

Figure 13.10. Paracetamol (**8**) and its toxic quinoneimine metabolite (**9**).

logical significance. Such a mechanism is believed to account for the carcinogenicity of a number of thioamides.

Other elements besides carbon, nitrogen, and sulfur can undergo metabolic redox reactions. The direct oxidation of oxygen atoms in phenols and alcohols is well documented for some substrates. Thus, the oxidation of secondary alcohols by some peroxidases can yield a hydroperoxyde and ultimately a ketone. Some phenols are known to be oxidized by cytochrome P450 to a semiquinone and ultimately to a quinone. A classical example is that of the anti-inflammatory drug paracetamol (**8**) (Fig. 13.10; acetaminophen), a minor fraction of which is oxidized by CYP2E1 to the highly reactive and toxic quinoneimine **9**.

Additional elements of limited significance in medicinal chemistry that are able to enter

redox reactions include silicon, phosphorus, arsenic, and selenium (Fig. 13.11). Note however that the enzymology and mechanisms of these reactions are insufficiently understood. For example, a few silanes have been shown to yield silanols *in vivo* (reaction 1). The same applies to some phosphines, which can be oxygenated to phosphine oxides by monooxygenases (reaction 2).

Arsenicals have received some attention because of their therapeutic significance. Both inorganic and organic arsenic compounds display an As(III)-As(V) redox equilibrium in the body. This is shown with the arsine-arsine oxide and arsenoxide-arsonic acid equilibria (reactions 3-A and 3-B and reactions 4-B and 4-C, respectively). Another reaction of interest is the oxidation of arseno compounds to arsenoxides (reaction 4-A), a reaction of importance in the bioactivation of a number of chemotherapeutic arsenicals.

The biochemistry of organoselenium compounds is of some interest. For example, a few selenols have been seen to oxidize to selenenic acids (reaction 5-A) and then to seleninic acids (reaction 5-B).

2.6 Reactions of Oxidative Cleavage

A number of oxidative reactions presented in the previous sections yield metabolic interme-

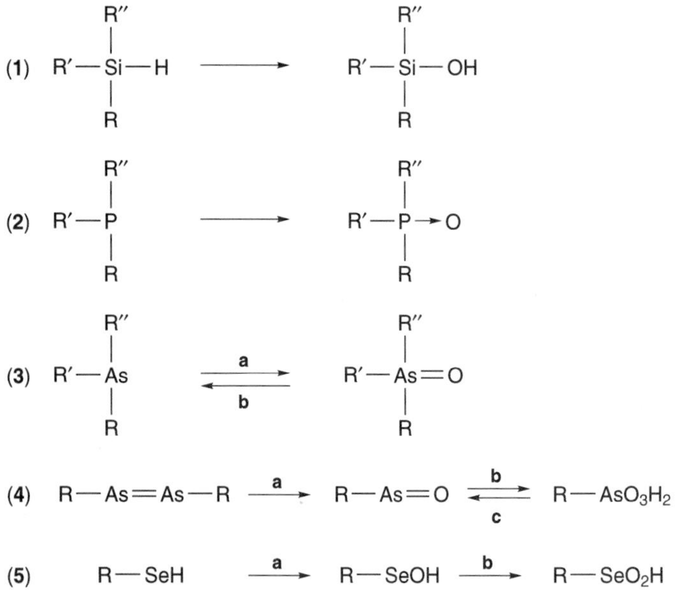

Figure 13.11. Some selected reactions of oxidation and reduction involving silicon, phosphorus, arsenic, and selenium in xenobiotic compounds.

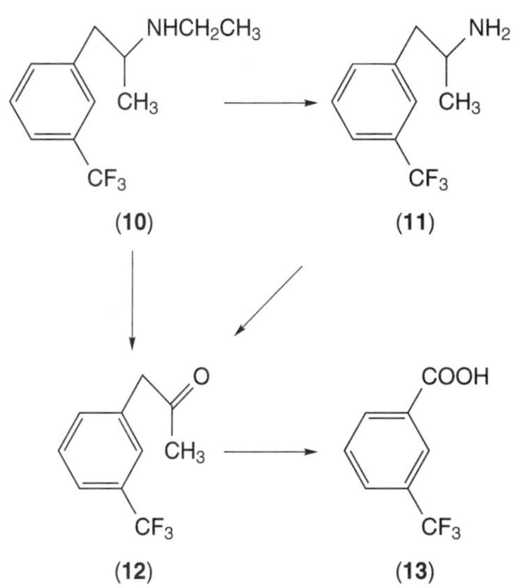

Figure 13.12. Fenfluramine (**10**), norfenfluramine (**11**), (*m*–trifluoromethyl)phenylacetone (**12**), and *m*–trifluoromethylbenzoic acid (**13**).

diates that readily undergo postenzymatic cleavage of a C-X bond (X being an heteroatom). As briefly mentioned, reactions 4-A and 4-B in Fig. 13.3 represent important metabolic pathways that affect many drugs. When X = N (by far the most frequent case), the metabolic reactions are known as *N*-demethylations, *N*-dealkylations, or deaminations, depending on the moiety being cleaved. Consider for example fenfluramine (**10**) (Fig. 13.12), which is *N*-deethylated to norfenfluramine (**11**), an active metabolite, and deaminated to (*m*-trifluoromethyl)phenylacetone (**12**), an inactive metabolite that is further oxidized to *m*-trifluoromethylbenzoic acid (**13**).

When X = O or S in reaction 4 (Fig. 13.3), the metabolic reactions are known as *O*-dealkylations or *S*-dealkylations, respectively. *O*-demethylation is a typical case of the former reaction. And when X = halogen in reactions

5-A and 5-B (Fig. 13.3), loss of halogen can also occur and is known as oxidative dehalogenation.

The reactions of oxidative C-X cleavage discussed above result from carbon hydroxylation and are catalyzed by cytochrome P450. However, *N*-oxidation reactions followed by hydrolytic C-N cleavage can also be catalyzed by cytochrome P450 (e.g., reactions 5-E and 5-H in Fig. 13.7). The sequence of reactions 5-A and 5-E in Fig. 13.7 is of particular interest because it is the mechanism by which monoamine oxidase deaminates endogenous and exogenous amines.

2.7 Reactions of Hydration and Hydrolysis

Hydrolases catalyze the addition of a molecule of water to a variety of functional moieties. Thus, epoxide hydrolase hydrates epoxides to yield *trans*-dihydrodiols (reaction 1-B in Fig. 13.5). This reaction is documented for many arene oxides, particularly metabolites of aromatic compounds, and epoxides of olefins. Here, a molecule of water has been added to the substrate without loss of a molecular fragment, therefore the use of the term "hydration" sometimes found in the literature.

Reactions of hydrolytic cleavage (hydrolysis) are shown in Fig. 13.13. They are frequent for organic esters (reaction 1), inorganic esters such as nitrates (reaction 2) and sulfates (reaction 3), and amides (reaction 4). These reactions are catalyzed by esterases, peptidases, or other enzymes, but nonenzymatic hydrolysis is also known to occur for sufficiently labile compounds under biological conditions of pH and temperature. Acetylsalicylic acid, glycerol trinitrate, and lidocaine are three representative examples of drugs undergoing extensive cleavage of the organic ester, inorganic ester, or amide group, respectively. The reaction is of particular significance in the activation of ester prodrugs (Section 5.2).

(1) $R-COO-R' \longrightarrow R-COOH + R'-OH$

(2) $R-ONO_2 \longrightarrow R-OH + HNO_3$

(3) $R-OSO_3H \longrightarrow R-OH + H_2SO_4$

(4) $R-CONHR' \longrightarrow R-COOH + R'-NH_2$

Figure 13.13. Major hydrolysis reactions involving esters (organic and inorganic) and amides.

3 CONJUGATION REACTIONS

3.1 Introduction

As defined in the Introduction, conjugation reactions (also infelicitously known as phase II reactions) result in the covalent binding of an endogenous molecule or moiety to the substrate. Such reactions are of critical significance in the metabolism of endogenous compounds, as witnessed by the impressive battery of enzymes that have evolved to catalyze them. Conjugation is also of great importance in the biotransformation of xenobiotics, involving parent compounds or metabolites thereof (3).

Conjugation reactions are characterized by a number of criteria:

1. They are catalyzed by enzymes known as transferases.
2. They involve a cofactor that binds to the enzyme in close proximity to the substrate and carries the endogenous molecule or moiety to be transferred.
3. The endogenous molecule or moiety is highly polar (hydrophilic), and its size is comparable with that of the substrate.

It is important from a biochemical and practical viewpoint to note that these criteria are neither sufficient nor necessary to define conjugations reactions. They are not sufficient, because in hydrogenation reactions (i.e., typical reactions of functionalization) the hydride is also transferred from a cofactor (NADPH or NADH). And they are not necessary, because all the above criteria suffer from some important exceptions mentioned below.

3.2 Methylation

3.2.1 Introduction. Reactions of methylation imply the transfer of a methyl group from the cofactor S-adenosyl-L-methionine (SAM) (**14**). As shown in Fig. 13.14, the methyl group in SAM is bound to a sulfonium center, giving it a marked electrophilic character and explaining its reactivity. Furthermore, it becomes pharmacokinetically relevant to distinguish methylated metabolites in which the positive charge has been retained or lost.

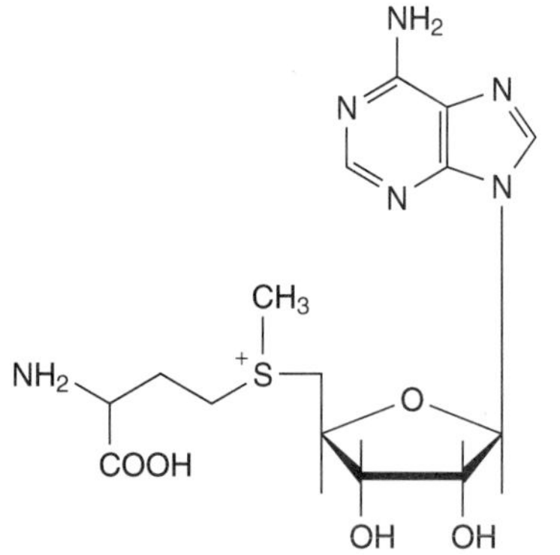

Figure 13.14. S–adenosyl–L–methionine (**14**).

A number of methyltransferases are able to methylate small molecules (46, 47). Thus, reactions of methylation fulfill only two of the three criteria defined above, because the methyl group is small compared with the substrate. The main enzyme responsible for O-methylation is catechol O-methyltransferase (EC 2.1.1.6; COMT), which is mainly cytosolic but also exists in membrane-bound form. Several enzymes catalyze reactions of xenobiotic N-methylation with different substrate specificities, e.g., nicotinamide N-methyltransferase (EC 2.1.1.1), histamine methyltransferase (EC 2.1.1.8), phenylethanolamine N-methyltransferase (noradrenaline N-methyltransferase; EC 2.1.1.28), and nonspecific amine N-methyltransferase (arylamine N-methyltransferase, tryptamine N-methyltransferase; EC 2.1.1.49) of which some isozymes have been characterized. Reactions of xenobiotic S-methylation are mediated by the membrane-bound thiol methyltransferase (EC 2.1.1.9) and the cytosolic thiopurine methyltransferase (EC 2.1.1.67) (3).

The above classification of enzymes makes explicit the three types of functionalities undergoing biomethylation, namely hydroxy (phenolic), amino, and thiol groups.

3.2.2 Methylation Reactions. Figure 13.15 summarizes the main methylation reactions seen in drug metabolism. O-Methylation is a common reaction of compounds containing a catechol moiety (reaction 1), with a usual regioselectivity for the *meta* position. The sub-

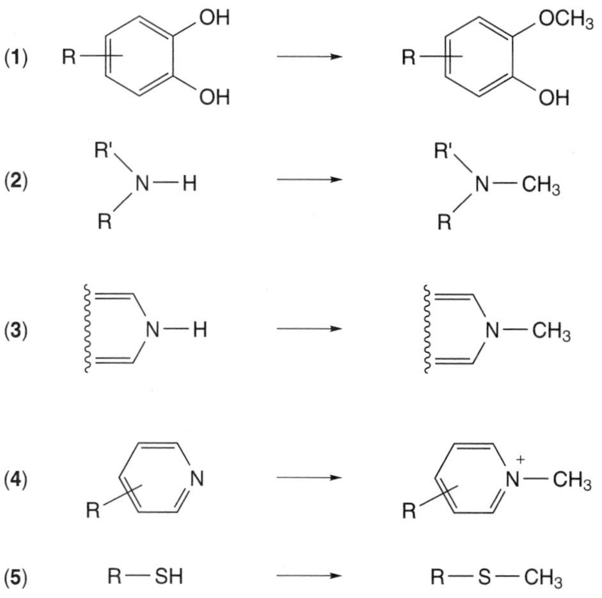

Figure 13.15. Major methylation reactions involving catechols, various amines, and thiols.

strates can be xenobiotics, particularly drugs, with L-DOPA being a classic example. More frequently, however, O-methylation occurs as a late event in the metabolism of aryl groups, after they have been oxidized to catechols (reactions 1, Fig. 13.5). This sequence was seen for example in the metabolism of the anti-inflammatory drug diclofenac (**15**) (Fig. 13.16), which in humans yielded 3'-hydroxy-4'-methoxy-diclofenac as a major metabolite with a very long plasma half-life (48).

Three basic types of N-methylation reactions have been recognized (reactions 2–4, Fig. 13.15). A number of primary amines (e.g., amphetamine) and secondary amines (e.g., tetrahydroisoquinolines) have been shown to be *in vitro* substrates of amine N-methyltransferase, whereas some phenylethanolamines and analogs are methylated by phenylethanolamine N-methyltransferase (reaction 2). However, such reactions are seldom of significance *in vivo*, presumably because of effective oxidative N-demethylation. A comparable situation involves the N—H group in an imidazole ring (reaction 3), exemplified by histamine (49). A therapeutically relevant example is that of theophylline (**16**) whose N(9)-methylation is masked by N-demethylation in adult but not newborn humans.

N-Methylation of pyridine-type nitrogen atoms (reaction 4, Fig. 13.15) seems to be of greater *in vivo* pharmacological significance than reactions 2 and 3 for two reasons. First, the resulting metabolites, being quaternary amines, are more stable than tertiary or secondary amines toward N-demethylation. And second, these metabolites are also more polar

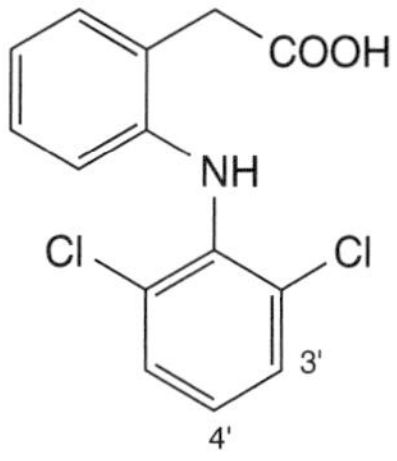

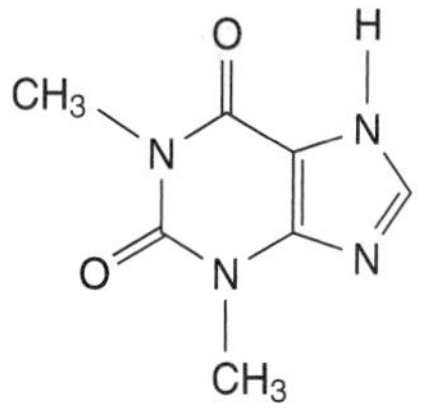

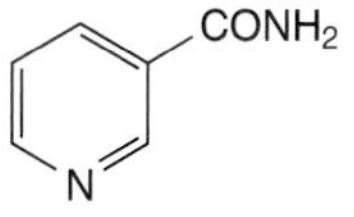

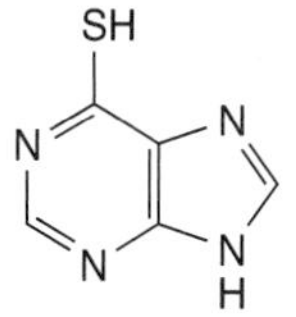

Figure 13.16. Diclofenac (**15**), theophylline (**16**), nicotinamide (**17**), and 6–mercaptopurine (**18**).

than the parent compounds, in contrast to the products of reactions 2 and 3. Good substrates are nicotinamide (**17**), pyridine, and a number of monocyclic and bicyclic derivatives (49).

S-Methylation of thiol groups (reaction 5) is documented for such drugs as 6-mercaptopurine (**18**) and captopril (50). Other substrates are metabolites (mainly thiophenols) resulting from the S—C cleavage of (aromatic) glutathione and cysteine conjugates (see below). Once formed, such methylthio metabolites can be further processed to sulfoxides and sulfones before excretion (i.e., reactions 2-A and 2-C in Fig. 13.9).

From Fig. 13.15, it is apparent that methylation reactions can be subdivided into two classes:

1. Those where the substrate and the product have the same electrical state; a proton in the substrate being exchanged for a positively charged methyl group (reactions 1–3 and 5).

2. Those where the product has acquired a positive charge, namely becomes a pyridine–type quaternary ammonium (reaction 4).

3.3 Sulfation

3.3.1 Introduction. Sulfation reactions consist of a sulfate being transferred from the cofactor 3'-phosphoadenosine 5'-phosphosulfate (**19**) (PAPS; Fig. 13.17) to the substrate under catalysis by a sulfotransferase. The three criteria of conjugation are met in these reactions. Sulfotransferases, which catalyze a variety of physiological reactions, are soluble enzymes that include aryl sulfotransferase (phenol sulfotransferase; EC 2.8.2.1), alcohol sulfotransferase (hydroxysteroid sulfotransferase; EC 2.8.2.2), amine sulfotransferase (arylamine sulfotransferase; EC 2.8.2.3), estrone sulfotransferase (EC 2.8.2.4), tyrosine-ester sulfotransferase (EC 2.8.2.9), steroid sulfotransferase (EC 2.8.2.15), and cortisol sulfotransferase (glucocorticosteroid sulfotransferase; EC 2.8.2.18). Among these enzymes, the former three are of particular significance in the sulfation of xenobiotics. Recent advances in the molecular biology of these enzymes has led to the recognition of three

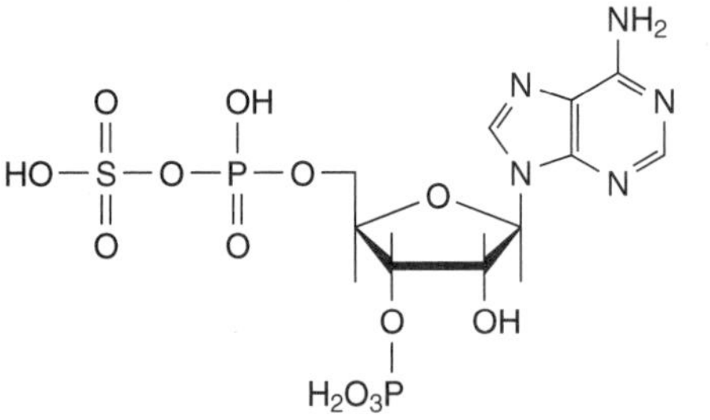

Figure 13.17. 3'–Phosphoadenosine 5'–phosphosulfate (**19**) (PAPS).

human phenol sulfotransferases, the thermostable, phenol-preferring SULT1A1 and SULT1A2 and the thermolabile, monoamine-preferring SULT1A3 (3, 51, 52).

The sulfate moiety in PAPS is linked to a phosphate group by an anhydride bridge whose cleavage is exothermic and supplies enthalpy to the reaction. The electrophilic —OH or —NH— site in the substrate will react with the leaving SO_3^- moiety, forming an ester sulfate or a sulfamate (Fig. 13.18). Some of these conjugates are unstable under biological conditions and will form electrophilic intermediates of considerable toxicological significance.

3.3.2 Sulfation Reactions. Sulfoconjugation of alcohols (reaction 1 in Fig. 13.18) leads to metabolites of different stabilities. Endogenous hydroxysteroids (i.e., cyclic secondary alcohols) form relatively stable sulfates, whereas some secondary alcohol metabolites of allylbenzenes (e.g., safrole and estragole) form highly genotoxic carbocations (53). Primary alcohols, e.g., methanol and ethanol, can also form sulfates whose alkylating capacity is well known (54). Similarly, polycyclic hydroxymethylarenes yield reactive sulfates believed to account for their carcinogenicity.

In contrast to alcohols, phenols form stable sulfate esters (reaction 2). The reaction is usually of high affinity (i.e., rapid), but the limited availability of PAPS restricts the amounts of conjugate being produced. Typical drugs undergoing limited sulfation include paracetamol (**8**) (Fig. 13.10) and diflunisal (**20**) (Fig. 13.19).

Aromatic hydroxylamines and hydroxylamides are good substrates for some sulfotransferases and yield unstable sulfate esters

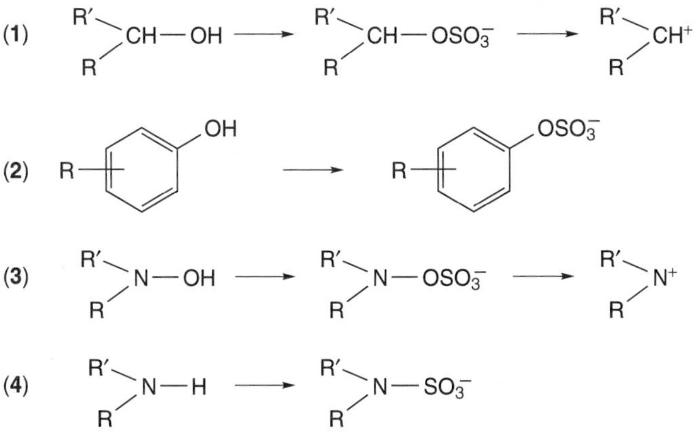

Figure 13.18. Major sulfation reactions involving primary and secondary alcohols, phenols, hydroxylamines and hydroxylamides, and amines.

(reaction 3 in Fig. 13.18). Indeed, heterolytic N-O cleavage produces a highly electrophilic nitrenium ion. This is a mechanism believed to account for part or all of the cytotoxicity of arylamines and arylamides (e.g., phenacetin). In contrast, significantly more stable products are obtained during *N*-sulfoconjugation of amines (reaction 4). Alicyclic amines, and pri-

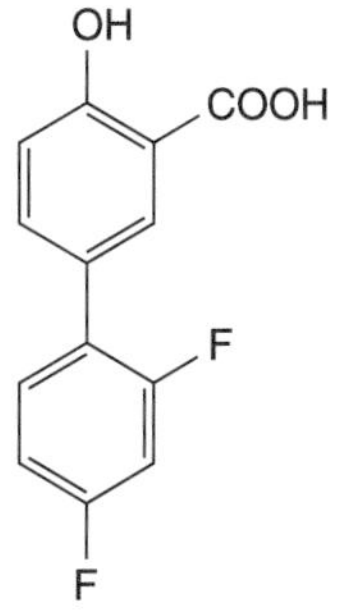

(20)

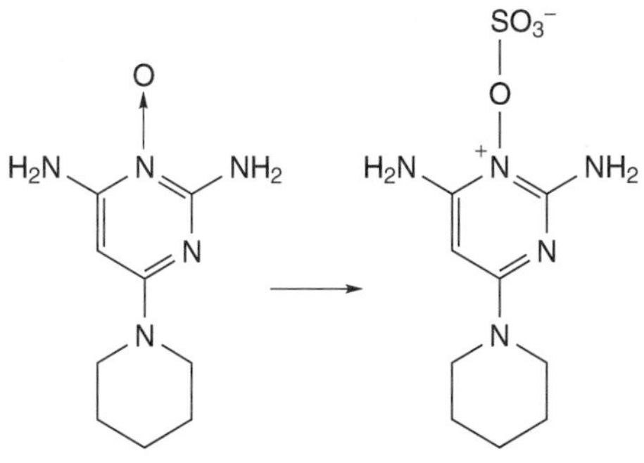

(21)   (22)

Figure 13.19. Diflunisal (**20**), minoxidil (**21**), and its *N,O*–sulfate ester (**22**).

mary and secondary alkyl- and aryl-amines, can all yield sulfamates (55). The significance of these reactions in humans is still poorly understood.

An intriguing and very rare reaction of conjugation occurs for minoxidil (**21**) (Fig. 13.19), an hypotensive agent also producing hair growth. This drug is an *N*-oxide, and the actual active form responsible for the different therapeutic effects is the *N,O*-sulfate ester (**22**) (56).

3.4 Glucuronidation and Glucosidation

3.4.1 Introduction. Glucuronidation is a major and very frequent reaction of conjugation. It involves the transfer to the substrate of a molecule of glucuronic acid from the cofactor uridine-5′-diphospho-α-D-glucuronic acid (**23**) (UDPGA; Fig. 13.20). The enzyme catalyzing this reaction is known as UDP-glucuronyltransferase (UDP-glucuronosyltransferase; EC 2.4.1.17, UDPGT) and consists of a number of proteins coded by genes of the *UGT* superfamily. The human UDPGT known to metabolize xenobiotics is the product of two gene families, *UGT1* and *UGT2*. These enzymes include UGT1A1 (bilirubin UDPGTs) and several UGT1A, as well as numerous phenobarbital-inducible or constitutively expressed UGT2B (57–61).

In addition to glucuronidation, this section briefly mentions glucosidation, a minor metabolic reaction seen for a few drugs. Candidate enzymes catalyzing this reaction could be

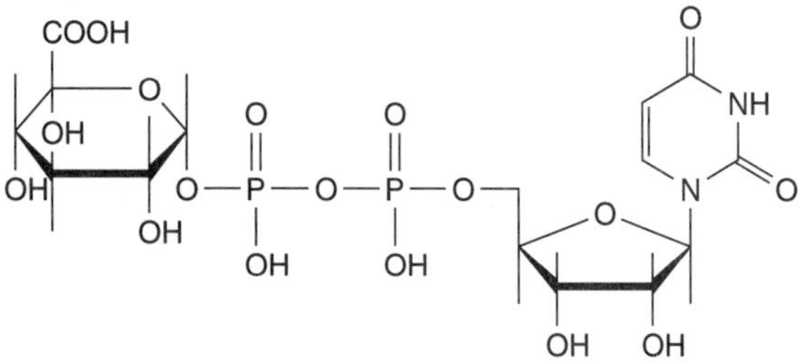

Figure 13.20. Uridine–5′–diphospho–α–D– glucuronic acid (**23**) (UDPGA).

(**23**)

phenol β-glucosyltransferase (EC 2.4.1.35), arylamine glucosyltransferase (EC 2.4.1.71), and nicotinate glucosyltransferase (EC 2.4.1.196).

3.4.2 Glucuronidation Reactions. Glucuronic acid exists in UDPGA in the 1α-configuration, but the products of conjugation are β-glucuronides. This is because the mechanism of the reaction is a nucleophilic substitution with inversion of configuration. Indeed, and as shown in Fig. 13.21, all functional groups able to undergo glucuronidation are nucleophiles, a common characteristic they share despite their great chemical variety. As a consequence of this diversity, the products of glucuronidation are classified as *O*-, *N*-, *S*-, and C-glucuronides.

O-Glucuronidation is shown in reactions 1–5 (Figure 13.21). A frequent metabolic reaction of phenolic xenobiotics or metabolites is their glucuronidation to yield polar metabolites excreted in urine and/or bile. *O*-Glucuronidation is often in competition with *O*-sulfation (see above), with the latter reaction predominating at low doses and the former at high doses. In biochemical terms, glucuronidation is a reaction of low affinity and high capacity, wheras sulfation displays high affinity and low capacity. A typical drug undergoing extensive glucuronidation is paracetamol (**8**) (Fig. 13.10). Another major group of substrates are alcohols: primary, secondary, or tertiary (reaction 2, Fig. 13.21). Medicinal examples include chloramphenicol and oxazepam. Another important example is that of morphine, which is conjugated on its phenolic and secondary alcohol groups to form the 3-*O*-glucuronide (a weak opiate antagonist) and the 6-*O*-glucuronide (a strong opiate agonist), respectively (62).

An important pathway of *O*-glucuronidation is the formation of acyl-glucuronides (reaction 3). Substrates are arylacetic acids (e.g., diclofenac) (**15**) (Fig. 13.16) and aliphatic acids (e.g., valproic acid). Aromatic acids are seldom substrates; a noteworthy exception is diflunisal (**20**) (Fig. 13.19), which yields both the acyl and phenolic glucuronides. The significance of acyl glucuronides has long been underestimated, perhaps because of analytical difficulties. Indeed, these metabolites are quite reactive, rearranging to positional isomers and binding covalently to plasma and seemingly also tissue proteins (63). Thus, acyl glucuronide formation cannot be viewed solely as a reaction of inactivation and detoxification.

A special class of acyl glucuronides are the carbamoyl glucuronides (reaction 4 in Fig. 13.21). A number of primary and secondary amines have been found to yield this type of conjugate, whereas, as expected, the intermediate carbamic acids are not stable enough to be characterized. Carvedilol (**24**) (Fig. 13.22) is one drug exemplifying the reaction, in addition to forming an *O*-glucuronide on its alcohol group and a carbazole-*N*-linked glucuronide (see below) (64). Much remains to be understood concerning the chemical and biochemical reactivity of carbamoyl glucuronides.

Hydroxylamines and hydroxylamides may also form *O*-glucuronides (reaction 5, Fig. 13.21). Thus, a few drugs and a number of aromatic amines are known to be *N*-hydroxylated and then *O*-glucuronidated. The glucuronidation of N—OH groups competes with *O*-sulfation, but the reactivity of *N*-*O*-glucuronides to undergo heterolytic cleavage and form nitrenium ions does not seem to be well characterized.

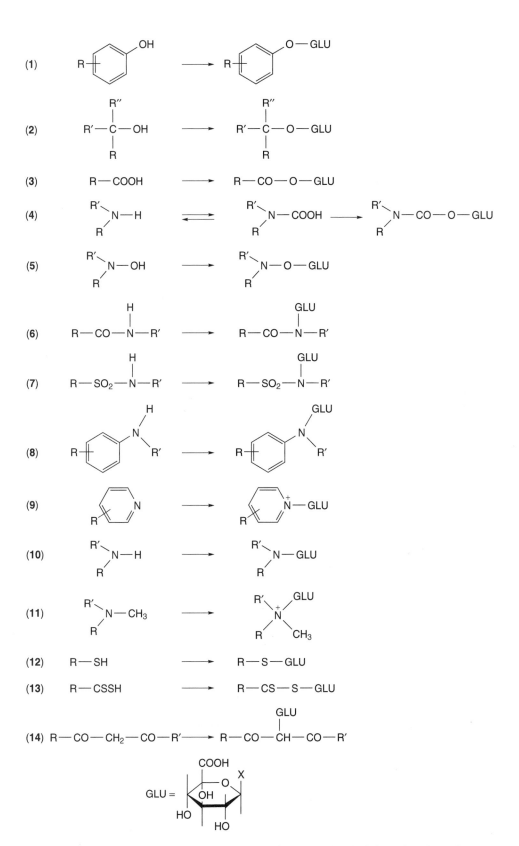

Figure 13.21. Major glucuronidation reactions involving phenols, alcohols, carboxylic acids, carbamic acids, hydroxylamines and hydroxylamides, carboxamides, sulfonamides, various amines, thiols, dithiocarboxylic acids, and 1,3–dicarbonyl compounds.

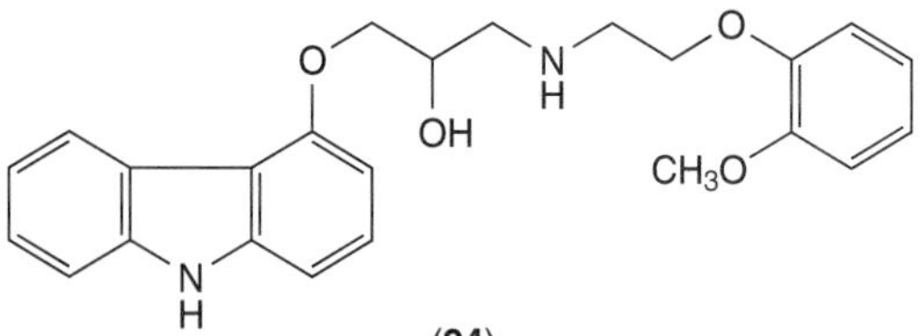

Figure 13.22. Carvedilol (**24**), phenytoin (**25**), and sulfadimethoxine (**26**).

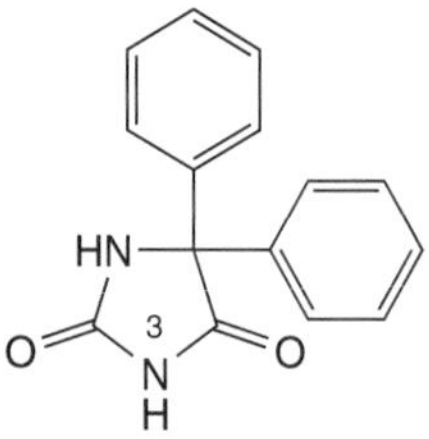

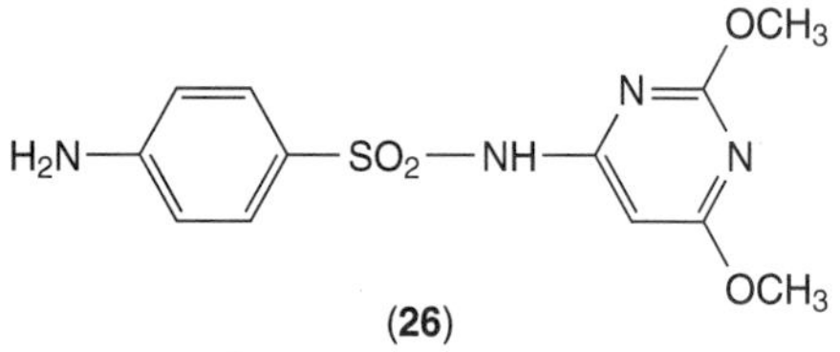

Second in importance to *O*-glucuronides are the *N*-glucuronides formed by reactions 6–11 in Fig.13.21: amides (reactions 6 and 7), amines of medium basicity (reactions 8 and 9), and basic amines (reactions 10 and 11). The *N*-glucuronidation of carboxamides (reaction 6) is exemplified by carbamazepine (**5**) (Fig. 13.6) and phenytoin (**25**) (Fig. 13.22). In the latter case, *N*-glucuronidation was found to occur at N (3). The reaction has special significance for sulfonamides (reaction 7) and particularly antibacterial sulfanilamides such as sulfadimethoxine (**26**) (Fig. 13.22), because it produces highly water-soluble metabolites that show no risk of crystallizing in the kidneys.

N-Glucuronidation of aromatic amines (reaction 8, Fig. 13.21) has been observed in only a few cases (e.g., conjugation of the carbazole nitrogen in carvedilol) (**24**). Similarly, there are a number of observations that pyridine-type nitrogens and primary and secondary basic amines can be *N*-glucuronidated (reactions 9 and 10, respectively). As far as human drug metabolism is concerned, another reaction of significance is the *N*-glucuronidation of lipophilic, basic tertiary amines containing one or two methyl groups (reaction 11) (65, 66).

More and more drugs of this type (e.g., antihistamines and neuroleptics), are found to undergo this reaction to a marked extent in humans, e.g., cyproheptadine (**27**) in Fig. 13.23.

Third in importance are the *S*-glucuronides formed from aliphatic and aromatic thiols (reaction 12 in Fig. 13.21) and from dithiocarboxylic acids (reaction 13) such as diethyldithiocarbamic acid, a metabolite of disulfiram. As for C-glucuronidation (reaction 14), this reaction has been seen in humans for 1,3-dicarbonyl drugs such as phenylbutazone and sulfinpyrazone (**28**) (Fig. 13.23).

3.4.3 Glucosidation Reactions. A few drugs have been observed to be conjugated to glucose in mammals (67). This is usually a minor pathway in the cases where glucuronidation is possible. An interesting medicinal example is that of some barbiturates such as phenobarbital, which yield the *N*-glucoside.

3.5 Acetylation and Acylation

All reactions discussed in this section involve the transfer of an acyl moiety to an acceptor group. In most cases, an acetyl is the acyl moiety being transferred, while the acceptor group may be an amino or hydroxy function.

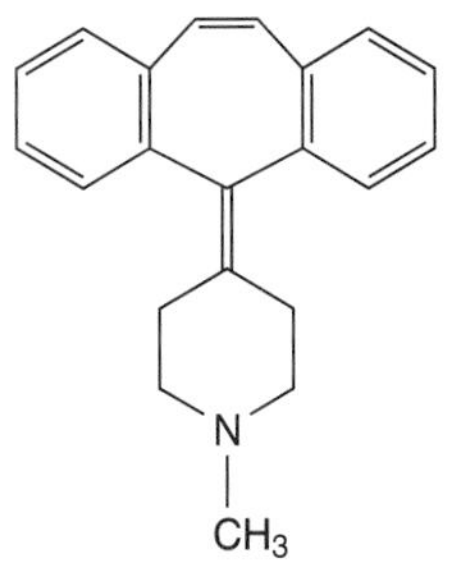

(27)

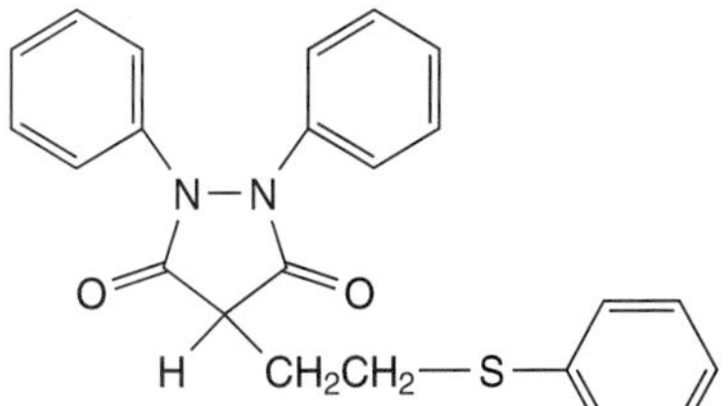

(28)

Figure 13.23. Cyproheptadine (**27**) and sulfinpyrazone (**28**).

3.5.1 Acetylation Reactions. The major enzyme system catalyzing acetylation reactions is arylamine *N*-acetyltransferase (arylamine acetylase; EC 2.3.1.5; NAT). Two enzymes have been characterized, NAT1 and NAT2; the latter has two closely related isoforms NAT2A and NAT2B whose levels are considerably reduced in the liver of slow acetylators (68, 69). The cofactor of *N*-acetyltransferase is acetylcoenzyme A (CoA-*S*-Ac, **29** with R = acetyl) (Fig. 13.24) where the acetyl moiety is bound by a thioester linkage.

Two other activities, aromatic-hydroxylamine *O*-acetyltransferase (EC 2.3.1.56) and *N*-hydroxyarylamine *O*-acetyltransferase (EC 2.3.1.118), are also involved in the acetylation of aromatic amines and hydroxylamines (see below). Other acetyltransferases exist, e.g., diamine *N*-acetyltransferase (putrescine acetyltransferase; EC 2.3.1.57) and aralkylamine *N*-acetyltransferase (serotonin acetyltransferase; EC 2.3.1.87), but their involvement in xenobiotic metabolism does not seem to be documented.

The substrates of acetylation, schematized in Fig. 13.25, are mainly amines of medium basicity. Very few basic amines (primary or secondary) of medicinal interest have been reported to form *N*-acetylated metabolites (reaction 1), and when they did, the yields were low. In contrast, a large variety of primary aromatic amines are *N*-acetylated (reaction 2). Thus, several drugs such as sulfonamides and *para*-aminosalicylic acid (**30**) (PAS; Fig. 13.26) are acetylated to large extents, not to mention various carcinogenic amines such as benzidine.

Arylhydroxylamines can also be acetylated, but the reaction is one of *O*-acetylation (reaction 3-A in Fig. 13.25). This is the reaction formally catalyzed by EC 2.3.1.118 with acetyl-CoA acting as the acetyl donor; the *N*-hydroxy metabolites of a number of arylamines are known substrates. The same conjugates can be formed by intramolecular *N,O*-acetyl transfer, when an arylhydroxamic acid (an *N*-aryl-*N*-hydroxyacetamide) is substrate of, e.g., EC 2.3.1.56 (reaction 3-B). In addition, such an arylhydroxamic acid can transfer its acetyl moiety to an acetyltransferase, which can then acetylate an

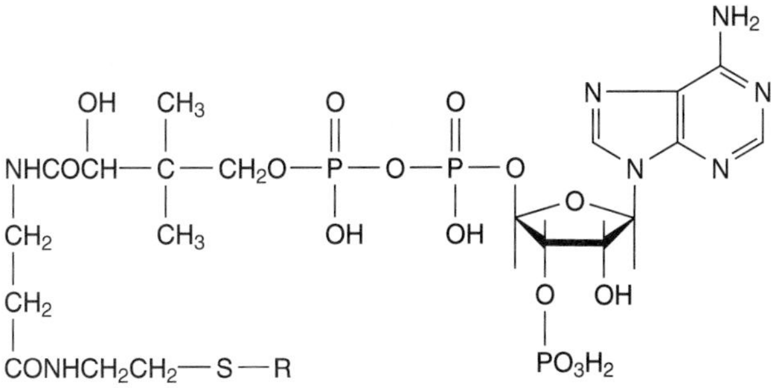

(29)

Figure 13.24. Acetylcoenzyme A (**29**). R, acetyl.

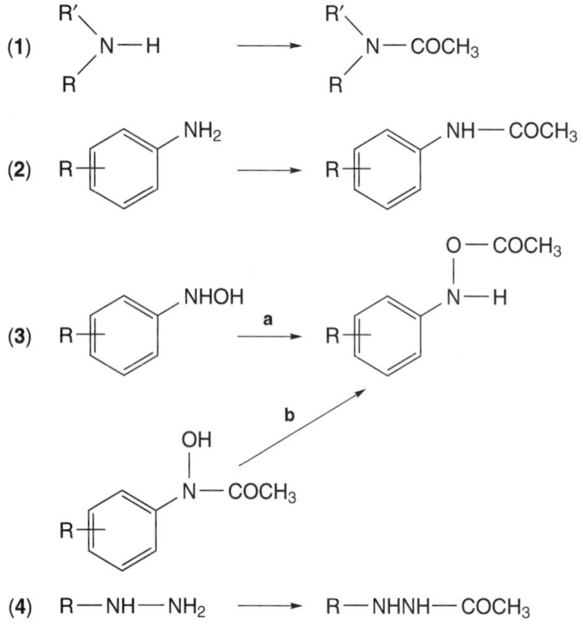

Figure 13.25. Major acetylation reactions involving aliphatic amines, aromatic amines, arylhydroxylamines, hydrazines, and hydrazides.

arylamine or an arylhydroxylamine (intermolecular N,N- or N,O-acetyl transfer). Because N-acetyltransferase (EC 2.3.1.5) can catalyze these various reactions, there is some doubt as to the

Figure 13.26. *para*–Aminosalicylic acid (**30**), isoniazid (**31**), and hydralazine (**32**).

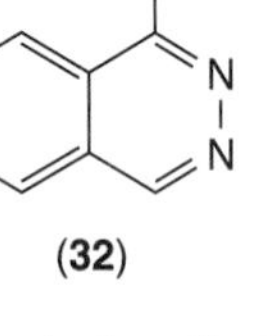

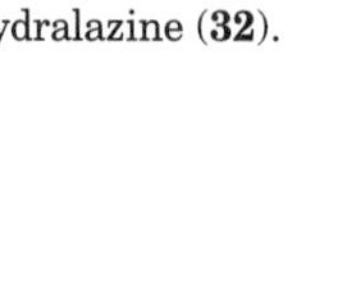

individuality of EC 2.3.1.56 and EC 2.3.1.118 in mammals (70).

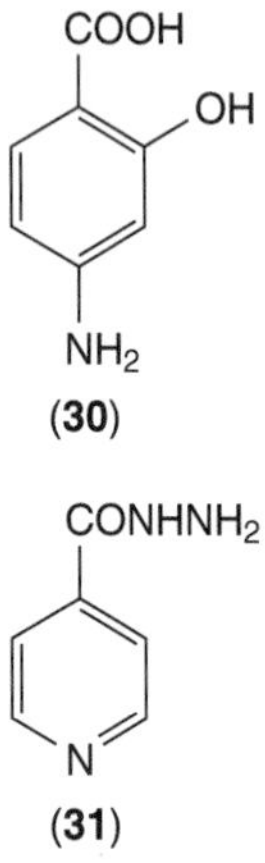

Besides amines, other nitrogen-containing functionalities undergo N-acetylation; hydrazines and hydrazides are particularly good substrates (reaction 4, Fig. 13.25). Medicinal examples include isoniazid (**31**) (Fig. 13.26) and hydralazine (**32**).

3.5.2 Other Acylation Reactions. A limited number of studies have shown N-formylation to be a genuine route of conjugation for some arylalkylamines and arylamines, particularly polycyclic aromatic amines. There is evidence to indicate that the reaction is catalyzed by arylformamidase (EC 3.5.1.9) in the presence of N-formyl-L-kynurenine.

A very different type of reaction is represented by the conjugation of xenobiotic alcohols with fatty acids, yielding highly lipophilic metabolites accumulating in tissues. Thus, ethanol and haloethanols form esters with palmitic acid, oleic acid, linoleic acid, and linolenic acid; enzymes catalyzing such reactions are cholesteryl ester synthase (EC 3.1.1.13) and fatty-acyl-ethyl-ester synthase (EC 3.1.1.67) (71). Larger xenobiotics such as tetrahydrocannabinols and codeine are also acy-

Table 13.7   Metabolic Consequences of the Conjugation of Xenobiotic Acids to Coenzyme A (CoA-SH)

R-COOH → R-CO-S-CoA → •
- Hydrolysis
- Formation of amino acid conjugates
- Formation of hybrid triglycerides
- Formation of phospholipids
- Formation of cholesteryl esters
- Formation of bile acid esters
- Formation of acyl-carnitines
- Protein acylation
- Unidirectional chiral inversion of arylpropionic acids (profens)
- Dehydrogenation and β-oxidation
- 2-Carbon chain elongation

lated with fatty acids, possibly by sterol *O*-acyltransferase (EC 2.3.1.26).

3.6   Conjugation with Coenzyme A and Subsequent Reactions

3.6.1 Conjugation with Coenzyme A. The reactions described in this section all have in common the fact that they involve xenobiotic carboxylic acids forming an acyl-CoA metabolic intermediate (**29**) (Fig. 13.24, R = xenobiotic acyl moiety). The reaction requires ATP and is catalyzed by various acyl-CoA synthetases of overlapping substrate specificity, e.g., acetate-CoA ligase (acetyl-CoA synthetase; EC 6.2.1.1), butyrate-CoA ligase [fatty acid thiokinase (medium chain); EC 6.2.1.2], long-chain-fatty-acid-CoA ligase [fatty acid thiokinase (long chain); acyl-CoA synthetase; EC 6.2.1.3], benzoate-CoA ligase (EC 6.2.1.25), and phenylacetate-CoA ligase (EC 6.2.1.30).

The acyl-CoA conjugates thus formed are seldom excreted, but they can be isolated and characterized relatively easily in *in vitro* studies. They may also be hydrolyzed back to the parent acid by thiolester hydrolases (EC 3.1.2). In a number of cases, such conjugates have pharmacodynamic effects and may even represent the active forms of some drugs, e.g., hypolipidemic agents. In the present context, the interest of acyl-CoA conjugates is their further transformation by a considerable variety of pathways (Table 13.7) (72). The most significant routes are discussed below.

3.6.2 Formation of Amino Acid Conjugates. Amino acid conjugation is a major route for a number of small aromatic acids and involves the formation of an amide bond between the xenobiotic acyl-CoA and the amino acid. Glycine is the amino acid most frequently used for conjugation (reaction 1 in Fig. 13.27), whereas a few glutamine conjugates have been characterized in humans (reaction 2). The enzymes catalyzing these transfer reactions are various *N*-acyltransferases, for example, glycine *N*-acyltransferase (EC 2.3.1.13), glutamine *N*-phenylacetyltransferase (EC 2.3.1.14), glutamine *N*-acyltransferase (EC 2.3.1.68), and glycine *N*-benzoyltransferase (EC 2.3.1.71). In addition, other amino acids can be used for conjugation in various animal species, e.g., alanine and taurine, as well as some dipeptides (3).

The xenobiotic acids undergoing amino acid conjugation are mainly substituted benzoic acids. In humans for example, hippuric acid and salicyluric acid (**33**) (Fig. 13.28) are the major metabolites of benzoic acid and salicylic acid, respectively. Similarly, *m*-trifluoromethylbenzoic acid (**13**) (Fig. 13.12), a major metabolite of fenfluramine (**10**), is excreted as the glycine conjugate. Phenylacetic acid derivatives can yield glycine and glutamine conjugates. Some aliphatic acids give glycine or taurine conjugates.

3.6.3 Formation of Hybrid Lipids and Sterol Esters. Incorporation of xenobiotic acids into lipids forms highly lipophilic metabolites that may burden the body as long-retained residues. In the majority of cases, triacylglycerol analogs (reaction 3 in Fig. 13.27) or cholesterol esters (reaction 4) are formed. The enzymes catalyzing such reactions are *O*-acyltransferases, including diacylglycerol *O*-acyltransferase (EC 2.3.1.20), 2-acylglycerol *O*-acyltransferase (EC 2.3.1.22), and sterol *O*-acyltransferase (cholesterol acyltransferase; sterol-ester synthase; EC 2.3.1.26; ACAT). Some phospholipid analogs, as well as some esters to the 3-hydroxy group of biliary acids, have also been characterized (73, 74).

The number of drugs and other xenobiotics that are currently known to form glyceryl or cholesteryl esters is very limited, but should increase as a result of increased awareness of researchers. One telling example is that of ibu-

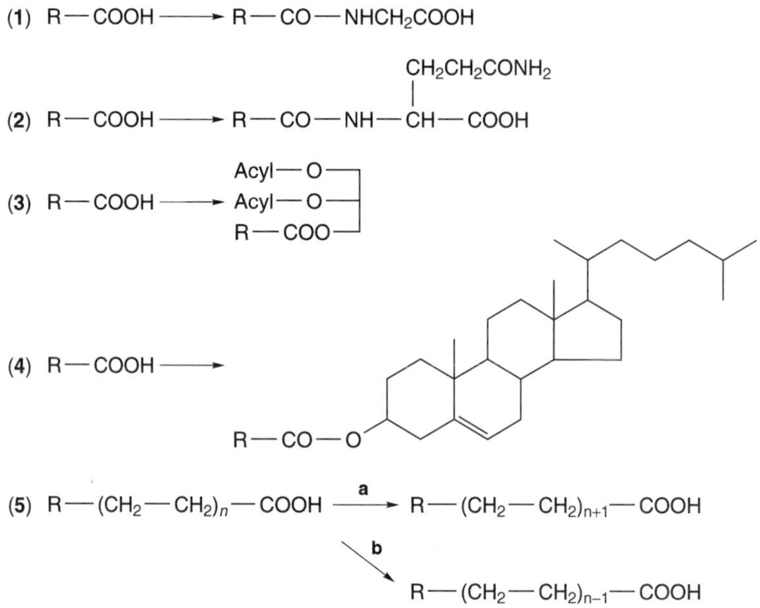

(1) R—COOH ⟶ R—CO—NHCH$_2$COOH

(2) R—COOH ⟶ R—CO—NH—CH—COOH

with CH$_2$CH$_2$CONH$_2$ substituent on the CH

(3) R—COOH ⟶ Acyl—O / Acyl—O / R—COO

(4) R—COOH ⟶

R—CO—O

(5) R—(CH$_2$—CH$_2$)$_n$—COOH $\xrightarrow{a}$ R—(CH$_2$—CH$_2$)$_{n+1}$—COOH

$\xrightarrow{b}$ R—(CH$_2$—CH$_2$)$_{n-1}$—COOH

Figure 13.27. Metabolic reactions involving acyl–CoA intermediates of xenobiotic acids, namely conjugations and two–carbon chain lengthening or shortening. Other products of β–oxidation are shown in Fig. 13.30.

profen (**34**) (Fig. 13.28), a much used anti-inflammatory drug whose (R)-enantiomer forms hybrid triglycerides detectable in rat liver and adipose tissue.

3.6.4 Chiral Inversion of Arylpropionic Acids. Ibuprofen (**34**) (Fig. 13.28) and other arylpropionic acids (i.e., profens) are chiral drugs existing as the (+)-(S) eutomer and the (−)-

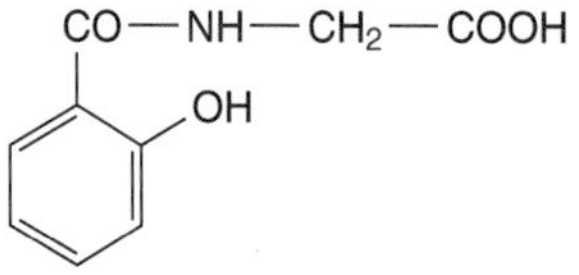

CO—NH—CH$_2$—COOH

OH

(**33**)

CH$_3$

CH—COOH

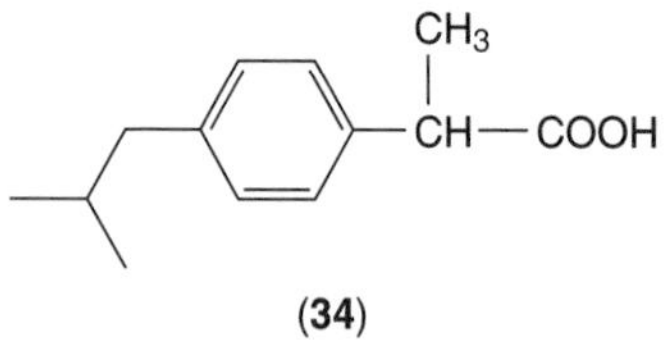

(**34**)

Figure 13.28. Salicyluric acid (**33**) and ibuprofen (**34**).

(R) distomer. These compounds undergo an intriguing metabolic reaction such that the (R)-enantiomer is converted to the (S)-enantiomer, whereas the reverse reaction is negligible. This unidirectional chiral inversion is thus a reaction of bioactivation, and its mechanism is now reasonably well understood (Fig. 13.29) (75).

The initial step in the reaction is the substrate stereoselective formation of an acyl-CoA conjugate with the (R)-form but not with the (S)-form. This conjugate then undergoes a reaction of epimerization possibly catalyzed by methylmalonyl-CoA epimerase (EC 5.1.99.1), resulting in a mixture of the (R)-profenoyl- and (S)-profenoyl-CoA conjugates. The latter can then be hydrolyzed as shown in Fig. 13.29 or undergo other reactions such as hybrid triglyceride formation (see below).

3.6.5 β-Oxidation and 2-Carbon Chain Elongation. In some cases, acyl-CoA conjugates formed from xenobiotic acids can also enter the physiological pathways of fatty acids catabolism or anabolism. A few examples are known of xenobiotic alkanoic and arylalkanoic

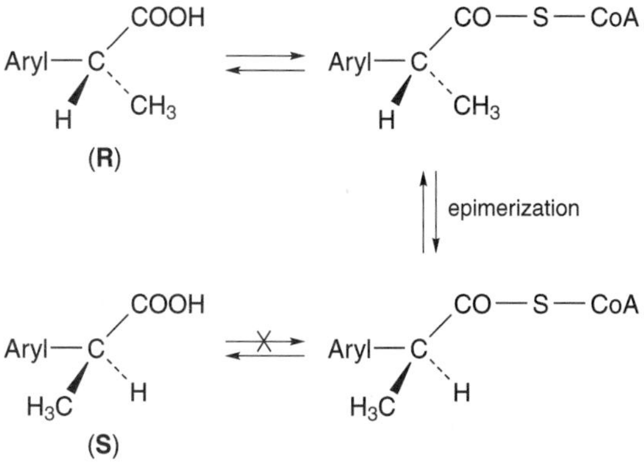

Figure 13.29. Mechanism of unidirectional chiral inversion of profens.

acids undergoing two-carbon chain elongation, or two-, four-, or even six-carbon chain shortening (reactions 5-A and 5-B in Fig. 13.27). In addition, intermediate metabolites of β-oxidation may also be seen, illustrated by valproic acid (**35**) (Fig. 13.30). Approximately 50 metabolites of this drug have been characterized; they are formed by β-oxidation, glucuronidation, and/or cytochrome P450-catalyzed dehydrogenation or oxygenation. Figure 13.30 shows the β-oxidation of valproic acid seen in mitochondrial preparations (76, 77). The resulting metabolites have also been found in unconjugated form in the urine of humans or animals dosed with the drug.

3.7 Conjugation and Redox Reactions of Glutathione

3.7.1 Introduction. Glutathione (**36**) (Fig. 13.31, GSH) is a thiol-containing tripeptide of major significance in the detoxification and toxification of drugs and other xenobiotics (78). In the body, it exists in a redox equilibrium between the reduced form (GSH) and an oxidized form (GS-SG). The metabolism of glutathione (i.e., its synthesis, redox equilibrium, and degradation) is quite complex and involves a number of enzymes (79, 80).

Glutathione reacts in a variety of manners. First, the nucleophilic properties of the thiol (or rather thiolate) group make it an effective conjugating agent, emphasized in this section. Second, and depending on its redox state, glutathione can act as a reducing or oxidizing agent (e.g., reducing quinones, organic nitrates, peroxides and free radicals, or oxidiz-

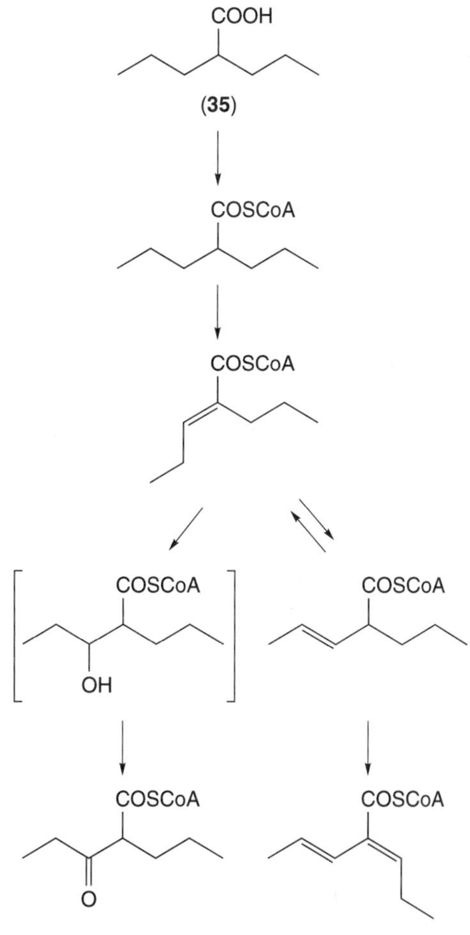

Figure 13.30. Mitochondrial β–oxidation of valproic acid (**35**).

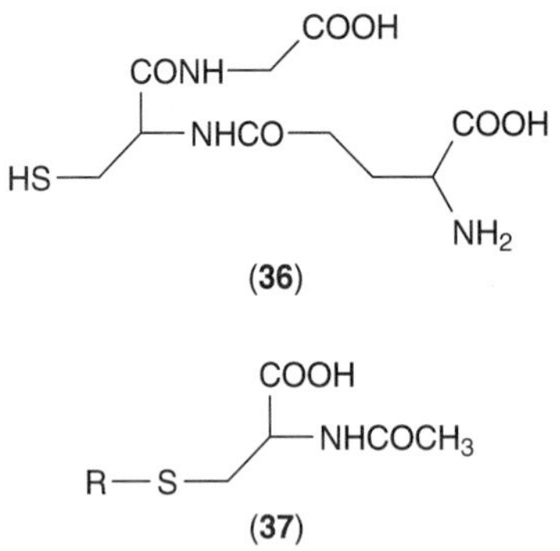

Figure 13.31. Glutathione (36) and N–acetylcys-teine conjugates (37).

ing superoxide). Another dichotomy exists in the reactions of glutathione, because these can be enzymatic (e.g., conjugations catalyzed by glutathione S-transferases and peroxide reductions catalyzed by glutathione peroxidase) or nonenzymatic (e.g., some conjugations and various redox reactions).

The glutathione S-transferases (EC 2.5.1.18; GSTs) are multifunctional proteins coded by a multigene family. They can act as enzymes as well as binding proteins. These enzymes are mainly localized in the cytosol as homodimers and heterodimers, and exist as four classes in mammals. The human enzymes comprise the following dimers: A1-1, A1-2, A2-2, A3-3 (alpha class), M1a-1a, M1a-1b, M1b-1b, M1a-2, M2-2, M3-3 (mu class), P1-1 (pi class), T1-1 (theta class), and a microsomal enzyme (MIC) (81, 82). The GST A1-2, A2-2, and P1-1 display selenium-independent glutathione peroxidase activity, a property also characterizing the selenium-containing enzyme glutathione peroxidase (EC 1.11.1.9). The GST A1-1 and A1-2 are also known as ligandin when they act as binding or carrier proteins, a property also displayed by M1a-1a and M1b-1b. In the latter function, these enzymes bind and transport a number of active endogenous compounds (e.g., bilirubin, cholic acid, steroid and thyroid hormones, and hematin), as well as some exogenous dyes and carcinogens.

The nucleophilic character of glutathione is due to its thiol group (pK_a 9.0) in its neutral form and even more to the thiolate form. In fact, an essential component of the catalytic mechanism of glutathione transferase is the marked increase in acidity (pKa 6.6) experienced by the thiol group during binding of glutathione to the active site of the enzyme (82). As a result, GSTs transfer glutathione to a very large variety of electrophilic groups; depending on the nature of the substrate, the reactions can be categorized as nucleophilic substitutions or nucleophilic additions. And with compounds of sufficient reactivity, these reactions can also occur nonenzymatically (14, 82, 83).

Once formed, glutathione conjugates (R-SG) are seldom excreted as such (they are best characterized in vitro or in the bile of laboratory animals), but usually undergo further biotransformation before urinary or fecal excretion. Cleavage of the glutamyl moiety [by glutamyl transpeptidase (EC 2.3.2.2)] and of the cysteinyl moiety [by cysteinylglycine dipeptidase (EC 3.4.13.6) or aminopeptidase M (EC 3.4.11.2)] leave a cysteine conjugate (R-S-Cys) that is further N-acetylated by cysteine-S-conjugate N-acetyltransferase (EC 2.3.1.80) to yield an N-acetylcysteine conjugate (37) (Fig. 13.31, R-S-CysAc). The latter type of conjugates are known as mercapturic acids, a name that clearly indicates that they were first characterized in urine. This however does not imply that the degradation of unexcreted glutathione conjugates must stop at this stage, because cysteine conjugates can be substrates of cysteine-S-conjugate β-lyase (EC 4.4.1.13) to yield thiols (R-SH). These in turn can rearrange as discussed below or be S-methylated and then S-oxygenated to yield thiomethyl conjugates ($R—S—CH_3$), sulfoxides ($R—SO—CH_3$), and sulfones ($R—SO_2—CH_3$).

3.7.2 Reactions of Glutathione. The major reactions of glutathione, both conjugations and redox reactions, are summarized in Fig. 13.32. Reactions 1 and 2 are nucleophilic additions and substitutions to sp^3-carbons, respectively, whereas reactions 3–8 are nucleophilic substitutions or additions at sp^2-carbons, sometimes accompanied by a redox reaction. Reactions at nitrogen or sulfur atoms are shown in reaction 9–11.

The first reaction in Fig. 13.32 is nucleophilic addition to epoxides (reaction 1-A) to yield a nonaromatic conjugate. This is followed by several metabolic steps (reaction 1-B), leading to an aromatic mercapturic acid. This is a frequent reaction of metabolically produced arene oxides (Fig. 13.5), documented for naphthalene and numerous drugs and xenobiotics containing a phenyl moiety. Note that the same reaction can also occur readily for epoxides of olefins (not shown in Fig. 13.32).

An important pathway of substitution at sp^3-carbons is represented in reaction 2-A, followed by the production of mercapturic acids (reaction 2-B). Various electron-withdrawing leaving groups (X in reaction 2) may be involved, which are either of xenobiotic (e.g., halogens) or metabolic origin (e.g., a sulfate group). Such a reaction occurs for example at the -CHCl$_2$ group of chloramphenicol and at the NCH$_2$CH$_2$Cl group of anticancer alkylating agents.

The reactions at sp^2-carbons are quite varied and complex. Addition to activated olefinic groups (e.g., α,β-unsaturated carbonyls) is shown in reaction 3. A typical substrate is acrolein (CH$_2$═CH—CHO). Quinones (*ortho*- and *para*-) and quinoneimines react with glutathione by two distinct and competitive routes, namely nucleophilic addition to form a conjugate (reaction 4-A), and reduction to the hydroquinone or aminophenol (reaction 4-B). A typical example is provided by the toxic quinoneimine metabolite (**9**) (Fig. 13.10) of paracetamol (**8**). Because in most cases quinones and quinoneimines are produced by the biooxidation of hydroquinones and aminophenols, respectively, their reduction by GSH can be seen as a futile cycle that consumes reduced glutathione. As for the conjugates produced by reaction 4-A, they may undergo reoxidation to *S*-glutathionylquinones or *S*-glutathionylquinone imines of considerable reactivity. These quinone or quinoneimine thioethers are known to undergo further GSH conjugation and reoxidation (84).

Haloalkenes are a special group of substrates of GS-transferases because they may react with GSH either by substitution (reaction 5-A) or by addition (reaction 5-B). Formation of mercapturic acids occurs as for other glutathione conjugates, but in this case, *S*-C cleavage of the *S*-cysteinyl conjugates by the renal β-lyase (reactions 5-C and 5-D) yields

thiols of significant toxicity because they rearrange by hydrohalide expulsion to form highly reactive thioketenes (reaction 5-E) and/or thioacyl halides (reactions 5-F and 5-G) (85).

With good leaving groups, nucleophilic aromatic substitution reactions also occur at aromatic rings containing additional electron-withdrawing substituents and/or heteroatoms (reaction 6-A). As for the detoxification of acyl halides with glutathione (reaction 7), a good example is provided by phosgene (O═CCl$_2$), an extremely toxic metabolite of chloroform that is inactivated to the diglutathionyl conjugate O═C(SG)$_2$.

The addition of glutathione to isocyanates and isothiocyanates has received some attention because of its reversible character (reaction 8) (86). Substrates of the reaction are xenobiotics such as the infamous toxin methyl isocyanate, whose glutathione conjugate behaves as a transport form able to carbamoylate various macromolecules, enzymes, and membrane components. The reaction is also of interest from a medicinal viewpoint because anticancer agents such as methylformamide seem to work by undergoing activation to isocyanates and then to the glutathione conjugate.

The reaction of *N*-oxygenated drugs and metabolites with glutathione may also have toxicological and medicinal implications. Thus, the addition of GSH to nitrosoarenes, probably followed by rearrangement, forms sulfinamides (reaction 9), which have been postulated to contribute to the idiosyncratic toxicity of a few drugs such as sulfonamides (87). As for organic nitrate esters such as nitroglycerine and isosorbide dinitrate, the mechanism of their vasodilating action is now believed to result from their reduction to nitric oxide (NO). Thiols, and particularly glutathione, play an important role in this activation. In the first step, a thionitrate is formed (reaction 10-A) whose *N*-reduction may occur by more than one route. For example, a GSH-dependent reduction may yield nitrite (reaction 10-B), which undergoes further reduction to NO; *S*-nitrosoglutathione (GS—NO) has also been postulated as an intermediate.

The formation of mixed disulfides between GSH and a xenobiotic thiol (reaction 11) has been observed in a few cases, e.g., with captopril.

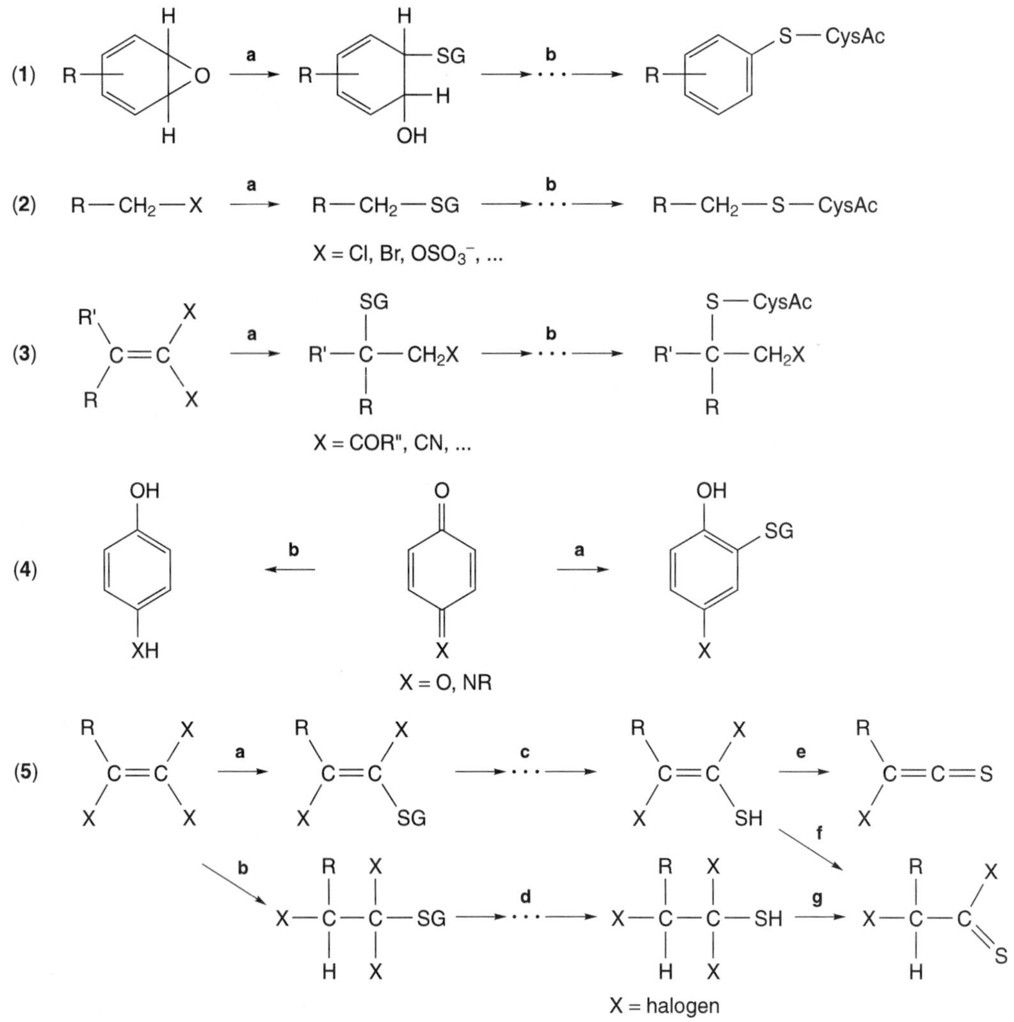

Figure 13.32. Major reactions of conjugation of glutathione, sometimes accompanied by a redox reaction.

Finally, glutathione and other endogenous thiols (including albumin) are able to inactivate free radicals (e.g. $R^•$, $HO^•$, $HOO^•$, $ROO^•$) and have thus a critical role to play in cellular protection (88). The reactions involved are highly complex and incompletely understood; the simplest are as follows.

$$GSH + X^• \rightarrow GS^• + XH$$

$$GS^• + GS^• \rightarrow GSSG$$

$$GS^• + O_2 \rightarrow GS—OO^•$$

$$GS—OO^• + GSH \rightarrow GS—OOH + GS^•$$

3.8   Other Conjugation Reactions

Sections 3.2–3.7 present the most common and important routes of xenobiotic conjugation, but these are by far not the only ones. A number of other routes have been reported whose importance is at present restricted to a few exogenous substrates, or which have re-

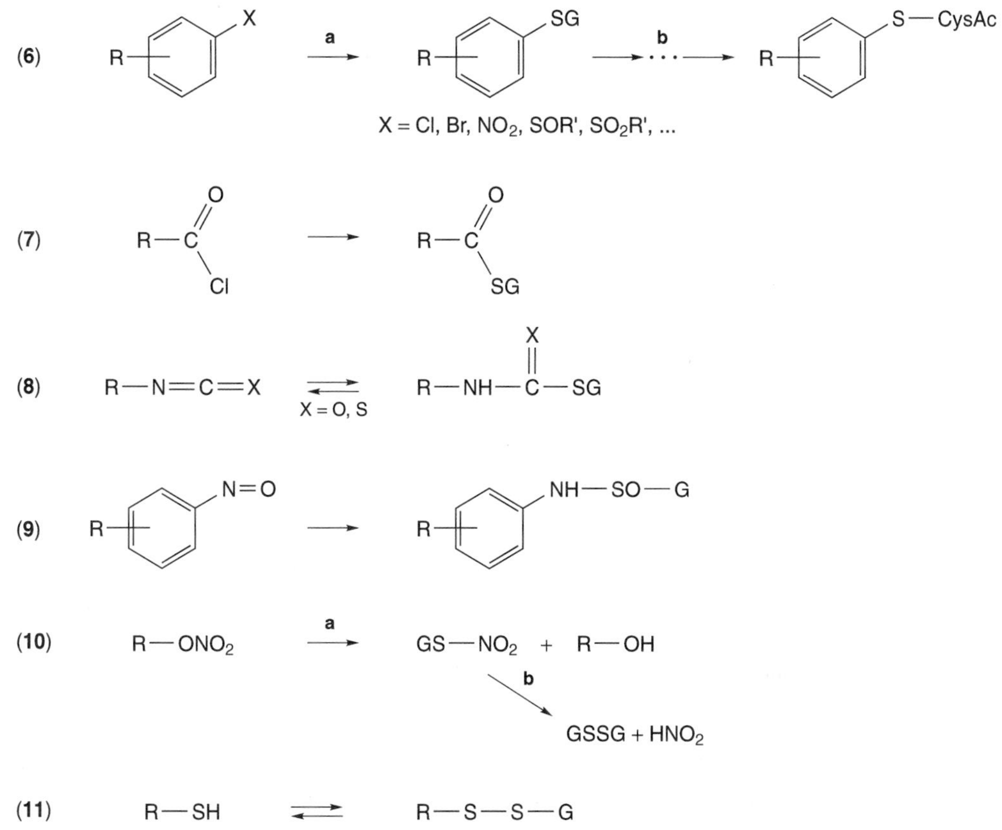

$$X = Cl, Br, NO_2, SOR', SO_2R', ...$$

Figure 13.32. (Continued.)

ceived only limited attention (89, 90). In this section, two pathways of pharmacodynamic significance will be discussed, phosphorylation and carbonyl conjugation (Fig. 13.33). In both cases, the xenobiotic substrates belong to narrowly defined chemical classes.

Phosphorylation reactions are of great significance in the processing of endogenous compounds and macromolecules. It is therefore astonishing that relatively few xenobiotics are substrates of phosphotransferases (e.g., EC 2.7.1), forming phosphate esters (reaction 1 in Fig. 13.33). The phosphorylation of phenol is a curiosity observed by some. In contrast, a number of antiviral nucleoside analogs are known to yield the mono-, di-, and triphosphates *in vitro* and *in vivo*, e.g., zidovudine (**38**) (Fig. 13.34, AZT), 2', 3'-dideoxycytidine, and 9-(1,3-dihydroxy-2-propoxymethyl)guanine. These conjugates are active forms of the drugs and are, in particular, incorporated in the DNA of virus-infected cells (91).

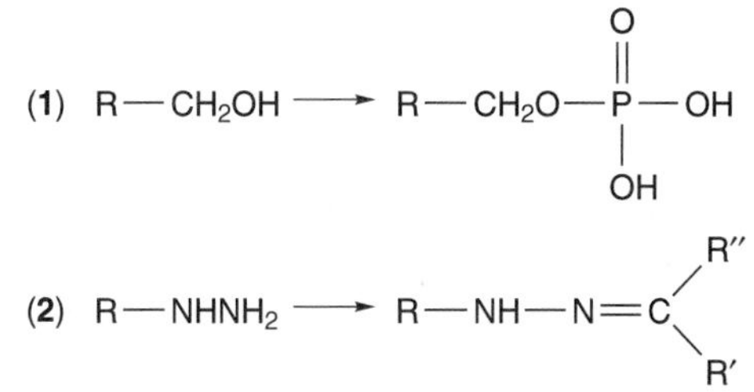

The second pathway of conjugation to be presented here is the reaction of hydrazines with endogenous carbonyls (reaction 2) (92). This reaction occurs nonenzymatically and involves a variety of carbonyl compounds, including aldehydes (mainly acetaldehyde) and ketones (e.g., acetone, pyruvic acid

Figure 13.33. Additional conjugation reactions discussed in Section 3.8, namely formation of phosphate esters and of hydrazones.

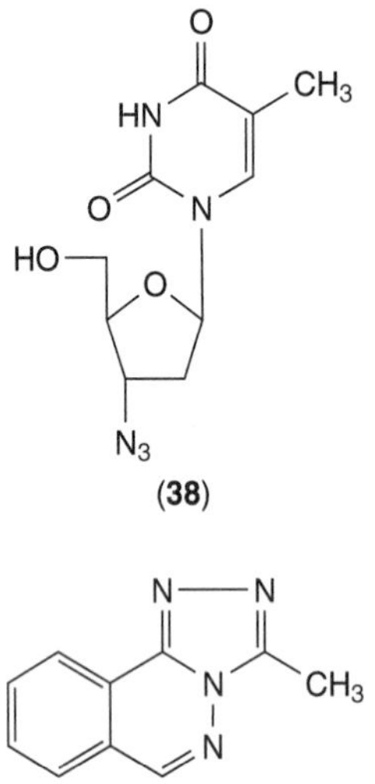

(38)

(39)

Figure 13.34. Zidovudine (**38**) and methyltriazolophthalazine (**39**).

CH_3—CO—COOH, and α-ketoglutaric acid
HOOC—CH_2CH_2—CO—COOH). The products thus formed are hydrazones, which may be excreted as such or undergo further transformation. Isoniazid (**31**) (Fig. 13.26) and hydralazine (**32**) (Fig. 13.26) are two drugs that form hydrazones in the body. For example, the reaction of hydralazine with acetyldehyde or acetone is a reversible one, meaning that the hydrazones are hydrolyzed under biological conditions of pH and temperature. In addition, the hydrazone of hydralazine with acetaldehyde or pyruvic acid undergoes an irreversible reaction of cyclization to another metabolite, methyltriazolophthalazine (**39**) (Fig. 13.34).

4 BIOLOGICAL FACTORS INFLUENCING DRUG METABOLISM

A variety of physiological and pathological factors influence xenobiotic metabolism and hence the wanted and unwanted activities associated with a drug. It can be helpful to distinguish between interindividual and intraindividual factors that can influence the capacity of an individual to metabolize drugs (Table 13.8). The interindividual factors are viewed as remaining constant throughout the

Table 13.8   Biological Factors Affecting Xenobiotic Metabolism

Interindividual Factors
Animal species
Genetic factors (genetic polymorphism)
Gender
Intraindividual Factors
Physiological changes
age
biological rhythms
pregnancy
Pathological changes
disease
stress
External influences
nutrition
enzyme induction by xenobiotics
enzyme inhibition by xenobiotics

life span of an organism and are the expression of its genome. In contrast, the intraindividual factors may vary depending on time (age or even time of day), pathological states, or external factors (nutrition, pollutants, drug treatment). Some of the factors will be discussed and exemplified below, others are just mentioned here and in Table 13.8.

4.1   Interindividual Factors

4.1.1 Animal Species. The use of different species as a surrogate for investigating and predicting the metabolism of new compounds in humans had been used throughout the 20th century even though it was recognized in the mid 1950s that wide species variation existed in the metabolism of a single drug (1, 2, 93). Currently, during the early phase of drug discovery there has been a steady switch from obtaining metabolism data with animal tissues and whole animals to screening studies with human cDNA expressed enzymes, cell fractions (microsomes), and cells. Although metabolism studies in animals have decreased in importance in drug design (unless developing a drug for that species), a greater understanding of the molecular, genetic, and physiological differences between humans and animal models are still required for correctly interpreting pharmacological safety studies required by regulatory agencies (94).

Increasingly, drug development is centered around the binding or inhibition of a human enzyme or receptor that has been cloned and isolated for use in a high-throughput drug development program. Before initiating studies in humans, the lead compound(s), often with a novel mechanism of action, must be evaluated for safety and toxicity in two mammalian species (one rodent, one non-rodent) (95). This information is used for the initial studies in humans, and it is important that the metabolism component of these animal studies are predictive for humans, which may or may not be true. A reported example would be the development of zidovudine (AZT) for treatment of patients with HIV (94). The initial animal studies in rats and dogs (two species commonly used) suggested that AZT did not undergo significant metabolism, and extrapolation of the data in animals to humans suggested that dosing every 12 h would be possible. When the first human subjects were given AZT, 70–80% of the dose was glucuronidated. Although the glucuronide conjugates were considered "inert" in relation to toxicity, the rate of glucuronidation was so rapid that initially AZT had to be given every 4 h to maintain a therapeutic drug level. Although it is well recognized that no two species will have the same complement of metabolizing enzymes, a detailed understanding at the molecular and physiological level in numerous species is needed so that the appropriate animal model(s) are used during drug development.

A new approach in understanding toxicity related to xenobiotics is the introduction of genetically manipulated mice, referred to as knockout or null mice. These knockout mice enable the use of an *in vivo* system that lacks specific metabolism enzyme(s), e.g., CYP, to determine if an observed toxicity in a rodent model might be caused by formation of an oxidative metabolite. An example would be the hepatotoxicity observed with acetaminophen in both mice and humans at high doses (96). The toxic metabolite, the electrophilic *N*-acetyl-benzoquinoneimine, seemed to be formed from CYP2E1 and CYP1A2 based on microsomal metabolism studies. CYP2E1 and CYP1A2 knockout mice confirmed this interpretation; the mice showed increased resis-

tance to the hepatotoxic effects of acetaminophen (97, 98). This animal model has suggested that the variability observed for hepatotoxicity of acetaminophen in humans may be associated with the level of these CYP enzymes. Studies of this type with knockout mice are helpful in understanding why some individuals that lack certain biotransformation enzyme(s) may or may not respond to a given drug.

4.1.2 Genetic Factors—Polymorphism in Metabolism. Drugs can undergo metabolism through a number of metabolic pathways. However, problems associated with the oxidative metabolism by the CYP1-4 enzymes listed in Table 13.5 are most frequently associated with clinical problems. As a new drug is being developed, its interaction with the CYP pathways are studied to determine if it acts as a substrate, an inhibitor, or an inducer of these enzymes. If the candidate drug is a substrate for one or more CYP enzymes, its metabolism is characterized through enzyme kinetics studies to determine K_m and V_{max} using a "bank" of human liver microsomes. Ultimately, the data obtained with the microsomes are confirmed using recombinant or cDNA-expressed CYP enzymes (most are commercially available) (99).

Table 13.9 is a partial list of drugs shown to be metabolized by specific CYP enzymes. This characterization is referred to as phenotyping and has become so important, especially for the CYP enzymes, that regulatory agencies expect it to be done on any newly introduced drug (100). By extracting structure-metabolism relationships (SMRs; Section 5.1) from databases of substrates and/or inhibitors, it has been possible in a number of cases to define some common structural features of such compounds (Table 13.9) (101–103).

After the specific CYP(s) involved in the metabolism of a drug are known, how does this information relate to the overall metabolism of a drug in a diverse population? The ability for an individual to metabolize a drug is dependent on the form (genotype), location, and amount of enzyme present. The amount of a specific metabolic enzyme present in the general population seems to be highly variable. Shown in Table 13.6 is also the variability of

Table 13.9   Examples of Substrates, Inhibitors, and Inducers Reported for Major Human Drug-Metabolizing CYP450 Enzymes (adapted from Refs. 27, 28, 98, 101–103)[a]

CYP1A2	
Common features of substrates	Planar polyaromatic/heterocyclic amines and amides, neutral or basic, lipophilic, with one putative H-bond donating site
Model substrates	Caffeine (N3-demethylation), 7-ethoxyresorufin (O-deethylation), phenacetin (O-deethylation), (R)-warfarin (C6-hydroxylation)
Substrates	Acetaminophen, acetanilide, aminopyrine, antipyrine, aromatic amines, chlorzoxazone, clozapine, 17β-estradiol, flutamide, imipramine, lidocaine, (S)- and (R)-mianserin, (S)- and (R)-naproxen, phenacetin, propafenone, tacrine, tamoxifen, theophylline, trimethadone, verapamil, (R)-warfarin, zolpidem
Inhibitors	7,8-Benzoflavone, ciprofoxacin, ellipticine, flavonoids (apigenin, naringenin, quercetin, etc.), fluoxamine, flurafylline, α-naphthoflavone, phenacetin
Inducers	Benzo[a]pyrene, charcoal-broiled beef, cigarette smoke, cruciferous vegetables, dihydralazine, 3-methylcholanthrene
CYP2A6	
Common features of substrates	Molecules of relatively low molecular weight, including ketones and nitrosamines
Model substrate	Coumarin (C7-hydroxylation)
Substrates	Cyclophosphamide, 1,3-butadiene, N,N-dimethylnitrosamine, halothane, paraxanthine, phenothiazines, sevoflurane, valproic acid, zidovudine
Inhibitors	Coumarin, diethyldithiocarbamate, 8-methoxypsoralen (methoxsalen)
Inducers	Dexamethasone
CYP2B6	
Common features of substrates	Nonplanar molecules, many with V-shaped structure
Model substrates	7-Benzyloxyresorufin (O-dealkylation), 7-pentoxyresorufin (O-dealkylation)
Substrates	Atrazine, bupropion, cyclophosphamide, dextromethorphan, dextromethorphan, diazepam, dibenzo[a,h]anthracene, halothane, lidocaine, (S)-mephenytoin, (S)- and (R)-mianserin, nicotine, testosterone
Inhibitors	Diethyldithiocarbamate, methoxsalen, orphenadrine
Inducers	Clofibrate, dexamethasone, mephenytoin, phenobarbital, phenytoin
CYP2C8	
Model substrate	Tolbutamide (p-methylhydroxylation)
Substrates	Arachidonic acid, benzo[a]pyrene, carbamazepine, clozapine, cyclophosphamide, (S)- and (R)-ibuprofen, omeprazole, paclitaxel, progesterone, retinoic acid, sulfadiazine, temazepam, testosterone, tienilic acid, trimethoprim, (R)-warfarin, zidovudine
Inhibitors	Sulfafenazole
Inducers	Rifampicin

Table 13.9   (*Continued*)

CYP2C9	
Common features of substrates	Neutral or acidic molecules with lipophilic site of oxidation a discrete distance from H-bond donor group or possibly anionic group
Model substrates	Diclofenac (C4'-hydroxylation), phenytoin (C4'-hydroxylation), tolbutamide (*p*-methyl-hydroxylation)
Substrates	Acetaminophen, arachidonic acid, benzo[a]pyrene, clozapine, cyclophosphamide, dapsone, dorzolamide, (*R*)-fluoxetine, hexobarbital, (*S*)- and (*R*)-ibuprofen, lauric acid, losartan, (*S*)- and (*R*)-naproxen, *N*-nitrosodimethylamine, ondansetron, phenylbutazone, piroxicam, 9-*cis*-retinoic acid, sulfadiazine, sulfamethoxazole, tetrahydrocannabinol, torasemide, trimethoprim, (*S*)-warfarin, zidovudine
Inhibitors	Cimetidine, fluconazole, fluvastatin, isoniazid, lovastatin, pravastatin, simvastatin, sulfaphenazole, tienilic acid, tolbutamide, warfarin
Inducers	Barbiturates, rifampicin
CYP2C19	
Common features of substrates	No clear features
Model substrates	(*S*)-Mephenytoin (C4'-hydroxylation), omeprazole (C5-hydroxylation)
Substrates	Amitriptyline, barbiturates, chlorproguanil, citalopram, clomipramine, clozapine, cyclophosphamide, diazepam, hexobarbital, imipramine, pentamidine, phenobarbital, phenytoin, propranolol, quinine, (*S*)- and (*R*)-warfarin, zidovudine
Inhibitors	Cimetidine, fluconazole, fluoxetine, fluoxamine, ketoconazole, (*S*)-mephenytoin, omeprazole, tranylcypromine
Inducers	Rifampicin
CYP2D6	
Common features of substrates	A basic nitrogen 5–7 Å from the lipophilic site of metabolism, which is in general on or near an aromatic system
Model substrates	Bufurolol (C1'-hydroxylation), debrisoquine (C4-hydroxylation), dextromethorphan (O-demethylation), sparteine (C-oxidations)

the various CYP enzymes as compiled by Rendic and DiCarlo (27, 28). Various studies using different methods to quantify levels of CYP enzymes in the liver come up with different abundance and ranges, but they are all consistent in showing that these levels are highly variable (104). It seems from this data that the CYP3A and CYP2C are the more abundant and vary the least, whereas the less abundant vary the most. One also notes that CYP2D6 is present in a comparatively small concentration. However, it can account for up to 80% of the *in vivo* metabolism of some drugs; an example is the O-demethylation of dextromethorphan (105). A lower variability is seen for the conjugating II enzymes, e.g., glutathione S-transferase (1.3- to 3.3-fold), UDP-glucuronyltransferase (3.1- to 10-fold), *N*-acetyltransferase (3- to 18-fold), phenol sulfotransferase (2- to 13-fold), steroid sulfotransferase (2.5- to 8-fold), and catechol O-methyltransferase (4- to 7-fold) (104). Overall, the levels and activi-

Table 13.9 (*Continued*)

Substrates	Ajmaline, alprenolol, amiflamine, amphetamine, aprindine, captopril, chlorpheniramine, cinnarizine, citalopram, clomipramine, clozapine, codeine, desipramine, dolasteron, encainide, flecainide, fluoxetine, fluphenazine, haloperidol, hydrocordone, imipramine, loratidine, methoxyphenamine, 3,4-methylenedioxymethamphetamine, metoprolol, mexiletine, (*S*)- and (*R*)-mianserin, nifedipine, olanzapine, omeprazole, oxycodone, perhexiline, phenformin, propaphenone, propranolol, remoxipride, ritonavir, saquinavir, selegiline, tamsulosin, timolol, tomoxetine, tramadol, trifluperidol, zolpidem
Inhibitors	Ajmaline, ajamlicine, flecainide, fluoxetine, isoniazid, ketoconazole, lobeline, nefazodone, paroxetine, (*R*)-propaphenone, quinidine, ritonavir, "statins," yohimbine, venlafaxine
Inducers	None known
CYP2E1	
Common features of substrates	Small linear or cyclic molecules (molecular weight < 200) including many industrial chemicals; otherwise relatively featureless
Model substrates	Aniline (C4-hydroxylation), chlorzoxazone (C6-hydroxylation), *p*-nitrophenol (aromatic hydroxylation)
Substrates	Acetaminophen, alcohols, benzene, benzo[a]pyrene, 1,3-butadiene, caffeine, clozapine, chlorzoxazone, dapsone, *N*,*N*-dimethylnitrosamine, enflurane, ethanol, felbamate, fluoxetine, halothane, halogenated alkanes, isoflurane, methylformamide, nitrosamines, odansetrone, phenobarbital, sulfadiazine, styrene, theophylline, toluene
Inhibitors	3-Amino-1,2,4-triazole, cimetidine, diethyldithiocarbamate, dihydrocapsaicin, dimethysulfoxide, disulfiram, ethanol, 4-methylpyrazole, phenylisothiocyanate
Inducers	Benzene, ethanol, isoniazid
CYP3A4	
Common features of substrates	Large to very large lipophilic molecules, with position of metabolism determined by chemical reactivity

ties of the CYP and conjugating II enzymes are believed to be influenced by both genetic and environmental factors.

Pharmacogenetics or the study of the genetic differences in drug metabolizing enzymes was needed to explain why a seemingly small percentage of subjects/patients receiving a drug would encounter unexpected side effects at low doses. When metabolism is primarily dependent on the transformation by a single metabolic enzyme, genetic variability can lead to an increased risk of an adverse drug reaction or therapeutic failure with that drug.

Based on the number of genes and frequency of polymorphism, it has been estimated that approximately two-thirds of the proteins in humans will have an amino acid difference between unrelated individuals. If one uses as an example the over 50 human CYP450s identified, this implies that approximately 30 of these enzymes will have changes in amino acids that may affect protein sequence and possibly substrate specificity

Table 13.9 *(Continued)*

Model substrates	Alprazolam (C4-hydroxylation), lidocaine (*N*-deethylation), erythromycin (*N*-demethylation), midazolam (C1'-hydroxylation), testosterone (C6β-hydroxylation)
Substrates	Acetaminophen, aldrin, alfentanil, amiodarone, aminopyrine, amitriptyline, amprenavir, androstenedione, antipyrine, astemizole, benzphetamine, budesonide, carbamazepine, celecoxib, chlorpromazine, chlorzoxazone, cisapride, clarithromycin, clozapine, cocaine, codeine, cortisol, cyclophosphamide, cyclosporin, dapsone, delavirdine, dextromethorphan, digitoxin, diltiazem, diazepam, erythromycin, 17β-estradiol, ethinylestradiol, etoposide, felbamate, fentanyl, flutamide, hydroxyarginine, ifosphamide, imipramine, indinavir, ketoconazole, lansoprazole, loratidine, losartan, lovastatin, (*S*)-mephenytoin, methadone, mianserin, miconazole, mifepristone, nelfinavir, nevirapine, nicardipine, nifedipine, odansetron, omeprazole, orphenadrine, proguanil, propafenone, quinidine, quinine, rapamycin, retinoic acid, ritonavir, saquinavir, selegiline, serindole, sufentanil, sulfinpyrazone, tacrolimus, tamoxifen, tamsulosin, taxol, teniposide, terfenadine, tetrahydrocannabinol, theophylline, toremifene, triazolam, trimethadone, trimethoprim, troleandomycin, verapamil, warfarin, zatosetron, zolpidem, zonisamide
Inhibitors	Amprenavir, cimetidine, clotrimazole, delavirdine, 6,7-dihydroxybergamottin, ethinylestradiol, fluoxetine, gestodene, indinavir, itraconazole, ketoconazole, miconazole, nelfinavir, nicardipine, ritonavir, saquinavir, troleandomycin, verapamil
Inducers	Carbamazepine, dexamethasone, glutethimide, nevirapine, phenobarbital, phenytoin, pregnenolone-16α-carbonitrile, progesterone, rifabutin, rifampin, ritonavir, St. John's Wort, sulfadimidine, sulfinpyrazone, troglitazone, troleandomycin

[a]Substrates may also be inhibitors, but inhibitors do not need to be substrates.

(106). Mutations that cause a disease are rare and often lead to early death. Genetic polymorphisms are more common and are officially defined as occurring in more than 1% of the population. Some of the clinically important polymorphisms associated with metabolism enzymes include CYP2C9, CYP2D6, CYP2C19, FMO, plasma pseudocholinesterase, *N*-acetyltransferase, thiopurine methyltransferase, and UDP-glucuronysyltransferase (107–113).

The polymorphisms can range from a change in a single nucleotide (referred to as SNPs) with two different alleles and a heterozygosity approaching 50%, to numerous nucleotide changes and alleles and low heterozygosity. The two alleles in an individual at a given locus (i.e., genotype) influence the kinetics of drug metabolism. The variant alleles can differ by modification at one or more point mutations, gene deletion, or gene duplication or multiduplication. As a result, the coded enzyme can have its activity unaffected, decreased, or increased. Depending on the enzyme system, occurrence of homozygous or heterozygous trait in relation to the allele, the activity associated with the allele, and the extent to which the alleles have been duplicated, an individual can can be classified as ultra-rapid metabolizer (duplicate copies of high activity alleles), extensive metabolizer (homozygous in high activity alleles), intermediate metabolizer (heterozygous in high activity/low activity alleles), and poor metabolizer (homozygous in low activity alleles).

An example consistent with this form of classification is the CYP2D6 O-demethylation of dextromethorphan (99). The number and complexity of CYP2D6 mutations are large, with over 70 allelic modifications being documented. Only a few allelic variants are common (CYP2D6*1, *2, *3, *4, *5, *6, *9, and *10) and only five alleles are functional (CYP2D6*1, *2, *9, *10, and *17), with activity comparable with the wild-type (K_m/V_{max} comparable with CYP2D6*1). The following distribution of metabolizers was observed in a population for the O-demethylation of dextromethorphan: ultrarapid (1–10%), extensive (75–85%), intermediate (10–15%), and poor (1–10%). Although it might be desirable to design drugs that would not be metabolized by the CYP2D6 pathway, many CNS drugs display the pharmacophoric pattern of CYP2D6 substrates, namely an appropriately substituted aromatic ring, a 2–3 carbon spacer, a protonated amine, and an oxidizable substituent 5–7 Å from the latter (Table 13.9).

4.1.3 Genetic Factors—Polymorphism in Absorption, Distribution, and Excretion. Initially, many clinical studies attempted to use drug metabolism to explain variability in pharmacokinetic (drug disposition) and pharmacodynamic (drug response) effects. It is now becoming increasingly evident that adsorption, distribution, and excretion also play a key role in the variability observed in drug and metabolite levels and drug response, and that phenotypic differences in both metabolism and drug transport are needed to explain variability observed in drug disposition (114). P-Glycoprotein (P-gp), a protein coded by the *MDR1* gene, is responsible for actively transporting many structurally different compounds from inside to outside cells against a concentration gradient (114). P-gp and CYP3A4 are coexpressed in the same cells and the combination of these systems are important in drug disposition, as observed for cyclosporin (114–116).

Given the existence of polymorphisms for both P-gp and CYP3A4, as well as in other transporters involved in absorption, distribution, and excretion, it is a challenge for any drug to be safely given within a relatively small dosage range to a majority of patients. With the advent of modern DNA methods, efforts are being made to develop DNA microchips capable of "fingerprinting" an individual's likely response to a drug. It seems to be a reasonable objective to phenotype the major systems involved in absorption, distribution, metabolism, and excretion to minimize adverse drug effects and improve drug therapy.

4.1.4 Ethnic Differences. Ethnic differences are basically genetic, with different proportions of "normal" and slow metabolizers observed in different populations. However, the influence of environmental factors such as nutrition and lifestyle cannot be excluded. An early example was the phenotyping of slow and fast acetylators of isoniazid. Slow acetylators (primarily caused by decreased activity of the NAT2 enzyme) exhibited a pronounced ethnic difference, e.g., 60% slow acetylators in Europe, ~50% in Africa, ~15% in China and Japan, and ~5% in Canadian Inuit (Eskimos) (117). It is now recognized that ethnicity and polymorphism are involved in a number of phase I and phase II enzymes, including CYP, FMO, methyltransferases, sulfotransferases, glucuronyltransferases, and acetylation.

An example of genetic variation within the CYP family would be with the CYP2C9 variants (CYP2C9*1, *2, *3). The allelic variants differ in their affinity (K_m) and/or intrinsic clearance (V_{max}/K_m) for CYP2C9 substrates such as warfarin (e.g., CYP2C9*2 exhibits decreased 6- and 7-hydroxylation of (S)-warfarin). The CYP2C9*2 variant is observed in black (~3%) and white (~10%) populations but is absent or rare in East Indians (99). Before the introduction of a new drug, studies are done to identify pharmacokinetic parameters and appropriate doses. However, because of ethnic differences, these studies may or may not be directly transposable to another country. The increased documentation of genetic polymorphisms as well as dietary and environmental chemicals that can modify the levels of various metabolizing enzymes make the importance of patient phenotyping increasingly important.

4.1.5 Gender Differences. Gender differences in drug pharmacokinetics has been observed for a limited number of drugs (118). It was logical to propose that this difference in pharmacokinetics may be caused by a difference in drug-metabolizing enzymes. In animal studies, primarily rats and mice, differences in functionalization (CYP) and conjugation (glucuronidation and sulfation) enzymes have been identified (119). For example, male rats have 10–30% more CYP enzymes compared with females, and they metabolize drugs such as indinavir and tolbutamide much faster. Male-specific (CYP2C11 and CYP2C18) and female-specific (CYP2C12 and CYP2C19) enzymes are found in the liver and seem to be under regulation of the sex hormones (120). These enzymes exhibit differences in substrate preferences that should contribute to the difference in metabolism of some drugs.

Sex-dependent differences in humans have been seen in the pharmacokinetics of a limited number of drugs. A lower clearance or higher plasma levels of certain drugs are seen for women compared with men, e.g., diazepam, thiothixene, rifampicin, gemcitabine, chloramphenicol, tetracycline, and erythromycin (119). However, metabolism alone is often not the reason for the differences in pharmacokinetics, because sex-specific differences in absorption, bioavailability, and distribution are well documented. There is evidence that CYP2C19, CYP2D6, and possibly CYP2E1 activities may be higher in men than women. Differences in drug metabolism have been seen for mephobarbital (CYP2C19), mephenytoin (CYP2C19), desipramine (CYP2D6), clomipramine (CYP2D6), propranolol (CYP2D6), and chlorzoxazone (CYP2E1) (118). For example, clomipramine metabolism, which is mediated by both CYP2C19 and CYP2D6, is significantly higher in men than in women (121). Metabolism also seems to be related to the sex-specific differences in the pharmacokinetics seen for acetaminophen (122) and propranolol (123), in which women form less of the glucuronide conjugate than men.

Oral contraceptives have been shown to increase the clearance of acetaminophen because of higher glucuronidation activity in women taking oral contraceptives compared with controls. In contrast, less is known about the effect of changes in hormone status in men on drug metabolism. Oral anabolic steroids have been shown to reduce the half-life and increase the clearance of antipyrine in men (124). Although some sex-specific differences have been observed, the mechanism by which this occurs is unknown. These differences could be a result of differences in sex steroid levels as seen with the rat, but proof in humans is lacking. Basically, because of the confounding effects of age, diet, and physiological factors, interpretation of the sex-dependent differences in drug pharmacokinetics in humans is very complex and the role of metabolism remains to be clarified.

4.2 Intraindividual Factors

4.2.1 Age. The pharmacokinetic and pharmacodynamic effects of a drug can be influenced by age, and drug metabolism plays an important role in understanding the differences observed. Differences between the levels of metabolism enzymes for the fetal and neonatal (first 4 weeks postpartum) liver versus the adult liver have been observed in both animal and human studies (125). At birth, total CYP levels are approximately 30% of adult levels and glucuronidation activity is at 10–30% of adult levels. Interestingly, sulfotransferase activity in neonates seems to be comparable with that in adults.

The best documented enzyme system for which decreased activity is observed in humans relates to the maturation of the CYP enzymes. In the fetal liver the CYPs 2C, 1A, 2A, 2B, 3A4, 3A5, and possibly 2E are absent or present at extremely low levels, whereas CYP3A7 is actively synthesized (126, 127). At birth, some of the CYP enzymes (2C, 2A, 2B, 3A4, 3A5, and 2E) begin to increase in concentration for the first week, and then remain relatively constant. In contrast, the CYP1A2 enzyme seems to be expressed much later, with levels that increase during the first

3 months. An example of the clinical importance of this maturation process would be the N-demethylation of diazepam by CYP2C19. *In vitro* data based on immunoblotting and enzymatic activity indicated that the level of CYP2C19 increases for the first week after birth then remains relatively constant at approximately 30% of the level observed in adults for a year. When urine specimens were taken after diazepam administration, formation of N-desmethyldiazepam was very low in infants that were 1–2 days of age, the amount of metabolite increased after 1 week of age, and then its formation remained stable up to 5 years (128). It is clear that drug biotransformation, especially for the CYP enzymes, is dependent on maturation of drug metabolizing enzymes, and that metabolism by neonates, children, and adults may be significantly different.

The elderly (usually defined as individuals over 65 years of age) also exhibit a difference in the pharmacokinetics of some drugs that seem to be related to drug metabolism. However, the origin of such differences is quite complex and is often related to other physiological changes such as reduction in liver blood flow and liver size (129, 130). Recent studies indicate that CYP450 activity is decreased by 30% in healthy elderly subjects compared with younger subjects, especially in the levels of CYP1A2 (131, 132). However, other studies have seen no relationship between aging and CYP activities on content in human liver (133, 134). Other enzymes such as gastric alcohol dehydrogenase have lower activity in older people, making them more sensitive to the CNS depressant effects of ethanol (135). As the number of older individuals increase in the population, especially those ages 80–90 years, adaptations in drug therapy and posology as related to metabolism will become increasingly important (136).

4.2.2 Biological Rhythms. There is increasing evidence that drug metabolism may undergo some daily variations as a result of a diurnal (circadian) rhythm. It is well established that nearly all physiological functions and parameters can vary with the time of day, e.g., heart rate, blood pressure, hepatic blood flow, urinary pH, and plasma concentrations

of hormones, other signal molecules, glucose, and plasma proteins (137). Drugs that have shown a pharmacokinetic difference in humans during the first part of the active/wake cycle include ethanol, nicotine, and theophylline (138, 139). For example, sustained-release theophylline dosed in the evening exhibited lower peak drug concentrations and a longer time to maximal concentration than when dosed in the morning (140). That the changes in blood levels of the drug were caused by metabolism rather than variations in absorption and excretion were not readily obvious (141–143). However, consistent with concepts concerning the induction of metabolizing enzymes by endogenous substances, a study in male rats found that a high hepatic microsomal 7-alkoxycoumarin O-dealkylase activity observed during the active cycle of the rat was related to the diurnal secretion of growth hormone (144). Although the circadian rhythm is not important in the initial stages of drug design, it is important to recognize that these daily fluctuations may influence the absorption, distribution, metabolism, and excretion (ADME) profile on which the formulation of a new drug is based.

4.2.3 Disease. A number of diseases may influence drug metabolism, although no generalization seems in sight. First and above all, diseases of the liver will have a direct effect on enzymatic activities in this organ, e.g., cirrhosis, hepatitis, jaundice, and tumors. Diseases of extrahepatic organs such as the lungs and the heart may also affect hepatic drug metabolism by influencing blood flow, oxygenation, and other vital functions. Renal diseases have a particular significance because of a decrease of the intrinsic xenobiotic-metabolizing activity of the kidneys and a decreased rate of urinary excretion (145).

Particular attention has been paid to infectious and inflammatory diseases, which have been associated with differential expression of various CYP enzymes causing either an increase (CYP2E1, CYP4A) or a decrease (CYP2C11) (146, 147). Animal models for inflammation or infection have been shown to result in decreased drug clearance with a reduced microsomal content of CYP enzymes and reduced drug metabolism. In humans, septic shock and infections (e.g., *Streptococcus pneumoniae*, influenza B virus, *Plasmodium falciparum*), influenza vaccine, bacterial endotoxin, cytokines, and interferon-α have all been associated with reduced drug clearance and reduced metabolism. An example is provided by the elevated levels of theophylline and the resulting toxicity that were observed in children with influenza B infections (148). The mechanism by which this occurs is not clear at this time, but nitric oxide produced during inflammation has been implicated. Nitric oxide causes a decrease in CYP450 expression, a decrease in CYP450 enzyme activity, and an increased degradation of damaged enzymes. Because interferons and various cytokines are used or being developed for cancer treatment, and because the latter usually includes multidrug treatment, it becomes increasingly important to understand which factors can influence drug metabolism in a variety of disease states.

4.2.4 Enzyme Inhibition. Interference with the metabolism of a drug by inhibition of the primary enzyme involved in its metabolism can lead to serious side effects or therapeutic failure. Inhibitors of the CYP enzymes are considered to be the most problematic, being the primary cause of drug–drug and drug–food interactions. Listed in Table 13.9 are some inhibitors of specific CYP enzymes, although inhibition by these compounds may not be comparable in all phenotypes (149).

As mentioned earlier, the screening for CYP inhibition is currently done at the earliest stages of drug development. Prospective drugs can be classified as potent ($IC_{50} < 1 \mu M$), moderate ($1 \mu M < IC_{50} < 10 \mu M$), or weak ($IC_{50} > 10 \mu M$) inhibitors of CYP with preference for further drug development given to weak or moderate inhibitors (150). It should be noted that a number of drugs such as quinidine (CYP2D6 microsomal IC_{50} $\sim$0.2 μM) and ketoconazole (CYP3A4 microsomal IC_{50} $\sim$15 nM) would be classified as potent inhibitors but are still clinically useful. The mechanism by which inhibition of the CYP enzymes occur can be organized into four general categories (151):

1. Competitive inhibition caused by two drugs competing for the binding site in the enzyme, e.g., (S)-omeprazole inhibiting the oxidation of diazepam, phenytoin, or (R)-warfarin.

2. Noncompetitive inhibition caused by slow reversible binding of lipophilic ligands to the heme iron (e.g., imidazole-based drugs such as cimetidine and antifungals), or to reversible binding to an allosteric site.

3. Noncompetitive, slowly reversible inhibition by *in situ* formation of a carbene ligand (as seen with methylenedioxybenzene derivatives) or a nitroso ligand (as seen with some amines).

4. Noncompetitive, irreversible inhibition requiring *in situ* metabolism to a reactive intermediate (mechanism-based inactivators, also known as suicide substrate inhibitors).

All four mechanisms can modify drug metabolism in clinical situations. An example of the first mechanism is the effect of (S)-omeprazole on the N-demethylation of diazepam (152). CYP2C19 is primarily responsible for the N-demethylation of diazepam, and the C'5-hydroxylation and 5-O-demethylation of (S)-omeprazole. When subjects were given diazepam after 5 days on (S)-omeprazole, the AUC of diazepam was 81% higher, its elimination half-life was increased from 43 to 86 h, and a 17% lower AUC for N-desmethyldiazepam was observed when compared with controls. For diazepam, these changes were not considered to be clinically significant, but for drugs with smaller a therapeutic margin this degree of inhibition could be important.

The second mechanism of inhibition occurs with the slow reversible ligand binding of a nucleophilic electron pair to the heme iron of the CYP. These types of interactions can occur with anilines, phenols, and several of the nitrogen- or oxygen-containing aromatic heterocycles (pyridines and imidazoles, etc.), provided the rest of the molecule is sufficiently lipophilic. Whether inhibition occurs with a specific compound depends on the electronic and steric effects of substituents close to the electron pair. Imidazoles provide a good example of how placement of a substituent can modify inhibitory activity. Placing a lipophilic substituent at the 1- or 4(5)-position of the imidazole allows the least sterically hindered nitrogen to act as a sixth ligand to the prosthetic heme iron. This leads to effective inhibition of the CYP enzyme. In contrast, a 2-substituted or 4,5-disubstituted imidazole provides enough steric hindrance to hinder or prevent inhibition (153). In general, heterocycles are frequent bioisosteric replacements for a benzene or naphthalene ring to improve water solubility or receptor interaction through hydrogen bonding. At times it may be necessary to compromise optimal receptor binding associated with a heterocycle to minimize interaction with the CYP enzymes through strategic placement of substituent(s) near a nucleophilic electron pair.

The third and fourth mechanisms occur by *in situ* metabolic activation of the inhibitor to a metabolite that binds reversibly to the heme iron (third mechanism) or irreversibly (covalent binding) to the prosthetic heme or the apoprotein (151, 154). Reversible binding is seen with lipophilic drugs containing a methylenedioxyphenyl group being oxidized to a carbene (155, 156), or an amine that can undergo hydroxylation and further oxidation to a nitroso intermediate (157). An example would be erythromycin, its derivatives, and its analogs. It has been proposed that the dimethylamino group present on the desoxamine sugar undergoes N-demethylation, and then further oxidation to the relatively lipophilic nitroso metabolite. This metabolite forms a relatively stable complex with CYP3A4 in the reduced (ferrous) state and inhibition occurs.

Mechanism-based irreversible inhibition occurs when a reactive metabolic intermediate is formed *in situ* that can (1) bind covalently with the prosthetic heme through N-alkyl/arylation (e.g., secobarbital), (2) alkylate the apocytochrome (e.g., chloramphenicol or 2-ethynylnaphthalene), or (3) cause destruction of the prosthetic heme to products that irreversibly bind to the apocytochrome (e.g. CCl_4) (158). These mechanism-based inactivators have primarily been designed and used for the selective inhibition of specific CYP enzymes and elucidating the mechanism of P450 reactions. Some drugs (e.g., aromatase inhibitors)

that are selective inhibitors of one or few CYP enzymes have been used in human cancer therapy by blocking key steps in the production and metabolism of steroid hormones. They have found application in the treatment of steroid-sensitive cancers, such as prostate, breast, and adrenal carcinomas (159). In such cases, CYP enzymes become drug targets and the drug-enzyme interaction is a pharmacodynamic rather than a pharmacokinetic process.

It is important to differentiate a moderate or weak irreversible inhibitor from a competitive inhibitor during the earliest stages of drug development. This is accomplished by determining the apparent IC_{50} of the test compound toward a standard substrate (e.g., CYP2D6-mediated O-demethylation of dextromethorphan) with and without a preincubation period of the enzyme with the test compound (160). Generally, preclinical candidates that are either potent competitive inhibitors ($IC_{50} < 1\ \mu M$) or are metabolized to reactive metabolites are dropped from further development.

4.2.5 Enzyme Induction. The term enzyme induction is used in a general sense to indicate that the amount and activity of a metabolizing enzyme has increased following exposure to a drug or chemical. In contrast to inhibition, induction is generally a slow process that is either caused by an increase in the synthesis of the enzyme or a decrease in its degradation. A drug that causes induction may increase its own metabolism (autoinduction, e.g., ethanol induction of CYP2E1) or increase metabolism of another drug with no effect on its own metabolism (e.g., rifampin induction of CYP2C19). Clinically, drug induction is considered less of a problem than drug inhibition, but there are some marked exceptions. For example, potent inducing agents such as antiepileptics can significantly modify the pharmacokinetic behavior of other drugs, e.g., warfarin. Another reason to avoid potent inducing agents during drug development is the positive relationship observed in rodent oncogenicity studies between administration of strong inducers and the occurrence of liver tumors, although this same relationship has not been seen in humans (161).

Primary interest has been directed toward the CYP enzymes in which five classes of inducing agents have been identified. Representative compounds for each class are as follows (98):

- 3-methylcholanthrene (rat CYP1A1, CYP1A2)
- phenobarbital (rat CYP2B2, CYP2A1, CYP2C6, CYP3A2)
- pregnenolone-16α-carbonitrile (rat CYP3A1, CYP3A2)
- clofibric acid (rat CYP4A1, CYP4A2. CYP4A3)
- isoniazid (rat CYP2E1)

The mechanisms of CYP induction have been studied for decades in rats and other laboratory animals, but only during the last 10 years has major progress been made. Although induction can occur because of stabilization of the enzyme, the underlying mechanism is not well understood. In contrast, the mechanisms underlying increased expression of the various metabolizing enzymes involve transcription factors and their regulating enzymes (162). The following cytosolic receptors seem to bind with specific inducing agents:

- the aryl hydrocarbon receptor or AhR, which binds polycyclic aromatic hydrocarbons and β-naphthoflavone
- the human pregnane-activated X receptor, PXR, which binds rifampin and pregnenolone-16α-carbonitrile
- the constitutively active or androstane receptor, CAR, which is activated by androstanolol and phenobarbital
- the peroxisome proliferator-activated receptor, PPARα, which binds clofibric acid

The following is a simplified sequence of events that leads to induction (163):

1. binding of an inducing agent to a receptor (AhR, PXR, CAR, PPARα)
2. heterodimerization with another receptor [Arnt or retinoic X receptor (RXR)] and transport into the nucleus
3. interaction with a response enhancer within the nucleus and transmission of signal to the CYP promoter region on the gene
4. change in chromatin structure

5. increased synthesis of mRNA followed by newly synthesized protein

Table 13.9 lists commonly recognized inducing drugs and chemicals for the CYP enzymes, showing also that some inducers act on multiple enzymes. Although ethical restraints have restricted detailed studies in humans, in rats, phenobarbital is recognized as primarily increasing levels of CYP2B1 and CYP2B2 by more than 20-fold. In addition, it increases the level of CYP2A1, CYP2C6, CYP3A2 levels two- to fourfold as well as a 50–300% increase in cytochrome b_5, NADPH-cytochrome P450 reductase, epoxide hydrolase, aldehyde dehydrogenase, glutathione S-transferase, and UDP-glucuronyltransferase. It is likely that other enzymes and response factor enhancers and inhibitors involved in this induction sequence have yet to be identified (162, 164–166). This is based on the observation that many structurally diverse chemicals, for which a single receptor would be highly unlikely, have been shown to cause comparable induction. This can be seen with phenobarbital in which a comparable induction can be caused by glutethimide, phenytoin, loratidine, doxylamine, griseofulvine, chlorpromazine, ketoconazole, chlordane, dieldrin, butylated hydroxytoluene, octamethylcyclotetrasiloxane, and 2,2′,4,4′,4,4′-hexachlorobiphenyl. As our understanding of the underlying mechanisms by which chemicals regulate the levels of the metabolic enzymes improves, an explanation may arise for the ethnic-, age-, disease-, and sex-related differences in drug pharmacokinetics that cannot currently be explained by genetic differences.

5 DRUG METABOLISM AND THE MEDICINAL CHEMIST

Having presented the various reactions of drug metabolism and the biological factors influencing them, we now turn our attention to topics of marked interest to medicinal chemists, namely SMRs, the prediction of drug metabolism, prodrug design, and toxophoric groups.

5.1 SMRs

5.1.1 Introduction. Two classes of factors will influence qualitatively (how? and what?) and quantitatively (how much? and how fast?) the metabolism of a given xenobiotic in a given biological system. The first class is that of the biological factors discussed in Section 4. The second class of factors are the various molecular properties that will influence a metabolic reaction, most notably configuration, electronic structure, and lipophilicity. When speaking of a metabolic reaction, however, we should be clear about what is meant. Indeed, enzyme kinetics allows a metabolic reaction to be readily decomposed into a binding and a catalytic phase. Despite its limitations, Michaelis-Menten analysis offers an informative approach for assessing the binding and catalytic components of a metabolic reaction (167). Here, the Michaelis constant K_m represents the affinity, with V_{max} being the maximal velocity, k_{cat} the turnover number, and K_m/V_{max} the catalytic efficiency. As we will see, this distinction makes much sense in SMRs.

5.1.2 Chirality and Drug Metabolism. The influence of stereochemical factors in xenobiotic metabolism is a well-known and abundantly reviewed phenomenon (168–174). Because metabolic reactions can produce more than one response (i.e., several metabolites), two basic types of stereoselectivity are seen, substrate stereoselectivity and product stereoselectivity (see Section 1.3). Substrate stereoselectivity occurs when stereoisomers are metabolized (1) differently (in quantitative and/or qualitative terms) and (2) by the "same" biological system under identical conditions. Substrate stereoselectivity is a well-known and abundantly documented phenomenon under *in vivo* and *in vitro* conditions. In fact, it is the rule for many chiral drugs, ranging from practically complete to moderate. A complete absence of substrate enantioselectivity has seldom been seen.

Michaelis-Menten analysis suggests that the molecular mechanism of substrate enantioselectivity can occur in the binding step (different affinities, K_m), in the catalytic step (different reactivities, V_{max}), or in both. There

are cases in which stereoselective metabolism is of toxicological relevance (173). This situation can be illustrated with disopyramide and mianserin, which undergo substrate stereoselective oxidation. Disopyramide (**40**) (Fig. 13.35), a chiral antiarrhythmic agent, is marketed as the racemate. Although disopyramide is generally well tolerated, several cases of hepatic toxicity have been described. *In vitro* studies using rat hepatocytes revealed a considerably higher cytotoxicity of the (S)-enantiomer, assessed by leakage of lactate dehydrogenase and morphological changes. The biotransformation of disopyramide involves mono-N-dealkylation (which is stereoselective for the (S)-enantiomer) and aromatic oxidation (which is stereoselective for the (R)-enantiomer). The low cytotoxicity of the N-dealkylated metabolites in rat hepatocytes leads to the hypothesis that toxicity is mainly a result of aromatic oxidation (175).

The metabolism of the antidepressant drug mianserin (**41**) (Fig. 13.35) revealed analogies and differences with disopyramide. Indeed, aromatic oxidation in human liver microsomes occurred more readily for the (S)-enantiomer, whereas N-demethylation was the major route for the (R)-enantiomer. At low drug concentrations, cytotoxicity toward human mononuclear leucocytes was caused by (R)-mianserin more than by (S)-mianserin and showed a significant correlation with N-demethylation (176). Thus, the toxicity of mianserin seemed associated with N-demethylation rather than with aromatic oxidation, in contrast to disopyramide. The chemical nature of the toxic intermediates was not established, but a comparison between disopyramide and mianserin emphasizes that no *a priori* expectation should influence toxicometabolic studies.

Product stereoselectivity occurs when stereoisomeric metabolites are generated (1) differently (in quantitative and/or qualitative terms) and (2) from a single substrate with a suitable prochiral center or face. Examples of metabolic pathways producing new centers of chirality in substrate molecules include ketone reduction, reduction of carbon–carbon double bonds, hydroxylation of prochiral methylenes, oxygenation of tertiary amines to N-oxides, and oxygenation of sulfides to sulfoxides. Product stereoselectivity may be a re-

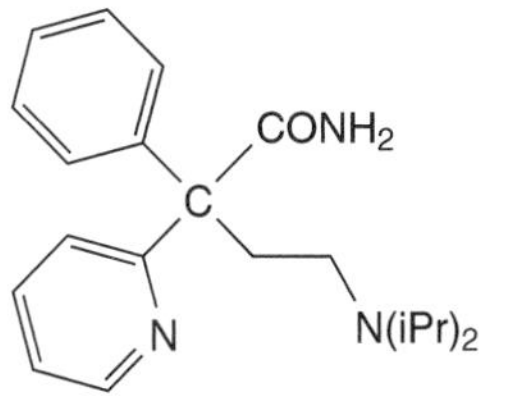

(40)

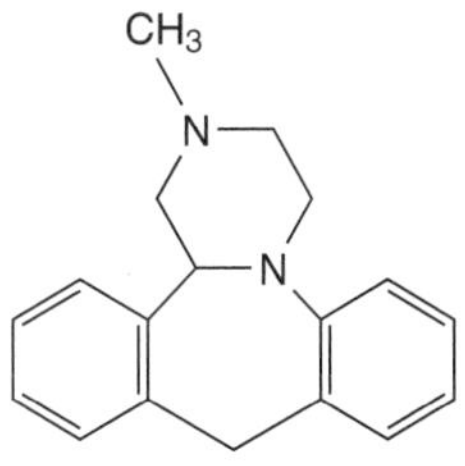

(41)

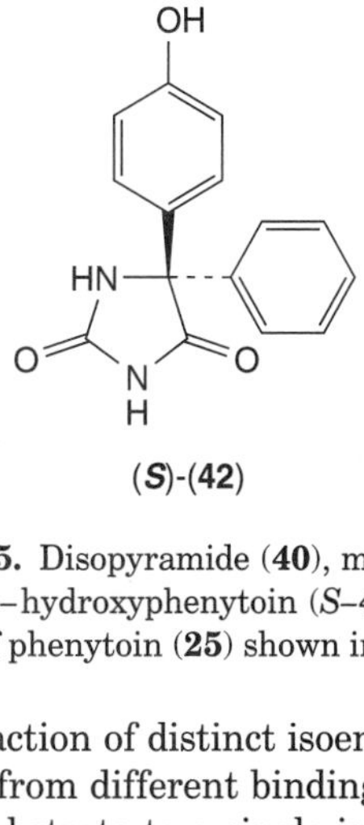

(S)-(42)

Figure 13.35. Disopyramide (**40**), mianserin (**41**), and (*S*)–*para*–hydroxyphenytoin (*S*–**42**), the chiral metabolite of phenytoin (**25**) shown in Fig. 13.22.

sult of the action of distinct isoenzymes, or it may result from different binding modes of a prochiral substrate to a single isoenzyme. In this case, each productive binding mode will bring another of the two enantiotopic or diastereotopic target groups in the vinicity of the catalytic site, resulting in diastereoisomeric enzyme–substrate complexes. Thus, the ratio of products depends on the relative probability of the two binding modes. In addition, the catalytic step, which involves diastereoisomeric transition states, may also influence or control product selectivity.

The possible toxicological consequences of product selective metabolism can be illustrated with the antiepileptic phenytoin (**25**) (Fig. 13.22). This compound is prochiral by

virtue of a center of prochirality featuring two enantiotopic phenyl rings. This drug is known to induce gingival overgrowth in chronically treated patients. In the absence of surgical removal (gingivectomy), this overgrowth can reach severe stages. Significant levels of phenytoin and its *para*-hydroxylated metabolite were found in the gingiva of chronically treated patients, whereas in experimental animals both the drug and its metabolite were found to induce gingival overgrowth at concentrations comparable with those achieved in patients. Gingival toxicity is postulated to result from covalent binding of a reactive metabolite formed during phenyl oxidation. In humans and most experimental animals except the dog, the *para*-hydroxylated metabolite has predominantly the (*S*)-configuration (**42**) (Fig. 13.35), with an *S/R* ratio in humans of approximately 10/1 (177). This implies an enzymatic discrimination of the two enantiotopic groups with a marked recognition of the *pro*-S ring.

It is thus clear that substrate and product stereoselectivities are the rule in the metabolism of stereoisomeric and stereotopic drugs. But if the phenomenon is to be expected *per se*, it is not trivial to predict which enantiomer will be the preferred substrate of a given metabolic reaction, or which enantiomeric metabolite of a prochiral drug will predominate. This is because of the fact that the observed stereoselectivity of a given reaction will depend both on molecular properties of the substrate and on enzymatic factors (e.g., the stereoelectronic architecture of the catalytic site in the various isozymes involved). In fact, substrate and product stereoselectivities are determined by the binding mode(s) of the substrate and by the resulting topography of the enzyme–substrate complex (178). Only molecular modeling (see below) can help us predict stereoselectivities in drug metabolism.

5.1.3 Qualitative Relations Between Metabolism and Lipophilicity. Comparing the overall metabolism of numerous drugs clearly reveals a global relation with lipophilicity (167). Indeed, there exist some highly polar xenobiotics known to be essentially resistant to any metabolic reaction, e.g., saccharin, disodium cromoglycate, and zanamivir. Furthermore, many *in vivo* metabolic studies have demonstrated a dependence of biotransformation on lipophilicity, suggesting a predominant role for transport and partitioning processes. A particularly illustrative example is offered by antiasthmatic chromone-2-carboxylic acids (179), whose pharmacokinetic behavior in the rat revealed the opposite influences of log D on renal clearance (which decreased with increasing lipophilicity) and metabolic clearance (which increased with increasing lipophilicity). Also, the more lipophilic β-blockers are extensively if not completely metabolized (e.g., propranolol), whereas the more hydrophilic ones undergo biotransformation for only a fraction of the dose (e.g., atenolol).

This global trend is in line with the Darwinian rationale for xenobiotic metabolism, which is believed to have evolved in an animal–plant "warfare," with herbivores, adapting to the emergence of protective chemicals (e.g., alkaloids) in plants (180).

The exception to the global and direct relation between extent of metabolism and lipophilicity is offered by the vast number of human-made, highly lipophilic polyhalogenated xenobiotics that now pollute our entire biosphere. Such compounds include polyhalogenated insecticides (e.g., DDT), polyhalogenated biphenyls, and dioxins, which have a strong propensity to accumulate in adipose tissues. In addition, these compounds are highly resistant to biotransformation in animals caused, in part, by their very high lipophilicity and in part by the steric shielding against enzymatic attack provided by the halo substituents.

5.1.4 Quantitative Relations Between Metabolism and Lipophilicity. Given its spectral characteristics, cytochrome P450 is of particular interest to investigate the binding step and the molecular factors that influence it. Indeed, substrate binding to CYP can be detected as a type I difference ultraviolet (UV) spectrum (peak at 385–390 nm, trough at ~420 nm) and quantified as K_s (in M units). This productive mode of binding is to be contrasted with unproductive ligand binding (peak at 425–430 nm, trough at ~390 or 410 nm), whereby a nitrogen or oxygen atom in the ligand interacts with the Fe(III) atom in

the heme (6). Of relevance here is the fact that the substrate binding mode to microsomal cytochrome P450 has repeatedly been shown to be related to lipophilicity. This is exemplified by an extensive study of the type I binding affinity of 50 model compounds to hamster liver microsomes. Linear relations were found between K_s and log D_{oct} (at pH 7.4) (181). These relations incorporated monocyclic hydrocarbons, bicyclic hydrocarbons, higher hydrocarbons, homologous carbamates, fatty acids, and fatty acyl methyl esters, respectively, with r^2 values in the range 0.90–0.99. The fact that all compounds could not be fitted into a single regression indicates that other properties in addition to lipophilicity influenced affinity, presumably electronic and steric factors.

When the results of Michaelis-Menten analyses are examined for QSMRs, it is often found that lipophilicity correlates with K_m but not with V_{max} or k_{cat}. This is documented for ester hydrolysis and oxidation of various chemical series by monooxygenases and other oxidases (167). Depending on the explored property space, the relationships between K_m and lipophilicity are linear or parabolic. Such results indicate that when relations are found between rate of metabolism and lipophilicity, the energy barrier of the reaction is largely similar for all compounds in the series, allowing lipophilicity to become the determining factor. A relevant example is provided by the CYP2D6-catalyzed oxidation of a series of fluorinated propranolol derivatives. Their K_m values spanned a 400-fold range and showed a weak but real correlation with the distribution coefficient. The same was true for the catalytic efficiency k_{cat}/K_m, but only because the k_{cat} values spanned a narrow sixfold range (182).

5.1.5 The Influence of Electronic Factors. Stereoelectronic properties may influence the binding of substrates to enzymatic active sites in the same manner as they influence the binding of ligands to receptors. For example, the K_m values of the above-mentioned propranolol derivatives were highly correlated with their basicity (pK_a), confirming that an ionic bond plays a critical role in the binding of amine substrates to CYP2D6 (182).

Table 13.10   A Classification of *In Silico* Methods to Predict Biotransformation

A) Local methods: Methods applicable to series of compounds with "narrow" chemical diversity and/or to
biological systems of "low" complexity
    QSAR (linear, multilinear, multivariate, . . .).
      affinities, relative rates, . . . (depending on the physicochemical properties considered)
    3D-QSAR (CoMFA, Catalyst, GRID/GOLPE, . . .)
      substrate behavior, relative rates, inhibitor behavior, . . .
    Molecular modeling and docking
      ligand yes/no (substrate? inhibitor?), regioselectivity, . . .
    Quantum mechanical (MO) methods (ab initio, semi-empirical)
      regioselectivity, mechanisms, relative rates, . . .
B) Global methods: Methods applicable to series of compounds with "broad" chemical diversity (*and, in
the future, to biological systems of "high" complexity*)
    Databases (Metabolite, Metabolism, . . .)
      nature of metabolites, reactive/adduct-forming metabolites, . . .
    Expert systems and their databases (MetabolExpert, METEOR, . . .)
      nature of metabolites, metabolic trees, reactive/adduct-forming metabolites, *relative importance
of these metabolites depending on biological factors,* . . .

Electronic properties are of particular interest in SMRs because they control the cleavage and formation of covalent bonds characteristic of a biotransformation reaction (i.e., the catalytic step). Correlations between electronic parameters and catalytic parameters obtained from *in vitro* studies (e.g., V_{max} or k_{cat}) allow a rationalization of substrate selectivity and some insight into reaction mechanism.

An example of a quantitative SMR study correlating electronic properties and catalytic parameters is provided by the glutathione conjugation of *para*-substituted 1-chloro-2-nitrobenzene derivatives (183). The values of log k_2 (second order rate constant of the nonenzymatic reaction) and log k_{cat} (enzymatic reaction catalyzed by various glutathione transferase preparations) were correlated with the Hammett resonance σ^- value of the substrates, a measure of their electrophilicity. Regression equations with positive slopes and r^2 values in the range 0.88–0.98 were obtained. These results quantitate the influence of substrate electrophilicity on nucleophilic substitutions mediated by glutathione, be they enzymatic or nonenzymatic.

Quantum mechanical calculations may also shed light on SMRs, revealing correlations between rates of metabolic oxidation and energy barrier in cleavage of the target C—H bond (184).

5.1.6 3D-QSMRs and Molecular Modeling. Sections 5.1.4 and 5.1.5 have exemplified

quantitative structure-activity relationships (QSARs) methods as applied to metabolism, i.e., QSMRs. Here, lipophilicity and electronic parameters are being used as independent variables and a metabolic parameter (often assessing affinity) as the dependent variable. Almost always, correlation equations of this type have been obtained for limited and/or closely related series of substrates, implying a very narrow exploration of structural diversity space. Also, the metabolic parameters were obtained from relatively simple biological systems. When it comes to predicting biotransformation, such QSMR correlations are typically local methods (Table 13.10) of low or negligible extrapolative capacity.

Three-dimensional (3D) methods are also of value in SMRs, namely 3D-QSARs and the molecular modeling of xenobiotic-metabolizing enzymes (Table 13.10). Indeed, they represent a marked progress in predicting the metabolic behavior (be it as substrates or inhibitors) of novel compounds. An important restriction, however, is that they can be applied to single enzymes only.

3D-QSMR methods yield a partial view of the binding/catalytic site of a given enzyme as derived from the 3D molecular fields of a series of substrates or inhibitors (the training set). In other words, they yield a "photographic negative" of such sites and will allow a quantitative prediction for novel compounds structurally related to the training set. Two popular methods in 3D-QSARs are Comparative

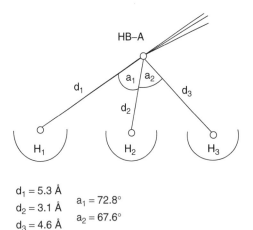

$d_1 = 5.3$ Å
$d_2 = 3.1$ Å $a_1 = 72.8°$
$d_3 = 4.6$ Å $a_2 = 67.6°$

Figure 13.36. Simplified 3D pharmacophoric model of CYP2B6 substrates as obtained with the Catalyst algorithm (186). The main features characteristic of substrates are three hydrophobic moieties (H1, H2, and H3) and a H–bond acceptor group (HB–A).

Molecular Field Analysis (CoMFA) and Catalyst. An early metabolic study using CoMFA involved the monoamine oxidase-catalyzed oxidation of MPTP analogs (185). A more recent study using the Catalyst algorithm described the pharmacophoric features characteristic of CYP2B6 substrates (186), a simplified representation of which is given in Fig. 13.36. Pharmacophoric features of inhibitors have also been obtained for a number of CYP enzymes using 3D-QSAR methods (187–189).

The molecular modeling of xenobiotic-metabolizing enzymes affords another approach to rationalize and predict drug–enzyme interactions (for reviews and illustrated examples see Refs. 102, 190). The methodology of molecular modeling is explained in detail elsewhere in this work and will not be presented here. Suffice it to say that its application to drug metabolism was made possible by the crystallization and X-ray structural determination of the first bacterial cytochromes P450 in the mid-1980s (see Ref. 191). Complexes of CYP101 with docked camphor (its substrate) or an inhibitor shed further light on the binding/catalytic site of this enzyme. As more and more amino acid sequences of mammalian (i.e., membrane-bound) cytochromes P450 became available, their tertiary structure was

modeled by homology superimposition with the experimentally determined CYP101 template or other bacterial CYPs (homology modeling). Known substrates and inhibitors were then docked *in silico*, affording pharmacophoric models. Given the assumptions made in homology modeling and the lack of accurate scoring functions, such pharmacophoric models cannot give quantitative affinity predictions. However, they can afford fairly reliable yes/no answers as to the affinity of test set compounds, and in favorable cases may also predict the regioselectivity of metabolic attack (Table 13.10) (192, 193).

The pharmacophoric models of a large number of mammalian and mostly human CYPs are now available (102, 190, 192–195), as well as other xenobiotic-metabolizing enzymes such as DT-diaphorase and glutathione *S*-transferases (196, 197). The first crystallization and X-ray structural elucidation of a mammalian cytochrome P450, CYP2C5 (198), is a breakthrough that removes the set of assumptions inherent in homology modeling and will thus improve the predictive power of molecular modeling.

Quantum mechanical methods are also classified as local in Table 13.10. Here, a word of caution is necessary, because such methods are in principle applicable to any chemical system. However, they cannot handle more than one metabolic reaction or catalytic mechanism at a time, and as such can only predict metabolism in simple biological systems, in contrast to the global methods presented in the next section. An example illustrating the power of QSMRs based on quantum mechanical calculations is provided by a study of catechol *O*-methyltransferase (COMT) (199). The *O*-methylation of a large number of catechol derivatives was investigated in the presence of recombinant soluble human COMT. A good correlation existed between V_{max} values and the Hammet σ_{p-} constant, such that the turnover rate decreased with increasing ionization of a phenolic group. The V_{max} values were also correlated with an index of the molecular electrostatic potential (MEP) of the anion. The K_m values were correlated with both lipophilicity and the MEP index. The mechanistic interpretation of these correlations was that increased stabilization of the catecholate anion led to

Table 13.11 Goals of a Global Metabolic Prediction, Classified by Increasing Difficulty

Goal 1: A list of all reasonable metabolites of a given compound

Goal 2: Same as above, organized in a metabolic tree

Goal 3: Same as above, plus a warning for reactive/adduct-forming metabolites

Goal 4: Same as above, plus a probability of formation based on molecular factors, acting as a filter against improbable metabolites

Goal 5: Same as above, plus a probability of formation under different biological conditions

stabilization of the Michaelis complex at the expense of the transition state complex.

5.1.7 Global Expert Systems to Predict Biotransformation. While medicinal chemists are not usually expected to possess a deep knowledge of the mechanistic and biological factors that influence drug metabolism, they will find it quite useful to have a sufficient understanding of SMRs to be able to predict reasonable metabolic schemes. A qualitative prediction of the biotransformation of a novel xenobiotic should allow (1) the identification of all target groups and sites of metabolic attack, (2) the listing of all possible metabolic reactions able to affect these groups and sites, and (3) the organization of the metabolites into a metabolic tree (goals 1 and 2 in Table 13.11). This is the information summarized in Sections 2 and 3. The next desirable feature would be a warning for potentially reactive/adduct-forming metabolites (goal 3).

Given the available information, there exists an ever increasing interest in expert systems hopefully able to meet goals 1–3 in Table 13.11. A few systems of this type are now available (200–202), examples of which are named in Table 13.10. Such systems will make correct qualitative or even semiquantitative predictions for a number of metabolites, but the risk of false positives must be taken very seriously. This is because of the great difficulty in meeting goal 4 (Table 13.11), devising efficient filters to remove unlikely metabolites, based on the molecular properties of substrates. Meeting goal 5, namely taking biological factors

(tissues, animal species, etc.) into account, is still far away.

Updating and validating such algorithms and their databases are also critical aspects. At this time, the European Cooperation in the Field of Scientific and Technical Research (Project COST B15) had begun an independent evaluation of existing expert systems used in the *in silico* prediction of ADME properties, with a view of publishing a consensus paper.

5.2 Modulation of Drug Metabolism by Structural Variations

5.2.1 Overview. Many examples of the SMRs discussed above involve overall molecular properties such as configuration, conformation, electronic distribution, or lipophilicity. An alternative means of modulating metabolism is by structural modifications of the substrate at its target site, a direct approach whose outcome is often more predictable than that of altering molecular properties by structural changes not involving the reaction center. Globally, structural variations at the reaction center can aim either at decreasing or even suppressing biotransformation, or at promoting it by introducing labile groups. Metabolic switching is a combination of the two goals, the aim being to block metabolism in one part of the molecule and to promote it in another.

Inertness toward biotransformation can often be observed for highly hydrophilic or lipophilic compounds. But high polarity and high lipophilicity tend to be avoided by drug designers because they may result in poor bioavailability and very slow excretion, respectively. However, metabolic stabilization can be achieved more conveniently by replacing a labile group with another, less or nonreactive moiety, provided this change is not detrimental to pharmacological activity (203). Classical examples include the following:

- introducing an N-t-butyl group to prevent N-dealkylation
- inactivating aromatic rings toward oxidation by substituting them with strongly electron-withdrawing groups (e.g., $-CF_3$, $-SO_2NH_2$, $-SO_3^-$)

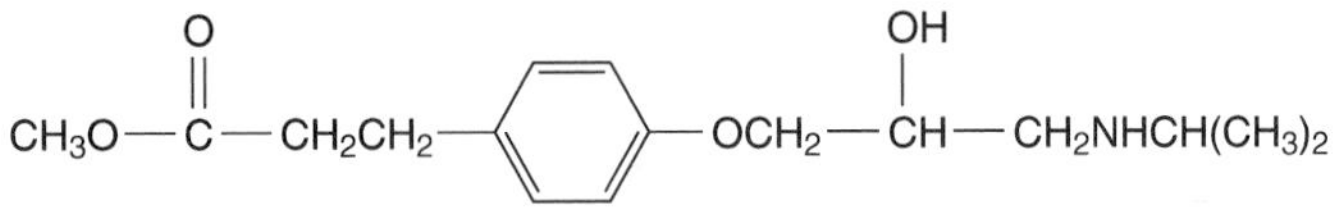

Figure 13.37. Esmolol (**43**).   (**43**)

• replacing a labile ester linkage with an amide group
• more generally protecting the labile moiety by steric shielding

In some cases, metabolic stabilization presents advantages such as the following:

• longer half-lives
• decreased possibilities of drug interactions
• decreased inter- and intrapatient variability
• decreased species differences
• decreased number and significance of active metabolites

Nevertheless, drawbacks cannot be ignored, e.g., too long half-lives and a risk of accumulation.

In contrast to metabolic stabilization, metabolic switching is a versatile means of deflecting metabolism away from toxic products to enhance the formation of therapeutically active metabolites and/or to obtain a suitable pharmacokinetic behavior.

Metabolic promotion can be achieved by introducing a functional group of predictable metabolic reactivity, for example, an ester linkage. This concept enjoys considerable success in the design of prodrugs, as discussed separately below. Another approach rendered possible by metabolic promotion is the design of "soft" drugs (204). The concept of soft drugs, which are defined as "biologically active compounds (drugs) characterized by a predictable *in vivo* metabolism to nontoxic moieties, after they have achieved their therapeutic role," has led to valuable therapeutic innovations such β-blockers with ultrashort duration of action. Examples of the latter are [(arylcarbonyl)oxy]propanolamines and esmolol (**43**) (Fig. 13.37). In both cases, esterase-mediated hydrolysis produces metabolites that are inactive due respectively to the loss of the side-chain or to a high polarity of the *para*-substituent.

For the sake of fairness, it must also be mentioned that the design of soft drugs is not without limitations. Whereas emphasis is placed on the predictability of their metabolism, this predictability is qualitative more than quantitative as a result of the many biological factors that influence their biotransformation. A similar limitation also applies to many prodrugs, as discussed in the following sections.

5.2.2 Principles of Prodrug Design. Prodrugs are defined as therapeutic agents that are inactive *per se* but are predictably transformed into active metabolites (205). As such, prodrugs must be contrasted with soft drugs, which as explained above, are active *per se* and yield inactive metabolites. And in a more global perspective, prodrugs and soft drugs seem to be the two extremes of a continuum of possibilities where both the parent compound and the metabolite(s) contribute to a large or small proportion to the observed therapeutic response.

Prodrug design aims at overcoming a number of barriers to a drug's usefulness (Table 13.12). Based on these and other considerations, the major objectives of prodrug design can be listed as follows:

• improved formulation (e.g., increased hydrosolubility)
• improved chemical stability
• improved patient acceptance and compliance
• improved bioavailability
• prolonged duration of action
• improved organ selectivity
• decreased side effects
• marketing considerations, "me-too" or "me-better" drugs

The successes of prodrug design are many, and a large variety of such compounds have

Table 13.12   Prodrugs: A Concept to Overcome Barriers to Drug's Usefulness (modified from Ref. 206)

Pharmaceutical barriers
 Insufficient chemical stability
 Poor solubility
 Inacceptable taste or odour
 Irritation or pain
Pharmacokinetic barriers
 Insufficient oral absorption
 Marked presystemic metabolism
 Short duration of action
 Unfavorable distribution in the body
Pharmacodynamic barriers
 Toxicity

Table 13.13   Complementary Viewpoints When Discussing Prodrugs

Chemical classification? (overlapping classes)
 Bioprecursors
 Classical carrier-linked prodrugs
 Site-specific chemical delivery systems
 Macromolecular prodrugs
 Drug–antibody conjugates
Mechanisms of activation? (may operate
  simultaneously)
 Enzymatic
  → biological variability
  → difficult optimization
 Nonenzymatic
  → no biological variability
  → easier optimization
Tissue/organ selectivity?
 Due to tissue-selective activation of classical
   prodrugs
 Produced by site-specific chemical delivery
   systems
Toxic potential?
 Of a metabolic intermediate (for bioprecursors)
 Of the carrier moiety or a metabolite thereof
Gain in therapeutic benefit?
 "*Post hoc*" design (prodrugs of established
   drugs)
  The gain can range from modest to marked
 "*Ad hoc*" design (a labile group is an initial
   specification)
  The gain is usually marked

proven their therapeutic value. When discussing this multidisciplinary field of medicinal chemistry, several complementary viewpoints can be adopted, namely a chemical classification, the nature of activation (enzymatic or nonenzymatic), the tissue selectivity, the possible production of toxic metabolites, and the gain in therapeutic benefit (Table 13.13).

In a chemical perspective, it may be convenient to distinguish between carrier-linked prodrugs, i.e., drugs linked to a carrier moiety by a labile bridge, and bioprecursors, which do not contain a carrier group and are activated by the metabolic creation of a functional group (207). A special group of carrier-linked prodrugs are the site-specific chemical delivery systems (204, 208). Macromolecular prodrugs are synthetic conjugates of drugs covalently bound (either directly or through a spacer) to proteins, polypeptides, polysaccharides, and other biodegradable polymers (209). A special case is provided by drugs coupled to monoclonal antibodies.

Prodrug activation occurs enzymatically, nonenzymatically, or sequentially (enzymatic step followed by nonenzymatic rearrangement). As much as possible, it is desirable to reduce biological variability; hence, the particular interest of nonenzymatic reactions of hydrolysis or intramolecular catalysis as discussed in the next section.

The problem of tissue or organ selectivity (targeting) is another important aspect of prodrug design. Many unsuccessful and a few successful attempts have been made to achieve

organ-selective activation of prodrugs (see below).

The toxic potential of metabolic intermediates, of the carrier moiety or of a fragment thereof should never be neglected. For example, some problems may be associated with formaldehyde-releasing prodrugs such as N- and O-acyloxymethyl derivatives or Mannich bases (see below). Similarly, arylacetylenes assayed as potential bioprecursors of anti-inflammatory arylacetic acids proved to be highly toxic because of the formation of intermediate ketenes (see pathway 3 in Fig. 13.5).

The gain in therapeutic benefit provided by prodrugs is a question that knows no general answer. Depending on both the drug and its prodrug, the therapeutic gain and when it can be characterized may be marked or modest. But as suggested in Table 13.13, a trend is apparent when comparing *post hoc*– and *ad hoc*–designed prodrugs. *Post hoc* design implies well-accepted drugs endowed with useful

Table 13.14   Examples of Common and Less Common Carrier Groups Used in Prodrug Design

Carrier groups linked to a hydroxy group (R-OH)
   Esters of simple or functionalized aliphatic carboxylic acids, e.g., R-O-CO-R'
   Esters of carbamic acids, e.g., R-O-CO-NR'R''
   Esters of amino acids (e.g., lysine), e.g., R-O-CO-CH(NH$_2$)R'
   Esters of ring-substituted aromatic acids, e.g., R-O-CO-aryl
   Esters of derivatized phosphoric acids, e.g., R-O-PO(OR')(OR'')
   (Acyloxy)methyl or (acyloxy)ethyl ethers, e.g., R-O-CH$_2$-O-CO-R' or
     R-O-CH(CH$_3$)-O-CO-R'
   (Alkoxycarbonyloxy)methyl or (alkoxycarbonyloxy)ethyl ethers, e.g., R-O-CH$_2$-O-CO-O-R' or
     R-O-CH(CH$_3$)-O-CO-O-R'
   O-glycosides
Carrier groups linked to a carboxylic group (R-COOH)
   Esters of simple alcohols or phenols, e.g., R-CO-O-R'
   Esters of alcohols containing an amino or amido function, e.g., R-CO-O-(CH$_2$)$_n$-NR'R'' or
     R-CO-O-(CH$_2$)$_n$-CO-NR'R'' or R-CO-O-(CH$_2$)$_n$-NH-COR'
   (Acyloxy)methyl or (acyloxy)ethyl esters of the type R-CO-O-CH$_2$-O-CO-R' or R-CO-O-CH(CH$_3$)-O-CO-R'
   Hybrid glycerides formed from diacylglycerols, e.g., R-CO-O-CH(CH$_2$-O-CO-R')$_2$
   Esters of diacylaminopropan-2-ols, e.g., R-CO-O-CH(CH$_2$-NH-COR')$_2$
   N,N-dialkyl hydroxylamine derivatives, e.g., R-CO-O-NR'R''
   Amides of amino acids (e.g., glycine), e.g., R-CO-NH-CH(R')-COOH
Carrier groups linked to an amino or amido group (RR'-NH)
   Amides formed from simple or functionalized acyl groups, e.g., RR'N-CO-R''
   Amides cleaved by intramolecular catalysis (with accompanying cyclization of the carrier moiety)
   Alkyl carbamates, e.g., RR'N-CO-O-R''
   (Acyloxy)alkyl carbamates, e.g., RR'N-CO-O-CH(R'')-O-CO-R'''
   (Phosphoryloxy)methyl carbamates, e.g., RR'N-CO-O-CH$_2$-O-PO$_3$H$_2$
   N-(acyloxy)methyl or N-(acyloxy)ethyl derivatives, e.g., RR'N-CH$_2$-O-CO-R'' or RR'N-CH(CH$_3$)-O-CO-R''
   N-Mannich bases, e.g., RR'N-CH$_2$-NR''R'''
   N-(N,N-dialkylamino)methylene derivatives of primary amines, e.g., RN=CH-NR'R''
   N-α-hydroxyalkyl derivatives of peptides
   Imidazolidinone derivatives of peptides
   Oxazolidines of ephedrines and other 1-hydroxy-2-aminoethane congeners

qualities but displaying some unwanted property that a prodrug form should ameliorate. The gain in such cases is usually modest, yet real, but may be marked if good targeting is achieved.

In contrast, *ad hoc* design implies active compounds suffering from some severe drawback (e.g., high hydrophilicity restricting bioavailability), which prevents therapeutic use. Here, a prodrug form may prove necessary, and its design will be integrated into the iterative process of lead optimization. A high therapeutic gain is obviously expected.

5.2.3 Chemical Aspects of Prodrug Design. Numerous carrier-linked prodrugs have been prepared from drugs containing an adequate functional group. Without aiming at comprehensiveness, a list of common and less common carrier groups is given in Table 13.14 for

three types of functional groups that are frequent sites of derivatization to form prodrugs. Thus, drugs containing an alcoholic or phenolic group can be conveniently derivatized to esters or labile ethers. Drugs containing a carboxylic group can form a variety of esters and amides, and their SMRs are documented in many studies. Drugs containing an NH group (i.e., amides, imides, and amines) are amenable to derivatization to a variety of prodrugs (210).

The reactivity of these prodrugs varies considerably. Whereas most of them are activated enzymatically, others are activated nonenzymatically, as discussed in the following paragraphs. Enzymatic hydrolysis followed by chemical breakdown is another possibility (e.g., carbamic acids formed from carbamates).

As mentioned in the previous section, it is often desirable to reduce biological variability by designing prodrugs whose activation occurs

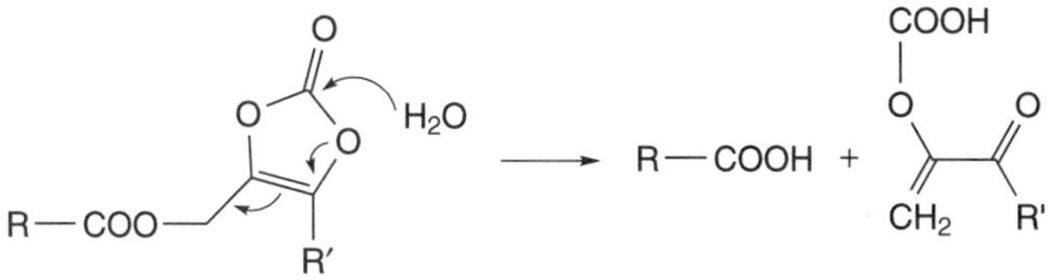

Figure 13.38. Mechanism of non-enzymatic breakdown of (2–oxo–1,3–dioxol–4–yl)methyl esters (212).

purely or predominantly by a nonenzymatic mechanism (211). Prodrugs of this type include the following:

- (2-oxo-1,3-dioxol-4-yl)methyl esters
- Mannich bases
- oxazolidines
- esters with a basic side-chain able to catalyze intramolecular hydrolysis
- esters and amides undergoing intramolecular nucleophilic cyclization-elimination

The first and last type are of particular interest and will be exemplified here. 5-Substituted (2-oxo-1,3-dioxol-4-yl)methyl esters have found a variety of applications as prodrugs of carboxylic acids. Their activation is mainly nonenzymatic, because they break down chemically under physiological conditions of pH and temperature as shown in Fig. 13.38 (212). The nature of the 5-substituent influences the rate of hydrolytic ring opening.

Activation by intramolecular cyclization-elimination occurs in specifically designed prodrugs of phenols, alcohols, and amines (213). A number of design strategies exist to achieve such mechanisms, their general chemical principle being shown in Fig. 13.39. In such a schematic representation, the carrier moiety is a side-chain attached to the drug by a carbonyl group (i.e., by an ester or amide function) and containing a nucleophilic group symbolized by Nu. The latter directly attacks the carbonyl in a reaction of nucleophilic substitution whose outcome is cyclization of the carrier moiety and elimination of the drug molecule.

Known intramolecular nucleophiles are basic amino, acidic amido, carboxylate, and hydroxy groups. Thus, the radiation sensitizer 5-bromo-2'-deoxyuridine (45) (Fig. 13.40) was derivatized with diamino acids to obtain prodrugs shown as (44) in Fig. 13.40 (with R = H or cyclohexyl) (214). These prodrugs were stable in acidic solutions. Under physiological conditions of pH and temperature, the reaction of cyclization proceeded cleanly according to reaction A with a half-life of 23 or 30 min for the two prodrugs (44) having R = H or cyclohexyl, respectively. In human plasma under the same conditions, the half-lives were longer, suggesting protection from breakdown by binding to proteins. In rat plasma, the half-life of the second compound was 5 min, indicating that enzymatic hydrolysis by rat plasma hydrolases (reaction B in Fig. 13.40) is possible in some cases.

5.2.4 Multistep Prodrugs. A number of prodrugs do not undergo a single, direct reaction of activation, but need two or more steps to liberate the active metabolite. Such a strategy may seem a source of needless complications,

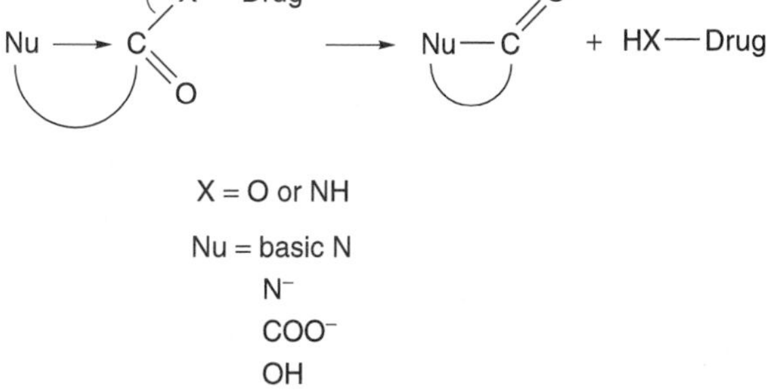

X = O or NH
Nu = basic N
N⁻
COO⁻
OH

Figure 13.39. General reaction scheme for the intramolecular activation of prodrugs by cyclization–elimination (modified from Ref. 213).

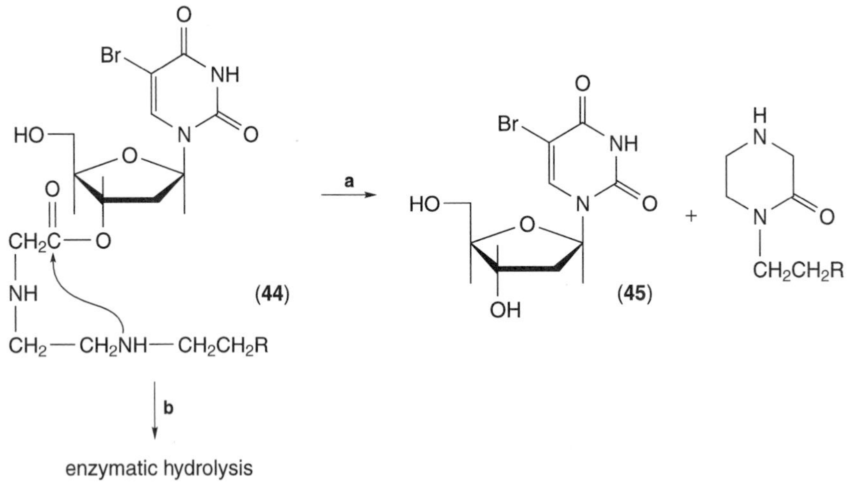

Figure 13.40. Activation of basic ester prodrugs of 5–bromo–2′–deoxyuridine (**45**) by cyclization of the pro–moiety (reaction a) and by enzymatic hydrolysis (reaction b) (214).

yet enough examples exist to show that facing these additional difficulties may be rewarding.

A nice example of this approach has been reported for two-step prodrugs of pilocarpine (**47**) (Fig. 13.41), aimed at improving ocular delivery (215). The prodrugs were lipophilic diesters of pilocarpic acid (**46**). The first activation step was enzymatic regiospecific O-acyl hydrolysis to remove the acyl carrier (reaction A). In a second step, intramolecular nucleophilic substitution-elimination led to loss of the alcohol carrier and to ring closure to pilocarpine (reaction B). Efficient enzymatic hydrolysis was seen in various ocular tissue preparations, confirming the potential of these diesters as prodrugs for ocular delivery.

Another type of two-step prodrug has been designed in recent years in which the reaction shown in Fig. 13.39 (intramolecular attack and elimination of the active metabolite) must be preceded by the metabolic unmasking of the nucleophile. Here like in the pilocarpine prodrugs (Fig. 13.41), the nucleophilic group is not immediately available for cyclization-elimination, but remains latent until liberated metabolically. The analogy stops here, however, because in the case of pilocarpine prodrugs, the eliminated moiety was a second carrier and not the drug as discussed now.

A number of promoieties have been designed to undergo the two-step elimination of a drug (216–218). Coumarinic acid is one such promoiety. A number of model amines were derivatized with coumarinic acid, their general structure being shown as (**48**) in Fig. 13.42. These model prodrugs undergo initial hydrolysis, followed by facile lactonization to coumarin and release of the active amine (reactions A and B in Fig. 13.42). A particular interest of this approach is its potential to design peptide prodrugs with greatly improved permeation properties (218). Here, both ends of the peptide are linked to the coumarinic acid promoiety (**49**) (Fig. 13.43). Enzymatic hydrolysis cleaves the ester bridge, allowing the phenolic group to be liberated to react intramolecularly. This second step is identical to that shown in Fig. 13.42.

As mentioned in Table 13.13, tissue or organ selectivity is one of the objectives of prodrug design. For example, much interest exists in dermal delivery (219) and brain penetration (208). In this context, site-selective chemical delivery systems might evoke the "magic bullets" of drug design, their selectivity being based on some enzymatic or physicochemical characteristic of a given tissue or organ. For example, the selective presence of cysteine conjugate β-lyase in the kidney suggests that this enzyme might be exploited for delivery of sulfhydryl drugs to this organ (220). Brain-selective dihydropyridine carriers have been extensively investigated (204, 208). A large variety of drugs (e.g., neurophar-

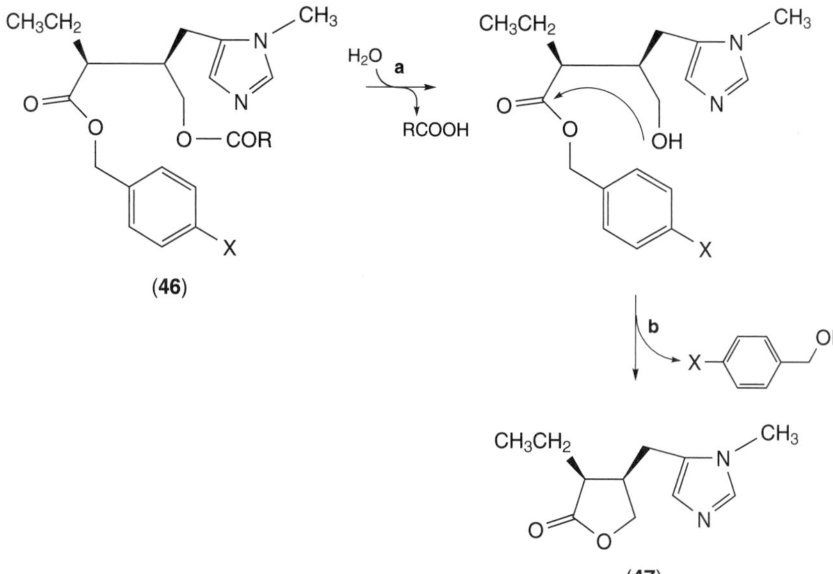

Figure 13.41. Two–step activation of pilocarpine prodrugs (215). The prodrugs are diesters of pilocarpic acid (**46**). Enzymatic hydrolysis (reaction a) cleaves the acyl carrier group. The product is a monoester of pilocarpic that undergoes cyclization to pilocarpine (**47**) upon intramolecular nucleophilic attack (cyclization) and elimination of the alcohol carrier (reaction b).

macological agents, steroid hormones, chemotherapeutic agents) have been coupled to dihydropyridine carriers, resulting in improved and sustained brain delivery in experimental animals.

Capecitabine (**50**) (Fig. 13.44) is a recently marketed site-selective multistep prodrug of the antitumor drug 5-fluorouracil (5-FU) (**53**) (221). The prodrug is well absorbed orally and is hydrolyzed by liver carboxylesterase. The resulting metabolite is a carbamic acid that spontaneously decarboxylates to 5'-deoxy-5-fluorocytidine (**51**). The enzyme cytidine deaminase, which is present in the liver and

tumors, transforms 5'-deoxy-5-fluorocytidine into 5'-deoxy-5-fluorouridine (**52**). Transformation into 5-FU (**53**) is catalyzed by thymidine phosphorylase and occurs selectively in tumor cells.

Capecitabine is of great interest in the context of this chapter. Clinically, it was first approved for the cotreatment of refractory metastatic breast cancer. Its therapeutic spectrum now includes metastatic colorectal cancer, and there are hopes that it might broaden further as positive results of new clinical trials become available. Capecitabine thus affords an impressive gain in therapeutic benefit com-

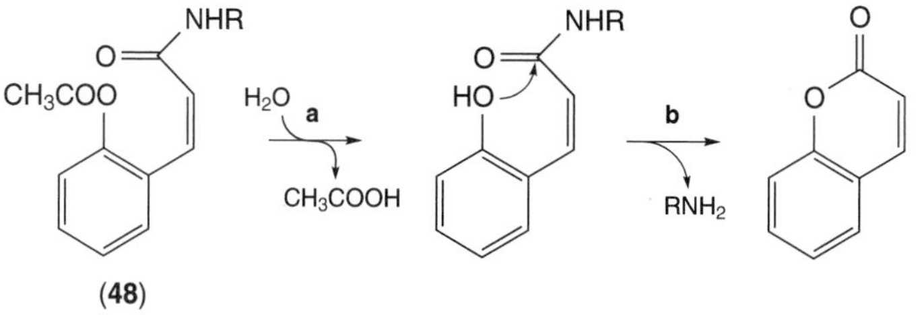

Figure 13.42. Amides of coumarinic acid *O*–acetyl ester (**48**) as prodrugs of active amines activated by cyclization–elimination in a two–step sequence. Reaction a is the hydrolase–catalyzed hydrolysis of the carboxylate moiety, followed by an intramolecular nucleophilic substitution with elimination of the active amine (reaction b) (216).

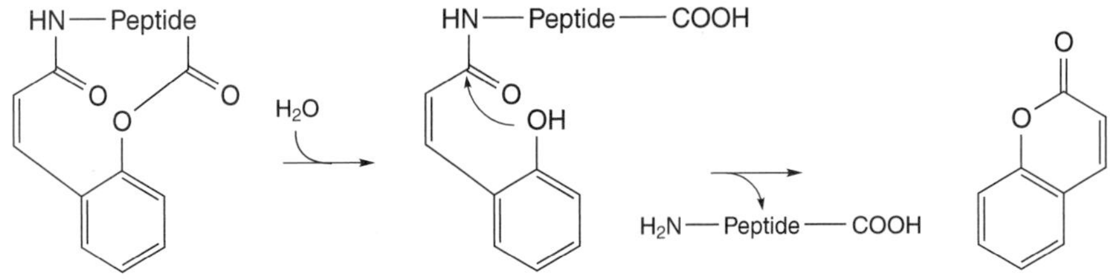

Figure 13.43. Two–step prodrugs of peptides derivatized with coumarinic acid (**49**) (218). Activation occurs by enzymatic cleavage of the ester bond, followed by intramolecular nucleophilic attack by the phenolic group on the carbonyl, resulting in cyclization to coumarin and elimination of the peptide.

pared with 5-FU because of its oral bioavailability and a relatively selective activation in and delivery to tumors.

5.3 The Concept of Toxophoric Groups

Biotransformation to reactive metabolic intermediates is one of the major mechanisms by which drugs exert toxic effects and particularly chronic toxicity. Other causes of (often acute) toxicity include all unwanted or exaggerated pharmacological responses. These various pharmacokinetic and pharmacodynamic mechanisms can be studied by structure-toxicity relationship methods (222, 223) and are thus amenable to rational corrective steps within the framework of drug design.

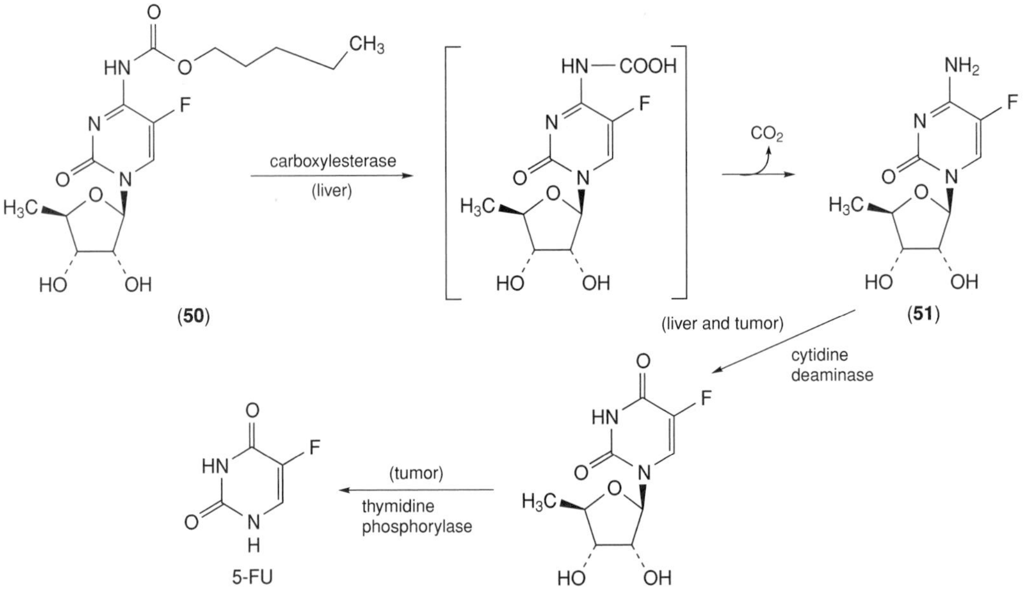

Figure 13.44. Metabolic activation of capecitabine (**50**), a site–selective multistep prodrug of the antitumor drug 5–fluorouracil (5–FU) (**53**). Following oral absorption, the prodrug is hydrolyzed by liver carboxylesterase to a carbamic acid that spontaneously decarboxylates to 5′–deoxy–5–fluorocytidine (**51**). The latter is transformed into 5′–deoxy–5–fluorouridine (**52**) by cytidine deaminase present in the liver and tumors. The third activation step occurs selectively in tumor cells and involves the transformation to 5–FU (**53**), catalyzed by thymidine phosphorylase (221).

Table 13.15 Major Toxophoric Groups and Their Metabolic Reactions of Toxification

Functionalization reactions
  Some aromatic systems that can be oxidized to epoxides, quinones, or quinonimines (Fig. 5, reaction 1)
  Ethynyl moieties activated by cytochrome P450 (Fig. 5, reaction 3)
  Some halogenated alkyl groups that can undergo reductive dehalogenation (Fig. 3, reaction 7)
  Nitroarenes that can be reduced to nitro anion-radicals, nitrosoarenes, nitroxides, and hydroxylamines (Fig. 7, reaction 4)
  Some aromatic amides that can be activated to nitrenium ions (reaction 3 in Fig. 7, followed by reaction 3 in Fig. 18)
  Some thiocarbonyl derivatives, particularly thioamides, that can be oxidized to S,S-dioxide (sulfene) metabolites (Fig. 9, reaction 3)
  Thiols that can form mixed disulfides (Fig. 9, reaction 1)
Conjugation reactions
  Some carboxylic acids that can form reactive acylglucuronides (Fig. 21, reaction 3)
  Some carboxylic acids that can form highly lipophilic conjugates (Fig. 27, reactions 3 and 4)

Table 13.16 Macromolecular Adducts as a Class of Toxic Metabolites (21)

Their formation may be
  non-enzymatic
  enzymatic
  post-enzymatic
The covalent bond may be
  strong (C—N, C—O, C—S)
  of medium energy (S—S)
The target macromolecule may be
  soluble
  membrane-bound
The fate of macromolecular adducts is largely unknown
Macromolecular adducts may be toxic due to
  loss of the macromolecule's original functions
  antigenic activity
  toxic breakdown products

the above that the presence of a toxophoric group necessarily implies toxicity. Reality is far less gloomy, as only potential toxicity is indicated. Given the presence of a toxophoric group in a compound, a number of factors will operate to render the latter either toxic or nontoxic.

1. The molecular properties of the substrate will increase or decrease its affinity and reactivity toward toxification and detoxification pathways.
2. Metabolic reactions of toxification are always accompanied by competitive and/or sequential reactions of detoxification that compete with the formation of the toxic metabolite and/or inactivate it. A profusion of biological factors control the relative effectiveness of these competitive and sequential pathways (Section 4).
3. The reactivity and half-life of a reactive metabolite control its sites of action and determine whether it will reach sensitive sites (18).
4. Dose, rate, and route of entry into the organism are all factors of known significance.
5. Above all, there exist essential mechanisms of survival value that operate to repair molecular lesions, remove them immunologically, and/or regenerate the lesioned sites (Fig. 13.45) (21).

For a reaction of toxification to occur at all, a target moiety must be present that has been termed a toxogenic, toxicophoric, or toxophoric group (Table 13.15) (see Refs. 224–226). Major functionalization reactions that activate toxophoric groups include oxidation to electrophilic intermediates or reduction to nucleophilic radicals (which may be followed by oxygen reduction to superoxide, hydroxyl radical, and other reactive oxygen species). The electrophilic or nucleophilic intermediates may then react with bio(macro)molecules, producing critical or noncritical lesions. Table 13.16 lists some characteristics of macromolecular adducts as a class of toxic metabolites (21). The toxic potential of radicals is particularly noteworthy. Of more recent awareness is the fact that some conjugation reactions may also lead to toxic metabolites, either reactive ones or long-retained residues.

A drug designer worthy of the name must be conversant with toxification reactions and toxophoric groups (Table 13.15) (6, 21, 227). However, it would be wrong to conclude from

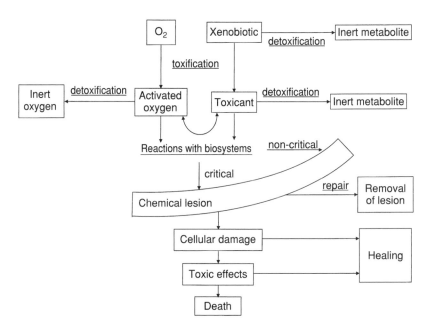

Figure 13.45. General scheme placing of toxification reactions (vertical arrows) and of detoxification (horizontal arrows) in a broader biological context (21).

In conclusion, the presence of a toxophoric group is not a sufficient condition for observable toxicity, a sobering and often underemphasized fact. Nor is it a necessary condition because other mechanisms of toxicity exist, for example, the acute toxicity characteristic of many solvents.

6   CONCLUDING REMARKS

Drug discovery and development are becoming more complex by the day, with physicochemical, pharmacokinetic, and pharmacodynamic properties being screened and assessed as early and as simultaneously as possible. But the real challenge lies with the resulting deluge of data, which must be stored, analyzed, and interpreted.

Storing, retrieving, and analyzing data calls for a successfull synergy between humans and algorithmic machines, in other words, between human and artificial intelligence. However, making sense of the data, interpreting their analyses, and rationally planning subsequent steps is an entirely different issue. The difference is a qualitative one, and it is the difference between information and knowledge.

This chapter is about both. Our first objective in writing it was obviously to supply as much useful information as would fit in the allocated pages. By useful information, we mean structured data as exemplified by the classification of metabolic reactions (Sections 2 and 3) and biological factors affecting them (Section 4).

Information becomes knowledge when it is connected to a context from which it receives meaning. This treatise is about medicinal chemistry, and indeed medicinal chemistry is the context of our chapter and of all others. By discussing the connection between drug metabolism and the medicinal chemistry context (Section 5), we have tried to be true to our second objective, which was to present medicinal chemists with meaningful information. Had this chapter been written for a treatise of molecular biology, much of the basic information would have been the same, but the context and the chapter's meaning would have been different. This also tells us that medicinal chemistry itself needs a context to acquire meaning, the context of human welfare to which responsible medicinal chemists are proud to contribute (228).

REFERENCES

1. B. Testa and P. Jenner, *Drug Metabolism. Chemical and Biochemical Aspects*, Dekker, New York, 1976.

2. P. Jenner and B. Testa, Eds., *Concepts in Drug Metabolism. Parts A and B*. Dekker, New York, 1980.

3. G. J. Mulder, Ed., *Conjugation Reactions in Drug Metabolism*, Taylor & Francis, London, 1990.

4. R. B. Silverman, *The Organic Chemistry of Drug Design and Drug Action*, Academic Press, San Diego, 1992.

5. F. C. Kauffman, Ed., *Conjugation-Deconjugation Reactions in Drug Metabolism and Toxicity*, Springer Verlag, Berlin, 1994.

6. B. Testa, *The Metabolism of Drugs and Other Xenobiotics—Biochemistry of Redox Reactions*, Academic Press, London, 1995.

7. P. R. Ortiz de Montellano, Ed., *Cytochrome P450. Structure, Mechanism, and Biochemistry*, 2nd ed., Plenum Press, New York, 1996.

8. T. F. Woolf, Ed., *Handbook of Drug Metabolism*, Dekker, New York, 1999.

9. B. Testa and J. M. Mayer, *Hydrolysis in Drug and Prodrug Metabolism—Chemistry, Biochemistry, and Enzymology*, Wiley VHCA, Zurich, in press.

10. C. Ioannides, Ed., *Enzyme Systems that Metabolise Drugs and Other Xenobiotics*, Wiley, Chichester, UK, 2002.

11. B. Testa in B. Testa, Ed., *Advances in Drug Research*, vol. **13**, Academic Press, London, 1984, pp. 1–58.

12. F. J. Di Carlo, *Drug Metabol. Rev.*, **13**, 1–4 (1982).

13. B. Testa, *Trends Pharmacol. Sci.*, **8**, 381–383 (1987).

14. B. Testa, *Drug Metabol. Rev.*, **13**, 25–50 (1982).

15. J. Caldwell, *Drug Metabol. Rev.*, **13**, 745–777 (1982).

16. H. G. Neumann in B. Testa, Ed., *Advances in Drug Research*, vol. **15**, Academic Press, London, 1986, pp. 1–28.

17. J. A. Timbrell, *Principles of Biochemical Toxicology*, 2nd ed., Taylor & Francis, London, 1991.

18. J. R. Gillette, *Drug Metabol. Rev.*, **14**, 9–33 (1983).

19. L. P. Balant, H. Roseboom, and R. M. Guntert-Remy in B. Testa, Ed., *Advances in Drug Research*, vol. **19**, Academic Press, London, 1990, pp. 1–138.

20. G. Gaviraghi, R. J. Barnaby, and M. Pellegatti in B. Testa, H. van de Waterbeemd, G. Folkers, and R. Guy, Eds., *Pharmacokinetic Optimization in Drug Research - Biological, Physicochemical and Computational Strategies*, Wiley VHCA, Zurich, 2001, pp. 3–14.

21. B. Testa in M. M. Campbell and I. S. Blagbrough, Eds, *Medicinal Chemistry into the Millenium*, Royal Society of Chemistry, Cambridge, 2001, pp. 345–356.

22. T. Walle, U. K. Walle, and L. S. Olanoff, *Drug Metabol. Disposit.*, **13**, 204–209 (1985).

23. Nomenclature Committee of the International Union of Biochemistry and Molecular Biology, *Enzyme Nomenclature 1992*. Academic Press, San Diego, 1992. Available online at http://www.chem.qmul.ac.uk/iubmb/enzyme/ and http://www.biochem.ucl.ac.uk/bsm/enzymes/.

24. F. J. Gonzalez, *Pharmacol. Therap.*, **45**, 1–38 (1990).

25. D. R. Nelson, T. Kamataki, D. J. Waxman, F. P. Guengerich, R. W. Estabrook, R. Feyereisen, F. J. Gonzalez, M. J. Coon, I. C. Gunsalus, O. Gotoh, K. Okuda, and D. W. Nebert, *DNA Cell. Biol.*, **12**, 1–51 (1993).

26. D. R. Nelson, L. Koymans, T. Kamataki, J. J. Stegeman, R. Feyereisen, D. J. Waxman, M. R. Waterman, O. Gotoh, M. J. Coon, R. W. Estabrook, I. C. Gunsalus, and D. W. Nebert, *Pharmacogenetics*, **6**, 1–42 (1996).

27. S. Rendic and F. J. Di Carlo, *Drug Metabol. Rev.*, **29**, 413–580 (1997).

28. S. Rendic, *Drug Metabol. Rev.* **34**, 83–48 (2002). Available online at http://drnelson.utmem.edu/human.P450.seqs.html and http://www.gentest.com.

29. G. K. Chambers, *Gen. Pharmacol.*, **21**, 267–272 (1990).

30. G. Duester, J. Farrés, M. R. Felder, R. S. Holmes, J. O. Höög, X. Parés, B. V. Plapp, S. J. Yin, and H. Jörnvall, *Biochem. Pharmacol.*, **58**, 389–395 (1999).

31. R. L. Felsted and N. R. Bachur, *Drug Metabol. Rev.*, **11**, 1–60 (1980).

32. H. P. Wirth and B. Wermuth, *FEBS Lett.*, **187**, 280–282 (1985).

33. R. Lindahl, *Crit. Rev. Biochem. Molec. Biol.*, **27**, 283–335 (1992).

34. J. M. Jez, T. G. Flynn, and T. M. Pennings, *Biochem. Pharmacol.*, **54**, 639–647 (1997).

35. V. Vasiliou, A. Bairoch, K. F. Tripton, and D. W. Nebert, *Pharmacogenetics*, **9**, 421–434 (1999).

36. P. A. Cossum, *Biopharm. Drug Disposit.*, **9**, 321–336 (1988).

37. E. Kyburz, *Drug News Perspect.*, **3**, 592–599 (1990).

38. C. Beedham, *Drug Metabol. Rev.*, **16**, 119–156 (1985).

39. B. A. Callingham, A. Holt, and J. Elliott, *Biochem. Soc. Transact.*, **19**, 228–233 (1991).

40. C. E. Castro, *Pharmacol. Therap.*, **10**, 171–189 (1980).

41. T. E. Eling and J. F. Curtis, *Pharmacol. Therap.*, **53**, 261–273 (1992).

42. J. P. Uetrecht, *Drug Metabol. Rev.*, **24**, 299–366 (1992).

43. K. R. Korzekwa, W. F. Trager, K. Nagata, A. Parkinson, and J. R. Gillette, *Drug Metabol. Disposit.*, **18**, 974–979 (1990).

44. B. Rambeck, T. May, and U. Juergens, *Therap. Drug Monit.*, **9**, 298–303 (1987).

45. J. W. Gorrod and A. Raman, *Drug Metab. Rev.*, **20**, 307–339 (1989).

46. M. Fujioka, *Int. J. Biochem.*, **24**, 1917–1924 (1992).

47. R. M. Weinshilboum, D. M. Otterness, and C. L. Szumlanski, *Ann. Rev. Pharmacol. Toxicol.*, **39**, 19–52 (1999).

48. J. W. Faigle, I. Böttcher, J. Godbillon, H. P. Kriemler, E. Schlumpf, W. Schneider, A. Schweizer, H. Stierlin, and T. Winkler, *Xenobiotica*, **18**, 1191–1197 (1988).

49. P. A. Crooks, C. S. Godin, L. A. Damani, S. S. Ansher, and W. B. Jakoby, *Biochem. Pharmacol.*, **37**, 1673–1677 (1988).

50. R. M. Weinshilboum, *Pharmacol. Therap.*, **43**, 77–90 (1989).

51. C. N. Falany, *Trends Pharmacol. Sci.*, **12**, 255–259 (1991).

52. R. B. Raftogianis, T. C. Wood, and R. M. Weinshilboum, *Biochem. Pharmacol.*, **58**, 605–616 (1999).

53. R. S. Tsai, P. A. Carrupt, B. Testa, and J. Caldwell, *Chem. Res. Toxicol.*, **7**, 73–76 (1994).

54. J. E. Manautou and G. P. Carlson, *Xenobiotica*, **22**, 1309–1319 (1992).

55. K. Iwasaki, T. Shiraga, K. Noda, K. Tada, and H. Noguchi, *Xenobiotica*, **16**, 651–659 (1986).

56. K. D. Meisheri, G. A. Johnson, and L. Puddington, *Biochem. Pharmacol.*, **45**, 271–279 (1993).

57. B. Burchell, D. W. Nebert, D. R. Nelson, K. W. Bock, T. Iyanagi, P. L. M. Jansen, D. Lancet, G. J. Mulder, J. R. Chowdhury, G. Siest, T. R. Tephly, and P. I. Mackenzie, *DNA Cell Biol.*, **10**, 487–494 (1991).

58. J. O. Miners and P. I. Mackenzie, *Pharmacol. Therap.*, **51**, 347–369 (1991).

59. G. J. Mulder, *Ann. Rev. Pharmacol. Toxic.*, **32**, 25–49 (1992).

60. P. I. Mackenzie, I. S. Owens, B. Burchell, K. W. Bock, A. Bairoch, A. Bélanger, S. Fournel-Gigleux, M. Green, D. W. Hum, T. Iyanagi, D. Lancet, P. Louisot, J. Magdalou, J. R. Chowdhury, J. K. Ritter, H. Schachter, T. R. Tephly, K. F. Tipton, and D. W. Nebert, *Pharmacogenetics*, **7**, 255–269 (1997).

61. A. Radominska-Pandya, P. J. Czernik, J. M. Little, E. Battaglia, and P. I. Mackenzie, *Drug Metabol. Rev.*, **31**, 817–899 (1999).

62. P. A. Carrupt, B. Testa, A. Bechalany, N. El Tayar, P. Descas, and D. Perrissoud, *J. Med. Chem.*, **34**, 1272–1275 (1991).

63. H. Spahn-Langguth and L. Z. Benet, *Drug Metabol. Rev.*, **24**, 5–48 (1992).

64. W. H. Schaefer, A. Goalwin, F. Dixon, B. Hwang, L. Killmer, and G. Kuo, *Biol. Mass Spectrom.*, **21**, 179–188 (1992).

65. H. Luo, E. M. Hawes, G. McKay, E. D. Korchinski, and K. K. Midha, *Xenobiotica*, **21**, 1281–1288 (1991).

66. U. Breyer-Pfaff, D. Fischer, and D. Winne, *Drug Metabol. Disposit.*, **25**, 340–345 (1997).

67. B. K. Tang, *Pharmacol. Therap.*, **46**, 53–56 (1990).

68. D. A. Price Evans, *Pharmacol. Therap.*, **42**, 157–234 (1989).

69. D. M. Grant, M. Blum, and U. A. Meyer, *Xenobiotica*, **22**, 1073–1081 (1992).

70. S. S. Mattano, S. Land, C. M. King, and W. W. Weber, *Molec. Pharmacol.*, **35**, 599–609 (1989).

71. H. K. Bhat and G. A. S. Ansari, *Chem. Res. Toxicol.*, **3**, 311–317 (1990).

72. K. Waku, *Biochim. Biophys. Acta.*, **1124**, 101–111 (1992).

73. J. Caldwell and B. G. Lake, *Biochem. Soc. Trans.* **13**, 847–862 (1985).

74. P. F. Dodds, *Life Sci.*, **49**, 629–649 (1991).

75. J. M. Mayer and B. Testa, *Drugs Future*, **22**, 1347–1366 (1997).

76. S. M. Bjorge and T. A. Baillie, *Drug Metabol. Disposit.*, **19**, 823–829 (1991).

77. J. Hulsman, *Pharm. Weekly Sci. Ed.*, **14**, 98–100 (1992).

78. D. J. Reed, *Ann. Rev. Pharmacol. Toxicol.*, **30**, 603–631 (1990).

79. A. Meister, *J. Biol. Chem.*, **263**, 17205–17208 (1988).

80. H. Sies, *Free Rad. Biol. Med.*, **27**, 916–921 (1999).

81. S. Tsuchida and K. Sato, *Crit. Rev. Biochem. Molec. Biol.*, **27**, 337–384 (1992).

82. R. N. Armstrong, *Chem. Res. Toxicol.*, **10**, 2–18 (1997).

83. B. Ketterer, *Drug Metabol. Rev.*, **13**, 161–187 (1982).

84. T. J. Monks and S. S. Lau, *Crit. Rev. Toxicol.*, **22**, 243–270 (1992).

85. W. Dekant, S. Vamvakas, and M. W. Anders, *Drug Metabol. Rev.*, **20**, 43–83 (1989).

86. T. A. Baillie and J. G. Slatter, *Acc. Chem. Res.*, **24**, 264–270 (1991).

87. A. E. Cribb, M. Miller, J. S. Leeder, J. Hill, and S. P. Spielberg, *Drug Metabol. Disposit.*, **19**, 900–906 (1991).

88. D. Ross, *Pharmacol. Therap.*, **37**, 231–249 (1988).

89. P. Jenner and B. Testa, *Xenobiotica*, **8**, 1–25 (1978).

90. B. Testa in D. J. Benford, J. W. Bridges, and G. G. Gibson, Eds., *Drug Metabolism—from Molecules to Man*, Taylor & Francis, London, 1987, pp. 563–580.

91. E. De Clercq in B. Testa, Ed., *Advances in Drug Research*, vol. **17**, Academic Press, London, 1988, pp. 1–59.

92. J. P. O'Donnell, *Drug Metabol. Rev.* **13**, 123–159 (1982).

93. B. N. LaDu, H. G. Mandel, and E. L. Way, *Fundamentals of Drug Metabolism and Drug Disposition*, Williams and Wilkins, Baltimore, 1971.

94. J. M. Collins, *Chem. Biol.-Interact.*, **134**, 237–242 (2001).

95. M3 Nonclinical Safety Studies for the Conduct of Human Clinical Trials for Pharmaceuticals, U.S. Department of Health and Human Services, FDA, CDER, CBER, 1997.

96. M. Treinen-Moslen in C. D. Klaassen, Ed., *Casarett and Coull's Toxicology—The Basic Science of Poisons*, 6th ed., McGraw-Hill, New York, 2001, pp. 471–489.

97. S. S. Lee, J. T. Buters, and T. Pineau, *J. Biol. Chem.*, **271**, 12063–12067 (1996).

98. A. Parkinson in C. D. Klaassen, Ed., *Casarett and Coull's Toxicology—The Basic Science of Poisons*, 6th ed., McGraw-Hill, New York, 2001, pp. 133–224.

99. H. G Xie, R. B. Kim, A. J. J. Wood, and C. M. Stein, *Ann. Rev. Pharmacol. Toxicol.*, **41**, 815–850 (2001).

100. A. D. Rodrigues, *Biochem. Pharmacol.*, **57**, 465–480 (1999).

101. D. A. Smith, S. M. Abel, R. Hyland, and B. C. Jones, *Xenobiotica*, **28**, 1095–1128 (1998).

102. D. F. V. Lewis, M. Dickins, P. J. Eddershaw, M. H. Tarbit, and P. S. Goldfarb, *Drug Metabol. Drug Interact.*, **15**, 1–49 (1999).

103. S. A. Smith, M. J. Ackland, and B. C. Jones, *Drug Discov. Today* **2**, 479–486.

104. K. R. Iyer and M. W. Sinz, *Chem.-Biol. Interact.*, **118**, 151–169 (1999).

105. A. Yu and R. L. Haining, *Drug Metabol. Disposit.*, **29**, 1514–1520 (2001).

106. D. R. Nelson, *Arch. Biochem. Biophys.*, **369**, 1–10 (1999).

107. A. K. Daly, *J. Mol. Med.*, **73**, 539–553 (1995).

108. U. A. Meyer, *Lancet*, **356**, 1667–1671 (2000).

109. R. H. Tukey and C. P. Strassburg, *Ann. Rev. Pharmacol. Toxicol.*, **40**, 581–616 (2000).

110. D. W. Hein, M. A. Doll, A. J. Fretland, M. A. Leff, S. J. Webb, G. H. Xiao, U. S. Devanaboyina, N. A. Nangju, and Y. Feng, *Cancer Epidemiol. Biomark. Prev.*, **9**, 29–42 (2000).

111. B. Ketterer, *Chem. Biol. Interact.*, **138**, 27–42 (2001).

112. H. Glatt, *Chem. Biol. Interact.*, **129**, 141–170 (2000).

113. R. M. Weinshilboum, D. M. Otterness, and C. L. Szumlanski, *Ann. Rev. Pharmacol. Toxicol.*, **39**, 19–52 (1999).

114. A. Ayrton and P. Morgan, *Xenobiotica*, **31**, 469–497 (2001).

115. J. Van Asperen, O. Van Tellingen, and J. H. Beijnen, *Pharmacol. Res.*, **37**, 429–435 (1998).

116. U. Fuhr, *Drug Safety*, **18**, 252–272 (1998).

117. W. Kalow and L. Bertilsson in B. Testa and U. A. Meyer, Eds., *Advances in Drug Research*, vol. **25**, Academic Press, London, 1994, pp. 1–53.

118. E. Tanaka, *J. Clin. Pharmacol. Therap.*, **24**, 339–346 (1999).

119. C. A. Mugford and G. L. Kedderis, *Drug Metabol. Rev.*, **30**, 441–498 (1998).

120. S. Bandiera, *Can. J. Physiol. Pharmacol.*, **68**, 762–768 (1990).

121. M. Gex-Fabry, A. E. Balant-Gorgia, L. P. Balant, and G. Garrone, *Clin. Pharmacokinet.*, **19**, 241–255 (1990).

122. D. R. Abernethy, M. Divoll, D. J. Greenblatt, and B. Ameer, *Clin. Pharmacol. Therap.*, **31**, 783–790 (1982).

123. T. Walle, K. Walle, R. S. Mathur, Y. Y. Palesch, and E. C. Conradi, *Clin. Pharmacol. Therap.*, **56**, 127–132 (1994).

124. S. G. Johnson, J. P. Kampmann, E. P. Bennet, and F. S. Jorgensen, *Clin. Pharmacol. Therap.*, **20**, 233–237 (1976).

125. P. J. Gow, H. Ghadrial, R. A. Smallwood, D. J. Morgan, and M. S. Ching, *Pharmacol. Toxicol.*, **88**, 3–15 (2001).

126. T. Cresteil, P. Beaune, P. Kremers, C. Celier, F. P. Guengerich, and J. P. Leroux, *Eur. J. Biochem.*, **151**, 345–350 (1985).

127. D. Lacroix, M. Sonnier, A. Moncion, G. Cheron, and T. Cresteil, *Eur. J. Biochem.*, **247**, 625–634 (1997).

128. J. M. Treluyer, G. Gueret, G. Cheron, M. Sonnier, and T. Cresteil, *Pharmacogenetics*, **7**, 441–452 (1997).

129. R. J. Cody, *Drugs Aging*, **3**, 320–334 (1993).

130. K. Woodhouse and H. A. Wynne, *Drugs Aging*, **2**, 243–255 (1992).

131. E. A. Sotaniemi, A. J. Arranto, O. Pelkonen, and M. Pasanen, *Clin. Pharmacol. Therap.*, **61**, 331–339 (1997).

132. R. E. Vestal, *Cancer*, **80**, 1302–1310 (1997).

133. D. L. Schmucker, K. W. Woodhouse, R. K. Wang, H. Wynne, O. F. James, M. McManus, and P. Kremers, *Clin. Pharmacol. Therap.*, **48**, 365–374 (1990).

134. T. Shimada, H. Yamazaki, M. Mimura, Y. Inui, and F. P. Guengerich, *J. Pharmacol. Exp. Therap.*, **270**, 414–423 (1994).

135. H. K. Seitz, G. Egerer, U. A. Simanowski, R. Waldherr, R. Eckey, D. P. Agarwal, H. W. Goedde, and J. P. von Wartburg, *Gut*, **34**, 1433–1437 (1993).

136. S. M. Lichtman and J. A. Skirvin, *Oncology*, **14**, 1743–1755 (2000).

137. B. Lemmer, *Sem. Perinatol.*, **24**, 280–290 (2000).

138. R. P. Sturtevant, F. M. Sturtevant, J. E. Pauly, and L. E. Scheving, *Int. J. Clin. Pharmacol.*, **16**, 594–599 (1978).

139. J. M. Gries, N. Benowitz, and D. Verotta, *Clin. Pharmacol. Therap.*, **60**, 385–395 (1996).

140. P. H. Scott, E. Tabachnik, S. MacLeod, J. Correia, C. Newth, and H. Levison, *J. Pediatr.*, **99**, 476–479 (1981).

141. J. G. Moore and E. Englert, *Nature*, **226**, 1261–1262 (1970).

142. B. Krauer and L. Dettli, *Chemotherapy*, **14**, 1–6 (1969).

143. B. Bruguerolle and B. Lemmer, *Life Sci.*, **52**, 1809–1824 (1993).

144. T. Furukawa, S. Manabe, T. Watanabe, S. Sharyo, and Y. Mori, *Toxicol. Appl. Pharmacol.*, **161**, 219–224 (1999).

145. P. Jenner and B. Testa in P. Jenner and B. Testa, Eds., *Concepts in Drug Metabolism. Part B*, Dekker, New York, 1981, pp. 423–513.

146. H. Iber, M. B. Sewer, T. B. Barclay, S. R. Mitchell, T. Li, and E. T. Morgan, *Drug Metabol. Rev.*, **31**, 29–41 (1999).

147. E. T. Morgan, *Drug Metabol. Rev.*, **29**, 1129–1188 (1997).

148. M. J. Kraemer, C. T. Furukawa, J. R. Koup, G. G. Shapiro, W. E. Pierson, and C. W. Bierman, *Pediatrics*, **69**, 476–480 (1982).

149. J. H. Lin and A. Y. H. Yu, *Ann. Rev. Pharmacol. Toxicol.*, **41**, 535–67 (2001).

150. R. E. White, *Ann. Rev. Pharmacol. Toxicol.*, **40**, 133–57 (2000).

151. B. Testa and P. Jenner, *Drug Metabol. Rev.*, **12**, 1–117 (1981).

152. T. Anderson, M. Hassan-Alin, G. Hasselgren, and K. Rohss, *Clin. Pharmacokinet.*, **40**, 523–537 (2001).

153. M. Murray, *Drug Metabol. Rev.*, **18**, 55–81 (1987).

154. R. R. Ortiz de Montellano in G. G. Gibson, Ed., *Progress in Drug Metabolism*, vol. **11**, Taylor and Francis, London, 1988, pp. 99–148.

155. M. Mesnil, B. Testa, and P. Jenner, *Biochem. Pharmacol.*, **37**, 3619–3622 (1988).

156. C. F. Wilkinson and M. Murray, *Drug Metabol. Rev.*, **15**, 897–917 (1984).

157. C. Bensoussan, M. Delaforge, and D. Mansuy, *Biochem. Pharmacol.*, **49**, 591–602 (1995).

158. J. R. Halpert, F. P. Guengerich, J. R. Bend, and M. A. Correia, *Tox. Appl. Pharmacol.*, **125**, 163–175 (1994).

159. J. P. Van Wauwe and P. A. J. Janssen, *J. Med. Chem.*, **32**, 2231–2239 (1989).

160. L. V. Favreau, J. R. Palamanda, C. C. Lin, and A. A. Nomeir, *Drug Metabol. Disposit.*, **27**, 436–439 (1999).

161. J. Whysner, P. M. Ross, and G. M. Williams, *Pharmacol. Therap.*, **71**, 153–191 (1996).

162. R. N. Hines, Z. Luo, T. Cresteil, X. Ding, R. A. Prough, J. L. Fitzpatrick, S. L. Ripp, K. C. Falkner, N. L. Ge, A. Levine, and C. J. Elferink, *Drug Metabol. Disposit.*, **29**, 623–633 (2001).

163. D. J. Waxman, *Arch. Biochem. Biophys.*, **369**, 11–23 (1999).

164. J. P. Whitlock Jr., *Ann. Rev. Pharmacol. Toxicol.*, **39**, 103–125 (1999).

165. L. C. Quattrochi and P. S. Guzelian, *Drug Metabol. Disposit.*, **29**, 615–622 (2001).

166. T. Sueyoshi and M. Negishi, *Ann. Rev. Pharmacol. Toxicol.*, **41**, 123–43 (2001).

167. B. Testa, P. Crivori, M. Reist, and P. A. Carrupt, *Perspect. Drug Discov. Design.*, **19**, 179–211 (2000).

168. P. Jenner and B. Testa, *Drug Metabol. Rev.*, **2**, 117–184 (1973).

169. T. Walle and U. K. Walle, *Trends Pharmacol. Sci.*, **7**, 155–158 (1986).

170. B. Testa and J. M. Mayer in E. Jucker, Ed., *Progress in Drug Research*, vol. **32**, Birkhäuser, Basel, 1988, pp. 249–303.

171. B. Testa, *Biochem. Pharmacol.*, **37**, 85–92 (1988).

172. F. Jamali, R. Mehvar, and F. M. Pasutto, *J. Pharm. Sci.*, **78**, 695–715 (1989).

173. J. Mayer and B. Testa in P. G. Welling and L. P. Balant, Eds., *Pharmacokinetics of Drugs*, Springer-Verlag, Berlin, 1994, pp. 209–231.

174. J. M. Mayer and B. Testa, *Int. J. BioChromatog.*, **5**, 297–312 (2000).

175. P. Le Corre, D. Ratanasavanh, D. Gibassier, A. M. Barthel, P. Sado, R. Le Verge, and A. Guillouzo in A. Guillouzo, Ed., *Liver Cells and Drugs*, vol. **164**, Colloque Inserm, Paris, 1988, pp. 321–324.

176. R. J. Riley, C. Lambert, N. R. Kitteringham, and B. K. Park, *Br. J. Clin. Pharmacol.*, **27**, 823–830 (1989).

177. T. C. Butler, K. H. Dudley, D. Johnson, and S. B. Roberts, *J. Pharmacol. Exp. Therap.*, **199**, 82–92 (1976).

178. K. Sugiyama and W. F. Trager, *Biochemistry*, **25**, 7336–7343 (1986).

179. D. A. Smith, K. Brown, and M. G. Neale, *Drug Metabol. Rev.*, **16**, 365–388 (1985).

180. F. J. Gonzalez and D. W. Nebert, *Trends Genet.*, **6**, 182–186 (1990).

181. K. A. S. Al-Gailany, J. B. Houston, and J. W. Bridges, *Biochem. Pharmacol.*, **27**, 783–788 (1978).

182. A. L. Upthagrove and W. L. Nelson, *Drug Metabol. Disposit.*, **29**, 1377–1388 (2001).

183. R. Morgenstern, G. Lundqvist, V. Hancock, and J. W. DePierre, *J. Biol. Chem.*, **263**, 6671–6675 (1988).

184. K. R. Korzekwa, J. Grogan, S. DeVito, and J. P. Jones in Snyder R. et al., Eds., *Biological Reactive Intermediates V*, Plenum Press, New York, 1996, pp. 361–369.

185. C. Altomare, P. A. Carrupt, P. Gaillard, N. El Tayar, B. Testa, and A. Carotti, *Chem. Res. Toxicol.*, **5**, 366–375 (1992).

186. S. Ekins, G. Bravi, B. J. Ring, T. A. Gillespie, J. S. Gillespie, M. Vandenbranden, S. A. Wrighton, and J. H. Wickel, *J. Pharmacol. Exp. Therap.*, **288**, 21–29 (1999).

187. S. Ekins, G. Bravi, S. Binkley, J. S. Gillespie, B. J. Ring, J. H. Wickel, and S. A. Wrighton, *Pharmacogenetics*, **9**, 477–489 (1999).

188. S. Ekins, G. Bravi, S. Binkley, J. S. Gillespie, B. J. Ring, J. H. Wickel, and S. A. Wrighton, *Drug Metabol. Disposit.*, **28**, 994–1002 (2000).

189. A. Poso, J. Gynther, and R. Juvonen, *J. Comp.-Aided Mol. Design.*, **15**, 195–202 (2001).

190. A. M. ter Laak and N. P. E. Vermeulen in B. Testa, H. van de Waterbeemd, G. Folkers, and R. Guy, Eds., *Pharmacokinetic Optimization in Drug Research—Biological, Physicochemical and Computational Strategies*, Wiley VHCA, Zurich, 2001, pp. 551–588.

191. T. L. Poulos, B. C. Finkel, and A. J. Howard, *J. Mol. Biol.*, **195**, 687–700 (1987).

192. M. J. de Groot, M. J. Ackland, V. A. Horne, A. A. Alex, and B. C. Jones, *J. Med. Chem.*, **42**, 4062–4070 (1999).

193. G. M. Keseru, *J. Comp.-Aided Mol. Design.*, **15**, 649–657 (2001).

194. G. D. Szklarz and J. R. Halpert, *J. Comp.-Aided Mol. Design.*, **11**, 265–272 (1997).

195. F. de Rienzo, F. Fanelli, M. C. Menziani, and P. G. De Benedetti, *J. Comp.-Aided Mol. Design.*, **14**, 93–116 (2000).

196. S. Chen, K. Wu, D. Zhang, M. Sherman, R. Knox, and C. S. Yang, *Mol. Pharmacol.*, **56**, 272–278 (1999).

197. M. J. de Groot and N. P. E. Vermeulen, *Drug Metabol. Rev.*, **29**, 747–799 (1997).

198. P. A. Williams, J. Cosme, V. Sridhar, E. F. Johnson, and D. E. McRee, *Mol. Cell*, **5**, 121–131 (2000).

199. P. Lautala, I. Ulmanen, and J. Taskinen, *Mol. Pharmacol.*, **59**, 393–402 (2001).

200. P. W. Erhardt, Ed., *Drug Metabolism—Databases and High-Throughput Testing During*

Drug Design and Development, International Union of Pure and Applied Chemistry and Blackwell Science, London, 1999.

201. D. R. Hawkins, *Drug Disc. Today*, **4**, 466–471 (1999).

202. B. L. Podlogar, I. Muegge, and L. J. Brice, *Curr. Op. Drug Disc. Dev.*, **4**, 102–109 (2001).

203. E. J. Ariëns and A. M. Simonis in J. A. Keverling Buisman, Ed., *Strategy in Drug Research*, Elsevier, Amsterdam, 1982, pp. 165–178.

204. N. Bodor in B. Testa, Ed., *Advances in Drug Research*, vol. **13**. Academic Press, London, 1984, pp. 255–331.

205. B. Testa and J. Caldwell, *Med. Res. Rev.*, **16**, 233–241 (1996).

206. V. J. Stella, W. N. A. Charman, and V. H. Naringrekar, *Drugs*, **29**, 455–473 (1985).

207. C. G. Wermuth in G. Jolles and K. R. H. Wooldridge, Eds., *Drug Design: Fact or Fantasy?*, Academic Press, London, 1984, pp. 47–72.

208. N. Bodor, *Ann. NY Acad. Sci.*, **507**, 289–306 (1987).

209. R. Duncan, *Anti-Cancer Drugs*, **3**, 175–210 (1992).

210. I. H. Pitman, *Med. Res. Rev.*, **1**, 189–214 (1981).

211. B. Testa and J. M. Mayer, *Drug Metabol. Rev.*, **30**, 787–807 (1998).

212. W. S. Saari, W. Halczenko, D. W. Cochran, M. R. Dobrinska, W. C. Vincek, D. C. Titus, S. L. Gaul, and C. S. Sweet, *J. Med. Chem.*, **27**, 713–717 (1984).

213. D. Shan, M. G. Nicolaou, R. T. Borchardt, and B. Wang, *J. Pharm. Sci.*, **86**, 765–767 (1997).

214. W. S. Saari, J. E. Schwering, P. A. Lyle, S. J. Smith, and E. L. Engelhardt, *J. Med. Chem.*, **33**, 2590–2595 (1990).

215. H. Bundgaard, E. Falch, C. Larsen, G. L. Mosher, and T. J. Mikkelson, *J. Pharm. Sci.*, **75**, 775–783 (1986).

216. B. Wang, H. Zhang, and W. Wang, *Bioorg. Med. Chem. Lett.*, **6**, 945–950 (1996).

217. G. M. Pauletti, S. Gangwar, B. Wang, and R. T. Borchardt, *Pharm. Res.*, **14**, 11–17 (1997).

218. W. Wang, J. Jiang, C. E. Ballard, and B. Wang, *Curr. Pharm. Design.*, **5**, 265–287 (1999).

219. S. Y. Chan and A. Li Wan Po, *Int. J. Pharmaceut.*, **55**, 1–16 (1989).

220. I. Y. Hwang and A. A. Elfarra, *J. Pharmacol. Exp. Therap.*, **251**, 448–454 (1989).

221. Y. Tsukamoto, Y. Kato, M. Ura, I. Horii, H. Ishitsuka, K. Kusuhara, and Y. Sugiyama, *Pharm. Res.*, **18**, 1190–1202 (2001).

222. D. E. Hathway, *Molecular Aspects of Toxicity*, The Royal Society of Chemistry, London, 1984.

223. M. T. D. Cronin, *Farmaco.*, **56**, 149–151 (2001).

224. E. J. Ariëns, *Drug Metabol. Rev.*, **15**, 425–504 (1984).

225. P. J. van Bladeren, *Trends Pharmacol. Sci.*, **9**, 295–299 (1988).

226. N. R. Pumford and N. C. Halmes, *Ann. Rev. Pharmacol. Toxicol.*, **37**, 91–117 (1997).

227. B. Testa and J. Mayer, *Farmaco.*, **53**, 287–291 (1998).

228. B. Testa, *Pharm. News*, **3**, 10–12 (1996).

CHAPTER FOURTEEN

Metabolic Considerations in Prodrug Design

Luc P. Balant
Clinical Research Unit
Department of Psychiatry
University Hospitals of Geneva
Geneva, Switzerland

Contents

Burger's Medicinal Chemistry and Drug Discovery
Sixth Edition, Volume 2: Drug Development
Edited by Donald J. Abraham
ISBN 0-471-37028-2   © 2003 John Wiley & Sons, Inc.

1 INTRODUCTION

The term *prodrug* was first introduced by Albert in 1958 (1) to describe compounds that undergo biotransformation before exhibiting their pharmacological effects. Since then, many papers have been published on this subject (2–8). Basically, prodrug design constitutes an area of drug research that is concerned with the optimization of drug delivery. This may be "systemic delivery" after oral or topical administration or "delivery to the site of action," independently of the route of administration. The rationale for prodrug design is that a molecule with optimal structural configuration and physicochemical properties for eliciting the desired pharmacological action and the expected therapeutic effect does not necessarily possess the best molecular form and properties for its delivery at the receptor sites. By attachment of a pro-moiety to the active moiety, a prodrug is formed that is designed to overcome the barrier that hinders the optimal use of the active principle (9). Usually, the use of the term prodrug implies a covalent link between an "active moiety" and a "carrier moiety," but some authors also use this term to characterize some form of salts of the active principle. In this review, only covalently bound moieties are considered.

Prodrugs also occur in the organism. As an example of an endogenous prodrug, proinsulin is synthesized in the pancreas to be released as its active moiety insulin and an inactive propeptide. To some extent, the bioactivation of neurotransmitters could also be termed prodrug design by nature and there are many other examples in the realm of mammalian endogenous compounds. Prodrugs derived from plant sources also exist. As an example, codeine activation to morphine is essential for its analgesic effect (Section 5.2).

Finally, prodrugs were first synthesized before the concept was introduced by Albert. As an example, heroin is to some extent an "accidental" prodrug. As a matter of fact, morphine crosses the blood–brain barrier relatively slowly. Diacetylmorphine (heroin) has a greater lipid solubility, and thus transfers into the brain, where hydrolysis converts heroin to morphine. The latter is ultimately responsible for the pharmacological effect (10). This is a good example of a prodrug that has found unintended use outside the field of medicine, although it has also been used to treat chronic pain.

The present review concentrates on prodrugs designed by pharmaceutical chemists to overcome biopharmaceutical or pharmacokinetic (essentially metabolic) problems encountered with active principles. Although this review does not consider the chemical methods used to synthesize prodrugs, the reader more specifically interested in chemical synthesis can find relevant bibliography in references presented in the following sections. The investigations presented in this review have been selected not because they lead to a commercialized drug product, but to illustrate the types of problems pharmaceutical chemists have tried to solve and the metabolic, kinetic, or biopharmaceutical rationale behind these attempts. Although very little published information is available on pharmacokinetic methods useful in the study of prodrugs and on the regulatory requirements needed for prodrug development, the authors have tried to develop some general ideas that might be useful in this regard.

1.1 The Pharmacokinetic Point of View

A prerequisite for the design of safe drugs is knowledge about the various metabolic reactions that xenobiotics and endogenous compounds undergo in the organism. Because pharmacological activity depends on molecular structure, the medicinal chemist is restricted in the choice of functional groups for the design of new drugs. Often he finds or she encounters a situation where a structure has adequate pharmacologic activity but has an inadequate pharmacokinetic profile (i.e., absorption, distribution, metabolism, and excretion). This is because the compounds synthesized by the medicinal chemist are usually screened on the basis of *in vitro* testing and because pharmacology and pharmacokinetic departments in the pharmaceutical industry often do not collaborate at the early stage of drug development. It is only later, when the new compound is tested in animals or in humans, that pharmacokinetic disadvantages become obvious. In many cases the compound is discarded, but in some instances biopharmaceutics can help solve the problem if drug behavior in the gastrointestinal tract is involved or if the elimination half-life is too short. In the first case, innovative pharmaceutical formulations or new routes of administration can be devised. In the second case slow-release formulations may help overcome the problem.

More difficult is the situation where a drug should be available as a solution but its solubility is too low to allow easy parenteral or ocular administration. In such cases, the prodrug approach may be appropriate, as discussed briefly below (Section 4.1). The same applies to the improvement in the half-life of unstable parenterals (Section 4.2). Similarly, gastrointestinal membrane or transdermal passage as well as hepatic first-pass metabolism are difficult to change by biopharmaceutical approaches, as being inherent in the molecular structure of the active moiety. To improve the pharmacokinetic properties of a molecule displaying such problems, one must modify its chemical structure. This can be done by the design of a prodrug or by the synthesis of an analog structure with improved pharmacokinetic properties. The latter case falls outside the scope of the present review. Accordingly, the following sections are mainly concerned with prodrugs.

Even if, according to the accepted definition, the couple "prodrug-and-its-active-derivative" should not be considered as the pair "parent-compound-and-metabolite," from a pharmacokinetic point of view the two entities are strictly identical. Accordingly, the equations used for the determination of the kinetics of metabolites (active or not) are to be used (11–14). As for metabolite kinetics, a clear appreciation of the pharmacokinetics of a substance generated *in vivo* is possible only if the fraction of metabolite formed from its precursor is known. In most cases, prodrug design tends to produce substances that are totally bioactivated by one chemical route. This is, however, not necessarily the case and adequate provision must be made for such deviations from the ideal case when planning and interpreting the results in pharmacokinetic and relative bioavailability investigations.

1.2 The Biopharmaceutical Point of View

In the present chapter, the adjective *biopharmaceutical* is used in the context of biopharmacy, that is, for matters related to pharmaceutical formulations. Accordingly, the meaning is fundamentally different from the substantive *biopharmaceutical(s)* used to denote active substances developed by biotechnology. This substantive is not used in this chapter.

From a biopharmaceutical point of view, prodrugs can be defined as precursors of active principles that can be used to modify a variety of both pharmaceutical and biological properties including, as discussed above, modification of the pharmacokinetics of the drug *in vivo* to improve absorption, distribution, metabolism, and excretion (ADME). This may lead to improvement of bioavailability by increased aqueous solubility, increase of drug product stability, enhancement of patient acceptance, and compliance by minimizing taste and odor problems, elimination of pain on injection, and decrease of gastrointestinal irritation (6). It is usually accepted that a prodrug should not possess any relevant pharmacological activity. In this context the term *prodrug* is used essentially when chemical modifications are deliberately introduced to improve an unsatisfactory situation encountered when the active compounds are administered per se. The classical example is the synthesis in 1899 of aspirin, in an attempt to improve therapy with salicylic acid.

The present review deals essentially with enhancement of gastrointestinal availability of drugs through prodrug administration, but some other aspects of the prodrug concept are also discussed briefly.

1.3   The Regulatory Point of View

Presently there is no clear opinion about the requirements for the development of a prodrug. In general, regulatory agencies are reluctant to register this type of product. Of particular concern is the fact that toxicological studies might not be relevant for human use of the drug because of differences in the rate and/or extent of formation of the active moiety.

This problem was raised by Aungst et al. (15) in 1987, concerning the extrapolation of data from animal models to humans. Discussing the results of their pharmacokinetic studies of nalbuphine prodrugs, a narcotic agonist/antagonist, they stated: "Animal models should be similar to man with regard to: (*i*) nalbuphine disposition (bioavailability) after oral dosing; and (*ii*) prodrug hydrolysis rates." It is evident that because of great interspecies variability in xenobiotic metabolism (well known from a toxicological point of view), this goal is quite difficult to achieve. Accordingly, animal experiments should be analyzed with caution and confirmatory experiments in humans must be performed. It could even be argued that for some active principles it would be wiser to conduct all experiments in humans. As an example of interspecies differences, it was found (16) that the pivaloyloxyethyl ester of methyldopa was essentially hydrolyzed presystematically to pivalic acid and methyldopa, whereas the succinimidoethyl derivative was readily hydrolyzed in the rat, but not in the dog. Similarly, it was found that for dyphylline prodrugs, the relative rates

of active moiety release were 1.3–13 times faster in rabbit plasma than in human plasma (17).

Nomenclature is also a subject of confusion, although of much lesser importance than the metabolic aspects discussed in the previous paragraph. As examples of potential confusion some terms used in Europe are briefly discussed. A similar confrontation of definitions could also be made for other countries. The nomenclature adopted in 1965 (18) by the European Communities presents no problem for prodrugs:

- *Proprietary medicinal product*: Any ready-prepared medicinal product placed on the market under a special name and in a special pack.
- *Medicinal product*: Any substance or combination of substances presented for treating or preventing disease in human beings or animals.
- *Substance*: Any matter irrespective of origin which may be human, animal, vegetable, chemical.

In its Notice on "Studies of Prolonged-Action Forms in Man" dated 1987 (19) the EC introduces the term of *active principle* presented in "pharmaceutical forms." This concept may apply to prodrugs, but only indirectly because the substance contained as such in the pharmaceutical form is not exactly the active principle. Similarly, in the Council recommendation of 1987 on Investigation on Bioavailability (20), the terms *active drug ingredient* or *therapeutic moiety* of a drug are used to define the chemical species that should be measured. Finally, in a newer definition of bioavailability, the term *active substance* is used (21).

In most situations, the terminology can be adapted to the characterization of the pharmacokinetics of prodrugs. Some authors use the terminology of a "prodrug" being bioactivated to the "drug" or even to the "parent drug," which is a rather misleading nomenclature. To avoid confusion with the term *drug product*, often used to describe the "pharmaceutical form," the terms *prodrug* and *active moiety* are generally used in the present review. There are, however, situations in which great care must be taken to analyze the potential consequences of imprecise definitions of these concepts. Section 8.4, on bioavailability assessment of prodrugs, is a good example of such potential problems.

Finally, there is one type of prodrug that may create potential problems at the regulatory level and it is represented by all prodrugs aiming at a prolongation of the duration of action of an active substance. As an example, it is not totally clear from the available guidelines how a prodrug administered intravenously as a solution with prolonged-release properties should be compared to the "drug" given by the same route of administration. In these cases, it is hoped that scientific common sense will prevail over strict adherence to guidelines that were not written with such particular cases in mind.

2 CHEMICAL BOND

Prodrug design and synthesis is an important field of pharmaceutical chemistry. This review, however, is centered around biopharmaceutical aspects of prodrugs. From a pharmacokinetic point of view it is important to understand the nature of the chemical bond linking the active moiety to its "carrier moiety," and the nature of the "carrier moiety." Knowledge of the nature of the chemical bond may help to explain the nature of the biotransformation process and its location in specific tissues or cells. The study of the fate in the body of the "carrier" moiety is particularly important from a safety point of view and should be investigated just as thoroughly as the active moiety. Clearly, in some cases, such as the esters of methanol or ethanol, the fate of the released carrier moiety is well known, and no extra study is needed during drug development. In other cases, additional pharmacokinetic investigations may be necessary.

A basic requirement for prodrug design is naturally the adequate reconversion of the prodrug to the active moiety in vivo. This prodrug–drug conversion may take place before absorption (e.g., in the gastrointestinal system), during absorption (e.g., in the gastrointestinal wall or in the skin), after absorption,

or at the specific site of drug action. Because the prodrug is usually inactive, it is important that the conversion be essentially complete because intact prodrug represents unavailable drug. However, the rate of conversion depends on the specific goal of prodrug design. As two opposite examples, a prodrug designed to overcome low solubility for an intravenous formulation should be converted very rapidly to the active moiety after injection. Conversely, if the objective of the prodrug is to produce a sustained drug action through rate-limiting conversion, the rate of conversion should not be too fast. Depot neuroleptics designed for once-a-month intramuscular administration are an interesting case. As discussed in Section 3.3, the rate-limiting step is their diffusion from the oily depot to the bloodstream, the conversion in plasma to the active moiety being very fast, as for most ester prodrugs.

This clearly indicates that the nature of the bond between the carrier and the active moiety plays a major role in prodrug design and that pharmacokinetic considerations are of utmost importance in this context. In the following paragraphs, some typical kinds of chemical bonds are discussed. It is not intended here to present an overview on this subject, but only to give some examples of how chemistry and pharmacokinetics interact in the field of prodrugs.

For a more detailed review of xenobiotic conversion in the body, see the chapter by B. Testa in the present volume (22) and a review by the late H. Bundgaard (9), who had been so active in this field and whose work inspired parts of this chapter.

In any case, the necessary conversion or activation of prodrugs in the body can take place by a variety of chemical or enzymatic reactions that, as stated before, must be selected to achieve the optimal conversion site and rate. There are, however, some limitations to this choice, given that the active moiety must have adequate chemically functional groups for the attachment of the carrier moiety. The most common prodrugs are those requiring a hydrolytic cleavage mediated by enzymatic catalysis, but reductive and oxidative reactions have also been used for the in vivo regeneration of the active moiety. Besides usage of the various enzyme systems of the body to carry out the necessary activation of pro-drugs, the buffered and relatively constant value of the physiological pH may be useful in triggering the release of a drug from a pro-drug. In these cases, the prodrugs are charac-terized by a high degree of chemical lability at pH 7.4, while preferably exhibiting a higher stability at other pHs.

2.1 Esters

As stated earlier, the rational design of biolog-ically reversible drug derivatives is based on the ability of the host tissue to regenerate the active moiety. This is frequently accomplished through the mediation of an enzyme system. Enzymes considered important to orally ad-ministered prodrugs are found in the gut wall, liver, and blood. In addition, enzyme systems present in the gut microflora may be impor-tant in metabolizing the prodrug before it reaches the intestinal cells (23). Because of the wide variety of esterases present in the target tissues for oral prodrug regeneration, it is not surprising that esters are the most numerous prodrugs designed when gastrointestinal ab-sorption is considered. The consequence, how-ever, of this multiplicity of esterases is that prodrug stability or instability must be tested not only in plasma but also in the presence of gut, liver, or brain esterases, as shown, for ex-ample, in a study of a series of prodrug esters including various acylated acetaminophen and acetylaminobenzoate compounds (24).

By appropriate esterification of molecules containing a hydroxyl or carboxyl group, it is possible to obtain derivatives with almost any desirable hydro- or lipophilicity as well as *in vivo* lability. Some aliphatic or aromatic esters are not sufficiently labile *in vivo* to ensure an adequate rate and extent of prodrug conver-sion. For example, simple alkyl or aryl esters of penicillins are not hydrolyzed to the active free penicillin acid *in vivo* and therefore have no therapeutic potential (22). The reason is to be found in the highly sterically hindered en-vironment about the carboxyl group in the penicillin molecule, which makes enzymatic attack very difficult. This shortcoming can be overcome by preparing a double ester type, in which the terminal ester grouping is less steri-cally hindered. The first step in the hydrolysis of such an ester is enzymatic cleavage of the

terminal ester bond with formation of a highly unstable hydroxymethyl ester, which rapidly dissociates to the acidic drug. This approach has been successfully used in pivampicillin (**1**)

(**1**)

to improve the oral bioavailability of ampicil-lin (Section 3.2). Other approaches used to overcome this problem are based on the use of carrier moieties that give highly labile esters. For example, glycolamide esters have been synthesized in this context (25).

Chloramphenicol prepared as its palmitate ester is an interesting study example with re-spect to polymorphism. This ester, synthe-sized to avoid the bitterness of the active moi-ety (see below), exists as four polymorphs, three crystalline forms and an amorphous one. Polymorphs A and B have been found in commercial preparations but only form B leads to satisfactory blood levels. This effect was first attributed to the lower solubility of the "inactive" form A (26), but it was later proved that the higher susceptibility of form B to esterases was, in fact, involved (27).

In some cases such as that of terbutaline (28), it was observed that improved absorption of the diester ibuterol was accompanied by shorter effect duration and a slightly in-creased extent of first-pass metabolism. Other prodrugs were thus synthesized in an attempt to improve resistance to first-pass hydrolysis. Preliminary results with bambuterol, (**2**), an *N*-isostere of ibuterol, displayed improved hy-drolytic stability. This might result from the introduction of the nitrogen atom, which re-duced the reactivity of the acyl function, and/or dimethylcarbamate being a cholinest-erase inhibitor, and thus bambuterol hydroly-sis could be partially inhibited because of re-

$$(CH_3)_2NCOO$$

$$
\begin{array}{c}
\text{OH} \\
\mid \\
\text{CHCH}_2\text{NHC(CH}_3)_3
\end{array}
$$

$$(CH_3)_2NCOO$$

(2)

versible inhibition of the esterases responsible for its own degradation.

2.2 Mannich Bases

Mannich base prodrugs are said to enhance the delivery of their parent drugs through the skin because of enhanced water solubility and enhanced lipid solubility. Those more polar prodrugs are also more effective in improving topical delivery than prodrugs that have been designed to incorporate only lipid-solubilizing groups into the structure of the parent drug (29). Mannich base prodrugs are regenerated by chemical hydrolysis (Section 2.8) without enzymatic catalysis (30). Various other chemical approaches can be used to achieve increased skin permeability by enhancing both water solubility and lipid solubility (31).

2.3 Macromolecular Prodrugs

Proteins (such as antibodies and lipoproteins), liposomes, synthetic polymers, or polysaccharides (such as dextran and inulin) are various types of macromolecules used as drug delivery systems. Polymers are used extensively in these systems, including nanoparticles, microcapsules, laminates, matrices, and microporous powders. In all these delivery systems, the drug is merely dispersed or incorporated into the system without the formation of a covalent bond between the drug and the polymer. The present chapter discusses only those polymeric drugs in which an active moiety is covalently bound to a polymeric backbone. As a complement, recent reviews of drug–polymer conjugates as a potential for improved chemotherapy are available (32, 33) and contain extensive bibliography. It must also be mentioned that some pharmaceutical chemists have combined a prodrug with another drug delivery system, such as liposomes or

polymer conjugates, as exemplified for 5-fluoro-2′-deoxyuridine (34).

Although the terms *macromolecular prodrug* or *polymeric prodrug systems* are relatively recent, the interest in the temporary covalent binding of pharmacologically active compounds to macromolecules for sustained release has been recognized for many years as a potential method for controlled drug delivery and has received increasing attention (35).

As discussed below, as early as 1975, Ringsdorf (36) drew attention to the need for tailor-made macromolecular prodrugs with a view to attaching a homing device or directing units, solubilizing units, and using a biostable or biodegradable backbone. Besides monoclonal antibodies used in conjunction with anticancer drugs, other macromolecules such as polyvinylic or polyacrylic, polysaccharidic, and poly(α-amino acid) backbones have been mostly used. Theoretically, the main advantage of synthetic polymers over naturally occurring macromolecules is that they can be tailored to meet individual requirements.

As typical examples, one may mention dextran, soluble starch, or hydroxyethyl starch-based ester prodrugs of naproxen (37, 38) or dextran-based prodrugs of mitomycin C (39, 40). Other polysaccharides have been tested for their usefulness as transport groups for therapeutic agents, such as starch for acetylsalicylic acid (41, 42), inulin for procainamide (43, 44), agarose for mitomycin (45) or for adriamycin (46), soluble starch for nicotinic acid (47) or for salicylate (48), cellulose for insulin (49), and hydroxypropyl cellulose for estrone and testosterone (40, 51). However, clinical experience with such prodrugs is still too limited for an adequate judgment on this aspect of macromolecular prodrugs.

Many other polymers have been tested as backbones for prodrugs. As examples, one may list poly-L-(glutamic acid) for clonidine controlled release (52), hyaluronic acid for hydrocortisone and derivatives (53–55), or the synthetic polyanionic polymer pyran for the antitumor, antibacterial, and antiviral agent muramyl dipeptide (56). Hyaluronic acid and pyran are interesting in the sense that they add a new dimension to the macromolecular prodrug approach. Hyaluronic acid is believed to be potentially useful for the control of infection or inflammation in the eye after surgery and to promote wound healing; its use as a prodrug backbone could also transform it into an active ingredient in some indications (56). Similarly, pyran has an effect on the immune system and it has been postulated that, combined with specific active principles, it could also play an active role in the effects of the drug (55). In these two cases, the prodrug would give rise to two active moieties, with possibly complementary therapeutic activities. Other polymers have also been tested, as exemplified by fluorouracil derivatives of organosilicon (57) or polyacrylates (58).

Recently, drug–antibody conjugates have become realities in the context of drug transport, drug targeting, and cancer therapy. An aspect of macromolecular prodrug synthesis that still seems to present a technical problem is the development of selective antitumor drug–protein conjugates with a covalent and reversible linkage for stability in biological fluids, together with adequate release at the site of action. As a matter of fact, a major problem is the rapid clearance of such complexes from the bloodstream because they are frequently recognized as foreign materials by cells of the reticuloendothelial system.

From a theoretical point of view, Ringsdorf (36) proposed a model for macromolecular prodrugs. The backbone must contain three essential units:

1. A device for controlling the physicochemical properties of the entire macromolecule, which mainly controls the hydrophilic–lipophilic balance, the electronic charge, and the solubility of the system.
2. The active moiety, which must be covalently bound to the polymer and must remain attached to it until the macromolecule reaches the desired site of action. The active molecule must be detached from the parent polymer at the site of action, the release taking place by hydrolysis or by specific enzymatic cleavage of the drug–polymer bond. In many cases, the active moiety is attached to the polymer through a spacer molecule, which is an amino acid or other simple molecule. The choice of the spacer molecule is of crucial impor-

tance for the generation of adequate releasing characteristics.
3. The "homing device," which should guide the entire drug–polymer conjugate to the target tissue. Presently, antibodies are mainly used for this purpose, although other approaches are possible. For example, given that a variety of cell systems are known to possess cell surface lectins with well-defined sugar specificity, glucosylated polymers could be of interest as potential cell-specific homing devices.

The choice of an active moiety for use in this type of systems is based on three criteria (59):

1. Only potent substances can be used because there is a restriction on the amount of drug that can be administered.
2. The active moiety must have a functional group by which it can bind with the polymer backbone directly or by means of a spacer molecule.
3. The active substance must be sufficiently stable and should not be excreted in its conjugate form until it is released at the desired site.

As can be seen from this list of criteria, macromolecular prodrugs represent a true challenge for the pharmaceutical chemist, particularly because very little is known about the behavior of xenobiotic macromolecules in the human body. Not only are the pharmacokinetic properties of such substances not well investigated, but their potential to provoke immunological reactions represents an additional difficulty. Such properties depend not only on the type of polymer but also on its size. This introduces an additional pharmacokinetic variable. As an example, small peptides tend to be easily taken up by the hepatocytes; larger molecules are filtered by the glomerulus and taken up by the tubular cells where they are catabolyzed; and still larger molecules, although filtered, are not taken up in the tubules, whereas moderately large molecules are not filtered at all. As a consequence, it is probable that much more basic research on the kinetics of these compounds is needed before the pharmaceutical chemists can select on scientifically sound bases the most adequate "backbone" molecule.

2.4 Prodrug Derivatives of Peptides

A major obstacle to the application of peptides as clinically useful drugs is their poor delivery characteristics. Most peptides are rapidly metabolized by proteolysis at most routes of administration and they possess short biological half-lives because of rapid metabolism (60, 61). In addition, they are generally nonlipophilic compounds that show poor water solubility and biomembrane penetration characteristics, which lead to poor absorption and low availability at their potential site of action. A possible approach to solve these delivery problems may be derivatization of the bioactive peptide to produce prodrugs or transport forms that possess enhanced physicochemical properties (9). Thus derivatization may, on one hand, protect small peptides against degradation by enzymes present at the mucosal barrier and, on the other hand, render hydrophilic peptides more lipophilic and hence facilitate their absorption (62). In recent years, significant progress has been made in developing prodrug approaches for improvement of water solubility, stability, and membrane permeability of peptides (63). For improving water solubility, the focus has been on the bioreversible introduction of ionizable functional groups to peptides, which helps to increase the polarity and thus water solubility of the peptide drugs. For improving stability, efforts have focused on stabilizing peptides against exopeptidase-mediated hydrolysis by bioreversibly masking the terminal carboxyl and/or amino groups. For improving permeability through biological barriers, recent efforts have focused on both improving peptide lipophilicity, to facilitate its passive permeation through biological membranes, and conjugation of a peptide to a carrier that allows for the active transport of the peptide–carrier conjugate.

Approaches have also been developed for improving peptide drug delivery to the central nervous system (64). As a matter of fact, the microvasculature of the central nervous system (CNS) is characterized by tight junctions between the endothelial cells and, thus, behaves as a continuous lipid bilayer that prevents the passage of polar and lipid-insoluble substances such as peptides. Highly active enzymes expressed in the morphological components of the microcirculation also represent a metabolic component that contributes to the homeostatic balance of the CNS. Peptides generally cannot enter the brain and spinal cord from the circulating blood because they are highly polar, lipid insoluble, and metabolically unstable. Active transport systems exist for only very few of them in the membranous barrier separating the systemic circulation from the interstitial fluid of the CNS. This blood–brain barrier is, therefore, the major obstacle to peptide-based drugs that are potentially useful against diseases affecting the brain and spinal cord. Chemical-enzymatic (prodrug and chemical delivery/targeting system) and biological carrier–based approaches have been developed to overcome the blood–brain barrier for these highly active and versatile molecules that are very attractive as a future generation of neuropharmaceuticals.

Such an approach has, for example, been proposed for thyrotropin-releasing hormone (THR, 3), a hypothalamic tripeptide that reg-

(3)

ulates the synthesis and secretion of thyrotropin from the anterior pituitary gland. THR is a potential drug for the management of neurological disorders. Its clinical use is greatly hampered by its rapid metabolism, leading, after parenteral administration, to a half-life of a few minutes. Moreover, its very low lipophilicity limits its ability to penetrate the blood–brain barrier. The prodrug approach has been tested and found potentially useful to improve the pharmacokinetic characteristics of THR (65, 66), particularly in protecting it against cleavage by carboxypeptidase A. The same approach has been used to protect peptides against degradation by another pancreatic proteolytic enzyme, α-chymotrypsin (67). In the same line of research, Gln-Leu-Pro-Gly, a progenitor sequence for the thyrotropin-releasing hormone (TRH) analog [Leu(2)]TRH [pGlu-Leu-Pro-NH(2)], was covalently and bioreversibly modified on its N- and C-termini to create a lipoidal brain-targeting system for the TRH analog (68). The mechanism of targeting and the recovery of the parent peptide at the target site involve several enzymatic steps. Because of the lipid insolubility of the peptide pyridinium conjugate obtained after one of these reactions, one of the rudimentary steps of brain targeting (i.e., trapping in the CNS) can be accomplished. The design also includes spacer amino acid(s) to facilitate the posttargeting removal of the attached modification.

The release of the TRH analog in the brain is orchestrated by a sequential metabolism through the use of esterase/lipase, peptidyl glycine alpha-amidating monooxygenase (PAM), peptidase cleavage, and glutaminyl cyclase. In addition to *in vitro* experiments to prove the designed mechanism of action, the efficacy of brain targeting for [Leu(2)]TRH administered in the form of chemical-targeting systems containing the embedded progenitor sequence was monitored by the antagonistic effect of the peptide on the barbiturate-induced anesthesia (measure of the activational effect on cholinergic neurons) in mice, and considerable improvement was achieved over the efficacy of the parent peptide upon using this paradigm. Other precursors have also been synthesized and successfully tested in animals (69).

Another area of considerable interest is the use of the prodrug concept for better kinetic properties of angiotensin-converting enzyme inhibitors (Section 3.2) or fibrinolytic enzymes (70).

2.5   Peptide Esters of Drugs

In the preceding section prodrugs of peptides were considered. They should not be confused with peptide esters of drugs. In the latter case, one forms α-amino acid or related short-chained aliphatic amino esters, for example, as a useful means of increasing the aqueous solubility of drugs containing a hydroxyl group

[e.g., with the aim of developing improved preparations for parenteral administration (71, 72)]. Ideally, such prodrugs should possess high water solubility at the pH of optimum stability and sufficient stability in aqueous solution to allow long-term storage (>2 years) of ready-to-use solutions and yet they should be converted quantitatively and rapidly *in vivo* to the active moiety. Considering these desirable properties of the prodrugs, the use of α-amino acids or related esters is not without problems. Although they are generally readily hydrolyzed by plasma enzymes (73, 74), they exhibit poor stability in aqueous solutions, as exemplified with esters of metronidazole (74, 75), acyclovir (76), corticosteroids (77, 78), and paracetamol (79, 80). One of the solutions to overcome such problems is the use of adequately selected spacer groups (Section 4.1).

Valaciclovir (81) and famciclovir (82, 83) are prodrugs, respectively, of the anti-herpesvirus drugs acyclovir (or aciclovir) and penciclovir. Valaciclovir is the L-valyl ester of the active antiviral component acyclovir, which is rapidly formed after oral administration, together with the essential amino acid L-valine. The systemic availability of the prodrug is three to five times greater than that of the active moiety (84, 85). Because acyclovir is often administered by the intravenous route for the treatment and suppression of the less susceptible herpes viral diseases (84), it is hoped that valaciclovir may ultimately succeed acyclovir as a first-line treatment for genital herpes or herpes zoster. Famciclovir is not an amino acid–type prodrug, but the dipropionyl prodrug of the *in vivo* penultimate metabolite 6-deoxypenciclovir (86). Famciclovir has been developed because of the limited oral absorption of penciclovir, which has not been commercialized as such for this reason (87). Famciclovir displays a reasonable systemic availability after oral administration (88).

2.6   Amine Prodrugs

The presence of a primary amine group in a drug can affect its physicochemical and biological properties in different ways (89). For example, drugs containing a primary amine group can undergo intra- or intermolecular aminolysis reactions, leading to reactive and/or potentially toxic products. When the primary amine group is present in a molecule with another ionizable functionality, like a carboxylic acid group, the molecule can display poor aqueous solubility and liposolubility attributed to the zwitterionic nature of the molecule in the physiological pH range, thereby potentially limiting its dissolution rate and/or its passive permeability. This is exemplified by the fact that after oral administration, the systemic availability of many peptides is low, in part also because of their enzymatic lability (see Section 2.4).

In addition, the terminal free amino acid groups are recognition sites for proteolytic enzymes like aminopeptidase and trypsin present in the gastrointestinal tract lumen, the brush border region, and the cytosol of the intestinal mucosa cells. For all these reasons the prodrug approach has been advocated (89, 90) for the improvement of the *in vivo* behavior of active principles containing primary amine groups. However, many attempts to impart "ester characteristics" (see Section 2.1) to amines have met with limited success and other approaches such as the synthesis of pro-prodrugs for amide prodrugs have been described (90). The pro-prodrug, for example, is designed to be stable chemically and is biologically converted to a prodrug. The latter is then chemically activated to the active amine group–containing substance. If the rate of the chemical reaction is sufficiently rapid, the biological reaction will become the rate-determining process in the overall activation. A derivative with these qualities has been said to possess an "enzymatic trigger" (91).

2.7 Lipidic Peptides

Lipidic prodrugs, also called drug–lipid conjugates, have the drug covalently bound to a lipid moiety, such as a fatty acid, a diglyceride, or a phosphoglyceride (92). Drug–lipid conjugates have been prepared to take advantage of the metabolic pathways of lipid biochemistry, allowing organs to be targeted or delivery problems to be overcome. Endogenous proteins taking up fatty acids from the bloodstream can be targeted to deliver the drug to the heart or liver. For glycerides, the major advantage is the modification of the pharmacokinetic behavior of the drug. In this case,

one or two fatty acids of a triglyceride are replaced by a carboxylic drug. Lipid conjugates exhibit some physicochemical and absorption characteristics similar to those of natural lipids. Nonsteroidal, anti-inflammatory drugs such as acetylsalicylic acid, indomethacin, naproxen, and ibuprofen were linked covalently to glycerides to reduce their ulcerogenicity. Mimicking the absorption process of dietary fats, lipid conjugates have also been used to target the lymphatic route (e.g., L-dopa, melphalan, chlorambucil, and GABA). Based on their lipophilicity and resemblance to lipids in biological membranes, lipid conjugates of phenytoin were prepared to increase intestinal absorption, whereas glycerides or modified glycerides of L-dopa, glycine, GABA, thiorphan, and *N*-benzyloxycarbonylglycine were designed to promote brain penetration. In phospholipid conjugates, antiviral and antineoplasic nucleosides were attached to the phosphate moiety. The long alkyl side-chains may also have the additional effect of protecting a labile parent drug from enzymatic attack, thereby enhancing metabolic stability. Lipidic prodrugs display advantages and drawbacks; in particular, one must mention the potential pharmacological activity of the fatty acid itself.

The α-amino acids with long hydrocarbon side-chains, the so-called lipidic amino acids and their homo-oligomers, the lipidic peptides, represent a class of compounds that combine structural features of lipids with those of amino acids. Several uses of lipidic amino acids and peptides have been proposed. Of particular interest is their potential use as a drug delivery system (93). The lipidic amino acids and peptides could be covalently conjugated to or incorporated into poorly absorbed peptides and drugs, to enhance the passage of the pharmacologically active compounds across biological membranes. Because of their bifunctional nature, the lipidic amino acids and peptides have the capacity to be chemically conjugated to drugs with a wide variety of functional groups. The linkage between drug and lipidic unit may either be biologically stable (i.e., a new drug is formed) or possess biological or chemical instability (i.e., the conjugate is a prodrug). In either case, the resulting conjugates would be expected to possess a high degree of membranelike character, which may be sufficient to facilitate their passage across membranes.

2.8 Chemical Hydrolysis

A serious drawback of prodrugs requiring chemical (i.e., nonenzymatic) release of the active moiety is the inherent lability of these compounds, thereby raising some stability–formulation problems, at least in cases of solution preparations. Such problems may sometimes be overcome by using an approach involving pro-prodrugs or double prodrugs, where use is made of an enzymatic-release mechanism before the spontaneous chemical reaction. It is then possible to overcome the stability problem of the prodrug.

3 GASTROINTESTINAL ABSORPTION

3.1 Improvement of Gastrointestinal Tolerance

The gastrointestinal lesions produced by the acid nonsteroidal anti-inflammatory (NSAI) agents are generally believed to be caused by two different mechanisms: a direct contact mechanism on the gastrointestinal mucosa and a generalized systemic action appearing after absorption, which can be demonstrated after intravenous administration. The relative importance of these mechanisms may vary from drug to drug. Acetylsalicylic acid and most newer NSAI drugs are carboxylic acids. Temporary masking of the acid function has been proposed as a promising means of reducing gastrointestinal toxicity resulting from the direct mucosal contact mechanism (94–101).

Several attempts have been made to develop bioreversible derivatives of acetylsalicylic acid (102–104). They nicely illustrate the potential problems encountered with prodrugs of prodrugs. A major difficulty in the design of aspirin prodrugs is the great enzymatic lability of the acetyl ester in aspirin, derivatized at its carboxyl group. Therefore, a prerequisite for any true aspirin prodrug is for the masking group to cleave faster than the acetyl ester moiety; otherwise, the derivatives will behave as prodrugs of salicylic acid.

3.2 Increase in Systemic Availability

Until recently, it seemed that the most fruitful area of derivatization was the improvement of passive drug absorption through epithelial tissue. Accordingly, an immense number of prodrugs featuring the addition of a hydrophobic group have been prepared to improve their gastrointestinal absorption. As early as 1975, Sinkula and Yalkowsky (23) listed a number of drug derivatives used as modifiers of absorption, and the list has steadily increased since then. It is in the field of poorly absorbed penicillin derivatives that the most successful prodrugs have been produced and commercialized. They include esters (Section 2.1) of ampicillin [e.g., bacampicillin, pivampicillin (**1**), and talampicillin], mecillinam (pivmecillinam), and carbenicillin (carfecillin: for urinary tract infections only). As an example, plasma concentrations of ampicillin attained with these esters are up to five times higher than those seen after oral ampicillin (105–107). Other antibiotics for which the prodrug approach has been tested include cefotiam (108) and erythromycin (109). Oseltamivir is also an ethyl ester prodrug and its active metabolite is a selective inhibitor of influenza virus neuraminidase (110). The systemic availability of the active metabolite is only about 5% and it is increased to 80% in its prodrug form. The biotransformation occurs through intestinal and hepatic esterases.

Furosemide (**4**) provides an interesting ex-

(4)

ample of the use of the prodrug concept for a substance presenting biopharmaceutical problems (111). This loop diuretic is only incompletely (40–60%) absorbed after oral administration (112) and, in addition, the systemic availability (both in terms of rate and extent of absorption) shows a high degree of inter- and intraindividual variability. Various possible reasons for the relatively low systemic availability have been considered such as acid-catalyzed degradation of the drug in the stomach, first-pass metabolism in the gut wall or in the liver, dissolution-limited absorption, and site-limited or site-specific absorption, although no firm explanation has been given (112). Data obtained in rats appear to indicate that the occurrence of a site-related absorption from the stomach or the upper part of the gastrointestinal tract is the most likely explanation for the incomplete and variable absorption pattern of furosemide. On the other hand, because of the short duration of the diuretic effect of conventional furosemide tablets or capsules, various slow-release preparations have been developed. A disadvantage of these preparations is, however, that their relative bioavailability is reduced to about 50% compared to that of conventional tablets. This is why the prodrug approach has been proposed (111) to improve the oral bioavailability characteristics of furosemide drug products.

Acyclovir (5) is an interesting example of

(5)

prodrug or pro-prodrug design for improved gastrointestinal absorption. This antiherpetic agent exhibits great selectivity in its antiviral action through conversion to the active triphosphorylated species by virtue of virus-specific thymidine kinase (113, 114). Acyclovir is thus a prodrug exhibiting site-specific conversion to the active moiety (Section 5.2); however, it suffers from poor oral systemic availability, with only 15% to 20% of an oral dose being absorbed in humans (115, 116). This can most likely be ascribed to the poor water solubility and low lipophilicity of the compound. 6-Deoxyacyclovir (desiclovir) has been found to be a promising prodrug with improved oral

absorption (117). The compound is 18 times more water soluble than acyclovir and is also more lipophilic. Its systemic availability is about 75% in healthy humans (118). From a formal point of view 6-deoxyacyclovir is a pro-prodrug (Section 5.4).

Prodrug design has also been successful in the area of angiotensin-converting enzyme (ACE) inhibitors, enalapril (6) being an ester

(6) R = C2H5
(7) R = H

prodrug of enalaprilat, (7), which has not been developed as such, but immediately commercialized as its prodrug. Enalaprilat binds tightly to the angiotensin-converting enzyme, yet is transported with low efficacy by the peptide carrier in the gastrointestinal tract. The prodrug enalapril has a higher apparent affinity for the carrier. This indicates that the reason for good oral absorption of enalapril is that it makes enalaprilat more peptidelike rather than more nonpolar (119). Dicarboxylic acid ACE inhibitors in development or commercialized based on esterification of the same carboxyl group include perindoprilat, ramiprilat, cilazaprilate, and benazeprilate.

Mixed triglycerides formed by coupling of drugs to diglycerides exhibit physicochemical properties (120) and absorption characteristics (121) similar to those of natural triglycerides, resulting in a different pharmacokinetic and/or pharmacodynamic profile compared to that of the unmodified drug. Such an approach has been used to improve the oral systemic availability of phenytoin (122).

The concept of bioreversible chemical modification has been proposed to diminish the gastrointestinal and/or hepatic first-pass metabolism of drugs with high extraction ratios. This approach has been tried for a number of drugs such as methyldopa (16, 123, 124), dopamine (125), etilefrine (126), L-dopa (127),

terbutaline (128), salicylamide (129), 5-flu-
orouracil (130–132), progesterone (133), β-es-
tradiol (134), naltrexone (135, 136), N-acetyl-
cysteine (137), and peptides (138).

For some prodrugs a specific problem may
arise if presystemic isomerization to an inactive compound occurs. As an example, ester
prodrugs of the delta-3-isomers of cephalosporins may isomerize to the microbiologically inactive delta-2-isomer (139), and it is important
for clinical effectiveness that, after reaching
the blood, hydrolysis of the ester group proceeds much more rapidly than isomerization
of the delta-3 double bond.

Another interesting approach is the use of
peptide carrier systems to improve intestinal
absorption (140). The concept is based on the
relatively recent finding that the intestinal
mucosal cell peptide transporter has a relatively broad specificity. These findings suggest
that this transporter could serve for polar prodrugs and analogs of di- and tripeptides. The
general scheme is that a polar drug with a low
membrane permeability is converted into a
prodrug that is transported by the peptide carrier into the mucosal cell. This prodrug may
still be very polar because what is required is
the correct structural features for the peptide
carrier–mediated transport. The prodrug
can still possess a high aqueous solubility in
the gastrointestinal lumen. After membrane
transport, the prodrug is subsequently hydrolyzed by a mucosal cell cytosolic enzyme.

3.3 Sustained-Release Prodrug Systems

Sustained-release products have traditionally
been one of the major areas of biopharmaceutical research. However, some sustained-release preparations available on the market for
many years are in fact derived from prodrug
design. The prodrug approach has been applied, for example, for depot neuroleptics. Esterification of the active substance with decanoic acid yields a very lipophilic prodrug
that is dissolved in Viscoleo. Intramuscular injection creates an oily depot from which the
prodrug slowly diffuses into the systemic circulation, where esterases quickly release the
active substances. These depot forms allow
the drug to be given only once or twice a
month. It is thus not a slow bioactivation of
the therapeutic moiety that is involved in the

present case, but a slow release from an oily
depot. The synthesis of the esterified precursor of neuroleptics permitted a new approach
to long-term treatment of schizophrenia.
Compounds available in depot prodrug formulations include fluphenazine (**8**), flupenthixol,

(**8**) R = H
(**9**) R = $C_9H_{19}CO^-$

and zuclopentixol (141, 142). It must, however, be stressed that prodrug design alone
was not sufficient for depot neuroleptics. A
specific solvent (a vegetable oil) and a specific
route of administration (intramuscular) had
to be combined with the prodrug concept because of the release of the active moiety from
the decanoate esters (**9**) is almost instantaneous when the prodrug reaches the systemic
circulation. These examples are excellent il-
lustrations of successful prodrugs that have
had a major impact on therapy by the modification of pharmacokinetic and biopharmaceutical characteristics of well-established
agents. Because of the lability of ester bonds in
biological fluids, ester prodrugs are usually
not designed (with the exceptions mentioned
above) for sustained-release systems, although such an approach has been tried with
zidovudine in an attempt to reduce side effects
in the treatment of AIDS (143).

Other slow-release systems have been devised on the basis of prodrug design. They rely
more specifically on the slow delivery of the
active moiety from a prodrug to which it is
covalently bound. As an example, a catechol
monoester of L-dopa has been shown, in rats,
to show longer duration of plasma L-dopa concentrations, as evaluated from "mean residence time" (144). Similarly, carbamate prodrugs have been used for a dopamine agonist
early in drug development (145). Classically,
sustained-release systems that use polymer
matrices rely on release of the drug by either
diffusion through the matrix, erosion of the

matrix, or a combination of both. The prodrug concept can also be applied for such systems, for example, by covalently binding the drug to a biodegradable polymer. This approach has been tested for naltrexone (146) and prazosin (147), but these prodrugs are not intended for oral administration. In these systems, two physicochemical processes may influence drug release. First, drug is released from the backbone polymer by hydrolysis (enzyme and/or acid–base catalyzed). Second, the free drug diffuses through the polymer matrix. If the rate of hydrolysis is lower than the rate of diffusion, it will govern the release rate. The fact that this rate-limiting process occurs at the boundary between unaffected nonswollen and previously swollen, degraded, or permeated material implies particular pharmaceutical designs (see also Section 2.3).

Another procedure that has been followed is synthesizing a monomer with the active substance, and then polymerizing this new molecule. Chloramphenicol has thus been attached to a methacrylic derivative by an acetal function, and then copolymerized with 2-hydroxyethyl-methacrylate (148). The copolymer can then, in addition, be dispersed into a biocompatible nondegradable polymer such as Eudragit, playing the role of a polymer matrix (149).

Ion-exchange systems have found many medicinal applications. As a controlled-release system, ion-exchange resins offer the benefit that release of a complexed drug is initiated by an influx of competing ions from the gastrointestinal tract. However, in general, release rates from unmodified resins are too great for adequate control of delivery and diffusional barriers to delay drug egress are employed in some preparations. Homologous series of O-n-acyl propranolol prodrugs have been used to study the influence of physicochemical properties of the drug on its release from resinates (150, 151).

3.4 Improvement of Taste

Although not directly related to the physiological process of absorption, it may happen that oral drugs with a markedly bitter taste may lead to poor compliance if administered as a solution, a syrup, or an elixir. The prodrug approach has been used in this context for chloramphenicol, clindamycin, or metronidazole (5). 3-Hydroxymorphinan opioid analgesics or antagonists are good examples of the potential of prodrug design because of their low oral bioavailability attributed to high first-pass metabolism and bitter taste (152). Oral administration is limited for most of these compounds and systemic availability can be substantially improved by sublingual dosing. However, discomfort ascribed to bitter taste is a common complaint from patients administered morphine or similar compounds buccally. Many prodrugs of these analgesics or their antagonists do not taste bitter and may represent an attractive alternative to injections when the oral route is not adequate because of low systemic availability.

3.5 Diminishing Gastrointestinal Absorption

Colon-specific drug delivery has potential in the treatment of a variety of colonic diseases such as colitis, colon cancer, radiation-induced colitis, and irritable bowel syndrome (153). The overall concept is to deliver an active ingredient as a prodrug having the ability to remain unabsorbed in the upper gastrointestinal tract and subsequently to release active substance, for example, by the action of bacterial glycosidases in the large intestine. Such an approach has been attempted for steroidal anti-inflammatory drugs such as dexamethasone (154, 155). This type of prodrug is interesting from a pharmacokinetic point of view, given that diminished systemic availability must be shown together with an increased local availability as compared to the active moiety administered per se. Thus specific colon delivery (156) and stability in various locations of the gastrointestinal tract (157) should be tested. Animal and *in vitro* models for such studies include "conventional," colitic, and germ-free rats (156, 157) or the guinea pig (158) as used for the study of dexamethasone β-D-glucosides or β-D-glucuronides. *In vitro* models are also used to investigate colonic membrane transfer, which include different intestinal preparations and monolayer cultures. These cell cultures can express a variety of the characteristic morphological, cytochemical, and transport features found in the intestinal enterocytes.

4 PARENTERAL ADMINISTRATION

Although it is not common to prepare pro-drugs for parenteral administration, some situations may occur in which this might be useful in increasing the amount of active moiety that will reach the systemic circulation.

4.1 Increased Aqueous Solubility

Examples in which increased "pharmaceutical availability" is important include glucocorticoids [such as dexamethasone (**10**) or methyl-

(**10**) R = H

(**11**) R = P—ONa (phosphate group with O and ONa)

prednisolone] used at high doses for emergency treatment of shock or other life-threatening situations. Because their water solubility is very low, these drugs are generally administered in the form of water-soluble esters as hemisuccinates and phosphates (**11**) (159–161). A similar approach has been used for allopurinol (162) and acyclovir (163), to make them suitable for parenteral or rectal administration as a solution. However, as simple as parenteral administration may seem compared to oral administration, the situation is often more complicated than expected. As an example, fetindomide is a potential pro-drug of mitindomide, an antitumor agent with poor solubility in water and in most pharmaceutically acceptable solvents (164). The pro-drug was designed so that mitindomide would be released *in vivo* by the loss of two molecules of phenylalanine and formaldehyde. It was observed that *in vitro* formaldehyde exerts a cat-

alytic effect on fetindomide hydrolysis (164). This study shows that the choice of a spacer group (in this case formaldehyde) is not always without important practical consequences. Another potential problem described for chloramphenicol sodium succinate is incomplete hydrolysis, leading to a "systemic availability" that is less than expected (165, 166).

Spacer groups also play an important role for the design of peptide esters with adequate solubility and stability (Section 2.5). Spacer groups are also important for other types of prodrugs designed to increase water solubility. Dextran-linked, water-soluble prodrug esters of metronidazole (167) are a good example of the importance of the spacer. The antitrichomal activity measured *in vitro* on *Trichomonas vaginalis* seemed to be correlated with the release rate of the active moiety, metronidazole, and with the hydrophobicity of the spacer. Metronidazole ester prodrugs have also been designed for formulating a parenteral solution to be administered by a single injection instead of repeated infusions. This form should be useful for rapid onset of action in the treatment of anaerobic bacteria (168).

4.2 Improvement in the Shelf Life of Parenterals

For a prodrug, the true utilization time has been defined (169) as the time during which the total concentration of prodrug and drug equals or exceeds 90% of the original prodrug concentration. It has consequently been argued that the storage of active ingredients in solutions as prodrugs might produce advantages (170). If it is assumed that a solution of the parent drug is useful until its concentration decreases to 90% of its initial value ($t_{90\%}$), a prodrug utilization time can be defined as the time during which administration of the prodrug provides a bioavailable dose of drug equal to or better than the parent drug at $t_{90\%}$ (61). Under optimal conditions, $t_{90\%}$ of the prodrug might be rather longer than $t_{90\%}$ for the active moiety stored as such. Caution must be used, however, because this concept is valid only if the prodrug degrades to the active principle and if the given definition is considered valid for the drug product under scrutiny.

A problem arises when a prodrug in solution prematurely converts to the drug through chemical hydrolysis during storage and simultaneously to an inactive product through a degradation pathway similar to that of the parent drug. For example, hydrolysis of ester prodrugs of penicillins may yield the corresponding penicillins while simultaneously forming inactive β-lactam degradation products.

5 DISTRIBUTION

5.1 Tissue Targeting

An interesting approach of specific drug delivery to the liver has been attempted with glutathione as a dextran conjugate for the potential treatment of hepatic poisoning (171). This prodrug was developed with the aim of overcoming both the poor hepatic uptake of glutathione and its rapid renal degradation into constituent amino acids. Similarly, the cholesterol-lowering agents lovostatin and simvastatin are prodrugs that show a liver-specific uptake and bioactivation after oral administration (172, 173).

Tissue targeting may also be directed at tumors through the use of monoclonal antibodies, as briefly mentioned earlier. Finally, brain targeting can also be attempted with the prodrug approach, as exemplified by estradiol by use of a redox-based chemical delivery system (174).

It is possible that for such types of drugs, new pharmacokinetic approaches would have to be derived to correctly describe their behavior in the body, particularly if one is interested in evaluating the degree of tissue targeting.

5.2 Activation at the Site of Action

Several examples of prodrugs are found in the purine and pyrimidine analogs that substitute for natural nucleotides and inhibit nucleic acid formation. For example, 5-fluorouracil is essentially harmless to mammalian host and tumor cells. Upon administration, the drug is subject to one of two opposing metabolic fates (10). Inactivation and elimination are accomplished by catabolism (about 80% of the dose) and by urinary excretion of unchanged drug

(about 5–20%). On the other hand, cytotoxicity in host and tumor cells occurs only after anabolism in actively proliferating cells. Because of a marked first-pass metabolism, oral administration leads to highly variable systemic availability and is thus an unsuitable and unreliable mode of therapy. It thus seems possible to improve oral delivery of this drug by use of the prodrug approach (130, 131). This has been attempted with 5'-deoxy-5-fluorouridine, a sugar derivative on the market for the oral route. It is a pro-prodrug if 5-fluorouracil is considered a prodrug. The pro-prodrug has a half-life of about 10–30 min in patients and is transformed mostly into 5-fluorouracil. It is possible that the therapeutic index of 5-fluorouracil might be increased by the prodrug approach. It seems that doxifluridine is selectively activated to 5-fluorouracil in sensitive tumor cells as opposed to bone-marrow cells (175). Should this be conclusively demonstrated, it would be an interesting example of increased target organ availability attributed to prodrug derivatization.

The kinetics of 5-fluorouracil and its metabolites are essentially nonlinear. Therefore it is extremely difficult to build models that would correctly describe the cascade of nonlinear transformations that are observed, starting from drug absorption to its transformation into the active moiety. More recently, capecitabine has been commercialized. It is a fluorpyrimidine carbamate available for oral administration. Concentrations of 5-fluorouracil in some tumors are higher than those in the adjacent healthy tissues. The tumor preferential activation of capecitabine to 5-fluorouracil is explained by tissue differences in the activity of cytidine deaminase and thymidine phosphorylase. It is interesting to note that the last of the three metabolic steps leading to 5-fluorouracil is the formation of 5'-deoxy-5-fluoridine. Capecitabine is thus a pro-prodrug (176–178).

The case of the antiherpes agent acyclovir was discussed previously (Section 3.2). This is also a prodrug with site-specific activation. A similar situation has been exploited in cancer chemotherapy because of the increased secretion of the plasminogen activator urokinase by various tumors and metastases as compared to normal cells. This may be used for the synthesis of urokinase-labile prodrugs of anticancer agents. In this case the transport or pro-moiety must be carefully selected so that the active moiety is specifically released by urokinase at an appropriate rate. In addition, the chemical properties of the active moiety per se must be such as to affect adequately the susceptibility of the carrier-active substance bond to be cleared by the enzyme (179) because if chemical properties of the active moiety are such as to render the prodrug resistant to urokinase action, the rate and extent of delivery in the target cell may then be inappropriate for adequate cancer therapy.

Omeprazole (**12**), a potent antiulcer agent,

(12)

is an effective inhibitor of gastric acid secretion, given that it is an inhibitor of the gastric H^+,K^+-ATPase (unlike acid secretion inhibitors such as cimetidine). The enzyme H^+,K^+-ATPase is responsible for gastric acid production and is located in the secretory membranes of the parietal cells. Omeprazole itself is not an active inhibitor of this enzyme, but is transformed within the acid compartments of the parietal cell into the active inhibitor, close to the enzyme.

Orally administered amines do not cross the blood–brain barrier, but neutral amino acids such as α-methyldopa are transported into the brain by a specific carrier system. α-Methyldopa is subsequently concentrated in neuronal cells, where it becomes a substrate in the catecholamine biosynthesis and is transformed into α-methylnoradrenaline (180).

Codeine activation to morphine by demethylation is controlled by the activity of the polymorphic cytochrome CYP2D6 (181). The clinical relevance of this observation derives from the fact that about 10% of Caucasians lack this metabolic pathway. Because poor metabolizers cannot activate this widely used drug, for

them codeine is an ineffective analgesic (182). A similar observation has been made for the metabolic activitation of hydrocodone to hydromorphone (183).

These substances metabolized within the CNS illustrate well one of the difficulties that may be encountered with prodrug kinetics. The prodrug may follow an ADME pattern perfectly well described by its systemic availability, volume of distribution, and both hepatic and renal clearances, but still have a pharmacological effect whose dependency on blood kinetics is only indirect. This is particularly true if the drug must diffuse through the blood–brain barrier, and is then metabolized by enzyme systems different from those found in the liver.

5.3   Reversible and Irreversible Conversion

The natural conversion of an inactive stereoisomer to an active one is another case in which no *a priori* intention existed when the compound was synthesized. Ibuprofen is a good example. From a pharmacokinetic point of view, one may consider that the reversibly formed metabolite represents a compartment for the parent compound, and adequate provision must be taken when analyzing this type of data.

Sulindac (**13**) is an interesting case of re-

(13)

versible metabolism to an active derivative by reduction to a sulfide. It is also metabolized irreversibly to an inactive sulfone (184, 185). *In vitro* anti-inflammatory test systems considered relevant to *in vivo* efficacy and correlations between drug concentration and biological effect *ex vivo* suggest that sulindac is a prodrug. Reoxidation of the sulfide to the parent form occurs before elimination, but the sulfide has a long half-life and prolonged duration of action. There is strong evidence that both liver (186) and gut flora (187) contribute to sulfide generation.

Although many alkylating agents are administered in the active form, cyclophosphamide (**14**) is a prodrug activated in the liver.

(**14**)

Initial conversion is to 4-hydroxycyclophosphamide, which is in spontaneous equilibrium with its tautomeric form, aldophosphamide. These function as inactive "transport" molecules (188). Subsequent spontaneous elimination of acrolein from aldophosphamide generates the active moiety, phosphoramide mustard. Because of its unique activation mechanism, numerous bioreversible prodrugs of phosphoramide mustard have been investigated in an attempt to improve its therapeutic index (189).

It is evident that modeling of such compounds is difficult and that great care must be taken not to generate models that have too many degrees of freedom or that cannot be solved in a univocal way.

5.4 The Double Prodrug Concept for Drug Targeting

Drug targeting may present particular problems that cannot be solved by a "simple" prodrug design (i.e., one transport plus one active moiety). For example, a prodrug designed to promote site-specific delivery through a target-specific cleavage mechanism (e.g., attributed to a specific enzyme activity) may not be successful if it is not able to reach the target tissue. For optimal activity, both conditions should be fulfilled at the same time. A promising solution to this problem is the double prodrug concept (91). Drug targeting may be achieved by preparing a prodrug form with good properties for transport of a prodrug that exhibits a site-specific bioactivation. Pilo-

carpine is a good example of an active moiety for which the double prodrug approach may be useful to overcome unfavorable pharmacokinetic properties (Section 7.1). Acyclovir is another example for which the double prodrug design might be useful (Section 3.2).

5.5 Prodrug Activation Enzymes in Cancer Gene Therapy or Gene-Directed Enzyme Prodrug Therapy (GDEPT)

GDEPT therapy of cancer is a novel approach, with the potential to selectively eradicate tumor cells, while sparing normal tissue from damage (190). Among the broad array of genes that have been evaluated for tumor therapy, those encoding prodrug activation enzymes are especially appealing because they directly complement ongoing clinical chemotherapeutic regimes (191, 192). These enzymes can activate prodrugs that have low inherent toxicity by use of both bacterial and yeast enzymes, or enhance prodrug activation by mammalian enzymes. The general advantage of the former is the large therapeutic index that can be achieved. The interest of the latter is the non-immunogenicity (supporting longer periods of prodrug activation) and the fact that the prodrugs will continue to have some efficacy after transgene expression is extinguished. Essentially, these prodrug activation enzymes mediate toxicity through disruption of DNA replication, which occurs at differentially high rates in tumor cells compared with most normal cells. In cancer gene therapy, vectors target delivery of therapeutic genes to tumor cells, in contrast to the use of antibodies in antibody-directed prodrug therapy. Vector targeting is usually effected by direct injection into the tumor mass or surrounding tissues, although the efficiency of gene delivery is usually low. Thus, it is important that the activated drug is able to act on nontransduced tumor cells. This bystander effect may require cell-to-cell contact or be mediated by facilitated diffusion or extracellular activation to target neighboring tumor cells. Effects at distant sites are believed to be mediated by the immune system, which can be mobilized to recognize tumor antigens by prodrug-activated gene therapy. Prodrug-activation schemes can be combined with each other and with other treatments, such as radiation, in a synergistic manner. Use of prodrug wafers for intratumoral drug activation and selective permeabilization of the tumor vasculature to prodrugs and vectors should further increase the value of this new therapeutic modality.

5.6 Antibody-Directed Enzyme Prodrug Therapies (ADEPT)

The activation of specially designed prodrugs by antibody–enzyme conjugates targeted to tumor-associated antigens is another form of the use of antibodies in cancer therapy (192, 193). It is different from the drug–antibody conjugates described in Section 2.3, but is similar in its concept to GDEPT described in Section 5.5.

5.7 Virus-Directed Enzyme Prodrug Therapy (VDEPT)

The virus-directed enzyme prodrug therapy (VDEPT) anticancer gene therapy strategy relies on the use of viral vectors for the efficient delivery to tumor cells of a "suicide gene" encoding an enzyme that ideally converts (as for the other approaches) a nontoxic prodrug to a cytotoxic agent (194).

5.8 Antitumor Prodrug by Use of Cytochrome P450 (CYP) Mediated Activation

An ideal cancer chemotherapeutic prodrug is completely inactive until metabolized by a tumor-specific enzyme, or by an enzyme that is only metabolically competent toward the prodrug under physiological conditions unique to the tumor. Human cancers, including colon, breast, lung, liver, kidney, and prostate, are known to express cytochrome P450 (CYP) isoforms, including 3A and 1A subfamily members (195). This raises the possibility that tumor CYP isoforms could be a focus for tumor-specific prodrug activation. Different approaches have been tested, including identification of prodrugs activated by tumor-specific polymorphic CYPs, use of CYP gene-directed enzyme prodrug therapy (Section 5.5), and CYPs acting as reductases in hypoxic tumor regions.

5.9 Radiation-Activated Prodrugs

Bioreductive drugs are designed to be activated by enzymatic reduction in hypoxic re-

gions of tumors, but activation of these drugs is not always fully suppressed by oxygen in normal tissues (196). A further limitation is that bioreductive drug activation depends on suitable reductases being expressed in the hypoxic zone. As an alternative approach, prodrugs may be reduced in hypoxic regions and thereby activated by ionizing radiation rather than by enzymes. This strategy is theoretically attractive, but design requirements for such radiation-activated cytotoxins are challenging. In particular, the reducing capacity of radiation at clinically relevant doses is small, thus necessitating the development of prodrugs capable of releasing very potent cytotoxins efficiently in hypoxic tissue.

6 TRANSDERMAL ABSORPTION

Metabolism of drugs by the skin is gaining interest because of its pharmacokinetic, pharmacological, therapeutic, and toxicological implications. The metabolic capacity of the skin can be exploited in the field of dermal drug delivery through the use of prodrugs (197). The prodrug approach in dermal drug delivery must clearly separate optimization of systemic delivery from delivery to the dermis for the treatment of skin diseases. The physicochemical attributes needed to reach these two objectives are clearly different.

The prodrug approach in dermal drug delivery has been the subject of numerous investigations, and various drugs have been considered such as vidarabine (198), aspirin (199), theophylline (29, 200, 201), purine analogs (202–205), 5-fluorouracil (200, 201, 204), indomethacin (206–208), dithranol (209), lonapalene (210), mitomycin C (211, 212), metronidazole (213, 214), 5-fluorocytosine (215), nicorandil (216), or zidovudine (217, 218). These investigations have shown that, although the effect is generally relatively small, drug derivatives can be made to permeate the skin more readily than the parent compound. However, guidelines for enabling optimal dermal delivery by use of the prodrug approach are still lacking. Few reports on prodrugs have dealt with derivatives consisting of homologous series (219) and few systematic evaluations of their physicochemical properties relevant for penetration of human skin have been published (220). Nevertheless, reviews of prodrug design for topical application are available (221, 222).

7 OCULAR ABSORPTION

A major challenge in ocular therapeutics is improving the poor local availability of topically applied ophthalmic drugs (223). Such low availability, often less than 10% of the applied dose, is largely attributable to precorneal processes that rapidly remove drugs from their absorption site and to the existence of the corneal structure designed to restrict passage of xenobiotics (224, 225). During the past decades, considerable effort has been devoted to prolonging precorneal drug retention through vehicle manipulations (viscosity, polymeric inserts), with the hope of enhancing ocular availability. Thus far, it seems that such methods have only resulted in moderate success. This is probably attributable to the modest improvement in ocular drug availability as exemplified by viscous solutions and by the lack of patient acceptance of, for example, inserts (223).

As stated by Lee (223), once the mechanisms have been better understood, by which topically applied drugs are ocularly absorbed, it would become possible to attempt other approaches to enhance corneal drug permeability:

- Modification of the *integrity of the corneal epithelium* transiently by use of penetration enhancers.
- Modification of the *physicochemical characteristics of the drug product* through ion-pair formation.
- Modification of the *physicochemical characteristics of the active principle* by prodrug derivatization.

However, thus far, only the prodrug approach seems to have met objectives that allow commercialization of such ophthalmic products.

7.1 Some Typical Examples

Prodrugs have been designed to improve corneal absorption. This approach has been applied with epinephrine (226–230), terbutaline (231), various prostaglandins (232), phenylephrine (233–235), and pilocarpine (236–241). For some pilocarpine derivatives the double prodrug approach has been used to overcome eye irritation and improve on poor water solubility (240, 241) (Section 5.4).

As discussed by Bundgaard (9), the latter drug is an interesting example of the potential of the double prodrug approach, given that it presents significant delivery problems. Its ocular availability is low and the elimination of the drug from its site of action in the eye is fast, resulting in a short duration of action. Furthermore, undesirable side effects such as myopia and miosis frequently occur as a result of noncorneal absorption or transient peaks of high drug concentration in the eye. To be useful, a potential prodrug should exhibit a higher lipophilicity than that of pilocarpine to enable efficient penetration through the corneal membrane. It should possess sufficient aqueous solubility and stability for formulation as eyedrops, it should be converted to the active parent drug within the cornea or once the corneal barrier has been passed, and it should lead to controlled release and hence prolonged duration of action of pilocarpine. Various diesters of pilocarpic acid (15) have

(15)

been shown to possess these desirable attributes (237–239). The compounds are highly stable in aqueous solution at pH 5–6, but readily converted quantitatively to pilocarpine (16) in the eye, through a sequential process involving enzymatic hydrolysis followed by spontaneous ring closure of the intermediate pilocarpic acid monoesters. The latter derivatives were originally developed as prodrug forms, but they suffered from limited

(16)

stability in aqueous solutions. This stability problem was solved by forming double prodrug pilocarpic acid esters. Because of their blocked hydroxyl group, these compounds are unable to undergo cyclization to pilocarpine in the absence of hydrolytic enzymes.

β-Adrenergic receptor blockers are widely used in the treatment of glaucoma. Their therapeutic usefulness may be limited by a relatively high incidence of cardiovascular and respiratory side effects. These arise as a result of absorption of the topically applied drug into the systemic circulation and are essentially the same as those seen with oral drug administration. A potentially useful approach for decrease in the systemic absorption of topically applied β-adrenergic receptor blockers, thereby diminishing their adverse effects, may be the development of prodrugs with improved corneal absorption characteristics (242, 243). This approach has been variably successful with nadolol (244, 245), propranolol (246), and timolol (247–254).

Timolol has been particularly well studied in the context of prodrug design for a reduction of systemic availability of ocularly applied xenobiotics (223). This aim is achieved either by reducing the instilled dose in proportion to the degree of enhancement in corneal drug absorption (248, 249) or by increasing the lipophilicity of the prodrugs, thereby impeding their absorption into the systemic circulation without negatively affecting their ocular absorption (242). The basis for the second method is the differential lipophilic characteristics of the membranes responsible for ocular absorption (cornea) and systemic absorption (conjunctival and nasal mucosa). Both methods have been found to be feasible in the case of timolol. It has thus been shown that it is possible to design prodrugs that are poorly absorbed into the bloodstream and yet well absorbed into the eye.

An even more extreme situation is found with carbonic anhydrase inhibitors such as acetazolamide, ethoxyzolamide, and methazolamide, which are useful for the treatment of glaucoma. Because of their limited aqueous solubility or unfavorable lipophilicity, they are not active when given topically to the eye and must be given orally or parenterally. Systemic side effects severely limit this mode of therapy and, consequently, numerous investigations are presently under way to find a new carbonic anhydrase inhibitor that readily penetrates the cornea or to prepare a prodrug with adequate water solubility and lipophilicity combined with the ability to be reconverted to the parent sulfonamide after corneal passage (255–257).

As stated by Lee (223), although prodrugs hold great promise in improving ocular delivery, they have yet to be considered routinely for improving the physicochemical characteristics of potent active principles originally developed for systemic use because in many cases such active substances show unacceptable side effects when they reach the systemic circulation after ocular application. The key of a new approach would be to take advantage of the ability of prodrugs to enhance the ocular potency of a drug candidate originally designed for systemic use, but which has been, in a second step, deliberately chosen for ophthalmology because of its lack of systemic potency. This would be an attempt to achieve relative "oculoselectivity." Therefore, the salient feature of the proposed approach is to select a less potent drug candidate to minimize the incidence of systemic side effects and then to offset this loss in potency by enhancing its ocular absorption through the use of prodrug design.

7.2 Ocular "Pharmacokinetics"

For all routes of administration, the absorption, distribution, and elimination kinetics are important for obtaining the desired therapeutic effect. For drugs expected to act in the eye after ocular absorption, pharmacokinetic parameters are difficult to obtain in both animals and humans. Accordingly, it has been proposed to rely essentially on pharmacodynamic measurements by use of a specific biological response after topical administration (258, 259). For example, the apparent absorption and elimination kinetics in the human iris have been estimated for several commonly used drugs, such as tropicamide (260, 261), pilocarpine (261, 262), homatropine (263), and phenylephrine (264, 265), through use of the measurement of pupil-response vs. time profiles. Such an approach is based on the assumption that the extent of the mydriasis is an instantaneous response to the quantity of the active principle residing in the biophase (iris dilator muscle) so that the time course of the pupil response directly reflects the change of the drug in the iris.

As reported by Chien and Schoenwald (259), for phenylephrine, a drug used during cataract surgery and ophthalmoscopic examination, the situation may be more complicated. As a matter of fact, a rebound miosis can be observed in some patients (266–268). A subsequent instillation of phenylephrine in these patients may then result in a reduction of mydriasis, suggesting that mydriatic tolerance may develop in the iris muscle. As a consequence, mydriasis measurement may not accurately reflect the pharmacokinetic behavior of the active principle. To test this hypothesis, the authors (259) compared pharmacokinetic and pharmacodynamic parameters obtained in the rabbit eye after topical instillation of phenylephrine and one of its prodrugs. The study conclusively showed that the kinetic parameters of phenylephrine estimated from its mydriasis profile did not accurately reflect the kinetics of drug distribution in the iris. These parameters also varied with the instillation of phenylephrine solution or prodrug suspensions. Such results clearly indicate that the study of the local kinetics of a drug or a prodrug after topical administration is a quite difficult task and that dynamic results obtained in humans should be extrapolated with care if pharmacokinetic deductions are made from this type of data.

8 PHARMACOKINETIC AND BIOPHARMACEUTICAL ASPECTS

8.1 Compartmental Approach

If one considers the one-compartment, open-body model with first-order input and output,

the concentration–time curve for the parent compound can be described by the Bateman function. For most drugs the absorption rate constant is significantly larger than the elimination rate constant. As a consequence, the shape of the concentration vs. time curve will, after some time, depend only on the elimination rate constant. During the postabsorptive phase, it will therefore be possible to estimate the "apparent half-life of elimination," which is related to the first-order elimination rate constant. On the other hand, if elimination is faster than absorption, the rate constant obtained from the slope of the terminal portion of the curve will be the absorption rate constant and not the elimination rate constant. This is called a flip-flop situation, and may be observed with drugs that are very rapidly eliminated or with dosage forms that slowly release the drug in an apparent first-order fashion.

The pharmacokinetic properties of metabolites are particularly interesting in this respect. At one time it was generally assumed that metabolite elimination was faster than formation because metabolites were considered to be more easily eliminated from the body than the parent compound. This assumption may be true when polar compounds such as glucuronides, sulfates, or glycine conjugates are the major biotransformation products, but need not be true when drugs are, for example, acetylated or hydroxylated. Nevertheless, a basic difference usually exists between parent drug and metabolite kinetics: for metabolites, elimination is normally faster than formation, whereas for most drugs elimination is slower than gastrointestinal absorption. The consequence of this reversal is that metabolites often show flip-flop kinetics.

For prodrugs, the situation is basically different because, in most cases, the chemical bond between transport moiety and active moiety is designed as labile to avoid, for example, urinary elimination of the prodrug, which would diminish the amount of active compound available at the site of action. However, this is not an inviolate rule, particularly if prodrug design is used to provide "slow formation" of the active moiety.

The pharmacokinetic models used to describe prodrug and active moiety kinetics are similar to those used for metabolites (11). It may thus often be difficult to calculate individual rate constants, when prodrug transformation occurs both on first pass and after reaching the systemic circulation. Even if complicated models are of little probable help for data analysis, they may be useful to determine the relative influence of the different rate constants on the bahavior of the prodrug and the active moiety. For example, comparing simulated data obtained with different models may allow evaluation of the impact of prodrug excretion on the AUC of the active moiety in function of the prodrug transformation rate.

8.2   Clearance Approach

The clearance concept can successfully be applied to analyze metabolite kinetics in blood (13). It can, by analogy, be used for the kinetics of the active moiety after prodrug administration. As with the compartmental approach, care must be taken when prodrug transformation occurs during first pass and/or prodrug excretion is important after it reaches the systemic circulation. Here again, some differences with metabolite kinetics occur in the function of two different aspects:

1. Metabolite kinetics of classical drugs are not always studied after their administration to humans, whereas the active moiety of a prodrug should always be well investigated from a pharmacokinetic point of view. This means that the basic pharmacokinetic parameters (11) are estimated and that they can be used to extract relevant information when the prodrug is administered.
2. Metabolite formation is usually the rate-limiting step, whereas this is not the case for the active moiety of a prodrug.

It must be stated that these two considerations also hold for the compartmental approach, although in the latter case, the danger of calculating artifactual parameters is more critical.

8.3   Study Design for Kinetic Studies in Humans

Ideally, the kinetics of the prodrug and the active moiety should be studied in the same patients or healthy volunteers. It is not possible to provide generally valid guidelines for study design because, as described in the previous sections, many factors may influence the analysis of the data, and consequently the way in which data must be collected. For example, it is certainly very different to study a prodrug such as bacampicillin with a very fast release of ampicillin after oral administration, or the decanoic ester of haloperidol given intramuscularly as a depot preparation in an oily vehicle.

Planning pharmacokinetic studies for topically applied prodrugs is even more problematic, and it must be realized that even kinetic studies of topically applied drugs (i.e., dermal, ocular, rectal, etc.) are not yet clearly delineated. In particular, analytical sensitivity often represents a major obstacle for the conduct of properly designed pharmacokinetic investigations (11).

Finally, it must be remembered that a prodrug intended for oral administration may show perfectly adequate release performances under normal conditions, but be unreliable (i.e., not transformed to the active moiety) in liver disease. This may be seen if regeneration occurs enzymatically in the liver. The prodrug should, accordingly, be tested in this pathological situation before marketing.

8.4   Bioavailability Assessment

The concept of bioavailability was developed in the early 1960s, when it was realized that the same active drug ingredient, in the same dose but formulated in different pharmaceutical products, might not have the same therapeutic and/or toxicological properties, even if the two formulations were administered according to the same dosage regimen. Since that time, numerous efforts have been made to establish definitions and guidelines for experimental protocols, specification of analytical methods, calculation of target pharmacokinetic parameters, statistical procedures, and clinical relevance of bioavailability studies. Over the years it has become evident that there are some active substances or particular situations to which the definitions are only partly applicable. A workshop was organized in 1989 to highlight some of those cases and to discuss the applicability of the current definitions, with special reference to their use in drug registration documents. A position paper presented the conclusions and consensus reached by participants from academia, the pharmaceutical industry, and regulatory authorities in a number of countries (269).

It was recognized that currently available and recognized definitions of bioavailability based on the concept of "active drug ingredient" or "therapeutic moiety" render mandatory consideration of the measurement of pharmacologically active metabolites. In this context two cases were considered:

1. An active drug ingredient is metabolized to active metabolites.
2. An inactive prodrug is biotransformed to active metabolites.

In the latter situation, two typical cases were discussed.

Case A. When a prodrug is pharmacologically inactive and is subject to an important presystemic transformation to the active moiety, it is correct to administer the active moiety by the intravenous route as the reference pharmaceutical form, as long as its behavior (CL, V_d) is identical for both modes of administration. The ratio of dose-normalized AUCs after oral and intravenous administration is the absolute bioavailability of the active moiety. This is, for example, the procedure that was used with ampicillin and its prodrugs bacampicillin and pivampicillin (107).

Case B. When a prodrug is pharmacologically inactive, subject to important presystemic and/or systemic biotransformation to the active moiety, and intended for both oral and intravenous administration, two practical situations may occur: the active moiety is available for parenteral administration or the active moiety cannot be administered intravenously to humans for practical, safety, or ethical reasons.

In the first situation, the procedure described in Case A is applicable. In the second

situation, it is necessary to administer the parent prodrug compound intravenously as the reference compound. The ratio of the dose-normalized AUCs of the parent prodrug indicates its absolute bioavailability; this parameter is often close to zero and of limited clinical relevance because, by definition, the prodrug is not active. The absolute bioavailability of the active moiety cannot usually be calculated, unless it is demonstrated that the prodrug is entirely biotransformed to the active moiety, or if the amount of parent drug transformed to other metabolites can be calculated for both routes of administration. Even if the absolute bioavailability of the active metabolite cannot be calculated in the strict sense of pharmacokinetic definitions, clinically relevant conclusions about the behavior of the active moiety after intravenous and oral administration may nevertheless be obtained from analysis of its concentration–time curves. For example, if these curves are similar (AUC, t_{max}, C_{max}) after both modes of administration, it is probable that therapeutic effects will be similar. In addition, if AUCs are equivalent, it can be stated that equivalent amounts of the active moiety are made available to the body after intravenous and oral administration (269).

8.5 Comparison of Animal and Human Data

As briefly discussed in Section 1.3, it is often difficult to compare pharmacokinetic data of prodrugs obtained in animals and humans. A basic problem is raised by the fact that the enzymatic system responsible for the bioactivation of the prodrug is not identical in different mammal species, including humans (270). Usually, the next stage after testing for purity and stability of the new compound involves the assessment of whether bioavailability of the parent molecule is increased after administration of the prodrug by gavage to laboratory animals. The species selection has very often been made according to the amount of information available in those particular laboratories and in the literature. It is a process that can be dishearteningly misleading. Increasing the range of animal species does not always lead to a better ability to predict bioavailability in humans. Hydrolysis studies are also important at this stage of drug development to ensure that any novel prodrug will hydrolyze in human tissues, as expected from animal data, and to clarify about why a particular prodrug is not performing as expected in animals. After selection of a prodrug candidate, it is essential to determine where and how rapidly hydrolysis takes place in the animal species to be used for safety evaluation before the first bioavailability studies in humans.

Another, complicating factor is when the metabolism of the active moiety also differs in different animals. As an example, methylprednisolone showed nonlinear kinetics in the rat (161), but linear kinetics in humans (160). It then becomes very difficult to analyze the respective advantages of different prodrugs tested in different animal species. Great care should thus be laid on proper and complete study design to allow for such comparisons, as shown for example, with a series of ester prodrugs of valproic acid (271) in dogs, of 5-fluorouracil prodrugs in the intestine of rabbits (272), or diacid angiotensin-converting enzyme inhibitors (273).

In view of these difficulties, it should be envisaged to test prodrug approaches at a very early stage of drug development if particular pharmacokinetic problems of specially interesting active moieties are detected. If liver metabolic activities are involved, it might be useful to consider the use of liver microsome preparations as shown for new dopaminergic compounds (274) and to include human microsomes to gain some insight into the situation potentially encountered in humans.

8.6 Validity of Classical Pharmacokinetic Concepts for Prodrug Design

As seen in the previous sections, classical pharmacokinetic concepts can be used for the study of prodrugs, despite the fact that they have essentially been derived from theoretical considerations related to the situation where the parent compound is the active moiety. As for metabolite kinetics, difficulties arise when the prodrug is not totally biotransformed to the "drug." There is one situation, however, in which it is questionable whether these concepts are fully appropriate, that is, drug targeting. For example, if a prodrug is very specifically targeted to one type of cells or to one organ system, it might be necessary to define a

"targeting index" to describe distribution of the prodrug and the active moiety, given that a concept such as the apparent volume of distribution might only poorly describe the distribution properties of the therapeutic agent. The same restriction may apply to the concepts of systemic availability, systemic clearance, and apparent half-life of elimination. This does not mean that classical pharmacokinetic concepts would become invalid because they are robust and well validated, but that some creativity is probably necessary to improve the descriptive power of pharmacokinetics to allow comparisons of "drug" and prodrug or between prodrugs. Similar considerations also apply to the concepts basic to biopharmaceutical evaluations. If the necessity to define new pharmacokinetic parameters for the description of targeted prodrugs is accepted, it will be of utmost importance to define only concepts that can be quantified based on experimental data. It is possible that development of imaging techniques such as PET or NMR scan will give an impetus to these foreseeable developments in pharmacokinetics.

A problem requiring special attention is the stereoselective *in vivo* activation of prodrugs derived from racemic mixtures, as exemplified by the stereoselective hydrolysis of *O*-acetylpropranolol (275), for which it was also found that the selectivity of plasma enzyme urase differs from that of liver and intestine enzymes.

9 SOME CONSIDERATIONS FOR PRODRUG DESIGN

9.1 Rationale of Prodrug Design

The design of prodrugs in a rational manner requires, as stated by Bundgaard (9), that the underlying causes that necessitate or stimulate the use of the prodrug approach be defined and clearly understood. It may then be possible to identify the means by which the difficulties can be overcome. The rational design of the prodrug can thus be divided into three basic steps:

1. Identification of the drug delivery problem.
2. Identification of the physicochemical properties required for optimal delivery.

3. Selection of a prodrug derivative that has the proper physicochemical properties and that will be cleaved in the desired biological compartment.

In this context it must be accepted that a very close collaboration is needed between the pharmaceutical chemists active in drug synthesis and those working in the area of xenobiotic metabolism. This is particularly important if more targeted prodrugs are designed in function of enzymes available at the right place, in the right amount, and with the right prodrug specificity.

Levodopa is an interesting case of this type of collaboration. Levodopa is a metabolic precursor of dopamine used for the treatment of Parkinson's disease for more than 30 years. The prodrug allows dopamine to be produced in the central nervous system in adequate quantities, given that dopamine itself passes the blood–brain barrier only with difficulty. During the lifetime of levodopa it became apparent that its dosage should be reduced to diminish side effects such as nausea or heart rhythm disturbances caused by metabolic transformation of levodopa in the periphery. One way to reach this goal is the association of benzerazide, an inhibitor of dopa decarboxylase that reduces peripheral dopamine concentrations without loss of therapeutic activity in the central nervous system. The present formulation of levodopa contains benzerazide at one-fourth of the levodopa dosage. A further improvement of the therapy with levodopa is the coadministration of tolcapone, an inhibitor of catechol-*O*-methyl-transferase. This decreases the metabolic pathway to 3-*O*-methyldopa and increases the overall therapeutic efficacy and tolerability (276, 277). In addition to these "pharmacodynamic" changes, the pharmaceutical formulation of levodopa has been continuously optimized.

9.2 Practical Considerations

In the rational design and synthesis of prodrugs, several factors should be considered before starting the development of a new compound intended for large-scale production (278):

- The chemical intermediates or modifiers should be available in a high state of purity at reasonable cost.
- Complicated synthetic schemes should be avoided and purification steps should be efficient without markedly increasing production costs. The production should be easy to scale up from the benchmark to industrial production.
- The prodrug should be stable in bulk form. This is of particular importance for substances like esters, which are likely to be degraded in the presence of even trace amounts of moisture.
- The *in vivo* lability should be sufficient to permit release of the active moiety at a rate adequate to ensure its therapeutic activity. Regeneration can be either chemical (pH effects) or/and enzymatic.
- The prodrug and the "carrier moiety" should be nontoxic. Relatively "safe" moieties include amino acids, short to medium length alkyl esters, and some of the macromolecules described previously.
- The pharmacokinetics of the active moiety should be well documented before starting prodrug synthesis, and, at a later stage, prodrug kinetics should be thoroughly investigated in humans.
- The biopharmaceutical consequences for prodrug formulations should be carefully evaluated.
- Finally, the prodrug should present some clinically relevant advantage over the active principle administered directly. In this context, it must be remembered that modification of one pharmacokinetic property frequently alters other properties of the drug molecule and caution must thus be exercised when embarking on a program of this nature.

10 CONCLUSIONS

Although prodrug design started more than 30 years ago and many reviews have been written on this subject, very little information is available in official guidelines or pharmacokinetic textbooks on the regulatory requirements or data analysis for this type of compounds. The present review is an attempt to gather and confront available information on the subject.

Some basic problems have, however, been left untouched. For example, the difficulty of extrapolating data from animals to humans encountered during toxicokinetic and toxicologic studies with drugs is amplified with prodrugs because not only might the metabolism of the active moiety differ, but also its availability from the prodrug. As a matter of fact, there is presently no published rationale for the conduct of animal and human pharmacokinetic programs during prodrug research and development.

We concluded a review on prodrugs (7) by quoting the question asked in 1985 (279) by Stella et al.: "Do prodrugs have advantages in clinical practice?" Our opinion was that

> Today, the answer is certainly YES in some particular cases, but for many drugs this aspect of drug design has received no clear and satisfactory solution. The main reason for this situation is that most prodrugs have been synthesized starting from valuable and well known drugs. As a consequence, the potential advantage of the new chemical entity over its "seasoned precursor" has often been only marginal. It is thus important that in the future, drug design of new chemical entities should incorporate "delivery and/or targeting components" from the earliest stages of research and development. This strategy might help substances too toxic, or unable to show adequate pharmacologic effects in their basal form to go through primary and secondary screening, before successfully reaching human testing. It is evident that if such an approach were to become an integral part of basic drug design and not just a hindsighted attempt to solve problems associated with older drugs, it would also be necessary to develop new biopharmaceutical and pharmacokinetic approaches to tackle the new challenges.

After 15 years, the authors still believe that this is a valid statement.

After this review, which focused more on pharmacokinetic aspects than on chemical synthesis, we can conclude that, indeed, additional thinking on new ways to approach the toxicokinetic and the clinical pharmacokinetics of prodrugs and their active moiety is of paramount importance if prodrug design is to remain (or to become?) an important part for research and development of new therapeutic agents. In parallel, great efforts must be undertaken to better understand the molecular basis of xenobiotic metabolism. It should then be easier to synthesize compounds that would show the most appropriate physicochemical characteristics.

REFERENCES

1. A. Albert, *Nature*, **182**, 421 (1958).
2. T. Higuchi and V. Stella, *Pro-drugs as Novel Drug Delivery Systems*, American Chemical Society, Washington, DC, 1975.
3. E. B. Roche, *American Pharmaceutical Association/Academy of Pharmaceutical Sciences, Symposium*, Washington, DC, (1977).
4. A. A. Sinkula and E. B. Roche, Eds., *American Pharmaceutical Association Academy of Pharmaceutical Sciences, Symposium*, Washington, DC, 1977, pp. 1, 17.
5. W. I. Higuchi, A. Kusai, J. L. Fox, N. A. Gordon, N. F. H. Ho, C. C. Hsu, D. C. Baker, and W. M. Shannon in T. J. Roseman and S. Z. Mansdorf, Eds., *Controlled Release Delivery Systems*, Marcel Dekker, New York, 1983, p. 43.
6. D. G. Waller and C. F. George, *Br. J. Clin. Pharmacol.*, **28**, 497 (1989).
7. L. P. Balant, E. Doelker, and P. Buri, *Eur. J. Drug. Metab. Pharmacokinet.*, **15**, 143 (1990).
8. L. P. Balant, E. Doelker, and P. Buri in A. Rescigno and A. K. Thakur, Eds., *New Trends in Pharmacokinetics*, Plenum Press, New York, 1991, p. 281.
9. H. Bundgard, *Drugs Future*, **16**, 443 (1991).
10. C. E. Inturrisi, B. M. Mitchell, K. M. Foley, M. Schulz, S. Seung-Uon, and R. W. Houde, *N. Engl. J. Med.*, **310**, 1213 (1984).
11. L. P. Balant and J. McAinsh in P. Jenner and B. Testa, Eds., *Concepts in Drug Metabolism*, Part A, Marcel Dekker, New York, 1980, p. 311.
12. M. Gibaldi and D. Perrier, *Pharmacokinetics*, 2nd ed., Marcel Dekker, New York, 1982, 494 pp.
13. M. Rowland and T. N. Tozer, *Clinical Pharmacokinetics*, Lea & Febiger, Philadelphia, 2nd ed., 1989, 546 pp.
14. L. Benet in M. E. Wolff, Ed., *Burger's Medicinal Chemistry and Drug Discovery*, 5th ed., John Wiley & Sons, New York, 1996.

15. B. J. Aungst, M. J. Myers, E. G. Shami, and E. Shefter, *Int. J. Pharm.*, **38**, 199 (1987).

16. S. Vickers, C. A. H. Duncan, H. G. Ramjit, M. R. Dobrinska, C. T. Dollery, H. J. Gomez, H. L. Leidy, and W. C. Vincek, *Drug Metab. Dispos.*, **12**, 242 (1984).

17. H. P. Huang and J. W. Ayres, *J. Pharm. Sci.*, **77**, 104 (1988).

18. Council directive 65/65/EEC, *Official J. Eur. Comm.*, **22**, 2, (1965).

19. Commission of the European Communities-*Note for Guidance*, (III/1962/87), 1987.

20. Council Recommendation 87/176/EEC, *Official J. Eur. Comm.*, L 73 (1987).

21. Commission of the European Communities-*Note for Guidance*, (CPMP/EWP/QWP/1401/98), 1991.

22. B. Testa in D. J. Abraham, Ed., *Burger's Medicinal Chemistry and Drug Discovery*, 6th ed., John Wiley & Sons, New York, 2003.

23. A. A. Sinkula and S. H. Yalkowsky, *J. Pharm. Sci.*, **64**, 181 (1975).

24. H. Seki, T. Kawaguchi, and T. Higuchi, *J. Pharm. Sci.*, **77**, 855 (1988).

25. N. M. Nielsen and H. Bundgaard, *J. Pharm. Sci.*, **77**, 285 (1988).

26. A. Aguiar and J. E. Zelmer, *J. Pharm. Sci.*, **58**, 983 (1969).

27. A. Burger, *Sci. Pharm.*, **45**, 269 (1977).

28. O. A. T. Olsson and L. A. Svensson, *Pharm. Res.*, **1**, 19 (1984).

29. K. B. Sloan and N. Bodor, *Int. J. Pharm.*, **12**, 299 (1982).

30. M. Johansen and H. Bundgaard, *Arch. Pharm. Chem. Sci.*, **9**, 40 (1981).

31. K. G. Siver and K. B. Sloan, *J. Pharm. Sci.*, **79**, 66 (1990).

32. R. Duncan, *Anticancer Drugs*, **3**, 175 (1992).

33. F. Kratz, U. Beyer, and M. T. Schutte, *Crit. Rev. Ther. Drug Carrier Syst.*, **16**, 245 (1999).

34. T. Kawaguchi, A. Tsugane, K. Higashide, H. Endoh, et al., *J. Pharm. Sci.*, **81**, 508 (1992).

35. M. Vert, *Crit. Rev. Ther. Drug Carrier Syst.*, **2**, 291, 327 (1986).

36. H. Ringsdorf, *J. Polym. Sci. Polym. Symp.*, **51**, 135 (1975).

37. E. Harboe, C. Larsen, M. Johansen, and H. P. Olesen, *Int. J. Pharm.*, **53**, 157 (1989).

38. C. Larsen, *Int. J. Pharm.*, **51**, 223 (1989).

39. K. Sato, K. Itakura, K. Nishida, Y. Takakura, M. Hashida, and H. Sezaki, *J. Pharm. Sci.*, **78**, 11 (1989).

40. T. Fujita, Y. Yasuda, Y. Takakura, M. Hashida, and H. Sezaki, *J. Controlled Release*, **11**, 149 (1990).

41. K. Kratzl and E. Kaufmann, *Monatschr. Chem.*, **92**, 371 (1961).

42. K. Kratzl, E. Kaufmann, O. Kraupp, and H. Stormann, *Monatschr. Chem.*, **92**, 378 (1961).

43. J. P. Remon, R. Duncan, and E. Schacht, *J. Controlled Release*, **1**, 47 (1984).

44. E. Schacht, E. Ruys, J. Vermeersch, and J. P. Remon, *J. Controlled Release*, **1**, 33 (1984).

45. T. Kojima, M. Hashida, S. Muranishi, and H. Sezaki, *Chem. Pharm. Bull.*, **26**, 1818 (1978).

46. T. R. Tritton, G. Yee, and L. B. Wingaard, *Fed. Proc.*, **42**, 284 (1983).

47. L. Puglisi, V. Caruso, R. Paoletti, P. Ferruti, and M. C. Tanzi, *Pharmacol. Res. Commun.*, **8**, 379 (1976).

48. A. Havron, B. Z. Weiner, and A. Zilkha, *J. Med. Chem.*, **17**, 770 (1974).

49. M. Singh, P. Vasudevan, T. J. M. Sinha, A. R. Ray, M. M. Misro, and K. Guha, *J. Biomed. Mater. Res.*, **15**, 655 (1981).

50. S. Yolles, J. F. Morton, and M. F. Sartori, *J. Polym. Sci. Polym. Chem. Ed.*, **17**, 4111 (1979).

51. S. Yolles, *J. Parenter. Drug Assoc.*, **32**, 188 (1987).

52. X. Li, N. W. Adams, D. B. Bennet, and S. W. Kim, *Proc. Int. Symp. Controlled Release Bioact. Mater.*, **17**, 244 (1990).

53. H. N. Joshi, V. J. Stella, and E. M. Topp, *Proc. Int. Symp. Controlled Release Bioact. Mater.*, **17**, 244 (1990).

54. L. Goei, E. Topp, V. Stella, et al. in R. M. Ottenbrite and E. Chiellini, Eds., *Polymers in Medicine—Biomedical and Pharmaceutical Applications*, Technomic Publishing, Lancaster, PA, 1992, p. 85.

55. L. Goei, E. Topp, V. Stella, et al. in R. M. Ottenbrite and E. Chiellini, Eds., *Polymers in Medicine—Biomedical and Pharmaceutical Applications*, Technomic Publishing, Lancaster, PA, 1992, p. 93.

56. T. M. T. Turk, S. Webb, and R. M. Ottenbrite in R. M. Ottenbrite and E. Chiellini, Eds., *Polymers in Medicine—Biomedical and Pharmaceutical Applications*, Technomic Publishing, Lancaster, PA, 1992, p. 167.

57. T. Ouchi, K. Hagita, M. Kwashima, T. Inoi, and T. Tashiro, *J. Controlled Release*, **8**, 141 (1988).

58. T. Ouchi, H. Fujie, S. Jokei, Y. Sakamoto, H. Chikashita, T. Inoi, and O. Vogl, *J. Polym. Sci. Polym. Chem. Ed.*, **24**, 2059 (1986).

59. N. N. Joshi, *Pharm. Technol.*, **12**, 118 (1988).

60. M. J. Humphrey and P. S. Ringrose, *Drug Metab. Rev.*, **17**, 283 (1986).

61. V. H. L. Lee and A. Yamamoto, *Adv. Drug. Deliv. Rev.*, **4**, 171 (1990).

62. H. Bundgaard in S. S. Davis, L. Illum, and E. Tomlinson, Eds., *Delivery Systems for Peptide Drugs*, Plenum Press, New York, 1986, p. 49.

63. W. Wang, J. Jiang, C. E. Ballard, and B. Wang, *Curr. Pharm. Des.*, **5**, 265 (1999).

64. L. Prokai, *Prog. Drug Res.*, **51**, 95 (1998)

65. H. Bundgaard and J. Møss, *Pharm. Res.*, **7**, 885 (1990).

66. J. Møss and H. Bundgaard, *Int. J. Pharm.*, **74**, 67 (1991).

67. A. H. Kahns and H. Bundgaard, *Pharm. Res.*, **8**, 1533 (1991).

68. L. Prokai, K. Prokai-Tatrai, X. Ouyang, H. S. Kim, W. M. Wu, A. Zharikova, and N. Bodor, *J. Med. Chem.*, **42**, 4563 (1999).

69. S. H. Yoon, J. Wu, W. M. Wu, L. Prokai, and N. Bodor, *Bioorg. Med. Chem.*, **8**, 1059 (2000).

70. F. Markwardt, *Pharmazie*, **8**, 521 (1989).

71. H. Bundgaard in H. Bundgaard, Ed., *Design of Prodrugs: Bioreversible Derivatives for Various Functional Groups and Chemical Entities*, Elsevier, Amsterdam/New York, 1985, pp. 1, 92.

72. E. Jensen and H. Bundgaard, *Int. J. Pharm.*, **71**, 117 (1991).

73. H. Bundgaard, C. Larsen, and P. Thorbek, *Int. J. Pharm.*, **18**, 67 (1984).

74. H. Bundgaard, C. Larsen, and E. Arnold, *Int. J. Pharm.*, **18**, 79 (1984).

75. M. J. Cho and L. C. Haynes, *J. Pharm. Sci.*, **74**, 883 (1985).

76. L. Colla, E. De Clercq, R. Busson, and H. Vanderhaeghe, *J. Med. Chem.*, **26**, 602 (1983).

77. M. Kawamura, R. Yamamoto, and S. Fujisawa, *J. Pharm. Soc. Jpn.*, **91**, 863 (1971).

78. K. Johnson, G. L. Amidon, and S. Pogany, *J. Pharm. Sci.*, **74**, 87 (1985).

79. E. Jensen, E. Falch, and H. Bundgaard, *Acta Pharm. Nord.*, **3**, 31 (1991).

80. I. M. Kovach, I. H. Pitman, and T. Higuchi, *J. Pharm. Sci.*, **70**, 881 (1981).

81. C. M. Perry and D. Fauls, *Drugs*, **52**, 754 (1996).

82. R. L. Jarvest, D. Sutton, and R. A. Vere Hodge, *Pharm. Biotechnol.*, **11**, 313 (1998).

83. D. K. Kim, N. Lee, Y. W. Kim, K. Chang, G. J. Im, W. S. Choi, and K. H. Kim, *Bioorg. Med. Chem.*, **7**, 419 (1999).

84. J. Soul-Lawton, E. Seaber, N. On, R. Wootton, P. Rolan, and J. Posner, *Antimicrob. Agents Chemother.*, **39**, 2759 (1995).

85. H. Steingrimsdottir, A. Gruber, C. Palm, G. Grimfors, M. Kalin, and S. Eksborg, *Antimicrob. Agents Chemother.*, **44**, 207 (2000).

86. M. R. Rashidi, J. A. Smith, S. E. Clarke, and C. Beedham, *Drug Metab. Disp.*, **25**, 805 (1997).

87. R. A. Vere Hodge, D. Sutton, M. R. Boyd, M. R. Harnden, and R. L. Jarvest, *Antimicrob. Agents Chemother.*, **33**, 1765 (1989).

88. K. S. Gill and M. J. Wood, *Clin. Pharmacokinet.*, **31**, 1 (1996).

89. V. H. Naringrekar and V. J. Stella, *J. Pharm. Sci.*, **79**, 138 (1990).

90. K. L. Amsberry, A. E. Gerstenberger, and R. T. Borchardt, *Pharm. Res.*, **8**, 455 (1991).

91. H. Bundgaard, *Adv. Drug Deliv. Rev.*, **3**, 39 (1989).

92. D. M. Lambert, *Eur. J. Pharm. Sci.*, **11** (Suppl. 2), S15–S27 (2000).

93. I. Toth, R. A. Hughes, M. R. Munday, C. A. Murphy, P. Mascagni, and W. A. Gibbons, *Int. J. Pharm.*, **68**, 191 (1991).

94. V. Cioli, S. Putzolu, V. Rossi, and C. Corradino, *Toxicol. Appl. Pharmacol.*, **54**, 332 (1980).

95. H. Bundgaard and N. M. Nielsen, *Int. J. Pharm.*, **43**, 101 (1988).

96. F. J. Perisco, J. F. Pritchard, M. C. Fisher, K. Yorgey, S. Wong, and J. Carson, *J. Pharmacol. Exp. Ther.*, **247**, 889 (1988).

97. L. Gu, J. Dunn, and C. Dvorak, *Drug Dev. Ind. Pharm.*, **15**, 209 (1989).

98. A. H. Kahns, P. B. Jensen, N. Mork, and H. Bundgaard, *Acta Pharm. Nord.*, **1**, 327 (1989).

99. M. C. Venuti, J. M. Young, P. J. Maloney, D. Johnson, and K. McGreevy, *Pharm. Res.*, **6**, 867 (1989).

100. V. R. Shanbhag, A. M. Crider, R. D. Gokhale, A. Harpalani, and R. M. Dick, *J. Pharm. Sci.*, **81**, 149 (1992).

101. V. K. Tammara, M. M. Narurkar, A. M. Crider, and M. A. Khan, *Pharm. Res.*, **10**, 1191 (1993).

102. H. Bundgaard, N. M. Nielsen, and A. Buur, *Int. J. Pharm.*, **44**, 151 (1988).

103. M. Hundewadt and A. Senning, *J. Pharm. Sci.*, **80**, 545 (1991).

104. H. Tsunematsu, E. Ishida, S. Yoshida, and M. Yamamoto, *Int. J. Pharm.*, **68**, 77 (1991).

105. D. A. Leight, D. S. Reeves, K. Simmons, A. L. Thomas, and P. J. Wilkinson, *Br. Med. J.*, **1**, 1378 (1976).

106. M. Rozencweig, M. Staquet, and J. Klastersky, *Clin. Pharmacol. Ther.*, **19**, 592 (1976).

107. M. Ehrnebo, S. O. Nilsson, and L. O. Boreus, *J. Pharmacokinet. Biopharm.*, **7**, 429 (1979).

108. Y. Yoshimura, N. Hamaguchi, and T. Yashiki, *Int. J. Pharm.*, **38**, 179 (1987).

109. M. Marvola, S. Nykänen, and M. Nokelainen, *Pharm. Res.*, **8**, 1056 (1991).

110. G. He, J. Massarella, and P. Ward, *Clin. Pharmacokinet.*, **37**, 471 (1999).

111. N. Mørk, H. Bundgaard, M. Shalmi, and S. Christensen, *Int. J. Pharm.*, **60**, 163 (1990).

112. M. Hammarlund-Udenaes and L. Z. Benet, *J. Pharmacokinet. Biopharm.*, **17**, 1 (1989).

113. G. B. Elion, P. A. Furman, J. A. Fyfe, P. De Miranda, L. Beauchamp, and H. J. Schaeffer, *Proc. Natl. Acad. Sci. USA*, **74**, 5716 (1977).

114. H. J. Schaeffer, L. Beauchamp, P. De Miranda, G. B. Elion, D. J. Bauer, and P. Collins, *Nature*, **272**, 583 (1978).

115. R. B. Van Dyke, J. D. Connor, C. Wyborny, M. Hintz, and R. E. Keeney, *Am. J. Med.*, **73**, 172 (1982).

116. P. De Miranda and M. R. Blum, *J. Antimicrob. Chemother.*, **12** (Suppl. B), 29 (1983).

117. T. A. Krenitsky, W. W. Hall, P. De Miranda, L. M. Beauchamp, H. J. Schaeffer, and P. D. Whiteman, *Proc. Natl. Acad. Sci. USA*, **81**, 3209 (1984).

118. B. G. Petty, R. J. Whitley, S. Liao, H. C. Krasny, L. E. Rocco, L. G. Davis, and P. S. Lietman, *Antimicrob. Agents Chemother.*, **31**, 1317 (1987).

119. D. I. Friedman and G. L. Amidon, *Pharm. Res.*, **6**, 1043 (1989).

120. J. R. Deverre, A. Gulik, Y. Letourneux, P. Courvreur, and J. P. Benoit, *Chem. Phys. Lipids*, **59**, 75 (1991).

121. A. Garzon-Aburbeh, J. H. Poupaert, M. Claesen, and P. Dumont, *J. Med. Chem.*, **29**, 687 (1986).

122. G. K. E. Scriba, *Pharm. Res.*, **10**, 1181 (1993).

123. S. Vickers, C. A. Duncan, S. D. White, G. O. Breault, R. B. Royds, P. J. De Schepper, and K. F. Tempero, *Drug Metab. Dispos.*, **6**, 646 (1978).

124. M. R. Dobrinska, W. Kukovetz, E. Beubler, H. L. Leidy, H. J. Gomez, J. Demetriades, and J. A. Bolognese, *J. Pharmacokinet. Biopharm.*, **10**, 587 (1982).

125. K. Murata, K. Noda, K. Kohno, and M. Samejima, *J. Pharm. Sci.*, **78**, 812 (1989).

126. J. Wagner, H. Grill, and D. Henschler, *J. Pharm. Sci.*, **69**, 1423 (1980).

127. N. Bodor, K. B. Sloan, T. Higuchi, and K. Sasahara, *J. Med. Chem.*, **20**, 1435 (1977).

128. Y. Hörnblad, E. Ripe, P. O. Magnusson, and K. Tegnes, *Eur. J. Clin. Pharmacol.*, **10**, 9 (1976).

129. M. D'Souza, R. Venkataramanan, A. D'Mello, and P. Niphadkar, *Int. J. Pharm.*, **31**, 165 (1986).

130. A. Buur and H. Bundgaard, *J. Pharm. Sci.*, **75**, 522 (1986).

131. H. Sasaki, T. Takahashi, J. Nakamura, J. Konishi, and J. Shibasaki, *J. Pharm. Sci.*, **75**, 676 (1986).

132. A. Buur and H. Bundgaard, *Int. J. Pharm.*, **36**, 41 (1987).

133. K. Basu, D. O. Kildsig, and A. K. Mitra, *Int. J. Pharm.*, **47**, 195 (1988).

134. M. A. Hussain, B. J. Aungst, and E. Shefter, *Pharm. Res.*, **5**, 44 (1988).

135. M. A. Hussain, C. A. Koval, M. J. Myers, E. G. Shami, and E. Shefter, *J. Pharm. Sci.*, **76**, 356 (1987).

136. M. A. Hussain and E. Shefter, *Pharm. Res.*, **5**, 113 (1988).

137. A. H. Kahns and H. Bundgaard, *Int. J. Pharm.*, **62**, 193 (1990).

138. A. Buur and H. Bundgaard, *Int. J. Pharm.*, **46**, 159 (1988).

139. A. N. Saab, L. W. Ditter, and A. A. Hussain, *J. Pharm. Sci.*, **77**, 906 (1988).

140. J. P. F. Bai, M. Hu, P. Subramanian, H. I. Mosberg, and G. L. Amidon, *J. Pharm. Sci.*, **81**, 113 (1992).

141. A. E. Balant-Gorgia and L. P. Balant, *Clin. Pharmacokinet.*, **13**, 65 (1987).

142. A. E. Balant-Gorgia, L. P. Balant, and A. Andreoli, *Clin. Pharmacokinet.*, **25**, 217 (1993).

143. T. Kawaguchi, K. Ishikawa, T. Seki, and K. Juni, *J. Pharm. Sci.*, **79**, 531 (1990).

144. M. Ihara, Y. Tsuchiya, Y. Sawasaki, A. Hisaka, H. Takehana, K. Tomimoto, and M. Yano, *J. Pharm. Sci.*, **78**, 525 (1989).

145. I. Den Daas, P. De Boer, P. G. Tepper, H. Rollema, and A. S. Horn, *J. Pharm. Pharmacol.*, **43**, 11 (1991).

146. N. Negishi, D. B. Bennett, C. S. Cho, S. Y. Jeong, W. A. R. Van Heeswijk, J. Feijen, and S. W. Kim, *Pharm. Res.*, **4**, 305 (1987).

147. X. Li, N. W. Adams, D. B. Bennett, J. Feijen, and S. W. Kim, *Pharm. Res.*, **8**, 527 (1991).

148. J. C. Meslard, L. Yean, F. Subira, and J. P. Vairon, *Makromol. Chem.*, **187**, 787 (1986).

149. N. Chafi, J. P. Montheard, and J. M. Vergnaud, *Int. J. Pharm.*, **45**, 229 (1988).

150. W. J. Irwin and K. A. Belaid, *Int. J. Pharm.*, **46**, 57 (1988).

151. W. J. Irwin and K. A. Belaid, *Int. J. Pharm.*, **48**, 159 (1988).

152. M. A. Hussain, B. J. Aungst, C. A. Koval, and E. Shefter, *Pharm. Res.*, **5**, 615 (1988).

153. A. Rubinstein, *Crit. Rev. Ther. Drug Carrier Syst.*, **12**, 101 (1995).

154. D. R. Friend, S. Phillips, A. McLeod, and T. N. Tozer, *J. Pharm. Pharmacol.*, **43**, 353 (1991).

155. B. Haeberlin and D. R. Friend, *Proc. Int. Symp. Controlled Release Bioact. Mater.*, **19**, 287 (1992).

156. B. Haeberlin, L. Empey, R. Fedorak, H. Nolen III, and D. R. Friend, *Proc. Int. Symp. Controlled Release Bioact. Mater.*, **20**, 174 (1993).

157. B. Haeberlin, H. Nolen III, W. Rubas, and D. R. Friend, *Proc. Int. Symp. Controlled Release Bioact. Mater.*, **20**, 172 (1993).

158. D. R. Friend, S. Phillips, A. McLeod, and T. N. Tozer, *Proc. Int. Symp. Controlled Release Bioact. Mater.*, **18**, 564 (1991).

159. P. Rohdewald, H. Möllmann, J. Barth, J. Rehder, and H. Derendorf, *Biopharm. Drug Dispos.*, **8**, 205 (1987).

160. H. Möllmann, P. Rohdewald, J. Barth, M. Verho, and H. Derendorf, *Biopharm. Drug Dispos.*, **10**, 453 (1989).

161. A. N. Kong and W. J. Jusko, *J. Pharm. Sci.*, **80**, 409 (1991).

162. H. Bundgaard, E. Jensen, E. Falch, and S. B. Pedersen, *Int. J. Pharm.*, **64**, 75 (1990).

163. H. Bundgaard, E. Jensen, and E. Falch, *Pharm. Res.*, **8**, 1087 (1991).

164. F. Sendo, C. Riley, and V. J. Stella, *Int. J. Pharm.*, **45**, 207 (1988).

165. M. C. Nahata and D. A. Powell, *Clin. Pharmacol. Ther.*, **30**, 368 (1981).

166. J. T. Burke, W. A. Wargin, R. J. Sherertz, K. L. Sanders, M. R. Blum, and F. A. Sarubbi, *J. Pharmacokinet. Biopharm.*, **6**, 601 (1982).

167. H. Vermeersch, J. P. Remon, D. Permentier, and E. Schacht, *Int. J. Pharm.*, **60**, 253 (1990).

168. E. Jensen, H. Bundgaard, and E. Falch, *Int. J. Pharm.*, **58**, 143 (1990).

169. M. A. Schwartz and W. L. Hayton, *J. Pharm. Sci.*, **61**, 906 (1972).

170. N. A. T. Nguyen, L. M. Mortada, and R. E. Notari, *Pharm. Res.*, **5**, 288 (1988).

171. Y. Kaneo, T. Tanaka, Y. Fujihara, H. Mori, and S. Iguchi, *Int. J. Pharm.*, **44**, 265 (1988).

172. D. E. Duggan, I. W. Chen, W. F. Bayne, R. A. Halpin, C. A. Duncan, M. S. Schwartz, R. J. Stubbs, and S. Vickers, *Drug Metab. Dispos.*, **17**, 166 (1989).

173. S. Vickers, C. A. Duncan, I. W. Chen, A. Rosegay, and D. E. Duggan, *Drug Metab. Dispos.*, **18**, 138 (1990).

174. M. H. Rahimy, J. W. Simpkins, and N. Bodor, *Pharm. Res.*, **7**, 1061 (1990).

175. L. J. Schaaf, B. R. Dobbs, I. R. Edwards, and D. G. Perrier, *Eur. J. Clin. Pharmacol.*, **34**, 439 (1988).

176. J. S. de Bono and C. J. Tvelves, *Invest. New Drugs*, **19**, 41 (2001).

177. J. G. Kuhn, *Ann. Pharmacother.*, **35**, 217 (2001).

178. B. Reigner, K. Blesch, and E. Weidekamm, *Clin. Pharmacokinet.*, **40**, 85 (2001).

179. P. Kurtzhals and C. Larsen, *Acta Pharm. Nord.*, **1**, 269 (1989).

180. W. M. Pardridge, *Physiol. Rev.*, **63**, 1481 (1983).

181. P. Dayer, J. Desmeules, T. Leemann, and R. Striberni, *Biochem. Biophys. Res. Commun.*, **152**, 411 (1988).

182. J. Desmeules, P. Dayer, M. P. Gascon, and M. Magistris, *Clin. Pharmacol.*, **45**, 122 (1989).

183. S. V. Otton, M. Schadel, S. W. Cheung, H. L. Kaplan, U. E. Busto, and E. M. Sellers, *Clin. Pharmacol. Ther.*, **54**, 463 (1993).

184. D. E. Duggan, L. E. Hare, C. A. Ditzler, B. W. Lei, and K. C. Kwan, *Clin. Pharmacol. Ther.*, **21**, 326 (1977).

185. D. E. Duggan, K. F. Hooke, R. M. Noll, H. B. Hucker, and C. G. Van Arman, *Biochem. Pharmacol.*, **27**, 2311 (1978).

186. K. Tatsumi, S. Kitimura, and S. Yanada, *Biochem. Biophys. Acta*, **747**, 86 (1983).

187. H. A. Strong, N. J. Warner, A. G. Renwick, and C. F. George, *Clin. Pharmacol. Ther.*, **38**, 387 (1985).

188. M. Colvin, R. B. Brundrett, M. N. N. Kan, I. Jardine, and C. Feneslau, *Cancer Res.*, **36**, 1121 (1976).

189. C. H. Kwon, *Arch. Pharm. Res.*, **22**, 533 (1999).

190. O. Greco and G. U. Dachs, *J. Cell Physiol.*, **187**, 22 (2001).

191. M. Aghi, M. Hochberg, and X. O. Breakefield, *J. Genet. Med.*, **148** (2000).

192. I. Niculescu-Duvaz, F. Friedlos, D. Niculescu-Duvaz, L. Davies, and C. J. Springer, *Anticancer Drug Des.*, **14**, 517 (1999).

193. K. N. Syrigos and A. A. Epenetos, *Anticancer Res.*, **19**, 605 (1999).

194. J. I. Grove, P. F. Searle, S. J. Weedon, N. K. Green, I. A. McNeish, and D. J. Kerr, *Anticancer Drug Des.*, **14**, 461 (1999).

195. L. H. Patterson, S. R. McKeown, T. Robson, R. Gallagher, S. M. Raleigh, and S. Orr, *Anticancer Drug Des.*, **14**, 473 (1999).

196. W. R. Wilson, M. Tercel, R. F. Anderson, and W. A. Denny, *Anticancer Drug Des.*, **13**, 663 (1998).

197. S. Y. Chan and A. Li Wan Po, *Int. J. Pharm.*, **55**, 1 (1989).

198. C. D. Yu, J. L. Fox, N. F. H. Ho, and W. I. Higuchi, *J. Pharm. Sci.*, **68**, 1341 (1979).

199. T. Loftsson and N. Bodor, *J. Pharm. Sci.*, **70**, 750 (1981).

200. K. B. Sloan, S. A. M. Koch, and K. G. Silver, *Int. J. Pharm.*, **21**, 251 (1984).

201. K. B. Sloan, E. F. Sherertz, and R. G. McTiernan, *Int. J. Pharm.*, **44**, 87 (1988).

202. B. Møllgaard, A. Hoelgaard, and H. Bundgaard, *Int. J. Pharm.*, **12**, 153 (1982).

203. K. B. Sloan, M. Hashida, J. Alexander, B. Bodor, and T. Higuchi, *J. Pharm. Sci.*, **72**, 372 (1983).

204. K. G. Siver and K. B. Sloan, *Int. J. Pharm.*, **48**, 195 (1988).

205. A. N. Saab, K. B. Sloan, H. D. Beall, and R. Villanueva, *J. Pharm. Sci.*, **79**, 1099 (1990).

206. K. B. Sloan, S. Selk, J. Haslam, L. Caldwell, and R. Shaffer, *J. Pharm. Sci.*, **73**, 1734 (1984).

207. K. B. Sloan, *Adv. Drug Deliv. Rev.*, **3**, 67 (1989).

208. F. P. Bonina, L. Montenegro, P. De Capraris, E. Bousquet, and S. Tirendi, *Int. J. Pharm.*, **77**, 21 (1991).

209. W. Wiegrebe, A. Retzow, E. Plumier, N. Ersoy, A. Garbe, H. P. Faro, and R. Kunert, *Arzneim.-Forsch.*, **34**, 48 (1984).

210. M. F. Powell, A. Magill, A. R. Becker, R. A. Kenley, S. Chen, and G. C. Visor, *Int. J. Pharm.*, **44**, 225 (1988).

211. M. Hashida, E. Mukai, T. Kimura, and H. Sezaki, *J. Pharm. Pharmacol.*, **37**, 542 (1985).

212. E. Mukai, K. Arase, M. Hashida, and H. Sezaki, *Int. J. Pharm.*, **25**, 95 (1985).

213. H. Bundgaard, A. Hoelgaard, and B. Møllgaard, *Int. J. Pharm.*, **15**, 285 (1983).

214. M. Johansen, B. Møllgaard, P. K. Wotton, C. Larsen, and A. Hoelgaard, *Int. J. Pharm.*, **32**, 199 (1986).

215. S. A. M. Koch and K. B. Sloan, *Int. J. Pharm.*, **35**, 243 (1987).

216. K. Sato, K. Sugibayashi, and Y. Morimoto, *Int. J. Pharm.*, **43**, 31 (1988).

217. T. Seki, T. Kawaguchi, and K. Juni, *Pharm. Res.*, **7**, 948 (1990).

218. T. Seki, T. Kawaguchi, K. Juni, K. Sugibayashi, and Y. Morimoto, *Proc. Int. Symp. Controlled Release Bioact. Mater.*, **18**, 513 (1991).

219. R. P. Waranis and K. B. Sloan, *J. Pharm. Sci.*, **77**, 210 (1988).

220. D. A. W. Bucks, *Pharm. Res.*, **1**, 148 (1984).

221. V. H. L. Lee and V. H. K. Li, *Adv. Drug Deliv. Rev.*, **3**, 1 (1989).

222. K. B. Sloan, Ed., *Prodrugs—Topical and Ocular Drug Delivery*, Marcel Dekker, New York, 1992.

223. V. H. L. Lee, *J. Controlled Release*, **11**, 79 (1990).

224. V. H. L. Lee and J. R. Robinson, *J. Ocular Pharmacol.*, **2**, 67 (1986).

225. H. Sasaki, K. Yamamura, T. Mukai, K. Nishida, J. Nakamura, M. Nakashima, and M. Ichikawa, *Crit. Rev. Ther. Drug Carrier Syst.*, **16**, 85 (1999).

226. A. I. Mandell, F. Stentz, and A. E. Kitabchi, *Ophthalmology*, **85**, 268 (1978).

227. D. A. McLure, T. Higuchi, and V. J. Stella, Eds., *Pro-drugs as Novel Drug Delivery Systems*, American Chemical Society, Washington, DC 1975, p. 224.

228. M. A. Hussain and J. E. Truelove, *J. Pharm. Sci.*, **65**, 1510 (1976).

229. N. Bodor and G. Visor, *Exp. Eye Res.*, **38**, 621 (1984).

230. N. Bodor and G. Visor, *Pharm. Res.*, **1**, 168 (1984).

231. T. L. Phipps, D. E. Potter, and J. M. Rowland, *J. Ocular Pharmacol.*, **2**, 225 (1986).

232. L. Z. Bito, *Exp. Eye Res.*, **38**, 181 (1984).

233. D. S. Chien and R. D. Schoenwald, *Biopharm. Drug Dispos.*, **7**, 453 (1986).

234. R. D. Schoenwald, J. C. Folk, V. Kumar, and J. G. Pier, *J. Ocular Pharmacol.*, **3**, 333 (1987).

235. R. D. Schoenwald and D. S. Chien, *Biopharm. Drug Dispos.*, **9**, 527 (1988).

236. H. Bundgaard, E. Falch, C. Larsen, G. L. Mosher, and T. J. Mikkelson, *J. Med. Chem.*, **28**, 979 (1985).

237. H. Bundgaard, E. Falch, C. Larsen, and T. J. Mikkelson, *J. Pharm. Sci.*, **75**, 36 (1986).

238. H. Bundgaard, E. Falch, C. Larsen, G. L. Mosher, and T. J. Mikkelson, *J. Pharm. Sci.*, **75**, 775 (1986).

239. G. L. Mosher, H. Bundgaard, E. Falch, C. Larsen, and T. J. Mikkelson, *Int. J. Pharm.*, **39**, 113 (1987).

240. T. Järvinen, P. Suhonen, S. Auriola, J. Vepsäläinen, A. Urtti, and P. Peura, *Int. J. Pharm.*, **75**, 249 (1991).

241. P. Suhonen, T. Järvinen, P. Rytkönen, P. Peura, and A. Urtti, *Pharm. Res.*, **8**, 1539 (1991).

242. H. Bundgaard, A. Buur, S. C. Chang, and V. H. L. Lee, *Int. J. Pharm.*, **33**, 15 (1986).

243. H. Sasaki, D. S. Chien, and V. H. L. Lee, *Pharm. Res.*, Suppl. 5, S98 (1988).

244. E. Duzman, C. C. Chen, J. Anderson, M. Blumenthal, and H. Twizer, *Arch. Ophthalmol.*, **100**, 1916 (1982).

245. E. Duzman, N. Rosen, and M. Lazar, *Br. J. Ophthalmol.*, **67**, 668 (1983).

246. A. Buur, H. Bundgaard, and V. H. L. Lee, *Int. J. Pharm.*, **42**, 51 (1988).

247. S. C. Chang, H. Bundgaard, A. Buur, and V. H. L. Lee, *Invest. Ophthalmol. Vis. Sci.*, **28**, 487 (1987).

248. H. Bundgaard, A. Buur, S. C. Chang, and V. H. L. Lee, *Int. J. Pharm.*, **46**, 77 (1988).

249. S. C. Chang, H. Bundgaard, A. Buur, and V. H. L. Lee, *Invest. Ophthalmol. Vis. Sci.*, **29**, 626 (1988).

250. S. C. Chang, D. S. Chien, H. Bundgaard, and V. H. L. Lee, *Exp. Eye Res.*, **46**, 59 (1988).

251. D. E. Potter, D. J. Shumate, H. Bundgaard, and V. H. L. Lee, *Curr. Eye Res.*, **7**, 755 (1988).

252. H. Sasaki, D. S. Chien, K. Lew, H. Bundgaard, and V. H. L. Lee, *Invest. Ophthalmol. Vis. Sci.*, Suppl. 29, 83 (1988).

253. H. Sasaki, D. S. Chien, H. Bundgaard, and V. H. L. Lee, *Pharm. Res.*, Suppl. 5, S164 (1988).

254. D. S. Chien, H. Sasaki, H. Bundgaard, A. Buur, and V. H. L. Lee, *Pharm. Res.*, **8**, 728 (1991).

255. A. Bar-Llan, N. I. Pessah, and T. H. Maren, *J. Ocular Pharmacol.*, **2**, 109 (1986).

256. J. D. Larsen, H. Bundgaard, and V. H. L. Lee, *Int. J. Pharm.*, **47**, 103 (1988).

257. J. D. Larsen and H. Bundgaard, *Int. J. Pharm.*, **51**, 27 (1989).

258. D. M. Maurice and S. Mishima in M. L. Sears, Ed., *Pharmacology of the Eye*, Springer-Verlag, New York, 1984, pp. 19–116.

259. D. S. Chien and R. D. Schoenwald, *Pharm. Res.*, **7**, 476 (1990).

260. V. F. Smolen and R. D. Schoenwald, *J. Pharm. Sci.*, **63**, 1582 (1974).

261. S. Yoshida and S. Mishima, *Jpn. J. Ophthalmol.*, **19**, 121 (1975).

262. M. Sugaya and S. Nagataki, *Jpn. J. Ophthalmol.*, **22**, 127 (1978).

263. H. D. Gambill, K. N. Ogle, and T. P. Kearns, *Arch. Ophthalmol.*, **77**, 740 (1967).

264. S. Mishima, *Invest. Ophthalmol. Vis. Sci.*, **21**, 504 (1981).

265. S. Matsumoto, T. Tsuru, M. Araie, and Y. Komuro, *Jpn. J. Ophthalmol.*, **26**, 338 (1982).

266. Y. Mitsui and Y. Takagi, *Arch. Ophthalmol.*, **65**, 626 (1961).

267. N. J. Haddad, N. J. Moyer, and F. C. Riley Jr., *Am. J. Ophthalmol.*, **70**, 729 (1970).

268. S. M. Meyer and F. T. Fraunfelder, *Ophthalmology*, **87**, 1177 (1980).

269. L. P. Balant, L. Z. Benet, H. Blume, et al., *Eur. J. Clin. Pharmacol.*, **40**, 123 (1991).

270. L. Mizen and G. Burton, *Pharm. Biotechnol.*, **11**, 345 (1998).

271. S. Hadad, T. B. Vree, E. Van Der Kleijn, and M. Bialer, *J. Pharm. Sci.*, **81**, 1047 (1992).

272. A. Buur, A. Yamamoto, and V. H. L. Lee, *J. Controlled Release*, **14**, 43 (1990).

273. J. D. Stuhler, H. Cheng, and B. R. Dorrbecker, *J. Pharm. Sci.*, **81**, 1071 (1992).

274. K. T. Hansen, P. Faarup, and H. Bundgaard, *J. Pharm. Sci.*, **80**, 793 (1991).

275. K. Takahashi, S. Tamagawa, J. Haginaka, H. Yasuda, T. Katagi, and N. Mizuno, *J. Pharm. Sci.*, **81**, 226 (1992).

276. U. E. Gasser, K. Jorga, C. Crevoisier, S. E. Hovens, and P. L. van Giersbergen, *Eur. Neurol.*, **41**, 206 (1999).

277. H. Baas, F. Zehrden, R. Selzer, R. Kohnen, J. Loetsc, and S. Harder, *Clin. Pharmacokinet.*, **40**, 383 (2001).

278. A. A. Sinkula, *Pro-drugs as Novel Drug Delivery Systems*, American Chemical Society, Washington, DC, 1975, pp. 116, 153.

279. V. J. Stella, W. N. A. Charman, and V. H. Naringrekar, *Drugs*, **29**, 445 (1985).

CHAPTER FIFTEEN

Retrometabolism-Based Drug Design and Targeting

NICHOLAS BODOR
Center for Drug Discovery
University of Florida
Gainesville, Florida
and
IVAX Research, Inc.
Miami, Florida

PETER BUCHWALD
IVAX Research, Inc.
Miami, Florida

Contents

Burger's Medicinal Chemistry and Drug Discovery
Sixth Edition, Volume 2: Drug Development
Edited by Donald J. Abraham
ISBN 0-471-37028-2   © 2003 John Wiley & Sons, Inc.

1 INTRODUCTION

Undeniably, drug research was the main scientific factor driving the considerable medical progress of the last century (1, 2). Medicinal chemistry has witnessed significant changes as a result of advancements in the elucidation of the molecular-biochemical mechanisms of drug action and also the result of other developments such as combinatorial chemistry and high-throughput screening. Despite all these, rational drug design, which would allow the development of effective pharmaceutical agents with minimal side effects, on as rational a basis as possible, is still an elusive goal. As already noticed by Paul Ehrlich, the first significant personality in medicinal and pharmaceutical chemistry, successful research still needs the four Gs (in German): *"Glück, Geduld, Geschick und Geld"* (that is, luck, patience, skill, and money). In fact, the situation seems to have worsened as a result of increasing regulation and the increased complexity of

drug research, which has arguably hampered true innovation. Despite exponentially increasing R&D expenditures, the number of launched new chemical entities (NCEs) has essentially stagnated in the last 25–30 years (3).

A main reason behind the alarmingly low success rate of most design processes is that, although pharmacologial potency is increased, toxicity concerns are often ignored. Most new therapeutic agents designed to bind to a specific receptor or found to have high activity ultimately are discarded when unacceptable toxicity or unavoidable side effects are encountered in later stages of the development. However, this should not be surprising for a number of reasons. Side effects are usually closely related to the intrinsic receptor affinity responsible for the desired activity. Furthermore, although in most cases the desired response should be localized to some organ or cell, the corresponding receptors are often distributed throughout the body. Finally, for

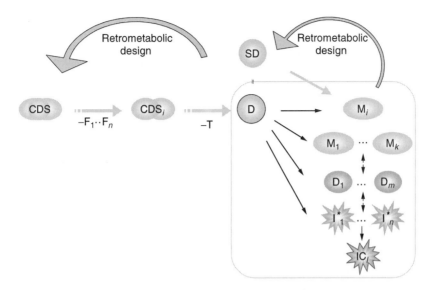

Figure 15.1. The retrometabolic drug design loop, including chemical delivery system (CDS) design and soft drug (SD) design. A schematic representation of possible metabolic pathways for drugs in general is also included (see text for details).

most drugs, metabolism generates multiple metabolites that can have a qualitatively or quantitatively different type of biological activity, including enhanced toxicity.

Consequently, drug design should focus not on increasing activity alone, but on increasing the therapeutic index (TI). This index reflects the degree of selectivity or margin of safety. It is usually defined as the ratio between the median toxic dose (TD_{50}) and the median effective dose (ED_{50}):

$$TI = TD_{50}/ED_{50} \qquad (15.1)$$

One must also take into account that metabolic conversion of a drug (D) can generate analog metabolites ($D_1 \cdots D_m$) that have structures and activities similar to the original drug but have different pharmacokinetic properties, other metabolites ($M_1 \cdots M_k$) including inactive ones (M_i), and potential reactive intermediates ($I_1^* \cdots I_n^*$) that can be responsible for various kinds of cell damage by forming toxic species (IC_l) (Fig. 15.1). Because all these compounds may be present simultaneously and in various time-dependent concentrations, in a rigorous approach, toxicity (T) has to be described as a combination of intrinsic toxicity/selectivity [$T_i(D)$] and toxicities attributed to various metabolic products:

$$T(D) = T_i(D) + T(D_1 \cdots D_m)$$
$$+ T(M_1 \cdots M_k) + T(I_1^* \cdots I_n^*) \qquad (15.2)$$

Therefore, as we have often argued, targeting and metabolism considerations should be an integral part of any drug design process, to ensure that the desired new chemical entity is designed with targeting and a preferred metabolic route in mind. That is, the molecule should be designed so that it has site specificity and selectivity built into its molecular structure. Site-specific delivery and site-specific action, if achievable and sufficient for efficacy, might alleviate undesired effects arising from intrinsic toxicity. By designing and predicting the major metabolic pathways, formation of undesired toxic, active, or high-energy intermediates might be avoidable. Surprisingly, the importance of early integration of metabolism, pharmacokinetic, and general physicochemical considerations in the drug design process has been clearly recognized in industrial settings only during the mid-1990s (4, 5), despite numerous publications describing these concepts and methodologies in the early 1980s (6–12).

2   RETROMETABOLIC DRUG DESIGN

2.1   Principles

Retrometabolic drug design approaches represent systematic methodologies that integrate structure-activity and structure-metabolism relationships and are aimed to design safe, locally active compounds with an improved therapeutic index (13–15). They include two distinct methods (Fig. 15.1). One is the design of *soft drugs* (6–12, 15, 16). Soft drugs are new, active therapeutic agents, often isosteric/isoelectronic analogs of a lead compound, with a chemical structure specifically designed to allow predictable metabolism into inactive metabolites after exerting the desired therapeutic effect. The other is the design of *chemical delivery systems* (CDS) (15, 17–23). A CDS is defined as a biologically inert molecule that requires several steps in its conversion to the active drug and that enhances drug delivery to a particular organ or site.

Although both approaches involve chemical modifications to obtain an improved therapeutic index and both require enzymatic reactions to fulfill drug targeting, the principles of soft drug and CDS design are distinct. Whereas the CDS is inactive by definition and sequential enzymatic reactions provide the differential distribution and drug activation, soft drugs are active therapeutic agents designed to be easily metabolized into inactive species. In an ideal situation, for a CDS, the drug is present at the site and nowhere else in the body because enzymatic processes produce the drug only at the site, whereas for a soft drug, the drug is present at the site and nowhere else in the body because enzymatic processes destroy the drug at those sites. Whereas CDSs are designed to achieve drug targeting at a selected organ or site, soft drugs are designed to afford a differential distribution that can be regarded as reversed targeting.

2.2   Terminology

These drug design approaches were designated as *retrometabolic* to emphasize that

metabolic pathways are designed going backward compared to actual metabolic processes in a manner somewhat similar to Corey's retrosynthetic analysis, in which synthetic pathways are designed going backward compared to actual synthetic laboratory operations.

2.2.1   Soft Drug versus Hard Drug. The soft drug terminology was originally introduced in 1976 (6, 24–26) to contrast Ariëns's theoretical, drug design concept of nonmetabolizable, hard drugs. Hard drugs do not undergo any metabolism and, hence, avoid the problems caused by reactive intermediates or active metabolites. For example, certain strongly lipophobic drugs, such as enalaprilat, lisinopril, cromolyn, and bisphophonates (e.g., alendronate), are not metabolized *in vivo*, and they can be regarded as hard drug examples (4).

Unfortunately, additional confusion was created with the introduction of the *antedrug* terminology by Lee in 1982 (27). The definition of the "antedrug" term (27, 28) is essentially identical to that of the soft drug (6, 24–26) and was introduced considerably later. Because the *ante-* prefix, which is very similar in meaning to the *pro-* prefix (e.g., prior to), implies the conceptual opposite of a soft drug (i.e., an inactive agent that has to be activated by metabolism), we suggest dropping the "antedrug" terminology altogether.

2.2.2   Soft Drug versus Prodrug. The difference between the *soft drug* and *prodrug* concepts should also be clarified because confusion related to these two different terms is still frequent (29). Prodrugs are pharmacologically inactive compounds that result from transient chemical modifications of biologically active species (30–35). This concept was introduced by Albert in 1958 (36), and the rationale for prodrug design is that the structural requirements needed to elicit a desired pharmacological action and those needed to provide optimal delivery to the targeted receptor sites may not be the same. Hence, a chemical change is introduced to improve some deficient physicochemical property, such as membrane permeability or water solubility, or to overcome some other problem, such as rapid elimination, bad taste, or a formulation difficulty. After administration, the prodrug, by virtue of its improved characteristics, is more systemically and/or locally available than the parent drug. However, before exerting its biological effect, the prodrug must undergo chemical or biochemical conversion to the active form. Therefore, in theoretical terms, the prodrug and the soft drug concepts are opposite to each other. Whereas prodrugs are, ideally, inactive by design and are converted by a predictable mechanism to the active drug, soft drugs are active *per se* and are designed to undergo a predictable and controllable metabolic deactivation.

2.2.3 Chemical Delivery System versus Prodrug.

A number of effects, such as poor selectivity, poor retention, and the possibility for reactive metabolites, may often conspire to decrease the therapeutic index of drugs masked as prodrugs. Chemical delivery systems (CDSs) were developed starting in the early 1980s to address these challenges (15, 17–23). The CDS concept evolved from the prodrug concept, but became essentially differentiated by the introduction of targetor moieties and by the employment of multistep activation. In addition to functional moieties, which are also contained by prodrugs and are included to provide protected or enhanced overall delivery, CDSs also contain targetor moieties responsible for targeting, site-specificity, and lock-in. CDSs are designed to undergo sequential metabolic conversions, disengaging the modifier functions and finally the targetor, after this moiety fulfills its site- or organ-targeting role.

3 SOFT DRUGS

As mentioned, soft drugs are new, active therapeutic agents, often isosteric/isoelectronic analogs of a lead compound, with a chemical structure specifically designed to allow predictable metabolism into inactive metabolites after exerting the desired therapeutic effect (6–12, 15, 16). Consequently, soft drugs are new therapeutic agents obtained by building into the molecule, in addition to the activity, an optimized deactivation and detoxification route. The desired activity is generally local, and the soft drug is applied or administered at

or near the site of action. Therefore, in most cases, they produce pharmacological activity locally, but their distribution away from the site results in a prompt metabolic deactivation that prevents any kind of undesired pharmacological activity or toxicity (see Fig. 15.9 below for an illustration).

In soft drug design, the goal is not to avoid metabolism, but rather to control and direct it. Inclusion of a metabolically sensitive moiety into the drug molecule makes possible the design and prediction of the major metabolic pathway and avoids the formation of undesired toxic, active, or high-energy intermediates. If possible, inactivation should take place as the result of a single, low-energy and high-capacity step that yields inactive species subject to rapid elimination. The most critical metabolic pathways are mediated by oxygenases. Because oxygenases exhibit not only interspecies but also interindividual variability and are subject to inhibition and induction (37), and because the rates of hepatic monooxygenase reactions are at least two orders of magnitude lower than the slowest of the other enzymatic reactions (38), it is usually desirable to avoid oxidative pathways as well as these slow, easily saturable oxidases. This suggest that the design of soft drugs should be based on scaffolds inactivated by hydrolytic enzymes. Rapid metabolism can be more reliably carried out by these ubiquitously distributed esterases. Not relying exclusively on metabolism or clearance by organs such as liver or kidney is an additional advantage because blood flow and enzyme activity in these organs can be seriously impaired in critically ill patients.

3.1 Enzymatic Hydrolysis

Carboxylic ester–containing chemicals are very efficiently hydrolyzed into the respective free acids by a class of enzyme designated as carboxylic ester hydrolases (EC 3.1.1). Unfortunately, because these widely occurring enzymes exhibit broad and overlapping substrate specificity toward esters and amides, and because, in many cases, their exact physiological role remains unclear, their classification is difficult (39–45). According to an older but still used classification system (46), the more important subclasses include carboxyl-

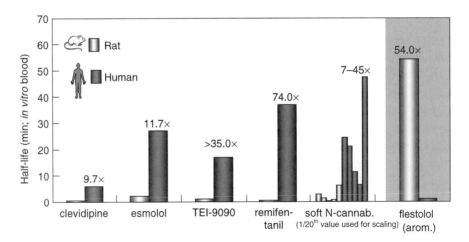

Figure 15.2. The interspecies variability of hydrolytic half-lives illustrated by rat versus human *in vitro* blood data.

esterase, EC 3.1.1.1 (carboxylic-ester hydrolase, ali-esterase, B-esterase, monobutyrase, cocaine esterase, etc.); arylesterase, EC 3.1.1.2 (A-esterase, paraoxonase); acetylcholinesterase, EC 3.1.1.7 (choline esterase I); and cholinesterase, EC 3.1.1.8 (choline esterase II, pseudocholinesterase, butyryl-choline esterase, benzoylcholinesterase, etc.).

Esterase activity not only depends on the substrate but also shows strong interspecies, interindividual, and interorgan variability (39, 45, 47, 48). For example, aliphatic esters tend to be metabolized much faster by rodents (rats, guinea pigs) than by humans, as illustrated by the *in vitro* blood data for clevidipine (49), esmolol (50, 51), isocarbacyclin methyl ester (TEI-9090) (52), remifentanil (53), soft cannabinoid analogs (54), and the aromatic ester-containing flestolol (55) in Fig. 15.2. Therefore, compared to the usual problems related to the extrapolation of animal test results to humans (56, 57), preclinical evaluation of ester-based soft drugs might be even more challenging and animal data less predictive of human clinical trial results.

Nevertheless, ester structures are of sufficient chemical stability to provide the shelf life required for drugs. One of the earliest therapeutic agents that made use of the advantages attainable by introduction of an ester moiety into the structure was etomidate (Amidate) (**1**) (Fig. 15.3). This is a unique short-acting nonbarbiturate hypnotic agent discovered in

1964 (58). It is eliminated by ester hydrolysis in plasma and liver (59). Etomidate is a potent intravenous (i.v.) hypnotic agent with a very rapid onset of action. However, its acid metabolite is inactive, and the duration of hypnosis after etomidate administration can be very short (<5 min) (60). Therefore, the therapeutic index of etomidate (18.0–32.0) is considerably larger than that of other hypnotic agents,

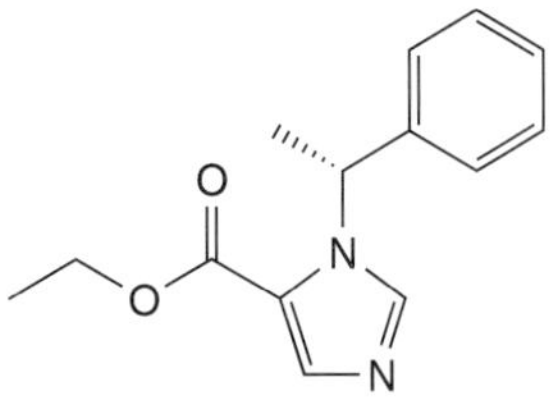

(1)

etomidate

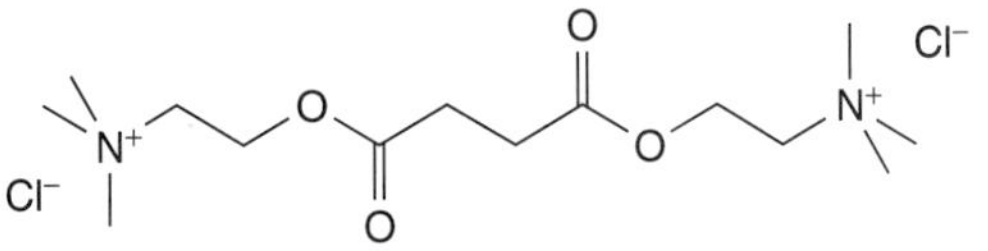

(2)

succinylcholine (chloride)

Figure 15.3. Etomidate (**1**), a short-acting hypnotic agent, and succinylcholine (**2**), a short-acting depolarizing neuromuscular agent, were early examples of compounds with short-acting and safe character provided by esterase mediated hydrolysis.

such as thiopental (2.5–4.3), methohexital (4.9–11.7), or propanidid (4.4–10.0) (60).

Short-acting ester-containing neuromuscular drugs, such as succinylcholine (Anectine, **2**) (Fig. 15.3) and mivacurium chloride (Mivacron) (61) designed to undergo hydrolysis by human plasma cholinesterase, also exploit similar principles, to ensure fast and spontaneous recovery upon cessation of administration. Hence, their durations of action are only 6–8 and 12–18 min, respectively, and the corresponding elimination routes remain functional even in renal failure.

3.2   Soft Drug Approaches

The soft drug concept was first introduced in 1976 (6), and reiterated in 1980 (24–26). A total of five major approaches have been identified (7, 8, 11, 12, 15, 16):

1. *Inactive metabolite–based soft drugs*: active compounds designed starting from a known (or hypothetical) inactive metabolite of an existing drug by converting this metabolite into an isosteric/isoelectronic analog of the original drug in such a way as to allow a facile, one-step controllable metabolic conversion after the desired therapeutic role has been achieved back to the very inactive metabolite from which the design started.

2. *Soft analogs*: close structural analogs of known active drugs that have a specific metabolically sensitive moiety built into their structure to allow a facile, one-step controllable deactivation and detoxication after the desired therapeutic role has been achieved.

3. *Active metabolite–based soft drugs*: metabolic products of a drug resulting from oxidative conversions that retain significant activity of the same kind as the parent drug. If activity and pharmacokinetic considerations allow it, the drug of choice should be the metabolite at the highest oxidation state that still retains activity.

4. *Activated soft drugs*: a somewhat separate class derived from nontoxic chemical compounds activated by introduction of a specific group that provides pharmacological activity. During expression of activity, the inactive starting molecule is regenerated.

5. *Pro-soft drugs*: inactive prodrugs (chemical delivery forms) of a soft drug of any of the above classes including endogenous soft molecules. They are converted enzymatically into the active soft drug, which is subsequently enzymatically deactivated.

Out of these approaches, the *inactive metabolite–based* and the *soft analog* approaches have been the most useful and successful strategies for designing safe and selective drugs. Both of these approaches focus on designing compounds that have a moiety that is susceptible to metabolic, preferentially hydrolytic, degradation built into their structure. This allows a one-step controllable decomposition into inactive, nontoxic moieties as soon as possible after the desired role is achieved and avoids other types of metabolic routes. Of course, judicious combination of *de novo* (e.g., receptor-based) design principles with general soft drug design principles can also result in *de novo* soft drugs.

3.3   Inactive Metabolite–Based Soft Drugs

This is a very versatile method for developing new and safe drugs, and it is indeed the soft drug design strategy that proved the most successful until now. The design starts from a known or designed inactive metabolite of a drug used as lead compound. Starting from the structure of this inactive metabolite, novel structures are designed that are isosteric and/or isoelectronic analogs of the drug from which the lead inactive metabolite was derived (*isosteric/isoelectronic analogy*). These new structures are designed to yield the starting inactive metabolite in one metabolic step (*metabolic inactivation*) without any other metabolic conversion (*predictable metabolism*). The specific binding and transport properties, as well as the metabolic rates of the new soft drugs can be controlled by structural modifications (*controlled metabolism*).

How much freedom one has for structural modifications in designing the new structures from the selected inactive metabolite depends on whether restrictive pharmacophore regions are involved in the formation of the inactive metabolite. Obviously, if they are not involved, there is more freedom for structural modifications, and one can deviate considerably from the requirement of the isosteric/isoelectronic analogy. Inactive metabolite–based soft drugs can also be obtained by starting, not from an actual isolated and identified inactive metabolite, but from a designed useful inactive metabolite (a hypothetical inactive metabolite). To illustrate the general principles, a number of drug classes will be reviewed, starting with those that already resulted in marketed drugs.

3.3.1 Soft β-blockers.

Because in this class inactive metabolite–based soft drugs can be obtained by introducing the hydrolytically sensitive functionality at a flexible pharmacophore region, there is considerable freedom for structural modifications. Consequently, transport and metabolism properties can be controlled more easily.

Fig. 15.4 compares the metabolism of the well-known β-blocker metoprolol (3) with that of the soft drugs (8) designed starting from one of its inactive metabolites (7). Metoprolol is extensively metabolized by the hepatic monooxygenase system both at the more restrictive β-amino alcohol pharmacophore region (resulting in 4) and at the more flexible pharmacophore region *para* to the phenol ring (resulting in 5 and 6) (62–64). Two of these metabolites, α-hydroxymetoprolol (5) and O-demethylmetoprolol (6), have selective β_1-blocker activity, but are 5–10 times less potent than metoprolol (62). The main metabolites detected are, however, the acids (4) and (7), and they are devoid of β-adrenoceptor activity or toxicity (LD$_{50}$ in mice is greater than 500 mg/kg i.v.) (62). Hence, the phenylacetic acid derivative (7) can be used as starting point for an inactive metabolite approach. By esterification of (7) and by introduction of some additional flexibility in the design (e.g., $n = 0$ or 2), a number of soft β-blockers structures (8) can be obtained with different receptor binding, transport, rate of cleavage, and metabolic properties.

3.3.1.1 Adaprolol.

If membrane transport (lipophilicity) and relative stability are important for pharmacological activity, then the R group of (8) should be relatively lipophilic and impart ester stability. From the various soft β-blockers developed in our laboratory ($n = 1$),

adaprolol, the adamantane ethyl derivative (10), was selected as a potential candidate as a topical antiglaucoma agent (65–69). Adaprolol produces prolonged and significant reduction of intraocular pressure (IOP), but it hydrolyzes relatively rapidly (66, 67). Therefore, local activity can be separated from undesired systemic cardiovascular/pulmonary activity, a characteristic much sought after in the search for antiglaucoma therapy (70). Following unilateral ocular treatment with adaprolol, no effects are produced in the contralateral eye because of systemic inactivation.

Several clinical studies of adaprolol maleate have already been completed, and no severe or clinically significant medical events have been reported. A double-masked comparison of adaprolol and timolol performed on 67 ocular hypertensive patients with IOP greater than 21 mmHg demonstrated that intraocular pressure was significantly reduced throughout the study in all treatment groups. Adaprolol reduced IOP by about 20%, whereas timolol reduced IOP by 25–30% (16). In patients over 70 years old, the IOP-reducing effects of 0.2% adaprolol and 0.5% timolol were statistically indistinguishable after 10 days of application (16). On the other hand, timolol reduced the systolic blood pressure with statistical significance, whereas neither of the adaprolol concentrations tested demonstrated such change. Timolol also showed a trend, although not statistically significant, to reduce the heart rate, whereas pulse was conserved in both adaprolol treatment groups. Hence, adaprolol has a safer cardiovascular profile than that of timolol, especially in the population over 70 years old.

If systemic ultrashort-action is the objective, then R groups that make (8) susceptible to rapid hydrolysis should be used. With such agents that have short half-lives ($\sim$15 min), steady-state plasma concentrations and readily adjustable effects can be rapidly achieved on intravenous administration and infusion. Also, drug effects can be rapidly eliminated by termination of the infusion. For example, a number of methyl-thiomethyl and related esters (8) ($n = 1$, R = CH_2SCH_3, CH_2SOCH_3, $CH_2SO_2CH_3$) were found as ultrashort acting (71). When injected intravenously, these compounds hydrolyzed ex-

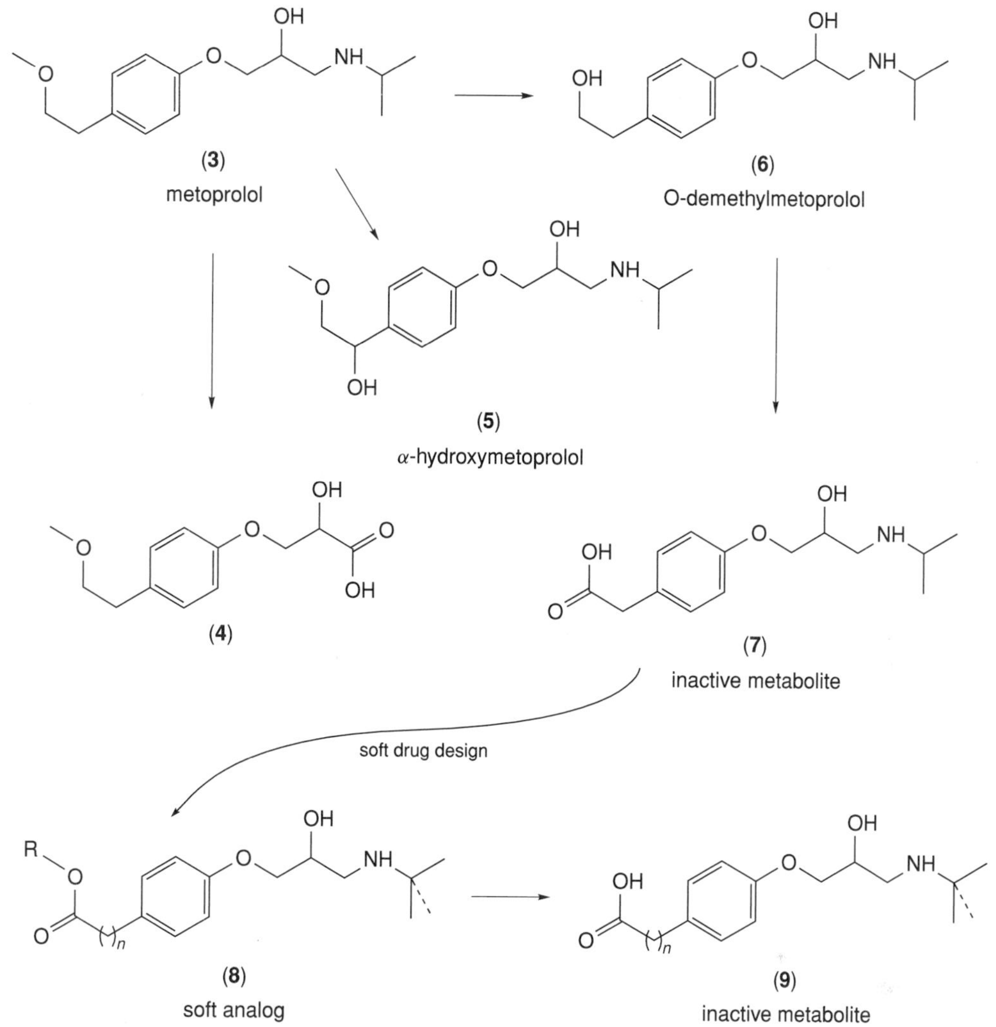

Figure 15.4. Comparison of the metabolism of metoprolol (**3**) with that of the soft drugs (**8**) designed starting from one of its inactive metabolites (**7**). The dashed line in structures (**8**) and (**9**) denotes the possibility of having either isopropyl- or *tert*-butyl–substituted amines.

tremely fast, much faster than simple alkyl esters. Many other compounds within this family of structures (**8**) (with $n = 0$ or 2), have been developed and tested in different laboratories, and a number of them have proved quite successful.

3.3.1.2 Esmolol. Esmolol (Brevibloc, **11**) (Fig. 15.5) is an ultrashort-acting (USA) β-blocker that relies on rapid metabolism by serum esterases (72, 73). By the late 1970s, it was already known that insertion of an ester moiety between the aromatic ring and the

β-amino alcohol side chain (74) or at the more remote *para* position (75) might not significantly affect β-blocking activity. It was also shown that the acid metabolites (**9**) ($n = 0, 1$) are devoid of β-adrenoceptor activity (62, 75). Somewhat later, following a systematic search of different β-blocker series that contained ester moieties inserted at different positions, esmolol was selected as the best candidate for development (72, 76–78).

For compounds of the general structure (**8**) in these series, duration of action decreased as

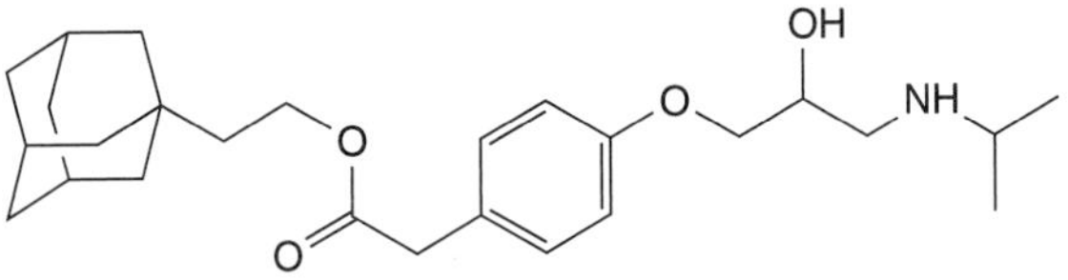

(10)

adaprolol

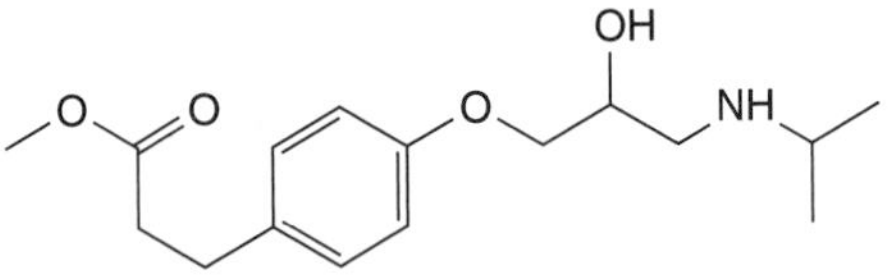

(11)

esmolol

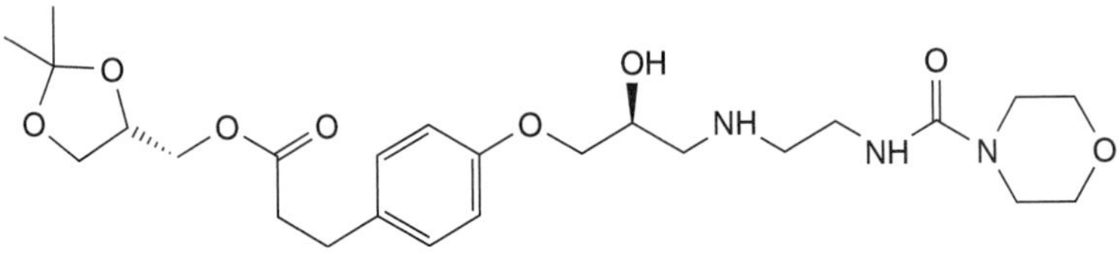

(12)

landiolol

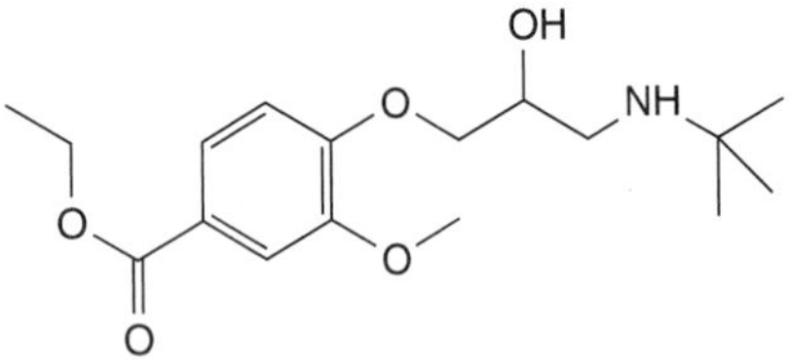

(13)

vasomolol

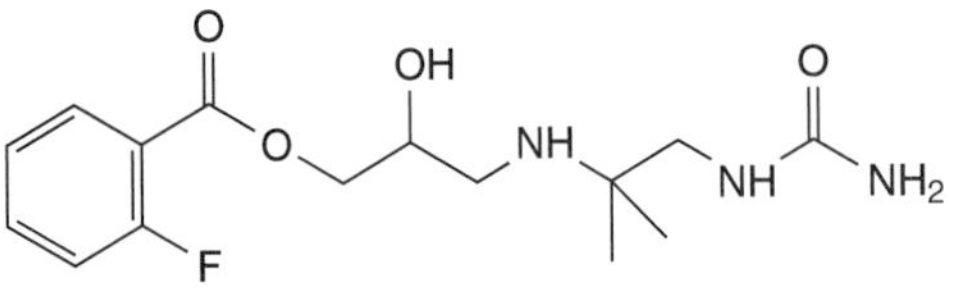

(14)

flestolol

Figure 15.5. Ester-containing β-blocker structures.

n increased from 0 to 2, probably because more rapid hydrolysis as steric hindrance decreased. Esmolol ($n = 2$, R = CH$_3$) was the fourth β-blocker to be approved by the FDA (1986) for intravenous clinical use, but it differs from the previous three (propranolol, metoprolol, labetalol) because its pharmacological effects dissipate within 15–20 min after stopping the drug (79). The elimination half-life of esmolol (5–15 min) is indeed considerably shorter than that of propranolol (2–4 h) (79). Its acid metabolite formed by hydrolysis of the ester group (**9**) ($n = 2$) has a relatively long half-life (3.7 h), but it is indeed inactive. Its β-adrenoceptor antagonist potency is about 1500 times lower than that of esmolol, and it is unlikely to exert any clinically significant effects during esmolol administration (79).

The *in vitro* human blood half-life of esmolol is around 25–27 min (50, 51). As a nice confirmation of the soft drug design principles, the presence of other ester-containing drugs, such as acetylcholine, succinylcholine, procaine, or chloroprocaine, have been shown to have no effect on this hydrolytic half-life, and consequently no metabolic interactions are to be expected (50).

3.3.1.3 Landiolol. Landiolol (ONO-1101, **12**) (Fig. 15.5) is a more recently developed

USA β-blocker with improved cardioselectivity (80–83). Modification of the R ester group of esmolol did not afford compounds superior to esmolol in β-blocking activity, duration of action, or cardioselectivity. However, additional modifications resulted in landiolol that, compared to esmolol, has a ninefold increased *in vivo* β-blocking activity and an eightfold increased *in vitro* β_1/β_2 cardioselectivity (255 versus 32) (80). Its structure (**12**) contains a morpholinocarbonylamino moiety and has S-configured hydroxyl and ester functions. It has proved to be a potent β-antagonist with effects that are removed quickly by washout (81–83). Landiolol was found to have an elimination half-life (2.3–4.0 min, *in vivo*, human) shorter than that of any other β-blocker (83).

3.3.1.4 Vasomolol, Flestolol, and Other Structures.
Vasomolol (**13**) was developed along quite similar design principles. It is a vanilloid-type USA β_1-adrenoceptor antagonist that has vasorelaxant activity and is devoid of intrinsic sympathomimetic activity (84). As with other similar agents, vasomolol infusion was characterized by a rapid onset of action. Steady state of β-blockade was attained within 10 min after initial infusion, and a rapid recovery from blockade occurred after discontinuation of the infusion (84).

Because relatively early in these studies, insertion of an ester moiety between the aromatic ring and the β-amino alcohol side chain was found to preserve reasonable β-blocking activity (74, 78), a variety of such structures were also synthesized and investigated. Half-lives in blood and liver suggested that ester hydrolysis is the major pathway for the inactivation of these [(arylcarbonyl)-oxy]propanolamines. A bulky 2-CH_3 substituent prevented the hydrolysis of the ester, but the 2-F substituent, which offered minimum steric hindrance but maximum electron-withdrawing effect, was a promising aromatic substituent for short action (78). Flestolol (**14**, ACC-9089) (Fig. 15.5), a compound with such a 2-F substituent, was selected for further toxicological evaluation and clinical study. Interestingly, an investigation of the metabolism of flestolol and other esters found polymorphic rates of ester hydrolysis in New Zealand white rabbit blood and cornea (55). About 30% of the animals studied were found as "slow" metaboliz-

ing ($t_{1/2}$ = 17 min, *in vitro*, blood) and about 70% were found as "fast" metabolizing ($t_{1/2} <$ 1 min). No such bimodal distribution of esterase activity was found in blood from rats, dogs, and humans or in the aqueous humor and iris–ciliary body complex of rabbits (55).

Another β-antagonist, which has a structure similar to the general structure (**8**) ($n =$ 2) but contains a reversed ester in its *para*-positioned side chain (L-653,328) (85), was also claimed to be a soft drug (70). The ocular instillation of 2% of this drug to human volunteers resulted in a reduction in IOP; however, this was less than that elicited by 0.5% timolol. Nevertheless, in contrast to timolol, there was no evidence of systemic β-adrenoceptor blockade (86). In addition, cumulative concentrations of L-653,328 up to 4% did not cause bronchoconstriction in asthmatic patients (87). Despite these, L-653,328 cannot be considered a true soft drug because its hydrolytic cleavage releases an active alcohol, L-652,698. In fact, L-653,328 was originally designed as the acetate ester prodrug of this active alcohol, and both the ester L-653,328 and the alcohol L-652,698 have modest β-receptor blocking activity. The K_i value for displacement of ^{125}I-iodocyanopindolol binding to β_1-binding sites in membrane fractions of rabbit left ventricle is somewhat smaller for the ester: 3.1 versus 5.7 μM, and the more lipophilic ester causes somewhat better IOP-lowering (85). As it turns out, this case represents neither an ideal prodrug nor an ideal soft drug design. The lack of systemic effects is attributed to the rapid oxidation of the alcohol in the systemic circulation into inactive carboxylic acid metabolites (86).

3.3.1.5 Soft Bufuralol Analogs.
Bufuralol (**15**) is a potent, nonselective β-adrenoceptor antagonist with β_2 partial agonist properties. Its effectiveness in the treatment of essential hypertension is probably the result of a favorable balance of β-blockade and β_2-agonist–mediated vasodilation. Bufuralol undergoes complex metabolism in humans, including stepwise oxidation toward an acid metabolite (**18**) through the corresponding hydroxy (**16**) and keto (**17**) intermediates (Fig. 15.6). These intermediates are still active (88, 89) and have different (interestingly, longer) elimination half-lives (90). Not only does a differential me-

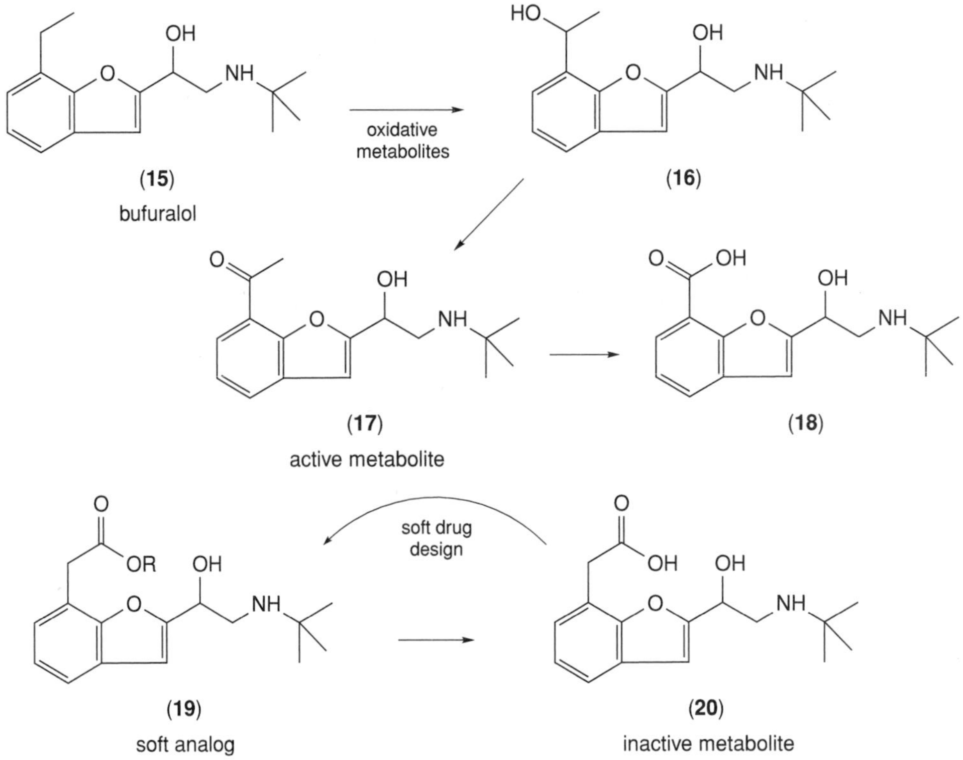

Figure 15.6. Metabolism of bufuralol (**15**) produces oxidative, active metabolites (**16**), (**17**) that lead to a final inactive acid metabolite (**18**). Starting from a designed (hypothetical) inactive metabolite (**20**), a series of inactive metabolite–based soft compounds (**19**) were designed.

tabolism of the two enantiomers occur, but differences attributed to genetic polymorphism are also encountered (91).

A soft drug approach may help avoid these problems, and recently, a number of ester-containing soft drug candidates (**19**) were synthesized and tested for β-antagonist activity by recording ECG and intra-arterial blood pressure in rats (92). This is an example of a hypothetical inactive metabolite–based approach, given that the retrometabolic design starts not from an actual, major metabolite, but from a hypothetical one (**20**). Nevertheless, this metabolite was confirmed to be inactive during the study: it did not decrease the heart rate significantly. Although in the isoproterenol-induced tachycardia model bufuralol at an i.v. dose of 1 mg/kg diminished heart rate for at least 2 h, the effects of equimolar soft drugs lasted for only 10–30 min (Fig. 15.7). The effects of the four most active compounds (**19**, R

= methyl, ethyl, isopropyl, *tert*-butyl) on resting heart rate and mean arterial pressure were also evaluated in comparison to esmolol following infusion (Fig. 15.7). These new soft drugs produced effects that were similar to that produced by esmolol both in magnitude and time course, and the corresponding infusion rates were 10 times smaller.

3.3.2 Soft Opioid Analgetics. Several short-acting 4-anilidopiperidine opioids, such as fentanyl (**21a**), sufentanil (**21b**), or alfentanil, have been introduced into anesthetic practice because they showed advantages over morphine. They do not cause significant histamine release and have shorter durations of action than that of morphine. Therefore, they can provide greater cardiovascular stability and less persistent postoperative respiratory depression. However, their terminal half-lives in humans are still longer than desired; for ex-

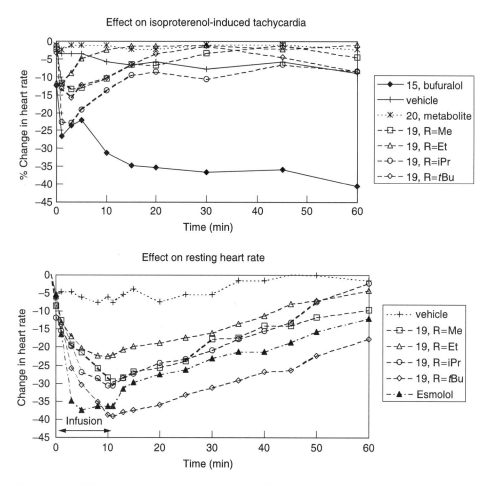

Figure 15.7. Effects of bufuralol-related soft drugs (**19**) on isoproterenol-induced tachycardia (i.v. bolus of 3.8 μmol/kg; vehicle 10% DMSO in 30% hydroxypropyl β-cyclodextrin) and resting heart rate (i.v. infusion, 2 μmol/kg/min, R = Et: 4 μmol/kg/min, esmolol: 20 μmol/kg/min; vehicle 0.9% NaCl) in rats. Data represent the means of at least three animals; error bars were omitted for better visibility.

ample, even the shortest-acting alfentanil has $t_{1/2} \approx 70$–90 min. This can result in drug accumulation and prolonged durations of action after multiple bolus injections or infusion. In addition, hepatic dysfunction may result in prolonged retention because the elimination of these compounds relies on hepatic metabolism (93).

A hypothetical inactive metabolite–based soft drug approach proved useful in solving these problems. A first attempt to incorporate ester and carbonate moieties into structure (**21**) (Fig. 15.8) at the R_2 side-chain (R_2 = —CH_2O_2CR', —CH_2O_2COR') produced potent analgetics, but durations of action were

still longer than desired (94). However, another design based on esterification of the hypothetical inactive metabolite (**23**) yielded remifentanil (**22a**, Ultiva), a unique ultra-short-acting opioid analgesic (53, 95–97).

Even if there is no evidence for the opioid analgetics represented by (**21**) (Fig. 15.8) to metabolize into the acids (**23**) (93), structures of type (**22**) can represent possible soft, short-acting compounds susceptible to hydrolytic inactivation. During the synthesis and pharmacological evaluation of a number of such compounds (53), it was established that the carfentanil piperidine (R_2 = CO_2CH_3) provided more potent analgetics than the fenta-

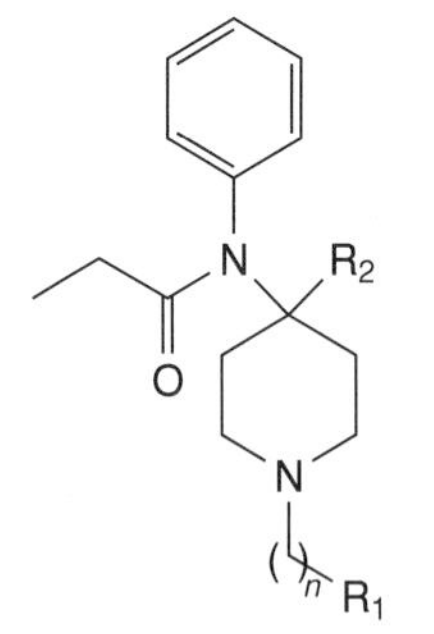

		n	R$_1$	R$_2$
(21a)	fentanyl	2	phenyl	H
(21b)	sufentanil	2	2-thienyl	CH$_2$OCH$_3$
(21c)	carfentanil	2	phenyl	CO$_2$CH$_3$

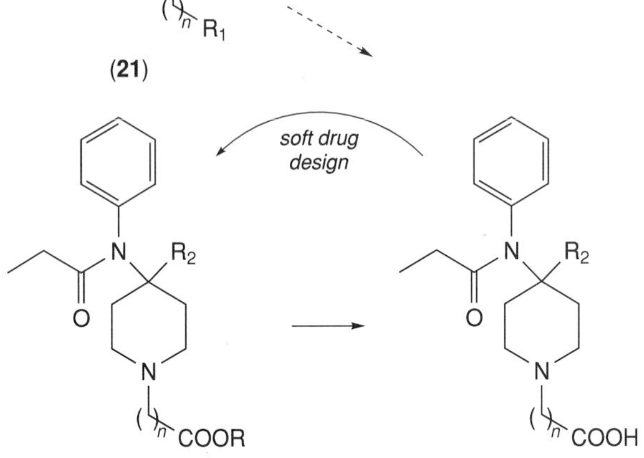

(21)

soft drug design

(22)

(23)

		n	R	R$_2$
(22a)	remifentanil	2	methyl	CO$_2$CH$_3$

Figure 15.8. Design of soft opioid analgetics (**22**) based on the hypothetical inactive metabolite (**23**).

nyl nucleus (R$_2$ = H). A separation (n) of two methylene units between the piperidine nitrogen and the ester function was found as optimal for added potency and decreased duration of action (53). Durations of actions, as measured *in vivo* by a classic rat tail withdrawal assay (98), ranged from extremely short to long (5–85 min), depending on the substitution of the alkyl portion of the ester (R). One of these compounds, remifentanil (**22a**) (Fig. 15.8), was approved by the FDA in 1996 for

clinical use as an ultrashort-acting opioid analgetic (Ultiva) during general anesthesia and monitored anesthesia care. Remifentanil has a half-life of 37 min in human whole blood (*in vitro*) and is nearly quantitatively converted to the corresponding acid (**23**). Furthermore, the carboxylic acid (**23**) was indeed found to be approximately 1000 times less potent in the *in vitro* guinea pig ileum assay and 350 times less potent in the *in vivo* rat tail withdrawal reflex model than its parent drug (**22**). Remifentanil

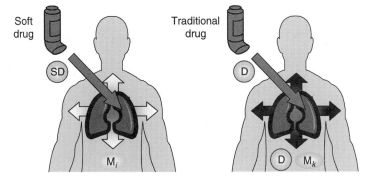

Figure 15.9. Illustration of the difference between soft (SD) and traditional (D) drugs for inhaled corticosteroids. For soft drugs, the designed-in metabolism rapidly deactivates any fraction that reaches the systemic circulation; hence, the local effect is accompanied by no or just minimal side effect.

is roughly equivalent in potency to fentanyl in the guinea pig ileum assay (EC_{50} values of 2.4 and 1.8 nM, respectively), and its effect can be antagonized by naloxone, an opiate antagonist (95). In 24 patients undergoing elective inpatient surgery, its terminal half-life ranged from 10 to 21 min, whereas that of its major metabolite (**23**) ($n = 2$, $R_2 = CO_2CH_3$) ranged from 88 to 137 min (97).

In the case of remifentanil it was also proved that, as predicted by the basic principles used in soft drug design, the possibility of drug interactions could be minimized by building metabolic considerations into the structure. Clearance, volume of distribution, and terminal half-life data indicated that co-administration of esmolol has no significant ($P < 0.05$) effect on the pharmacokinetics (or pharmacodynamics) of remifentanil in rats, despite both drugs being metabolized by non-specific esterases (99, 100).

3.3.3 Soft Corticosteroids.

One of the most active fields for soft drug design was the field of anti-inflammatory corticosteroids. Traditional corticosteroids are very useful drugs, but they are known to have multiple adverse effects that seriously limit their usefulness. Even if corticosteroids are most often applied only topically, significant portions of the topically applied drugs (e.g., lung, nasal mucosa, gastrointestinal tract, eye, or skin) reach the general circulatory system. Consequently, the resulting systemic side effects, such as adrenal suppression, effects on bone and growth, skin thinning and easy bruising, or increased risk of cataracts and glaucoma, together with local side effects, such as oral candidiasis or dysphonia, limit their use. Corticosteroids are also subject to different oxidative and/or reductive metabolic conversions, and formation of various steroidal metabolites can lead to undesirably complex situations. A considerable number of attempts were aimed to improve this situation, and soft drug approaches are particularly well suited for this purpose (Fig. 15.9).

There is a frequent misconception regarding soft drugs, particularly soft steroids, that has to be clarified. Often, the soft nature is associated with fast hydrolytic degradation, but this is not necessarily so. If hydrolysis is too rapid, then only weak activity may be ob-tained. The desired increase of the therapeutic index can be achieved only if the drug is sufficiently stable to reach the receptor sites at the target organ and to produce its desired effect, but the free, non-protein-bound drug undergoes facile hydrolysis to avoid unwanted, systemic side effects. To successfully separate the desired local activity from systemic toxicity, an adequate balance between intrinsic activity, solubility/lipophilicity, tissue distribution, protein binding, and rate of metabolic deactivation has to be achieved. In the case of slow, sustained release to the general circulatory system from the delivery site, even a relatively slow hydrolysis could result in a very low, almost steady-state systemic concentration.

3.3.3.1 Loteprednol Etabonate. Loteprednol etabonate (**26**) (Figs. 15.10 and 15.11) is an active corticosteroid that lacks serious side effects and that received final FDA approval in 1998 as the active ingredient of two ophthalmic preparations, Lotemax and Alrex (101, 102). Currently, it is the only corticosteroid approved by the FDA for use in all inflammatory and allergy-related ophthalmic disorders, including inflammation after cataract surgery, uveitis, allergic conjunctivitis, and giant papillary conjunctivitis. Loteprednol etabonate resulted from a classic inactive metabolite–based soft drug approach (103–114).

Hydrocortisone (**24**) (R_1, R_2, R_3, X_1, $X_2 = H$, no Δ^1; Fig. 15.10) undergoes a variety of oxidative and reductive metabolic conversions (115). Oxidation of its dihydroxyacetone side-chain leads to formation of cortienic acid (**25**) through a 21-aldehyde (21-dehydrocortisol) and a 21-acid (cortisolic acid). Cortienic acid is an ideal lead for the inactive metabolite approach because it lacks corticosteroid activity and is a major metabolite excreted in human urine. To obtain active compounds, the important pharmacophores found in the 17α and 17β side-chains had to be restored. Suitable isosteric/isoelectronic substitution of the α-hydroxy and β-carboxy substituents with esters or other types of functions should restore the original corticosteroid activity and also incorporate hydrolytic features to help avoid accumulation of toxic levels. More than 120 of such soft steroids (**24**) that resulted from modifications of the 17β-carboxyl function and the 17α-hydroxy function together with other

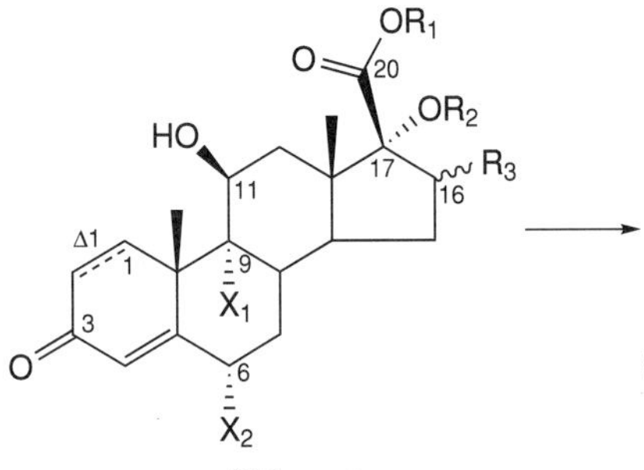

(**24**)

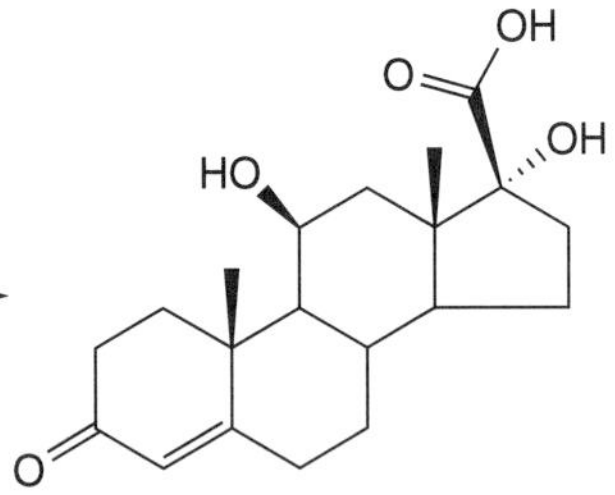

(**25**)

cortienic acid
(inactive metabolite)

R_1 = alkyl, haloalkyl, etc.;
R_2 = alkyl, alkoxyalkyl, COOalkyl, etc.;
R_3 = H, α- or β-CH_3;
X_1, X_2 = H, F;
Δ^1 = double bond

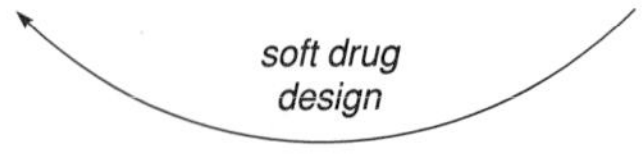

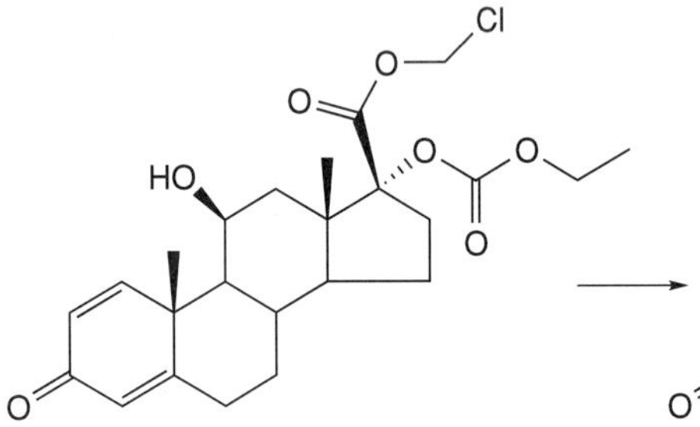

(**26**)
loteprednol
etabonate

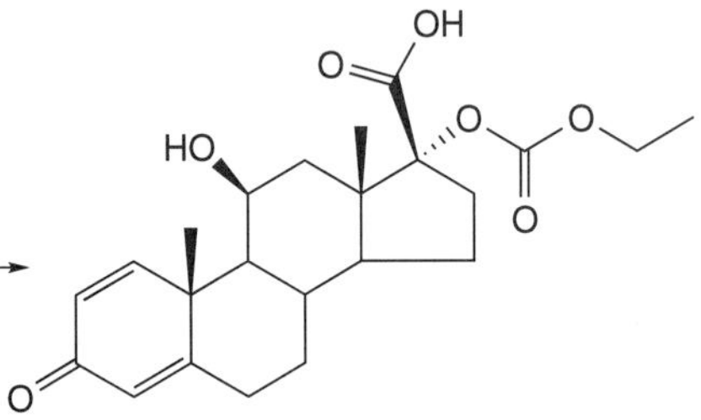

(**27**)

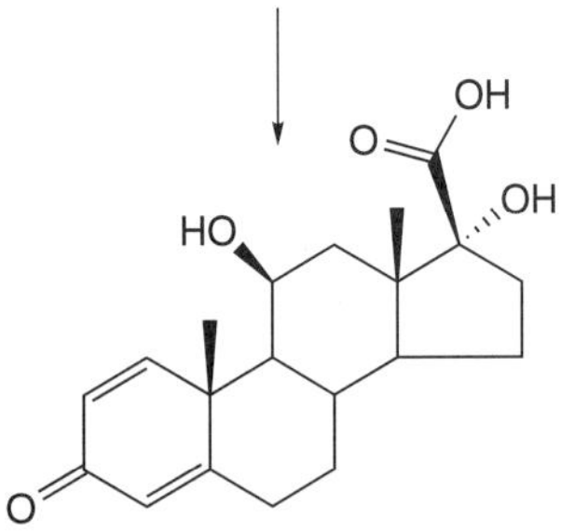

(**28**)

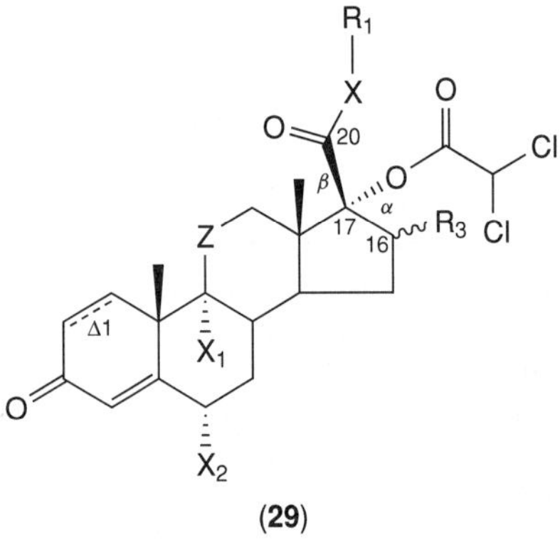

(**29**)

R_1: C_1-C_4 alkyl, C_1-C_4 alkoxy, C_1-C_4 alkylthio
R_3: H, α- or β-CH_3, α- or β-OH, $=CH_2$
X_1: H, F, Cl
X_2: H, F, Cl, $-CH_3$
X: O, S
Z: C=O, β-CHOH, β-CHCl
Δ^1: double bond (present or absent)

Figure 15.10. The inactive metabolite–based design and the metabolism of loteprednol etabonate (**26**). The general structure of the new 17α-dichloroacetate ester soft steroids is also included (**29**).

(a)

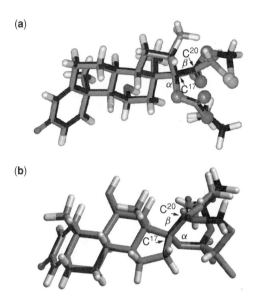

(b)

Figure 15.11. Overlapping pharmacophore structures of corticosteroids. (a) Clobetasol propionate (in gray) and the soft corticosteroid loteprednol etabonate (**26**) (in black). The view is from the α side, from slightly below the steroid ring system. (b) Loteprednol etabonate (in black) and a 17α-dichloroester soft steroid (in gray). This view is from the β side, from above the steroid ring system. See color insert.

changes intended to enhance corticosteroid activity [introduction of Δ^1, fluorination at $6\alpha(X_2)$ and/or $9\alpha(X_1)$, methylation at 16α or 16$\beta(R_3)$] have been synthesized at Otsuka Pharmaceutical Co. (Japan).

A haloester in the 17β position and a novel carbonate (105) or ether (116) substitution in the 17α position were found as critical functions for activity. Incorporation of 17α carbonates or ethers was preferred over 17α esters, to enhance stability and to prevent formation of mixed anhydrides that might be produced by reaction of a 17α ester with a 17β acid functionality. Such mixed anhydrides were assumed toxic and probably cataractogenic. A variety of 17β esters were synthesized. Because this position is an important pharmacophore that is sensitive to small modifications, the freedom of choice was relatively limited. For example, although chloromethyl or fluoromethyl esters showed very good activity, the chloroethyl or α-chloroethylidene derivatives demonstrated very weak activity.

Simple alkyl esters also proved virtually inactive. Consequently, the 17β-chloromethyl ester was held constant and 17α-carbonates with different substituents on the steroid skeleton were varied for further investigation. Loteprednol etabonate, and some of the other soft steroids, provided a significant improvement of the therapeutic index determined as the ratio between the anti-inflammatory activity and the thymus involution activity (114, 117, 118). Furthermore, binding studies using rat lung cytosolic corticosteroid receptors showed that some of the compounds approach and even exceed the binding affinity of the most potent corticosteroids known.

Loteprednol (**26**) was selected for development based on various considerations including the therapeutic index, availability, synthesis, and "softness" (the rate and easiness of metabolic deactivation). Early studies in rabbits (106, 109) and rats (110) demonstrated that, consistent with its design, (**26**) is indeed active, is metabolized into its predicted metabolites (**27, 28**) (Fig. 15.10), and these metabolites are inactive (105). The pharmacokinetic profile of loteprednol indicated that, when absorbed systemically, it is rapidly transformed to the inactive metabolite (**27**) and eliminated from the body mainly through the bile and urine (110, 111, 113). It did not affect the intraocular pressure in rabbits (109), an observation confirmed later in various human studies (Fig. 15.12) (119). Consistent with the soft nature of this steroid, systemic levels or effects cannot be detected even after chronic ocular administration (120).

Clinical studies proved that it is a safe and effective treatment for contact lens–associated giant papillary conjunctivitis (GPC), seasonal allergic conjunctivitis, postoperative inflammation, or uveitis (101, 102). Based on promising results from animal studies (112–114), loteprednol etabonate is also being developed for treatment of asthma, rhinitis, colitis, and dermatological problems.

3.3.3.2 17α-Dichloroester Soft Steroids. Recently, a new class of soft steroids with 17α-dichloroester substituent has been identified (**29**) (Fig. 15.10) (121). This is a unique design: no known corticosteroid contains halogen substituents at the 17α position. Nevertheless, the pharmacophore portions of these steroids,

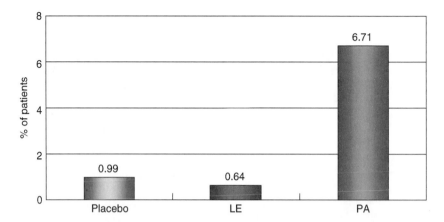

Figure 15.12. Pooled data showing the percent of patients with IOP elevation greater than 10 mmHg among patients not wearing contact lenses and treated for more than 28 days. The number of patients within each group was as follows: placebo, $n = 304$; loteprednol etabonate (LE), $n = 624$; prednisolone acetate (PA), $n = 164$ (119).

including the halogen atoms at 17α, can be positioned so as to provide excellent overlap with those of the traditional corticosteroids (Fig. 15.11b). Dichlorinated substituents seem required for activity and sufficiently soft nature, and two justifications seem likely. First, with dichlorinated substituents, one of the Cl atoms will necessarily point in the direction needed for pharmacophore overlap, but with monochlorinated substituents, steric hindrance will force the lone Cl atom to point away from this desired direction. Second, whereas dichloro substituents increase the second-order rate constant k_{cat}/K_M of enzymatic hydrolysis in acetate esters by a factor of about 20 compared to that of the unsubstituted ester, monochloro substituents do not cause any change (122).

Contrary to loteprednol-type soft steroids (Fig. 15.10), in this class of soft steroids, hydrolysis primarily cleaves not the 17β-positioned, but the 17α-positioned ester. Nevertheless, the corresponding metabolites are also inactive. Selected members of this class have shown better receptor binding than that of loteprednol etabonate; were proven as effective as, or even more effective than, budesonide in various asthma models; and, in agreement with their soft nature, were found as having low toxicity in animal models and in human clinical trials (121, 123).

3.3.3.3 Fluocortin Butyl. Fluocortin butyl (**30**) (Fig. 15.13) is an anti-inflammatory steroid obtained in one of the early approaches aimed at integrating ester moieties into steroid structures. Metabolism studies on fluocortolone revealed a number of oxidative and reductive metabolites in human urine (124), including fluocortolone-21-acid, an inactive metabolite. Synthesis and pharmacological evaluation of its different ester derivatives yielded fluocortin butyl (**30**, Vaspit, Novoderm, Varlane), the butyl ester of a C-21 carboxy steroid (125–128). The ester is an anti-inflammatory agent of rather weak activity, and any portion absorbed systemically following topical application is hydrolyzed into inactive species.

However, the low intrinsic activity of this steroid hindered its widespread use. The glucocorticoid receptor affinity and the topical anti-inflammatory potency of fluocortin butyl (**30**) are severalfold lower than those of dexamethasone (126). Fluocortin butyl ameliorated allergic rhinitis at daily doses of 2–8-mg divided into two to four daily inhalations (129, 130), but it did not protect against bronchial obstruction in bronchial provocation tests, even at 8-mg doses divided into four daily inhalations, in contrast to a 10 times lower dose of beclomethasone dipropionate (131).

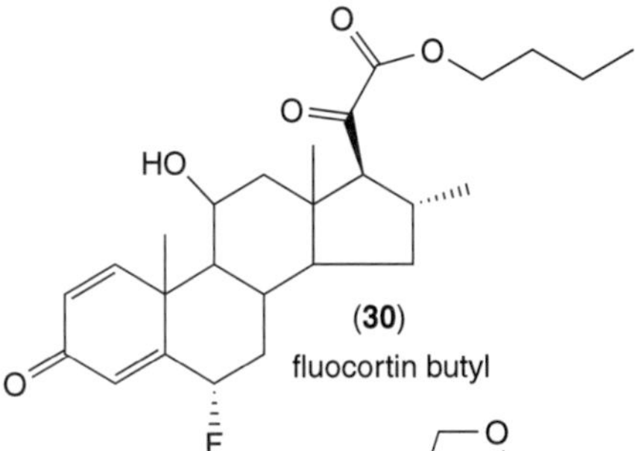

(**30**)
fluocortin butyl

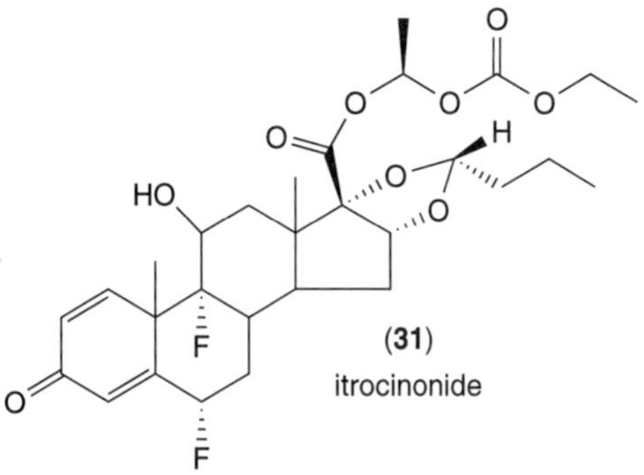

(**31**)
itrocinonide

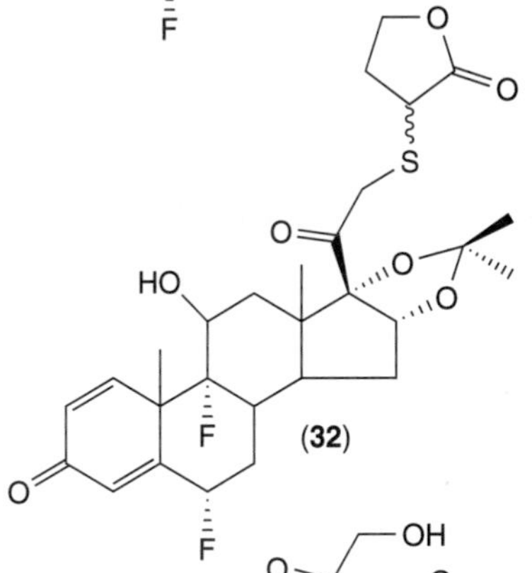

(**32**)

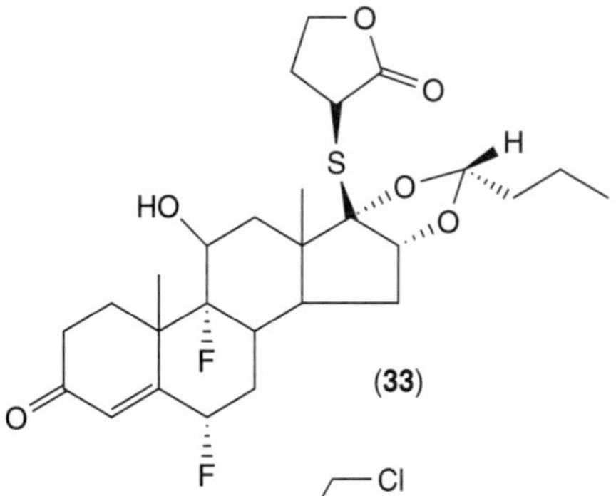

(**33**)

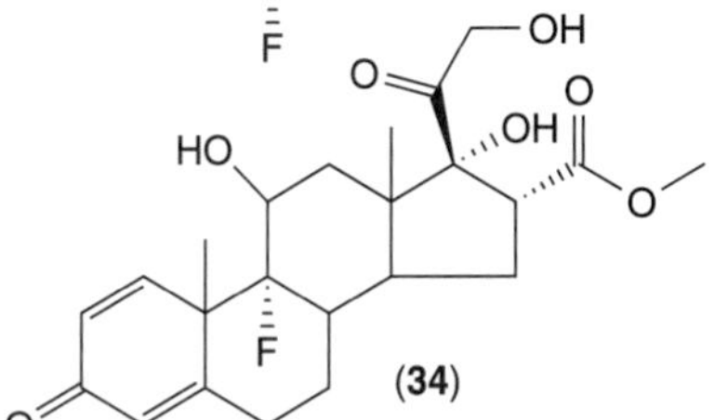

(**34**)
FP16CM

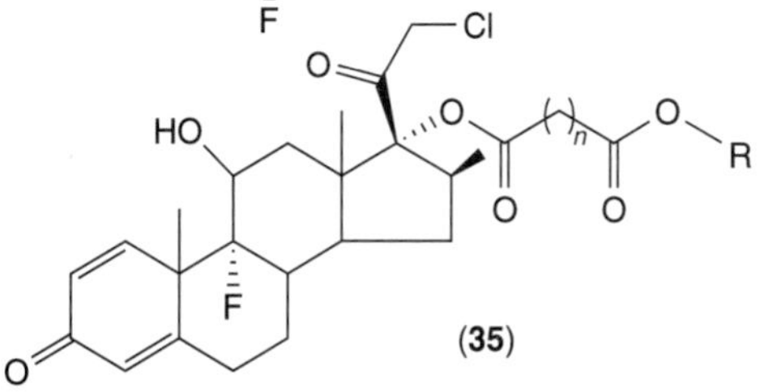

(**35**)

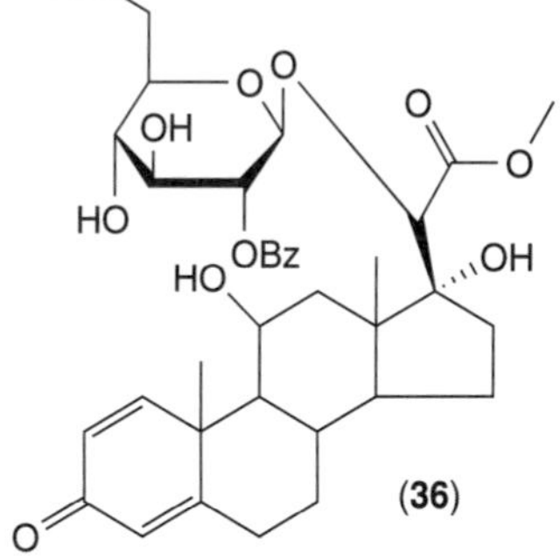

(**36**)

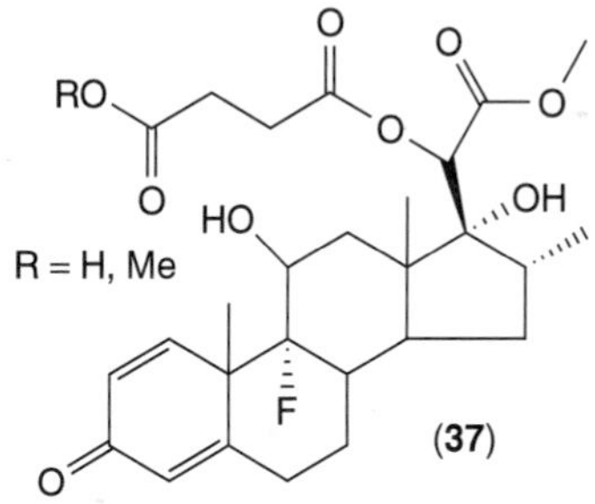

(**37**)

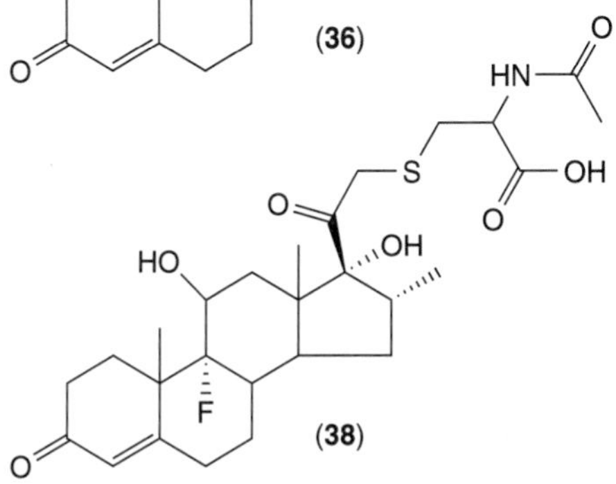

(**38**)

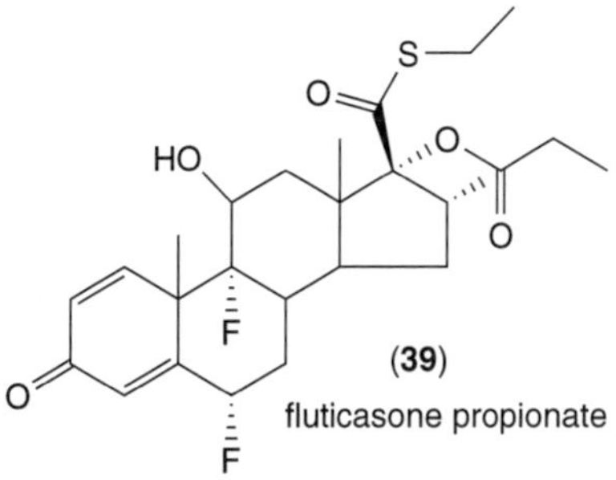

(**39**)
fluticasone propionate

Figure 15.13. Corticosteroid structures discussed in the text.

3.3.3.4 Itrocinonide.

A soft steroid series containing 17β-methyl-carbonate ester moieties susceptible toward hydrolysis was also developed during the 1980s (132). Receptor affinities varied significantly with the substituents and with the stereochemistry of the chiral center at the 17β ester. The selected double-fluorinated compound, itrocinonide (31) (Fig. 15.13), had a receptor affinity similar to that of budenoside and a sufficiently rapid rate of *in vitro* hydrolysis ($t_{1/2}$ = 30 min in human blood at 37°C) (132). The *in vivo* potency of itrocinonide was less than that of budesonide, but, in agreement with soft drug design principles, the ratio between its anti-inflammatory efficacy in airways/lung and its systemic steroid activity (i.e., thymus involution or plasma cortisol suppression) was much better than the corresponding ratio for budesonide. It also had very good systemic tolerance in human volunteers and asthmatics. Because of its short plasma half-life (~30 min, which is about one-fifth that of budenoside and fluocortin butyl), itrocinonide lacked measurable systemic glucocorticoid activity. In patients with asthma or seasonal rhinitis, itrocinonide administered as a dry powder formulation did exert some antiasthmatic and antirhinitic efficacy, although these effects were not sufficient enough to compete with the efficacy of current inhaled steroids (132).

3.3.3.5 Glucocorticoid γ-Lactones.

Recently, various γ-lactone derivatives including 21-thio derivatives of fluocinolone acetonide with γ-lactones and cyclic carbonates (e.g., 32) (133) and sulfur-linked γ-lactones incorporated at the 17β position (e.g., 33) (134) were also explored. For these compounds, human serum paraoxonase (E.C. 3.1.8.1) was claimed to be the metabolizing enzyme (133), which is of interest because this enzyme has a much lower activity in lung tissue than in plasma, and thus it can provide improved site-specific activity for inhaled compounds. Contrary to the corresponding esters, 21-thio—linked lactones were stable in human lung S9 preparation ($t_{1/2}$ > 480 min for 32), but rapidly hydrolyzed in human plasma ($t_{1/2}$ < 1 min) (133). The rate of hydrolysis was also rapid in plasma for the 17-linked lactones possessing a sulfur in the α-position of the butyrolactone group ($t_{1/2}$ < 5 min), whereas C-linked lactones were

stable (134). Among the compounds of this series, (33) showed promising topical anti-inflammatory activity in the rat ear edema model and much lower systemic effects than those of budesonide in the thymus involution test.

3.3.3.6 Other Corticosteroid Designs.

Various, mostly prednisolone-based ester derivatives, were synthesized and investigated in a series of attempts (designated as antedrug designs) (27, 28, 135–143). They were aimed to improve the local-to-systemic activity ratio of anti-inflammatory steroids and may be considered as based on hypothetical inactive metabolites. Studied compounds include ester derivatives of steroid 21-oic acids (27), a number of 16α-carboxylate analogs (e.g., 34) (28, 135–137, 140), 6-carboxylate analogs (138), and (16α,17α-d) isoxazolines derivatives (139, 141, 142). Some of these compounds were found to have relatively low activity, similar to that of hydrocortisone or prednisolone, and they also achieved some, but not very significant, improvement in the local-to-systemic activity ratio. Relative binding affinities (RBA, considering $RBA_{dexamethasone}$ = 100, IC_{50} = 7.3 n*M*) of two such 9α-fluorinated steroids for the cytosolic glucocorticoid receptor are 7.9 and 3.7 for FP16CM (34) and its 21-acetate derivative FP16CMAc, respectively (28). For comparison, loteprednol etabonate (26), a nonfluorinated soft steroid, has an RBA value of around 200 (105).

Other groups also made attempts to separate local and systemic effects by integrating moieties susceptible to rapid, nonhepatic metabolism within the corticosteroid structure. One of the more successful attempts explored 17α-(alkoxycarbonyl)alkanoate analogs (35) of clobetasol propionate (144). Again, this can be considered as a hypothetical inactive metabolite—based approach, and a corresponding metabolite (35) (*n* = 2, R = H) has indeed been shown to be inactive. Esters that were susceptible to rapid hydrolysis exhibited good separation of topical anti-inflammatory to systemic activity. The study also indicated the existence of an optimal volume for the 17α side chain. For example, the methyl succinate derivative (35) (*n* = 2, R = methyl) showed as potent topical anti-inflammatory activity as clobetasol propionate, but a dramatically reduced thymolytic activity. Therefore, the corresponding therapeutic index was increased more than 130-fold compared to that of clobetasol propionate. It has to be mentioned, however, that for this compound (**35**), as well as for the glucocorticoid γ-lactones (**32**, **33**), active compounds may be formed from the inactive metabolite, for example, in the case of (**35**) by chemical cleavage of the succinate ester to the active clobetasol.

Another effort involved the design, synthesis, and testing of a colon-targeted pro-soft drug (**36**) for possible oral treatment of ulcerative colitis (145). These C-20 oxyprednisolonate 21-esters contain glucopyranosyl ethers to render the pro-soft drugs hydrophilic and thus poorly absorbable in the small intestine. Removal of the glucopyranosyl ethers releases the corresponding active soft drug. This process is mediated by colonic bacteria within the colonic lumen, as demonstrated *in vivo* after administration in the jejunum of guinea pigs. In the systemic circulation, degradation of the C-21 esters rapidly releases the inactive acid metabolites. Interestingly, the half-lives in guinea pig plasma for the two different ester stereoisomers were quite different, being 2.6 and 166.8 min for the 20*R*-dihydroprednisolonate and 20*S*-dihydroprednisolonate, respectively. Somewhat later, steroid-17-yl methyl glycolates with succinyl group at C-20 derived from prednisolone and dexamethasone (**37**) were also investigated by the same group (146). In fact, this is again a pro-soft drug–type approach, as first an active compound, the 21-ester, is released, and then this is further metabolized into an inactive metabolite.

In another separate study, three series of compounds were synthesized in which sulfur-containing amino acids were incorporated into the steroidal structure at the 21 position (147). The rationale for this, which the authors considered as being more or less along the principles of soft drug design, was that sulfur-containing compounds have shown, generally, a good cutaneous distribution as well as relative rapid biotransformation and fast elimination, with the oxidized metabolites being inactive in most cases. The selected most promising compound of this series was

(**38**), for which *in vivo* results showed a local activity about 10 times less and a systemic activity about 970 times less than that of dexamethasone.

Finally, before closing this section, it should be mentioned that some other steroid drugs such as fluticasone propionate (**39**), tipredane, or butixocort 21-propionate are often and erroneously called soft drugs (148, 149). Thiol ester corticosteroids such as fluticasone have been shown to be metabolized in the liver by oxidative cleavage of the thiol ester bond and not by hydrolysis in the plasma (150). Consequently, even if fluticasone propionate itself lacks oral activity because high hepatic first-pass metabolism to the corresponding (inactive) 17-carboxylic acid, it has systemic effects if given subcutaneously (151). Fluticasone propionate was found to have a terminal half-life of 7.7–8.3 h in 12 healthy male subjects after inhaled administration of 500, 1000, and 2000 µg of drug using a metered-dose inhaler. In these subjects it produced dose-related cortisol suppression; the highest administered dose of fluticasone resulted in cortisol concentrations that were lower than the limit of detection (152). The slow elimination of fluticasone led to accumulation during repeated dosing. This accumulation may explain the marked decrease in plasma cortisol seen during treatment with fluticasone propionate within the clinical dose range (153). Furthermore, it is a highly lipophilic steroid and it shows an increased terminal half-life after inhalation, which usually is an indication of slow, rate-limiting absorption ("flip-flop pharmacokinetics") (154).

3.3.4 Soft Estrogens. Estrogens represent another group of steroids in which soft drug approaches can provide new therapeutic agents with a beneficial separation of local and systemic effects. Menopause-related estrogen depletion is more than likely associated with a variety of symptoms that range the gamut from vasomotor complaints to cognitive deficits. Estrogen administration [hormone replacement therapy (HRT)] is known to alleviate most of these symptoms, but because of an association with increased risk for cancer, stroke, and other metabolic diseases, such therapies are either not recommended for, or

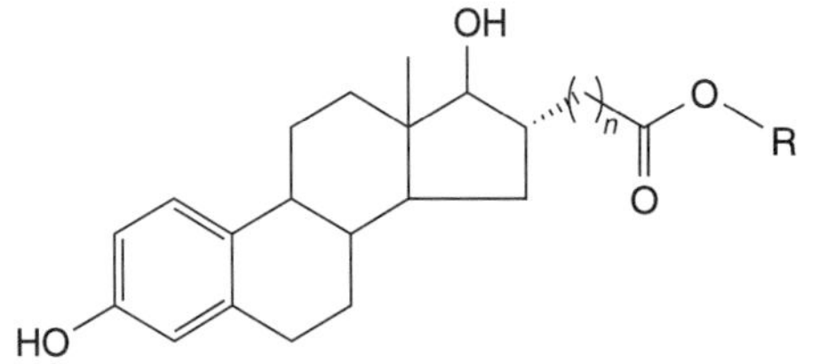

(40)

n = 0, 1, 2

Figure 15.14. Soft estrogens designed for the treatment of vaginal dyspareunia.

avoided by, many women (155, 156). Vaginal dyspareunia is a common disease affecting a large proportion of menopausal women (around 40% within 10 years of the onset of menopause), and topical application of estrogen has been used for treatment. Locally active soft estrogens with reduced systemic activity may provide a therapeutic alternative, and a series of estradiol-16α-carboxylic acid esters (**40**) (Fig. 15.14) were synthesized and examined recently for this purpose (157). Whereas none of the acids (**40**) (R = H, n = 0, 1, 2) showed significant estrogen receptor binding, the esters did. For them, receptor binding decreased with increasing n (Fig. 15.14) or branching of the alcohol portion (R = isopropyl, neopentyl), but not with increasing length of the alcohol chain (R = methyl—butyl). The rate of hydrolysis in rat hepatic microsomes increased with increasing chain length (methyl to butyl) and was especially high for fluorinated alcohol chains (e.g., R = CH_2CHF_2). Three of the most promising compounds (**40**, n = 0, R = CH_3, CH_2CH_3, CH_2CH_2F) were also tested for systemic and local action in rodent *in vivo* models. All of them, and especially the fluoroethyl ester, showed good separation of local and systemic estrogenic action.

3.3.5 Soft β_2-Agonists. β_2-Agonists represent an important class of drugs in the therapy of asthma because of their β_2-receptor mediated bronchodilating activity (158). Compounds such as terbutaline (**41**) (Fig. 15.15), fenoterol, or salbutamol are chemical analogs of epinephrine and are β_2-selective agents. These agents, including the longer acting for-

moterol or salmeterol, are most frequently taken by aerosol. The majority of the drug administered this way is swallowed, and only about 10% of the dose reaches the lung directly. Therefore, there is a great potential to produce unwanted side effects such as tachycardia or skeletal muscle tremor. Again, a soft drug approach might yield viable solutions. Incorporation of a metabolically labile ester group into such structures has been attempted both in the nitrogen substituent (159) and on the aryl system (160) (Fig. 15.15).

The activity of compounds (**43**) (n = 0 or 1, R = CH_3) or (**44**) (R = CH_3) surpassed that of terbutaline (**41**) or isoprenaline. The corresponding carboxylic acids (**43** or **44**, R = H) are essentially devoid of β_2-agonist activity (they are 1–4 orders of magnitude less potent); therefore, their use as inactive metabolites in the design of more potent esters is justified (Fig. 15.15) (159, 160). Compound (**44**) with R = CH_3 (ZK 90.055) was selected for further pharmacological and toxicological evaluation. Consistent with the design, it rapidly hydrolyzed in the presence of guinea pig liver homogenate and showed good *in vivo* bronchospasmolytic activity when given by inhalation. Meanwhile, it was almost inactive on oral administration and had no effect on the heart rate of guinea pigs following inhalation of up to 10-fold the dose that was active in the bronchospasm experiment (160).

Soft β-agonist structures were also investigated as possible soft antipsoriatic agents (161). Both soft drugs and pro-soft drugs were synthesized as models of topical antipsoriatic β-adrenergic agonists. The structure of the soft drugs was similar to that of (**43**) (Fig. 15.15, n = 1, R = CH_3, CH_2CH_3), and the pro-soft drugs were obtained by esterification of the phenolic functions. In the presence of porcine liver carboxyesterase, the pivaloyl ester groups of the prodrug underwent rapid hydrolysis ($t_{1/2}$ = 8.4 min) to release the soft drug, which then also underwent hydrolysis ($t_{1/2}$ = 456 min) to the inactive carboxylate anion (161). The soft drug was a full β-agonist on the guinea pig tracheal preparation producing a maximal response similar to that achieved with isoprenaline. The pro-soft drug produced only slowly developing responses at high con-

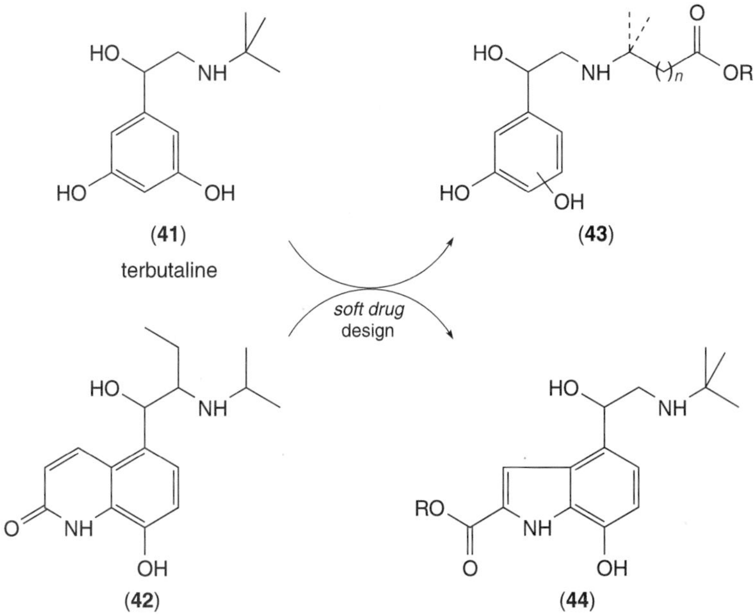

Figure 15.15. Soft β_2-agonists (**43**, **44**) designed based on lead structures such as terbutaline (**41**) or procaterol (**42**).

centrations (>10 μM) and had better transport properties across a silicone membrane.

3.3.6 Soft Insecticides/Pesticides. The very same concepts used for soft drug design can be extended to the design of less toxic commercial chemical substances, provided that adequate structure–activity relationship (SAR) and structure–metabolism relationship (SMR) data of analogous substances can be gathered (*soft chemical* design). The following two examples are not actual designs based on such principles, but observations made in hindsight. Nevertheless, they illustrate the possibilities inherent to such approaches. Also, they provide examples for the design of environmentally safe, nontoxic chemicals (green chemistry) (162).

3.3.6.1 Chlorobenzilate. One instance in which these principles have been used (unintentionally) for the design of nonpharmaceutical products is chlorobenzilate (**52**), an ethyl ester—containing analog of dicofol (**46**) and dichlorodiphenyltrichloroethane (DDT, chlorophenothane, **45**) (Fig. 15.16). DDT was the first chemical that revolutionized pest control,

and it was also used to control typhus and malaria. Its insecticide properties were discovered in 1939 (163), and it was widely used as a pesticide in the United States, although it was banned in 1972 for all but essential public health use and a few minor uses. The decision was prompted by the prospect of ecological imbalance from continued use of DDT, by the development of resistant strains of insects, and by suspicions that it causes a variety of health problems, including cancer. DDT undergoes complex *in vivo* metabolism, including oxidation (**46**, **47**), iterative dehydrohalogenation/reduction cycles (**48–50**), and hydrolysis (**51**) (Fig. 15.16) (164). The acid metabolite (**51**) is of low toxicity; can be excreted as a water-soluble species; and is, indeed, a major metabolite detected in feces and urine. Therefore, it is an ideal lead compound for a formal inactive metabolite approach (Fig. 15.16).

Not surprisingly, the corresponding ethyl ester, ethyl-4,4′ -dichlorobenzilate (chlorobenzilate, **52**), is also active as a pesticide, but has much lower carcinogenicity than that of either DDT or dicofol (kelthane, **46**). For ex-

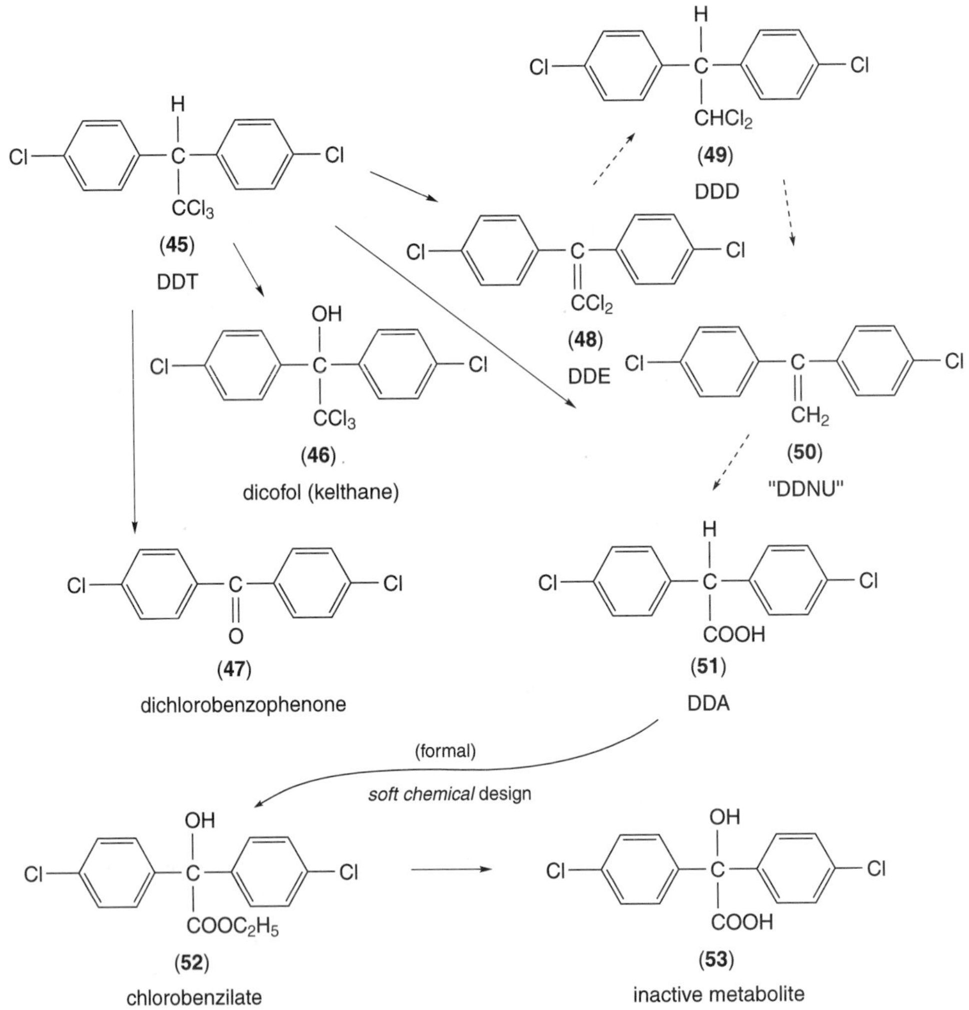

Figure 15.16. Chlorobenzilate (**52**) can be regarded as a soft chemical obtained using an inactive metabolite–based approach based on the metabolism of DDT (**45**).

ample, the carcinogen concentration deter-mined in mice is 6000 mg/kg for chlorobenzi-late compared to 10 and 264 mg/kg for DDT and dicofol, respectively (165, 166). The oral median lethal dose (LD_{50}) for female rats is 1220 mg/kg for chlorobenzilate compared to 118 and 1000 mg/kg for DDT and dicofol, re-spectively (167). The ethyl ester moiety appar-ently functions similarly to that of the trichlo-romethyl group of DDT in restoring pesticidal activity. However, because in exposed subjects the labile ethyl ester group enables rapid me-

tabolism to the free, nontoxic carboxylic acid, chlorobenzilate is considerably less toxic than DDT.

3.3.6.2 Malathion. Malathion (**54**) (Fig. 15.17) is an excellent example to illustrate an additional, not yet sufficiently explored aspect of the design of soft chemicals. As mentioned, it is desirable to design soft chemicals deacti-vated by carboxylesterases. For soft chemicals intended to be used as pesticides, in addition to the usual advantages of soft drug design, the differential distribution of these enzymes

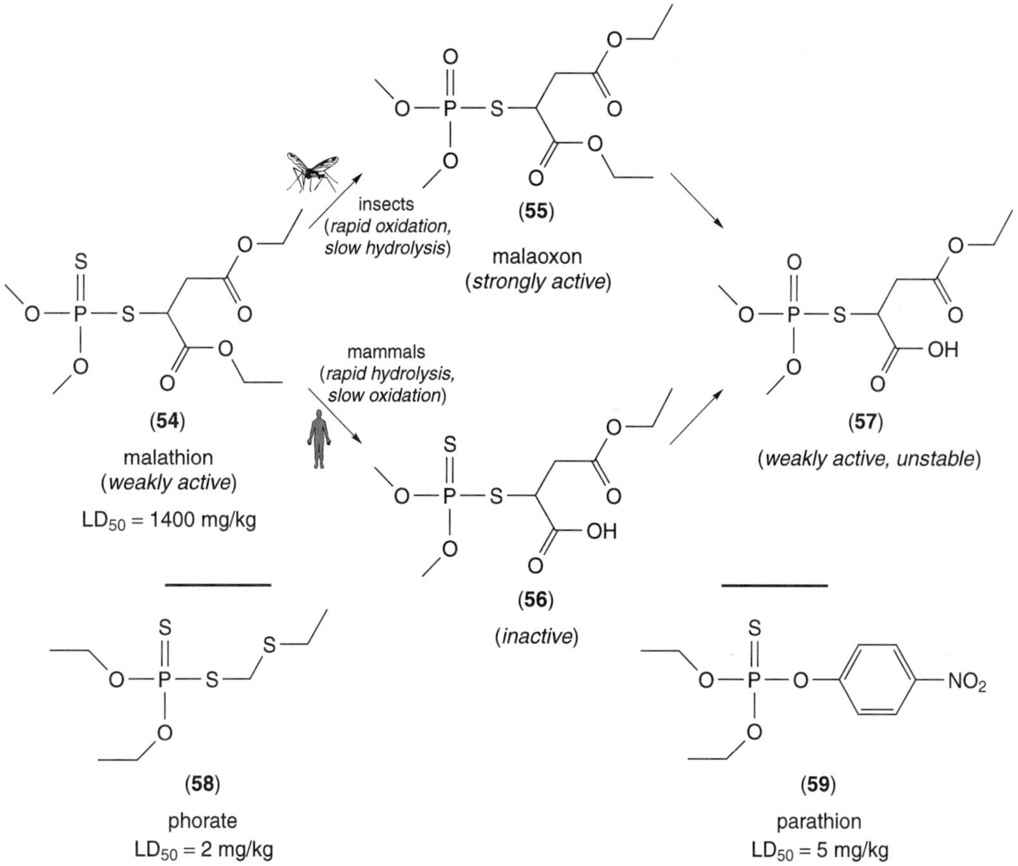

Figure 15.17. Malathion (**54**), its oxidative activation to malaoxon (**55**), a much more active cholinesterase inhibitor, and its deactivation by carboxylesterases. Oxidative activation is the dominant pathway in insects, but hydrolytic deactivation is the dominant pathway in mammals. Because of its hydrolytic deactivation in mammals, malathion is much less toxic than other organophosphates such as phorate (**58**) or parathion (**59**).

between vertebrates and insects may also provide selectivity based on metabolism. An elegant example is provided by malathion (**54**), a widely used organophosphate insecticide (Fig. 15.17). Malathion is detoxified through a variety of metabolic pathways, one of the most prominent of which is the hydrolysis of one of its two ethyl carboxylester groups. The carboxylesterase that hydrolyzes and thereby detoxifies malathion is widely distributed in mammals, but only sporadically in insects, where in some rare cases it is responsible for insecticide resistance (168, 169). In the meantime, insects seem to possess a very active oxidative enzyme system that transforms malathion (**54**) into malaoxon

(**55**), a much more active cholinesterase inhibitor (Fig. 15.17). Probably, all insects and all vertebrates possess both an esterase and an NADPH-dependent oxidase system, but the balance of action of these two systems varies from one organism to another and provides selectivity of action. A similar mechanism may provide considerable selectivity for other soft chemicals to be designed and may result in safer, *soft* insecticides, for example, in the parathion family. These compounds are not susceptible to such deactivation mechanisms and, consequently, have unacceptably high mammalian toxicities. For example, acute oral LD_{50} values in male rats are 2 and 5 mg/kg for phorate (**58**) and parathion (**59**), respectively, compared to 1400 mg/kg for malathion (**54**).

3.3.7 Soft Anticholinergics: Inactive Metabolite–Based Approach.

Soft anticholinergics provide a good illustration for the flexibility and potential of the general soft drug design concept. Our work in this area resulted in two entirely different classes of soft anticholinergics. *Inactive metabolite–based* classes (**64**, **69**) (Fig. 15.18), which are discussed here, were obtained by using methylatropine (**60**) (170–175), *N*-methylscopolamine (**61**) (176, 177), or glycopyrrolate (**62**) (178) as lead. The *soft analog* class (**73**), which is discussed in the following section, contains soft quaternary analogs (26, 179).

Anticholinergics are competitive inhibitors of acetylcholine and have many useful clinical effects, such as mydriasis/cycloplegia, prevention of motion sickness, local antisecretory activity, and others. However, their use is limited by a number of side effects, such as dry mouth, photophobia, irritability, disorientation, hallucinations, and cardiac arrhythmia (180). For example, their local antisecretory activity was long thought to be beneficial for inhibiting eccrine sweating (perspiration) (181, 182). A wide range of anticholinergic agents, such as atropine, scopolamine, or their quaternary ammonium salts, are known to inhibit perspiration, but because they produce many side effects, they cannot be used as antiperspirants. A locally active soft analog may again represent a viable solution.

To obtain such compounds that have high local, but practically no systemic activity, different series of soft anticholinergics based on methylatropine, *N*-methylscopolamine, glycopyrrolate, or propantheline were designed in our laboratories (183). Several new molecules synthesized were found to be potent anticholinergics both *in vitro* and *in vivo*, but in contrast to their hard analogs they had no systemic anticholinergic activity following topical administration. For example, methylatropine- or methylscopolamine-derived phenylmalonic acids (**64**, $n = 0$) (170–173, 175, 176) and phenylsuccinic acids (**64**, $n = 1$) (174, 177) served as useful hypothetical inactive metabolites for the design of soft anticholinergics. This *inactive metabolite–based* approach exploits the idea that the benzylic hydroxy function in methylatropine (**60**) or methylscopolamine (**61**) could be oxidized to the carboxylic acids (**65**) ($n = 0$), which are hypothetical (180, 184, 185) inactive metabolites. Esterification of this carboxy function to afford soft drug series (**64**) may restore activity and meanwhile ensure facile, hydrolytic deactivation.

Soft anticholinergic esters of this kind (**64**) showed good intrinsic activity, as indicated by the pA_2 value of 7.85 of tematropium (**66**) compared with 8.29 for atropine. However, *in vivo* activities were much shorter than those for the "hard" atropine. Accordingly, when equipotent mydriatic concentrations of atropine and tematropium (**66**) were compared following ocular administration, the same maximal mydriasis was obtained, but the area under the curve (mydriasis vs. time) for the soft compound was only 11–19% of that for atropine (170, 172). This is consistent with the facile hydrolytic deactivation of the soft drug. Similarly, the cardiovascular activity of compound (**66**) showed ultrashort duration. The effect of (**66**) on the heart rate and its ability to antagonize the cholinergic cardiac depressant action induced by acetylcholine injection or by electrical vagus stimulation was determined in comparison with atropine (sulfate) and methylatropine (nitrate). A dose of 1 mg/kg of atropine or methylatropine could completely abolish the bradycardia induced by acetylcholine injection or by electrical vagus stimulation for more than 2 h following i.v. injection. On the other hand, similar doses of tematropium exerted antimuscarinic activity for only 1–3 min following i.v. injection. Even a 10-fold increase in its dose to 10 mg/kg did not lead to any significant prolongation of the duration of anticholinergic activity.

As a further variation, the corresponding ester analogs of methylscopolamine were also investigated (176). For example, the ethyl ester was shorter acting than even tropicamide, and, consistent with the soft drug approach, the untreated eye did not show any mydriasis, as opposed to administration of methscopolamine. Other structures like the cyclopentyl derivative (**67**) (PCMS-2) (175) or different phenylsuccinic analogs of methylatropine (174) and methylscopolamine (177) were also investigated. Compound (**67**) was equipotent

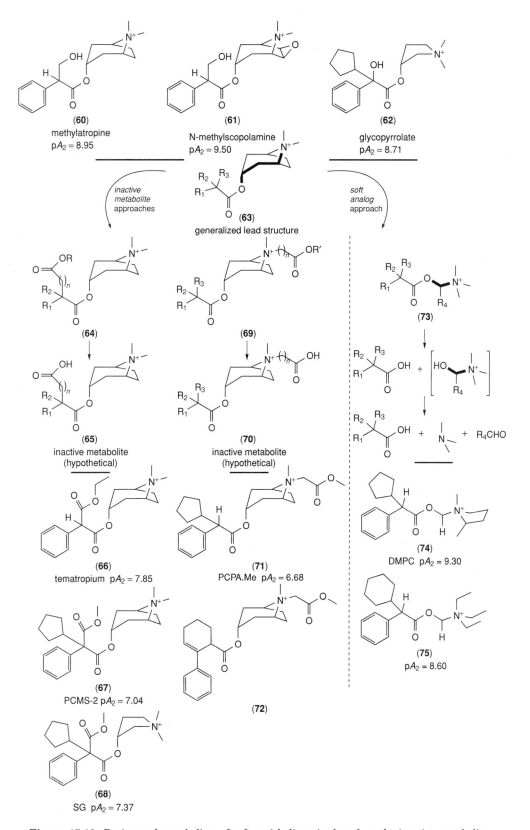

Figure 15.18. Design and metabolism of soft anticholinergics based on the inactive metabolite–based approach (**64**, **69**; substitutions at two different positions) and the soft analog approach (**73**). pA_2 values shown are for *in vitro* anticholinergic activity determined by guinea pig ileum assay with carbachol as agonist.

to atropine in protecting against carbachol-induced bradycardia in rats, but its duration of action was again significantly shorter (15–30 min vs. more than 2 h) (175). Similar ester analogs (**68**) of glycopyrrolate (**62**) were also explored (178).

For soft anticholinergics, the inactive metabolite–based approach can also yield a different class of compounds in which the hydrolytically labile ester-containing side-chain is attached to the quaternary nitrogen head (**69**). A few such compounds derived from methylatropine (**71**) (186) or glycopyrrolate (187) have been recently synthesized and tested. Furthermore, compounds such as (**72**) have also been explored in an attempt to obtain soft anticholinergics with muscarinic receptor subtype selectivity (188). Hence, a lead compound (LG50643) was selected that has been shown to be a potent and selective antagonist for the M_3 receptor subtype (189), and it was derivatized following the same procedure to obtain soft compounds such as **72**. Receptor-binding studies on cloned muscarinic receptors indicated that these soft anticholinergics have reasonable activity (pK_i values of 7.5–8.9) and that two of them show muscarinic receptor subtype selectivity (M_3/M_2) (188). Consistent with their soft nature, these compounds were short acting and were rapidly eliminated from plasma.

3.4 Soft Analogs

Compounds classified as soft analogs are close structural analogs of known active drugs (lead compounds), but they have a moiety that is susceptible to metabolic, preferentially hydrolytic, degradation built into their structure. The built-in metabolism should be the major, and preferentially, the only metabolic route for drug deactivation (*predictable metabolism*), and the rate of the predictable metabolism should be controllable by structural modifications (*controlled metabolism*). The predicted metabolism should not require enzymatic processes leading to highly reactive intermediates, and the products resulting from the metabolism should be nontoxic and have no significant biological or other activities (*metabolic inactivation*). Finally, the metabolically weak spot should be located within the molecule so that the overall physical, phys-

icochemical, steric, and complementary properties of the soft analog are very close to those of the lead compound (*isosteric/isoelectronic analogy*).

3.4.1 Soft Anticholinergics: Soft Quaternary Analogs.
As mentioned, in addition to the inactive metabolite–based classes of soft anticholinergics, an entirely different class containing soft quaternary analogs has been explored. Structural differences between "hard" and "soft" anticholinergics of this kind are relatively small, but nonetheless profound, as illustrated on the right side of Fig. 15.18. *Soft analog* anticholinergic structures were obtained by shortening the bridge of two or three carbon atoms separating the quaternary head and ester function of traditional, "hard" anticholinergics illustrated by the generalized structure (**63**) to just one carbon separation as shown in structure (**73**). This allows facile hydrolytic deactivation by way of a short-lived intermediate to the corresponding acid, tertiary amine, and aldehyde, all inactive as anticholinergics, as shown in Fig. 15.18.

At the time of the design, a separation of at least two carbon atoms between the ester oxygen and the quaternary nitrogen of such anticholinergic structures was thought to be critical for effective receptor binding. Nevertheless, several compounds of type (**73**) were found to be at least as potent as atropine (26). For example, (**74**) was equipotent with atropine in various anticholinergic tests, but it was very short acting after i.v. injection. Therefore, when applied topically to humans, it produced high local antisecretory activity but no systemic toxicity. In a more recent study, similarly designed soft analogs of propantheline have also been investigated as potential antiperspirants/antiulcerative agent (179).

3.4.2 Soft Antimicrobials
3.4.2.1 Cetylpyridinium Analogs: Soft Quaternary Salts.
The very first soft analogs designed were "soft quaternary salts" represented by the generalized structure (**79**), intended for antimicrobial use (6, 190). Similar to the previously described anticholinergics structures, these substances undergo a facile hydrolytic cleavage process through a very short-lived intermediate to deactivate

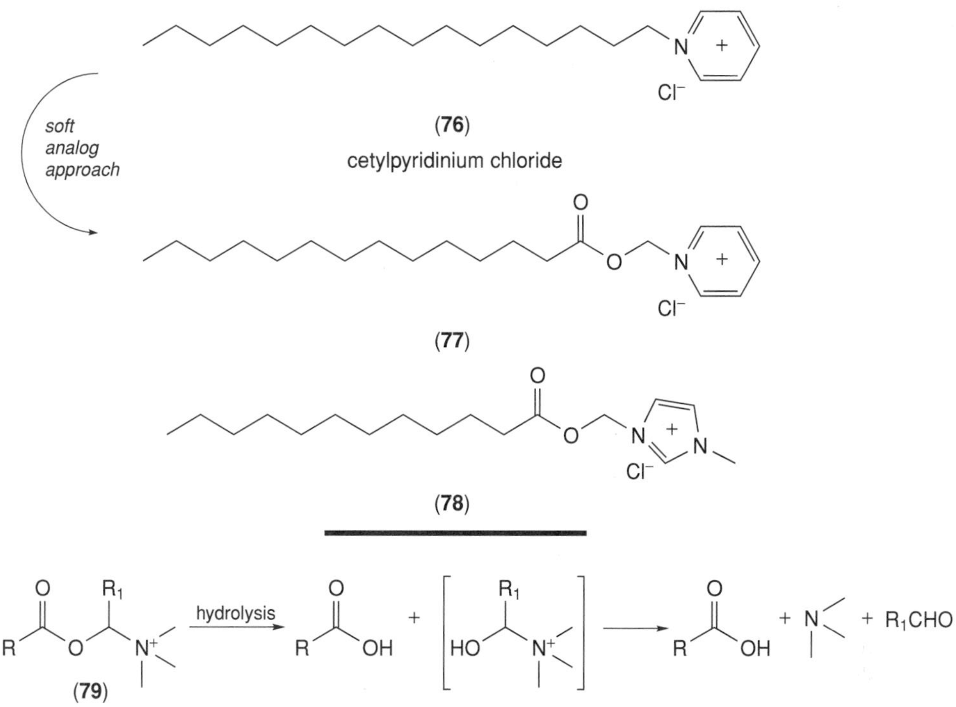

Figure 15.19. Cetylpyridinium chloride (**76**) and soft analog antimicrobials (**77**, **78**). The general hydrolytic deactivation mechanism of soft quaternary salts (**79**) through a very short-lived intermediate to an acid, an amine, and an aldehyde is also shown.

and form an acid, an amine, and an aldehyde, as shown in Fig. 15.19. This mechanism was initially designed to develop a prodrug of aspirin, the synthesis of which, however, was unsuccessful (29).

The simplest example of useful true soft analogs (Fig. 15.19) is provided by the isosteric analogs (**77**, **78**) of cetylpyridinium chloride (**76**). Cetylpyridinium, a known "hard" quaternary antimicrobial agent, needs several oxidative (generally β-oxidation) steps to lose its surface-active, antimicrobial properties. The quaternary salts represented by (**76**) and (**77**) are very similar: both contain side chains that are essentially 16 atoms in length. Their physicochemical properties are also very similar. For example, their critical micelle concentrations determined by a molecular light-scattering method are 1.3×10^{-4} and $1.7 \times 10^{-4}\,M$, respectively (24). *Hard* and *soft* compounds possess comparable antimicrobial activity as measured by their contact germicidal efficiency, but *soft* compounds undergo facile hy-

drolytic cleavage, leading to their deactivation. Because of this, the *soft* (**78**) is about 40 times less toxic than the *hard* (**76**): the corresponding oral LD_{50} values for white Swiss male mice are 4110 and 108 mg/kg, respectively.

3.4.2.2 Long-Chain Esters of Betaine and Choline. Another, unrelated effort was directed toward the development of long-chain esters (**80**, **83**) of betaine (**85**) or choline (**82**) as soft antimicrobial agents (Fig. 15.20) (191–195). These compounds are ester-containing structural analogs of amphiphilic quaternary ammonium compounds, which, similar to cetylpyridinium (**76**), are surface active substances known for their membrane-disruptive and antimicrobial activities. Contrary to the previous design, however, hydrolysis here is not followed by additional, fast degradation, but results in well-investigated and common compounds, such as choline (**82**), betaine (**85**), and fatty acids (**81**) (Fig. 15.20). For example, the alkanoylcholines were found active

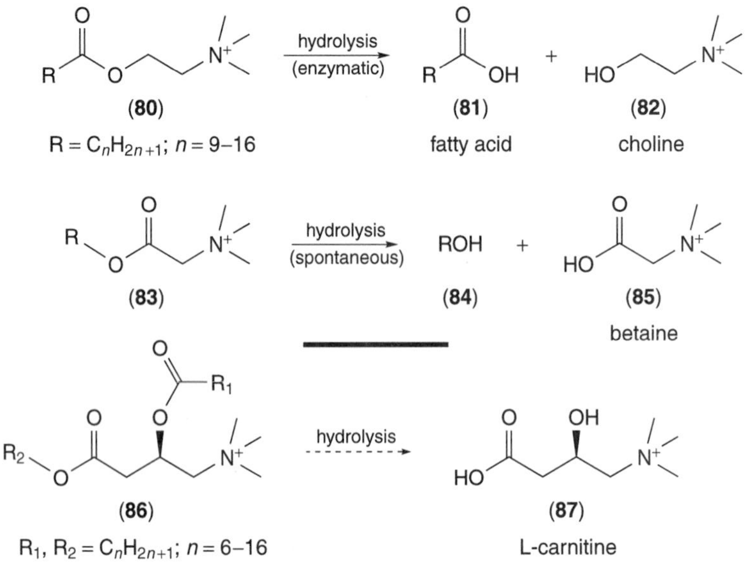

Figure 15.20. Soft quaternary ammonium antimicrobials designed as long-chain ester of choline, betaine, or L-carnitine.

against gram-negative and gram-positive bacteria, as well as yeasts (192). Activity increased with increasing chain length and was similar to that of the stable, *hard* quaternary ammonium compounds of similar length such as hexadecyltrimethylammonium bromide. Considerable differences in the binding affinity of compounds with different hydrocarbon chains at different concentrations to *Candida albicans* were observed, and they seemed related to the critical micelle concentration of the compounds (193).

3.4.2.3 L-Carnitine Esters. Another class of *soft* broad-spectrum antimicrobials devoted to curing dermatological infections was designed based on quaternary ammonium L-carnitine esters (**86**) (Fig. 15.20) (196). The series, particularly members characterized by alkyl chains with a total of 16–18 carbons, showed good activity against a wide range of bacteria, yeasts, and fungi. They also showed low *in vitro* cytotoxicity and good *in vivo* dermal tolerance. However, the decomposition of these compounds was not analyzed in detail. Judicious esterification (R_2) of the carboxy function is required for antimicrobial activity, but a free hydroxy group at position 3 of the carnitine skeleton does not annihilate activity.

From the soft drug design point of view, it is also important to note that the common constituent of all these compounds, L-carnitine (**87**), has no pharmacological effects for doses of up to 15 g/day.

3.4.3 Soft Antiarrhythmic Agents

3.4.3.1 ACC-9358 Soft Analogs. ACC-9358 (**88**) (Fig. 15.21) is an orally active class Ic antiarrhythmic agent that underwent clinical trials and for which a number of soft analogs (**89**) were synthesized and tested (197). Replacement of the formanilide function of ACC-9385 with alkyl esters resulted in compounds with similar antiarrhythmic activity. Esters attached directly to the aromatic ring (**89**, $n = 0$) of the bis(aminomethyl)phenol moiety were resistant to hydrolysis in human blood, but distancing the ester (**89**, R = —CH$_2$CH(CH$_3$)$_2$) from the aromatic ring by one, two, and three methylene units afforded *soft* compounds with human blood half-lives of 8.7, 25.9, and 2.0 min, respectively (197). As in most other cases, branching on the carbon attached to the oxygen atom of the alkoxy functionality tended to inhibit ester hydrolysis. The antiarrhythmic activity, as measured *in vitro* in the guinea pig right atrium, of a number of acid metabolites

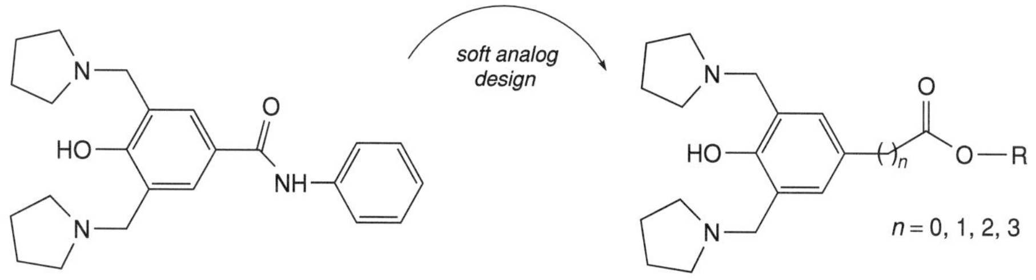

Figure 15.21. Soft analogs (**89**) of ACC-9358 (**88**), an orally active class Ic antiarrhythmic agent that underwent clinical evaluation. Soft analogs such as (**90**) and (**91**) showed good activity and sufficiently short half-life in human blood.

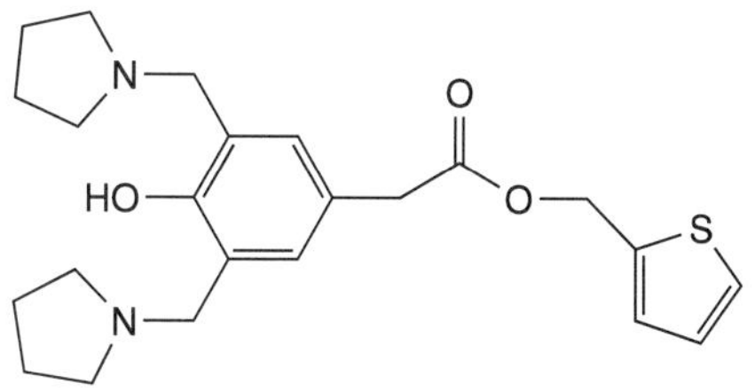

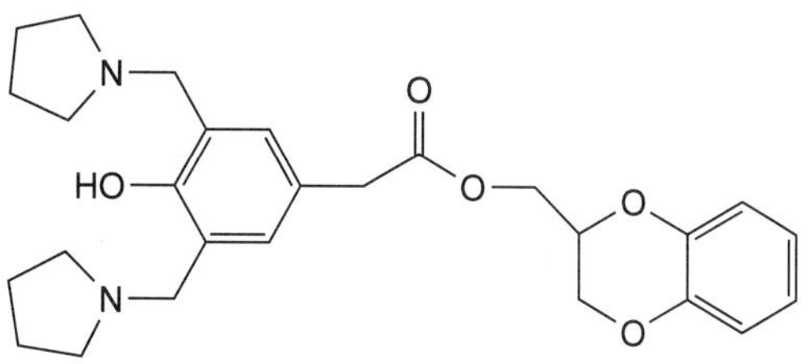

(**89**, R = H, n = 0, 1, 2) was indeed significantly less than that of the corresponding ester soft drugs.

In a more recent work, additional esters (**89**) derived from aromatic and heterocyclic alcohols were investigated to improve the lipophilic character and enhance the *in vivo* potency, biodistribution, and duration profile (198). A number of these esters showed consistent ability to convert acetylstrophanthidin-induced arrhythmias in guinea pig right atria to normal sinus rhythm with an ED_{50} value of less than 10 μg/mL. Based on their shorter half-life in human blood, esters (**90**) ($t_{1/2}$ = 3.5 min) and (**91**) ($t_{1/2}$ = 7.1 min) were selected for *in vivo* evaluation. They both demonstrated greater potency than that of lidocaine in the 24-h Harris dog model and equal potency to lidocaine in the oubaine-intoxicated dog model (198). In addition, the lipophilicity of these compounds was lower than that of lidocaine, suggesting a lower ability to penetrate the blood—brain barrier and, thus, lower CNS liability. Considering all these observations, (**91**) was chosen as a potential development

candidate because it possessed the most desirable pharmacological and pharmacokinetic profile. Some analogs where the ester moiety was attached to the second aromatic ring of ACC-9385 (197) and some monoaminomethylene-appended analogs were also explored (198).

3.4.3.2 Amiodarone Soft Analogs. Among currently available antiarrhythmic agents, amiodarone (**92**), a structural analog of thyroid hormone, has electrophysiological effects that most closely resemble those of an ideal drug. However, amiodarone is highly lipophilic, is eliminated extremely slowly, and has unusually complex pharmacokinetic properties, frequent side effects, and clinically significant interactions with many commonly used drugs (199). Consequently, despite its high efficacy, amiodarone is used only for life-threatening ventricular antiarrhythmias that are refractory to other drugs.

An active soft analog may solve many of these problems, and because amiodarone has a butyl side chain, its structure is well suited for such a design (Fig. 15.22). A number of possi-

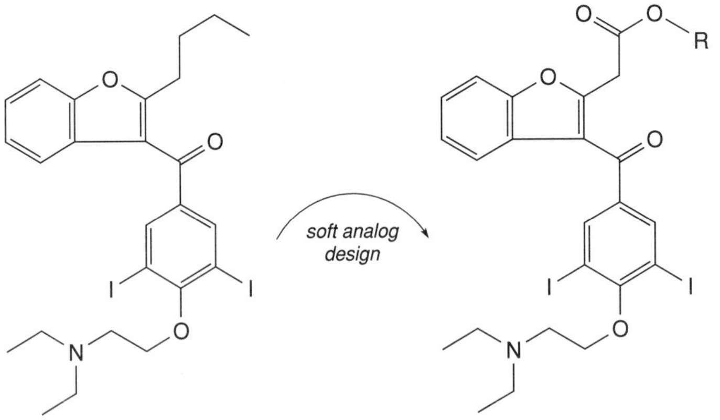

(92) amiodarone

(93)

(93a) R = Me (ATI-2001)
(93b) R = Et (ATI-2010)
(93c) R = iPr (ATI-2064)
(93d) R = iBu
(93e) R = *s*Bu (ATI-2042)
(93f) R = nPe (ATI-2054)
(93g) R = cyclohexyl
(93h) R = (R)menthyl
(93i) R = (R)endobornyl
(93j) R = methyladamantane

Figure 15.22. Amiodarone (**92**) and possible soft analogs (ATI-2000 series, **93**), in which the butyl side chain at position 2 of the benzofurane moiety of amiodarone is replaced with an ester-containing side chain to allow facile hydrolytic degradation.

ble soft analogs (ATI-2000 series, **93**) were synthesized and tested for activity and duration of action (200–203). For example, the electrophysiological effects of the first investigated analog, the methyl ester ATI-2001 (**93**, R = CH_3), were found to be even greater than those of amiodarone in guinea pig isolated heart, and, in agreement with the soft drug design principles, they were more readily reversible. At equimolar concentration (1 μM), the soft analog caused significantly greater slowing of heart rate, depression of atrioventricular and intraventricular conduction, and prolongation of ventricular repolarization than did amiodarone. However, unlike amiodarone, its effects were significantly reversed during washout of the drug.

Because the half-life of ATI-2001 in human plasma was found to be only 12 min (203), which may be too short to allow long-term management of cardiac arrhythmias, esters with longer or more branched side chains were also examined. These modifications were found to markedly alter the magnitude and time course of the induced electrophysiological effects, and the *sec*-butyl and isopropyl esters were considered to merit further investigation (202). In agreement with the principles of soft drug design, the common acid metabolite (**93**, R = H) was found to have no electrophysiological activity (202). Taken together, these findings suggest that such soft drugs may prove to be a valuable addition to current antiarrhythmic therapy, although further studies are needed.

These compounds represent a possible example of an orally active soft drug. Even if the structure contains an ester moiety to allow enzymatic hydrolysis, it is possible to maintain activity for ester-containing drugs after oral administration. Indeed, many ester-containing drugs are orally administered. A study of the butyl ester prodrug of indomethacin in rats also showed that hydrolysis of the ester bond is mainly carried out in the circulatory system and the bond is barely hydrolyzed in the intestinal tract (204).

The above examples illustrate the generally applicable isosteric-type soft analog design, where an ester or a reversed ester function replaces two neighboring methylene groups. In some of these cases, when sufficient structural variability is introduced, the distinction between a soft analog and a (real or

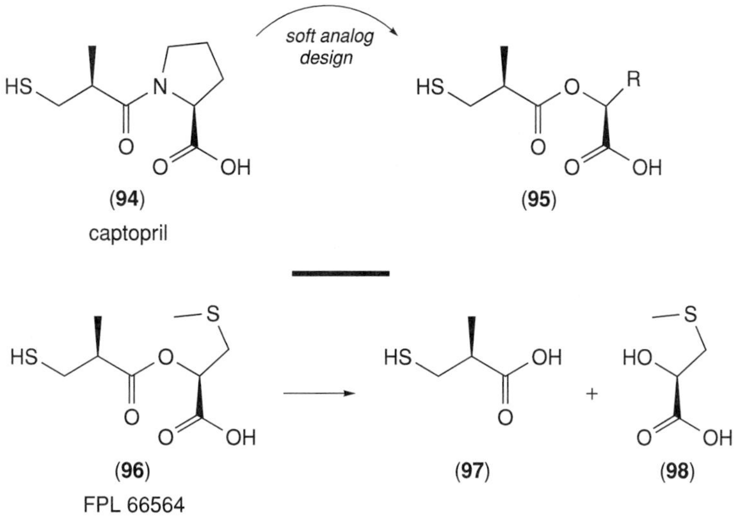

Figure 15.23. Design and metabolism of ultrashort-acting ACE inhibitors (**95**) based on soft analogs of captopril (**94**). Hydrolytic cleavage of these compounds results in small, hydrophilic metabolites (e.g., **97**, **98**) that have no ACE inhibitory activity.

hypothetical) inactive metabolite–based design may become somewhat blurred. For example, larger esters can no longer be regarded as strict structural analogs, but they can be regarded as esters of the (hypothetical) inactive acid metabolite (**93**) (R = H).

3.4.4 Soft Angiotensin-Converting Enzyme (ACE) Inhibitors.

ACE inhibitors, such as captopril (**94**) or enalapril, are widely accepted vasodilators in chronic heart failure. However, their use in acute conditions is restricted because of the prolonged duration of their effect. Again, a *soft analog* approach may provide an active, ultrashort-acting (USA) ACE inhibitor that may represent a viable solution. In work based on these ideas, a number of captopril analogs (**95**) (Fig. 15.23) with the proline amide bond replaced by esters susceptible to hydrolytic cleavage were investigated (205). Whereas no captopril hydrolysis was observed in human blood, a number of soft analogs, especially those with thioalkyl substituents, were degraded in a sufficiently fast manner. Potency, as measured on purified rabbit lung ACE, could be further increased with larger substituents, but the hydrolysis of these compounds became unacceptably slow. Soft analog (**96**) (FPL 66564) was selected as a

potential drug candidate because it showed the required balance of ACE inhibition potency (5.7 nM) and degradation rate (human blood $t_{1/2}$ = 14 min). As required by general soft analog design principles, its hydrolytic products (**97**, **98**) are without ACE inhibitory activity and, being small hydrophilic molecules, should not present any clearance problems. The *in vivo* testing of compound (**96**) showed a dose-dependent inhibition of angiotensin I pressor response in the anesthetized rat, but the effect rapidly dissipated to baseline levels on termination of the i.v. infusion (205).

3.4.5 Soft Dihydrofolate Reductase (DHFR) Inhibitors.

A series of esters was synthesized recently as possible dihydrofolate reductase (DHFR) inhibitors that are susceptible toward hydrolytic degradation (148, 206–208). DHFR is involved in the reduction of dihydrofolate into tetrahydrofolate, and reduced folates are important cofactors in the biosynthesis of nucleic acids and amino acids. Hence, DHFR inhibitors can limit cellular growth. Consequently, classical DHFR inhibitors such as methotrexate (**99**) (Fig. 15.24) or nonclassical DHFR inhibitors such as trimetrexate (**100**) have shown antineoplastic or antiprotozoal activity, and they might also be useful in the treatment of inflammatory bowel disease (IBD), rheumatoid arthritis, psoriasis, and asthma. A number of compounds composed of a bicyclic aromatic unit connected by an ester-containing bridge to another aromatic ring (e.g., **101**, **102**) have been synthesized and investigated as possible therapeutic agents against *Pneumocystis carinii* pneumonia and IBD with increased safety. Substitution of the methyleneamino bridge common to antifolates with an ester-containing bridge (Fig. 15.24) retained DHFR-inhibitory activity. The best ester-containing inhibitors were about 10 times less potent inhibitors than trimetrexate and piritrexim in the DHFR assay, but provided slightly better pcDHFR selectivity index, which was defined as the ratio of IC_{50} (rat liver DHFR) to IC_{50} (*P. carinii* DHFR). Furthermore, the hydrolytic metabolites were all poor inhibitors. *In vitro* hydrolysis using human and rat tissues or available esterases were relatively slow for most of the esters. Human and rat liver fractions were more active than human duodenal mucosa and human blood leukocytes in hydrolyzing the compounds. Contrary to (**101**), effective *in vivo* hydrolysis and a favorable pharmacokinetic profile could be demonstrated for (**102**) (207). Finally, (**101**) exhibited good anti-inflammatory activity in a colitis model in mice (208), but showed unsatisfactory results in a rat arthritic model (206).

3.4.6 Soft Cannabinoids. The most important pharmacologically active member of the cannabinoid family is Δ^9-tetrahydrocannabinol (THC, **103**) (Fig. 15.25), the main active constituent of marihuana, a psychoactive agent used for thousands of years. Cannabinoids have many potential therapeutic benefits, one of the more interesting of which is the reduction of intraocular pressure (IOP) (209). They are known to produce significant and dose-dependent IOP-lowering activity, even if applied topically (209, 210). However, despite apparently doing this by a mechanism different from that of other antiglaucoma drugs (211), their therapeutic potential is diminished because this effect could not be separated from strong CNS and cardiovascular effects.

As in other cases, a topically applied, soft drug susceptible to metabolic inactivation may afford separation between the local, desired effect and the unwanted, systemic side effects. Consequently, new soft cannabinoids (**105**) were synthesized and evaluated (54, 212). They are structural analogs of SP-1 (**104**), a nitrogen-containing cannabinoid derivative that has been shown earlier to have IOP-lowering activity (210, 211, 213). In agreement with the SAR hypothesis used for the design, all the compounds that were successfully synthesized by using a Pechmann condensation had IOP-lowering activity, but the common acid metabolite of the soft analogs (**105**, R = H) was inactive. Activities were evaluated in a number of *in vivo* experiments after both i.v. and topical administration in rabbits. The results obtained were somewhat equivocal because of the extremely low aqueous solubility of the compounds tested, but they were in agreement with a soft drug that is active and is rapidly inactivated. For example, when administered i.v. in rabbits, the ethyl-substituted soft analog (**105**) (R = C_2H_5, R' = C_6H_5) produced IOP-lowering activity that was parallel in both eyes, lasted for only 15 min, and had a maximum of 18 $\pm$ 3% ($P <$ 0.005). When dissolved in emulphor (EL-719 PF618; PEG 40 castor oil) and administered topically, the same soft analog produced an IOP-lowering effect that lasted longer and was significant ($P < 0.05$) at $t = 1$ h. In a first evaluation (212), an indirect-response E_{max} PK/PD model (214) connecting plasma concentration and IOP-lowering effect explained well the experimental results obtained in rabbits following i.v. administration (Fig. 15.26).

3.4.7 Soft Ca^{2+} Channel Blockers
3.4.7.1 Perhexiline Analogs. Perhexiline is a Ca^{2+} channel blocker that is effective in the treatment of angina pectoris, the most common symptom of chronic ischemic heart disease, but is of limited use because of side effects such as hepatotoxicity, weight loss, and peripheral neuropathy. Given that these undesirable effects are related to the slow metabolism and accumulation of perhexiline, soft analogs may represent a conceivable alternative. A class of analogs with an amide moiety inserted as a possible enzymatically labile cen-

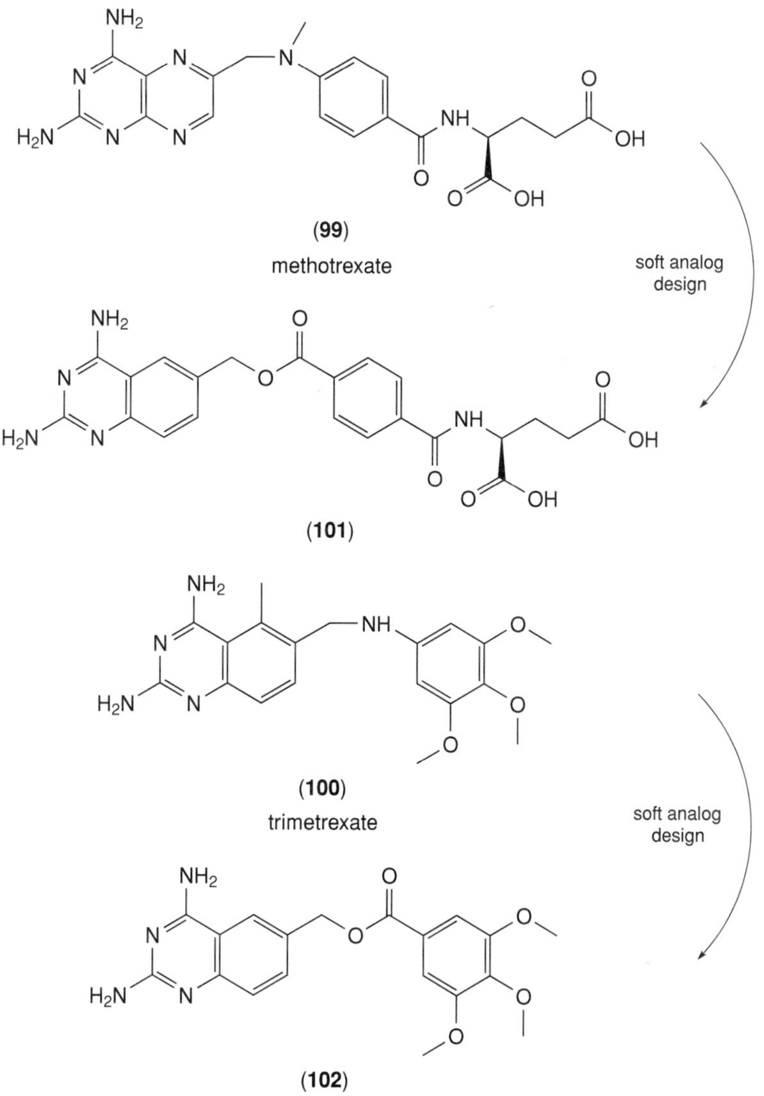

Figure 15.24. Design of soft analogs (**101**, **102**) of methotrexate (**99**) and trimetrexate (**100**) as possible DHFR inhibitors.

ter was investigated for perhexiline (215–217). Most of these newly designed compounds proved to be more active calcium antagonists on depolarized pig coronary artery than perhexiline and produced concentration-related coronary dilation in the perfused guinea pig heart. Some of them also provided promising *in vivo* results. However, metabolic degradation of amide moieties may not be fast enough to provide adequate metabolic lability and

may take place only with the involvement of cytochrome P450.

3.4.7.2 Clevidipine. Clevidipine (**106**) (Fig. 15.27) is an ultrashort-acting, soft calcium channel blocker currently under clinical development for i.v. use in the reduction and control of blood pressure in cardiac surgical procedures (49, 218–220). It is a dihydropyridine-type calcium channel antagonist, structurally related to felodipine (**109**), but with an

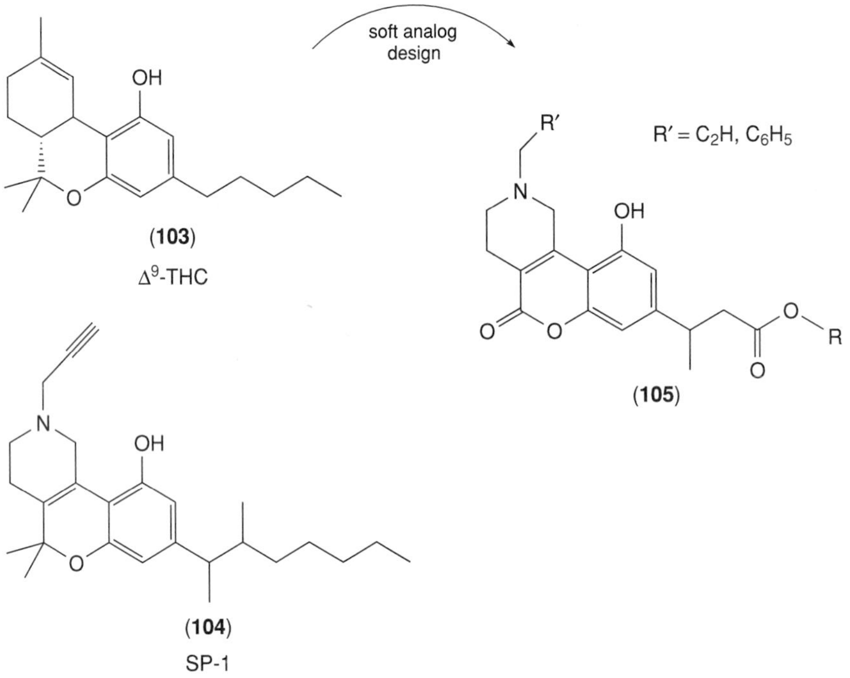

Figure 15.25. Synthesized soft cannabinoid analog structures (**105**) and representative lead structures used for the design (Δ^9-THC, **103**; SP-1, **104**).

additional, less-hindered ester moiety introduced into its structure through an acyloxyalkyl-type substitution. Contrary to the two hindered esters already present in the felodipine structure, this ester is susceptible to enzymatic degradation and leads to an acidic, inactive metabolite (**108**) through an unstable intermediate (**107**) (Fig. 15.27) (49). Such acyloxyalkyl-type double esters have been extensively used in prodrug design to trigger sufficiently fast, substitution-controllable degradations (31). This design resulted in sufficiently fast degradation (e.g., *in vitro* human blood $t_{1/2}$ of 5.8 min) and formation of an equivalent amount of the corresponding primary metabolite, which has been shown to be devoid of any vasodilating effect in both animal and human experiments (49, 218). Clevidipine is a high-clearance drug with a small volume of distribution, which results in very short half-lives. It appears suitable for blood pressure control, for example, after coronary artery bypass grafting (219).

Clevidipine has two possible enantiomers, and hydrolysis half-lives were only about 10%

different for the R and S isomers. A slight stereoselectivity in the extensive plasma binding of clevidipine (free fractions of 0.43% and 0.32% for S- and R-clevidipine, respectively) might be the reason for this difference (49, 220). Hence, from a pharmacokinetic point of view, there seems to be no advantage in using a single enantiomer instead of a racemic mixture. The hydrolytic half-life of clevidipine was also investigated in blood from pseudocholinesterase-deficient volunteers because the homozygous atypical allele for the corresponding gene has been reported to occur with a frequency of about 1 in 3500 for Caucasians. In the small number of subjects studied, *in vitro* half-life increased by only about 50% (49).

3.5 Active Metabolite–Based Soft Drugs

Active metabolite–based soft drugs are metabolic products of a drug resulting from oxidative conversions that retain significant activity of the same kind as the parent drug (7, 10–12). Most drugs undergo stepwise metabolic degradation to yield intermediates and

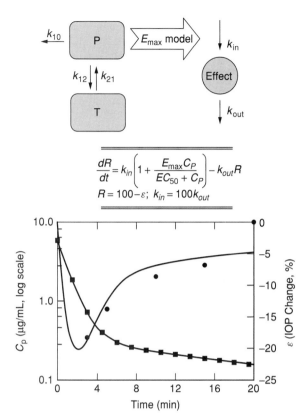

$$\frac{dR}{dt} = k_{in}\left(1 + \frac{E_{max}C_P}{EC_{50} + C_P}\right) - k_{out}R$$

$$R = 100 - \varepsilon; \quad k_{in} = 100k_{out}$$

Figure 15.26. PK/PD relationship between the concentration C_P (■) and IOP-lowering effect ε (●) of soft cannabinoid **105** in rabbits following i.v. administration. Curves represent the best fit of the data obtained with the indirect response model used as a first evaluation and characterized by the equations shown and the following parameters: EC_{50} = 1.2 μg/mL; E_{max} = 40%; k_{out} = 0.75. Notation: R, response; k_{in}, zero-order influx rate constant of the process assumed to cause the IOP-lowering effect and assumed to be stimulated by the drug; k_{out}, corresponding first-order efflux rate constant; ε, measured effect (IOP change %); E_{max}, maximum stimulatory effect of the drug on the influx rate; C_P, plasma concentration obtained from a two-compartment pharmacokinetic model (P, plasma; T, tissue) characterized by first-order rate constants k_{10}, k_{12}, and k_{21}; EC_{50}, plasma concentration causing 50% of the E_{max} effect on k_{in}.

structural analogs (D_1...D_m) that often have activity similar to that of the original drug molecule (D) (Fig. 15.1). These general oxidative metabolic transformations put a burden on the saturable and slow oxidative enzyme system, and result in compounds that have different selectivity, binding, distribution, and elimination properties. Therefore, at any given time, a mixture of active components is present with continuously changing relative concentration. This can result in complex, almost uncontrollable situations, making safe and effective dosing almost impossible.

In agreement with the basic soft drug design principles, judicious selection of an active metabolite can yield a potent drug that will undergo a one-step deactivation process, given that it is already at the highest oxidation state. For example, if sequential oxidative metabolic conversion of a drug takes place, such as the quite common hydroxyalkyl → oxo → carboxy sequence, in which the carboxy function is generally the inactive form, some previous ox-

idative metabolite (preferably the one just before deactivation) could be the best choice for a drug. Despite numerous publications suggesting the utility of active metabolites for drug design purposes (7, 10–12), the idea was long neglected and only recently began to generate interest. Nevertheless, there are examples of active metabolites used as a source of new drug candidates because of better safety profiles; for example, oxyphenbutazone, the active *p*-hydroxy metabolite of phenylbutazone; oxazepam, the common active metabolite of chlordiazepoxide, halazepam, chlorazepate, and diazepam; or pravastatin, the hydroxylated structure derived from mevastatin (compactin).

Bufuralol (**15**) (Fig. 15.6), the nonselective β-adrenoceptor antagonist with $β_2$ partial agonist properties discussed earlier, also provides a good illustration. As mentioned, bufuralol undergoes complex metabolism in humans, including stepwise oxidation toward an acid metabolite (**18**) through the corre-

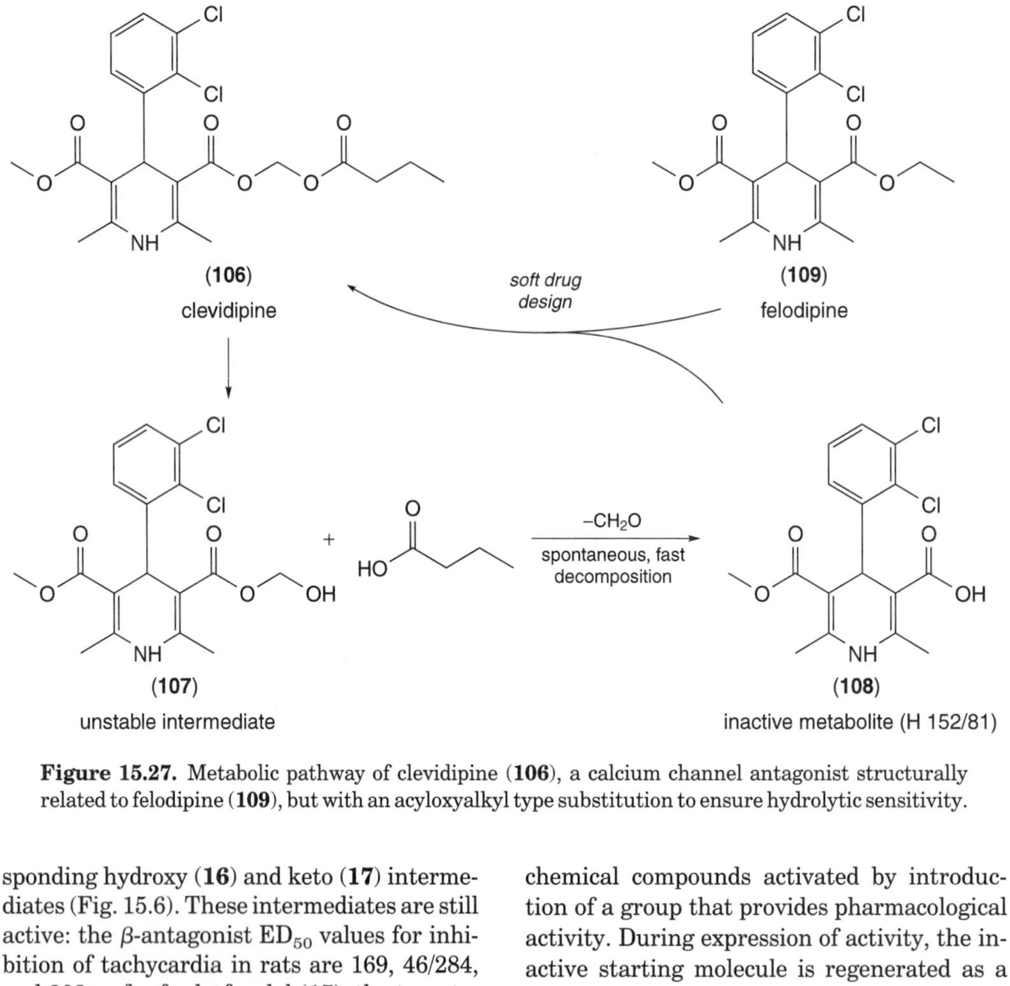

Figure 15.27. Metabolic pathway of clevidipine (**106**), a calcium channel antagonist structurally related to felodipine (**109**), but with an acyloxyalkyl type substitution to ensure hydrolytic sensitivity.

sponding hydroxy (**16**) and keto (**17**) intermediates (Fig. 15.6). These intermediates are still active: the β-antagonist ED_{50} values for inhibition of tachycardia in rats are 169, 46/284, and 203 μg/kg for bufuralol (**15**), the two stereoisomers of (**16**), and (**17**), respectively (88, 89). They also have different and long elimination half-lives: biological half-lives are 4, 7, and 12 h for bufuralol (**15**), (**16**), and (**17**), respectively (90). Differential metabolism of the two enantiomers and differences ascribed to genetic polymorphism also occur (91). According to the principle of active metabolite design, the ketone (**17**), the highest active oxidative metabolite, should be the drug of choice (11). This compound still retains most of the activity, but is deactivated by one-step oxidation to structure (**18**).

3.6 Activated Soft Drugs

Activated soft compounds are not analogs of known drugs, but are derived from nontoxic chemical compounds activated by introduction of a group that provides pharmacological activity. During expression of activity, the inactive starting molecule is regenerated as a result of a hydrolytic process. An example of activated soft compound is provided by N-chloramine antimicrobials. During an effort to identify locally active antimicrobial agents of low toxicity, N-chloramines based on amino acids, amino alcohol esters, and related compounds were developed. These compounds (Fig. 15.28), particularly those derived from α-disubstituted amino acid esters and amides (221, 222), serve as a source of positive chlorine (Cl^+), which was assumed to be primarily responsible for antimicrobial activity. However, when the chlorine is lost, before or after penetration through microbial cell walls, the nontoxic initial amine is regenerated (Fig. 15.28).

The antimicrobial activity of chloramines was known before, but chemical instability

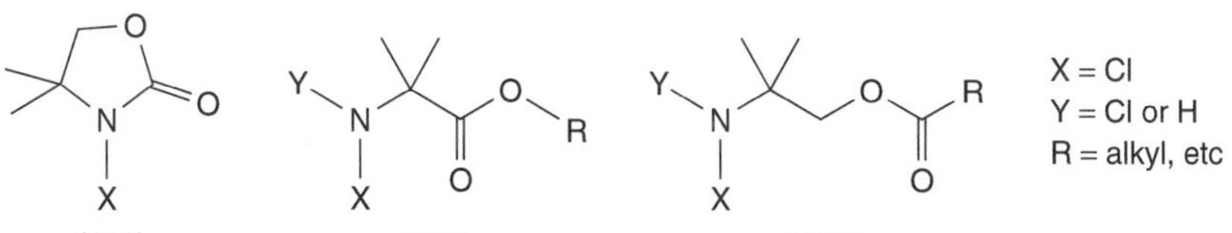

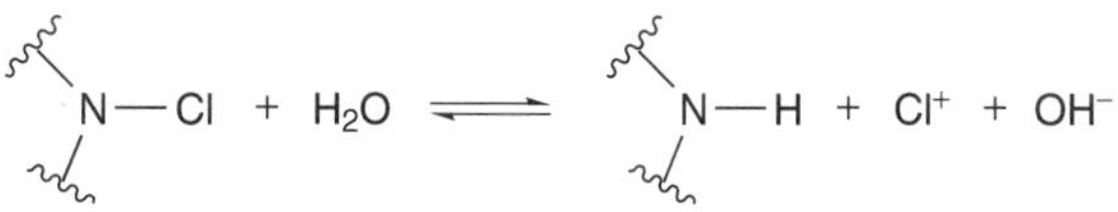

Figure 15.28. General structures of representative low chlorine potential, soft *N*-chloramine antimicrobials (**110–112**), and the mechanism of their decomposition.

and high reactivity limited their widespread use. After establishing the mechanism of their decomposition (221), it was realized that stable *N*-chloramines that have much lower "chlorine potential" and are much less corrosive could be obtained if the α-carbons lack hydrogen (221–224). Structures (**110**), (**111**), and (**112**) in Fig. 15.28 illustrate some of these low chlorine potential, soft chloramines. Structure (**110**) proved to be a particularly effective bactericide in a laboratory water treatment plant (225). It is exceptionally stable in water and in dry storage (226, 227); it is nontoxic in chicken's drinking water at 200 mg/L, and it detoxifies aflatoxin (228). The mechanism of action of these compounds involves inhibition of bacterial growth by inhibiting DNA, RNA, and protein synthesis (229). When the chlorine interacts with —SH-containing enzymes, the original precursor is regenerated. Thus, the predictable and controllable metabolism of soft chloramines gives a definite advantage over hard, lipophilic, aromatic antimicrobials containing C—Cl bonds.

Another example of activated soft compound is provided by the class of soft alkylating agents prepared during the development of soft quaternary salts (25). These compounds are α-halo esters and are relatively mild alkylating agents. Their low alkylating potential should allow transport to tumor cells without indiscriminate alkylation. In addition, being activated esters, the circulating free part can be deactivated by esterases. These should result in a better separation of desired activity and unwanted toxicity. One of these compounds, chloromethyl hexanoate,

was found to have anticancer activity in the P388 lymphocytic leukemia test (25).

3.7 Pro-Soft Drugs

As their name implies, pro-soft drugs are inactive prodrugs (chemical delivery forms) of a soft drug of any of the above classes including endogenous soft molecules. They are converted enzymatically into the active soft drug, which is subsequently enzymatically deactivated. Soft drugs, as any other drug, can be the subject of prodrug design resulting in pro-soft drugs, but, as results from their definitions, one cannot conceive a "soft prodrug." Two examples of pro-soft drugs have already been discussed: a colon-targeted pro-soft drug (**36**) for possible oral treatment of ulcerative colitis (145) and a possible pro-soft antipsoriatic β-adrenergic agonists (161).

Derivatives of natural hormones and other biologically active agents such as neurotransmitters have well-developed mechanisms for their disposition and, therefore, can be considered *natural soft drugs* (e.g., hydrocortisone, dopamine). Because their metabolism is usually fast and their transport is specific, they can become useful as drugs only if sustained, local, or site-specific delivery is developed for them. Compounds designed for such purposes can also be considered pro-soft drugs.

A possible example for sustained chemical release at the site of application for hydrocortisone (**116**), a natural soft drug, is shown in Fig. 15.29. The 4,5-unsaturated 3-ketone group, being essential for binding and activity, is a good target for modification. Spirothiazolidine derivatives (**113**) were selected (230) be-

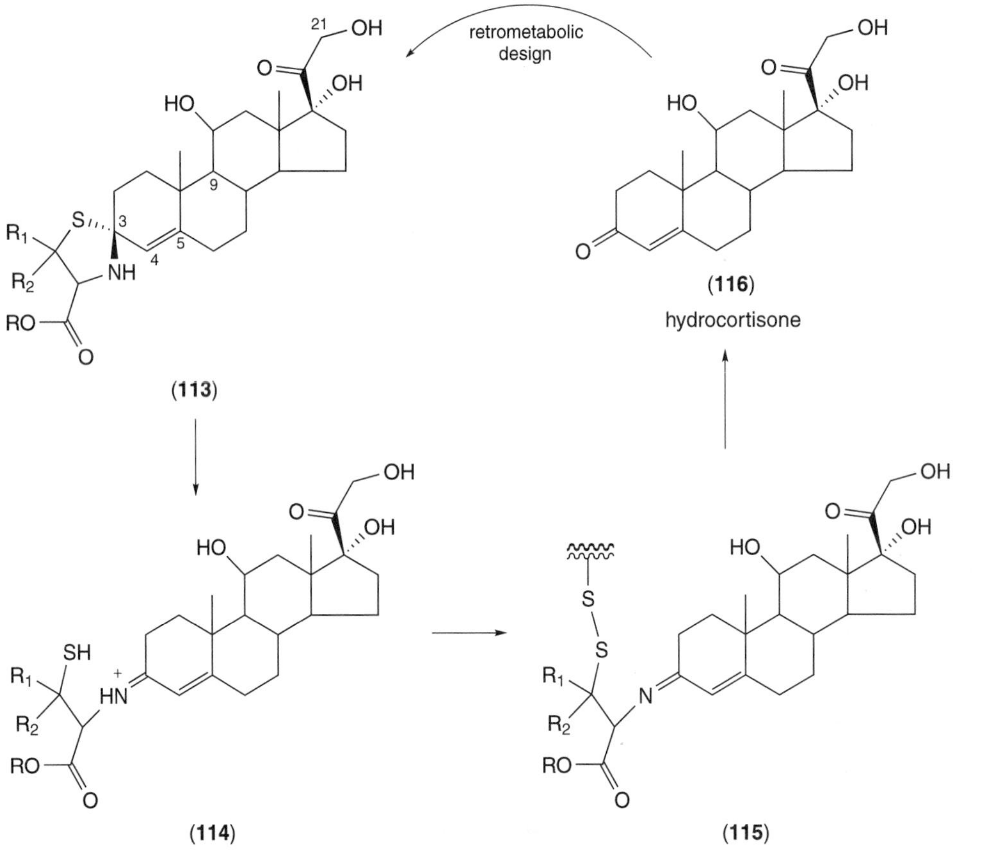

Figure 15.29. Hydrocortisone (**116**) can be regarded as a natural soft drug, and a spirothiazolidine derivative (**113**) serves as a pro-soft drug for controlled release. Opening of the thiazolidine ring (**114**) is followed by disulfide bridging (**115**) to trap the steroid at the site of application.

cause they should be subject to biological cleavage of the imine formed after spontaneous cleavage of the carbon—sulfur bond (231). Spirothiazolidines of hydrocortisone acetate were about three to four times more potent than hydrocortisone derivatives when tested for topical anti-inflammatory activity. Meanwhile, they delivered significantly less hydrocortisone transdermally than did either the unmodified hydrocortisone or hydrocortisone 21-acetate. These results are consistent with a local tissue binding through a disulfide bridge, as suggested in Fig. 15.29. The opening of the spirothiazolidine ring of (**113**) as shown in structure (**114**) allows trapping of the steroid with a disulfide bridging (**115**) at the site of application. Hydrolysis releases the active component (**116**) only locally (232). A similar

behavior of the thiazolidine of progesterone confirmed this local binding in the skin (230).

3.8 Computer-Aided Design

Because of the considerable flexibility of retrometabolic drug design, for certain lead compounds a large number of possible soft structural analogs can be designed by applying the various "soft transformation rules." Finding the best drug candidate among them may prove tedious and difficult. Fortunately, computer methods developed to calculate various molecular properties, such as molecular volume, surface area, charge distribution, polarizability, aqueous solubility, partition and permeability coefficient, and even hydrolytic lability, make possible more quantitative design (233–244). The capabilities of quantita-

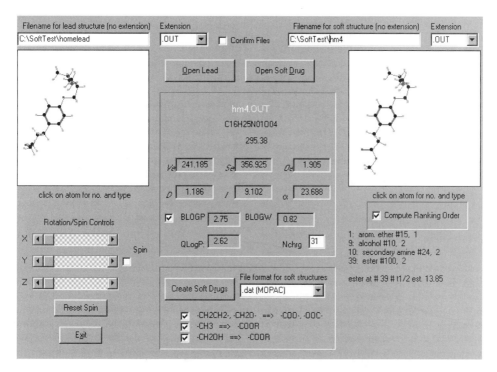

Figure 15.30. Windows 95/98 interface of the computer-aided soft drug design program.

tive design have been further advanced by developing expert systems that combine the various structure-generating rules of soft drug design with the developed predictive softwares to provide a ranking order based on isosteric/isoelectronic analogy (16, 245–247). The interface of a recent version of the corresponding computer program is shown in Fig. 15.30.

The overall approach is general in nature and can be used starting with essentially any lead. The system can provide full libraries of possible new "soft" molecular structures and an analogy-based ranking of these candidates, making possible more thorough and more quantitative design. Because candidates that are unlikely to have reasonable activity can be eliminated ahead of synthesis and experimental testing, considerable savings are achieved in laboratory time and expense.

3.8.1 Structure Generation. The expert system designed to aid the generation and ranking of novel soft drug candidates has been described in the literature in detail (16, 245–

247). As an important part, it contains the rules to transform certain substructures of the lead compound according to the principles of retrometabolic design. For the soft drug approach, the two most successful strategies, the inactive metabolite–based and the soft analog strategies, were implemented. The program can generate common oxidative metabolites of the lead compound; for example, it can find $—CH_3$, $—CH_2OH$, or other alkyl groups in the lead and replace them by $—COOH$ or $—(CH_2)_n—COOH$ groups, or it can find phenyl ($— C_6H_5$) groups and replace them by phenol ($—C_6H_4OH$) groups, and so forth. Accordingly, the computer can then generate new soft drugs by derivatization of the "oxidized" metabolite to regenerate active soft drugs. The program can also generate soft analogs by looking for the presence of neighboring methylene/hydroxymethylene groups and by replacing them in an isosteric/isoelectronic fashion with corresponding esters $—O—CO—$, reversed esters $—CO—O—$, or other functions.

3.8.2 Candidate Ranking. Because in certain cases a large number of analog structures may be designed, it is desirable to have some prediction regarding their activities and/or toxicities. Structures that are better isosteric/isoelectronic analogs of the lead are also more promising drug candidates. The present expert system provides a ranking order based on the closeness of calculated properties to those of the lead compound using fully optimized geometries obtained from advanced semiempirical AM1 calculations (248) for all the compounds that are of interest.

Because most new structures are close structural analogs (often exact isomers), it is important to include as many and as relevant properties as possible. At present, four parameter categories are used with equal weight to describe isosteric/isoelectronic relations: molecular size/shape descriptors (V, volume; S, surface area; O, ovality); electronic properties (D, dipole moment; α, average polarizability; I, ionization potential); predicted solubility/partition properties (log W, log $P_{o/w}$); and atomic charge distribution on the unchanged portions. Each of these parameters can be obtained from the optimization output, and they all play important roles in determining binding and transport properties; they should give a relatively good description of the isosteric/isoelectronic analogy. Because all these properties are measured in different units and vary over different ranges, ranking factors (RF) were introduced for each property, and a weighted average is used for the final ranking (16, 245–247). Following synthesis and experimental testing of a few best candidates, further refinements are possible; for example, experimental results might suggest different weighting for the properties used in ranking. Illustrative examples were also provided in details in the reviews mentioned (16, 245–247).

3.8.3 Hydrolytic Lability. Hydrolytic lability is also of considerable relevance; therefore, a quantitative structure-metabolism relationship (QSMR) study was undertaken recently to identify a structure-based QSMR equation that is not limited to congener series (241). An equation accounting for 80% of the variability of the *in vitro* human blood log $t_{1/2}$ of seven different drug classes containing a total of 67

compounds was obtained by introducing the inaccessible solid angle Ω_h as a measure of steric hindrance. The final equation used to estimate log $t_{1/2}$, in addition to $\Omega hO=$, includes the AM1-calculated charge on the carbonyl carbon ($q_{C=}$) and a calculated log octanol/water partition coefficient (**QLogP**) (237, 240, 243) as parameters:

$$\log t_{1/2} = -3.805 + 0.172\Omega_h^{O=}$$
$$- 10.146q_{C=} + 0.112\textbf{QLogP}$$
$$n = 67, \quad r = 0.899, \quad (15.3)$$
$$\sigma = 0.356, \quad F = 88.1$$

In general agreement with previous results, we found steric effects as having the most important influence on the rate of enzymatic hydrolysis. Lipophilicity, as measured by **QLogP** and some of the electronic parameters, such as the charge on the carbonyl C ($q_{C=}$), also proved informative, but to a much lesser degree. Half-lives were found to increase with increasing steric hindrance around the ester moiety, as measured by the inaccessible solid angle Ω_h. An important novelty was the finding that the rate of metabolism as measured by log $t_{1/2}$ seems to be more strongly correlated with the steric hindrance of the carbonyl sp^2 oxygen ($\Omega hO = :r2 = 0.58$, $n = 79$) than with that of the carbonyl sp^2 carbon as measured by the inaccessible solid angle ($\Omega hC = : r2 = 0.29$). This provides additional evidence for the important, possibly even rate-determining role played by hydrogen bonding at this oxygen atom in the mechanism of this enzymatic reaction. The corresponding fully computerized predictive method was integrated within the expert system (247).

3.8.4 Illustration: Esmolol. A formal computer-aided soft drug design that uses homometoprolol (**117**) as the lead compound provides a good illustration of the process of computerized structure-generation and candidate ranking. Starting with structure (**117**), four different soft drug analogs are generated by oxidizing the methoxymethyl function to its corresponding carboxylic acid type metabolites and converting them into various esters

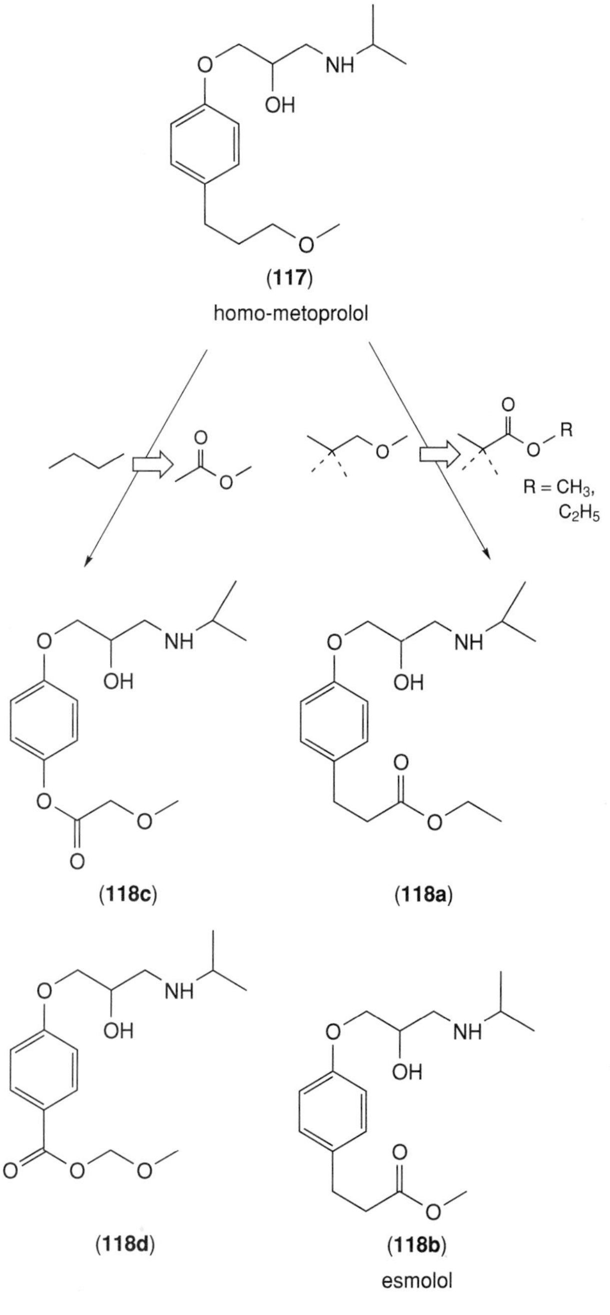

(117)
homo-metoprolol

R = CH₃,
C₂H₅

R = CH_3, C_2H_5

(118c)   (118a)

(118d)   (118b)

esmolol

Figure 15.31. Illustration of the formal soft drug design process that uses homo-metoprolol (**117**) as lead to generate four soft drug analogs including esmolol (**118b**).

(R methyl or ethyl) (**118a**, **118b**), respectively, by replacing neighboring methylene groups with —O—CO— or —CO—O— functions (**118c**, **118d**), as shown in Fig. 15.31. Properties are then calculated based on AM1-optimized structures. Table 15.1 summarizes the analogy-ranking in order of decreasing overall analogy together with the half-lives estimated using the present program (Fig. 15.30). Charge distributions were compared for atoms in the aromatic ring and the β-amino alcohol side-chain (including hydrogens). The

Table 15.1   Ranking Order and Predicted Hydrolytic Half-Lives for Homo-Metoprolol (117) Analogs 118 (Fig. 15.31)

Compound	$RF_S{}^a$	RF_E	RF_P	RF_Q	RF	Predicted $t_{1/2}$ (min)
118b	1.00	3.30	2.12	1.00	1.85	13.9
118c	13.12	2.59	9.94	18.99	11.16	21.3
118a	36.94	5.37	4.56	1.47	12.09	14.7
118d	17.11	16.18	13.88	15.89	15.76	1.8

^{a}Ranking factors (RF) were computed as described in the references mentioned. Smaller values indicate better isosteric/isoelectronic analogy.

obtained results clearly show (**118b**) as an outstanding best analog in practically all categories. The nice aspect of this hindsight design is that structure (**118b**) corresponds to esmolol (Brevibloc), an ultrashort-acting (USA) β-blocker discussed earlier (**11**) (Fig. 15.5).

4   CHEMICAL DELIVERY SYSTEMS

Chemical delivery system (CDS) approaches provide novel, systematic methodologies for targeting active biological molecules to specific target sites or organs based on predictable enzymatic activation (15, 21, 23). CDSs are inactive chemical derivatives of a drug obtained by one or more chemical modifications. They provide site-specific or site-enhanced delivery through sequential, multistep enzymatic and/or chemical transformations. The newly attached moieties are monomolecular units that are in general smaller than, or of similar size as, the original molecule. Hence, the *chemical delivery system* term is used here in a stricter sense: they do not include systems in which the drug is covalently attached to large "carrier" molecules.

In a general formalism, the bioremovable moieties introduced can be classified into two categories. Targetor (T) moieties are responsible for targeting, site specificity, and lock-in. Modifier functions ($F_1 \cdots F_n$) serve as lipophilizers, protect certain functions, or fine-tune the necessary molecular properties to prevent premature, unwanted metabolic conversions. CDSs are designed to undergo sequential metabolic conversions, first disengaging the modifier functions and then the targetor moiety, after it fulfilled its site- or organ-targeting role. These transformations provide targeting

(differential distribution) of the drug by either exploiting site-specific transport properties, such as those provided by the presence of a blood–brain barrier (BBB), or by recognizing specific enzymes found primarily, exclusively, or at higher activity at the site of action.

Three major CDS classes have been identified:

1. *Brain-targeting (enzymatic physical-chemical–based) CDSs*: exploit site-specific traffic properties by sequential metabolic conversions that result in considerably altered transport properties.
2. *Site-specific enzyme-activated CDSs*: exploit specific enzymes found primarily, exclusively, or at higher activity at the site of action.
3. *Receptor-based transient anchor-type CDSs*: provide enhanced selectivity and activity through transient, reversible binding at the receptor.

4.1   Brain-Targeting (Enzymatic Physical-Chemical–Based) CDSs

4.1.1   Design Principles. Brain-targeting chemical delivery systems represent the most developed CDS class. To obtain such a CDS, the drug is chemically modified to introduce the protective function(s) and the targetor (T) moiety. Upon administration, the resulting CDS is distributed throughout the body. Predictable enzymatic reactions convert the original CDS by removing some of the protective functions and modifying the T moiety, leading to a precursor form (T^+–D), which is still inactive, but has significantly different physicochemical properties (Fig. 15.32). Whereas these intermediates are continuously elimi-

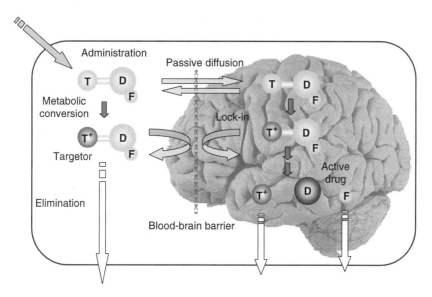

Figure 15.32. Schematic representation of the sequential metabolism employed by brain-targeting CDSs to allow targeted and sustained delivery of the active drug D.

nated from the "rest of the body," at the targeted site, which has to be delimited by some specific membrane or other distribution barrier, efflux–influx processes are altered and provide a specific concentration. Consequently, release of the active drug only occurs at the site of action.

For example, the blood–brain barrier (BBB) can be regarded as a biological membrane that is in general permeable to lipophilic compounds, but not to hydrophilic molecules (Fig. 15.33). In most cases, these transport criteria apply to both sides of the barrier. The BBB is a unique membranous barrier that tightly segregates the brain from the circulating blood (249, 250). Capillaries of the vertebrate brain and spinal cord lack the small pores that allow rapid movement of solutes from circulation into other organs; these capillaries are lined with a layer of special endothelial cells that lack fenestrations and are sealed with tight junctions. Tight epithelium, similar in nature to this barrier, can also be found in other organs (skin, bladder, colon, lung) (251). It is now well established that these endothelial cells, together with perivascular elements such as astrocytes and pericytes, constitute the BBB. In brain capillaries, intercellular cleft, pinocytosis, and fenestrae

are virtually nonexistent; exchange must pass transcellularly. Therefore, only lipid-soluble solutes that can freely diffuse through the capillary endothelial membrane may passively cross the BBB. In general, such exchange is overshadowed by other nonspecific exchanges. Despite the estimated total length of 650 km and total surface area of 12 m² of capillaries in human brain covered by only about 1 mL of total capillary endothelial cell volume, this barrier is very efficient and makes the brain practically inaccessible for lipid-insoluble compounds, such as polar molecules and small ions. The BBB also has an additional, enzymatic aspect: solutes crossing the cell membrane are subsequently exposed to degrading enzymes present in large numbers inside the endothelial cells that contain large densities of mitochondria, metabolically highly active organelles. Furthermore, active transport can significantly alter both inward and outward transport for compounds that are substrates of the corresponding transporters.

Therefore, if a lipophilic compound enters the brain and is converted there into a hydrophilic molecule, it will become "locked in": it will no longer be able to come out (Fig. 15.32). With such a system, targeting is assisted because the same conversion, as it takes place in

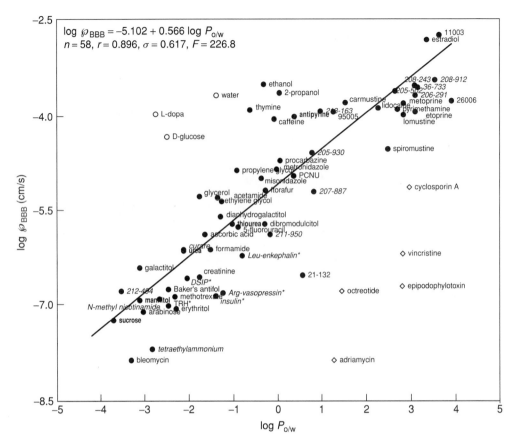

Figure 15.33. *In vivo* log permeability coefficient of rat brain capillaries (log $\wp_{BBB}$) as a function of log octanol/water partition coefficient (log $P_{o/w}$), the most commonly used measure of lipophilicity. For compounds in italic, the log distribution coefficient measured at physiological pH was used (log $D_{o/w}$). Values denoted with a star are for guinea pig. The strong deviants below the line that are denoted with diamonds are known substrates for P-glycoprotein, a multidrug transporter that actively removes them from the brain.

the rest of the body, accelerates peripheral elimination and further contributes to brain targeting. In principle, many targetor moieties are possible for a general system of this kind (252–257), but the one based on the 1,4-dihydrotrigonelline ↔ trigonelline (coffearine) system, in which the lipophilic 1,4-dihydro form (T) is converted *in vivo* to the hydrophilic quaternary form (T$^+$) (Fig. 15.34), proved the most useful. This conversion takes place easily everywhere in the body because it is closely related to the ubiquitous NAD(P)H ⇌ NAD(P)$^+$ coenzyme system associated with numerous oxidoreductases and cellular respiration (258, 259). Because oxidation takes

place with direct hydride transfer (260) and without generating highly active or reactive radical intermediates, it provides a nontoxic targetor system (261). Furthermore, it was shown (262) that the trigonellinate ion formed after cleavage of the CDS undergoes rapid elimination from the brain, most likely by involvement of an active transport mechanism that eliminates small organic ions; therefore, the T$^+$ moiety formed during the final release of the active drug D from the charged T$^+$–D form will not accumulate within the brain (262, 263).

Although the charged T$^+$–D form is locked behind the BBB into the brain, it is easily elim-

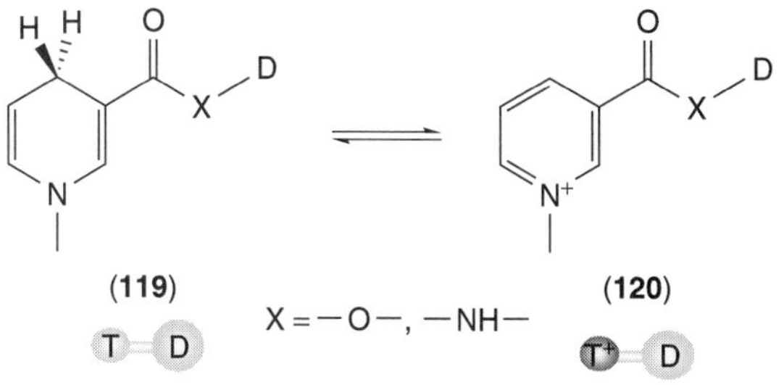

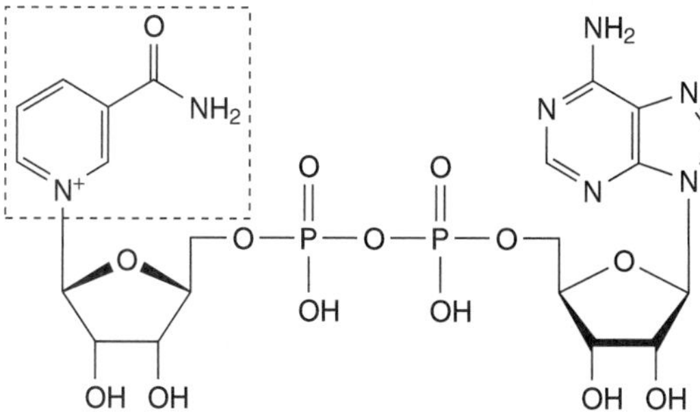

(121)
NAD⁺ nicotinamide adenine dinucleotide

Figure 15.34. The trigonellinate ⇌ 1,4-dihy-drotrigonellinate redox system used in brain-targeting CDSs exploits the analogy with the ubiquitous $NAD(P)^+ \rightleftharpoons NAD(P)H$ coenzyme system to convert the lipophilic 1,4-dihydro form (T, **119**) into the hydrophilic quaternary form (T^+, **120**).

inated from the body as a result of the acquired positive charge, which enhances water solubility. After a relatively short time, the delivered drug D (as the inactive, locked-in T^+–D) is present essentially only in the brain, providing sustained and brain-specific release of the active drug. It has to be emphasized again that the system not only achieves delivery to the brain but it also achieves preferential delivery, which means brain targeting. This should allow smaller doses and reduce peripheral side effects. Furthermore, because the "lock-in" mechanism works against the concentration gradient, the system also provides more prolonged effects. Consequently, these CDSs can be used not only to deliver compounds that otherwise have no access to the brain but also to retain lipophilic compounds within the brain, as it has indeed been achieved, for example, with a variety of steroid hormones.

The CDS approach has been explored with a wide variety of drug classes: biogenic amines: phenylethylamine (17, 264, 265), tryptamine (266, 267); steroid hormones: testosterone (257, 268–270), progestins (271), progesterone (272), dexamethasone (273, 274), hydrocortisone (257), estradiol (275–293); anti-

infective agents: penicillins (294–297), sulfonamides (298); antivirals: acyclovir (21, 299), trifluorothymidine (300, 301), ribavirin (302–304), Ara-A (302), deoxyuridines (305–307), 2′-F-5-methylarabinosyluracil (308); antiretrovirals: zidovudine (AZT) (309–323), 2′,3′-dideoxythymidine (263), 2′,3′-dideoxycytidine (324), ganciclovir (325), cytosinyl-oxathiolane (326); anticancer agents: lomustine (CCNU) (327, 328), HECNU (328), chlorambucil (329); neurotransmitters: dopamine (330–334), GABA (335, 336); nerve growth factor (NGF) inducers: catechol derivatives; anticonvulsants: GABA (336), phenytoin (339), valproate (340), stiripentol (341), felodipine (342); monoamine oxidase (MAO) inhibitors: tranylcypromine (343, 344); antidementia drugs (cholinesterase inhibitors): tacrine (345, 346); antioxidants: LY231617 (347); chelating agents: ligands for technetium complexes (348); nonsteroidal anti-inflammatory drugs (NSAIDs): indomethacin (349), naproxen (349); anesthetics: propofol (350); nucleosides: adenosine (351); and peptides: tryptophan (352, 353), Leu-enkephalin analog (354, 355), thyrotropin-releasing hormone (TRH) analogs (356–359), kyotorphin analogs (22, 360), and S-adenosyl-L-methionine (SAM) (361). Selected examples are summarized in the following chapters together with representative physicochemical properties, metabolic pathways, and pharmacological data.

4.1.2 Site-Targeting Index and Targeting-Enhancement Factors.

In most cases, drugs are intended to exert their action at some selected site, but they are delivered systemically, and only a small fraction of the dose reaches the intended target site (e.g., organ or cell). Drug targeting or drug delivery systems (including most prodrugs) are specifically designed to improve this situation. To assess the site-targeting effectiveness of drugs or drug delivery systems in general from a pharmacokinetic perspective, one can define a *site-targeting index* (STI) as the ratio between the area under the curve (AUC) for the concentration of the drug itself at the targeted site and that at a systemic site (e.g., blood or plasma):

$$STI = AUC_{target}/AUC_{blood} \qquad (15.4)$$

This index gives an accurate measure on how effectively the active therapeutic agent is actually delivered to its intended site of action. For example, as data in Table 15.2 indicate, some drugs (e.g., AZT and benzylpenicillin) are very ineffective in penetrating the BBB and reaching brain tissues; the AUCs of their brain concentrations are less than 1% of the corresponding blood values.

To assess the effectiveness of a drug delivery system compared to the drug itself, one can define a *site-exposure enhancement factor* (SEF) to measure the change of the AUC of the active drug at the target site after administration of the delivery system compared to administration of the drug alone:

$$SEF = AUC_{target}^{Delivery\ System}$$
$$\div AUC_{target}^{Drug\ Alone} \qquad (15.5)$$

Of course, this definition assumes equivalent doses. For slightly different doses, linearity between dose and AUC is a reasonable assumption, and the ratio of dose-normalized AUCs (AUC/dose) can be used in the above definition. Because a good delivery system not only increases exposure to the active agent at the target site, but also decreases the corre-

sponding systemic exposure, a *targeting-enhancement factor* (TEF) can be introduced that measures the relative improvement in the STI produced by administration of the delivery system compared to administration of the drug itself:

$$TEF = STI^{Delivery\ System}/STI^{Drug\ Alone} \qquad (15.6)$$

TEF as defined above is the most rigorous measure that can be used to quantify the improvement in (pharmacokinetic) targeting produced by a delivery system. It compares not just concentrations, but concentrations along a time period, and it compares actual, active drug concentrations both at target and systemic sites. If (*1*) the targeting or delivery ("carrier") moiety produces no toxicity of its own, (*2*) the therapeutic effect is linearly related to AUC_{target}, and (*3*) side effects are linearly related to AUC_{blood} (assumptions that are mostly satisfied), then TEF also represents the enhancement produced in the therapeutic index (TI) defined by Equation 15.1: $TI^{Delivery\ System}/TI^{Drug\ Alone}$. As it turns out, TEF, as defined here, represents the same ratio as a drug-targeting index used in a theoretical pharmacokinetic treatment (362).

Table 15.2 presents such AUC-based SEF and TEF values for brain-targeting CDSs in different species with available *in vivo* experimental data. In a number of cases, CDSs produced a very substantial increase of the targeting effectiveness, often more than an order of magnitude.

4.1.3 Zidovudine (AZT)–CDS.

One of the many devastating consequences of the infection by human immunodeficiency virus type 1 (HIV-1), the causative agent of AIDS, is an encephalopathy with subsequent dementia. To adequately treat AIDS dementia, antiviral agents must reach the CNS in therapeutically relevant concentrations, but many potentially useful drugs cannot penetrate into brain tissue. Zidovudine (azidothymidine, AZT; **122**) (Fig. 15.35) was the first drug approved for the treatment of AIDS. This modified riboside has been shown to be useful in improving the neuropsychiatric course of AIDS encephalopathy in a few patients, although the doses required

Table 15.2 Site-Targeting Indices (STI), Site-Exposure, and Targeting Enhancement Factors (SEF, TEF) of the Brain-Targeting CDSs with Available _In Vivo_ Data

Compound Species (reference)	Dose (μmol/kg)	Time (h)	D only AUC (μg × h/mL) Brain	Blood	STI	CDS AUC (μg × h/mL) Brain	Blood	STI	SEF Eq. 15.5	TEF Eq. 15.6
AZT										
Mouse (322)[a]	187	12	1.21	26.64	**0.045**	11.28	25.38	0.444	9.32	9.79
Rat (314)	130	6	0.23	26.33	**0.009**	7.09	9.67	0.733	31.51	85.84
Rabbit (311)	64	3	11.22	37.92	**0.296**	26.98	34.07	0.792	2.40	2.68
AZDU										
Mouse (322)[a]	197	12	2.09	25.83	**0.081**	11.43	25.79	0.443	5.47	5.48
Ganciclovir										
Rat (325)[b]	80	6	0.66	10.46	**0.063**	3.61	1.42	2.542	5.44	40.05
Benzylpenicillin, S = Me										
Rat (296)	60	1	0.20	6.05	**0.033**	0.74	4.56	0.162	3.70	4.91
Dog (297)[c]	30	4	0.16	24.52	**0.007**	5.07	11.75	0.431	31.69	66.13
Benzylpenicillin, S = Et										
Rat (296)	60	1	0.20	6.05	**0.033**	2.12	2.95	0.719	10.60	21.74
Dog (297)[c]	30	4	0.16	24.52	**0.007**	0.48	0.61	0.787	3.00	120.59
Dexamethasone										
Rat (273)	20	6	1.36	19.48	**0.070**	1.69	8.97	0.188	1.24	2.70

[a]Brain and serum concentrations monitored.
[b]Brain and plasma concentrations monitored.
[c]CSF and blood concentrations monitored.

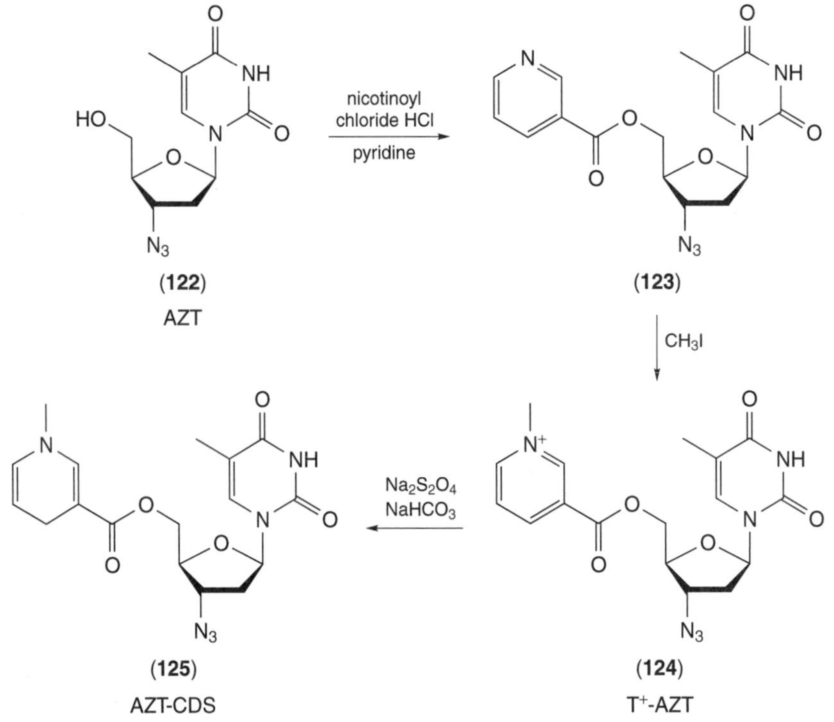

Figure 15.35. Synthesis of AZT–CDS (**125**) (310). Because AZT (**122**) contains a primary alcohol at the 5′-position, attachment of the targetor through an ester bond can be relatively easily carried out, and it illustrates one of the simplest possible synthetic routes for brain-targeting CDSs.

for this improvement usually precipitated severe anemia.

Because AZT–CDS (**125**) is relatively easily accessible synthetically (Fig. 15.35), following the increasing occurrence of AIDS, AZT–CDS was investigated in a number of laboratories to enhance the access of AZT to brain tissues (309–323). AZT–CDS is a crystalline solid, which is stable at room temperature for several months when protected from light and moisture. It is relatively stable in pH 7.4 phosphate buffer, but rapidly oxidizes in enzymatic media. In addition, T⁺–AZT (**124**), the depot form of AZT, was shown to gradually release the parent compound.

The ability of AZT–CDS to provide increased brain AZT levels has been demonstrated in a number of different *in vivo* animal models. The corresponding AUC data for mice (322), rats (314), and rabbits (311) are presented in Table 15.2. In all species, AZT–CDS

provided substantially increased brain-targeting of AZT. For example, the targeting-enhancement factor of the AUC-based site-targeting index was 86 in rats. It is interesting to note that the relatively large difference in the TEF values for different species is mainly attributable to the variability of the brain–STI of AZT itself; AZT–CDS gave very similar values in all three species.

In rabbits, the brain/blood ratio never exceeded 1.0 after AZT administration, indicating that the drug is always in higher concentration in the blood than in the brain, but it approached 3.0 at 1 h postdosing with AZT–CDS. Similarly, in dogs, AZT levels were 46% lower in blood, but 175–330% higher in brain after AZT–CDS administration compared to parent drug administration (315). Furthermore, *in vitro* experiments found AZT–CDS not only more effective in inhibiting human immunodeficiency virus (HIV) replication than AZT itself, presumably attributable to hydrolysis, but also less toxic to host lymphocytes (314, 317, 318). In conclusion, a number of different studies found an improved delivery to the brain and a decreased potential for dose-limiting side effects, suggesting that AZT–CDS may become useful in the treatment of AIDS encephalopathy and the related dementia.

4.1.4 Ganciclovir–CDS. Enhanced brain delivery with a CDS approach has also been achieved for ganciclovir (**126**) (Fig. 15.36), another antiviral agent that may be useful in treating human encephalitic cytomegalovirus (CMV) infection (325). CMV infections are common and usually benign; however, human CMV occurs in 94% of all patients suffering with AIDS, and it has been implicated as a deadly cofactor. Although positive results were achieved in the treatment of the associated retinitis, the treatment of encephalitic cytomegalic disease is far less successful with ganciclovir. A brain-targeted CDS (**127**) (Fig. 15.36) was, therefore, investigated, and improved organ selectivity was demonstrated. As shown in Fig. 15.36 and Table 15.2, an almost fivefold increase in the relative brain exposure to ganciclovir and a simultaneous seven times decrease in blood exposure was achieved with the CDS in rats.

4.1.5 Benzylpenicillin–CDS. The poor CNS penetration of β-lactam antibiotics, including penicillins and cephalosporins, often makes the treatment of various bacterial infections localized in brain and cerebrospinal fluid difficult. Furthermore, because these antibiotics return from the CNS back to the blood (active transport may be involved), simple prodrug approaches did not provide real solutions. Consequently, a number of CDSs have been synthesized and investigated (294–297). Because in these structures the targetor moiety had to be coupled to an acid moiety, a spacer function had to be inserted between the targetor and the drug. This renders additional flexibility to the approach, and several possibilities were investigated. Benzylpenicillin–CDS was one of the first cases in which the influence of such functions on the stability and delivery properties of CDSs could be investi-

gated. Systematic studies of the effects of various modifications have been performed since then (254, 355).

In rats, i.v. administration of benzylpenicillin–CDS with an ethylene 1,2-diol spacer gave an AUC-based brain-exposure enhancement factor for benzylpenicillin of around 11 (Table 15.2) (296). From the several coupling possibilities investigated, diesters of methylene diol and ethylene 1,2-diol proved worthy of further investigation, but ester–amide combinations did not prove successful. Consequently, the diesters have also been investigated in rabbits and dogs (297).

Both in rabbits and in dogs, i.v. administration of equimolar doses of CDSs provided benzylpenicillin levels in brain and cerebrospinal fluid (CSF) that were substantially higher and more prolonged than in the case of parent drug administration. In this study, the diester of methylene diol provided more significant increase in benzylpenicillin brain exposure (SEF = 32); the diester of ethylene 1,2-diol gave better targeting, but released lower drug concentrations because of the formation of an inactive hydroxyethyl ester. Because dogs were not euthanized, only blood and CSF data were collected (Table 15.2), and these indicated better CSF penetration in dogs than in rabbits.

Although parenteral injection of benzylpenicillins may sometimes cause undesired side effects, such as seizures, the high CNS levels obtained in dogs and rabbits after CDS administration were not accompanied by any toxic side effects (297). Similarly, although the delivered compound exhibited toxicity in the rotorod test at 0.5 h following a 300 mg/kg i.p. dose, felodipine–CDS displayed no toxicity at either 0.5 or 4 h after a 100 mg/kg i.p. dose or 0.5 h for a 300 mg/kg i.p. dose (342). Some toxicity, which is not surprising considering the toxicity of the delivered compound itself, was observed at 4 h for the 300 mg/kg CDS dose.

4.1.6 Estradiol–CDS. Among all CDSs, estradiol–CDS (Estredox) is in the most advanced investigation stage (phase I/II clinical trials). Estrogens are endogenous hormones that produce numerous physiological actions and are among the most commonly prescribed drugs in the United States, mainly for HRT in

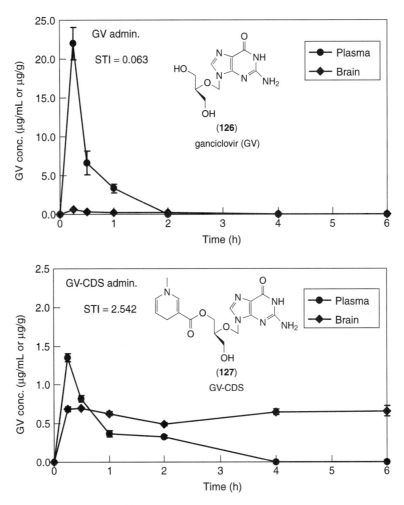

Figure 15.36. Ganciclovir (GV, **126**) concentrations in brain and plasma as a function of time after an i.v. dose of 80 μmol/kg of GV (**126**) or GV–CDS (**127**). Data are means $\pm$ SEM in rats.

postmenopausal women and as a component of oral contraceptives (363). They are lipophilic steroids that are not impeded in their entry to the CNS; hence, they can readily penetrate the BBB and achieve high central levels after peripheral administration. However, estrogens are poorly retained within the brain. Therefore, to maintain therapeutically significant concentrations, frequent or sustained (transdermal) doses have to be administered. Constant peripheral exposure to estrogens has been related, however, to a number of pathological conditions including cancer, hypertension, and altered metabolism (155, 156, 364–

366). Furthermore, the Heart and Estrogen/Progestin Replacement Study (HERS) suggested that HRT does not affect the incidence of coronary heart disease (CHD) (367, 368). Because estrogen is suspected to lead to a 30% increase in the risk of breast cancer, rising to 50% if it is taken for more than 10 years (369), there is less justification now for a systemic HRT. In industrialized nations, the average woman spends about a third of her life in the postmenopausal stage (menopause occurs at an average age of 52); hence, whether to take HRT or not is an important decision. Because the CNS is the target site for many estrogenic actions (370), brain-targeted delivery may provide safer and more effective agents in many cases.

Estrogen CDSs could be useful in reducing the secretion of luteinizing hormone-releasing hormone (LHRH) and, hence, in reducing the secretion of luteinizing hormone (LH) and gonadal steroids. As such, they could be used to achieve contraception and to reduce the growth of peripheral steroid-dependent tumors, such as those of the breast, uterus, and prostate, and to treat endometriosis. They also could be useful in stimulating male and female sexual behavior, and in the treatment of menopausal vasomotor symptoms ("hot flashes") (371). Other potential uses are in neuroprotection (372), in the reduction of body weight, or in the treatment of Parkinson's disease (373), depression, and various types of dementia, including Alzheimer's disease (AD) (155, 374, 375). AD, which still has no specific cure, results in progressively worsening symptoms that range from memory loss to declining cognitive ability. It affects an estimated 10% of the population over 65 years of age and almost 50% of that over 85 years of age (376). Findings on the effect of estrogen in AD are somewhat conflicting (377, 378). Nevertheless, estrogen produces many effects (370) that may delay the onset of AD, prevent AD, or improve the quality of life in AD, and therapeutic effects may require sustained and sufficiently high estradiol levels (379, 380).

Estradiol (E_2, **130**) (Fig. 15.37) is the most potent natural steroid. It contains two hydroxy functions: one in the phenolic 3 position and one in the 17 position. With these synthetic handles, three possible CDSs can be designed attaching the targetor at the 17-, at the 3-, or at both positions. Attachment at either position, but especially at the 17 position, should greatly decrease the pharmacological activity of E_2, given that these esters are known not to interact with estrogen receptors (381).

Since its first synthesis in 1986 (275), E_2–CDS (**128**) has been investigated in several models (276–286, 288–293). *In vitro* studies with rat organ homogenates as the test matrix indicated half-lives of 156.6, 29.9, and 29.2 min for (T at the 17 position) in plasma, liver, and brain homogenates, respectively (275).

Thus, E_2–CDS is converted to the corresponding quaternary form (T^+–E_2) (**129**) faster in the tissue homogenates than in plasma. This is consistent with the hypothesis of a membrane-bound enzyme, such as the members of the NADH transhydrogenase family, acting as oxidative catalyst. These studies also indicated a very slow production of E_2 from T^+–E_2, suggesting a possible slow and sustained release of estradiol from brain deposits of T^+–E_2.

To detect doses of E_2–CDS (**128**), T^+–E_2 (**129**), and E_2 (**130**) of physiological significance, a selective and sensitive method was needed. Therefore a precolumn-enriching HPLC system was used (282) that allowed accurate detection in plasma samples and organ homogenates with limits of 10, 20, and 50 ng/mL or ng/g for T^+–E_2, E_2–CDS, and E_2, respectively. This study proved that in rats, E_2 released from the T^+–E_2 intermediate formed after i.v. E_2–CDS administration has an elimination half-life of more than 200 h and brain E_2-levels are elevated four to five times longer after administration than after simple estradiol treatment (Fig. 15.38) (282). Proving effective targeting, another study also found that steroid levels between 1 and 16 days after E_2–CDS treatment were more than 12-fold higher in brain samples than in plasma samples (286). Studies in orchidectomized rats proved that a single i.v. injection of E_2–CDS (3 mg/kg) suppressed LH secretion by 88, 86, and 66% relative to DMSO controls at 12, 18, and 24 days, respectively, and that E_2 levels were not elevated relative to the DMSO control at any sampling time (278). A single i.v. administration of doses as low as 0.5 mg/kg to ovariectomized rats induced prolonged (3–6 weeks) pharmacological effects as measured by LH suppression (276, 278, 286), reduced rate of weight gain (279, 284–286), or, in castrated male rats, reestablishment of copulatory behavior (277). A large number of other very encouraging results have been obtained in various animal models and phase I/II clinical trials (Fig. 15.39); most of them have been reviewed previously (21, 283, 288, 289, 291). Clinical evaluations suggest a potent central effect with only marginal elevations in systemic estrogen levels; therefore, E_2–CDS may become

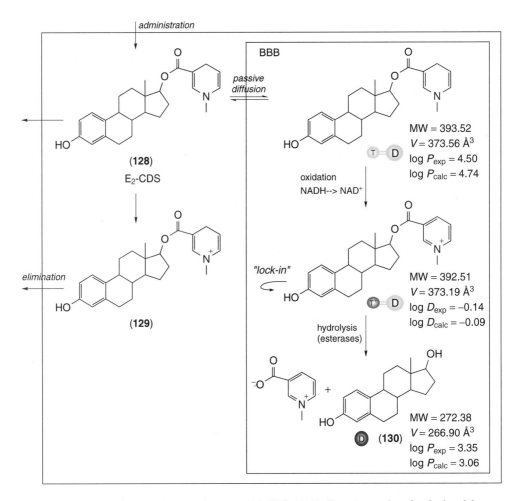

Figure 15.37. *Lock-in* mechanism for estradiol–CDS (**128**). Experimental and calculated logarithms of the octanol–water partition (log P) and distribution (log D) coefficients are shown to illustrate the significant changes in partition properties. The lipophilic CDS (log $P > 4$) can easily cross the BBB, whereas the hydrophilic intermediate (log $D < 0$) is no longer able to come out, providing a sustained release of the active estradiol (**130**).

a useful and safe therapy for menopausal symptoms or for estrogen-dependent cognitive effects.

Recently, E_2–CDS also was shown to provide very encouraging neuroprotective effects. In ovariectomized rats, pretreatment with E_2–CDS decreased the mortality caused by middle cerebral artery occlusion (MCAO) from 65% to 16% (292). E_2–CDS reduced the area of ischemia by 45–90% or 31%, even if administered after MCAO by 40 or 90 min, respectively. Another recent study provided evidence that treatment with E_2–CDS can protect cholin-

ergic neurons in the medial septum from lesion-induced degeneration (293).

4.1.7 Cyclodextrin Complexes. The same physicochemical characteristics that allow successful chemical delivery also complicate the development of acceptable pharmaceutical formulations. Increased lipophilicity allows partition into deep brain compartments, but also confers poor aqueous solubility. The oxidative lability, which is needed for the lock-in mechanism, and the hydrolytic instability, which releases the modifier functions or the

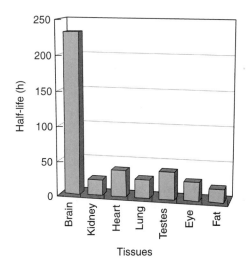

Figure 15.38. Elimination half-lives in various tissues for the T^+–E_2 (**129**) formed after i.v. administration of 38.1 μmol/kg E_2–CDS (**128**) in rats.

active drug, combine to limit the shelf life of the CDS. Cyclodextrins may provide a possible solution. They are torus-shaped oligosaccharides that contain various numbers of α-1,4-linked glucose units (6, 7, and 8 for α-, β-, and γ-cyclodextrin, respectively). The number of units determines the size of a conelike cavity into which various compounds can include and form stable complexes (382–385). Formation of the host–guest inclusion complex generally involves only the simple spatial entrapment of the guest molecule without formation of covalent bonds.

The corresponding inclusion complex with 2-hydroxypropyl-β-cyclodextrin (HPβCD) solved essentially all problems with E_2–CDS (386). This modified cyclodextrin was selected based on its low toxicity observed using various administration routes and based on the fact that alkylation or hydroxyalkylation of the glucose oligomer can disrupt hydrogen bonding and provide increased water solubility for the compound and for its inclusion com-

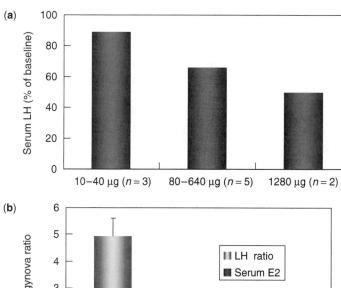

Figure 15.39. (a) Effect of various grouped doses of E_2–CDS (**128**) on mean LH suppression relative to baseline values in postmenopausal human volunteers. (b) The effect of i.v. or buccal E_2–CDS relative to oral Progynova (estradiol valerate) on LH suppression or E_2 serum elevation in a study of 18 postmenopausal volunteers. Data are collected from all dosing groups and are corrected for baseline and dose. E_2–CDS was significantly more active as measured by the LH suppression, yet had less of a tendency to elevate serum E_2 levels.

plexes as well (387–390). Indeed, the aqueous solubility of E_2–CDS was enhanced about 250,000-fold in a 40% (w/v) HPβCD solution (from 65.8 ng/mL to 16.36 mg/mL). The phase solubility diagram indicated that a 1:1 complex forms at low HPβCD concentration but a 1:2 complex occurs at higher HPβCD concentrations. The stability of E_2–CDS was also significantly increased, allowing formulation in an acceptable form. The rate of ferricyanide-mediated oxidation, a good indicator of oxidative stability, was decreased by a factor of about 10, and shelf life was increased about fourfold, as indicated by t_{90} and t_{50} values in a temperature range of 23–80°C (386). The cyclodextrin complex even provided better distribution by preventing retention of the solid material precipitated in the lung.

Similarly promising results were obtained for testosterone–CDS (270), lomustine–CDS (328), and benzylpenicillin–CDS (391). For the latter, aqueous solubility was enhanced about 70,000-fold in a 20% (w/v) HPβCD solution (from 50–70 ng/mL to 4.2 mg/mL), and stability was also increased.

4.1.8 Molecular Packaging.

More recently, the CDS approach has been extended to achieve successful brain deliveries of a Leu-enkephalin analog (354, 355), thyrotropin-releasing hormone (TRH) analogs (356–359), and kyotorphin analogs (22, 360). Neuropeptides, peptides that act on the brain or spinal cord, represent the largest class of transmitter substance and have considerable therapeutic potential (22, 392, 393). The number of identified endogenous peptides is growing exponentially; recently, almost 600 sequences for neuropeptides and related peptides have been listed (394). However, delivery of peptides through the BBB is an even more complex problem than delivery of other drugs because they can be rapidly inactivated by ubiquitous peptidases (395–398). Therefore, for a successful delivery, three issues have to be solved simultaneously: (1) enhance passive transport by increasing the lipophilicity, (2) ensure enzymatic stability to prevent premature degradation, and (3) exploit the lock-in mechanism to provide targeting.

The solution described here is a complex molecular packaging strategy, in which the peptide unit is part of a bulky molecule dominated by lipophilic modifying groups that direct BBB penetration and prevent recognition by peptidases (399). Such a brain-targeted packaged peptide delivery system contains the following major components: the redox targetor (T); a spacer function (S), consisting of strategically used amino acids to ensure timely removal of the charged targetor from the peptide; the peptide itself (P); and a bulky lipophilic moiety (L), attached through an ester bond or sometimes through a C-terminal adjuster (A) at the carboxyl terminal to enhance lipid solubility and to disguise the peptide nature of the molecule (see Fig. 15.41 below for an illustration).

To achieve delivery and sustained activity with such complex systems, it is very important that the designated enzymatic reactions take place in a specific sequence. On delivery, the first step must be the conversion of the targetor to allow for lock-in. This must be followed by removal of the L function to form a direct precursor of the peptide that is still attached to the charged targetor. Subsequent cleavage of the targetor-spacer moiety finally releases the active peptide.

4.1.8.1 Leu-Enkephalin Analog.

Since the discovery of endogenous peptides with opiate activity (400), many potential roles have been identified for these substances (401, 402). Possible therapeutic applications could extend from treatment of opiate dependency to numerous other CNS-mediated dysfunctions. Opioid peptides are implicated in regulating neuroendocrine activity (403–405), motor behavior (406), seizure threshold (407, 408), feeding and drinking (409), mental disorders (401, 410), cognitive functions (401, 411), cardiovascular functions (412, 413), gastrointestinal functions (401, 414), and sexual behavior and development (402). However, analgesia remains their best known role, and it is commonly used to evaluate endogenous opioid peptide activity. Whereas native opioid peptides could alter pain threshold after intracerebroventricular (i.c.v.) dosing, they were ineffective after systemic injection. It was reasonable, therefore, to attempt a brain-targeted CDS approach.

The first successful delivery and sustained release was achieved for Tyr-D-Ala-Gly-Phe-D-

Leu (DADLE), an analog of leucine-enkephalin, a naturally occurring linear pentapeptide (Tyr-Gly-Gly-Phe-Leu) that binds to opioid receptors (Fig. 15.40) (354). In rat brain tissue samples collected 15 min after i.v. CDS-administration, electrospray ionization mass spectrometry clearly showed the presence of the locked-in T^+–D form at an estimated concentration of 600 pmol/g of tissue. The same ion was absent from the sample collected from the control animals treated with the vehicle solution only. To optimize this delivery strategy, an effective synthetic route was established for peptide CDSs, and the role of the spacer and the lipophilic functions was investigated (355). Four different CDSs were synthesized by a segment-coupling method. Their i.v. injection produced a significant and long-lasting (>5 h) response in rats monitored by classic (98) tail-flick latency test. Compared with the delivered peptide itself, the packaged peptide and the intermediates formed during the sequential metabolism had weak affinity to opioid receptors. The antinociceptive effect was naloxone-reversible and methylnaloxonium-irreversible. Because quaternary derivatives such as methylnaloxonium are unable to cross the BBB, these techniques (355, 415) are used to prove that central opiate receptors are responsible for mediating the induced analgesia. It could be concluded, therefore, that the peptide CDS successfully delivered, retained, and released the peptide in the brain.

The efficacy of the CDS package was strongly influenced by modifications of the spacer (S) and lipophilic (L) components, proving that they have important roles in determining molecular properties and the timing of the metabolic sequence. The bulkier cholesteryl group used as L showed a better efficacy than that of the smaller 1-adamantaneethyl, but the most important factor for manipulating the rate of peptide release and the pharmacological activity turned out to be the spacer (S) function: proline as spacer produced more potent analgesia than did alanine.

4.1.8.2 TRH Analogs. A similar strategy was used (356–359) for the CNS delivery of a thyrotropin-releasing hormone (TRH, Glp-His-Pro) analog (Glp-Leu-Pro) in which histidine (His^2) was replaced with leucine (Leu^2) to

dissociate CNS effects from thyrotropin-releasing activity (416). Such compounds can increase extracellular acetylcholine (ACh) levels, accelerate ACh turnover, improve memory and learning, and reverse the reduction in high-affinity choline uptake induced by lesions of the medial septal cholineric neurons (417, 418). Therefore, as reviewed in the literature (419, 420), these peptides are potential agents for treating motor neuron diseases (421), spinal cord trauma (422), or neurodegenerative disorders such as Alzheimer's disease (423) and may also have a potential cytoprotective role (424).

Because the peptide that has to be delivered has no free —NH$_2$ and —COOH termini, a precursor sequence (Gln-Leu-Pro-Gly) that will ultimately produce the final TRH analog was packaged for delivery. Therefore, two additional steps had to be included in the metabolic sequence: one in which the C-terminal adjuster glycine is cleaved by peptidyl glycine α-amidating monooxygenase (PAM) to form the ending prolinamide, and one in which the N-terminal pyroglutamyl is formed from glutamine by glutaminyl cyclase (356). In summary, the following sequential biotransformation has to take place after delivery to the brain: first, lock-in by oxidation of the dihydrotrigonellyl (T) to the corresponding pyridinium salt, then removal of cholesterol, oxidation of glycine by peptidyl glycine α-amidating monoxygenase to prolineamide, cleavage of the targetor-spacer combination, and, finally, cyclization of glutamine by glutaminyl cyclase to pyroglutamate.

Selection of a suitable spacer proved also important for the efficacy of TRH–CDSs as measured by the decrease in the barbiturate-induced sleeping time in mice, an interesting and well-documented effect of this neuropeptide (425, 426). After i.v. administration of 15 μmol/kg, the Leu^2-TRH analog itself produced only a slight decrease of 17 ± 7% compared to the vehicle control. Equimolar doses of the CDS compounds with Pro, Pro-Pro, and Pro-Ala spacers produced statistically significant ($p < 0.05$) decreases of 47 ± 6%, 51 ± 7%, and 56 ± 4%, respectively (358). Treatment with a TRH–CDS also significantly improved memory-related behavior in a passive-avoid-

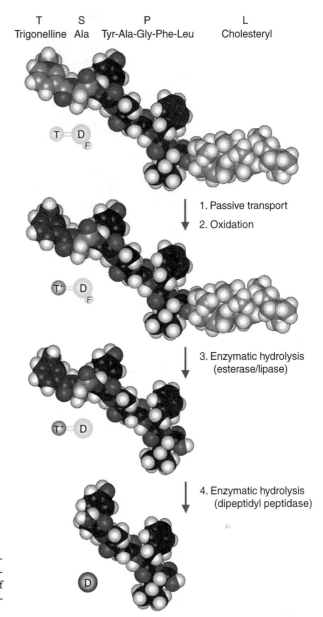

T   S   P   L
Trigonelline Ala   Tyr-Ala-Gly-Phe-Leu   Cholesteryl

1. Passive transport
2. Oxidation

3. Enzymatic hydrolysis
   (esterase/lipase)

4. Enzymatic hydrolysis
   (dipeptidyl peptidase)

Figure 15.40. CPK structures illustrating the sequential metabolism of the molecular package used for brain delivery of a leucine enkephalin analog. See color insert.

ance paradigm in rats bearing bilateral fimbrial lesions without altering thyroid function (357).

Because the amide precursor was found to be susceptible to deamination by TRH-deaminase, a side-reaction process competitive with the designed-in cleavage of the spacer—Gln peptide bond, another analog of TRH was also examined (359). The Pro-Gly C-terminal was replaced by pipecolic acid (Pip), to obtain a TRH analog that is active in the carboxylate form and, hence, to eliminate the need for formation of the amide. The CDS was prepared by a 5 + 1 segment-coupling approach, and the cholesteryl ester of pipecolic acid was prepared using either Fmoc (fluorenylmethyloxycarbonyl) or Boc (*tert*-butyloxycarbonyl) as protection. Pharmacological activity was assessed by antagonism of barbiturate-induced sleeping time in mice, and the pipecolic acid analog was found to be even somewhat more effective than the previous package.

4.1.8.3 Kyotorphin Analogs.

Finally, a kyotorphin analog (Tyr-Lys) that has activity similar to kyotorphin itself (427) was also successfully delivered (22, 360). Kyotorphin (Tyr-Arg) is an endogenous neuropeptide that exhibits analgesic action through the release of endogenous enkephalin, and its analgesic activity is about fourfold greater than that of Met-enkephalin (428). It also has other effects that may prove useful, such as seizure-protective effects (429) or effects on nitric oxide synthase (NOS) (430–432). Because this peptide contains a free amine residue, and because preliminary studies indicated that such ionizable functions might prevent successful delivery, this free-amine functionality was additionally "packaged" by attachment of a Boc group (131) (Fig. 15.41). It was assumed that this attachment is bioconvertible, and the lock-in mechanism will allow time for its enzymatic removal.

Because during the synthesis of this CDS package, α- and ϵ-amine functions had to be differentially functionalized, combined liquid phase Fmoc and Boc chemistry was employed. With this synthetic method at hand, a number of various spacers were explored for this delivery system (proline, proline-alanine, and proline-proline), and the double proline (131) (Fig. 15.41) provided the best pharmacological effect among them. The corresponding CDS showed good activity on the rat tail-flick latency test even with a single proline spacer. Activity was already significant at a 3 μmol/kg (1.0 mg/kg) dose and leveled off at about 22 μmol/kg. This represents an improvement of about two orders of magnitude compared to the approximately 200 mg/kg dose necessary to observe activity when the peptide itself is given intravenously. Several intermediates and building blocks were also studied, but only administration of the whole molecular package produced significant pharmaceutical response, confirming that only the designed metabolic sequence as a whole is effective in delivering peptides across the BBB.

4.1.8.4 Brain-Targeted Redox Analogs (BTRA).

Kyotorphin analogs also served for the design and evaluation of a novel and conceptually different method that can provide brain-targeted activity for peptides containing amino acids with basic side chains (lysine, arginine) (360). Within this approach, the targetor T moiety, which by conversion to its charged T^+ form is responsible for the lock-in mechanism, is not attached to the peptide from outside, but it is integrated within novel *redox* amino acids building blocks that replace the original basic amino acid of the active peptide. Consequently, isosteric/isoelectronic analogy between the original and the redox side chain and not cleavage of the redox targetor is expected to provide activity. Figure 15.41 compares the structure of the classical brain-targeted CDS (131) with the structure of the novel brain-targeted redox analog (BTRA) (132) used for the present kyotorphin analog. The ϵ-amine function of lysine can be directly replaced by the nicotinamide function by use of the Zincke procedure to form pyridinium salts (353, 360, 433–435). As shown in Fig. 15.42, both the CDS and the BTRA produced significant and prolonged analgesic effect in rats, and both could be reversed by naloxone, demonstrating again that central opiate receptors must be responsible for mediating the induced analgesia. Such novel amino acids may provide both effective replacement and targeting for other lysine- or arginine-containing peptides.

In conclusion, molecular packaging of peptides is a rational drug design approach that achieved the first documented noninvasive brain delivery of these important biomolecules in pharmacologically significant amounts. Because drug delivery is the potential Achilles' heel of biotechnology's peptide drug industry (436), this approach may represent an important step toward future generations of high-efficiency neuropharmaceuticals obtained from biologically active peptides.

4.2 Site-Specific Enzyme-Activated CDS

These CDSs exploit enzymatic conversions that take place primarily, exclusively, or at higher activity at the site of action as a result of differential distribution of certain enzymes. If such enzymes can be found, their use in a

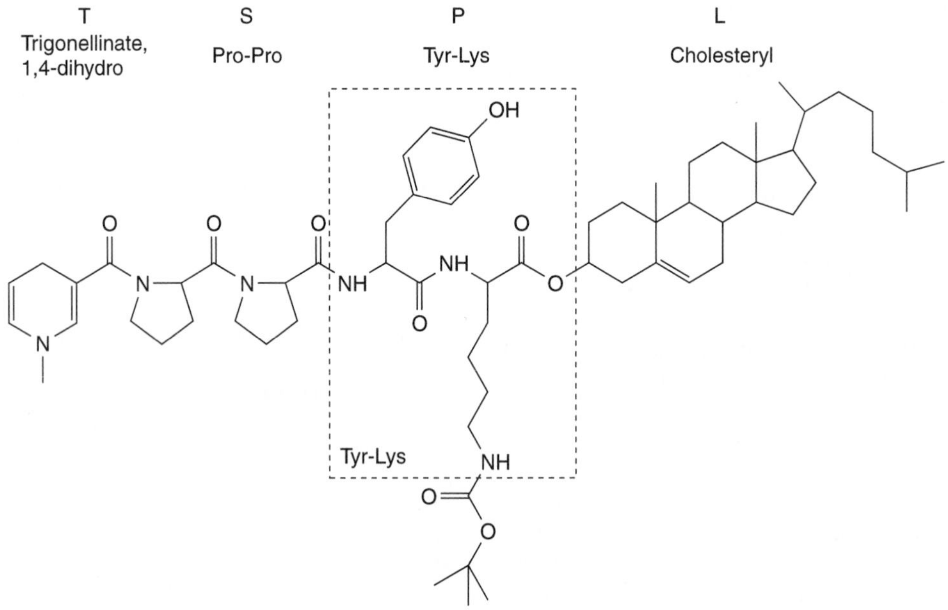

(**131**)
YK-CDS

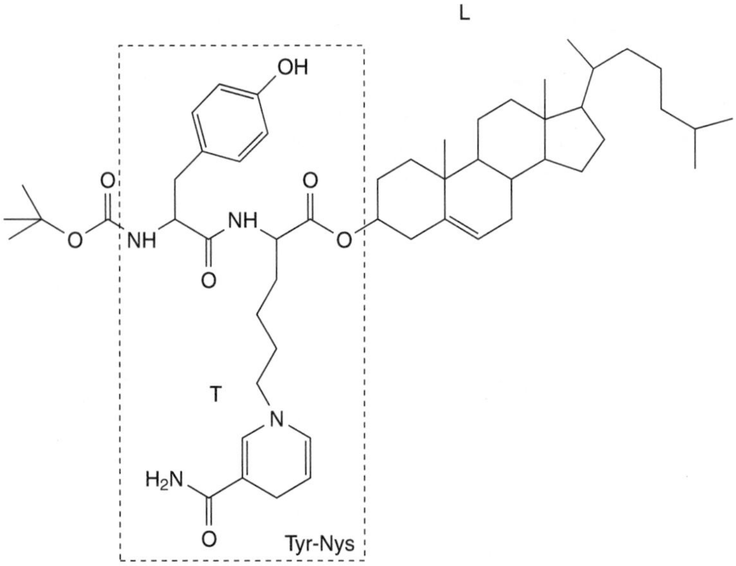

(**132**)
BTRA

Figure 15.41. Delivery of kyotorphin analogs was achieved using both using a chemical delivery system (**131**) and a brain-targeted redox analog (**132**) approach.

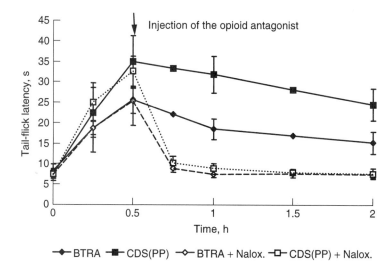

Figure 15.42. Reversal of the analgesia produced by 22.3 μmol/kg YK–CDS (**131**) and BTRA (**132**) administered i.v. by naloxone (6.1 μmol/kg, s.c.) administered 30 min later. Data represents means $\pm$ SE of six rats for each group.

strategically designed system can lead to high site specificity. The targetor moiety and the eventual protective function(s) are introduced by chemical modifications that also involve important pharmacophore functionalities. Following administration, the resulting CDS is distributed throughout the body. It is continuously eliminated from the "rest of the body" without producing any pharmacological effects. However, at the targeted site, where the necessary enzymes are found primarily, exclusively, or at higher activity, predictable enzymatic reactions convert the original CDS to the active drug.

Certainly, the system depends on the existence of such enzymes. Successful site- and stereospecific delivery of intraocular pressure (IOP)-reducing β-adrenergic blocking agents to the eye was achieved with a retrometabolic design (118, 437–442), based on a general metabolic process that appears to apply to all lipophilic ketone precursors of β-amino alcohols. Contrary to the free catechol-containing adrenalone, lipophilic esters of adrenalone were effectively reduced within the eye to yield epinephrine by way of the corresponding esters. However, oral administration did not result in any formation of epinephrine (443). Hence, the ketone function may act as a targetor that can be reduced to produce the β-amino alcohol pharmacophore necessary for activity. Because the approach could be extended to other β-agonists (444, 445), the pos-

sibility arose to produce ophthalmically useful β-adrenergic antagonists within the eye from the corresponding ketones. However, because ketones of the phenol ether–type β-blockers decompose in aqueous solutions to form the corresponding phenols, they are not good bioprecursors to produce β-amino alcohols. To stabilize them, they were converted to oximes, which need to undergo a facile enzymatic hydrolysis to the bioreducible ketone. Although the oximes are much more stable than the ketones, their aqueous stability still does not provide an acceptable shelf life [e.g., even at pH 4.5, which can be considered the lowest acceptable pH limit for ophthalmic vehicle solutions and where the oxime is more stable, $t_{90\%}$ at room temperature of a 1% w/v solution is only 44 ± 5 days for alprenoxime (**136**) (440)]. Additional and significant stabilization was obtained, however, by using methoximes, methylethers of the corresponding oximes. For example, for the methoxime analog of alprenolol, no significant decomposition was observed during storage at room temperature within 1 year at pH values around 6.5 (440).

To summarize, in these CDSs, a β-amino oxime or alkyloxime function (**133**) (Fig. 15.43) replaces the corresponding β-amino alcohol pharmacophore part of the original molecules (**135**). These oxime or alkyloxime derivatives exist in alternative Z (syn) or E (anti) configuration. They are enzymatically hydrolyzed within the eye by enzymes located in the

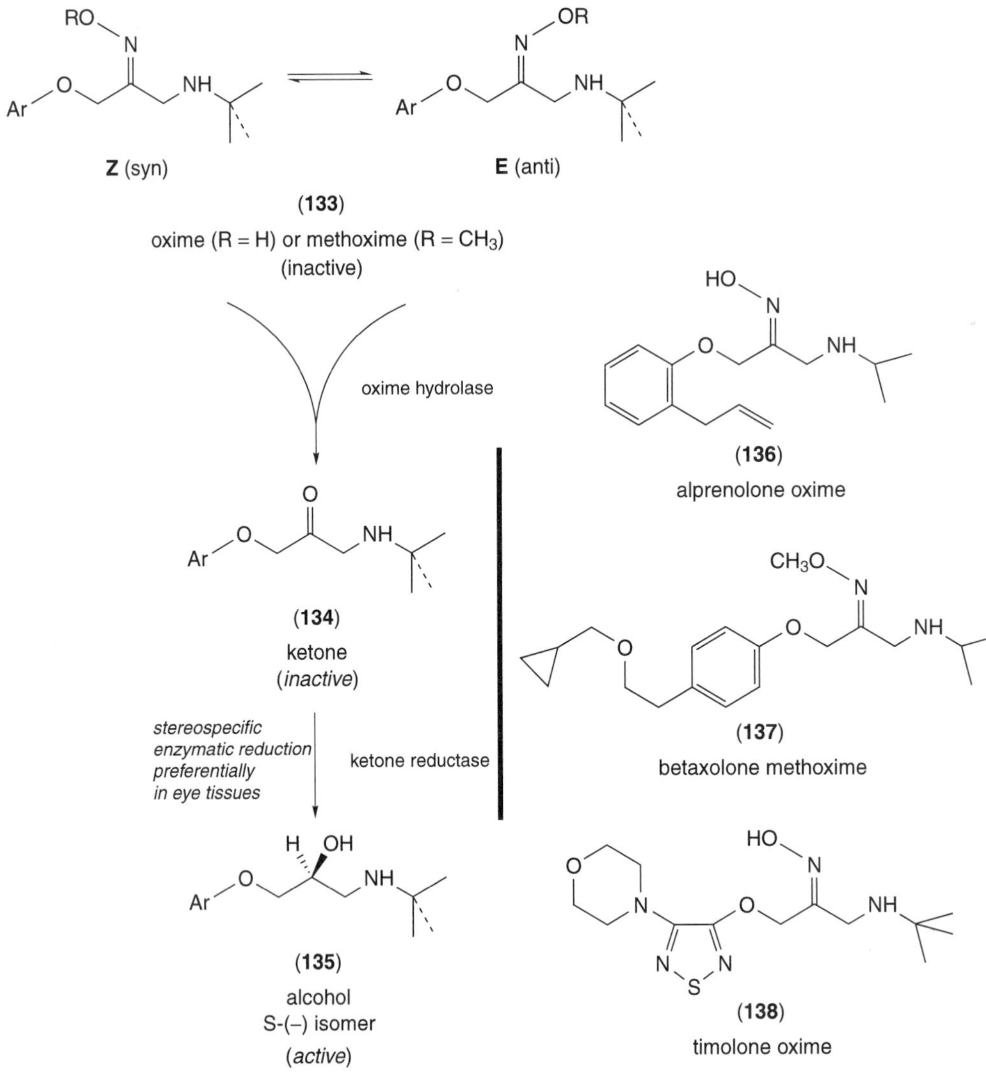

Figure 15.43. Sequential activation of oximes in the site- and stereospecific delivery of β-adrenergic antagonists to the eye. The original oximes or methoximes (**133**) and the intermediate ketones (**134**) are inactive; they are enzymatically converted into the active S-(−)-β-adrenergic blocker alcohols (**135**) in a site- and stereospecific manner.

iris–ciliary body, and subsequently, reductive enzymes also located in the iris–ciliary body produce only the active S-(−) stereoisomer of the β-blocker (20). A variety of such oxime and methoxime analogs of alprenolol (**136**), betaxolol (**137**), propranolol, and timolol (**138**) were synthesized and tested (437, 440–442, 446–448). They were all shown to undergo the predicted specific activation within, and only within, the eye. The highest concentrations of

the active β-antagonists were observed in the iris–ciliary body. Following i.v. injection, the oxime disappeared rapidly from the blood, and at no time could the corresponding alcohol be detected (437). This indicates that the required enzymatic hydrolysis—reduction activation sequence, which occurs in the eye, does not take place in the systemic circulation.

The oxime and methoxime analogs of alprenolol (**136**) (440, 447) and betaxolol (**137**)

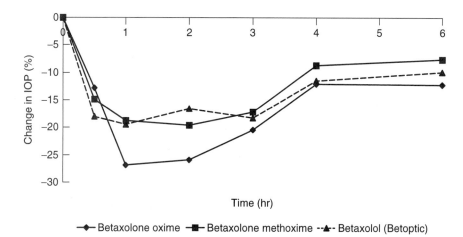

Figure 15.44. IOP reducing activities of betaxolol (Betoptic) and its oxime and methoxime analogs in New Zealand albino rabbits. Doses of 100 μL solutions formulated in saline with 0.01% EDTA and 0.01% benzalkonium chloride (pH 6.0) were administered in one eye (1.84–1.86 mg). Data are from Alcon Laboratories.

(442) were found to produce significant and long-lasting reduction in the IOP of rabbits following uni- or bilateral administration (Fig. 15.44). In most cases, the novel analogs produced more pronounced and longer-lasting effects than did the parent compounds and were also less irritating. On the other hand, the i.v. bolus injection of alprenoxime led to only insignificant transient bradycardia, and no activity was found after oral or topical administration in rats and rabbits (447). Alprenolol in a similar dose produced sustained and significant bradycardia for more than 30 min. A study in dogs (448) confirmed that no significant cardiac electrophysiologic parameters are altered after systemic treatment with alprenoxime or its methyl ester analog, even at doses that far exceed that effective in reducing IOP. In the meantime, administration of alprenolol itself exerted profound effects on cardiac function consistent with β-blocking activity and resulting in changes of 30–140%. Alprenoxime also did not alter isoproterenol-induced tachycardia in dogs (448). In a similar manner and in contrast to betaxolol itself, the oxime or methoxime analogs of betaxolol had no effect on isoproterenol-induced tachycardia in Sprague–Dawley rats at doses up to 20 μmol/kg (442).

In conclusion, oxime or methoxime derivatives showed significant IOP-lowering activity, but even their i.v. administration did not produce the active β-blockers metabolically. Therefore, they are void of any cardiovascular activity, a major drawback of classical antiglaucoma agents. Methoxime derivatives provide sufficient stabilization for acceptable shelf life.

4.3 Receptor-Based Transient Anchor-Type CDS

This last class of CDSs is based on formation of a reversible covalent bond between the delivered entity and some part of the targeted receptor site to enhance selectivity and activity. The targetor moiety of these CDSs is expected to undergo a modification that allows transient anchoring of the active agent. This class is less well developed than the previous ones, but it may become of increasing importance with the accumulation of receptor-related information.

The possibilities of a receptor-based CDS for naloxone, a pure opiate antagonist, were explored because opiate addiction is a very serious problem and long-lasting antagonists are highly desirable. The presence of an essential —SH group near the binding site in opioid receptors was long suspected (449), given that the agonist-binding capacity is destroyed by —SH reagents like N-ethylmaleimide. Mutagenesis and other studies (450, 451) are be-

ginning to elucidate the mechanism of this inhibition of binding at opioid receptors,
receptors that are members of the G-protein—
coupled receptor family. There is also evidence
that Leu-enkephalin or even morphine analogs containing activated sulfhydryl groups
can form mixed disulfide linkages with opioid
receptors. The modified agonist caused receptor activation that persisted following extensive washing, was naloxone-reversible, and returned after naloxone was washed away (452).
Chlornaltrexamine, an irreversible alkylating
affinity label on the opiate receptor, can maintain its antagonist effect for an astonishing
period of 2–3 days, proving that covalent binding can increase the duration of action at these
receptors.

Our investigation focused on a more reversible, spirothiazolidine-based system. Spirothiazolidine derivatives were selected because they should be subject to biological
cleavage of the imine formed after spontaneous cleavage of the carbon—sulfur bond (231),
and because they have been explored earlier in
delivery systems for controlled-release endogenous agents like hydrocortisone or progesterone (Fig. 15.29) (230, 232). Given that modifications at the 6-keto position of naloxone are
possible without loss of activity, the spirothiazolidine ring was attached at this site in a
manner similar to that done for hydrocortisone (Fig. 15.29). Opening of this spirothiazolidine ring allows oxidative anchoring of the
delivered agent to nearby —SH groups
through disulfide bridging. The blocking of
the receptor is reversible; endogenous —SH
compounds presumably will regenerate the
—SH group of the receptor. If an adequate
conformation can be obtained, stronger and
longer binding is expected at the opiate receptors. Indeed, IC_{50} values of naloxone-6-spi-
rothiazolidine for these receptors indicated increased affinities (15). The increase was
especially significant for the γ-receptors
known to be involved, among others, in sedative analgesia (IC_{50} values of 3.9 and 60.0 nM
for naloxone–CDS and naloxone, respectively). The intrinsic activity for guinea pig
ileum of naloxone–CDS also showed an almost
25-fold increase compared with that of naloxone. Because other G-protein–coupled receptors (e.g., TRH, D_1 and D_2 dopamine, vaso-

pressin, follicle-stimulating hormone, and
cannabinoid receptors) are also sensitive to
sulfhydryl reagents (451), there is considerable potential for this CDS class if modifications of the corresponding ligands at adequate
sites can be performed.

In addition, because the earlier evaluations
of spirothiazolidine-based systems indicated
an enhanced deposition to lung tissue, the possibility of using similar mechanisms in developing other lung tissue targeting CDSs
seemed worth exploring. Lipoic acid, a nontoxic coenzyme for acyl transfer and redox reactions, was investigated as targetor moiety
for selective delivery to lung tissue because of
its ability to also form disulfide linkages (453).
The corresponding CDSs for chlorambucil,
an antineoplastic agent, and cromolyn, a
bischromone used in antiasthmatic prophylaxis, were built by way of an ester linkage. In
vitro kinetic and in vivo pharmacokinetic
studies showed that the respective CDSs were
sufficiently stable in buffer and biological media; hydrolyzed rapidly into the respective active parent drugs; and, relative to the underi-
vatized parent compounds, they significantly
enhanced delivery and retention of the active
drug in lung tissue. For example, following administration to rats, about 20% of the administered dose was found in lung with cromolyn–
CDS; with parent drug administration, only
about 0.2% was found for the same period of
up to 30 min. Compared with the administration of parent drug, chlorambucil–CDS also
produced an over 20-fold increase in the
amount of chlorambucil delivered to rabbit
lung tissue after 30 min after administration.
Increased lipophilicity in these systems is also
likely to have a beneficial effect on the bioavailability of these highly hydrophilic compounds. These CDSs are designed to anchor
the drug entity in lung tissue and provide a
sustained release of active compound over a
protracted period of time.

5 CONCLUSIONS

The present chapter attempted to systematize
and summarize several novel metabolisn-
based drug design approaches collected under
the umbrella of retrometabolic drug design

and targeting. These approaches incorporate two major classes: soft drugs and chemical delivery systems. They achieve their drug-targeting roles in opposite ways, but they have in common the basic concept of designed metabolism to control drug action and targeting. These approaches are general in nature, can be applied to essentially all drug classes, and are based on specific design rules whose applications can be enhanced by specific computer programs. To illustrate the concepts, many specific examples from a variety of fields were presented, including already marketed drugs that resulted from the successful application of such design principles.

REFERENCES

1. J. Le Fanu, *The Rise and Fall of Modern Medicine*, Carrol & Graf, New York, 1999.

2. J. Drews, *Science*, **287**, 1960–1964 (2000).

3. B. G. Reuben in C. G. Wermuth, Ed., *The Practice of Medicinal Chemistry*, Academic Press, London, 1996, pp. 903–938.

4. J. H. Lin and A. Y. Lu, *Pharmacol. Rev.*, **49**, 403–449 (1997).

5. H. van de Waterbeemd, D. A. Smith, K. Beaumont, and D. K. Walker, *J. Med. Chem.*, **44**, 1313–1333 (2001).

6. N. Bodor in E. B. Roche, Ed., *Design of Biopharmaceutical Properties through Prodrugs and Analogs*, Academy of Pharmaceutical Sciences, Washington, DC, 1977, pp. 98–135.

7. N. Bodor in J. A. K. Buisman, Ed., Strategy in Drug Research. Proceedings of the 2nd IUPAC-IUPHAR Symposium on Research, Noordwijkerhout, The Netherlands, Elsevier, Amsterdam, 1982, pp. 137–164.

8. N. Bodor, *Trends Pharmacol. Sci.*, **3**, 53–56 (1982).

9. N. Bodor, *Adv. Drug Res.*, **13**, 255–331 (1984).

10. N. Bodor, *Med. Res. Rev.*, **3**, 449–469 (1984).

11. N. Bodor, *CHEMTECH*, **14**, 28–38 (1984).

12. N. Bodor in R. Dulbecco, Ed., *Encyclopedia of Human Biology*, Vol. **7**, Academic Press, San Diego, 1991, pp. 101–117.

13. N. Bodor, *Adv. Drug Deliv. Rev.*, **14**, 157–166 (1994).

14. N. Bodor in R. Rapaka, Ed., *Membranes and Barriers: Targeted Drug Delivery*, Vol. **154**, NIDA Research Monograph Series, NIDA, Rockville, MD, 1995, pp. 1–27.

15. N. Bodor and P. Buchwald, *Pharmacol. Ther.*, **76**, 1–27 (1997).

16. N. Bodor and P. Buchwald, *Med. Res. Rev.*, **20**, 58–101 (2000).

17. N. Bodor, H. H. Farag, and M. E. Brewster, *Science*, **214**, 1370–1372 (1981).

18. N. Bodor, *Methods Enzymol.*, **112**, 381–396 (1985).

19. N. Bodor, *Ann. N. Y. Acad. Sci.*, **507**, 289–306 (1987).

20. N. Bodor and L. Prokai, *Pharm. Res.*, **7**, 723–725 (1990).

21. N. Bodor and M. E. Brewster in R. L. Juliano, Ed., *Targeted Drug Delivery*, Vol. **100**, Handbook of Experimental Pharmacology, Springer-Verlag, Berlin, 1991, pp. 231–284.

22. N. Bodor and P. Buchwald, *Chem. Br.*, **34**, 36–40 (1998).

23. N. Bodor and P. Buchwald, *Adv. Drug Deliv. Rev.*, **36**, 229–254 (1999).

24. N. Bodor, J. J. Kaminski, and S. Selk, *J. Med. Chem.*, **23**, 469–474 (1980).

25. N. Bodor and J. J. Kaminski, *J. Med. Chem.*, **23**, 566–569 (1980).

26. N. Bodor, R. Woods, C. Raper, P. Kearney, and J. Kaminski, *J. Med. Chem.*, **23**, 474–480 (1980).

27. H. J. Lee and M. R. I. Soliman, *Science*, **215**, 989–991 (1982).

28. A. S. Heiman, D.-H. Ko, M. Chen, and H. J. Lee, *Steroids*, **62**, 491–499 (1997).

29. N. Bodor in H. Bundgaard, Ed., *Design of Prodrugs*, Elsevier, Amsterdam, 1985, pp. 333–354.

30. V. J. Stella in T. Higuchi and V. J. Stella, Eds., *Prodrugs as Novel Drug Delivery Systems*, American Chemical Society, Washington, DC, 1975, pp. 1–115.

31. H. Bundgaard, Ed., *Design of Prodrugs*, Elsevier Science, Amsterdam, 1985.

32. N. Bodor and J. J. Kaminski, *Annu. Rep. Med. Chem.*, **22**, 303–313 (1987).

33. L. P. Balant and E. Doelker in M. E. Wolff, Ed., *Burger's Medicinal Chemistry and Drug Discovery*, Vol. **1**, Wiley–Interscience, New York, 1995, pp. 949–982.

34. C. G. Wermuth, J.-C. Gaignault, and C. Marchandeau in C. G. Wermuth, Ed., *The Practice of Medicinal Chemistry*, Academic Press, London, 1996, pp. 671–696.

35. V. J. Stella, Ed., *Adv. Drug Deliv. Rev.* **19**, 111–330 (1996).

36. A. Albert, *Nature*, **182**, 421–427 (1958).

37. J. R. Gillette, *Drug Metab. Rev.*, **10**, 59–87 (1979).

38. G. J. Mannering in B. Testa and P. Jenner, Eds., *Concepts in Drug Metabolism Part B*, Marcel Dekker, New York, 1981, pp. 53–166.

39. K.-B. Augustinsson, *Ann. N. Y. Acad. Sci.*, **94**, 844–860 (1961).

40. K. Krisch in P. D. Boyer, Ed., *The Enzymes*, Vol. **5**, Academic Press, New York, 1971, pp. 43–69.

41. E. Heymann in W. B. Jakoby, J. R. Bend, and J. Caldwell, Eds., *Metabolic Basis of Detoxication, Biochemical Pharmacology and Toxicology*, Academic Press, New York, 1982, pp. 229–245.

42. C. H. Walker and M. I. Mackness, *Biochem. Pharmacol.*, **32**, 3265–3569 (1983).

43. F. W. Williams, *Clin. Pharmacokinet.*, **10**, 392–403 (1985).

44. F.-J. Leinweber, *Drug Metab. Rev.*, **18**, 379–439 (1987).

45. T. Satoh and M. Hosokawa, *Annu. Rev. Pharmacol. Toxicol.*, **38**, 257–288 (1998).

46. International Union of Biochemistry, Nomenclature Committee, *Enzyme Nomenclature 1984. Recommendations of the Nomenclature Committee of the International Union of Biochemistry on the Nomenclature and Classification of Enzyme-Catalyzed Reactions*, Academic Press, Orlando, FL, 1984.

47. P. Buchwald, *Min. Rev. Med. Chem.*, **1**, 101–111 (2001).

48. P. Buchwald and N. Bodor, *Pharmazie*, **57**, 87–93 (2002).

49. H. Ericsson, B. Tholander, and C. G. Regårdh, *Eur. J. Pharm. Sci.*, **8**, 29–37 (1999).

50. C. Y. Quon and H. F. Stampfli, *Drug Metab. Dispos.*, **13**, 420–424 (1985).

51. C. Y. Quon, K. Mai, G. Patil, and H. F. Stampfli, *Drug Metab. Dispos.*, **16**, 425–428 (1988).

52. T. Minagawa, Y. Kohno, T. Suwa, and A. Tsuji, *Biochem. Pharmacol.*, **49**, 1361–1365 (1995).

53. P. L. Feldman, M. K. James, M. F. Brackeen, J. M. Bilotta, S. V. Schuster, A. P. Lahey, M. W. Lutz, M. R. Johnson, and H. J. Leighton, *J. Med. Chem.*, **34**, 2202–2208 (1991).

54. A. Buchwald, *Ph.D. Thesis*, University of Florida, Gainesville, 2001.

55. H. F. Stampfli and C. Y. Quon, *Res. Commun. Mol. Pathol. Pharmacol.*, **88**, 87–97 (1995).

56. W. R. Chappell and J. Mordenti, *Adv. Drug Res.*, **20**, 1–116 (1991).

57. D. B. Campbell, *Ann. N. Y. Acad. Sci.*, **801**, 116–135 (1996).

58. E. F. Godefroi, P. A. J. Janssen, C. A. M. van der Eycken, A. H. M. T. van Heertum, and C. J. E. Niemegeers, *J. Med. Chem.*, **8**, 220–223 (1965).

59. P. J. Lewi, J. J. P. Heykants, and P. A. J. Janssen, *Arch. Int. Pharmacodyn. Ther.*, **220**, 72–85 (1976).

60. P. A. J. Janssen, C. J. E. Niemegeers, K. H. L. Schellekens, and F. M. Lenaerts, *Arzneim.-Forsch.*, **21**, 1234–43 (1971).

61. J. J. Savarese, H. H. Ali, S. J. Basta, P. B. Embree, R. P. F. Scott, N. Sunder, J. N. Weakly, W. B. Wastila, and H. A. El-Sayad, *Anesthesiology.*, **68**, 723–732 (1988).

62. K. O. Borg, E. Carlsson, K.-J. Hoffmann, T.-E. Jönsson, H. Thorin, and B. Wallin, *Acta Pharmacol. Toxicol.*, **36**, (Suppl. V), 125–135 (1975).

63. C.-G. Regårdh and G. Johnsson, *Clin. Pharmacokinet.*, **5**, 557–569 (1980).

64. P. Benfield, S. P. Clissold, and R. N. Brogden, *Drugs*, **31**, 376–429 (1986).

65. N. Bodor, Y. Oshiro, T. Loftsson, M. Katovich, and W. Caldwell, *Pharm. Res.*, **3**, 120–125 (1984).

66. N. Bodor, A. El-Koussi, M. Kano, and M. M. Khalifa, *J. Med. Chem.*, **31**, 1651–1656 (1988).

67. N. Bodor and A. El-Koussi, *Curr. Eye Res.*, **7**, 369–374 (1988).

68. P. Polgar and N. Bodor, *Life Sci.*, **48**, 1519–1528 (1991).

69. N. Bodor, A. El-Koussi, K. Zuobi, and P. Kovacs, *J. Ocul. Pharmacol. Ther.*, **12**, 115–122 (1996).

70. M. F. Sugrue, *J. Med. Chem.*, **40**, 2793–2809 (1997).

71. H.-S. Yang, W.-M. Wu, and N. Bodor, *Pharm. Res.*, **12**, 329–336 (1995).

72. P. W. Erhardt, C. M. Woo, W. G. Anderson, and R. J. Gorczynski, *J. Med. Chem.*, **25**, 1408–1412 (1982).

73. P. W. Erhardt in P. W. Erhardt, Ed., *Drug Metabolism. Databases and High Throughput Testing During Drug Design and Development*, Blackwell Science, Oxford, 1999, pp. 62–69.

74. H. Tatsuno, K. Goto, K. Shigenobu, Y. Kasuya, H. Obase, Y. Yamada, and S. Kudo, *J. Med. Chem.*, **20**, 394–397 (1977).

75. J. P. O'Donnell, S. Parekh, R. J. Borgman, and R. J. Gorczynski, *J. Pharm. Sci.*, **68**, 1236–1238 (1979).

76. P. W. Erhardt, C. M. Woo, R. J. Gorczynski, and W. G. Anderson, *J. Med. Chem.*, **25**, 1402–1407 (1982).

77. P. W. Erhardt, C. M. Woo, W. L. Matier, R. J. Gorczynski, and W. G. Anderson, *J. Med. Chem.*, **26**, 1109–1112 (1983).

78. S.-T. Kam, W. L. Matier, K. X. Mai, C. Barcelon-Yang, R. J. Borgman, J. P. O'Donnell, H. F. Stampfli, C. Y. Sum, W. G. Anderson, R. J. Gorczynski, and R. J. Lee, *J. Med. Chem.*, **27**, 1007–1016 (1984).

79. P. Benfield and E. M. Sorkin, *Drugs*, **33**, 392–412 (1987).

80. S. Iguchi, H. Iwamura, M. Nishizaki, A. Hayashi, K. Senokuchi, K. Kobayashi, K. Sakaki, K. Hachiya, Y. Ichioka, and M. Kawamura, *Chem. Pharm. Bull.*, **40**, 1462–1469 (1992).

81. K. Muraki, H. Nakagawa, N. Nagano, S. Henmi, H. Kawasumi, T. Nakanishi, K. Imaizumi, T. Tokuno, K. Atsuki, Y. Imaizumi, and M. Watanabe, *J. Pharmacol. Exp. Ther.*, **278**, 555–563 (1996).

82. A. Kitamura, A. Sakamoto, T. Inoue, and R. Ogawa, *Eur. J. Clin. Pharmacol.*, **51**, 467–471 (1997).

83. H. Atarashi, A. Kuruma, M. Yashima, H. Saitoh, T. Ino, Y. Endoh, and H. Hayakawa, *Clin. Pharmacol. Ther.*, **68**, 143–150 (2000).

84. Y.-T. Lin, B.-N. Wu, J.-R. Wu, Y.-C. Lo, L.-C. Chen, and I.-J. Chen, *J. Cardiovasc. Pharmacol.*, **28**, 149–157 (1996).

85. M. F. Sugrue, P. Gautheron, J. Grove, P. Mallorga, M.-P. Viader, J. J. Baldwin, G. S. Ponticello, and S. L. Varga, *Invest. Ophthalmol. Vis. Sci.*, **29**, 776–784 (1988).

86. K. Bauer, F. Brunner-Ferber, L. M. Distlerath, E. A. Lippa, B. Binkowitz, P. Till, and G. Kaik, *Clin. Pharmacol. Ther.*, **49**, 658–664 (1991).

87. K. G. Bauer, F. Brunner-Ferber, L. M. Distlerath, E. A. Lippa, B. Binkowitz, P. Till, and G. Kaik, *Br. J. Clin. Pharmacol.*, **34**, 122–129 (1992).

88. T. C. Hamilton and V. Chapman, *Life Sci.*, **23**, 813–820 (1978).

89. P. J. Machin, D. N. Hurst, and J. M. Osbond, *J. Med. Chem.*, **28**, 1648–1651 (1985).

90. R. J. Francis, P. B. East, S. J. McLaren, and J. Larman, *Biomed. Mass. Spectrom.*, **3**, 281–285 (1976).

91. P. Dayer, T. Leemann, A. Kupfer, T. Kronbach, and U. A. Meyer, *Eur. J. Clin. Pharmacol.*, **31**, 313–318 (1986).

92. S.-K. Hwang, A. Juhasz, S.-H. Yoon, and N. Bodor, *J. Med. Chem.*, **43**, 1525–1532 (2000).

93. W. Meuldermans, J. Hendrickx, W. Lauwers, R. Hurkmans, E. Swysen, J. Thijssen, P. Timmerman, R. Woestenborghs, and J. Heykants, *Drug Metab. Dispos.*, **15**, 905–913 (1987).

94. J. A. Colapret, G. Diamantidis, H. K. Spencer, T. C. Spaulding, and F. G. Rudo, *J. Med. Chem.*, **32**, 968–974 (1989).

95. M. K. James, P. L. Feldman, S. V. Schuster, J. M. Bilotta, M. F. Brackeen, and H. J. Leighton, *J. Pharmacol. Exp. Ther.*, **259**, 712–718 (1991).

96. T. D. Egan, H. J. M. Lemmens, P. Fiset, D. J. Hermann, K. T. Muir, D. R. Stanski, and S. L. Shafer, *Anesthesiology*, **79**, 881–892 (1993).

97. C. L. Westmoreland, J. F. Hoke, P. S. Sebel, C. C. Hug, and K. T. Muir, *Anesthesiology*, **79**, 893–903 (1993).

98. F. E. D'Amour and D. L. Smith, *J. Pharmacol. Exp. Ther.*, **72**, 74–79 (1941).

99. S. H. Haidar, J. E. Moreton, Z. Liang, J. F. Hoke, K. T. Muir, and N. D. Eddington, *J. Pharm. Sci.*, **86**, 1278–1282 (1997).

100. S. H. Haidar, J. E. Moreton, Z. Liang, J. F. Hoke, K. T. Muir, and N. D. Eddington, *Pharm. Res.*, **14**, 1817–1823 (1997).

101. S. Noble and K. L. Goa, *BioDrugs*, **10**, 329–339 (1998).

102. J. F. Howes, *Pharmazie*, **55**, 178–183 (2000).

103. N. Bodor (to Otsuka Pharmaceutical Co.), Belgian Pat. 889,563 (Cl. CO7J), November 3, 1981.

104. N. Bodor and M. Varga, *Exp. Eye Res.*, **50**, 183–187 (1990).

105. P. Druzgala, G. Hochhaus, and N. Bodor, *J. Steroid Biochem.*, **38**, 149–154 (1991).

106. P. Druzgala, W.-M. Wu, and N. Bodor, *Curr. Eye Res.*, **10**, 933–937 (1991).

107. M. Alberth, W.-M. Wu, D. Winwood, and N. Bodor, *J. Biopharm. Sci.*, **2**, 115–125 (1991).

108. N. S. Bodor, S. T. Kiss-Buris, and L. Buris, *Steroids*, **56**, 434–439 (1991).

109. N. Bodor, N. Bodor, and W.-M. Wu, *Curr. Eye Res.*, **11**, 525–530 (1992).

110. N. Bodor, T. Loftsson, and W.-M. Wu, *Pharm. Res.*, **9**, 1275–1278 (1992).

111. G. Hochhaus, L.-S. Chen, A. Ratka, P. Druzgala, J. Howes, N. Bodor, and H. Derendorf, *J. Pharm. Sci.*, **81**, 1210–1215 (1992).

112. N. Bodor, T. Murakami, and W.-M. Wu, *Pharm. Res.*, **12**, 869–874 (1995).

113. N. Bodor, W.-M. Wu, T. Murakami, and S. Engel, *Pharm. Res.*, **12**, 875–879 (1995).

114. N. Bodor and P. Buchwald in R. P. Schleimer, P. M. O'Byrne, S. J. Szefler, and R. Brattsand, Eds., *Inhaled Steroids in Asthma: Optimizing Effects in the Airways, Lung Biology in Health and Disease*, Vol. **163**, Marcel Dekker, New York, 2002, pp. 541–564.

115. C. Monder and H. L. Bradlow, *Recent Progr. Horm. Res.*, **36**, 345–400 (1980).

116. P. Druzgala and N. Bodor, *Steroids*, **56**, 490–494 (1991).

117. N. Bodor in E. Christophers, A. M. Kligman, E. Schöpf, and R. B. Stoughton, Eds., *Topical Corticosteroid Therapy: A Novel Approach to Safer Drugs*, Raven Press, New York, 1988, pp. 13–25.

118. N. Bodor in H. van der Goot, G. Domány, L. Pallos, and H. Timmerman, Eds., *Trends in Medicinal Chemistry '88 Proceedings of the Xth International Symposium on Medicinal Chemistry*, Elsevier, Amsterdam, 1989, pp. 145–164.

119. G. D. Novack, J. Howes, R. S. Crockett, and M. B. Sherwood, *J. Glaucoma*, **7**, 266–269 (1998).

120. J. Howes and G. D. Novack, *J. Ocul. Pharmacol. Ther.*, **14**, 153–158 (1998).

121. N. Bodor, U.S. Pat. 5,981,517, November 9, 1999.

122. P. Barton, A. P. Laws, and M. I. Page, *J. Chem. Soc. Perkin Trans.* **2**, 2021–2029 (1994).

123. A. Miklós, Z. Magyar, É. Kiss, I. Novák, M. Grósz, M. Nyitray, I. Dereszlay, E. Czégeni, A. Druga, J. Howes, and N. Bodor, *Pharmazie*, **57**, 142–146 (2002).

124. E. Gerhards, B. Nieuweboer, G. Schulz, H. Gibian, D. Berger, and W. Hecker, *Acta Endocr.*, **68**, 98–126 (1971).

125. H. Laurent, E. Gerhards, and R. Wiechert, *J. Steroid Biochem.*, **6**, 185–192 (1975).

126. J. F. Kapp, H. Koch, M. Töpert, H.-J. Kessler, and E. Gerhards, *Arzneim.-Forsch.*, **27**, 2191–2202 (1977).

127. J. F. Kapp, B. Gliwitzki, P. Josefiuk, and W. Weishaupt, *Arzneim.-Forsch.*, **27**, 2206–2213 (1977).

128. R. Reckers, *Arzneim.-Forsch.*, **27**, 2240–2244 (1977).

129. T. F. Hartley, P. L. Lieberman, E. O. Meltzer, J. N. Noyes, D. S. Pearlman, and D. G. Tinkelman, *J. Allergy Clin. Immunol.*, **75**, 501–507 (1985).

130. H. A. Orgel, E. O. Meltzer, C. W. Bierman, E. Bronsky, J. T. Connell, P. L. Lieberman, R. Nathan, D. S. Pearlman, H. L. Pence, R. G. Slavin, et al., *J. Allergy Clin. Immunol.*, **88**, 257–264 (1991).

131. P. S. Burge, J. Efthimiou, M. Turner-Warwick, and P. T. Nelmes, *Clin. Allergy*, **12**, 523–531 (1982).

132. A. Thalén, P. H. Andersson, P. T. Andersson, B. Axelsson, S. Edsbäcker, and R. Brattsand in R. P. Schleimer, P. M. O'Byrne, S. J. Szefler, and R. Brattsand, Eds., *Inhaled Steroids in Asthma: Optimizing Effects in the Airways, Lung Biology in Health and Disease*, Vol. **163**, Marcel Dekker, New York, 2002, pp. 521–537.

133. K. Biggadike, R. M. Angell, C. M. Burgess, R. M. Farrell, A. P. Hancock, A. J. Harker, W. R. Irving, C. Ioannou, P. A. Procopiou, R. E. Shaw, Y. E. Solanke, O. M. P. Singh, M. A. Snowden, R. J. Stubbs, S. Walton, and H. E. Weston, *J. Med. Chem.*, **43**, 19–21 (2000).

134. P. A. Procopiou, K. Biggadike, A. F. English, R. M. Farrell, G. N. Hagger, A. P. Hancock, M. V. Haase, W. R. Irving, M. Sareen, M. A. Snowden, Y. E. Solanke, C. J. Tralau-Stewart, S. E. Walton, and J. A. Wood, *J. Med. Chem.*, **44**, 602–612 (2001).

135. I. B. Taraporewala, H. P. Kim, A. S. Heiman, and H. J. Lee, *Arzneim.-Forsch.*, **39**, 21–25 (1989).

136. A. S. Heiman, H. P. Kim, I. B. Taraporewala, and H. J. Lee, *Arzneim.-Forsch.*, **39**, 262–267 (1989).

137. H. M. McLean, M. A. Khalil, A. S. Heiman, and H. J. Lee, *J. Pharm. Sci.*, **83**, 476–479 (1994).

138. A. S. Heiman, D. Hong, and H. J. Lee, *Steroids*, **59**, 324–329 (1994).

139. T. Kwon, A. S. Heiman, E. T. Oriaku, K. Yoon, and H. J. Lee, *J. Med. Chem.*, **38**, 1048–1051 (1995).

140. K.-J. Yoon, M. A. Khalil, T. Kwon, S.-J. Choi, and H. J. Lee, *Steroids*, **60**, 445–451 (515–521) (1995.

141. M. A. Khalil, M. K. Maponya, D.-H. Ko, Z. You, E. T. Oriaku, and H. J. Lee, *Med. Chem. Res.*, **6**, 52–60 (1996).

142. D.-H. Ko, M. F. Maponya, M. A. Khalil, E. T. Oriaku, Z. You, and H. J. Lee, *Med. Chem. Res.*, **7**, 313–324 (1997).

143. H. J. Lee and D.-H. Ko, *Soc. Biomed. Res. Symp.*, **7**, 33–40 (1997).

144. H. Ueno, A. Maruyama, M. Miyake, E. Nakao, K. Nakao, K. Umezu, and I. Nitta, *J. Med. Chem.*, **34**, 2468–2473 (1991).

145. T. Kimura, T. Yamaguchi, K. Usuki, Y. Kurosaki, T. Nakayama, Y. Fujiwara, Y. Matsuda, K. Unno, and T. Suzuki, *J. Controlled Release*, **30**, 125–135 (1994).

146. T. Suzuki, E. Sato, H. Tada, and Y. Tojima, *Biol. Pharm. Bull.*, **22**, 816–821 (1999).

147. C. Milioni, L. Jung, and B. Koch, *Eur. J. Med. Chem.*, **26**, 947–951 (1991).

148. M. Graffner-Nordberg, K. Sjödin, A. Tunek, and A. Hallberg, *Chem. Pharm. Bull.*, **46**, 591–601 (1998).

149. P. J. Barnes, *Nature*, **402**, (Suppl.), B31–B38 (1999).

150. J. T. H. Ong, B. J. Poulsen, W. A. Akers, J. R. Scholtz, F. C. Genter, and D. J. Kertesz, *Arch. Dermatol.*, **125**, 1662–1665 (1989).

151. G. H. Phillipps, *Respir. Med.*, **84**, (Suppl. A), 19–23 (1990).

152. S. Rohatagi, A. Bye, C. Falcoz, A. E. Mackie, B. Meibohm, H. Möllmann, and H. Derendorf, *J. Clin. Pharmacol.*, **36**, 938–941 (1996).

153. L. Thorsson, K. Dahlström, S. Edsbäcker, A. Källén, J. Paulson, and J.-E. Wirén, *Br. J. Clin. Pharmacol.*, **43**, 155–161 (1997).

154. H. Derendorf, G. Hochhaus, B. Meibohm, H. Möllmann, and J. Barth, *J. Allergy Clin. Immunol.*, **101**, S440–S446 (1998).

155. R. A. Lobo, *Am. J. Obstet. Gynecol.*, **173**, 982–989 (1995).

156. V. Beral, E. Banks, G. Reeves, and P. Appleby, *J. Epidemiol. Biostat.*, **4**, 191–210 (1999).

157. D. C. Labaree, T. Y. Reynolds, and R. B. Hochberg, *J. Med. Chem.*, **44**, 1802–1814 (2001).

158. B. Waldeck, *Pharmacol. Toxicol.*, **77**, (Suppl. III), 25–29, 1995.

159. R. Albrecht and O. Loge, *Eur. J. Med. Chem.*, **20**, 51–55 (1985).

160. R. Albrecht, J. Heindl, and O. Loge, *Eur. J. Med. Chem.*, **20**, 57–60 (1985).

161. H. S. Gill, S. Freeman, W. J. Irwin, and K. A. Wilson, *Eur. J. Med. Chem.*, **31**, 847–859 (1996).

162. N. Bodor, *CHEMTECH*, **25**, 22–32 (1995).

163. P. Müller, Ed., *DDT: The Insecticide Dichlorodiphenyltrichloroethane and Its Significance*, Birkhäuser Verlag, Basel, Switzerland, 1955.

164. R. D. O'Brien, *Insecticides: Action and Metabolism*, Academic Press, New York, 1967.

165. S. K. Kashyap, S. K. Nigam, A. B. Karnik, R. C. Gupta, and S. K. Chatterjee, *Int. J. Cancer.*, **19**, 725–729 (1979).

166. International Agency for Research on Cancer (IARC), **30**, 73–101 (1983).

167. T. B. Gaines, *Toxicol. Appl. Pharmacol.*, **14**, 515–534 (1969).

168. K. A. Hassall, *The Biochemistry and Uses of Pesticides*, 2nd., Macmillan, London, 1990.

169. E. Hodgson and R. J. Kuhr, Eds., *Safer Insecticides: Development and Use*, Marcel Dekker, New York, 1990.

170. R. Hammer, K. Amin, Z. E. Gunes, G. Brouillette, and N. Bodor, *Drug Des. Deliv.*, **2**, 207–219 (1988).

171. N. Bodor, A. El-Koussi, and R. Hammer, *J. Biopharm. Sci.*, **1**, 215–223 (1990).

172. R. H. Hammer, W.-M. Wu, J. S. Sastry, and N. Bodor, *Curr. Eye Res.*, **10**, 565–570 (1991).

173. G. N. Kumar, R. H. Hammer, and N. S. Bodor, *Bioorg. Med. Chem.*, **1**, 327–332 (1993).

174. R. H. Hammer, E. Gunes, G. N. Kumar, W.-M. Wu, V. Srinivasan, and N. S. Bodor, *Bioorg. Med. Chem.*, **1**, 183–187 (1993).

175. A. Juhász, F. Huang, F. Ji, P. Buchwald, W.-M. Wu, and N. Bodor, *Drug. Dev. Res.*, **43**, 117–127 (1998).

176. G. N. Kumar, R. H. Hammer, and N. Bodor, *Drug Des. Discov.*, **10**, 11–21 (1993).

177. G. N. Kumar, R. H. Hammer, and N. Bodor, *Drug Des. Discov.*, **10**, 1–9 (1993).

178. F. Ji, F. Huang, A. Juhasz, W. Wu, and N. Bodor, *Pharmazie*, **55**, 187–191 (2000).

179. G. Brouillette, M. Kawamura, G. N. Kumar, and N. Bodor, *J. Pharm. Sci.*, **85**, 619–623 (1996).

180. T. Ali-Melkillä, J. Kanto, and E. Iisalo, *Acta Anaesthesiol. Scand.*, **37**, 633–642 (1993).

181. F. S. K. MacMillan, H. H. Reller, and F. H. Synder, *J. Invest. Dermatol.*, **43**, 363–377 (1964).

182. A. E. Lasser, *IMJ–Ill. Med. J.*, **131**, 314–317 (1967).

183. G. N. Kumar and N. Bodor, *Curr. Med. Chem.*, **3**, 23–36 (1996).

184. J. D. Gabourel and R. E. Gosselin, *Arch. Int. Pharmacodyn.*, **115**, 416–431 (1958).

185. R.E. Gosselin, J.D. Gabourel, and J. H. Wills, *Clin. Pharmacol. Ther.*, **1**, 597–603 (1960).

186. F. Huang, P. Buchwald, C. E. Browne, H. H. Farag, W.-M. Wu, F. Ji, G. Hochhaus, and N. Bodor, *AAPS Pharm. Sci.*, **3,** article 30 (2002).

187. F. Ji, W.-M. Wu, and N. Bodor, *Pharmazie*, **57**, 138–141 (2002).

188. F. Huang, Ph.D. Thesis, University of Florida, Gainesville, FL (1999).

189. G. D'Agostino, A. R. Renzetti, F. Zonta, and A. Subissi, *J. Pharm. Pharmacol.*, **46**, 332–336 (1994).

190. N. Bodor (to INTERx Research Corp., Lawrence, KS), U.S. Pat. 3,998,815, December 21, 1976.

191. M. Lindstedt, S. Allenmark, R. A. Thompson, and L. Edebo, *Antimicrob. Agents Chemother.*, **34**, 1949–1954 (1990).

192. B. Ahlström, M. Chelminska-Bertilsson, R. A. Thompson, and L. Edebo, *Antimicrob. Agents Chemother.*, **39**, 50–55 (1995).

193. B. Ahlström, M. Chelminska-Bertilsson, R. A. Thompson, and L. Edebo, *Antimicrob. Agents Chemother.*, **41**, 544–550 (1997).

194. B. Ahlström and L. Edebo, *Microbiology*, **144**, 2497–2504 (1998).

195. B. Ahlström, R. A. Thompson, and L. Edebo, *APMIS*, **107**, 318–324 (1999).

196. M. Calvani, L. Critelli, G. Gallo, F. Giorgi, G. Gramiccioli, M. Santaniello, N. Scafetta, M. O. Tinti, and F. De Angelis, *J. Med. Chem.*, **41**, 2227–2233 (1998).

197. D. M. Stout, L. A. Black, C. Barcelon-Yang, W. L. Matier, B. S. Brown, C. Y. Quon, and H. F. Stampfli, *J. Med. Chem.*, **32**, 1910–1913 (1989).

198. R. J. Chorvat, L. A. Black, V. V. Ranade, C. Barcelon-Yang, D. M. Stout, B. S. Brown, H. F. Stampfli, and C. Y. Quon, *J. Med. Chem.*, **36**, 2494–2498 (1993).

199. T. R. Vrobel, P. E. Miller, N. D. Mostow, and L. Rakita, *Progr. Cardiovasc. Dis.*, **31**, 393–426 (1989).

200. M. J. P. Raatikainen, C. A. Napolitano, P. Druzgala, and D. M. Dennis, *J. Pharmacol. Exp. Ther.*, **277**, 1454–1463 (1996).

201. M. J. P. Raatikainen, T. E. Morey, P. Druzgala, P. Milner, M. D. Gonzalez, and D. M. Dennis, *J. Pharmacol. Exp. Ther.*, **295**, 779–785 (2000).

202. T. E. Morey, C. N. Seubert, M. J. P. Raatikainen, A. E. Martynyuk, P. Druzgala, P. Milner, M. D. Gonzalez, and D. M. Dennis, *J. Pharmacol. Exp. Ther.*, **297**, 260–266 (2001).

203. A. Juhász and N. Bodor, *Pharmazie*, **55**, 228–238 (2000).

204. T. Ogiso, M. Iwaki, T. Tanino, T. Nagai, Y. Ueda, O. Muraoka, and G. Tanabe, *Biol. Pharm. Bull.*, **19**, 1178–1183 (1996).

205. A. J. G. Baxter, R. D. Carr, S. C. Eyley, L. Fraser-Rae, C. Hallam, S. T. Harper, P. A. Hurved, S. J. King, and P. Meghani, *J. Med. Chem.*, **35**, 3718–3720 (1992).

206. M. Graffner-Nordberg, J. Marelius, S. Ohlsson, Å. Persson, G. Swedberg, P. Andersson, S. E. Andersson, J. Åqvist, and A. Hallberg, *J. Med. Chem.*, **43**, 3852–3861 (2000).

207. M. Graffner-Nordberg, K. Kolmodin, J. Åqvist, S. F. Queener, and A. Hallberg, *J. Med. Chem.*, **44**, 2391–2402 (2001).

208. M. Graffner Nordberg, *Ph.D. Thesis*, Uppsala University, Uppsala, Sweden, 2001.

209. R. K. Razdan and J. F. Howes, *Med. Res. Rev.*, **3**, 119–146 (1983).

210. K. Green and K. Kim, *Proc. Soc. Exp. Biol. Med.*, **154**, 228–231 (1977).

211. K. Green, J. F. Bigger, K. Kim, and K. Bowman, *Exp. Eye Res.*, **24**, 189–196 (1977).

212. A. Buchwald, H. Derendorf, F. Ji, N. V. Nagaraja, W.-M. Wu, and N. Bodor, *Pharmazie*, **57**, 108–114 (2002).

213. H. G. Pars, F. E. Granchelli, R. K. Razdan, J. K. Keller, D. G. Teiger, F. J. Rosenberg, and L. S. Harris, *J. Med. Chem.*, **19**, 445–454 (1976).

214. B. Meibohm and H. Derendorf, *Int. J. Clin. Pharmacol. Ther.*, **35**, 401–413 (1997).

215. G. Marciniak, D. Decolin, G. Leclerc, N. Decker, and J. Schwartz, *J. Med. Chem.*, **31**, 2289–2296 (1988).

216. N. Decker, M. Grima, J. Velly, G. Marciniak, G. Leclerc, and J. Schwartz, *Arzneim.-Forsch./ Drug Res.*, **38**, (II), 905–908 (1988).

217. N. Decker, M. Grima, J. Velly, G. Marciniak, G. Leclerc, and J. Schwartz, *Arzneim.-Forsch./ Drug Res.*, **38**, (II), 1110–1114 (1988).

218. H. Ericsson, B. Tholander, J. A. Björkman, M. Nordlander, and C. G. Regårdh, *Drug Metab. Dispos.*, **27**, 558–564 (1999).

219. N. Kieler-Jensen, A. Jolin-Mellgard, M. Nordlander, and S. E. Ricksten, *Acta Anaesthesiol. Scand.*, **44**, 186–193 (2000).

220. H. Ericsson, J. Schwieler, B. O. Lindmark, P. Lofdahl, T. Thulin, and C. G. Regårdh, *Chirality*, **13**, 130–134 (2001).

221. J. J. Kaminski, N. Bodor, and T. Higuchi, *J. Pharm. Sci.*, **65**, 553–557 (1976).

222. J. J. Kaminski, N. Bodor, and T. Higuchi, *J. Pharm. Sci.*, **65**, 1733–1737 (1976).

223. J. J. Kaminski, M. M. Huycke, S. H. Selk, N. Bodor, and T. Higuchi, *J. Pharm. Sci.*, **65**, 1737–1742 (1976).

224. M. Kosugi, J. J. Kaminski, S. H. Selk, I. H. Pitman, N. Bodor, and T. Higuchi, *J. Pharm. Sci.*, **65**, 1743–1746 (1976).

225. H. D. Burkett, J. H. Faison, H. H. Kohl, W. B. Wheatley, S. D. Worley, and N. Bodor, *Water Res. Bull.*, **17**, 874–879 (1981).

226. S. D. Worley, W. B. Wheatley, H. H. Kohl, H. D. Burkett, J. H. Faison, J. A. Van Hoose, and N. Bodor in R. L. Jolley, W. A. Brungs, J. A. Cotruvo, R. B. Cumming, J. S. Mattice, and V. A. Jacobs, Eds., *Water Chlorination: Environmental Impact and Health Effects*, Vol. **4**, Ann Arbor Science Publishers, Ann Arbor, MA, 1983, pp. 1105–1113.

227. S. D. Worley, W. B. Wheatley, H. H. Kohl, H. D. Burkett, J. A. Van Hoose, and N. Bodor, *Ind. Eng. Chem. Prod. Res. Dev.*, **22**, 716–718 (1983).

228. E. C. Mora, H. H. Kohl, W. B. Wheatley, S. D. Worley, J. H. Faison, H. D. Burkett, and N. Bodor, *Poultry Sci.*, **61**, 1968–1971 (1982).

229. H. H. Kohl, W. B. Wheatley, S. D. Worley, and N. Bodor, *J. Pharm. Sci.*, **69**, 1292–1295 (1980).

230. N. Bodor and K. B. Sloan, *J. Pharm. Sci.*, **71**, 514–520 (1982).

231. W. M. Schubert and Y. Motoyama, *J. Am. Chem. Soc.*, **87**, 5507–5508 (1965).

232. N. Bodor, K. B. Sloan, R. J. Little, S. H. Selk, and L. Caldwell, *Int. J. Pharm.*, **10**, 307–321 (1982).

233. N. Bodor, Z. Gabanyi, and C.-K. Wong, *J. Am. Chem. Soc.*, **111**, 3783–3786 (1989).

234. N. Bodor, A. Harget, and M.-J. Huang, *J. Am. Chem. Soc.*, **113**, 9480–9483 (1991).

235. N. Bodor and M.-J. Huang, *J. Pharm. Sci.*, **81**, 272–281 (1992).

236. N. Bodor and M.-J. Huang, *J. Pharm. Sci.*, **81**, 954–960 (1992).

237. N. Bodor and P. Buchwald, *J. Phys. Chem. B*, **101**, 3404–3412 (1997).

238. P. Buchwald and N. Bodor, *Proteins*, **30**, 86–99 (1998).

239. P. Buchwald and N. Bodor, *J. Phys. Chem. B*, **102**, 5715–5726 (1998).

240. P. Buchwald and N. Bodor, *Curr. Med. Chem.*, **5**, 353–380 (1998).

241. P. Buchwald and N. Bodor, *J. Med. Chem.*, **42**, 5160–5168 (1999).

242. P. Buchwald, *Perspect. Drug Disc. Des.*, **19**, 19–45 (2000).

243. P. Buchwald and N. Bodor, *J. Am. Chem. Soc.*, **122**, 10671–10679 (2000).

244. P. Buchwald and N. Bodor, *J. Pharm. Pharmacol.*, **53**, 1087–1098 (2001).

245. N. Bodor, P. Buchwald, and M.-J. Huang, *SAR QSAR Environ. Res.*, **8**, 41–92 (1998).

246. N. Bodor, P. Buchwald, and M.-J. Huang in J. Leszczynski, Ed., *Computational Molecular Biology, Theoretical and Computational Chemistry*, Vol. **8**, Elsevier, Amsterdam, 1999, pp. 569–618.

247. P. Buchwald and N. Bodor, *Pharmazie*, **55**, 210–217 (2000).

248. M. J. S. Dewar, E. G. Zoebisch, E. F. Healy, and J. J. P. Stewart, *J. Am. Chem. Soc.*, **107**, 3902–3909 (1985).

249. G. W. Goldstein and A. L. Betz, *Sci. Am.*, **255**, 74–83 (1986).

250. M. W. B. Bradbury, Ed., *Physiology and Pharmacology of the Blood–Brain Barrier*, Springer-Verlag, Berlin, 1992.

251. C. Crone in A. J. Suckling, M. G. Rumsby, and M. W. B Bradbury, Eds., *The Blood–Brain Barrier in Health and Disease*, Ellis Horwood, Chichester, UK, 1986, pp. 17–40.

252. N. Bodor and M. E. Brewster, *Pharmacol. Ther.*, **19**, 337–386 (1983).

253. T. Ishikura, T. Senou, H. Ishihara, T. Kato, and T. Ito, *Int. J. Pharm.*, **116**, 51–63 (1995).

254. E. Pop, *Curr. Med. Chem.*, **4**, 279–294 (1997).

255. G. Somogyi, S. Nishitani, D. Nomi, P. Buchwald, L. Prokai, and N. Bodor, *Int. J. Pharm.*, **166**, 15–26 (1998).

256. G. Somogyi, P. Buchwald, D. Nomi, L. Prokai, and N. Bodor, *Int. J. Pharm.*, **166**, 27–35 (1998).

257. N. Bodor, H. H. Farag, M. D. C. Barros, W.-M. Wu, and P. Buchwald, *J. Drug Target*, **10**, 63–71 (2002).

258. J. Rydström, J. B. Hoek, and L. Ernster in P. D. Boyer, Ed., *The Enzymes*, Vol. **13**, Academic Press, New York, 1976.

259. J. B. Hoek and J. Rydström, *Biochem. J.*, **254**, 1–10 (1988).

260. N. Bodor, M. E. Brewster, and J. J. Kaminski, *J. Mol. Struct.*, **206**, 315–334 (1990).

261. M. E. Brewster, K. S. Estes, R. Perchalski, and N. Bodor, *Neurosci. Lett.*, **87**, 277–282 (1988).

262. N. Bodor, R. G. Roller, and S. J. Selk, *J. Pharm. Sci.*, **67**, 685–687 (1978).

263. E. Palomino, D. Kessel, and J. P. Horwitz, *J. Med. Chem.*, **32**, 622–625 (1989).

264. N. Bodor and H. H. Farag, *J. Med. Chem.*, **26**, 313–318 (1983).

265. N. Bodor and A. M. Abdelalim, *J. Pharm. Sci.*, **74**, 241–245 (1985).

266. N. Bodor, T. Nakamura, and M. E. Brewster, *Drug Des. Deliv.*, **1**, 51–64 (1986).

267. N. Bodor, H. H. Farag, and P. Polgar, *J. Pharm. Pharmacol.*, **53**, 889–894 (2001).

268. N. Bodor and H. H. Farag, *J. Pharm. Sci.*, **73**, 385–389 (1984).

269. N. Bodor and A. M. Abdelalim, *J. Pharm. Sci.*, **75**, 29–35 (1986).

270. W. R. Anderson, J. W. Simpkins, M. E. Brewster, and N. Bodor, *Drug Des. Del.*, **2**, 287–298 (1988).

271. M. E. Brewster, K. S. Estes, and N. Bodor, *Pharm. Res.*, **3**, 278–285 (1986).

272. M. E. Brewster, M. Deyrup, K. Czako, and N. Bodor, *J. Med. Chem.*, **33**, 2063–2065 (1990).

273. W. R. Anderson, J. W. Simpkins, M. E. Brewster, and N. Bodor, *Neuroendocrinology*, **50**, 9–16 (1989).

274. T. Siegal, F. Soti, A. Biegon, E. Pop, and M. E. Brewster, *Pharm. Res.*, **14**, 672–675 (1997).

275. N. Bodor, J. McCornack, and M. E. Brewster, *Int. J. Pharm.*, **35**, 47–59 (1987).

276. J. W. Simpkins, J. McCornack, K. S. Estes, M. E. Brewster, E. Shek, and N. Bodor, *J. Med. Chem.*, **29**, 1809–1812 (1986).

277. W. R. Anderson, J. W. Simpkins, M. E. Brewster, and N. Bodor, *Pharmacol. Biochem. Behav.*, **27**, 265–271 (1987).

278. K. S. Estes, M. E. Brewster, J. W. Simpkins, and N. Bodor, *Life Sci.*, **40**, 1327–1334 (1987).

279. K. S. Estes, M. E. Brewster, and N. S. Bodor, *Life Sci.*, **42**, 1077–1084 (1988).

280. W. R. Anderson, J. W. Simpkins, M. E. Brewster, and N. Bodor, *Endocr. Res.*, **14**, 131–148 (1988).

281. M. E. Brewster, K. S. Estes, and N. Bodor, *J. Med. Chem.*, **31**, 244–249 (1988).

282. G. Mullersman, H. Derendorf, M. E. Brewster, K. S. Estes, and N. Bodor, *Pharm. Res.*, **5**, 172–177 (1988).

283. J. Howes, N. Bodor, M. E. Brewster, K. Estes, and M. Eve, *J. Clin. Pharmacol.*, **28**, 951, Abstr. 181 (1988).

284. J. W. Simpkins, W. R. Anderson, R. Dawson Jr., E. Seth, M. Brewster, K. S. Estes, and N. Bodor, *Physiol. Behav.*, **44**, 573–580 (1988).

285. J. W. Simpkins, W. R. Anderson, R. Dawson Jr., and N. Bodor, *Pharm. Res.*, **6**, 592–600 (1989).

286. D. K. Sarkar, S. J. Friedman, S. S. C. Yen, and S. A. Frautschy, *Neuroendocrinology*, **50**, 204–210 (1989).

287. W. J. Millard, T. M. Romano, N. Bodor, and J. W. Simpkins, *Pharm. Res.*, **7**, 1011–1018 (1990).

288. M. E. Brewster, J. W. Simpkins, and N. Bodor, *Rev. Neurosci.*, **2**, 241–285 (1990).

289. K. S. Estes, P. M. Dewland, M. E. Brewster, H. Derendorf, and N. Bodor, *Pharm. Ztg. Wiss.*, **136**, 153–158 (1991).

290. M. E. Brewster, M. S. M. Bartruff, W. R. Anderson, P. J. Druzgala, N. Bodor, and E. Pop, *J. Med. Chem.*, **37**, 4237–4244 (1994).

291. K. S. Estes, M. E. Brewster, and N. Bodor, *Adv. Drug Deliv. Rev.*, **14**, 167–175 (1994).

292. J. W. Simpkins, G. Rajakumar, Y.-Q. Zhang, C. E. Simpkins, D. Greenwald, C. J. Yu, N. Bodor, and A. L. Day, *J. Neurosurg.*, **87**, 724–730 (1997).

293. O. Rabbani, K. S. Panickar, G. Rajakumar, M. A. King, N. Bodor, E. M. Meyer, and J. W. Simpkins, *Exp. Neurol.*, **146**, 179–186 (1997).

294. E. Pop, W.-M. Wu, E. Shek, and N. Bodor, *J. Med. Chem.*, **32**, 1774–1781 (1989).

295. E. Pop, W.-M. Wu, and N. Bodor, *J. Med. Chem.*, **32**, 1789–1795 (1989).

296. W.-M. Wu, E. Pop, E. Shek, and N. Bodor, *J. Med. Chem.*, **32**, 1782–1788 (1989).

297. W.-M. Wu, E. Pop, E. Shek, R. Clemmons, and N. Bodor, *Drug Des. Deliv.*, **7**, 33–43 (1990).

298. M. E. Brewster, M. Deyrup, K. Seyda, and N. Bodor, *Int. J. Pharm.*, **68**, 215–229 (1991).

299. V. Venkatraghavan, E. Shek, R. Perchalski, and N. Bodor, *Pharmacologist*, **28**, 145 (1986).

300. K. Rand, N. Bodor, A. El-Koussi, I. Raad, A. Miyake, H. Houck, and N. Gildersleeve, *J. Med. Virol.*, **20**, 1–8 (1986).

301. A. El-Koussi and N. Bodor, *Drug Des. Deliv.*, **1**, 275–283 (1987).

302. P. G. Canonico, M. Kende, and B. Gabrielsen, *Adv. Virus. Res.*, **35**, 271–312 (1988).

303. M. Bhagrath, R. Sidwell, K. Czako, K. Seyda, W. Anderson, N. Bodor, and M. E. Brewster, *Antiviral Chem. Chemother.*, **2**, 265–286 (1991).

304. M. Deyrup, R. Sidwell, R. Little, P. Druzgala, N. Bodor, and M. E. Brewster, *Antiviral Chem. Chemother.*, **2**, 337–355 (1991).

305. K. W. Morin, L. I. Wiebe, and E. E. Knaus, *Carbohydr. Res.*, **249**, 109–116 (1993).

306. K. W. Morin, E. E. Knaus, and L. I. Wiebe, *J. Labelled Compd. Radiopharm.*, **35**, 205–207 (1994).

307. J. Balzarini, K. W. Morin, E. E. Knaus, L. I. Wiebe, and E. De Clercq, *Gene Ther.*, **2**, 317–322 (1995).

308. E. Pop, W. Anderson, J. Vlasak, M. E. Brewster, and N. Bodor, *Int. J. Pharm.*, **84**, 39–48 (1992).

309. M. E. Brewster, R. Little, V. Venkatraghavan, and N. Bodor, *Antiviral Res.*, **9**, 127 (1988).

310. R. Little, D. Bailey, M. E. Brewster, K. S. Estes, R. M. Clemmons, A. Saab, and N. Bodor, *J. Biopharm. Sci.*, **1**, 1–18 (1990).

311. M. E. Brewster, W. Anderson, and N. Bodor, *J. Pharm. Sci.*, **80**, 843–845 (1991).

312. E. Pop, M. E. Brewster, W. R. Anderson, and N. Bodor, *Med. Chem. Res.*, **2**, 457–466 (1992).

313. E. Pop, Z. Z. Liu, J. Vlasak, W. Anderson, M. E. Brewster, and N. Bodor, *Drug Del.*, **1**, 143–149 (1993).

314. Y. Mizrachi, A. Rubinstein, Z. Harish, A. Biegon, W. R. Anderson, and M. E. Brewster, *AIDS*, **9**, 153–158 (1995).

315. M. E. Brewster, W. R. Anderson, A. I. Webb, L. M. Pablo, D. Meinsma, D. Moreno, H. Derendorf, N. Bodor, and E. Pop, *Antimicrob. Agents Chemother.*, **41**, 122–128 (1997).

316. P. T. Torrence, J. Kinjo, K. Lesiak, J. Balzarini, and E. DeClerq, *FEBS Lett.*, **234**, 135–140 (1988).

317. S. R. Gogu, S. K. Aggarwal, S. R. S. Rangan, and K. C. Agrawal, *Biochem. Biophys. Res. Commun.*, **160**, 656–661 (1989).

318. S. K. Aggarwal, S. R. Gogu, S. R. S. Rangan, and K. C. Agrawal, *J. Med. Chem.*, **33**, 1505–1510 (1990).

319. R. H. Lupia, N. Ferencz, S. K. Aggarwal, K. C. Agrawal, and J. J. L. Lertora, *Clin. Res.*, **38**, 15A (1990).

320. R. H. Lupia, N. Ferencz, J. J. L. Lertora, S. K. Aggarwal, W. J. George, and K. C. Agrawal, *Antimicrob. Agents Chemother.*, **37**, 818–824 (1993).

321. J. Gallo, F. Boubinot, D. Doshi, J. Etse, V. Bhandti, R. Schinazi, and C. K. Chu, *Pharm. Res.*, **6**, S161 (1989).

322. C. K. Chu, V. S. Bhadti, K. J. Doshi, J. T. Etse, J. M. Gallo, F. D. Boudinot, and R. F. Schinazi, *J. Med. Chem.*, **33**, 2188–2192 (1990).

323. J. M. Gallo, J. T. Etse, K. J. Doshi, F. D. Boudinot, and C. K. Chu, *Pharm. Res.*, **8**, 247–253 (1991).

324. E. Palomino, D. Kessel, and J. P. Horwitz, *Nucleosides Nucleotides*, **11**, 1639–1649 (1992).

325. M. E. Brewster, K. Raghavan, E. Pop, and N. Bodor, *Antimicrob. Agents Chemother.*, **38**, 817–823 (1994).

326. M. Camplo, A. S. Charvet Faury, C. Borel, F. Turin, O. Hantz, C. Trabaud, V. Niddam, N. Mourier, J. C. Graciet, J. C. Chermann, and J. L. Kraus, *Eur. J. Med. Chem.*, **31**, 539–546 (1996).

327. K. Raghavan, E. Shek, and N. Bodor, *Anticancer Drug Des.*, **2**, 25–36 (1987).

328. K. Raghavan, T. Loftsson, M. E. Brewster, and N. Bodor, *Pharm. Res.*, **9**, 743–749 (1992).

329. N. Bodor, V. Venkatraghavan, D. Winwood, K. Estes, and M. E. Brewster, *Int. J. Pharm.*, **53**, 195–208 (1989).

330. N. Bodor and J. W. Simpkins, *Science*, **221**, 65–67 (1983).

331. N. Bodor and H. H. Farag, *J. Med. Chem.*, **26**, 528–534 (1983).

332. J. W. Simpkins, N. Bodor, and A. Enz, *J. Pharm. Sci.*, **74**, 1033–1036 (1985).

333. F. A. Omar, H. H. Farag, and N. Bodor, *J. Drug Target.*, **2**, 309–316 (1994).

334. V. Carelli, F. Liberatore, L. Scipione, M. Impicciatore, E. Barocelli, M. Cardellini, and G. Giorgioni, *J. Controlled Release*, **42**, 209–216 (1996).

335. W. R. Anderson, J. W. Simpkins, P. A. Woodard, D. Winwood, W. C. Stern, and N. Bodor, *Psychopharmacology*, **92**, 157–163 (1987).

336. P. A. Woodard, D. Winwood, M. E. Brewster, K. S. Estes, and N. Bodor, *Drug Des. Deliv.*, **6**, 15–28 (1990).

337. A. Kourounakis, N. Bodor, and J. Simpkins, *Int. J. Pharm.*, **141**, 239–250 (1996).

338. A. Kourounakis, N. Bodor, and J. Simpkins, *J. Pharm. Pharmacol.*, **49**, 1–9 (1997).

339. E. Shek, T. Murakami, C. Nath, E. Pop, and N. S. Bodor, *J. Pharm. Sci.*, **78**, 837–843 (1989).

340. E. Pop and N. Bodor, *Epilepsy Res.*, **13**, 1–16 (1992).

341. A. V. Boddy, K. Zhang, F. Lepage, F. Tombret, J. G. Slatter, T. A. Baillie, and R. H. Levy, *Pharm. Res.*, **8**, 690–697 (1991).

342. S. H. Yiu and E. E. Knaus, *J. Med. Chem.*, **39**, 4576–4582 (1996).

343. E. Pop, K. Prókai-Tátrai, W. Anderson, J.-L. Lin, M. E. Brewster, and N. Bodor, *Eur. J. Pharmacol.*, **183**, 1909 (1990).

344. K. Prókai-Tátrai, E. Pop, W. Anderson, J.-L. Lin, M. E. Brewster, and N. Bodor, *J. Pharm. Sci.*, **80**, 255–261 (1991).

345. M. E. Brewster, C. Robledo-Luiggi, A. Miyakeb, E. Pop, and N. Bodor in E. M. Meyer, J. W. Simpkins, and J. Yamamoto, Eds., *Novel*

Approaches to the Treatment of Alzheimer's Disease, Advances in Behavioral Biology, Vol. **36**, Plenum, New York, 1989, pp. 173–183.

346. E. Pop, K. Prókai-Tátrai, J. D. Scott, M. E. Brewster, and N. Bodor, *Pharm. Res.*, **7**, 658–664 (1990).

347. E. Pop, F. Soti, W. R. Anderson, J. A. Panetta, K. S. Estes, N. S. Bodor, and M. E. Brewster, *Int. J. Pharm.*, **140**, 33–44 (1996).

348. D. Bailey, R. Perchalski, M. Bhagrath, E. Shek, D. Winwood, and N. Bodor, *J. Biopharm. Sci.*, **2**, 205–218 (1991).

349. M. J. Phelan and N. Bodor, *Pharm. Res.*, **6**, 667–676 (1989).

350. E. Pop, W. Anderson, K. Prókai-Tátrai, J. Vlasak, M. E. Brewster, and N. Bodor, *Med. Chem. Res.*, **2**, 16–21 (1992).

351. W. Anderson, E. Pop, S.-K. Lee, N. Bodor, and M. Brewster, *Med. Chem. Res.*, **1**, 74–79 (1991).

352. E. Pop, W. Anderson, K. Prókai-Tátrai, M. E. Brewster, M. Fregly, and N. Bodor, *J. Med. Chem.*, **33**, 2216–2221 (1990).

353. E. Pop, K. Prókai-Tátrai, M. E. Brewster, and N. Bodor, *Org. Prep. Proc. Int.*, **26**, 687–690 (1994).

354. N. Bodor, L. Prokai, W.-M. Wu, H. H. Farag, S. Jonnalagadda, M. Kawamura, and J. Simpkins, *Science*, **257**, 1698–1700 (1992).

355. K. Prokai-Tatrai, L. Prokai, and N. Bodor, *J. Med. Chem.*, **39**, 4775–4782 (1996).

356. L. Prokai, X.-D. Ouyang, W.-M. Wu, and N. Bodor, *J. Am. Chem. Soc.*, **116**, 2643–2644 (1994).

357. L. Prokai, X. Ouyang, K. Prokai-Tatrai, J. W. Simpkins, and N. Bodor, *Eur. J. Med. Chem.*, **33**, 879–886 (1998).

358. L. Prokai, K. Prokai-Tatrai, X. Ouyang, H.-S. Kim, W.-M. Wu, A. Zharikova, and N. Bodor, *J. Med. Chem.*, **42**, 4563–4571 (1999).

359. S.-H. Yoon, J. Wu, W.-M. Wu, L. Prokai, and N. Bodor, *Bioorg. Med. Chem.*, **9**, 1059–1063 (2000).

360. P. Chen, N. Bodor, W.-M. Wu, and L. Prokai, *J. Med. Chem.*, **41**, 3773–3781 (1998).

361. L. Prokai, K. Prokai-Tatrai, and N. Bodor, *Med. Res. Rev.*, **20**, 367–416 (2000).

362. C. A. Hunt, R. D. MacGregor, and R. A. Siegel, *Pharm. Res.*, **3**, 333–344 (1986).

363. C. L. Williams and G. M. Stancel in J. G. Hardman and L. E. Limbird, Eds., *Goodman & Gilman's The Pharmacological Basis of Therapeutics*, McGraw-Hill, New York, 1996, pp. 1411–1440.

364. N. M. Kaplan, *Annu. Rev. Med.*, **29**, 31–40 (1978).

365. K. Fotherby, *Contraception*, **31**, 367–394 (1985).

366. J. D. Yager and J. G. Liehr, *Annu. Rev. Pharmacol. Toxicol.*, **36**, 203–232 (1996).

367. G. Wells and D. M. Herrington, *Drugs Aging*, **15**, 419–422 (1999).

368. T. W. Meade and M. R. Vickers, *J. Epidemiol. Biostat.*, **4**, 165–190 (1999).

369. S. Aldridge, *Magic Molecules: How Drugs Work*, Cambridge University Press, Cambridge, 1998.

370. B. S. McEwen and S. E. Alves, *Endocr. Rev.*, **20**, 279–307 (1999).

371. G. V. Upton, *J. Reprod. Med.*, **29**, 71–79 (1984).

372. C. Behl and D. Manthey, *J. Neurocytol.*, **29**, 351–358 (2000).

373. C. Leranth, R. H. Roth, J. D. Elswoth, F. Naftolin, T. L. Horvath, and D. E. Redmond Jr., *J. Neurosci.*, **20**, 8604–8609 (2000).

374. A. Maggi and J. Perez, *Life Sci.*, **37**, 893–906 (1985).

375. K. Yaffe, G. Sawaya, I. Lieberburg, and D. Grady, *JAMA*, **279**, 688–695 (1998).

376. L. Gopinath, *Chem. Br.*, **34**, 38–40 (1998).

377. S. G. Haskell, E. D. Richardson, and R. I. Horwitz, *J. Clin. Epidemiol.*, **50**, 1249–64 (1997).

378. V. W. Henderson, *Neurology*, **48**, S27–S35 (1997).

379. S. Asthana, S. Craft, L. D. Baker, M. A. Raskind, R. S. Birnbaum, C. P. Lofgreen, R. C. Veith, and S. R. Plymate, *Psychoneuroendocrinology*, **24**, 657–677 (1999).

380. S. Asthana, L. D. Baker, S. Craft, F. Z. Stanczyk, R. C. Veith, M. A. Raskind, and S. R. Plymate, *Neurology*, **57**, 605–612 (2001).

381. L. Janocko, J. M. Lamer, and R. B. Hochberg, *Endocrinology*, **114**, 1180–1186 (1984).

382. W. Saenger, *Angew. Chem. Int. Ed. Engl.*, **19**, 344–362 (1980).

383. J. Szejtli, *Cyclodextrins and their Inclusion Complexes*, Akadémiai Kiadó, Budapest, 1982.

384. J. S. Pagington, *Chem. Br.*, **23**, 455–458 (1987).

385. J. Szejtli, *Chem. Rev.*, **98**, 1743–1753 (1998).

386. M. E. Brewster, K. E. Estes, T. Loftsson, R. Perchalski, H. Derendorf, G. Mullersman, and N. Bodor, *J. Pharm. Sci.*, **77**, 981–985 (1988).

387. J. Pitha and J. Pitha, *J. Pharm. Sci.*, **74**, 987–990 (1985).

388. J. Pitha, J. Milecki, H. Fales, L. Pannell, and K. Uekama, *Int. J. Pharm.*, **29**, 73–82 (1986).

389. A. Yoshida, H. Arima, K. Uekama, and J. Pitha, *Int. J. Pharm.*, **46**, 217–222 (1988).

390. M. E. Brewster, K. S. Estes, and N. Bodor, *Int. J. Pharm.*, **59**, 231–243 (1990).

391. E. Pop, T. Loftsson, and N. Bodor, *Pharm. Res.*, **8**, 1044–1049 (1991).

392. A. J. Kastin, R. H. Ehrensing, W. A. Banks, and J. E. Zadina in E. R. de Kloet, V. M. Wiegant, and D. de Wied, Eds., *Neuropeptides and Brain Function*, Progress in Brain Research, Vol. **72**, Elsevier, Amsterdam, 1987, pp. 223–234.

393. C. B. E. Nemeroff, *Neuropeptides in Psychiatric and Neurological Disorders*, Johns Hopkins University Press, Baltimore, 1988.

394. C. H. V. Hoyle, *Neuropeptides: Essential Data*, Wiley, Chichester, 1996.

395. W. A. Banks and A. J. Kastin, *Adv. Exp. Med. Biol.*, **274**, 59–69 (1990).

396. J. Brownlees and C. H. Williams, *J. Neurochem.*, **60**, 793–803 (1993).

397. A. Ermisch, P. Brust, R. Kretzschmar, and H.-J. Rühle, *Physiol. Rev.*, **73**, 489–527 (1993).

398. R. Oliyai and V. J. Stella, *Annu. Rev. Pharmacol. Toxicol.*, **33**, 521–544 (1993).

399. N. Bodor and L. Prokai in M. Taylor and G. Amidon, Eds., *Peptide-Based Drug Design: Controlling Transport and Metabolism*, American Chemical Society, Washington, DC, 1995, pp. 317–337.

400. J. Hughes, T. V. Smith, H. W. Kosterlitz, L. Fothergill, B. A. Morgan, and H. R. Morris, *Nature*, **258**, 577–579 (1975).

401. G. A. Olson, R. D. Olson, and A. J. Kastin, *Peptides*, **6**, 769–791 (1985).

402. G. A. Olson, R. D. Olson, and A. J. Kastin, *Peptides*, **13**, 1247–1287 (1992).

403. R. J. Bicknell, *J. Endocr.*, **107**, 437–446 (1985).

404. M. J. Millan and A. Herz, *Int. Rev. Neurobiol.*, **26**, 1–83 (1985).

405. S. S. C. Yen, M. E. Quigley, R. L. Reid, J. F. Ropert, and N. S. Cetel, *Am. J. Obstet. Gynecol.*, **152**, 485–493 (1985).

406. R. Sandyk, *Life Sci.*, **37**, 1655–1663 (1985).

407. H. Frenk, *Brain Res. Rev.*, **6**, 197–210 (1983).

408. F. C. Tortella, J. B. Long, and J. W. Holaday, *Brain Res.*, **332**, 174–178 (1985).

409. C. A. Baile, C. L. McLaughlin, and M. A. Della-Fera, *Physiol. Rev.*, **66**, 172–234 (1986).

410. C. Schmauss and H. M. Emrich, *Biol. Psychiatry.*, **20**, 1211–1231 (1985).

411. I. Izquierdo and C. A. Netto, *Ann. N. Y. Acad. Sci.*, **444**, 162–177 (1985).

412. J. W. Holaday, *Annu. Rev. Pharmacol. Toxicol.*, **23**, 541–594 (1983).

413. M. W. Johnson, W. E. Mitch, and C. S. Wilcox, *Prog. Cardiovasc. Dis.*, **27**, 435–450 (1985).

414. F. Porreca and T. F. Burks, *J. Pharmacol. Exp. Ther.*, **227**, 22–27 (1983).

415. A. J. Kastin, M. A. Pearson, and W. A. Banks, *Pharmacol. Biochem. Behav.*, **40**, 771–774 (1991).

416. T. Szirtes, L. Kisfaludy, É. Pálosi, and L. Szporny, *J. Med. Chem.*, **27**, 741–745 (1984).

417. E. M. Santori and D. E. Schmidt, *Regul. Pept.*, **1**, 69–74 (1980).

418. Y. Itoh, T. Ogasawara, T. Mushiroi, A. Yamazaki, Y. Ukai, and K. Kimura, *J. Pharmacol. Exp. Ther.*, **271**, 884–890 (1994).

419. J. A. Kelly, *Essays Biochem.*, **30**, 133–149 (1995).

420. G. W. Bennett, T. M. Ballard, C. D. Watson, and K. C. Fone, *Exp. Gerontol.*, **32**, 451–469 (1997).

421. G. G. Yarbrough, *Life Sci.*, **33**, 111–118 (1983).

422. A. I. Faden, R. Vink, and T. K. McIntosh, *Ann. N. Y. Acad. Sci.*, **553**, 380–384 (1989).

423. A. H. Mellow, T. Sunderland, R. M. Cohen, B. A. Lawlor, J. L. Hill, P. A. Newhouse, M. R. Cohen, and D. L. Murphy, *Psychopharmacology*, **98**, 403–407 (1989).

424. A. Horita, M. A. Carino, J. Zabawska, and H. Lai, *Peptides*, **10**, 121–124 (1989).

425. A. Horita, M. A. Carino, and J. R. Smith, *Pharmacol. Biochem. Behav.*, **5**, (Suppl. 1), 111–116 (1976).

426. A. Horita, M. A. Carino, and H. Lai, *Fed. Proc.*, **45**, 795 (1986).

427. K. Rolka, E. Oslisok, J. Krupa, M. Kruszinsky, L. Baran, E. Przegalinska, and G. Kupryszewski, *Pol. J. Pharmacol. Pharm.*, **35**, 473–480 (1983).

428. H. Takagi, H. Shiomi, H. Ueda, and H. Amano, *Nature*, **282**, 410–412 (1979).

429. L. S. Godlevsky, A. A. Shandra, I. I. Mikhaleva, R. S. Vastyanov, and A. M. Mazarati, *Brain Res. Bull.*, **37**, 223–226 (1995).

430. T. Arima, Y. Kitamura, T. Nishiya, H. Takagi, and Y. Nomura, *Neurosci. Lett.*, **212**, 1–4 (1996).

431. T. Arima, Y. Kitamura, T. Nishiya, T. Taniguchi, H. Takagi, and Y. Nomura, *Neurochem. Int.*, **30**, 605–611 (1997).

432. J. Y. Summy-Long, V. Bui, S. Gestl, E. Koehler-Stec, H. Liu, M. L. Terrell, and M. Kadekaro, *Brain Res. Bull.*, **45**, 395–403 (1998).

433. T. Zincke, *Ann. Chem.*, **330**, 361–374 (1903).

434. H. Lettré, W. Haede, and E. Ruhbaum, *Ann. Chem.*, **579**, 123–132 (1953).

435. Y. Génisson, C. Marazano, M. Mehmandoust, D. Gnecco, and B. C. Das, *Synlett*, 431–434 (1992).

436. B. M. Wallace and J. S. Lasker, *Science.*, **260**, 912–913 (1993).

437. N. Bodor, A. El-Koussi, M. Kano, and T. Nakamuro, *J. Med. Chem.*, **31**, 100–106 (1988).

438. A. El-Koussi and N. Bodor, *Int. J. Pharm.*, **53**, 189–194 (1989).

439. N. Bodor, *Adv. Drug Deliv. Rev.*, **16**, 21–38 (1995).

440. L. Prokai, W.-M. Wu, G. Somogyi, and N. Bodor, *J. Med. Chem.*, **38**, 2018–2020 (1995).

441. N. Bodor, H. H. Farag, G. Somogyi, W.-M. Wu, M. D. C. Barros, and L. Prokai, *J. Ocul. Pharmacol.*, **13**, 389–403 (1997).

442. H. H. Farag, W.-M. Wu, M. D. C. Barros, G. Somogyi, L. Prokai, and N. Bodor, *Drug Des. Discov.*, **15**, 117–130 (1997).

443. N. Bodor and G. Visor, *Exp. Eye Res.*, **38**, 621–626 (1984).

444. I. K. Reddy and N. Bodor, *J. Pharm. Sci.*, **83**, 450–453 (1994).

445. I. K. Reddy, S. R. Vaithiyalingam, M. A. Khan, and N. S. Bodor, *J. Pharm. Sci.*, **90**, 1026–1033 (2001).

446. A. Simay, L. Prokai, and N. Bodor, *Tetrahedron*, **45**, 4091–4102 (1989).

447. N. Bodor and A. El-Koussi, *Pharm. Res.*, **8**, 1389–1395 (1991).

448. P. Polgar and N. Bodor, *Life Sci.*, **56**, 1207–1213 (1995).

449. J. R. Smith and E. J. Simon, *Proc. Natl. Acad. Sci. USA*, **77**, 281–284 (1980).

450. D. B. Joseph and J. M. Bidlak, *Eur. J. Pharmacol.*, **267**, 1–6 (1996).

451. M. Shahrestanifar, W. W. Wang, and R. D. Howells, *J. Biol. Chem.*, **271**, 5505–5512 (1996).

452. K. Kanematsu, R. Naito, Y. Shimohigashi, M. Ohno, T. Ogasawara, M. Kurono, and K. Yagi, *Chem. Pharm. Bull.*, **38**, 1438–1440 (1990).

453. M. Saah, W.-M. Wu, K. Eberst, E. Marvanyos, and N. Bodor, *J. Pharm. Sci.*, **85**, 496–504 (1996).

CHAPTER SIXTEEN

Drug Discovery: The Role of Toxicology

JAMES B. MOE
JAMES M. MCKIM, JR.
Pharmacia Corporation
Kalamazoo, Michigan

Contents

Burger's Medicinal Chemistry and Drug Discovery
Sixth Edition, Volume 2: Drug Development
Edited by Donald J. Abraham
ISBN 0-471-37028-2   © 2003 John Wiley & Sons, Inc.

609

1 INTRODUCTION

1.1 History

Before the 1980s, toxicologists' principal role in the pharmaceutical industry was the assessment of the safety of drug candidates that had already entered the formal drug development process. In this role, they generated and interpreted data to provide an essential risk-benefit analysis that related to the predicted safety of drug candidates when administered to the target human or veterinary patient populations. Data generated from animals treated with the drug candidate were those drawn from antemortem and postmortem observations and laboratory analyses including hematology, serum chemistry analysis, urinalysis, gross pathology, and histopathology. *In vitro* testing was largely limited to the Ames test or other genetic toxicity assays. The scientists who performed these tests were mainly general toxicologists, pathologists, and genetic toxicologists.

With the scientific advances that were introduced to target highly specific tissue receptors and pharmacological targets, the toxicologists' role, of necessity, expanded from a largely empiric approach to one of investigating the adverse effects linked to the agonism or antagonism of these specific targets. The scientific specialties involved in toxicology also expanded to include molecular toxicology, safety pharmacology, cell biology, and immunopathology. As the field of toxicology expanded, it became apparent that this diversity of scientific skills could contribute to candidate selection and optimization in the drug discovery effort, providing valuable information that chemists could use in establishing structure-activity relationships (SAR) for new classes of compounds. This information was useful in rational drug design, and success rates improved as these candidates were moved into development. As a result of these trends, whereas toxicologists continued their contributions in traditional risk assessment and problem solving, there was increased emphasis on predicting and preventing toxicity in drug discovery and development.

In this chapter, the contributions of state-of-the-art toxicology to the discovery process will be described. It is not the intent of this chapter to attempt to cover the entire field of toxicology; that is well covered in standard textbooks (1, 2). The role of toxicology in drug development and regulatory-driven risk assessment is well established, directed by guidelines such as those published from the International Conferences on Harmonization (e.g., Proceedings of The Fourth International Conference On Harmonization) and will not be covered to any great extent.

1.2 Overview of the Drug Discovery and Development Process

Modern drug discovery starts with the identification of a pharmacologic target that is hypothetically the primary cause of disease (3). Potential targets include host cell genes, receptors, signaling systems, organelles, and biochemicals such as enzymes (4, 5). Additionally, an element of a disease modifying process, such as an inflammatory mediator, may be a target. Biological processes required for propagation of infectious agents have also proven to be therapeutically useful targets; examples include protease and reverse transcriptase of the human immunodeficiency virus (HIV). Common to all targets selected as therapeutic opportunities is the hypothesis that some type of pathogenetic linkage exists to the disease-causing process, rather than to specific signs, symptoms, or effects. Validation of the therapeutic usefulness of targets is achieved through study of *in vitro* systems, animal models of disease, or clinical trials. Selection of potential lead compounds that interact with validated targets is often achieved through application of high throughput screening (HTS), which facilitates rapid screening of thousands of potential lead compounds using very small quantities of test substance (6). Through HTS, a compound library may yield a subset of leads that bind or otherwise react with the target, indicating therapeutic potential. This subset, which may still include hundreds of leads, is subjected to a process of lead selection and optimization to arrive at a smaller group of perhaps 10 leads that have the best attributes in terms of specificity and potency of interactions with the target. This smaller group of leads can be tested in a number of *in vitro* metabolism, pharmaceutics, and toxicology tests to determine their relative probability of success in later development. At this point, limited, single dose *in vivo* metabolism, bioavailability, and toxicity studies are conducted, first in rodents and later in non-rodents. Based on the results of these preliminary *in vivo* studies, more extensive, repeated-dose, range-finding *in vivo* studies are performed to help characterize the relationship of adverse effects to systemic exposure and to identify target organs for expression of toxicity.

If the collective results of the toxicology studies indicate that one or more leads have an appropriate toxicity profile to progress as development candidates, more definitive studies, conducted according to Good Laboratory Practices (GLP) standards, are conducted in rodents and non-rodents to support initial clinical testing in humans. As development proceeds from exposure in healthy human volunteers to proof-of-concept studies for tolerance and efficacy in small numbers of patients (phase I and phase II trials) to large clinical (phase III) trials, nonclinical toxicity testing continues. Depending on the intended clinical indication and duration of treatment, animal chronic toxicity studies as long as 1 year and carcinogenicity studies as long as 2 years in duration may be needed to support approval and registration by governmental regulatory authorities.

1.3 Changing Paradigm in Drug Discovery and Toxicology Problem Prevention versus Problem Solving

Over the past decade, discovery scientists have refined and extended HTS and decreased their reliance on whole-animal (*in vivo*) pharmacology models, which had previously been used to select chemicals as potential candidates for development as therapeutic agents. This has resulted in dramatic reductions in the number of research animals used, as well as in time and human resources required to generate an extensive database related to a specific therapeutic target. To leverage these advances and

more effectively develop pharmaceutical products, the nonclinical development sciences (pharmaceutics/formulation development, metabolism, and toxicology) were required to become involved at an earlier stage in the discovery process.

In recent years, toxicologists have become integral members of discovery teams, participating in the therapeutic target identification, chemical lead selection, candidate optimization, and early product characterization processes (7). In this role, toxicologists make important contributions by identifying toxicity liabilities at earlier stages of the discovery process, providing essential SAR information that medicinal chemists and other discovery scientists can use to validate therapeutic targets and design new molecular templates.

1.4 Role of Toxicology for Improving Speed and Success Rates in Drug Development

In the pharmaceutical industry, more than one-third of the new drug candidates that enter preclinical and clinical evaluations for safety are dropped from development because of unanticipated toxicity. This represents a tremendous loss of time and resources, which results in increased cycle time, increased development costs, and significant delays in the availability of beneficial drugs to clinicians and their patients. Therefore, to reduce the number of failures in late development caused by safety issues, the potential toxic liability of new chemical entities is now being evaluated much earlier in the discovery process.

Toxicologists contribute to speed and success rates in drug discovery and later in the development phases by detecting and characterizing adverse event liability as early as possible through timely application of appropriate predictive methods. Through increased use of *in vitro* technology and other predictive methods, the toxicologist improves selection and optimization of molecular templates as drug candidates. Use of improved predictive methods enables early characterization of potential toxicity before initiation of extensive *in vivo* studies. Use of *in vivo* systems consumes more test compound and requires more time, and should be reserved for candidates that are most likely to progress to development. All efforts should be made to detect adverse effects

Figure 16.1. Generic screening funnel. It is important to develop a basic set of physical-chemical and biological data for new chemical entities early in the discovery funnel. Compounds are screened for activity against a specified therapeutic target. Agents with a high activity toward the desired target are termed "Hits," and these hit chemicals are subjected to a series of *in vitro* screens to evaluate some of the basic parameters that help define a good drug. Compounds with the highest probability of success, based on the results of the *in vitro* screens, are tested in animals to show proof of concept and to obtain basic pharmacokinetic parameters. At this point, there may be one or two compounds remaining, and these must be evaluated for toxic liability in 14-day repeated-dose rodent and non-rodent toxicity studies, followed by a 28-day study conducted under Good Laboratory Practices (GLP).

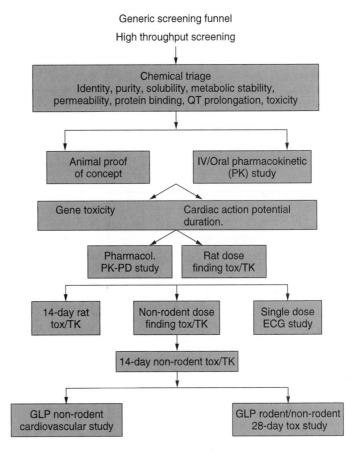

before the drug candidate has entered development, which involves more extensive, regulatory-driven GLP studies in animals or clinical trials in human subjects. One example of a model for achieving these goals is depicted in Fig. 16.1.

2 MECHANISMS OF ADVERSE EFFECTS

The traditional scientific approach to pharmaceutical safety assessment was based largely on the concept of xenobiotic-mediated direct injury to the whole animal, organ, tissue, or cell manifested by death; altered function; or modified growth rates. As scientific knowledge advanced, it became apparent that a variety of mechanisms were involved including direct injury to cellular macromolecules (e.g., DNA, RNA, proteins) or organelles (e.g., mitochondria, endoplasmic reticulum), adaptive responses (e.g., induction or inhibition of cyto-

chrome P450 isoenzymes), or pharmacologically mediated response. The following sections briefly describe some examples of each of these mechanisms. More detailed information is provided in standard toxicology textbooks and current literature.

2.1 Direct Cellular Injury

Toxicity is often discussed in terms of organ systems. Compounds can cause renal or liver damage, pancreatic injury, or neurological effects. The tissue of a specific organ is comprised of cells that have taken on the morphology and phenotype that characterize the organ system. Therefore, a renal toxin must ultimately produce an adverse effect on cells that make up the kidney. Consequently, investigations into the mechanisms of organ toxicity must focus on how chemicals produce adverse effects by perturbing cellular and subcellular processes essential for the survival of the cell.

Identification of subcellular changes help elucidate the mechanism of toxicity, which is considerably more informative in the field of drug development than simply knowing that a compound causes liver or kidney damage.

The integrity of cells is maintained by the cell membrane, the basic structure of which is a lipid bilayer interspersed with large globular proteins. Some of these proteins function as pumps that control osmotically active molecules such as sodium, potassium, and calcium. This osmotic balance is critical to the survival of the cell and hence the tissue to which the cell belongs. The cell membrane surrounds the cytosol, which contains the nucleus, mitochondria, endoplasmic reticulum, golgi apparatus, peroxisomes, and lysosomes. In addition, there are many proteins in the cytosol not contained within organelles that perform a number of important biochemical functions, which are critical to the survival and functionality of the cell. The combined actions of cytosolic enzymes (glycolysis) and mitochondria (citric acid cycle) provide energy in the form of adenosine triphosphate (ATP) used to fuel many cellular functions, including the manufacturing of new proteins through transcription and translation. Some organs, such as brain, muscle, and heart, have cells with additional features that give these tissues their unique functional characteristics. Thus, each tissue type is composed of many cells that use important mechanisms required to sustain health and to define the functionality of the organ.

It is important to remember that all drugs and chemicals will produce adverse effects if the dose is high enough and the exposure long enough. Therefore, the dose determines whether the effect observed will be beneficial or adverse. This concept is important to remember while reading the discussions that follow as many drugs are used as examples of adverse effects. Figure 16.2 is a diagrammatic representation of the many different cellular targets discussed below.

Toxicity may be an extension of the desired pharmacology, or it may involve an entirely different mechanism. New chemicals can be successfully developed as drugs if the dose that produces the desired therapeutic effect is sufficiently smaller than the one that produces toxicity. To develop an understanding of a compound's effective dose range versus its toxic dose range, it is necessary to conduct dose-response experiments that measure the desired and toxic responses over several doses. Once generated, these data can be used to develop numerical values that relate benefit to risk. The therapeutic index (TI) is defined by the ratio of the dose that causes 50% mortality (LD_{50}) to the dose that produces the desired effect in 50% of the animals (ED_{50}). The greater the TI, the safer the drug. The major disadvantage of using LD_{50} and ED_{50} values is that the shape of the dose-response curve is not taken into consideration because only the median point of the dose-response curve is used. One approach used in the pharmaceutical industry to address this deficiency is the ratio of the no observed adverse effect level (NOAEL) to the pharmacologically effective level (ED_{50}). This ratio is sometimes referred to as the margin of safety (MOS). The relationship between the dose-response curves is shown in Fig. 16.3.

2.1.1 Cell Membrane–Associated Toxicity. Cellular toxicity occurs when exogenous chemicals interact with the cell and disrupt the essential biochemical processes that maintain homeostasis. There are many potential sites where exogenous chemicals could disrupt normal cellular function, resulting in cell injury and ultimately cell death. The more essential the function is to the health of the cell, the more severe the toxicity. For example, compounds that directly interact with the cell membrane may perturb its ability to control the movement of ions and maintain the osmotic balance between the interior and exterior of the cell. In addition, intercellular communication may be reduced or blocked by compounds that perturb the physical and chemical properties of the membrane environment.

Membrane-active compounds can elicit adverse effects in three primary ways. First, compounds that are positively charged at physiologic pH may alter membrane surface potential in a manner proportional to their concentration in the membrane. This in turn could have an adverse effect on membrane function. Second, the chemical may absorb

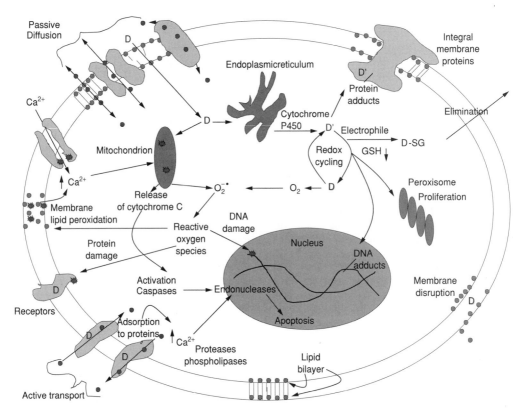

Figure 16.2. Mechanisms of cellular toxicity. Tissues are comprised of cells, and each cell is defined by its cell membrane. The cell membrane is composed of a lipid bilayer, which contains proteins that function as ion channels and receptors. Compounds that disrupt the membrane environment can directly or indirectly alter the normal function of these proteins. In each cell, there are numerous subcellular organelles, all of which are potential targets for toxicity. Cytochrome P450 enzymes in the endoplasmic reticulum may metabolize drugs that enter the cell. Metabolism has one of two effects on the drug's potential toxicity: (1) it may reduce toxicity by eliminating parent compound, or (2) it may increase toxicity by generating a reactive (electrophilic) metabolite. Drugs may inhibit critical functions in mitochondria or damage DNA in the nucleus, which can lead to cell death by apoptosis or necrosis.

into membrane proteins, thereby disrupting their normal function. Third, the compounds may incorporate in the lipid phase of the membrane, resulting in pronounced effects on membrane dynamics and structure. A characteristic of many β-adrenergic blocking drugs (exaprolol, alprenolol, propranolol, and metipranolol) is their ability to nonspecifically disrupt many membrane functions simultaneously. It is likely that these agents are partitioning in the lipid phase and disrupting membrane structure, which in turn disrupts the membrane proteins associated with the effects observed (8). The antitumor drug valino-

mycin increases membrane permeability to potassium (9), which results in cell death. The antifungal agent amphotericin B is highly lipophilic and easily absorbs into membrane lipids where it interacts with membrane sterols (cholesterol in mammals and ergosterol in fungi) to create channels in the membrane, which allows small molecules to be lost, eventually leading to cell death (10). Membrane toxicity associated with compounds such as those discussed above can be reduced by preventing the compound from interacting with the cell membrane. One way of achieving this is to package the membrane-active drug into

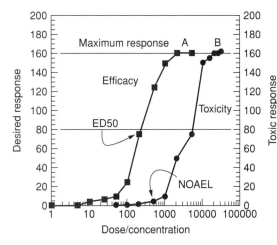

Figure 16.3. Dose-response data. Potential drugs are evaluated for efficacy, potency, and toxicity by examining the relationship between dose and response. The response may be the desired effect (efficacy curves) or a measure of adverse effects (toxicity curves). In most cases the amount of drug required to produce the desired effect is lower than the amount required to produce toxicity. Compound A is more potent than B. NOAEL, no-observed-adverse effect level.

membrane-vesicles called liposomes. This prevents membrane interaction, maintains efficacy, and significantly reduces toxicity. It should be clear from the preceding discussion that many drugs and chemicals have the ability to interact with and disrupt the function of cell membranes, which can lead to cell death.

2.1.2 Subcellular Targets of Toxicity. Several drugs and chemicals elicit their adverse effects by interacting with subcellular targets. Once in the intracellular milieu, there are many potential sites for a drug or chemical to interact with to produce toxicity. Macromolecules such as proteins, lipids, and nucleic acids are all potential targets. Subcellular organelles such as the mitochondria, lysosomes, endoplasmic reticulum, and nucleus are also potential sites of toxicity. One of the most essential processes in terms of cell health is cellular respiration and the production of ATP. Drugs or chemicals that inhibit mitochondrial oxidative phosphorylation or β-oxidation of fatty acids and subsequently the production of ATP and the ratio of nicotinamide-adenine dinucleotide phosphate to its reduced form (NAD/NADH)-reducing equivalents cause cell death in a relatively short period of time. The aminoglycoside antibiotics, gentamicin, neomycin, kanamycin, and streptomycin, inhibit mitochondrial respiration, resulting in renal and hepatic toxicity (11). Rotenone, oligomycin, and antimycin A are examples of compounds that inhibit specific complexes in oxidative phosphorylation, resulting in cell death.

2.1.3 Proteins as Targets for Toxicity. Cytotoxicity can result from the parent or unmodified drug or chemical or a reactive metabolite formed *in vivo*. For an excellent review on metabolism and the formation of reactive intermediates, the reader is referred to Ref. 12. Reactive intermediates, e.g., electrophiles, can undergo covalent protein binding. Although it is possible that some protein binding merely reflects the scavenging of electrophilic intermediates, recent evidence suggests that covalent protein binding can define the mechanism of toxicity. The formation of protein adducts can result in at least three outcomes: (1) altered protein function, (2) altered immune recognition, and (3) redistribution of proteins and enzyme inhibition (13). It is now clear that covalent protein binding can determine the target organ of toxicity. The application of immunological detection systems to this issue has provided the data to support this conclusion. Diclofenac is a nonsteroidal anti-inflammatory drug that produces hepatotoxicity in a small number of humans (~15%). Although the exact mechanisms underlying diclofenac hepatotoxicity are unclear, evidence indicates that covalent protein binding could explain the observed toxicity (14).

When administered at high doses, the analgesic acetaminophen (APAP) is hepatotoxic and has been shown to covalently modify a select group of proteins. The modified proteins include microsomal glutamine synthase, mitochondrial glutamate dehydrogenase, aldehyde dehydrogenase, carbamyl phosphate synthetase I, cytosolic selenium-binding protein, and nuclear lamin A. Inhibition of these enzymes is important in elucidating the mechanisms of APAP toxicity. To develop a causal relationship between the formation of specific protein adducts and an observed toxicity, it is necessary to purify, sequence, and identify the adduct proteins and then show that the observed toxicity is a result of alteration of the enzyme (14–18).

2.1.4 Nucleic Acids as Targets for Toxicity.

Many compounds and their metabolites interact with nucleic acids to form covalent adducts or noncovalent associations, such as intercalation, that can have significant effects on cell health. Covalent modification of DNA bases occurs when molecules with electrophilic properties react with the nucleotides in nucleophilic centers. Noncovalent DNA modifications occur when molecules intercalate into the DNA strand. There are more than 15 potential nucleophilic sites where electrophilic compounds can form adducts with nucleotide bases. In addition, the phosphorous in the phosphodiester bond is also susceptible to modification. There are two basic classes of electrophiles that interact with nucleophilic centers in DNA. The first group is characterized by the presence of an atom carrying a partial or full positive charge. The second group carries one or more unpaired electrons in its outer orbital. Group 1 electrophiles are formed metabolically, usually by cytochrome P450 enzymes, whereas group 2 electrophiles are free radicals formed by homolytic fission of a covalent bond or by accepting or losing an electron.

Examples of compounds that form covalent bonds with DNA bases include diethylnitrosamine, which forms O^4 adducts with deoxyadenosine, and O^6 and N^7 adducts with deoxyguanine and aromatic amines, which react with DNA to form O^6 and N^2 adducts with deoxyguanine. If these mutations occur in genes that control cell division, they can lead to cell transformation and ultimately the de-

velopment of cancer. Compounds that have noncovalent interactions include actinomycin D and mitamycin, which intercalate into the DNA strand. These interactions profoundly affect DNA replication and transcriptional processes (19–21).

Compounds that form DNA adducts or DNA association complexes can cause errors in DNA replication, leading to mutations and inhibit the transcription of genes. DNA adducts and DNA-drug complexes may be harmless in nonreplicating cells and lethal in cells undergoing division. The cells within tumors are characterized by uncontrolled cell replication. Many antitumor drugs are designed to kill replicating cells by alkylating DNA. Unfortunately, other dividing cell populations, such as the cells of bone marrow, are also susceptible to these drugs. It is important to remember that in addition to direct chemical-mediated alkylation of DNA, many compounds induce the formation of reactive oxygen species such as hydroxyl radical, superoxide anion, and hydrogen peroxide. These highly reactive oxygen species can also modify DNA by forming 8-oxodeoxyygaunosine, which as been associated with aging and carcinogenesis (22). Several endogenous and exogenous compounds undergo metabolism to reactive intermediates known as quinones. Quinones formed from benzene, aromatic hydrocarbons, estrogens, and catecholamines alkylate DNA or proteins directly or undergo redox cycling to form reactive oxygen species (22).

2.1.5 Lipids as Targets for Toxicity.

Lipids are a diverse group of biomolecules present in eukaryotic organisms that are characterized by being water-insoluble but highly soluble in organic solvents. There are many different lipid molecules in eukaryotic organisms. These include phospholipids, cholesterol, glycolipids, sphingolipids, glycerol, triacylglycerol, diacylglycerol, and free fatty acids. In addition, there are specialized lipids used for transport known collectively as lipoproteins and include low density lipoproteins (LDL), high density lipoproteins (HDL), and chylomicrons. There are several bioactive lipids, including diacylglycerol, phosphatidic acid, lysolipids, arachodonic acid, and ceramide (23).

These unique molecules have many important biological roles. They can serve as a source of energy, signal molecules, and components of various cell membranes. The primary lipids in biological membranes are phospholipids, cholesterol, and glycolipids (24).

Phospholipids can originate from glycerol or from sphingosine. Phosphoglycerides are phospholipids made from glycerol. They are composed of a three-carbon glycerol backbone, two fatty acid chains (usually 16 or 18 carbons in length), and a phosphorylated alcohol. The balance of lipids *in vivo* is essential to normal development and general health. This is evident by the lethality of lipid storage diseases and disease processes such as atherosclerosis (24).

A disruption of lipid structure or composition caused by drugs and chemicals can have significant adverse effects on cell function and overall homeostasis. As discussed above, many compounds indirectly alter lipid structures through noncovalent interactions. Hydrophobic drugs absorb into the lipids of cells and indirectly disrupt membrane permeability, the function of transport proteins, and membrane potential (25). The metabolism of many drugs and chemicals result in the formation of reactive intermediates. An example of this can be seen with quinone compounds, which undergo redox cycling with their semiquinone radicals, leading to the formation of reactive oxygen species and the initiation of lipid peroxidation. The degradation of lipids through lipid peroxidation is a significant mechanism of lipid-mediated toxicity (26). Lipid peroxidation causes the breakdown of critical lipids in the cell membrane and the subsequent release bioactive molecules such as lipid hydroperoxides, chain-cleavage products, and polymeric materials (27). The important role that lipids play in maintaining the health of cells and organs is evident by the fact that the oxidation of lipids has been implicated in the pathogenesis of atherothrombosis, the development of certain types of cancer, aging, and acute cytotoxicity (27–32).

2.2 Adaptive Response

Changes in organ weight or histopathological appearance, regarded as toxic effects, often represent an adaptive response in structure or function of the cell to the test substance. Accurate interpretation of the nature and cause of structural and functional changes is essential in assessing early drug candidates for potential toxicity in later preclinical studies and safety in clinical trials. Electron microscopic examination of tissues is useful in differentiating injury to the cell from adaptive change and in correlating structural changes in the subcellular organelles (peroxisomes, endoplasmic reticulum, and mitochondria) with functional adaptation to xenobiotics or physiological demand (33).

There are several examples of compounds that can produce changes in rodents that are not relevant to anticipated effects in humans. The antiseizure drug, phenobarbital, produces a well characterized sequence of biochemical and morphological changes in rats, which includes liver and thyroid enlargement and a pleiotrophic expression of cytochrome P450 enzymes (CYP2B1/2B2, CYP3A1/2, epoxide hydrolase, and uridine diphosphate (UDP)-glucuronosyltransferase) (34). Enlargement of these organs is caused by increased numbers of cells (hyperplasia) and increased volume of individual cells (hypertrophy). Hypertrophy of liver cells is at least partially attributable to proliferation of endoplasmic reticulum. In addition, phenobarbital exposure causes hepatic and thyroid tumors in 2-year rodent carcinogenicity studies (34–36). Investigations into the mechanisms underlying the rodent biochemical changes revealed that thyroid tumors are the result of an adaptive response to an induction in hepatic metabolizing enzymes that reduce circulating thyroid hormone levels. This causes a compensatory increase in thyroid stimulating hormone (TSH), which hyperstimulates the thyroid to increase the release of thyroxin (T4/T3). The hyperstimulation of the thyroid causes thyroid hypertrophy and hyperplasia and ultimately thyroid neoplasia. This adaptive response is the underlying cause of thyroid tumors in 2-year rodent studies (37). This sequence of events has not been documented to occur in humans. Although phenobarbital clearly increases the incidence of hepatic tumors in rodents, a correlation between phenobarbital and hepatic tumors has not been shown in patients with epilepsy who were prescribed phenobarbital for the treatment of seizures associated with this disease (38–42). The research that identified the disconnect between rodent effects and anticipated effects in humans has allowed several drugs and chemicals described as phenobarbital-like (loratadine, oxazepam, doxylamine succinate, and phenobarbital), with respect to the biochemical and morphological changes, to be approved for consumer use (37, 43–47).

In the middle to late 1980s, there were a number of reports of compounds with low systemic toxicity producing nephropathy and renal tumors in male rats (48). The renal lesions were characterized by an increase in a protein known as $\alpha 2\mu$-globulin. Compounds that produce this syndrome are nongenotoxic, produce renal tumors in male rats but not in female rats, and are associated with hyaline droplets forming in the proximal segment of the renal tubule. Tumors typically develop after repeated exposure to high concentrations of compounds. The synthesis of $\alpha 2\mu$-globulin is controlled by androgen and occurs exclusively in the liver of male rats. Under normal circumstances, $\alpha 2\mu$-globulin is excreted from the liver, filtered by the glomerulus in the kidney, reabsorbed in the proximal tubule, and then degraded by lysosomal enzymes. It is now known that exposure to some chemicals can cause binding of small hydrocarbon molecules to $\alpha 2\mu$-globulin. The $\alpha 2\mu$-globulin-hydrocarbon structure can still be filtered by the glomerulus and is absorbed in the proximal tubular cells; however, it cannot be degraded by lysosomal proteases. Consequently, it accumulates in the kidney and eventually takes on the form of hyaline droplets. The presence of these hyaline droplets stimulates cellular hyperplasia, which over time acts as a tumor promoter. This sequence of events has been determined to be male-rat specific with no relevance to safety in humans (49, 50).

Peroxisome proliferation (PP) occurs in rodents treated with agents that modify lipid metabolism through interaction with the peroxisome proliferator activator receptor (PPAR)-α in the nucleus of cells (51). When treated for long periods of time with drugs that cause PP, rodents commonly develop liver tumors, whereas a similar association

has not been established in humans. The basis for the apparent difference in susceptibility to hepatic tumorigenesis is not understood, but may relate to the greater expression of PPAR-α in the liver of rodents compared with humans (52).

The endoplasmic reticulum (ER) contributes to a variety of functions, including protein synthesis and detoxification of xenobiotics, and disturbance of these functions can be linked to cell injury (53). Prolonged exposure to ethanol induces proliferation of the ER of in the liver and enhanced activity of cytochrome P450 enzymes, which in turn can result in production of hepatotoxic or carcinogenic metabolites (54).

Mitochondria are involved in key functions essential to cell survival, notably those related to energy supply (55). Size and number of mitochondria increase with physiological demand in response to a variety of stimuli, including increased contractile activity of skeletal muscle (56). Injury from exposure to xenobiotics can result in structural changes (swelling, increase in number) in mitochondria that share some morphological features with those resulting from increased physiological demand (57).

2.3 Pharmacologically Mediated Effects

Pharmacologically mediated adverse effects can be placed into one of three categories: (1) primary pharmacodynamic (PD) effects, (2) secondary PD effects, and (3) those either not linked to the primary PD effect or caused by an unknown mechanism. Primary pharmacodynamic adverse effects are those which represent an overexpression of the primary, intended pharmacological effect of the xenobiotic. An example of an adverse primary PD effect occurs when an overdose of insulin, a critical hormone involved in the homeostasis of blood glucose, is administered. A deficiency of insulin results in diabetes mellitus and is expressed by hyperglycemia. Overdosing of insulin results in hypoglycemia, which can have immediate and severe clinical consequences.

Adverse secondary PD effects are those caused by a recognized pharmacological effect that is different from, but physiologically linked to, the primary, or desired PD effect. An example of this is hyperprolactinemia, which is associated with psychotherapeutic agents that antagonize the activity of dopamine-2 (D2) receptors in the brain (58). When D2 antagonists are administered to rodents in long-term toxicity studies, prolactin levels in the plasma are increased, which in turn stimulate cellular proliferation in the mammary gland and increase the incidence of mammary tumors. The directly causal relationship of hyperprolactinemia with breast cancer in rodents has not been established in humans (59).

Adverse pharmacological effects not recognized as being mechanistically linked to the primary PD effect occur with many therapeutic agents. Several categories of noncardiac drugs, including antipsychotics and antidepressants, cause alteration of cardiac electrophysiology, expressed as prolongation of the QT interval of the QRST wave complex in electrocardiograms of human patients and animals (60). Prolongation of the QT interval has been associated with increased incidence of life-threatening irregularities in cardiac function. Because these adverse effects can be separated from the primary PD effect, early mechanistic investigations and judicious screening facilitate SAR-driven chemistry, which reduces the likelihood of these liabilities appearing later in nonclinical studies or clinical trials.

3 *IN VIVO* TOXICITY TESTING

Contemporary scientific and regulatory standards require the use of well-validated animal models for fully defining the toxicity profile of drug products. Traditionally, the selection of new therapeutic entities was achieved mainly through tests in animals, limiting the numbers of compounds that could be tested, consuming excessive resources, and returning results relatively slowly. Although the emerging *in vitro* methods described in this chapter for selecting drug candidates and characterizing their toxicity increase the efficiency of these processes, *in vivo* testing is still a valuable and widely used tool in drug discovery and early development. *In vivo* determination of the adverse effect liability of new therapeutic entities is achieved through collaboration with

discovery and metabolism scientists in conducting serum chemistry, hematology, histopathology, and pharmacokinetic examinations on animals used in primary pharmacologic studies. These examinations provide valuable preliminary insight by relating primary pharmacodynamic effects and potential toxic liabilities to systemic exposure of the drug candidate and its metabolites.

A guiding principle in use of *in vivo* testing during the early stages of drug discovery and development is to obtain as much information as early as possible, using the following: an appropriate species, fewest possible animals, and smallest possible amount of test substance. Important factors in the selection of test animal species include scientific considerations such as pharmacologic responsiveness to the test compound, metabolic profile (often suggested by prior *in vitro* testing), bioavailability, and pharmacokinetics. Practical factors such as body size, cost, ease of handling, and availability must also be taken into consideration. In the lead optimization and early toxicity characterization phases of drug discovery, when quantities of test compound are often limited, mice or other small rodents are the species of choice. In designing the studies, dose levels and duration of dosing are key considerations. Acute testing, involving administrations of various, single, ascending doses of test compound to small groups of animals, is often adequate to compare the general tolerability of a set of early drug candidates and can provide general information regarding the toxicity profile, especially if adverse events include convulsions or other profound clinical changes. To achieve these goals, it is necessary to achieve adequate systemic exposure, approximating and exceeding the concentrations that result in primary pharmacodynamic effects. Simple formulations, such as aqueous solutions or suspensions, are administered by the route that is most direct and well tolerated and achieves the desired plasma concentrations for exposure. Parenteral (intravenous, intraperitoneal, or subcutaneous) injection may be the preferable route of administration even if the ultimate goal is an orally administered product.

Information gained from early *in vivo* studies is essential in determining the doses to be used in more definitive follow-up studies. Critical information gained from the first, or early follow-up, *in vivo* studies include an indication of the maximum tolerated dose (MTD), principal target organs for expression of toxicity, and hematologic or other laboratory changes that might serve as biomarkers of toxicity. Whenever possible, plasma, serum, and urine should be collected from test animals for bioanalysis and correlation of adverse effects with systemic exposure. From correlation of toxicity with systemic exposure, general estimates of NOAEL and lowest observed effect level (LOEL) can be made. Because the scope and design of these early studies are limited, risk assessment to support early trials in human subjects must be based on results from the longer-term and more definitive toxicology studies (i.e., carried out according to GLP), which are key components of registration application documents.

4   EMERGING TECHNOLOGIES FOR ASSESSING TOXICITY

Application of emerging technologies allows the optimization of potency and efficacy while monitoring potential adverse interactions of new chemicals with biological systems. In addition, it is possible to develop structure activity versus structure toxicity relationships, develop *in vitro* margins of safety, estimate a therapeutic index, and use *in vitro* data to estimate *in vivo* toxicity.

To provide meaningful toxicity data to discovery chemists during the early phase of lead identification; the test systems must be robust and allow data to be delivered in a timely manner so that the information can be incorporated into the decision-making process. At this early time in discovery, compound availability is usually limited, which means that the *in vitro* systems employed must have low compound requirements. Moreover, these test systems must be amenable to HTS formats, such as 96- or 384-well cell culture, and technically simple for rapid turnaround and successful integration with robotics platforms.

New technologies with applications to *in vitro* toxicity screening are being developed and marketed at an astounding rate. The dis-

cussions that follow highlight some of the more significant technology platforms.

4.1   *In Silico* Toxicology

In recent years, there has been an increased interest, by regulatory agencies, pharmaceutical companies, and chemical companies, in the development and use of computer-aided computational methods to predict toxicity. The objective has been the development of software systems that can predict the potential toxicity of a new compound based on compound structure characteristics and historical information stored in government and industrial databases. This approach has been fueled by a desire to reduce animal usage, reduce the cost of drug development, evaluate structure-related toxicity in a "virtual" environment even before a new compound has been synthesized, and improve the downstream success of new drugs. Thus, the term *in silico* toxicology can be defined as the "application of computer technology and information processing to analyze, model, and estimate chemical toxicity based on SAR." Additional terms used to describe the same process include e-TOX and Com Tox (61).

In silico approaches focus on many important endpoints of toxicity; however, the ability to predict carcinogenic or mutagenic properties has been the primary focus of initial efforts. This was most likely because a large amount of toxicity data for a wide range of chemicals was readily available in government and corporate databases. In addition, these endpoints have a high level of importance assigned to them by the regulatory agencies and the public.

The e-Tox programs currently available were developed based on one of two approaches: (*1*) expert systems (ES) based on rules or toxicity and (*2*) ES based on statistical or correlative methods. Both approaches attempt to evaluate the chemical itself, the biological activity associated with the structure, and the algorithm that describes the relationship between the two. In general, ES consist of data and rule acquisition, a reasoning system, and a rule generator. Thus, an ES can reason with expert knowledge, explain its reasoning, and integrate new knowledge into an existing database (62).

Two examples of ES rules-based systems that are commercially available include Deductive Estimation of Risk from Existing Knowledge (DEREK) and Oncologic. Both systems make predictions based on a series of rules contained in their database. These rules were developed by experts who reviewed the toxicity associated with single chemicals or groups of chemicals with similar structures (e.g., congeneric chemicals) to develop structure toxicity relationships (STR). The systems have been expanded to include known or hypothetical mechanisms of toxicity associated with the congeneric chemicals. Therefore, DEREK contains rules, established by experts, for predicting chemical toxicity (carcinogenicity) for a panel of toxicological endpoints. The system can also provide estimates on metabolite formation and mechanisms associated with the predicted toxicity. DEREK makes qualitative, not quantitative, predictions based on structural properties known to produce toxicity. Although the system was developed around carcinogenic compounds, it has been expanded to include mutagenicity, skin sensitization, irritation, reproductive toxicity, and neurotoxicity. In comparison, Oncologic only contains rules for estimating carcinogenic potential (61–66).

In contrast to the rules-based approach, statistical/correlative systems such as Toxicity Prediction by Komputer Assisted Technology (TOPKAT), Computer Automated Structure Evaluation (CASE), and MULTICASE, an improved version of CASE, depend on complex algorithms, based on existing knowledge obtained from a large noncongeneric or heterogeneous group of chemicals, to build the STR or SAR used for predicting toxicity (66–73).

4.2 Profiling Gene Expression, Proteins, and Metabolites

The terminology used in the "Omics" technologies may be confusing at first glance and therefore requires some explanation.

In general, the genome refers to the cell's DNA, DNA sequence, and chromatin organization. Transcriptome describes the cells complement of mRNA produced by the transcription of DNA sequences or genes, and the proteome is all cellular protein translated from the mRNA. The term metabonome refers to endogenous metabolites of cellular processes in tissues, blood, bile, or urine.

4.2.1 Genomics. Significant advances in DNA sequencing and automation technologies have allowed scientists to embark on projects aimed at obtaining the entire genomic sequence of various species including humans. These endeavors gave rise to the field of genomics, which refers to research aimed at characterizing the DNA sequence and gene structure of a particular organism. It is believed that this new "gene" knowledge may allow scientists to predict an individual's risk of developing certain diseases or how they will respond to drug therapy (pharmacogenomics/pharmacogenetics) (74) or environmental chemical exposure.

Efforts in genomics have given rise to new technology platforms that allow scientists to evaluate and monitor gene transcription, translation, and functionality. Functional genomics describes the technologies employed to evaluate the biological significance of specific genes. Evaluation of gene expression and the effects of chemicals on gene expression profiles in the past have focused either on one gene or on a small family of genes. Examples would include the cytokines, cytochrome P450s, and endocrine receptors. Some of the technologies available for studying the effects of new drugs or chemicals on these genes included Northern blots and PCR/differential display for mRNA levels, Western blots for protein levels, and biochemical assays for enzyme activity. New analytical technologies resulting from the genomics effort, such as DNA microarrays, have allowed for studying the expression profile of the entire genome rather than studying the expression profile of a single gene or a small set of genes.

DNA microarrays can be used in two general ways: (1) a DNA-based comparison of genomic content and (2) the RNA-based comparison of gene expression. This methodology enables the expression patterns of literally thousands of genes to be analyzed in a single experiment. DNA microarrays are prepared by immobilizing the gene sequences of interest on a solid support (e.g., glass or nylon membrane). Typically this is accomplished using cDNA or oligonucleotides containing the

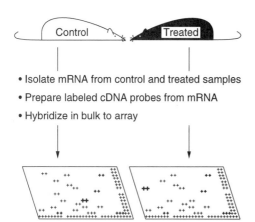

• Isolate mRNA from control and treated samples
• Prepare labeled cDNA probes from mRNA
• Hybridize in bulk to array

Figure 16.4. Analysis of multiple genes with DNA microarrays.

coding sequence for each gene of interest. Changes in gene expression levels (suppression or induction) are determined by extracting mRNA from cells or tissues exposed to various new chemical entities. The extracted mRNA is labeled and hybridized to the immobilized gene sequences. Quantification is accomplished with a phosphoimager equipped with software that allows the operator to evaluate a large amount of data in a timely manner (Fig. 16.4).

The field of genomics and the advent of technologies like DNA microarrays have already made a significant impact on the pharmaceutical industry. Over the past 3 decades, the use of antibiotics has increased. Unfortunately, the resistance of microorganisms to the current battery of drugs has also increased. To keep up with the rate that current drugs are becoming ineffective, new targets must be identified and this information used to develop novel classes of drugs. The old process of identifying new anti-infectives involved testing natural products or derivatives of known compounds *in vitro*. This approach has not enabled pharmaceutical companies to develop novel classes of antibiotics in a time frame that can keep up with the increasing number of resistant microbes and the subsequent demand for new drugs. Improved DNA sequencing technologies developed in the field of genomics has led to the genomic sequencing of more than 20 bacterial strains. With this information, it is possible to identify bacterial

genes essential to the microbe. Once identified, these become new targets for drug design and may lead to a new generation of structurally unique antimicrobial drugs (75–78). This is an example of how functional genomics can be integrated into the drug discovery process.

Toxicogenomics is the application of genomic-based technologies to the field of toxicology. By comparing gene and protein expression profiles obtained from test systems exposed to new chemical entities with profiles obtained from controls, it is possible to identify gene/protein expression profiles that are predictive of subcellular targets and organ-specific toxicity. This information can then be used to develop *in vitro* screens for specific types of toxicity. A current limitation of this approach in toxicology is the accurate interpretation of extremely large quantities of expression data. How much induction is adverse? How much suppression is adverse? Which genes in fact lead to adverse effects in the cell? The answers to these questions may be obtained as research continues; however, an immediate application of this technology might be to identify the mechanisms of indirect toxicity. This would allow the analysis to focus on a smaller group of genes that respond to chemical insult and ultimately lead to cytotoxicity (79, 80). Another approach would be to evaluate a large number of compounds with similar mechanisms of toxicity in the gene expression assays and look for similar patterns of expression. Recent studies (81) have shown that this relationship between mechanism of toxicity and gene expression profiles may indeed be correct.

All cells in an organism carry the same complement of DNA (genome), and yet cells from different tissues have dramatically different functions. The genes expressed in the neuronal cells of the brain are not the same as those expressed in the liver. Therefore, the genome carries information about what a cell might be. In comparison, the functionality of a cell is defined by its complement of proteins (proteome). Proteins represent what the cell actually is in terms of identity and function. Thus, although a great deal of information can be obtained from gene expression profiling, the functional result of those changes cannot be known without evaluating the protein products of those genes.

4.2.2 Proteomics. Proteomics is the technology that allows the evaluation of protein expression profiles. The methodology of proteomics can be broken down into four general phases: separation, detection, analysis, and identification. The separation of proteins is done by two-dimensional gel electrophoresis (2D-gels). This involves the separation of proteins first by their isoelectric point (pI) and then by their molecular weight. The separated proteins are visualized with stains such as Coomasie Brilliant Blue or silver stains. This technology has been used for many years, and the details of running 2D-gels can be found in any number of excellent references (82). The 2D-gels are evaluated by image analysis to analyze the effect of time and treatment on the protein expression profiles. The image analysis software enables multiple gels to be compared and a differential expression pattern to be obtained.

An ideal approach to unknown chemicals is to evaluate alterations in gene expression and then correlate these to changes in the relative abundance or function of the proteins coded by those same genes.

4.2.3 Metabonomics. Metabonomics is similar to genomics and proteomics in that it identifies changes in the profiles of small endogenous molecules that are the normal products of various metabolic pathways. Application to the field of toxicology is based on the premise that a chemical insult disrupts normal metabolic pathways, resulting in a change in the metabolite profiles measured in blood, urine, and other biofluids. The goal is to gain an overall understanding of how a chemical affects an organism, and in some instances, to predict the site of *in vivo* toxicity (e.g., liver or kidney). Biofluids are collected from control and treated animals, and the metabolite profile analyzed by nuclear magnetic resonance (NMR). The results are processed using pattern recognition algorithms to evaluate differences between control and treated patterns (83–86).

4.2.4 Branched DNA Analysis of Gene Expression. Branched DNA (bDNA; Quantagene Technology, Emeryville, CA) is a method for the quantitation of gene expression (mRNA) that achieves sensitivity by amplifying the detection signal not the target itself. This maintains the relative abundance of target sequences to other gene transcripts in the transcriptome. The system is unique in that it does not require extensive RNA isolations, gels, or autoradiography, and the assay can achieve sensitivities similar to those obtained with PCR. The assay can be performed in 96-well cell culture plates, and multiple targets can be assessed in the same 96-well plate. The bDNA assay is performed by preparing multiple oligonucleotide probes per target gene of interest. Two types of oligonucleotide probes are prepared for each target of interest. These are known as capture extenders (CEs) and label extenders (LEs). One section of the CE probe consists of a sequence complementary to the target mRNA, whereas another section contains a sequence complimentary to a synthetic nucleotide attached to the bottom of the Quantagene 96-well plate. Cells are lysed to release total RNA, and an aliquot of this is transferred to the Quantagene plate. The CE probes are added to capture the mRNA of interest. The probe-mRNA complex binds to the nucleotide sequence attached to the plate. The well is washed to remove all unbound mRNA. The mRNA of interest (target) remains in the plate bound by the capture extender probes. The second set of probes, known as the LE probes, is added to the plate. This probe has a section with a sequence complementary to the target and a section with a sequence complementary to a nucleotide on which is built the signal amplification branched oligonucleotide DNA molecules (bDNA amplifier). Label probes conjugated with alkaline phosphatase are added and allowed to hybridize to the bDNA amplifier complex. A chemiluminescence substrate is added for detection (Fig. 16.5). A software system supplied by Quantagene called Probe Designer allows the design of CEs and LEs that have minimal cross-reactions with other mRNAs (87, 88).

Step 1: Lyse cells to access target mRNA and hybridize LE and CE probes

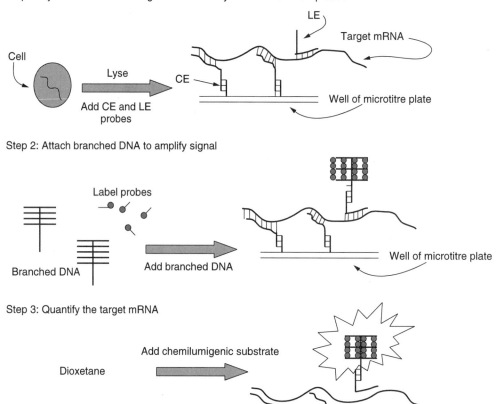

Step 2: Attach branched DNA to amplify signal

Step 3: Quantify the target mRNA

Figure 16.5. Branched DNA (bDNA) methodology.

4.3 *In Vitro* Cell-Based Toxicity Screening

4.3.1 Multi-Endpoint Analysis. Until recently, *in vitro* evaluations were typically done
using primary cell cultures and a limited number of doses to monitor changes in one or two
endpoints. The new paradigm is to evaluate
toxicity much earlier in the discovery process.
To be useful to discovery teams, these evaluations must be done with a small amount of
compound, require no more than 1 or 2 weeks
to turn data around, provide information on a
new compound's toxicity relative to other
chemicals in class, enable the prioritization of
new leads based on some parameter of toxicity
(e.g., the concentration that produces a half-

maximal response or TC_{50}), allow the development of STRs, provide information on mechanism and subcellular targets, and provide
information on *in vivo* toxicity.

Achieving these many objectives has required not one, but several, biochemical endpoints capable of evaluating multiple biochemical processes which are essential to the
health of the cell. The Tox-Cluster Assay System (89) incorporates assays that monitor mitochondrial function, mitogenesis, energy
status, cell death (necrotic/apoptotic), and oxidative stress with an immortalized cell line
derived from rat liver. These endpoints are
monitored over exposure concentrations that
range from 0.1 to 300 μM. By combining the

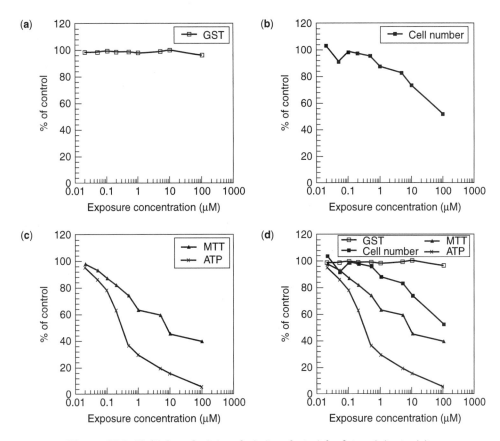

Figure 16.6. Multiple endpoint analysis (tox-cluster) for determining toxicity.

concentration response profiles obtained for each endpoint, it is possible to provide a more complete toxicity profile.

An example, using the insecticide rotenone (89, 90), of the information that can be obtained from this system is shown in Fig. 16.6. Rotenone was initially exposed to cells for 24 h. This resulted in a sharp reduction in all parameters. To gain more information on the mechanisms underlying cell death, the experiment was repeated for 6 h, as shown above. If membrane integrity [glutathione S-transferase (GST)] had been used to assess toxicity, the compound would have appeared safe. If cell number had been the endpoint used, there would have been a small reduction in cells, but the reason for the loss of cells (death or reduced rate of proliferation) would not be known. When the common markers of 3-[4,5-dimethylthiazol-2-yl]-2,5-diphenyltetrazolium bromide (MTT) reduction and ATP levels

were used, a steep dose-dependent reduction in both parameters was observed. Again, if these were the only assays used, it would not be possible to ascertain the reason for the reduction. MTT and ATP are general markers of mitochondrial function, which could have been reduced by acute cell death, inhibition of mitochondrial function, or direct inhibition of the enzymes responsible for synthesis or degradation. When all parameters are combined, it is possible to deduce that the reduction in MTT and ATP was most likely the result of direct interaction with mitochondria resulting in a loss of cellular ATP. Cell replication requires a large amount of energy; therefore, the reduction in cell number was caused by a reduced rate of cell replication. Indeed, the mechanism of rotenone toxicity is inhibition of oxidative phosphorylation. Multiple experiments comparing the *in vitro* data to *in vivo* toxicity showed excellent correlation.

Clearly, no system can predict all mechanisms of toxicity. Toxicity related to the production of active metabolites, the immune system, the endocrine system, or the CNS are examples of toxic mechanisms that would not be detected.

This multi-parameter approach provides reliable information to drug discovery teams on the relative toxicity of new compounds in a class, toxicity relative to similar drugs already on the market, STRs, subcellular targets, identification of the mechanism of adverse effects, and plasma concentrations where toxicity would be expected to occur *in vivo*.

4.3.2 High Content Screening.

High content screening (HCS) is a relatively new technology platform that incorporates fluorescent probes to monitor biochemical events, a cell-imaging system, and bioinformatics in a single instrument (Cellomics Inc., Pittsburgh, PA). The system provides a mechanism to evaluate a complete biological profile of a single cell by monitoring multiple endpoints simultaneously (91, 92). The system uses high resolution imaging to quantify several different targets identified by using fluorescent probes specific for each biochemical event of interest. These endpoints are quantified on discrete fluorescence channels in a near simultaneous fashion (Fig. 16.7). Although HCS has gained popularity in the area of drug discovery where it has been used for identifying novel therapeutic targets and for assessing drug effects on these targets, HCS also holds a great deal of promise for the field of toxicology. For example, HCS has been used to monitor the potential of new drugs to induce apoptosis. To accomplish this, fluorescent probes were designed to monitor specific biochemical events indicative of apoptosis. These included mitochondrial mass and membrane potential, caspase 3 and caspase 8, F-actin, and nuclear fragmentation. In this study, three different cell lines (MCF-7, L929, and BHK) were exposed to the apoptotic agents staurosporine and taxol. HCS enabled the detection and quantification of changes occurring in intact cells and showed that both compounds induced apoptosis. In addition, these studies provided significant insight into the differential effects of time and dose in different cell types (91). It is not difficult to envision the design and use of many different probes to monitor specific cellular events, which could be potential targets of toxicity. The small space requirements, its ability to monitor multiple events, and its potential for rapid throughput makes HCS technology extremely promising.

4.3.3 Mouse and Human Embryonic Stem Cells in Toxicology.

One of the most interesting and potentially valuable discoveries in developmental biology has been the ability to isolate and culture embryonic stem cell lines. Before it is possible to understand the research potential of these cell lines, it is necessary to first understand what embryonic stem cells are and how they are obtained. Stem cells are characterized by the ability to self-replicate for long periods of time and maintain the capacity to differentiate into one of many different cell types (pluripotent). In the preimplantation embryo, early cells do not resemble the differentiated cells of adult organs. In the normal environment, fertilization occurs in the oviduct. Over the next few days as the embryo moves down the oviduct into the uterus, it undergoes several cellular divisions, resulting in cells known as blastomeres. Blastomeres are not committed to becoming any particular cell type and thus the preimplantation embryo maintains a great deal of plasticity. After about 5 days of development, the outer layer of cells, known as the trophectoderm, separates from the inner cell mass (ICM) to become the placenta. The cells of the ICM maintain their pluripotent properties prior to implantation. However, after implantation, the ICM cells become restricted in their ability to differentiate. If the ICM cells are removed from their *in vivo* environment and cultured under appropriate *in vitro* conditions, the cells maintain their ability to proliferate and to differentiate into any number of possible cell types. These totipotent cells are known as the embryonic stem cell line (93–96).

The successful derivation of embryonic stem cell lines from the mouse blastocyst was accomplished in 1981, and the potential value of these unique cells was realized (97, 98). In 1998, the first human-derived embryonic stem cell line was successfully developed from the

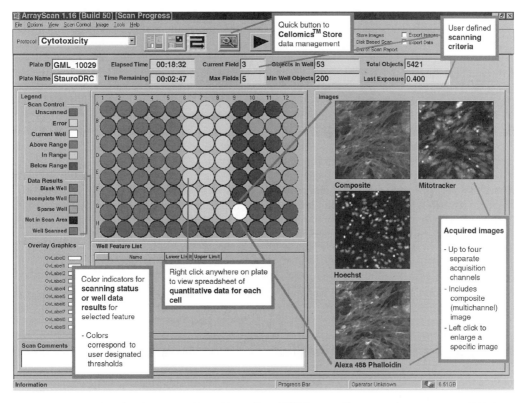

Figure 16.7. Graphical user interface of ArrayScan HCS System. Staurosporine treatment of BHK cells in 96-well microplate. Drug concentrations increase from left to right (controls in first two columns). Well color coding reflects (1) well status (current well being scanned is shown in white whereas unscanned wells are shown in grey) and (2) well results (pink wells have insufficient cell number, blue wells are below user-defined threshold range, red wells are above range, and green wells are in range for the selected cell parameter). Images correspond to current well, with three monochrome images reflecting three respective fluorescence channels, and composite (three-channel) color overlay in upper left panel. Nuclei are identified by the Hoechst dye (blue), microtubules by fluorescently labeled phalloidin (green), and mitochondria by Mitotracker stain (red). See color insert.

blastocyst of preimplantation embryos produced by *in vitro* fertilization (96). Because embryonic stem cells can proliferate for prolonged periods and maintain the potential to differentiate into specific cell types, they hold great promise for several areas of medical research. They offer the possibility of an unlimited source of tissue for transplant, and this technology alone may provide significant treatments for heart disease, diabetes, Parkinson's disease, and leukemia. Human embryonic stem cell lines may provide a new *in vitro* model that will provide information on infertility, spontaneous abortions, and other developmental problems. This model may also provide human-derived tissues that could be

used in the development of new drugs and as a model to evaluate potential toxicity much earlier in the drug discovery process.

Because human embryonic stem cell lines are derived from a fertilized human embryo, many moral, ethical, and political issues have slowed research requiring human embryonic stem cell lines (99–101). Continued research is essential to fully understanding the potential benefits of these unique cells.

4.3.4 Screening for Cardiotoxicity. Early detection of potential for adverse cardiovascular effects has become increasingly important with the recognition of the clinical risks associated with increased prolongation of the QT segment of the electrocardiogram of patients taking noncardiac drugs (60). Advances in electronics have facilitated remote recording of physiologic effects of drug candidates in unrestrained, instrumented rodents and larger animals. Whereas these *in vivo* measurements are valuable in detecting adverse cardiovascular effects and are an essential component of the toxicology testing program for drug candidates, they occur relatively late in the candidate selection process, require larger amounts of test substance, are expensive, and can lack sensitivity. *In vitro* tests can overcome many of the disadvantages of *in vivo* cardiovascular testing as a screening tool and contribute valuable information in establishing chemical SARs.

Tests for action potential duration (APD) are conducted using isolated heart tissue (Purkinje fibers, papillary muscle, or ventricle) from donor animals in a perfusion solution that contains test substance (102). Multiple preparations of the APD test system can be obtained from one animal, and test conditions can be manipulated to better define effects related to the test substance. Prolongation of the QT segment is associated with blockade of the inward rectifier potassium (Ikr) channel in cardiac tissue (103, 104). The Ikr channel is encoded by the human ether-a-gogo–related gene (HERG), which has become an important screening tool for potential QT prolongation (105). Whereas the *in vitro* tests for APD prolongation and Ikr blockade have advanced the ability to screen for QT prolongation, there are a number of technical limitations, and the results do not necessarily correlate with occurrence of adverse cardiac effects in patients. Therefore, these tests are regarded as a means of identifying the presence of a potential risk. Results of all *in vitro* and *in vivo* cardiovascular tests must be considered and integrated in developing a complete risk assessment (106).

5 SUMMARY AND CONCLUSIONS

Advances in identification of therapeutic targets and medicinal chemistry to interact with these targets have driven changes in the way that toxicologists contribute to the discovery and development of drugs. The emphasis is increasingly on identifying cellular and molecular alterations that are mechanistically related to expressions of toxicity. Identification of mechanisms of toxicity has improved understanding of adverse effects of therapeutic agents and resulted in more relevant assessment of risk versus benefit. Additionally, improved mechanistic understanding has enabled use of *in vitro* methodologies that facilitate prediction of toxic liabilities and contribute to structure activity decisions in the synthesis of new therapeutic molecules. The teamwork between medicinal chemists and toxicologists working in this new paradigm will be a key factor in driving advances in discovery of new therapeutic agents.

REFERENCES

1. M. A. Gallo in C. D. Klaassen, Ed., *Casarett and Doull's Toxicology: The Basic Science of Poisons,* McGraw Hill, New York, 1996, pp. 1–35.

2. R. M. Diener in G. I. Sipes, C. A. McQueen, and A. J. Gandolfi, Eds., *Comprehensive Toxicology,* Pergamon, New York, 1997.

3. J. Drews, *Arzneimittelforschung,* **45,** 934–939 (1995).

4. J. M. Herz, W. J. Thomsen, and G. G. Yarbrough, *J. Recept. Signal Transduct. Res.,* **17,** 671–776 (1997).

5. A. Heguy, A. A. Stewart, J. D. Haley, D. E. Smith, and J. G. Foulkes, *Gene Expr.,* **4,** 337–344 (1995).

6. B. A. Kenny, M. Bushfield, D. J. Parry-Smith, S. Fogarty, and J. M. Treherne, *Prog. Drug Res.,* **51,** 245–269 (1998).

7. G. L. Cockerell, C. S. Aaron, and J. B. Moe, *Toxicol. Pathol.,* **27,** 477–478 (1999).

8. K. Ondrias, A. Stasko, V. Jancinova, and P. Balgavy, *Mol. Pharmacol.,* **31,** 97–102 (1987).

9. R. L. Juliano, S. Daoud, H. J. Krause, and C. W. Grant, *Ann. N Y Acad. Sci.,* **507,** 89–103 (1987).

10. R. L. Juliano, C. W. Grant, K. R. Barber, and M. A. Kalp, *Mol. Pharmacol.,* **31,** 1–11 (1987).

11. J. M. Weinberg, P. G. Harding, and H. D. Humes, *Arch. Biochem. Biophys.,* **205,** 232–239 (1980).

12. A. Parkinson, in C. D. Klasassen, Ed., *Casarett and Doull's Toxicology: The Basic Science of Poisons,* McGraw Hill, New York, 1996, pp. 113–187.

13. S. D. Cohen and E. A. Khairallah, *Drug. Metab. Rev.*, **29**, 59–77 (1997).

14. D. Bjorkman, *Am. J. Med.*, **105**, 17S–21S (1998).

15. J. B. Bartolone, R. B. Birge, S. J. Bulera, M. K. Bruno, E. V. Nishanian, S. D. Cohen, and E. A. Khairallah, *Toxicol. Appl. Pharmacol.*, **113**, 19–29 (1992).

16. S. J. Bulera, R. B. Birge, S. D. Cohen, and E. A. Khairallah, *Toxicol. Appl. Pharmacol.*, **134**, 313–320 (1995).

17. N. R. Pumford, B. M. Martin, and J. A. Hinson, *Biochem. Biophys. Res. Commun.*, **182**, 1348–1355 (1992).

18. N. R. Pumford, T. G. Myers, J. C. Davila, R. J. Highet, and L. R. Pohl, *Chem. Res. Toxicol.*, **6**, 147–150 (1993).

19. M. C. Poirier, *Mutat. Res.*, **288**, 31–38 (1993).

20. D. E. Muscarella and S. E. Bloom, *Biochem. Pharmacol.*, **53**, 811–822 (1997).

21. D. W. Hein, M. A. Doll, A. J. Fretland, K. Gray, A. C. Deitz, Y. Feng, W. Jiang, T. D. Rustan, S. L. Satran, and T. R. Wilkie Sr, *Mutat. Res.*, **376**, 101–106 (1997).

22. J. L. Bolton, M. A. Trush, T. M. Penning, G. Dryhurst, and T. J. Monks, *Chem. Res. Toxicol.*, **13**, 135–160 (2000).

23. N. D. Ridgway, D. M. Byers, H. W. Cook, and M. K. Storey, *Prog. Lipid Res.*, **38**, 337–360 (1999).

24. L. Stryer, in *Biochemistry*, W. H. Freeman and Co., New York, 1995, pp. 263–273.

25. A. A. Spector and M. A. Yorek, *J. Lipid Res.*, **26**, 1015–1035 (1985).

26. J. L. Bolton, E. Pisha, F. Zhang, and S. Qiu, *Chem. Res. Toxicol.*, **11**, 1113–1127 (1998).

27. H. Esterbauer, *Am. J. Clin. Nutr.*, **57**, 779S–786S (1993).

28. T. Y. Aw, *Free Radic. Res.*, **28**, 637–646 (1998).

29. T. J. Lyons, *Am. J. Cardiol.*, **71**, 26B–31B (1993).

30. A. W. Girotti, *J. Photochem. Photobiol.*, *B13*, 105–118 (1992).

31. T. Ramasarma, *Indian J. Biochem. Biophys.*, **27**, 269–274 (1990).

32. E. Cadenas and H. Sies, *Adv. Enzyme Regul.*, **23**, 217–237 (1985).

33. P. J. Goldblatt and W. T. Gunning III, *Ann. Clin. Lab. Sci.*, **14**, 159–167 (1984).

34. J. Whysner, P. M. Ross, and G. M. Williams, *Pharmacol. Ther.*, **71**, 153–191 (1996).

35. E. Thorpe and A. I. Walker, *Food Cosmet. Toxicol.*, **11**, 433–442 (1973).

36. G. M. Williams, *Ann. NY Acad. Sci.*, **349**, 273–282 (1980).

37. R. J. Griffin, C. N. Dudley, and M. L. Cunningham, *Fundam. Appl. Toxicol.*, **29**, 147–154 (1996).

38. J. Clemmesen and S. Hjalgrim-Jensen, *Ecotoxicol. Environ. Saf.*, **1**, 457–470 (1978).

39. S. B. Shirts, J. F. Annegers, W. A. Hauser, and L. T. Kurland, *J. Natl. Cancer Inst.*, **77**, 83–87 (1986).

40. J. H. Olsen, J. D. Boice Jr., J. P. Jensen, and J. F. Fraumeni Jr., *J. Natl. Cancer Inst.*, **81**, 803–808 (1989).

41. J. H. Olsen, J. D. Boice Jr., and J. F. Fraumeni Jr., *Br. J. Cancer*, **62**, 996–999 (1990).

42. J. H. Olsen, H. Wallin, J. D. Boice Jr., K. Rask, G. Schulgen, and J. F. Fraumeni Jr., *Cancer Epidemiol. Biomarkers Prev.*, **2**, 449–452 (1993).

43. A. Parkinson, R. P. Clement, C. N. Casciano, and M. N. Cayen, *Biochem. Pharmacol.*, **43**, 2169–2180 (1992).

44. J. A. Skare, V. A. Murphy, R. C. Bookstaff, G. A. Thompson, M. A. Heise, Z. D. Horowitz, J. H. Powell, A. Parkinson, and J. V. St Peter, *Arch. Toxicol.* Suppl., **17**, 326–340 (1995).

45. J. M. McKim Jr., P. C. Wilga, G. B. Kolesar, S. Choudhuri, A. Madan, L. W. Dochterman, J. G. Breen, A. Parkinson, R. W. Mast, and R. G. Meeks, *Toxicol. Sci.*, **41**, 29–41 (1998).

46. J. M. McKim Jr., S. Choudhuri, P. C. Wilga, A. Madan, L. A. Burns-Naas, R. H. Gallavan, R. W. Mast, D. J. Naas, A. Parkinson, and R. G. Meeks, *Toxicol. Sci.*, **50**, 10–19 (1999).

47. J. M. McKim Jr., G. B. Kolesar, P. A. Jean, L. S. Meeker, P. C. Wilga, R. Schoonhoven, J. A. Swenberg, J. I. Goodman, R. H. Gallavan, and R. G. Meeks, *Toxicol. Appl. Pharmacol.*, **172**, 83–92 (2001).

48. EPA, *alpah2μ-Globulin: Association with Chemically Induced Renal Toxicity and Neoplasia in the Male Rat*, U.S. Environmental Protection Agency, Washington, DC, 1991.

49. J. Caldwell, and J. J. Mills in *General and Applied Toxicology*, Groves Dictionaries Inc., New York, 1999, pp. 128–129.

50. W. G. Flamm and L. D. Lehman-McKeeman, *Regul. Toxicol. Pharmacol.*, **13**, 70–86 (1991).

51. N. Latruffe and J. Vamecq, *Biochimie*, **79**, 81–94 (1997).

52. F. J. Gonzalez, J. M. Peters, and R. C. Cattley, *J. Natl. Cancer Inst.*, **90**, 1702–1709 (1998).

53. W. Paschen and J. Doutheil, *Acta Neurochir. Suppl.*, **73**, 1–5 (1999).

54. C. S. Lieber, *Physiol. Rev.*, **77**, 517–544 (1997).

55. D. B. Zorov, B. F. Krasnikov, A. E. Kuzminova, M. Y. Vysokikh, and L. D. Zorova, *Biosci. Rep.*, **17**, 507–520 (1997).

56. R. A. Butow and E. M. Bahassi, *Curr. Biol.*, **9**, R767–R769 (1999).

57. M. E. Sobaniec-Lotowska, *Exp. Toxicol. Pathol.*, **49**, 225–232 (1997).

58. M. J. Reymond and J. C. Porter, *Horm. Res.*, **22**, 142–152 (1985).

59. B. K. Vonderhaar, *Endocr. Relat. Cancer*, **6**, 389–404 (1999).

60. F. De Ponti, E. Poluzzi, and N. Montanaro, *Eur. J. Clin. Pharmacol.*, **56**, 1–18 (2000).

61. E. J. Matthews, R. D. Benz, and J. F. Contrera, *J. Mol. Graph Model*, **18**, 605–615 (2000).

62. E. Benfenati and G. Gini, *Toxicology*, **119**, 213–225 (1997).

63. R. Benigni, *Mutat. Res.*, **334**, 103–113 (1995).

64. C. A. Marchant, *Environ. Health Perspect.*, **104** (Suppl 5), 1065–1073 (1996).

65. J. E. Ridings, M. D. Barratt, R. Cary, C. G. Earnshaw, C. E. Eggington, M. K. Ellis, P. N. Judson, J. J. Langowski, C. A. Marchant, M. P. Payne, W. P. Watson, and T. D. Yih, *Toxicology*, **106**, 267–279 (1996).

66. A. M. Richard, *Mutat. Res.*, **400**, 493–507 (1998).

67. A. M. Richard, *Mutat. Res.*, **305**, 73–97 (1994).

68. K. Enslein, V. K. Gombar, and B. W. Blake, *Mutat. Res.*, **305**, 47–61 (1994).

69. D. F. Lewis, *Regul. Toxicol. Pharmacol.*, **20**, 215–222 (1994).

70. V. K. Gombar, K. Enslein, and B. W. Blake, *Chemosphere*, **31**, 2499–2510 (1995).

71. E. J. Matthews and J. F. Contrera, *Regul. Toxicol. Pharmacol.*, **28**, 242–264 (1998).

72. G. Klopman and M. Tu, *J. Med. Chem.*, **42**, 992–998 (1999).

73. M. J. Prival, *Environ. Mol. Mutagen*, **37**, 55–69 (2001).

74. B. B. Spear, M. Heath-Chioizz, and J. Huff, *Trends Mol. Med.*, **7**, 201–204 (2001).

75. C. P. Gray and W. Keck, *Cell Mol. Life Sci.*, **56**, 779–787 (1999).

76. P. M. Selzer, S. Brutsche, P. Wiesner, P. Schmid, and H. Mullner, *Int. J. Med. Microbiol.*, **290**, 191–201 (2000).

77. C. E. Barry III, M. Wilson, R. Lee, and G. K. Schoolnik, *Int. J. Tuberc. Lung Dis.*, **4**, S189–S193 (2000).

78. D. J. Payne, P. V. Warren, D. J. Holmes, Y. Ji, and J. T. Lonsdale, *Drug Discov. Today*, **6**, 537–544 (2001).

79. W. D. Pennie, J. D. Tugwood, G. J. Oliver, and I. Kimber, *Toxicol. Sci.*, **54**, 277–283 (2000).

80. M. R. Fielden and T. R. Zacharewski, *Toxicol. Sci.*, **60**, 6–10 (2001).

81. J. F. Waring, R. Ciurlionis, R. A. Jolly, M. Heindel, and R. G. Ulrich, *Toxicol. Lett.*, **120**, 359–368 (2001).

82. S. D. Patterson, *Curr. Opin. Biotechnol.*, **11**, 413–418 (2000).

83. J. K. Nicholson, J. C. Lindon, and E. Holmes, *Xenobiotica*, **29**, 1181–1189 (1999).

84. E. Holmes, J. K. Nicholson, and G. Tranter, *Chem. Res. Toxicol.*, **14**, 182–191 (2001).

85. E. Holmes, A. W. Nicholls, J. C. Lindon, S. C. Connor, J. C. Connelly, J. N. Haselden, S. J. Damment, M. Spraul, P. Neidig, and J. K. Nicholson, *Chem. Res. Toxicol.*, **13**, 471–478 (2000).

86. D. G. Robertson, M. D. Reily, R. E. Sigler, D. F. Wells, D. A. Paterson, and T. K. Braden, *Toxicol. Sci.*, **57**, 326–337 (2000).

87. S. Bushnell, J. Budde, T. Catino, J. Cole, A. Derti, R. Kelso, M. L. Collins, G. Molino, P. Sheridan, J. Monahan, and M. Urdea, *Bioinformatics*, **15**, 348–355 (1999).

88. F. S. Nolte, *Adv. Clin. Chem.*, **33**, 201–235 (1998).

89. J. M. McKim, P. C. Wilga, D. K. Petrella, J. F. Pregenzer, R. K. Patel, and G. L. Cockerell, *The Toxicologist*, **60**, 306 (2001).

90. D. D. Saunders and C. Harper in A. W. Hayes, Ed., *Principles and Methods of Toxicology*, 1994, p. 398.

91. K. U. Schallreuter, J. Moore, J. M. Wood, W. D. Beazley, D. C. Gaze, D. J. Tobin, H. S. Marshall, A. Panske, E. Panzig, and N. A. Hibberts, *J. Investig. Dermatol. Symp. Proc.*, **4**, 91–96 (1999).

92. R. N. Ghosh, Y. T. Chen, R. DeBiasio, R. L. DeBiasio, B. R. Conway, L. K. Minor, and K. T. Demarest, *Biotechniques*, **29**, 170–175 (2000).

93. M. Schuldiner, O. Yanuka, J. Itskovitz-Eldor, D. A. Melton, and N. Benvenisty, *Proc. Natl. Acad. Sci. USA*, **97**, 11307–11302 (2000).

94. J. Itskovitz-Eldor, M. Schuldiner, D. Karsenti, A. Eden, O. Yanuka, M. Amit, H. Soreq, and N. Benvenisty, *Mol. Med.*, **6**, 88–95 (2000).

95. M. F. Pera, B. Reubinoff, and A. Trounson, *J. Cell Sci.*, **113**, 5–10 (2000).

96. J. A. Thomson, J. Itskovitz-Eldor, S. S. Shapiro, M. A. Waknitz, J. J. Swiergiel, V. S. Marshall, and J. M. Jones, *Science,* **282**, 1145–1147 (1998).

97. M. J. Evans and M. H. Kaufman, *Nature,* **292**, 154–156 (1981).

98. G. R. Martin, *Proc. Natl. Acad. Sci. USA,* **78**, 7634–7638 (1981).

99. J. A. Robertson, *Nat. Rev. Genet.,* **2**, 74–78 (2001).

100. D. M. Trepagnier, *J. LA State Med. Soc.,* **152**, 616–624 (2000).

101. J. R. Meyer, *J. Med. Ethics,* **26**, 166–170 (2000).

102. M. M. Adamantidis, D. L. Lacroix, J. F. Caron, and B. A. Dupuis, *J. Cardiovasc. Pharmacol.,* **26**, 319–327 (1995).

103. Y. G. Yap and A. J. Camm, *Clin. Exp. Allergy,* **29** (Suppl 3), 174–181 (1999).

104. M. Taglialatela, P. Castaldo, A. Pannaccione, G. Giorgio, A. Genovese, G. Marone, and L. Annunziato, *Clin. Exp. Allergy,* **29** (Suppl 3), 182–189 (1999).

105. J. S. Mitcheson, J. Chen, M. Lin, C. Culberson, and M. C. Sanguinetti, *Proc. Natl. Acad. Sci. USA,* **97**, 12329–12333 (2000).

106. P. Champeroux, E. Martel, C. Vannier, V. Blanc, J. Y. Leguennec, J. Fowler, and S. Richard, *Therapie,* **55**, 101–109 (2000).

CHAPTER SEVENTEEN

Drug Absorption, Distribution, and Elimination

LESLIE Z. BENET
BEATRICE Y. T. PEROTTI
University of California
Department of Pharmacy
San Francisco, California

LARRY HARDY
Aurigene Discovery Technologies
Lexington, Massachusetts

Contents

Burger's Medicinal Chemistry and Drug Discovery
Sixth Edition, Volume 2: Drug Development
Edited by Donald J. Abraham
ISBN 0-471-37028-2   © 2003 John Wiley & Sons, Inc.

1 OVERVIEW OF PRODUCT DEVELOPMENT ISSUES

Although many compounds come through drug discovery every year, only a rare few can actually make their way through product development and pass the scrutiny of the U.S. Food and Drug Administration (FDA). With some ups and downs in marketing, these newly approved pharmaceutical products eventually land on the shelves in pharmacies, where they are dispensed to patients upon presentation of the proper prescriptions.

Why are so many compounds "killed" in the product development stage? Because the bottom line question in drug development asks, "Is this drug safe and effective in the human body?" many of these compounds simply do not prove to be safe and effective in the human body.

To answer this safety and efficacy question satisfactorily, development scientists must address issues from many areas. Some of the common questions are:

- What route of administration will be used to deliver the drug?
- How much of the drug gets into the systemic circulation and to the site of action?
- Where in the body will the drug distribute?
- How long does the drug remain intact in the body?
- How does the drug get eliminated from the body?
- Is the administered compound metabolized into other entities?
- Is the pharmacological effect derived from the administered compound or/and from metabolites?
- How much drug in the body is needed to elicit the pharmacological effect?
- Can patients develop allergic or toxic reactions toward the administered compound or its metabolites?
- How much drug in the body is needed to initiate such adverse reactions?

- What is the mechanism for such allergic reactions?
- Are these reactions reversible?
- Is the drug or the metabolites carcinogenic, teratogenic, or mutagenic? If so, what is the underlying mechanism?

2 DEFINING DRUG ABSORPTION, DISPOSITION, AND ELIMINATION THROUGH USE OF PHARMACOKINETICS

Initial progress in product development rests heavily on a thorough understanding of the absorption, distribution, and elimination characteristics of a drug. Absorption, distribution, and elimination processes begin when a dose is administered and may govern the appearance of any therapeutic effect (Fig. 17.1). Pharmacokinetics is used to quantitate these processes. Apart from its usefulness in explaining the safety and toxicity assessment data that will be discussed later in the chapter, pharmacokinetic analysis is used primarily to design appropriate dosing regimens and also to quantitatively define drug disposition (2).

The term *pharmacokinetics* can be defined as what the body does to the drug. The parallel term, *pharmacodynamics*, can be defined as what the drug does to the body. A fundamental hypothesis of pharmacokinetics is that a relationship exists between a pharmacologic or toxic effect of a drug and the concentration of the drug in a readily accessible site of the body (e.g., blood). This hypothesis has been documented for many drugs (3, 4), although for some drugs no clear relationship has been found between pharmacologic effect and plasma or blood concentrations.

This chapter focuses on the conceptual approach to pharmacokinetics, its use as a tool in therapeutics, and in defining drug disposition. The more common mathematical modeling-exponential equation-data analysis approach to pharmacokinetics will not be discussed here. These mathematical techniques are necessary to determine the important pharmacokinetic parameters that are to be discussed; however, for medicinal chemists who need to

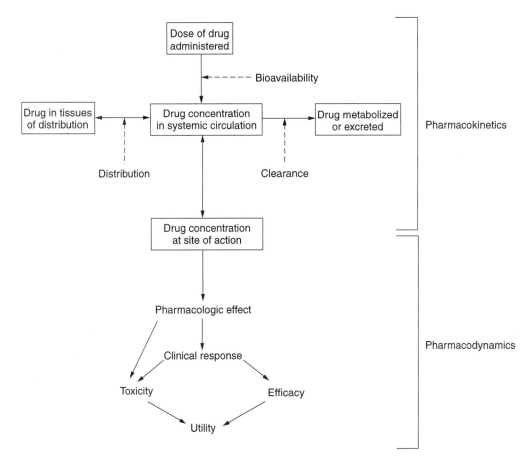

Figure 17.1. Schematic representation of the dose-response relationship of a drug (1). (Reproduced courtesy of the author and Appleton & Lange.)

interact with other pharmaceutical scientists in the drug development process, a conceptual approach to pharmacokinetics will be more useful. Nevertheless, to even gain a conceptual understanding of pharmacokinetics and its application to drug development, some basic simple equations, as discussed in this chapter, are necessary. This approach should allow the medicinal chemist to appreciate why modification of a drug molecule may lead to important changes in drug disposition.

3 IMPORTANT PHARMACOKINETIC PARAMETERS USED IN DEFINING DRUG DISPOSITION

Among the many pharmacokinetic parameters (Table 17.1) that can be determined when defining a new drug substance, clearance (CL) is the most important defining drug disposition. Bioavailability (F) and volume of distribution (V) are also of primary importance when pharmacokinetics is used to define drug absorption.

3.1 Clearance

Clearance (CL) is the measure of the ability of the body to eliminate a drug. Initially, clearance will be looked at from a physiological point of view. Figure 17.2 depicts how a drug is removed from the systemic circulation when it passes through an eliminating organ. The rate of presentation of a drug to a drug elimination organ is the product of organ blood flow (Q) and the concentration of drug in the arterial blood entering the organ (C_A). The rate of exit

Table 17.1 Ten Critical Pharmacokinetic and Pharmacodynamic Parameters in Drug Development in Order of Importance[a]

1. Clearance
2. Effective concentration range
3. Extent of availability
4. Fraction of the available dose excreted unchanged
5. Blood/plasma concentration ratio
6. Half-life
7. Toxic concentration
8. Extent of protein binding
9. Volume of distribution
10. Rate of availability

[a]Ref. 5 reproduced courtesy of L. Z. Benet and Plenum Press.

of a drug from the drug-eliminating organ is the product of the organ blood flow (Q) and the concentration of the drug in the venous blood leaving the organ (C_V). By mass balance, the rate of elimination (or extraction) of a drug by a drug-eliminating organ is the difference between the rate of presentation and the rate of exit.

$$\text{Rate of presentation} = QC_A \quad (17.1)$$

$$\text{Rate of exit} = QC_V \quad (17.2)$$

$$\begin{aligned}\text{Rate of elimination} &= QC_A - QC_V \\ &= Q(C_A - C_V)\end{aligned} \quad (17.3)$$

The extraction ratio (ER) of an organ can be defined as the ratio of the rate of elimination to the rate of presentation.

Extraction ratio

$$= (\text{rate of elimination}/ \quad (17.4a)$$
$$\text{rate of presentation})$$

$$\begin{aligned}ER &= Q(C_A - C_V)/QC_A \\ &= (C_A - C_V)/C_A\end{aligned} \quad (17.4b)$$

The maximum possible extraction ration is 1.0 when no drug emerges into the venous blood upon presentation to the eliminating organ (i.e., $C_V = 0$). The lowest possible extraction ratio is zero, when all the drug passing through the potential drug-eliminating organ appears in the venous blood (i.e., $C_V = C_A$). Drugs with an extraction ratio of more than 0.7 are by convention considered as high extraction ratio drugs, whereas those with an extraction ratio of less than 0.3 are considered as low extraction ratio drugs.

The product of organ blood flow and extraction ration of an organ represents a rate at which a certain volume of blood is completely cleared of a drug. This expression defines the organ clearance (CL_{organ}) of a drug.

$$CL_{organ} = QER = Q(C_A - C_V)/C_A \quad (17.5)$$

From Equations 17.3 and 17.5, one can see that clearance is a proportionality constant between rate of elimination and the arterial drug concentration.

At steady state, by definition, rate in equals rate out. Rate is given by the dosing rate ($Dose/\tau$, i.e., dose divided by dosing interval τ) multiplied by the drug availability F, whereas

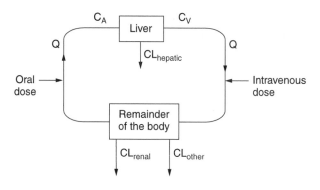

Figure 17.2. Schematic representation of the concentration-clearance relationship.

rate out is the rate of elimination (clearance multiplied by the systemic concentration C).

$$\text{Rate in} = \text{rate out} \quad (17.6)$$

$$F(Dose/\tau) = CL(C) \quad (17.7)$$

When Equation 17.7 is integrated over time from zero to infinity, Equation 17.8 results.

$$F(Dose) = CL(AUC) \quad (17.8)$$

where AUC is the area under the concentration time curve and F is the fraction of dose available to the systemic circulation (see Bioavailability in Section 3.2).

Clearance may be calculated as the available dose divided by the area under the systemic drug concentration curve.

$$CL = \frac{F(Dose)}{AUC} \quad (17.9)$$

The maximum value for organ clearance is limited by the blood flow to the organ (i.e., extraction ratio is 1). The average blood flow to the kidneys is approximately 66 L/h and to the liver, approximately 81 L/h. Clearance can occur in many sites in the body and is generally additive. Elimination of a drug may occur as a result of processes occurring in the liver, the kidney, and other organs. The total systemic clearance will be the sum of the individual organ clearances.

$$CL_{\text{total}} = CL_{\text{hepatic}} + CL_{\text{renal}} + CL_{\text{other}} \quad (17.10)$$

Among the many organs that are capable of eliminating drugs, the liver, in general, has the highest metabolic capability. Drug molecules in blood are bound to blood cells and plasma proteins such as albumin and alpha$_1$-acid glycoprotein. Yet only unbound drug molecules can pass through hepatic membranes into hepatocytes, where they are metabolized by hepatic enzymes or transported into the bile. Thus to be eliminated, drug molecules must partition out of the red blood cells and dissociate from plasma proteins to become unbound or free drug molecules. Because unbound drug molecules are free to parti-

tion into and out of the blood cells and hepatocytes, there is an equilibrium of free drug concentration between the blood cells, the plasma, and the hepatocytes. The ratio between unbound drug concentration and the total drug concentration constitutes the fraction unbound (f_{u}).

Fraction unbound =

unbound drug concentration/

total drug concentration

or

$$f_{\text{u}} = C_{\text{u}}/C \quad (17.11)$$

Because an equilibrium exists between the unbound drug molecules in the blood cells and the plasma, the rate of elimination of unbound drugs is the same in the whole blood as in the plasma at steady state. Thus,

$$CL_{\text{p}}C_{\text{p}} = CL_{\text{b}}C_{\text{b}} = CL_{\text{u}}C_{\text{u}} \quad (17.12)$$

where the subscripts p, b, and u refer to plasma, blood, and unbound, respectively.

From the material presented so far, one may intuitively imagine that hepatic drug clearance will be influenced by hepatic blood flow, fraction unbound, and intrinsic clearance; that is, the intrinsic ability of the organ to clear unbound drug. The simplest model that describes hepatic clearance in terms of these physiological parameters is the well-stirred model (6). Assuming instantaneous and complete mixing, the well-stirred model states that hepatic clearance (with respect to blood concentration) is

$$CL_{\text{hep}} = \frac{Q_{\text{hep}}f_{\text{u}}CL_{\text{int}}}{Q_{\text{hep}} + (f_{\text{u}}CL_{\text{int}})} \quad (17.13)$$

Note that f_{u} is calculated from unbound and total concentration in whole blood.

Equation 17.5 advises that hepatic clearance is the product of hepatic blood flow and hepatic extraction ratio. Therefore, as shown in Equation 17.13, the hepatic extraction ratio is

$$ER_{\text{hep}} = \frac{f_u CL_{\text{int}}}{Q_{\text{hep}} + (f_u CL_{\text{int}})} \quad (17.14)$$

By examining Equations 17.13 and 17.14, one finds that for drugs with a high extraction ratio (i.e., ER approaches 1.0), $f_u CL_{\text{int}}$ is much greater than Q_{hep}, and clearance approaches Q_{hep}. In other words, the clearance for a high extraction ratio drug, imipramine for example, is perfusion rate limited. For drugs with a low extraction ratio (i.e., ER approaches zero), Q_{hep} is much greater than $f_u CL_{\text{int}}$, and clearance is approximated by $f_u CL_{\text{int}}$. An example of a low extraction ratio drug is acetaminophen.

The intrinsic ability of an organ to clear a drug is directly proportional to the activity of the metabolic enzymes in the organ. Such metabolic processes, both *in vitro* and *in vivo*, are characterized by Michaelis-Menten kinetics:

$$\text{Rate of metabolism} = V_{\text{MAX}} C/(K_M + C) \quad (17.15)$$

in which V_{MAX}, the maximum rate at which metabolism can proceed, is proportional to the total concentration of enzyme. K_M is the Michaelis-Menten constant corresponding to the drug concentration that yields one-half of the maximum rate of metabolism. Dividing both sides of Equation 17.15 by the systemic concentration (C) yields

$$\text{Rate of metabolism}/C = CL_{\text{metabolism}}$$
$$= V_{\text{MAX}}/(K_M + C) \quad (17.16)$$

Because identification of a saturable process occurs only for low extraction ratio compounds (i.e., $CL_{\text{metabolism}} = f_u CL_{\text{int}}$), the relationship between classical enzyme kinetics and pharmacokinetics is revealed:

$$f_u CL_{\text{int}} = V_{\text{MAX}}/(K_M + C) \quad (17.17)$$

Because V_{MAX} and K_M can be obtained from *in vitro* metabolism experiments, development scientists may reasonably predict the *in vivo* clearance parameter of a low or intermediate extraction ratio drug from *in vitro* data, by use of Equations 17.13 and 17.17. By use of appropriate scaling factors, the *in vivo* clearance parameter in humans can be ap-

proximated from *in vitro* metabolism data from other species (7, 8).

The kidneys are also important drug-eliminating organs. Renal clearance (CL_r) is a proportionality term between urinary excretion rate and systemic concentration. By integrating Equation 17.18a over time from zero to infinity, one obtains renal clearance, which is the ratio of the total amount of the drug excreted unchanged to the area under the systemic concentration curve. The total amount of drug excreted unchanged can be measured experimentally or can be calculated from the dose, if the fraction of the available dose excreted unchanged (f_e) is known.

$$CL_r = \text{Urinary excretion rate}/C \quad (17.18a)$$
$$= \text{Total amount of drug}$$
$$\text{excreted unchanged}/AUC \quad (17.18b)$$
$$= f_e[F(Dose)]/AUC \quad (17.18c)$$

It is obvious from the preceding arguments that the conceptual approach to clearance requires measurements of blood concentrations. Yet, bioanalytical measurements are often carried out in plasma because of the ease of sample handling. However, measuring the blood to plasma ratio allows one to convert clearance values, determined in plasma, to their corresponding blood values by use of Equation 17.12.

3.2 Bioavailability

The *bioavailability* of a drug product through various routes of administration is defined as the fraction of unchanged drug that is absorbed intact and then reaches the site of action; or the systemic circulation following administration by any route. For an intravenous dose of a drug, bioavailability is defined as unity. For drugs administered by other routes of administration, bioavailability is often less than unity. Incomplete bioavailability may be attributed to a number of factors that can be subdivided into categories of dosage-form effects, membrane effects, and site of administration effects. Obviously, the route of administration that offers maximum bioavailability is the direct input at the site of action for which the drug is developed. This arrangement may be difficult to achieve because the site of action is not known for disease states and, in other cases, the site of action may be completely inaccessible, even when the drug is placed into the blood stream. The most commonly used route is oral administration. However, orally administered drugs may decompose in the fluids of the gastrointestinal lumen, or be metabolized as they pass through the gastrointestinal membrane. In addition, once a drug passes into the hepatic portal vein, it may be cleared by the liver before entering into the general circulation. The loss of drug as it passes through drug-eliminating organs for the first time is known as the *first-pass effect*.

The fraction of an oral dose available to the systemic circulation, considering both absorption and the first-pass effect, can be found by comparing the ratio of AUCs after oral and intravenous dosing.

$$F = AUC_{oral}/AUC_{i.v.} \qquad (17.19)$$

If an assumption is made that all of a drug dose is absorbed through the gastrointestinal tract intact, and that the only extraction takes place at the liver, then the maximum bioavailability (F_{max}) is

$$F_{max} = 1 - ER_{hep} \qquad (17.20)$$

From Equations 17.5 and 17.20, one can derive the following relationship for F_{max}:

$$F_{max} = 1 - (CL_{hep}/Q_{hep}) \qquad (17.21)$$

For high extraction ratio drugs, where CL_{hep} approaches Q_{hep}, F_{max} will be small. For low extraction ratio drugs, Q_{hep} is much greater than CL_{hep} and F_{max} will approximate one.

Recently, bioavailability and clearance data obtained from a crossover study of cyclosporine kinetics before and after rifampin dosing revealed a new understanding of drug metabolism and disposition of the compound (9). Healthy volunteers were given cyclosporine, intravenously and orally, before and after CYP3A (P450 3A) enzymes were introduced by rifampin. As expected, the blood clearance and cyclosporine increased from 0.31 to 0.42 L

h^{-1} kg^{-1} as a result of the induction of the drug's metabolizing enzymes (i.e., an increase in V_{max} in Equation 17.15). There was no change in volume of distribution, but there was a dramatic decrease in bioavailability from 27% to 10% in these individuals.

A decrease in bioavailability is to be expected because cyclosporine undergoes some first-pass metabolism as it goes through the liver after oral dosing. However, if one predicts on the basis of pharmacokinetics what the maximum bioavailability [as calculated by Equation 17.21] would be before and after rifampin dosing, the maximum bioavailability would decrease from 77% to 68%. Thus, there would be an expected cyclosporine bioavailability decrease of approximately 12% just on the basis of clearance changes resulting from inducing CYP3A enzymes in the liver. In fact, there was a bioavailability decrease of 60%. Furthermore, bioavailability was significantly less than would be predicted at maximum. Although some of that lower bioavailability was attributed to formulation effects, the discrepancy between the theoretical maximum bioavailability and the achievable bioavailability of cyclosporine remained a question.

However, on the basis of new findings during the last several years about the high prevalence of CYP3A isozymes in the gut, significant metabolism of cyclosporine in the gut as well as in the liver was speculated. This hypothesis can consistently explain the *significantly lower* bioavailability that would be predicted, even if all of the drug could be absorbed into the blood stream. This finding, particularly quantification of the magnitude of gut metabolism (more than two-thirds of the total metabolism for an oral dose of cyclosporine occurs in the gut), would not have been realized had pharmacokinetics not been used in the analysis of the given data.

3.3 Volume of Distribution

The volume of distribution (V) relates the amount of drug in the body to the concentration of drug in the blood or plasma, depending on the fluid measured. At its simplest, this relationship is defined by Equation 17.22:

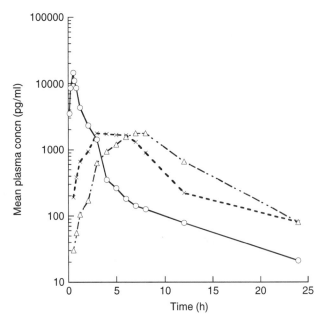

Figure 17.3. Mean serum IFN-α concentrations after a single 36×10^6 U dose as an intravenous infusion ($\bigcirc$) or an intramuscular ($\times$) or subcutaneous ($\triangle$) injection (10). (Reproduced courtesy of the author and *Clin. Pharmacol. Ther.*)

$$V = \text{amount of drug in body}/C \quad (17.22)$$

For a normal 70-kg man, the plasma volume is 3 L, the blood volume is 5 L, the extracellular fluid outside the plasma is 12 L, and the total body water is approximately 42 L. However, many classical drugs exhibit volumes of distribution far in excess of these known volumes. The volume of distribution for digoxin is about 700 L, which is approximately 10 times greater than the total body volume of a 70-kg man. This example serves to emphasize that the volume of distribution does not represent a real volume. Rather, it is an *apparent* volume that should be considered as the size of the pool of body fluids that would be required if the drug were equally distributed throughout all portions of the body. In fact, the relatively hydrophobic digoxin has a high apparent volume of distribution because it distributes predominantly into muscle and adipose tissue, leaving only a very small amount of drug in the plasma in which the concentration of drug is measured.

At equilibrium, the distribution of a drug within the body depends on binding to blood cells, plasma proteins, and tissue components. Only the unbound drug is capable of entering and leaving the plasma and tissue compartments. A memory aid for this relationship can be summarized by the expression

$$V = V_p + V_{TW} f_u / f_{u,T} \quad (17.23)$$

where V_p is the volume of plasma, V_{TW} is the aqueous volume outside plasma, f_u is the fraction unbound in plasma, and $f_{u,T}$ is the fraction unbound in tissue. Thus, a drug that has a high degree of binding to plasma proteins (i.e., low f_u) will generally exhibit a small volume of distribution. Unlike plasma protein binding, tissue binding of a drug cannot be measured directly. Generally, this parameter is assumed to be constant unless indicated otherwise.

In Equation 17.22, the body is considered as a single homogeneous pool of body fluids as described above for digoxin. For most drugs, however, two or three distinct pools of distribution space appear to exist. This condition results in a time-dependent decrease in the measurable blood or plasma concentration, which reflects distribution into other body pools independent of the body's ability to eliminate the drug. Figure 17.3 describes mean serum IFN-α concentrations after a 40-min intravenous infusion as well as after intramuscular and subcutaneous injections of the same dose. Note the logarithmic biphasic nature of the mean plasma concentration-time curve after the intravenous infusion. This biphasic nature represents both the distribution and elimination processes.

Generally, comparisons of volume of distribution are made by use of a parameter designated as the volume of distribution at steady state (V_{SS}), which reflects the sum of the volumes of all the pools into which a drug may distribute. V_{SS} can be calculated from the area under the moment curve ($AUMC$) and the area under the curve (AUC) as defined by Benet and Galeazzi (11):

$$V_{SS} = Dose_{i.v.} AUMC/AUC^2 \quad (17.24)$$

$AUMC$ can be calculated from areas under a plot of concentration·time vs. time. Both AUC and $AUMC$ may be calculated from the coefficients and exponents of the equations used to describe the multicompartment nature of drug kinetics as depicted in Fig. 17.3. These concepts will be revisited following a discussion of half-life.

4 IMPORTANT PHARMACOKINETIC PARAMETERS USED IN THERAPEUTICS

In addition to the three parameters, clearance (CL), bioavailability (F), and volume of distribution (V) discussed previously, a fourth parameter, half-life ($t_{1/2}$), is also crucial in therapeutics. The decreasing order of importance of these four parameters is clearance, bioavailability, half-life, and volume. Clearance defines the dosing rate, bioavailability defines dose adjustment, and half-life defines the dosing interval. Volume of distribution defines the loading dose.

4.1 Dosing Rate

As discussed in Equation 17.6 in Section 3.1, rate of presentation equals rate of exit at steady rate. Whereas rate of presentation is the product of bioavailability and dosing rate, rate of exit is the product of clearance and average (steady-state) concentration. By replacing the average concentration with the target concentration, the dosing rate can be computed with known values of bioavailability and clearance.

$$\text{Rate in} = \text{rate out} \quad (17.6)$$

Bioavailability × dosing rate

$$= \text{Clearance} \quad (17.7a)$$

× average concentration

$$F(\text{dosing rate}) = CL(C_{\text{target}}) \quad (17.7b)$$

4.2 Dose Adjustment

There is often a therapeutic concentration range of drug in the blood that is necessary to elicit a clinical effect without causing drug toxicity. The boundaries of this range are set by the minimum effective concentration (MEC) and the minimum toxic concentration (MTC). For example, to maintain a steady rate of presentation of 100 mg/min, one can administer either a completely bioavailable intravenous (i.v.) infusion at 100 mg/min or a sustained release oral dosage form with 50% bioavailability at 200 mg/min. As shown in Equations 17.7a and 17.7b, the actual dosing rate depends on the bioavailability of the dosage form. For proper dosing adjustment, the bioavailability (F) of a given dosage form stands as a must-know parameter in therapeutics.

4.3 Dosing Interval

Half-life ($t_{1/2}$) is an extremely useful kinetic parameter in terms of therapeutics, given that this parameter, together with therapeutic index, helps define the dosing interval at which drugs should be administered. By definition, half-life is the time required for 50% of the drug remaining in the body to be eliminated. In three and one-third half-lives, 90% of the dose would have been eliminated. Table 17.2 shows the percentage of dose lost in different numbers of half-lives. If the dosing interval is

Table 17.2 Percentage of Dose Lost as a Function of Number of Half-Lives

Number of Half-Lives	Dose Lost (%)
1	50
2	75
3	87.5
3.3	90
4	93.8
5	96.9

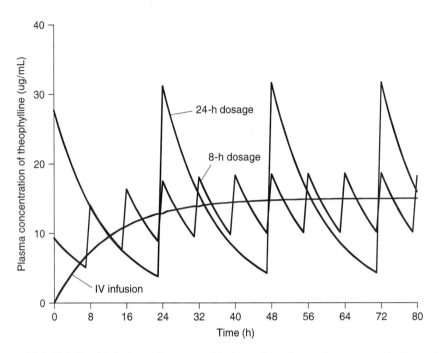

Figure 17.4. Relationship between frequency of dosing and maximum plasma concentrations when a steady-state plasma level of 15 μg/mL is desired. The time course of plasma concentrations for 43.2 mg/h intravenous infusion, and 8-hourly 340 mg oral dose, and a 24-hourly 1020 mg oral dose are depicted. (Reproduced courtesy of the author and Appleton & Lange.)

long relative to the half-life, large fluctuations in drug concentration will occur. On the other hand, if the dosing interval is short relative to half-life, significant accumulation will occur. The half-life parameter also allows one to predict drug accumulation within the body and quantifies the approach to plateau that occurs with multiple dosing and constant rates of infusion. Conventionally, three and one third half-lives are used as the time required to achieve steady state under constant infusion. The concentration level achieved after this time is already 90% of the steady-state concentration and, clinically, it is difficult to distinguish a 10% difference in concentrations.

After determining the extent of oral drug bioavailability, half-life is probably the next most important parameter in terms of deciding the appropriateness of a drug for further development. Drugs with very short half-lives create problems in maintaining steady-state concentrations in the therapeutic range. Thus, a successful drug product with a short half-life will require a dosage form that allows

a relatively constant prolonged input. Drugs with a very long half-life are favored in terms of efficacy considerations; however, this long half-life may be a negative characteristic in terms of toxicity considerations. Figure 17.4 illustrates the importance of half-life in defining the dosing interval.

Figure 17.4 depicts the relationship between the frequency of theophylline dosing and the plasma concentration time course when a steady-state theophylline plasma level of 15 μg/mL is desired. The smoothly rising curve shows the plasma concentration achieved with an i.v. infusion of 43.2 mg/h to a patient exhibiting an average theophylline clearance of 0.69 mL min^{-1} kg^{-1}. The steady-state theophylline plasma level achieved is midway within the therapeutic concentration range of 10–20 μg/mL. The figure also depicts the time courses for 8-hourly administration of 340 mg and 24-hourly administration of 1020 mg, assuming that these doses are administered in an immediate release dosage form. In each case the mean steady-state concentration is 15 μg/mL. However, the peak to trough ratio and the concentrations achieved with the once-daily dosing (i.e., 1020 mg) of a rapidly released formulation would result in concentrations in the toxic range, exceeding 20 μg/mL for certain periods of time, as well as concentrations less than 10 μg/mL, for significant periods during each dosing interval. In contrast, the 8-hourly dosing, which is approximately equivalent to the half-life of theophylline, shows a twofold range in peak to trough, and theophylline levels stay within the therapeutic plasma concentration range.

Although half-life is a very important parameter in therapeutics for defining the dosing interval, half-life can be a very misleading parameter when one is attempting to use pharmacokinetics as a tool in defining drug disposition. As depicted in Equation 17.25, half-life varies as a function of the two physiologically related parameters, volume and clearance.

$$t_{1/2} = 0.693(V/CL) \qquad (17.25)$$

Half-life has little value as an indicator of the processes involved in either drug elimination or distribution. Yet, early studies of drug pharmacokinetics in disease states have relied on drug half-life as the sole measure of alteration in drug disposition. Disease states can affect the physiologically related parameters, volume of distribution and clearance; thus, the derived parameter, half-life, will not necessarily reflect the expected change in drug elimination.

As clearance decreases, because of a disease process, half-life would be expected to increase. However, this reciprocal relationship is exact only when the disease does not change the volume of distribution. For example, as Klotz and coworkers have shown, the increase in half-life of diazepam with age does not result from a decrease in clearance but rather results from an increase in volume as the patient ages (12). Clearance, a measure of the body's ability to eliminate the drug, does not significantly decrease with age for diazepam. However, when volume increases, less drug is in the blood flowing to the liver, and elimination can occur only for those molecules that

come into contact with the liver. Thus the time that the drug remains in the body is increased. Another example of how half-life is not a good predictor of mechanisms of elimination is given by studies with the oral antihypoglycemic tolbutamide. The half-life of tolbutamide decreases in patients with acute viral hepatitis; that is, the drug appears to be eliminated faster by a diseased liver, the exact opposite from what one might expect. Here, the disease appears to decrease protein binding in both plasma and tissues, causing no change in volume of distribution but an increase in free fraction in the plasma, which results in an increase in total clearance and, subsequently, a decrease in half-life (13).

Equation 17.25 describes the half-life relationship for a drug that appears to follow one-compartment body kinetics; that is, when the body is considered to be a single homogeneous pool of body fluids. However, many drugs appear to exhibit multiple distribution pools and therefore may have multiple half-lives (as was depicted in Fig. 17.3 for IFN-α). Drugs with multiple half-lives are usually reported in the literature as having "distribution" and "terminal elimination" half-lives. Defining the "relevant" half-life in such situations has been addressed by Benet and coworkers (2, 14, 15).

Consider the situation in which a drug is best described by a two-pool model, as has been suggested for IFN-α by Wills et al. (10). The data in the Wills study were recalculated to represent the equation describing the concentration (ng/mL) of the drug after a 228 μg dose of IFN-α as a function of time (h), as given in Equation 17.26.

$$C = 14.13e^{-1.04t} + 0.545e^{-0.136t} \qquad (17.26)$$

The disposition constants in the exponents of Equation 17.26 correspond to half-lives of 0.667 h (40 min) and 5.1 h. In the interferon literature, the 5.1-h half-life is generally referred to as the mean elimination half-life, whereas the 40-min half-life, if mentioned at all, is generally referred to as a distribution half-life. These representations may not be accurate; the relevance of a particular half-life may be defined in terms of the fraction of the clearance that is related to each half-life. Note in Equation 17.8 that clearance is inversely related to area under the drug concentration time curve (AUC). When the equation describing the time course of drug concentrations requires more than one exponential term, this circumstance can be represented by Equation 17.27. For an n-compartment pharmacokinetic model,

$$C = L_1e^{-\lambda_1 t} + L_2e^{-\lambda_2 t} + \cdots + L_ne^{-\lambda_n t} \quad (17.27)$$

AUC can be calculated as the ratio of coefficients and exponents as in Equation 17.28.

$$AUC = L_1/\lambda_1 + L_2/\lambda_2 + \cdots + L_n/\lambda_n \quad (17.28)$$

Calculating AUC for IFN-α as described in Equation 17.26 yields

$$AUC \text{ (ng h}^{-1}\text{ mL}^{-1}) = \begin{array}{cc} 13.59 & + & 4.01 \\ (77\%) & & (23\%) \end{array}$$

Note that 77% of the AUC relates to the coefficient and exponent for the 40-min half-life, which suggests that the 5.1-h half-life is in fact a minor contributor to the prediction of steady-state concentrations of IFN-α. The importance of these fractional areas can be observed if Equation 17.7 is rearranged to predict steady-state concentrations (C_{SS}):

$$C_{SS} = F(Dose)/(\tau CL) \quad (17.7c)$$

Now, given that CL can be defined as given in Equation 17.9 as the relationship between an available single dose and AUC, Equation 17.7c becomes

$$C_{SS} = [F(Dose)]AUC/\tau[F(Dose)] \quad (17.29)$$

Thus, the ability to correctly predict C_{SS} is dependent on the accuracy of the measurement of AUC. If the 5.1-h half-life for IFN-α is ignored, the data of Wills et al. suggest that the value for AUC, and therefore the steady-state concentration, will be underestimated by only 23%, given that this longer half-life represents a relatively small fraction of the total IFN-α clearance (10). This error in drug concentrations could probably be ignored with confidence because a 23% difference may often be within analytical error for protein drugs, as well as within the day-to-day variability in a particular patient.

The above calculations assume that drug concentrations are important in defining the efficacy or toxicity of IFN-α. If this is true, the clinician can safely ignore the 5.1-h half-life in patients with normal elimination characteristics because little change in steady-state drug levels will be observed. However, it may be that the response, particularly toxicity, is related to the amount of drug in the body, rather than the systemic concentration. The amount in the body at steady state (A_{SS}) is the product of the systemic concentration and the steady-state volume of distribution:

$$A_{SS} = C_{SS}V_{SS} \quad (17.30)$$

As can be seen from Equation 17.24, the accurate calculation of V_{SS} requires an understanding of how AUMC is related to the fractional area of each half-life. The complications of this relationship will not be discussed in this introductory chapter but the correct estimation of V_{SS} is always significantly determined by the terminal half-life.

More recently, Benet has described so-called multiple dosing half-lives, the half-life for a drug that is equivalent to the dosing interval to choose so that plasma concentrations (Equation 17.31) or amounts of drug in the body (Equation 17.32) will show a 50% drop during a dosing interval at a steady state. These parameters are defined in terms of the mean residence time in the central compartment ($MRTC$) and the mean residence time in the body (MRT).

$$t_{1/2_{MD}}{}^{plasma} = 0.693MRTC \quad (17.31)$$

$$t_{1/2_{MD}}{}^{amount} = 0.693MRT \quad (17.32)$$

MRTC in a one-compartment body model is the inverse of the rate constant for elimination. In a multiple-compartment model, where the multiple dosing plasma half-life is useful, MTRC is given by the volume of the central compartment where drug concentrations are measured divided by clearance. MRT in Equation 17.32 is the ratio of AUMC/AUC.

4.4 Loading Dose

For drugs with long half-lives, the time to reach steady state is substantial. In these instances, it may be desirable to administer a loading dose that promptly raises the concentration of a drug in plasma to the projected steady-state value. In theory, only the amount of the loading dose needs to be computed, not the rate of its administration. To a first approximation, this is true. The amount of drug required to achieve a given steady-state concentration in the plasma is the amount of drug that must be in the body when the desired steady state is reached. For intermittent dosage schemes, the amount is that at the average concentration. The volume of distribution is the proportionality factor that relates the total amount of drug in the body to the concentration in the plasma. If a loading dose is administered to achieve the desired steady-state concentration, then

Loading dose

= amount in the body at steady state

$$= C_{P,SS} V_{SS} \quad (17.33)$$

For most drugs, the loading dose can be given as a single dose by the chosen route of administration. However, for drugs that follow complicated a multicompartment model, the distribution phase cannot be ignored in the calculation of the loading dose. If the rate of absorption is rapid relative to distribution (this circumstance is always true for i.v. bolus administration), the concentration of drug in plasma that results from an appropriate loading dose can initially be considerably higher than desired. Severe toxicity may occur, albeit transiently. This may be particularly important, for example, in the administration of antiarrhythmic drugs, where an almost immediate toxic response is obtained when plasma concentrations exceed a particular level. Thus, although estimation of the amount of the loading dose may be quite correct, the rate of administration can be crucial in preventing excessive drug concentrations, and slow administration of an i.v. drug (over minutes rather than seconds) is almost always wise.

5 HOW IS PHARMACOKINETICS USED IN THE DEVELOPMENT PROCESS?

5.1 Preclinical Development

The pharmacokinetics of a new molecular entity must be defined in the animal species used in preclinical drug development. This information serves as a portion of the required data necessary for submission of an IND (investigational new drug) application. Generally, the absorption, distribution, metabolism, and excretion of a new molecular entity is characterized in a small animal such as a rodent (usually the rat) and in a large animal (usually the dog and/or monkey). The drug should be characterized in terms of its clearance, volume of distribution, bioavailability, and half-life. In addition, the linearity of the drug pharmacokinetics over the doses anticipated for use in toxicology studies must be determined. Where nonlinearities are observed, the Michaelis-Menten parameters are characterized. These preclinical pharmacokinetics data may be used in an initial prediction of the disposition characteristics in humans. Theoretically, intrinsic clearance or clearance and volume of distribution may be scaled up from the preclinical animal species to predict parameters in humans. At this time, these approaches have not been as successful as one would hope. With the rapid development in the understanding of the metabolic isozymes involved in drug metabolism, and their conservation of characteristics across animal species, it is anticipated that scale-up procedures from animal data may be more readily used in the future.

5.2 Pharmacokinetics in Initial Phase I Human Studies

Phase I human studies are designed to evaluate the absorption, distribution, metabolism, and excretion characteristics of a new molecular entity in humans. Except for potentially toxic drugs, as used in life-threatening diseases such as cancer and AIDS, phase I studies are usually undertaken in healthy volunteers. Here, the pharmacokinetic characteristics of the drug are defined as in the preclinical studies. As described previously, the extent of availability after oral dosing and the half-life of the drug are two of the critical parameters used by the drug industry in making decisions as to whether further study of the drug is justified. For example, if first-pass elimination of a drug is extensive, as defined by Equation 17.21, the drug manufacturer must make a decision as to whether it is economically feasible to pursue further studies of this drug in humans. If the drug is unique in its pharmacodynamic characteristics, or is useful in treating life-threatening conditions, where intravenous dosing will be acceptable to the medical community, then companies may pursue further development, even though they anticipate that oral bioavailability may be low and consequently highly variable from patient to patient. Similarly, as discussed previously, drugs with very short half-lives will be useful only if a controlled-release dosage form of the drug can be developed to maintain necessary systemic concentrations without excessive dosing throughout the day. As described earlier in the section on preclinical development, Phase I studies are also useful to determine whether a drug exhibits nonlinear kinetics and whether clearance may change during multiple dosing, as a result of either induction or inhibition of elimination pathways. Another important part of Phase I evaluation is the effect of meals on drug bioavailability, both in terms of the extent and the rate of availability. The studies carried out in initial Phase I evaluation are used to predict the appropriate dose and the drug disposition characteristics that might be expected in patient populations.

5.3 Pharmacokinetics in Phase II and Phase III Studies

Once the drug disposition characteristics of a new molecular entity are defined, the drug must also be evaluated in the patient populations that will receive the drug. Phase II studies, the first studies in patients, are designed to select the appropriate dosage and conditions to be used in the pivotal multicenter large-scale clinical studies (phase III) required to prove safety and efficacy. Phase II and phase III studies do not routinely involve detailed studies designed to characterize the drug's pharmacokinetics. Rather, on the basis of the information gained in the Phase I stud-

ies, investigators predict the systemic concentration time course that would be observed in patients receiving the drug. Generally, a few plasma samples are taken during phase II and phase III studies, which are then used in a "population pharmacokinetic model" to determine whether the characteristics of the drug in the patient population differ from those determined in healthy volunteers. This knowledge, of course, is important to accurately predict the appropriate dosage regimen to be administered to the patient population, and to define the labeling for the dosage form to be marketed in the future.

5.4 Late Phase I Pharmacokinetic Studies

Once a drug sponsor has obtained information in patient populations through phase III studies that a drug is safe and effective, a number of so-called phase I studies are then carried out to complete the pharmacokinetic package of information to be supplied to regulatory agencies. These studies include evaluating the effects of disease states on the drug pharmacokinetics. For example, the effect of decreased renal function, the effect of hepatic disease, the effects of age and gender, and the evaluation of potential interactions with other drugs that the patients may be taking are carried out during the late phase I studies. These studies are characterized as phase I because they are usually carried out in subjects who do not have the disease for which the drug has been prescribed, but rather have a particular characteristic that the regulatory agencies and the company wish to have evaluated.

Often, the final dosage form to be marketed by the drug manufacturer is not exactly coincident with the dosage form that was used in the efficacy and safety studies in phase III patient populations. Thus, the drug sponsor must carry out a bioequivalence study between these dosage forms to ensure the regulatory agencies that the product to be marketed is equivalent, in terms of the active ingredients, to the dosage form used in the pivotal phase III studies proving efficacy and safety.

5.5 Pharmacokinetics in Phase IV

Phase IV constitutes studies that are carried out after regulatory approval and commercial sales of the product. In some cases, regulatory agencies may request the pharmaceutical manufacturer to carry out a drug interaction or a disease interaction study after approval of the drug for marketing. These studies are used to ensure that the labeling and the dosage recommendations under particular conditions are accurate. In addition, the pharmaceutical manufacturer may need to change the manufacturing processes for the dosage form after the drug is on the market. Under these conditions, bioequivalence studies would be required by the regulatory agencies to ensure that the new product is equivalent to that previously marketed by the company.

6 APPLICATIONS IN AREAS OTHER THAN DRUG DISCOVERY

In addition to the wide application of pharmacokinetics in the drug discovery process and in clinical pharmacology, pharmacokinetics has also been found to be extremely useful in environmental science. In toxicological studies, such as pesticide exposure, water contaminants exposure, and air-borne carcinogens exposure, pharmacokinetics becomes a valuable tool in evaluating the safety level of such compounds in humans or domestic animals.

REFERENCES

1. L. Z. Benet in B. G. Katzung, Ed., *Basic and Clinical Pharmacology*, Appleton & Lange, Norwalk, CT, 1992, pp. 35–48. (For updated coverage see: N. H. G. Holford in B. G. Katzung, Ed., *Basic and Clinical Pharmacology*, 6th ed., Appleton & Lang, Norwalk, CT, 1995, pp. 35–50.)

2. L. Z. Benet, *Eur. J. Resp. Dis.*, **65**, 45–61 (1984).

3. L. Z. Benet, J. R. Mitchell, and L. B. Sheiner in A. G. Gilman, T. W. Rall, A. S. Nies, and P. Taylor, Eds., *Goodman and Gilman's The Pharmacological Basis of Therapeutics*, Pergamon Press, Elmsford, NY, 1990, pp. 3–32. (For updated coverage see: G. R. Wilkinson in J. G. Hardman, L. E. Limbird, and A. G. Gilman, Eds., *Goodman and Gilman's The Pharmacological Basis of Therapeutics*, 10th ed., McGraw-Hill, New York, 2001, pp. 3–29.)

4. L. Z. Benet and R. L. Williams in A. G. Gilman, T. W. Rall, A. S. Nies, and P. Taylor, Eds., *Goodman and Gilman's The Pharmacological Basis*

of Therapeutics, Pergamon Press, Elmsford, NY, 1990, pp. 1650–1735. (For updated coverage see: E. M. Ross and T. P. Kenakin in J. G. Hardman, L. E. Limbird, and A. G. Gilman, Eds., *Goodman and Gilman's The Pharmacological Basis of Therapeutics*, 10th ed., McGraw-Hill, New York, 2001, pp. 31–43.)

5. L. Z. Benet in A. Yacobi, J. P. Skelly, V. P. Shag, and L. Z. Benet, Eds., *Integration of Pharmacokinetics, Pharmacodynamics and Toxicokinetics in Rational Drug Development*, Plenum Press, New York, 1993, pp. 115–123.

6. K. S. Pang and M. Rowland, *J. Pharmcokinet. Biopharm.*, **5**, 625–653 (1977).

7. H. Boxenbaum, *J. Pharmacokinet. Biopharm.*, **8**, 165–170 (1980). [For updated coverage see: H. Boxenbaum and C. DiLea, *J. Clin. Pharmacol.*, **35**, 957–966 (1995).]

8. J. Mordenti, S. A. Chen, J. A. Moore, B. L. Ferraiolo, and J. D. Green, *Pharm. Res.*, **8**, 1351–1359 (1991). [For updated coverage see: (a) T. Lave, P. Coassolo, and B. Reigner, *Clin. Pharmacokinet.*, **36**, 211–231 (1999); (b) R. S. Obach, *Curr. Opin. Drug Discov. Dev.*, **4**, 36–44 (2001); (c) K. A. Bachmann and R. Ghosh, *Curr. Drug Metab.*, **2**, 299–314 (2001).]

9. M. F. Hebert, J. P. Roberts, T. Prueksaritanont, and L. Z. Benet, *Clin. Pharmacol, Ther.*, **52**, 452–457 (1992). [For updated coverage see: (a) K. E. Thummel and G. R. Wilkinson, *Annu. Rev. Pharmacol. Toxicol.*, **38**, 389–430 (1998); (b) S. D. Hall, K. E. Thummel, P. B. Watkins, K. S. Lown, L. Z. Benet, M. F. Paine, R. R. Mayo, D. K. Turgeon, D. G. Bailey, R. J. Fontana, and S. A. Wrighton, *Drug Metab. Dispos.*, **27**, 161–166 (1999).]

10. R. J. Wills, S. Dennis, H. E. Spiegel, D. M. Gibson, and P. I. Nadler, *Clin. Pharmacol. Ther.*, **35**, 722–727 (1984).

11. L. Z. Benet and R. L. Galeazzi, *J. Pharm. Sci.*, **68**, 1071–1074 (1979).

12. U. Klotz, G. R. Avant, A. Hoyumpa, S. Schenker, and G. R. Wilkinson, *J. Clin. Invest.*, **55**, 347–359 (1975). [For updated coverage see: D. J. Greenblatt, *J. Clin. Psychiatry*, **54**, 8–13 and 55–56 (1993).]

13. R. L. Williams, T. F. Blaschke, P. J. Meffin, K. L. Melmon, and M. Rowland, *Clin. Pharmacol. Ther.*, **21**, 301–309 (1977). [For updated coverage see: L. Z. Benet and B. A. Hoener, *Clin. Pharmacol. Ther.*, **71**, 115–121 (2002).]

14. B. L. Ferraiolo and L. Z. Benet in R. T. Borchard, A. J. Repta, and V. J. Stella, Eds., *Directed Drug Delivery: The Multi-disciplinary Problem*, Humana Press, Clifton, NJ, 1985, pp. 13–33.

15. C. A. Gloff and L. Z. Benet, *Adv. Drug Deliv. Rev.*, **4**, 359–386 (1990).

CHAPTER EIGHTEEN

Physicochemical Characterization and Principles of Oral Dosage Form Selection

Gregory E. Amidon
Xiaorong He
Michael J. Hageman
Pharmacia Corporation
Kalamazoo, Michigan

Contents

Burger's Medicinal Chemistry and Drug Discovery
Sixth Edition, Volume 2: Drug Development
Edited by Donald J. Abraham
ISBN 0-471-37028-2　© 2003 John Wiley & Sons, Inc.

1 INTRODUCTION

Although thousands of compounds, really hundreds of thousands in today's pharmaceutical industry, are synthesized and evaluated every year, very few make it to clinical testing and fewer still make it to the market. The reasons for failure are many. One thing is always true though: a marketable dosage form with the desired drug delivery properties must be developed to commercialize a product! Because of the challenges associated with drug discovery and development, the opportunity to identify and develop a safe and effective product benefits greatly from the integration of pharmacology, chemistry, toxicology, metabolism, clinical research, and formulation development. Every discovery team will benefit by keeping this in mind. The ability to identify a suitable dosage form can make or break a product. The dosage form must achieve the desired concentration at the desired site (often considered the blood) for the desired duration. Furthermore, the dosage form must be robust and manufacturable!

To initiate formulation development activities—that is, the identification of a "marketable" product—important physical and chemical properties (physicochemical properties) as well as permeability properties need to be determined. Evaluation of these properties early in the discovery process can help discovery teams not only to select which templates to pursue but also to identify the most promising leads. This information is also valuable "downstream" as decisions regarding dosage form design are being made. The focus of this chapter is primarily on those physicochemical properties that are most important in the evaluation of discovery leads and provides the information and guidance needed as dosage form development is started and progresses. It will become apparent that the dosage forms that make sense to consider are dictated to a large extent by physicochemical properties. Throughout this chapter there is an emphasis on the tools and principles that will be helpful to the medicinal chemist during the drug discovery process, with particular emphasis on oral dosage forms.

It must be emphasized that, throughout the discovery and development process, collaboration between the drug discovery experts, medicinal chemistry scientists, and clinical development and drug delivery scientists is extremely important in identifying lead compounds that have the best chance of surviving the development process and lead to a marketable product. In some ways, drug discovery and development is a bit like a crapshoot; the best we can do is load the dice in our favor to improve our odds of finding successful compounds by understanding and applying sound scientific principles.

2 SOLID FORM SELECTION

Usually, for oral drug delivery, crystalline solids are preferred. This is especially true for solid dosage forms such as tablets or capsules, where the solid form of the active ingredient is often a critical component in determining dosage form manufacturing, performance, and stability. Ideally, though not always, the most thermodynamically stable form is used because it will generally provide the greatest physical and chemical stability. Therefore, early identification and selection of the solid form to be used in development becomes paramount, given its direct impact on physicochemical and drug delivery attributes.

Many of the physicochemical properties discussed in this chapter are, in fact, dependent on the solid form. Aqueous solubility, hygroscopicity, and chemical stability are three obvious examples where very large differences may exist between solid forms. It is therefore valuable to begin a rigorous process of identifying solid forms early in the discovery process. This information is valuable as discovery efforts proceed. As lead compounds are identified, a systematic approach to the synthesis of crystalline salts should take place where possible and appropriate by use of different solvents and counterions (1). Once salts of interest are identified, a systematic approach to the crystallization, identification, and characterization of all solid forms including salts should be undertaken. This often requires a systematic approach to crystallization, often from a variety of solvents, to identify polymorphs as well as alternative pseudopolymorphs such as solvates (2, 3). Selection of the right solid form or salt can allow the formulation scientist to design the dosage form with optimal physicochemical and drug-release properties. A thorough understanding of the polymorphs, pseudopolymorphs, hydrates, solvates, salt forms, and amorphous forms maximizes the opportunity to understand, control, and predict the behavior of a compound in the solid state, identify the appropriate dosage forms to consider, and develop a marketable product.

This section provides a brief overview of solid form considerations. The physicochemical characterization described in this chapter can, in effect, be applied to each of the forms that have been identified and isolated because each solid form will have a unique set of physicochemical properties. Careful consideration of these properties will inevitably lead to the

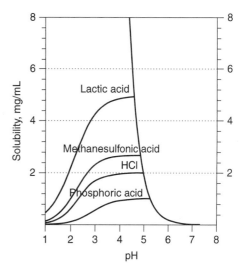

Figure 18.1. pH-solubility profile of different acids of a free base.

identification of better lead compounds and forms with which to enter development.

2.1 Salts

The preparation of salts is frequently undertaken to improve the physicochemical properties of an ionizable compound. Most often, improvement of solubility and dissolution rate is desired. However, improvement in crystallinity (e.g., melting point), stability, or hygroscopicity may be possible (1, 2). Figure 18.1 shows the solubility of several salts of terfenadine and the free base as a function of pH. Development of salts, particularly soluble salts of insoluble compounds, is not without challenge, however. Complete characterization of salt forms is needed (4). If the salt is very soluble, precipitation of the insoluble free base (or free acid) may occur *in vivo* under physiological pH conditions.

Systematic screening for salts is advantageous at an early stage to identify the most desirable salts from the perspective of solubility, dissolution rate, solid state stability, hygroscopicity, toxicity, and drug delivery. In particular, consideration of the toxicological properties of the counterion is needed. For some counterions, toxicological concerns may exist because of the high doses needed during drug safety evaluation, whereas the quantity

Table 18.1   Selected Counterions Suitable for Salt Formation

	Anions	Cations

Preferred
    Universally accepted by regulatory agencies with no significant
        limit on quantities
	Citrate[a]
	Hydrochloride[a]
	Phosphate[a]
		Calcium[a]
		Potassium[a]
		Sodium[a]

Generally accepted
    Generally accepted but with some limitation on quantities
        and/or route of administration
	Acetate[a]
	Besylate
	Gluconate
	Lactate
	Malate
	Mesylate[a]
	Nitrate[a]
	Succinate
	Sulfate[a]
	Tartrate[a]
	Tannate
	Tosylate
		Choline
		Ethylenediamine
		Tromethamine

Suitable for use
    More limited approval with some limitation on quantities
        which are acceptable in safety studies or human use
        and/or route of administration
	Adipate
	Benzoate
	Cholate
	Edisylate
	Hydrobromide
	Fumarate
	Maleate[a]
	Napsylate
	Pamoate
	Stearate
		Arginine
		Glycine
		Lycine
		Magnesium
		Meglumine

[a]Commonly used.

of the counterion present in the marketed dosage form may not be problematic. Table 18.1 is a summary of the most commonly used counterions for pharmaceutical salt formation. A review of U.S. Food and Drug Administration (FDA) approved drugs shows over 60 different counterions have been used (1).

2.2   Polymorphs

Polymorphs differ in solid crystalline phase structure (crystal packing) but are identical in the liquid and vapor states (5). As such, different polymorphs may exhibit very different properties. In effect, two different polymorphs of the same molecule may be as different in their physicochemical properties as the crystals of two different compounds. Properties such as melting point, solubility, dissolution rate, stability, hygroscopicity, and density can all vary with polymorph. In fact, it is a safe bet that multiple polymorphs will exist of a crystalline form and a large number of pharmaceutical compounds have been shown to crystallize in multiple polymorphic forms. It is therefore appropriate and even "mandatory" to actively pursue polymorph identification (2, 6). The first form isolated in the medicinal chemistry laboratory is not necessarily the most stable, soluble, or desirable form for testing or development. Identification of the most stable form, which is generally the form that is most desirable for development, is critical to the successful development of a stable solid dosage form. It is also valuable to identify and characterize the other metastable polymorphic forms that may have more desirable properties. Even if a metastable crystalline form is ultimately chosen for development, an understanding of the properties of the thermodynamically stable form is critical to the development of a dosage form with maximum stability.

2.3 Pseudopolymorphs

Pseudopolymorphs are not strictly polymorphs because they differ from each other in the solid crystalline phase, such as through the incorporation of solvents (solvates) or water (hydrates). Pseudopolymorphs may also be suitable for development and should be identified and evaluated during solid form screening activities.

2.3.1 Hydrates. Generally, hydrates are considered appropriate pseudopolymorphs for development. Many drugs are marketed as hydrates, presumably because they are either the most stable form at "typical" relative humidities or because they offer other drug delivery advantages. Hydrates often, though not always, are less soluble in water than the corresponding anhydrous form. If the hydrate is less soluble, it often crystallizes when the anhydrous form is suspended in water and allowed to equilibrate.

An important consideration for hydrates is the humidity range in which interconversion of the anhydrous form and hydrate occurs. In theory, exposing anhydrous or hydrous forms to different relative humidities and evaluating the solid form and moisture content present at "equilibrium" can characterize this. However, the interconversion rate between anhydrous and hydrate forms is often slow. A more rapid and efficient method is to slurry anhydrous and hydrate forms in solvents of varying water activity (7, 8). The slurry facilitates and speeds conversion to the most stable form at the specified water activity and temperature. Characterization of the solid form at equilibrium in the slurry allows a more precise determination of the relative humidity (RH; i.e., water activity in the vapor phase) at which conversion occurs. Appropriate selection of excipients, manufacturing processing conditions, packaging, and storage conditions may minimize changes in form that can occur for such compounds in a dosage form. However, the development needs and regulatory burdens to demonstrate adequate physical stability (e.g., no conversion to another form in the dosage form) may be undesirable (6) and it is often best to avoid these forms if possible.

As a general rule, anhydrous forms that do not convert to the hydrate below 75% RH (at equilibrium) are likely to exhibit adequate physical and chemical stability in oral solid dosage forms. Adequate manufacturing and packaging can be designed to protect most oral solid dosage forms from exposure to >75% RH. Conversely, hydrates that do not convert to the anhydrous form until the relative humidity drops below about 20% (at equilibrium) are also likely to exhibit adequate physical stability in solid dosage forms.

2.3.2 Solvates. Generally, solvates are undesirable as a final form because the solvent is frequently unacceptable for human use. However, the formation of solvates can frequently occur during drug synthesis, and understanding their physicochemical properties is a key to proper control and crystallization of the desired solid form. Desolvated solvates can retain the structure of the solvate even after the solvent is removed by drying or vacuum. Small changes in lattice parameters may occur and the remaining desolvated solvate form tends to be less ordered (2). Under some circumstances, a desolvated solvate may be used as an intermediate in the preparation of the final solid form.

2.4 Amorphous Solids

Although development of oral dosage forms most often takes advantage of crystalline solid forms, amorphous solids have occasionally been used and offer some unique opportunities to overcome limitations associated with crystalline forms. This is sometimes necessary if no crystalline forms can be identified. Amorphous solids may be prepared by solvent evaporation, freeze-drying, or coprecipitation processes. Because of their higher energy state, amorphous solids may be used to improve solubility or dissolution rate. However, caution and careful formulation is required to ensure that no undesirable form changes occur within the dosage form that could compromise physical or chemical properties of the product throughout its shelf life. Amorphous solids are generally quite hygroscopic, less chemically stable than crystalline forms, and difficult to handle.

Table 18.2 Important Physical, Chemical, and Biological Properties for Oral Drug Delivery

Physical Properties	Chemical Properties	Biological Properties
Polymorphic form(s)	Ionization constant (pK_a)	Membrane permeability
Crystallinity	Solubility product (K_{sp}) of salt forms	Gut metabolism
Melting point	Chemical stability in solution	First-pass metabolism
Particle size, shape, surface area	Chemical stability in solid state	Systemic metabolism
Density	Photolytic stability	
Hygroscopicity	Oxidative stability	
Aqueous solubility as a function of pH	Incompatibility with formulation additives	
Solubility in organic solvents	Complexation with formulation additives	
Solubility in presence of surfactants (e.g., bile acids)		
Dissolution rate		
Wettability		
Partition coefficient (octanol-water)		

2.5 Nonsolid Forms

Nonsolid forms of active pharmaceutical ingredients are rarely used in oral dosage forms primarily because of difficulties in isolation and in carrying out weighing and processing steps commonly used in pharmaceutical manufacturing. It is possible, however, to incorporate liquids directly into liquid formulations such as syrups and in liquid-filled capsule preparations. However, this approach rarely offers any advantage over the use of a solid form and should be avoided if at all possible through the use of appropriate solid form crystallization or salt selection studies.

3 PHYSICOCHEMICAL PROPERTY EVALUATION

Measurement of key physicochemical properties of the most relevant solid forms of lead compounds early in the discovery process can help discovery teams not only to select which templates to pursue but also to identify promising leads. This information is particularly valuable "downstream" as decisions regarding dosage form design are being made. Some of the key physical, chemical, and biological properties that should be of interest to both the discovery team and the formulation scientist are listed in Table 18.2. Of these, the physicochemical properties that have a clear impact on the feasibility of oral formulation development are: melting point, partition coefficient, aqueous solubility, biological membrane permeability, hygroscopicity, ionization constants, solution stability, and solid state stability. These are discussed in greater detail in the following sections. It should be kept in mind that many of these properties are dependent on the solid form, and complete characterization of each of the most relevant solid forms is needed to provide a complete physicochemical picture.

Many of the physicochemical properties of interest are dependent on the solid form and, unfortunately, successful prediction of polymorphic forms is inexact. This, in combination with the fact that prediction of physicochemical properties is also very challenging, makes *ab initio* prediction very difficult and imprecise. However, some discussion of predictive tools is included in this chapter. A general comment regarding *ab initio* prediction is that "order of magnitude" predictions may be possible once some basic physicochemical information is available. However, the complexity and diversity of the chemistry space make reliable predictions across a broad spectrum of chemical structures very difficult. It is not surprising then that physicochemical predictions across more narrowly defined chemical spaces (e.g., chemical or therapeutic classes) can be more reliable and useful. Drug delivery, formulation, and computational chemistry experts will likely be able to provide a useful perspective on opportunities to take advantage of such *ab initio* predictions within the chemistry space that discovery teams often operate.

3.1 Melting Point

Melting point is defined as the temperature at which the solid phase exists in equilibrium with its liquid phase. As such, the melting point is a measure of the "energy" required to overcome the attractive forces that hold the crystal together. Melting point determination is of great value and can successfully be accomplished by any of several commonly used methods, including visual observation of the melting of material packed in a capillary tube (Thiele arrangement), hot-stage microscopy, or other thermal analysis methods such as differential scanning calorimetry (DSC). Careful characterization of thermal properties such as that possible with DSC provides the investigator with an opportunity to assess and quantify the presence of impurities as well as the presence or interconversion of polymorphs and pseudopolymorphs. Melting points and the energetics of desolvation can also be evaluated, as can the enthalpies of fusion for different solid forms.

As a practical matter, low melting materials tend to be more difficult to handle in conventional solid dosage forms. Melting points below about 60°C are generally considered to be problematic and melting points greater than 100°C are considered desirable. Temperatures in conventional manufacturing equipment such as high shear granulation equipment, fluid bed granulation, and drying as well as production tablet machines can exceed 40–80°C. Although amorphous solids do not have a distinct melting point, they undergo softening as temperatures approach the glass-transition temperature. Furthermore, common handling procedures (e.g., weighing, processing) can be difficult for low melting materials. Alternative dosage forms (liquid type) may be required for liquid or low melting materials. A comparison of melting points of polymorphs also provides a perspective on the relative stability of polymorphic forms (5). For monotropic polymorphs, the highest melting polymorph is the most stable at all temperatures, whereas for enantiotropic polymorphs, the highest melting polymorph is not necessarily the most stable at all conditions (5).

Ab initio prediction of melting point is not currently very practical because there is no general relationship yet that relates melting points of compounds to chemical structure. Some success has been achieved for small data sets of hydrocarbons and substituted aromatics (9). Yalkowsky and coworkers (10–12) have had some success in the use of a group contribution and molecular geometry approach to predict the melting points of aliphatic compounds. To date, melting point predictions for these limited data sets are in the range of ±35°C. The melting point of organic molecules is primarily controlled by the intermolecular forces (van der Waals, dipolar forces, hydrogen bonds) and molecular symmetry (11). Computational tools to predict polymorphs and melting points have been used to a limited extent thus far (13). Greater molecular symmetry, which determines how efficiently molecules will pack in a crystal, and the presence of hydrogen donor groups both significantly increase intermolecular interactions in the solid state and increase melting point.

3.2 Partition Coefficient

The partition coefficient is defined for dilute solutions as the molar concentration ratio of a single, neutral species between two phases at equilibrium:

$$P = [A]_o/[A]_w \qquad (18.1)$$

Usually the logarithm (base 10) of the partition coefficient ($\log P$) is used because partition coefficient values may range over 8–10 orders or magnitude. Indeed, the partition coefficient, typically the octanol-water partition coefficient, has become a widely used and studied physicochemical parameter in a variety of fields, including medicinal chemistry, physical chemistry, pharmaceutics, environmental science, and toxicology. Although P is the partition coefficient notation generally used in the pharmaceutical and medicinal chemistry literature, environmental and toxicological sciences have more traditionally used the term K or K_{ow}.

One of the earliest applications of oil/water partitioning to explain pharmacological activity was the work of Overton (14) and Meyer (15) over a century ago, which demonstrated that narcotic potency tended to increase with oil/water partition coefficient. The estimation and application of partition coefficient data to drug delivery began to grow rapidly in the 1960s (16–21), to become one of the most widely used and studied physicochemical parameters in medicinal chemistry and pharmaceutics.

Selection of the octanol-water system is often justified in part because, like biological membrane components, octanol is flexible and contains a polar head and a nonpolar tail. Hence, the tendency of a drug molecule to leave the aqueous phase and partition into octanol is viewed as a measure of how efficiently a drug will partition into and diffuse across biological barriers such as the intestinal membrane. Although the octanol-water partition coefficient is, by far, most commonly used, other solvent systems such as cyclohexane-water and chloroform-water systems offer additional insight into partitioning phenomena.

Partition coefficients are relatively simple to measure, at least in principle. However, the devil is in the details and certain aspects demand sufficient attention that rapid throughput methodologies have not yet been successfully developed to cover a broad range of partitioning. Several recent reviews of experimental methods provide an abundance of practical information on the accurate determination of partition coefficients (22, 23). Indeed, some of the motivation to develop reliable predictions of partition coefficient lies in the fact that measurement is often time-consuming and challenging (24).

With the widespread application of lipophilicity and partitioning to biophysical processes, a wide variety of tools is currently available to estimate partition coefficient. Several recent reviews of programs and methods that are commercially available have been published (25–28). Predictive methods may be broken down into the following basic approaches to partition coefficient estimation: (1) group contribution methods, by use of molecular fragments; (2) group contribution, by use of atom-based contributions; (3) conformation-dependent or molecular methods; (4) combined fragment and atom-based methods; and (5) other physicochemical methods.

As mentioned earlier, partition coefficient refers to the distribution of the neutral species. For ionizable drugs in which the ionized species does not partition into the organic phase, the apparent partition coefficient (D) can be calculated from the following:

Acids:
$$\log D = \log P - \log(1 + 10^{(\mathrm{pH}-\mathrm{p}K_a)}) \quad (18.2)$$

Bases:
$$\log D = \log P - \log(1 + 10^{(\mathrm{p}K_a-\mathrm{pH})}) \quad (18.3)$$

Critical reviews of computational methods are available in the literature (24–28). Each computational method has strengths and weaknesses, although it is important to keep in mind that any computational tool is only as good as its database and that extrapolation to compound structures that lie outside that data set is risky. In general, predictive methods can be viewed as providing the best estimates for the chemistry spaces used to develop the model. For medicinal chemists, this means that the greatest success is likely to be achieved with the development of specific relationships for the class of compounds of interest. Measurement of representative compounds within a therapeutic class may very likely allow more accurate prediction of the properties of the entire class.

3.3 Aqueous Solubility

At its simplest, the importance of aqueous solubility in determining oral absorption can be seen from the following equation describing the flux of drug across the intestinal membrane:

$$\mathrm{Flux} = P_m(C_i - C_b) \quad (18.4)$$

where P_m is the intestinal membrane permeability, C_i is the aqueous drug concentration (un-ionized) in the intestine, and C_b is the portal blood concentration (un-ionized).

If solid drug is present in the intestine, the concentration in the intestinal tract may approach or equal its aqueous solubility, S, if dissolution and release of drug from the dosage form is sufficiently rapid ($C_i \approx S$). From

Equation 18.4, then, it is apparent that the flux of drug across the intestine is directly proportional to its aqueous solubility. For drugs that have high intestinal membrane permeability (P_m) aqueous solubility may be the limiting factor to adequate drug absorption. Generally, only the un-ionized species is absorbed; thus, for ionizable compounds, the concentration of the un-ionized form should be considered in Equation 18.4.

Generally, drug solubility is determined by adding excess drug to well-defined aqueous media and agitating until equilibrium is achieved. Appropriate temperature control, solute purity, agitation rate, and time as well as monitoring of the solid phase at equilibration are needed to ensure that high quality solubility data are obtained (29). In particular, it is important to evaluate the suspended solid form present at equilibrium because conversion to another solid form (e.g., polymorph, pseudopolymorph, hydrate, salt) may occur during equilibration. If a form change has occurred, the measured solubility is more likely representative of the solubility of the final form present rather than the starting material. Efforts to develop high throughput screening methods to measure or classify solubility have recently been undertaken (30) with some success, although they generally suffer from being a dynamic measure (i.e., not equilibrium) and increased variability attributed to higher throughput.

A number of different techniques have been proposed for estimating aqueous solubility. They can broadly be classified as (1) methods based on group contributions, (2) techniques based on experimental or predicted physicochemical properties (e.g., partition coefficient, melting point), (3) methods based on molecular structure (e.g., molar volume, molecular surface area, topological indices), and (4) methods that use a combination of approaches (29, 31, 32). Although all of the methods have some theoretical basis, their use in predicting aqueous solubility is largely empirical. Detailed discussions may be found in the literature of the fundamentals of solubility measurement and prediction (29, 30). Each approach has advantages and has been successfully applied to a variety of classes of compounds to develop and test the accuracy

of solubility predictions. Usually, approaches that are developed from structurally related analogs yield more accurate predictions (32).

Aqueous solubility is determined, in a simple sense, by the interaction of solute molecules in the crystal lattice, interactions in solution, and the entropy changes as the solute passes from the solid phase to the solution phase. Accordingly, the pioneering work of Yalkowsky and Valvani (33) successfully estimated the solubility of rigid short-chain nonelectrolytes with the following equation:

$$\log(S) = -\log(P) - 0.01(\text{MP}) + 1.05$$

$$(18.5)$$

where S is molar solubility, P is the octanol-water partition coefficient, and MP is the melting point in °C.

The Yalkowsky-Valvani equation provides insight into the relative importance of crystal energy (melting point) and lipophilicity (partition coefficient). This semiempirical approach has subsequently been applied to and refined for a variety of solutes and classes of compounds (33–37). From Equation 18.5, one can see that the octanol-water partition coefficient is a significant predictor of aqueous solubility. A 1 log unit change in aqueous solubility can be expected for each log unit change in partition coefficient. By comparison, a melting point change of 100°C is required to have the same 1 log unit change on solubility. The Yalkowsy-Valvani and similar equations can be used to predict aqueous solubility often within a factor of 2, by use of predicted partition coefficients and measured melting points.

Aqueous solubility prediction continues to be an active area of research, with a wide variety of approaches being applied to this important and challenging area. To date, group contribution approaches as well as correlation with physicochemical properties (partition coefficient) appear to be the most promising (29, 32). It is important to keep in mind that correlations that are developed from structurally related analogs would consistently yield more accurate predictions.

3.4   Dissolution Rate

Aqueous solubility can also play a critical role in the rate of dissolution of drug and hence release from dosage forms. The dissolution rate of a solute from a solid was shown by Noyes and Whitney (38) to be

$$\text{Dissolution rate} = (A/V)(D/h)(S - C)$$

$$(18.6)$$

where A is the drug surface area; V is the volume of dissolution medium; D is the aqueous diffusion coefficient; h is the "aqueous diffusion layer thickness," which is dependent on viscosity and agitation rate; S is the aqueous drug solubility at the surface of the dissolving solid; and C is the concentration of drug in the bulk aqueous phase.

From the Noyes-Whitney equation, the dissolution rate is seen to be directly proportional to the aqueous solubility (S) as well as the surface area (A) of drug exposed to the dissolution medium. It is common practice, especially for low solubility drugs, to increase the dissolution rate by increasing the surface area of a drug through particle size reduction. If the drug surface area is too low, the dissolution rate may be too slow and absorption may become dissolution rate limited. For high solubility drugs, the dissolution rate is generally fast enough that a high drug concentration is achieved in the lumen and extensive particle size reduction is not needed. Synthesis of high solubility salts of weak acids or bases is commonly undertaken to facilitate rapid dissolution in the gastrointestinal (GI) tract. On the basis of theoretical considerations (39), as a rough "rule of thumb," if the particle diameter (in micrometers) is less than the aqueous solubility (in $\mu g/mL$), further particle size reduction is probably not needed to achieve conventional immediate release dissolution profiles.

3.5   Permeability

Lead compounds generated in today's pharmaceutical research environment, frequently through the use of high throughput screen programs, often have unfavorable biopharmaceutical properties. These compounds are generally more lipophilic, less soluble, and of higher molecular weight (30). Indeed, permeability, solubility, and dose, referred to as the *triad* (30), determine whether a drug molecule can be developed into a commercially viable product with the desired properties. As described earlier by Equation 18.4, intestinal permeability can be critically important both in controlling the rate and extent of absorption and in achieving desired plasma levels.

It is often assumed that the major factors determining transport of drugs across the intestinal membrane are molecular size and hydrophobicity. Although true to some extent, this is a simplistic view of drug transport across a complex biological barrier (40). There are essentially four mechanisms by which drugs may cross the intestinal membrane: passive diffusion, carrier-mediated transport, endocytosis, and paracellular transport. Of these, passive diffusion across the intestinal membrane follows Fick's law (Equation 18.4) and is often the major mechanism for low molecular weight, lipophilic compounds. Transport is proportional to intestinal membrane permeability. Passive transport depends to a large extent on three interdependent physicochemical parameters: lipophilicity, polarity, and molecular size (41–43). Maximizing passive absorption generally involves optimizing these properties. Molecules, most commonly hydrophilic in nature, may also pass through the tight junctions that exist between adjacent epithelial cells. However, tight junctions are estimated to constitute only 0.1% of the surface area of the intestine (44) and so limit this mechanism. Some compounds are reported to open tight junctions, thereby increasing paracellular transport.

Epithelial cells are also known to contain P-glycoprotein (P-gp) efflux pumps that serve to pump drugs out of the cells and back into the lumen against a concentration gradient. The small intestine is particularly rich in P-gp pumps and this mechanism has been shown to limit the oral absorption of a variety of molecules such as cyclosporin, digoxin, ranitidine, and cytotoxic drugs (40, 45, 46). The role of efflux pumps in drug absorption is a topic of great interest and research and its importance may be greater than is currently recognized (46).

Although physicochemical descriptors of drug molecules are generally not adequate to

Table 18.3   Rule of Five[a]

Molecular weight > 500	
Log P > 5	
Number of H-bond acceptors > 10	(sum of nitrogens and oxygens in molecule)
Number of H-bond donors > 5	(sum of OHs and NHs in molecule)

[a]Ref. 30.

precisely predict oral bioavailability, there is certainly value to the medicinal chemist to understand some of the basic molecular properties that influence permeability. Of these, passive transport in particular depends to a large extent on three interdependent physicochemical parameters: lipophilicity, polarity, and molecular size (41, 42, 47–50).

Drug lipophilicity is widely used as a predictor of membrane permeability, given that partitioning of drug into the lipophilic epithelial cells is a necessary step for passive diffusion. Chief among the measures of lipophilicity is the octanol-water partition coefficient discussed earlier in greater detail. However, lipophilicity alone is inadequate to accurately predict bioavailability, and this is no surprise considering the complicated multifaceted nature of transport across the intestinal membrane (51). Even when limiting predictions to drugs that are absorbed by passive diffusion, only an approximate relationship between lipophilicity and permeability is observed (41). However, excessively high lipophilicity is clearly a detriment to efficient absorption, probably because it is reflected in a very low aqueous solubility. Also, highly lipophilic drugs may become sequestered in the cell, with little improvement in permeability across the membrane. In a recent study by Lipinski and coworkers (48), only about 10% of the compounds that entered late-stage clinical testing (phase II) had a log P value greater than 5.

Additional factors that appear to influence permeability are polarity and molecular weight. An excessive number of hydrogen bond donors and hydrogen bond acceptors have been correlated to decreased permeability. Lipinski and others (48, 52, 53) also concluded that molecules with more than five hydrogen bond donors (number of NH + number of OH) and more than 10 hydrogen bond acceptors (number of nitrogens + number of oxygens) are not common in compounds that have reached later-stage clinical development and this is likely the result of decreased absorption. Finally, they conclude that molecular weight also appears to be a factor. Very few compounds with a molecular weight greater than 500 proceed very far in development. On the basis of these observations, Lipinski and coworkers defined the "Rule of Five" as described in Table 18.3 as a reasonable rule-based guideline to consider (30). If molecules exceed two or more of the "limits," the medicinal chemist should be concerned that oral absorption may be a significant problem.

With the difficulties associated with accurate estimation of permeability based only on physicochemical properties, a variety of methods of measuring permeability have been developed and used, among which are: (1) cultured monolayer cell systems, such as Caco-2 or MDCK; (2) diffusion cell systems that use small sections of intestinal mucosa between two chambers; (3) in situ intestinal perfusion experiments performed in anesthetized animals such as rats; and (4) intestinal perfusion studies performed in humans (40, 54–62). All of these methods offer opportunities to study transport of drug across biological membranes under well-controlled conditions. Caco-2 monolayer systems in particular have become increasingly commonly used in recent years and human intestinal perfusion methods are also becoming more commonly available. Correlations between Caco-2 permeability and absorption in humans have been developed in several laboratories (63–72). As shown in Fig. 18.2, a correlation between absorption in humans and Caco-2 permeability is obtained in each laboratory but with a displacement of the curve (73). A comparison of the variability in Caco-2 permeability values from different laboratories demonstrates that direct comparison of results between laboratories must be done with caution. For this reason it is neces-

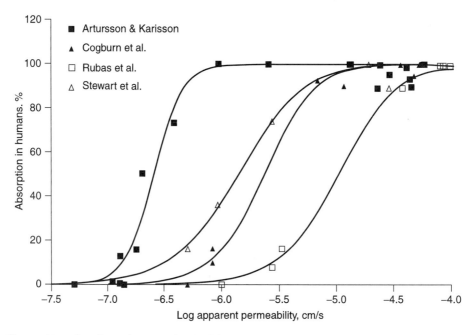

Figure 18.2. Correlation between absorbed fraction in humans after oral administration and permeability in Caco-2 monolayers obtained from four different laboratories. [Reprinted with permission of the publisher (73).]

sary to use a set of reference compounds to accurately characterize a drug molecule as poorly or highly permeable. Generally the best correlation to the *in vivo* situation is obtained for drugs absorbed by passive transport, but transport by different mechanisms may also be characterized by the use of these *in vitro* and *in situ* methods, depending on selected cell lines or *in situ* models.

3.6 Ionization Constant

Knowledge of acid-base ionization properties is essential to an understanding of solubility properties, partitioning, complexation, chemical stability, and drug absorption. The ionized molecule exhibits markedly different properties from the corresponding un-ionized form.

For weak acids, the equilibrium between the free acid (HA) and its conjugate base (A$^-$) is described by the following equilibrium equation:

$$HA = H^+ + A^- \qquad (18.7)$$

and the corresponding acid dissociation constant (K_a) is given by

$$K_a = [H^+][A^-]/[HA] \qquad (18.8)$$

By definition, pK_a is described as

$$pK_a = -\log(K_a) = pH + \log[HA]/[A^-] \qquad (18.9)$$

Corresponding equations for a weak base (B) and its conjugate acid (BH$^+$) are described by the following:

$$BH^+ = B + H^+ \qquad (18.10)$$

$$K_a = [B][H^+]/[BH^+] \qquad (18.11)$$

$$pK_a = pH + \log[BH^+]/[B] \qquad (18.12)$$

Of particular interest to the medicinal chemist and formulation scientist is the impact of pK_a on apparent aqueous solubility and partitioning.

Taking a weak base as an example, the total aqueous solubility (S_T) is equal to the sum of the ionized ([BH$^+$]) and un-ionized species ([B]) concentrations in solution.

$$S_T = [B] + [BH^+] \qquad (18.13)$$

$$S_T = [B] + [B][H^+]/K_a = [B](1 + [H^+]/K_a) \qquad (18.14)$$

Generally the un-ionized form, in this case the free base, is the less soluble species in water. The solubility of the un-ionized free base form is defined as the intrinsic solubility (S_b).

Assuming that the solution is saturated with respect to free base at all pH values, Equation 18.14 can be expressed as

$$S_T = S_b(1 + [H^+]/K_a) \qquad (18.15)$$

or,

$$S_T = S_b(1 + 10^{(pK_a - pH)}) \qquad (18.16)$$

The corresponding solubility equation for a weak acid with an intrinsic solubility of S_a is given by

$$S_T = S_a(1 + K_a/[H^+]) \qquad (18.17)$$

and,

$$S_T = S_a(1 + 10^{(pH - pK_a)}) \qquad (18.18)$$

Typical solubility profiles are shown in Fig. 18.3 for a weak acid and a weak base and several significant conclusions and implications are worth pointing out. Taking the free base as an example once again, at pH values greater than the pK_a, the predominant form present in solution is the un-ionized form; hence the total solubility is essentially equal to the intrinsic solubility. At pH = pK_a, the drug is 50% ionized and the total solubility is equal to twice the intrinsic solubility. As the pH drops significantly below the pK_a, a rapid increase in total solubility is observed because the percentage ionized is dramatically increasing. In fact, for each unit decrease in pH, the total aqueous solubility will increase 10-fold, as seen in Fig. 18.3. The total solubility will continue to increase in such a manner as long as the ionized form continues to be soluble. Such dramatic increases in solubility as a function of pH demonstrate the importance of control-

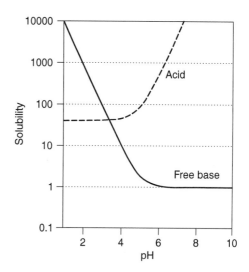

Figure 18.3. pH-solubility profile for a free base and acid.

ling solution pH and also offer the formulation scientist a number of possible opportunities to modify dosage form and factors leading to oral absorption properties.

For weak acids, one will observe a rapid increase in total solubility as the pH exceeds the pK_a because, in this case, the ionized conjugate base concentration will increase with increasing pH. Often, for weak acids and bases, the medicinal chemist and formulation scientist must understand the solubility properties of both the un-ionized species and its corresponding conjugate form because each may limit solubility. In this regard, the work of Kramer and Flynn (74) is particularly instructive. As seen in their work, the free base solubility curve is as predicted at high pH. In this pH range, the free base form is the least soluble form and limits the total solubility, as predicted by equation (18.16). Also shown in their work (similar to Fig. 18.1) is the solubility curve for the corresponding salt. At low pH, it is the solubility of the salt that limits the total solubility. From this solubility curve, one can correctly conclude which solid form will exist at equilibrium as a function of pH. This basic principle is of significance *in vivo*, for example, because one might imagine dosing patients with a soluble salt, which could rapidly dissolve in the low pH of the stomach, but as the drug in the gastric contents entered the intestine where solution pH approaches neutral, precipitation of the free base could occur. Such changes have been proposed as an explanation for the poor bioavailability of highly soluble salts of weak bases.

There are currently a number of software packages that allow for reasonably accurate pK_a estimates based on a variety of approaches, including the application of linear free energy relationships based on group contributions, chemical reactivity, and calculated atomic charges. In addition to predictive tools, a variety of reliable methods for measuring pK_a values are available, including tritrametric and spectroscopic methods as well as aqueous solubility curve measurements as a function of pH.

3.7 Hygroscopicity

Moisture uptake is a significant concern for pharmaceutical powders. Moisture has been shown to have a significant impact, for example, on the physical, chemical, and manufacturing properties of drugs, excipients, and formulations. It is also a key factor in decisions related to packaging, storage, handling, and shelf life, and successful development requires a sound understanding of hygroscopic properties. Moisture sorption isotherms relate the equilibrium water content of the solid material to the atmospheric relative humidity (i.e., water activity in the vapor phase) to which it is exposed. Isotherms can yield an abundance of information regarding the physical state of the solid and the conditions under which significant changes may occur. Conversion from an anhydrous form to a hydrated form may be observed when the relative humidity exceeds a critical level and moisture content rapidly increases in the solid. Quantitative measurement of moisture content also provides valuable information on the type of hydrate that formed.

Measurement of moisture uptake is typically done by either of two general methods. The classical approach involves equilibration of solid at several different humidities and the subsequent determination of water content either by gravimetric or analytical methods such as Karl Fischer titration or loss on drying. Moisture adsorption or desorption may be measured by use of this method and the pro-

cess is effective but tedious and time-consuming. A relatively recent development is the use of automated controlled-atmosphere systems in conjunction with an electronic microbalance (75–77). Such systems can generate an atmosphere with well-controlled humidity passing over a sample (often only a few milligrams are needed) and weight change is monitored. Such systems can be programmed to carry out a series of humidity increments to generate the adsorption curve and/or a series of decrements to generate moisture desorption. In this way hysteresis may be observed as well as any phase or form changes that are associated with moisture sorption. Examples of moisture sorption curves are shown in Figs. 18.4 and 18.5.

Dynamic moisture sorption, in particular, provides an excellent opportunity to study solid form conversion; Fig. 18.6 depicts a typical sorption curve of an antiarrhythmic compound that shows the conversion of an anhydrate to a monohydrate. Moisture uptake by the anhydrous form is very small on the moisture uptake curve until a critical humidity of about 70% is achieved. At this point, rapid moisture uptake occurs and a hydrate form containing 10% moisture is generated. Subsequent reduction in the humidity (desorption) shows the hydrate to remain until approximately 5% RH, when it spontaneously converts to the anhydrous form. It is important to recognize, however, that conversion between solid forms is very time dependent. The relative humidities at which conversion was seen in Fig. 18.6 are significantly dependent on the length of time the solid material was equilibrated. For the material shown in Fig. 18.6, conversion from the anhydrous to the hydrate "at equilibrium" will occur somewhere between 10 and 70% RH. More precise determination of the critical humidity at which conversion occurs may be determined as described in Section 2.3.1.

Prediction of moisture sorption is not currently possible, although it is certainly influenced by crystal structure, amorphous components, and solubility. In general, water adsorption to the surface of crystalline materials will result in very limited moisture uptake. Only 0.1% water uptake would be predicted for monolayer coverage of a crystalline

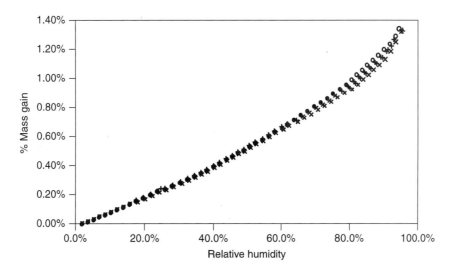

Figure 18.4. Moisture sorption as a function of relative humidity for an antiarrhythmic compound. (Reprinted from Ref. 75 with permission from Elsevier Science.)

material with an average particle size of 1 μm (78). Typically, pharmaceutical powders are in the range of 1–200 μm in diameter, so significant moisture uptake by powders is likely attributable to reasons other than simple surface adsorption. Amorphous regions tend to be much more prone to moisture sorption, and high moisture uptake is likely to reflect the presence of amorphous regions or form changes such as the formation of stochiometric hydrates or nonstochiometric hydrates (clathrates). Moisture sorption has, in fact, been used to quantitate the amorphous content of predominantly crystalline materials.

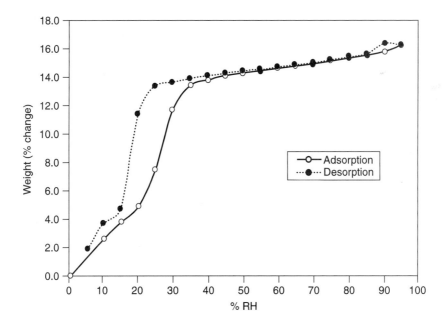

Figure 18.5. Moisture sorption as a function of relative humidity of a protein kinase inhibitor. [Reprinted with permission of the publisher (76).]

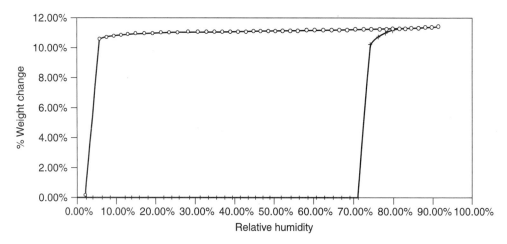

Figure 18.6. Moisture sorption as a function of relative humidity of an antiarrhythmic compound.

3.8 Stability

Both solution and solid state stability are key considerations for oral delivery. The drug molecule must be adequately stable in the dosage form to ensure a satisfactory shelf life. For oral dosage forms, it is generally considered that 2 years is the minimum acceptable shelf life. This allows sufficient time for the manufacture and storage of bulk drug, the manufacture of the dosage form, shipping, storage, and, finally, sale to and use by the consumer. Loss of potency is an obvious consideration and generally stability guidelines require that at least 90% of the drug remain at the end of the shelf life. More often though, shelf life is determined by the appearance of relatively low levels of degradation products. Although a 5–10% loss of drug perhaps may be considered acceptable, the appearance of a degradation product or impurity of unknown toxicity at a level of 0.1–1% will likely require identification or qualification. Detailed guidance regarding stability has been provided by regulatory agencies such as those in the FDA Guidance for Industry and the International Conference on Harmonization (ICH).

3.8.1 Solution Stability. Solution stability is important for oral products because the drug generally has to dissolve in the gastric or intestinal fluids before absorption. Residence time in the stomach varies between 15 min and several hours depending on fasting/fed state. In addition, the stomach is generally quite acid for a majority of subjects but may depend on disease state. In this context, stability under acid conditions over a period of several hours at 37°C is satisfactory, with no significant appearance of degradation products of unknown toxicity. Residence time in the small intestine, where the pH may range from 5–7, is approximately 3 h, whereas residence in the large intestine ranges up to 24 h. Stability studies for up to 24 h in the pH range of 5–7 at 37°C, with no significant appearance of degradation products of unknown toxicity, generally indicates that significant decomposition in the intestine will not occur.

Buffered aqueous solution stability studies are typically done at pH values of 1.2–2 and in the range of 5–7. A complete degradation rate profile can provide valuable information regarding the degradation mechanism and degradation products. A complete study and understanding of solution stability is particularly critical for aqueous and cosolvent solution formulations that may be developed for pediatric or geriatric populations. The medicinal chemist will likely have an excellent understanding of the possibility of acid or base catalysis and degradation in aqueous solutions. A close collaboration with the formulation scientist can ensure that a careful analysis and study of potential decomposition mechanisms are adequately investigated early

Table 18.4 The Biopharmaceutics Classification System[a]

Class I	Class II	Class III	Class IV
High solubility	Low solubility	High solubility	Low solubility
High permeability	High permeability	Low permeability	Low permeability

[a]Ref. 83.

in development. This will minimize the chances of surprises later in development.

3.8.2 Solid State Stability. Adequate solid state stability is often critical for many drugs because solid dosage forms (tablets, capsules) are generally the preferred delivery system. Stability of the drug in the dosage form for several years at room temperature is generally required. Unstable drugs may be developed, although the time and resources needed are generally much greater and the chances of failure far greater.

Accelerated stability studies are often carried out early in development on pure drug to assess stability and identify degradation products and mechanism. Testing at 50°C, 60°C, or even 70°C under dry and humid conditions (75% RH) for 1 month are often sufficient to provide an initial assessment. More quantitative assessments of drug and formulation stability are carried out to support regulatory filings and generally follow regulatory guidelines (79).

The field of solution and solid state stability is expansive, varied, and beyond the scope of this chapter. Stability studies described earlier at a variety of conditions provide the perspective and understanding needed to make meaningful predictions of long-term stability and shelf life (80, 81). Typically, solid state decomposition occurs either by zero-order or first-order processes. Arrhenius analysis and extrapolation to room temperature may provide additional confidence that the dosage form will have acceptable stability. Generally though, regulatory guidance allows for New Drug Applications to project shelf life on the basis of accelerated conditions, although data at the recommended storage temperature are generally required to support the actual shelf life of marketed products.

4 DOSAGE FORM DEVELOPMENT STRATEGIES

An important aspect of pharmaceutical formulation development is to "facilitate" drug absorption and to ensure that an adequate amount of drug reaches the systemic circulation. Most orally administered drugs enter systemic circulation by way of a passive diffusion process through the small intestine. This is easily seen from the following equation (82):

$$M = P_{\mathrm{eff}} A C_{\mathrm{app}} t_{\mathrm{res}} \qquad (18.19)$$

where the amount of drug absorbed (M) is proportional to the effective membrane permeability (P_{eff}), the surface area available for absorption (A), the apparent luminal drug concentration (C_{app}), and the residence time (t_{res}). Because it is difficult to alter or control surface area and residence time, formulation strategies often focus on enhancing either drug permeability across the apical membrane or drug concentration at the absorption site. Recently, a Biopharmaceutics Classification System (BCS) has been proposed as a tool to categorize compounds into four classes according to these two key parameters, solubility and permeability (83). Although the BCS does not address other important factors such as the drug absorption mechanism and presystemic degradation or complexation, it nonetheless provides a useful framework for identifying appropriate dosage forms and strategies to consider which may provide opportunities to overcome physicochemical limitations. It is within the BCS context that this chapter discusses dosage form development strategies that should be considered.

4.1 Biopharmaceutics Classification System

According to the BCS, compounds are grouped into four classes according to their solubility and permeability, as shown in Table 18.4 (83).

4.1.1 Solubility. In recent guidelines issued by the FDA, solubility within the BCS is defined as the "minimum concentration of drug, milligram/milliliter (mg/mL), in the largest dose strength, determined in the physiological pH range (pH 1–7.5) and temperature (37 ± 0.5°C) in aqueous media" (84). Drugs with a dose-to-solubility ratio of less than or equal to 250 mL are considered highly soluble; otherwise, they are considered poorly soluble. In other words, the highest therapeutic dose must dissolve in 250 mL of water at any physiological pH. Because the dissolution rate is closely tied to solubility (see Section 3.4), the FDA also provides dissolution criteria for immediate release (IR) products. A rapidly dissolving IR drug product should release no less than 85% of the labeled drug content within 30 min, by use of United States Pharmacopeia (USP) dissolution apparatus I at 100 rpm in each of the following media: (1) 0.1 N HCl or simulated gastric fluid (USP) without enzymes; (2) pH 4.5 buffer; and (3) pH 6.8 buffer or simulated intestinal fluid (USP) without enzymes. One needs to realize, however, that these compendial dissolution media may drastically underestimate *in vivo* performance of poorly soluble compounds, especially lipophilic compounds. Even though a lipophilic compound is poorly soluble in aqueous environment, it may be sufficiently soluble in the presence of bile salts and other native components of the GI tract. To better simulate physiological conditions of the GI tract, Dressman and coworkers recently proposed use of "biorelevant" media, to take into account the effect of composition, volume, and hydrodynamics of the luminal contents on drug dissolution and solubility (85).

4.1.2 Permeability. In the same guidance as mentioned earlier, permeability is defined "as the effective human jejunal wall permeability of a drug and includes an apparent resistance to mass transport to the intestinal membrane" (84). High permeability drugs are considered to be those "with greater than 90% oral absorption in the absence of documented instability in the gastrointestinal tract, or whose permeability attributes have been determined experimentally" (84). A list of compounds has been compiled to allow researchers to establish a correlation between *in vitro* permeability measurements and *in vivo* absorption. Accordingly, a drug with a human permeability greater than $2–4 \times 10^{-4}$ cm/s would be expected to have greater than 95% absorption (83). Permeability measurements in predictor models such as Caco-2 or *in situ* perfusions are generally related back to the reference compounds. A rough guide is that compounds with permeability greater than that of metoprolol are considered high permeability. Even though this BCS was designed to guide decisions with respect to *in vivo* and *in vitro* correlations and the need for bioequivalence studies, it can also be used to categorize the types of formulation strategies that might be pursued. Table 18.5 summarizes dosage form options for each biopharmaceutics class.

4.2 Class I: High Solubility and High Permeability

Compounds belonging to Class I are highly soluble and permeable. When formulated in an immediate-release dosage form, a Class I compound should rapidly dissolve and be well absorbed across the gut wall. However, absorption problems may still occur if the compound is unstable, forms an insoluble complex in the lumen, undergoes presystematic metabolism, or is actively secreted from the gut wall. Potential formulation strategies that may overcome such absorption barriers are discussed later in further detail.

In many cases, the challenge to formulate Class I drugs is not to achieve rapid absorption, but rather to achieve the target release profile associated with a particular pharmacokinetic and/or pharmacodynamic profile. In such cases, a controlled-release dosage form may be more desirable to tailor the blood profile to maintain the plasma concentration at a more sustained level (see Fig. 18.7). Controlled release technology is well established in the pharmaceutical industry, and there are at least 60 commercially available oral controlled-release products (86). Depending on the release mechanism, controlled release technology can be classified into four major categories: dissolution controlled, osmotically controlled, diffusion controlled, and chemically controlled dosage forms. Extensive reviews are available in this area (86, 87). It is

Table 18.5 Dosage Form Options Based on Biopharmaceutics Classification System

Class I: High Solubility, High Permeability	Class II: Low Solubility, High Permeability
• No major challenges for immediate release dosage forms • Controlled-release dosage forms may be needed to limit rapid absorption profile	Formulations designed to overcome solubility or dissolution rate problems • Salt formation • Precipitation inhibitors • Metastable forms • Solid dispersion • Complexation • Lipid technologies • Particle size reduction

Class III: High Solubility, Low Permeability	Class IV: Low Solubility, Low Permeability
Approaches to improve permeability • Prodrugs • Permeation enhancers • Ion pairing • Bioadhesives	• Formulation would have to use a combination of approaches identified in Class II and Class III to overcome dissolution and permeability problems • Strategies for oral administration are not really viable. Often use alternative delivery methods, such as intravenous administration.

Class V: Metabolically or Chemically Unstable Compounds[a]
Approaches to stabilize or avoid instability • Prodrugs • Enteric coating (protection in stomach) • Lipid vehicles (micelles or emulsions/microemulsions) • Enzyme inhibitor • Lymphatic delivery (to avoid first-pass metabolism) • Lipid prodrugs • P-gp efflux pump inhibitors

[a]Class V compounds do not belong to the Biopharmaceutics Classification System. Compounds in this class may have acceptable solubility and permeability, but can still pose significant absorption challenge if they undergo luminal degradation, presystemic elimination, or are effluxed by p-glycoproteins.

beyond the scope of this chapter to review all controlled release technology. However, it is still worth pointing out the considerations in designing oral controlled-release dosage forms, particularly those relevant to Class I compounds. These considerations include:

1. How long is the GI transit time?
2. Is there substantial colonic absorption?
3. What is the dose required?
4. Is there a safety concern if dose dumping occurs?

Dose dumping of Class I compounds, in particular, may cause more safety concerns than for other classes of compounds because Class I compounds are expected to be absorbed rapidly.

4.3 Class II: Low Solubility and High Permeability

Compounds belonging to Class II have high permeability but low aqueous solubility. In the past decade, an increasing number of Class II compounds have emerged from the discovery pipeline, thereby stimulating the development of a variety of dosage forms and drug delivery technologies. Major Class II technologies are designed to deal with poor solubility and dissolution characteristics and include

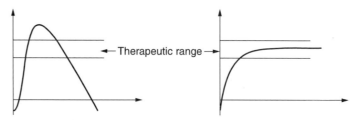

Figure 18.7. Modification of drug release profile to achieve maximum therapeutic effect. (Reprinted from Ref. 134 with permission of the publisher.)

salt formation, size reduction, use of metastable forms, complexation, solid dispersion, and lipid-based formulations. Despite the diversity of these technologies, the central theme of Class II formulation approaches remains the same: to enhance the drug dissolution rate and/or solubilize drugs at the absorption site to provide faster and more complete absorption. The following section discusses the advantages and limitations of major technologies for Class II compounds.

4.3.1 Salt Formation. Salt formulation is one of the most commonly used approaches to deal with Class II compounds as a way to enhance drug solubility and dissolution rate. Salt selection largely depends on pK_a. It is generally accepted that a minimum difference of 3 units between the pK_a value of the group and that of its counterion is required to form stable salts (88). Many other factors also influence salt selection, such as the physical and chemical characteristics of the salt, safety of the counterion, therapeutic indications, and route of administration. The main purpose of salt formation is to enhance the rate at which the drug dissolves. For this purpose, sodium and potassium salts are often first considered for weakly acidic drugs, whereas hydrochloride salts are often first considered for weakly basic drugs. A wide variety of counterions have been successfully used in pharmaceuticals and many of the most common ones are listed in Table 18.1.

Salt formation does have its limitations. It is not feasible to form salts of neutral compounds and it may be difficult to form salts of very weak bases or acids. Even if a stable salt can be formed, the salt may be hygroscopic, exhibit complicated polymorphism, or have poor processing characteristics. In addition, formulation of a stable and soluble salt may not be as straightforward as one would expect.

Conversion from salt to either free acid or base has been a common problem both *in vitro* and *in vivo*. Such conversion may cause surface deposition of a less soluble free acid or free base on a dissolving tablet and prevent further drug release. Even if a solid dosage form completely dissolves, the un-ionized form may still precipitate out in the lumen before absorption. In such cases, use of precipitation inhibitors may significantly improve the bioavailability of rapidly dissolving salts.

4.3.2 Precipitation Inhibition. As mentioned earlier, the central theme of Class II technologies is to enhance the drug dissolution rate or solubilization to provide more rapid and complete absorption. However, a significant increase in "free drug" concentration above equilibrium solubility results in supersaturation, which can lead to drug precipitation. This has been a common problem of many Class II technologies. The supersaturation (σ), which is the driving force for both nucleation and crystal growth, is frequently defined as follows:

$$\sigma = \ln(C/S) \qquad (18.20)$$

where C is the solution concentration and S is the solubility of the compound of interest at a given temperature. The crystallization rate generally increases with σ and decreases with viscosity of the crystallization medium. Certain inert polymers, such as hydroxypropyl methylcellulose (HPMC), polyvinylpyrrolidone (PVP), polyvinyl alcohol (PVA), and polyethylene glycol (PEG) are known to prolong the supersaturation of certain compounds from a few minutes to hours. It is suggested that these polymers increase the viscosity of the crystallization medium, thereby reducing the crystallization rate of drugs (89–91). In addition, these polymers may present a steric barrier to drug molecules and inhibit drug crystallization through specific intermolecular interactions on growing crystal surfaces. Polymers such as acacia, poloxamers, HPMC, and PVP have been shown to be adsorbed onto certain faces of host crystals, reduce the crystal growth rate of the host, and produce smaller crystals (92, 93).

In addition to polymers, synthetic impurities may also influence the crystallization rate of drugs. Extensive research has been carried out to examine effect of "tailor-made additives" on the crystallization of drugs (94). These additives are generally structurally tailored to resemble drug molecules and can be incorporated into the lattice of the drug to some extent. Upon incorporation, additives can impede further drug crystallization through specific host-additive interactions. In theory, potent additives can be designed to inhibit drug precipitation more effectively than can polymers. However, unless these additives are pharmaceutically acceptable, use of such additives in a formulation can raise safety and regulatory concerns.

4.3.3 Metastable Forms. The solid-state structure of drugs, such as the state of hydration, polymorphic form, and crystallinity have a significant effect on physicochemical properties, such as solubility and dissolution rate, which was discussed earlier in this chapter. In general, anhydrous forms, for example, dissolve faster and have higher solubility than that of hydrates in an aqueous environment. Although some studies have shown that hydrates of certain drugs dissolve faster than anhydrous forms, such studies may be complicated by phase transition between anhydrous to hydrated forms or differences in particle size and wettability between anhydrous and hydrated materials.

Polymorphic form also influences dissolution rate and solubility. By definition, metastable polymorphs should have higher solubility and faster dissolution rates than those of their more stable crystalline counterparts because they possess a higher Gibbs free energy. Generally, only a moderate enhancement of solubility and dissolution rate can be achieved through polymorphic modification, although

exceptions do exist. Greater increases can be achieved through the use of amorphous material, which is a noncrystalline solid that is metastable with respect to the crystalline form. Amorphous forms can be viewed as an extension of the liquid state below the melting point of the solid (95). In some cases, amorphous materials can significantly enhance the dissolution rate and solubility and lead to a three- to fourfold increase in bioavailability (96).

Common processing methods, such as freeze drying, spray drying, and milling, may partially or completely transform a crystalline material into amorphous forms. Because the solid-state structure can significantly impact bioavailability, it is important to control processing methods so that a pure form (or a mixture of forms with fixed ratio) is produced consistently. Even if a reproducible processing method is available, one still faces the inevitable challenge: metastable forms are destined to convert to thermodynamically more stable polymorphs with time. Polymorphic transformation is a kinetic issue that depends on many factors such as crystal defects, residual solvent, processing, and storage conditions. Amorphous forms are particularly sensitive to moisture level in the product as well as in the atmosphere because water significantly lowers the glass-transition temperature of the amorphous form and facilitates recrystallization. For this reason, there is often a reluctance to develop a metastable form of a drug unless there is enough confidence that the metastable form will not transform to the stable form during storage within a desirable shelf life. To overcome this problem, several approaches have been used to prolong the shelf life of metastable forms.

4.3.4 Solid Dispersion. Sekiguchi and Obi were the first to develop a solid dispersion method to enhance the bioavailability of a poorly water soluble drug (97). Their method involved melting a physical mixture of drug with hydrophilic carriers to form a eutectic mixture, in which the drug was present in a microcrystalline state. When a drug is homogeneously dispersed throughout the solid matrix, this type of formulation is also termed a "solid dispersion." However, a drug may not always be present in a "microcrystalline state." As later demonstrated by Goldberg et al. (98, 99), a certain fraction of drug might be molecularly dispersed in a carrier, thereby forming a solid solution. The key difference between a solid dispersion and a solid solution is that the former is a homogeneous physical mixture of components, whereas the latter is a molecular dispersion of one component in another. In the solid dispersion, each component still preserves its own crystal lattice, whereas in solid solution, there are no individual crystals of each component but, rather, the molecules are mixed together at the molecular level. A solid dispersion is generally considered to release drug as very fine colloidal particles upon contact with an aqueous environment, thereby enhancing the dissolution rate of poorly soluble drugs through increased surface area (100). Other factors that could lead to enhanced dissolution rate include possible creation of an amorphous drug as well as generally increased solubility and wettability of a drug in the solid dispersion matrix.

Despite the promises, complicated processing methods have limited commercial viability of solid dispersions. There are two common methods to produce solid dispersions. One is to melt drugs and hydrophilic carriers such as PEG, PVP, PVA, and HPMC or other sugars, followed by cooling and hardening of the melt. The other is to dissolve drug and carrier in a common solvent, followed by solvent evaporation. The melting technique often involves high temperature, which presents a challenge for processing thermal-labile compounds, whereas the cosolvent technique has its own problems. Because solid dispersion often uses hydrophilic carriers and hydrophobic drugs, it is difficult to find a common solvent to dissolve both components. Regardless of processing method, solid dispersion materials are often soft, waxy, and possess poor compressibility and flowability. This presents additional manufacturing challenges, especially during the scale-up process.

Physical instability and the preparation of reproducible material are two significant challenges in developing solid dispersions. It is not uncommon to produce wholly or partially amorphous drug during processing, which will eventually transform to a more stable crystalline form over time. The rate of transformation may be greatly influenced by storage conditions, formulation composition, and processing methods. So far, very few solid dispersion products have been marketed.

4.3.5 Complexation. It has been well established in the literature that complexation is an effective way to solubilize hydrophobic compounds. Nicotinimide is known to complex with aromatic drugs through π donor-π acceptor interaction (101). Similar π-π interaction also occurs between salts of benzoic acid or salicylic acid and drugs containing aromatic rings such as caffeine (102). Obviously, aromaticity is an important factor in this type of complexation. Unfortunately, from a safety perspective the use of these types of complexing agents for products is not really very viable.

Cyclodextrin (CD) exemplifies another type of complexation, that is, complexation through inclusion. Cyclodextrins are torus-shape "oligosaccharides composed of 6–8 dextrose units (α-, β-, and γ-cyclodextrins, respectively) joined through 1–4 bonds" (103). It has a lipophilic cavity with a 6.0- to 6.5-Å opening, which can form inclusion complexes by taking up a guest molecule into the central cavity. Formation of complexes alters the physicochemical properties such as solubility, dissolution rate, stability, and volatility of both drug molecules and CD molecules. A drug's solubility and dissolution rate usually increase when forming inclusion complex with CDs.

Inclusion complexation has also been used to stabilize, decrease the volatility, and ameliorate the irritancy and toxicity of drug molecules. In addition, modified CDs such as carboxymethyl derivatives (e.g., CME-β-CD) exhibit pH-dependent solubility, and therefore can be used in enteric formulations.

Many drugs interact most favorably with β-CD. Unfortunately, β-CD has the lowest solubility (1.8% in water at 25°C) among the three non-derivatized CDs, thereby limiting its solubilization capacity. To enhance the solubility of the β-CD and to improve its safety, derivatives of β-CD have been developed. Among them, hydroxypropyl-β-cyclodextrins (HP-β-CDs) and sulfobutylether-β-cyclodextrins (SBE-β-CDs, mainly as SBE7-β-CD, where 7 refers to the average degree of substitution) have captured interest in formulating poorly soluble drugs in immediate-release dosage forms. A disadvantage of going to these derivatized cyclodextrins as an excipient is higher cost.

It is well demonstrated in the literature that complexation with CD could significantly enhance bioavailability of poorly soluble compounds. Apart from bioavailability enhancement, other advantages of the use of cyclodextrin include modification of stability, ease of manufacturing, and reproducibility compared to other Class II strategies such as use of "higher energy forms" and solid dispersions. So far, at least 10 oral products containing CD have gained approval from the FDA (103).

A frequent consideration when forming inclusion complexes is how fast the drug is released from the complex *in vivo*. Stella and Rejewski showed that weakly to moderately bound drug, in fact, rapidly dissociates from CD upon dilution (103). For strongly bound drugs or when dilution is minimal, competitive displacement is important for rapid and complete dissociation. Although rapid reversibility of the complexation process is essential for drug absorption, it also poses the potential to reduce bioavailability. Excess free drug may precipitate upon dilution in the GI tract before absorption. If such precipitation is significant, it may prove valuable to combine drug/CD complexes with precipitation inhibitors.

Another potential concern with CD is safety because there is only limited experience with marketed products at this point. Although oral administration of CD is generally considered safe, it may cause increased elimination of bile acids and certain nutrients (104). In addition, CD may also cause membrane destabilization through its ability to extract membrane components such as cholesterol and phospholipids. In this regard, cyclodextrin may act as a permeation enhancer to enhance the mucosal permeation of the drug (105).

4.3.6 Lipid Technologies. It has been known for a long time that lipid-based formulations can significantly improve the bioavailability of hydrophobic drugs by facilitating drug dissolution, dispersion, and solubilization, either directly from administered lipids or through intraluminal lipid processing. However, the preference for solid dosage forms usually prevails because of physical and chemical instability associated with lipid formulations. It was not until recently that lipid technologies generated much interest, likely because of the increasing numbers of hydrophobic compounds emerging from discovery programs. Several common types of lipid-based dosage forms include lipid suspensions and solutions, micelle solubilization, microemulsions, macroemulsions (or emulsions), and liposomes.

4.3.6.1 Lipid Suspensions and Solutions. A typical lipid solution is composed of triglycerides or mixed glycerides and surfactants (106). Although lipid solutions are easy to formulate, they have limited solvent capacities except for very lipophilic drugs ($\log P > 4$) (107). Consequently, the design of unit dose lipid solutions is often not a practical approach, especially for high dose compounds. Further, a typical lipid solution may be poorly dispersible in water. In such cases, digestibility of the lipid formulation may be important to achieve good bioavailability because lipolysis is commonly believed to facilitate release of drug from colloidal solution, thereby leading to faster absorption (108). Nondigestible lipids such as mineral oil (liquid paraffin) and sucrose polyesters "can actually limit/reduce drug absorption by retaining a portion of the co-administered drug" (109). Studies have also shown that the bioavailability of a digestible lipid formulation tends to be higher than that of nondigestible formulations. However, given the complexity of lipid digestion, it may be difficult to interpret the effects of lipid vehicles. Common digestible lipids are dietary lipids (including glycerides, fatty acids, phospholipids, and cholesterol/cholesterol esters) and their synthetic derivatives. A few excellent reviews on how digestibility may influence bioavailability are available (108, 109).

Lipid suspensions are also known to enhance the bioavailability of hydrophobic drugs. Unlike lipid solutions, suspended drug needs to undergo additional dissolution before the absorption. Therefore, factors such as drug particle size and amount suspended may also influence the bioavailability.

Table 18.6 Physical Characteristics of Different Lipid Colloidal Systems[a]

	Micelles	Microemulsions	Emulsions	Liposome
Spontaneously obtained	Yes	Yes	No	No
Thermodynamically stable	Yes	Yes	No	No
Turbidity	Transparent	Transparent	Turbid	Transparent to turbid[b]
Typical size range	<0.01 μm	~ 0.1 μm or less	0.5–5 μm	0.025–25 μm
Cosurfactant used	No	Yes	No	No
Surfactant concentration	<5%	>10%	1–20%	0.5–20%
Dispersed phase concentration	<5%	1–30%	1–30%	1–30%

[a]Modified from ref. 135.
[b]Depending on droplet size.

4.3.6.2 Micelle Solubilization. The early interest in lipid formulations was initiated by the findings that coadministration of drug with food enhanced the bioavailability of many drugs. Intake of food stimulates secretion of bile salts into the duodenum, increasing bile salts concentration in the duodenum from a typical 1–4 mmol/L in the fasted state to 10–20 mmol/L in the fed state (110). It is hypothesized that micelles or mixed micelles formed by bile salts and digested lipids could significantly solubilize a hydrophobic drug, thereby enhancing drug absorption. This prompted investigation of the use of simple lipid solutions and suspensions for hydrophobic drugs discussed earlier. As summarized in Table 18.6, normal micelles are transparent and thermodynamically stable liquid solutions consisting of water and amphiphile. Micelles have low viscosity, long shelf life, and are easy to prepare. However, they have limited capacity to solubilize oil and hydrophobic drugs.

Small amounts of surfactants are often added to formulations to significantly improve drug wettability. Also, when added below their critical micelle concentration (CMC), the surfactant can adhere to the surface of the drug and reduce the interfacial tension between the drug and the dissolution medium. Above the CMC, surfactants form micelles that can solu-

bilize the drug. Micelle solubilization may either increase drug absorption, by increasing the amount of drug that is solubilized and available at the absorption surface (111), or it may, in some cases, reduce diffusion of the drug to the absorption surface and reduce absorption (112). See Table 18.7 for a list of common surfactants that are pharmaceutically acceptable.

4.3.6.3 Emulsions. Emulsions have much higher solvent capacity than micelles for hydrophobic materials. However, emulsions are metastable colloids that will phase separate over a period of time and form a two-phase system (i.e., oil phase and aqueous phase). Because of its physical instability, large energy input (usually mechanical mixing) is required to form an emulsion.

4.3.6.4 Microemulsions. Unlike emulsions, microemulsions are transparent and thermodynamically stable colloidal systems, formed under certain concentrations of surfactant, water, and oil (Fig. 18.8). The transparency is because the droplet size of the microemulsions is small enough (<100 nm) that they do not reflect light. Because of its thermodynamic stability, microemulsions may have long shelf lives and spontaneously form with gentle agitation. However, microemulsions are not infinitely stable upon dilution because dilution

Table 18.7 List of Common Surfactants That Are Pharmaceutically Acceptable

Nonionic	Polysorbates (Tweens), sorbitan esters (Spans), polyoxyethylene monohexadecyl ether
Anionic	Sodium lauryl sulfate, SLS or SDS, sodium docusate
Cationic	Quaternary ammonium alkyl salts such as hexadecyl trimethyl ammonium bromide (CTAB), didodcecylammonium bromide (DDAB)
Zwitterionic	Phospholipids

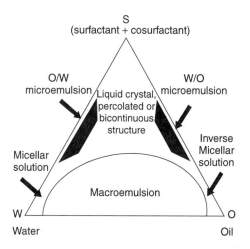

S
(surfactant + cosurfactant)

O/W
microemulsion

Liquid crystal, percolated or bicontinuous structure

W/O
microemulsion

Inverse Micellar solution

Micellar solution

Macroemulsion

W
Water

O
Oil

Figure 18.8. Hypothetical phase regions of microemulsion systems of oil (O), water (W), and surfactant + consurfactant (S). (Reprinted from Ref. 135 with permission of the publisher.)

changes the composition of the colloidal system. Microemulsions also have a high capacity for hydrophobic drugs that further adds to their attractiveness as a promising drug delivery system for poorly water soluble compounds.

4.3.6.5 Liposomes. Liposomes are a metastable colloidal system consisting of natural lipids and cholesterol. Unlike micelles, emulsions, and microemulsions, liposomes use ingredients that are part of biological membranes. Therefore, liposomes have relatively few problems with toxicity, unlike the exogenous surfactants present in other colloidal systems. Although liposomes have generated great interest in the past decades, oral administration of liposomes remains highly controversial, and thus is not discussed in detail.

4.3.6.6 Self-Emulsifying Drug Delivery Systems (SEDDS). SEDDS, closely related to microemulsions and emulsions, are isotropic lipid solutions typically consisting of a mixture of surfactant, oil, and drug that rapidly disperses to form fine emulsion droplets upon administration. If the droplet size is comparable to typical microemulsion droplet size, SEDDS become SMEDDS or a self-microemulsifying drug delivery system. One apparent advantage of SMEDDS and SEDDS over microemulsions is elimination or reduction of the aqueous phase, thereby significantly reducing the

dose volume. SEDDS often have a dose volume that is small enough to allow encapsulation into soft or hard gelatin capsules. In addition, use of SEDDS avoids or partially avoids common physical stability problems associated with emulsions.

Early work by Pouton demonstrated that a good SEDDS formulation could significantly enhance the dissolution and bioavailability of poorly soluble compounds (113). Pouton proposed two criteria to describe the efficiency of SEDDS formulation: (1) the rate of emulsification and (2) the particle size distribution of the resultant emulsion. An efficient SEDDS should produce fine dispersions (<1 μm) rapidly at a reproducible rate. Efficient SEDDS or SMEDDS have demonstrated their potential in delivering hydrophobic compounds. The most notable case is a SMEDDS formulation of cyclosporin A (Neoral), in which the formulation has significantly increased the bioavailability as well as decreased patient variability (114, 115).

Despite the potential of SEDDS and SMEDDS in oral delivery of poorly soluble drugs, few oral SEDDS formulations have been marketed so far. This is partly attributable to traditional preferences to develop a solid dosage form, and partly to inherent limitations associated with lipid products. One limitation of SEDDS systems in general is physical and chemical instability caused by undesirable interactions between drug and excipient or among excipients. Another major limitation of SEDDS is that many hydrophobic drugs may not have sufficient solubility in pharmaceutically acceptable lipids. It is a common misperception that poor water solubility means good lipid solubility. Although SEDDS may have relatively higher solubilization capacities than those of simple lipid solutions and micelles, most hydrophobic drugs are not very soluble in long-chain hydrocarbon oils. In general, many hydrophobic drugs ($2 < \log P < 4$) are more soluble in small/medium-chain oils such as Miglyol 812 than in long-chain oils. However, it is rare that the drug load of a SEDDS formula can exceed 30%.

High concentrations of surfactants in SEDDS also raise safety concerns, especially for drugs intended for chronic therapy. In addition, common lipid components such as fatty acids, glycerides, and several surfactants are known to act as absorption enhancers (116). There is a host of safety issues associated with absorption enhancers (for details, refer to Sections 4.4.2 and 4.4.3). Furthermore, formulating a SEDDS system is not a trivial exercise. It requires understanding of complicated phase behavior of a system, consisting of at lease four basic components: oil, drug, surfactant(s), cosurfactant(s), as well as water for microemulsion. It is hoped that research in this area will build a large enough database to help formulate SEDDS in the future.

4.3.7 Size Reduction. Particle size reduction is a common method to enhance the dissolution rate of poorly soluble drugs. The underlying principle is that the dissolution rate is directly proportional to the surface area, which increases with size reduction. The most common way to reduce particle size is through milling. There are several types of milling equipment including the cutter mill, revolving mill, hammer mill, roller mill, attrition mill, and fluid-energy mill. Equipment selection depends on target particle size distribution as well as characteristics of drugs. For example, the cutter mill is often used for fibrous material and product size is 180–850 μm, whereas the fluid-energy mill (sometimes referred to as micronizing mill) is often used for moderately hard and friable crystalline materials with typical size distribution created of 1–30 μm (117). Although size reduction generally enhances the dissolution rate of poorly soluble compounds, there is a critical threshold below which further reduction in particle size will not enhance absorption. For compounds that are extremely insoluble, the critical threshold may be in the nanoparticle range. In such cases, nanoparticle technology may come in handy. Refer to Section 3.4 on dissolution rate for further details on the importance and impact of particle size on dissolution rate.

Heat and mechanical impact generated during the milling process can cause both physical and chemical instability. During the milling process, the localized temperature may rise as high as 100°C and cause chemical degradation or physical conversion. Therefore, it is important to always evaluate the effect of milling on the physical and chemical

properties of drugs. In addition to stability problems, very small particle size powder often possesses poor flow properties and wettability. In general, size reduction of hydrophobic material increases the tendency for powder to aggregate in an aqueous environment. Powder aggregation reduces the effective surface area of a drug, thereby reducing the dissolution rate. Excipients such as surfactants, sugars, polymers, or other excipients may be added to a formulation to minimize aggregation. Small amounts of surfactants are often added to formulations to significantly improve drug wettability.

4.4 Class III: High Solubility and Low Permeability

The limiting factor for Class III compounds is the effective permeability across the GI tract. Given the difficulty of altering membrane permeability, Class II technologies are often used to formulate Class III compounds. The underlying principle for such substitution is that increasing the drug concentration in the GI tract should increase absorption of a drug, if it is absorbed through the passive diffusion process (the assumption is valid in most cases). However, Class II solubilization technologies may not significantly enhance drug absorption if the solubility and dissolution rate are high. Theoretically, the most effective way to enhance the absorption of Class III compound is to overcome the absorption rate-limiting barrier, permeation. The following section reviews some emerging technologies that have showed early promises to formulate Class III compounds.

4.4.1 Prodrugs. Poor membrane permeation is most commonly attributed to either low partitioning into the lipid membrane or low membrane diffusivity. The most direct solution is to: (*1*) modify a drug's structure to increase lipophilicity, (*2*) reduce molecular weight, or (*3*) remove hydrogen-bonding groups. Prodrugs are one way to structurally modify the active compound to improve membrane permeability and still maintain activity of the parent drug upon bioreversion. Successful prodrug approaches include an approved antihypertensive agent, fosinopril (an acyloxyalkyl prodrug of fosinoprilat), and various angiotensin-converting enzyme (ACE) inhibitors (118).

There are five important criteria in prodrug design: (*1*) adequate stability to the variable pH environment of the GI tract, (*2*) adequate solubility or solubilization mechanisms, (*3*) enzymatic stability to luminal contents as well as the enzymes found in the brush border membrane, (*4*) good permeability and adequate log *P*, and (*5*) the prodrug should revert to the parent drug either in the enterocyte or once absorbed into systemic circulation (118). Post-enterocyte reversion is more desirable because conversion in the enterocyte would also allow for back diffusion into the GI lumen.

Among these five criteria, knowledge of both the rate of bioreversion and the biological distribution of reconversion sites is often most critical for prodrug success. If bioreversion is fast and nonspecific, prodrug reversion may take place before the limiting barrier is overcome; if too slow, the prodrug may readily reach the site of action but not release enough parent drug to elicit a pharmacological response prior to clearance of the prodrug. Knowledge about the biological distribution of reconversion sites will help predict the location of active drug. Ideally, reconversion sites should coincide with the target site.

Unfortunately, it is often difficult to satisfy all five criteria simultaneously, among which control of reconversion rate proves especially difficult. In addition, increasing lipophilicity often reduces aqueous solubility, which makes it even more difficult to formulate. Perhaps one of the biggest concerns is that prodrugs are considered new chemical entities that require a new set of preclinical studies. Therefore, the prodrug approach is often less preferred if simpler formulation approaches are available. The incorporation of a prodrug strategy really should occur in very early preclinical evaluations. Prodrug approaches do offer the opportunity to expand intellectual property through patents.

4.4.2 Permeation Enhancers. Use of permeation enhancers is an alternative way to enhance drug permeation through the biological membrane by transiently altering the integrity of the mucosal membrane. Permeation enhancers may act at either the apical cell membrane (transcellular pathway) or the tight junctions between cells (paracellular pathway). There are several ways that permeation enhancers may interact with the cell membrane. Some fatty acids such as oleic acid have been found to disrupt the configuration of the lipid region (116, 119). Some enhancers such as salicylic acid may interact with membrane protein, which carries on important membrane functions. Medium chain monoglycerides may extract cholesterol out of the cell membrane (120). Chelators such as EDTA and some bile acids could chelate Ca^{2+} in the tight junction, which can lead to pore openings from 8 to 14 Å (121).

Because of its potential damage to the membrane, permeation enhancers elicit great safety concern, especially in chronic therapy. Some of these include: (*1*) potential tissue irritation and damage, (*2*) effect of the enhancer on structural integrity of the mucosal membrane, (*3*) reversibility of membrane perturbation, (*4*) long-term effect of continued exposure to the enhancer, and (*5*) potential to also enhance absorption of any potential harmful substances that are also present in the intestine. All these issues may have significant toxicity ramifications. Therefore, the FDA has not approved any permeation enhancer, although use of some common excipients that are reported to enhance absorption may be acceptable (Table 18.8).

4.4.3 Ion Pairing. Ion pairing has been proposed to enhance effective permeability of polar or hydrophilic drugs that exhibit poor permeability properties. In this approach, an ionizable drug is coadministered with an excess concentration of a counterion. In theory, the ionized drug will associate with the counterion and partition into the membrane as a more lipophilic ion pair. Although several animal studies reported moderate success with the ion-pairing approach, in most cases the formulation was directly administered onto the absorption surface. Therefore, these studies did not reflect the effect of dilution, dispersion, and other counterions in the GI tract on ion-pair dissociation (122). In fact, although the concept of ion pairing has been around for

Table 18.8 **Compounds Shown to Have Intestinal Absorption-Enhancing Effects**[a]

Classes	Examples
Bile salts	Sodium deoxycholate, sodium glycocholate, sodium taurocholate, and their derivatives
Surfactant	Polyoxyethylene alkyl ethers, polysorbate, sodium lauryl sulfate, dioctyl sodium sulfosuccinate
Fatty acids	Sodium caprate, oleic acid
Glycerides	Natural oils, medium chain glycerides, phospholipids, polyoxyethylene glyceryl esters
Acyl carnitines and cholines	Palmitoyl carnitine, lauroyl choline
Salicylates	Sodium salicylate, sodium methoxysalicylate
Chelating agents	EDTA
Swellable polymers	Starch, polycarbophil, chitosan
Others	Citric acid, cyclodextrin

[a]Modified from Ref. 116.

almost four decades, lack of preclinical evidence or commercial feasibility has limited further research and development of this approach.

4.4.4 Improving Residence Time: Bioadhesives. Increasing residence time at the absorption site could also enhance drug absorption. Bioadhesive drug delivery systems have been proposed as a means to increase GI tract residence time. The original concept of bioadhesion is to administer drug in a bioadhesive polymer matrix that adheres to mucosal membranes to prolong the residence time (12–24 h), thereby increasing the contact time between drug and the absorption site. Several review articles and books have extensively reviewed the concept of bioadhesive polymers (123–126). This approach seems to have lost its popularity because of disappointing animal and human data.

4.5 Class IV: Low Solubility and Low Permeability

Class IV compounds exhibit both poor solubility and poor permeability and pose tremendous challenges to formulation development. As a result, a substantial investment in dosage form development with no guarantee of success should be expected. Class IV compounds are rarely developed or reach the market as oral products. A combination of Class II and Class III technologies could be used to formulate Class IV compounds. However, redesigning drug molecules to enhance solubility

and/or permeability or searching for a non-oral route may be more likely to succeed.

4.6 "Class V": Other Absorption Barriers

Although the BCS provides a useful framework for recognizing solubility and permeability as two key parameters controlling absorption, additional "barriers" that limit drug absorption do exist beyond the scope of the BCS. Luminal complexation can reduce the free drug concentration available for absorption. Luminal degradation further degrades compounds such as proteins and peptides that are susceptible to intestinal enzymes or microorganisms. Presystemic elimination includes both traditional first-pass metabolism and also intestinal metabolism. The significance of intestinal drug metabolism is a relatively recent discovery, as evidenced by the high intestinal concentration of CYP3A4, which is present in the intestine at approximately 80–100% of the CYP3A4 concentration in the liver (127). In addition, P-gp, a membrane transporter, further reduces drug absorption by retrograding efflux of the drug into the intestinal lumen in the secretory direction (or basolateral-to-apical direction).

This section reviews potential formulation strategies addressing the above-mentioned issues. However, given the difficult nature of overcoming such absorption barriers and the limited knowledge in this area, most of the formulation strategies reviewed in this section are highly experimental and yet to be proved.

4.6.1 Luminal Degradation. Luminal degradation can be attributed to chemical decomposition in the aqueous intestinal environment or metabolism by luminal digestive enzymes or luminal microorganisms. Degradation in the acidic environment of the stomach is relatively easy to solve either by use of an enteric coating dosage form or by formulating with antacid agents (122). Chemical instability in the slightly acidic to neutral pH of the small intestine (pH range: 4–7) may be more difficult to solve, especially for molecules that cannot be "protected" from the aqueous environment by complexation or solubilization. Another difficult problem to solve is enzymatic degradation in the GI lumen. Enzymatic degradation, combined with poor permeability, has significantly limited the oral absorption of proteins and peptides. Prodrugs are one approach to protect the parent compound from enzymatic degradation. Lipid vesicles and micelles may also be able to shield their encapsulated contents from luminal degradation. Micelle formation, for example, has been shown to slow ester hydrolysis of benzoylthiamine disulfide, resulting in increased *in situ* and *in vivo* absorption (128). Water-in-oil microemulsions have also demonstrated some potential in delivering peptides and proteins, although at a very low capacity or efficiency (129). Upon aqueous dilution in the GI tract, water-in-oil (w/o) microemulsions can undergo phase separation or inversion that can cause dose dumping and expose the encapsulated water-soluble drug to luminal degradation. Therefore, in cases where drug absorption is significantly enhanced in w/o microemulsions, it may not be simply because the w/o microemulsion protects the drug from luminal degradation. One cannot rule out the possibility that certain lipid excipients may act as an absorption enhancer and increase absorption of proteins and peptides.

Coadministration of the drug with an enzymatic inhibitor may also protect the drug from luminal enzymatic degradation. The key to this approach is that the inhibitor needs to effectively protect the drug until the drug is dissolved and absorbed. This would require large amounts of inhibitor to overcome dilution in the lumen. For the best protection, the drug may be encapsulated with the inhibitor in lipid vesicles or polymeric membranes. However, this approach may raise serious safety concerns and would not be recommended as a general approach to overcome enzymatic degradation.

4.6.2 Presystemic Elimination (First-Pass Metabolism and Intestinal Metabolism). Presystemic elimination includes both intestinal metabolism and hepatic first-pass metabolism. The significance of the former is a relatively recent discovery. Among enzymes discovered in the human intestine, CYP3A4 is by far the most important enzyme to drug metabolism. The CYP3A4 intestinal concentrations are approximately 80–100% of that in the liver. Other enzymes, such as CYP3A5, CYP1A1, CYP2C8–10, CYP2D6, and CYP2E1, have also been identified in the small intestine, although their levels are significantly lower than that in the liver (127).

Few oral formulation approaches are available to overcome presystemic elimination. Although coadministration of drug with an enzymatic inhibitor could boost bioavailability, it is not recommended as a viable approach because it raises serious safety concerns. Intestinal lymphatic transport offers the possibility of avoiding hepatic first-pass metabolism. The intestinal lymphatics are the major absorption gateway for natural lipids, lipid derivatives, and cholesterol. However, only highly lipophilic compounds (log $P > 5$–6), such as lipid-soluble vitamins or xenobiotics, can gain significant access to the systemic circulation through the lymphatics. The vast majority of pharmaceutical compounds are not lipophilic enough and, when they are, their solubilities are extremely poor. Lipophilic prodrugs may be designed for the purpose of enhancing intestinal lymphatic drug delivery. The design sophistication varies from simple chemical modification (e.g., derivative compounds through simple ester or ether linkages) to sophisticated functional design, in which a "functionally based" promoiety is added to facilitate compound incorporation into the normal lipid-processing pathways. Comprehensive reviews in this area are available (130, 131). In general, the use of lipid prodrugs to target the lymphatics is a wide-open research area. Much needs to be done before one can assess the practicality of this approach.

4.6.3 P-gp Efflux Mechanism.

Enterocyte P-gp is an apically polarized efflux transporter that was first identified in multidrug-resistant cancer cells but later also found to be present in the intestinal brush border region. P-gp reduces drug absorption by actively transporting a drug in the secretory direction back into the intestinal lumen. Interestingly, P-gp appears to share a large number of substrates and inhibitors with CYP3A (127). Little is known about how to overcome the absorption barrier posed by the P-gp efflux pump. Certain nonionic surfactants such as Cremophor have been shown to inhibit the P-gp efflux pump *in vitro* (132). However, to counter GI dilution, a much higher amount of surfactant may be required to achieve a similar effect *in vivo*. Another approach for overcoming P-gp efflux is to coadminister the drug with a P-gp inhibitor. For example, docetaxel has very poor oral bioavailability, partly because of its affinity for the intestinal P-gp efflux pump and partly because of possible metabolism of docetaxel by cytochrome P450 in the gut and liver. In a recent clinical study with 14 patients, the mean oral bioavailability in patients taking docetaxel was only $8 \pm 6\%$, whereas the bioavailability of docetaxel in patients receiving both the drug and cyclosporine A (both a P-gp substrate and inhibitor) was $90 \pm 44\%$ (133). Although effective, this approach does raise a series of safety concerns. Some important questions include:

1. Given that the P-gp efflux pump and CYP3A share a large number of substrates, what is the effect of administrating a P-gp inhibitor on liver and gut metabolism?
2. If both P-gp and CYP3A are inhibited, what are the potential implications caused by potential toxic substances that are usually metabolized by CYP3A?
3. Is the inhibition transient or long-lasting, reversible or irreversible?
4. What is the impact of intersubject variability on P-gp inhibition?

These issues may have serious toxicity implications. Therefore, without fully understanding the mechanism of P-gp inhibitors or the natural role of P-gp transporters, this approach is too risky to be considered as a routine method to improve oral bioavailability.

Overall, many so-called Class V compounds face significant delivery challenges that cannot be easily overcome by traditional methods. Although some emerging Class V technologies may be able to overcome such challenges to a certain extent, these strategies are highly exploratory and are yet to be proved. One may be better off seeking a non-oral delivery route to overcome the absorption barriers posed by Class V compounds. Such delivery routes may include nasal, oral mucosal, or intravenous administration. Details regarding these delivery routes are beyond the scope of this chapter.

4.7 Excipient and Process Selection in Dosage Form Design

The preceding sections classify the formulation strategies on the basis of the solubility and permeability of drugs. Once a formulation strategy is identified, it is important to choose suitable excipients and processing methods to achieve the objective of a selected dosage form, in terms of both dosage form performance and manufacturability. Although it is beyond the scope of this chapter to delve into the details of formulation development, a brief overview of some considerations may provide a useful insight into the factors considered by the formulation scientist in developing formulations. Typically, excipients are added to drugs to allow for the manufacture of dosage forms that meet performance (e.g., drug release, stability) and manufacturing requirements. Common excipients for tablet formulations, for example, include tablet fillers, binders, disintegrants, wetting agents, glidants, and lubricants. In addition to selecting appropriate excipients, suitable processing steps and processing conditions must be identified. For tablet dosage forms, typical processing steps include mixing, granulation, sizing, and compression. The granulation process can produce product with improved performance and manufacturing properties and either dry granulation or wet granulation can be done, depending on whether a granulating fluid is used. For materials with appropriate physical, chemical, and mechanical properties, direct compression without the granulation step may be most appropriate and desirable.

Proper selection of excipients and processes will impact the performance of a dosage form. For example, a poorly soluble drug often tends to be poorly wettable, too. If the objective is to obtain a fast-dissolving and dispersing dosage form, inclusion of a wetting agent such as sodium lauryl sulfate or polysorbate 80 may be appropriate or even necessary. Processing methods may also significantly impact dosage form performance. For example, it may not be appropriate to wet granulate amorphous drug because water may lower the glass-transition temperature and facilitate recrystallization during or after processing. In other situations, wet granulation can be used to avoid potential segregation and content uniformity problems where there is a significant difference in particle size or bulk density between the drug and excipients. Overall, a wise selection of excipients and processes relies on a sound understanding of the physical, chemical, and mechanical properties of the drug and excipients. A formulation may be successfully scaled up and consistently meet performance and manufacturing requirements only when one fully understands the complex relationship between the drug, excipients, processing, and the desired dosage form performance criteria.

5 CONCLUSION

This chapter is composed of two parts. Part I provides an overview of physicochemical characterization and its relevance to drug delivery. Part II categorizes various dosage form options according to the well-established Biopharmaceutics Classification System (BCS). Physicochemical properties are closely linked to biopharmaceutical properties of drug candidates. The BCS captures this link by highlighting the important effects of solubility and permeability on drug absorption. Therefore, a sound understanding of how physicochemical properties may affect absorption is essential to make a smart choice of which drug candidate(s) to select and which of the wide variety of oral dosage formulation options to pursue.

6 ACKNOWLEDGMENT

The authors thank their colleague Michael Hawley for reviewing this manuscript and for his many helpful suggestions.

REFERENCES

1. S. M. Berge, L. D. Bighley, and D. C. Monkhouse, *J. Pharm. Sci.*, **66**, 1–19 (1977).

2. S. Bryn, R. Pfeiffer, M. Ganey, C. Hoiberg, and G. Poochikian, *Pharm. Res.*, **12**, 945–954 (1995).

3. S. Byrn, R. R. Pfeiffer, G. Stephenson, D. Grant, and W. B. Gleason, *Chem. Mater.*, **6**, 1148–1158 (1994).

4. K. R. Morris, M. G. Fakes, A. B. Thakur, A. W. Newman, A. T. Serajuddin, et al., *Int. J. Pharm.*, **105**, 209–217 (1994).

5. J. Haleblian and W. McCrone, *J. Pharm. Sci.*, **58**, 911–929 (1969).

6. Anonymous. Proceedings of the International Conference on Harmonization, Fed. Register, **62**, 62889–62910 (1997).

7. H. Zhu, C. Yuuen, and D. Grant, *Int. J. Pharm.*, **135**, 151–160 (1996).

8. H. Zhu and D. Grant, *Int. J. Pharm.*, **139**, 33–43 (1996).

9. A. R. Katritzky, U. Maran, M. Karelson, and V. S. Lobanov, *J. Chem. Inf. Comput. Sci.*, **37**, 913–919 (1997).

10. J. F. Krzyzaniak, P. B. Myrdal, P. Simamora, and S. H. Yalkowsky, *Ind. Eng. Chem. Res.*, **34**, 2530–2535 (1995).

11. P. Simamora and S. H. Yalkowsky, *Ind. Eng. Chem. Res.*, **33**, 1405–1409 (1994).

12. P. Simamora, A. H. Miller, and S. H. Yalkowsky, *J. Chem. Inf. Comput. Sci.*, **33**, 437–440 (1993).

13. P. Verwer and F. J. J. Leusen, *Reviews in Computational Chemistry*, Wiley-VCH, New York, 1998, pp. 327–365.

14. E. Overton, *Phys. Chem*, **8**, 189–209 (1891).

15. H. Meyer, *Arch. Exp. Pathol. Pharmakol.*, **42**, 109–118 (1899).

16. P. Buchwald and N. Bodor, *Curr. Med. Chem.*, **5**, 353–380 (1998).

17. C. Hansch, P. Maloney, T. Fujita, and R. Muir, *Nature*, **194**, 178–180 (1962).

18. C. Hansch and T. Fujita, *J. Am. Chem. Soc.*, **86**, 1616–1626 (1964).

19. C. Hansch, *Acc. Chem. Res.*, **2**, 232–239 (1969).

20. C. Hansch, R. Muir, T. Fujita, P. G. F. Maloney, and M. Streich, *J. Am. Chem. Soc.*, **85**, 2817–2825 (1963).

21. T. Fujita, J. Iwasa, and C. Hansch, *J. Am. Chem. Soc.*, **86**, 5175 (1964).

22. P. Taylor, *Comprehensive Medicinal Chemistry*, Pergamon Press, New York, 1990, pp. 241–294.

23. J. Sangster, *Octanol-Water Partition Coefficients: Fundamental and Physical Chemistry*, John Wiley & Sons, Chichester, UK, 1997.

24. A. J. Leo, *J. Pharm. Sci.*, **76**, 166–168 (1987).

25. A. J. Leo, *Chem. Pharm. Bull.*, **43**, 512–513 (1995).

26. R. Mannhold and K. Dross, *Quant. Struct.-Act. Relat.*, **15**, 403–409 (1996).

27. H. Vandewaterbeemd and R. Mannhold, *Quant. Struct.-Act. Relat.*, **15**, 410–412 (1996).

28. R. Mannhold, R. F. Rekker, C. Sonntag, A. M. Terlaak, K. Dross, and E. E. Polymeropoulos, *J. Pharm. Sci.*, **84**, 1410–1419 (1995).

29. S. Yalkowsky and S. Banerjee, *Aqueous Solubility: Methods of Estimation for Organic Compounds*, Marcel Dekker, New York, 1992.

30. C. A. Lipinski, F. Lombardo, B. W. Dominy, and P. J. Feeney, *Adv. Drug Deliv. Rev.*, **23**, 3–25 (1997).

31. S. H. Yalkowsky, G. L. Flynn, and G. L. Amidon, *J. Pharm. Sci.*, **61**, 983–984 (1972).

32. J. Huuskonen, *Combin. Chem. High Throughput Screen.*, **4**, 311–316 (2001).

33. S. H. Yalkowsky and S. C. Valvani, *J. Pharm. Sci.*, **69**, 912–922 (1980).

34. S. H. Yalkowsky, S. C. Valvani, and T. J. Roseman, *J. Pharm. Sci.*, **72**, 866–870 (1983).

35. S. C. Valvani, S. H. Yalkowsky, and T. J. Roseman, *J. Pharm. Sci.*, **70**, 502–507 (1981).

36. S. H. Yalkowsky, *J. Pharm. Sci.*, **70**, 971–973 (1981).

37. G. L. Amidon, S. H. Yalkowsky, and S. Leung, *J. Pharm. Sci.*, **63**, 1858–1866 (1974).

38. A. Noyes and W. Whitney, *J. Am. Chem. Soc.*, **19**, 930–934 (1897).

39. B. Rohrs and G. E. Amidon, Unpublished results.

40. L. Barthe, J. Woodley, and G. Houin, *Fundam. Clin. Pharmacol.*, **13**, 154–168 (1999).

41. G. Camenisch, J. Alsenz, H. Vandewaterbeemd, and G. Folkers, *Eur. J. Pharm. Sci.*, **6**, 313–319 (1998).

42. G. Camenisch, G. Folkers, and H. Vandewaterbeemd, *Int. J. Pharm.*, **147**, 61–70 (1997).

43. H. Vandewaterbeemd, G. Camenisch, G. Folkers, and O. A. Raevsky, *Quant. Struct.-Act. Relat.*, **15**, 480–490 (1996).

44. H. N. Nellans, *Adv. Drug Deliv. Rev.*, **7**, 339–364 (1991).

45. J. Hunter and B. H. Hirst, *Adv. Drug Deliv. Rev.*, **25**, 129–157 (1997).

46. L. Z. Benet, C. Y. Wu, M. F. Hebert, and V. J. Wacher, *J. Controlled Release*, **39**, 139–143 (1996).

47. C. A. Lipinski, *J. Pharmacol. Toxicol. Methods*, **44**, 235–249 (2000).

48. C. A. Lipinski, F. Lombardo, B. W. Dominy, and P. J. Feeney, *Adv. Drug Deliv. Rev.*, **46**, 3–26 (2001).

49. D. J. Livingstone, M. G. Ford, J. J. Huuskonen, and D. W. Salt, *J. Comput.-Aided Mol. Des.*, **15**, 741–752 (2001).

50. M. C. Desai, P. F. Thadeio, C. A. Lipinski, D. R. Liston, R. W. Spencer, and I. H. Williams, *Bioorg. Med. Chem. Lett.*, **1**, 411–414 (1991).

51. C. P. Lee, R. L. A. Devrueh, and P. L. Smith, *Adv. Drug Deliv. Rev.*, **23**, 47–62 (1997).

52. M. H. Abraham, H. S. Chadha, G. S. Whiting, and R. C. Mitchell, *J. Pharm. Sci.*, **83**, 1085–1100 (1994).

53. D. A. Paterson, R. A. Conradi, A. R. Hilgers, T. J. Vidmar, and P. S. Burton, *Quant. Struct.-Act. Relat.*, **13**, 4–10 (1994).

54. S. Winiwarter, N. M. Bonham, F. Ax, A. Hallberg, H. Lennernäs, and A. Karlen, *J. Med. Chem.*, **41**, 4939–4949 (1998).

55. H. Lennernäs, *J. Pharm. Sci.*, **87**, 403–410 (1998).

56. A. L. Ungell, S. Nylander, S. Bergstrand, A. Sjoberg, and H. Lennernäs, *J. Pharm. Sci.*, **87**, 360–366 (1998).

57. H. Lennernäs, *J. Pharm. Pharmacol.*, **49**, 627–638 (1997).

58. H. Lennernäs, S. Nylander, and A. L. Ungell, *Pharm. Res.*, **14**, 667–671 (1997).

59. U. Fagerholm, M. Johansson, and H. Lennernäs, *Pharm. Res.*, **13**, 1336–1342 (1996).

60. H. Lennernäs, K. Palm, U. Fagerholm, and P. Artursson, *Int. J. Pharm.*, **127**, 103–107 (1996).

61. H. Lennernäs, J. R. Crison, and G. L. Amidon, *J. Pharmacokinet. Biopharm.*, **23**, 333–337 (1995).

62. H. Lennernäs, *Eur. J. Pharm. Sci.*, **2**, 39–43 (1994).

63. P. Artursson, *Crit. Rev. Ther. Drug Carrier Syst.*, **8**, 305–330 (1991).

64. P. Artursson and J. Karlsson, *Biochem. Biophys. Res. Commun.*, **175**, 880–885 (1991).

65. P. Artursson and R. T. Borchardt, *Pharm. Res.*, **14**, 1655–1658 (1997).

66. B. H. Stewart, Y. Wang, and N. Surendran, *Ann. Rep. Med. Chem.*, **35**, 299–307 (2000).

67. B. H. Stewart, O. H. Chan, N. Jezyk, and D. Fleisher, *Adv. Drug Deliv. Rev.*, **23**, 27–45 (1997).

68. B. H. Stewart, O. H. Chan, R. H. Lu, E. L. Reyner, H. L. Schmid, H. W. Hamilton, B. A. Steinbaugh, and M. D. Taylor, *Pharm. Res.*, **12**, 693–699 (1995).

69. W. Rubas, N. Jezyk, and G. M. Grass, *Pharm. Res.*, **10**, 113–118 (1993).

70. N. Jezyk, W. Rubas, and G. M. Grass, *Pharm. Res.*, **9**, 1580–1586 (1992).

71. G. M. Grass, W. Rubas, and N. Jezyk, *FASEB J.*, **6**, A1002 (1992).

72. J. N. Cogburn, M. G. Donovan, and C. S. Schasteen, *Pharm. Res.*, **8**, 210–216 (1991).

73. P. Artursson, K. Palm, and K. Luthman, *Adv. Drug Deliv. Rev.*, **46**, 27–43 (2001).

74. S. F. Kramer and G. Flynn, *J. Pharm. Sci.*, **61**, 1896–1904 (1972).

75. M. S. Bergren, *Int. J. Pharm.*, **103**, 103–114 (1994).

76. G. L. Engel, N. A. Farid, M. M. Faul, L. A. Richardson, and L. L. Winneroski, *Int. J. Pharm.*, **198**, 239–247 (2000).

77. K. R. Morris, A. W. Newman, D. E. Bugay, S. A. Ranadive, A. T. Serajuddin, et al., *Int. J. Pharm.*, **108**, 195–206 (1994).

78. C. Ahlneck and G. Zografi, *Int. J. Pharm.*, **62**, 87–95 (1990).

79. U.S. FDA Guidance for Industry: *Stability Testing of Drug Substances and Drug Products*, U.S. Food and Drug Administration, Washington, DC, 1998.

80. K. Connors, G. Amidon, and V. Stella, *Chemical Stability of Pharmaceuticals: A Handbook for Pharmacists*, Wiley-Interscience, New York, 1986.

81. J. Carstensen and C. Rhodes, *Drug Stability: Principles and Practices*, Marcel Dekker, New York, 2000.

82. H. Lennernäs, *Pharm. Res.*, **12**, 1573 (1995).

83. G. L. Amidon, H. Lennernäs, V. P. Shah, and J. R. Crison, *Pharm. Res.*, **12**, 413 (1995).

84. Center for Drug Evaluation and Research (CDER), *Immediate Release Solid Oral Dosage Forms Scale-up and Postapproval Changes: Chemistry, Manufacturing, and Controls, In Vitro Dissolution Testing, and In Vivo Bioequivalence Documentation*, US Food and Drug Administration, Washington, DC, 1995.

85. J. B. Dressman and C. Reppas, *Eur. J. Pharm. Sci.*, **11**, S73–S80 (2000).

86. V. V. Ranade and M. A. Hollinger, *Oral Drug Delivery in Drug Delivery Systems*, CRC Press, Boca Raton/New York/London/Tokyo, 1996, pp. 166–167.

87. A. F. Kydonieus, *Treatise on Controlled Drug Delivery: Fundamentals, Optimization, Applications*, Marcel Dekker, New York, 1992.

88. R. J. Bastin, M. J. Bowker, and B. J. Slater, *Org. Proc. Res. Dev.*, **4**, 427 (2000).

89. H. Suzuki and H. Sunada, *Chem. Pharm. Bull.*, **46**, 482 (1998).

90. H. Suzuki and H. Sunada, *Chem. Pharm. Bull.*, **45**, 1688 (1997).

91. T. Loftsson, *Pharmazie*, **53**, 733 (1998).

92. E. Shefter, *Techniques and Solubilization of Drugs*, Marcel Dekker, New York, 1981, pp. 159–182.

93. A. Hasegawa, M. Taguchi, R. Suzuki, T. Miyata, H. Nakagawa, and I. Sugimoto, *Chem. Pharm. Bull.*, **36**, 4941 (1988).

94. I. Weissbuch, R. Popovitz-Biro, M. Lahav, and L. Leiserowitz, *Acta Crystallogr. B*, **51**, 115 (1995).

95. S. R. Byrn, R. P. Pfeiffer, and J. G. Stowell, *Solid-State Chemistry of Drugs*, SSCI, West Lafayette, IN, 1999, 249 pp.

96. H. M. Abdou, *Dissolution, Bioavailability and Bioequivalence*, Mack Publishing, Easton, PA, 1989, pp. 56–72.

97. K. Sekguchi and N. Obi, *Chem. Pharm. Bull.*, **9**, 866 (1961).

98. A. H. Goldberg, M. Gibaldi, and J. L. Kanig, *J. Pharm. Sci.*, **55**, 482 (1966).

99. A. H. Goldberg, M. Gibaldi, and J. L. Kanig, *J. Pharm. Sci.*, **55**, 487 (1966).

100. A. T. M. Serajuddin, *J. Pharm. Sci.*, **88**, 1058 (1999).

101. A. A. Rasool, A. A. Hussain, and L. W. Dittert, *J. Pharm. Sci.*, **80**, 387 (1991).

102. D. Hörter and J. B. Dressman, *Adv. Drug Del. Rev.*, **46**, 75 (2001).

103. V. J. Stella and R. A. Rajewski, *Pharm. Res.*, **14**, 556 (1997).

104. D. O. Thompson, *Crit. Rev. Ther. Drug Carrier Syst.*, **14**, 1–104 (1997).

105. R. A. Rajewski and V. J. Stella, *J. Pharm. Sci.*, **85**, 1142 (1996).

106. C. W. Pouton, *Eur. J. Pharm. Sci.*, **11**, S93 (2000).

107. C. W. Pouton, *Adv. Drug Del. Rev.*, **25**, 47 (1997).

108. K. J. MacGregor, J. K. Embleton, J. E. Lacy, E. A. Perry, L. J. Solomon, H. Seager, and C. W. Pouton, *Adv. Drug Del. Rev.*, **25**, 33 (1997).

109. A. J. Humberstone and W. N. Charman, *Adv. Drug Del. Rev.*, **25**, 103 (1997).

110. B. N. Sigh, *Clin. Pharmacokinet.*, **37**, 213 (1999).

111. G. E. Amidon, W. I. Higuchi, and N. F. H. Ho, *J. Pharm. Sci.*, **71**, 77–84 (1982).

112. S. Mall, G. Buckton, and D. A. Rawlins, *Int. J. Pharm.*, **131**, 41 (1994).

113. C. W. Pouton, *Int. J. Pharm.*, **27**, 335 (1985).

114. E. A. Mueller, J. M. Kovarik, J. B. Van Bree, W. Tetzloff, and K. Kutz, *Pharm. Res.*, **11**, 301–304 (1994).

115. J. M. Kovarik, E. A. Mueller, J. B. Van Bree, W. Tetzloff, and K. Kutz, *J. Pharm. Sci.*, **83**, 444–446 (1994).

116. B. J. Aungst, H. Saitoh, D. L. Burcham, S. M. Huang, S. A. Mousa, and M. A. Hussain, *J. Controlled Release*, **41**, 19 (1996).

117. E. L. Parrott, *The Theory and Practice of Industrial Pharmacy*, Lea & Febiger, Philadelphia 1986, 37 pp.

118. J. P. Krise and V. J. Stella, *Adv. Drug. Deliv. Rev.*, **19**, 287 (1996).

119. S. Muranishi, *Crit. Rev. Ther. Drug Carrier Syst.*, **7**, 1 (1990).

120. Y. Watanabe, E. J. van Hoogdalem, and A. G. de Boer, *J. Pharm. Sci.*, **77**, 847 (1988).

121. M. Tomita, M. Shiga, M. Hayashi, and S. Awazu, *Pharm. Res.*, **5**, 341 (1988).

122. B. J. Aungst, *J. Pharm. Sci.*, **82**, 979 (1993).

123. D. Duchene, F. Touchard, and N. A. Peppas, *Drug Dev. Ind. Pharm.*, **14**, 283 (1988).

124. A. J. Moës, *Crit. Rev. Ther. Drug Carrier Syst.*, **8**, 143 (1993).

125. N. A. Peppas and P. A. Buri, *J. Controlled Release*, **2**, 257 (1985).

126. V. M. Lenaerts and R. Gurny, *Bioadhesive Drug Delivery Systems*, CRC Press, Boca Raton, FL, 1990.

127. V. J. Wacher, L. Salphati, and L. Z. Benet, *Adv. Drug Deliv. Rev.*, **46**, 89 (2001).

128. I. Utsumi, K. Kohno, and Y. Takeuchi, *J. Pharm. Sci.*, **63**, 676–681 (1974).

129. P. P. Constanitinides, *Pharm. Res.*, **12**, 1561 (1995).

130. W. N. Charman and C. J. H. Porter, *Adv. Drug. Del. Rev.*, **19**, 149 (1996).

131. W. N. Charman, *J. Pharm. Sci.*, **89**, 967 (2000).

132. M. M. Nerurkar, P. S. Burton, and R. T. Borchardt, *Pharm. Res.*, **13**, 528–534 (1996).

133. M. M. Malingré, D. J. Richel, J. H. Beijnen, H. Rosing, F. J. Koopman, W. W. T. B. Huinink, M. E. Schot, and J. H. M. Schellens, *J. Clin. Oncol.*, **19**, 1160 (2001).

134. J. Devane, *Pharm. Technol.*, **22**, 68–80 (1998).

135. R. P. Bagwe, J. R. Kanicky, B. J. Palla, P. K. Patanhali, and D. O. Shah, *Crit. Rev. Ther. Drug Carrier Syst.*, **18**, 77 (2001).

CHAPTER NINETEEN

The FDA and Regulatory Issues

W. Janusz Rzeszotarski
Food and Drug Administration
Rockville, Maryland

Contents

Burger's Medicinal Chemistry and Drug Discovery
Sixth Edition, Volume 2: Drug Development
Edited by Donald J. Abraham
ISBN 0-471-37028-2 © 2003 John Wiley & Sons, Inc.

1 CAVEAT

The reader is reminded that all the information that is provided in this chapter is freely available on the Web from the government and other sources and subject to change. Instead of a bibliography, only the hyperlink sources are included in text, and the reader is advised to check them frequently.[1]

2 INTRODUCTION

"The problem is to find a form of association which will defend and protect with the whole common force the person and goods of each associate, and in which each, while uniting himself with all, may still obey himself alone, and remain as free as before." So wrote Jean Jacques Rousseau, Citizen of Geneva, in The Social Contract or Principles of Political Right (1762) (http://www.blackmask.com/books10c/socon.htm). His teachings were well known to the Founding Fathers. The Miracle at Philadelphia, the Constitutional Convention of May–September 1787, so gloriously described by Catherine Drinker Bowen, established federalism in the United States and provided for regulation of commerce between the states.

The progress of federalism was slow, and a trigger was needed. On the 25th of April 1846, Mexican troops crossed the Rio Grande to attack U.S. dragoons; this provided an excuse for the U.S.-Mexican war of 1846–1848. The state of medical support for the U.S. troops in Mexico was appalling. The drugs imported for them, counterfeited. In reaction to these events the U.S. Congress passed the Drug Importation Act of 1848, considered by many as the cornerstone of drug regulation in the United States. The act itself required U.S.

Customs Service (already in existence) inspection to stop entry of adulterated drugs from overseas. If one subscribes to George F. Will's precept: "We are not a democracy, we are a republic," the ensuing train of legislative endeavor provides for a fascinating story of continuous interaction between the governing and the governed.

Therefore, the Food and Drug Administration exists by the mandate of the U.S. Congress with the Food, Drug & Cosmetics Act (http://www.fda.gov/opacom/laws/fdcact/fdctoc.htm) as the principal law to enforce. The Act, based on it regulations developed by the Agency, constitutes the basis of the drug approval process (http://www.fda.gov/cder/regulatory/applications/default.htm). The name Food and Drug Administration is relatively new. In 1931 the Food, Drug, and Insecticide Administration, then part of the U.S. Department of Agriculture, was renamed the Food and Drug Administration (http://www.fda.gov).

3 CHRONOLOGY OF DRUG REGULATION IN THE UNITED STATES

The history of food and drug law enforcement in the United States and the consecutive modifications of the 1906 Act are summarized below (from http://www.fda.gov/cder/about/history/time1.htm).

1820 Eleven physicians meet in Washington, DC, to establish the U.S. Pharmacopeia, the first compendium of standard drugs for the United States.

1846 Publication of Lewis Caleb Beck's Adulteration of Various Substances Used in Medicine and the Arts helps document problems in the drug supply.

1848 Drug Importation Act passed by Congress requires U.S. Customs Service inspection to stop entry of adulterated drugs from overseas.

---

[1] The views expressed are my own and do not necessarily represent those of nor imply endorsement from the Food and Drug Administration or the U.S. Government.

1903 Lyman F. Kebler, M. D., Ph.C., assumes duties as Director of the Drug Laboratory, Bureau of Chemistry.

1905 Samuel Hopkins Adams' 10-part exposé of the patent medicine industry, "The Great American Fraud," begins in Collier's. The American Medical Association, through its Council on Pharmacy and Chemistry, initiates a voluntary program of drug approval that would last until 1955. To earn the right to advertise in AMA and related journals, companies submitted evidence, for review by the Council and outside experts, to support their therapeutic claims for drugs.

1906 The original Food and Drugs Act is passed by Congress on June 30 and signed by President Theodore Roosevelt. It prohibits interstate commerce in misbranded and adulterated foods and drugs. The Meat Inspection Act is passed the same day. Shocking disclosures of unsanitary conditions in meatpacking plants, the use of poisonous preservatives and dyes in foods, and cure-all claims for worthless and dangerous patent medicines were the major problems leading to the enactment of these laws.

1911 In U.S. versus Johnson, the Supreme Court rules that the 1906 Food and Drugs Act does not prohibit false therapeutic claims but only false and misleading statements about the ingredients or identity of a drug.

1912 Congress enacts the Sherley Amendment to overcome the ruling in U.S. versus Johnson. It prohibits labeling medicines with false therapeutic claims intended to defraud the purchaser, a standard difficult to prove.

1914 The Harrison Narcotic Act imposes upper limits on the amount of opium, opium-derived products, and cocaine allowed in products available to the public; requires prescriptions for products exceeding the allowable limit of narcotics; and mandates increased record keeping for physicians and pharmacists that dispense narcotics. A separate law dealing with marihuana would be enacted in 1937.

1933 FDA recommends a complete revision of the obsolete 1906 Food and Drugs Act. The first bill is introduced into the Senate, launching a 5-year legislative battle. FDA assembles a graphic display of shortcomings in pharmaceutical and other regulation under the 1906 act, dubbed by one reporter as the Chamber of Horrors and exhibited nationwide to help draw support for a new law.

1937 Elixir Sulfanilamide, containing the poisonous solvent diethylene glycol, kills 107 persons, many of whom are children, dramatizing the need to establish drug safety before marketing and to enact the pending food and drug law.

1938 The Federal Food, Drug, and Cosmetic Act of 1938 is passed by Congress, containing new provisions:

> Requiring new drugs to be shown safe before marketing; starting a new system of drug regulation.

> Eliminating the Sherley Amendment requirement to prove intent to defraud in drug misbranding cases.

> Extending control to cosmetics and therapeutic devices.

> Providing that safe tolerances be set for unavoidable poisonous substances.

> Authorizing standards of identity, quality, and fill-of-container for foods.

> Authorizing factory inspections.

> Adding the remedy of court injunctions to the previous penalties of seizures and prosecutions.

> Under the Wheeler-Lea Act, the Federal Trade Commission is charged to oversee advertising associated with products, including pharmaceuticals, otherwise regulated by FDA.

> FDA promulgates the policy in August that sulfanilamide and selected other dangerous drugs

must be administered under the direction of a qualified expert, thus launching the requirement for prescription only (non-narcotic) drugs.

1941 Insulin Amendment requires FDA to test and certify purity and potency of this life-saving drug for diabetes.

Nearly 300 deaths and injuries result from distribution of sulfathiazole tablets tainted with the sedative phenobarbital. The incident prompts FDA to revise manufacturing and quality controls drastically, the beginning of what would later be called good manufacturing practices (GMPs).

1945 Penicillin Amendment requires FDA testing and certification of safety and effectiveness of all penicillin products. Later amendments would extend this requirement to all antibiotics. In 1983 such control would be found no longer needed and was abolished.

1948 Supreme Court rules in U.S. versus Sullivan that FDA's jurisdiction extends to the retail distribution, thereby permitting FDA to interdict in pharmacies illegal sales of drugs—the most problematical being barbiturates and amphetamines.

1951 Durham-Humphrey Amendment defines the kinds of drugs that cannot be used safely without medical supervision and restricts their sale to prescription by a licensed practitioner.

1952 In U.S. versus Cardiff, the Supreme Court rules that the factory inspection provision of the 1938 FDC Act is too vague to be enforced as criminal law.

A nationwide investigation by FDA reveals that chloramphenicol, a broad-spectrum antibiotic, has caused nearly 180 cases of often fatal blood diseases. Two years later, the FDA would engage the American Society of Hospital Pharmacists, the American Association of Medical Record Librarians, and later the American Medical Association in a voluntary program of drug reaction reporting.

1953 Factory Inspection Amendment clarifies previous law and requires FDA to give manufacturers written reports of conditions observed during inspections and analyses of factory samples.

1955 HEW Secretary Olveta Culp Hobby appoints a committee of 14 citizens to study the adequacy of FDA's facilities and programs. The committee recommends a substantial expansion of FDA staff and facilities, a new headquarters building, and more use of educational and informational programs.

1962 Thalidomide, a new sleeping pill, is found to have caused birth defects in thousands of babies born in western Europe. News reports on the role of Dr. Frances Kelsey, FDA medical officer, in keeping the drug off the U.S. market, arouse public support for stronger drug regulation.

Kefauver-Harris Drug Amendments are passed to ensure drug efficacy and greater drug safety. For the first time, drug manufacturers are required to prove to the FDA the effectiveness of their products before marketing them. In addition, the FDA is given closer control over investigational drug studies, FDA inspectors are granted access to additional company records, and manufacturers must demonstrate the efficacy of products approved before 1962.

1963 Advisory Committee on Investigational Drugs meets, the first meeting of a committee to advise FDA on product approval and policy on an ongoing basis.

1965 Drug Abuse Control Amendments are enacted to deal with problems caused by abuse of depressants, stimulants, and hallucinogens.

1966 FDA contracts with the National Academy of Sciences/National Research Council to evaluate the effectiveness of 4000 drugs approved on the basis of safety alone between 1938 and 1962.

1968 FDA Bureau of Drug Abuse Control and the Treasury Department's Bureau of Narcotics are transferred to the Department of Justice to form the Bureau of Narcotics and Dangerous Drugs (BNDD), consolidating efforts to police traffic in abused drugs. A reorganization of BNDD in 1973 formed the Drug Enforcement Administration.

The FDA forms the Drug Efficacy Study Implementation (DESI) to incorporate the recommendations of a National Academy of Sciences investigation of effectiveness of drugs marketed between 1938 and 1962.

Animal Drug Amendments place all regulation of new animal drugs under one section of the Food, Drug, and Cosmetic Act—Section 512—making approval of animal drugs and medicated feeds more efficient.

1970 In Upjohn versus Finch, the Court of Appeals upholds enforcement of the 1962 drug effectiveness amendments by ruling that commercial success alone does not constitute substantial evidence of drug safety and efficacy.

The FDA requires the first patient package insert: oral contraceptives must contain information for the patient about specific risks and benefits.

The Comprehensive Drug Abuse Prevention and Control Act replaces previous laws and categorizes drugs based on abuse and addiction potential vis-à-vis therapeutic value.

1972 Over-the-Counter Drug Review initiated to enhance the safety, effectiveness, and appropriate labeling of drugs sold without prescription.

1973 The U. S. Supreme Court upholds the 1962 drug effectiveness law and endorses FDA action to control entire classes of products by regulations rather than to rely only on time-consuming litigation.

1976 Vitamins and Minerals Amendments ("Proxmire Amendments") stop the FDA from establishing standards limit-

ing potency of vitamins and minerals in food supplements or regulating them as drugs based solely on potency.

1982 Tamper-resistant packaging regulations issued by the FDA to prevent poisonings such as deaths from cyanide placed in Tylenol capsules. The Federal Anti-Tampering Act passed in 1983 makes it a crime to tamper with packaged consumer products.

1983 Orphan Drug Act passed, enabling FDA to promote research and marketing of drugs needed for treating rare diseases.

1984 Drug Price Competition and Patent Term Restoration expedites the availability of less costly generic drugs by permitting the FDA to approve applications to market generic versions of brand-name drugs without repeating the research done to prove them safe and effective.

At the same time, the brand-name companies can apply for up to 5 years of additional patent protection for the new medicines they developed to make up for time lost while their products were going through the FDA's approval process.

1987 The FDA revises investigational drug regulations to expand access to experimental drugs for patients with serious diseases with no alternative therapies.

1988 The Prescription Drug Marketing Act bans the diversion of prescription drugs from legitimate commercial channels. Congress finds that the resale of such drugs leads to the distribution of mislabeled, adulterated, subpotent, and counterfeit drugs to the public. The new law requires drug wholesalers to be licensed by the states; restricts reimportation from other countries; and bans sale, trade, or purchase of drug samples and traffic or counterfeiting of redeemable drug coupons.

1991 The FDA publishes regulations to accelerate reviews of drugs for life-threatening diseases.

1992 Generic Drug Enforcement Act imposes debarment and other penalties for illegal acts involving abbreviated drug applications.

Prescription Drug User Fee requires drug and biologics manufacturers to pay fees for product applications and supplements and other services. The act also requires the FDA to use these funds to hire more reviewers to assess applications.

1994 The FDA announces it could consider regulating nicotine in cigarettes as a drug, in response to a citizen's petition by the Coalition on Smoking and Health.

Uruguay Round Agreements Act extends the patent terms of U.S. drugs from 17 to 20 years.

1995 The FDA declares cigarettes to be "drug delivery devices." Restrictions are proposed on marketing and sales to reduce smoking by young people.

1997 Food and Drug Administration Modernization Act (FDAMA) re-authorizes the Prescription Drug User Fee Act of 1992 and mandates the most wide-ranging reforms in agency practices since 1938. Provisions include measures to accelerate review of devices, advertising unapproved uses of approved drugs and devices, health claims for foods in agreement with published data by a reputable public health source, and development of good guidance practices for agency decision-making. The fast track provisions are intended to speed up the development and the approval review process for . . . "drug intended for the treatment of a serious or life-threatening condition and it demonstrates the potential to address unmet medical needs for such a condition."

4   FDA BASIC STRUCTURE

Employing over 9000 employees, the FDA's structure reflects the tasks on hand and consists of a number of centers and offices.

Center for Biologics Evaluation and Research (CBER)

Center for Devices and Radiological Health (CDRH)

Center for Drug Evaluation and Research (CDER)

Center for Food Safety and Applied Nutrition (CFSAN)

Center for Veterinary Medicine (CVM)

National Center for Toxicological Research (NCTR)

Office of the Commissioner (OC)

Office of Regulatory Affairs (ORA)

As an agency of the U.S. Government, the FDA does not develop, manufacture, or test drugs. Drug approval is entirely based on sponsor's (manufacturer's) reports of a drug's studies so that the appropriate Center can evaluate its data. The evaluation of submitted data allows the Center reviewers (1) to establish whether the drug submitted for approval works for the proposed use, (2) to assess the benefit-to-risk relationship, and (3) to determine if the drug will be approved. The approval of low molecular weight molecular entities rests within the CDER authority (http://www.fda.gov/cder/) and is the subject of this chapter. An analogous center CBER regulates biological products like blood, vaccines, therapeutics, and related drugs and devices (http://www.fda.gov/cber/). The reader interested in other centers or aspects of FDA activities is advised to visit appropriate sites. In general outline, the drug approval process is divided into an Investigational New Drug (IND) Application Process (with its phases representing a logical and safe process of drug development); New Drug Approval, and the post-approval activities. For as long as an approved drug remains on the market, all aspects pertinent to its safety are under constant scrutiny by the FDA.

5   IND APPLICATION PROCESS

The tests carried out in the pre-clinical investigation of a potential drug serve the purpose to determine whether the new molecule has the desired pharmacological activity and is reasonably safe to be administered to humans in limited, early-stage clinical studies. Before any new drug under pre-clinical investigation is administered to patients to determine its value as a therapeutic or diagnostic, the drug's sponsor must obtain permission from the FDA through the IND process (http://www.fda.gov/cder/regulatory/applications). By definition, a sponsor is a person or entity who assumes responsibility for compliance with applicable provisions of the Federal Food, Drug, and Cosmetic Act (the FDC Act) and related regulations and initiates a clinical investigation. A sponsor could be an individual, partnership, corporation, government agency, manufacturer, or scientific institution. In a way, the IND is an exemption from the legal requirement to transport or distribute across state lines only drugs approved for marketing. Although not approved, the molecule has to conform to specific requirements under the Federal Food, Drug, and Cosmetic Act as interpreted by the Code of Federal Regulations (CFR). The CFR, a codification of the general and permanent regulations published in the Federal Register by the Executive departments and agencies, provides detailed information of requirements for each step of the approval process (http://www.access.gpo.gov/nara/cfr/waisidx_98/21cfr312_98.html). The Federal Register is the additional important source for information on what regulations FDA proposes and notices issues.

A sponsor wishing to submit an IND is assisted and guided by a number of regulatory mechanisms and documents created to secure the uniformity of applications and to guarantee consistency of the review process. The logical development of information and guidance is as follows: (1) from the Federal Food, Drug, and Cosmetic Act to (2) the Code of Federal Regulations to (3) the use of available guidance documents issued by the CDER/CBER and International Conference on Harmonization (ICH). In their review process the FDA reviewers also depend on Manuals of Policies and Procedures (MAPPs), which constitute approved instructions for internal practices and procedures followed by CDER staff. MAPPs are to help standardize the new drug review process and other activities and are available for the public (http://www.fda.gov/cder/mapp.htm).

5.1 Types of IND

The CFR does not differentiate between the "commercial" and "non-commercial," "research," or "compassionate" IND. The three general "types" of INDs below are often mentioned, but again the nomenclature used is not recognized by 21 CFR312.3. The term Commercial IND is defined in CDER's MAPP 6030.1 as: "An IND for which the sponsor is usually either a corporate entity or one of the institutes of the National Institutes of Health (NIH). In addition, CDER may designate other INDs as *commercial* if it is clear the sponsor intends the product to be commercialized at a later date" (http://www.fda.gov/cder/mapp/6030-1.pdf). The term Screening IND is defined in CDER's MAPP 6030.4 (http://www.fda.gov/cder/mapp/6030-4.pdf) as "A single IND submitted for the sole purpose of comparing the properties of closely related active moieties to screen for the preferred compounds or formulations. These compounds or formulations can then become the subject of additional clinical development, each under a separate IND."

The same goes for the fast track programs of the FDA originating from the section 112(b) "Expediting Study and Approval of Fast Track Drugs" of the Food and Drug Administration Modernization Act (FDAMA) of 1997. The FDAMA amendments of the Act are designed to facilitate the development and expedite the review of new drugs that are intended to treat serious or life-threatening conditions and that demonstrate the potential to address unmet medical needs (fast track products).

5.1.1 An Investigator IND. An investigator is an individual who conducts a clinical investigation or is a responsible leader of a team of investigators. Sponsor is a person who takes responsibility far and initiates a clinical investigation. Sponsor may be a person or an organization, company, university, etc. Sponsor-investigator is a physician who both initiates and conducts an investigation, and under whose immediate direction the investigational drug is administered or dispensed. A physician might submit a research IND to propose studying an unapproved drug, or an approved product for a new indication or in a new patient population. The investigator's name appears on the Investigational New Drug Application forms (Forms FDA 1571 and 1572) as the name of person responsible for monitoring the conduct and progress of clinical investigations.

5.1.2 Emergency Use IND.

Emergency Use IND of an investigational new drug (21 CFR 312.36) allows the FDA to authorize use of an experimental drug in an emergency situation that does not allow time for submission of an IND in accordance with the Code of Federal Regulations. It is also used for patients who do not meet the criteria of an existing study protocol or if an approved study protocol does not exist.

5.1.3 Treatment IND.

Treatment IND (21 CFR 312.34) is submitted for experimental drugs showing promise in clinical testing for serious or immediately life-threatening conditions while the final clinical work is conducted and the FDA review takes place. An immediately life-threatening disease means a stage of a disease in which there is a reasonable likelihood that death will occur within a matter of months or in which premature death is likely without early treatment. For example, advanced cases of AIDS, herpes simplex encephalitis, and subarachnoid hemorrhage are all considered to be immediately life-threatening diseases. Treatment INDs are made available to patients before general marketing begins, typically during phase III studies. Treatment INDs also allow FDA to obtain additional data on the drug's safety and effectiveness (http:// www.fda.gov/cder/handbook/treatind.htm).

5.2 Parallel Track

Another mechanism used to permit wider availability of experimental agents is the "parallel track" policy (Federal Register of May 21, 1990) developed by the U.S. Public Health Service in response to AIDS. Under this policy, patients with AIDS whose condition prevents them from participating in controlled clinical trials can receive investigational drugs shown in preliminary studies to be promising.

5.3 Resources for Preparation of IND Applications

As listed above, to assist in preparation of IND, numerous resources are available on the Web to provide the sponsor with (1) the legal requirements of an IND application, (2) assistance from CDER/CBER to help meet those requirements, and (3) internal IND review principles, policies, and procedures.

5.3.1 Pre-IND Meeting.

In addition to all documents available on the Web, under the FDAMA provisions and the resulting guidances (http://www.fda.gov/cder/fdama/default. htm), a sponsor can request all kinds of meetings with the FDA to facilitate the review and approval process. The pre-IND meetings (21 CFR 312.82) belong to type B meetings and should occur with the division of CDER responsible for the review of given drug therapeutic category within 60 days from when the Agency receives a written request. The list of questions and the information submitted to the Agency in the Information Package should be of sufficient pertinence and quality to permit a productive meeting.

5.3.2 Guidance Documents.

Guidance documents representing the Agency's current thinking on a particular subject can be obtained from the Web (http://www.fda.gov/ cder/guidance/index.htm) or from the Office of Training and Communications, Division of Communications Management (http://www. fda.gov/cder/dib/dibinfo.htm). One should remember that the guidance documents merely provide direction and are not binding on either part. A Guidance for Industry "Content and Format of Investigational New Drug Applications (INDs) for Phase I Studies of Drugs, Including Well-Characterized, Therapeutic, Biotechnology-derived Products" (http://www. fda.gov/cder/guidance/index.htm) is a place to start. This particular Guidance, based on 21CFR 312, provides a detailed clarification of CFR requirements for data and data presentation to be included in the initial phase I IND document, permitting its acceptance by the Agency for review.

5.3.3 Information Submitted with IND. To be acceptable for review by the FDA, the IND application must include the following groups of information.

5.3.3.1 Animal Pharmacology and Toxicology Studies. Pre-clinical data to permit an assessment as to whether the product is reasonably safe for initial testing in humans. Also included are the results of any previous experience with the drug in humans (often foreign use).

5.3.3.2 Manufacturing Information. Information pertaining to the composition, manufacturer, stability, and controls used for manufacturing the drug substance and the drug product. This information is assessed to ensure that the company can adequately produce and supply consistent batches of the drug.

5.3.3.3 Clinical Protocols and Investigator Information. Detailed protocols for proposed clinical studies to assess whether the initial-phase trials will expose subjects to unnecessary risks needs to be provided. In addition, information on the qualifications of clinical investigators, professionals (generally physicians) who oversee the administration of the experimental compound, to assess whether they are qualified to fulfill their clinical trial duties is needed. Finally, commitments to obtain informed consent from the research subjects, to obtain review of the study by an institutional review board (IRB), and to adhere to the investigational new drug regulations are also required.

Once the IND is submitted, the sponsor must wait 30 calendar days before initiating any clinical trials. During this time, the FDA has an opportunity to review the IND for safety to assure that research subjects will not be subjected to unreasonable risk.

5.4 The First Step, the Phase I IND Application

The content of the Phase I IND Application (http://www.fda.gov/cder/guidance/index.htm) must include the following:

A. FDA Forms 1571 (IND Application) and 1572 (Statement of Investigator)

B. Table of Contents

C. Introductory Statement and General Investigational Plan

D. Investigator's Brochure

E. Protocols

F. Chemistry, Manufacturing, and Control (CMC) Information

G. Pharmacology and Toxicology Information

H. Previous Human Experience with the Investigational Drug

I. Additional and Relevant Information

Ad C. It should succinctly describe what the sponsor attempts to determine by the first human studies. All previous human experience with the drug, other INDs, previous attempts to investigate followed by withdrawal, foreign marketing experience relevant to the safety of the proposed investigation, etc., should be described. Because the detailed development plans are contingent on the results of the initial studies, limited in scope and subject to change, that section should be kept as brief as possible.

Ad D. Before the investigation of a drug by participating clinical investigators may begin, the sponsor should provide them with an Investigator Brochure. The recommended elements of Investigator's Brochure are subject of ICH document ICH E6 (http://www.fda.gov/cder/guidance/iche6.htm) and should provide a compilation of the clinical and non-clinical data relevant to the study in human subjects. The brochure should include a brief description of the drug substance, summaries of pharmacological and toxicological effects, pharmacokinetics and biological disposition in animals, and if known, in humans.

Also included should be a summary of known safety and effectiveness in humans from previous clinical studies. Reprints of published studies may be attached. Based on prior experience or on related drugs, the brochure should describe possible risks and side effects and eventual precautions or need of special monitoring.

Ad E. Protocols for phase I studies need not be detailed and may be quite flexible compared with later phases. They should provide the following: (*1*) outline of the investigation, (*2*) estimated number of patients involved, (3) description of safety exclusions, (4) description of dosing plan, duration, and dose or method of determining the dose, and (5) specific detail elements critical to safety. Monitoring of vital signs and blood chemistry and toxicity-based stopping or dose adjustment rules should be specified in detail.

Ad F. Phase I studies are usually conducted with the drug substance of drug discovery origin. It is recognized that the synthetic methods may change and that additional information may be accumulated as the studies and development progress. Nevertheless, the application should provide CMC information sufficient to evaluate the safety of drug substance. The governing principle is that the sponsor should be able to relate the drug product proposed for human studies to the drug product used in animal toxicology studies. At issue is the comparability of the (im)purity profiles. Also addressed should be the issues of stability of the drug product and the polymorphic form of the drug substance as they might change with the change of synthetic methods. The CMC information section to be provided in the phase I application should consist of following sections.

1. CMC Introduction: Should address any potential human risks and the proposed steps to monitor such risks and describe the eventual chemical and manufacturing differences between batches used in animal and proposed for human studies.

2. Drug Substance:
   a. Brief description of the drug substance including some physicochemical characterization and proof of structure.
   b. The name and address of manufacturer.
   c. Brief description of manufacturing process with a flow chart and a list of reagents, solvents and catalysts.
   d. Proposed acceptable limits of assay and related substances (impurities) based on actual analytical results with the certificates of analysis for batches used in animal toxicological studies and stability studies and batches destined for clinical studies.

   e. Stability studies may be brief but should cover the proposed duration of the study. A tabular presentation of past stability studies may be submitted.

3. Drug Product:
   a. List of all components.
   b. Quantitative composition of the investigational drug product.
   c. The name and address of manufacturer.
   d. Brief description of manufacturing and packaging process.
   e. Specifications and methods assuring identification, strength, quality and purity of drug product.
   f. Stability and stability methods used. The stability studies should cover the duration of toxicologic and clinical studies.

4. Placebo (see part 3)

5. Labels and Labeling: Copies or mock-ups of proposed labeling that will be provided to each investigator should be provided.

6. A claim for Categorical Exclusion from submission or submission of Environmental Assessment. The FDA believes that the great majority of drug products should qualify for a categorical exclusion.

Ad G. The Pharmacology and Toxicology Information is usually divided into the following sections.

1. Pharmacology & Drug Distribution which should contain, if known: description of drug pharmacologic effects and mechanisms of action in animals and its absorption, distribution, metabolism and excretion.

2. Toxicology: Integrated Summary of toxicologic effects in animals and *in vitro*. In cases where species specificity may be of concern, the sponsor is encouraged to discuss the issue with the Agency. In the early phase of IND, the final fully quality-assured individual study reports may slow preparation and delay the submission of application. If the integrated summary is based on unaudited draft reports, the sponsor is required to submit an update by 120 days after the start of the human studies and identify the differences. Any new find-

ings discovered in preparation of final document affecting the patient safety must be reported to FDA in IND safety reports. To support the safety of human investigation the integrated summary should include:

a. Design of the toxicologic studies and deviations from it. The dates of trials. References to protocols and protocol amendments.

b. Systematic presentation of findings highlighting the findings that may be considered by an expert as possible risk signals.

c. Qualifications of individual who evaluated the animal safety data. That individual should sign the summary attesting that the summary accurately reflects the data.

d. Location of animal studies and where the records of the studies are located, in case of an inspection.

e. Declaration of compliance to Good Laboratory Practices (GLP) or explanation why the compliance was impossible and how it may affect the interpretation of findings.

3. Toxicology—Full Data Tabulation. Each animal toxicology study intended to support the safety of the proposed clinical study should be supported by a full tabulation of data suitable for detailed review. A technical report on methods used and a copy of the study protocol should be attached.

Ad H. Previous Human Experience with the Investigational Drug may be presented in an integrated summary report. The absence of previous human experience should be stated.

Ad I. Additional and Relevant Information may be needed if the drug has a dependence or abuse potential, is radioactive, or if pediatric safety and effectiveness assessment is planned. Any information previously submitted need not to be resubmitted but may be referenced.

Once the IND is submitted to FDA, an IND number is assigned, and the application is forwarded to the appropriate reviewing division. The reviewing division sends a letter to the

Sponsor-Investigator providing the IND number assigned, date of receipt of the original application, address where future submissions to the IND should be sent, and the name and telephone number of the FDA person to whom questions about the application should be directed. The IND studies shall not be initiated until 30 days after the date of receipt of the IND by the FDA. The sponsor may receive earlier notification by the FDA that studies may begin.

5.4.1 Phase I of IND. The initial introduction of an investigational new drug into humans may be conducted in patients, but is usually conducted in healthy volunteer subjects. Phase I studies are usually designed to obtain, in humans, sufficient information about the pharmacokinetics, pharmacological effects, and metabolism, the side effects associated with increasing doses, and perhaps, preliminary evidence on effectiveness of the drug. The information collected should permit the design of well-controlled, scientifically valid phase II studies. The studies might even attempt to study the structure-activity relationships and the mechanism of action. The total number of subjects in the phase I study may vary. Depending on intent, it is usually in the range of 20–80 and rarely exceeds 100. The phase lasts several months and 70% of investigated drugs pass that phase. Beginning with phase I studies, the CDER can impose a clinical hold (i.e., prohibit the study from proceeding or stop a trial that has started) for reasons of safety or because of a sponsor's failure to accurately disclose the risk of study to investigators. The review process, illustrated in Fig. 19.1, begins with the moment the IND application is assigned to individual reviewers representing various disciplines.

5.4.2 Phase II of IND. The initial (phase I) studies can be conducted in a group of patients, but most likely are conducted in healthy volunteers. In phase II, the early clinical studies of the effectiveness of the drug for a particular indication or indications are conducted in patients with the disease or condition. They are also used to determine the common short-term side effects and risks associated with the drug. The number of pa-

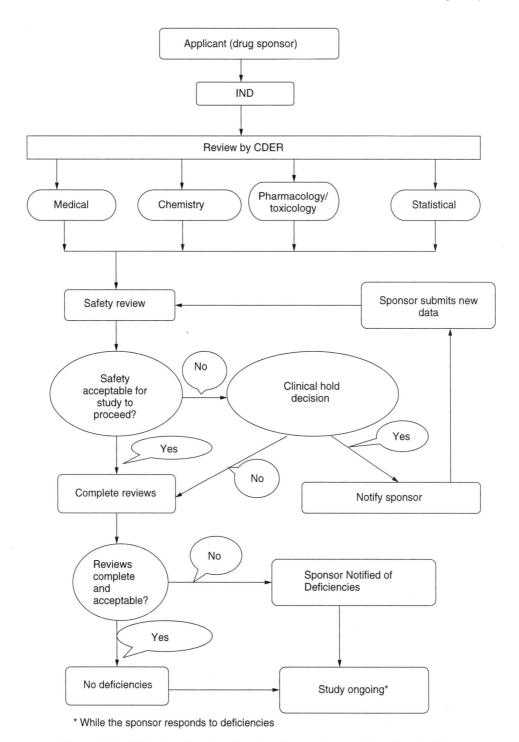

Figure 19.1. IND review flow chart from http://www.fda.gov/cder/handbook/ind.htm.

tients in phase II studies is still small and does not exceed several hundred. The studies that have to be well-controlled and closely monitored last several months to 2 years. Approximately 33% of drugs investigated pass that phase.

5.4.3 Phase III of IND. Phase III studies are expanded controlled and uncontrolled trials. They are performed after preliminary evidence suggesting effectiveness of the drug has been obtained in phase II and are intended to gather the additional information about effectiveness and safety that is needed to evaluate the overall benefit–risk relationship of the drug. Phase III studies also provide an adequate basis for extrapolating the results to the general population and transmitting that information in the physician labeling. Phase III studies usually include several hundred to several thousand people.

In both phases II and III, CDER can impose a clinical hold if a study is unsafe (as in phase I), or if the protocol is clearly deficient in design in meeting its stated objectives. Great care is taken to ensure that this determination is not made in isolation, but reflects current scientific knowledge, agency experience with the design of clinical trials, and experience with the class of drugs under investigation. Out of 100 drugs entering phase I, over 25 should pass phase III and go into the New Drug Application (NDA) approval process. According to FDA calculations (http://www. fda.gov/fdac/special/newdrug/ndd_toc.html), about 20% of drugs entering IND phase are eventually approved for marketing. The numbers agree with similar representation of Pharmaceutical Research and Manufacturers of America (PhRMA; http://www.phrma.org/ index.phtml?mode=web), showing that on the average, it takes 12–15 years and over $500 million to discover and develop a new drug. Out of 5000 compounds entering the preclinical research, only 5 go to IND and only 1 is approved (http://www.phrma.org/publications/ documents/factsheets//2001–03-01.210.phtml).

5.4.4 Phase IV of IND. 21 CFR 312 Subpart E provides for drugs intended to treat life-threatening and severely-debilitating illnesses. In that case, the end-of-phase I meet-

ings would reach agreement on the design of phase II controlled clinical trials. If the results of preliminary analysis of phase II studies are promising, a treatment protocol may be requested and when granted would remain in effect until the complete data necessary for marketing application are assembled. Concurrent with the marketing approval, the FDA may seek agreement to conduct post-marketing, phase IV studies (21CFR312.85).

5.5 Meetings with the FDA (http://www. fda.gov/cder/guidance/2125fnl.pdf)

Section 119(a) of the FDAMA of the 1997 Act directs the FDA to meet with sponsors and applicants, provided certain conditions are met, for the purpose of reaching agreement on the design and size of clinical trials intended to form the primary basis of an effectiveness claim in a NDA submitted under section 505(b) of the Act. These meetings are considered special protocol assessment meetings. All in all, there are three categories of meetings between sponsors or applicants for PDUFA products and CDER staff listed in the above guidance: type A, type B, and type C.

5.5.1 Type A. A type A meeting is one that is immediately necessary for an otherwise stalled drug development program to proceed. Type A meetings generally will be reserved for dispute resolution meetings, meetings to discuss clinical holds, and special protocol assessment meetings that are requested by sponsors after the FDA's evaluation of protocols in assessment letters. Type A meetings should be scheduled to occur within 30 days of FDA's receipt of a written request for a meeting from a sponsor or applicant for a PDUFA product.

5.5.2 Type B. Type B meetings are (*1*) pre-IND meetings (21 CFR 312.82), (*2*) certain end of phase I meetings (21 CFR 312.82), (*3*) end of phase II/pre-phase III meetings (21 CFR 312.47), and (*4*) pre-NDA/BLA meetings (21 CFR 312.47). Type B meetings should be scheduled to occur within 60 days of the Agency's receipt of the written request for a meeting.

5.5.3 Type C. A type C meeting is any meeting other than a type A or type B meeting, but it should be regarding the development

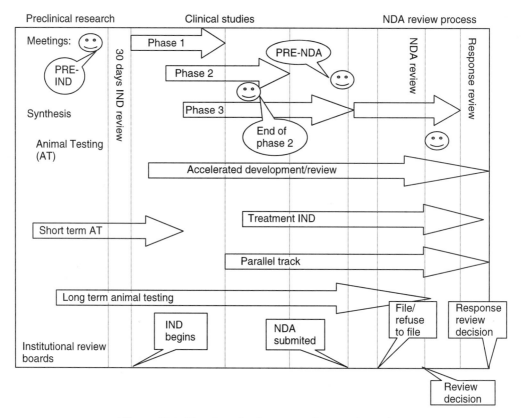

Figure 19.2. New drug development and approval process.

and review of a product in a human drug application as described in section 735 (1) of the Act.

6   DRUG DEVELOPMENT AND APPROVAL TIME FRAME

The development and approval process is presented in Fig. 19.2. In the preclinical phase, the sponsor conducts the short-term animal testing and begins more extensive long-term animal studies. It is advisable to meet with the appropriate division of CDER in a pre-IND meeting to clarify the content of an application. When a sufficient amount of necessary data is gathered into an IND document, the application is filed with the FDA. The Agency has 30 days from the date the document is received to review the IND application, request additional information, and reach the

decision of whether the phase I studies using human subjects can begin (see Fig. 19.1). Depending on the amount of information available or developed about the investigated drug, the phases of IND can overlap. There is no time limit on the duration of IND phases, and the time limits are simply determined by the results and economics. Approval of a drug doesn't end the IND process, which may continue for as long the sponsor intends to accumulate additional information about the drug, which may lead to new uses or formulation (see Fig. 19.2).

Accelerated development/review (*Federal Register*, April 15, 1992) is a highly specialized mechanism for speeding the development of drugs that promise significant benefit over existing therapy for serious or life-threatening illnesses for which no therapy exists. This process incorporates several novel elements aimed at making sure that rapid development and review is balanced by safeguards to protect both the patients and the integrity of the regulatory process.

6.1 Accelerated Development/Review

Accelerated development/review can be used under two special circumstances: when approval is based on evidence of the product's effect on a "surrogate endpoint," and when the FDA determines that safe use of a product depends on restricting its distribution or use. A surrogate endpoint is a laboratory finding or physical sign that may not be a direct measurement of how a patient feels, functions, or survives, but is still considered likely to predict therapeutic benefit for the patient.

The fundamental element of this process is that the manufacturers must continue testing after approval to demonstrate that the drug indeed provides therapeutic benefit to the patient (21CFR314.510). If not, the FDA can withdraw the product from the market more easily than usual.

6.2 Fast Track Programs

Fast Track Programs (http://www.fda.gov/cder/fdama/default.htm and http://www.fda.gov/cder/guidance/2112fnl.pdf), Section 112(b), of the Food and Drug Administration Modernization Act of 1997 (FDAMA) amends the Federal Food, Drug, and Cosmetic Act (the Act) by adding section 506 (21 U.S.C. 356) and directing the FDA to issue guidance describing its policies and procedures pertaining to fast track products. Section 506 authorizes the FDA to take actions appropriate to facilitate the development and expedite the review of an application for such a product. These actions are not limited to those specified in the fast track provision but also encompass existing FDA programs to facilitate development and review of products for serious and life-threatening conditions.

The advantages of Fast Track consist of scheduled meetings with the FDA to gain Agency input into development plans, the option of submitting a New Drug Application in sections, and the option of requesting evaluation of studies using surrogate endpoints (see Accelerated Approval). "The Fast Track desig-

nation is intended for products that address an unmet medical need, but is independent of Priority Review and Accelerated Approval. An applicant may use any or all of the components of Fast Track without the formal designation. Fast Track designation does not necessarily lead to a Priority Review or Accelerated Approval" (http://www.accessdata.fda.gov/scripts/cder/onctools/accel.cfm).

6.3 Safety of Clinical Trials

The safety and effectiveness of the majority of investigated, unapproved drugs in treating, preventing, or diagnosing a specific disease or condition can only be determined by their administration to humans. It is the patient that is the ultimate premarket testing ground for unapproved drugs. To assure the safety of patients in clinical trials, the CDER monitors the study design and conduct of clinical trials to ensure that people in the trials are not exposed to unnecessary risks. The information available on the Web refers the sponsors and investigators to the necessary CFR regulations and guidances. The most important parts of CFR regulating clinical trials are as follows.

1. Financial disclosure section under 21 CFR 54. This covers financial disclosure for clinical investigators to ensure that financial interests and arrangements of clinical investigators that could affect reliability of data submitted to the FDA in support of product marketing are identified and disclosed by the sponsor (http://www.fda.gov/cder/about/smallbiz/financial_disclosure.htm).

2. Parts of 21 CFR 312 that include regulations for clinical investigators (http://www.fda.gov/cder/about/smallbiz/CFR.htm#312.60 and further):

   312.60 General Responsibilities of Investigators

   312.61 Control of the Investigational Drug

   312.62 Investigator Record Keeping and Record Retention

   312.64 Investigator Records

   312.66 Assurance of Institutional Review Board (IRB) Review

312.68 Inspection of Investigator's Records and Reports

312.69 Handling of controlled Substances

312.70 Disqualification of a Clinical Investigator

The important part of any clinical investigation is the presence and activity of an Institutional Review Board (http://www.fda.gov/oc/ohrt/irbs/default.htm), a group that is formally designated to review and monitor biomedical research involving human subjects. An IRB has the authority to approve, require modifications in (to secure approval), or disapprove research. This group review serves an important role in the protection of the rights and welfare of human research subjects. An IRB review is to assure, both in advance and by periodic review, that appropriate steps are taken to protect the rights and welfare of humans participating as subjects in the research. To achieve that, IRBs use a group process to review research protocols and related materials (e.g., informed consent documents and investigator brochures) to ensure protection of the rights and welfare of human subjects.

7 NDA PROCESS (HTTP://WWW.FDA. GOV/CDER/REGULATORY/ APPLICATIONS/NDA.HTM)

By submitting the NDA application to the FDA, the sponsor formally proposes to approve a new drug for sale and marketing in the United States. The information on the drug's safety and efficacy collected during the animal and human trials during the IND process becomes part of the NDA application. The review process of the submitted NDA (Fig. 19.3) is expected to answer the following questions:

1. Is the new drug safe and effective in its proposed use(s)? Do the benefits of the drug outweigh the risks?
2. Is the proposed labeling (package insert) of the drug appropriate and complete?
3. Are the manufacturing and control methods adequate to preserve drug's identity, strength, quality, and purity?

As for IND, the preparation of NDA submission is based on existing laws and regulations and is guided by various guidance documents representing the Agency's current thinking on particular subjects to be included in the NDA documentation. The preparation of NDA submission is based on the Federal Food, Drug, and Cosmetic Act (http://www.fda.gov/opacom/laws/fdcact/fdctoc.htm), as amended, which is the basic drug law in the United States. Its interpretation is provided by the Code of Federal Regulations: 21CFR 314—Applications for FDA Approval to Market a New Drug or an Antibiotic Drug and is available in PDF format at http://www.fda.gov/opacom/laws/fdcact/fdctoc.htm. Further help in understanding of the NDA process is obtained from the available online guidances (http://www.fda.gov/cder/guidance/index.htm) and CDER Manuals of Policies and Procedures (MAPPs; http://www.fda.gov/cder/mapp.htm). The list of guidances is particularly long and needs constant monitoring because some of them may be updated or withdrawn. Many of them address the format and content of the application to assure uniformity and consistency of the review process and decision-making. Particularly useful are the following MAPPs (in PDF):

6050.1—Refusal to Accept Application for Filing from Applicants in Arrears

7211.1—Drug Application Approval 501(b) Policy

7600.6—Requesting and Accepting Non-Archivable Electronic Records for New Drug Applications

7.1 Review Priority Classification

Under the Food and Drug Administration Modernization Act (FDAMA), depending on the anticipated therapeutic or diagnostic value of the submitted NDA, its review might receive a "Priority" (P) or "Standard" (S) classification. The designations "Priority" (P) and "Standard" (S) are mutually exclusive. Both original NDAs and effectiveness supplements receive a review priority classification, but manufacturing supplements do not. The basics of classification, discussed in CDER's MAPP 6020.3 (http://www.fda.gov/cder/mapp/6020-3.pdf), are listed below.

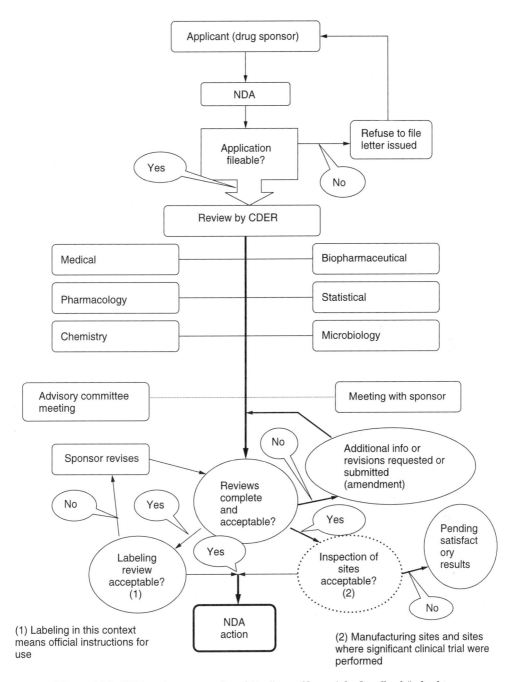

Figure 19.3. NDA review process from http://www.fda.gov/cder/handbook/index.htm.

7.2 P—Priority Review

The drug product, if approved, would be a significant improvement compared with marketed products [approved (if such is required), including non-"drug" products/therapies] in the treatment, diagnosis, or prevention of a disease. Improvement can be demonstrated by (1) evidence of increased effectiveness in treatment, prevention, or diagnosis of disease; (2) elimination or substantial reduction of a treatment-limiting drug reaction; (3) documented enhancement of patient compliance; or (4) evidence of safety and effectiveness of a new subpopulation. (The CBER definition of a priority review is stricter than the definition that CDER uses. The biological drug, if approved, must be a significant improvement in the safety or effectiveness of the treatment diagnosis or prevention of a serious or life-threatening disease).

7.3 S—Standard Review

All non-priority applications will be considered standard applications. The target date for completing all aspects of a review and the FDA taking an action on the "S" application (approve or not approve) is 10 months after the date it was filed. The "P" applications have the target date for the FDA action set at 6 months.

7.4 Accelerated Approval (21CFR Subpart H Sec. 314.510)

Accelerated Approval is approval based on a surrogate endpoint or on an effect on a clinical endpoint other than survival or irreversible morbidity. The CFR clearly states that the FDA . . . "may grant marketing approval for a new drug product on the basis of adequate and well-controlled clinical trials establishing that the drug product has an effect on a surrogate endpoint that is reasonably likely, based on epidemiologic, therapeutic, pathophysiologic, or other evidence, to predict clinical benefit or on the basis of an effect on a clinical endpoint other than survival or irreversible morbidity. Approval under this section will be subject to the requirement that the applicant study the drug further, to verify and describe its clinical benefit, where there is uncertainty as to the relation of the surrogate endpoint to clinical benefit, or of the observed clinical benefit to ultimate outcome. Post-marketing studies would usually be studies already underway. When required to be conducted, such studies must also be adequate and well-controlled. The applicant shall carry out any such studies with due diligence." Therefore, an approval, if it is granted, may be considered a conditional approval with a written commitment to complete clinical studies that formally demonstrate patient benefit.

8 U.S. PHARMACOPEIA AND FDA

The USP Convention is the publisher of the *United States Pharmacopeia* and *National Formulary (USP/NF)*. These texts and supplements are recognized as official compendia under the Federal Food, Drug & Cosmetic Act (FD&C Act). As such, their standards of strength, quality, purity, packaging, and labeling are directly enforceable under the adulteration and misbranding provisions without further approval or adoption by the FDA (http://www.usp.org/frameset.htm; http://www.usp.org/standards/fda/jgv_testimony.htm).

The Federal Food, Drug, and Cosmetic Act §321(g) (1) states: "The term "drug" means (A) articles recognized in the official United States Pharmacopoeia, official Homeopathic Pharmacopoeia of the United States, or official National Formulary, or any supplement to any of them; and . . . " (http://www.mlmlaw.com/library/statutes/federal/fdcact1.htm). That statement and additional arguments evolving from it may lead to a misapprehension that the USP and the FDA are at loggerheads over the authority to regulate the quality of drugs marketed in the United States. Nothing could be further from the truth. The harmonious collaboration of the FDA with many of the USP offices and committees may serve as a model of interaction between a federal agency and a non-governmental organization such as USP.

CDER's MAPP 7211.1 (http://www.fda.gov/cder/mapp/7211–1.pdf) establishes policy applicable to drug application approval with regard to official compendial standards and Section 501(b) of the Act: "When a USP monograph exists and an ANDA/NDA application is submitted to the Agency, reviewers are not to approve regulatory methods/specifications (i.e., those which must be relied upon for enforcement purposes) that differ from those in the USP, unless a recommendation is being or has been sent to the USPC through Compendial Operations Branch (COB) to change the methods/specifications. Direct notification to the U.S. Pharmacopeial Convention, Inc. by applicants does not absolve reviewers of their obligation to notify COB. Each Office within the Center should determine its own standard operating procedures under the policy decision."

9 CDER FREEDOM OF INFORMATION ELECTRONIC READING ROOM

The 1996 amendments to the Freedom of Information (FOI) Act (FOIA) mandate publicly accessible "electronic reading rooms" (ERRs) with agency FOIA response materials and other information routinely available to the public, with electronic search and indexing features. The FDA (http://www.fda.gov/foi/electrr.htm) and many centers (http://www.fda.gov/cder/foi/index.htm) have their ERRs on the Web.

The International Conference on Harmonization of Technical Requirements for Registration of Pharmaceuticals for Human Use (ICH; http://www.ifpma.org/ich1.html) brings together the regulatory authorities of European Union (EU-15), Japan, and the United States, and experts from the pharmaceutical industry in these three regions. The purpose is to make recommendations on ways to achieve greater harmonization in the interpretation and application of technical guidelines and requirements for product registration to reduce or obviate the need to duplicate the testing carried out during the research and development of new medicines. The objective of such harmonization is a more economical use of human, animal, and material resources, and the elimination of unnecessary delay in the global development and availability of new medicines while maintaining safeguards on quality, safety, and efficacy, and regulatory obligations to protect public health.

A series of guidances have been issued (such as http://www.fda.gov/cder/guidance/4539Q.htm) that provide recommendations for applicants preparing the Common Technical Document for the Registration of Pharmaceuticals for Human Use (CTD) for submission to the FDA.

10 CONCLUSION

The ever-expanding field of medicinal chemistry and the heterogeneity of treatment approaches require constant vigilance to maintain the balance between the safe and the novel. The process of new drug evaluation to determine the risk/benefit quotient is affected by many conflicting factors. Nobody, be it the inventor, the generic, the regulatory, the physician, or the patient, is immune from temptation. The principles of social contract as it was written by Jean Jacques Rousseau and accepted by John Adams still apply. The globalization of pharmaceutical industry in G-7 countries and the harmonization of regulatory process between the United States, EU-15, and Japan will have a profound impact on how and how many of the new drugs are going to be developed in the future. The reader is advised to stay familiar with the Web and attuned to the FDA and PhRMA pages.

CHAPTER TWENTY

Intellectual Property in Drug Discovery and Biotechnology

Richard A. Kaba
Timothy P. Maloney
James P. Krueger
Rudy Kratz
Julius Tabin
Fitch, Even, Tabin & Flannery
Chicago, Illinois

Contents

Burger's Medicinal Chemistry and Drug Discovery
Sixth Edition, Volume 2: Drug Development
Edited by Donald J. Abraham
ISBN 0-471-37028-2   © 2003 John Wiley & Sons, Inc.

1 INTRODUCTION

Intellectual property is the branch of law that protects and, indeed, encourages the creation of certain products of the human mind or intellect. This chapter is intended to provide a basic understanding and appreciation of intellectual property law, especially as it relates to patents, trademarks, and trade secrets, in the United States and, to a lesser extent, in the remainder of the world (1). Issues and concerns particularly related to the drug discovery and development process are emphasized.

By making effective use of the legal protection afforded by the intellectual property laws in the United States and elsewhere, the drug developer can protect its investment, enhance the value of the technology being developed, and earn a profit sufficient to allow and encourage further research into improving existing drugs and therapies as well as developing new drugs and therapies. By better understanding these intellectual property laws, the drug developer, together with experienced intellectual property counsel, can develop an effective intellectual property strategy. In this

way, new and emerging technologies as well as new drug discoveries can be identified, managed, and protected as an integral part of an organization's research and development activities to create a strong intellectual property portfolio.

The rewards flowing from the development of a strong intellectual property portfolio can be significant. A patent allows the patent holder to exclude others from making, using, offering for sale, or selling the patented invention during the term of the patent. A carefully crafted intellectual property portfolio (including pending patent applications, issued patents, trademarks, and trade secrets) can also serve many other purposes. It can be used defensively to prevent others from patenting the invention. It can present legitimate barriers to competitors attempting to enter a new field. It can allow time for recouping investments and establishing market position and identity. It can be used to generate revenues through licensing arrangements or outright sales of the patents, trademarks and associated goodwill, or trade secrets. It can be useful in obtaining outside financing or entering into shared research arrangements, joint ventures, or cross-licensing arrangements. In many instances, a start-up biotechnology company's only marketable asset is its intellectual property. A carefully crafted and maintained patent portfolio can be an especially beneficial asset when seeking outside funding or negotiating an agreement with a large, well-established, and well-funded partner.

The application of intellectual property law to the field of drug discovery and biotechnology presents unique and challenging issues for the individual researcher, the research organization or company, and the intellectual property counsel. These issues arise mainly because of the fast-developing nature of the drug discovery and biotechnology field, the enormous investment in time and money required in the current regulatory climate to develop a new drug or treatment process and bring it to the marketplace, and the opportunity derived from the "biotechnology revolution" to achieve rapid breakthroughs in the health care area with the potential for substantial economic rewards. How the industry meets these challenges, and how the legal system evolves and adapts to this rapidly changing field, will significantly affect the development of the burgeoning biotechnology-pharmaceutical industry and the health care system in general. How well individual companies or research organizations protect their intellectual property will determine, to a significant degree, who will survive and prosper.

The development of a drug or treatment process from its conception until its introduction in the marketplace generally requires 6 to 10 years, sometimes even more. This delay is generally attributed to the time required for research and development, pilot plant studies, scale-up studies, animal studies, clinical studies, obtaining the necessary regulatory approvals [e.g., from the U.S. Food and Drug Administration (FDA)], marketing studies, and the like. A successful drug development program can cost hundreds of millions of dollars. The ability to protect that human and economic investment has become an increasingly important factor in the drug discovery and development process. A business organization, whether a start-up company or an established pharmaceutical giant, often cannot justify the

necessary investment if its intellectual property cannot be reasonably protected. Without such protection, a so-called free rider could offer the same or very similar drug or treatment process based on the developer's own research data at a significantly lower cost. Without the ability to protect and recover one's investment and earn a profit, drug discovery and development, for all practical purposes, could be carried out or sponsored only by governments or large nonprofit organizations. This would severely limit the number of persons generating new ideas, decrease the number of new drugs entering the marketplace, and increase the time required for the development of new drugs or treatment processes.

Intellectual property law—that body of law that includes patents, trademarks, trade secrets, and copyrights—provides the framework and mechanism by which investment in intellectual property can be protected. The drug discovery process, especially as it has developed in response to federal regulation and the recent biotechnology revolution, faces new and difficult challenges and issues within the field of intellectual property law. The biotechnology-pharmaceutical industry must recognize and understand these challenges and issues to take advantage of the protection now offered and to be prepared to adapt to modifications that may be made in intellectual property law in the future.

Patents are generally considered the strongest form of protection available for intellectual property and, therefore, should be the cornerstone of the intellectual property protection strategy. An effective patent strategy or program must first identify new and emerging technologies and inventions. A significant part of this program is educating researchers and other employees about the importance of protecting intellectual property and providing mechanisms and incentives to encourage them to bring forward their ideas and innovations for appropriate evaluation. Next, the patent strategy should provide a mechanism to evaluate the inventions and determine whether to file a patent application on a given invention and, if so, when and where to file throughout the world. It must also determine whether and when to update pending patent applications when new information and data become available. This will generally require a careful case-by-case evaluation of each invention, including the likelihood of patentability and success in the ongoing research and FDA approval processes. Unfortunately, such decisions must almost always be made without complete data or information and long before concrete assessments and estimates can be made concerning the ultimate technological and economic success of the invention.

Drug discovery technology has become increasingly complex and multidisciplinary. It is increasingly difficult for meaningful research to be carried out by individuals or even small research teams. Rather, large multidisciplinary teams bringing wide-ranging expertise to bear on a given problem are generally needed to stay ahead of the competition. The existence and requirements of such teams may have a significant effect on the patentability of drugs and treatment processes.

In addition to being new or novel, a product or process to be patentable must not have been obvious at the time the invention was made to "a person having ordinary skill in the art to which said subject matter pertains" (2). Just who is a person of ordinary skill in the drug discovery area? Clearly a person of ordinary skill in the art of drug discovery is at least a highly skilled individual, probably with an advanced degree. Does that person have a Master's or Ph.D. degree in the field to which the invention is most closely related? Or does that person have advanced-level knowledge in more than one field associated with the invention? If so, in how many fields? Is the person of ordinary skill a single individual or a mythical person having the combined knowledge and skill of a multidisciplinary team in which each member possesses ordinary skill in a specific art? These questions remain for the U.S. Patent and Trademark Office (PTO), and ultimately the courts, to decide. Whatever the ultimate resolution, this determination will dramatically affect the patentability of drug-related inventions.

Drug discovery has accelerated over the past several decades because of the continuous and phenomenally rapid advance of the underlying technology. The amount of information and data in the literature is enormous and is growing at an increasing rate, as evidenced by the successful sequencing of the human genome. What is unobvious today to one of "ordinary skill in the art" may well be obvious tomorrow, next week, or next month. This rapidly expanding body of technical information dramatically increases the pressure to seek patent protection as early as possible, oftentimes before the invention is fully developed and its ramifications and significance are fully known.

Another unique aspect of the drug discovery industry is that the majority of research is directed toward a relatively limited number of well-known target diseases or disorders (e.g., AIDS and cancer) and enabling technologies (e.g., receptors) that are useful in drug discovery. Given the importance of these diseases, disorders, and enabling technologies, and the potentially huge economic rewards, many research groups and organizations have turned their resources toward these relatively few targets (3). Although one hopes that this intense competition will lead to near-term breakthroughs in new drugs, methods of treatment, and cures, the intense competition makes it more difficult to protect inventions made along the way. Also, because of the large number of groups working and filing patent applications in the same or closely related biotechnology research areas, the number of potential invention priority contests (i.e., patent interferences; see Section 3.7 below) in the PTO is likely to be significantly higher than that in other technologies. The increased possibility of interferences in the area of drug discovery and treatment processes contributes to the pressure to file as quickly as possible.

Throughout most of the world outside the United States, patents are granted to the first to file rather than the first to invent. In such countries, the failure to file quickly can result in loss of valuable patent rights. Moreover, if others independently make the same invention and obtain a patent, the inventor who files late may be prevented from using the very technology upon which vast sums and significant human resources were spent.

The changing pathway for drug discovery also influences the way in which inventors or their assignees interact with the patent system (4). Historically, drug discovery and drug development were carried out by large pharmaceutical companies. More and more, the basic discovery and initial stages of drug development are being carried out by university research teams and start-up or relatively small biotechnology companies. These groups generally do not have the internal economic resources to seek worldwide patent protection or to carry a new drug or therapy through the clinical stages. Outside funding, strategic alliances, or licensing arrangements are usually necessary as the research and development progresses. In seeking funding or other business arrangements, researchers are generally required to disclose at least basic business and/or technical information. On the other hand, extreme care should be taken to prevent public disclosure of inventions before the appropriate patent applications are filed. The United States has a 1-year grace period in which a patent application can be filed after the first public disclosure, public use, or offer for sale of the subject matter of the invention. In most other countries, however, there is an "absolute novelty" requirement—public disclosure of the subject matter of the invention anywhere in the world before filing the patent application will likely preclude foreign patent protection (see Section 2.3). It is critical, therefore, that secrecy be maintained until the initial patent application is filed covering the invention. Once the initial patent application is filed (usually in the country where the invention is made or developed), corresponding applications can be filed within 1 year in most other countries under prevailing international agreements (i.e., the Paris Convention; see Section 5.1 below), claiming the benefit of the filing date of the initial application.

Secrecy may be very difficult to maintain if one is seeking outside funding. Most potential investors demand significant business and technical details before making the desired investment. To the extent possible, however, the amount of technical information provided should be strictly limited and its use and dissemination carefully controlled. Confidentiality agreements are helpful in maintaining secrecy and are highly recommended when seeking private funding or joint research arrangements. Public funding and offerings, which trigger the disclosure requirements of the Securities Act, the Securities Exchange Commission (SEC), and state Blue Sky laws, present even more difficult problems. Public offerings, when possible, should be delayed until patent applications are filed because public disclosure, even if accidental or in violation of a confidentiality agreement, can preclude patent protection in most countries of the world. Once again, there is great pressure for filing patent applications as quickly as possible.

The requirements set forth in the patent law and as dictated by procedures for obtaining research funding, both in the United States and the rest of the world, also strongly encourage filing a patent application covering a new drug or treatment process as soon as possible. In many cases, it may be desirable or even necessary to file the patent application before complete data are available. For example, a patent application covering a protein for which only a partial DNA coding sequence has been determined may be filed in the United States, to establish an early priority date for the invention. This would permit such preliminary information to be disclosed with reduced risk of losing valuable patent rights. Once the sequence is complete, a continuation-in-part (CIP) application including the additional data may be filed in the United States (and, if appropriate, the original application abandoned). The new material added in the CIP application receives the priority date of the actual CIP filing date. In some cases, it may be desirable to file several CIP applications as new data and/or discoveries are made. Using such an approach, however, has risks. In the United States a patent application must be "enabling"; that is, it must provide a "written description of the invention, and of the manner and process of making and using it, in such full, clear, concise, and exact terms as to enable any person skilled in the art to which it pertains, or with which it is most nearly connected, to make and use the same" (5). If a patent application is filed too early, before sufficient data are available to allow an enabling disclosure, the application may be rejected or any resulting patent may be invalid. In the protein example, if it is determined that a complete DNA sequence is required for enablement, the first filed patent application would not be legally sufficient; the CIP application containing the full sequence, however, would be enabling.

One must take into account that the United States generally has a more stringent enablement requirement than that of many other countries. Thus, in the DNA sequencing example above, a partial sequence may be sufficient in other parts of the world. In such countries, a patent application having only a partial sequence and an earlier filing date may have priority over a second patent application filed by another party where the second application is based on a U.S. application, the filing of which was delayed because the full sequence was not yet complete (6).

This chapter presents a discussion of provisional and utility patents, trademarks, and trade secrets and emphasizes their use in protecting intellectual property in the drug discovery and biotechnology areas. Other forms of intellectual property protection will be mentioned briefly. This chapter cannot, of course, provide the reader with sufficient detail to allow him or her to protect drug-related technology effectively and comprehensively. It is imperative to obtain competent legal counsel specializing in the area of intellectual property and technology transfer, preferably counsel with the appropriate drug research and biotechnology experience, as early in the research and development process as possible. This chapter should enable the reader to communicate and interact more effectively with counsel as they jointly fashion, within the ongoing research and development process, the appropriate legal protection for the particular technology.

2 PATENT PROTECTION AND STRATEGY

The U.S. Constitution provides that "Congress shall have [the] power . . . [t]o promote the progress of science and useful arts by securing for limited times to authors and inventors the exclusive right to their respective writings and discoveries" (7). In exercising that power, Congress has established a system for granting utility patents (8), design patents, and plant patents. Utility patents protect the structural and functional aspects of products

or processes and are granted for a term ending 20 years from the initial filing date of the utility patent. Design patents protect the ornamental design or aspect of a useful product and are granted for a term of 14 years. Plant patents grant the right to exclude others from reproducing, selling, or using an asexually reproduced plant variety for a term of 17 years. Certain sexually reproduced plant varieties can be protected under the Plant Variety Protection Act of 1970 for a term of 18 years; plant variety "certificates" under this program are issued by the U.S. Department of Agriculture. Although design and plant patents can, in some cases, be an important part of a company's patent portfolio, this chapter will concentrate on utility patents.

Utility patents (hereinafter *patents*) are generally considered the strongest form of legal protection for intellectual property. They grant the patent holder a "legal monopoly" on the invention for 20 years after the initial filing date in the United States. During the term of the patent, the patent holder can prevent others from making, using, offering for sale, or selling the patented invention in the United States or importing the patented invention into the United States. In exchange for this limited right to exclude, the patentee must fully disclose the invention to the public; at the end of the patent term, the invention is dedicated to the public.

Patent protection is limited geographically. For the most part, a U.S. patent does not provide legal protection from, or prevent, an act occurring outside the United States (and its territories and possessions), although that same act would fall within the scope of patent protection if carried out in the United States (the one exception is discussed in Section 4). The same is generally true for other countries. Thus, a comprehensive patent strategy should take into account the possibility of obtaining patent protection in all countries where an invention will be exploited (i.e., sold, manufactured, or used).

The U.S. patent system is designed to protect new and nonobvious products and processes. It protects the application of ideas and laws of nature; it does not protect the ideas or laws of nature themselves. Thus, Einstein's $E = mc^2$ equation would not have been patentable, even though Einstein or others might have obtained patent protection for a nuclear power plant, a nuclear rocket engine, or the myriad other products and processes derived from this basic principle. The idea or basic principle itself is available for all to use and develop.

In the United States, the 20-year term (9) of a patent is measured from its earliest nonprovisional filing date for which benefit is claimed. Largely because of the time involved in the FDA approval process and associated clinical trials, a considerable portion of the patent term can elapse before a patented drug can be sold in the marketplace. Thus, the period effectively available for the drug developer to recoup its investment and earn a profit can be considerably shorter than the 20-year patent term.

Beginning in the early 1980s, Congress has taken steps to significantly strengthen and improve the patent system. These steps include adding a reexamination process (10), authorizing the formation of the U.S. Court of Appeals for the Federal Circuit to hear appeals from the PTO and in patent-infringement cases, and providing for the extension of the patent term for certain drug-related inventions and for certain delays in the PTO. The patent system now provides substantially better and more consistent protection to technology in general than in the past and is now an even more attractive mechanism for the drug discovery and biotechnology industry to protect its intellectual property. Patent terms may be extended up to 5 years for qualifying drugs, medical devices, food additives, and methods "primarily us[ing] recombinant DNA technology in the manufacture of the product," to compensate for certain, but not all, delays in the regulatory and approval process (11). The patent term may be extended only if the approval of the first commercial use of the patented product occurs during the original patent term and the extension is applied for within 60 days of the approval. Given the limited time in which to apply for the extension, it is important that patent counsel be informed when approval of a drug or medical device is granted, so that the application for term extension can be filed within the required time period. The patent term extension process is jointly administered by the PTO and the Food and Drug Administration (FDA). The PTO determines whether a patent qualifies for the extension, and the FDA determines the allowable extension term.

2.1 Patent Strategy

Pharmaceutical and other high-technology industries are increasingly global in nature. More and more, competition in global markets requires effective global protection for intellectual property. As barriers to trade decrease and the legal mechanisms for protecting intellectual property are strengthened around the world, such protection will become even more important. It is an important part of any drug discovery organization's intellectual property strategy to determine how best to protect its intellectual property throughout its global market.

The simplest strategy would be to file patent applications for each invention in each and every country having a patent system. Such a strategy could, of course, be prohibitively expensive and in most cases less than cost-effective. A cost-benefit strategy should be applied to each particular invention and its uses. It is essential to evaluate the market potential of the invention around the world and the ability to control that market based on patent protection in key countries. For countries with interrelated markets, it may be possible to protect technology effectively in one country with a patent in another.

Appropriate technical, business, marketing, and legal personnel should be involved in the decision-making process so that all relevant factors can be considered in determining whether to file a patent application on a given invention, and, if so, when and where to file. The relevant factors will, of course, vary from case to case, as will their relative importance. For example, a start-up company interested in developing a single drug or family of drugs may have a strong interest in obtaining as comprehensive patent coverage as possible. Unfortunately, such a start-up company may not have the resources necessary to seek such comprehensive patent coverage. A pioneer invention (i.e., one that breaks new ground or provides an important technical breakthrough and will likely dominate a particular industry segment) generally warrants wider patent protection than an invention that provides only an incremental improvement in an existing technology. Patent protection around the world is further discussed in Section 5.

Improvements of previously patented inventions deserve special consideration in developing a patent strategy. Patents covering even relatively minor improvements can be important elements in expanding and extending protection of a basic technology. If the inventor or assignee of the improved invention is also the holder of the patent of the basic invention, there are generally two options. The improvement can either be kept as a trade secret or patented. Although a trade secret potentially has an unlimited lifetime (i.e., until secrecy is lost), the actual lifetime is likely to be much shorter, especially in the drug industry, where detailed FDA disclosures are required. However, if the improvement is only an obvious variation of a basic invention, reliance on trade secret law may be the only option.

Generally, patent protection of the improvement is the preferred option. A patent for an improved drug or process may allow for additional patent protection for commercially significant embodiments past the term of the basic patent on the original invention. Obtaining such improvement patents will also make it more difficult for competitors to penetrate or expand into a market. "Driving stakes in the ground" in the form of improvement patents all around the basic or core invention makes it more difficult for any potential competitor to carve out a niche in the market.

In most cases where the developer of the improvement does not hold the patent on the basic or core invention, keeping the improvement as a trade secret is not a realistic option unless the patent covering the basic invention is due to expire in a relatively short time. Therefore, seeking patent protection is generally the best option. Assume the basic drug X is protected by a patent held by Company A and that Competitor B develops and patents a significantly improved drug formulation Y containing drug X. Company A can continue to market drug X but cannot offer drug formulation Y. Competitor B cannot offer either drug X alone or in the form of formulation Y. Competitor B may wish simply to license the improvement to Company A and collect revenues through a license. Or Competitor B can use its patent position on formulation Y as leverage in seeking access to the market. In many cases, Company A and Competitor B will agree to cross-license each other so that each can offer the improved formulation Y. Thus, for a competitor seeking to enter a market otherwise closed by another's patent position, improvement patents can provide valuable leverage.

Several differences between U.S. patent law and the patent law of almost every foreign country also significantly affect patent strategy, especially the determination of when to file a patent application in the United States. Some of the most important of these include the rules for determining priority and the requirement of "absolute novelty" essentially everywhere except the United States. These differences are discussed next.

2.2 First to Invent versus First to File

In our world of rapidly advancing technology, and particularly for very active research areas such as drug discovery and biotechnology, investigators at varying locations are often working in the same general area, often on the same specific research topic, and frequently discover essentially the same invention within a very short time of each other. Thus, the issue often arises as to which of two (or more) inventors is entitled to a patent on a contemporaneously discovered invention.

U.S. patent law establishing entitlement to a patent in the case of essentially simultaneous invention is different from the law of substantially all other countries throughout the world. Nearly all countries award the patent to the first party who files a patent application (i.e., the first-to-file rule) (12). The United States, however, follows the first-to-invent rule whereby, at least in theory, the first to invent is generally entitled to the patent, even though he or she was not the first to file a patent application. Thus, it is possible that one party (i.e., the first to invent) who loses the race to the PTO may be entitled to patent protection in the United States, whereas another party (i.e., the first to file) may be entitled to patent protection for the same invention in most other countries. This possibility increases the incentive for a party to file a patent application covering the invention as quickly as possible.

The U.S. PTO, on discovering that two or more parties have copending patent applications or a patent application and a recently issued patent claiming the same invention, may set up an interference proceeding to determine which party is the first inventor of the subject matter. Such a determination is not straightforward. In an attempt to make the procedure as predictable as possible, a great number of rules (both substantive and procedural) have been adopted by the PTO to govern the proceedings and the gathering of the evidence necessary to establish the facts surrounding the making of the invention by each party. These rules give the party who was the first to file (the senior party) significant substantive and procedural advantages that significantly increase the senior party's chances of prevailing in the interference proceeding.

Generally, in the United States, the party who is first to "reduce an invention to practice" is given priority and awarded the patent, unless another party who reduced the invention to practice at a later date can prove that he or she was the first to conceive the invention and worked diligently to reduce it to practice from a time before the other party's date of conception (13). Reduction to practice may be an "actual reduction to practice" (physically making or carrying out the invention) or a "constructive reduction to practice" (filing a patent application). Therefore, in the United States, at least in theory, the filing date of an U.S. patent application may not control the outcome of the priority contest between parties who each actually reduced the invention to practice. As just noted, however, the party who files first has certain practical advantages in the interference proceedings.

Currently there is considerable interest in the world community for the United States to harmonize its laws with those of the rest of the industrialized world and adopt the first-to-file system. Although efforts have been made in the United States to adopt the first-to-file rule, there has been considerable resistance to such a change. Adoption of a first-to-file rule does not appear likely in the near future. Should the United States ever adopt such a rule, it will likely insist that other major industrial countries enact changes in their laws to favor true international protection of patentable subject matter. The adoption of such a first-to-file rule in the United States may have only a relatively small effect on the drug discovery field, especially for large corporate entities and others involved in the global marketplace because they already have significant incentives to file patent applications as quickly as possible. The adoption of the first-to-file rule in the United States would initially appear to have a significant effect on individual inventors or small organizations who may be interested almost entirely in the U.S. market or who do not have adequate resources for quickly developing inventions or filing patent applications. However, such individuals and small organizations are generally at a significant disadvantage in any interference proceeding simply because of the cost involved. Such individual inventors and small organizations may not, therefore, be as deeply affected in a practical sense by a first-to-file rule as one might first imagine.

2.3 Absolute Novelty

In most countries, a public disclosure of an invention before filing a patent application precludes obtaining patent protection for the invention. This is in contrast to the United States, where an applicant has 1 year after publication, public use, or offer of sale of the subject matter of the invention in which to file a patent application. The effect of such a public disclosure is the same whether it is made by the inventor or by another (14). Thus, if patent protection outside the United States is desired, a patent application must be filed in at least one Paris Convention country (see Section 5.1) before the public disclosure, followed by the filing of the corresponding applications in other countries within 1 year. Public disclosure within the convention year does not adversely affect any later filed applications filed within the convention year.

Valuable patent rights can be lost because of early disclosure of the patentable technology. Such loss can be especially damaging to an organization involved in drug discovery because of the global market for drug and drug-related technology. Because of required public disclosure related to the FDA approval process, the patent rights associated with drug discovery are especially at risk through premature disclosure. Disclosure of technical information should be closely monitored and controlled as part of a comprehensive intellectual property program. All employees, including research, medical, technical, and business personnel, should be carefully educated in regard to the confidential nature of technical information and the consequences of premature disclosure. An essential component of such a program is an evaluation procedure for all articles, abstracts, seminars, or presentations before actual submission and/or presentation. In addition to preventing premature disclosures, such an evaluation program can aid in educating personnel on the importance of protecting confidential information.

An intellectual property committee responsible for reviewing and approving all disclosures containing technical information, including FDA submissions, is highly recommended. In cases where disclosure is a potential problem, the committee should, if possible, delay publication until the appropriate patent applications are filed. To avoid delays in FDA submissions and scientific or technical publications, such submissions or publications should be reviewed by the committee as early as possible, to allow for sufficient time to prepare and file any necessary patent applications (15). Because of the importance of FDA submissions and the amount of technical data involved, it may be desirable to have at least one individual responsible for FDA matters included on the committee to ensure proper coordination between the functions.

The need for monitoring and controlling technical information does not end with the filing of the initial patent application. Disclosures relating to an invention claimed in an earlier patent application may also contain information concerning new inventions or improvements of the earlier claimed invention, which may be the subject of later filed applications. In addition, control of the disclosure of inventions contained in a patent application may allow for filing patent applications in countries requiring absolute novelty after expiration of the convention year, should that become necessary or desirable. Failures to file

such applications within the convention year may be intentional (i.e., too expensive or perceived lack of technical merit) or accidental. If funds become available, or technical merit is established, or the accident discovered, applications can be filed after the convention year if there has not been a public disclosure. Any application filed after the convention year, however, cannot rely on the earlier priority date; its effective filing date will be the actual filing date in the specific country. Loss of the priority date can be significant, given that additional prior art may be available against the application.

2.4 Patent Term

The U.S. Congress has recently and significantly changed U.S. patent law to conform it to international trade negotiations, which resulted in the General Agreement on Trade and Tariffs (GATT) (16). Before June 8, 1995, patents granted in the United States had patent terms of 17 years measured from the date that the patent was granted. The GATT implementing legislation changed the term of patents based on applications filed on or after June 8, 1995. The change to patent term applies only to utility and plant patents; it does not apply to design patents.

All patents resulting from new applications, continuation applications, continuation-in-part applications, and divisional applications filed on or after June 8, 1995, have a 20-year term. The patent term begins on the issue date of the patent and ends 20 years after the filing date of the earliest nonprovisional application relied on (17). If an applicant claims the benefit of an earlier application filing date, the earliest application filing date relied on will control, and the patent term will end 20 years from that date (18). Further, an international application filed under the PCT and designating the United States has the effect of a national application regularly filed in the PTO, so that the filing date of such an international application would be used for the purpose of calculating the 20-year term. The filing dates for foreign applications or provisional applications do not start the clock running on the 20-year term; the 20-year term is measured from the filing date of the regular U.S. application.

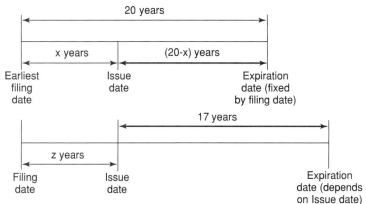

Figure 20.1. 17- versus 20-year patent term.

Because the 20-year term starts to run as of its priority date, the 20-year term can be, and often will be, significantly shorter than the 17-year term. The 17-year term (lower time line) and 20-year term (upper time line) are illustrated in Fig. 20.1. For any patent resulting from an application filed after June 8, 1995, the patent term will be less than 17 years whenever the patent issues more than 3 years after the priority date (i.e., the patent term is less than 17 when x is greater than 3 in Fig. 20.1).

The 20-year patent term is likely to have its biggest impact in cases where a series of continuing applications are filed to obtain a patent, whether those continuing applications are continuation applications, continuation-in-part applications, or divisional applications. The scenario set forth in Table 20.1 illustrates the effect of GATT on patent term in a hypothetical case with several continuing applications.

As shown in Table 20.1, a parent application was filed on April 1, 1989, and a restriction requirement was issued by the PTO on June 1, 1990, objecting that the application is directed to more than one invention. Assume a first divisional application directed to one of the inventions was filed on June 7, 1995 (1 day before the effective date of the new 20-year patent term provision) and a second divisional application directed to another invention was filed June 8, 1995 (the effective date of the 20-year patent term provision). Although both divisional applications issue on the same date (July 1, 1998), the expiration of patent term differs significantly. For the first divisional application, the patent term is 17 years from the issue date because the 17-year term is longer than the 20-year term (as measured from April 1, 1989 priority date); hence, this patent will expire on July 1, 2015. For the second divisional application, the patent term expires 20 years from the filing date of the first application (i.e., the priority date of April 1, 1989); this patent will expire on April 1, 2009. Hence, the second divisional application has a patent term of only 10 years, 9 months versus a patent term of 17 years for the first divisional application. In this hypothetical case, a difference of only one day in the filing dates would have resulted in a loss of over 6 years of the

Table 20.1 Effect of GATT on the Patent Term of Continuation Applications

Parent Application Filed:	April 1, 1989	
Restriction Requirement:	June 1, 1990	
	First	Second
Divisional Filed:	June 7, 1995	June 8, 1995
Issue Date:	July 1, 1998	July 1, 1998
Expiration Date:	July 1, 2015	April 1, 2009
Term:	17 years	10 years, 9 months

Table 20.2   Example of Resetting Patent Term

U.S. Patent 5,376,293
- Issued 12/27/94
- Based on CIP application filed on 9/9/93 claiming benefit of parent application filed on 9/14/92
- 17-year term = 12/27/94 to 12/27/11
- 20-year term = 12/27/94 to 9/14/12

patent term for the second divisional application. Congress recognized this potential loss of patent term relating to continuing application practice and provided several transitional provisions that attempt, at least for certain applications pending before June 8, 1995, to avoid this potential loss (19).

2.4.1 Effect on Issued Patents and Pending Applications. The GATT provisions extend the term of certain already issued U.S. patents to the greater of 17 years from the date of issue, or 20 years from the priority date of the original application. Hence, patents "in force" on June 8, 1995 and patents resulting from applications filed before June 8, 1995 will have the longer of the 17- or 20-year terms. The patent term will be "reset" for patents "in force" on June 8, 1995, which had a pendency period of less than 3 years (20). The effect of a reset patent term is shown in Table 20.2.

For the patent in Table 20.2, the 17-year term is measured from the issuance date of the patent; the 20-year term is measured from the filing date of the parent application. Given that in this case the 20-year term is longer, the patent term will be reset to September 14, 2012, thereby providing the patentee with a longer patent term (by about 9.5 months) than originally expected. Although not required in the present example, one must also take into account any terminal disclaimers submitted in the various applications in calculating such "reset" patent terms (21).

Remedies for infringement for such "reset" patents are generally the same as those available during the original term of the patent (i.e., the 17-year term) (22). The effect of such "reset" patents on existing licenses will, at least in the first instance, be determined by negotiations between the parties.

2.4.2 Patent Term Extension. Congress recognized that, under the new 20-year patent term, delays in the PTO could reduce the period of exclusivity afforded by a patent and provided provisions relating to patent term extension to at least moderate this effect. To that end, at least some of the potentially lost term in some cases can be restored. These patent term extension provisions should be especially important to the biotechnology and pharmaceutical industries because their inventions are routinely subject to extended delays in the prosecution in the PTO as well as delays caused by the Food and Drug Administration (FDA) approval process. The extensions are independent from other previously available extensions because of premarketing regulatory review of the patented product after the patent issues (23). The extensions provided for FDA regulatory review can be granted for up to a maximum of 5 years; there is no cap on extensions available for PTO delay.

Delays during examination that qualify for patent term extension include secrecy orders, involvement in interferences, and successful appeals to the Board of Patent Appeals and Interferences or to the federal courts. The first two qualifying delays (secrecy orders and interferences) are generally initiated by the PTO. Only the third qualifying delay, relating to successful appeals, involves a process that can be routinely initiated by the patent applicant or assignee. It is important to note that these are the only bases for extensions of time for delays in the PTO for patents filed before May 29, 2000. Thus, for example, if the PTO simply loses such an application—and thereby delays the prosecution while the file is sought—the applicant cannot regain the time lost to the patent terms, even though the loss may be completely attributed to PTO error.

Amended term extension provisions, in effect for any patent filed on or afer May 29, 2000, address delays occasioned by the failure of the PTO to take certain actions within prescribed time periods. The extension is on a day-for-day basis, that is one day for each day after the prescribed period until the action is taken. An extension is also available when the PTO fails to issue a patent within 3 years of the U.S. filing date, not including time consumed by secrecy orders, appeals, or delays requested by the applicant. Although doing away with a previous 5-year limitation on the amount of term extension, the amended provisions do impose other limitations. The extension may not exceed the actual number of days that issuance was delayed, may not adjust the term beyond an expiration date specified in a terminal disclaimer, and is reduced by the time the applicant failed to use "reasonable efforts" to prosecute the application.

The term extensions relating to delays in the PTO are automatically determined by the PTO (24). Such term extensions are granted only to the extent that periods for the various causes do not overlap. Thus, for example, if an application remained under a secrecy order for 2 years and was involved in a successful appeal for 3 years, 1 year of which overlapped with the secrecy order period, the patentee would be entitled to, at most, a 4-year extension.

2.4.3 Patent Prosecution. The changes relating to patent terms in the U.S. GATT legislation have fundamentally changed U.S. prosecution practice. The potential loss of patent term has a significant impact on patent prosecution strategy and dramatically changes the patent prosecution environment from a relatively stable one to one of a much more dynamic nature as a consequence of the tension between enforceability of the patent and length of the patent term. With the 17-year term, it was not uncommon (especially where commercial development of the invention was many years in the future) to delay (or at least, not rush) the actual issuance of the patent because the full 17-year term was available whenever the patent issued. Because the time factor was normally not critical, one could carefully attempt to obtain the best possible claim coverage without regard for the time required to convince the PTO of the patentability of the invention. Under the 20-year term, however, delays in the PTO effectively reduce the period of enforceability of any resulting patents. Thus, it is generally in the applicant's best interest to speed up prosecution in the PTO to obtain the longest possible patent term. One must not, however, lose sight of the overall objective: a valid patent with a reasonable patent term and claims providing reasonable protection for the invention. Simply

speeding up prosecution of the patent to obtain a longer term without careful consideration of the strength of the resulting patent is not a recommended approach. An invalid patent with a long term, even the full 20 years, has little or no value; a patent with a long term and claims that are too narrow to provide adequate protection may have very little value; and a valid patent with only a very short term may have little or no value. The potential reduction of the patent term may especially affect the chemical, pharmaceutical, and biotechnology industries where pendency of applications is typically long, restriction requirements are frequent, and regulatory delay in other agencies may reduce the effective patent term (25). In the new dynamic environment presented by the 20-year patent term, the key objective is to obtain valid patents having reasonable breadth in as reasonably short a time period as possible. Efforts must be made to speed up prosecution within the PTO without impairing patent validity or claim coverage.

To achieve this objective (i.e., a valid patent with reasonable claims and a reasonable term), patent prosecution should be modified to speed up, to the extent possible and reasonable, overall patent prosecution, including, for example, the filing of divisional, continuation, and continuation-in-part applications, and responses to office actions, while maintaining the high quality of the resulting product. For example, an applicant's response to restriction requirements should be modified relative to the current procedure (26). In the pharmaceutical field, applications typically include claims to a family of new compounds, methods for making them, and one or more therapeutic uses. These types of claims are often considered by the PTO to be separate inventions and are subject to multiple restriction requirements. Before the enactment of the GATT legislation, it was possible to obtain successive patents to each group of claims, each of which had a 17-year term from the date of grant. Thus, under the old system, the overall period of protection could be effectively extended by filing the divisional application just before the parent application issued. This technique is no longer effective under the 20-year patent term because the term of each divisional application will end at the same time as the parent case (i.e., 20 years after the earliest priority date). In most cases, consideration should be given to filing divisional applications as early as possible in response to a restriction requirement to avoid potential loss of patent term in the divisional applications. Thus, under the new system, the simultaneous prosecution of multiple divisional applications is important to preserve the maximum patent term for each divisional application (27).

Before the enactment of the GATT legislation, applications that may have been an improvement or modification to an invention on which an application was already pending, were filed as continuation-in-part applications claiming the benefit of the filing date of the earlier filed application. Claims of a continuation-in-part application are entitled to the parent application filing date if the claimed subject matter is disclosed in the parent application in the manner provided by 35 U.S.C. § 112, first paragraph, and if other requirements of 35 U.S.C. § 120 are satisfied. With the enactment of the GATT provisions, the patent term of a continuation-in-part application will end 20 years from the filing date of its parent application. Hence, where the invention to be claimed is based on new matter that is patentably distinct from the parent application, it may be appropriate to file a new application rather than a continuation-in-part application. By doing so, the new application will have its 20-year term calculated from its filing date rather than the priority date of the earlier application.

Under the new 20-year term rules, shortening the time between filing an application and issuance of the patent increases the patent term. Thus, it is generally in the applicant's best interest to speed up the prosecution process so long as the validity of the claims is not compromised or the scope of the claims unduly limited. Speeding up the examination process may also be beneficial in making decisions concerning patent applications outside the United States and, if the patentability analysis is favorable, in seeking outside funding. If the applicant meets certain conditions, he or she can attempt to speed up the PTO process by filing a petition with the PTO to have the application declared "special" (28). If special status is granted, the PTO should provide expedited examination of the application.

For most applications, special status is probably not available. Timely and effective communications with the patent Examiner and responding fully to Official Actions from the PTO as quickly as possible are techniques that are likely to help speed prosecution. Before the enactment of GATT, an applicant responding to a first office action might make only limited amendments to claims and extensively argue patentability of the pending claims in an attempt to get broad claim coverage. Moreover, in many cases, it was desirable to delay the actual issuance of the patent, especially in the face of lengthy FDA approval procedures, because the 17-year term would begin only when the patent actually issued. However, under the new 20-year term, cases generally should be placed in the best possible position for allowance as quickly as possible in light of the first office action and the known prior art to minimize the prosecution period (29). Speeding up the process in this manner does require close cooperation between the attorney prosecuting the application and the client. In appropriate cases, the attorney should provide draft responses to the client as quickly as reasonably possible; likewise, clients should provide their input regarding the draft response as quickly as reasonably possible.

Another method of speeding up the prosecution process—and one that is highly recommended—involves the use of the interview process. In many cases, examiners can more quickly appreciate the invention, and its patentability, when the invention is presented orally either by a telephonic or personal interview. Such interviews are especially helpful after the first office action (assuming the claims are rejected) because the case for patentability can be presented in light of the art and arguments presented by the Examiner. In many instances, the attendance of the inventor and/or presentation of demonstrations can be very helpful. Such interviews help define the issues of concern to the examiner more clearly and allow the next response (assuming agreement is not reached) to be more focused. Often the Examiner may indicate amendments during the interview that would be considered favorably. In some cases, agreement can be reached. Even if agreement cannot be reached, the attorney generally has a much better idea of the Examiner's concerns and the Examiner generally has a much better understanding of the invention. Further prosecution is likely to be much more focused and directed. Indeed, narrowing the issues under consideration using interviews generally can avoid addressing many issues in writing and can, therefore, limit the potential effect of prosecution history estoppel during any subsequent litigation.

In appropriate cases, after final rejection or a second rejection on the merits, the applicant should consider going directly to the appeal process rather than filing continuation applications to continue prosecution before the PTO. Because the time taken for an appeal, if that appeal is ultimately successful, can be used to extend the 20-year patent term up to 5 years, it is likely that the appeal process will become even more important (30). In fact, in suitable cases, it will be recommended that the appeal be initiated as soon as possible. Thus, it is generally recommended that the claims be placed in the best possible form for appeal when responding to the first office action. Once again, a personal interview after the first office action can be very helpful in placing the claims in the best position for allowance or, if necessary, for appeal. If not satisfied with the Board's decision, the applicant may appeal that decision either to the Court of Appeals for the Federal Circuit based on the record before the PTO or to a federal district court for a *de novo* review. If the examiner's position is overturned, the Court of Appeals for the Federal Circuit or the district court can order the PTO to issue the patent. However, appeal to either the Court of Appeals for the Federal Circuit or a federal district court destroys the secrecy of the application as well as that of the record of the proceedings within the PTO and, thus, destroys any trade secrets that may have been contained therein (assuming they have not already been lost through publication of the U.S. or other applications).

2.5 Publication of Patent Applications

Recently, Congress provided for publication of U.S. patent applications filed on or after November 29, 2000 (31). Applications filed on or after this date, with some exceptions, will be published 18 months (32) after the earliest filing date for which priority is sought. This change brings the U.S. system in line with most other countries with regard to publication of pending applications.

Congress provided several exceptions to the publication requirement. Applications that are not pending at the end of the 18-month period should not be published (33). Applications that are undergoing national security review or that have been subject to a secrecy order should not be published (34). Provisional applications will not be published (35). In addition, an applicant can request nonpublication if the U.S. application is not to be filed in another country or under a multilateral international agreement that provides for publication. To qualify for this exemption, the application must certify at the time the application is filed that the invention disclosed in the application has not been and will not be the subject of an application filed in another country or under a multilateral international agreement that provides for publication (36). In cases where the U.S. application contains more extensive information or description than that of the corresponding applications filed in another country or under a multilateral international agreement that provides for publication, the applicant may file a redacted copy of the application removing the new information or description for publication (37).

Of course, publication will destroy any trade secrets contained in the application. In practice, however, the new publication rules do not significantly affect an applicant's ability to protect trade secrets contained in the application because publication effectively occurs, assuming the proper certification is made, only if the application would have been published in another country or under a multilateral international agreement that provides for publication.

Publication will provide an opportunity to more easily track competitors and emerging technologies and will normally provide the public with knowledge of the invention at an earlier time, thereby advancing the state of the art. In addition, the publication will qualify as prior art as of its publication date. Thus, inventions for which patents never issue will become part of the state of the art and be available to the public.

3   REQUIREMENTS FOR PATENTS

To obtain a patent on an invention in the United States, the inventor or inventors must, as the initial step, file a patent application describing the invention in such terms as to teach one of ordinary skill in the art how to make and use the invention and claiming the subject matter that the inventor (or inventors) regards as the invention. The subject matter of the claimed invention must be within the statutory classes of patentable inventions. In addition, the claimed invention must have utility and be both new and nonobvious. The requirements for patentability (especially as to what constitutes patentable subject matter) can vary considerably throughout the world.

This section addresses the requirements for patentability in the United States and, to a much lesser extent, variations encountered in a few representative countries. The actual patenting procedure in the U.S. PTO are also briefly discussed.

3.1   Patentable Subject Matter in the United States

In the United States, patentable subject matter includes "any new and useful process, machine, manufacture, or composition of matter, or any new and useful improvement thereof" (38). An invention must be claimed so as to fit within one of the four statutory classes of inventions: process, machine, manufacture, and composition of matter. A process is essentially the means to achieve a desired end; for example, a method to synthesize a drug or a method of using a drug to treat a specific condition. The other patentable classes are basically the end products themselves. These four classifications are broad and generally encompass the vast majority of technological advances. Examples of generally nonpatentable subject matter include laws of nature or abstract ideas, products of nature, algorithms [a physical process using an algorithm can, however, be patented (39)], and printed materials (40).

The subject matter of most inventions clearly falls within one of the four statutory classes. But for cases where the issue of patentable subject matter is raised, the line between what is patentable and nonpatentable often cannot be clearly drawn and must be evaluated on a case-by-case basis. Consider "products of nature": compounds occurring in nature normally cannot be patented. So a naturally occurring drug collected, for example, from a particular plant species is not generally patentable. However, if through human intervention the naturally occurring drug is produced in a purer or more concentrated form not naturally occurring, the new form of the drug may be patentable. Even if the actual drug is not patentable, a new and unobvious use for that drug may be patentable; or, the combination of the drug with other active ingredients may be patentable. Likewise, a new method of preparing, concentrating, or purifying the drug (even if the drug is exactly the same as the naturally occurring drug) may also be patentable.

Patentable pharmaceutical inventions generally and broadly include drugs, diagnostics, intermediates, drug formulations, dosage forms, methods of treatments, kits containing the drug or diagnostic, methods of preparing the drug or diagnostic, and the like. Microorganisms, plasmids, cell lines, DNA, animals, and other biological materials have been found patentable if they are the result of human intervention or manipulation. Initially, the PTO resisted granting patents on living organisms on the basis that the claimed invention was a product of nature. The U.S. Supreme Court in the landmark case of *Diamond v. Chakrabarty* (41), however, held that new lifeforms (i.e., bacteria altered genetically to digest crude oil) can be patentable subject matter. In 1987 the PTO formally issued a notice indicating that it "considers nonnaturally occurring non-human multicellular living organisms, including animals, to be patentable subject matter" (42). It is now clearly established that nonnaturally occurring biological materials, including microorganisms, plants, and animals, can be protected by patents in the United States (43).

Sequences of human genes and their regulatory regions, either whole or partial, can be patented. Portions of genes such as expressed sequence tags (EST) or single nucleotide polymorphisms (SNP) can also be patented (44). Such partial sequences are potentially importance in the detection and screening of diseased genes. The U.S. PTO allows patenting of EST and SNP partly as a result of their utility in tracing ancestry, parentage, gene mapping, and screening for predisposition to genetic disease or environmental insults (45).

In the United States, it has been presumed that patents on human being are precluded by the U.S. Constitution (46). However, the definition of "human being" has become increasingly unclear with the advancement of cloning technologies. For example, a laboratory animal into which a single human gene has been transferred would not be regarded as human by most definitions of the term in common usage, and should not thereby be excluded from patent protection. Conversely, a human being whose somatic cells contain a single non-human gene, or multiple non-human genes, introduced for therapeutic purposes would still be considered human and thereby not patentable (47). The question of how many characteristics may be transplanted before an animal is considered a "human being" may become an important consideration that will impact patentability (48).

In addition to fitting into one of the statutory classes, a patentable invention must also possess utility (i.e., it must provide some useful function to society). The invention must be operable and accomplish some function that is not clearly illegal (49). A chemical compound for which the only known use is to make another compound, which does not have a known use, is not patentable; if, however, the final product is useful, then the starting material is also useful and should have sufficient utility for patentability (50). The patent specification must disclose at least one nontrivial utility for the invention. Thus, at least one practical utility for a new composition of matter (e.g., a new drug) must be disclosed. Where possible, however, it is generally recommended that several utilities are disclosed to reduce the risk that the PTO will find the invention lacking utility. Composition of matter claims covering the drug will generally protect all uses of the drug, even ones not disclosed and ones discovered

after the patent issues. The utility of the drug does not have to be patentable in its own right. Nor does the utility have to be developed or discovered by the inventor of the composition of matter; the inventor merely has to disclose, and sometimes provide proof of, the utility.

Generally, if the utility disclosed in the patent specification is easily understood and is consistent with known scientific laws, the PTO will not question it. If, however, the disclosed utility clearly conflicts with general scientific principles or is incredible on its face (e.g., a perpetual motion machine), the PTO will presume the invention lacks utility and will require strong evidence supporting operability (51). Thus, for example, it is likely that an invention asserting, as the only utility, a complete cure for leukemia, other cancer, or AIDS will require an especially strong showing of effectiveness (52).

For inventions involving pharmaceuticals and methods of treatment, the PTO often requires a relatively high level of proof for the disclosed utility. A utility likely to be deemed incorrect or unbelievable by one skilled in the relevant art in view of the contemporary knowledge of that art will require adequate proof. The proof required can generally be based on clinical data, *in vivo* data, *in vitro* data, or combinations thereof, where such data would convince one skilled in the art. The data may be included in the patent specification as filed or provided (or, if appropriate, supplemented) in a later submitted declaration or affidavit (53).

The data necessary to support utility will, of course, vary as the relevant art advances. For example, proving the effectiveness of a new drug, which is the first of a new class of drugs for a particular disease, will require more convincing evidence than a later, but still new and nonobvious, drug of the same class. The PTO has recently indicated that

> if the utility relied on is directed solely to the treatment of humans, evidence of utility, if required, must generally be clinical evidence . . . although animal tests may be adequate where the art would accept these as appropriately correlated with human utility or where animal tests are coupled with other evidence, including clinical evidence and a structural

similarity to compounds marketed commercially for the same indicated uses. If there is no assertion of human utility or if there is an assertion of animal utility, operativeness for use on standard test animals is adequate for patent purposes (54).

Thus, the data required to show utility for human use are generally of the amount and type that those skilled in the art would consider acceptable for extrapolation to *in vivo* human effectiveness. Clinical data, although not required, will generally be preferred and, where available, should normally be provided. For the PTO to require clinical testing in all cases of pharmaceutical inventions, however, would have a decidedly detrimental effect on the industry and its research efforts and, most likely, would be inconsistent with the overall goals of the patent system.

3.2 Patentable Subject Matter Outside the United States

Patentable subject matter in many countries (especially for health-related and biotechnology inventions) is often significantly restricted compared to practices in the United States. Some countries do not allow plants or animals to be patented. The United States, Japan, and the European Patent Office currently allow genetically engineered animals to be protected by patent (55). Many countries do not allow pharmaceuticals or methods of medical treatment to be patented. Each country will, of course, have its own specific limitations and exceptions for patentable subject matter (56). It is not possible in the present chapter to provide an even limited discussion of such patentable subject matter. Moreover, any details provided could very well be out of date in a relatively short time. However, a few examples for selected countries (57) are helpful to illustrate the variations in patentable subject matter:

1. *Australia*. Medicines that are mixtures of known ingredients are generally not patentable (58).
2. *Canada*. Plants and methods of medical treatment are generally not patentable (59).

3. *France*. Surgical or therapeutic treatment of humans and animals or diagnostic methods for humans and animals are generally not patentable (however, the first use of a known composition or substance for carrying out these methods may be patentable); plants and animals and biological processes for producing them (except microbiological processes and products) are generally not patentable (60).
4. *Germany*. Plants, animals, and biological processes for producing plants or animals (except microbiological processes and products) are not patentable; surgical, therapeutic, and diagnostic methods applied to humans or animals are generally not patentable, although the products used in these processes may be patentable (61).
5. *Italy*. Plants and animals as well as the methods of producing them (except microbiological processes) are generally not patentable; surgical, therapeutic, and diagnostic methods applied to humans and animals are generally not patentable, although compositions for carrying out these methods may be; pharmaceuticals are patentable, but the compounding of medicine in a pharmacy is not (62).
6. *Sweden*. Methods for treating humans and animals are generally not patentable, although products used in these methods may be patentable; plants and animals and processes (except microbiological processes and products) for producing them are generally not patentable (63).
7. *United Kingdom*. Animals and plants and nonmicrobiological processes for producing them are generally not patentable; surgical, therapeutic, and diagnostic methods for humans and animals are generally not patentable, although compositions for use in such methods may be patentable (64).
8. *Japan*. Generally similar to patentable subject matter in the United States (65).

Applicants interested in seeking worldwide protection for inventions, especially for the drug discovery and biotechnology industries, must take these differences in patentable subject matter into account. Even in jurisdictions where specific classes of inventions cannot be patented, it is often possible to claim the invention in a manner so as to be within patentable subject matter. For example, many countries that do not allow drugs to be patented may allow claims directed to a method of making the drug. Likewise, many countries that do not allow patent claims directed at methods for medical treatment of humans may allow claims directed to devices or compositions for carrying out the treatment processes. Generally, locally accredited patent counsel in the relevant country is retained to aid in the prosecution of the application because of the need for knowledge and understanding of substantive and procedural patent law in the relevant country. One of the tasks of such local counsel is to adapt the legal definition of the invention to local law and practice, especially in regard to patentable subject matter.

3.3 Patent Specification

The specification is that part of a patent application in which the inventor describes and discloses his or her invention in detail. In the United States, the specification "shall contain a written description of the invention, and of the manner and process of making and using it, in such full, clear, concise, and exact terms as to enable any person skilled in the art to which it pertains, or with which it is most nearly connected, to make and use the same, and shall set forth the best mode contemplated by the inventor of carrying out his [or her] invention" (66). Furthermore, the specification "shall conclude with one or more claims particularly pointing out and distinctly claiming the subject matter the applicant regards as his [or her] invention" (67). Thus, the specification must meet four general requirements: (1) provide a written description; (2) provide sufficient detail to teach persons of ordinary skill how to make and use the invention; (3) reveal the best mode of making and using the invention known by the inventor at the time the application is filed; and (4) provide at least one claim covering the applicant's invention (68). Each of these requirements is discussed in turn below. Almost without exception, the requirements for a legally sufficient specification in other countries are less stringent than those in the United States.

Thus a patent specification that is legally sufficient in the United States is usually sufficient for filing in other countries, with only relatively minor modification to conform it to specific national regulations and practices of the relevant country (69).

3.3.1 Written Description. The claimed invention must be the invention described in the specification. In other words, the claims must be "supported" by the specification as originally filed. Thus, the claims cannot encompass or contain more or different elements, steps, or compositions than are described in the specification. The claims cannot be amended or new claims added during prosecution of the patent application in the PTO unless the portion added is found within or supported by the specification as originally filed. The test for whether amended or new claims are supported is generally "whether one skilled in the art, familiar with the practice of the art at the time of the filing date, could reasonably have found the 'later' claimed invention in the specification as [originally] filed" (70).

Although new matter cannot be introduced into the specification or claims by amendment during prosecution (71), not all additions to the specification constitute new matter. For example, an applicant can generally supplement the specification by relying on well-known principles, prior art, and "inherency" (72). New matter, however, can be added by filing a CIP application claiming a priority date of the earlier filed application for the subject matter disclosed in the original application. The effective filing date for the newly added subject matter (i.e., the new matter) is the actual filing date of the CIP application. Nonetheless, it is generally recommended that the specification as originally filed be as complete as possible.

Care should be taken in preparing the patent specification to consider fully the ramifications of the invention and the possibility of extending or expanding the scope of the invention. Thus, if a specific new drug is developed, consideration should be given to structural and functional analogs. For example, in many cases, the effectiveness of a drug will not be significantly affected by replacing a methyl group with an alkyl group containing, for example, two to six carbon atoms. Such variations, if allowed by the technology and the prior art, can considerably expand the scope of the claimed invention and the scope of protection afforded by the patent (73).

3.3.2 Enablement.

In the United States, the specification, as filed, must teach one of ordinary skill in the art how to make and use the invention without undue experimentation. Enablement is essentially what the inventor gives to the public in exchange for the exclusive right afforded by the patent. The public must be able to understand the invention based on the specification to build upon the invention and develop new technology. Furthermore, the public is entitled to a complete description of the invention and the manner of making and using it so that the public can practice the invention after the patent expires. A specification that fails to teach one of ordinary skill in the art how to make and use the invention is not legally sufficient.

The enablement requirement does not mandate that each and every detail of the invention be included or that the specification be in the form of a detailed "cookbook" with every step specified to the last detail (74). Rather, the skilled artisan must be able to practice the invention without "undue experimentation." Generally, "a considerable amount of experimentation is permissible, if it is merely routine, or if the specification . . . provides a reasonable amount of guidance with respect to the direction in which the experimentation should proceed" (75). Although the acceptable amount of experimentation will vary from case to case, the factors normally considered by the PTO and the courts in determining the level of permissible experimentation include the following: (1) quantity of experimentation required, (2) amount of direction or guidance provided by the specification, (3) presence or absence of working examples, (4) nature of the invention, (5) state of the prior art, (6) relative skill of workers in the art, (7) predictability or unpredictability of the art, and (8) breadth of the claims (76). The level of acceptable experimentation will vary as the state of the art advances and the level of skill in the art increases.

Working examples are one factor in determining whether the specification is enabling. They are not required if the specification otherwise teaches one skilled in the art how to practice the invention without undue experimentation. However, in appropriate cases and if drafted with sufficient detail, such examples provide a relatively easy and straightforward way in which to make the specification enabling. So-called paper examples [i.e., examples describing work that has not actually been carried out (77)] can be used to satisfy the enablement requirement if the level of predictability for the art and invention is sufficiently high.

Pharmaceutical patents, especially those involving microorganisms or other biological material, can raise significant enablement issues. These issues arise whether the microorganism or other biological material is simply used in an invention sought to be patented or is itself the subject matter sought to be patented. If the microorganism is known and readily available or can be prepared using a procedure described in the specification, the enablement requirement is fulfilled. Otherwise the applicant may need to take additional steps to comply with the enablement requirement. In such case, this requirement can normally be satisfied by the deposit of a microorganism or other biological material in a depository that meets PTO requirements. The deposit must be made in "a depository affording permanence of the deposit and ready accessibility thereto by the public if a patent is granted" (78). Once the patent is granted, all restrictions on the availability of the deposit must be irrevocably removed. The PTO has indicated that permanent availability is

> satisfied if the depository is contractually obligated to store the deposit for a reasonable time after expiration of the enforceable life of the patent. The [PTO] will not insist on any particular period after expiration of the enforceable life of the patent. The enforceable life of the patent for this purpose is considered to be seventeen years plus six (6) years to cover the statute of limitations. Any deposit which is made under the Budapest Treaty will be for a term acceptable to the [PTO], unless the thirty years from the date of deposit will expire before the end of the enforceable life of the patent. With this one ex-

ception, any deposit made under the Budapest Treaty will meet all of the requirements for a suitable deposit except that assurances must also be provided that all restrictions on the availability to the public of the deposited microorganism or other biological material will be irrevocably removed upon the granting of the patent (79).

The submitter will generally have a continuing responsibility to replace the deposit during the enforceable life of the patent should the deposit become nonviable or otherwise unavailable from the depository.

The Budapest Treaty enables an applicant to make a deposit of a microorganism or other biological material in a single "international cell depository authority" and thereby satisfy the enabling requirements of the signatory nations to the treaty (and, generally, nonsignatory countries as well) (80). Where a deposit is required and patent protection outside the United States will be sought, a deposit under the Budapest Treaty will generally be recommended to minimize the number of deposits required. Deposits made under the Budapest Treaty are generally available to certain "certified parties" 18 months after the priority date (i.e., after the foreign patent application is published) (81). For purposes of U.S. patent law, deposits can be made outside of the Budapest Treaty so long as the conditions required by the PTO are ensured. Under U.S. law, deposits are generally not required to be released to the public until the patent issues (82). Should the applicant later determine that foreign protection will be sought, a non-Budapest Treaty deposit can generally be converted to a Budapest Treaty deposit. The major depository in the United States is the American Type Culture Collection (ATCC) in Manassas, Virginia. The ATCC is an "international cell depository authority" under the Budapest Treaty; the ATCC also accepts non-Budapest Treaty deposits.

Deposits can also provide a convenient mechanism whereby the patent owner can monitor individuals and organizations obtaining samples of the deposit during the life of the patent. Most depositories will provide the submitter with notice of sample requests (in some cases a relatively small fee may be required).

Such information can be helpful in enforcing patent rights. Especially in cases where the deposited material is patented, the patent holder may wish to consider a simple "letter license," allowing the requester to use the deposited material only in a certain manner (e.g., research purposes only) and requiring the requester to report any commercial uses of, or derived from, the deposited material.

3.3.3 Best Mode. The specification must "set forth the best mode contemplated by the inventor of carrying out his [or her] invention" (83). The inventor cannot hide or conceal the best physical mode of making or using the invention. The purpose of this requirement "is to restrain inventors from applying for patents while at the same time concealing from the public preferred embodiments of their inventions" (84). Questions of failure to meet the best mode requirements are rarely raised during prosecution of a patent in the PTO. Best mode issues are more often raised during interference proceedings or during litigation to enforce the patent.

The inventor must disclose the best mode known to him or her at the time the application was filed. If the inventor was aware of a better mode at the time of filing the application but did not disclose it, the entire patent is invalid and no claims are enforceable. However, a better mode of carrying out the invention discovered after the filing date need not be disclosed. Inventors should be carefully advised about the importance of the best mode requirement and of the necessity of informing patent counsel of any improvements made or considered before the actual filing date. Of particular concern is the time period between preparation of the application and its filing date; any potential improvements made during this period must be carefully evaluated to determine whether they must be included in the specification to satisfy the best mode requirement. Any improvements made after the application is filed—especially those made shortly after the filing date—should be carefully documented in case it ever becomes necessary to prove they were made after the filing date.

Generally, the best mode requirement focuses on the inventor's state of mind at the time the application was filed. It is the best mode contemplated by the inventor, not anyone else, for carrying out the invention. It is generally immaterial whether the failure to include the best mode in the specification was intentional or accidental. The subjective test is whether the inventor knew of a better mode at the time of the filing date and, if so, whether it is *adequately* disclosed in the specification (85). In some cases, deposit of biological materials may also be required to satisfy the best mode requirement. It is possible to include the best mode in the specification but to describe the best mode so poorly as to effectively conceal it. Because the potential penalty for failure to meet the best mode requirement is invalidation of the entire patent, considerable care should be taken to identify the best mode contemplated by the inventor and to describe it carefully and fully.

3.3.4 Claims. The specification must "conclude with one or more claims particularly pointing out and distinctly claiming the subject matter which the applicant regards as his [or her] invention" (86). Claims are the numbered paragraphs (each only one sentence long) at the end of the patent specification. The claims define the metes and bounds of the exclusive right granted by the patent. Each claim defines a separate right to exclude others from making, using, offering for sale, or selling embodiments within the scope of the specific claim.

The claims are critically important. They require careful drafting, preferably by highly qualified and experienced patent counsel, to cover the invention as broadly and comprehensively as appropriate. Normally the claims should be drafted as broadly as the prior art and the specification allow. Allowable claim scope will, of course, depend in large part on the state of the art. Inventions in a crowded art (i.e., technology with a relatively large amount of closely related prior art) can generally only be claimed relatively narrowly. Inventions for which there is relatively little prior art can be claimed more broadly. Generally, one attempts to draft one or more independent claims that describe the invention as broadly as the prior art will allow and then narrower, dependent claims specific to pre-

ferred embodiments, with dependent claims of intermediate scope in between. A dependent claim refers to an earlier claim and incorporates by reference all the limitations of the earlier claim, and adds additional limitations. Thus, the scope of the claimed invention generally varies from the broader independent claim or claims to the narrower, more restricted dependent claims. Should a court later find, for example, that the broader claims are obvious over the prior art, the narrower claims, with added limitations or elements, may still be valid and enforceable.

Normally it is also preferred that the invention be claimed so as to fit into as many of the statutory classifications as possible. For example, for a newly discovered drug, one might attempt to claim a method of making the drug, the drug itself, formulations containing the drug, and a method of using the drug to treat one or more conditions. For a new diagnostic, one might attempt to claim a method of making the diagnostic, the diagnostic itself, a method of using the diagnostic, and a kit containing the diagnostic. Other aspects of claims are discussed below.

3.4 Provisional Applications

Effective June 8, 1995, inventors may file a new type of patent application—the provisional application—in the U.S. PTO. Provisional applications may offer significant advantages to at least some patent applicants. However, if not used carefully, the provisional application may, in some instances, lead to the loss of significant patent rights.

The provisional application was intended to provide a domestic priority document and thereby "level the playing field" for domestic patent applicants relative to their non-United States counterparts. Applicants who have filed previously in countries outside the United States have been able to, and still can under the current statute, use their earlier-filed foreign patent application to secure a priority date for the United States application (if the U.S. application is filed within 1 year of the first-filed foreign application) (87). A domestic patent applicant applying for a U.S. patent could not, before June 8, 1995, obtain a priority date earlier than the actual filing date of his or her first-filed U.S. application. Under the

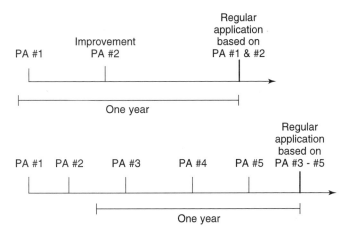

Figure 20.2. Multiple provisional applications.

new system, any applicant (domestic or foreign) may file a provisional application in the United States (88). If the regular application (i.e., nonprovisional application) is filed within 12 months of the provisional application and claims benefit of the provisional application, the filing date for the provisional application can be used as the priority date for the regular application (89). A provisional application can provide an earlier priority date for a domestic applicant just as an earlier-filed foreign application can for a foreign applicant.

The filing of the provisional application does not trigger the 20-year patent term. Rather, only the later-filed regular application starts the 20-year period (90). Thus, by filing a provisional application, followed within 12 months by a regular application claiming priority from the provisional application, an applicant can effectively obtain a patent term of 21 years (as measured from the date of the provisional application) (91).

The priority date of the provisional application provides protection for the regular application against prior art dated after the provisional filing date and/or statutory bar events occurring after the provisional filing date (92). Moreover, the provisional filing date can also be used in interference proceedings to obtain procedurally advantageous senior status and/or to prove an earlier invention date.

The provisional application can be especially useful in cases where an invention is undergoing active development. This use of provisional applications is illustrated in Fig. 20.2.

The top portion of Fig. 20.2 demonstrates a case in which the invention is fully developed at the time the first provisional application 1 (PA #1) is filed but a significant improvement is developed within the year. The significant improvement is then included in PA #2. When the regular application is filed within 1 year of PA #1, the benefits of both PA #1 and PA #2 are claimed. In this manner, both the basic invention (PA #1) and the improvement (PA #2) have the earliest possible priority dates (93).

The bottom portion of Figure 20.2 demonstrates a case in which the invention is developed over a period greater than 1 year. At each point of a significant advancement, a provisional application is filed (i.e., PA #1 through PA #5). Close to the end of the 1-year period, as measured from PA #1, one should consider whether the regular application should be filed at that time. Generally, if at the 1-year anniversary it is determined that the earliest provisional application filed was fully enabled (and, thus, provides the benefit of an earlier filing date), it is recommended (and probably required if foreign applications are to be filed) that a regular application be filed and that regular application claim benefit of all earlier provisional applications. In that case, the regular application would be filed shortly after PA #5 (or perhaps in place of it) in the bottom portion of Fig. 20.2 and the benefits of all earlier provisional applications will be claimed.

In some cases, however, one may not be ready to file the regular application at the 1-year anniversary of PA #1. For example, the invention may still be undergoing development and not be at a stage where a decision can be made as to whether to maintain the invention as a trade secret. Or, it may be determined that the earliest filed provisional applications were not enabled (94) and will not, therefore, provide the benefit of an earlier priority date (95). Or, the development may take an unexpected turn, or the actual invention may have changed so dramatically, that the earliest provisional applications are simply not relevant to the invention disclosed and claimed in the regular application. The regular application can be filed at any time during the development process and claim the benefit of all provisional applications filed within the preceding year. In the example illustrated, the regular application would claim benefit of PA #3 through PA #5. The earlier provisional applications (PA #1 and PA #2) would have become abandoned at their respective 1-year anniversaries and would remain secret (96). If the invention were fully developed earlier, one or more of these earlier provisional applications could have been claimed. In this manner, the earliest priority date can be obtained for each significant development in the inventive process. Of course, if the development is complete within 1 year of the first provisional application, then all provisional applications should be claimed when the regular application is filed.

3.4.1 Statutory Requirements. The statutory requirements for a provisional application are somewhat less stringent than those for a regular application. (The requirements for a regular application were not changed by GATT.) A provisional application, however, must fully comply with the requirements of the first paragraph of 35 U.S.C. § 112 to provide a priority date (97). That is, the provisional application must include a written description of the invention in sufficient detail to allow or enable one of ordinary skill in the art to make and use the invention and it must include the "best mode" known to the inventors as of the provisional application filing date. Drawings, if required by the nature of the invention, must also be included. In addition, the provisional application must include a cover sheet identifying it as a provisional

application, a list of the "inventors" of the disclosed subject matter, and payment of the required filing fee. Provisional application filing fees are relatively modest; currently they are $160 for a large entity and $80 for a small entity.

Provisional applications do not require patent claims (98) or a declaration signed by the named "inventors" (99). Each named inventor must, however, have made a "contribution" to the subject matter disclosed in the provisional application (100). A provisional application cannot claim the benefit of any earlier application (foreign or domestic) (101). In some cases, a regular application may be converted to a provisional application (102). The provisional application is not examined in the PTO and is maintained in secrecy within the PTO. One year after its filing date, the provisional application is automatically abandoned. A regular application must be filed within 1 year of the provisional application to claim benefit of the provisional application's filing date (103). In other words, the applicant has 1 year in which to decide whether to go forward with a regular application if the benefit of the priority date is desired. Should a regular application not be filed, the provisional application will not be published. Thus, in appropriate cases, the applicant might maintain the invention disclosed in the abandoned provisional application as a trade secret. The regular application could also be filed after the abandonment (assuming no statutory bars), although the benefit of the provisional application filing date could not be claimed.

The provisional application can effectively give an applicant 2 years after a statutory bar (e.g., public use or sale of the invention) in which to decide whether to file a regular application. A statutory bar can be avoided if the provisional application is filed within 1 year of the statutory bar event (104). If the regular application claiming benefit of the provisional application is filed within 1 year of the provisional application, it is entitled to benefit of the provisional application's filing date. Thus, the regular application could be filed up to 2 years after the otherwise disqualifying event.

3.4.2 Best Mode. The provisional application must disclose the best mode known to the inventors as of the filing date (105). If the regular application has a "substantive content" different from the provisional application, the best mode must be updated as of the time of the filing of the regular application (106). Thus, in such cases, any improvements (if they form part of the best mode) developed between the filing dates of the provisional and the regular applications must be disclosed in the regular application.

In some cases, the use of a provisional application may allow an applicant to seek patent protection on a basic invention and to keep a later-developed improvement as a trade secret. If the regular application has the same "substantive content" as the provisional application, the PTO has acknowledged that updating the best mode may not be required. Thus, by filing a "complete" provisional application (with a full set of claims) (107), an applicant appears to have the options of keeping later-developed improvements as trade secrets or including (and protecting) them in the later-filed regular application. A regular application that has the same "substantive content" as that of the provisional application would be analogous to a regular continuation application in which the best mode need not be updated (108). Of course, the Court of Appeals for the Federal Circuit has yet to specifically speak to this issue. Because of this uncertainty, we still recommend in most cases that the regular application be updated to include any later-developed best mode and to claim the improvements.

3.4.3 Concerns. The requirements for a provisional application appear deceptively simple. Indeed, the PTO initially urged that the provisional application would provide a quick and inexpensive method for inventors to obtain a priority date because claims were not required and, supposedly, the written description need not be as detailed as a regular application (109). Almost immediately, however, it was recognized that the use of the provisional application might not be as easy as first suggested (110). Nonetheless, it is also evident that the new provisional application offers significant benefits to the patent applicant. It es-

tablishes an early priority date, effectively tacks an extra year onto the 20-year patent term, has a relatively low filing fee, provides an easy way to avoid filing at least some incomplete regular applications [e.g., one without a signed declaration (111)], and, in at least some cases, may allow an applicant to keep later-developed improvements as trade secrets.

Failure to fully understand and appreciate the requirements and limitations of the provisional application can, at least in some cases, result in loss of valuable patent rights. As noted above, the provisional application disclosure must adequately enable the invention under 35 U.S.C. § 112. Only material adequately disclosed in the provisional application should be entitled to the priority date of the provisional application. The initial provisional application should thus be as complete as possible. Material added in the regular application to "fill out" or complete the disclosure of the provisional application may not be able to rely on the date of the provisional application. Claims in the regular application supported only by the disclosure added in the regular application may, therefore, be unpatentable over prior art dated between the priority date and the filing date of the regular application or because of statutory bars the provisional application was thought to overcome. Thus, filing a "barebones" provisional application is generally not recommended.

A barebones provisional application may be useful when it is necessary to prepare and file a provisional application as quickly as possible (e.g., to avoid an imminent statutory bar); even then, however, as complete a provisional application as time permits would still be preferred. Such a barebones provisional application might also be used to advantage in relatively simple mechanical inventions where "a picture may be worth a thousand words." To be effective in establishing a priority date, however, the barebones provisional application must still enable one of ordinary skill in the art to make or use the invention and must contain the best mode known to the inventors as of its filing date.

Enablement-type issues may not arise during prosecution of the patent application; thus an applicant may obtain a patent based on the provisional application's priority date. During enforcement of the patent, however, a potential infringer will almost certainly argue that the relevant claims are not supported by the provisional application disclosure and are, therefore, not entitled to its filing date. An applicant with an otherwise strong and valuable patent could find that his or her patent is rendered valueless because of an attempt to save money by filing the barebones provisional application. Even if the potential infringer is not successful, this type of defense may cause the patent holder to expend considerable time and expense during litigation to counter these arguments; from the patent holder's perspective, these resources may be better spent on other issues in the litigation.

Another potential problem is related to the claims, or lack thereof, in provisional applications. Claims are not required in the provisional application. Claim drafting is critical to patent protection and can be an expensive part of the preparation of the patent application. An applicant may wish to avoid claims in the provisional application to save money (or at least postpone such expenditures), especially if he or she can personally draft the remainder of the application. But such an approach can be, we believe, hazardous. First, most patent professionals drafting an application begin with the claims (or at least the broad independent claims). The claims help in more fully understanding the invention and guide the drafting of the written description of the invention. Once drafted, the claims provide a framework for the specification and help to ensure that all claimed features of the invention are fully and adequately described in the application. A provisional application drafted without carefully considering the claims runs the significant risk of not providing adequate support to the later-drafted claims that the applicant ultimately wishes to include in the later-filed regular application.

Lack of claims in the provisional application may also cause difficulty in naming the inventors. The "inventors" named in a provisional application are those who "contributed" to the disclosure. At this time, it is not clear who these "contributors" may actually be. It is possible that some so-called contributors may not be inventors of later-drafted claims. Indeed, individuals not even associated with the invention or the assignee (e.g., competitors who have published articles or issued patents that are incorporated into the provisional application disclosure by reference) could be considered as "contributors" and, thus according to the statute, required to be named as "inventors" (112). Although these individuals (as well as other "contributors" and noninventors of the claimed invention) should not be named as inventors in the regular application, their inclusion on the provisional application may cause significant problems. For example, an individual from outside the organization named on the provisional application would have no incentive to assign the provisional application to the organization, thereby making sale or transfer of the technology more difficult. An employee named as an inventor in a provisional application but then not listed as an inventor on the regular application can also present problems. If that employee has left the company after the filing of the provisional application (perhaps under strained circumstances), he or she (or a later potential infringer) may argue that the individual was removed only because of their departure and, thus, the removal was improper. Even if there are no legal challenges, employee morale could suffer when individuals are first listed and then removed as inventors.

3.4.4 Recommendations. Although we must wait for the Federal Circuit to address these issues, there appears to be at least one approach for using provisional applications that avoids, or at least minimizes, these potential problems. Applicants using provisional applications should, in our opinion, file provisional applications having as complete a disclosure as possible, even to the extent of including claims. Ideally, this provisional application should be so complete that it can be filed without any changes as the regular application within the 1-year period. If the regular application is a virtual copy of the provisional application, a potential infringer cannot reasonably argue that the claims were not supported in the provisional application and, therefore, not entitled to the priority date (at least to any greater extent than the regular application). Moreover, including claims in the provisional application may simplify the naming of inventors. It seems reasonable in such a case that the named "contributors" of the provisional application could be selected based only on the claimed invention (rather than the much broader material disclosed). Thus, by including claims, the inventors of the provisional application could be selected by the same standards as inventors for the regular application. The Federal Circuit must, of course, ultimately decide whether such an approach is acceptable.

It is important to appreciate that the filing of a provisional application also starts the convention year under the Paris Convention for foreign applications. Thus, an applicant who files a regular application claiming benefit of a provisional application must also file the appropriate foreign applications within that same 1-year period. An applicant using a provisional application as a priority document must be alert so that his or her foreign applications are filed before the end of the convention year. Failure to file the appropriate foreign applications within the 1-year period may cause the loss of the priority date for any foreign applications and, perhaps, the loss of foreign patent rights. Docketing systems should be modified to incorporate provisional applications and their effects on foreign filing and other dates.

3.4.5 Applicants from Countries Other Than the United States. Although provisional applications are designed to provide a priority document for domestic applicants, foreign applicants may use provisional applications to their advantage. A U.S. patent claiming benefit of an earlier-filed foreign application has a prior art effect against other applicants only as of its U.S. filing date, not as of its foreign priority date. In other words, the foreign priority date provided a "shield" but not a "sword" against prior art for the foreign applicant (113). It is possible (but until the Federal Circuit has spoken, not certain) that the provisional application can be used by foreign applicants to create an earlier patent-defeating date (i.e., the filing date of the provisional application). A foreign applicant could file a provisional application in the United States as quickly as possible (based on the export regulations of the relevant coun-

try) after filing his or her foreign application. Although the provisional application cannot claim the benefit of the foreign application, the regular application (filed within the convention year) could claim benefit of both the foreign application and the provisional application. The effective date of the regular application against prior art (and statutory bars) would be the foreign priority date; the effective date as an offensive "sword" against later-filed U.S. applications (i.e., as prior art) would likely be the filing date of the provisional application. Thus, for a relatively low cost, a foreign applicant may be able to significantly increase the effect of his or her application as prior art against other patent applicants in the United States.

3.5 Invention Must Be New and Unobvious

In the United States, an invention to be patentable must be both new (114) and unobvious (115) over the prior art. Generally the prior art is the body of existing technological information (i.e., the state of the art) against which the invention is evaluated. Prior art may include other patents from anywhere in the world, printed publications or printed patent applications from anywhere in the world, U.S. patent applications that eventually issue as patents, public use or offer for sale in the United States of the subject matter embodying the invention, and, depending on the circumstances, unpublished and unpatented research activities of others in the United States. These types of prior art are defined in § 102 of the Patent Statute.

Issued U.S. patents are effective as prior art as of their filing dates. Patents from other countries, as well as publications from anywhere in the world, are effective as prior art in the United States as of their actual publication dates. Admissions by the inventor as to the content or status of the prior art, even if later shown to be incorrect, are also part of the prior art for that invention. Care should be taken, therefore, to ensure that admissions against interest are not made in the patent application or during prosecution of the application before the PTO. Unless it is absolutely clear that a given document qualifies legally as prior art against the invention, it should not be referred to as "prior art" or otherwise admitted to be "prior art."

3.5.1 Novelty under 35 U.S.C. § 102. The determination of novelty of an invention is generally straightforward: the issue is whether the invention is old or new. If a single prior art reference shows identically every element of the claimed invention (i.e., anticipation), the invention is not novel and thus not patentable. The anticipating reference, however, must be enabling. In other words, it must teach one of ordinary skill in the art how to make and practice the invention. Thus, the inclusion of only the name or chemical structure of a compound in a reference, without providing a method of making the compound, will generally not be considered to have anticipated a later patent claiming that compound (116). On the other hand, an anticipating prior art reference is not required to provide a use or utility for a compound (117). Thus, a reference that teaches how to make a specific compound but does not disclose a use will still prevent a later inventor, who discovers a use, from claiming the compound itself (118). The later inventor may be able to claim the use or a process taking advantage of a newly discovered property of the compound, if such use or process is new and nonobvious. In some cases, especially where new uses have been discovered, the effective amount required or the mode of administration of, for example, a drug may provide sufficient novelty for patentability.

An invention may also not be patentable under § 102 if certain events (so-called statutory bars) occur more than 1 year before the patent application is filed in the PTO. The statutory bars are designed to encourage the inventor to file his or her patent application in a timely manner. For example, a written description of the invention, a public use, or offer of sale of the subject matter of the invention in the United States more than 1 year before the filing date of the application bars the invention from being patented. An inventor thus has a 1-year grace period after such a public disclosure in which to file a patent application in the United States.

Except for the United States, most other countries require absolute novelty. A written description, public use, or offer of sale that actually discloses the invention would, in most cases, prevent the inventor from obtaining patent protection outside of the United States. If worldwide patent protection is important, as is likely in almost all drug-related inventions, one should file a patent application covering the invention before public disclosure. Potential public disclosures should be carefully screened and, where appropriate, controlled to ensure that valuable patent rights are not lost.

3.5.2 Obviousness under 35 U.S.C. § 103. Even if the invention is novel under § 102, the invention may not be patentable under § 103 if "the differences between the subject matter sought to be patented and the prior art are such that the subject matter as a whole would have been obvious at the time the invention was made to a person having ordinary skill in the art to which the subject matter pertains" (119). The determination of obviousness under § 103 is considerably more difficult than the determination of novelty under § 102. Obviousness is determined using a three-part inquiry: (1) a determination of the scope and content of the prior art; (2) a determination of the differences between the prior art and the claimed invention; and (3) a determination of the level of ordinary skill in the art (120).

Relevant and applicable prior art is that which is pertinent to a determination of obviousness of the claimed invention under § 103. Prior art for determining obviousness is not limited to the narrow technical area of the invention. The relevant prior art is normally found in the general technical field (or fields) to which the claimed invention is directed as well as analogous fields to which one of ordinary skill in that field or fields would reasonably turn or use to solve the problem. Factors used to evaluate the level of ordinary skill often include (1) educational level of the inventor, (2) nature of problems generally encountered in the art, (3) solutions provided by the prior art for these problems, (4) rate of innovation in the art, (5) level of sophistication of the art, and (6) educational level of active workers in the art (121). These factors, when applied to the drug discovery and development field, suggest a very high level of ordinary skill in this art.

Additionally, so-called secondary considerations or "objective indices of nonobviousness" are often used in determining whether an invention is obvious (122). Secondary considerations that can be used to support nonobviousness and patentability include, for example, commercial success, a long-standing problem that resisted solution until the invention, failure of others to solve the problem, unexpected results, copying of the invention by others in the art, and initial skepticism in the art concerning the success or value of the invention. Such secondary considerations provide evidence about how others in the art viewed the advance provided by the claimed invention and are often used in arguments presented to the PTO by the applicant in support of patentability. For such secondary considerations to be useful in establishing patentability, there must be a direct relationship between the secondary factor and the merits of the claimed invention. Therefore, to be useful in demonstrating nonobviousness, commercial success should be predominately the result of the benefits and merits of the invention rather than, for example, advertising or marketing considerations (123). Independent and essentially simultaneous development of the invention by others in the art to solve the same or very similar problem is one secondary consideration that supports the conclusion that the invention may be obvious.

Against this factual background of the three-pronged test and secondary considerations (if any), the obviousness or nonobviousness of the invention is determined. The decision maker must step back in time and determine whether a person of ordinary skill would have found the invention as a whole obvious at the time the invention was made. The Court of Appeals for the Federal Circuit has offered the following guidelines for this determination:

> [T]he following tenets of patent law ... must be adhered to when applying § 103: (1) the claimed invention must be considered as a whole ... [because] though the difference between [the] claimed invention and prior art may seem slight, it may also have been the key to advancement of the art ... ; (2) the references

must be considered as a whole and suggest the desirability and thus the obviousness of making the combination; (3) the references must be viewed without the benefit of hindsight vision afforded by the claimed invention; [and] (4) "ought to be tried" is not the standard with which obviousness is determined (124).

In the PTO, obviousness rejections are normally based on the combination of two or more prior art references. However, to support an obviousness rejection, the prior art references must do more than simply provide the elements of the claimed invention when combined. The references must also provide the motivation for the person of ordinary skill to combine the references in a way so as to achieve the claimed invention (125).

The standard for determining obviousness is not the same as "obvious to try." Thus, a claim cannot be properly rejected on obviousness grounds simply because all the parameters or numerous possibilities in the prior art reference could be varied or tried to arrive at the successful combination of the claimed invention, unless the references also provided directions to the successful combination. Nor can a claim be properly rejected because it might be obvious to try a new technology or approach that offers promise for solving a particular problem unless the prior art provides more than general guidance as to the form of the invention or methods to achieve it (126). Nor can a claim be properly rejected using hindsight to reconstruct the invention from the prior art using the applicant's own patent specification (127).

Once a case of *prima facie* obviousness is made out, the burden shifts to the applicant to rebut obviousness by showing that the invention had unexpected or surprising results. Unexpected results for a drug or drug formulation can include properties superior to those of prior art compounds including, for example, greater pharmaceutical activity than would have been predicted from the prior art, greater effectiveness, reduced toxicity or side effects, effectiveness at lower dosage, greater site-specific activity, and the like, as well as properties that the prior art compounds do not have. The unexpected results can be included in the patent application itself or presented by affidavit or declaration during prosecution of the patent application. If such unexpected results cannot be shown in cases of *prima facie* obviousness, the invention is not patentable. On the other hand, if a case of *prima facie* obviousness is not made out, it is not necessary to offer rebuttal evidence in the form of unexpected or surprising results for patentability.

3.5.2.1 Composition of Matter Claims. Prior art compounds that are structurally similar to the claimed compound or drug may render the claimed compound obvious and therefore unpatentable. But "[a]n assumed similarity based on a comparison of formulae must give way to evidence that the assumption is erroneous" (128). Recently, the Court of Appeals for the Federal Circuit reaffirmed this standard for *prima facie* obviousness as applied to composition of matter claims:

> This court . . . reaffirms that structural similarity between claimed and prior art subject matter, proved by combining references or otherwise, where the prior art gives reasons or motivation to make the claimed compositions, creates a *prima facie* case of obviousness, and that the burden (and opportunity) then falls on an applicant to rebut that *prima facie* case. Such rebuttal or argument can consist of a comparison of test data showing that the claimed compositions possess unexpectedly superior properties or properties that the prior art does not have (129).

The court also affirmed that *prima facie* obviousness does not require, given structural similarity, the same or similar utility between the claimed compositions and the prior art composition (130).

Structural similarity to support a case of *prima facie* obviousness is often found in cases of homologous series of compounds, isomers, steroisomers, esters and corresponding free acids, and the like. Bioisosterism, which may be of particular interest to the medicinal and pharmaceutical researcher, can also give rise to *prima facie* obviousness for both compounds and methods of using such compounds (131). Bioisosterism recognizes that substitution of an atom or group of atoms for another atom or group of atoms having similar size, shape, and electron density generally provides compounds having similar biological activity.

For example, the Court of Appeals for the Federal Circuit found that the use of amitriptyline for treating depression was obvious in view of the closely related antidepressant imipramine. Imipramine differs structurally from amitriptyline by replacement of the unsaturated carbon in the center ring of amitriptyline with a nitrogen atom. Another prior art reference showed that chloropromazine (a phenothiazine derivative) and chloroprothixene (a 9-aminoalkylene-thioxanthene derivative) had similar biological properties. Chloropromzine and chloroprothixene differ in the same manner as imipramine and amitriptyline (i.e., an unsaturated carbon vs. nitrogen in the central ring structure) and are also closely related (but not as closely as imipramine and amitriptyline). In fact, this prior art reference "concluded that, when the nitrogen atom located in the central ring of the phenothiazine compound is interchanged with an unsaturated carbon atom as in the corresponding 9-amionalkylene-thioxanthene compound, the pharmacological properties of the thioxanthene derivative resemble very strongly the properties of the corresponding thiazines" (132).

As indicated above, the applicant has the opportunity to rebut the case of *prima facie* obviousness by presenting evidence showing unexpected or surprising results. In cases where it is known that the compound to be claimed is structurally similar to known compounds, it is generally preferred that the data supporting patentability be included in the application as filed. Even if the *prima facie* case cannot be overcome, it may be possible in some cases to obtain at least some patent protection by claiming a method of making the compound or a method of using the compound.

3.5.2.2 Process Claims. There are generally two types of process or method claims: (1) claims directed to a method of synthesizing or otherwise transforming a chemical or biological material and (2) claims directed to a method of using particular compounds to achieve a desired result. The same basic obviousness standard is generally applied to process claims as is applied to other claims, including composition of matter claims. The Court of Appeals for the Federal Circuit in *In re Durden* held that the use of novel starting materials and/or the production of novel products (even if the products are patentable in their own right) in a known process for making a compound may make the process new but does not necessarily make it unobvious (133). The PTO interpreted *In re Durden* as effectively holding that an old process could never become unobvious through the use of novel starting materials and/or the creation of novel products. The Court of Appeals for the Federal Circuit later suggested that *In re Durden* merely "refused to adopt an unvarying rule that the fact that nonobvious starting materials and nonobvious products are involved *ipso facto* makes the process obvious" and added:

> The materials used in a claimed process, as well as the results obtained therefrom, must be considered along with the specific nature of the process, and the fact that new or old, obvious or nonobvious, materials are used or result from the process are only factors to be considered, rather than conclusive indicators of the obviousness or nonobviousness of a claimed process (134).

The patentability of process claims, whether method of making or method of using, should be determined in the same way as any other claim, that is, by applying § 103 to determine whether the invention as a whole is obvious or unobvious.

3.6 Procedure for Obtaining Patents in the U.S. PTO

Obtaining a patent in the United States involves an *ex parte* procedure solely between the inventor or the inventor's assignee and the PTO (135). The proceedings between the applicant and the PTO are conducted in writing. The proceedings themselves and the written record of the proceedings, often called the file or prosecution history, are carried out and maintained in secret until the application is published or the patent actually issues. Pending utility patent applications are now published 18 months after their priority date unless, at the time of filing, the applicant requests that the application not be published and certifies that it has not been and will not be filed in any country outside of the United States that provides for publication (136). If and when the application is published or the patent actually issues, the secrecy of the proceeding ends and the entire written record is open to the public. Should a patent not issue, the record will be maintained in secret (137). Thus, assuming corresponding patent applications were not filed in other countries and not published in the United States, an abandoned U.S. patent application will remain secret and may be maintained as a trade secret. If a foreign patent application has been filed, however, retention of any trade secrets contained therein is possible only until the time the foreign application is published (i.e., 18 months after the priority date).

For inventions developed in the United States, the patenting process generally begins with filing a patent application meeting the statutory requirements in the PTO. A flow chart generally illustrating the U.S. patenting procedure is shown in Fig. 20.3. Once an application is filed in the United States, the applicant or assignee has 12 months in which to file patent applications throughout the rest of the world. Normally the PTO will not have undertaken the initial examination of the application before the expiration of the 12-month period. Thus, one normally does not have the benefit of the PTO's initial patentability review before decisions must be made concerning filings elsewhere in the world.

Because the proceedings in the PTO are *ex parte* in nature, the applicant has a "duty of candor" to be forthcoming in dealing with the PTO throughout the prosecution of the patent application. This duty of candor includes providing the PTO with all information that may be material to the examination of applicant's patent application (e.g., potential prior art) of which the applicant is aware. The duty of candor does not require the applicant to carry out a literature or prior art search. However, if the applicant or the applicant's representative is aware of relevant prior art, it must be disclosed. Normally, such relevant prior art is provided to the PTO in an Information Disclosure Statement (IDS). The duty to disclose relevant prior art continues throughout the prosecution of the application. Therefore, if the applicant later becomes aware of relevant prior art, including prior art found by patent

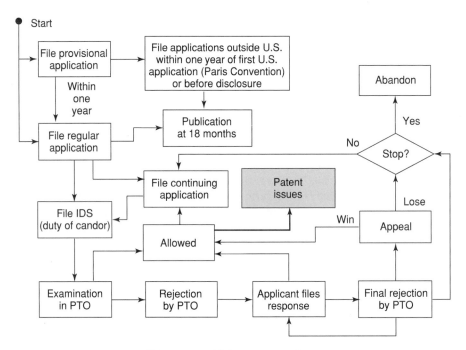

Figure 20.3. U.S. patent procedure.

offices outside the Untied States during prosecution of the corresponding applications filed in other countries, that information must be provided to the PTO. Failure to comply with this duty of candor can result in any patent issuing from the application being held unenforceable.

Upon receipt of a patent application, the PTO assigns the application a serial number and filing date. Shortly thereafter and if disclosure of the content of the application would not be detrimental to the national security, the PTO issues a foreign filing license, which allows the application to be filed in other countries (138). Thereafter, patent applications based on the United States patent application can be filed elsewhere in the world.

In some instances, the first response the applicant has from the PTO is a restriction requirement. The filing fee entitles the applicant to examination of one invention (e.g., a drug itself, a method of making the drug, and a method of using the drug may be considered separate inventions). If the application claims more than one invention, the PTO can issue a restriction requirement, thereby separating the claims into groups corresponding to the various inventions and requiring the applicant to choose (i.e., elect) one of the groups of claims for examination (139). The other, nonelected inventions can be refiled as divisional applications while relying on the priority date of the original application. Such divisional applications must be filed before the original application is abandoned or issues as a patent. Such divisional applications will expire 20 years from the filing date of the earliest filed nonprovisional application for which benefit has been claimed (140).

A patent examiner at the PTO examines the patent application to determine whether it meets the statutory requirements. This includes a search by the examiner of the prior art. Using the prior art uncovered in the search and information submitted by the applicant, the examiner evaluates the claims and determines whether the claimed invention is novel and nonobvious. Although rare, especially in drug discovery and biotechnology areas, the examiner may find the claims as submitted to be patentable. The patent would, on payment of the appropriate issue fees, then proceed to issue.

More likely, the examiner will initially find at least some of the claims to be anticipated or obvious over the prior art and will inform the applicant of this finding in an office action, along with the rationale for the rejection or rejections. The applicant has a limited time [normally 3 months (141)] in which to respond to the office action. The applicant can choose, especially if the examiner's arguments appear persuasive, to abandon the application and, if the application has not been and will not be published, retain the invention as a trade secret. The applicant can attempt to amend some or all of the claims to overcome the rejections and/or present arguments demonstrating why the examiner has improperly rejected the claims. The applicant can also present, if appropriate, additional information or data, including information relating to secondary considerations, demonstrating the novel and nonobvious nature of the invention. Such information (or data) is normally presented in an affidavit or declaration.

The examiner, on reconsideration of the claims in light of the applicant's response (and any supporting information or data), can repeat the rejections, submit new rejections based on other prior art or reasoning, or withdraw the rejections wholly or in part. If the examiner withdraws all the rejections and allows the claims, the patent will issue upon payment of the issue fee. If the examiner allows some claims but continues the rejection of other claims, the applicant has several options. The applicant could contest the rejections of the claims once again in the pending application. Or, the rejected claims could be canceled and the application passed to issue with the allowed claims. After a final rejection, any rejected claims can be appealed to the Board of Patent Appeals and Interferences in the PTO. It is often preferred in cases having both allowed and rejected claims to allow a patent to issue with the allowed claims and file a continuation application with the rejected claims before the parent application actually issues. By effectively separating the allowed claims from the rejected claims, the risk that the Board of Patent Appeals and Interferences will undermine the decision on patentability of the already allowed claims on appeal of the rejected claims is significantly reduced.

If the applicant is not successful in overcoming the examiner's rejections and the examiner makes the rejections final, several options remain. Again the applicant may simply abandon the application and, if the application has not been and will not be published, retain the invention as a trade secret. Or, the applicant can refile the application as a divisional, continuation, or continuation-in-part application and continue prosecution in the PTO. The applicant may also appeal the examiner's rejection to the Board of Patent Appeals and Interferences within PTO. If not satisfied with the Board's decision, the applicant may appeal that decision either to the Court of Appeals for the Federal Circuit based on the record before the PTO or to a federal district court for a *de novo* review. If the examiner's position is overturned, the Federal Circuit or the district court can order the PTO to issue the patent. Appeal to either the Federal Circuit or a federal district court destroys the secrecy of the application as well as that of the record of the proceedings within the PTO and thus destroys any trade secrets that may have been contained therein.

3.7 Interference

With the first-to-invent system in the United States, it is sometimes necessary to determine which of two or more inventors (or groups of inventors) first invented the subject matter that is claimed in common by the parties. Interferences are the proceedings within the PTO for making such determinations. These proceedings, which are overseen by senior examiners within the PTO, are ultimately decided by the Board of Patent Appeals and Interferences in the PTO. The party who first conceives an invention and first reduces it to practice will normally be awarded priority and will be awarded the U.S. patent (142). This is not the case, however, if another party, who reduced the invention to practice at a later date, can prove that he or she was the first to conceive the invention and proceeded diligently to reduce it to practice from a time before the other party's date of conception. The diligence of the first to reduce the invention to practice is normally immaterial in the priority contest.

Patent interferences are possible between two or more copending applications or between a pending application and a recently issued patent (143). In the case of copending applications, an applicant will often be notified by the examiner that his or her application appears to be in allowable condition but that prosecution is being suspended for consideration of the potential declaration of an interference. In cases involving a pending application and an issued patent, the examiner may cite the issued patent as a reference and suggest that the applicant may wish to "copy" claims from the patent to provoke an interference, or the applicant may become aware of the issued patent and attempt to amend his or her application (i.e., "copy" claims) to contain claims from the issued patent to provoke an interference. In the case of a pending application and an issued patent, claims from the issued patent must be copied by the applicant within 1 year of the issue date of the patent to provoke an interference. Moreover, if the applicant has an effective filing date less than 3 months after the effective filing date of the issued patent, a significantly reduced showing is generally required by the applicant to justify the interference (144). If the PTO determines that both parties have allowable claims directed to the same subject matter, a formal Declaration of Interference is issued.

An interference is a complex, multistage, *inter partes* procedure designed to determine which party has priority with respect to the patentable subject matter that is disclosed and claimed by two or more inventors. The interference normally proceeds through the following stages: (*1*) declaration of interference by the PTO, (*2*) motion period, (*3*) filing of preliminary statements, (*4*) discovery, (*5*) testimony period, (*6*) final hearing, (*7*) decision by Board of Patent Appeals and Interferences, and (*8*) appeal and court review. Each of these stages is governed by complex procedural and substantive rules with many potential pitfalls for the unwary. Failure to follow these rules carefully can result in an adverse ruling or, ultimately, judgment against the party violating them.

The party with the earlier priority date is designated the senior party and is presumed to be the first inventor. The other party is desig-

nated the junior party and has the burden of proving an earlier date of invention, generally by a preponderance of the evidence. If, however, the junior party's application was filed after the issuance of the senior party's patent, an earlier invention date must be proven beyond a reasonable doubt. Throughout the interference proceeding, the senior party retains significant procedural and substantive advantages.

Once an interference has been declared by the PTO, the parties are given an opportunity to redefine the interfering subject matter (i.e., the counts of the interference that are similar in form to patent claims) and, if possible, to assert an earlier effective filing date based on a related U.S. patent application or a corresponding patent application in another country. In redefining the interfering subject matter, each party will generally seek to amend the counts in a manner more favorable to itself (i.e., consistent with and supported by his or her evidence concerning conception and reduction to practice). The parties, if appropriate, may also raise issues that are not directly related to the dates of conception or reduction to practice. For example, one party may allege that the other party derived the interfering subject matter from someone else or that the other party's application has deficiencies that render the interfering subject matter unpatentable to that party. Either party can also argue that the subject matter in question is simply not patentable to anyone (i.e., effectively that neither party is entitled to a patent) (145). Generally, however, the most significant issues relate to establishing the respective dates of conception and reduction to practice and, where appropriate, diligence in reducing the invention to practice.

During the interference, parties can rely on inventive activities occurring in the United States and, with certain limitations, a North American Free Trade Agreement Implementation Act (NAFTA) country or a World Trade Organization (WTO) country to prove a date of invention before the actual filing date (146). The parties cannot rely on inventive activities within non-NAFTA or non-WTO countries to establish a date of invention; in such cases the parties will be limited to their foreign priority date as the date of invention. Thus, in an interference proceeding, activities within the United States, independent of the date, can be used to establish an earlier date of invention. Additionally, a party can establish a date of invention based on inventive activities in a NAFTA member country on or after December 8, 1993 or in a WTO member country on or after January 1, 1996 (147).

Thus, for applications filed after January 1, 1996, applicants from other countries (at least those carrying out inventive activities in NAFTA and WTO countries) are, at least in theory, on equal footing with U.S. applicants in proving earlier dates of inventions in an interference. In many cases, however, applicants and researchers from other countries will need to modify their methods of research record keeping to take advantage of this provision regarding foreign activities. Applicants from other countries must also understand, and adapt their operations to, the types of proofs and evidence necessary to prove dates of conception, diligence, and actual reduction to practice under U.S. law, as discussed below.

In adopting legislation regarding inventive activities outside the United States, Congress recognized that many countries do not provide levels or opportunities for discovery during judicial proceedings similar to those in the United States. Thus, an applicant from another country could potentially gain a significant advantage over his or her U.S. counterpart in an interference proceeding if the applicant from another country could use U.S. discovery procedures to gain information regarding the U.S. applicant's position while at the same time resisting such discovery based on his or her country's limited or otherwise inadequate discovery proceedings.

To address this potential problem, Congress has provided that "appropriate inferences" should be drawn when evidence regarding a date of invention cannot be obtained from a party (and perhaps even a third party) in a NAFTA or WTO country to the extent "such information could be made available in the United States" (148). "Appropriate inferences" are not defined and thus, depending on the circumstances, may range from mere "slaps on the wrist" to loss of rights in the proceeding. It is likely that the type of discovery that "could be made available" will depend

on the type of proceeding involved. Thus, the effect of this requirement could, for example, be very different in an interference proceeding before the PTO (where the scope of discovery is limited) and a later appeal before the federal courts (where the scope of discovery is much broader). Because of these uncertainties, the practice effects of this provision are unclear.

Once the PTO has declared the interference, each party must file a preliminary statement by a date set by the PTO. Facts alleged in the preliminary statement must be proved later in the proceedings. The preliminary statement must provide the dates of invention each party will rely on. A party intending to rely on an invention date earlier than his or her filing date must allege the earlier invention date and identify facts supporting the earlier date in the preliminary statement. The parties are held strictly in their proofs to any dates alleged in the preliminary statement. Thus a party able to introduce evidence showing an invention date earlier than alleged in the preliminary statement will still be held to the alleged date. The preliminary statement, therefore, should be carefully prepared and allege only dates and facts that the party can prove by clear and convincing evidence. A party relying on dates of conception and actual reduction to practice for the invention date must generally prove such dates by corroborated evidence. The testimony of an inventor or coinventor must be corroborated as it relates to priority of invention.

The preliminary statements are placed under seal and provided to the opposing party only at a later time set by the PTO. In the initial stages of the proceeding, neither party is aware of the alleged invention date of the opposing party or whether the opposing party has even alleged an invention date earlier than its filing date. After filing the preliminary statements, the parties generally undertake discovery and testimony to develop their cases. Once all evidence periods have expired, each party presents the formal record of evidence on which it wishes to rely. After the parties have submitted formal briefs enumerating their legal arguments and, where appropriate, rebuttal arguments, and after any oral arguments, the interference is decided by a board of three senior PTO examiners. The losing party may appeal the decision to the United States Court of Appeals for the Federal Circuit based on the interference record or bring a *de novo* civil action in an appropriate federal district court.

Conception generally is considered to occur when the inventor forms a definite perception of the complete invention sufficient to allow one of ordinary skill in the art to understand the concept and reduce it to practice without further inventive steps. Reduction to practice can be either constructive (i.e., by filing a patent application meeting the statutory requirement) or actual (i.e., by making and testing the invention to demonstrate that it yields the desired result). Therefore, to achieve an actual reduction to practice for a pharmaceutical, there must be both an available process for making the chemical compound (i.e., the drug) and testing to establish its utility. To be useful in an interference proceeding, such evidence generally must be corroborated by one or more persons who are not inventors. Circumstantial evidence can also be used to help establish an actual reduction to practice. Often the best corroborative evidence is provided by laboratory notebooks used to record ideas and experiments associated with the invention. Preferably, laboratory notebooks should be signed and dated daily by the investigators actually doing the work and diligently witnessed in writing by at least one noninventor coworker who has read and understood the record. The witness should be able to understand the significance of the experiments being witnessed. Generally, the witness should not be actively involved in the work, so as to reduce the risk that the witness will later be found to be a coinventor. If a witness is later determined to be a coinventor, he or she cannot corroborate the work. For particularly important inventions, two witnesses should be considered so that even if one witness is effectively disqualified (e.g., one witness is later found to be a coinventor), there still remains a corroborating witness. Without independent corroborating evidence of the events surrounding the invention, including conception and the reduction to practice and, possibly, diligence, a junior party will most likely lose the priority contest even if he or she was the first to invent. Should a party need to prove

diligence, a corroborated written record showing almost continuous activity from a time before the other party's date of conception up to the time of that party's own reduction to practice is generally required. If a party fails to work on the project for several successive weeks without adequate explanation, such inactivity will often be sufficient to destroy the case for diligence. The activity of the inventor in developing another invention does not normally constitute an adequate excuse for such a period of inactivity.

A comprehensive intellectual property program should, therefore, give special attention both to implementing acceptable record-keeping procedures (including procedures for witnessing notebooks and other records of invention in a timely manner) and to ensuring that these record-keeping procedures are followed. Once again, the intellectual property committee is the logical focus for this task.

Interferences are long, costly, and complex procedures that are laden with procedural pitfalls for both the senior party and, especially, the junior party. The senior party is heavily favored to be awarded the right to a patent on the contested invention (149). Based on the advantage of the senior party in such interference proceedings, patent applications should be filed in the United States as quickly as possible to increase the probability of achieving senior status in any interference that may be declared.

3.8 Correction of Patents

Issued patents are often found to contain errors of varying degrees. Minor errors (e.g., clerical or typographical errors and erroneous inclusion or exclusion of an inventor), if made without deceptive intent, can usually be corrected through a Certificate of Correction issued by the PTO at the request of the inventor or assignee (150). Generally, a mistake is not minor if its correction would materially affect the scope or meaning of the patent claims.

Patents that are wholly or partly inoperative or invalid may, in some cases, be corrected by reissuing the patent. Reissue patents are generally sought because (1) the claims in the original patent are either too narrow or too broad, (2) the specification contains inaccuracies or errors, (3) priority from a patent application filed in another country was not claimed or was claimed improperly, or *(4)* reference to a prior copending application was not included or was improperly made.

To obtain a reissue patent, the patent owner must establish that *(1)* the patent is considered "wholly or partly inoperative or invalid" because of a "defective specification or drawing" or because the inventor claimed "more or less than he [or she] had a right to claim," *(2)* the defect arose "through error without any deceptive intention," *(3)* new matter is not introduced into the specification, and *(4)* the claims in the reissue application meet the legal requirements for patentability (151). An application for a reissue patent that contains claims enlarging the scope of the original patent must be filed within 2 years of the date of grant of the original patent. The term of a reissue patent is the unexpired portion of the original patent; a reissue patent cannot extend the duration of the original patent. An infringer may have a personal defense of intervening rights to continue an otherwise infringing activity if the activity or preparation for the activity took place before the grant of the reissue patent (152). The reissue process is often used to "clean up" a patent before embarking on litigation to enforce that patent.

A patent owner or any third party can seek a review of an issued patent by the PTO on the basis of prior art not considered by the PTO during the original examination (153). Such a reexamination procedure, although technically not a mechanism for correcting a patent, allows the PTO to determine the correctness of the original patent grant in light of the newly presented prior art. A third party requesting reexamination has a limited, but potentially significant, opportunity to present written arguments against patentability in response to the applicant's initial statement concerning patentability.

Recent changes in rules governing reexamination proceedings allow a third party to request and participate in a reexamination proceeding using either *ex parte* or *inter partes* proceedings. Using ex parte proceedings, the third-party requestor is limited to responding to an applicant's initial statement concerning patentability (if filed) (154); no other input is

allowed by the third party. Under *inter partes* proceedings, the third party has the opportunity to respond to arguments and amendments made by the patent owner throughout the reexamination (155). The third party in the *inter partes* proceeding is limited to issues raised by the patent owner or PTO and cannot raise new issues. The third party in an *inter partes* proceeding is, however, barred from later asserting invalidity in a civil action on any ground that was raised, or could have been raised, during the reexamination proceeding (156). Because of this estoppel effect, *inter partes* reexamination proceedings are not likely to be used extensively by third parties.

The original patent is not presumed valid during the reexamination proceeding. The PTO can reaffirm the original grant, substitute a new grant by allowing new or amended claims, or withdraw the original grant. The legal presumption of validity of the patent is not strengthened by the successful completion of the reexamination proceeding. However, the practical effect of the reexamination proceeding may considerably strengthen the patent in the eyes of a judge or jury in later litigation. Courts and juries will generally look with particular favor on a patent that has twice been found patentable by the PTO (i.e., once during the original prosecution in the PTO and then again during the reexamination proceedings). On the other hand, the reexamination proceeding provides an accused or a potential infringer the possibility of invalidating a patent outside of the court system at a considerably reduced cost. An accused infringer should, of course, carefully consider the risks and implications of the PTO upholding the patent grant through the reexamination process before embarking on such a process because of the limited involvement of the accused infringer in the process and the strengthened validity resulting if the PTO finds the original grant sustained. Considering the potential risk to third parties that the patent might actually be strengthened and considering their limited involvement in the reexamination proceeding, it is not generally recommended that third parties request reexamination unless the prior art is believed, to a high level of certainty, to render the patent invalid (157).

4 ENFORCEMENT OF PATENTS

Fundamentally, a patent provides the right to exclude others from practicing the invention covered by (i.e., "claimed in") the patent. Enforcement of patents is limited both temporally and geographically. A patent can be enforced only against acts occurring during its term (158). Patent term is discussed in detail in Section 2.4 above.

4.1 Geographic Scope

Generally, a patent can be enforced only within the country in which it was granted. There is one exception to the general principle that a U.S. patent does not provide legal protection against acts outside its geographical borders. Under U.S. law, the importation, use, sale, or offer for sale of a nonpatented product in the United States is an act of infringement, if the nonpatented product was made outside the United States by a process patented in the United States (159). Comparable laws exist in many other countries. Thus, a U.S. patent covering a new and novel process for producing known products or drugs cannot be circumvented by manufacturing it outside the United States and then importing the product or drug into the United States. The U.S. patent holder cannot, of course, use its U.S. patent to prevent the use of that process overseas or the importation of the resulting products into other countries.

The unique situation in Europe deserves mention. In this time of economic globalization, the members of the European Union have taken measures to stop cross-border infringements involving more than one country. After grant, a European patent becomes a bundle of national patents in those countries designated by the patent owner. In cases of patent infringement in several of these countries, it is possible to sue all infringers in one proceeding. Under jurisdictional rules of the 1973 Brussels Convention, the patent owner may sue several joint infringers residing in different countries in one action, provided that at

least one of them is a resident of the country where the action is brought (160). Thus, assuming the patentee has a national patent in a particular European country, it is possible in the case of multicountry infringements to enforce the patent through another country's court system.

4.2 The Trial and Appellate Courts

In the United States, patent infringement is a federal cause of action. To enforce a patent, the patent holder normally brings suit against the alleged infringer in a federal district court. A patent dispute may also be initiated by a party seeking to have the patent declared invalid, unenforceable, or not infringed in response to having been threatened under the patent. Actions involving U.S. patents are conducted like all other civil actions in the federal courts, and are governed by the Federal Rules of Civil Procedure and the Federal Rules of Evidence. Most federal judges have no formal training in the sciences or patent law, and patent cases represent only a small percentage of federal cases. Because patent cases are sparsely distributed among the district court judges, the typical judge has very little experience handling the numerous complex issues involved (161).

The United States is unique in using juries to decide commercial civil trials, including patent disputes. The right to a trial by jury is enshrined in the Seventh Amendment to the U.S. Constitution and has been exercised readily by patent litigants. The incidence of jury demands in patent cases has grown from about 3% in the late 1960s to more than 50% in recent years (162). The jury's role is to listen to the evidence presented at trial to determine the facts, and to determine disputed issues by applying the patent law to those facts. When the evidence presents underlying disputed fact issues, the jury plays a crucial role in determining patent validity, infringement, and damages.

Two other forums for resolving patent infringement suits also deserve mention. The United States Court of Federal Claims has exclusive jurisdiction over patent infringement claims made against the federal government (163). The United States International Trade Commission ("ITC") has authority to preclude importation of products covered by a U.S. patent or made abroad by a process covered by a U.S. patent (164). Significant procedural and substantive differences exist between these forums and the district courts. For example, there is no right to trial by jury in the Court of Federal Claims, nor is injunctive relief available against the government or suppliers to the government (165). ITC investigations are conducted by ITC staff, whose determinations must consider factors, such as the public welfare, competitive economic conditions, and the interests of U.S. consumers, which do not necessarily align with the parties' interests (166). Detailed treatment of these two tribunals is beyond the scope of this chapter.

The 94 federal judicial districts are organized into 12 regional circuits, each of which has its own court of appeals. With the Federal Courts Improvement Act ("FCIA") of 1982 (167), the U.S. Congress created the Court of Appeals for the Federal Circuit and vested it with exclusive jurisdiction over patent cases. One of its objectives in doing so was to promote uniformity in the area of patent law and diminish the uncertainty created by inconsistent application of the patent law among the regional circuits (168). The Federal Circuit also hears appeals of patent matters from the PTO, the Court of Federal Claims, and the ITC, as well as appeals of a variety of non-patent matters. Matters involving issues of patent law make up a substantial portion of the Court's docket, and the Court is considered to have specialized expertise in the patent area.

A party may petition the U.S. Supreme Court to accept an appeal from an adverse decision of the Federal Circuit. The Supreme Court has discretionary control over its docket. It generally declines to hear the vast majority of cases it is asked to decide, and will only rarely take a patent case. Thus, the Federal Circuit is effectively the court of last resort for most patent litigants.

4.3 The Parties

4.3.1 Plaintiffs. A patent owner or an assignee of all rights in a patent may bring suit. The assignee must possess the assigned patent rights at the time that the suit is filed (169).

Where a patent is co-owned by multiple persons, all must jointly bring suit (170). Corporations must therefore make certain that each of the joint inventors execute an assignment document to avoid subsequent difficulties in bringing an infringement action. A transfer of less than the entire patent, an undivided share of the patent, or all rights in the patent in a geographical region of the U.S. is ordinarily deemed a license, and conveys no right to sue. A licensee may sue in its own name for patent infringement only if he owns "all substantial rights" under the patent, such that the license operates as a "virtual" assignment. In certain circumstances, a licensee with less than all substantial rights may have standing to sue as a coplaintiff (171). License provisions expressly granting the licensee a right to sublicense without the licensor's approval, to exclude others from using the technology, to file suit for infringement, and to defend the licensor's interest in such a suit may confer standing to a licensee as a coplaintiff (172).

4.3.2 Defendants. Section 271 of the Patent Code provides a remedy against those who make, use, sell, offer for sale the patented invention in the United States, or import the patent invention into the United States (173). Those who actively induce infringement by knowingly and actively aiding another's direct infringement can also be held liable (174). A remedy may also be had for contributory infringement against one who sells a material part of the invention that he knew was particularly adapted to infringe the patent and is not a staple item of commerce suitable for noninfringing use. To prove inducement or contributory infringement, a patentee must prove at least one act of direct infringement (175). The establishment of both active inducement and contributory infringement requires knowledge or awareness of the infringement of the patent by the infringer. As opposed to the case of direct infringement, a truly "innocent infringer" cannot be liable for either actively inducing infringement or for contributory infringement. It is not necessary that the direct infringer be named as a party, although doing so often simplifies the task of obtaining evidence sufficient to prove direct infringement.

A patentee has a cause of action against those who supply components of a patented product in the United States intending that the components be assembled into an infringing product abroad (176). A patentee may also sue one who imports, offers to sell or uses in the United States a product made by a patented method (177). Those who purchase a patented product from the patent owner or a licensee receive an automatic implied license to use and resell the product (178). The lawful owner of a patented device may also repair it or replace worn parts so long as this does not amount to recreating the patented device (179).

For a composition of matter claim covering a drug, the manufacturer who prepares the drug using any process, the drugstore that sells it, and the patient who takes it may each be liable as direct infringers; the doctor who prescribes it may be liable by actively inducing infringement by the patient. For a claim covering a process for making a specific drug, both the manufacturer who makes the drug in the United States using the patented process and the manufacturer who makes the drug outside the United States using the patented process and sells the drug in the United States would be liable as infringers. Manufacturing the drug by another process would not, of course, infringe the process claim (assuming the process used and the patented process are not equivalent under the doctrine of equivalents). For a claim covering a method of treatment using a specific drug, the ultimate user might be liable as a direct infringer; the manufacturer, druggist, and doctor might by liable if they actively induce infringement by the patient. For a claim covering a process of making a specific drug that employs an ingredient capable of use only in the process, the users of the process may be liable as direct infringers and the manufacturer of the ingredient who sells it to the users of the process may be liable as a contributory infringer. Thus, from the patent holder's viewpoint, it is desirable to have as many different types of claims in the patent to protect against infringement.

There is a special patent infringement remedy for a patent owner when a party files an Abbreviated New Drug Application ("ANDA") or a "paper NDA." Both ANDAs and paper NDAs permit companies to seek approval to market a generic drug before the patent coverage on the product has expired without the costly studies required for a so-called pioneer drug. It is an act of infringement to file an ANDA or paper NDA for a generic drug that is the same as the "pioneer drug" previously approved. By defining this somewhat artificial act of infringement, Congress provided a mechanism for early resolution of infringement disputes presented by the generic formulation. Monetary damages are not available until actual commercial manufacture and sale of the generic drug.

Pioneer drug applicants are required to provide the FDA with the identity and expiration date of any patent that claims the drug for which FDA approval is requested, or a method of using such drug (180). ANDAs and paper NDAs must certify that: (1) no such patents have been identified by the original pioneer drug applicant; (2) the identified patents have expired; (3) the date that patents will expire; or (4) that the patents are invalid or will not be infringed by the making, use, or sale of the generic drug (181). This certification determines the date on which FDA approval of the ANDA or paper NDA may be made effective and when marketing may begin. If the first or second certification is made, approval can be immediate. If the third certification is made, approval is delayed until the patents expire. If the applicant makes the fourth certification, the generic applicant must provide the original pioneer drug applicant a detailed statement supporting the assertion of noninfringement or invalidity. The original pioneer drug applicant may then bring suit for infringement within 45 days, causing delay of the approval until the earlier of 30 months or a court ruling that the patent is not infringed or is invalid.

4.3.3 Declaratory Judgment Plaintiffs. A party sufficiently threatened with an infringement suit may file an action in a district court seeking to have the patent declared invalid or not infringed. The court's jurisdiction over such actions derives from the Federal Declaratory Judgments Act of 1934 (182). A declaratory judgment action must be based on an "actual controversy" between the parties. The

party filing suit must show (1) that it has or is prepared to produce the accused product or use the accused method, and (2) that the patentee's conduct created an objectively reasonable apprehension that the accused infringer would eventually be sued if it continued its activity (183). An express charge of infringement easily satisfies the "actual controversy" requirement. Accordingly, the patentee should consider omitting such charges when notifying others of patent rights, and express instead a desire to discuss a potential license. However, statements falling short of an express infringement charge can give rise to declaratory judgment jurisdiction if the totality of circumstances create a reasonable apprehension of suit (184). A manufacturer not directly charged by the patentee may bring such a suit based on threats made against its customers.

4.4 Determining Infringement

When speaking of infringement, it is the claims of the patent that are infringed. The claims set the legal bounds of the technical area in which the patent holder can prevent others from making, using, selling or offering to sell, or importing the patented invention; in other words, the claims define the metes and bounds of the exclusive right granted by the patent. Each claim defines a separate right; some claims may be infringed, whereas others are not. Claims also define the bounds outside of which others can operate without infringing the patent.

The process for determining patent infringement has two main steps. First, the claims of the patent are interpreted to determine their scope and meaning. Next, the properly interpreted claims are compared to the accused product or process (185). Patent claim interpretation is performed by the judge based on the words of the claims themselves interpreted in light of the written description and drawings, and the record of proceedings in the PTO. Words in a claim are interpreted to have their ordinary meaning in the applicable field unless they were given a special meaning in the written specification or during prosecution (186).

The terms *comprising, consisting essentially*, and *consisting* are used in patent claims to link the preamble and the limitations of the claimed invention; these "terms of art" have special meanings in patent claims and can dramatically affect the scope or coverage of a patent claim. For example, a claim directed to a "composition of matter *comprising* compound X" would be infringed by a formulation containing compound X regardless of the presence of other components; an infringing composition must contain the listed element and can contain any other elements. A claim directed to a "composition of matter *consisting essentially* of compound X and compound Y" would be infringed by a formulation containing both compounds X and Y and other components that do not materially affect the basic and novel characteristic of the invention. A claim directed to a "composition of matter *consisting of* compound X and compound Y" would be infringed by a formulation containing only compounds X and Y (and normal impurities and the like); an infringing composition in this case must have only the listed elements and no others.

In the United States, patent infringement can occur either by literal infringement or under the doctrine of equivalents. Literal infringement of a patent requires that each and every limitation or element of at least one claim be found in the composition, device, or process as claimed. If a single limitation is not present, there is no literal infringement. However, infringement may still be found under the doctrine of equivalents where the differences between the claimed invention and the accused composition, device, or process are "insubstantial." Evidence that the accused composition, device, or process performs "substantially the same function, in substantially the same way, to achieve the same result" as the composition, device, or process defined in the patent claim may also support a finding of equivalency, although this test is more suitable for analyzing mechanical devices than chemical compositions or processes.

The doctrine of equivalents gives a court the ability to effectively expand the scope of a patent claim beyond the literal language of the claim. Otherwise, a party who copies a patent's inventive concept, but makes some trivial or obvious change, would avoid legal infringement. The doctrine of equivalents is a flexible concept. Pioneer patents will normally be entitled to a broader range of equivalents than patents claiming a relatively small advance over the prior art. As a technology develops and matures (i.e., becoming a "crowded art"), claims generally will be entitled to a smaller range of equivalents.

Balanced against this need to protect against invention pirating is the public's interest in knowing with reasonable certainty what a given patent covers. The doctrine of equivalents creates uncertainty as to the scope of patent rights and may thereby discourage legitimate efforts to design around existing patents. Important technical advancements often are the result of such design-around activities. A proper balance between these competing considerations is crucial to a properly functioning patent system. The courts have struggled to formulate and apply the doctrine in a manner that achieves this difficult goal. There are important limitations on the doctrine of equivalents intended to prevent patent rights from being extended too broadly. First, the doctrine is applied on an element-by-element basis. Thus, for each claim limitation there must be a corresponding element in the accused product or process that is identical or equivalent (i.e., only insubstantially different) to that limitation for there to be infringement by equivalents. Stated differently, if a theory of equivalence would effectively read a claim limitation out of the claim altogether, then there is no infringement.

In 1995 the Federal Circuit addressed basic issues concerning the doctrine of equivalents and held that (1) the determination of equivalents is a fact issue to be resolved by the jury; (2) the test focuses on the substantiality of differences between the claimed and accused products or processes, although other factors such as evidence of copying or designing around may also be relevant; and (3) equivalency is determined objectively from the perspective of a hypothetical person of ordinary skill in the relevant art (187). Two years later, the Supreme Court confirmed equivalents infringement as a viable doctrine, but expressed concern that the doctrine had become "unbounded by the patent claims" (188). The Supreme Court did little to clarify the proper test of equivalency, but required a "special vigilance against allowing the concept of equivalence to eliminate completely any [claim] elements" (189).

Further, a composition, device, or process cannot infringe a patent under the doctrine of equivalents where the broader interpretation of the patent claims necessary to cover the accused device or process would also cover "prior art" (i.e., the state of the technology before the filing of the patent application) (190).

A third important limitation on the doctrine of equivalents is "prosecution history estoppel," which prevents a patent owner from obtaining coverage under the doctrine of equivalents of subject matter that was relinquished during prosecution of the patent application. The purpose of prosecution history estoppel is to protect the notice function of the claims. Courts traditionally adopted a "flexible bar" approach under which the subject matter surrendered was determined by what a reasonable competitor would likely conclude from reviewing the prosecution file. However, in its recent *Festo* decision, the Court of Appeals for the Federal Circuit ruled that when a claim limitation is narrowed during prosecution, application of the doctrine of equivalents to that claim element is completely barred (191). This complete bar approach would eliminate the public's need to speculate as to the subject matter surrendered by a claim amendment. The *Festo* decision also confirmed that prosecution history estoppel applies to all narrowing amendments made for any reason related to patentability, even those made for purposes of clarification and not in response to a prior art rejection (192). Arguments made during prosecution, without a claim amendment, can also create an estoppel if they demonstrate a surrender of subject matter (193).

Although the *Festo* decision would lend certainty to the determination of patent scope, there is concern that it represents an overly rigid approach that will, in effect, severely restrict patent rights. The Supreme Court recently vacated the Federal Circuit's *Festo* decision and returned the case to the Federal Circuit for further consideration. The claim scope actually surrendered with the amendment will, however, apparently remain an important consideration in applying the doctrine of equivalents and prosecution history estoppel. However, the Supreme Court shifted the burden of proof on the issue of prosecution history estoppel from the alleged infringer to the patent owner. According to the Supreme Court, the patent owner should bear the burden of overcoming both of the following presumptions: (1) that amendments which narrow a patent claim, for any reason, create prosecution history estoppel; and (2) that all coverage between the originally proposed claim and the issued claim was surrendered by the patent owner. Further developments in this area of the law should be expected. In light of the most recent *Festo* rulings, however, the broadest claims should be drafted in a manner (i.e., significantly more narrowly) to reduce the likelihood that amendments relating to patentability will be required during prosecution.

4.5 Defenses to Infringement

In a patent infringement suit, the alleged infringer may assert that the accused product or process does not infringe the patent claims either literally or under the doctrine of equivalence. The burden is on the patentee to show that the claims cover the alleged infringing device or process. The alleged infringer may also attempt to show that the patentee is estopped from expanding the claims under the doctrine of equivalents sufficiently to cover the alleged infringing device or process, either because of admissions or arguments presented during the prosecution of the patent before the PTO or because the broader reading of the claims would encompass and cover prior art devices or processes. Noninfringement of the patent claims by the accused device or process is frequently raised as a defense in patent infringement suits.

The validity of the patent can also be raised as a defense. Patents are presumed valid. Thus, the alleged infringer must prove the patent is invalid by a clear and convincing standard of proof. Validity generally involves three components: novelty, utility, and nonobviousness. For a given claim, if the court finds that a claim lacks any one of the three attributes, that claim is invalid and cannot be infringed by the alleged infringer or anyone else. In other words, a holding by court that a patent claim is invalid prevents the patentee

from asserting the specific claim against anyone. Other claims in the patent may, however, remain valid and enforceable.

The validity of a patent claim is often attacked by a challenger presenting prior art that, in the challenger's opinion, anticipates the patent claims or renders them obvious. If the prior art offered by the challenger is not materially different from that considered by the PTO during examination of the patent, the challenger will have a difficult time showing that the patent is not valid; in effect, the challenger must convince the judge or jury that the PTO failed to do its job properly in granting the patent. If the challenger can present prior art that was not before the PTO examiner, and especially if the new prior art is more relevant to the invention than the prior art before the examiner, the challenger will have an easier time proving invalidity. In important cases, the challenger may be willing to go to great lengths to find additional prior art, including conducting extensive literature searches, hiring technical experts, interviewing other researchers in the field, and even posting rewards for helpful information.

A challenger may also attempt to invalidate a patent by showing sales, offers of sale, or public use of the invention more than 1 year before the effective filing date of the patent application. Evidence of such activity by the patentee or some other party is often not known to the PTO and, if available, may thus pose a serious challenge to patent validity. In other cases, a challenger may also attempt to invalidate a patent by showing that the specification is not enabling or that it fails to disclose the best mode of carrying out the invention that was known at the time the application was filed. These defenses are used less often and are considered less likely to persuade a jury at trial.

The challenger may also assert that the patentee acted improperly before the PTO in obtaining the patent. This inequitable conduct or fraud defense is raised in many patent infringement suits. The process of obtaining a patent is an *ex parte* proceeding between the applicant and the PTO. Because representations made by the applicant are generally accepted at face value, the courts have established a relatively high standard of candor and conduct for the applicant in the dealings with the PTO. A challenger asserting inequitable conduct will attempt to prove that the applicant was not entirely forthcoming during the *ex parte* proceedings before the PTO. Inequitable conduct might include, for example, an applicant failing to bring relevant and material prior art known to the applicant to the attention of the PTO, attempting to "bury" an especially relevant prior art document in a large collection of seemingly less relevant prior art, falsifying data, or otherwise misleading the PTO. The challenger must show by clear and convincing evidence that the alleged mischaracterization or other inequitable conduct was "material" and that the applicant acted with the required intent to mislead or deceive the PTO.

The materiality and intent requirements of inequitable conduct are interrelated. For example, a high level of materiality can create an inference that the conduct was willful, and a specific showing of willfulness can reduce the level of materiality required. Most often, information is considered material if there is a reasonable likelihood that a reasonable examiner would have considered it important in deciding whether to allow the patent. The necessary "intent, " which can be shown by circumstantial evidence, is usually characterized as an intent to deceive the PTO. Simple negligence or an erroneous judgment made in good faith will generally not support a finding of the necessary intent (194). Gross negligence may support a finding of the required intent. A specific showing of wrongful intent clearly provides the necessary intent (195). In contrast to invalidity based on the other criteria discussed, a finding of unenforceability based on inequitable conduct affects the entire patent, not just the claims to which the inequitable conduct may specifically relate. It may also render related continuing patents unenforceable.

There is no general "experimental use" exemption or defense against a charge of infringement (196). There is, however, a so-called clinical trial exemption that is of special interest to the drug discovery and development industry. This clinical trial exemption generally provides that it is not infringement to make, use, or sell a patented invention "solely for uses reasonably related to the development and submission of information under a Federal law which regulates the manufacture, use, or sale of drugs or veterinary biological products" (197). Thus, a third party may use certain patented drug-related devices, compositions, or processes during the life of the patent to develop data and information reasonably necessary for use in, say, the FDA approval process. This exemption is generally intended to allow a manufacturer to obtain FDA approval during the life of the patent to be able to offer a generic drug as quickly as possible after the patent on the drug expires. If the manufacturer is forced to delay clinical trials and similar data-generating activity until the patent term expires, the FDA approval could be delayed until well after the end of the patent term, thereby effectively extending the patent term and depriving the public of the alternative product. This provision and the patent term extension provision for drug-related inventions are generally designed to ensure that the patentee obtains the full measure of protection normally offered by a patent, but no more.

4.6 Remedies for Infringement

If a patent claim is found infringed and not invalid, the court "shall award . . . damages adequate to compensate for the infringement but in no event less than a reasonable royalty . . . together with interest and costs" (198). The purpose of such compensatory damages is to return to the patentee the value of the loss associated with the unauthorized use of the patented item or process. Where the patentee or exclusive licensee is exploiting the patent and lost or diverted sales can be demonstrated that, but for the infringement, would have gone to the patentee or exclusive licensee, lost profits (rather than a reasonable royalty) might be the appropriate estimate of damages.

4.6.1 Lost Profits. To prove lost profits, the patentee must show: (1) demand for the patented product; (2) ability to meet the demand; (3) absence of noninfringing substitutes; and (4) the amount of profit that would have been earned (199). However, this is not the exclusive test, and the patentee may find other ways to prove with reasonable probability that it would have made all or some of the infringing sales (200).

Lost profits may be proven based on diverted sales, eroded prices, or increased costs. In a two-supplier market, it is presumed that the patentee would have made the infringer's sales or charged higher prices without the infringing competition (201). In markets with more than two competitors, market shares and growth rate projections may be considered (202). Price erosion considerations can become significant to the damage calculation if the patentee can show it could have charged higher prices.

In *Rite-Hite v. Kelley* (203), the Court of Appeals for the Federal Circuit held that lost profits on unpatented products that directly compete with the infringing product are recoverable if the lost sales were reasonably foreseeable. The patentee in *Rite-Hite* made a product covered by its patent and a higher priced unit that was not covered. The allegedly infringing product was intended to compete with the unpatented unit. The Federal Circuit held that the patentee was entitled to recover profits it lost on diverted sales of the unpatented unit as compensation for the infringement.

A patentee may also recover damages based on unpatented components sold with a patented product, either on a lost profit or royalty theory. *Rite-Hite* further clarified that doing so requires proof that the unpatented components function together with the patent product in some manner to produce a desired result. Unpatented components sold with the patented product only for convenience or business advantage may not be included in the damage calculation (204). However, the infringer who proceeds with the infringing activity, reasoning that the worst-case outcome is having to pay a royalty on its sales, may be unpleasantly surprised when ordered to pay damages based on lost profits of the patentee's unpatented products and accessories.

4.6.2 Reasonable Royalty. If lost profits cannot be shown, the patentee is entitled to a reasonable royalty, which is calculated as the amount a reasonable person would have been willing to pay as a royalty in a hypothetical negotiation at the time infringement began.

Some of the factors that may be considered in determining the reasonable royalty include the infringer's expected profit, whether the license would have been exclusive, actual established royalties under the patent or in the particular field, and the parties' bargaining positions at the time the infringement began (205).

4.6.3 Enhanced Damages. Infringement can range from unknowing or accidental to deliberate or reckless disregard of the patentee's legal rights. For "willful infringement," damages can be increased up to three times the actual damages (206) and may, in exceptional cases, include an award of attorney fees (207). Willful infringement can be found where an infringer knew of the patent and lacked a reasonable legal justification for the infringing actions. The potential infringer with notice of the patent has an affirmative duty to determine whether any of its actions would constitute infringement. This duty can generally be discharged by obtaining competent legal advice concerning the potential infringement. A failure to seek such advice, however, may not by itself be sufficient for a finding of willful infringement, nor will obtaining such advice automatically prevent a finding of willful infringement. The risk of up to triple damages in the case of willful infringement, and attorney fees in exceptional cases, are designed to deter infringement and encourage would-be infringers to evaluate their actions carefully in light of the claims of relevant patents.

4.6.4 Injunctive Relief. In addition to money awards, the court may grant an injunction to "prevent the violation of any right secured by patent, on such terms as the court deems reasonable" (208). Preliminary injunctions, by which the alleged infringer is required or forbidden to do certain specified acts before a full trial on the merits of the case, serve to preserve the status quo in the market during the suit. These are rarely granted. To obtain a preliminary injunction, the patentee must show (*1*) a reasonable likelihood of success on the merits, (*2*) irreparable harm to the patentee if the injunction is not granted, (*3*) hardships favoring the patentee, and (*4*) that the public interest will be favored by granting the injunction (209). Such a showing is often difficult to make at the early stages of the proceeding.

In contrast to preliminary injunctions, permanent injunctions are routinely granted after a trial court has found the patent to be valid and infringed. Such injunctions generally forbid the infringer from continuing the infringing acts during the remaining term of the patent, or may be worded more precisely to forbid the use of specific compositions, processes, design features, and the like.

4.7 Commencement of Proceedings

A suit to enforce or attack a patent may be filed against a given defendant in any district court in which requirements of personal jurisdiction and venue are met. The rules governing venue in patent cases seek to prevent the defendant from having to litigate in an inconvenient forum. Generally, the proper venue for a patent infringement defendant is any district where the defendant resides or has committed acts of infringement and has an established place of business (210).

The requirement that the court have power to exercise personal jurisdiction over the defendant depends on whether the defendant has established some minimum contact with the state in which the district court is located and, if so, whether the assertion of jurisdiction over the defendant is fair and reasonable in light of all of the circumstances (211). All of the defendant's contacts with the forum state, such as offices, employees, bank accounts, product sales, advertising, telephone listings, and the like, may be considered. Even a non-U.S. company with no United States operations may be subject to jurisdiction if it places an infringing product into distribution channels that it reasonably knows will lead to product sales in the forum state (212).

The complaint that commences the action must identify the parties, state the basis of the court's jurisdiction, and contain a "short and plain statement showing that the pleader is entitled to relief" (213). In a patent infringement complaint, the statement will normally allege that the plaintiff owns or otherwise has rights to enforce the patent, that the defendant is infringing the patent by making, using,

selling, offering for sale, or importing products embodying the patented invention, and that the infringement has damaged plaintiff. The plaintiff may request preliminary and/or final injunctive relief and an accounting for damages.

The defendant must respond to the patent complaint by either filing an answer or bringing motions challenging the complaint on one or more procedural grounds. An answer must admit or deny each of the plaintiff's allegations, and raise any "affirmative defenses." Affirmative defenses are matters outside the plaintiff's main case that are in the nature of avoidance, such as invalidity of any patent claim in suit, unenforceability of the patent, laches, and implied license (214). The most common motions filed in lieu of an answer in patent cases are those challenging the court's jurisdiction over the defendant, requesting transfer to a more convenient district court, or alleging that an indispensable party with an interest in the dispute has not been joined.

4.8 Discovery

Extensive and invasive discovery is a hallmark of U.S. patent litigation. The rules relating to discovery encourage full disclosure by the parties to avoid potentially unfair surprises at trial (215). The scope of permissible discovery is extremely broad, and allows a party to seek from its opponent or a nonparty any information that is not protected by a recognized privilege and is either relevant to an issue in the case or appears "reasonably calculated" to lead to admissible relevant evidence (216). The patentee will seek evidence regarding the accused product or process, the profitability of the infringing activities, the accused infringer's awareness of and intent with regard to the patent, and the accused infringer's defenses. Discovery into the latter subject may encompass noninfringement defenses, prior art defenses, and equitable defenses such as undue delay. Persons requested to provide such information may include the management, research, engineering, and accounting personnel of the accused company, retained experts, customers and suppliers, former employees, and even competitors. The accused infringer will generally seek evidence regarding the development of the invention and prosecution of the patent application, details of the patentee's commercial product or process, prior art to the asserted patent, the patentee's awareness of uncited prior art, and the profitability of the patented product or process. Discovery requests for such information may be directed to the inventors, the patentee's own engineering, marketing and financial personnel, the patent attorney involved in prosecution, and other companies with prior related technology. Obtaining discovery from a nonparty requires counsel to issue a subpoena requesting specific information or a deposition, which the district court may quash or modify if the subpoena request is overbroad or otherwise unreasonable (217).

Two important limitations on discovery of relevant information in patent actions are the attorney-client privilege and the work product immunity doctrine. The former protects confidential communications between lawyer and client for the purpose of securing legal advice. The latter protects documents prepared by or for counsel in connection with litigation, and counsel's mental impressions.

The five discovery procedures commonly employed in patent actions are requests for production, interrogatories, requests to admit, depositions, and inspections. Production requests are used to obtain tangible evidence such as documents, drawings, photographs, product samples, and digitally stored data. A party may also request to inspect and test the accused or patented product or process. Interrogatories are written questions posed to the opposing party that must be answered in writing under oath, subject to any valid objections. The written answers may be used as evidence at trial. Requests to admit seek written admissions that certain facts are correct. Although used somewhat sparingly in patent cases, requests to admit can be effective tools for resolving issues, such as the genuineness of important documents or dates of important events, before trial.

The deposition process permits counsel to obtain sworn testimony from witnesses before trial. Depositions enable counsel to learn more about the underlying facts and events, explore an opponent's case, identify additional evidence and witnesses, and obtain admissions regarding important facts. Deposition testimony may be admissible at trial if the witness is unavailable to testify in person (218). Depositions are the primary mechanism of obtaining testimony from nonparty witnesses who reside outside the geographic limits of the court's trial subpoena power. A deposition transcript may also be used to contradict the testimony of a deponent who later testifies at trial. The deposition of a party may be used at trial for any purpose (219). Depositions are also important opportunities for each party to assess the strengths and weakness of their case and the credibility and demeanor of key witnesses.

The discovery rules also require the parties to identify each expert witness they will call at trial and provide a report summarizing the expert's credentials and opinions (220).

4.9 Summary Judgment

A court may dispose of an action without conducting a trial where the evidence presents no genuinely disputed issues of fact. A party may seek "summary judgment" at any time, although such requests are most commonly made after the completion of discovery. Courts are frequently requested to resolve patent cases in the context of summary judgment motions. If complex factual issues exist, and expert testimony is required to explain the technology to the court, such motions should be denied (221). However, when the facts are not disputed, and an issue such as infringement depends only on a question of law, summary judgment becomes an effective way to resolve cases more quickly and efficiently.

Rulings commonly made on summary judgment motions include invalidity due to anticipation (i.e., lack of novelty), no literal infringement because of the absence of a claim element, or infringement when all elements are present. Where the court finds no literal infringement, the appropriateness of granting summary judgment of no infringement under the doctrine of equivalents depends on whether one of the legal limits on the doctrine applies, such as prosecution history estoppel or the prior art. Issues such as obviousness and unenforceability attributed to inequitable conduct often involve fact issues precluding the use of summary judgment.

4.10 Trial

The trial of the patent action that is not set-
tled or resolved by the court involves the presentation of evidence, submission of the case
to the trier of fact, followed by the entry of
judgment. The purpose of the trial is to decide
facts based on the testimony of witnesses and
examination of evidence. In a jury trial, factual issues are determined by the jury and legal issues are decided by the judge (222). If a
jury is not requested, the judge also performs
the fact-finding role. The judge presides over
the trial to ensure that the process is fair to
the litigants. In carrying out this obligation,
the judge will resolve the parties' disputes regarding the conduct of the trial, rule on objections to the admissibility of evidence or improper questions asked of witnesses, instruct
the jury as to the findings it must make and
the governing law, and control the conduct of
the attorneys. Before going to trial, all parties
will educate the judge with a trial brief outlin-
ing their case and what they intend to prove at
trial.

Another aspect of the court's role relates to
expert testimony. A trial judge has broad discretion to consider whether a witness is qualified as an expert, and whether the expert's
knowledge will assist the jury. In response to
concerns about the prevalence of "junk science" in federal civil cases, the Supreme Court
recently determined that this requires a preliminary assessment of whether the expert's
reasoning and methodology is scientifically
valid (223). Technical or financial expert analysis that does not satisfy accepted standards in
the expert's particular field is subject to exclusion.

Each party may present live witness testimony, deposition testimony, documents, and
demonstrative evidence such as charts, photographs, videos, and physical objects. Given the
potential for jury confusion, trial counsel must
carefully plan the sequence and manner of
presenting the evidence that will aid the court
and jury in understanding the complex technical and legal issues involved and lead the decision maker to the desired result. Generally,
counsel will attempt to present a simple,
straightforward case focusing on a few key

themes that the judge and jury can understand (224).

After the presentation of evidence by both
parties, trial counsel delivers closing arguments to the fact finder after which a decision
is reached. Certain post-trial motions are
available to the losing party, the most common
of which is a motion to set a jury's verdict aside
as unreasonable in view of the evidence (225).
Either party may appeal to the Court of Appeals for the Federal Circuit.

4.11 Alternative Dispute Resolution

Suits to enforce or defeat patents are invariably a component of larger business objectives, and the tactics employed must derive
from the underlying business strategies. In
most countries, patent litigation is a very expensive endeavor (226). Absent unique circumstances, litigants in the United States pay
their own attorney fees and costs. In many
other countries, the prevailing party may recover some or all of its fees and costs. Legal
and expert witness fees, travel expenses, document expenses, and court fees can mount
quickly.

Numerous other factors necessarily enter
into the strategic planning of the patent owner
and accused infringer. In the United States,
strong remedies are available to the victorious
patent owner. The accused infringer faces not
only the risk of being enjoined from continuing the commercial activities in question, but
may also be required to pay compensatory
damages, as well as enhanced damages and
attorney fees. The financial consequences of
an adverse decision of infringement are thus
potentially catastrophic. A legal victory for the
accused infringer can be a hollow one if it
spends all of its profits defending its right to
continue its commercial activities.

On the other hand, because patent invalidity is an absolute defense to an infringement
charge, the patent owner risks investing substantial litigation expense and effort only to
emerge from trial with a valueless, invalid
patent and no means of preventing competitors from practicing the once-protected technology.

Patent legal proceedings, which can last
years, can seriously disrupt business planning
and strategy as well as divert human and eco-

nomic resources away from the drug discovery and development process. The extremely high cost of litigation generally and especially in complex patent cases, coupled with unpredictability of outcome, the considerable time periods involved, and the likelihood of appeals resulting in even further delays, may increase the desire of both parties to reach an acceptable settlement before or even during trial. A settlement reached on reasonable business terms can, in many cases, provide a more favorable and satisfactory outcome for both parties.

In many cases, alternative dispute resolution processes (e.g., negotiation, mediation, arbitration, neutral expert fact-finding, and the like) can offer significant benefits over traditional litigation for resolving patent-related disputes, including infringement (227). Such benefits include, for example, faster resolution of the disputed issues, the ability to tailor the process to the needs of the parties, the ability to select fact-finders or decision makers with the educational and technical backgrounds suitable for the technology, generally lower cost, increased predictability of outcome, and a finite and definite resolution of the dispute. Alternative dispute resolution processes are likely to be used with increasing frequency in patent-related cases.

5 WORLDWIDE PATENT PROTECTION

Because the marketplace for the products from drug discovery and development has become global in nature, worldwide patent protection has become increasingly important. Seeking patent protection in many countries throughout the world can be very expensive. The cost of obtaining patent protection should be weighed against the benefits derived from patent protection on a country-by-country basis. The countries most often chosen for patenting purposes include the United States, Canada, the European Community (usually designating at least Germany, France, United Kingdom, Italy, and Sweden), Australia, and Japan. For some specific products or processes and marketing considerations, other countries may be as important as, if not more important than, those just listed. Pharmaceuti-

cal companies, for example, may wish to file patent applications in most or all countries where they (or their subsidiaries, affiliates, or licensees) are likely to produce and/or market a new drug.

In most cases, it is simply too expensive to attempt to file patent applications in a majority of the countries of the world, much less in every country. The evaluation of where to file should be undertaken on a case-by-case basis, taking into account the technology itself and the marketplace. In many cases, it may be possible to obtain adequate patent protection without seeking patent coverage in a large number of countries. For example, if there are interrelated markets, a patent in one country often can offer practical and effective protection (but not legal, enforceable protection) against infringing acts in another country or countries. Thus, for example, a competitor may be discouraged from offering a drug in Canada, if that drug cannot be offered in the United States because of the existence of a blocking U.S. patent. In effect, the United States and Canada (and to an increasing extent, Mexico) form a single North American market (228). Taking such market considerations into account throughout the world, it may be possible, depending on the specific technology, to obtain practical worldwide patent protection through patents in only a relatively few countries.

5.1 International Agreements

Having determined where to file, one faces the task of filing patent applications in the appropriate countries. International agreements have made this process much easier. Although it is possible to file separate patent applications in each of the countries selected, this procedure is rarely used. Rather, the procedures of various international intellectual property treaties can be used to simplify this administrative task considerably. The principal international treaty governing patents is the Paris Convention for the Protection of Industrial Property (229), which has approximately 162 signatory member nations. A patent application filed in any member nation creates a priority date for applications filed within 12 months (i.e., the convention year) in other convention nations. Thus, an applicant can file a patent application in the United States and then file separate patent applications in other member countries within the ensuing 12 months. Such applications have an effective filing date, which is the same as the filing date of the U.S. application (230). For non-Paris Convention countries (e.g., Taiwan), patent applications must be filed directly in the national patent offices. Patent applications in non-Paris Convention countries must rely on their actual filing date in the particular country. The filing procedure using the Paris Convention still requires an application to be filed in every country in which protection is sought and is, therefore, still unwieldy if a large number of countries are selected.

The European Patent Convention (EPC) (231) allows for the filing of a single patent application designating selected member European countries that, following prosecution before and issuance by the European Patent Office, becomes effective as national patents in the designated countries. Currently, 24 European nations (listed in Table 20.3) are members of the EPC. In addition, six other countries, although not members, can be reached through the EPC procedure (232). A patent issued by EPC is not a true European patent; rather it is a grant of separate national patents in the member countries designated by the applicant, each of which is enforceable under the laws of the country in which it was granted. Although applications can be filed in individual countries, the use of the single EPC application has been widely accepted as a convenient and less expensive mechanism to obtain coverage when seeking patent protection in four or more of the European member countries. If protection is desired in only a few member countries (i.e., three or less), national applications in the individual countries may be a less expensive alternative. The decision of whether to use a single EPC application or whether to file directly in individual countries depends on many factors and should be decided on a case-by-case basis.

Since its adoption in 1978, the Patent Cooperation Treaty (PCT) (233) has become an increasingly important and useful mechanism to obtain patent protection throughout the world. Currently the PCT has about 115 mem-

Table 20.3 Members of the European Patent Convention (as of mid-2002)

EPC Member States	
Austria	Italy
Belgium	Liechtenstein
Bulgaria	Luxembourg
Czech Republic	Monaco
Cyprus	Netherlands
Denmark	Portugal
Estonia	Slovak Republic
Finland	Spain
France	Sweden
Germany	Switzerland
Greece	Turkey
Ireland	United Kingdom

ber states, including the United States, Canada, Japan, Australia, and most European countries. The member states are listed in Table 20.4. The PCT allows the filing of a single international application that has the same effect as if separate applications were filed in each designated country. The PCT does not create an international patent and does not modify the substantive requirements for patentability in the member countries. It simply reduces the effort and resources necessary to file the patent application initially in multiple countries at the same time, thus proving an effective mechanism for filing an international application (especially as the convention year draws to a close and it becomes necessary to file applications very quickly).

Typical practice for an applicant based in the United States might involve filing an application in the U.S. PTO, and then filing applications in the desired countries throughout the world within the convention year. Alternatively, the procedures offered by the EPC and/or PCT could be used for filing such applications. Filing a PCT application directly in the local national patent office (assuming it is a PCT-receiving office) designating most, if not all, of the PCT countries is increasingly becoming the preferred practice both in the United States and the remainder of the world. In the United States, this procedure would involve filing a PCT application in the U.S. PTO that designates the desired PCT countries. Because of the increasing importance of the role of the PCT in obtaining patent protection, it is

Table 20.4 Contracting States of the Patent Cooperation Treaty (as of late-2001)

	PCT Contracting States
Africa	Algeria, Benin, Burkina Faso, Cameroon, Central African Republic, Chad, Congo, Côte d'Ivoire, Equatorial Guinea, Gabon, Gambia, Guinea, Guinea-Bissau, Kenya, Lesotho, Liberia, Madagascar, Malawi, Mali, Mauritania, Morocco, Mozambique, Niger, Oman, Senegal, Sierra Leone, South Africa, Sudan, Swaziland, Togo, Tunisia, Uganda, United Republic of Tanzania, Zambia, and Zimbabwe
Americas	Antigua and Barbuda, Barbados, Brazil, Canada, Columbia, Costa Rica, Cuba, Dominica, Ecuador, Grenada, Mexico, Saint Lucia, Trinidad and Tobago, and the United States of America
Asia/Pacific	Armenia, Azerbaijan, Australia, China, Democratic People's Republic of Korea, Georgia, India, Indonesia, Israel, Japan, Kazakhstan, Kyrgyzstan, Mongolia, New Zealand, Philippines, Republic of Korea, Singapore, Sri Lanka, Tajikistan, Turkmenistan, United Arab Emirates, Uzbekistan, and Viet Nam
Europe	Albania, Austria, Belarus, Belgium, Belize, Bosnia and Herzegovina, Bulgaria, Croatia, Cyprus, Czech Republic, Denmark, Estonia, Finland, France, Georgia, Germany, Greece, Hungary, Iceland, Ireland, Italy, Latvia, Liechtenstein, Lithuania, Luxembourg, Monaco, Netherlands, Norway, Poland, Portugal, Republic of Moldova, Romania, Russian Federation, Slovakia, Slovenia, Spain, Sweden, Switzerland, The former Yugoslav Republic of Macedonia, Turkey, Ukraine, United Kingdom, and Yugoslavia

important to have at least a basic understanding of PCT procedures to capitalize on the advantages it offers.

5.2 PCT Patent Practice

A single PCT application can be filed in a PCT-receiving office (the U.S. PTO is one such receiving office) designating all or only certain member states. For example, a U.S. applicant could file a PCT patent application in the English language in the U.S. PTO and designate the appropriate PCT member states (234) in which protection is desired. A flow chart generally illustrating typical PCT practice is shown in Figure 20.4. Once the PCT application is filed, an international search is performed by the International Search Authority, through a national patent office or intergovernment alliance (e.g., the U.S. PTO or the European Patent Office). The application is published 18 months after the effective filing date (i.e., the priority date). Therefore, 18 months after the priority date any trade secrets or other technical information contained in the application are disclosed to the public.

The PCT also provides for an optional international preliminary examination related to the patentability of the claimed invention.

Although the results of the preliminary examination are not binding on individual member states, the results of preliminary examination as well as the international search report can offer significant insight and assistance for an applicant in determining the likelihood of ultimately obtaining patent protection and, thus, whether to proceed with the application in the individual designated states.

The applicant must elect whether to go forward in some or all of the designated countries (i.e., enter the national stage in the elected countries) within a certain time period after the priority date. Until April 1, 2002, this election must be made within 20 months of the priority date. The election can be delayed a further 10 months by entering so-called Chapter II proceedings and payment of additional fees. Effective April 1, 2002, the PCT Assembly has decided to change the time limit for entering the national stage in elected countries (235). At that time and assuming individual countries of interest have modified their local law to adopt these PCT changes, an applicant will have up to 30 months within which to enter the national stage without making an earlier election and without paying additional fees. The Assembly decided to make this change because many applicants were en-

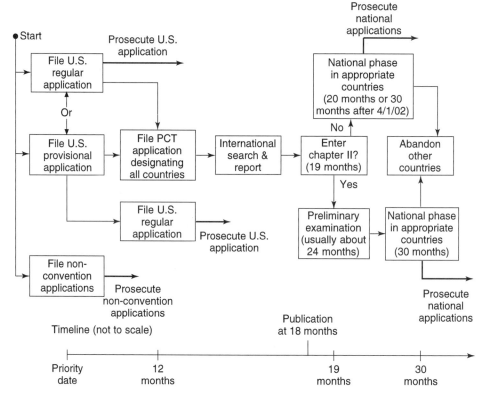

Figure 20.4. PCT practice.

tering Chapter II proceedings solely to delay entry into the national stage. Chapter II proceedings, including examination, still remain available.

By delaying the final selection of the elected countries up to 30 months after the priority date, the applicant can postpone significant fees associated with entering the national stages, including the costs associated with preparing the required foreign translations. This delay may, at least in some cases, allow the applicant to have a better understanding of the patentability, marketability, and/or commercial potential of the invention. If the viability or importance of the invention has decreased, the number of countries where the application will be pursued can be appropriately reduced (even to zero) for a significant cost savings. The PCT process, therefore, allows additional time to consider the appropriate countries in which to pursue foreign patent protection. It is important to remem-

ber when considering the PCT process that, although the number of PCT countries in which patent protection is sought can be reduced, the PCT countries in which the national stage can be entered is limited to those listed or selected in the original PCT filing.

After the PCT application has been searched and published, it is transferred to the patent offices of the individual countries designated by the applicant for entry into the national stage. In the national stage, the individual countries proceed to grant or reject the application in accordance with their specific domestic laws.

5.3 Other Aspects of Patent Laws in Other Countries

The U.S. patent system differs in a number of ways from many other national patent systems. Some of these differences are discussed above, including patentable subject matter, priority of invention, and absolute novelty.

Applicants evaluating whether and where to seek patent protection throughout the world should also be aware of working requirements and compulsory licensing requirements that are included in various forms in the patent laws of many countries. Applicants should also be aware of, and perhaps use in appropriate cases, opposition proceedings in certain countries whereby anyone, including competitors, can oppose the granting of a patent and, thereby, become involved in the patenting process.

In the United States there is no requirement to use a patented invention. The patent holder, if he or she so chooses, may prevent others from making, using, selling, or importing the invention during the entire term of the patent, regardless of whether the patent holder practices or uses the invention. In contrast, many other countries have working requirements and/or compulsory license provisions.

A typical working provision provides that a patentee can lose patent rights if (1) the patentee does not use the invention or discovery within the relevant country within a fixed time period (usually 1 to 4 years) after the grant of the patent or (2) the patentee ceases the use of the invention or discovery for a fixed time period (usually 1 to 3 years consecutively) unless the patentee can justify the cause of the inaction. Thus in many countries the patent owner must either use the invention or run the risk of losing certain rights otherwise granted by the patent. A sufficient use must be determined on a case-by-case basis in light of each specific country's working provisions. Use or manufacture on a commercial scale is sufficient in literally all cases to satisfy such working provisions. Where commercial use is not possible, production on a more limited scale with offers of sale of the product may be acceptable. Such uses should be carefully documented in the event the working of the invention is contested and must be proven. Where it is not possible to use the invention on even a limited scale, it may be possible to satisfy the working provision by offering licenses to parties within the country who would be reasonably interested in such licenses or by advertising the availability of such licenses in appropriate local or regional

media. In most cases, the working of the invention by a licensee satisfies the working requirement. In some cases, and especially where countries have entered into agreements granting reciprocal rights, working in another country may satisfy the working requirement. Generally, the greater the demand for the patented invention within the country, the more difficult it will be to justify manufacture outside the country.

Additionally, in many countries, if the invention is not adequately worked within the country or if the public interest so requires, the law provides for and requires the patent holder to grant licenses under the patent upon application. Generally, such compulsory licenses are not available until 4 years after the filing of the application or 3 years after the grant of the patent, whichever is later. In many countries, compulsory licenses can be granted, regardless of any working requirement, when it is in the public interest. Inventions relating to food products, pharmaceuticals, and health-related products can fall within this public interest provision. In general, the royalty from such a compulsory license is agreed on by the parties. Should the parties not reach agreement as to an acceptable royalty, the compulsory licensing provisions generally provide that the royalty will be determined by the government. In general, the royalty set by the government to be paid by a domestic organization to a foreign patent holder is likely to be lower than the royalty determined in an arms-length negotiation between the licensor and licensee.

Both the working requirements and compulsory license provisions vary considerably from country to country. Thus, the laws of each relevant country must be reviewed to determine the potential effects of such provisions on specific inventions and patents. Such effects should also be taken into account in determining where to seek patent protection. Even in cases for which it is unlikely that the working requirement can be satisfied and that compulsory licenses might be required, it is still generally advisable to seek patent protection. It is possible that no one will actually seek such a compulsory license. Even if a compulsory license is granted, the patent owner may be able to object on the grounds that the delay in working was unavoidable for economic reasons and thereby frustrate or delay the grant of a compulsory license for an extended period of time. The granting of a compulsory license might, of course, provide a reasonable return in the form of royalty payments. Any royalty may be better than letting a competitor freely practice the invention. Finally, especially if the grant of the compulsory license is contested, the patent holder will generally have a number of years in which to develop the market for himself or herself before the compulsory license is finalized. A competitor coming into the market at that later time may find the market more difficult to penetrate.

For many years, prosecution of a U.S. patent application was conducted in secrecy, but this rule of secrecy has been changed (236). Before this change, third parties were generally unaware that a patent application had even been filed, much less what, if any, claims would be granted until the patent actually issued. Now, patent applications in the Untied States are generally published 18 months after the priority date of the application. Publication may be avoided under certain circumstances, as discussed in Section 2.5. This change in U.S. prosecution practice conforms to the practice of most other countries. In nearly all other countries, a patent application is published 18 months after the priority date.

U.S. prosecution practice, however, still differs from that of many other countries following publication of an application or grant of a patent. In many other countries, once the patent application has been allowed by a national patent office or the patent has been granted, the application or patent may be subject to an opposition proceeding. In general, such oppositions give third parties an opportunity to bring additional prior art or other factors affecting the grant of the patent to the attention of the pertinent national patent office. Oppositions generally must be filed within a fixed time (normally 1–12 months) after publication for opposition. Once filed, however, the opposition proceeding itself may take several years and include a lengthy appeal process from the national patent office's decision. In many cases, the opposition procedure has made it even more expensive to obtain patents, especially important patents, in such countries as Japan and many European countries (e.g., Germany, Netherlands, and the Scandinavian countries). In some countries, the patent is not granted or issued and is not enforceable until the opposition proceedings are complete and the national patent office actually issues the patent. In Europe, this potential problem has been alleviated to some degree by the advent of the EPC. Although the EPC has a 9-month period in which an opposition can be brought, the opposition does not forestall the issuance of the individual national patents that, once issued, are immediately enforceable against a third-party infringer who may be opposing the patent grant in the European Patent Office. There is some movement globally to prevent or at least reduce the ability of a competitor to stall interminably the issuance of key patents until late in the patent term (usually 20 years as measured from the priority date). Of course, the desirability of curtailing oppositions and their effect will depend on which side of the fence one is on. For a competitor, delaying or even preventing the grant of such a patent may be important.

In the United States, the reexamination proceeding allows a limited form of opposition for issued patents. During a reexamination, the claims of an issued patent are examined in light of prior art that was not considered by the PTO in the original prosecution. In the past, the involvement of third parties in reexamination proceedings was strictly limited, but United States practice has been changed to allow the participation of third parties in reexamination proceedings. The reexamination process is discussed in more detail in Section 3.8 above.

6 TRADEMARKS

Although laws regulating and protecting trademarks were mainly developed in the twentieth century, trademarks have been in use in one form or another (e.g., artisan's "potter's marks") for at least 4000 years (237). The motivation for placing such a potter's mark on products probably arose from the artisan's pride in his work and the desire of individual artisans to take credit for what he had produced. For them, their marks were a means of identifying and distinguishing their products from similar works by other artisans. That concept still remains the primary function of a trademark and generally is the basis for trademark law and protection throughout the world.

Trademarks are the words, names, slogans, pictures, symbols, or designs that are used to identify the source or origin of goods. Similarly, service marks are the words, names, slogans, pictures, symbols, or designs that are used to identify and distinguish services. Trademarks and service marks are generally governed by the same legal principles. Throughout this section, references to *trademarks* or *marks* will generally include and apply to service marks as well. Trademarks perform four basic functions: (*1*) identify one seller's goods or services and distinguish them from the goods or services of others, (*2*) indicate a common source of goods or services, (*3*) indicate a certain level of quality of the goods or services, and (*4*) assist in selling the goods or services (e.g., advertising). Trademarks do not provide protection for the underlying goods or services. That is the role of patents or trade secrets.

The ability to identify and distinguish the goods of one party from those of another is an essential prerequisite and function for a trademark. Contemporary trademark laws generally do not recognize property right in a trademark per se. Rather, the "property" in which a trademark owner may claim a legitimate and protectable interest is the goodwill of the business symbolized by the trademark. It is the value of that goodwill that establishes the value and worth of the trademark. Thus a competitor should be precluded from misappropriating another's goodwill by using the trademark that symbolizes that goodwill.

In the United States, the Lanham Act (238) as amended provides the framework for registration of trademarks in the PTO and for claims for infringement of federally registered trademarks as well as related claims for specific acts of unfair competition committed "in commerce." The Trademark Counterfeiting Act of 1984 (239) provides several remedies against parties who counterfeited federally registered trademarks, including *ex parte* seizures and criminal sanctions.

6.1 Trademarks as Marketing Tools

Trademarks are uniquely suited to facilitate the marketing of drugs. The value of a trademark is measured by the goodwill that is generated from sales of products using that trademark. Goodwill, however, is dependent on the consumer's favorable perception of the goods. If the customer is satisfied with the product, he or she is likely to form a favorable impression as to products bearing the same trademark, and may look for the same mark when purchasing similar products in the future (i.e., "brand loyalty"). Such brand loyalty is difficult to establish where products are essentially interchangeable and compete mainly through price. Generally, however, it is relatively easy to establish brand loyalty for drugs and other health-related products, especially when they are patented. During the patent term for a new drug, the consumer's ability to select an alternate medication may be limited or nonexistent. Consequently, patent protection can aid in the creation of brand loyalty, which may continue after the patent expires. Furthermore, successful medications act positively and directly upon the patient; they heal, relieve pain, lower blood pressure, ease breathing, or otherwise improve the health and comfort of the patient. Each successful use may result in a positive psychological reinforcement as to the importance and value of that particular medication. That value can translate directly into goodwill. If that goodwill can be successfully linked to a trademark in the mind of the consumer or prescribing doctor, the marketing potential for that trademark can be significantly increased.

Unlike a patent, the duration of a trademark is not limited to a fixed term. Hence, a trademark can be a valuable tool to help protect and maintain a market long after patent protection has expired. Individuals accustomed to buying a trademarked product during the life of a patent (when the patent owner or his licensee may be the only one offering the product) often will continue to seek that same product after the expiration of the patent. Such product and brand loyalty, along with its associated goodwill (if carefully developed and nurtured) can represent a significant obstacle for market entry and/or penetration by others, even after the patent expires. Efforts to establish and promote the trademark undertaken during the exclusive period offered by the patent may pay dividends well into the future. Thus, in addition to seeking patent protection, drug companies should, at a very early stage, develop a marketing strategy to maximize and link the potential goodwill of a new drug with a particular trademark. Although generic drugs offered after the drug patent expires may cost less, many patients will request, and many doctors will prescribe, the trusted product they used in the past and can identify by its associated trademark.

6.2   Selection of Trademarks

Before bringing a new drug to the marketplace, careful consideration should be given to the selection of an appropriate trademark. Normally, a trademark that will not only identify and distinguish the drug in the marketplace but also secure a market advantage for the manufacturer and/or seller will be preferred. Some words, however, can be more successfully developed as trademarks than others because of their distinctiveness. There is a "spectrum of distinctiveness" for potential trademarks. At one end is the arbitrary or fanciful word, in the middle are suggestive or descriptive words relating to the product or its characteristics, and at the other end is the generic name of the product itself. An arbitrary or fanciful word is ideally suited for use as a trademark. A generic name cannot be used as a trademark for the product that it describes. Such a generic name, however, might function as a trademark for other, unrelated products. Thus, words such as *mustang, jaguar,* and *cougar* can function as trademarks for automobiles, notwithstanding their recognized meanings. For a new drug, however, it may be preferable to coin a new term or word as opposed to appropriating an existing word.

A suggestive term (i.e., one that suggests but does not directly describe something associated with the goods or services) may function as a trademark without further evidence of the public's recognition of its status as a trademark. A descriptive term (i.e., one that

directly describes something associated with the goods or services) cannot function as a trademark until and unless it is established that, in addition to its descriptive meaning, it has also acquired a "secondary meaning" (i.e., a new meaning attached to the term relating to its use as a trademark, which identifies and distinguishes the goods of a particular manufacturer or merchant) (240).

For example, generic names of animals such as mustang, jaguar, and cougar are successful as trademarks for cars because such names have connotative meanings that *suggest* that the characteristics associated with these animals might also be found in the cars themselves (e.g., speed, agility, power). However, if instead of merely suggesting the qualities or characteristics of a product, a word actually *describes* the qualities or characteristics, then it cannot be recognized as a trademark by the PTO until it has been shown to have acquired secondary meaning. Thus, *coupe, sedan, convertible,* and *fastback* are not likely candidates for acquiring secondary meaning and trademark status for automobiles because they clearly describe types of automobiles. Such terms should be free to be used by all. Other descriptive terms, however, may become legitimate trademarks by acquiring secondary meaning.

Whether a particular term is suggestive or descriptive is a gray area in trademark law. In attempting to distinguish between suggestive and descriptive terms, the PTO has indicated that

> Suggestive marks are those which require imagination, thought or perception to reach a conclusion as to the nature of the goods or services. Thus a suggestive term differs from a descriptive term, which immediately tells something about the goods or services. Suggestive marks, like fanciful and arbitrary marks, are registerable on the Principal Register without proof of secondary meaning. . . . The great variation in facts from case to case prevents the formation of specific rules for specific fact situations. Each case must be decided on its own merits (241).

The acquisition of secondary meaning for a descriptive mark usually requires a calculated effort by the mark's owner to establish a specific relationship or recognition between the mark and the particular goods or service. Thus the mark's owner can attempt through marketing techniques to form an association in the relevant marketplace between the mark and the specific goods or services for which the mark is used. For example, Owens-Corning Fiberglass Corporation employed an advertising and marketing campaign (using the Pink Panther cartoon character) for its pink-colored insulation to establish that color as an identification of its insulation. This effort to strengthen the public's recognition and association of the desired mark with the product was so successful that Owens-Corning was able to establish secondary meaning and obtain a federal registration for the color pink for its insulation product (242). Proof of secondary meaning, however, is not always easy to acquire. The Court of Appeals for the Federal Circuit held that several million dollars of advertising and many millions of dollars of sales under a particular term were insufficient to prove that the term has acquired secondary meaning in the marketplace (243). The court stressed the need for consumer surveys to measure whether the sales and advertising have been successful in creating secondary meaning for a particular mark.

Descriptive and suggestive terms are generally not good candidates as trademarks for drugs. Because a new drug often represents a new discovery, it is useful to select or develop a new word or name to identify this product to the public. The selection process of a new trademark for a new drug (or any new product) must, however, take into account the fact that the Lanham Act prohibits certain items from ever being registered as trademarks. Registered trademarks cannot include, for example, "immoral, deceptive, or scandalous" matter, state or national flags or similar insignia, names of living individuals (except by written consent) or a deceased U.S. president during the life of his widow (except by written consent), "merely descriptive or deceptively misdescriptive" names, "primarily geographically descriptive or deceptively misdescriptive" names, and surnames (244). The selection process should also take into account, especially if registrations in other countries are desired, the possible meaning of any potential mark in those other countries and languages. For example, an arbitrary and fanciful word in the English language may have a specific (perhaps negative, scandalous, or bizarre) meaning in another country or language.

A mark cannot be registered "which so resembles a mark registered in the Patent and Trademark Office, or a mark or trade name previously used in the United States by another and not abandoned, as to be likely, when used on or in connection with the goods of the applicant, to cause confusion, or to cause mistake, or to deceive" (245). Once a tentative choice of one or more potential trademarks is made, it is prudent, if not essential, to conduct a trademark search to protect the expected investment in the chosen mark. Normally, an initial computer search of the records in the PTO should be made to locate any similar marks that are registered or for which registration is pending before the PTO. This search of the PTO files is a relatively fast and inexpensive procedure by which possibly insurmountable problems (e.g., so-called knock-out marks, which likely prevent registration) can be identified. If very similar marks are not found in the PTO search, a more comprehensive search, including databases containing state trademark registrations and common law marks, is generally undertaken to provide additional security that the mark in question is available for adoption. Preferably an outside search agency conducts the search using databases containing registered and unregistered names and marks in use throughout the United States. If the proposed trademark is intended for use on products or services that may be marketed in countries outside the United States, the search should be expanded to include foreign trademark applications and registrations in such foreign countries. If no closely similar marks are found (or, in some cases, if such marks are used on very different types of goods), one may proceed to register the new mark in the PTO and, if appropriate, seek registrations in other countries as well.

6.3 Registration Process

In the United States, trademark rights arise from use of a mark, rather than from its registration in the PTO (246). However, registration in the PTO provides significant advantages to a trademark owner. Furthermore, marks can be registered based only on a bona fide intent to use the mark with a particular product or a specific service (247). Under the intent-to-use filing provisions, however, it is still necessary to establish actual commercial use of the mark before the actual registration will issue. One important benefit of the intent-to-use process is the recognition of the date of filing of an intent-to-use application as the priority date, on which trademark rights begin, rather than the date the mark is first used commercially on the product or for the service. The intent-to-use provisions can be especially useful in marketing a new drug because they provide a method for obtaining protection for a trademark before the drug actually enters the marketplace.

A trademark application must include both the mark for which registration is sought and a description of the goods and/or services upon which it is to be used. Once an application for registration of a mark has been filed in the PTO, the examination process is essentially the same whether the application is based on actual use or intent to use. A trademark examiner in the PTO reviews the application to determine whether it meets the statutory requirements. The examiner may issue a refusal to register the mark based on a failure to meet one or more of the statutory requirements. The applicant is given an opportunity to present arguments or to amend the application to overcome these objections. If the examiner finds the application to be in order, the mark will be published for opposition in the PTO's *Official Gazette*. Anyone who believes he or she may be injured by the registration of the mark may file a Notice of Opposition within 30 days of the date of publication unless an extension of time is granted.

If an opposition is not filed, an application based on actual use will mature into an issued registration or an intent-to-use application will be allowed. For an allowed intent-to-use application, the applicant has 6 months within which to establish actual use of the mark. If actual use of the mark is not established within this 6-month period, the applicant may request, upon payment of required fees, extensions of time (in 6-month increments up to a total of five extensions). If after 36 months from the date of allowance, actual

use of the mark in commerce has not been made, the application will be deemed abandoned. Thereafter, a new application for registration of the same mark may be filed, but the earlier priority date will be lost.

The initial term of a registered mark is 10 years. Between the 5th and 6th years of the initial term, the registrant must file an affidavit or declaration stating that the mark is still in use with the goods specified in the registration. Failure to timely file this affidavit or declaration will result in the cancellation of the registration. If the registrant files an affidavit or declaration confirming, in addition to the mark still being in use, that it has been in continuous use for the preceding 5 years, the registration becomes "incontestable" (248).

Trademark rights in the United States arise from use and not from the registration of the mark. However, registration of a mark in the PTO provides the mark's owner many significant procedural advantages. First, a certificate of registration of a mark on the principal register constitutes "*prima facie* evidence of the validity of the registration . . . and the registrant's exclusive right to use the registered mark in commerce on or in connection with the goods and services specified in the certificate" (249). The registrant's rights are not limited geographically to the locations in the United States where the mark has actually been used. The registrant's "exclusive" rights extend throughout the United States (and territories and possessions). Moreover, if the registration has become "incontestable" based on a consecutive 5 years of continuous use after registration, then the registrant's "exclusive rights" are incontestable except upon certain limited grounds specified in the Lanham Act (250) and the incontestable registration constitutes "conclusive evidence" of the registrant's "exclusive rights" in the registered mark (251). The registration is also "constructive notice of the registrant's claim of ownership" throughout the United States. Constructive notice means that, after a mark has been registered, others have legal notice of the registrant's trademark rights. even if they are not aware of the registration (252). Constructive notice prevents later users of a mark from relying on the common law defense of "good faith" adoption of the mark. To take advantage of this constructive notice provision, the registrant must use the mark with the proper registration notation (253).

These procedural advantages are of considerable importance should the registrant attempt to enforce the trademark rights in court. The federal courts, rather than state courts, have original jurisdiction over causes of action arising from federally registered marks (254). Furthermore, a registrant may claim an earlier priority date based on an intent to use application. The ability to claim an earlier priority date may be critical in a contest between rival claimants of the same or similar marks. A single day of priority over a junior party has been held sufficient to sustain a senior party's rights and require the junior party to discontinue further use of his or her mark (255).

Registration also provides a relatively simple mechanism to prevent the importation of goods bearing a counterfeit or infringing copy of the mark into the United States. Upon the filing of a certified copy of a trademark registration with the U.S. Customs Service and payment of the required fee, the Customs Service will undertake to exclude entry into the United States of products bearing an infringing copy of the Registrant's mark (256). The presumed ability to determine infringement (i.e., that the alleged infringing mark is confusingly similar to the registered mark) by direct visual inspection allows for the simplified procedure.

Finally, registration of a trademark entitles the registrant to both injunctive and monetary relief for trademark infringement. Under the Lanham Act, monetary relief can include (1) the profits that an infringer has earned by marketing products under the infringing trademark and (2) any damages that have been sustained by the registrant from the infringement (257). Moreover, "[i]n assessing profits [of the infringer] the plaintiff [registrant] shall be required to prove defendant's sales only; defendant must prove all elements of costs or deduction claimed" (258). Similarly, the court may increase the award of damages to the registrant by three times the amount of damages proven at trial (259) and, in exceptional cases, the court may award attorney fees to the prevailing party (260). The

court may also order "that all labels, signs, prints, packages, wrappers, receptacles, and advertisements in the possession of the defendant bearing the registered mark . . . shall be delivered up and destroyed" (261).

In an action for trademark infringement, the issue is whether a mark used by the defendant with particular goods or services is likely to cause confusion, mistake, or deception of the public. Generally, the best evidence of likelihood of confusion is that which shows substantial actual confusion. However, actual confusion is not necessary to prove infringement of a federally registered trademark: "It has been said that the most successful form of copying is to employ enough points of similarity to confuse the public with enough points of difference to confuse the courts" (262). The registration of a trademark in the PTO considerably strengthens trademark rights and makes them considerably easier to enforce.

6.4 Oppositions and Cancellations

Once the mark is published in the *Official Gazette*, anyone believing that he or she will be injured by the issuance of the registration may file a Notice of Opposition within 30 days following publication. Additionally, anyone who believes he or she may be injured by maintenance of a registration on the Principal Register may file a Petition for Cancellation of that registration. Once either a Notice of Opposition or Petition for Cancellation is filed, the application is transferred to the Trademark Trial and Appeal Board for a determination of the merits of the opposition or cancellation request. Opposition and cancellation proceedings are, in many ways, procedurally similar to a court trial. The only issues before the Board, however, are whether an applicant should be allowed to register the mark or whether a registrant can maintain the registration for the mark. Either party can appeal the Board's decision to the Court of Appeals for the Federal Circuit on the record established before the Board or to a federal district court for a *de novo* review of the Board's decision.

6.5 Preserving Trademark Rights through Proper Use

Trademark rights can be diminished and even destroyed through improper use. A trademark is to be used as an adjective modifying the generic name of a product. The trademark owner should never use the trademark as the name of the product itself. The trademark owner should also attempt to prevent others from doing so.

Once established, the trademark must be protected and its usage carefully monitored and controlled. A trademark (even if arbitrary or fanciful when first coined) can be lost if it becomes a descriptive or generic name for the product itself and thus fails to identify the particular product offered by the trademark owner. The original *aspirin* trademark lost its status as a trademark in the United States when it lost its distinctiveness in identifying the particular Bayer Company product. The general public came to regard this term as identifying the type of drug rather than distinguishing the product of Bayer. Consequently, the term *aspirin* was held to have fallen into the public domain in the United States and not the exclusive property of Bayer (263).

In addition to proper use through its own advertising and marketing, the trademark owner should police the use of the mark by others. For example, the trademark owner may wish to subscribe to a service that will search various media databases on a routine basis for improper uses of the mark. Or, the trademark owner can carry out the search on its own. When improper uses are found, a letter or other notification can be sent explaining why the use is improper and requesting proper usage. Depending on the degree of misuse and the importance of the mark, other mechanisms, including, for example, advertising and marketing campaigns, might be used. Documentation of such policing should be maintained in the event it becomes necessary during litigation or cancellation proceedings to prove that the trademark owner acted in a proper and prudent manner to protect the trademark.

When used, the mark should be set apart, preferably in bold type, from the other words around it. If the mark has not been registered in the PTO, it should be followed whenever possible by the designation "TM"; if it has

been registered, it should be followed by the proper registration notice:

> [A] registrant of a mark registered in the Patent and Trademark Office, may give notice that his mark is registered by displaying with the mark the words "Registered in U.S. Patent and Trademark Office" or "Reg. U.S. Pat. & Tm. Off." or the letter R enclosed within a circle, thus ®; and in any suit for infringement under this Act by such a registrant failing to give such notice of registration, no profits and no damages shall be recovered under the provisions of this Act unless the defendant had actual notice of the registration (264).

6.6 Worldwide Trademark Rights

It is impossible here to discuss in any detail the plethora of laws of other countries regarding the establishment and protection of trademark rights, but a few comments may be in order. As noted above, when selecting a trademark it is prudent to verify whether the desired mark has some meaning in another language or sounds similar to a word in another language. In a few instances, trademark owners have discovered to their chagrin that an English or made-up word used as a trademark is so similar to a word in another language having a negative or otherwise inappropriate connotation that it is impossible to use the trademark in some countries. It is far better to discover such problems before significant resources are expended to develop the goodwill associated with that mark.

In the global marketplace, trademark protection normally will be required in various countries throughout the world. However, like patent protection, trademark protection is limited geographically. Determining the appropriate countries in which to seek trademark protection should be an integral part of the intellectual property strategy. Because of the generally lower cost of obtaining trademark protection, it may be advantageous to obtain such protection in a larger number of countries than where one might seek patent protection. As a general rule of thumb, trademark protection should be perfected, at a minimum, in every foreign country in which patent protection is sought. Similarly, if licensing in other countries is contemplated, registration of the mark in those countries is advised. Licensees are likely to be interested in both the patent and trademark rights. It may also be appropriate to obtain trademark registrations in countries having potentially significant markets, even though patent protection will not be sought and/or licenses will not be granted. It is prudent to ensure that a well-known drug can be exported into such countries under its trademark, and that someone else does not acquire the rights to market a drug or other health-related product under that trademark.

When and to what extent trademark rights should be acquired in countries throughout the world must necessarily be determined on a case-by-case basis. In countries outside the United States, trademark rights generally arise from registration rather than from use of the mark. It is therefore prudent to file applications for registration of a selected mark as soon as practicable in those countries where substantial use of the trademark is anticipated. Therefore, when a drug is ready to be announced to the public under a particular trademark, even if it is not anticipated that sales under the mark will begin for quite some time, the strategy for establishing trademark rights in other countries should generally already be in place. Unfortunately, it is not uncommon for unrelated parties, upon learning that a new product will be marketed under a particular mark, to attempt to register the mark in at least some countries in advance of the originator of the trademark. The unrelated party could, for example, hold the mark for ransom or transfer the rights to the mark to a local company for marketing similar products in that country under the mark. Although redress may, in some instances, be achieved through the courts in appropriate countries, this can be expensive and may involve many years of litigation. Thus, in considering the development of trademark rights for new drugs in foreign countries, an "ounce of prevention" may be worth considerably more than several "pounds of cure."

6.7 Other Rights under the Lanham Act

The Lanham Act in § 43(a) also creates a federal cause of action for "false designation of origin and false description of goods" (265).

Courts have construed § 43(a) as regulating any act or representation that might cause the purchasing public mistakenly to believe that a product originating from one manufacturer or merchant originated from some other manufacturer or merchant; § 43(a) now forms the basis of the federal law of unfair competition.

Under § 43(a), some characteristics of a product can be protected from imitation by others. Thus, for example, the color, shape, and size of pills have been held to be protectable. The court allowed a drug company, even after the patent on the drug expired, to market the particular drug exclusively in capsules of a particular color, shape, and size (266). Other drug companies could, of course, market generic forms of the drug after the patent had expired, but not in the format in which the public had come to know and recognize the product. Thus, the exclusive right to market a drug in capsules of the same color shape and size that the public has come to associate with "the authentic product" can be a protectable property interest and a valuable marketing tool. Section 43(a) is not restricted to registered trademarks. Rather, it embraces the broad panoply of "trade dress" for a product (i.e., the total visual combination of elements in which a product or service is packaged and offered to the public). The development of valuable rights that may be protected under this section depends primarily on how the product is marketed and presented to the public.

The protection offered by § 43(a) was expanded further by the Trademark Revision Act of 1988 to cover false representations made in regard to another's goods. Section 43(a) provides that any person who

> in commercial advertising or promotion, misrepresents the nature, characteristics, qualities or geographical origins of his or her *or another person's goods* or services or commercial activities shall be liable in a civil action by any person who believes that he or she is or is likely to be damaged by the act (267).

Thus, false or misleading statements about one's own goods or another's goods can give rise to a cause of action.

7 TRADE SECRETS

Information that has value because it is not generally known (e.g., business and technical, patentable and nonpatentable) can be protected as a trade secret against discovery by improper means or through breach of confidence. To be protected as a trade secret, the information must in fact be "secret" (i.e., not generally known by the industry). The duration of a trade secret is the length of time the information is kept secret. Public disclosure of the information by the trade secret owner or anyone else results in the loss of the protection offered by the trade secret.

The practical role of trade secrets in the drug development and discovery industry may, of course, be significantly limited by the public disclosure of information and data required by the FDA approval process, publication of pending patent applications, and the large number of groups working in this area, who may independently discover the secret. Information or data that the FDA publicly discloses or otherwise makes available to the public loses its status as a trade secret. However, where trade secret protection is not available because of the inability to maintain secrecy, patent protection may be the only viable form of protection available.

Although trade secret law may provide only limited and short-term protection for technical and business information in the drug discovery and development area, such protection can be a valuable component of an overall intellectual property strategy. Trade secret protection can be especially important in protecting technology at its earliest stages of development (e.g., before publication of pending patent applications and/or release of data and information by the FDA). Of course, trade secrets can be used to protect technical or business information that is not disclosed to the FDA (or, if disclosed, not released to the public by the FDA) or to the public through, for example, published patent applications or other publications. In such cases, trade secrets can provide a viable alternative and/or adjunct to obtaining patent protection, especially where the patentability of an invention is in doubt. Like patents and trademarks, trade se-

crets and general technical know-how can be sold outright or licensed.

Relying too heavily on trade secret protection to protect intellectual property, however, has significant limitations and risks. The cornerstone of a trade secret is secrecy. Once secrecy is lost (regardless of how it is lost), the protection afforded by trade secret law is lost and competitors are free to use the technology. To reiterate, the duration of a trade secret is the length of time the information is kept secret. Furthermore, even if secrecy is maintained, the technology can be used by others as long as it is discovered in a fair and honest manner (e.g., independent discovery or reverse engineering). Thus, it may be proper for a competitor to analyze a new drug and, based on the information obtained, seek FDA approval to market the drug (assuming there are no blocking patents and the drug sample was obtained properly). Public disclosure of the information by the trade secret owner or others (even by one who improperly obtained and/or disclosed the information) effectively terminates the protection offered by the trade secret. Therefore, the scope of protection available and risks associated with trade secrets must be carefully considered and evaluated, and procedures defined and implemented to protect and maintain the required secrecy, before significant reliance is placed on trade secret protection as an alternative to patent protection. This is especially true for the drug discovery and development industry, where detailed technical disclosures to the FDA are generally required. Ideally, patent protection and trade secret protection should be carefully coordinated to provide maximum protection.

7.1 Trade Secret Definition

Trade secret protection is generally governed by state law; thus, its definition can vary from state to state. One common definition provides that a trade secret consists "of any formula, pattern, device, or compilation of information that is used in one's business, and that gives him an opportunity to obtain an advantage over competitors who do not know or use it" (268). Another definition, which a significant number of states (269) have adopted in some version, is provided by the Uniform Trade Secrets Act: "[I]nformation, including a formula, pattern, compilation, program, device, method, technique, or process that: (1) derives independent economic value, actual or potential, from not being generally known to, and not being readily ascertainable by proper means by, other persons who can obtain economic value from its disclosure or use, and (2) is the subject of efforts that are reasonable under the circumstances to maintain its secrecy" (270). Therefore, trade secrets generally can include formulae for chemical compounds and drugs; processes for manufacturing, treating, and preserving materials; patterns and designs for a machine or other device; computer software; business strategies and plans; customers lists; and similar business and technical information having economic value.

7.2 Requirements for Trade Secret Protection

For a protectable trade secret to exist, generally it must meet four interrelated criteria: (1) it must be the proper subject matter for a trade secret (i.e., it must fall within the type of information protectable as a trade secret), (2) it must not generally be known in the trade (i.e., it must be a secret), (3) it must be of commercial value to the holder, and (4) it must be treated and maintained as a secret. Although the trade secret must be secret, novelty in the patent sense is not required. Thus, an obvious improvement in a drug manufacturing process, which could not be protected by a patent, could be retained as a trade secret so long as it is not known to others in the industry (assuming the other criteria are met). The third requirement, commercial value, is generally met if knowledge or use of the trade secret by the holder provides some competitive advantage. The fourth criterion essentially requires that the trade secret holder treat the information in an appropriate manner (i.e., reasonable measures must be taken to keep and maintain the information as a secret). The efforts to maintain secrecy will vary with the information and the financial resources of the organization. At a minimum, such reasonable efforts should include limiting access to the information to key employees who have a need to know, having employees sign confidentiality agreements, and alerting employees about the status of the sensitive information that is considered to be a trade secret (e.g., appropriate labeling of documents as *confidential*, storing such documents in a secure manner, and marking process areas that are off-limits). Disclosure to outsiders should be limited to that necessary for business reasons and should be carefully controlled; such disclosure generally should be through confidentiality agreements.

In addition to the general criteria above, courts have used a number of specific factors in determining the existence of a trade secret. Some of these factors include (1) the extent to which the information is known outside of one's business; (2) the extent to which it is known by employees and others involved in the business; (3) the extent of measures taken to guard the secrecy of the information; (4) the value of the information to the trade secret holder and, potentially, to competitors; (5) the amount of effort or money expended in developing the information; and (6) the ease or difficulty with which the information could be properly acquired or duplicated by others (271).

In setting up a program to protect trade secrets, these factors should be carefully considered to maximize the probability of a court later finding that a protectable trade secret does in fact exist. For example, documents containing trade secrets should be labeled *confidential* or with a similar notation. Overuse of a *confidential* stamp, however, should be avoided. If all documents are routinely labeled *confidential* without regard to the trade secret content, a court might later determine that employees were not properly informed of the trade secrets or that trade secrets were treated no differently than other information. Moreover, if all documents are marked *confidential*, employees may not treat the trade secrets with the appropriate care, thereby increasing the risk that actual secrecy will be lost. In some instances, several classifications of information with varying degrees of control might be appropriate.

It is generally desirable to have a comprehensive and documented program for protection of trade secrets. This program can be invaluable in maintaining a competitive advantage in the marketplace, as well as providing a means to demonstrate to a court that protectable trade secrets existed and were treated in the appropriate manner. Such a program should have a mechanism for identifying trade secrets and then protecting and maintaining them. This trade secret program can form an integral component of an overall security program to maintain patentable inventions as secrets until the appropriate patent applications are filed. Such a program can be implemented through an intellectual property committee responsible for general intellectual property matters.

7.3 Enforcement of Trade Secrets

Trade secrets generally protect only against wrongful disclosure or discovery of information by competitors or others. Thus, one might have a cause of action against, say, an employee who leaks information to a competitor, or against a competitor who discovers a trade secret through improper industrial espionage (e.g., bribing an employee to disclose a trade secret or by breaking into a computer system or facility), or against one who improperly obtains and/or uses a trade secret (e.g., misrepresentation or breach of an implied or express confidentiality agreement). In addition, in appropriate cases, one might bring suit to prevent the improper disclosure of a trade secret (272). For example, a key employee, who resigns to join a competitor, might be enjoined from disclosing trade secrets of his or her former employee to the new employer.

Not all means of discovering a trade secret are actionable. For example, it is acceptable to learn of the trade secret by independent discovery, by reverse engineering, or by evaluation of products or data available publicly. Thus, for example, a trade secret holder would not have a cause of action against a competitor who independently discovers the trade secret. In addition, one who properly obtains the trade secret without any obligation to maintain the trade secret in confidence is free to use it and, if desired, disclose it to the public, thereby destroying the original trade secret status. Indeed, one who independently and properly discovers an invention held as trade secret by another may be able to obtain a patent covering the invention. In such a case, the potential rights of the patentee and the

trade secret holder relative to each other appear to remain an unresolved question (273).

Remedies for misappropriation of a trade secret can include actual and punitive damages as well as injunctive relief. An injunction may only prevent the wrongdoer from using the trade secret information for a fixed length of time. Some courts will limit the length of the injunction to the estimated time it would take a hypothetical competitor to discover the trade secret by reverse engineering (a so-called lead-time injunction). Only the wrongdoer may be prevented from using the trade secret. Other competitors, as well as the general public, are generally free to use the trade secret to the extent that it has been publicly disclosed.

Although trade secrets potentially offer protection for an unlimited duration (i.e., so long as secrecy is maintained), in practice the time of protection is often relatively brief. One estimate is that most trade secrets have an average life expectancy of about 3 years (274). Because of the intense competition, employee mobility, and FDA disclosure requirements, the lifetime of an average trade secret in the drug discovery and development industry may be even shorter.

Even within such a short lifetime, however, trade secrets remain a useful adjunct to patent protection. For example, trade secret protection can be used to protect an invention before filing a patent application and while the application is pending before the PTO up until the time of publication. Trade secrets may also be used to protect later improvements in patented processes or materials that do not, in themselves, warrant filing separate patent applications.

7.4 Relationship of Trade Secrets and Patents

Trade secret and patent protection have coexisted in the United States for more than 200 years. The U.S. Supreme Court in 1974 made it clear that federal patent law does not preempt state trade secret law (275). Nonetheless, the disclosure requirement of patent law and the secrecy requirement of trade secret law are often in conflict. The Patent Statute requires a patent specification that teaches one of ordinary skill in the art how to make and use the invention and that discloses the best mode of carrying out the invention known to the inventor as of the application filing date. Any trade secrets disclosed in the patent specification lose their status as trade secrets once the patent application is made public (published or issued as a patent). Failure to disclose a trade secret in an application where the trade secret is necessary for enablement or best mode considerations will result in an invalid patent.

Although the issue is easily stated, it is considerably more difficult in practice to determine which trade secrets relating to an invention must be disclosed. Clearly an applicant should not attempt to obtain patent protection for an invention while seeking to keep the commercial embodiment (the best mode) as a trade secret. Yet as noted, the best mode requirement relates to the applicant's knowledge at the time the application is filed. Improvements made before the filing date may be required to be included in the original application. Such improvements made after the filing date, even if they constitute a better method of practicing the invention, can be retained as trade secrets. Improvements made after the filing date of the original application may, however, have to be disclosed in subsequent CIP applications adding new matter to the specification (276).

Generally, it is not necessary to disclose trade secrets that are related to the invention but are not required for its operation and are not related to the best mode of operation known to the applicant as of the filing date of the patent application. Of course, by not disclosing related trade secrets in a patent application, one runs the significant risk that a court may later hold the patent invalid or unenforceable for failure to provide an enabling specification or to disclose the best mode. In a close case, it may be preferable to err on the side of disclosing more than the required minimum. After all, trade secrets, for the most part, have only a limited lifetime, especially if disclosed in submissions to the FDA or other federal agencies.

7.5 Freedom of Information Act

So-called sunshine-type laws, including state and federal Freedom of Information Acts (FOIAs) and state right-to-know laws, can significantly impact the ability of pharmaceutical and drug discovery companies to retain the secrecy required for viable trade secrets. The general purpose of these laws is to increase the openness of governmental processes and decision making. Yet release of information by the government under FOIA can destroy valuable trade secrets rights. Some of the information submitted to government agencies will be routinely released to the public as part of the functioning of the agencies. Other information may be released based on specific requests by members of the public.

The federal FOIA mandates disclosure of official information of the administrative agencies of the federal executive branch (including, e.g., FDA and EPA) unless the information falls within one of the nine statutory exemptions (277). FOIA provides that "each agency, upon any request for records that (i) reasonably describes such records, and (ii) is made in accordance with published rules stating the time, place, fees (if any), and procedures to be followed, shall make the records promptly available to any person" (278). Thus, any member of the public, including domestic or foreign competitors, can obtain records through the FOIA. Such records can include drug-related submissions to FDA and identifications of new chemical compounds submitted to EPA.

Under FOIA, the burden of proof for withholding information is on the government agency having possession of the information. Potentially a government agency can rely on two FOIA exemptions to withhold trade secret-type information. FOIA exemption 3 generally allows an agency to withhold information exempted from disclosure by another statute, "provided that such statute (A) requires that the matters be withheld from the public in such a manner as to leave no discretion on the issue, or (B) establishes particular criteria for withholding or refers to particular types of matters to be withheld" (279). This provision, taken together with the Trade Secrets Act (280), may provide a basis for exempting trade secrets from disclosure under FOIA. Exemption 4 provides that the disclosure requirements of FOIA do not apply to "trade secrets and commercial or financial information obtained from a person and privileged or confidential" (281). On their face, these exemptions appear to provide considerable protection against public disclosure for trade secrets disclosed to government agencies such as FDA and EPA. The courts, especially the U.S. Court of Appeals for the District of Columbia, have tended to read the exemptions narrowly. Furthermore, the U.S. Supreme Court has held that FOIA exemptions are permissive rather than mandatory:

> FOIA by itself protects the submitters' interest in confidentiality only to the extent that this interest is endorsed by the agency collecting the information. Enlarged access to governmental information undoubtedly cuts against the privacy concerns of nongovernmental entities, and as a matter of policy some balancing and accommodation may well be desirable. We simply hold here that Congress did not design the FOIA Exemptions to be mandatory bars to disclosure (282).

Therefore, an agency retains the discretion to disclose information that falls within the exemption. Agencies may tend to grant more liberal disclosure simply to avoid lawsuits by requesters seeking to compel disclosure (283).

The North American Free Trade Agreement (NAFTA) may also significantly affect trade secrets that are disclosed to the FDA or other regulatory agencies for product approval. Under the Treaty, member countries are required to maintain data submitted for product approval to governmental or administrative agencies confidential where the data involved "considerable effort" unless disclosure is needed to protect the public interest (284). Such data cannot be used by competitors for a reasonable period of time (i.e., defined as not less than 5 years from date of product approval) after submission (285). The full effect of NAFTA on trade secrets and their potential disclosure under FOIA must await the adoption of statutes and regulation implementing the Treaty. Drug discovery organizations would be well advised to kept abreast of developments associated with, or resulting from, NAFTA that relate to trade secrets in general and, more specifically, the FDA's drug approval procedures. Where appropriate, such

organizations should modify their treatment of trade secret information accordingly.

The details and nuances of the law governing FOIA disclosure are beyond the scope of this chapter. However, any submitter of information and data, especially information and data involving drug discovery and development relating to the public health, should realize (and perhaps expect) that government agencies may at some time release information or data either to the public at large or to individuals or organizations that submit specific requests. This possibility increases the importance and significance of patent protection, if appropriate, relative to trade secret protection. The intellectual property strategy devised for the drug discovery and development organization should take this factor into account.

7.6 Trade Secret Protection Outside the United States

It is impossible here to discuss in any detail the protection afforded to trade secrets in other countries. Moreover, any details provided could be out of date in a relatively short time as new cases are decided or new statutes are adopted. A few general comments, however, may be in order.

Protection for trade secrets in other countries varies dramatically, ranging from essentially none or very little up to, and even exceeding, the level of protection provided in the United States. Therefore, local counsel in the relevant country should be consulted in the event that trade secret or related issues arise. A few examples for selected countries, however, can be helpful to illustrate the scope of variations for the protection for trade secrets (286).

1. *Australia.* In Australia, trade secret protection is based on common law (both English and Australian). The subject matter of the trade secret must relate to a trade. Portions of the trade secret may be known, but the overall result must not be known or achievable to the public. Protection can be based on breach of an express or implied contract or breach of a confidential relationship. Remedies can include injunctions

(preliminary or permanent); damages; profits; and in appropriate cases, destruction of property embodying the trade secret.

2. *France.* Trade secret protection per se does not exist in France. However, the combined protection afforded to manufacturing secrets, commercial secrets, and "know-how" is similar to trade secret protection. To be eligible for such protection, the information must not be known by others in France (287). Manufacturing secrets must actually be used, or be ready for immediate use, in industry; thus, ongoing research and development information or data may not qualify as a manufacturing secret. Protection is also afforded to know-how; technical information that will ultimately be used in industry may be protected as know-how. Commercial secrets include commercial and financial information. Remedies for improper use include damages, injunctions, and specific performance.

3. *Germany.* In Germany, protection is generally afforded to "industrial and commercial secrets" that relate to a business enterprise. The information must not be generally known or available and the holder of the information must intend to maintain its secrecy. Remedies include damages, criminal sanctions, and injunctions.

4. *Italy.* Legal protection is generally afforded to commercial and industrial secrets. To be protected, the secret must meet the following requirements: (*1*) it must not be known or readily available to competitors or others in Italy; (*2*) there must be an objective, justifiable economic reason for maintaining secrecy; and (*3*) the holder must have taken adequate steps to maintain secrecy. Manufacturing secrets must be connected with a manufacturing or production activity; commercial secrets relate to other activities of the organization. Remedies include damages, injunctions, declaratory relief, and removal of the effects of unfair competition.

5. *Japan.* In Japan, there is generally no protection afforded trade secrets. However, expressed contracts or agreements not to dis-close specific information will generally be enforced. Remedies for violating such agreements include damages and injunctive relief.

6. *Mexico.* Currently, the protection afforded trade secrets in Mexico is relatively limited and weak. The situation is likely to change because of the recent adoption of NAFTA. NAFTA, covering the United States, Canada, and Mexico, requires each country, at a minimum, to provide "adequate and effective protection and enforcement of" intellectual property rights, specifically including trade secret rights. The actual effect of NAFTA must, of course, await its actual implementation in the member countries.

7. *United Kingdom.* Although there appears to be no generally accepted definition of trade secrets in the UK, protection is generally afforded to commercial, industrial, and scientific information that is not generally known and that is capable of industrial or commercial application, where the holder has acted in a manner consistent with an intent to keep the information secret. Remedies include injunction, damages, accounting and profits, and inspection of the defendant's premises for materials relating to the trade secret.

Even in countries that do not afford significant protection through laws directly covering trade secrets, protection may be possible through contracts or agreements to protect confidential information disclosed to other parties. Preferably and wherever possible, any required disclosure of the trade secrets or other confidential information to third parties (e.g., to employees, potential business partners, vendors, and the like)—whether in the United States or elsewhere—should be made using confidentiality or secrecy agreements between the parties. Improper disclosure of such trade secrets or information likely would be a breach of contract. In countries not recognizing trade secrets or offering little trade secret protection, redress for improper disclosure of a trade secret could be sought under contract law. In countries offering significant protection for trade secrets, redress could be sought under contract law and/or trade secret law.

Trade secret holders interested in using their trade secrets throughout the world must protect their secrecy of the relevant information in each and every country in which it is used. The loss of secrecy anywhere in the world can affect, and often destroy, the trade secret around the world. Perhaps even higher safeguards could be maintained in countries that do not offer adequate trade secret protection because such countries might provide ideal havens for individuals or organizations seeking to discover trade secrets for their own use or for sale to others. Even where redress may be obtained in local courts for improper use or disclosure of a trade secret, such litigation can be expensive, and it is unlikely that a damage award could be obtained that would reasonably compensate the trade secret holder for loss of his or her trade secret throughout the world.

8   OTHER FORMS OF PROTECTION

Other forms of protection for intellectual property that are available in the United States include copyrights, statutory invention registrations, and design patents. These methods of protecting intellectual property generally have only limited applicability to drug discovery and related technology. Copyrights, for example, protect a work of authorship; they do not protect inventions such as new drugs, diagnostic assays, or methods of treatment. Thus, although the copyright owner may prevent others from making copies of, for example, a published article, manual, pamphlet, or computer program, he or she cannot prevent others from using the ideas or data contained therein. Copyrights can be useful in drug discovery by protecting printed materials such as advertising, manuals, pamphlets, computer programs, and the like. For instance, a computer program useful in DNA sequencing can be protected by copyright, even though the ideas contained in the computer program and the actual DNA sequences determined using the program cannot be protected by copyright. A copyright is created

once the work of authorship is produced in any tangible form. Although not required, registering the copyright in the Copyright Office of the Library of Congress provides certain advantages should it become necessary to enforce the copyright (288).

Limited protection can also be provided by the statutory invention registration (SIR) administered by the PTO (289). This procedure provides for a patentlike document that officially and affirmatively places the invention in the public domain for defensive purposes. The SIR is essentially a defensive publication that can be used when the inventor does not wish to obtain patent rights, yet wishes to be free of any later patents by others claiming the same invention. The SIR is treated in the same manner as a patent for defensive purposes, that is, both as prior art and as establishing a constructive reduction to practice in interference proceedings. The SIR cannot be used to prevent others from making, using, offering to sell, selling, or importing the disclosed invention or inventions. The publication of the invention in a trade or scientific publication or other media has essentially the same defensive effect as a SIR. However, the effective date of a publication is generally the actual date of publication; for the SIR, the effective date is the filing date (290). The SIR procedure is often used to disclose work done at federal research agencies for which agency does not wish to seek patent protection. Anyone, however, can use the SIR procedure. The SIR program should be considered as an alternative to publication in a technical journal when one has determined not to seek patent protection for a specific invention, but wishes to prevent others from obtaining patents covering that invention (291).

Design patents can be used to protect the ornamental design or aspect of a useful product (292). A design patent grants the holder the exclusive right to exclude others from making, using, and selling designs closely resembling the patented design. To be eligible for protection in this manner, the design must be novel, nonobvious, and ornamental. For example, a design patent could cover an ornamental packaging design for a drug or diagnostic kit or an ornamental design for a pill or capsule. The term of a design patent is

14 years from the date of grant. Thus, where possible and appropriate (e.g., for a unique pill or capsule design), trademark protection may be preferred because its duration is generally limited only by continued use (see section 6). Design patents do not have claims like utility patents. Rather, a design patent contains one or more drawings that define the scope of protection. The drawings are compared to the appearance of an alleged infringing product. Infringement is found "if, in the eye of an ordinary observer, giving such attention as a purchaser usually gives, two designs are substantially the same, if the resemblance is such as to deceive such an observer, inducing him to purchase one supposing it to be the other, the first one patented is infringed by the other" (293). Generally, design patents are governed by the same rules for validity as utility patents. However, they cover very different aspects of a given product: a utility patent relates to the functional aspects, whereas a design patent relates only to the ornamental aspects of an article. Thus, it is possible to have a utility patent and a design patent covering the same product. For a drug manufacturer, design patents might be used to help establish brand loyalty and recognition in a manner similar to trademarks.

9 CONCLUSION

Careful use of the intellectual property system, especially the patent system, in the United States and elsewhere in the world can enable those in the drug discovery and biotechnology industry to protect the fruits of their labor. By making effective use of the legal protection afforded by the intellectual property laws, a drug developer can protect its investment, enhance the value of its technology, and earn a profit sufficient to allow further research into improving existing drugs and therapies as well as developing new drugs and therapies. Indeed, by providing such protection, the intellectual property system seeks to encourage the development of new and useful technology and products. For any industry, attention to the protection of intellectual property at the earliest stages of its development is of the utmost importance. This espe-

cially applies to the drug discovery and biotechnology industry, however, because of the rapidly developing nature of the technology and the FDA submission requirements. The formation of an intellectual property committee that oversees the protection process and offers overall guidance in the development of the intellectual property strategy, on a continuing basis, is highly recommended. The organization and the intellectual property committee should work closely with qualified legal counsel, preferably patent counsel skilled in the relevant technology, to fashion internal mechanisms for protecting the technology, for seeking the appropriate legal protection, and for developing appropriate enforcement policies. In this manner, one can obtain the appropriate legal protection that the particular technology demands and deserves.

ENDNOTES

1. For following up on specific points, more detail is available in several excellent treatises devoted to intellectual property. See, e.g., D. Chisum, Patents (1990); I. P. Cooper, Biotechnology and the Law (2001); J. Rosenstock, The Law of Chemical and Pharmaceutical Invention (1993); M. Adelman, Patent Law Perspectives (2001); K. Burchfiel, Biotechnology and the Federal Circuit (1995); P. Rosenberg, Patent Law Fundamentals (1989); S. Ladas, Patents, Trademarks and Related Rights: National and International Protection (1975); E. B. Lipscomb, Lipscomb's Walker on Patents (3d ed.; 1986); J. McCarthy, Trademarks and Unfair Competition (2001); M. F. Jager, Trade Secret Law (2001); R. M. Milgrim, Milgrim on Trade Secrets (2001).

2. 35 U.S.C.A. § 103 (West Supp. 2001); see also Standard Oil Co. v. American Cyanamid Co. 774 F.2d 448, 454 (Fed. Cir. 1985).

3. Congress has attempted to alleviate this problem somewhat with the Orphan Drug Act, which provides, with some limitations, an exclusive 7-year right to market a drug for treatment of a disease affecting less than 200,000 individuals or for which there is "no reasonable expectation" that the developmental costs of the drug can be recovered through sales in the United States. 21 U.S.C.A. §§ 360aa–360ee (West Supp. 2001).

4. In the United States, patent applications must be filed in the name of the inventors. Inventors

may assign their rights in the patent application and any patents that may issue therefrom to third parties (i.e., assignees).

5. 35 U.S.C.A. § 112 (West 1984).

6. An export license is generally required to export technology developed in the United States. 35 U.S.C.A. § 184 (West Supp. 2001); 37 C.F.R. § 5.11 (2001); see also 22 C.F.R. parts 120–130 (1993) (International Traffic in Arms Regulations of the Department of State); 15 C.F.R. part 700 (2001) (Export Administration Regulations of the Department of Commerce); 10 C.F.R. part 810 (2001) (Foreign Atomic Energy Activities Regulations of the Department of Energy). Thus, a patent application for an invention made in the United States generally cannot first be filed in another country with a lesser enabling requirement unless the appropriate foreign filing license is obtained.

7. U.S. Const. art. I, § 8, cl. 8.

8. Applicants can now file a provisional application (described in more detail in section 3.4). Such a provisional application cannot be converted into a utility patent application or mature into a utility patent. A utility patent application can be filed within 1 year of the provisional application and claim benefit of the provisional application filing date; such a utility patent application can, of course, mature into a utility patent.

9. The United States has adopted a 20-year patent term for all applications filed on or after June 8, 1995. Before this time, the patent term was 17 years as measured from the actual issue date. For patents issuing from applications filed before June 8, 1995, the patent term is either 17 years as measured from the issue date or 20 years measured from the priority date, whichever is longer. The 20-year term is discussed in more detail in section 2.4.

10. The reexamination process was modified in 1999, thereby allowing a larger role to third parties in the process. See section 3.8.

11. 35 U.S.C.A. § 156 (West Supp. 2001).

12. Generally, most countries provide an exception to the first-to-file rule in cases of derivation (i.e., where the first to file, who is not actually an inventor, learns of the invention from the actual inventor who files second).

13. 35 U.S.C.A. § 102(g) (West Supp. 2001).

14. A notable exception to this general rule is found in Canada. Generally, an inventor has a 1-year grace period from the date of a public disclosure in which to file a Canadian application if the public disclosure is made by the in-

ventor or by a person who obtained the information from the inventor. There is no grace period (i.e., absolute novelty applies) if the disclosure is made by someone other than the inventor who did not obtain the information from the inventor. E. Hanellin (Ed.), Patents Throughout the World C-4, C-5 (2001).

15. The FDA submission will generally not be released to the public by the FDA upon receipt. Rather, at some later time FDA may make the information publicly available. However, one should not rely on the FDA delaying disclosure if valuable patent rights are at stake. See also section 7.5.

16. The Uruguay Round Agreements Act of 1994, Pub. L. No. 103-465, 108 Stat. 4809 (1995); 59 Fed. Reg. 63,951–63,966 (1994) (proposed rules); 60 Fed. Reg. 20,195–20,231 (1995) (codified at 37 C.F.R. parts 1 and 3) (final rules).

17. The end of the patent term under the 17- or 20-year term can be advanced by failure to pay maintenance fees or by a terminal disclaimer. The end of the patent term can be extended in some cases by the patent term extension provisions. See section 2.4.2.

18. By not claiming benefit of the earlier filed application in a later filed application, the shortening of the patent term can be avoided. This strategy is not generally recommended, given that events occurring between filing dates of the earlier application and the later filed application could bar the granting of a valid patent. In many cases, the patentee may not become aware of such events until the time he or she attempts to enforce the patent. This strategy could be used, however, in the case of a continuation-in-part application where the newly claimed subject matter depends only on the newly added subject matter included in the continuation-in-part application and is patentably distinct from the subject matter disclosed in the earlier filed application. See also section 2.4.

19. The first transitional provision allows an applicant to avoid filing a continuing application in a case under final rejection if (1) the case has been pending for at least 2 years as of June 8, 1995; and (2) the required fee is paid. 37 C.F.R. § 1.129(a) (2001). The second transitional provision allows an applicant to avoid filing a divisional application to prosecute a restricted invention if (1) the case has been pending for at least 3 years as of June 8, 1995; (2) the restriction requirement was issued by the PTO after April 8, 1995; and (3) the required fee for each invention is paid. 37 C.F.R. § 1.129(b) (2001).

This second transitional provision is also not available if the PTO did not issued a restriction requirement before April 8, 1995 because of actions of the applicant. For example, if the applicant suspended examination or filed continuation applications and abandoned the parent applications, before the first office action, the Examiner would not have had an opportunity to issue a restriction requirement. In such cases, a restriction requirement would be appropriate and this transitional provision will not apply.

20. The phrase "in force" is not a term of art in patent law and is not defined in the implementing legislation. The most likely meaning is that such a patent has not expired as of June 8, 1995. Some patentee will almost certainly attempt to argue that his or her patent, although expired on June 8, 1995, was still "enforceable" on June 8, 1995, and, thus, "in force" on June 8, 1995, and entitled to the longer of the 17- or 20-year terms. A suit for infringement can be brought after expiration of a patent, although recovery is limited to the 6 years before the claim. 35 U.S.C.A § 286 (West 1984).

21. 35 U.S.C.A. § 154(c)(1) (West Supp. 2001).

22. There are limits on the remedies available for acts or activities that "were commenced or for which substantial investment was made before" June 8, 1995 and that became infringing because of a "reset" patent term. In such cases, payment of an "equitable remuneration" will allow for the continuation of such acts. 35 U.S.C.A. § 154(c)(2) (West Supp. 2001). The statute does not define "equitable remuneration."

23. 35 U.S.C.A. § 156 (West Supp. 2001).

24. In contrast, the premarket extension period is not automatically determined by either the PTO or the regulating agency (e.g., FDA). The patent holder must apply to the PTO for such a premarket extension (along with the specific time period requested) within a specific time period (60 days) after regulatory approval. 37 C.F.R. §§ 1.710–1.760 (2001).

25. The effective "marketable" patent life (i.e., the time left in the patent term once the product is on the market) for a biotechnology drug product is estimated to be about 11 years. Feisee, "Study Finds Biotech Patents Short-Changed on Term Length," BioNews 5 (October/November 2001).

26. One possibility regarding restriction requirements is to argue before the PTO that the requirement in a particular case is not appropriate. Generally, however, arguments suggesting that restriction requirements are improper are often difficult to win. Moreover, the time involved in contesting a restriction requirement will delay the ultimate grant of any patents and, therefore, reduce the effective patent term. Generally, therefore, it is not recommended that restriction requirements be contested.

27. Under the current system, an applicant may question whether even to file such divisional applications. If broad and adequate protection can be provided in the parent case, it appears that divisional applications may not be necessary. For example, if broad composition of matter claims can be obtained in the parent application, method of use claims presented in a divisional application may offer little additional protection and offer no additional patent term extension. Little additional protection is provided because the broad composition of matter claims will cover all uses of the compound, including a new use. In the event, however, that the composition of matter claims are later found to be invalid or unenforceable, the method of use claims could prove to be valid and enforceable. Thus, there are strong arguments that such divisional applications should be filed even under the current system.

28. 37 C.F.R. § 1.102 (2001). An application can be advanced out of turn for examination if the Commissioner of Patents and Trademarks believes advancement is justified. Suitable reasons for advancement include, for example, applicant's health or age or that "the invention will materially enhance the quality of the environment or materially contribute to the development or conservation of energy resources." A more significant showing will generally be required for advancement based on other reasons.

29. Recently the Court of Appeals for the Federal Circuit ruled that when a claim limitation is narrowed during prosecution, application of the doctrine of equivalents to that claim element is completely barred. Festo Corp. v. Shoketsu Kinzoku Kogyo Kabushiki Co., Ltd., 234. 3d 558 (Fed. Cir. 1999), vacated and remanded, 122 S. Ct. 1831 (2002). Although the U.S. Supreme Court has vacated this decision stand, it could still have a significant impact on claim drafting; more specifically, claims would initially be drafted significantly more narrowly so as to avoid the need to amend claims during prosecution. See Section 4.4 for more details regarding Festo's potential impact.

30. If the issuance of a patent is delayed because of appellate review by the Board of Patent Appeals and Interferences or by a federal court, and the appeal is successful, the patent term is extended for a period of the appeal (up to 5 years). The extension is reduced by any time attributable to appellate review before the expiration of 3 years from the filing date of the application and for the period of time the applicant did not act with "due diligence." 35 U.S.C.A. § 154(b) (West Supp. 2001). See also Section 2.4.

31. American Inventors Protection Act of 1999, Pub. L. No. 106-113, 113 Stat. 1501 (1999); 65 Fed. Reg. 17946–17971 (2000) (proposed rules); 65 Fed. Reg. 57,024–57,061 (2000) (codified at 37 C.F.R. parts 1 and 5) (final rules).

32. Applicants can request publication earlier than the end of the 18-month period. 37 C.F.R. § 1.129 (2001).

33. 37 C.F.R. § 1.211(a)(1) (2001).

34. 37 C.F.R. § 1.211(a)(2) (2001).

35. 37 C.F.R. § 1.211(b) (2001).

36. 37 C.F.R. § 1.213 (2001). If an application directed to the invention is filed in another country or under a multilateral international agreement that provides for publication, the applicant has 45 days in which to notify the PTO; failure to notify can result in abandonment of the U.S. application.

37. 37 C.F.R. § 1.217 (2001).

38. 35 U.S.C.A. § 101 (West 1984).

39. Algorithms are procedures or formulas for solving mathematical problems. An algorithm cannot be patented itself. A physical process using algorithms can, however, be patented. See, e.g., Arrhythmia Research Technology, Inc. v. Corazonix Corp., 958. 2d 1053 (Fed. Cir. 1992) (method and apparatus claims relating to the analysis of electrocardiographic signals employing an algorithm were patentable).

40. Recently, the Federal Circuit held that claims directed to business methods may constitute patentable subject matter. State Street Bank & Trust Co. v. Signature Fin. Group, Inc., 149. 3d 1368, 1373–1377 (Fed. Cir. 1998).

41. Diamond v. Chakrabarty, 447 U.S. 303 (1980).

42. Donald J. Quigg, Notice: "Animals-Patentability" (April 7, 1987) (notice issued by Assistant Secretary and Commissioner of Patents and Trademarks). The PTO Notice reaffirmed that an "article of manufacture or composition of matter will not be considered patentable unless given a new form, quality, properties or combination not present in the original article existing in nature." The PTO Notice also added that a "claim directed to or including within its scope a human being will not be considered to be patentable subject matter."

43. See, e.g., Ex parte Hibberd, 227 U.S.P.Q. 443 (Bd. Pat. App. & Int. 1985) (corn plant with increased level of tryptophan was patentable); U.S. Patent No. 4,736,866 (April 12, 1988) (first animal patent; transgenic mouse with cancer causing gene); U.S. Patent No. 5,183,949 (Feb. 2, 1993) (rabbit infected with HIV-1 virus).

44. Doll, 280 Science 689–690 (1998).

45. Idem.

46. The Thirteenth Amendment to the Constitution provides that "Neither slavery nor involuntary servitude, except as punishment for crime whereof the party shall have been duly convicted, shall exist within the United States." U.S. Const. amend. XIII.

47. A claim directed to a method of treatment comprising transferring somatic cells containing such genes might be patentable.

48. See also Daniel, "Of Mice and 'Manimal': The Patent and Trademark Office's Latest Stance Against Patent Protection for Human-Based Inventions," 7 J. Intell Prop. L. 99 (1999).

49. An invention that can be used for both "useful" and illegal or fraudulent purposes may be patented. Only an invention that could only be used for an illegal or fraudulent purpose would be rendered unpatentable because of the utility requirement. For example, a process for making tobacco only appear to be of higher quality or grade and having no other function or utility (i.e., the sole utility being to deceive the public) was not patentable because of lack of utility. Rickard v. Du Bon, 103 F. 868 (2d Cir. 1900).

50. See Brenner v. Manson, 383 U.S. 519, 528–536 (1966).

51. See, e.g., In re Langer, 503. 2d 1380, 1391–1392 (C.C.P.A. 1974); Newman v. Quigg, 877. 2d 1575, 1581 (Fed. Cir. 1989), cert. denied, 495 U.S. 932 (1990).

52. See, e.g., In re Jolles, 628. 2d 1322, 1326–1327 (C.C.P.A. 1980); Ex parte Kranz, 19 U.S.P.Q.2d 1216, 1218–1219 (Bd. Pat. App. & Int. 1991); Ex parte Balzarini, 21 U.S.P.Q.2d 1892, 1897 (Bd. Pat. App. & Int. 1991).

53. See generally Manual of Patent Examining Procedure, § 608.01(p) (2001).

54. Idem. (emphasis in original; citations omitted).

55. European Patent Handbook, 2nd ed., Matthew Bender, New York, September 2001.

56. For a detailed description of patent law and procedures in countries throughout the world, see E. Hanellin (Ed.), Patents Throughout the World (2001).

57. The countries discussed here are those that are usually recommended for foreign filings. See section 5.

58. Australian Intellectual Property Office, "Australian Patents for: Microorganisms; Cell Lines; Hybridomas; Related Biological Materials and Their Use; and Genetically Manipulated Organisms" (2001) (available: www.ipaustralia.gov.au).

59. Canadian Intellectual Property Office, "A Guide to Patents" (January 2000) (available: http://cipo.gc.ca).

60. North, "The U.S. Expansion of Patentable Subject Matter: Creating a Competitive Advantage for Foreign Multinational Companies?," 18 B. U. Intl. L. J. 111 (2000).

61. Idem.

62. Idem.

63. Idem.

64. United Kingdom Patent Office, "European Directive (98/44/EC) on the Legal Protection of Biotechnological Inventions" (1998) (available: www.patent.gov.uk).

65. Hiraki, "Problems Regarding the Patentability of Genomics and Scope of Protection of ESTs in Japan," CASRIP Newsletter (2000).

66. 35 U.S.C.A. § 112 (first paragraph) (West 1984).

67. 35 U.S.C.A. § 112 (second paragraph) (West 1984).

68. The requirements for a legally sufficient specification in most countries in the remainder of the world are generally less stringent than in the United States. Thus, a patent application that is legally sufficient in the United States is usually sufficient for filing in other counties, with only relatively minor modifications to conform it to local regulations and practices.

69. See, e.g., Helfgott, "A 'Global' Patent Application," 74 J. Pat. & Trademark Off. Soc. 26 (1992).

70. Texas Instruments v. International Trade Commission, 871 F.2d 1054, 1062 (Fed. Cir. 1989).

71. 35 U.S.C.A. § 132 (West 1984).

72. See, e.g., In re Wands, 858 F.2d 731 (Fed. Cir. 1988); Kennecott Corp. v. Kyocera International, Inc., 835 F.2d 1419, 1422 (Fed. Cir. 1987).

73. Such variations may fall within the scope of a claim under the doctrine of equivalents. See section 4.4. It is, however, easier to prove literal infringement. Moreover, an original claim of properly expanded scope may have its own scope expanded, based on the doctrine of equivalents.

74. In re Gay, 309 F.2d 769, 774 (C.C.P.A. 1962) ("Not every last detail is to be described, else patent specifications would turn into production specifications, which they were never meant to be.").

75. In re Jackson, 217 U.S.P.Q. 804, 807 (Bd. Pat. App. & Int. 1986).

76. Ex parte Forman, 230 U.S.P.Q. 546, 547 (Bd. Pat. App. & Int. 1986); In re Wands, 858. 2d 731, 737 (Fed. Cir. 1988).

77. Past tense—which might imply that the work was actually carried out—should not be used for paper examples. It should be clear from the text of the example that the work described was not carried out.

78. Manual of Patent Examining Procedure, § 608.01(p) (2001).

79. Idem.

80. Budapest Treaty on the International Recognition of Microorganisms for the Purposes of Patent Procedure of April 28, 1977, reprinted in E. Hanellin (Ed.), Patents Throughout the World, App. C (2001). As of October 15, 2001, 52 nations (including, for example, the United States, Japan, Australia, and most European countries) were contracting states of the Budapest Treaty.

81. The "certified parties" entitled to obtain a deposit under the Budapest Treaty are determined by the laws and regulations of the country or countries in which a patent application is filed. See Rule 11, Regulations Under the Budapest Treaty on the International Recognition of Microorganisms for the Purposes of Patent Procedure of April 28, 1977, reprinted in E. Hanellin (Ed.), Patents Throughout the World, App. C (2001).

82. 37 C.F.R. § 1.808 (2001).

83. 35 U.S.C.A. § 112 (first paragraph) (West 1984).

84. DeGeorge v. Bernier, 768 F.2d 1318, 1324 (Fed. Cir. 1985).

85. Spectra-Physics, Inc. v. Coherent, Inc., 827 F.2d 1524, 1536 (Fed. Cir. 1987).

86. 35 U.S.C.A. § 112 (second paragraph) (West 1984).

87. 35 U.S.C.A. § 119 (West Supp. 2001).

88. 35 U.S.C.A. § 111(b) (West Supp. 2001).

89. 35 U.S.C.A. § 119(e)(1) (West Supp. 2001).

90. 35 U.S.C.A. § 154(a)(3) (West 1995).

91. Although the ultimate patent cannot be enforced until the actual issue date, the provisional application allows the applicant to use the "patent pending" label on marketed products or promotional materials.

92. 35 U.S.C.A. § 119(e)(1) (West Supp. 2001); 35 U.S.C.A. § 102 (West Supp. 2001).

93. Of course, one could file the regular application containing the improvement at either the time of PA #2 in Figure 20.2 or at the 1-year date. If filed at the time of PA #2, the 20-year term would begin at that date and, thus, the effective patent term would be reduced. If filed at the end of the 1-year period, the improvement would effectively lose its early priority date and may, therefore, be exposed to additional prior art during examination.

94. As detailed in section 3.4, a provisional application must meet the requirements of 35 U.S.C. § 112 by providing a written description that "enables" one of ordinary skill in the art to make and use the invention.

95. Such a case could arise, for example, during gene sequencing. The initial sequencing data could provide the basis for an early provisional application with subsequent sequencing data being included in later provisional applications. Although it is hoped that the early sequence data would enable the gene and thus provide an early priority date, it may be determined later (before the 1-year anniversary) that the early data was not sufficient for enablement purposes. Thus, one could postpone the filing of the regular application until after the 1-year anniversary of the first provisional application because it would not provide any added benefit.

96. This same strategy can also be used where there is some question as to whether the earlier provisional applications are fully enabled. Thus, if there is some concern whether PA #1 is enabled, claiming benefit of a series of provisional applications (the later ones that are assumed to be fully enabled) allows the earliest priority date to be obtained for the regular application.

97. 35 U.S.C.A. § 111(b)(1) (West Supp. 2001).

98. 35 U.S.C.A. § 111(b)(2) (West Supp. 2001).

99. 35 U.S.C.A. § 111(b)(8) (West Supp. 2001).

100. 60 Fed. Reg. 20, 195, 20,222 (1995).

101. 35 U.S.C.A. § 111(b)(7) (West Supp. 2001).

102. 35 U.S.C.A. § 111(b)(6) (West Supp. 2001). A regular application can be converted to a provisional application within the first year; the effective date of the provisional application would be the original filing date of the regular application. If a regular application is filed (before the end of the original 1-year period), the applicant could take advantage of the delay in calculating the beginning of the 20-year term. Some have suggested that this procedure could be used to obtain a preview of the Examiner's views of the invention by filing a regular application, obtaining the first office action, and then converting the regular application to a provisional application, and refiling the regular application (perhaps modified to reflect the results of the first office action). Prosecution on the second-filed regular application could then begin anew. This method relies on receiving the first office action within 1 year of the regular application filing date. It seems unlikely, especially given the large number of new applications filed before the June 8, 1995 GATT deadline, that many first office actions will be received within the first year, thereby rendering this method ineffective. Moreover, if such a practice became widespread, the PTO (or Examiners on their own) might choose to delay most first office actions until at least 1 year after the filing date.

103. 35 U.S.C.A. § 119(e)(1) (West Supp. 2001).

104. 35 U.S.C.A. § 102 (West Supp. 2001).

105. 35 U.S.C.A. § 111(b)(1)(A) (West Supp. 2001).

106. 60 Fed. Reg. 20,195, 20,209 (1995).

107. A "complete" provisional application is intended to mean an application that could have been filed, without changes, as a regular application.

108. Transco Prods. Inc. v. Performance Contracting, Inc., 38 F.3d. 551, 557–559, 32 U.S.P.Q.2d 1077, 1082–1083 (Fed. Cir. 1994).

109. 59 Fed. Reg. 63,951, 63,952 (1994); 60 Fed. Reg. 20,195, 20,196 (1995).

110. See, e.g., 60 Fed. Reg. 20,195, 20,201–20,202 (1995).

111. A regular application can be filed without signatures of the inventors. The PTO then issues a Notice to File Missing Parts requiring a signed declaration to be filed within a certain time period (normally 1 to 2 months with extensions of time possible). In the meantime, however, the 20-year period has already begun. A provisional application can be filed without signatures. If desired, the regular application could be filed as soon as the inven-

tor's signature on the declaration is obtained or could be delayed until close to the 1-year period.

112. 60 Fed. Reg. 20,195, 20,222 (1995). Of course, carried to the extreme, a "contributor" could be anyone who helped develop the background area of the invention. It seems very unlikely that Congress intended all such "contributors" to be named on a provisional application. The issue becomes where to draw the line in naming inventors.

113. In re Hilmer, 359 F.2d 859, 149 U.S.P.Q. 480 (C.C.P.A. 1966); In re Hilmer 165 U.S.P.Q. 255 (C.C.P.A. 1970).

114. 35 U.S.C.A. § 102 (West Supp. 2001).

115. 35 U.S.C.A. § 103 (West Supp. 2001).

116. If the method of making the compound is clearly within the skill in the art, such a reference could place the compound within the public domain. In such a case, the reference could anticipate a claim to the compound.

117. In re Schoenwald, 964 F.2d 1122 (Fed. Cir. 1992).

118. In re Spada, 911 F.2d 705 (Fed. Cir. 1990).

119. 35 U.S.C.A. § 103 (West Supp. 2001).

120. Graham v. John Deere Co., 383 U.S. 1, 17 (1966).

121. Environmental Designs, Ltd. v. Union Oil Co., 713 F.2d 693, 696–697 (Fed. Cir. 1983).

122. Graham v. John Deere Co., 383 U.S. at 17–18.

123. Pentec, Inc. v. Graphic Controls Corp., 776 F.2d 309, 315–316 (Fed. Cir. 1985).

124. Hodosh v. Block Drug Co., 786 F.2d 1136, 1143 n.5 (Fed. Cir. 1986) (citations omitted).

125. See, e.g., In re Dow Chem. Co., 837 F.2d 469, 473 (Fed. Cir. 1988).

126. In re O'Farrell, 853 F.2d 894, 903 (Fed. Cir. 1988).

127. See. e.g., Graham v. John Deere Co., 383 U.S. at 36 (courts should "resist the temptation to read into the prior art the teachings of the invention in issue"); Uniroyal, Inc. v. Rudkin-Wiley Corp., 837 F.2d 1044, 1051 (Fed. Cir. 1988); W. L. Gore & Assoc., Inc. v. Garlock, Inc., 721 F.2d 1540, 1553 (Fed. Cir. 1983).

128. In re Papesch, 315 F.2d 381, 391 (C.C.P.A. 1963).

129. In re Dillon, 919 F.2d 688, 692–693 (Fed. Cir. 1990) (en banc).

130. In re Dillon, 919 F.2d at 693.

131. See, e.g., In re Merck & Co., 800 F.2d 1091, 1096–1098 (Fed. Cir. 1986) ("bioisosterism was commonly used by medicinal chemists be-

fore 1959 in an effort to design and predict drug activity"); Imperial Chem. Industr. v. Danbury Pharmacal, 745 F. Supp. 998 (D. Del. 1990) (bioisosterism theory used to invalidate patent under 35 U.S. C. § 103).

132. In re Merck & Co., 800 F.2d at 1095.

133. In re Durden, 763 F.2d 1406, 1410 (Fed. Cir. 1985).

134. In re Dillon, 919 F.2d at 695 (dictum).

135. The inventor or assignee is normally represented by a patent agent or attorney in proceedings before the PTO. Such agents or attorneys are admitted to practice before the PTO upon passing a bar examination given by the PTO. The inventor may, however, represent himself or herself.

136. American Inventors Protection Act of 1999 (AIPA) (P. L. 106-113) and 37 C.F.R. § 1.211(a) (2001).

137. The written record of an abandoned application is open for public inspection if a later U.S. patent formally relies on, or is formally related to, the abandoned application.

138. 35 U.S.C.A. § 184 (West Supp. 2001); 37 C.F.R. §§ 5.11–5.25 (2001). Unless the PTO notifies the applicant otherwise, the applicant can file foreign applications 6 months after the U.S. filing date without a foreign filing license.

139. Applicants can also attempt to convince the examiner that the restriction requirement is improper.

140. Before changes implemented under GATT, divisional applications potentially allowed the applicant to obtain several patents in series on the same basic concept and thereby extend patent protection beyond the 17-year term of the first patent to issue. This is no longer possible under the 20-year term.

141. Up to three additional months for responding normally can be obtained by payment of extension fees.

142. Such a party could lose priority if that party "abandoned, suppressed, or concealed" the invention. 35 U.S.C.A. § 102(g) (West Supp. 2001). For example, if the first party to conceive and reduce the invention to practice decided to keep the invention a trade secret and was spurred to file an application only upon learning of another's invention of the same subject matter, the first party's priority could be extinguished by its failure to file its patent application in a timely manner.

143. On occasion the PTO may inadvertently issue two patents that claim the same subject mat-

ter. In such cases, the PTO does not have jurisdiction to institute an interference proceeding. If one of the patentees brings an appropriate civil action in a federal district court, the court could determine which patentee is entitled to a patent claiming the subject matter.

144. In such cases, the PTO will normally accept an affidavit or declaration filed by the applicant alleging a basis upon which the applicant is entitled to a judgment relative to the patentee. If the applicant's effective filing date is more than 3 months after the effective filing date of the patentee, the PTO will generally demand evidence and arguments as to why the applicant is prima facie entitled to judgment relative to the patentee. 37 C.F.R. § 1.608(b) (2001).

145. A party who cannot prove an earlier filing date may prefer that no one obtain a patent so he or she can practice the invention without restriction. See, e.g., Perkin v. Kwon, 886 F.2d 325 (Fed. Cir. 1989). Such a party (i.e., one who knows that priority cannot be proven) could also seek a favorable license (especially while the preliminary statements remain sealed) and thereby, at least to some degree, benefit from the protection offered by the other party's patent (if granted by the PTO). The other party may be willing to make concessions in the terms of such an agreement because it would effectively terminate the interference and eliminate the risk of losing the priority contest.

146. See 35 U.S.C.A. § 104 (West Supp. 2001).

147. Manual of Patent Examining Procedure, § 2138.02 (2001).

148. 35 U.S.C.A. § 104(a)(3) (West Supp. 2001).

149. In two-party interference proceedings between fiscal years 1992 and 1994, the senior and junior parties prevailed in about 62 and 28%, respectively; the remainder were split decisions. For proceedings decided by the Board, the senior and junior parties prevailed in about 53 and 32%, respectively; again, the remainder were split decisions. Of the 673 two-party interferences during this period, about 21% proceeded to a decision by the Board. Calvert & Sofocleous, "Interference Statistics for Fiscal Years 1992 to 1994," 77 J. Pat. & Trademark Off. Soc. 417 (1995).

150. 35 U.S.C.A. §§ 254–256 (West Supp. 2001).

151. 35 U.S.C.A. § 251 (West Supp. 2001); 37 C.F.R. §§ 1.171–1.179 (2001).

152. 35 U.S.C.A. § 252 (West Supp. 2001).

153. 35 U.S.C.A. §§ 301–307 (West 1984 & Supp. 2001).

154. Idem.

155. 35 U.S.C.A. §§ 311–318 (West Supp. 2001).

156. 35 U.S.C.A. § 315(c) (West Supp. 2001).

157. One exception to this general rule is where a third party is precluded from entering a market by a blocking patent that, in the opinion of the third party, is invalid based on prior art not considered by the PTO. By requesting a reexamination, the third party would hope to invalidate the patent and clear the path for its entry into the market. If unsuccessful, the third part may wish to move on to other business opportunities.

158. A suit alleging infringement during the term of a patent can generally be brought after the patent expires. Damages cannot, however, be recovered for infringing acts occurring more than 6 years before filing the complaint. 35 U.S.C.A. § 286 (West 1984).

159. 35 U.S.C.A. § 271(g) (West Supp. 2001).

160. Marshall, "The Enforcement of Patent Rights in Germany," 31 IIC 646 (2000).

161. See Tenth Annual Judicial Conference of the United States Court of Appeals for the Federal Circuit, 146 F.R.D. 205, 372 (1992) [Judge Cohen (E. D. Mich.)] ("[T]here's an underappreciation of the general lack of experience of district judges [regarding patent matters] . . . [D]istrict judges have to constantly learn and re-learn patent law. They simply cannot keep current with developments in the law.").

162. Hon. P. R. Michel and M. Rhyu, "Improving Patent Jury Trials," The Federal Circuit Bar Journal, Vol. 6, No. 2 (1997).

163. 28 U.S.C.A. § 1498 (West Supp. 2001).

164. 19 U.S.C.A. § 1337 (West Supp. 2001).

165. Trojan, Inc. v. Shat-R-Shield, Inc., 885 F.2d 854, 855–856 (Fed. Cir. 1989).

166. 19 U.S.C.A. § 1337(a) (West Supp. 2001).

167. Pub. L. No. 97-164, 96 Stat 25 (1990).

168. See H. R. Rep. 97-312, 97th Cong., 1st Sess. 23 ("[T]he central purpose [of the FCIA] is to reduce the widespread lack of uniformity and uncertainty of legal doctrine that exists in the administration of patent law.").

169. Mas-Hamilton Group v. LaGard, Inc., 156 F.3d 1206, 1210 (Fed. Cir. 1998). The Federal Circuit clarified in Mas-Hamilton Group that a party that had standing to sue for infringement when suit was filed does not lose standing merely because the right to sue for past

damages was separately assigned during the lawsuit. The plaintiff's right to seek injunctive relief and future damages at the time of filing suit were sufficient to confer standing. Idem. at 1211.

170. Willingham v. Lawton, 555 F.2d 1340, 1343 (6th Cir. 1977).

171. Enzo APA & Sons, Inc. v. Geapag A. G., 134 F.3d 1090, 1093 (Fed. Cir. 1998); Textile Prods., Inc. v. Mead Corp., 134 F.3d 1481, 1484 (Fed. Cir. 1998).

172. Textile Prods., supra, note 170 at 1484–1485.

173. 35 U.S.C.A. § 271 (West Supp. 2001).

174. 35 U.S.C.A. § 271(b) (West 1984).

175. Water Tech. Corp. v. Calco, Ltd., 850 F.2d 660, 668 (Fed. Cir. 1988).

176. 35 U.S.C.A. § 271(f) (West Supp. 2001).

177. 35 U.S.C.A. § 271(g) (West Supp. 2001).

178. Bandag, Inc. v. Al Bolser's Tire Stores, Inc., 750 F.2d 903 (Fed. Cir. 1984).

179. Aro Mfg. Co. v. Convertible Top Replacement Co., 365 U.S. 336, 344–346 (1961).

180. 21 U.S.C.A. § 355(b)(1) (West Supp. 2001).

181. 21 U.S.C.A. §§ 355(b)(2)(A) & 355 (j)(2)(A)(vii) (West Supp. 2001).

182. 28 U.S.C.A. § 2201 (West 1984).

183. Arrowhead Indus. Water, Inc. v. Ecolochem, Inc., 846 F.2d 731 (Fed. Cir. 1988).

184. EMC Corp. v. Norand Corp., 89 F.3d 807 (Fed. Cir. 1996).

185. Markman v. Westview Instruments, Inc., 52 F.3d 967 (Fed. Cir. 1995) (en banc), aff'd, 116 S. Ct. 1384 (1996).

186. Idem. at 979–980.

187. Hilton Davis Chemical Co. v. Warner-Jenkinson Co., Inc., 62. 2d 1352 (Fed. Cir. 1995).

188. Warner-Jenkinson Co., Inc. v. Hilton Davis Chemical Co., 520 U.S. 17, 28–29 (1997).

189. Idem. at 40.

190. See, e.g., Wilson Sporting Goods Co. v. David Geoffrey & Assoc., 904 F.2d 677, 684 (Fed. Cir.).

191. Festo Corp. v. Shoketsu Kinzoku Kogyo Kabushiki Co., Ltd., 234 F.3d 558 (Fed. Cir. 1999), vacated and remanded, 122 S. Ct. 1831 (2002).

192. Idem. at 566.

193. Idem. at 568.

194. However, such intent may be sufficient for a finding of willful infringement if the materiality of the particular prior art is especially high.

See, e.g., N. V. Akzo v. E. I. Du Pont de Nemours, 810. 2d 1148, 1153 (Fed. Cir. 1987).

195. See, e.g., Kingsdown Medical Consultants Ltd. v. Hollister Inc., 863 F.2d 867, 876 (Fed. Cir. 1988); Atlas Powder Co. v. E. I. Du Pont de Nemours, 750. 2d 1569, 1578 (Fed. Cir. 1984).

196. A very limited "experimental use" exception allows otherwise infringing acts if the acts are "for amusement, to satisfy idle curiosity or for strictly philosophical inquiry." Roche Products, Inc. v. Bolar Pharmaceutical Co., 733 F.2d 858, 863 (Fed. Cir. 1984). Commercial activities are very unlikely to fall within this exception.

197. 35 U.S.C.A. § 271(e) (West Supp. 2001). This exemption specifically excludes from its coverage "a new animal drug or veterinary biological product . . . which is primarily manufactured using recombinant DNA, recombinant RNA, hybridoma technology, or other processes involving site specific genetic manipulation techniques."

198. 35 U.S.C.A. § 284 (West Supp. 2001).

199. Gyromat Corp. v. Champion Spark Plug Co., 735 F.2d 549, 549 (Fed. Cir. 1984).

200. Carella v. Starlight Archery & ProLine Co., 804 F.2d 135, 141 (Fed. Cir. 1986).

201. Lam, Inc. v. Johns-Manville Corp., 718 F.2d 1056, 1065 (Fed. Cir. 1983).

202. State Indus., Inc. v. Mor-Flo Indus., Inc., 883 F.2d 1573 (Fed. Cir. 1989).

203. Rite-Hite Corp., 56 F.3d 1538 (Fed. Cir. 1995).

204. Idem. at 1550–1551.

205. See Fromson v. Western Litho Plate & Supply Co., 853 F.2d 1568, 1576–1578 (Fed. Cir. 1988); Georgia Pacific Corp. v. United States Plywood Corp., 318 F. Supp. 1116, 1120 (S.D.N.Y. 1970).

206. 35 U.S.C.A. § 284 (West Supp. 2001).

207. 35 U.S.C.A. § 285 (West 1984). An award of attorney fees might be appropriate for an especially willful and egregious infringer or against a patentee who litigated in bad faith or committed fraud during the prosecution of the patent in the PTO. See, e.g., Machinery Corp. of Am. v. Gullfiber AB, 774 F.2d 467, 472–473 (Fed. Cir. 1985).

208. 35 U.S.C.A. § 283 (West 1984).

209. Atlas Powder Co. v. Ireco Chemicals., 773 F.2d 1230, 1231 (Fed. Cir. 1985).

210. See 28 U.S.C.A. §§ 1391(c) & 1400(b) (West 1993).

211. See, e.g., Red Wing Shoe Co. v. Hockerson-Halberstadt, Inc., 148 F.3d 1355 (Fed. Cir. 1998).

212. See Beverly Hills Fan Co. v. Royal Sovereign Corp., 21 F.3d 1558 (Fed. Cir. 1994) (holding district court in Virginia had personal jurisdiction over Chinese corporation who manufactured accused product in Taiwan and sold to a New Jersey importer because it placed the products in a "stream of commerce," knowing the likely destination of some of the products was Virginia).

213. Fed. R. Civ. P. 8(a).

214. Fed. R. Civ. P. 8(c).

215. See United States v. Procter & Gamble Co., 356 U.S. 677, 682–683 (1958).

216. For example, an internal corporate memorandum that is inadmissible at trial, because of the hearsay rule against out-of-court statements, may nonetheless be discoverable given its potential to identify witnesses who can be called to testify directly at trial regarding the contents of the memorandum.

217. See Fed. R. Civ. P. 45.

218. The testimony is entered into the record by reading the transcript of questions and answers, or by playing the video of a videotaped deposition.

219. A corporation may be deposed through one or more designated representatives who speak on the corporation's behalf. Fed. R. Civ. P. 30(b)(6).

220. Fed. R. Civ. P. 26(a)(2)(B).

221. See P. M. Palumbo v. Don-Joy Co., 762 F.2d 969, 975 (Fed. Cir. 1985).

222. Examples of fact issues include disputes regarding the structure and operation of the accused device, the teachings of the prior art and level of skill in the art, and the amount of damages sustained on account of the infringement. Examples of legal issues include claim interpretation, and whether the prior art prevents infringement under the doctrine of equivalents.

223. Daubert v. Merrell Dow Pharmaceuticals, Inc., 509 U.S. 579 (1993).

224. As an example of simplifying the case for the jury, consider the patentee who has received a broad claim construction from the court, but must still demonstrate to the jury that the claims read on the accused product. Although the patent drawings are technically irrelevant to that issue, the patentee's counsel may use every available opportunity to show the jury that the accused product looks similar to the patent drawings to visually reinforce the infringement theory. The jury is more likely to understand the drawings than the scope of the claims.

225. Fed. R. Civ. P. 50(a).

226. See Blonder-Tongue Labs, Inc. v. University of Illinois Foundation, 402 U.S. 313, 334 (1971).

227. See, e.g., Brunda, "Resolution of Patent Disputes by Non-litigation Procedures," 15 AIPLA Q. J. 73 (1987); Arnold, "Alternative Dispute Resolution in Intellectual Property Cases," 9(1) AIPLA Selected Legal Papers 289 (July 1991).

228. The North American Free Trade Agreement (NAFTA) between the United States, Canada, and Mexico continues the trend toward such a single North American market. See generally R. M. Milgrim, 2 Milgrim on Trade Secrets § 9.07 (2001); Paul, Hastings, Janofsky & Walker, North American Free Trade Agreement: Summary and Analysis (1993).

229. Paris Convention of March 20, 1883, as revised at Stockholm on July 14, 1967, reprinted in E. Hanellin (Ed.), Patents Throughout the World, App. B (2001) (earlier revisions are also included).

230. Applications can be filed in Paris Convention countries after the 12-month period if there has been no public disclosure. However, the priority date of the earlier filed application cannot be claimed.

231. European Patent Convention of October 5, 1973 (as amended to 1979), reprinted in E. Hanellin (Ed.), Patents Throughout the World, App. B(1)(d) (2001).

232. These countries include Albania, Latvia, Lithuania, the former Yugoslav Republic of Macedonia, Romania, and Slovenia.

233. Patent Cooperation Treaty of June 19, 1970 (as amended to 1985), reprinted in E. Hanellin (Ed.), Patents Throughout the World, App. B (2001).

234. For European countries, an applicant can designate an EPC application (which in turn designates the desired member states) or separate national applications in the individual European countries.

235. World Intellectual Property Organization, PCT Newsletter 1 (October 2001).

236. See 35 U.S.C.A. § 122 (West Supp. 2001).

237. F. I. Schechter, The Historical Foundations of the Law Relating to Trademarks (1925). See also Diamond, "The Historical Development of Trademarks," 65 Trademark Rep. 265 (1975).

238. 15 U.S.C.A. §§ 1051–1127 (West 1963, 1976, 1982 & Supp. 2001).

239. 18 U.S.C.A. § 2320 (West Supp. 2001).

240. Secondary meaning is "acquired when the name and the business becomes synonymous in the public mind; and [it] submerges the primary meaning of the name . . . in favor of its meaning as a word identifying that business." Visser v. Macres, 137 U.S.P.Q. 492, 494 (Cal. Dist. Ct. App. 1963).

241. U.S. PTO, Trademark Manual of Examining Procedure §§ 1209.01(a) & (b) (2d ed.; 1993) (citations omitted).

242. In re Owens-Corning Fiberglas Corp., 774. 2d 1116, 1125 (Fed. Cir. 1992).

243. Braun Inc. v. Dynamic Corp. of Am., 975 F.2d 815, 826 (Fed. Cir. 1992).

244. 15 U.S.C.A. §§ 1052(a)–(c) & (e) (West 1997 & Supp. 2001).

245. 15 U.S.C.A. § 1052(d) (West Supp. 2001).

246. Registration of a mark is not necessary. A common law trademark can be created by using the mark in commerce. Generally, such a common law trademark affords protection to its owner only in the geographical area of actual use.

247. The Trademark Law Revision Act of 1988.

248. 15 U.S.C.A. §§ 1065 & 1115 (West Supp. 2001).

249. 15 U.S.C.A. § 1057(b) (West Supp. 2001).

250. 15 U.S.C.A. § 1065 (West Supp. 2001).

251. 15 U.S.C.A. § 1115 (West Supp. 2001).

252. 15 U.S.C.A. § 1072 (West 1997).

253. See 15 U.S.C.A. § 1111 (West 1997) (relevant portion quoted in text at note 262 infra).

254. 15 U.S.C.A. § 1121 (West Supp. 2001).

255. Walt Disney Productions v. Kusan, Inc., 204 U.S.P.Q. 284 (C.D.Cal. 1979).

256. 15 U.S.C.A. § 1124 (West Supp. 2001).

257. 15 U.S.C.A. § 1117 (West Supp. 2001). The registrant is also entitled to recover the "costs" of the litigation against an infringer.

258. Idem.

259. Idem.

260. Idem.

261. 15 U.S.C.A. § 1118 (West Supp. 2001).

262. Baker v. Master Printers Union, 34 F. Supp. 808, 811 (D.N.J. 1940).

263. Bayer Co. v. United Drug Co., 272 F. 505, 509–510 (2d Cir. 1921).

264. 15 U.S.C.A. § 1111 (West 1997).

265. 15 U.S.C.A. § 1125(a) (West Supp. 2001).

266. Ciba-Geigy Corp. v. Bolar Pharmaceutical Co., 547 F. Supp. 1095 (D.N.J. 1982), aff'd, 719. 2d 56 (3d Cir. 1983).

267. 15 U.S.C.A. § 1125(a)(2) (West Supp. 2001).

268. Restatement (First) of Torts, § 757 cmt. b (1939).

269. As of 2001, about 43 states and the District of Columbia had adopted the Uniform Trade Secrets Act or some version thereof. See R. M. Milgrim, Milgrim on Trade Secrets § 305[1] (2001).

270. Uniform Trade Secrets Act, § 1(4) (1985).

271. Restatement (First) of Torts, § 757 cmt. b (1939).

272. Uniform Trade Secrets Act, § 2 (1985).

273. See, e.g., Robbins, "The Rights of the First Inventor-Trade Secret User as Against Those of the Second Inventor-Patentee (Part I)," 61 J. Pat. Off. Soc. 574 (1979); Jorda, "The Rights of the First Inventor-Trade Secret User as Against Those of the Second Inventor-Patentee (Part II)," 61 J. Pat. Off. Soc. 593 (1979); R. M. Milgrim, 2 Milgrim on Trade Secrets § 8.02[2] (2001).

274. Jorda supra note 273. Of course, some trade secrets may have a considerable longer life. For example, the original formula for Coca-Cola(r) has remained a trade secret for well over 100 years.

275. Kewanee Oil Co. v. Bicron Corp., 416 U.S. 470, 491 (1974).

276. See, e.g., Transco Products Inc. v. Performance Contracting Inc., 38 F.3d 551 (Fed. Cir. 1994) (patentee not required to update best mode in a continuation application in which no new matter was added).

277. 5 U.S.C.A. § 552 (West Supp. 2001). Virtually every state has adopted a corresponding state version of FOIA.

278. 5 U.S.C.A. § 552(a)(3)(A) (West Supp. 2001).

279. 5 U.S.C.A. § 552(b)(3) (West Supp. 2001).

280. 18 U.S.C.A. § 1905 (West Supp. 2001).

281. 5 U.S.C.A. § 552(b)(4) (West Supp. 2001).

282. Chrysler Corp. v. Brown, 441 U.S. 281, 293 (1979).

283. A trade secret owner may seek, through a so-called reverse FOIA suit, to enjoin an agency from releasing information containing trade secrets. See generally R. M. Milgrim, 2 Milgrim on Trade Secrets § 6.02A[3] (2001).

284. North American Free Trade Agreement, art. 1711(5).

285. North American Free Trade Agreement, art. 1711(6).

286. The specific information on trade secrets, which follows for countries outside the United States, is generally taken from R. M. Milgrim,

4 Milgrim on Trade Secrets, App. R, R-49 (1993) [reprinting Prasinos, "International Legal Protection of Computer Programs," 26 Idea 173 (1984)]. The trade secrets laws of these countries have likely changed, perhaps significantly, since the publication of this article. Nonetheless, this information illustrates the variability expected in trade secret law throughout the world. Clearly one should seek guidance from local counsel in the relevant country before setting up a trade secret protection program or attempting to enforce a trade secret in a particular country.

287. Some countries (e.g., France and Italy) allow trade secret protection for information that is not known within their national boundaries, even if it is known in other countries. In this age of almost instant communication, computer databases, and the Internet, it is increasingly unlikely that information that is known elsewhere in the world can effectively remain a secret for any significant length of time in a given country (especially in highly industrialized countries).

288. See, e.g., 17 U.S.C.A. §§ 411–412 (West Supp. 2001).

289. 35 U.S.C.A. § 157 (West Supp. 2001).

290. Essentially the same effect can be obtained with an issued patent by simply not enforcing the patent or filing a disclaimer whereby the patent is dedicated to the public. 35 U.S.C.A. § 253 (West 1984); 37 C.F.R. § 1.321(a) (2001). Such dedication to the public does not affect the defensive aspects of the patent.

291. The time lag between submission and the actual publication date in a scientific journal, especially in peer-reviewed journals, can be significant. Such delays may allow for another inventor to file a patent application covering the invention before the actual publication and ultimately obtain patent protection.

292. 35 U.S.C.A. §§ 171–173 (West 1984 & Supp. 2001).

293. Gorham Co. v. White, 81 U.S. (14 Wall.) 511, 528 (1871).

Index

Terms that begin with numbers are indexed as if the number were spelled out; e.g., "3D models" is located as if it were spelled "ThreeD models."

Abbreviated New Drug Application, 742
Absorption, distribution, and elimination, 633–634
  pharmacokinetic applications in development process, 645–647
  pharmacokinetic parameters in drug disposition, 635–641
  pharmacokinetic parameters in therapeutics, 641–645
  polymorphism in, 472
Absorption, distribution, metabolism, and excretion (ADME), 31
  prodrugs, 501
  receptor targeting drugs, 320, 342–343
  transport proteins, 266
Absorption, distribution, metabolism, excretion, and toxicity (ADMET)
  transport proteins, 266
ABT-773
  structure, 194
AC4437, 176
Acacia
  for precipitation inhibition, 669
Acarbose, 205, 212
Accelerated Approval, in New Drug Approval, 700
Accelerated development/review, FDA, 687, 696–697
ACCELERATOR RTS, 49
ACC-9358 soft analogs, 562–563
ACE inhibitors
  combinatorial library of captopril analogs, 18, 20
  prodrug approach, 508, 511
  soft analogs, 565
Acetaminophen
  hepatic toxicity, 615–616

Acetylation
  role in metabolism, 456–459
Acetylcholine, 391. *See also* Nicotinic acetylcholine receptors
  gating of anion-selective channels, 360
  snail ACh-binding protein, 362–364
Acetylcholinesterase, 375, 538
Acetylsalicylic acid (aspirin)
  prodrugs, 505, 510
Acid-base ionization properties, 660–662
Actinomycin D
  reaction with nucleosides, 616
Action potential duration testing, 627
Active moiety, prodrugs, 500, 502–503
Active pharmaceutical ingredients, 408. *See also* Drug synthesis
  solid-state properties, 414
Active transport, 257–265
Acyclovir
  prodrug, 508, 511, 515–516
Acylation
  role in metabolism, 456–459
Adair equation, 299
Adaprolol, 540–541
Adaptive toxicological responses, 617–618
Adhesins, 262–263
ADME studies, *See* Absorption, distribution, metabolism, and excretion (ADME)
ADMET studies, *See* Absorption, distribution, metabolism, excretion, and toxicity (ADMET)
Adornments (in combinatorial libraries), 14–15
β_2-Adrenergic receptor
  recombinant DNA studies, 102
β-Adrenoreceptor antagonists, *See* β-Blockers

Adriamycin
  prodrugs, 505
Adulteration of Various Substances Used in Medicine and the Arts (Beck), 684
Adurazyme, 218–219
Aerosol administration
  phosphorothioate oligonucleotides, 134
Affirmative defenses, in patent infringement cases, 748
Agalsidase alfa, 211, 212
Age
  and metabolism, 474
Aggrecan, 238
Agitation
  in large-scale synthesis, 415–416
AIDS, *See* HIV protease inhibitors
Alamar Blue, 50
Alanine-scanning mutagenesis, 88–89
Albumin, *See* Serum albumin
Aldehydes
  residues in pseudopeptide arrays, 12
Aldurazyme, 219, 220
Alfentanil, 544
Algorithms
  nonpatentability, 718
Allosteric effectors, 296–313
  basis as therepeutic agents, 312–313
Allosteric proteins, 295–296
  allosteric transitions at molecular level, 305–312
  common features of, 304–305
  hemoglobin as example, 296, 301–306
  targets for drug discovery, 313–315, 314
  theories of, 296–301
Alpha$_1$-acid glycoprotein
  binding of drugs to, 637
Alpha-L-iduronidase, 219, 220
Alprenolol
  membrane function disruption, 614